Selected Art Conventions

PRINCIPLES OF HUMAN PHYSIOLOGY

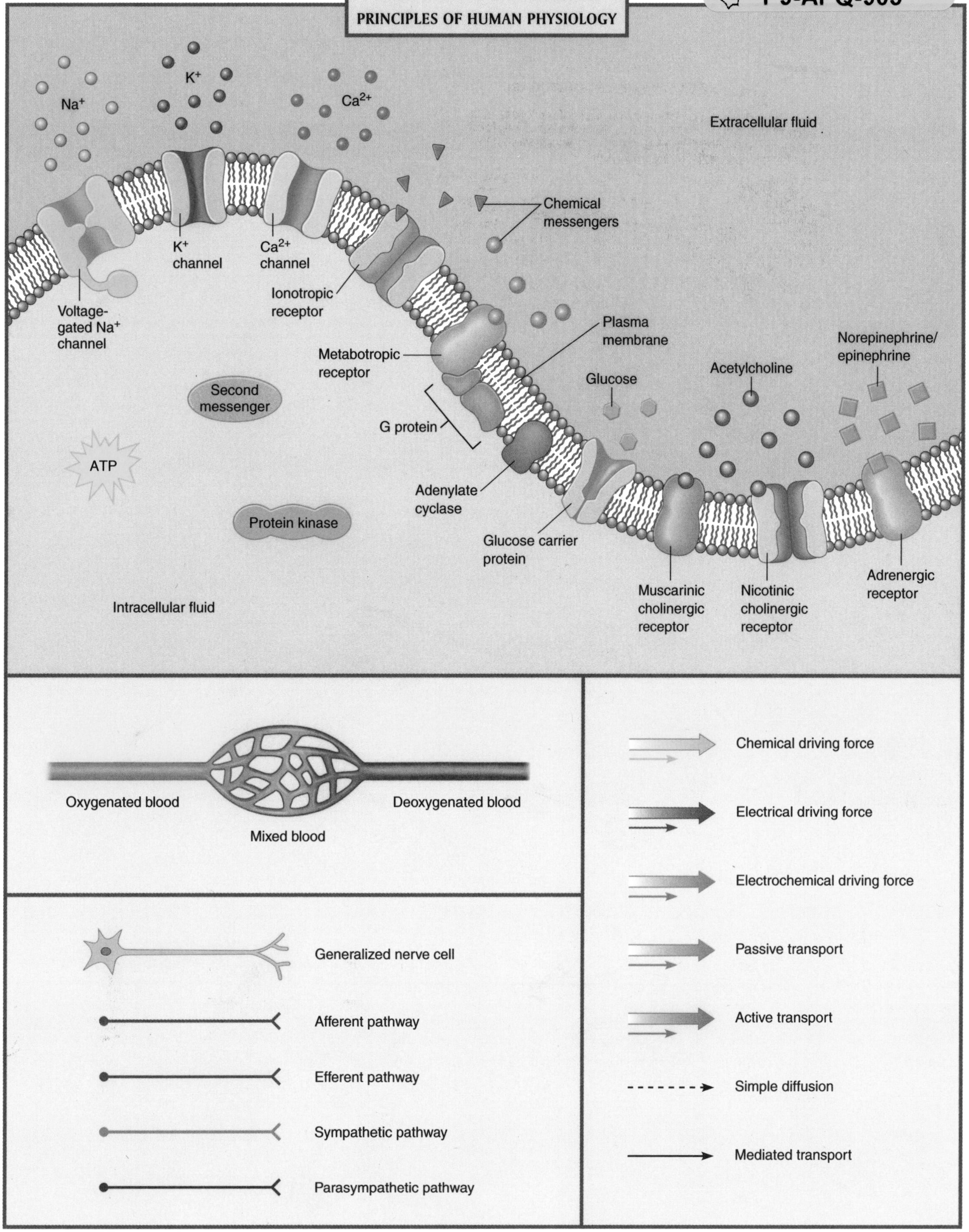

Na+
K+
Ca2+
Extracellular fluid
Chemical messengers
K+ channel
Ca2+ channel
Voltage-gated Na+ channel
Ionotropic receptor
Plasma membrane
Metabotropic receptor
Norepinephrine/ epinephrine
Acetylcholine
Glucose
Second messenger
G protein
ATP
Adenylate cyclase
Protein kinase
Glucose carrier protein
Adrenergic receptor
Muscarinic cholinergic receptor
Nicotinic cholinergic receptor
Intracellular fluid
Oxygenated blood
Deoxygenated blood
Mixed blood
Chemical driving force
Electrical driving force
Electrochemical driving force
Generalized nerve cell
Passive transport
Afferent pathway
Active transport
Efferent pathway
Simple diffusion
Sympathetic pathway
Mediated transport
Parasympathetic pathway

The Physiology Place

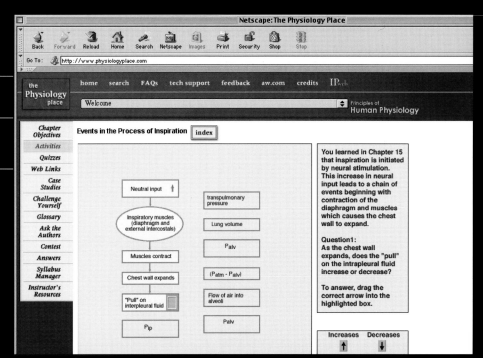

Log on.

Tune in.

Succeed.

To help you succeed in physiology, your professor has arranged for you to enjoy access to a great media resource, The Physiology Place. You'll find that this resource that accompanies your textbook will enhance your course materials.

What your system needs to use these media resources:

WINDOWS™
250 MHz Intel Pentium processor or greater
Windows 98, NT, 2000, XP
32 MB RAM installed
800 X 600 screen resolution, thousands of colors
Browser: Internet Explorer 5.0 or Netscape
    Communicator 4.7, 7.0
Plug-Ins: Shockwave Player 8, Flash Player 4
NOTE: Use of Netscape 6.0 and 6.1 are not recom-
    mended due to a known compatibility issue
    between Netscape 6.0 and 6.1 and the Flash
    and Shockwave plug-ins.

MACINTOSH™
233 MHz PowerPC
OS 9.2 or higher
32 MB RAM available
800 x 600 screen resolution, thousands of colors
Browser: Internet Explorer 5.0 or Netscape
    Communicator 4.7, 7.0
Plug-Ins: Shockwave Player 8, Flash Player 4
NOTE: Use of Netscape 6.0 and 6.1 are not recom-
    mended due to a known compatibility issue
    between Netscape 6.0 and 6.1 and the Flash
    and Shockwave plug-ins.

Got technical questions?
For technical support, please visit
www.aw.com/techsupport and complete
the appropriate online form. You can also
call our tech support hotline at
800-6-Pro-Desk (800-677-6337) Monday-
Friday, 8 a.m. to 5 p.m. CST.

Here's your personal ticket to success:

How to log on to www.physiologyplace.com:

1. Go to www.physiologyplace.com.
2. Click *Principles of Human Physiology*.
3. Click "Register."
4. Scratch off the silver foil coating below to reveal your pre-assigned access code.
5. Enter your pre-assigned access code exactly as it appears below.
6. Complete the online registration form

to create your own personal user Login Name and Password.
7. Once your personal Login Name and Password are confirmed by email, go back to www.physiologyplace.com, type in your new Login Name and Password, and click "Log In."

Your Access Code is:

Record your new login name and password on the back of this card.

Cut out this card and keep it handy. It's your ticket to valuable information. Students who purchase a used copy of this book may not have a valid access code. Students may purchase a subscription to this website at www.physiologyplace.com.

Important: Please read the License Agreement, located on the launch screen before using The Physiology Place. By using the web site, you indicate that you have read, understood, and accepted the terms of this agreement.

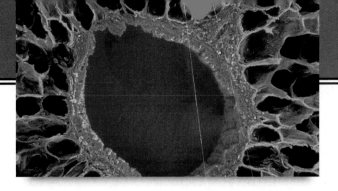

Principles of Human Physiology

WILLIAM J. GERMANN
UNIVERSITY OF DALLAS

CINDY L. STANFIELD
UNIVERSITY OF SOUTH ALABAMA

Contributors

Joseph G. Cannon
Kellett Chair in Allied Health Sciences
Medical College of Georgia
Chapter 23

Mary Jane Niles
University of San Francisco
Chapter 22

Benjamin
Cummings

San Francisco Boston New York
Cape Town Hong Kong London Madrid Mexico City
Montreal Munich Paris Singapore Sydney Tokyo Toronto

Publisher: Daryl Fox
Project Editor: Claire Brassert
Text Development Editor: Jane Tufts
Art Development Editor: Laura Southworth
Managing Editor: Wendy Earl
Associate Editor: Mary Ann Murray
Publishing Assistant: Marie Beaugureau
Online Associate Project Editor: Erin Joyce
Production Editor: Janet Vail
Art and Photo Coordinator: Kelly Murphy
Principal Artist: Rolf Swanson, Precision Graphics
Photo Researcher: Diane Austin
Cover Image: Tomo Narashima
Cover Design: F. tani Hasegawa
Copyeditor: Alan Titche
Manufacturing Buyer: Stacey Weinberger
Market Development Managers: Deirdre McGill, Cheryl Cechvala
Senior Marketing Manager: Lauren Harp

Photography and illustration credits appear on page 769.

Library of Congress Cataloging-in-Publication Data
Germann, William J.
     Principles of human physiology.William J. Germann, Cindy L. Stanfield;
contributors, Joseph G. Cannon, Mary Jane Niles.
       p. cm.
    Includes index.
    ISBN 0-8053-5662-2
    1. Human physiology. I. Stanfield, Cindy L. II. Title.

QP34.5 .G465 2001
612–dc21   2001047446

ISBN 0–8053–5622–2
1 2 3 4 5 6 7 8 9 10—QWV—05 04 03
www.aw.com/bc

To Laura and Ryan (W.J.G.)

To Jim Stanfield, Addie Thurston (mom), and all my brothers and sisters—Steve, Joe, Jim, Pat, Mike, Janice, and John (C.L.S.)

About the Authors

William J. Germann earned a Ph.D. in physiology at the University of Michigan. One of Germann's teachers at Michigan was Arthur Vander, whose commitment to education made a lasting impression and ultimately inspired him to pursue a career in undergraduate teaching. Germann currently teaches human physiology and a number of other biology courses at the University of Dallas in Irving, Texas. He believes that a teacher's job is not just to communicate information, but to inspire students to want to learn. As chair of the Biology Department, Germann spends much of his time advising premedical students and others interested in pursuing careers in healthcare, research, or teaching. Germann's research interest is in the field of electrophysiology and ion channel regulation in epithelial cells. He is a member of the Biophysical Society, the American Association of University Professors, and the Texas Association of Advisors for Health Professions.

Cindy L. Stanfield earned both a B.S. and a Ph.D in physiology at the University of California at Davis. She was exposed to and became fascinated by neurophysiology research as an undergraduate student. As a graduate student, Stanfield taught several physiology laboratory courses, and developed an interest in teaching. She currently teaches three undergraduate human physiology courses at the University of South Alabama, and continues with her neurophysiology research on sensory modulation. One of Stanfield's goals as a teacher is to expose students not only to the existing knowledge base, but also to the discovery of new knowledge through research experiences in the hopes that students will be excited about science. Stanfield is a member of the Society for Neuroscience, the International Association for the Study of Pain, the American Pain Society, Sigma Xi, and the Golden Key Honor Society. She lives in Mobile, Alabama with her husband, Jim, and their cats and dog.

Preface

In teaching physiology over the years, we have observed that students at the undergraduate level differ widely in their motivations for studying physiology. After college, some go on to graduate study in science or health-related professions, others obtain work in research labs or industry, and still others pursue careers in non-scientific fields. For all of these students, regardless of their backgrounds or aspirations, the study of physiology should be an immensely rewarding and valuable experience, and we hope that this text will help to make it that way.

We have noticed that most students eagerly approach physiology at the start of the course, but some run into difficulty with the subject and lose their motivation. Why do some students find physiology so challenging? One reason, perhaps, is that students are trained to memorize facts, whereas physiology requires the understanding of concepts. Another possibility is that physiology requires a broad background in other sciences, such as biology, chemistry, and physics. Some students are unprepared in one or more of these subjects, which makes their study of physiology more difficult, and often leads to disappointment and frustration. This text aims to make it as easy as possible for students to learn physiology while at the same time giving them a solid treatment of the subject. The book accomplishes this by providing tools to facilitate understanding, providing a beautiful, pedagogically sound art program to support the text, stimulating student interest in human physiology, integrating the subject material, and providing reinforcement of the most important concepts.

CLEAR AND PRECISE WRITING STYLE

Reviewers of this text tell us the one thing they want their students to do is grasp the central concepts of physiology. Unfortunately, physiology can be heavy on distracting jargon and "information overload." In addition, complex concepts may require lengthy explanation, causing students to lose track of where they are and how a topic relates to physiology as a whole. To aid students in understanding, our text employs an informal writing style with minimal jargon. In addition, we focus on core concepts and key information that students will find useful or interesting. Much effort has been expended to eliminate extraneous information, which tends to distract rather than inform.

MATH, PHYSICS, AND CHEMISTRY TOOLBOXES

A quick glance through the text reveals the strategic use of **Toolboxes (Figure 1)**. We created these boxes to untangle coverage of math, physics, and chemistry from the body of the text, but still keep these topics available to students who might need review in these subjects. Toolboxes give students what they need to know when they need to know it. To ensure clarity, concepts are explained using real examples.

TOOLBOX
EQUILIBRIUM POTENTIALS AND THE NERNST EQUATION

With knowledge of an ion's charge and its intracellular and extracellular concentrations, we can find the equilibrium potential of any ion using the Nernst equation:

$$E = \frac{61}{z} \log \frac{C_o}{C_i}$$

where E is the equilibrium potential, z is the charge (valence) of the ion, and C_o and C_i are the concentrations outside and inside the cell, respectively. In this form, the equation gives the value of the equilibrium potential in millivolts and assumes that the temperature is at or near the normal body temperature of 37°C.

Using typical intracellular and extracellular concentrations, we can find the equilibrium potential for sodium, E_{Na}, by making the appropriate substitutions, as follows:

$$E_{Na} = \frac{61}{1} \log \frac{145 \text{ mM}}{15 \text{ mM}} = 60.1 \text{ mV} \approx 60 \text{ mV}$$

In similar fashion, we can find the equilibrium potential for potassium, E_K:

$$E_K = \frac{61}{1} \log \frac{4 \text{ mM}}{140 \text{ mM}} = -94.2 \text{ mV} \approx -94 \text{ mV}$$

Note that in both cases, the valence is +1, so that the sign of the equilibrium potential depends solely on the direction of the concentration gradient. Note also that E_K is larger in magnitude than E_{Na}. This makes sense because a larger concentration gradient requires a larger membrane potential to balance it, and the K^+ gradient is larger than the Na^+ gradient. If a concentration gradient is small, intracellular and extracellular ion concentrations will be more nearly equal. If the concentrations are identical, the ratio C_o/C_i equals 1, making the equilibrium potential equal to zero. (Recall that the log of 1 is zero.) As the concentrations become more and more dissimilar, the ratio becomes either much larger or much smaller than 1, making the log term larger and more positive, or larger and more negative, respectively.

FIGURE 1 *Toolboxes* pull math, chemistry, and physics out of the text, but keep the information available to students who may need review.

BEAUTIFUL, PEDAGOGICALLY SOUND ART PROGRAM

Instructors tell us that the illustrations and photos in a book can greatly impact students' chances of understanding and retaining material. The art that accompanies the text was crafted specifically for this text. We carefully thought through each figure to make the art tell the story of the text and help students understand physiological concepts. Beautifully drawn **anatomy diagrams (Figure 2)** succinctly illustrate the structures that students need to know to understand the physiology. Numerous **flow charts (Figure 3)** visually describe processes, neatly summarizing cause and effect relationships, and showing negative feedback loops where applicable. Rectangular boxes in these charts show events such as changes in physiological variables, and ovals represent structures such as cells, organs, or tissues, that either cause or respond to these events.

Arrows pointing up or down show increases or decreases. Selected flow charts appear on the book's web site at **www.physiologyplace.com**, where students can test their understanding of physiological processes by completing the flowcharts online. **Consistent, pedagogically sound use of color and shapes** means students understand the art intuitively **(Figure 4)**. For example, consistently colored arrows denote active and passive transport throughout the art, so students come to recognize them as they work through the text.

Where possible, the text art matches up with the art in the media offerings, to provide a seamless path from text to media for students.

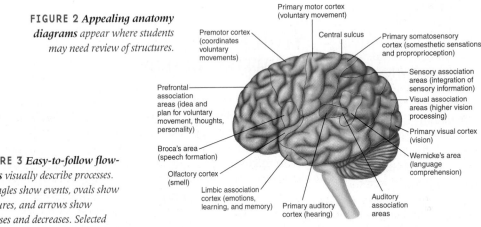

FIGURE 2 *Appealing anatomy diagrams appear where students may need review of structures.*

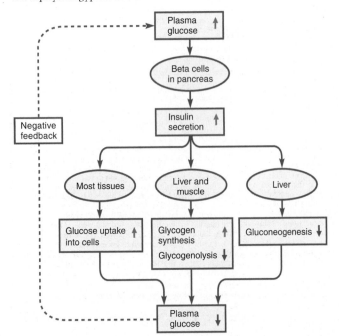

FIGURE 3 *Easy-to-follow flow-charts visually describe processes. Rectangles show events, ovals show structures, and arrows show increases and decreases. Selected charts appear as activities at www.physiologyplace.com.*

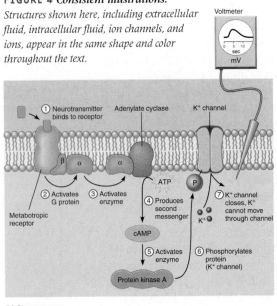

FIGURE 4 *Consistent illustrations.*
Structures shown here, including extracellular fluid, intracellular fluid, ion channels, and ions, appear in the same shape and color throughout the text.

(b) Slow response

STIMULATING STUDENT INTEREST

Instructors across the country have confirmed what we have observed in our classrooms: Students are stimulated to learn more when they understand how a subject relates to real life. To capitalize on this interest, we discuss topics of clinical relevance within the text where appropriate. In addition, **Discovery** boxes **(Figure 5)** discuss clinical applications of physiology (other than disease), new discoveries in physiology, physiological topics relevant to everyday life, or simply interesting facts about the system under discussion. **When It Goes Wrong** boxes **(Figure 6)** discuss certain disorders and treatments in more depth.

DISCOVERY
VAULTS AND CHEMOTHERAPY

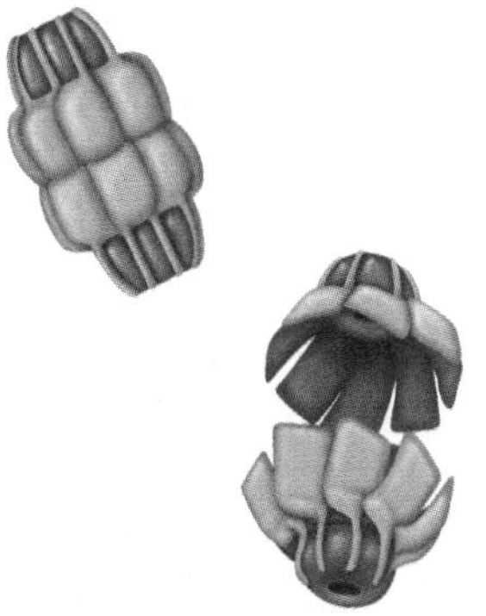

Vaults, shown in the accompanying figure, were discovered by researchers in the 1980s. Even though virtually every cell in the body has thousands of vaults, their function has still not been established. Current hypotheses suggest that vaults function in the transport of molecules, such as mRNA, from the nucleus to the cytoplasm, and that they may also have a role in protein synthesis.

In the late 1990s, researchers discovered th... ance t...

called *multidrug resistance*, have a greater concentration of vaults in their cells. However, this correlation does not prove that the increased number of vaults is the *cause* of the drug resistance; further studies are needed to determine if and how vaults influence drug resistance. In addition, a protein found in vaults, called *major vault protein* or MVP, is elevated in these patients. In the future, MVP may be used as a marker to identify patients who...

FIGURE 5 *Discovery boxes reveal relationships between concepts presented in different chapters, and discuss interesting topics and emerging discoveries in further detail.*

WHEN IT GOES WRONG
TREATING DEPRESSION

Most people feel a little blue now and then, and most people have had experiences that made them extremely sad, such as the death of a loved one. At such a time, people may describe themselves as being depressed. But clinical depression is much more than feeling sad, and generally it is not induced by an event.

Depression has many symptoms, including lack of energy, abnormal eating habits (too much or too little), and/or difficulty sleeping or sleeping too much. Often the person feels worthless and may be preoccupied with thoughts of suicide. A depressed person has difficulty functioning in society. The cause(s) of depression and its symptoms are not well understood, but depression is an illness associated with biochemical changes in the brain.

Much evidence suggests that depression is associated with deficiencies in the biogenic amines serotonin and norepinephrine. Indeed, one of the side effects associated with medications that decrease biogenic amines—such as the drug reserpine, used to treat high blood pressure—is depression. Therefore, pharmacological treatment strategies often seek to increase biogenic amine concentrations in the brain.

One class of antidepressants is *monoamine oxidase inhibitors*. Monoamine oxidase is the enzyme that breaks down biogenic amines, including norepinephrine and serotonin. Because these antidepressants inhibit their degradation, these neurotransmitters remain in the synaptic cleft for a longer period of time; the effect is similar to having increased the release of these neurotransmitters. Monoamine oxidase inhibitors, including phenelzine (Nardil™) and isocarboxazid (Marplan™), have been used successfully to treat many cases of clinical depression.

In many neural circuits, monoamine oxidase is located inside the presynaptic neuron, where it degrades neurotransmitters that have been actively transported back into the cell that released them (reuptake). In these cells, inhibition of monoamine oxidase is believed to increase the amount of neurotransmitter that is packaged into synaptic vesicles. The result is that more neurotransmitter is released in response to a given stimulus.

A relatively new class of commonly used antidepressants (first approved by the FDA in 1987) is *selective serotonin reuptake inhibitors (SSRIs)*. By decreasing the reuptake of serotonin into the cell that released it, SSRIs selectively increase the amount of serotonin present at the synaptic cleft. SSRIs are more specific than monoamine oxidase inhibitors because they only affect serotonergic synapses. SSRIs include fluoxetine (Prozac™) and paroxetine (Paxil™).

Another class of commonly used antidepressants is the *tricyclics*, named for the presence of three carbon rings. Although tricyclics are the oldest of the antidepressants (first used in the 1950s), their mechanisms of action are the least understood. Several hypotheses on the mechanisms have been proposed, most of which involve alterations of activity at adrenergic or serotonergic synapses. Tricyclics include imipramine (Tofranil™), amitriptyline (Elavil™), and desipramine (Norpramin™).

Does it surprise you that even though a drug has been around for about 50 years, the mechanism of its action is unknown? This is actually a fairly common occurrence, as drugs are often used with little knowledge of their mechanism of action. Even the therapeutic benefits of monoamine oxidase inhibitors and SSRIs for depression are not fully understood, for even though these drugs immediately decrease monoamine oxidase activity or serotonin reuptake, respectively, they do not affect the depression until they have been taken for several weeks. Much research remains to be done on the mechanisms of action of antidepressants and other medications.

FIGURE 6 *When It Goes Wrong boxes focus on disease conditions related to the topics of each chapter, and connect the topics to the real world.*

INTEGRATING SUBJECT MATERIAL

One of the challenges and rewards of studying physiology is understanding how concepts tie together. To help students understand the overall function of individual systems, we summarize topics in **Orientation Charts (Figure 7A)**. These charts show how the topics presented in a chapter, or series of chapters, fit together. They clarify how organ systems work as a whole to perform a function, such as delivering oxygen to and removing carbon dioxide from tissues. Where applicable, **partial orientation charts (Figure 7B)** summarize chunks of information in the chapters. The large orientation charts pull together all of the pieces from the smaller charts.

Single organ systems do not function alone; therefore, at the end of each system chapter, or series of chapters, **Systems Integration** charts **(Figure 8)** show how each organ system influences the others.

Finally, to present physiology as an integrated whole, the text concludes with a chapter on the body's physiological response to exercise. Exercise affects virtually all aspects of body function and is a topic of interest for many students making it an ideal "capstone" for the entire book. Chapter 1 concludes with a story that follows two fictional characters, Bill and Jane, as they run a marathon race from beginning to end. In each chapter, **Exercise Links (Figure 9)** discuss the relevance of specific topics to exercise in the context of the marathon story. Thus, the marathon story serves as a thread that wends its way through the entire book, guiding the students toward the story's culmination in the final chapter.

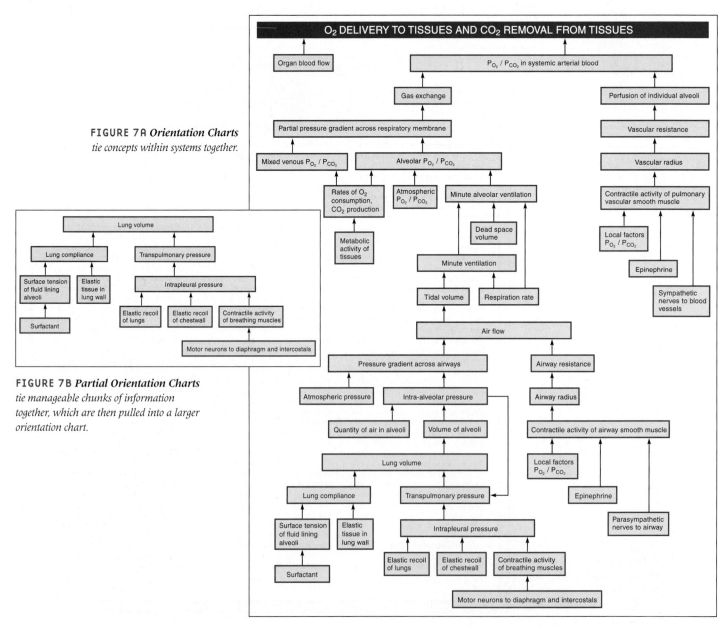

FIGURE 7A Orientation Charts
tie concepts within systems together.

FIGURE 7B Partial Orientation Charts
tie manageable chunks of information together, which are then pulled into a larger orientation chart.

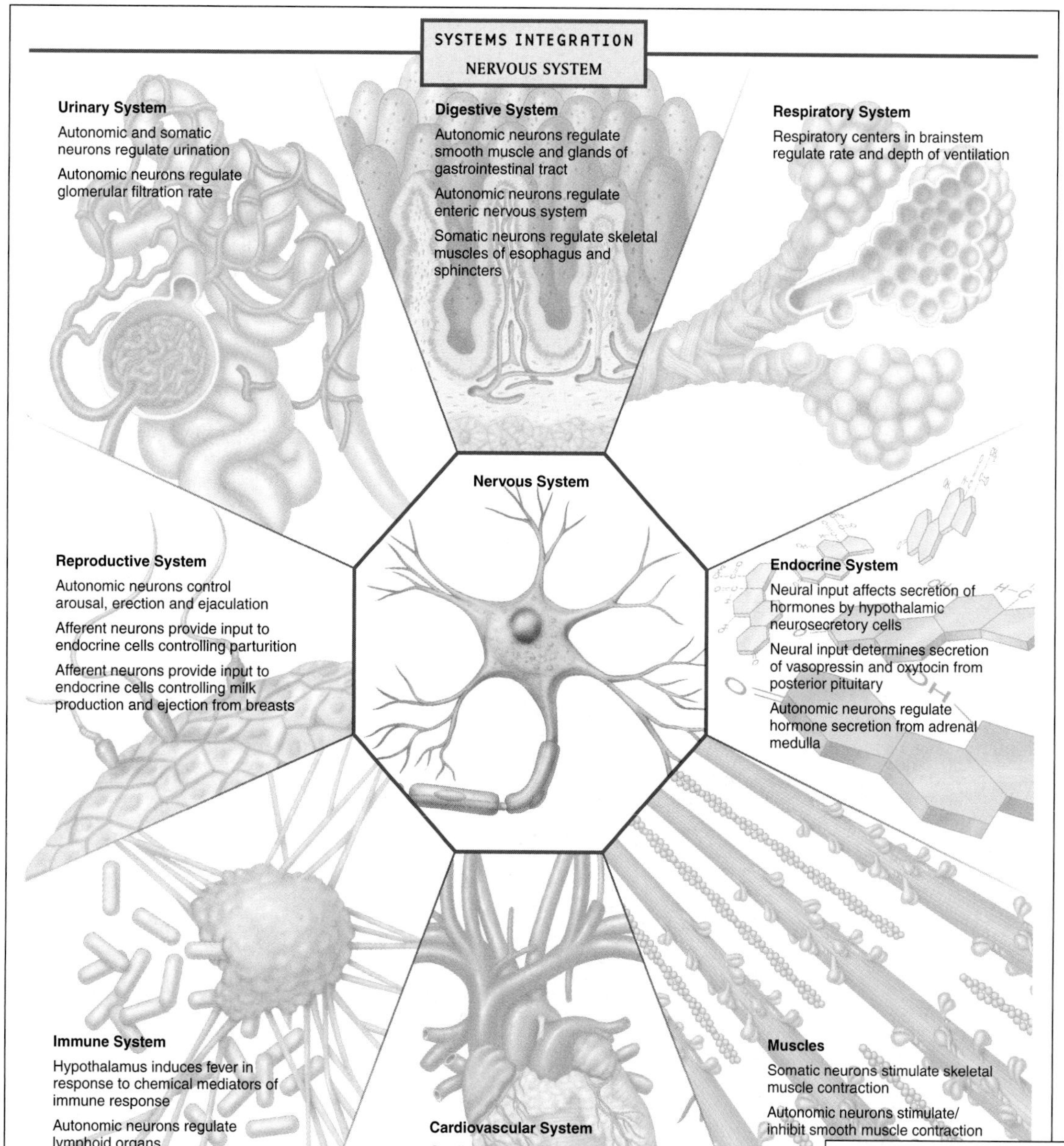

Urinary System

Autonomic and somatic neurons regulate urination

Autonomic neurons regulate glomerular filtration rate

Digestive System

Autonomic neurons regulate smooth muscle and glands of gastrointestinal tract

Autonomic neurons regulate enteric nervous system

Somatic neurons regulate skeletal muscles of esophagus and sphincters

Respiratory System

Respiratory centers in brainstem regulate rate and depth of ventilation

Nervous System

Reproductive System

Autonomic neurons control arousal, erection and ejaculation

Afferent neurons provide input to endocrine cells controlling parturition

Afferent neurons provide input to endocrine cells controlling milk production and ejection from breasts

Endocrine System

Neural input affects secretion of hormones by hypothalamic neurosecretory cells

Neural input determines secretion of vasopressin and oxytocin from posterior pituitary

Autonomic neurons regulate hormone secretion from adrenal medulla

Immune System

Hypothalamus induces fever in response to chemical mediators of immune response

Autonomic neurons regulate lymphoid organs

Stress stimulates CRH secretion (and thus, cortisol secretion) through neural pathways

Cardiovascular System

Cardiovascular centers in brainstem regulate heart beat and blood vessel radius through autonomic neurons

Muscles

Somatic neurons stimulate skeletal muscle contraction

Autonomic neurons stimulate/inhibit smooth muscle contraction

Autonom... decrease... cardiac m...

FIGURE 8 *Systems Integration charts at the end of each set of system chapters reinforce how each body system connects to the others.*

Exercise Link

Although a person does not consciously sense blood pH or the other individual inputs from visceral receptors, there may be times when we are able to sense the overall message being sent by these receptors. In the later stages of the marathon, Jane and especially Bill experienced two sensations that we call fatigue. One was a reduction in the responsiveness of their leg muscles because metabolic fuels were running low and waste products were building up in ways that hampered normal function. The other sensation is variously called "central" or "psychological" fatigue. Although physiologists do not fully understand this second type of fatigue, it is suspected that visceral receptor inputs, along with influences of body temperature and blood glucose levels, feed back on the motor cortex, reducing the drive to exercise. In addition, these inputs touch on the edge of our consciousness to produce those hard-to-describe feelings we associate with fatigue.

FIGURE 9 *Exercise Links tie each chapter to a marathon story introduced in Chapter 1 and discussed in detail in Chapter 23.*

MEDIA

Benjamin Cummings has long been a leader in technology with fun, innovative software programs that truly teach conceptual material. We developed the text with this in mind. You'll find that the media offerings described here will help students take their learning a step beyond what is in the book. Animations and tutorials bring the concepts in the book to life and help students with different learning styles. The following illustrations show, for example, how **glomerular filtration** is covered in multiple ways: in a figure in the text **(Figure 10)** in a case study at *The Physiology Place* **(Figure 11)**, in a tutorial in *InterActive Physiology®* **(Figure 12)** and in an experiment on *PhysioEx™* **(Figure 13)**. Quizzes allow students to test their knowledge as they work through the material online or on CD-ROM.

Glomerular filtration. (a) *Glomerular filtration pressure is the result of four Starling forces: (1) hydrostatic pressure in the glomerular capillaries (P_{GC}), (2) hydrostatic pressure in Bowman's capsule (P_{BC}), (3) oncotic pressure in the glomerular capillaries (π_{GC}), and (4) oncotic pressure in Bowman's capsule (π_{BC}). The net filtration pressure is 16 mm Hg.* **(b)** *The filtration fraction is the proportion of renal plasma that is filtered into Bowman's capsule. The normal filtration fraction is 20%.*

If proteins leaked out of the glomerular capillaries (which would decrease π_{GC} and increase π_{BC}), what would happen to the glomerular filtration pressure and to the glomerular filtration rate?

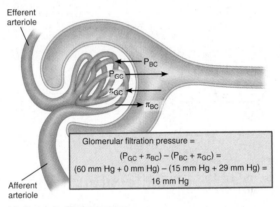

Efferent arteriole

P_{BC}
P_{GC}
π_{GC}
π_{BC}

Glomerular filtration pressure =
$(P_{GC} + \pi_{BC}) - (P_{BC} + \pi_{GC}) =$
$(60 \text{ mm Hg} + 0 \text{ mm Hg}) - (15 \text{ mm Hg} + 29 \text{ mm Hg}) =$
16 mm Hg

Afferent arteriole

(a) Glomerular filtration pressure

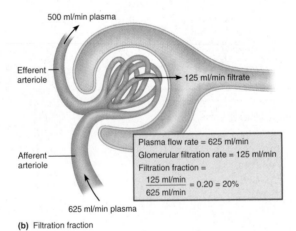

500 ml/min plasma

Efferent arteriole

125 ml/min filtrate

Plasma flow rate = 625 ml/min
Glomerular filtration rate = 125 ml/min
Filtration fraction =
$\dfrac{125 \text{ ml/min}}{625 \text{ ml/min}} = 0.20 = 20\%$

Afferent arteriole

625 ml/min plasma

(b) Filtration fraction

Both would increase.

FIGURE 10 *A figure from the discussion on glomerular filtration illustrates the concept in the text. Note the figure question, which challenges students to take their studying one step further.*

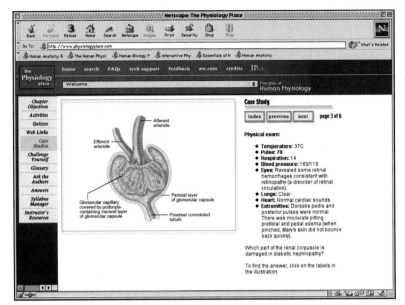

FIGURE 11 *With challenging activities, flow chart exercises, and case studies,* **The Physiology Place** *is the perfect site for online learning. In this case, students apply what they have learned about renal physiology to the real world. Visit* www.physiologyplace.com *for a demo. A subscription to* **The Physiology Place** *comes free with the purchase of a new copy of* **Principles of Human Physiology.**

FIGURE 12 **InterActive Physiology®** **(IPweb™)** *has already helped thousands of students understand and visualize complex physiological processes and thus succeed in this challenging course. Animations, tutorials, and quizzing help students master difficult concepts. IP brings the topic of glomerular filtration alive with animations that clarify processes. Here the animation shows autoregulation of GFR during different states of activity. References to IP are found in the book's chapter summaries. Every new copy of* **Principles of Human Physiology** *includes an IP Sampler CD-ROM. The full* **InterActive Physiology®** *program can be packaged with the text for an additional charge.*

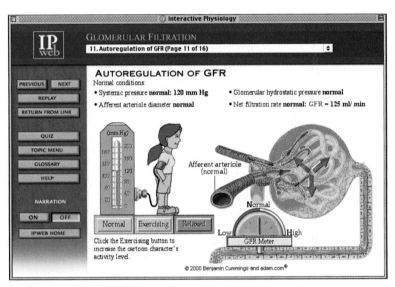

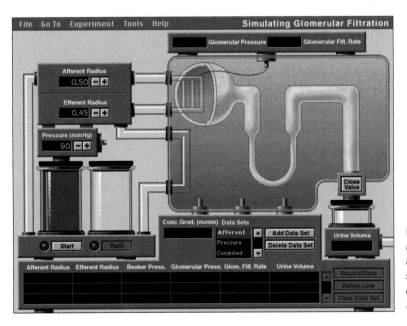

FIGURE 13 **PhysioEx™** *offers an alternative lab environment in which to master difficult concepts. Computerized simulations allow students to repeat lab experiments as needed to achieve a sound understanding of physiological processes.*

STUDY TOOLS REINFORCE
KEY CONCEPTS LEARNED

To learn effectively, students need to know they are on the right track as they are studying. Five study tools help students test their knowledge and go back to review material when necessary: Study Hints, Quick Tests, Chapter Summaries, End-of-Chapter Exercises, and Figure Questions. Each chapter opens with **Study Hints (Figure 14)**, a short list of topics discussed in previous chapters that students should know in order to get the most out of the chapter at hand. **Quick Tests (Figure 15)**, short sets of review questions at the ends of chapter sections, let students check their understanding of material before tackling new material. Each chapter closes with a detailed but concise **Chapter Summary (Figure 16)** and a complete bank of **End-of-Chapter Exercises (Figure 17)**, which include multiple-choice, objective, and essay questions. Multiple choice and objective questions help students to test their knowledge. Essay questions require students to synthesize material or to explain involved concepts. **Figure Questions (Figure 10)** appear with selected figures throughout the text. Long a hallmark of Benjamin Cummings texts, these questions ask students to take their studying one step further. Answers are provided at the bottom of the page.

STUDY HINTS

1. Lipid bilayer structure, p. 35

2. Membrane proteins, p. 36

3. ATP, p. 77

FIGURE 14 *Study Hints reference topics in previous chapters that students need to understand fully in order to get the most out of the current chapter.*

Quick Test 3.4

1. Is the oxidation of glucose catabolic or anabolic? Does it release energy or require it?

2. Where does the energy for ATP synthesis come from? When ATP is broken down, what is the released energy used for?

3. When energy from glucose oxidation is used to make ATP, only a certain fraction of the released energy is used for this purpose. What happens to the rest of the released energy?

FIGURE 15 *Quick Tests provide strategic checkpoints in the text to help students test their comprehension of the material and reread the previous section if necessary before moving forward. Answers to the Quick Tests are available online at www.physiologyplace.com.*

CHAPTER SUMMARY

The Autonomic Nervous System, p.312

There are two main branches of the efferent nervous system: the autonomic nervous system and the somatic nervous system. Table 10.4 (p. 312)compares the properties of the two branches of the autonomic nervous system with those of the somatic nervous system. The autonomic nervous system includes the parasympathetic and sympathetic nervous

adrenal medulla, stimulating the release of the hormone epinephrine. All preganglionic neurons contain the neurotransmitter acetylcholine. The parasympathetic postganglionic neurons also contain the neurotransmitter acetylcholine, but most sympathetic postganglionic neurons contain the neurotransmitter norepinephrine. The receptors for acetylcholine on postganglionic neurons are nicotinic

activity include the brainstem, hypothalamus, and limbic system.

> **IP** Nervous II, Synaptic Transmission, pages 3–11
>
> **IP** Nervous II, Ion Channels: pages 4–5, 7*
>
> **IP** Nervous I, The Membrane Potential, pages 3–6; 11

The Somatic Nervous System, p.323

The somatic division of the efferent

FIGURE 16 *Chapter Summaries pull together the main points of each chapter. IP (InterActive Physiology®) references point students to relevant pages in this award-winning software, where students can find lively animation and interactive quizzing to help them review topics further.*

EXERCISES

FIGURE 17 *End-of-Chapter Exercises allow students to self-test on the material covered in the chapter.*

Multiple-Choice Questions

1. Suppose that the electrochemical force for anion X (X^-) acts to move the anion out of the cell. If a neurotransmitter binding to its receptor opened channels for X^- on the postsynaptic cell, then the response would
   a) be an EPSP.
   b) be an IPSP.
   c) be stabilization of the membrane.
   d) not occur.

2. Suppose that all the calcium could be

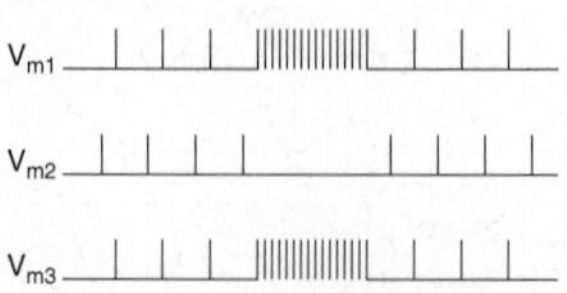

V_{m1} ————

V_{m2} ————

V_{m3} ————

3. From this voltage tracing, one can conclude that
   a) the stimulation depolarizes the cell membrane.

8. A fast EPSP is produced by
   a) the opening of sodium-selective channels.
   b) the opening of potassium-selective channels.
   c) the opening of channels selective for both sodium and potassium.
   d) the opening of calcium-selective channels.

9. The enzyme that catalyzes the synthesis of acetylcholine is
   a) adenylate cyclase.

Acknowledgments

Writing a textbook is a huge undertaking—rewarding, challenging, at times frustrating, and ultimately indescribable; it is something that one must experience first-hand in order to appreciate. Throughout the creation of this book, we benefited from the expertise, hard work, and encouragement of many editors, reviewers, artists, production people, instructors, and students who guided us along the way. We extend to them our deepest and heartfelt thanks, for without them this book would not exist.

Special thanks go to Mary Jane Niles (University of San Francisco) and Joseph G. Cannon (Medical College of Georgia). Mary Jane wrote Chapter 22, The Immune System, and masterfully made this complex subject easily intelligible and a joy to read about. Joe wrote the marathon story at the end of Chapter 1, the Exercise Link boxes that appear in each chapter, and Chapter 23, The Whole Body: Integrated Physiological Responses to Exercise. The innovative and entertaining marathon story and links provide a thread that runs through the entire book, and integrate many concepts into a coherent whole. We think Mary Jane and Joe's work has improved the book immeasurably.

If the experience of writing has taught us anything, it is that the creation of a book is a team effort. We were fortunate to have a winning team of publishing professionals at Benjamin Cummings to help us. To begin, we want to acknowledge our publisher, Daryl Fox, who first signed one of us (W.J.G.) to write the book and then stood behind the project the entire way. Two remarkable editors helped us transform our vision into the final product: Jane Tufts and Laura Southworth. With great intelligence, insight, warmth, and humor, Jane, our Text Development Editor, proved not only to be an invaluable colleague (and sometimes a daunting taskmaster as well!), but also a valued friend. From the time we began work on this project, Jane never lost sight of the book's ultimate aim, which is to educate students. Laura Southworth, our Art Development Editor, took our crude pencil sketches and transformed them into beautiful, concise illustrations that are effective teaching tools. She also ensured the art was consistent throughout the text and integrated the art style with that of the *InterActive Physiology®* CDs. Her quick-wittedness and sly sense of humor made her a pleasure to work with and some hard days of drudgery a lot more bearable. Our Project Editor, Claire Brassert, enlisted one of us (C.L.S.) as an author, managed the book project through development and into production, and hired the supplements authors. Claire, who has been with us from the beginning, steered the project through some rough spots, and we appreciate her stubborn "stick-to-it-iveness." Mary Ann Murray, Associate Editor, and Marie Beaugureau, Publishing Assistant, both cheerfully and tirelessly prepped manuscripts and coordinated the supplements for the book. Erin Joyce, Online Associate Project Editor, managed the development and production of the book's website, The Physiology Place, and the IP Sampler CD-ROM that accompanies the text. Deirdre McGill and Cheryl Cechvala worked as market development managers on the book and headed up the class testing program for the book, which yielded valuable feedback from students and instructors across the country. Janet Vail, Production Editor, moved the manuscript through production, kept track of countless details, and made sure every last piece of the book came together. Kelly Murphy, Art and Photo Coordinator, shepherded the art through production and carefully checked each piece. Photo Researcher, Diane Austin, secured photos for the book. The staff at Precision Graphics created the beautiful artwork for the book. In particular, we want to thank the lead artist, Rolf Swanson, whose talent and tireless dedication to this project made the art program possible. Alan Titche, Copyeditor, tirelessly and mercilessly scrutinized the manuscript. Although writing will always be hard work, the people at Benjamin Cummings made the job easier.

We wish to thank—and congratulate—our "comrades in arms," the following people who helped us by writing the supplemental materials: Christine Iltis, who authored the Instructor's Guide; Dave Kurjiaka, who wrote the Test Bank; and Cory Etchberger, Margaret Nordlie, and Pat Munn, who wrote the Study Guide. We've been extremely fortunate to have had the help of many insightful reviewers, instructors, and students whose names appear on the following pages and who shared ideas with us in focus group meetings or who read parts of the text. You can rest assured that we value your comments and considered each of them with the utmost seriousness. Thank you.

Cindy Stanfield thanks her husband Jim, whose support and patience made the project possible for her. She would also like to thank several of her colleagues at the University of South Alabama who supported her endeavors and provided valuable feedback on the development of the book. Stephen Itaya, Julio Turrens, Vicki Barrett, Michael Spector, and Jim Stubbs all reviewed sections of the book and answered questions related to their areas of expertise. Clara Massey was kind enough to provide ECG traces, and Ian Campbell (University of California at Davis) provided EEG records. The efforts of all these individuals, and others, helped to make the book stronger.

William Germann extends gratitude to the University of Dallas for giving him a year-long sabbatical and for providing financial support so that he could free up time to write during the first formative years of this book. He also gives many thanks to his colleagues in the Department of Biology, who supported him over the years and who also took up some of the slack during his absence: Marcy Brown-Marsden, Frank Doe, David Pope, and Cathy Shupe. Most of all, he thanks Joe Cannon, whom he has known since the days they were both graduate students in the Department of Physiology at the University of Michigan, for superlative contributions to the book. Nothing he can say could adequately express his gratitude to and admiration for someone who has turned out to be such a great physiologist.

In closing, we hope that students will find this text truly helps in studying physiology, and that it reflects our enthusiasm for the subject and our dedication to helping students learn. We welcome your comments and suggestions for future editions.

William J Germann   *Cindy Lou Stanfield*

William Germann and Cindy Stanfield
Applied Sciences
Benjamin Cummings
1301 Sansome St.
San Francisco, CA 94111

REVIEWERS

Thomas Adams
Michigan State University

Brian Adrian
Grand Valley State University

David Anderson
Pennsylvania State University

Patricia D. Ashby
Eastern New Mexico University

Albert Baccari
Montclair Community College

Ralph Barclay
Wayne State University

Debra Barnes
Contra Costa College

Janis Beaird
University of West Alabama

Cynthia Beck
George Mason University

Dale Benos
University of Alabama at Birmingham

Joseph Berger
Springfield College

David G. Bernard
University of Texas, Arlington

Ron Beumer
Armstrong Atlantic State University

Eric Bittman
University of Massachusetts, Amherst

Sunny Boyd
University of Notre Dame

Mark Bracken
Utah Valley State College

Edward Brandt
Shenandoah University

Virginia Brooks
Oregon Health Sciences University

William Brothers
San Diego Mesa College

Nishi Bryska
University of North Carolina, Charlotte

Michael Canute
Hawaii Pacific University

John Capeheart
University of Houston

Andrew N. Clancy
Georgia State University

William Cliff
Niagara University

David Cochrane
Tufts University

Frank Corotto
North Georgia College and State University

Charles Costa
Eastern Illinois University

Joseph Crivello
University of Connecticut

Leon Cuervo
Florida International University

Dan Deaver
Advanced Inhalation Research

Russell DiFiore
Pasadena City College

Leon Dorosz
San Jose State University

Gary Dudley
University of Georgia

Jean Pierre Dujardin
Ohio State University

John Dziak
Community College of Allegheny County

Steven Eiger
Montana State University

Yasir El-Sharif
City University of New York, College of Staten Island

Cory Etchberger
Longview Community College

Kenneth Etzel
University of Pittsburgh

Bridget Falkenstein
Sierra College

Richard Falvo
Southern Illinois University

Ralph E. Ferges
Palomar College

Philip J. Fernandez
Grand Canyon University

Milton Fingerman
Tulane University

Kathleen A. Fitzpatrick
Merrimack College

George Fortes
University of California, San Diego

Frank Frisch
Chapman University

Irja Galvan
Western Oregon University

Greg Garman
Centralia College

Nicholas R. Geist
Sonoma State University

Debabrata Ghosh
Texas Southern University

Alice Gibb
Northern Arizona University

Roger Gilchrist
University of Alabama, Birmingham

Bruce Gladden
Auburn University

Jack Goldberg
University of California, Davis

Todd Gordon
Kansas City Kansas Community College

Sheldon R. Gordon
Oakland University

Tamara Greco
Eastern Michigan University

Jean Grassman
City University of New York, Brooklyn College

Michael Griffin
Angelo State University

Charles J. Grossman
Xavier University

Michael Guinan
University of California, Davis

Douglas C. Gula
Miami University

David L. Hammerman
Long Island University

Dennis C. Haney
Furman University

John P. Harley
Eastern Kentucky University

Janet L. Haynes
Long Island University

Laura Hebert
Angelina College

James Herman
Texas A & M University

Tina Hines
University of Pittsburgh

James Hoffman
Diablo Valley College

Charles W. Holliday
Lafayette College

Michael J. Holmes
University of Pittsburgh

Sandra Hsu
University of San Francisco

Mark Hubley
Washington College

Kerry Hull
Bishops University

Christine Iltis
Salt Lake Community College

William Jackson
Western Michigan University

Paul Jarrell
Pasadena City College

Najma Javed
Ball State University

J. Kelly Johnson
University of Kansas

Penny Knoblich
Minnesota State University, Mankato

Nuran Kumbaraci
Steven's Technical Institute

David Kurjiaka
Ohio University

David C. Lennartz
Mount St. Mary's College

John Lepri
University of North Carolina, Greensboro

James Long
Boise State University

John Lovell
Kent State University

Jennifer Lundmark
California State University, Sacramento

Ronald Lynch
University of Arizona

Duncan MacDougall
McMaster University

Jennifer Marcinkiewicz
Kent State University

Christel Marschall
Lansing Community College

Theresa Martin
College of San Mateo

Todd McBride
California State University, Bakersfield

Jaqueline S. McLaughlin
Pennsylvania State University, Berks/Lehigh Valley

Katherine Mechlin
Wright State University

Robert Meckler
City College of San Francisco

Essie Meisami
University of Illinois, Urbana-Champaign

Jon S. Miller
Northern Illinois University

John W. Mills
Clarkson University

Jeanne Mitchell
Truman State University

Stacia Moffett
Washington State University

Ann Motekaitis
California State University, Sacramento

David P. Muehleisen
University of Utah

Donald Mulcare
University of Massachusetts, Dartmouth

Patricia Munn
Longview Community College

Ruth K. Nash
College of Marin

Linda R. Nichols
Santa Fe Community College

Gemma Niermann
University of California, Berkeley

Colleen Nolan
St. Mary's University

Margaret Nordlie
University of Mary

Howard J. Normile
Wayne State University

Brian J. Norris
California State University, San Marcos

Nathan Norris
West Valley College

Bruce O'Gara
Humboldt State University

Angela Ottava
Michigan State University

T. Lon Owen
Northern Arizona University

M.J. Parkes
University of Birmingham, UK

Shawn Pearcy
Wayne State College

John Perry
Oklahoma City Community College

Dickson Phiri
Mesa College

John Pigage
University of Colorado

Steven Price
Virginia Commonwealth University

Richard Puzdrowski
University of Houston, Clear Lake

David Quadagno
Florida State University

Ralph Reiner
College of the Redwoods

David Rivers
Loyola College

Mary Anne Rokitka
State University of New York, Buffalo

Barry Rothman
San Francisco State University

Jane Rudolph
Salt Lake Community College

Elizabeth M. Rust
University of Michigan

Allen Sanborn
Barry University

Brian Shmaefsky
Kingwood College

Lynne Schneider
University of Vermont

Thomas Secrest
Austin Community College

Arnold Sillman
University of California, Davis

C. Wayne Simpson
University of Missouri, Kansas City

James W. Small
Rollins College

Davis Smith
San Antonio College

George Snow
Mount St. Mary's College

Michael Stamper
College of West Virginia

Philip Stephens
Villanova University

Mary Lynne Stephanou
Santa Monica College

James Stockand
University of Texas, San Antonio

Kevin Strang
University of Wisconsin

James A. Strauss
Pennsylvania State University

Alan F. Sved
University of Pittsburgh

Karen Swearingen
Mills College

Steven Swoap
Williams College

Sherry Tamone
Sonoma State University

Deborah Taylor
Kansas City Kansas Community College

Carl Thurman
University of Northern Iowa

H. Ti Tien
Michigan State University

Paola S. Timiras
University of California, Berkeley

Tom Tomasi
Southwest Missouri State University

Mark Tomita
City University of New York, Brooklyn College

Joseph Tupper
Syracuse University

David R. Wade
Southern Illinois University

Christine K. Wade
University of Wyoming

Tracy Wagner
Washburn University

Bruce Wainman
McMaster University

Tim Wakefield
John Brown University

Benjamin Walcott
State University of New York, Stony Brook

Jin-Hui Wang
University of Kansas

Stephen Warburton
New Mexico State University

David Washington
University of Central Florida

James Watrous
St. Joseph's University

Ralph E. Werner
The Richard Stockton College of New Jersey

Eric P. Widmaier
Boston University

Nancy Williams
Pennsylvania State University, Fayette

F. Ray Wilson II
Baylor University

Darla Wise
Concord College

Nicholas J. D. Wright
Eastern New Mexico University

Marianna J. Zamlauski-Tucker
Ball State University

FOCUS GROUPS

Faculty

Allen Bedford
Bryn Athyn College

Mary Bober
Santa Monica College

Denise Cavener
Santa Monica College

Theresa Chen
College of Notre Dame

Richard Connett
Monroe Community College

Paul Emerick
Monroe Community College

Ana Escadon
L.A. Harbor College

Michael Guinan
University of California, Davis

David Hammerman
Long Island University

Robert Hyde
San Jose State University

William Jackson
Western Michigan University

Vasily Kolchenko
Pace University

David Lennartz
Mount St. Mary's College

Eileen Lynd-Balta
St. John Fisher College

Theresa Martin
College of San Mateo

Robert Meckler
City College of San Francisco

James Murphy
Monroe Community College

Ruth Nash
College of Marin

Nathan Norris
West Valley College

Dougald Scott
Cabrillo College

Rachel Simons
Monroe Community College

Gary Skuse
Rochester Institute of Technology

Michael Sneary
San Jose State University

Edmund Tong
Wheaton College

Carlene Tonini
College of San Mateo

Mary Jo Witz
Monroe Community College

Student

Bryn Athyn College

Francis Darkwah
Nicola Echols
Ruth Homber
Shondra Perry
Shada Sullivan

College of San Mateo

Deborah Bilafer
Nidia Medina
Robin Garoutte
Colleen Mahoney
Manolo Cacella
Angela Stratton
Maribeth King
Nina Breiz
Jim Tupas
Noel Segali
Helen Murphy
Steven Fulton

Texas A&M University

Mallika Anand
Ben Alexander
Katie Jackson
Nick Campbell
Andrew Svetlick

CLASS TESTERS

Magdalene Abraham
Mississippi Delta Community College

Ateegh Al-Arabi
Johnson County Community College

William Andresen
William Rainey Harper College

Gwen Bachman
University of Nebraska

Floyd Banks
Chicago State University

Cynthia Beck
George Mason University

Charles Biles
East Central University

Edward Bilsky
University of Northern Colorado

William Boyko
Sinclaire Community College

Mark Bracken
Utah Valley State College

Andrew Browe
Indiana University of Pennsylvania

Elaine S. Brubacher
University of Redlands

Maureen Burton
University of South Dakota

Michael Canute
Hawaii Pacific University

John Capeheart
University of Houston Downtown

Chun-fan Chen
Florida International University

Chris Chistopher
Santa Rosa Junior College

Craig Clifford
Northeastern State University

Jonathan Day
California State University, Chico

Nick Despo
Thiel College

Kathryn Dickson
California State University, Fullerton

Christopher Dunbar
Brooklyn College

Cory Etchberger
Longview Community College

Robert Farrell
Pennsylvania State University

Michael Ferrari
University of Arkansas

Charles Foster
State University of New York, Potsdam

Rhonda Gamble
Mineral Area College

Jorge Golowasch
Rutgers University

Tamara Greco
Eastern Michigan University

Joseph Gregorek
Gannon University

Michael Griffin
Angelo State University

John Harley
Eastern Kentucky University

Elizabeth Hayes
University of Wisconsin, Fond du Lac

Lisa Hays
Evergreen Valley College

James Herman
Texas A & M University

Stephan Hourdez
Pennsylvania State University

Najma Javed
Ball State University

Clyde E. Johnson
Southern University at Baton Rouge

Robert David Jones
Adelphi University

Henry Kayongo-Male
South Dakota State University

Peter King
Francis Marion University

Gopal Krishna
Moberly Area Community College

Jill Kruper
Murray State University

Nuran M. Kumbaraci
Stevens Institute of Technology

Gavin Lawson
Bridgewater College

Curtis Lee
Dallas Baptist University

John Lepri
University of North Carolina, Greensboro

David Mallory
Marshall University

Theresa Martin
College of San Mateo

Bill Mathena
Kaskaskia College

Vikki McCleary
University of North Dakota School of Medicine

Kerry McDonald
University of Missouri

Jackie McLaughlin
Pennsylvania State University, Berks-Lehigh Valley College

Robert Meckler
City College of San Francisco

Robb Moats
University of Florida, Jacksonville

Suzanne Moshier
University of Nebraska, Omaha

Donald J. Mulcare
University of Massachusetts, Dartmouth

Greg Mullen
University of Northern Colorado

Patricia Munn
Longview Community College

Mahalashmi Nagarajan
Wilberforce University

Colleen Nolan
St. Mary's University Texas

Marie O'Rourke
College of St. Mary

Robert Okazaki
Weber State University

Gibson Oriji
William Paterson University

Sujata Patel
University of Chicago

Shawn Pearcy
Wayne State College

Dave Pendergast
State University of New York, Buffalo—South Campus

Judy Penn
Shoreline Community College

Lisa Perrino
University of Maryland

Mary Perry
Allan Hancock College

Larry A. Reichard
Maple Woods Community College

Tricia A. Reichert
Colby Community College

Carmen Rexach-Zellhoefer
Merced College

Barbara Roller
Florida International University

Jane Rudolph
Salt Lake Community College

Arleen Sawitzke
Salt Lake Community College

Brian Shmaefsky
Kingwood College

Linda Smith
Stockton State College

John R. Steele
Ivy Tech State College

Mary Lynne Stephanou
Santa Monica College

Cathryn Stevens
Louisiana State College

Bonnie Tarricone
Ivy Tech State College

Phillip Taylor
Glenville State College

Carl Thurman
University of Northern Iowa

James Toepfer
Youngstown State University

Edmund Tong
Wheaton College

William Trotter
Des Moines Area Community College

Dana Vaughan
University of Wisconsin, Oshkosh

Warren Wade
Lousiana College

Tracy Wagner
Washburn University

Curt Walker
Dixie College

Wade Warren
Louisiana College

David Washington
University of Central Florida

John Water
Pennsylvania State University

Cheryl Watson
Central Connecticut State University

Flora Watson
South Carolina State University, Stanislaus

Allison Wilson
Benedict University

Hing-Sing Yu
University of Texas

Henry H. Ziller
Southeastern Louisiana State University

ART BOARD MEMBERS

Andrew Clancy
Georgia State University

Cathy Stevens
Louisiana State University

David Wilson
Miami University

Balwant Khatra
California State University, Long Beach

Peter Bannerman
Children's Hospital of Philadelphia

Robert Mitchell
Pennsylvania State University

Stephen Dodd
University of Florida

Thomas Adams
Michigan State University

John McReynolds
University of Michigan

Judith Gibber
Columbia University

Robert Moats
University of North Florida

Scott Hadley
Grand Valley State University

Scott Moye-Rowley
University of Iowa

Brief Contents

1 Introduction to Physiology 1

2 The Cell: Structure and Function 21

3 Cell Metabolism 62

4 Cell Membrane Transport 100

5 Chemical Messengers and the Endocrine System 131

6 Nerve Cells and Electrical Signaling 170

7 Synaptic Transmission and Neural Integration 201

8 The Nervous System: Central Nervous System 223

9 The Nervous System: Sensory Systems 260

10 The Nervous System: Autonomic and Motor Systems 311

11 Muscle Physiology 333

12 The Cardiovascular System: Cardiac Function 369

13 The Cardiovascular System: Blood, Blood Flow, and Blood Pressure 403

14 The Cardiovascular System: Regulation of Function 442

15 The Respiratory System: Breathing Mechanics 469

16 The Respiratory System: Gas Exchange and Regulation of Breathing 495

17 The Urinary System: Renal Function 532

18 The Urinary System: Fluid and Electrolyte Balance 564

19 The Digestive System 601

20 The Endocrine System: Regulation of Energy Metabolism and Growth 639

21 The Reproductive System 670

22 The Immune System 708

23 The Whole Body: Integrated Physiological Responses to Exercise 741

Answers to Odd-Numbered Multiple-Choice and Objective Questions 763

Credits 769

Glossary 771

Index 789

Contents

1 Introduction to Physiology   1

Organization of the Body   2
  Cells, Tissues, Organs, and Organ Systems • The Overall Body Plan: A Simplified View

Homeostasis: A Central Organizing Principle of Physiology   8

When It Goes Wrong: Heat Exhaustion and Heat Stroke   9

  Negative Feedback Control in Homeostasis • Homeostasis in Action: Thermoregulation

Challenging Homeostasis: A Study in Organ Systems Integration During Exercise   15
  The Hierarchy of Resource Allocation in Response to Activity • Bill and Jane's Marathons

Chapter Summary   19
Exercises   20

2 The Cell: Structure and Function   21

Biomolecules   22
  Carbohydrates • Lipids • Proteins • Nucleotides and Nucleic Acids

Cell Structure   34
  Structure of the Plasma Membrane • Structure of the Nucleus • Contents of the Cytosol • Structure of Membranous Organelles • Structure of Nonmembranous Organelles

Discovery: Vaults and Chemotherapy   42

Cell-to-Cell Adhesions   43
  Tight Junctions • Desmosomes • Gap Junctions

General Cell Functions   46
  Metabolism • Cellular Transport • Intercellular Communication

Protein Synthesis   48
  The Role of the Genetic Code • Transcription • Translation • Destination of Proteins • Post-Translational Processing and Packaging of Proteins • Regulation of Protein Synthesis • Protein Degradation

Cell Division   56
  Replication of DNA • The Cell Cycle

When It Goes Wrong: Cancer   59

Chapter Summary   59
Exercises   60

3 Cell Metabolism   62

Types of Metabolic Reactions   63
  Hydrolysis and Condensation Reactions • Phosphorylation and Dephosphorylation Reactions • Oxidation-Reduction Reactions

Metabolic Reactions and Energy   65
  Energy-Releasing and Energy-Requiring Reactions • Energy Changes in Reactions • Activation Energy

Reaction Rates   70
  Factors Affecting the Rates of Chemical Reactions • The Role of Enzymes in Chemical Reactions

Glucose Oxidation: The Central Reaction of Energy Metabolism   77
  ATP: The Medium of Energy Exchange • ATP in Action

Stages of Glucose Oxidation: Glycolysis, the Krebs Cycle, and Oxidative Phosphorylation   81

Glycolysis • The Krebs Cycle • Oxidative Phosphorylation • Summary of Glucose Oxidation • Conversion of Pyruvate to Lactic Acid

When It Goes Wrong: Lactose Intolerance   90

Energy Storage and Use: Metabolism of Carbohydrates, Fats, and Proteins   91

Glycogen Metabolism • Gluconeogenesis: Formation of New Glucose • Fat Metabolism • Protein Metabolism

Chapter Summary   96
Exercises   97

4   Cell Membrane Transport   100

Factors Affecting the Direction of Transport   101

Passive Transport versus Active Transport • Driving Forces Acting on Molecules

Factors Affecting the Rate of Transport   108

Factors Affecting Rates of Passive Transport • Factors Affecting Rates of Active Transport

Simple Diffusion: Passive Transport Through the Lipid Bilayer   111

The Basis for Simple Diffusion • Factors Affecting Membrane Permeability in Simple Diffusion

Facilitated Diffusion: Passive Transport Through Membrane Proteins   113

Transport Proteins in Facilitated Diffusion • Factors Affecting Membrane Permeability in Facilitated Diffusion

Active Transport   115

Primary Active Transport • Secondary Active Transport • Pumps and Leaks: How Active and Passive Transport Mechanisms Coexist in Cells

Osmosis: Passive Transport of Water Across Membranes   119

Discovery: Determining the Osmotic Pressure of a Solution   120

Osmolarity • Osmotic Pressure • Tonicity

Epithelial Transport: Movement of Molecules Across Two Membranes   124

Epithelial Structure • Epithelial Solute Transport • Epithelial Water Transport

When It Goes Wrong: Cystic Fibrosis   127

Chapter Summary   128
Exercises   129

5   Chemical Messengers and the Endocrine System   131

Intercellular Communication   132

Direct Communication Through Gap Junctions • Indirect Communication via Chemical Messengers

Chemical Messengers   133

Functional Classification of Chemical Messengers • Chemical Classification of Messengers

Discovery: Antihistamines   136

Synthesis and Release of Chemical Messengers • Transport of Messengers

Signal Transduction Mechanisms   140

Properties of Receptors • Signal Transduction Mechanisms for Intracellular Receptor-Mediated Responses • Signal Transduction Mechanisms for Membrane-Bound Receptor-Mediated Responses

Long-Distance Communication via the Nervous and Endocrine Systems   151

When It Goes Wrong: Bacterial Infections that Attack G Proteins   152

Endocrine Glands   152

Primary Endocrine Organs • Secondary Endocrine Organs

Hormone Actions at the Target Cell   161

Control of Hormone Levels in Blood • Hormone Interactions

Chapter Summary   167
Exercises   168

6   Nerve Cells and Electrical Signaling   170

Overview of the Nervous System   171
Cells of the Nervous System   172

Discovery: Neurogenesis   173

Neurons • Glial Cells

Electrical Signals in Neurons   177

Resting Membrane Potential • Gated Channels • Electrical Signaling Through Changes in Membrane Potential • Graded Potentials

When It Goes Wrong: Neurotoxins   185

Neural Integration

Action Potentials   187

Ionic Basis of an Action Potential • The Role of Voltage-Gated Ion Channels in Action Potentials • Refractory Periods • Propagation of Action Potentials

Chapter 6 continues

Chapter 6 continued

The Basis of Neural Stability   197
Chapter Summary   198
Exercises   199

7 Synaptic Transmission and Neural Integration   201

Types of Synapses   202
Chemical Synapses   202

*Functional Anatomy of Chemical Synapses • Signal
Transduction Mechanisms at Chemical Synapses • Excitatory
Synapses*

Discovery: Synaptic Vesicles   206

Inhibitory Synapses

Neural Integration   210

Summation • Frequency Coding

Presynaptic Modulation   212

Presynaptic Facilitation • Presynaptic Inhibition

Neurotransmitters: Structure, Synthesis, and
Degradation   214

*Acetylcholine • Biogenic Amines • Amino Acid
Neurotransmitters • Neuroactive Peptides*

When It Goes Wrong: Treating Depression   218

Other Neurotransmitters

Chapter Summary   220
Exercises   221

8 The Nervous System: Central Nervous System   223

General Anatomy of the Central Nervous System   224

*Protective Structures • Blood Supply to the Central Nervous
System • The Blood-Brain Barrier • Gray Matter and White
Matter*

When It Goes Wrong: Stroke   230

The Spinal Cord   231
The Brain   237

Cerebral Cortex • Subcortical Nuclei

Discovery: The Story of Phineas Gage   243

Integrated CNS Function: Reflexes   244

Stretch Reflex • Withdrawal and Crossed-Extensor Reflexes

Integrated CNS Function: Voluntary Motor Control   247

*Neural Components for Smooth Voluntary Movements
• Cortical Control of Voluntary Movement • The Control of*

*Posture by the Brainstem • The Role of the Cerebellum in
Motor Coordination • The Basal Nuclei in Motor Control*

Integrated CNS Function: Language   251
Integrated CNS Function: Sleep   251

*Functions of Sleep • Sleep-Wake Cycles • Electrical Activity
During Wakefulness and Sleep*

Integrated CNS Function: Emotions and Motivation   254
Integrated CNS Function: Learning and Memory   255

Learning • Memory • Plasticity in the Nervous System

Chapter Summary   258
Exercises   259

9 The Nervous System: Sensory Systems   260

General Principles of Sensory Physiology   261

Receptor Physiology • Sensory Pathways • Sensory Coding

The Somatosensory System   270

*Somatosensory Receptors • The Somatosensory Cortex
• Somatosensory Pathways • Pain Perception*

When It Goes Wrong: Insensitivity to Pain   275

Vision   277

*Anatomy of the Eye • The Nature and Behavior of Light Waves
• Accommodation • Clinical Defects in Vision • Regulating the
Amount of Light Entering the Eye • The Retina • Photo-
transduction • Adaptation to Light and Dark • Neural
Processing in the Retina*

Discovery: Color Vision   289

*Neural Pathways for Vision • Parallel Processing in the Visual
System • Depth Perception*

The Ear and Hearing   291

*Anatomy of the Ear • The Nature of Sound Waves • Sound
Amplification in the Middle Ear • Signal Transduction for
Sound • Neural Pathways for Sound*

The Ear and Equilibrium   299

*Anatomy of the Vestibular Apparatus • The Semicircular Canals
and the Transduction of Rotation • The Utricle and Saccule and
the Transduction of Linear Acceleration • Neural Pathways for
Equilibrium*

Taste   303

*Anatomy of Taste Buds • Signal Transduction in Taste • Neural
Pathway for Taste*

Olfaction   306

*Anatomy of the Olfactory System • Olfactory Signal
Transduction • Neural Pathway for Olfaction*

Chapter Summary   308
Exercises   309

The Nervous System: Autonomic and Motor Systems 311

The Autonomic Nervous System 312
Dual Innervation in the Autonomic Nervous System • Anatomy of the Autonomic Nervous System • Autonomic Neurotransmitters and Receptors • Autonomic Neuroeffector Junctions • Regulation of Autonomic Function

The Somatic Nervous System 323
Anatomy of the Somatic Nervous System • The Neuromuscular Junction

When It Goes Wrong: Myasthenia Gravis 328

Discovery: Curare 329

Chapter Summary 331
Exercises 331

Muscle Physiology 333

Skeletal Muscle Structure 334
Structure at the Cellular Level • Structure at the Molecular Level

The Mechanism of Force Generation in Muscle 338
The Sliding-Filament Model • The Crossbridge Cycle: How Muscles Generate Force • Excitation-Contraction Coupling: How Muscle Contractions Are Turned On and Off • Muscle Cell Metabolism: How Muscle Cells Provide ATP to Drive the Crossbridge Cycle

The Mechanics of Skeletal Muscle Contraction 344
The Twitch • Factors Affecting the Force Generated by Individual Muscle Fibers

When It Goes Wrong: Tetanus 349

Regulation of Force Generated by Whole Muscles • Velocity of Shortening

Types of Skeletal Muscle Fibers 253
Differences in Speed of Contraction: Fast-Twitch Fibers and Slow-Twitch Fibers • Differences in the Primary Mode of ATP Production: Glycolytic Fibers and Oxidative Fibers • Slow Oxidative, Fast Oxidative, and Fast Glycolytic Fibers

Skeletal Muscles at Work 357
Attachment of Muscles to Bones • Bones as Levers: Implications for Muscle Contraction

Other Muscle Types 360
Smooth Muscles • Cardiac Muscle

Chapter Summary 366
Exercises 367

The Cardiovascular System: Cardiac Function 369

An Overview of the Major Components of the Cardiovascular System and Their Functions 369
Blood • Blood Vessels • The Heart

The Path of Blood Flow Through the Heart and Vasculature 374

Electrical Activity of the Heart 377
The Conduction System of the Heart • Spread of Excitation Through the Heart Muscle • The Ionic Basis of Electrical Activity in the Heart • Recording the Electrical Activity of the Heart with an Electrocardiogram

The Cardiac Cycle 386
The Pump Cycle • Atrial and Ventricular Pressure • Aortic Pressure • Ventricular Volume • Heart Sounds

Cardiac Output and Its Control 392
Autonomic Input to the Heart • Factors Affecting Cardiac Output: Changes in Heart Rate • Factors Affecting Cardiac Output: Changes in Stroke Volume • Integration of Factors Affecting Cardiac Output

Chapter Summary 400
Exercises 401

The Cardiovascular System: Blood, Blood Flow, and Blood Pressure 403

The Composition of Blood 404
Plasma • Formed Elements

Platelets and Hemostasis 407
Vascular Spasm • Platelet Plug • Formation of a Clot

The Structure and Function of Blood Vessels 411
Arteries • Arterioles • Capillaries • Venules • Veins

Patterns of Blood Flow Within the Cardiovascular System 417
Parallel Flow • Independent Regulation of Blood Flow in the Systemic Circuit

Physical Laws Governing Blood Flow and Blood Pressure 420
Pressure Gradients in the Cardiovascular System • Resistance in the Cardiovascular System

Discovery: Permeability and Resistance 423

Discovery: The Role of Turbulence in Blood Pressure Measurement 425

Relating Pressure Gradients and Resistance in the Systemic Circulation

Chapter 13 continues

Chapter 13 continued

Factors Affecting Flow and Distribution of Blood to Organs   427

   Determinants of Organ Blood Flow: Mean Arterial Pressure and Organ Resistance • Determinants of Mean Arterial Pressure: Heart Rate, Stroke Volume, and Total Peripheral Resistance

How Changes in Central Venous Pressure Affect Blood Flow to Organs   430

   The Skeletal Muscle Pump • The Respiratory Pump • Blood Volume • Venomotor Tone

Movement of Fluid Across Capillary Walls   432

When It Goes Wrong: Heart Failure   434

   Forces Driving the Movement of Fluid Into and Out of Capillaries • Factors Affecting Filtration and Absorption Across Capillaries • The Lymphatic System

Chapter Summary   439
Exercises   440

 The Cardiovascular System: Regulation of Function   442

Extrinsic Control of Cardiovascular Function: Regulation of Mean Arterial Pressure   443

   Neural Control of Mean Arterial Pressure

When It Goes Wrong: Hypertension   450

   Hormonal Control of Mean Arterial Pressure • Control Initiated by Cardiac and Venous Baroreceptors (Volume Receptors) • Integrated Neural and Hormonal Control of Cardiovascular Function: Restoration of Mean Arterial Pressure Following Hemorrhage

Intrinsic Control of Cardiovascular Function: Regulation of Blood Flow Distribution to Organs   455

   Why Intrinsic Control? • Local Factors that Control Vascular Resistance

Other Cardiovascular Regulatory Processes   461

   Respiratory Sinus Arrhythmia • Chemoreceptor Reflexes • Thermoregulatory Responses • Responses to Exercise

Chapter Summary   467
Exercises   467

The Respiratory System: Breathing Mechanics   467

Overview of Respiratory Function   470
Anatomy of the Respiratory System   470

   Upper Airways • The Respiratory Tract • Structures of the Thoracic Cavity

Forces for Pulmonary Ventilation   475

   Pulmonary Pressures • Mechanics of Breathing

Factors Affecting Pulmonary Ventilation   484

   Lung Compliance • Airway Resistance

Clinical Significance of Respiratory Volumes and Air Flows   488

   Lung Volumes and Capacities • Pulmonary Function Tests

When It Goes Wrong: Chronic Obstructive Pulmonary Disease   490

   Alveolar Ventilation

Chapter Summary   493
Exercises   494

The Respiratory System: Gas Exchange and Regulation of Breathing   495

Overview of the Pulmonary Circulation   496
Diffusion of Gases   497

   Partial Pressure of Gases • Solubility of Gases in Liquids

Exchange of Oxygen and Carbon Dioxide   500

   Gas Exchange in the Lungs • Gas Exchange in Respiring Tissue

When It Goes Wrong: Pulmonary Edema   504

   Determinants of Alveolar P_{O_2} and P_{CO_2}

Transport of Gases in the Blood   505

   Oxygen Transport in the Blood

When It Goes Wrong: Anemia   510

   Carbon Dioxide Transport in the Blood

Central Regulation of Ventilation   514

   Neural Control of Breathing by Motor Neurons • Generation of Breathing Rhythm in the Brainstem • Peripheral Input to Respiratory Centers

Control of Ventilation by Chemoreceptors   518

   Chemoreceptors • Chemoreceptor Reflexes

Local Regulation of Ventilation and Perfusion   521

   Ventilation–Perfusion Ratios • Local Control of Ventilation and Perfusion

The Respiratory System in Acid-Base Homeostasis   522

   Acid-Base Disturbances in Blood • The Role of the Respiratory System in Acid-Base Balance

Discovery: The Effects of High Altitude   525

Chapter Summary   529
Exercises   530

The Urinary System: Renal Function   532

Functions of the Urinary System   533
Anatomy of the Urinary System   534

Structures of the Urinary System • Macroscopic Anatomy of the Kidney • Microscopic Anatomy of the Kidney • Blood Supply to the Kidney

Basic Renal Exchange Processes   538

Glomerular Filtration • Reabsorption • Secretion

Regional Specialization of the Renal Tubules   549

Mechanisms of Sodium Reabsorption in the Renal Tubule

Discovery: Dialysis   550

Nonregulated Reabsorption in the Proximal Tubule • Regulated Reabsorption and Secretion in the Distal Tubule and Collecting Duct • Water Conservation in the Loop of Henle

Excretion   553

Excretion Rate • Clearance • Micturition

Chapter Summary   561
Exercises   562

 The Urinary System: Fluid and Electrolyte Balance   564

The Concept of Balance   565

Factors Affecting the Plasma Composition • Solute and Water Balance

Water Balance   567

Osmolarity and the Movement of Water • Water Reabsorption in the Proximal Tubule • Role of the Medullary Osmotic Gradient in Water Reabsorption • Water Reabsorption in the Distal Tubule and Collecting Duct

Sodium Balance   575

The Effects of Aldosterone

Discovery: Don't Drink the Water   576

Atrial Natriuretic Peptide
Potassium Balance   579

Renal Handling of Potassium Ions • Regulation of Potassium Secretion by Aldosterone

Calcium Balance   580

Renal Handling of Calcium Ions • Hormonal Control of Plasma Calcium Concentrations

Interactions Between Fluid and Electrolyte Regulation   581
Acid-Base Balance   583

Sources of Acid-Base Disturbances

When It Goes Wrong: Kidney Disease   586

Defense Mechanisms Against Acid-Base Disturbances • Compensation for Acid-Base Disturbances

Chapter Summary   598
Exercises   599

 The Digestive System   601

Overview of Digestive System Function   602
Functional Anatomy of the Digestive System   602

The Digestive Tract • The Accessory Glands

Digestion and Absorption of Nutrients and Water   614

Carbohydrates • Proteins • Lipids • Vitamins • Minerals and Water

General Principles of Gastrointestinal Regulation   620

Discovery: Lipoproteins and Plasma Cholesterol   621

When It Goes Wrong: Ulcers   622

Neural and Endocrine Pathways of Gastrointestinal Control • Regulation of Food Intake

Gastrointestinal Secretion and Its Regulation   624

Saliva Secretion • Acid and Pepsinogen Secretion in the Stomach • Secretion of Pancreatic Juice and Bile • Rates of Fluid Movement in the Digestive System

Gastrointestinal Motility and Its Regulation   628

Electrical Activity in Gastrointestinal Smooth Muscle • Chewing and Swallowing • Gastric Motility • Motility of the Small Intestine • Motility of the Colon

Chapter Summary   636
Exercises   637

The Endocrine System: Regulation of Energy Metabolism and Growth   639

An Overview of Whole Body Metabolism   640

Adenosine Triphosphate (ATP) • Anabolism • Regulation of Metabolic Pathways

Energy Intake, Utilization, and Storage   641

Absorption, Utilization, and Storage of Energy in Carbohydrates • Absorption, Utilization, and Storage of Energy in Proteins • Absorption, Utilization, and Storage of Energy in Fats

Energy Balance   643

*Energy Input • Energy Output • Metabolic Rate • Negative
and Positive Energy Balance*

Energy Metabolism During the Absorptive and
Postabsorptive States   645

*Metabolism During the Absorptive State • Metabolism During
the Postabsorptive State*

Regulation of Absorptive and Postabsorptive
Metabolism   649

*The Role of Insulin • The Role of Glucagon • Effects of
Epinephrine and Sympathetic Nervous Activity on Metabolism*

Hormonal Regulation of Growth   654

Effects of Growth Hormone

When It Goes Wrong: Diabetes Mellitus   657

Other Hormones that Affect Growth

Thyroid Hormones   661

*Synthesis and Secretion of Thyroid Hormones • Actions of
Thyroid Hormones*

Glucocorticoids   663

*Factors Affecting Secretion of Glucocorticoids • Actions of
Glucocorticoids • The Role of Cortisol in Stress Response
• Effects of Abnormal Glucocorticoid Secretion*

Chapter Summary   668
Exercises   669

21 The Reproductive System   670

An Overview of Reproductive Physiology   671

*The Role of Gametes in Sexual Reproduction • Gene Sorting
and Packaging in Gametogenesis: Meiosis • Components of the
Reproductive System • Events Following Fertilization
• Patterns of Reproductive Activity Over the Human Life Span*

The Male Reproductive System   676

*Functional Anatomy of the Male Reproductive Organs
• Hormonal Regulation of Reproductive Function in Males
• Sperm and Their Development • The Sexual Response in
Males*

The Female Reproductive System   683

Functional Anatomy of the Female Reproductive Organs

When It Goes Wrong: Erectile Dysfunction   685

*Ova and Their Development • The Sexual Response in Females
• The Menstrual Cycle • Long-Term Hormonal Regulation of
Female Reproductive Function*

Discovery: Birth Control Methods   694

Fertilization, Implantation, and Pregnancy   695

*Events of Fertilization • Early Embryonic Development and
Implantation • Later Embryonic and Fetal Development
• Hormonal Changes During Pregnancy*

Parturition and Lactation   701

Events of Parturition • Lactation

Chapter Summary   705
Exercises   706

22 The Immune System   708

Anatomy of the Immune System   709

Leukocytes • Lymphoid Tissues

Organization of the Body's Defenses   713

Nonspecific Defenses • Immune Responses

Humoral Immunity   721

*The Role of B Lymphocytes in Antibody Production • Antibody
Function in Humoral Immunity*

Cell-Mediated Immunity   724

*Roles of T Lymphocytes in Cell-Mediated Immunity • Helper T
Cell Activation • Cytotoxic T Cell Activation: The Destruction of
Virus-Infected Cells and Tumor Cells*

Immune Responses in Health and Disease   728

*Generating Immunity: Immunization • Roles of the Immune
System in Transfusion and Transplantation*

When It Goes Wrong: Multiple Sclerosis   731

Immune Dysfunctions

**Discovery: Gene Therapy for Severe Combined
Immunodeficiency Disease   734**

Chapter Summary   738
Exercises   739

23 The Whole Body: Integrated Physiological Responses to Exercise   741

Principles of Physiological Integration   742

*The Metabolic Demands of Exercise • How Physiological
Function Is Increased*

The Start: Transition from Rest to Exercise   742

*Autonomic Nervous System Input: Fight or Flight • Energy
Sources and Mobilization: Different Fuels for Different
Needs • Cardiovascular Adjustments: Anticipating the
Needs • Respiratory Responses: Recruiting Reserve
Capacities • Temperature Regulation*

The Long Haul: Almost Steady State   746

Hormonal Control of Energy Metabolism • Cardiovascular Control: Factors Influencing Blood Pressure • Respiration: The Effort of Breathing • The Gastrointestinal Tract: The Forgotten System • Thermoregulation: Intensifying the Competition for Blood • Fluid Balance: The Weak Link in Steady State

The Decline to the End: "The Wall" or the Finish Line?   753

The Prime Directive: Protect the Brain • Fatigue: The Early Warning System • Tactical and Biological Influences on Fatigue • Conscious Decisions Versus Autonomic Control

The Aftermath   755

Syncope: The Autonomic System's Final Veto Power • Water Intoxication • Postexercise Fever • Delayed-Onset Muscle Soreness • Effect of Marathon Running on Susceptibility to Sickness

Chapter Summary   758
Exercises   759

Answers to Odd-Numbered Multiple-
   Choice and Objective Questions   763
Credits   769
Glossary   771
Index   789

1

Introduction to Physiology

OBJECTIVES

■ Name the four major types of cells in the human body, and describe their defining characteristics.

■ Define the following terms: *total body water, intracellular fluid, extracellular fluid, plasma, interstitial fluid, internal environment.*

■ Define *homeostasis* and explain its significance to the function of the body.

■ Describe the role of negative feedback in homeostasis, and define the following terms: *sensor, effector, input, output, integrating center, set point, error signal, regulated variable.*

CHAPTER OUTLINE

Organization of the Body   2

Homeostasis: A Central Organizing Principle
    of Physiology   8

Challenging Homeostasis: A Study in Organ
    Systems Integration During Exercise   15

Above: Electron micrograph of a microglia cell

Physiology, the science of body function, may be a child of human reason, but it was undoubtedly conceived in passion—our natural desire to understand ourselves and the universe we inhabit. How do our eyes and ears allow us to perceive the world around us? What makes our muscles move? How do we get energy from the food we eat? These and other questions occur to virtually everyone at one time or another. For some people these questions persist beyond mere curiosity and become the driving force for a lifetime of study. It is because of the efforts of such persons that the science of physiology exists today.

Like many other fields of study, physiology has a simple definition but is almost impossible to describe succinctly. Simply put, physiology attempts to explain the workings of the body using established principles from physics and chemistry. This description may seem clear enough, but the same could also be said of other biological sciences such as biochemistry and cell biology. It is difficult to say what makes physiology distinctive because it, like many other fields of scientific study, lacks distinct boundaries. To explain how the body works, physiologists use tools from many different fields, including biochemistry, cell biology, genetics, physics, and even engineering. In physiology, there is something for everyone. If you enjoy thinking about things on a cellular or molecular level, for example, you will find topics such as membrane transport or nerve cell signaling especially interesting. If you like thinking on a larger, organ-level or "whole-body" scale, the study of cardiovascular or respiratory function might be your cup of tea. Most physiologists would say that the best thing about studying physiology is that you get to explore *all* of these things.

Because nearly everyone is curious about how the human body works, there is a very good chance that studying physiology will be one of your most satisfying academic experiences. The study of physiology offers you not only the satisfaction of knowing something about the workings of the body, but it also may provide you a deep, perhaps even profound, understanding of it. You will also come to realize that physiology, like the other sciences, is not just a collection of well-worn facts but rather a "work in progress." You will recognize that there are significant gaps in our understanding of how the body works, and you will also see that much of our current understanding will be subject to change as new discoveries are made.

Regardless of your background or current affinities, your study of physiology will broaden your scientific interests and will widen the scope of your outlook. You will begin to see the "big picture," understanding body function not as a collection of unrelated phenomena but as a connected whole. You might even discover something else—that physiology is beautiful. Most of us who have decided to make it our life's work think so.

ORGANIZATION OF THE BODY

If you have ever spent time examining a detailed anatomical chart or model of the human body, you have seen that it is an exceedingly complex and intricate structure. Despite the complexity of its structure, however, there is an underlying simplicity to the function of the human body.

To a physiologist, perhaps the most interesting thing about the body is that its operation can be explained in terms of a relatively small set of easily understood principles. For this reason, a physiologist's approach to studying the body is to strip away all unnecessary details so that the essentials—that is, the unifying themes and principles—can be seen more clearly. To get an idea of what this means, consider Figure 1.1, which shows four kinds of cells found in the brain. Although these cells are all clearly different, you can see that they can be classified into a relatively small number of categories on the basis of similarities in their morphology (shape). This is fortunate, because the brain contains billions of such cells and it is unlikely that any two are exactly alike! When you consider the function of these cells, however, the similarities among them extend even further, allowing them to be grouped into just one category: All these cells (and many others throughout the body) are specialized to transmit information in the form of electrical signals from one body location to another. Because of these similarities, all these cells are classified as *neurons* (or *nerve cells*).

If the body's underlying simplicity is one of physiology's major themes, another is the degree of interaction among its various parts. Although each of the body's **cells,** the smallest living units, is independently capable of carrying out its own basic life processes, the various types of cells are specialized to perform different functions important to the operation of the body as a whole, and for this reason all the cells ultimately depend on each other for their survival. Similarly, the body's organs are also specialized to perform certain tasks vital to the operation of other organs. You know, for example, that your cells need oxygen to live and that oxygen is delivered to your cells by the bloodstream, but consider some of the many things that must occur in order to ensure that oxygen delivery is sufficient to meet the cells' needs. Oxygen is carried in the bloodstream by cells called *erythrocytes,* which are manufactured by *bone marrow,* a tissue found inside certain bones. To ensure that adequate numbers of erythrocytes are present in the blood, the synthesis of these cells is regulated by a hormone called *erythropoietin,* which is secreted by the kidneys. To ensure adequate blood flow to the body's tissues, the heart must pump a sufficient volume of blood every minute, and for this reason the rate and force of its contractions are regulated by the nervous system. To ensure that the blood carries enough oxygen, the lungs must take in sufficient quantities of air, which requires the control of breathing muscles (such as the diaphragm) by the nervous system.

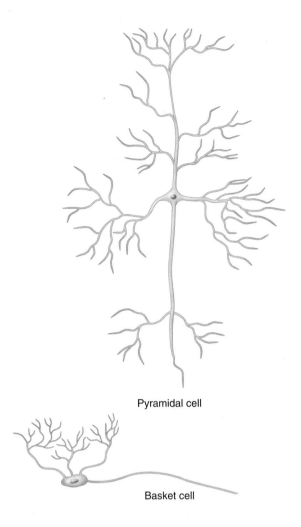

Pyramidal cell

Basket cell

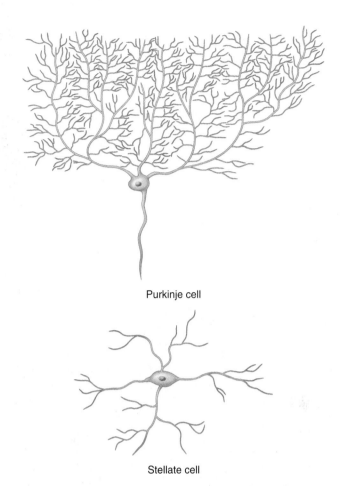

Purkinje cell

Stellate cell

FIGURE 1.1 **Four kinds of cells found in the brain.**

Finally, to provide the energy necessary to drive these and other processes, the gastrointestinal system breaks ingested food down into smaller molecules, which are absorbed into the bloodstream and distributed to cells throughout the body.

This example shows that proper body function requires not only that each part be able to carry out its own particular function, but also that the parts be able to work together in a coordinated manner. To help you better understand how the body's parts work together, the remainder of the chapter outlines broad principles pertaining to body function in general; the functions of specific organs and organ systems are the topics of later chapters.

Cells, Tissues, Organs, and Organ Systems

Although over 200 distinguishable kinds of cells are present in the body, there are only four major types of cells: (1) *neurons,* (2) *muscle cells,* (3) *epithelial cells,* and (4) *connective tissue cells.* Representative cells belonging to each of these cell types are shown in Figure 1.2. These classifications are very broad and are based primarily on functional differences. There are other more rigorous ways to classify cells based on anatomical distinctions and embryological origins.

As mentioned previously, nerve cells or **neurons** (Figure 1.2a) are specialized to transmit information in the form of electrical signals. For this purpose, neurons typically possess a number of branches that function either to receive signals from or transmit signals to other cells. Certain neurons, such as those in the eyes that respond to light or those in the skin that respond to touch, receive information from the outside environment and allow us to perceive the world through our senses. Other neurons relay signals to muscles, glands, and other organs, enabling the control of movement, hormone secretion, and other bodily functions. Still other neurons, such as those in the brain, are involved in information processing, enabling us to conceptualize, remember, formulate plans of action, and experience emotion.

Muscle cells (Figure 1.2b) are generally elongated in shape and are specialized to respond to certain signals (such as neural input) by contracting, thereby generating mechanical force and movement. These cells are found in the muscles of the arms, legs, and other body parts whose movements are under voluntary control, but they are also found in structures not under voluntary control, such as internal organs (the heart and stomach, for example) and blood vessels. The flexing of an arm, the pumping of

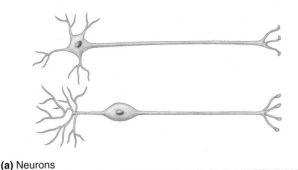

(a) Neurons

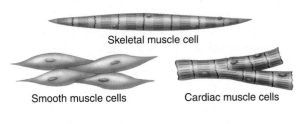

Skeletal muscle cell

Smooth muscle cells   Cardiac muscle cells

(b) Muscle cells

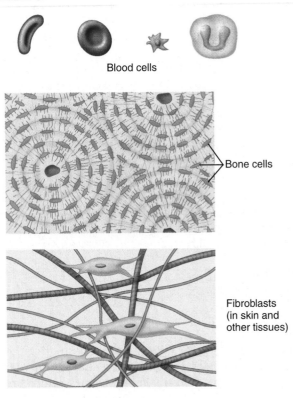

Blood cells

Bone cells

Fibroblasts
(in skin and
other tissues)

(d) Connective tissue cells

Basement
membrane

Basement
membrane

(c) Epithelial cells

FIGURE 1.2 **Major cell types in the human body. (a)** *Neurons.*
(b) *Muscle cells.* **(c)** *Epithelial cells.* **(d)** *Connective tissue cells.*

blood by the heart, and the mixing of food in the stomach
are all examples of muscle cells in action.

Epithelial cells are found in tissues called **epithelia**
(singular: **epithelium**), which consist of a continuous,
sheetlike layer of cells in combination with a thin underlying layer of noncellular material called a *basement membrane* (Figure 1.2c). Depending on the epithelium in
question, the cell layer may be one cell thick or several
cells thick, and the cells may vary in shape from short
and flattened in some cases to tall and oblong in others.
In all cases, however, cells are joined closely together to
form a barrier that prevents material on one side of the
epithelium from mixing freely with material on the other
side. Appropriately, epithelia are found wherever body
fluids must be kept separate from the external environment, such as the skin surface or the lining of the lungs.
Epithelia are also found in the linings of hollow organs
such as the stomach, intestines, and blood vessels, where
they separate fluids in the interior cavity from the surrounding body fluids. The interior cavity of a hollow organ or vessel is generally referred to as the **lumen.** (In
blood vessels and chambers of the heart, the epithelial
lining is referred to as an *endothelium.*)

Certain epithelial cells are specialized to transport
specific materials, such as inorganic ions, organic molecules, or water, from one location to another. One example occurs in the lining of the stomach, where cells
transport acid (hydrogen ions) into the lumen of the

stomach to aid in the digestion of food. Another example
is found in the lining of the intestine, where cells transport nutrients and water from the lumen of the intestine
into the bloodstream.

The last remaining major cell type, **connective tissue
cells,** is the most diverse. This cell type includes blood cells,
bone cells, fat cells, and many other kinds of cells that
might seem at first to have little in common in terms of
structure or function (Figure 1.2d).

In a narrow sense, the term *connective tissue* refers to any
structure whose primary function is to provide physical
support for other structures, to anchor them in place, or to
link them together. Familiar examples of connective tissue
structures are *tendons,* which anchor muscles to bones; *ligaments,* which connect bones together; and the elastic tissue
in the skin that gives it its toughness and flexibility. Another example of a connective tissue is the bones themselves, which provide direct or indirect support for all of the
body's structures. In most cases, connective tissue consists

of widely scattered cells embedded in a mass of noncellular material called the *extracellular matrix,* which contains a dense meshwork of proteins and other large molecules. Among the most important constituents of the extracellular matrix are the long, fibrous proteins *elastin* (which gives the tissue elasticity) and *collagen* (which gives the tissue tensile strength, the ability to resist stretching).

In a broader sense, the term *connective tissue* encompasses fluids such as the blood and lymph, which do not provide structural support like other connective tissue but instead serve to "connect" the various parts of the body together by providing avenues of communication. The blood, for example, delivers oxygen from the lungs to the rest of the body's tissues and carries hormones from the glands that secrete them to the tissues that respond to them. Similarly, the lymph carries water and other materials that leak out of blood vessels throughout the body and returns them to the blood.

It is a general rule that cells of a given type tend to group in the body with cells of the same type. Nerve cells, for example, are always found in conjunction with other nerve cells, and epithelial cells are always joined together with other epithelial cells. Any such collection of cells performing similar functions is referred to as a **tissue.** (The term *tissue* is also used more loosely to refer to any of the materials of which the body is composed.) Generally, when two or more tissues combine to make up structures that perform particular functions, those structures are called **organs.** The heart, for example, is an organ whose primary function is to pump blood. Although composed mostly of muscle tissue, it also contains nervous tissue (the endings of nerves that control the heartbeat), epithelial tissue (which lines the heart's chambers), and connective tissue (which makes up the heart's valves and other tissues that hold the muscle fibers together).

The various organs are organized into **organ systems,** collections of organs that work together to perform certain functions. An example is the *cardiovascular system,* whose function is to deliver blood to all the body's tissues. The cardiovascular system includes the heart, blood vessels, and the blood (which is not an organ, but rather a tissue). Another organ system is the *gastrointestinal system,* whose function is to break food down into smaller molecules (a process called *digestion*) and to transport these molecules into the bloodstream. This organ system includes the mouth, salivary glands, esophagus, stomach, intestines, liver, gallbladder, and pancreas. In some organ systems (for example, the cardiovascular and gastrointestinal systems), the organs are physically connected together. In other cases the organs are disconnected and more widely scattered. This is true of the *endocrine system,* which encompasses all the glands in the body that secrete hormones, and the *immune system,* which protects the body from invading microorganisms and other foreign materials.

Although the concept of an organ system is simple in principle, the distinction between one organ system and another is not always clear-cut—many organs perform functions that are integral to more than one organ system.

A prime example is the pancreas, which is considered to be part of both the digestive system, because it secretes fluid and digestive enzymes into the intestines, and the endocrine system, because it secretes certain hormones.

The Overall Body Plan: A Simplified View

When physiologists attempt to understand and explain body functions, they usually try to reduce the body's complexity to its essential elements so that unifying themes and principles can be seen more easily. This tendency to simplify is nowhere more apparent than in Figure 1.3, which shows a physiologist's "minimalist" view of the human body. This figure does not look anything like a real body: Not only is it the wrong shape, but it is simplistic and seems to be missing some parts. The gastrointestinal system, for example, is drawn as a straight tube that extends through the body from one end to the other, while the lungs are shown as a single hollow sac. The body's intricate network of blood vessels is shown as a simple loop, and the heart, which pumps blood around this loop, as just a box. Different cell types, such as nerve, muscle, and connective tissue cells, are drawn to look alike and are given the generic label "cells." Furthermore, the kidneys are shown simply as a single blind-ended tubule that leads to the outside. (In reality, each kidney is composed of large numbers of such tubules, which produce urine.)

Despite its lack of realism—or more correctly, because of it—Figure 1.3 highlights an important concept that might not be so readily apparent from a more realistic picture of the body: that the interior of the body and the body's external environment are separated by a single layer of epithelial tissue.

The Body's External Environment

Figure 1.3 clearly shows the layer of epithelial tissue that separates the **external environment** from the interior of the body. This epithelial barrier includes not only the skin, but also the linings of the lungs, gastrointestinal system, and kidney tubules. This means that when air enters the lungs or food enters the stomach, these materials are still actually *outside* the body because they are on the external side of this epithelial barrier. Figure 1.3 also shows that this barrier is continuous; that is, there is no real separation between the outer surface of the skin and the inside surfaces of the lungs, gastrointestinal system, and kidney tubules! They are all part of the same "fabric," if you will.

The Body's Internal Environment

In order to live, cells must take in oxygen and nutrients from their surroundings and release carbon dioxide and other waste products into their surroundings. The ultimate source of oxygen and nutrients, and the ulti-

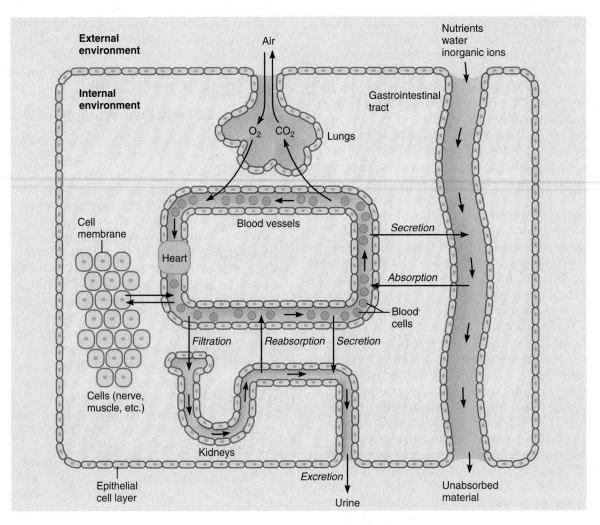

FIGURE 1.3 A highly simplified view of the overall plan of the human body.
Flows of material are indicated by arrows.

repository for discarded waste products (including carbon dioxide), is the external environment. Note in Figure 1.3, however, that most of the body's cells are not able to exchange materials directly with the external environment because they are not in direct contact with it. Instead, cells receive oxygen and nutrients from the bloodstream, which also carries carbon dioxide and waste products away from cells. Notice also, however, that most cells are not in *direct* contact with the blood, but are surrounded instead by a separate fluid that exchanges materials with the blood. Because this fluid constitutes the immediate environment of most of the body's cells, it is called the **internal environment.** (The term *internal environment* also applies to the fluid in the bloodstream that surrounds blood cells.)

The Exchange of Materials Between the External and Internal Environments To do its job, the blood must obtain oxygen, nutrients, and other needed materials such as water from the external environment and must release carbon dioxide and other unneeded materials into

it. Figure 1.3 shows that the exchange of material between the blood and the external environment occurs in different places, including the lungs, the gastrointestinal tract, and the kidneys.

In the lungs, oxygen enters the bloodstream from the air that is breathed in, while carbon dioxide exits the bloodstream and is expelled in the air that is breathed out. In the gastrointestinal tract, the water, inorganic salts, and nutrients obtained from digested food are transported from the lumen to the bloodstream, a process referred to as *absorption*. To aid in the digestion of food, stomach acid and fluids rich in enzymes are transported into the lumen from the bloodstream, a process called *secretion*. Unabsorbed materials (plus bacteria and cellular debris) remain in the gastrointestinal tract and are ultimately eliminated from the body as feces.

In the kidneys, fluid from the bloodstream first enters tubules via a mechanism known as *filtration*. Then as this fluid travels along the length of the tubules, needed materials (including water, inorganic salts, and nutrients) are selectively transported back into the bloodstream, a

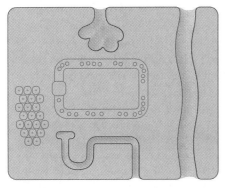

(a) Total body water (TBW)

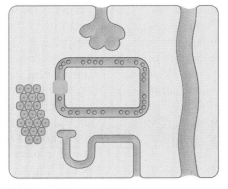

(b) Intracellular fluid (ICF)

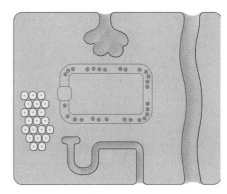

(c) Extracellular fluid (ECF)

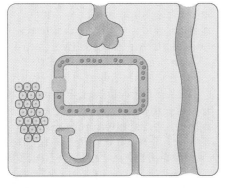

(d) Plasma

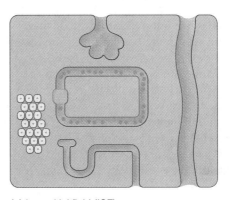

(e) Interstitial fluid (ISF)

FIGURE 1.4 Body fluid compartments. *The various fluid compartments are indicated by shading in several simplified body plans similar to that shown in Figure 1.3.* **(a)** *Total body water.* **(b)** *Intracellular fluid.* **(c)** *Extracellular fluid.* **(d)** *Plasma.* **(e)** *Interstitial fluid.*

Which of the following does not include plasma: total body water, extracellular fluid, or interstitial fluid?

process known as *reabsorption.* At the same time, unneeded materials are selectively transported from the bloodstream into the tubules, a process called *secretion.* The fluid that eventually reaches the ends of the tubules constitutes the urine, which is eliminated from the body (a process called *excretion*). Materials contained in the urine include cellular waste products as well as excess salts and water that are not needed by the body.

Body Fluid Compartments The most abundant substance in the body is *water,* which acts as a solvent for the great variety of solutes found in body fluids. These solutes include small molecules such as inorganic ions, sugars, and amino acids, and large molecules such as proteins. Figure 1.3 shows that the interior of the body is divided into separate compartments (which are filled with fluid) by barriers of different types, including epithelial tissues and *cell membranes,* which separate the contents of cells from their surroundings. Although these compartments are physically separated, they are still able to exchange materials with each other because the barriers that separate them are *permeable;* that is, they permit molecules to pass through them. These barriers let certain types of

molecules through more easily than others and even exclude some molecules entirely. Thus, it is more accurate to say that cell membranes and epithelial tissues are *selectively permeable* or *semipermeable.*

The volume of water contained in all the body's compartments is **total body water (TBW),** the total volume of fluid enclosed within the outer epithelial layer (Figure 1.4a). For a person weighing 70 kilograms (150 pounds), the volume of TBW is 42 liters, which accounts for about 60% of total body weight. Total body water includes water present in fluid located inside cells, called **intracellular fluid** or **ICF** (Figure 1.4b), and fluid located outside cells, called **extracellular fluid** or **ECF** (Figure 1.4c). In the body plan diagram the volume of ECF relative to ICF is highly exaggerated, for in reality about two-thirds of TBW is in the ICF, and only about one-third is in the ECF. (Note as well that *extracellular fluid* is synonymous with the *internal environment.*)

ICF and ECF are separated by cell membranes and differ significantly in composition. Intracellular fluid contains many proteins and is relatively rich in potassium,

Interstitial fluid

for example, while extracellular fluid contains few proteins and is relatively rich in sodium. These differences in composition are necessary for the proper functioning of cells and are made possible because the permeability of cell membranes to many solutes is relatively low. These membranes are virtually impermeable to proteins and other large solutes and have a limited permeability to sodium, potassium, and many other small solutes. (Oxygen, carbon dioxide, and certain small solutes can pass through cell membranes easily, however.)

Of the total volume of extracellular fluid, about 20% is found in the blood, while the remainder is found outside the blood. The portion that is present in the blood (that is, the liquid, noncellular part of the blood) is **plasma** (Figure 1.4d). The portion that is present outside the blood and that bathes most of the cells in the body is called the **interstitial fluid** or **ISF** (Figure 1.4e). Unlike ECF and ICF, plasma and interstitial fluid are very similar in composition; the only major difference is that plasma is relatively rich in proteins, which are scarce in interstitial fluid. The similarity in composition between plasma and interstitial fluid is due to the fact that the walls of the smallest and most numerous blood vessels, called *capillaries,* are highly permeable to most solutes except proteins.

Quick Test 1.1

1. Name the four major types of cells in the body and describe their defining characteristics.

2. Define the following terms: *cell, tissue, organ, organ system.*

3. Why is extracellular fluid referred to as the body's internal environment?

4. Give the proper term for each of the following: (a) all the water that is contained in the body, (b) fluid that is contained within cells, (c) fluid that is located outside cells, (d) fluid that is located outside cells and found in the blood, and (e) fluid that is located outside cells and found outside the blood.

HOMEOSTASIS: A CENTRAL ORGANIZING PRINCIPLE OF PHYSIOLOGY

When the heart of an amphibian or a reptile, such as a frog or a turtle, is removed from the animal and placed in a beaker of Ringer's solution (which is similar in composition to interstitial fluid), it will continue to beat for over an hour, even if the temperature and other conditions are not carefully controlled. However, to find a human heart that sits in a beaker and beats under the same conditions, you would have to watch a B-grade science fiction movie! The reason that it is not possible to keep a human heart alive for an indefinite

period of time in a beaker of fluid is that heart cells, and most other cells in the human body, are extremely sensitive to changes in the conditions in their immediate environment. (Amphibian and reptile cells are also sensitive to such changes, but to a lesser degree.) When an organ or tissue is removed from the body, the cells' environment begins to change immediately because blood flow to the tissue is interrupted. For instance, the concentration of oxygen in the tissue begins to fall and the concentration of carbon dioxide begins to rise. Even when human tissue is kept in an incubator under controlled laboratory conditions, it is difficult to duplicate the cells' normal environment precisely, and so most tissues will eventually die. This is the reason why a heart cannot be removed from an organ donor and stored indefinitely in a laboratory until a suitable recipient is found. In fact, individuals who need a heart must usually wait months or years until a suitable donor is found, and then they must receive the heart within hours.

Given our cells' sensitivity to changing conditions, how can the body tolerate the widely varying conditions in the external environment? After all, humans can live in very hot climates, such as the tropics, and also in much colder climates. We can live at sea level, where oxygen is plentiful, or in the mountains, where the oxygen concentration in air is lower. We can live in the dryness of a desert or in the extreme humidity of a rain forest. How can the body adapt to such a variety of conditions? The answer is that the body has all sorts of regulatory mechanisms that work to keep conditions in the *internal environment* relatively constant despite any changes that might occur in the external environment. This maintenance of constant conditions in the internal environment is known as **homeostasis.** As you progress through this book, you will discover that the concept of homeostasis looms large as a central organizing principle that runs through many different subject areas in physiology.

To say that the internal environment is regulated to remain constant means that the *composition, temperature,* and *volume* of extracellular fluid do not change significantly under normal conditions. The extracellular fluid is normally kept at a temperature near 37°C or 98.6°F (normal body temperature), and concentrations of many solutes (oxygen, carbon dioxide, sodium, potassium, calcium, and glucose, for example) are also kept relatively steady. The ability to maintain such constancy is important because the body is continually faced with potentially disrupting changes that can originate either in the external environment or within the body itself. When the environment warms up or you begin to exercise, for example, your body temperature rises. In either case, the rise in body temperature activates regulatory mechanisms that work to reduce body temperature and bring it back down toward normal. As you study physiology, you will see that the body is able to maintain constant conditions in the internal environment only through the efforts of different organ systems working together.

If you have ever tried to perform heavy manual labor on a hot summer day or have overexerted yourself in an athletic contest, you may have experienced weakness and dizziness as a result. If your exertions were severe or the weather was very hot, you may have even collapsed and lost consciousness momentarily. If this has happened to you, then you have experienced *heat exhaustion.* Heat exhaustion is a consequence of the body's effort to regulate its temperature—in particular, its efforts to get rid of excess heat. When the body must get rid of a large quantity of heat, massive quantities of sweat can be produced, leading to a significant reduction in blood volume. In addition, blood flow to the skin increases markedly, which diverts blood from other areas of the body. Together, these changes produce a reduction in blood pressure, which reduces blood flow to the brain and precipitates the symptoms just described.

A far more serious condition is *heat stroke,* in which the body's temperature rises out of control due to failure of the thermoregulatory system. Extreme overexertion or high environmental temperatures can overwhelm the body's capacity for getting rid of heat. When this happens, the body's temperature rises in spite of its thermoregulatory efforts. As the temperature continues to rise, the brain begins to malfunction. Delirium sets in, followed by a loss of consciousness. Eventually the brain's thermoregulatory centers begin to fail. When this occurs, the brain inappropriately stops sending signals to the sweat glands that tell them to secrete. As a result, sweat production comes to a halt, which compromises the body's ability to get rid of heat and causes the temperature to rise even faster. Furthermore, the increased body temperature causes the body's metabolic processes to speed up, so that heat production rises, which only adds to the problem. If left

untreated, this spiral of events leads inexorably to death.

The skin of individuals experiencing heat stroke has a flushed appearance (due to increased blood flow) but will also be dry (due to the absence of sweat). These signs make it easy to distinguish heat stroke from heat exhaustion, in which sweating is profuse and the skin is flushed and wet. If someone is experiencing heat stroke or is in danger of doing so, immediate medical attention is of the utmost importance. Often, a person's life can be saved by immersing them in ice water, which reduces the body temperature quickly to within the range at which normal thermoregulation is possible. Assuming that the elevated temperatures have not caused permanent damage to the brain's thermoregulatory centers, regulatory mechanisms can then take over.

Even though homeostatic regulatory mechanisms work to resist changes in the internal environment, every regulatory system has its limitations, even when it is undamaged by disease or trauma and is functioning normally. For example, body temperature can be maintained close to normal only so long as environmental temperatures are not too extreme and other stresses placed on the regulatory system are not too great. However, severe exercise or high environmental temperatures can cause body temperature to begin to rise out of control, with potentially fatal consequences (When It Goes Wrong: Heat Exhaustion and Heat Stroke, above). In fact, it is generally true that failure of any system to maintain homeostasis ultimately gives rise to signs and symptoms of disease because such failure adversely affects the function of organ systems.

Negative Feedback Control in Homeostasis

Because body temperature is not free to vary but is instead regulated to stay within relatively narrow limits, it is referred to as a **regulated variable.** Plasma concentrations of potassium, sodium, and calcium are also regulated variables because they are kept constant by homeostatic regulatory mechanisms. Most homeostatic regulatory mechanisms follow the same pattern: If a regulated variable increases, the system responds by making it decrease; if it decreases, the system responds by making it increase. Systems behaving in this manner are said to operate by **negative feedback.**

A familiar example of a negative feedback system is the cruise control in a car, which operates to keep the speed of the car steady at a certain desired point (Figure 1.5a and b). If a car running on level ground starts up

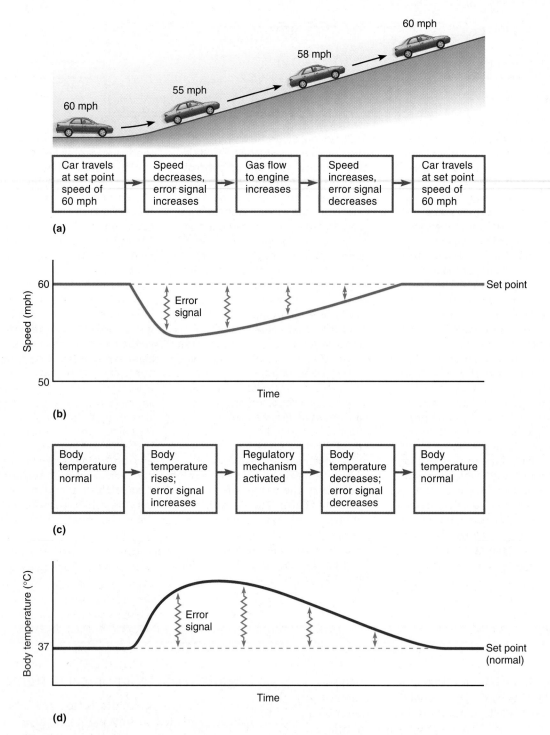

FIGURE 1.5 Negative feedback control of a regulated variable. (a) *Events occurring as the speed of an automobile is regulated by a cruise control mechanism.* **(b)** *A graph showing changes in the automobile's speed as it climbs a hill. The dashed line represents the set point speed; vertical arrows indicate the error signal.* **(c)** *Events occurring as body temperature is adjusted to normal following an initial rise.* **(d)** *A graph showing changes in body temperature.*

a hill, the car will begin to slow down. When the control mechanism detects a difference between the actual speed of the car and the desired speed, it feeds more gasoline to the engine, and the car's speed increases. When the car's speed reaches the desired speed, the system "throttles back" to maintain that speed. As long as the car's actual

speed does not differ from the desired speed, the system makes no further adjustments to the flow of gasoline.

Like the cruise control in a car, most homeostatic regulatory mechanisms make adjustments only when they detect a difference between the actual value of the regulated variable and the normal "desired" value, called

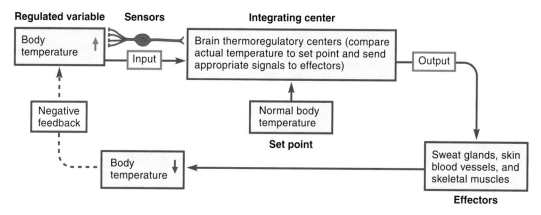

FIGURE 1.6 A negative feedback loop. *This feedback loop operates in the control of body temperature, as described in the text. Up and down arrows within boxes indicate increases and decreases, respectively. The dashed line indicates that the response of the system affects the input.*

the **set point.** Any difference between the actual value and the set point constitutes an **error signal.** Because these mechanisms normally work to bring the regulated variable closer to the set point, they ultimately function to make error signals as small as possible. An error signal that arises when body temperature rises above the normal set point of 37°C, for example, activates regulatory mechanisms that will eventually bring the temperature back down to near 37°C (Figure 1.5c and d).

To operate properly, a homeostatic regulatory mechanism must have a means of detecting the regulated variable. This is accomplished through the action of **sensors,** cells (often neurons) that are sensitive to the variable in question. For instance, certain blood vessels contain cells called *chemoreceptors* that are sensitive to concentrations of oxygen and carbon dioxide in the blood, and in the brain and other parts of the body there are neurons called *thermoreceptors* that are sensitive to temperature. Typically, the sensors relay signals (called **input**) to an **integrating center** (often a particular set of neural circuits in the brain), which compares the regulated variable to the set point and orchestrates the appropriate response. In response to the input it receives, the integrating center relays signals (called **output**) to the cells, tissues, or organs that bring about the final response. These cells, tissues, or organs are called **effectors.**

Now that we are familiar with the basic elements of homeostatic control mechanisms, we can see where the term *negative feedback* comes from, using temperature regulation as an example. To monitor body temperature, thermoreceptors send input to the brain's *thermoregulatory center,* the region responsible for temperature regulation. When the temperature rises above the set point, the thermoregulatory center sends output to effectors, which bring about a fall in body temperature. Because this fall in temperature is detected by the thermoreceptors and relayed back to the thermoregulatory center, the output of the system is effectively *fed back* into the system's input, forming what is known as a *feedback loop* (Figure 1.6). It is

called *negative* feedback because the response of the system (the fall in temperature) is opposite in direction to the change that set it in motion (a rise in temperature).

Negative feedback is important because it lets a regulatory system "know" whether or not it is making appropriate adjustments to the regulated variable. Once a rise in body temperature triggers a compensatory lowering of the temperature, for example, the brain's thermoregulatory center must be kept informed of the actual body temperature on a continual basis. Otherwise, thermoregulatory mechanisms might continue to lower the temperature below the set point. To illustrate principles of negative feedback in a more concrete fashion, mechanisms of body temperature control are discussed in greater detail in the next section.

In addition to negative feedback systems, a few *positive feedback* systems are also important in physiology. In **positive feedback,** the response of the system goes in the *same* direction as the change that sets it in motion. In females, for example, the pituitary gland (a small gland located at the base of the brain) secretes a hormone called *luteinizing hormone* (LH) that stimulates the ovaries to secrete hormones called *estrogens,* which regulate reproductive function. Under certain conditions, a rise in the plasma estrogen concentration can trigger an increase in the secretion of LH. This stimulates estrogen secretion, which enhances LH secretion even more, leading to further estrogen secretion, and so on. The result is a rapid rise in plasma LH, known as the *LH surge,* which triggers ovulation. Unlike negative feedback, which minimizes changes in physiological variables, positive feedback is useful in certain physiological systems because it allows a variable to change rapidly in response to a stimulus.

Even though a variable may change rapidly in positive feedback, it does not increase indefinitely or spiral out of control. This is because some factor always acts to terminate the positive feedback loop either by removing the original stimulus or limiting the system's ability to respond to that stimulus. During an LH surge, for example, the LH concentration rises rapidly to a peak and then begins to fall because the surge triggers ovulation, which temporarily inhibits the ovaries' ability to secrete estrogens. The resulting fall in plasma estrogen levels removes the stimulus that caused LH secretion to rise in the first place, thereby allowing LH levels to fall.

Homeostasis in Action: Thermoregulation

When nighttime comes to the desert, the snakes, lizards, and insects that were once active in the noonday sun begin to sink into a state of relative torpor; many are barely able to move at all. This change occurs because the falling temperature causes these animals' bodies to cool, which slows down biochemical reactions and other metabolic processes. By contrast, humans and other mammals are less affected by changes in the ambient temperature because they have the ability to maintain their body temperature within a fairly narrow range (a process called **thermoregulation**). Animals with this ability are said to be *homeothermic,* whereas those lacking this ability are *poikilothermic.*

All animals (and other living things) produce heat as a by-product of metabolism, but homeothermic animals are able to control body temperature by regulating the rates at which heat is produced and lost from the body. By increasing the rate of heat production and/or decreasing the rate of heat loss, body temperature is made to rise; by decreasing heat production and/or increasing heat loss, body temperature is made to fall.

However, there are limits to what these thermoregulatory mechanisms can accomplish. Prolonged exposure to very cold temperatures can cause body temperature to fall below the set point, a condition known as **hypothermia.** This happens, for example, to those unfortunate enough to be trapped in mountain snowstorms or swept from a boat into icy waters. Such misfortunes can quickly lead to stupor, loss of consciousness, multiple organ failure, and ultimately death. In contrast, overexertion or exposure to very high environmental temperatures can cause the body temperature to rise above the set point, a condition called **hyperthermia.** If severe enough it can lead to loss of consciousness, convulsions, respiratory failure, and death. Adverse effects begin to appear when body temperature approaches 41°C (106°F); a temperature of 43°C (109°F) or higher is usually fatal.

Mechanisms of Heat Transfer Between the Body and the External Environment

Under most conditions, the body loses heat to the environment because the surrounding temperature is normally lower than body temperature. When the rate of heat loss equals the rate of heat generation, body temperature does not change. Generally speaking, heat loss occurs via three different mechanisms: (1) *radiation,* (2) *conduction,* and (3) *evaporation.*

In **radiation,** thermal energy is transferred from the body to the environment in the form of electromagnetic waves. It is a general law of physics that all objects emit and absorb these waves, though to varying degrees. When an object is warmer than its surroundings, it emits more energy than it absorbs and loses heat. By contrast, if an object is cooler than its surroundings, it absorbs more energy than it emits and gains heat. If you have ever felt heat from a fire when you were standing several feet away, it is because you were absorbing radiated heat energy.

Conduction is the transfer of thermal energy between objects that are in direct contact with each other. As in radiation, heat is always transferred from the warmer object to the cooler object. When you touch cold metal or cold water, for example, you feel colder because thermal energy is transferred directly from your skin to the surrounding medium.

In **evaporation,** heat is lost from an object through the evaporation of water from its surface. When water evaporates from your body, the process that converts it from liquid form to gaseous form absorbs thermal energy. Some evaporation occurs through the skin, the lining of the lungs, and other moist surfaces such as the lining of the mouth, and is known as *insensible water loss* because it occurs on a continual basis without you being aware of it. Your body also loses water through the evaporation of *sweat,* a salt-containing solution secreted by numerous small *sweat glands* in the skin. Unlike insensible water loss, which is unavoidable, sweat production is regulated according to the body's needs. When increased heat loss is desirable, sweat production is increased. As a result, more water evaporates from the skin surface, which carries away more thermal energy from the body.

When the environmental temperature is warmer than body temperature, radiation and conduction actually transfer heat *into* the body. Because this transfer only adds to the heat generated by the body itself, how can the body regulate its temperature under these conditions? The answer is that the body relies solely on *evaporation* to carry heat away and therefore increases the production of sweat. Sweating cools the body under these conditions because water continues to evaporate even when it is cooler than its surroundings (assuming that the humidity of the surroundings is not too high).

When you go out in cold weather and the wind is blowing, you feel much colder than you would if the air were still because heat loss is accelerated by the movement of cool air across the skin surface. Under still conditions, the air that is closer to your skin warms up due to the absorption of heat from the body's surface. This warmer air forms a kind of "blanket" around you that slows down the rate of heat loss by conduction. Because the air in this protective layer contains moisture that has evaporated from the skin, it also tends to be moister than the surrounding air. The presence of this moisture near the skin reduces the rate of evaporative heat loss. When the surrounding air is moving, however, the thickness of this protective "blanket" of air is reduced, and conductive and evaporative heat loss both increase—which is why a fan helps to cool you on a hot summer's day. (The movement of air across your skin is an example of *convection,* in which heat is transferred from one place to another by a moving fluid.)

The Components of the Body's Thermoregulatory System

The body's thermoregulatory efforts are coordinated by a number of different centers in the brain, the most important of which is in the *hypothalamus,* a region located in the base of the brain just above the pituitary gland. Input to these brain centers comes from thermoreceptors located within the brain itself, the spinal cord, and other internal organs. These receptors, known as *central thermoreceptors,* are important because they monitor the temperature deep within the body (known as the **core temperature**). Other thermoreceptors, called *peripheral thermoreceptors,* are located in the skin, whose temperature is usually well below core temperature and is more variable. These receptors, which are of less importance in thermoregulation, enable us to sense the temperature of the environment. This ability is indirectly important in temperature regulation because it allows us to compensate for changes in environmental temperature by making behavioral adjustments, such as dressing appropriately or avoiding extreme temperatures altogether.

Output from brain thermoregulatory centers is transmitted by neurons to various effectors that vary the rate of heat production or loss. The effectors include *sweat glands,* which control evaporative heat loss by increasing or decreasing sweat secretion; *blood vessels in the skin,* which control conductive and radiative heat loss by increasing or decreasing blood flow to the skin; and *skeletal muscles,* which control heat production through *shivering*—bursts of rapid, involuntary contractions that generate heat as a metabolic by-product. When the core temperature changes, brain thermoregulatory centers attempt to compensate by sending appropriate commands to these effectors. As long as the outside temperature is within a narrow range called the *thermoneutral zone* (25–30°C), alteration of skin blood flow is usually sufficient to adjust the core temperature to normal. Outside this range, the other effector responses are called into play.

The Thermoregulatory Response to a Drop in Temperature: A Negative Feedback System in Operation

Suppose the temperature of the body's surroundings drops from 30°C to 15°C (Figure 1.7a). As the body's

rate of heat loss increases, the core temperature gradually drops below the normal set point, from 37°C to an eventual minimum of 36°C (Figure 1.7b). As this is occurring, information concerning the body's temperature is relayed by central thermoreceptors to the brain's thermoregulatory centers, which detect the error signal and orchestrate the following responses (Figure 1.7c):

1. *Decreased sweat production:* If the sweat glands are active initially, sweat production decreases or even stops altogether, which reduces evaporative heat loss.

2. *Decreased blood flow to skin:* Certain blood vessels in the skin constrict in response to neural commands, leading to a decrease in blood flow to the skin. Because the blood normally carries heat from the warmer deep regions of the body to the cooler outer regions, this reduction in flow causes skin temperature to decrease. As a result, rates of radiative and conductive heat loss decrease.

3. *Stimulation of shivering:* Skeletal muscles begin to shiver, which generates heat. This heat adds to the heat generated by other metabolic processes, and the body's overall rate of heat production increases.

The reduction in heat loss coupled with the increase in heat production raises core temperature. If these compensatory changes are able to restore the balance between heat production and loss, body temperature eventually returns to near its normal value and stays there (see Figure 1.7b). But no negative feedback system is perfect, and thermoregulatory mechanisms cannot hold body temperature absolutely constant. Instead, the temperature undergoes small variations and fluctuates above and below the set point; such fluctuations are normal and occur in all physiological variables.

In contrast to the events depicted in Figure 1.7, a rise in the surrounding temperature will cause a rise in core temperature, which triggers the opposite responses—increased sweat production, increased blood flow to the skin, and a decrease or complete cessation of shivering. The resulting increase in heat loss and decrease in heat production should act to reduce the core temperature, returning it to near normal.

Fever, an elevation of body temperature that frequently occurs during illness, does *not* result from a failure of the thermoregulatory system. Rather, fever results from a resetting of the set point that causes the system to raise the body's temperature in a controlled fashion. Fever is most often triggered in response to bacterial or viral infection, which stimulates certain blood cells (*white blood cells*) to proliferate and secrete various chemical substances. One or more of these substances act on thermoregulatory centers in the brain to increase the set point and therefore act as *pyrogens* (fever-inducers). Because the resulting temperature elevation stimulates an increase in the rate of many immune responses against invading microorganisms, fever is actually considered beneficial because it enhances the body's ability to defend itself.

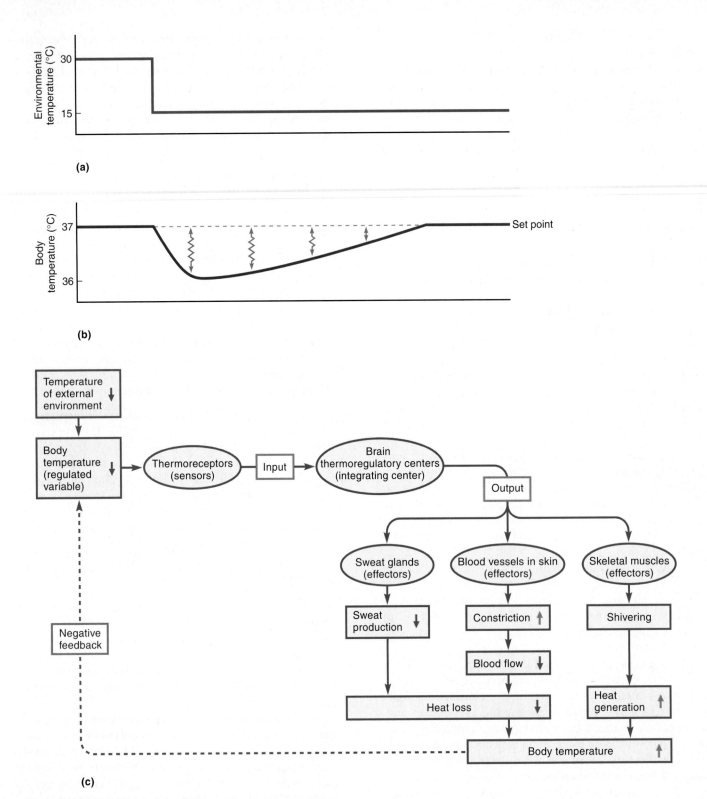

FIGURE 1.7 Thermoregulation. (a) *A graph of environmental temperature over time showing a drop from 30°C to 15°C.* **(b)** *A graph of body temperature over time showing the set point temperature of 37°C (dashed line) and error signal (vertical arrows).* **(c)** *Events occurring in the body in response to a drop in environmental temperature.*

When body temperature is constant, how do rates of heat generation and heat loss compare?

They are equal.

Quick Test 1.2

1. Define *homeostasis* and explain how it enables the body to adapt to changes in its environment.

2. Explain how negative feedback works to maintain homeostasis, and define the following terms: *regulated variable, set point, error signal, integrating center, sensor, effector.*

3. Explain the difference between positive feedback and negative feedback.

4. Name three types of effectors that are important in thermoregulation, and describe how each responds to a drop in body temperature.

CHALLENGING HOMEOSTASIS: A STUDY IN ORGAN SYSTEMS INTEGRATION DURING EXERCISE

The science of physiology progresses according to the belief that the function of the body can be understood only by examining the functions of its parts. As a natural consequence of this belief, physiologists tend to shift the focus of their investigations to objects on smaller and smaller scales—from organs and organ systems to cells, and then from cells to molecules. As a result, much of physiology today is scarcely distinguishable from biochemistry or molecular biology and bears little resemblance to the discipline of 50 years ago, when "physiologists" worked with animals in a laboratory and "biochemists" worked with test tubes containing molecules isolated from their organ of origin.

The new "molecular" approach to physiology has yielded invaluable insights into the workings of the body and has sparked a recent growth of knowledge that is nothing less than explosive. However, these gains come at a price: The current wealth of knowledge is too great to permit any one person a thorough understanding of all the body's functions. For this reason, physiologists must specialize and gain a firm grasp of their particular, narrower field. Thus, a world-renowned expert in cardiac electrophysiology, for example, might know relatively little about how the brain or stomach function and may not have thought much about these organs since the days when he or she was a student.

As a student of physiology, you will be faced with the same problem that professional physiologists face, although on a smaller scale. To learn how a particular organ works, you may have to spend a lot of time studying the functions of particular cells or systems of molecules inside those cells, and it may be difficult to assemble all these details into a coherent picture of the organ's function and its significance to the rest of the body.

As you study the nervous system, for example, you will learn that the signals transmitted by nerve cells result from the action of special cell membrane proteins called *ion channels,* of which there are several types. In examining how these channels work to generate signals, you will learn that channels regulate the flows of different ions, and that these flows affect the distribution of electric charges across the membrane. When trying to sort all of this out, it is easy to forget that you are actually learning about how signals are transmitted in the nervous system, a process that directly or indirectly influences the activity of virtually every tissue or organ in the body. Put another way, as you learn about ions flowing across membranes, you are really learning about something that bears on much of what you encounter throughout this book. This example illustrates that a thorough understanding of physiology requires not only the ability to focus in on details, but also the ability to "pull back" and see things in a larger perspective (that is, to integrate). Indeed, this is a prerequisite for understanding how the various organ systems work together to maintain homeostasis and to perform other body functions.

As you progress through this book, you will spend most of your time learning about specific organs and organ systems. Although these systems are described in separate chapters, in reality these systems operate concurrently and influence each other extensively. To reinforce your learning and help you integrate your knowledge of organ systems into a coherent picture of whole-body function, the final chapter examines the physiological adjustments that occur in exercise, which affects every organ system in the body. In particular, we will follow the changes that occur in the bodies of two runners, Bill and Jane, during the course of a 26.2-mile marathon, as described shortly.

The Hierarchy of Resource Allocation in Response to Activity

As we learned earlier, the flow of blood distributes resources to each part of the body by carrying oxygen and nutrients (such as glucose) into tissues and waste products away from tissues. Because the tissues' need for resources can change, however, blood flow must be adjusted to meet the needs of each organ. Throughout exercise and all other activities (even the "activities" of sleeping or sitting in a chair reading), there is a specific hierarchy that governs how resources get allocated to each organ: The brain has top priority. Maintaining stable blood and oxygen delivery to the brain requires that the heart also receives an unconditional allotment of blood. During injury or stress, blood flow to other organs can be sacrificed for the sake of the brain (and heart). But effective clearance of metabolic waste products remains a necessity, and therefore blood flow to the kidneys and liver cannot be interrupted by very much for very long. Owing to their low energy needs

TABLE 1.1 PHYSICAL CHARACTERISTICS OF THE TWO MARATHON RUNNERS		
	BILL	**JANE**
Age (years)	33	31
Height (cm)	177	166
Weight (kg)	72	57
Percent fat	16	22
Aerobic capacity (ml of O_2/kg/min)	55	51

at rest, blood flow to the skeletal muscles, digestive tract, and skin can be more severely restricted for longer periods of time.

Different types of exercise can elicit different patterns of physiological response. We will look at the physiological responses to running a marathon because it imposes large stresses on healthy individuals. People participating in a marathon can increase their energy expenditure (and hence their oxygen consumption and heat production) by 15-fold or more. In addition, they must maintain this increased metabolism for relatively long periods of time—anywhere between two and six hours, depending on their talents and training status. The runners either finish the 26.2-mile race, or they are forced to drop out along the way. The factors that determine how fast (or if) they finish will be described here.

Although we must wait until the final chapter for our full treatment of exercise physiology, we don't have to wait that long to begin our exploration of the subject. In fact, we can begin shortly by following the experiences of our two runners, who as we join them are beginning their preparations to run their marathon. As you read through the following story, you may not understand all the terms or statements in the narrative, but do not worry about that now. When you study later chapters in this textbook and learn about various physiological mechanisms, we will use features called *Exercise Links* to point out how particular mechanisms play a role in the experiences of our intrepid runners. At the end of the book you will be asked to read the story again. By then, all the terminology will be familiar, and you will have a more complete understanding of the events.

Before we consider their race experiences we turn to some relevant physical characteristics for Bill and Jane, displayed in Table 1.1. (Male and female runners were chosen as the subjects of this story to illustrate some physiological differences between the sexes that affect exercise performance.) Bill is taller and heavier because he has more muscle and a larger skeleton, but Jane has a higher percentage of body fat. Bill has a higher *aerobic capacity;* that is, he has a greater ability to use oxygen to generate energy when exercising at his maximal effort. (This measure is one indicator of physical fitness.) Aero-

bic capacity is measured as the volume of oxygen (ml of O_2) that is consumed per minute. Because a larger person usually has more cells that are using oxygen, regardless of physical fitness, the rate of oxygen consumption is adjusted to the person's body weight (per kilogram) when calculating aerobic capacity. Based simply on aerobic capacity, one might predict that Bill can run slightly faster than Jane for a prolonged period of time. Over shorter race distances ("10K" races, which are 10 kilometers or 6.2 miles) this is in fact true: Bill's best time is 43 minutes, and Jane's best is about a minute slower. But as we will see, aerobic capacity is only one of many factors that determine performance in a marathon.

Many integrated physiological processes will play out during the marathon, and several decisions by the two runners will have profound influences on these processes. For those of you who have participated in endurance events, some of these situations will be familiar. Now, let's examine Bill and Jane's marathon experiences.

Bill and Jane's Marathons

Our narrative is divided into three time periods. It begins not as the runners leave the starting line, but six months earlier, as Bill and Jane begin to train for the marathon.

Training

If Bill and Jane had decided impulsively to run a marathon without preparation, the chances that they might finish would be limited. Exercise causes depletion of

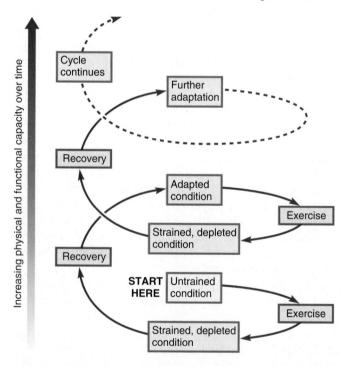

FIGURE 1.8 The upward spiral of exercise training. *Small increments of strain and debilitation resulting from each exercise session are followed by overcompensation during recovery, which leads to an adapted condition with higher physical and functional capacity than the previous condition.*

TABLE 1.2 ANATOMICAL AND FUNCTIONAL CHANGES RESULTING FROM TRAINING

ORGAN, TISSUE, OR STRUCTURE	ANATOMICAL CHANGES	FUNCTIONAL CHANGES
Heart	Size increases	Resting heart rate decreases; resting stroke volume increases, which increases pumping efficiency
Blood	Volume of blood and red cell mass increase	Oxygen-carrying capacity of blood increases
Skeletal muscle	Muscle mass, capillary density, and glycogen stores increase	Ability to develop useful power increases
Lungs	N/A	Ability of airways to widen in response to neural and hormonal influences increases, which increases ease of breathing during exercise
Bone	Density increases	Bone strength increases, which increases the body's ability to withstand the physical stresses of exercise
Endocrine and autonomic nervous systems	N/A	Sensitivity of tissues to insulin and pituitary hormones, and to autonomic input, increases, which enhances mobilization of energy stores during exercise
Various cells	N/A	Synthesis of stress proteins inside cells increases
Brain	N/A	Learning leads to improved technique, which decreases exertion; coping mechanisms decrease perceived exertion
Sweat glands	N/A	Production of more copious and more dilute sweat, which enhances the body's ability to cool itself

resources stored in the body (water, energy, and nutrients) and a certain amount of "wear and tear" on the muscles, tendons, and joints. During recovery from exercise, physiological processes not only restore the resources and repair the strained tissues, but also overshoot the original (untrained) condition by increasing the storage capacities and strengthening the tissues (adapted condition). Thus training comprises repetitive cycles of strain and depletion followed by a recovery process of overcompensation resulting in an upward spiral of increasing physical and functional capacity (Figure 1.8).

Without worrying about the actual physiological mechanisms yet, several aspects of training are apparent in Figure 1.8. First, each exercise session causes some debilitation; thus a person in training must be careful not to "overdo it" in any single session. Second, a recovery period is essential for the positive adaptations to occur;

thus proper rest intervals are as important as the exercise sessions themselves. A third issue is not obvious in the figure: As an individual reaches a new level of adaptation, the exercise stimulus needed to keep the spiral rising increases. The increased stimulus can be accomplished by increasing either the intensity or the duration of the exercise (or a little of both). After many months of training, Bill and Jane have developed the capacity and strength to complete a marathon with relatively little debilitation. As listed in Table 1.2, some organs and tissues actually grow larger in response to training, but many of the changes are the result of improved function rather than increased size.

Bill and Jane (as do many marathon trainees) reduce their training intensity in the week immediately preceding the race. With their goal so close at hand, they want to minimize the "debilitation" of training (and also want to avoid accidental injury after training so long).

Prelude to the Race

In the 24–48 hours just before the race, Bill and Jane "carbo-load" (eat lots of starchy and sweet foods such as bread, pasta, and fruits). These types of food are the source of the most rapidly used energy source—carbohydrate—and the runners are "topping off their fuel tanks" in preparation for the long journey ahead of them. Some runners, including Bill, drink plenty of coffee on the morning of the race. The caffeine in coffee is a stimulant, and Bill knew enough physiology to understand that caffeine also promotes use of carbohydrates as an energy source. Jane, however, knew an additional important fact—that caffeine adversely affects the body's ability to maintain proper water volumes (blood, for example, is mostly water). Jane drank less coffee. These and other decisions will have important consequences in each runner's marathon experience.

The Race

The race starts at 9 A.M. under clear skies and comfortable (68°F) temperatures. It is a popular race that draws several thousand runners. The "elite" runners are placed in a reserved area immediately behind the starting line; everyone else is asked to begin farther back, according to their expected finishing time. Bill and Jane are in the middle of the pack. When the gun goes off, only those up front are actually able to start running. Farther back, people are just too tightly packed; they begin with a slow shuffle and are not able to break into a running stride until the pack spreads out over the first mile. Bill is frustrated by this—in his excitement he wants to be running free, so he darts back and forth across the width of the road, bolting through every gap between runners ahead of him in order to advance. Jane is also running slower than she would like but decides to accept the pace. She knows that she has 26 miles ahead of her—and that there will be plenty of opportunity to make up for minutes lost at the beginning.

By about four miles out, the pack is well dispersed and everyone has settled into their race pace. Bill is about a half mile ahead of Jane and running at a faster pace. The first aid station, where runners can get water and other fluid-replacement beverages, is coming up. The station is crowded, so Bill would have to slow down to pick up something to drink. Because he drank a full liter of water 45 minutes before the race started, Bill does not feel the least bit thirsty, so he decides to skip this aid station. When Jane arrives at the aid station, she slows down to get a cup of water. In fact, she walks as she drinks it, knowing from experience that otherwise more of it splashes on her face than gets in her mouth. The next aid station two miles further along is less crowded, so Bill grabs a cup on the run and keeps going as he drinks. He is able to swallow about one gulp; the rest splashes on his face and down his shirt.

The next few miles are uneventful for Bill, except for a growing awareness of a distention in his bowels. He keeps running, hoping the sensations will pass, but instead the urgency to make a stop increases dramatically. He searches desperately for one of the portable facilities that race organizers have placed along the course for runners in his predicament. He finally finds one and relieves himself of a watery diarrhea. He has spent less than two minutes off the race course but begins running again with increased speed to make up for lost time. Jane has settled into a steady pace. She consistently drinks water at each aid station. For her, the middle portion of the race (miles 4 through 20) is relatively comfortable and uneventful.

By 11 A.M. the sun is almost overhead in the summer sky and the temperature has increased to 86°F. Bill is nearing the 18-mile mark about now. When he wipes his hand across his brow he feels salt encrusted along his temples. Although he was not aware of sweating very much, the presence of this salt suggests otherwise. He decides to make sure to take more water as each aid station presents itself. Bill still feels strong and decides that marathons are not as tough as they are hyped up to be.

By mile 20, however, Bill just wants the anguish to end—but he still has six more miles to go! Within two miles his legs have become leaden and do not seem to respond appropriately to his wishes. His stride no longer has any "spring" to it, but instead is "stiff-legged." His hips hurt, which causes further deterioration of his stride, and painful blisters are forming on his toes. Apart from these specific complaints, he experiences a general sense of fatigue and has a headache. His running pace deteriorates, and he must even walk at times. Meanwhile, Jane is fatigued, but her pace has not deteriorated to the same extent as Bill's. In fact, she catches up and ultimately passes Bill in the last three miles.

Jane finishes in 3 hours and 36 minutes. She is very tired, has a few blisters, and is glad to be finished, but otherwise she feels no unusual symptoms. Bill finishes 12 minutes later. He crosses the finish line, walks on a few paces, and then comes to a complete stop. As he stands there, he feels faint and his knees buckle. Fortunately, race volunteers catch him and ease him into a horizontal position in the shade just off the road.

After a while he feels a bit better. He sits up against a tree and drinks almost a full liter of water. Friends find him and take him home. About an hour later his head aches terribly, and he begins retching. After this episode subsides he falls asleep for a few hours. That evening he feels chilled and wraps himself in a blanket, even though it is a mild summer's evening.

Jane's muscles are somewhat sore the next day, but she is able to carry out normal activities with little difficulty. By contrast, Bill gasps every time he moves his tender and swollen legs. All movement requires a major, painful effort. His muscles feel stiffer and sorer than they felt after the race, and even his biceps are sore. These symptoms slowly resolve over the next few days, but it is a full week before Bill is totally pain free.

Organization of the Body, p. 2

The smallest living units are cells, which are specialized to carry out different functions in the body. Cells of specific types are organized into tissues, which are combined to make organs, which perform specific functions. Collections of organs that work together to perform certain tasks are organ systems. The body contains four major types of cells: (a) neurons, which are specialized to transmit electrical signals from one place to another; (b) muscle cells, which generate mechanical force and movement; (c) epithelial cells, which are organized into epithelia—sheet-like tissues found in the skin and the linings of hollow organs; and (d) connective tissue cells, which have diverse functions and are found in tendons, blood, lymph, and other connective tissue. The body's cells exchange material with fluid surrounding them, referred to as the body's internal environment. Materials are delivered to the internal environment and removed from it by the bloodstream, which exchanges materials with the external environment.

Internal barriers divide the body into several fluid-filled compartments. The total volume of water in all compartments is total body water (TBW), which includes intracellular fluid (ICF) and extracellular fluid (ECF). ICF and ECF differ markedly in composition and are separated by cell membranes. ECF includes plasma, the liquid component of blood, and interstitial fluid (ISF), which surrounds cells located outside the blood. Plasma and ISF are similar in composition and are separated by the epithelial tissues that line blood vessels.

IP Fluid and Electrolytes, Introduction to Body Fluids, pages 1–14

Homeostasis: A Central Organizing Principle of Physiology, p. 8

The body is able to tolerate changing conditions in the external environment through homeostasis, maintenance of constant conditions in the internal environment. To maintain homeostasis, regulatory mechanisms work to control certain physiological variables (regulated variables) so that variations are kept to a minimum. Most of these control mechanisms operate through negative feedback, in which changes in a regulated variable trigger responses that proceed in the opposite direction. Homeostatic regulatory mechanisms work to minimize error signals, differences between the actual value of a regulated variable and the normal or "desired" value (the set point). Homeostatic regulatory mechanisms include the following components: (a) sensors, which monitor the regulated variable; (b) an integrating center, which receives signals (input) from the sensors, compares the regulated variable to the set point, and orchestrates the appropriate response; and (c) effectors, which receive signals (output) from the integrating center and bring about the final response. Some physiological variables are regulated by positive feedback, in which a change in a variable triggers a response that proceeds in the same direction. An example of a homeostatic regulatory process is thermoregulation, which is achieved through adjustments in rates of metabolic heat production and heat loss. When body temperature falls below the set point (normally 37°C), a condition of hypothermia exists; when body temperature rises above the set point, it is hyperthermia. Normally, the body loses heat to its surroundings, which occurs via three routes: radiation, conduction, and evaporation.

The thermoregulatory system acts to maintain constant core temperature by using the following components: (a) thermoregulatory centers in the brain; (b) thermoreceptors, temperature-sensitive neurons that relay signals to the thermoregulatory centers; and (c) a variety of effectors, including sweat glands in the skin (which secrete sweat), skin blood vessels, which regulate blood flow through the skin, and skeletal muscles, which generate heat by shivering. A drop in body temperature below the set point triggers the following responses: (a) decreased sweat production, which reduces evaporative heat loss; (b) decreased blood flow to skin, which reduces conductive and radiative heat loss; and (c) shivering, which increases heat production. The combined effect of these effector responses acts to raise body temperature back toward the set point. A rise in body temperature above the set point triggers the opposite response.

Multiple-Choice Questions

1. When you touch a piece of cold metal, thermal energy travels into the metal from your skin. This is an example of
   a) radiation.
   b) conduction.
   c) convection.
   d) evaporation.

2. Which of the following best illustrates the concept of positive feedback?
   a) The secretion of acid by cells in the stomach lining is suppressed when the acidity of the stomach contents increases.
   b) A rise in blood pressure stimulates the elimination of water in the urine, which reduces the blood pressure.
   c) A severe increase in body temperature can trigger the shutdown of sweat glands, leading to further increases in temperature.
   d) An increase in the carbon dioxide concentration of the blood stimulates breathing, which increases the rate at which carbon dioxide is eliminated from the body.

3. When the temperature of the body's surroundings is higher than that of the body itself, the body loses heat through
   a) conduction.
   b) radiation.
   c) evaporation.
   d) none of the above.

4. The hormone aldosterone stimulates the reabsorption of sodium ions from the lumen of a kidney tubule. Based on your knowledge of the body's cell types, you can surmise that this hormone acts on
   a) neurons.
   b) muscle cells.
   c) epithelial cells.
   d) connective tissue cells.

5. Endothelial cells are found in
   a) muscle tissue.
   b) the lining of the stomach.
   c) the blood.
   d) the lining of blood vessels.

Objective Questions

1. The body's internal environment is synonymous with (extracellular fluid/intracellular fluid).

2. Maintenance of constant conditions in the internal environment is known as _____.

3. Plasma is extracellular fluid. (true/false)

4. A thermoreceptor is an example of a/an (sensor/effector).

5. In homeostasis, *all* physiological variables are regulated to stay constant. (true/false)

6. In hypothermia, body temperature is (above/below) the set point.

7. Sweating accounts for 100% of the body's evaporative heat loss. (true/false)

8. A decrease in blood flow to the skin tends to (raise/lower) body temperature.

9. The protein elastin is found in (epithelial/connective) tissue.

10. Plasma and interstitial fluid are identical in composition. (true/false)

Essay Questions

1. Draw as many parallels as you can between the mechanism of body temperature regulation and the control of room temperature by a thermostat. In your discussion, include a mention of set points, error signals, sensors, effectors, and negative feedback.

2. Draw a tree diagram illustrating the relationship between total body water, extracellular fluid, intracellular fluid, interstitial fluid, and plasma.

3. Suppose that molecules of glucose are ingested, absorbed into the bloodstream, and then converted inside muscle cells to carbon dioxide, which is eliminated through the lungs. Describe the path of these molecules as they travel through the body, being sure to mention the various barriers (epithelia or cell membranes) that must be crossed.

Find the answers to these exercises, and additional study tools, at the Physiology Place (www.physiologyplace.com).

The Cell: Structure and Function

OBJECTIVES

- Know the general characteristics of the four major classes of biomolecules (carbohydrates, proteins, lipids, and nucleic acids) and give a brief description of their functions in cells.

- Describe the structure and major functions of each of the following cellular components: plasma membrane, nucleus, ribosomes, rough endoplasmic reticulum, smooth endoplasmic reticulum, Golgi apparatus, mitochondria, lysosomes, peroxisomes, cytoskeleton.

- Define *transcription* and *translation,* and describe the role of each of the following in protein synthesis: DNA, genes, codons, genetic code, messenger RNA, transfer RNA, ribosomes, anticodons, rough endoplasmic reticulum.

- Explain how genetic information is stored in DNA, and how this information is passed on to other cells during mitosis.

- Briefly describe what happens to proteins following their synthesis, taking into account the different fates of membrane proteins, secreted proteins, and cytoplasmic proteins.

- Define *endocytosis* and *exocytosis,* and explain the primary difference between these and other mechanisms of cellular transport. Distinguish between receptor-mediated endocytosis and pinocytosis.

CHAPTER OUTLINE

Biomolecules   22

Cell Structure   34

Cell-to-Cell Adhesions   43

General Cell Functions   46

Protein Synthesis   48

Cell Division   56

Above: Electron micrograph of a cell

If you have ever taken a drop of pond water and looked at it under a microscope, then you've probably observed several single-celled organisms. These organisms survive by relatively simple processes. They take up nutrients from their environment, break them down, and convert them to usable energy or to substrates required to synthesize large molecules needed for life. They release waste products into the environment. They are motile in the water and respond to various types of stimuli. Lastly, these organisms can grow and reproduce.

The cells of your body are remarkably similar to these unicellular organisms. Your cells require nutrients for energy and for synthesizing needed molecules. Your cells generate waste products that must be eliminated. Some of your cells are motile, and most of your cells can grow and reproduce. However, unlike the unicellular organisms, your cells are part of a multicellular organism—you. As such, your cells must carry out coordinated activities, and different types of cells are specialized for certain functions, as discussed in Chapter 1. The ability of organ systems to carry out their functions depends on the cells that form them. Therefore, it is essential that you have a basic understanding of cellular physiology before you learn about the different organ systems. This chapter describes the basic structures and functions of cells.

BIOMOLECULES

In Chapter 1 you learned about the concept of homeostasis, and that physiology involves the study of how the organ systems work together to maintain homeostasis. To understand organ systems, we must understand their basic units: cells. To understand cells, we must first understand biomolecules.

Biomolecules are molecules that are synthesized by living organisms and contain carbon atoms (Toolbox: Atoms and Molecules). Carbon has four electrons in its outer shell and forms covalent bonds by sharing these electrons with other atoms, including other carbon atoms. Other atoms common to biomolecules include oxygen, hydrogen, and nitrogen. The ability of carbon to share four electrons and to form covalent bonds with other carbon atoms enables carbon-containing molecules to be large and complex. The carbon atoms are often arranged in chains or in a ring.

The four basic types of biomolecules are *carbohydrates, lipids, proteins,* and *nucleotides.* Some biomolecules are *polymers,* which consist of repeated subunits. For example, proteins are polymers of *amino acids.* The following sections discuss each class of biomolecule. Table 2.1 lists the functional groups commonly found in biomolecules.

Carbohydrates

Carbohydrates are composed of carbon, hydrogen, and oxygen in a ratio of 1:2:1, with the general chemical formula $(CH_2O)_n$. The chemical formula can also be written as $C_n(H_2O)_n$, which might be interpreted as *hydrated* carbons, or carbons surrounded by water—thus the name *carbohydrates.* However, the name can be deceiving because the carbons bond to hydroxyl groups (—OH) and hydrogens (—H), not to water molecules. The presence of several hydroxyl groups makes carbohydrates polar molecules, and therefore they readily dissolve in water (Toolbox: Polar Molecules and Hydrogen Bonds, p. 24).

On the basis of their molecular sizes, carbohydrates can be further classified into three major groups: (1) *monosaccharides,* (2) *disaccharides,* and (3) *polysaccharides.* **Monosaccharides** are the simple sugars, composed of a single unit (Figure 2.1a). The most common monosaccharide is **glucose,** which is an important source of energy for our cells. Glucose has the general formula $C_6H_{12}O_6$. Two other common monosaccharides, *fructose* and *galactose,* also have the general formula $C_6H_{12}O_6$. However, each is a distinct molecule because the atoms are arranged differently, giving each its own chemical properties. *Ribose* and *deoxyribose* are two other common monosaccharides. These molecules are important components of *nucleotides,* another class of biomolecules that are described later.

Disaccharides are carbohydrates formed by the covalent bonding of two monosaccharides (Figure 2.1b). Common examples of disaccharides include *sucrose,* which is composed of a glucose and a fructose joined together, and *lactose,* which is composed of a glucose and a galactose joined together. Sucrose is also known as table sugar, whereas lactose is a carbohydrate found in milk. Notice that the names of the monosaccharides and disaccharides all end in *-ose.*

Polysaccharides are carbohydrates formed by the covalent bonding of several monosaccharides. **Glycogen** (Figure 2.1c) is a polymer of glucose subunits and is found in animal cells. Several types of cells in the body can store glucose as glycogen and then later break glycogen down when they need glucose for energy. **Starch** is a polysaccharide found in plants. Humans consume starch in various plant food products; the digestion process makes the glucose subunits of starch available as energy sources. **Cellulose,** another polysaccharide found in plants, is consumed by humans, but we are unable to digest and absorb it. Therefore, cellulose, also known as

All matter is composed of fundamental units called *atoms,* which are on the order of one billionth of a centimeter in diameter. Atoms, in turn, are made up of three types of elementary particles: (1) *protons,* which each carry one unit of positive electrical charge; (2) *electrons,* which each carry one unit of negative electrical charge; and (3) *neutrons,* which carry no charge. Normally, an atom possesses equal numbers of protons and electrons, giving it a net charge of zero. The protons and neutrons are densely concentrated at the atom's center in a core called the *nucleus,* while the electrons travel in orbits or *shells* located at various distances from the nucleus.

Atoms are distinguished from one another on the basis of their *atomic number,* which equals the number of protons (and thus also the number of electrons) they possess. The atomic number also determines the chemical properties of an atom. Pure substances consisting entirely of atoms having the same atomic number are referred to as *elements.* Although over 100 elements are known, just four account for over 99% of all the atoms in the body: *hydrogen* (H), *carbon* (C), *nitrogen* (N), and *oxygen* (O), whose atomic numbers are 1, 6, 7, and 8, respectively.

Most substances are composed of two or more atoms linked or *bonded* together to form *molecules.* Molecules of water, the most abundant substance in the body, contain two hydrogen atoms and one oxygen and are designated by the formula H_2O. Carbon dioxide, a waste product generated by cells, contains two oxygen atoms and one carbon atom and is designated as CO_2. Most often, a molecule's atoms are held together by *covalent bonds,* which consist of pairs of electrons shared between adjacent atoms. Because all atoms of a given type possess the same number of sharable electrons, each element has a certain capacity for forming bonds. Hydrogen atoms form only a single bond, but oxygen, nitrogen, or carbon can form two, three, or four bonds, respectively:

$$H \cdot \qquad \cdot \overset{\cdot}{O} \cdot \qquad \cdot \overset{\cdot}{N} \cdot \qquad \cdot \overset{\cdot}{\underset{\cdot}{C}} \cdot$$

As long as each atom forms the correct number of bonds, they can be combined to form a virtually unlimited variety of molecules. Molecules of water, hydrogen (H_2), and methane or *natural gas* (CH_4) can be represented as follows:

$$H \overset{\cdot \cdot}{\underset{\overset{\cdot \cdot}{O}}{\,}} H \qquad H \colon H \qquad H \colon \overset{H}{\underset{H}{\overset{\cdot \cdot}{C}}} \colon H$$

Water   Hydrogen   Methane

Sometimes, atoms share two pairs of electrons, forming *double bonds.* This is illustrated in the following representation of a molecule of carbon dioxide:

$$O \colon\colon C \colon\colon O$$

Carbon dioxide

TABLE 2.1	**COMMON FUNCTIONAL GROUPS FOUND IN BIOMOLECULES**		

FUNCTIONAL GROUP	CHEMICAL FORMULA	STRUCTURE	CHEMICAL PROPERTY
Hydroxyl	—OH	—O—H	Polar
Sulfhydryl	—SH	—S—H	Polar
Phosphate	$-HPO_4^-$	$-O-\overset{\displaystyle O}{\overset{\|}{\underset{\underset{\textstyle O-}{\|}}{P}}}-OH$	Polar
Carboxyl	—COOH	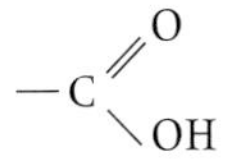 $-C\overset{\displaystyle O}{\underset{\textstyle OH}{\diagup}}$	Acid
Amino	$-NH_2$	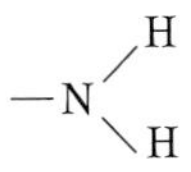 $-N\overset{\textstyle H}{\underset{\textstyle H}{\diagup}}$	Base

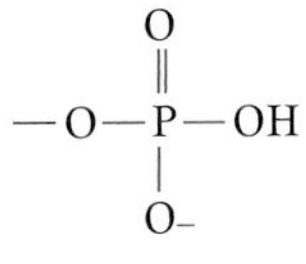

When two atoms are covalently bonded, they share electrons. However, this sharing may or may not be equal. Certain atoms, such as oxygen (O), nitrogen (N), or sulfur (S), hold onto electrons tightly. Therefore, when bonded to other atoms, they have a tendency to pull electrons away from the other atoms. As a result, each oxygen, nitrogen, or sulfur atom acquires a partial negative charge because it has slightly more than its "fair share" of electrons; that is, it has more electrons than are necessary to balance the positive charge of the protons in its nucleus. In constrast, the atom to which the oxygen, nitrogen, or sulfur is bonded is left with less than its "fair share" of electrons, which gives it a partial positive charge. This can be illustrated as follows for molecules in which oxygen, nitrogen, or sulfur is bound to hydrogen, with the R representing the remainder of the molecule and the δ representing a partial charge:

$$\overset{\delta^-}{R:\overset{}{O}:}\overset{\delta^+}{H} \qquad \overset{\delta^-}{R:\underset{\underset{\overset{|}{H}\,\delta^+}{}}{N}:}H \qquad \overset{\delta^-}{R:\overset{}{S}:}\overset{\delta^+}{H}$$

Bonds characterized by such unequal electron sharing are known as *polar bonds*. In constrast, in *nonpolar bonds* electrons are shared more or less equally, such that the atoms remain uncharged. Carbon to hydrogen bonds (C—H) are a very common example of nonpolar bonds found in the molecules of the body.

The presence of electrical charges within a polar molecule produces electrical forces. Direction of electrical forces follow a simple rule: *Opposite charges attract; like charges repel.* The positive region of one polar molecule is electrically attracted to the negative region of another polar molecule (or in some cases, the positive and negative regions can be within the same large polar molecule). This electrical attraction holds the two polar molecules together, forming a *hydrogen bond.* The polar bond is called a hydrogen bond because in a polar molecule it is usually a hydrogen that contains the partial positive charge. A hydrogen bond between two polar water molecules can be represented as follows:

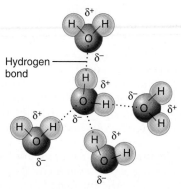

Even though hydrogen bonds are weak bonds and can easily be broken, they are important in determining the structure of large biomolecules such as proteins and nucleic acids, and in establishing the properties of substances in water. Water, the most abundant molecule in the body, is a polar molecule and a liquid at body temperature. How molecules behave in water is critical to cell function. Polar molecules are electrically attracted to the polar water molecules, and therefore they dissolve in water. Polar molecules are called hydrophilic ("water-loving") because of their ability to dissolve in water. Nonpolar molecules are called hydrophobic ("water-fearing") because they do not dissolve in water.

Although water is the most common solvent in the body, membranes are comprised of lipids, and how a substance behaves in lipids is also important to physiology. Nonpolar molecules dissolve in lipids and can permeate the phospholipid bilayer of membranes, and thus are called lipophilic ("lipid-loving"). Polar molecules cannot dissolve in lipids nor permeate the phospholipid bilayer of membranes, and thus are called lipophobic ("lipid-fearing"). The table below summarizes the nature of molecules held together by covalent bonds. Notice that hydrophilic molecules are lipophobic, and hydrophobic molecules are lipophilic.

ELECTRON SHARING IN COVALENT BOND	CLASS OF COVALENT BOND	PROPERTY IN WATER	PROPERTY IN LIPID
Equal sharing	Nonpolar	Hydrophobic	Lipophilic
Unequal sharing	Polar	Hydrophilic	Lipophobic

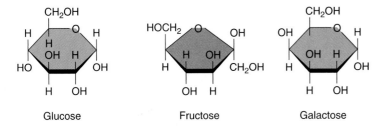

(a) Monosaccharides

Glucose Fructose Galactose

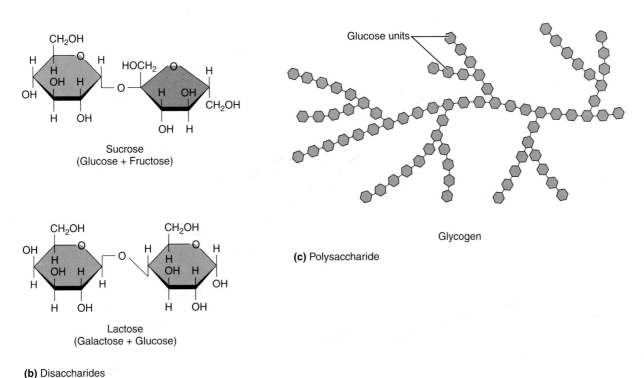

Sucrose
(Glucose + Fructose)

Lactose
(Galactose + Glucose)

(b) Disaccharides

Glucose units

Glycogen

(c) Polysaccharide

FIGURE 2.1 Carbohydrates. (a) *Examples of three mono-saccharides with the same chemical formula: $C_6H_{12}O_6$.* **(b)** *Examples of two disaccharides. Sucrose, or table sugar, consists of a glucose and a fructose. Lactose, a carbohydrate found in milk, consists of a glucose and a galactose.* **(c)** *A polysaccharide, which is a polymer of a monosaccharide. Glycogen is a glucose polymer found in animals.*

dietary fiber, passes through our gastrointestinal system. In addition to storing energy, polysaccharides are important components of cell membranes, a topic to be described later.

Lipids

Lipids are a diverse group of biomolecules that contain primarily carbon and hydrogen atoms linked together by nonpolar covalent bonds. Therefore, lipids generally are nonpolar molecules and do not dissolve in water. Most lipids also contain some oxygen and several contain phosphates ($-HPO_4^-$), which, depending on the structure, may provide areas of the molecule that are polar. A molecule that contains both polar and nonpolar regions is

called **amphipathic.** There are four main classes of lipids that vary both structurally and functionally: (1) *triglycerides,* (2) *phospholipids,* (3) *eicosanoids,* and (4) *steroids.*

Triglycerides

Triglycerides—what we typically call "fat"—are composed of two components: one glycerol molecule and three fatty acid molecules. **Glycerol** is a three-carbon carbohydrate (Figure 2.2a) that functions as the "backbone" of a triglyceride; **fatty acids** are long chains of carbon atoms with a carboxyl group ($-COOH$) at one end (Figure 2.2b and c). **Triglycerides** are formed by linking each of the three fatty acids to a different carbon in the glycerol backbone (Figure 2.2d).

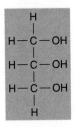

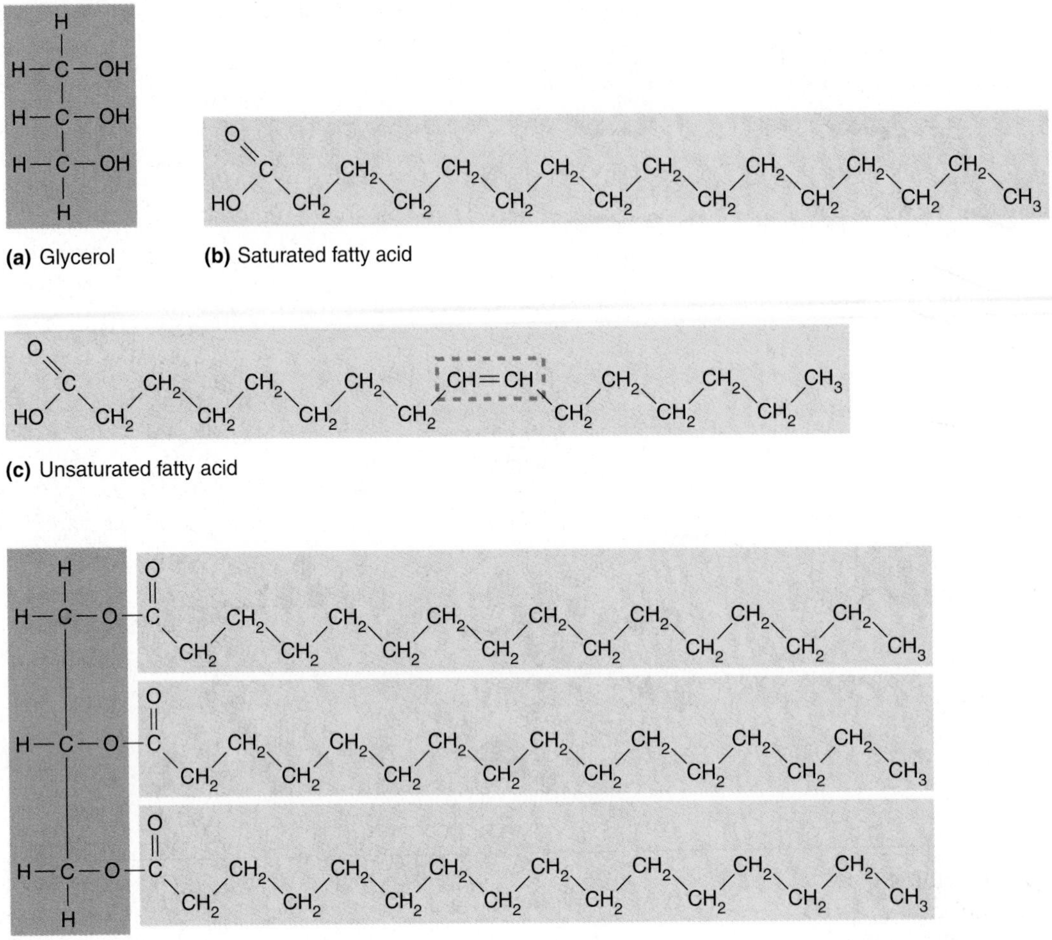

(a) Glycerol

(b) Saturated fatty acid

(c) Unsaturated fatty acid

(d) Triglyceride

FIGURE 2.2 Triglycerides. (a) *Glycerol, a three-carbon carbohydrate found in both triglycerides and phospholipids.* **(b)** *A fatty acid chain, which usually consists of an even number of carbon atoms with a carboxyl group (—COOH) on the end. This example is a saturated fatty acid because it has no double bonds between carbon atoms.* **(c)** *This fatty acid is unsaturated because of the presence of the double bond between carbons 9 and 10. Because it contains only one double bond, it is a monounsaturated fatty acid.* **(d)** *A triglyceride, which is composed of a glycerol backbone and three fatty acids.*

Each of the fatty acids in Figure 2.2b and 2.2c has 16 carbons. Which of the two fatty acids has more hydrogens?

Most fatty acids have an even number of carbon atoms, most commonly 16 or 18 carbons. An important feature of fatty acid chains is the number of double bonds between carbons. If there are no double bonds, then each carbon is linked to a maximum number of hydrogen atoms and is therefore *saturated* with hydrogen atoms. Thus, **saturated fatty acids** contain carbons linked only by single bonds (Figure 2.2b). In contrast, **unsaturated fatty acids** contain one or more pairs of carbons linked by double bonds and thus have fewer hydrogens per carbon (Figure 2.2c). A *monounsaturated fatty acid* contains exactly one double-bonded pair of carbons, whereas a *polyunsaturated fatty acid* contains more than one double-bonded pair of carbons. The degree of saturation determines important properties of a lipid, including some with significant clinical implications. For example, saturated fatty acids are implicated in the development of plaques that can clog arteries, which can lead to stroke or heart attack.

Triglycerides and fatty acids are nonpolar molecules due to the presence of nonpolar carbon-to-carbon and carbon-to-hydrogen bonds. They do not dissolve in water, but they readily dissolve in nonpolar solvents such as oil.

The fatty acid in Figure 2.2b

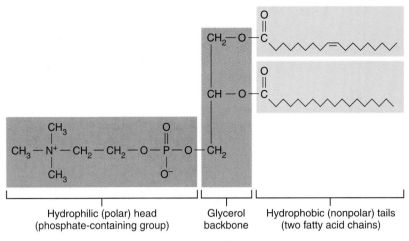
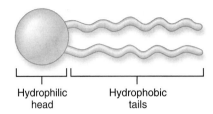

| Hydrophilic (polar) head (phosphate-containing group) | Glycerol backbone | Hydrophobic (nonpolar) tails (two fatty acid chains) |

(a) Phospholipid molecule (phosphatidyl choline)

| Hydrophilic head | Hydrophobic tails |

(b) Schematic representation of phospholipid molecule

FIGURE 2.3 Phospholipids. (a) *A phospholipid, which consists of a glycerol backbone linked to two fatty acid chains and a phosphate-containing group. The phospholipid shown here is* phosphatidyl choline. **(b)** *The standard way to schematically depict a phospholipid, emphasizing the hydrophilic head and hydrophobic tails of the molecule.*

Phospholipids

Phospholipids are lipids that contain a phosphate group. They are similar in structure to triglycerides in that a glycerol forms the backbone. However, instead of three fatty acids, a phospholipid contains two fatty acids and has a phosphate group attached to the third carbon of glycerol (Figure 2.3). The two fatty acids form the *tail* region of the phospholipid, which is nonpolar because of their long chains of carbon atoms. The phosphate group is generally attached to another chemical group, and together they form the *head* region of the phospholipid, which is polar. Therefore, phospholipids have both a polar region and a nonpolar region, making them amphipathic molecules.

The amphipathic property of phospholipids gives them unique behaviors in an aqueous or watery environment. The polar regions can dissolve in water, but the nonpolar regions cannot. Therefore, when phospholipids are placed in an aqueous environment, the polar regions face the water, and the nonpolar regions face each other. Phospholipids form two physiologically important structures when placed in an aqueous environment: phospholipid bilayers and micelles (Figure 2.4). In a phospholipid bilayer (Figure 2.4a), which is the core structure of cell membranes, the phospholipids are arranged into two parallel layers: The tails of parallel phospholipids face inward toward each other, and the heads face the outside, where they come into contact with the aqueous environment. A micelle (Figure 2.4b) is a spherical structure composed of a single layer of phospholipids; it functions in the transport of nonpolar molecules in an aqueous environment. The heads of the phospholipids face outward, where they come into contact with an aqueous environment; the tails face inward, forming a hydrophobic interior.

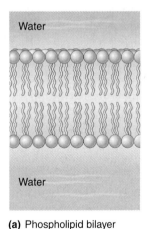
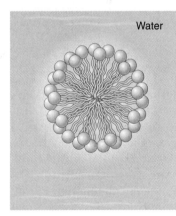

Water

Water

Water

(a) Phospholipid bilayer

(b) Micelle

FIGURE 2.4 Structures formed by phospholipids in an aqueous environment. (a) *A lipid bilayer, which consists of two sheets of phospholipids aligned such that the polar heads face the aqueous environment and the nonpolar tails face each other.* **(b)** *A micelle, a sphere formed by a single layer of phospholipids aligned such that the polar heads face the aqueous environment and the nonpolar tails face each other.*

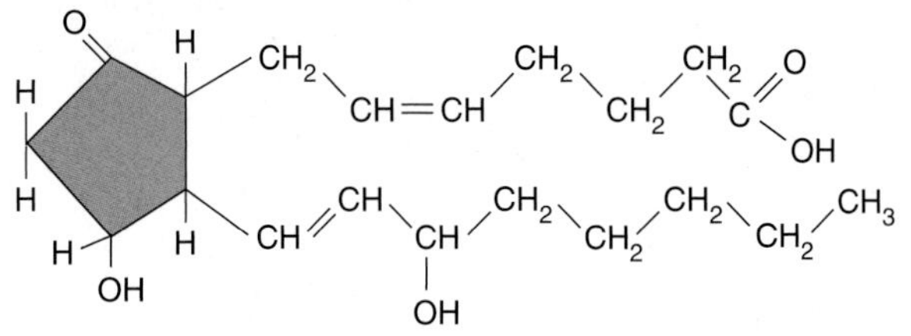

FIGURE 2.5 Eicosanoids. *In these modified 20-carbon fatty acids, the carbons in the middle of the carbon chain form a ring that causes the molecule to fold upon itself.*

Eicosanoids

Eicosanoids are modified fatty acids that function in intercellular communication. An eicosanoid is derived from a 20-carbon fatty acid and contains a five-carbon ring in the middle (Figure 2.5). This ring causes the molecule to fold upon itself, such that two chains of carbon atoms extend parallel to each other away from the ring. Eicosanoids include *prostaglandins, thromboxanes,* and *leukotrienes.*

Steroids

Steroids have a unique chemical structure consisting of three six-carbon rings and one five-carbon ring (Figure 2.6a). The most common steroid is cholesterol (Figure 2.6b). Due to the polar hydroxyl group on one end of the mostly nonpolar cholesterol molecule, it is a slightly amphipathic molecule. Cholesterol is an important component of the *plasma membrane,* the membrane surrounding cells. In addition, cholesterol is the precursor to all other steroids. Many steroids function as hormones, including *testosterone* (Figure 2.6c), *estradiol, cortisol,* and *calcitrol* (also called *1,25-dihydroxyvitamin D₃*).

Proteins

Proteins are a diverse group of biomolecules that carry out many cellular functions. **Proteins** are polymers of **amino acids,** relatively small biomolecules that contain an amino group and a carboxyl group (Figure 2.7a). Attached to the central carbon is an amino group, a carboxyl group, a hydrogen, and an R or residual group. There are 20 different R groups, and thus 20 different amino acids. The three amino acids shown in Figure 2.7b—alanine, tyrosine, and glutamate—have R groups with different chemical properties: nonpolar, polar, and charged, respectively. Although we will be discussing amino acids as components of proteins, they have other functions as well, including intercellular communication.

Polymers of amino acids are formed by joining two amino acids together by a *peptide bond;* these polymers are therefore called **polypeptides.** A peptide bond forms between the amino group of one amino acid and the carboxyl group of another amino acid by a *condensation reaction,* which is a reaction that releases water as two small molecules are joined together (Figure 2.8). Polypeptides vary in length from just two to several hundred amino acids; the name given to them differs based on the length or function. **Peptides** are short chains of amino acids, usually less than 50. Proteins consist of chains that generally are longer than 50 amino acids and often are hundreds of amino acids long.

The function of a protein is highly dependent on its three-dimensional structure or *conformation.* Protein structure can be described at as many as four different levels: *primary, secondary, tertiary,* and *quaternary structure* (Figure 2.9). The levels of protein structure can be likened to a telephone receiver's cord: Primary structure is the cord stretched out straight, secondary structure is the coiled cord, tertiary structure is the loops and bends in the coiled cord, and quaternary structure is the existence of two cords wrapped together. Now let's relate this analogy to the details of protein structure.

Primary protein structure simply refers to the sequence of amino acids, which is determined by peptide bonds within the peptide chain (see Figure 2.9a). Secondary protein structure is the folding pattern produced by hydrogen bonds between the hydrogen atom in the amino group of one amino acid and the oxygen atom in the carboxyl group of another amino acid in the same polypeptide. These hydrogen bonds fold the polypeptide into various shapes, such as α-helixes and β-pleated sheets (see Figure 2.9b), which contribute to the three-dimensional structure of proteins.

Tertiary protein structure is the folding produced by interactions between the R groups of different amino acids in the same polypeptide (see Figure 2.9c). Recall that the R groups have different chemical properties. The types of chemical interactions that can occur depend on the amino acids involved and include the following: (1) *hydrogen bonds,* (2) *ionic bonds,* (3) *van der Waals forces,* and (4) *covalent bonds.* Hydrogen bonds can form between polar R groups. Ionic bonds can form between ionized or charged R groups (Toolbox: Ions and Ionic Bonds, p. 31). Van der Waals forces are electrical attractions between the electrons of one atom and the protons of another atom. Covalent bonds form between the R groups of two cysteines. The R group of cysteine is a sulhydryl group, and the resulting covalent bond is called a *disulfide bridge.*

Quaternary protein structure exists only in proteins containing more than one polypeptide chain. An example is the protein hemoglobin (see Figure 2.9d), which functions in the transport of oxygen in the blood. Hemoglobin is a single protein consisting of four separate polypeptide chains. Hemoglobin, like all proteins, can function properly only when in the correct conformation.

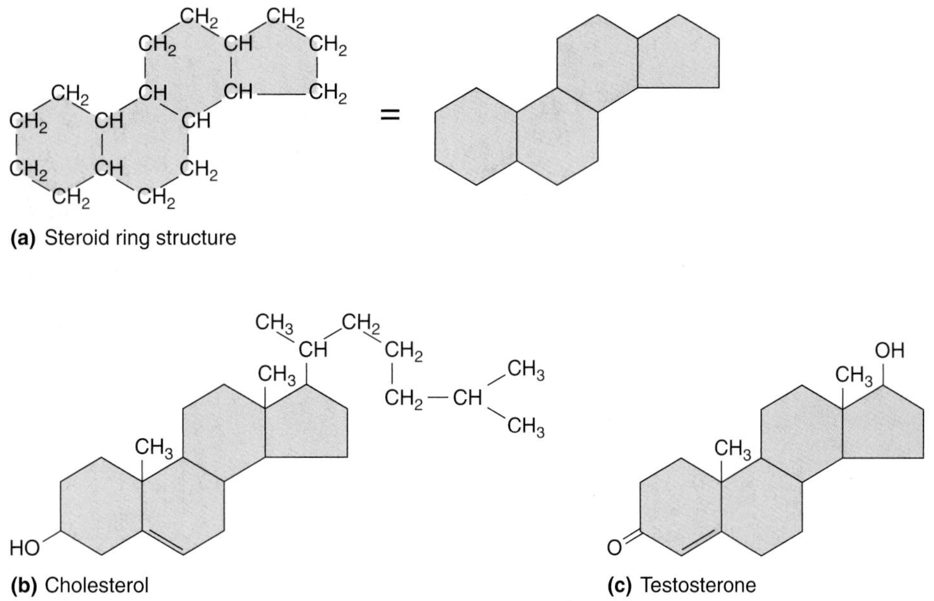

(a) Steroid ring structure

(b) Cholesterol

(c) Testosterone

FIGURE 2.6 Steroids. (a) *The basic structure of all steroids, which consists of three six-carbon rings and one five-carbon ring.* **(b)** *Cholesterol, the most common steroid and the precursor for all other steroids in the body. (c) Testosterone, an example of a steroid hormone.*

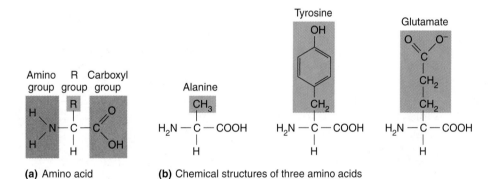

(a) Amino acid
(b) Chemical structures of three amino acids

FIGURE 2.7 Amino acids. (a) *The basic structure of an amino acid. The central carbon is bonded to an amino group, a carboxyl group, a hydrogen, and an R group.* **(b)** *Structures of three of the 20 amino acids. The different R group in each of these amino acids gives them different chemical properties: Alanine is nonpolar, tyrosine is polar, and glutamate is charged.*

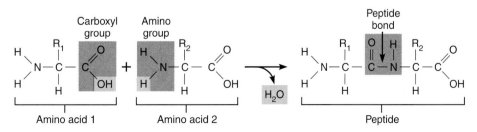

FIGURE 2.8 Formation of a peptide bond by a condensation reaction.
The peptide bond is formed between two amino acids, and water is released.

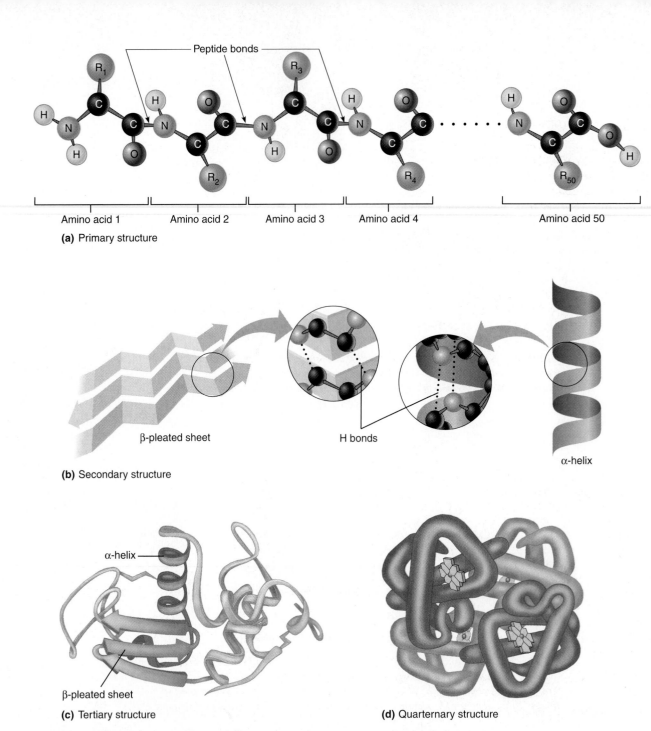

(a) Primary structure

(b) Secondary structure

β-pleated sheet

H bonds

α-helix

α-helix

β-pleated sheet

(c) Tertiary structure

(d) Quarternary structure

Peptide bonds

R₁ ... R₅₀ labels and atom labels within figure

Amino acid 1   Amino acid 2   Amino acid 3   Amino acid 4   Amino acid 50

FIGURE 2.9 Levels of protein structure. (a) *Primary protein structure, which is the sequence of amino acids.* **(b)** *Secondary protein structure, which is caused by hydrogen bonding between the amino hydrogen of one amino acid and the carboxyl oxygen of another amino acid. Common secondary structures include β-pleated sheets and α-helixes.* **(c)** *Tertiary protein structure, which is the folding pattern produced by interactions between the R groups of amino acids. The protein shown here is the enzyme* lysozyme. **(d)** *Quaternary protein structure, which is the arrangement of more than one polypeptide chain in a single protein. Shown here is* hemoglobin, *which consists of four polypeptide chains.*

Some atoms have a tendency to gain or lose electrons completely, in the process acquiring an excess or deficit of electrons. Particles such as this have net negative or positive charges, respectively, and are known as *ions.* Negatively charged ions are known as *anions;* positively charged ions are known as *cations.* When anions and cations are present in solids, they tend to form crystals in which the cations and anions are closely associated. A familiar example is *sodium chloride* (NaCl) or table salt, which contains sodium ions (Na^+) and chloride ions (Cl^-). Sodium ions are formed when sodium atoms lose an electron, producing an ion with 11 protons and ten electrons. Chloride ions are formed when chloride atoms gain an electron producing an ion with 17 protons and 18 electrons. This process occurs as follows:

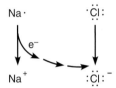

In a crystal of NaCl, the cations (Na^+) and anions (Cl^-) are held together by electrical forces of attraction due to their opposite charges. These forces are sometimes called *ionic bonds.* When ionic solids are dissolved in water, ionic bonds are disrupted, leaving cations and anions free to dissociate into separate particles. For sodium chloride this process can be illustrated as shown at right.

Solutions containing dissolved ions are described as being *electrolytic* because they are good conductors of electricity, and ionic substances are referred to as *electrolytes.* Body fluids are electrolytic and contain a number of small ions (known as *inorganic ions*), including sodium, potassium (K^+), calcium (Ca^{2+}), hydrogen (H^+), magnesium (Mg^+), chloride, sulfate (SO_4^{2-}), and bicarbonate (HCO_3^-).

Ionized chemical groups can also be found on certain types of biomolecules. Molecules containing significant numbers of polar bonds or ionized groups are described as being *polar* or *hydrophilic;* those possessing mostly nonpolar bonds and lacking ionized groups are described as being *nonpolar* or *hydrophobic.*

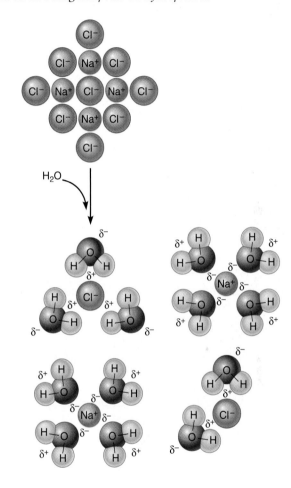

The three-dimensional conformation of proteins can be classified as either *fibrous* or *globular* (Figure 2.10). **Fibrous proteins** are generally extended, elongated strands that function in structure or contraction. Examples of fibrous proteins include *collagen,* a protein found in tendons and bone, and *tropomyosin,* a protein found in muscle cells. **Globular proteins** are coiled, foiled, irregular, and bulky. Among their many functions, they act as *chemical messengers* for intercellular communication, as *receptors* that bind chemical messengers, as *carrier proteins* that transport substances in blood or across membranes, and as *enzymes* that catalyze chemical reactions in the body. Examples of globular proteins include the oxygen binder *myoglobin,* the chemical messenger *growth hormone,* and a membrane transport protein called the Na^+/K^+ *pump.* Some proteins have both globular and fibrous components; one example is the muscle protein *myosin,* which has a globular head and fibrous tail. All the proteins mentioned in this discussion are discussed further in detailed descriptions of organ systems later in the book.

Some proteins have other classes of organic molecules attached to them. For example, *glycoproteins* have carbohydrates attached to the polypeptide chains,

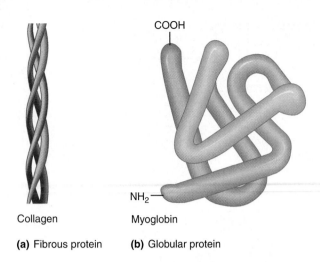

COOH

Collagen

NH₂

Myoglobin

(a) Fibrous protein

(b) Globular protein

Fibrous tail

Globular head

Myosin

(c) Mixed protein

FIGURE 2.10 Three-dimensional structures of proteins.
(a) *A fibrous protein, collagen, which consists of thin strands.* **(b)** *A globular protein, myoglobin.* **(c)** *A mixed protein, myosin, with a fibrous tail region and a globular head region.*

What types of chemical forces cause the myoglobin molecule to fold into its three-dimensional structure?

whereas *lipoproteins* have lipids attached to them. Glycoproteins are important components of the plasma membrane that surrounds cells; they also contribute to cell recognition, which is the ability of the immune system to recognize cells that are part of the body. Lipoproteins play an important role in the transport of lipids in blood.

Nucleotides and Nucleic Acids

Nucleotides have an important function in the transfer of energy within cells, and they form the genetic material of the cells. The basic structure of a nucleotide is shown in Figure 2.11a. **Nucleotides** contain one or more phosphate groups, a five-carbon carbohydrate, and a nitrogenous base. The carbohydrates found in nucleotides are ribose and deoxyribose. The nitrogenous bases in nucleotides include two classes: (1) *pyrimidines,* which contain a single carbon ring and include *cytosine, thymine,* and *uracil,* and (2) *purines,* which contain a double carbon ring and include *adenine* and *guanine.* Several nucleotides function in the exchange of cellular energy, including

adenosine triphosphate (ATP), *nicotinamide adenine dinucleotide* (NAD), *flavin adenine dinucleotide* (FAD), and *coenzyme A* (CoA). A nucleotide that has one phosphate, such as adenosine monophosphate (AMP), is called a *nucleotide monophosphate* (Figure 2.11b); similarly, a nucleotide with two phosphates, such as adenosine diphosphate (ADP), is a *nucleotide diphosphate,* and a nucleotide with three phosphates, such as ATP, is a *nucleotide triphosphate.*

Some nucleotides form a ring due to covalent bonding between an oxygen of the phosphate group, and a carbon of the carbohydrate group, as shown in Figure 2.12. These nucleotides, called *cyclic nucleotides,* include chemical messengers (molecules that function as signals for the cell to do something) inside cells such as *cyclic AMP* (cAMP) and *cyclic GMP* (cGMP).

Polymers of nucleotides include the **nucleic acids** that function in the storage and expression of genetic information: **deoxyribonucleic acid (DNA)** and **ribonucleic acid (RNA)** (Figure 2.13). DNA molecules are found in a cell's nucleus, where they store the genetic information. RNA molecules are found in both a cell's nucleus and its cytoplasm, and they are necessary for the expression of genetic information. Both DNA and RNA are polymers of nucleotides, with the phosphate group of one nucleotide linked to the carbohydrate of another, forming a chain with the bases sticking out to the sides. However, there are important differences in their structures, which we discuss next.

DNA (see Figure 2.13a) consists of two strands of nucleotides coiled together into a double helix. The carbohydrate in DNA is deoxyribose. The bases in DNA include adenine (A), guanine (G), cytosine (C), and thymine (T). The ends of the strands are labeled as 3' or 5', with 3' corresponding to the carbohydrate end and 5' corresponding to the phosphate end. The two strands of DNA are held together by hydrogen bonding between bases according to the **law of complementary base pairing,** which states that *whenever two strands of nucleic acids are held together by hydrogen bonds, G in one strand is always paired with C in the opposite strand, and A is always paired with T in DNA (or with U in RNA).* Cytosine and guanine form three hydrogen bonds between them, and adenine and thymine form two hydrogen bonds between them (see Figure 2.13b). Because of this pairing, the two strands of DNA are complementary to each other.

RNA (see Figure 2.13c) consists of a single strand of nucleotides with a 3' end and a 5' end. The carbohydrate in RNA is ribose. The bases in RNA include adenine, guanine, cytosine, and uracil. The bases of RNA also follow the law of complementary base pairing; guanine and cytosine form hydrogen bonds between them, as in DNA, but adenine forms hydrogen bonds with uracil instead of thymine. Even though RNA is single stranded, complementary base pairing is necessary to synthesize RNA from DNA (the RNA will be complementary to the DNA, except that thymine will be replaced with uracil), and to enable folding of a single RNA molecule upon itself.

Hydrogen bonds, ionic bonds, and van der Waals forces

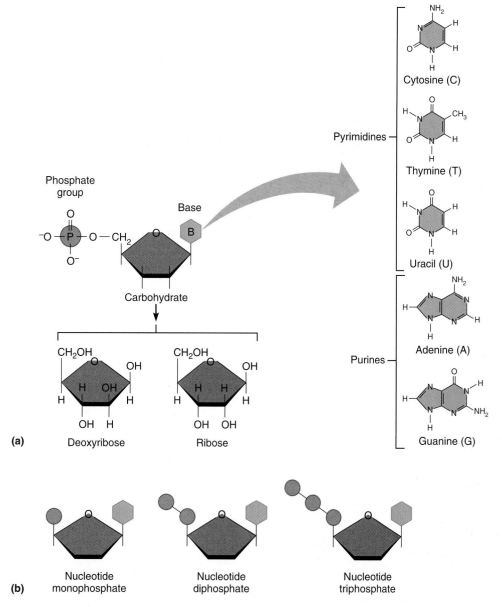

Cytosine (C)

Thymine (T)

Uracil (U)

Pyrimidines

Adenine (A)

Guanine (G)

Purines

Phosphate group

Base

Carbohydrate

CH₂OH

OH

H OH

H H

OH H

Deoxyribose

CH₂OH

OH

H H

H H

OH OH

Ribose

(a)

Nucleotide monophosphate

Nucleotide diphosphate

Nucleotide triphosphate

(b)

FIGURE 2.11 **Nucleotides. (a)** *The basic structure of a nucleotide. The carbohydrate in a nucleotide is either deoxyribose or ribose. The five possible bases include the pyrimidines, which contain a single ring, and the purines, which contain two rings.* **(b)** *A schematic representation of nucleotides. Nucleotide monophosphates contain a single phosphate group, nucleotide diphosphates contain two phosphate groups, and nucleotide triphosphates contain three phosphate groups.*

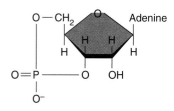

Cyclic AMP (cAMP)

FIGURE 2.12 **The cyclic nucleotide cyclic AMP (cAMP), a common messenger molecule.**

Q u i c k   T e s t   2 . 1

1. Are carbohydrates generally polar or nonpolar? What about triglycerides?

2. Name the two specialized structures phospholipids can form in an aqueous environment. What physiological functions are associated with these structures?

3. Name the subunits that make up the following polymers: glycogen, proteins, and nucleic acids. What are the general functions of each of these polymers?

4. Cholesterol is what type of biomolecule?

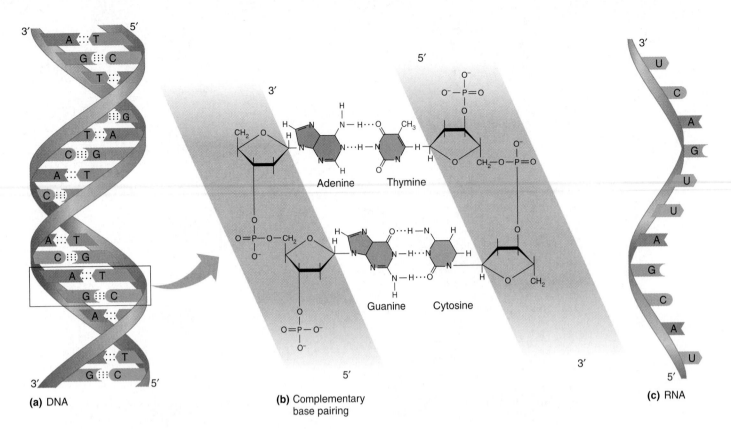

FIGURE 2.13 Nucleic acids. (a) *DNA, which consists of two strands of nucleotides held together by hydrogen bonding according to the law of complementary base pairing.* **(b)** *Complementary base pairing, with adenine (A) and thymine (T) forming two hydrogen bonds between them, and cytosine (C) and guanine (G) forming three hydrogen bonds between them.* **(c)** *RNA, which consists of a single strand of nucleotides.*

What type of chemical bond links the phosphates and carbohydrates together to form a chain?

CELL STRUCTURE

When put together in a certain way, the biomolecules form the basic units of life: cells. The human body contains more than 100 trillion cells that work together to maintain homeostasis. Remarkably, all of these cells derive from a single fertilized egg. During development, the cells differentiate into over 200 different types of cells, each with its own specialized function. Although cells can be specialized both anatomically and functionally, they generally have the same basic components. In this section we describe the basic components of a "typical" cell (Figure 2.14).

Each cell is bound by a **plasma membrane,** which separates the cell from the extracellular fluid. Inside the cell are two main components: the **nucleus,** which is a membrane-bound structure that contains the genetic information for the cell, and the **cytoplasm,** which includes everything inside the cell except the nucleus. The cyto-

plasm itself consists of two main components: the *cytosol,* or intracellular fluid, and the *organelles*. The **cytosol** is a gel-like fluid. The **organelles,** which are structures made up of a variety of biomolecules, carry out specific functions in the cell, much as the organs carry out specific functions in the body. There are *membranous organelles,* which are separated from the cytosol by one or more membranes, and *nonmembranous organelles,* which have no such boundary with the cytosol.

This section describes the anatomy of a cell; general functions of the various components of a cell are discussed in a subsequent section.

Structure of the Plasma Membrane

The cell contains several types of membrane that function as barriers between compartments. This section focuses on the structure of the plasma membrane, which separates the cell from its external environment. The structures of the *nuclear envelope,* which separates the nucleus

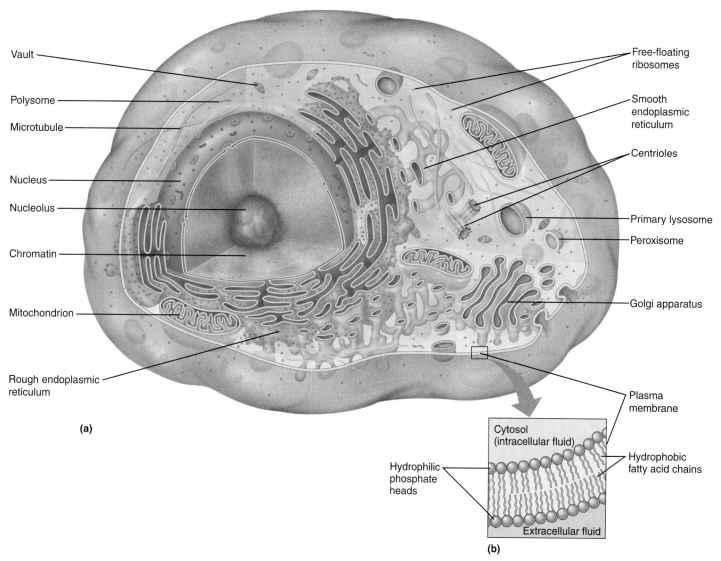

Vault

Polysome

Microtubule

Nucleus

Nucleolus

Chromatin

Mitochondrion

Rough endoplasmic
reticulum

Free-floating
ribosomes

Smooth
endoplasmic
reticulum

Centrioles

Primary lysosome

Peroxisome

Golgi apparatus

Plasma
membrane

(a)

Cytosol
(intracellular fluid)

Hydrophobic
fatty acid chains

Hydrophilic
phosphate
heads

Extracellular fluid

(b)

FIGURE 2.14 Structures of a typical cell. (a) *Three-dimensional view of the cell. The plasma membrane separates the cell from the extracellular fluid. The nucleus contains the genetic information. The cytoplasm contains the organelles and the cytoskeleton, both of which are surrounded by a fluid called cytosol. Organelles include mitochondria, endoplasmic reticulum, golgi apparatus, lysosomes, peroxisomes, ribosomes, and vaults.* **(b)** *Enlargement of a portion of the plasma membrane showing the lipid bilayer.*

from the cytoplasm, and those of the membranes around organelles, which separate the internal compartment(s) of the organelles from the cytosol, are similar to those of the plasma membrane.

Phospholipid Bilayer

The structure of the plasma membrane, described as a *fluid mosaic,* consists of phospholipids, proteins, and cholesterol (Figure 2.15). The phospholipids are arranged in a bilayer such that the hydrophilic heads face the aqueous environments inside the cell (the cytosol) and outside the cell (the interstitial fluid); the hydrophobic tails face each other. This bilayer forms the basic structure of the

membrane and provides a barrier to the movement of large polar molecules. Although water molecules are polar, they can generally cross the lipid bilayer because of their small size. The membrane is considered *fluid* because the phospholipids and the other molecules in the membrane are not linked together by chemical bonds and can therefore move about laterally, and even occasionally move from one side of the bilayer to the other. Cholesterol molecules are found within the lipid bilayer, where they interfere with hydrophobic interactions between phospholipid tails, which could cause crystallization of the bilayer and decrease the fluidity. Cholesterol also decreases the ability of water to cross the lipid bilayer.

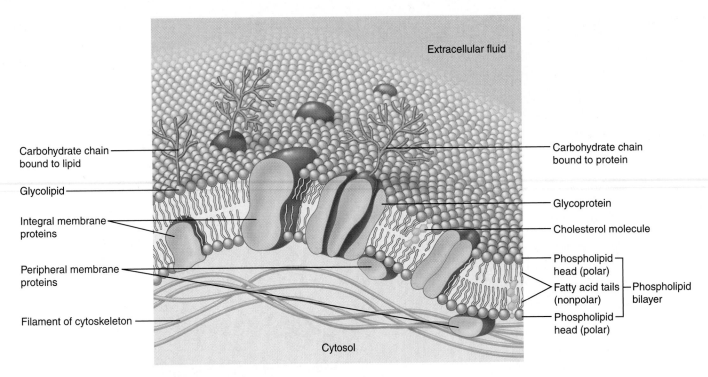

FIGURE 2.15 Plasma membrane. *The flexible plasma membrane is primarily a double layer of phospholipids, with the hydrophobic tails of the phospholipid buried within the bilayer, and the hydrophilic phosphate heads facing the cytosol and extracellular fluid. Integral membrane proteins are scattered throughout the bilayer. Peripheral membrane proteins are associated with the integral membrane proteins, primarily on the side facing the cytosol, where they function as part of the cytoskeleton. Cholesterol molecules are scattered throughout the lipid bilayer. Carbohydrates bound to membrane lipids or membrane proteins face the extracellular fluid. These lipids and proteins are called glycolipids and glycoproteins, respectively.*

Membrane Proteins

The plasma membrane is described as a *mosaic* because of the presence of proteins that are dispersed in the bilayer, like islands in a sea of phospholipids. There are two main classes of membrane proteins: *integral membrane proteins* and *peripheral membrane proteins* (see Figure 2.15). **Integral membrane proteins** are embedded within the lipid bilayer, and therefore they can be dissociated from the membrane only by physically disrupting the bilayer. Integral membrane proteins are amphipathic molecules that are in contact with both the lipid bilayer and the aqueous environment. The polar surface(s) of the protein face the aqueous environment, which could be the cytosol, extracellular fluid, or both. The nonpolar areas are embedded within the lipid bilayer.

Some integral membrane proteins are called **transmembrane proteins,** because they span the lipid bilayer, with surfaces exposed to both the cytosol and extracellular fluid. Often, a transmembrane protein crosses the membrane at several places. Transmembrane proteins include channels that allow ions to *permeate* (or cross) the membrane, and carrier proteins that transport molecules from one side of the membrane to the other. Other integral membrane proteins are located on only one side of the membrane. Some of the proteins facing the cytosol function as enzymes that catalyze chemical reactions in the cytosol or as a special class of proteins called G proteins (described later). Some of the proteins facing the extracellular fluid function as enzymes that catalyze reactions in the extracellular fluid or as receptor molecules that bind chemical messengers from other cells.

Peripheral membrane proteins are loosely bound to the membrane by associations with integral membrane proteins or phospholipids. Peripheral membrane proteins can be dissociated from the membrane and still leave the membrane intact. Most peripheral membrane proteins are located on the cytosolic surface of the plasma membrane and often function as part of a group of proteins that make up the *cytoskeleton.*

Membrane Carbohydrates

Also associated with the plasma membrane are carbohydrates, which are covalently bound to the membrane lipids or to proteins to form glycolipids or glycoproteins, respectively (see Figure 2.15). Plasma membrane carbohydrates are located only on the extracellular surface, where they have two primary functions: They form the *glycocalyx*, a protective layer that also functions in holding cells together (Figure 2.16), and they function in cell

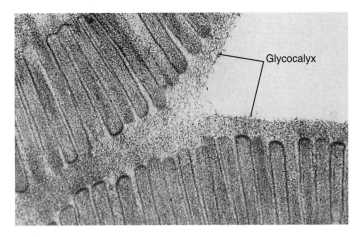

FIGURE 2.16 Electron micrograph of the plasma membrane showing the glycocalyx.

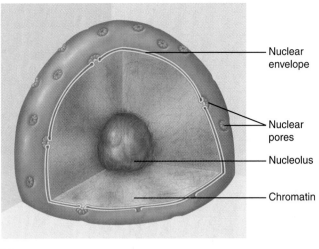

(a)

recognition; that is, they label the cell as a part of the body or as a distinct type of cell. Recognition is an important aspect of both the immune response and tissue growth.

Structure of the Nucleus

Most cells have a single nucleus that contains the genetic material of the cell, DNA. The DNA exists as thin threads called **chromatin,** except during cell division. The nucleus generally appears as a prominent spherical structure in the cell (Figure 2.17). Surrounding the nucleus is the **nuclear envelope,** which consists of two membranes. These membranes fuse intermittently, leaving gaps called **nuclear pores** that allow selective movement of molecules between the nucleus and the cytoplasm. Within the nucleus is a structure called the *nucleolus,* which is the site of synthesis of a type of RNA called *ribosomal RNA* (rRNA). The nucleus functions in the transmission and expression of genetic information. Encoded in the DNA is the genetic information for RNA and protein synthesis. The very important roles of DNA and RNA in protein synthesis are described later in this chapter.

Contents of the Cytosol

The cytosol is more than simply a fluid that bathes the organelles; it is a site of important chemical reactions and also a site for storage of molecules. Several enzymes that catalyze specific chemical reactions are located in the cytosol. Energy in the form of triglycerides or glycogen is stored in masses called **inclusions.** Molecules that are to be released from the cell, or **secreted,** are stored in membrane-bound sacs called **secretory vesicles.** In addition, the molecular composition of the cytosol is criti-

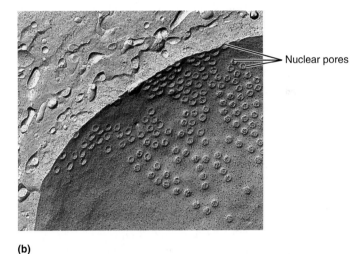

(b)

FIGURE 2.17 Nucleus. (a) *The nucleus is separated from the cytoplasm by the nuclear envelope. Pores in the nuclear envelope allow the movement of specific substances between the nucleus and cytoplasm. Located in the nucleus is the DNA that contains the genetic information of each individual. The DNA exists as thin threads called chromatin. Also within the nucleus is the nucleolus, the site of rRNA synthesis.* **(b)** *Electron micrograph of a freeze-fractured nuclear envelope showing nuclear pores.*

cal to the function of cells. For example, the ionic composition of the cytosol plays a crucial role in the activities of nerve and muscle cells.

Structure of Membranous Organelles

The membranes that surround certain organelles provide a barrier between the cytosol and the interior of the organelle, which creates compartments within the cytoplasm. In some cases, two membranes surround an organelle, creating compartments within the organelle. The structure of each membranous organelle is described next.

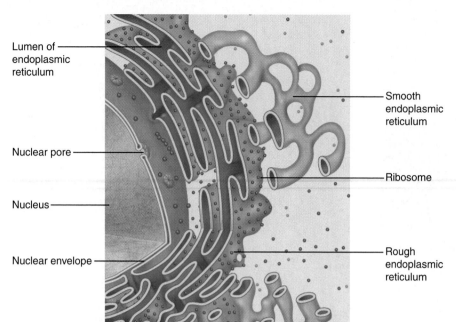

Lumen of endoplasmic reticulum

Nuclear pore

Nucleus

Nuclear envelope

Smooth endoplasmic reticulum

Ribosome

Rough endoplasmic reticulum

FIGURE 2.18 **Endoplasmic reticulum.** *The endoplasmic reticulum's membrane is continuous with the outer membrane of the nuclear envelope. The rough endoplasmic reticulum is closest to the nucleus and consists of flattened sacs with ribosomes attached to its external surface. The membrane of the rough endoplasmic reticulum is continuous with that of the smooth endoplasmic reticulum, which is tubular and lacks ribosomes.*

Endoplasmic Reticulum

The **endoplasmic reticulum** consists of an elaborate network of membrane enclosing an interior compartment called the lumen (Figure 2.18). There are two types of endoplasmic reticulum that differ in appearance and function: *rough endoplasmic reticulum* and *smooth endoplasmic reticulum.* The rough endoplasmic reticulum gets its name from its granular or "rough" appearance under magnification, due to the presence of **ribosomes,** which are complexes of rRNA and proteins that function in protein synthesis. In addition, rough endoplasmic reticulum looks like flattened sacs. Smooth endoplasmic reticulum, by contrast, consists of tubules and does not have ribosomes attached to it, giving it a "smooth" appearance. The membrane of the rough endoplasmic reticulum is continuous on one side with the outer membrane of the nuclear envelope, and on the other side with the smooth endoplasmic reticulum (see Figure 2.18).

The endoplasmic reticulum is important in the synthesis of several types of biomolecules. The rough endoplasmic reticulum is associated with the synthesis of proteins that will be secreted from the cell or incorporated into the plasma membrane, or are destined for another organelle. The smooth endoplasmic reticulum is the site of the synthesis of lipids, including triglycerides and steroids, and is a site for storage of calcium ions. In addition, the smooth endoplasmic reticulum is specialized in certain cells. In liver cells, for example, the smooth endoplasmic reticulum contains detoxification enzymes that degrade toxic substances in the blood.

Golgi Apparatus

The **Golgi apparatus** consists of membrane-bound flattened sacs called *cisternae* (Figure 2.19). The Golgi apparatus is closely associated with the endoplasmic reticulum on one side, called the *cis face,* although the membranes of the Golgi apparatus and endoplasmic reticulum are separate. The other side of the Golgi apparatus is called the *trans face.* The Golgi apparatus processes molecules synthesized in the endoplasmic reticulum, and prepares them for transport to their final location. The Golgi apparatus packages molecules into vesicles and directs the vesicles to the appropriate location. Some vesicles transport substances to intracellular sites, whereas secretory vesicles transport substances out of the cell.

Mitochondria

Mitochondria are bound by two membranes (Figure 2.20). The outer mitochondrial membrane separates the mitochondrion from the cytosol, whereas the inner mitochondrial membrane divides each mitochondrion into two compartments: the **intermembrane space,** the area between the two membranes, and the **mitochondrial matrix,** the innermost compartment. The inner mitochondrial membrane is a functional compartment in itself because it houses a series of proteins and other molecules called the *electron transport chain.* To increase the area for the electron transport system, the inner membrane is folded into tubules called **cristae.** Mitochondria are often called the "powerhouse" of the cell because most of the cell's usable energy form, ATP, is produced in these organelles.

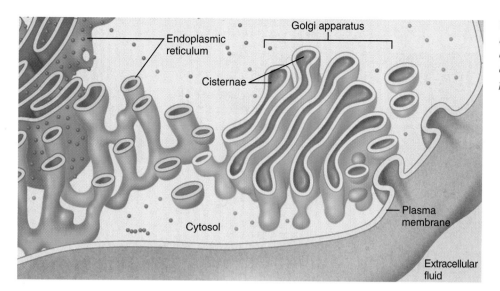

FIGURE 2.19 **Golgi apparatus.** *The Golgi apparatus consists of stacks of flattened sacs called cisternae. One side of the Golgi apparatus faces the endoplasmic reticulum; the other side faces the plasma membrane.*

The number of mitochondria per cell varies considerably among different cell types based on the particular cell's energy needs. For example, because muscle cells demand large amounts of energy they contain numerous mitochondria. By contrast, red blood cells, which function primarily in the transport of blood gases, contain no mitochondria.

Lysosomes

Lysosomes are small spherical organelles surrounded by a single membrane (Figure 2.21). Lysosomes contain enzymes that degrade intracellular debris and extracellular debris that has been taken into the cell. For example, after lysosomes fuse with old organelles that are no longer functioning, the enzymes inside the lysosomes then break down the organelles and usable components are recycled and waste products are eliminated from the cell. In the case of extracellular debris, cells can engulf extracellular particles by a process called *endocytosis,* in which the particles are enclosed within a vesicle and brought inside the cell. Lysosomes then fuse with the vesicle, enabling their enzymes to degrade the particles. The process of endocytosis is described in more detail later in this chapter.

Peroxisomes

Peroxisomes are spherical organelles that are slightly smaller than lysosomes and are surrounded by a single membrane. They function in the degradation of molecules such as amino acids, fatty acids, and toxic foreign matter. In the process of this degradation, they often produce hydrogen peroxide (H_2O_2), which in itself is toxic. One of the most notable characteristics of peroxisomes is

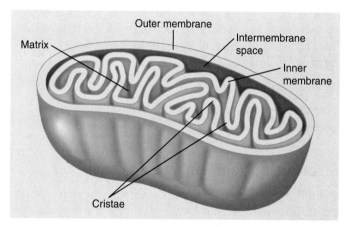

(a)

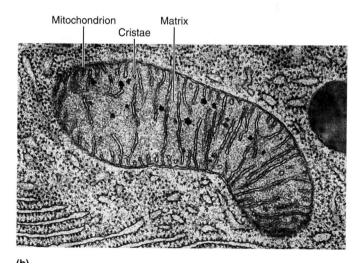

(b)

FIGURE 2.20 **Mitochondria.** **(a)** *Drawing of a mitochondrion. The outer mitochondrial membrane separates the mitochondrion from the cytosol, whereas the inner mitochondrial membrane separates the mitochondrion into an inner matrix and the intermembrane space. The inner membrane has numerous folds called cristae.* **(b)** *Electron micrograph of a mitochondrion.*

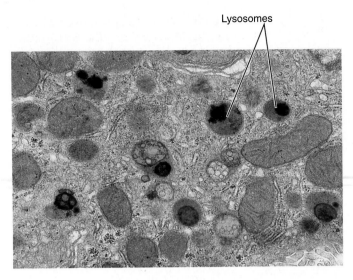

Lysosomes

FIGURE 2.21 Electron micrograph of lysosomes.

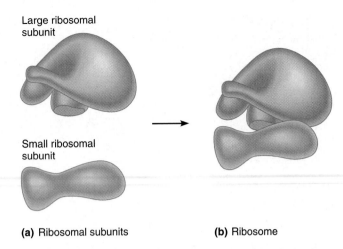

Large ribosomal subunit

Small ribosomal subunit

(a) Ribosomal subunits   **(b)** Ribosome

FIGURE 2.22 Ribosomes. *In order for protein synthesis to begin, the two free subunits in the cytosol* **(a)** *must come together to form a ribosome* **(b)**.

the presence of the enzyme *catalase,* which catalyzes the breakdown of hydrogen peroxide to form water and oxygen:

$$2 \ H_2O_2 \longrightarrow 2 \ H_2O + O_2$$

This reaction prevents the buildup of toxic hydrogen peroxide.

Structure of Nonmembranous Organelles

Even though nonmembranous organelles have no barrier separating them from the cytosol, they are still considered organelles because they consist of biomolecules organized into structures that perform specific functions within the cell.

Ribosomes

Ribosomes are dense granules composed of rRNA and proteins that function in protein synthesis. Each ribosome consists of a small subunit and a large subunit that are separate and located in the cytosol when the ribosome is not active (Figure 2.22a). During the process of protein synthesis (discussed later in this chapter), the two subunits come together in the cytosol to form a functional ribosome (Figure 2.22b). After protein synthesis is initiated, some ribosomes remain free in the cytosol, whereas others become attached to the rough endoplasmic reticulum. Proteins synthesized in association with free ribosomes can remain in the cytosol or enter a mitochondrion, the nucleus, or a peroxisome. Proteins synthesized in association with the rough endoplasmic reticulum will cross or enter a membrane, such as the plasma membrane or the membrane of an organelle.

Vaults

Vaults are the most recently discovered organelle, and their function is not fully understood. Vaults are barrel-shaped structures, approximately three times larger than ribosomes, that consist of two identical subunits. Vaults may be involved in the transport of molecules, such as mRNA, between the nucleus and the cytoplasm. Although little is known about them, vaults are receiving a lot of research attention because they may be involved in the development of resistance to chemotherapy in the treatment of cancer (Discovery: Vaults and Chemotherapy, p. 42).

Centrioles

Centrioles are short cylindrical structures consisting of bundles of protein filaments. Each cell has two centrioles oriented perpendicular to each other (see Figure 2.14). They function in directing the development of a structure called the *mitotic spindle* during cell division.

Cytoskeleton

The **cytoskeleton** is a flexible lattice of fibrous proteins, also called filaments, that gives the cell structure and support, much as the skeleton provides support for your body (Figure 2.23). The cytoskeleton is not rigid, nor does it have a fixed structure. The fibrous proteins that form it can disassemble and reassemble as necessary. Functions of the cytoskeleton include:

- Mechanical support and structure
- Intracellular transport of materials
- Suspension of organelles
- Formation of adhesions with other cells

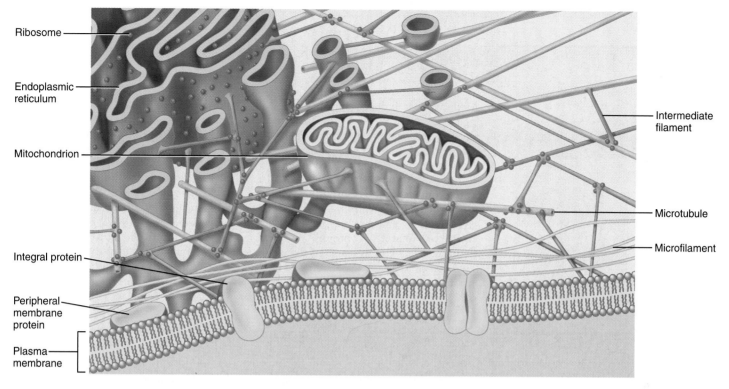

FIGURE 2.23 **Cytoskeleton.** *The cytoskeleton consists of three types of filaments that give the cell structural support and enable some degree of motility or contractility. Organelles are suspended in the cytosol by the filaments.*

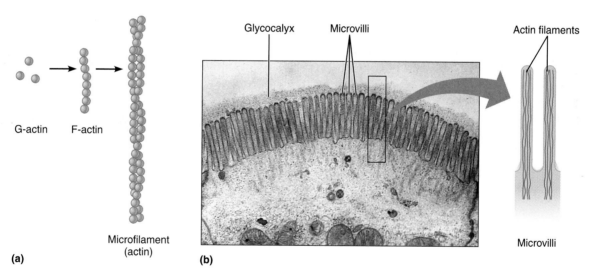

FIGURE 2.24 **Microfilaments. (a)** *An actin filament, the most common microfilament. Actin filaments consist of two helically arranged strands of F-actin, a fibrous protein formed of chains of the globular protein G-actin.* **(b)** *Electron micrograph of microvilli on the surface of an epithelial cell. The enlarged drawing shows a schematic of the arrangement of microfilaments that maintain the fingerlike microvilli.*

- Contraction
- Movement of certain cells

Several types of filaments form the cytoskeleton, including *microfilaments, intermediate filaments,* and *microtubules.* The classification of filaments is based on their diameters.

Microfilaments

Microfilaments have the smallest diameter among the fibrous proteins. One microfilament, called *actin* (Figure 2.24a), has several functions in cells, including muscle contraction, "amoeboidlike" movement of cells, and separation of the cytoplasm during cell division. Other

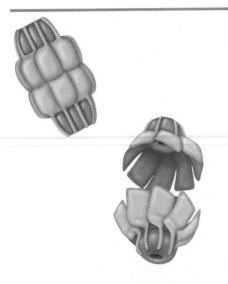
Vaults, shown in the accompanying figure, were discovered by researchers in the 1980s. Even though virtually every cell in the body has thousands of vaults, their function has still not been established. Current hypotheses suggest that vaults function in the transport of molecules, such as mRNA, from the nucleus to the cytoplasm, and that they may also have a role in protein synthesis.

In the late 1990s, researchers discovered that cancer patients who have a resistance to a wide range of chemotherapies, called *multidrug resistance,* have a greater concentration of vaults in their cells. However, this correlation does not prove that the increased number of vaults is the *cause* of the drug resistance; further studies are needed to determine if and how vaults influence drug resistance. In addition, a protein found in vaults, called *major vault protein* or MVP, is elevated in these patients. In the future, MVP may be used as a marker to identify patients who have resistance to chemotherapy.

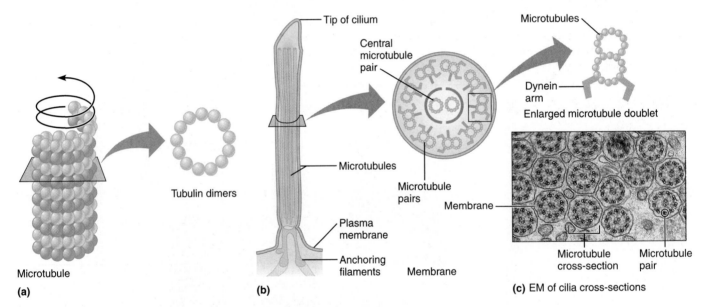

FIGURE 2.25 Microtubules. (a) *A microtubule, the largest of the cytoskeletal elements. Each is composed of a long hollow tubule of a globular protein called tubulin organized in dimers.* **(b)** *The arrangement of microtubules in cilia and flagella. Nine pairs of microtubules are arranged around a tenth central pair. The cross section shows the pairs of microtubules with their dynein arms.* **(c)** *Electron micrograph of cross sections through cilia, showing the ten pairs of microtubules.*

actin microfilaments provide structural support for special cell projections called **microvilli** (Figure 2.24b), which are often found in epithelial cells that are specialized for the exchange of molecules.

Intermediate Filaments

Intermediate filaments have a diameter between that of microfilaments and microtubules. Intermediate filaments tend to be stronger and more stable than microfilaments. Intermediate filaments called *keratin* are found in skin and hair cells. *Myosin* is an intermediate filament found in muscle cells, where it works with actin to produce contraction.

Microtubules

Microtubules, the filaments with the largest diameter, are composed of long hollow tubes of a spherical protein

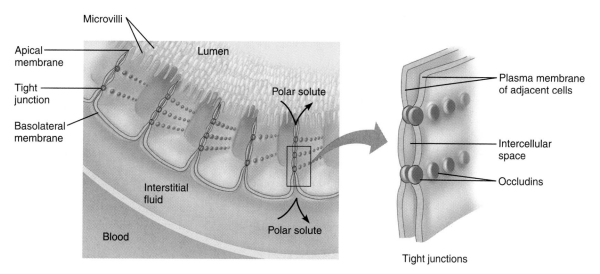

FIGURE 2.26 Tight junctions. *Tight junctions are commonly found in epithelial tissue, where they limit the movement of molecules through the interstitial space between epithelial cells. Proteins called occludins link two adjacent cells together such that they form nearly impermeable adhesions. Polar molecules that cannot cross lipid bilayers also generally cannot move between cells connected by tight junctions.*

called *tubulin* (Figure 2.25a). Microtubules provide strength to the cytoskeleton. They also form *spindle fibers,* which aid in the distribution of chromosomes during cell division, and assist with directional movement of vesicles and large biomolecules within the cytosol. In addition, microtubules are components of two motile structures: *cilia* and *flagella* (Figure 2.25b).

Cilia and flagella are hairlike protrusions from the cell surface that can move in a wavelike manner. Movement of cilia propels particles along hollow conduits in the body. In the respiratory tract, for example, cilia move mucus containing trapped inhaled particles out of the lungs and up toward the mouth, where it can be swallowed. A cilium consists of ten pairs of microtubules, with nine of the pairs surrounding a central pair. The paired microtubules are connected by proteins called *dynein arms* that help to generate the force needed for the microtubules to slide past each other; the sliding movement leads to the wavelike motion of a cilium. Flagella are similar to cilia in basic structure, but they are longer and are found in only one cell type in humans, the sperm. Flagella move sperm through the female reproductive tract toward the ovum. Table 2.2 summarizes the different components of the cell and their major functions. (When It Goes Wrong: Primary Ciliary Dyskinesia, www.physiologyplace.com, Challenge Yourself)

Quick Test 2.2

1. Where is the genetic information of a cell stored? What type of molecule stores this information?

2. Describe the basic structure of the plasma membrane. What are the functions of phospholipids and cholesterol? Name four functions of membrane proteins.

3. Describe the basic functions of each of the following organelles: rough endoplasmic reticulum, smooth endoplasmic reticulum, Golgi apparatus, ribosomes, mitochondria, lysosomes, and peroxisomes.

4. What are the three types of filaments found in the cytoskeleton? Name a special function of each type of filament.

CELL-TO-CELL ADHESIONS

In many tissues cells are held together by special membrane proteins called *cell adhesion molecules.* Other proteins function in more specialized cell adhesion, in which the purpose is more than merely holding the cells together. There are three main types of special junctions: *tight junctions, desmosomes,* and *gap junctions.*

Tight Junctions

Tight junctions are commonly found in epithelial tissue that is specialized for molecular transport. In **tight junctions,** integral membrane proteins called *occludins* fuse adjacent cells together to form a nearly *impermeable* barrier to the movement of substances between cells (Figure 2.26). Because of this barrier, polar solutes generally must cross the epithelial cell layer to go from one side of an

TABLE 2.2 SUMMARY OF CELL STRUCTURES
AND FUNCTIONS

CELL PART	STRUCTURE	FUNCTION
Plasma membrane	Lipid bilayer with scattered proteins and cholesterol molecules	Maintains boundary of cell and integrity of cell structure; embedded proteins serve multiple functions
Nucleus	Surrounded by double-layered nuclear envelope	Houses the DNA, which dictates cellular function and protein synthesis
Nucleolus	Dark oval structure inside the nucleus	Synthesis of ribosomal RNA
Cytosol	Gel-like fluid	Cell metabolism, storage
Membranous organelles		
Rough endoplasmic reticulum	Continuous with the nuclear envelope; flattened sacs dotted with ribosomes	Protein synthesis and post-translational processing
Smooth endoplasmic reticulum	Continuous with rough endoplasmic reticulum; tubular structure without ribosomes	Lipid synthesis and post-translational processing of proteins; transport of molecules from endoplasmic reticulum to Golgi apparatus
Golgi apparatus	Series of flattened sacs near the endoplasmic reticulum	Post-translational processing; packaging and sorting of proteins
Mitochondria	Oval-shaped, with an outer membrane and an inner membrane with folds called cristae that project into the matrix	ATP synthesis
Lysosomes	Granular, saclike; scattered throughout the cytoplasm	Breakdown of cellular and extracellular debris
Peroxisomes	Similar in appearance to lysosomes, but smaller	Breakdown of toxic substances
Nonmembranous organelles		
Vaults	Small, barrel-shaped	Unknown; possibly transport of molecules between nucleus and cytoplasm
Ribosomes	Granular organelles composed of proteins and RNA; located in cytosol or on surface of rough endoplasmic reticulum	Protein synthesis
Centrioles	Two cylindrical bundles of protein filaments that are perpendicular to each other	Direction of mitotic spindle development during cell division
Cytoskeleton	Composed of protein filaments, including microfilaments, intermediate filaments, and microtubules	Structural support of cell; cell movement and contraction

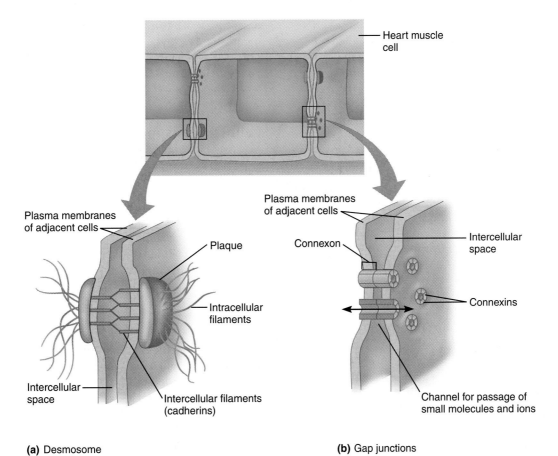

(a) Desmosome

(b) Gap junctions

FIGURE 2.27 Desmosomes and gap junctions, as shown in heart muscle.
(a) *Desmosomes, which are strong junctions between cells in tissues subjected to stress. Protein filaments extend from a plaque located near the plasma membrane where the two cells join together. Some of these filaments are located inside the cells (intracellular filaments), whereas other filaments called cadherins (intercellular filaments) extend into the intercellular space, linking the two cells together.* **(b)** *Gap junctions, which are areas of membrane containing proteins called connexons that form channels linking the cytoplasm of two adjacent cells. These channels allow ions and small molecules to pass between adjacent cells.*

epithelium to the other *(transepithelial transport),* rather than going around the cells *(paracellular movement).*

Epithelial cells often line hollow structures, such as the organs of the gastrointestinal tract or the tubules in the kidneys. The internal compartment of a hollow organ, the *lumen,* is separated from the cytosol of each epithelial cell by the *apical membrane.* The membrane facing the extracellular fluid is called the *basolateral membrane.* Epithelial cells lining the organs often have tight junctions that restrict or regulate the movement of molecules from the lumen of the organ into the blood, or from the blood into the lumen. Because of these tight junctions, the composition of the contents of the lumen can differ from that of the blood.

Desmosomes

Desmosomes are found in tissues subject to mechanical stress, such as those in the heart, uterus, and skin. A

desmosome is a filamentous junction between two adjacent cells that provides strength so that the cells do not tear apart when the tissue is subject to stress. Figure 2.27a shows the structure of desmosomes in the heart, where they are closely associated with another type of adhesion called *gap junctions* (described next). At the site of each desmosome is a *plaque* formed by glycoproteins clustered inside each cell. Extending from this plaque are both intracellular protein filaments and protein filaments called *cadherins* that cross the plasma membrane into the extracellular space. The cadherins are linked to intracellular filaments at the plaque.

Gap Junctions

Gap junctions, although found in a variety of tissue types, are most notable for their presence in smooth muscle and in the muscle of the heart, where the gap junctions allow the muscle to contract as a unit. **Gap junctions** are areas where two adjacent cells are connected by membrane proteins called *connexons* (Figure 2.27b). Each connexon is composed of six smaller membrane proteins called *connexins*. The connexons in the adjacent cells bind to each other, forming small channels that enable ions and small molecules to move between the two cells. Movement of ions between cells electrically couples the cells, such that if the electrical properties of one cell change, ions move from this cell to the adjacent cell through the gap junction, producing the same type of electrical change. This process allows the cells of the heart, for example, to act in a coordinated fashion so that they can contract in unison. Gap junctions also allow the passage of certain small molecules, including some chemical messengers such as the molecule cAMP. In this manner, a signal initiated in one cell can be transmitted to adjacent cells.

GENERAL CELL FUNCTIONS

Our cells have specialized functions that contribute to homeostasis. For example, muscle cells contract to generate force, and nerve cells use electrical impulses and chemical messengers for communication. Despite these special functions, most cells of the body carry out several general functions, which we discuss next. The specialized functions of cells are described in subsequent chapters.

Metabolism

All cells of the body carry out a variety of chemical reactions. **Metabolism** refers to all the chemical reactions that occur in the body. Metabolism can be divided into two classes of chemical reactions: (1) **anabolism,** the synthesis of large molecules from smaller molecules, which requires an input of energy, and (2) **catabolism,** the breakdown of large molecules to smaller molecules, which releases energy. Every second, thousands of anabolic and catabolic reactions are occurring in virtually every cell of the body. These reactions are highly regulated by the presence of enzymes. You will learn more about cell metabolism and enzymes in Chapter 3.

Exercise Link

The rates of all these anabolic and catabolic reactions are constantly changing (adapting) to environmental and behavioral circumstances. The energy required for exercise comes from the catabolism of glucose and triglycerides. In extreme exercise, such as running a marathon, even protein molecules are broken down for energy. When individuals are resting after a meal, anabolic processes are promi-

nent: New proteins are synthesized from dietary amino acids, glycogen is synthesized from dietary glucose, and triglycerides are formed from dietary fatty acids and stored in fat cells.

Cellular Transport

We have seen that the lipid bilayer is a barrier to hydrophilic molecules. However, cells require hydrophilic molecules to carry out their functions, and cells also release hydrophilic molecules into the extracellular environment. The process by which molecules move into and out of cells is called *membrane transport*. Next we examine the two types of cellular transport: movement across membranes and movement within membrane-bound compartments.

Transport of Molecules Across Membranes

Nonpolar molecules can permeate the plasma membrane and therefore will simply move across the membrane by diffusion, described in Chapter 4. The smallest of the polar molecules, such as water, can also permeate most plasma membranes. However, larger polar molecules cannot permeate the lipid bilayer; some of these molecules can cross plasma membranes only with the assistance of transmembrane proteins. For example, some molecules such as glucose and amino acids are transported across the membrane by *carrier proteins*. Other molecules, such as ions, can move through *ion channels*. Details regarding the functions of these proteins are described in Chapter 4.

Transport of Material Within Membrane-bound Compartments

Proteins and other *macromolecules* are too large to be transported across the plasma membrane by carrier proteins or to move through channels. Instead, movement of these macromolecules involves packaging the molecules into vesicles. Macromolecules enter cells through the formation of vesicles from the plasma membrane in a process called **endocytosis**. In **exocytosis,** macromolecules within cells are packaged into secretory vesicles, which then fuse with the plasma membrane and release their contents into the interstitial fluid. Both of these processes, which require the input of energy, are described next.

Endocytosis There are three forms of endocytosis: *phagocytosis, pinocytosis,* and *receptor-mediated endocytosis* (Figure 2.28). In all cases, extracellular fluid and sometimes particulate matter is brought into the cell by the formation of a vesicle.

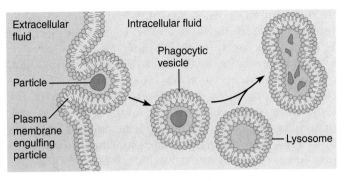

Extracellular fluid

Intracellular fluid

Phagocytic vesicle

Particle

Plasma membrane engulfing particle

Lysosome

(a) Phagocytosis

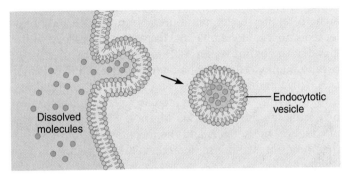

Dissolved molecules

Endocytotic vesicle

(b) Pinocytosis

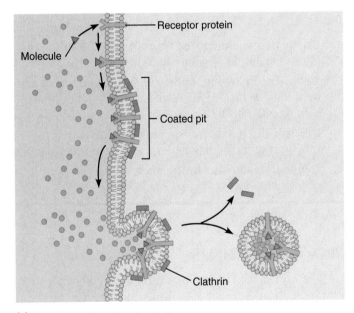

Receptor protein

Molecule

Coated pit

Clathrin

(c) Receptor-mediated endocytosis

FIGURE 2.28 **Endocytosis. (a)** *Phagocytosis. After the plasma membrane surrounds a particle in the extracellular fluid, the membrane pinches together to form a phagocytic vesicle around the particle, which then usually fuses with a lysosome. The enzymes of the lysosome degrade the particle.* **(b)** *Pinocytosis. The plasma membrane indents to form an endocytotic vesicle, containing extracellular fluid and dissolved solutes.* **(c)** *Receptor-mediated endocytosis. First, receptor proteins on the plasma membrane bind specific molecules, triggering endocytosis. The plasma membrane then indents around the molecules, forming a vesicle that pinches off of the plasma membrane and enters the cell.*

In **phagocytosis,** which means "cell-eating," a cell uses amoeboidlike movements of its plasma membrane to extend the membrane around a particle in the extracellular fluid (see Figure 2.28a). When the membrane completely surrounds the particle, the two sides of the plasma membrane pinch together to form a *phagocytic vesicle* (also called a *phagosome*) in the cytoplasm; the particle and some extracellular fluid are inside the vesicle. In this manner, the cell has *engulfed* the particle. Once inside the cell, the membrane of the phagocytic vesicle fuses with the membrane of a lysosome, forming a *phagolysosome,* which exposes the engulfed particle to the degradative enzymes of the lysosome. The enzymes break down the particle and usable components are recycled. Phagocytosis is common in *white blood cells,* which are responsible for removing foreign particles and bacteria from our bodies.

In **pinocytosis,** which means "cell-drinking," the plasma membrane indents, and its outer edges pinch together to form an *endocytotic vesicle* in the cytoplasm (see Figure 2.28b). Pinocytosis is a nonspecific process, and the contents of the resulting vesicle are extracellular fluid containing dissolved solutes.

Receptor-mediated endocytosis is similar to pinocytosis in that the plasma membrane indents to form the vesicle (see Figure 2.28c). Unlike pinocytosis, however, receptor-mediated endocytosis is specific. Proteins in the plasma membrane function as *receptors* that recognize and bind specific particles in the extracellular fluid. Binding of particles to receptors concentrates the particles to areas where endocytosis will occur. The area of plasma membrane that forms the vesicle is coated with proteins (called clathrin) on its cytosolic surface. The membrane indents in this area, forming what is called a **coated pit.** The coated pit becomes a coated vesicle containing the receptors and the particles bound to them. The protein coat rapidly leaves the vesicle, and the clathrin molecules are recycled. The now uncoated vesicle fuses with a lysosome, forming an *endolysosome.* The enzymes in the lysosome will then degrade the particles brought into the cell. The receptors are often recycled by exocytosis, which is described next.

Exocytosis Exocytosis is basically endocytosis in reverse: A vesicle inside the cell fuses with the plasma membrane and releases its contents into the extracellular fluid (Figure 2.29). Exocytosis has three functions: (1) to add components to the plasma membrane, (2) to recycle receptors removed from the plasma membrane by endocytosis, and (3) to secrete specific substances out of the cell and into the extracellular fluid.

The first two functions are related in that both add components to the plasma membrane. During exocytosis, whatever components are present in the vesicle membrane will be added to the plasma membrane. A cell can add certain proteins, phospholipids, or carbohydrates to the plasma membrane, or it can replace the membrane

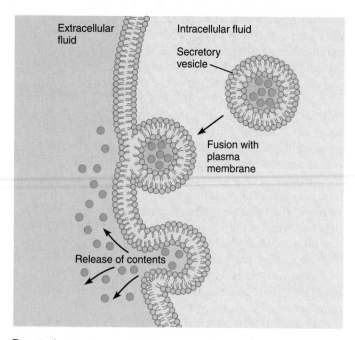

Extracellular fluid

Intracellular fluid

Secretory vesicle

Fusion with plasma membrane

Release of contents

Exocytosis

FIGURE 2.29 Exocytosis. *An intracellular vesicle fuses with the plasma membrane, and the contents of the vesicle are secreted from the cell.*

that is lost during endocytosis. In fact, endocytosis and exocytosis must be balanced in a cell; otherwise, the size of the plasma membrane will change.

The third function of exocytosis, the secretion of materials, can serve several different functions. Certain white blood cells secrete antibodies to fight infections. Most cells—in particular nerve cells and endocrine cells—secrete chemical messengers that communicate with other cells. Cells lining certain hollow ducts or passageways, such as the gastrointestinal tract or respiratory airways, secrete a sticky fluid called mucus, which acts as a protective coating. You will learn about many other examples of secretion in the chapters to come.

Transcytosis

Epithelial cells are often linked together by tight junctions. As a result, if a macromolecule is to move from the lumen to the interstitial fluid or vice versa, it must cross both the apical and basolateral membranes. Macromolecules cross epithelial cells by a process called **transcytosis** (Figure 2.30), which involves both endocytosis and exocytosis. During transcytosis, a large molecule is taken into the cell by endocytosis, but the endocytotic vesicle does not fuse with a lysosome. Instead, the vesicle travels to the opposite side of the cell and fuses with the plasma membrane to release its content by exocytosis.

Intercellular Communication

Communication is another function that all cells perform. A single cell in the body cannot exist in isolation. Cells

must communicate with each other to maintain an environment in which they can survive. Communication can be direct, as occurs through gap junctions. Often, however, communication involves the release from one cell of a chemical messenger that acts on a second cell. The two communicating cells can be adjacent to each other, or they can be in very different areas of the body. In general, messenger molecules are released from one cell and reach other cells, called *target cells,* which contain proteins that function as *receptors* for that specific messenger. The binding of the messenger molecule to its specific receptor triggers a response in the target cell.

In one example of such intercellular communication, a chemical messenger, antidiuretic hormone (ADH), is released into the blood by exocytosis from an endocrine gland in the brain. When ADH reaches its target cells—certain kidney cells—the hormone binds with its receptors. This binding stimulates the insertion of water pores into the kidney cells, which increases water conservation by decreasing the production of urine. The different types of chemical messengers and the mechanisms by which they produce a response in target cells are the topics of Chapter 5.

Quick Test 2.3

1. Briefly describe the three types of cell adhesions, and give an example of where each is found.

2. Define the following terms: *metabolism, anabolism,* and *catabolism.* Which requires an input of energy, anabolism or catabolism?

3. Describe the role of transmembrane proteins in the transport of molecules across the plasma membrane.

4. Define the following terms: *phagocytosis, pinocytosis, receptor-mediated endocytosis, exocytosis.*

PROTEIN SYNTHESIS

By now you should realize that proteins are very important biomolecules in our cells. Contained within the DNA are the codes for the synthesis of all the proteins in our cells. This code is called the **genetic code,** and it is universal among all animal species. Even though DNA is located in the nucleus, protein synthesis takes place in the cytoplasm. Therefore, the code must be transmitted to another molecule, called **messenger RNA** (mRNA), that can move into the cytoplasm. The general steps of protein synthesis are as follows:

1. DNA is *transcribed* according to the genetic code to form a complementary mRNA in the nucleus.

2. mRNA moves from the nucleus to the cytoplasm.

3. mRNA is *translated* by ribosomes to form the correct amino acid sequence of the protein in the cytoplasm.

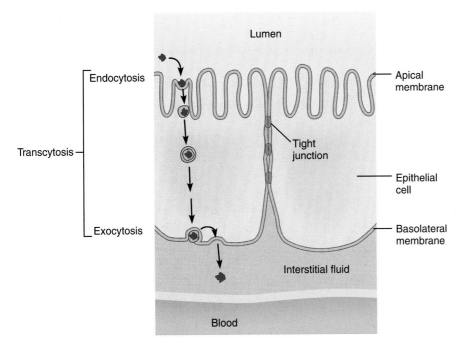

The Role of the Genetic Code

Recall that DNA consists of two strands of nucleotides held together by hydrogen bonding between complementary base pairs. The sequence of the four bases in DNA codes for the sequence of amino acids in proteins. The section of DNA that codes for a particular protein or proteins is called a **gene** (Figure 2.31a).

Twenty different amino acids are found in proteins. The sequence of the DNA base pairs must provide enough information to specify each amino acid in a polypeptide chain. A sequence of three base pairs, called a **triplet,** codes for a single amino acid (Figure 2.31b). Each base pair can be thought of as a letter of an alphabet that is used in words that are all three letters long and indicate a particular amino acid. With four bases that can take up three possible positions in a triplet, there are $4^3 = 64$ possible words in the genetic code. During transcription, which is described shortly, this code is passed on to mRNA, whose base sequence is complementary to DNA. Therefore, a DNA triplet is complementary to a three-base sequence in mRNA, called a **codon,** that codes for the same amino acid (Figure 2.31c).

Note that there are 64 possible codons, but only 20 amino acids. Thus, even though each codon codes for only one amino acid, some amino acids are coded for by more than one codon. For example, CCC always codes for the amino acid proline, but CCG also codes for proline. One codon, called an *initiator codon,* is found in every mRNA. This initiator codon, AUG, also codes for the amino acid methionine. There are three *termination codons* that indicate that the end of the protein code has occurred. Because these termination codons do not code for any amino acid, only 61 of the 64 codons actually code for amino acids.

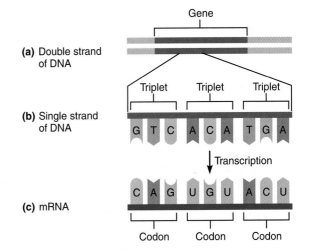

FIGURE 2.31 **The genetic code. (a)** *Genes, which are portions of DNA that code for a particular protein(s).* **(b)** *Triplets, the three-base sequences that code for the amino acid sequence in a gene.* **(c)** *Transcribed mRNA codons, which are complementary to the code in DNA triplets.*

Transcription

Transcription is the process in which RNA is synthesized using information contained in the DNA, and it occurs in the nucleus. Three types of RNA can be transcribed from DNA: (1) mRNA, (2) rRNA, and (3) *tRNA (transfer RNA).* Each type of RNA is involved in protein synthesis. This section describes the process of transcribing RNA from DNA, using mRNA as the example.

Figure 2.32 illustrates the process of transcription. The first step is the uncoiling of DNA and its separation into two strands at the site where transcription is occurring. One of the strands of DNA contains the gene and functions as a template for the synthesis of mRNA. The section of DNA

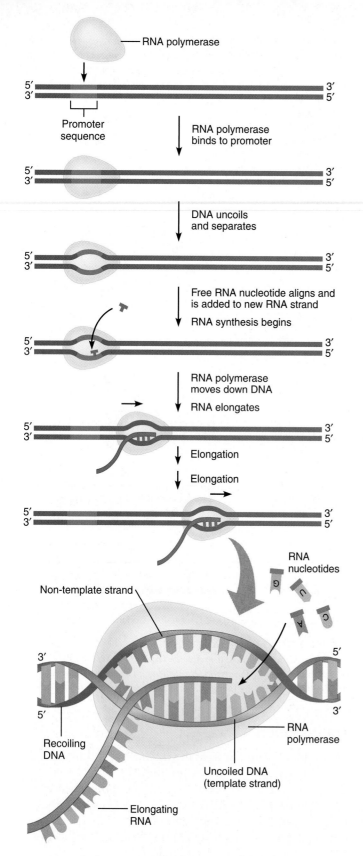

FIGURE 2.32 **Transcription.** *RNA polymerase binds to the promoter sequence of a DNA molecule, causing the DNA to uncoil. The two strands of DNA separate, exposing the bases for complementary base pairing with free nucleotides. The free nucleotides align, and RNA polymerase catalyzes the formation of a bond between them, generating an elongating strand of mRNA. The enlargement shows that the process of elongation occurs according to the law of complementary base pairing.*

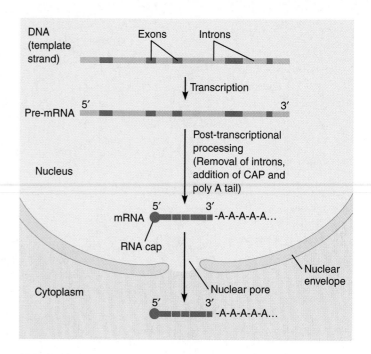

FIGURE 2.33 **Post-transcriptional processing.** *Following transcription, mRNA must undergo further processing in the nucleus. Transcribed introns must be removed, and the remaining exons spliced together. A chemical structure called a cap is added to the 5' end, and a poly A tail (several nucleotides with the adenine base) is added to the 3' end. After processing is complete, the mRNA molecule can move from the nucleus to the cytoplasm.*

with the gene is identified by the **promoter sequence,** a specific base sequence to which the enzyme *RNA polymerase* can bind. When RNA polymerase binds to the promoter sequence, it initiates the separation of DNA into two strands. This allows free ribonucleotides in the triphosphate form (ATP, UTP, GTP, and CTP) to align with the DNA template according to the law of complementary base pairing. (Recall that the base uracil in RNA replaces thymine in DNA.) RNA polymerase also catalyzes the formation of bonds between the aligned ribonucleotides, making a polynucleotide that will eventually be the single-stranded mRNA.

The mRNA undergoes *post-transcriptional processing* in the nucleus before the final mRNA product moves through the nuclear pores and enters the cytoplasm (Figure 2.33). Within a gene are regions of excess bases, called *introns,* that do not code for the amino acid sequence of proteins. Although these introns are transcribed in the initial mRNA molecule, they must be removed before mRNA leaves the nucleus. After introns are cut out of the initial mRNA molecule, the remaining coding segments, called *exons,* are joined together. Other aspects of post-transcriptional processing include addition of a chemical group called a *cap* to the 5' end, which is necessary for initiating translation, and addition of several adenine nucleotides called a poly A tail to the 3' end, which protects the mRNA from degradation in the cytoplasm. After processing is complete, the mRNA enters the cytoplasm, and translation of mRNA can occur.

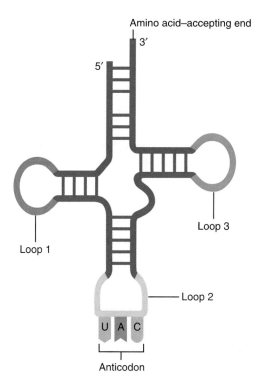

Amino acid–accepting end

FIGURE 2.34 Transfer RNA. *Hydrogen bonding between comple-mentary base pairs is responsible for the cloverleaf structure of transfer RNA. The tRNA binds a specific amino acid on one end, and it has a base sequence called an anticodon that recognizes the complementary mRNA codon.*

Translation

Translation is the process in which polypeptides are synthe-sized using mRNA codons as a template for the assembly of the correct amino acids along the sequence. Another type of RNA molecule, tRNA, carries the appropriate amino acid to the ribosome based on interactions with the mRNA codon. Due to complementary base pairing within its single strand of nucleotides, tRNA is shaped somewhat like a cloverleaf (Figure 2.34). The 3′ end of the tRNA molecule contains a binding site for a specific amino acid. Another region of the tRNA contains a base sequence, called the *an-ticodon,* that is complementary to the mRNA codon. It is the correct interaction between the codon on the mRNA and the anticodon on the tRNA that assures that the correct amino acid is added to the elongating polypeptide chain.

Translation occurs in the cytoplasm in association with ribosomes. Recall that ribosomes can be found free in the cytosol or attached to the rough endoplasmic retic-ulum. The ribosome aligns the mRNA with the tRNA carrying the appropriately coded amino acids, and con-tains the enzymes that catalyze the formation of peptide bonds between the amino acids, thus forming a polypep-tide. Ribosomes can hold two tRNA molecules at the same time in two distinct ribosomal regions identified as the *P site* and the *A site.*

Now that we have some knowledge of the compo-nents needed for translation, we turn to the process itself.

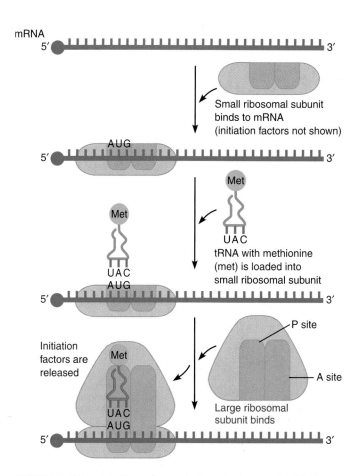

FIGURE 2.35 Initiation of translation. *In the cytosol, initiation factors bind to the cap region of a mRNA molecule and to the small ribosomal subunit, triggering the binding of the small ribosomal subunit to the mRNA. A tRNA with the anticodon complementary to the initiation codon of mRNA binds to the mRNA by the law of complementary base pairing. The large ribosomal subunit binds such that the first tRNA is located in the P site of the ribosome, and the initiation factors are released.*

The first step of translation is *initiation* (Figure 2.35). Ri-bosomes not actively involved in translation are found free in the cytosol as their individual subunits. To start translation, certain *initiation factors* bind to the cap that was added to the 5′ end of the mRNA molecule during post-transcriptional processing; other initiation factors form a complex with the small ribosomal subunit and a *charged tRNA* (tRNA bound to its amino acid) bearing the anticodon complementary to the initiation codon (AUG). Binding of the initiation factors triggers the binding of the small ribosomal subunit to the mRNA, such that the first initiation codon (AUG) found within the mRNA is aligned correctly for the start of translation. The large ribosomal subunit then binds, causing dissociation of the initiation factors and alignment of the first tRNA in the P site of the ribosome. A second charged tRNA molecule with the appropriate anticodon then enters the A site of the ribosome, and translation begins.

During the process of translation (Figure 2.36), an enzyme in the ribosome catalyzes the formation of a peptide bond between the two amino acids. The first

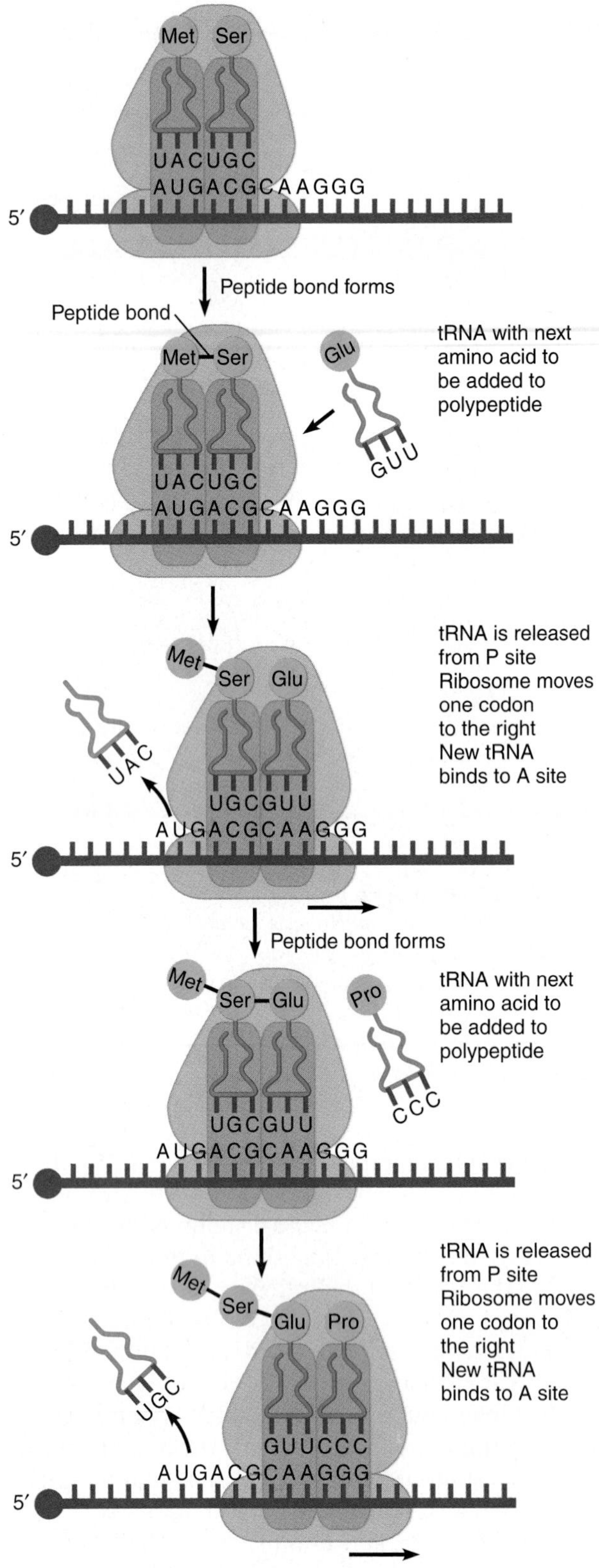

FIGURE 2.36 Translation. *Following initiation, tRNA molecules will occupy sites on the ribosome according to complementary base pairing between the anticodon of the tRNA and the codon of the mRNA. The tRNA carries the appropriate amino acid to be added to the polypeptide chain. Enzymes in the ribosome catalyze the formation of a peptide bond. The first tRNA molecule is released from the P site of the ribosome. The ribosome then moves down the mRNA molecule one codon, putting the second tRNA in the P site. A new tRNA occupies the A site, bringing with it the next amino acid to be added to the polypeptide chain.*

What is a DNA triplet code for the amino acid serine?

amino acid, which is always methionine, is released from the tRNA molecule, and the *free tRNA* (tRNA without its amino acid) leaves the P site on the ribosome. The ribosome then moves down the mRNA three bases (one codon), placing the second tRNA in the P site on the ribosome. The next charged tRNA enters the A site, bringing in the third amino acid to be added to the polypeptide. This process continues until a termination codon is reached on the mRNA. At that point, the polypeptide is released, and the ribosome and mRNA dissociate. More than one ribosome translates a given mRNA molecule at the same time, forming what is called a *polyribosome.*

Destination of Proteins

Proteins synthesized in a cell can be destined for the nucleus, cytosol, plasma membrane, or one of the organelles, or can be secreted from the cell into the extracellular fluid. Unless the final destination of a protein is to remain in the cytosol, the first functioning sequence of amino acids that is translated in a polypeptide chain is the **leader sequence.** The leader sequence is synthesized during translation by ribosomes free in the cytosol and determines the destination of the protein (Figure 2.37a). First, the leader sequence determines whether a protein will be synthesized in the cytosol or in association with the endoplasmic reticulum. Second, the leader sequence determines whether proteins synthesized in the cytosol will remain there (in which case there is no leader sequence) or enter a mitochondrion, a peroxisome, or the nucleus (Figure 2.37b). The leader sequence determines the fate of proteins by specific interactions with other proteins on either the membrane of an organelle or the nuclear membrane. How the leader sequence causes ribosomes to attach to the endoplasmic reticulum and where proteins synthesized by those ribosomes are destined is described shortly.

The process of protein synthesis in association with the endoplasmic reticulum is illustrated in Figure 2.38 on page 54. Protein synthesis starts in the cytosol, as previously described. If the leader sequence designates that synthesis is to occur in association with the endoplasmic reticulum, then the leader sequence and its associated ribosome bind to proteins on the endoplasmic reticulum

TGC

called *signal recognition proteins.* Once the ribosome is attached to the endoplasmic reticulum, protein synthesis continues, with the polypeptide chain passing through the membrane and entering the lumen of the rough endoplasmic reticulum as it is formed. The polypeptide chain travels from the lumen of the rough endoplasmic reticulum to the lumen of the smooth endoplasmic reticulum, and eventually to the Golgi apparatus by mechanisms described shortly. Along this route, another important process also occurs: *post-translational processing.*

Post-Translational Processing and Packaging of Proteins

Translation alone generally does not provide a functional protein. Changes in the initial polypeptide chain, called **post-translational processing,** must occur to produce the final protein product. Post-translational processing includes cleavage of some amino acids, which can occur anywhere along the route from the rough endoplasmic reticulum to the Golgi apparatus, and may even occur after packaging by the Golgi apparatus. For example, the leader sequence must be removed from the polypeptide; cleavage of other "excess" amino acids is also common. Other post-translational processing includes the addition of chemical groups such as lipids or carbohydrates to form lipoproteins and glycoproteins, respectively. Carbohydrates are added to the polypeptide in the endoplasmic reticulum and Golgi apparatus by a process called *glycosylation.*

From the lumen of the endoplasmic reticulum, the polypeptide must get to the Golgi apparatus (unless it is destined to remain in the endoplasmic reticulum), the lumen of which is not continuous with that of the endoplasmic reticulum. Therefore, the smooth endoplasmic reticulum packages the polypeptide into small *transport vesicles* that bud off of the smooth endoplasmic reticulum (see Figure 2.38). The membrane of the vesicle then fuses with the membrane of the Golgi apparatus, emptying its contents into the lumen of the Golgi apparatus at a Golgi cisterna close to the endoplasmic reticulum.

Once in the Golgi apparatus, the polypeptide undergoes further processing as it travels to the trans face on the opposite end of the Golgi apparatus. The Golgi apparatus sorts and packages proteins into vesicles that are targeted for their appropriate locations. The precise mechanism of how vesicles are directed to their targets is not fully understood, but certain chemicals function as markers for the sorting of proteins. Targets for proteins packaged by the Golgi apparatus include sites within the cell (lysosomes or Golgi apparatus) or the plasma membrane; alternatively, proteins may be secreted from the cell by exocytosis.

Figure 2.39 on page 55 summarizes the steps of protein synthesis and targeting.

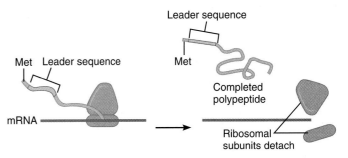

(a) Translation of leader sequence and polypeptide in cytosol

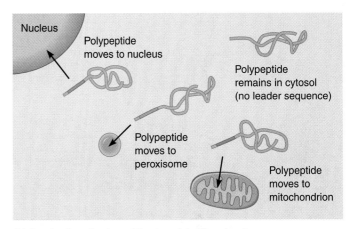

(b) Destination of polypeptides translated in cytosol

FIGURE 2.37 Targeting of proteins translated in the cytosol. **(a)** *The first amino acid of a new polypeptide chain is a methionine (met) because its codon is also the initiation codon. Another sequence synthesized early during translation is the leader sequence. Following translation, the ribosome, mRNA, and polypeptide separate.* **(b)** *The leader sequence determines the ultimate fate of a protein—whether it stays in the cytosol (in which case no leader sequence would have been translated), moves into a mitochondrion or peroxisome, or enters the nucleus.*

Regulation of Protein Synthesis

Cellular function can often be regulated by varying the amount of a certain protein in the cell. For example, by increasing the number of carrier proteins, the transport of a particular molecule can be increased. Protein synthesis can be regulated either at the level of transcription or at the level of translation.

Regulation of Transcription

Once synthesized and in the cytoplasm, mRNA is present in a functional form only briefly. Some mRNA binds to proteins in the cytoplasm that render it inactive until an appropriate chemical signal is received; other mRNA is rapidly degraded by enzymes in the cytosol. Since mRNA is necessary for protein synthesis, the rate of transcription affects the amount of protein in a cell. Transcription can be either turned on *(induced)* or turned off *(repressed),* depending on whether the cell requires more or less of the

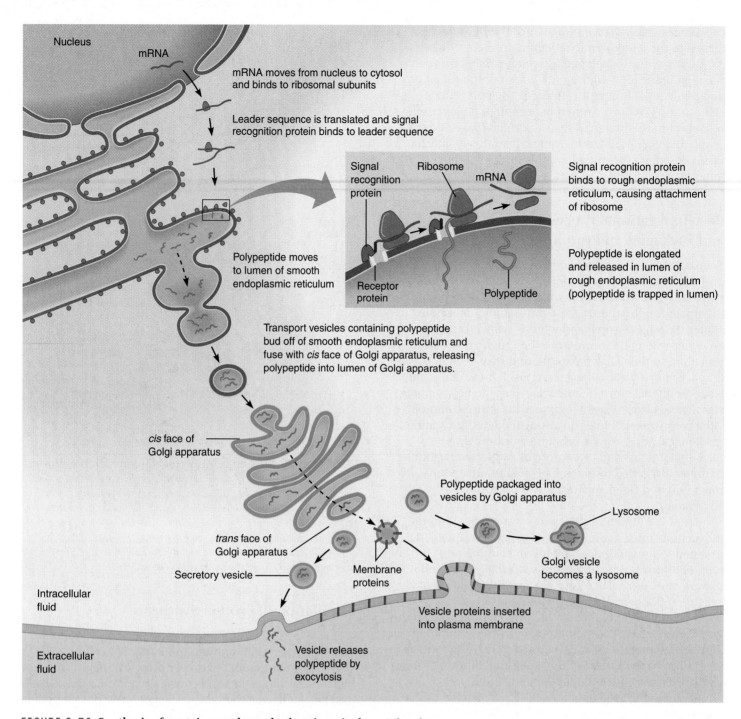

Nucleus

mRNA

mRNA moves from nucleus to cytosol
and binds to ribosomal subunits

Leader sequence is translated and signal
recognition protein binds to leader sequence

Signal
recognition
protein

Ribosome

mRNA

Signal recognition protein
binds to rough endoplasmic
reticulum, causing attachment
of ribosome

Receptor
protein

Polypeptide

Polypeptide is elongated
and released in lumen of
rough endoplasmic
reticulum (polypeptide is
trapped in lumen)

Polypeptide moves
to lumen of smooth
endoplasmic reticulum

Transport vesicles containing polypeptide
bud off of smooth endoplasmic reticulum and
fuse with *cis* face of Golgi apparatus, releasing
polypeptide into lumen of Golgi apparatus.

cis face of
Golgi apparatus

Polypeptide packaged into
vesicles by Golgi apparatus

Lysosome

trans face of
Golgi apparatus

Secretory vesicle

Membrane
proteins

Golgi vesicle
becomes a lysosome

Vesicle proteins inserted
into plasma membrane

Intracellular
fluid

Extracellular
fluid

Vesicle releases
polypeptide by
exocytosis

FIGURE 2.38 Synthesis of proteins on the endoplasmic reticulum. *When the
leader sequence causes the ribosome to attach to the endoplasmic reticulum, translation
continues, with the polypeptide forming in the lumen of the rough endoplasmic reticulum.
The polypeptide then moves to the lumen of the smooth endoplasmic reticulum, where it is
packaged into a transport vesicle. The membrane of the transport vesicle fuses with the
membrane of a cisterna on the* cis *face of the Golgi apparatus, releasing the polypeptide into
the lumen of the Golgi apparatus. The Golgi apparatus packages the polypeptide into a
vesicle and targets the vesicle to the appropriate location, which could be a lysosome or the
plasma membrane.*

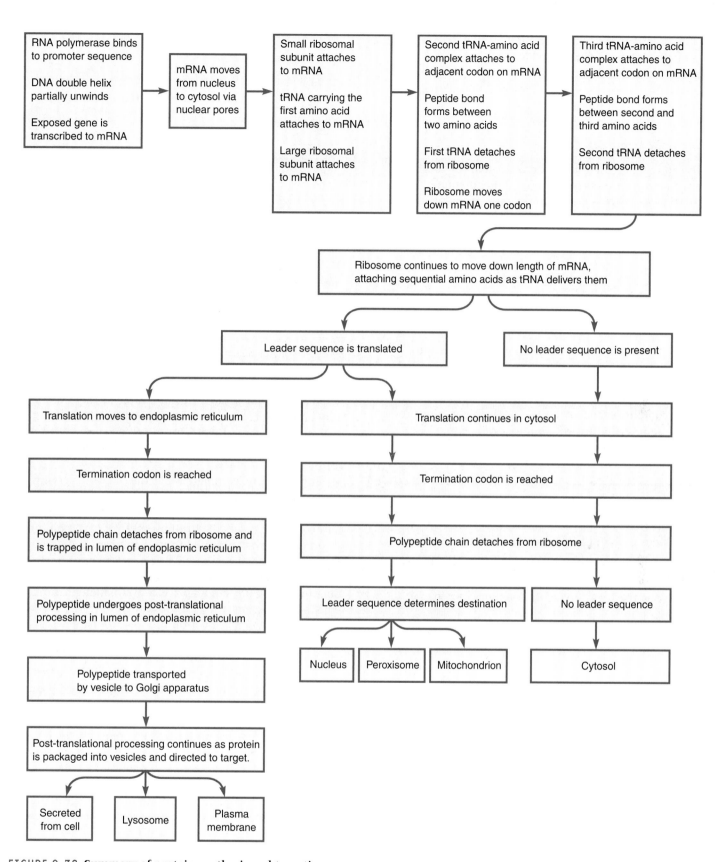

FIGURE 2.39 **Summary of protein synthesis and targeting.**

protein. Regulation of transcription typically occurs at the step in which RNA polymerase binds to the promoter sequence on the DNA. At this stage, certain molecules either enhance or decrease the ability of RNA polymerase to bind to the promoter sequence.

Regulation of Translation

Cells can regulate the translation of proteins by mechanisms that remain poorly understood. Regulation of translation typically occurs during initiation. For initiation to occur, at least 11 proteins, including initiation factors, must be present. Some of these proteins can be rendered active or inactive, thereby turning translation on or off.

Protein Degradation

Proteins do not exist indefinitely in the cytosol, but are continually degraded by enzymes called *proteases*. To prevent the nondiscriminant destruction of cytosolic proteins, those proteins to be destroyed are often tagged with a polypeptide called *ubiquitin*. Ubiquitin functions much like a leader sequence in that it directs the protein to a protein complex called a *proteosome*, which contains proteases that degrade the protein into small peptide fragments.

Quick Test 2.4

1. Define transcription and translation. Where in the cell does each of these processes occur?

2. Given the DNA triplet ATC, what is the mRNA codon that would be transcribed?

3. Describe the functions of RNA polymerase in transcription.

4. What is the name of the first part of a polypeptide chain that is translated from mRNA? What is its function?

CELL DIVISION

Most cells in the body have a limited life span, and therefore cell division is necessary to provide a continuing line of cells for the body. Cell division is also necessary for tissue growth, especially during the early years of life. There are two types of cell division: **mitosis,** which occurs in most cells, and **meiosis,** which occurs only in reproductive cells. Here we focus on mitosis; meiosis is described in Chapter 21.

Replication of DNA

Each cell in the body (except mature red blood cells) contains a copy of all the body's genes in its DNA. Therefore,

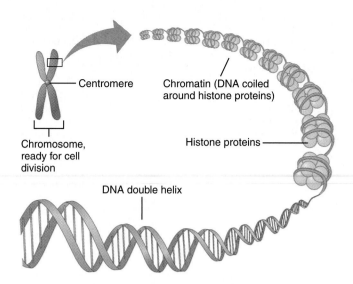

FIGURE 2.40 A chromosome. *A chromosome is a tightly wound and organized structure present just prior to cell division. At other times a cell's genetic material is present in the nucleus as chromatin, which consists of the DNA double helix wrapped around histone proteins.*

when a cell divides, DNA must be copied exactly so that each daughter cell receives a complete set of DNA. Before we can understand the process of cell division, we must understand the mechanism of copying DNA, called **replication.**

Together, one complete molecule of DNA carrying a specific set of genetic information, plus its associated proteins, make up a chromosome (Figure 2.40). Most human cells contain 23 pairs of chromosomes. Of each pair, one chromosome was inherited from the father, and one from the mother. Together, these 23 pairs of chromosomes form the **human genome.** (Discovery: The Human Genome, www.physiologyplace.com, Challenge Yourself.) Within the nucleus, the chromosomes are coiled around proteins called *histones*. Chromosomes are usually scattered throughout the nucleus in the form of beadlike structures called chromatin.

During DNA replication, each strand of DNA serves as a template for the synthesis of a new strand of DNA (Figure 2.41). The process is termed *semiconservative* because each new DNA molecule consists of a new strand plus an old strand (or conserved strand) that functioned as the template for the new strand. The process starts with the unwinding of DNA and then separation of the two strands.

Each strand of DNA acts as a template for adding the appropriate complementary nucleotide to the growing new strand. The nucleotides are present in their triphosphate form free in the nucleus (ATP, TTP, CTP, and GTP). When they align according to the law of complementary base pairing, the enzyme *DNA polymerase* catalyzes a reaction that adds the next nucleotide to the growing polynucleotide chain. In the process, two phosphates are cleaved from the nucleotide, providing some of the energy for the reaction. The end result of the process is two identical

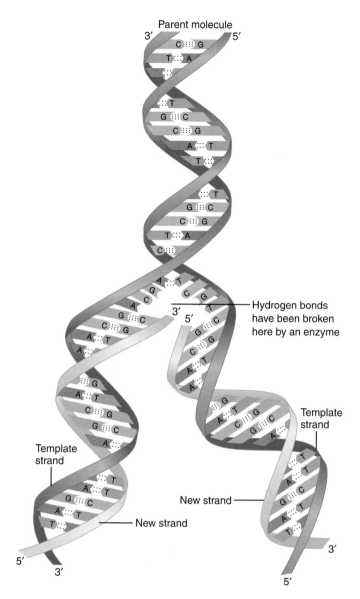

FIGURE 2.41 DNA replication. *The DNA double helix unwinds and the two strands separate, allowing the bases to align with free nucleotides. Each strand of DNA is a template for the synthesis of a new strand.*

copies of the DNA, one to pass on to each of two daughter cells during cell division.

Normally, DNA replication is precise; any errors that may have occurred are corrected by enzymes in the nucleus. However, alterations in the DNA, called *mutations*, can occur. Mutations within a single cell are usually not deleterious to the whole body. However, mutations of genes that control cell growth, called *proto-oncogenes*, can be very serious because growth of cells containing mutated proto-oncogenes is no longer regulated. Mutations such as this are responsible for cancer (When It Goes Wrong: Cancer, p. 59).

The Cell Cycle

The life cycle of a cell can be separated into two main divisions: *interphase* and *cell division.* During **interphase,** the

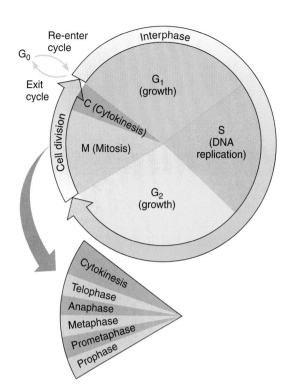

FIGURE 2.42 The cell cycle. *Most cells move through a continuous cycle that includes cell division. Interphase includes the periods between cell divisions and includes four phases: G0, G1, S, and G2. Mitosis and cytokinesis are the periods of actual cell division. Mitosis includes five phases: prophase, prometaphase, metaphase, anaphase, and telophase. Cytokinesis is the division of the cytoplasm. Some cells leave the cycle and enter G0, during which no activities related to cell division occur. Some of these cells remain in G0, whereas others reenter the cycle in response to a chemical signal.*

cell is carrying out its normal physiological functions; during cell division, the cell stops performing many of its normal physiological functions. Figure 2.42 shows the cell cycle. Notice that interphase is the period between the previous cell division and the beginning of the next cell division. The duration of interphase varies considerably among cell types. For example, cells lining the gastrointestinal tract undergo cell division every few days, whereas skeletal muscle cells never undergo cell division.

Interphase can be divided into four phases: *G0, G1, S, and G2.* The "G" stands for "gap," since little related to cell division is happening in the nucleus. During G0, the cell is carrying out its functions that are unrelated to cell division. Cells that do not undergo cell division, such as skeletal muscle cells, remain in G0 for their entire life. Other cells, such as certain white blood cells, require a signal to move from G0 to G1. During G1 the cell starts showing signs of an upcoming cell division, such as an increase in the rate of protein synthesis. In the S phase, DNA replication occurs, and the chromosomes now exist in pairs. During interphase, the centrioles duplicate so that they also exist in pairs at the start of cell division. During G2, rapid protein synthesis continues as the cell grows to about twice its original size. From G2 the cell enters cell division.

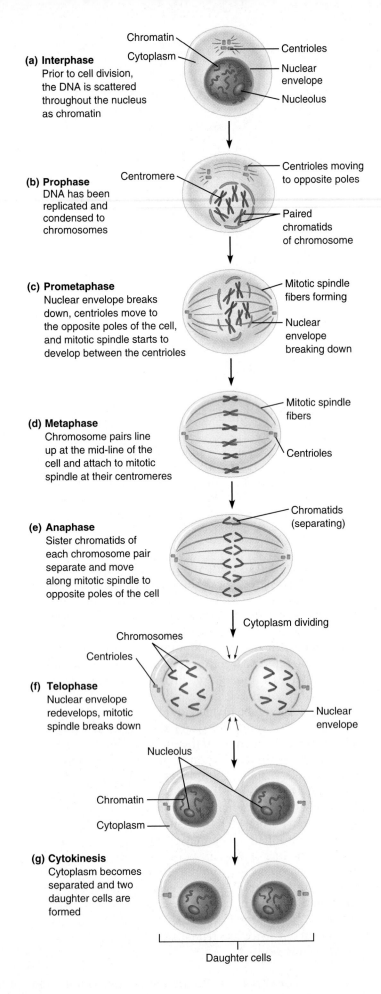

(a) Interphase
Prior to cell division, the DNA is scattered throughout the nucleus as chromatin

Chromatin
Cytoplasm
Centrioles
Nuclear envelope
Nucleolus

(b) Prophase
DNA has been replicated and condensed to chromosomes

Centromere
Centrioles moving to opposite poles
Paired chromatids of chromosome

(c) Prometaphase
Nuclear envelope breaks down, centrioles move to the opposite poles of the cell, and mitotic spindle starts to develop between the centrioles

Mitotic spindle fibers forming
Nuclear envelope breaking down

(d) Metaphase
Chromosome pairs line up at the mid-line of the cell and attach to mitotic spindle at their centromeres

Mitotic spindle fibers
Centrioles

(e) Anaphase
Sister chromatids of each chromosome pair separate and move along mitotic spindle to opposite poles of the cell

Chromatids (separating)

Cytoplasm dividing

Chromosomes
Centrioles

(f) Telophase
Nuclear envelope redevelops, mitotic spindle breaks down

Nuclear envelope

Nucleolus
Chromatin
Cytoplasm

(g) Cytokinesis
Cytoplasm becomes separated and two daughter cells are formed

Daughter cells

Cell division can be divided into two phases: *mitosis* and *cytokinesis* (see Figure 2.42). Mitosis includes five phases: (1) *prophase*, (2) *prometaphase*, (3) *metaphase*, (4) *anaphase*, and (5) *telophase*. Although the phases of mitosis are continuous and overlap, they can be identified by certain events (Figure 2.43).

1. Prophase: The chromatin pairs condense, such that the chromosomes are now visible under the light microscope. In this condensed and duplicated state, each individual of the chromosome pairs is called a **chromatid,** and the two chromatids in a chromosome are linked together at a structure called the *centromere*. Microtubules in the cell's cytoskeleton disassemble into their tubulin components, which will be used subsequently to form the mitotic spindle. The centriole pairs start moving to opposite poles of the cell, and the mitotic spindle develops between them.

2. Prometaphase: The nuclear envelope breaks down, the nucleolus is no longer visible, the centrioles are at opposite poles of the cell, and the chromosomes become linked at their centromeres to the spindle fibers.

3. Metaphase: The chromosomes are aligned in a plane at the middle of the cell.

4. Anaphase: The chromatid pairs separate, and the chromosomes move along the mitotic spindle toward opposite poles of the cell.

5. Telophase: New nuclear envelopes develop on the two sides of the cell, the chromosomes begin to decondense to chromatin, and the mitotic spindle begins breaking down.

Cytokinesis is the division of the cytoplasm. The cytoplasm divides by cleavage, which starts during the latter phases of mitosis (anaphase or telophase). The cleavage is accomplished by contraction of microfilaments (primarily actin) that form a ring around the midline of the cell, and the mitotic spindle disassembles. The two resulting daughter cells then enter interphase at G0.

Quick Test 2.5

1. Define *replication*. How does replication differ from transcription?

2. Name the four phases of interphase. During which phase does DNA replication occur?

3. Name the five stages of *mitosis*. During which stage does the nuclear envelope start to disintegrate?

FIGURE 2.43 Events in cell division. (a) *The appearance of the cell in interphase, which precedes cell division.* **(b–f)** *The major events in the five phases of mitosis.* **(g)** *The major events in cytokinesis.*

What type of filaments make up the mitotic spindle fibers?

Each year in the United States over 1 million individuals are diagnosed with cancer, and approximately 500,000 people die every year from this disease. Even though cancer can arise in nearly every type of body tissue and leads to illness in diverse ways, at least one theme is common to cancer: Cancerous cells divide uncontrollably and eventually displace the body's normally functioning cells and consume energy vital to normal body cells.

Cell growth and division is controlled by genes called *proto-oncogenes*. A mutation in a proto-oncogene can produce an *oncogene* that leads to abnormal cell growth and division. Mutations may occur spontaneously, or they may be induced by substances called *mutagens*. One example of a mutagen is DDT, which was used in pesticides in the 1960s. In some cases, including cancer, the mutated cells divide rapidly, passing the mutation on to their daughter cells, which then also divide rapidly. To rid the body of cancer, these mutated cells must be eliminated.

Most therapeutic strategies, including radiation therapy and chemotherapy, are directed at different stages of actively dividing cells. Because cancer cells divide rapidly, these therapies harm cancer cells more than they harm the normal, infrequently dividing cells of the body. Unfortunately, normal body cells that divide frequently are also affected by these therapeutic strategies. Thus because hair cells undergo rapid cell division, hair loss is one of the consequences of cancer therapy.

The role of genetics and of environmental agents (such as tobacco) in the initiation of abnormal cell growth and division is currently the topic of intense research. Researchers are learning to manipulate the body's immune system, hoping to halt the growth of cancerous cells by "tricking" the patient's own immunity into destroying the cancer. Sometimes a patient's immune system will not recognize cancer cells as foreign because the cancer cells are not different enough from normal cells to stimulate an immune reaction. At other times the immune system may recognize cancer cells, but its response is not strong enough to destroy the cancer. Various kinds of immunotherapies have been designed to help the immune system recognize cancer cells as distinctly foreign, and to strengthen the attack in the hopes that malignant cells will be destroyed.

CHAPTER SUMMARY

Biomolecules, p. 22

The four basic types of biomolecules are carbohydrates, lipids, proteins, and nucleotides. Carbohydrates are polar molecules and include monosaccharides, disaccharides, and polysaccharides. Lipids are generally nonpolar molecules and include triglycerides, phospholipids, eicosanoids, and steroids. The phospholipids are amphiphathic molecules. Proteins are polymers of amino acids. Nucleic acids are polymers of nucleotides and include DNA and RNA.

Cell Structure, p. 34

The plasma membrane separates the cell from the interstitial fluid. The plasma membrane is composed of phospholipids in a lipid bilayer and proteins and cholesterol embedded within the bilayer. Carbohydrates are found on the outer surface of the membrane as glycolipids or glycoproteins.

Inside the cell are the nucleus and cytoplasm. The nucleus contains the genetic material of the cell, DNA, and thus functions in the transmission and expression of genetic information. The cytoplasm consists of cytosol and organelles. The cytosol is a gel-like fluid that contains enzymes, glycogen and triglyceride stores, and secretory vesicles. The organelles include the rough and smooth endoplasmic reticulum, which function in protein and lipid synthesis, respectively; the Golgi apparatus, which processes molecules and prepares them for transport; mitochondria, which are the site of most ATP synthesis in the cells; lysosomes and peroxisomes, which degrade waste products; and ribosomes, which are the sites of protein synthesis.

The cytoskeleton gives the cells their shape, provides structural support, transports materials within the cell, suspends organelles, forms adhesions with other cells to form tissues, and causes contraction or movement in certain cells. The cytoskeleton consists of protein filaments including microfilaments, intermediate filaments, and microtubules.

Cell-to-Cell Adhesions, p. 43

Three specialized types of cell adhesions are commonly found between cells. Tight junctions are commonly found in epithelial tissue, where they form impermeable barriers to the movement of substances between cells. Desmosomes are strong junctions that provide resistance to mechanical stress. Gap junctions provide channels between two adjacent cells

that enable ions and small molecules to move from the cytosol of one cell to the cytosol of the other cell.

General Cell Functions, p. 46

Almost all cells of the body carry out several general functions, including metabolism, cellular transport of molecules across membranes, and intercellular communication.

Protein Synthesis, p. 48

Genes in the DNA contain the codes for protein synthesis. This code is transcribed to mRNA in the nucleus. The mRNA then moves to the cytoplasm where it is translated by ribosomes to synthesize proteins.

The leader sequence of a protein determines its fate. A protein will be synthesized either in the cytosol or in association with the endoplasmic reticulum. For those proteins synthesized in the cytosol, the leader se-

quence determines whether the protein will remain in the cytosol or enter a mitochondrion, a peroxisome, or the nucleus. For those proteins synthesized in association with the endoplasmic reticulum, the protein will eventually be packaged into vesicles by the Golgi apparatus and directed to the appropriate site in the cell or secreted from the cell.

Cell Division, p. 56

The cell cycle includes interphase and cell division. Interphase is the period between cell divisions; mitosis and cytokinesis make up the process of cell division. During the S phase of interphase, the DNA is replicated so that each daughter cell will receive an exact DNA copy. The five phases of mitosis are prophase, prometaphase, metaphase, anaphase, and telophase. Mitosis is followed by cytokinesis.

EXERCISES

Multiple-Choice Questions

1. Which of the following biomolecules is *not* a polymer?
   a) polysaccharide
   b) phospholipid
   c) protein
   d) nucleic acid

2. A fatty acid that has two double bonds between carbon atoms is called
   a) a saturated fatty acid.
   b) a monounsaturated fatty acid.
   c) a polyunsaturated fatty acid.
   d) an eicosanoid.

3. Which of the following molecules is *not* a component of a phospholipid?
   a) cholesterol
   b) glycerol
   c) phosphate
   d) fatty acid

4. Hydrogen bonding between the amino hydrogen of one amino acid and the carboxyl oxygen of another amino acid in a protein is responsible for
   a) primary protein structure.
   b) secondary protein structure.

   c) tertiary protein structure.
   d) quaternary protein structure.

5. Which of the following nucleic acids is *not* a pyrimidine?
   a) cytosine
   b) thymine
   c) uracil
   d) adenine

6. Glycogen and lipids are stored
   a) as inclusions in the cytosol.
   b) in storage vesicles.
   c) in secretory vesicles.
   d) in lysosomes.

7. Which organelle produces most of a cell's ATP?
   a) peroxisome
   b) Golgi apparatus
   c) mitochondrion
   d) smooth endoplasmic reticulum

8. Which cell-to-cell adhesion allows quick transmission of electrical signals between adjacent cells?
   a) desmosome
   b) gap junction
   c) tight junction
   d) all of the above

9. To initiate transcription of DNA, the enzyme RNA polymerase binds to the
   a) initiator codon.
   b) termination codon.
   c) promoter sequence.
   d) centromere.

10. The base sequence on tRNA that recognizes the mRNA codon by the law of complementary base pairing is called the
    a) tRNA codon.
    b) anticodon.
    c) amino codon.
    d) leader codon.

11. For proteins to be synthesized in association with the rough endoplasmic reticulum, the leader sequence must bind to a _____ on the membrane of the endoplasmic reticulum.
    a) transport vesicle
    b) secretory vesicle
    c) coated vesicle
    d) signal recognition protein

12. During what phase of mitosis is the mitotic spindle developing?
    a) prophase
    b) metaphase

c) anaphase
d) telophase

13. During what phase of mitosis do the chromosome pairs move to opposite poles of the cell?
   a) prophase
   b) metaphase
   c) anaphase
   d) telophase

Objective Questions

1. Monosaccharides are (polar/nonpolar/amphipathic) molecules.

2. Triglycerides are (polar/nonpolar/amphipathic) molecules.

3. The precursor molecule for all steroids is _____.

4. According to the law of complementary base pairing, in RNA, adenine base pairs with _____.

5. Transmembrane proteins are examples of (integral/peripheral) membrane proteins.

6. The _____ is a carbohydrate coating on the extracellular surface of the cell membrane.

7. The membrane of the smooth endoplasmic reticulum is continuous with the membrane of the Golgi apparatus. (true/false)

8. Microfilaments provide structural support for hairlike projections of the plasma membrane called _____.

9. The proteins that form gap junctions between two adjacent cells are called _____.

10. A section of DNA that codes for a specific protein is called a _____.

11. The mRNA codon that is transcribed from the DNA triplet ATC is _____.

12. More than one codon may code for a single amino acid. (true/false)

13. The first section of a polypeptide to be translated is called the _____ , and it is important in directing the translated protein to its final destination.

14. Replication of DNA occurs during (interphase/mitosis).

15. Division of the cytoplasm into two daughter cells during cell division is called _____.

Essay Questions

1. Name the four major classes of biomolecules. Describe the major function of each class.

2. Define the four levels of protein structure, and explain the chemical interactions responsible for each level.

3. Describe the structure of the plasma membrane. Name at least one function for each of the major components.

4. Certain cells in the adrenal gland synthesize and secrete lots of steroid hormones. What organelle would be abundant in these cells, and why?

5. Name the three special types of cell adhesions and list the special functions of each.

6. Define each of the following terms: *transcription, translation,* and *replication.* Where does each of these processes occur in a cell?

7. Describe the anatomical and functional relationship between the rough endoplasmic reticulum, smooth endoplasmic reticulum, and Golgi apparatus in protein synthesis.

Find the answers to these exercises, and additional study tools, at the Physiology Place (www.physiologyplace.com).

Cell Metabolism

OBJECTIVES

▪ Define metabolism and distinguish catabolic reactions from anabolic reactions.

▪ Describe what occurs in the following basic types of metabolic reactions: hydrolysis, condensation, phosphorylation, dephosphorylation, oxidation, reduction.

▪ Explain the relationship of energy change in a reaction to (1) the energies of the reactants and products and (2) the direction of the reaction.

▪ Define activation energy and describe its influence on the rates of reactions.

▪ Describe the general mechanism of enzyme action, and discuss the various factors that influence the rates of enzyme-catalyzed reactions.

▪ Describe the role of ATP in energy metabolism.

▪ Describe in general terms the events in the following stages of glucose metabolism: glycolysis, the Krebs cycle, and oxidative phosphorylation.

▪ Describe in general terms how the body is able to obtain energy from the breakdown of fats, proteins, and glycogen, and how it stores energy by synthesizing these compounds.

▪ Define the following terms: *glycolysis, glycogenesis, lipolysis, lipogenesis, proteolysis, gluconeogenesis.*

CHAPTER OUTLINE

Types of Metabolic Reactions   63

Metabolic Reactions and Energy   65

Reaction Rates   70

Glucose Oxidation: The Central Reaction of Energy Metabolism   77

Stages of Glucose Oxidation: Glycolysis, the Krebs Cycle, and Oxidative Phosphorylation   81

Energy Storage and Use: Metabolism of Carbohydrates, Fats, and Proteins   91

Above: Electron micrograph of a mitochondrion in a heart muscle cell

STUDY HINTS

1. Biomolecules, p. 22

2. Mitochondria, p. 38

3. Membrane structure, p. 34

Imagine a factory that manufactures hundreds of chemical compounds using only raw materials that can be found in any grocery store. This factory operates 24 hours a day and can speed up or slow down production on a moment's notice according to the demand for its products. The factory generates only environmentally safe waste products and can repair itself when the need arises. Sounds like a utopian dream of the future, right? Well, it's not: You carry billions of chemical factories that operate just like this in your own body—your cells.

Whenever you walk, run, read a book, eat a meal, or just relax, you are using energy harnessed from chemical reactions. In fact, every function of your body depends on the chemical reactions that occur within cells. Certain reactions release energy that can be used to perform work, as when you contract your muscles to lift a chair. Other reactions form the molecular building blocks of which cells are made, such as proteins and DNA.

The sum total of all chemical reactions that occur in cells is called **metabolism.** Although over a thousand different chemical reactions can be occurring in a cell at any given time, reactions follow certain patterns; thus they can be classified into a relatively small number of basic types, such as hydrolysis reactions, phosphorylation reactions, and oxidation-reduction reactions. In this chapter we focus on the types of reactions that provide the energy necessary to perform cellular work. First this energy must be extracted from biomolecules ingested in food; then it must be processed before it is either used for work or set aside in storage for future use. The set of reactions involved in energy storage and use (otherwise known as *energy exchange*) is called **energy metabolism.**

TYPES OF METABOLIC REACTIONS

Any chemical reaction involves the transformation of materials that enter the reaction, called **reactants,** into different materials, called **products.** In the reaction

$$A + B \rightarrow C + D$$

the reactants A and B are transformed into the products C and D. By convention, reactions are usually written with reactants on the left and products on the right. A reaction is said to "go forward" when it proceeds from left to right—that is, when reactants are being converted to products. Under certain conditions, some reactions may go in the opposite direction; in that case a reaction is said to go "in reverse."

A proper understanding of chemical reactions requires us to realize that even though we say a reaction goes forward or in reverse, it actually is *bidirectional,* meaning that it goes in both directions at the same time; that is, when reactant molecules are being converted to products, some product molecules are also being converted to reactants. To indicate this, reactions are often drawn with double arrows, as follows:

$$A + B \rightleftharpoons C + D$$

When we say that a reaction goes forward or in reverse, we are referring to the direction of the overall (net) reaction. Suppose, for example, that over a certain time interval some product molecules are converted to reactants, but an even greater number of reactant molecules are converted to products. In this case, the net reaction goes forward because more reactant molecules are converted to products than vice versa. Unless noted otherwise, we use the terms *forward* or *reverse* to refer to the direction of the *net* reaction.

Metabolic reactions are broadly classified as being either *catabolic* or *anabolic,* depending on whether the product molecules are larger or smaller in size than the reactant molecules. A reaction is **catabolic** if it involves the breakdown of larger molecules into smaller ones. Examples of catabolic reactions are the breakdown of proteins into amino acids and the breakdown of glycogen into individual glucose molecules, both of which occur when food is being digested. A reaction is **anabolic** if it involves the production of larger molecules from smaller reactants. Examples are the synthesis of proteins from amino acids and the synthesis of glycogen from glucose, both of which occur in cells following the absorption of ingested nutrients. The breakdown and synthesis of larger molecules are sometimes referred to as *catabolism* and *anabolism,* respectively.

In most cases, metabolic reactions are linked together in a series of steps such that the products of one reaction serve as the reactants in the next. Such a series of reactions is referred to as a **metabolic pathway,** which can be written either as

$$A + X \rightarrow B \rightarrow C \rightarrow D + Y$$

or as

$$A \underset{X}{\overset{}{\rightarrow}} B \rightarrow C \overset{Y}{\rightarrow} D$$

This particular pathway consists of a series of three reactions. In the first step, the starting material A reacts with X to form B, which is then transformed into C in the second reaction. C then enters the third reaction, which yields D and Y. The final products, D and Y, are referred to as **end-products** of the pathway. Substances in the middle of the pathway (B and C in this example) are called **intermediates.**

Later in this chapter we take a detailed look at a group of metabolic pathways involved in *glucose oxidation,* an important reaction because it supplies much of a cell's energy needs. Although each of the individual reactions that make up these pathways is unique, many have features in common and can be classified into one of a few general categories. In the following sections we examine three general types of metabolic reactions, which occur not only in glucose oxidation, but in all metabolic processes: (1) *hydrolysis and condensation* reactions, (2) *phosphorylation and dephosphorylation* reactions, and (3) *oxidation-reduction* reactions.

Hydrolysis and Condensation Reactions

When you eat a meal containing protein, your gastrointestinal tract breaks down the relatively large protein molecules into amino acids, which are considerably smaller and more easily absorbed into the bloodstream. This reaction, like most other catabolic reactions that occur in the body, is an example of *hydrolysis.* In **hydrolysis,** which means "splitting with water," water reacts with molecules, causing breakage of the bonds that link a molecule together. In this case, the bonds that are broken are the peptide bonds that link amino acids together to form proteins. The general form of a hydrolysis reaction is as follows:

$$A—B + H_2O \rightarrow A—OH + H—B$$

In this reaction, the molecule A—B is broken down into its constituent parts, A and B, as a result of the breaking of the bond that originally linked them together. In this process, a water molecule (H_2O) splits into two parts, a hydroxyl group (—OH) and a hydrogen (—H), one of which combines with A while the other combines with B. As a result, the bond between A and B is broken and replaced with new bonds.

The reverse of hydrolysis, called *condensation,* involves the joining together of two or more smaller molecules to form a larger one, as when amino acids are joined together to form proteins. In this process, water is generated as a product:

$$A—OH + H—B \rightarrow A—B + H_2O$$

Phosphorylation and Dephosphorylation Reactions

In many metabolic reactions, particularly those involved in energy exchange, a phosphate group is either added to a molecule or removed from it. Addition of a phosphate

group (abbreviated as P) is known as **phosphorylation** and can be written in an abbreviated fashion as follows:

$$A + P_i \rightarrow A—P$$

Here, the free phosphate group is abbreviated as P_i, for inorganic phosphate, which under physiological conditions exists mostly in the ionized forms HPO_4^{3-} or $H_2PO_4^{2-}$. The bond that is formed in this reaction (that is, the bond linking A to P) is known as a *phosphate bond.* Removal of a phosphate group, known as **dephosphorylation,** can be written as follows:

$$A—P \rightarrow A + P_i$$

Oxidation-Reduction Reactions

The removal of electrons from any molecule is called **oxidation;** the addition of electrons to a molecule is called **reduction.** Thus an *oxidation-reduction reaction* is any reaction in which electrons are removed from one reactant molecule and transferred to another. Once electrons have been removed from a molecule, that molecule is said to be *oxidized;* a molecule that has gained electrons (or that has not yet had its electrons removed) is said to be *reduced.*

Depending on the context, the term *oxidation* can be narrowly defined as the reaction of any molecule with oxygen. One example of oxidation in this sense is the combustion of paper or gasoline. A more physiologically relevant example is the oxidation of glucose ($C_6H_{12}O_6$), which occurs in cells as follows:

$$C_6H_{12}O_6 + 6\ O_2 \rightarrow 6\ CO_2 + 6\ H_2O$$

In this reaction, O_2 is the molecular form of oxygen, the form in which oxygen is present in the atmosphere, and CO_2 is carbon dioxide. This narrow definition of oxidation is consistent with the broader definition (removal of electrons) because an oxygen atom holds onto electrons very tightly and tends to pull them away from other atoms to which it is bonded, even though it doesn't remove these electrons completely.

The term *oxidation* is also used in reference to the removal of hydrogen atoms from a molecule, as shown in the following reaction:

$$HA—BH \rightarrow A\!=\!B + 2\ H$$

In this example, removal of two hydrogen atoms causes the single bond linking A and B to be replaced with a double bond. Even though hydrogen atoms (not electrons) are removed, the reaction is still considered oxidation because these atoms each carry an electron. Therefore, removal of these atoms involves the removal of electrons that originally belonged to the reactant molecule. (As a reflection of this, hydrogen atoms are sometimes referred to as *electrons* or *reducing equivalents* because they are equivalent to electrons.) Addition of hydrogens to a molecule, the reverse of the previous reaction, is called *reduction* and is shown as follows:

$$A{=}B + 2\,H \rightarrow HA{-}BH$$

Some oxidation or reduction reactions involve the removal or addition of actual electrons, which causes changes in a molecule's electrical charge. As we will see, this occurs in *oxidative phosphorylation*, a process occurring in mitochondria that is crucial to energy metabolism. At certain points in this process, electrons (e^-) are removed from pairs of hydrogen atoms, causing these atoms to become positively charged hydrogen ions (H^+), an example of oxidation:

$$H_2 \rightarrow 2\,H^+ + 2\,e^-$$

At other points in oxidative phosphorylation the reverse occurs, with electrons combining with hydrogen ions to form uncharged hydrogen atoms, an example of reduction:

$$2\,H^+ + 2\,e^- \rightarrow H_2$$

Quick Test 3.1

1. What is the primary distinction between an anabolic reaction and a catabolic reaction?
2. Define the following terms: *reactant, product, metabolic pathway, intermediate, end-product.*
3. Describe in general terms what occurs in hydrolysis, phosphorylation, and oxidation reactions.
4. For each of the reaction types in Question 3, give the term that applies to the opposite reaction.

METABOLIC REACTIONS AND ENERGY

The metabolic reactions that occur in the body perform many functions. One of the most important functions is enabling cells to transform raw materials from the environment into the great variety of substances from which the body is made. Metabolic reactions also provide us with **energy,** broadly defined as the capacity to perform work. The body most obviously performs work during movement. When you push a grocery cart, for example, you exert a certain amount of force against the cart and move it a certain distance. The amount of work you perform is the product of the force exerted and the distance moved (work = force × distance). However, the body performs work in other ways, too—even while at rest—and it must do so on a continual basis just to stay alive. The heart performs work when it pumps blood, the kidneys perform work when they form urine, and cells perform work when they multiply to repair damaged tissues. All this work is performed using energy derived from metabolic reactions. In this section we explore the nature of the relationship between chemical reactions and energy.

Energy-Releasing and Energy-Requiring Reactions

Although certain metabolic reactions release energy that can be used by cells to perform work, other metabolic reactions consume or require energy. As a general rule, reactions that release energy are catabolic. You know, for example, that you obtain energy from the food you eat, but before nutrients can be absorbed, the food must be digested. In this process, large molecules such as proteins and starch are broken down into smaller molecules such as amino acids and glucose. These nutrients are then absorbed into the bloodstream and taken up by cells throughout your body. To obtain energy from these nutrients, cells must break them down into still smaller molecules. Cells can also use amino acids, glucose, and other small nutrient molecules for other purposes, such as the synthesis of glycogen or cellular proteins, but these reactions are anabolic and do not release energy. Instead, energy must be put into them if they are to occur, which is true of anabolic reactions in general.

Energy Changes in Reactions

It is an inescapable truth of nature that chemical reactions are always accompanied by either the release of energy or the input of energy. In other words, all reactions involve energy exchange of one sort or another. (This exchange occurs because molecules possess energy; different types of molecules possess different amounts of energy.) When a reaction releases energy, it does so because the reactant molecules possess more energy than the product molecules; in this case, reactants enter the reaction with a certain amount of energy, and products come out of the reaction with less energy. According to the *first law of thermodynamics,* energy can be neither created nor destroyed but instead can only change form. Therefore, if the products come out of the reaction with less energy than the reactants had when they went in, then the "extra" energy must be released in one form or another during the course of the reaction:

$$\text{reactants} \rightarrow \text{products} + \text{energy}$$

The released energy might be heat or take some other form, and under certain conditions it can be used to perform work. A familiar example of an energy-releasing reaction is the combustion of gasoline. When gasoline is burned in the open air, it generates energy—heat and light—in the form of flames. When burned in the internal combustion engine of a car, some of this energy is used to perform the work of moving the car.

The energy change of any reaction (expressed as ΔE) is the difference between the energy of the products

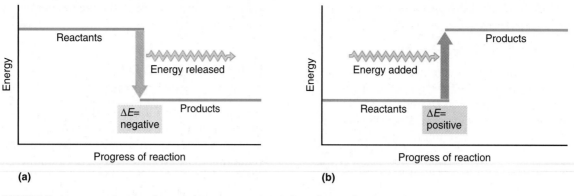

FIGURE 3.1 Energy changes in reactions. *Wavy arrows indicate the transfer of energy in the reaction.* **(a)** *When the energy of the reactants is higher than that of the products, the energy change is negative, and energy is released in the reaction.* **(b)** *When the energy of the products is higher than that of the reactants, the energy change is positive, and energy is added to the reaction.*

($E_{products}$) and the energy of the reactants ($E_{reactants}$): $\Delta E = E_{products} - E_{reactants}$. (The symbol Δ or delta is frequently used to denote a change in some quantity.) Because the products have less energy than the reactants in an energy-releasing reaction, the energy change ΔE in such a reaction is negative. This is illustrated in Figure 3.1a, where distance along the *y*-axis represents the amount of energy present in reactant or product molecules, and distance along the *x*-axis represents the progress of the reaction—that is, the transformation of reactants into products.

Figure 3.1b depicts the energy change that occurs in an energy-requiring reaction, which we can write as follows:

reactants + energy → products

In this case, the products possess more energy than the reactants, so that the energy change ΔE is positive. Because energy cannot be created, the "extra" energy acquired by the products must come from some source other than the reactant molecules themselves. Perhaps the most familiar example of an energy-requiring process is the boiling of water. To make water boil, energy must be put into it (by heating it on a stove, for instance).

The energy change ΔE is typically expressed in units of energy or heat, such as calories (cal) or kilocalories (kcal), but it may be given in joules (J) or kilojoules (kJ). A **calorie** is the amount of energy that must be put into 1 gram (or 1 ml) of water to raise its temperature by 1 degree centigrade (°C) under a standard set of conditions. One calorie is equivalent to 4.18 joules, and a kilocalorie or kilojoule is 1000 calories or joules, respectively. The amount of energy released or consumed in a reaction depends not only on the nature of the reactants and products involved, but also on the quantity of reactants consumed or products produced. As this quantity increases, the size of the energy change increases in direct proportion. For this reason, ΔE is commonly expressed in units of kcal/mole or some equivalent.

Kinetic Energy Versus Potential Energy

When we say that reactions release or require energy, or that reactant or product molecules possess a certain amount of energy, what do we mean? To answer this question, we first must realize that there are two basic types of energy: kinetic energy and potential energy. **Kinetic energy,** sometimes called *energy of motion,* is the type of energy possessed by moving objects. The kinetic energy of an object depends on the object's mass and how fast it is moving; kinetic energy increases when either of these factors increases. Atoms and molecules possess kinetic energy because they undergo *thermal motion,* a random movement or vibration that occurs at any temperature above absolute zero. This kinetic energy increases in direct proportion to the molecules' temperature and is often referred to as *thermal energy* or *heat.* In contrast, **potential energy** is not associated with motion but instead is the type of energy that can be stored and eventually converted to kinetic energy. (In other words, potential energy has the *potential* to become kinetic energy.) When we lift an object or push it up a hill, for example, we expend energy and perform work. At the same time, some of this energy is stored as potential energy because the potential energy of the object increases as it rises higher and higher above the ground or gains elevation. This potential energy can be converted to kinetic energy by allowing the object to fall to the ground or roll down the hill (Figure 3.2).

Now we can see that when we speak of the energy change in a reaction or of the energy of reactant or product molecules, we are talking about *potential* energy. In energy diagrams such as Figure 3.1, it is assumed that reactant and product molecules are at the same temperature (that is, have the same kinetic energy), so any difference in energy between reactants and products must reflect differences in potential energy. In the next section we see that the energy change in a reaction not only determines whether energy is released or required, but also the direction in which the reaction proceeds.

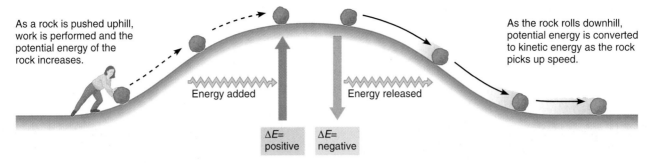

As a rock is pushed uphill, work is performed and the potential energy of the rock increases.

Energy added

Energy released

As the rock rolls downhill, potential energy is converted to kinetic energy as the rock picks up speed.

$\Delta E =$ positive

$\Delta E =$ negative

FIGURE 3.2 **Potential energy and kinetic energy.**

How the Energy Change of a Reaction Determines its Direction

We know from experience that chemical reactions have a natural tendency to run in particular directions. For instance, once you light a piece of paper, it burns until it is reduced to ashes (provided there is sufficient oxygen), but you never see the combustion of paper stop and go in reverse, such that ashes are reconstituted into paper. The reason for this is that the burning of paper is an energy-releasing reaction, and *energy-releasing reactions always proceed spontaneously in the forward direction.* (To say that a process occurs spontaneously means that it occurs on its own—that is, without any energy being put into it.) To reconstitute ashes into paper would require the paper-burning reaction to go in reverse, which would require energy rather than release it. This does not happen because *energy-requiring reactions do not go forward spontaneously; they go forward only when energy is put into them.* (In fact, energy-requiring reactions will spontaneously go in reverse if energy is not put into them.)

These rules reflect another fundamental truth of nature: Systems have a natural tendency to go from states of higher potential energy to states of lower potential energy. We know, for example, that a rock will roll downhill spontaneously but not uphill (not unless it has momentum to begin with), because the potential energy of the rock decreases when it rolls downhill; to roll uphill, its potential energy would have to increase. Likewise, an energy-releasing reaction (reactants → products + energy) goes forward spontaneously because the potential energy decreases as high-energy reactants are transformed into lower-energy products. In contrast, an energy-requiring reaction (reactants + energy → products) does not go forward spontaneously because the potential energy increases as low-energy reactants are converted to higher-energy products. (Strictly speaking, it is the *free energy* change of a reaction, not the *potential energy* change, that determines its direction, but free energy is a kind of chemical potential energy.)

Given the energy-releasing nature of catabolic reactions and the energy-requiring nature of anabolic reactions, we can surmise that catabolic reactions should occur spontaneously in cells, whereas anabolic reactions should occur only when energy is put into them. Ana-bolic reactions are able to occur in cells because certain cellular mechanisms couple these reactions with catabolic reactions. Such coupling harnesses the energy that is released from catabolic reactions and uses it to drive anabolic reactions.

A reaction is said to be in **equilibrium** when on balance it is going neither forward nor in reverse, such that there is no net reaction direction. Under these conditions, the conversions of reactant molecules to product molecules, and of product molecules to reactant molecules, occur at the same rate. (A reaction is at equilibrium when the reactants and products have the same energy, such that the energy change ΔE is zero.)

The Law of Mass Action

The energy of reactant or product molecules is determined by many different factors, including the numbers and types of chemical bonds that are present, the nature of the solvent in which the molecules are dissolved, and the molecules' *concentrations.* In general, as the concentration of molecules increases, the energy of those molecules increases (Toolbox: Solutions and Concentrations, p. 69). Because of this connection between concentration and energy, a given reaction can be made to go in either direction—forward or in reverse—by raising or lowering the concentrations of reactants relative to products. This phenomenon is commonly referred to as the **law of mass action:** An increase in the concentration of reactants relative to products tends to push a reaction forward, and an increase in the concentration of products relative to reactants tends to push a reaction in reverse.

The law of mass action is illustrated in Figure 3.3 using a model consisting of two water reservoirs connected by a tube, with the left-hand reservoir representing "reactants" and the right-hand reservoir representing "products." The difference between the reservoirs' heights indicates differences in energy between reactants and products based on their molecular properties. The volume of the water in the reservoirs represents reactant and product concentrations. The direction of water flow in the tube represents the direction of the reaction.

In Figure 3.3a, the left-hand reservoir is above the right-hand reservoir, indicating a reaction that normally

An energy-releasing reaction goes forward spontaneously...

...but can be forced to go in reverse if the concentration of products is increased.

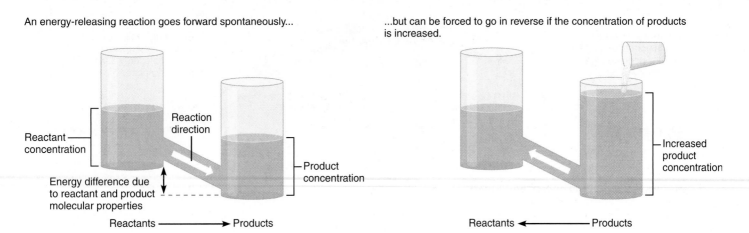

(a)

An energy-requiring reaction goes in reverse spontaneously...

...but can be forced to go forward if the concentration of reactants is increased.

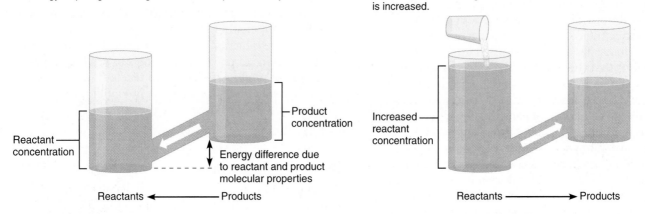

(b)

FIGURE 3.3 The law of mass action. *In this model, in which the volume of water in the reservoirs represents the concentrations of products and reactants, energy-releasing reactions can be forced to go in reverse when product concentrations are increased relative to reactant concentrations* **(a)**, *and energy-requiring reactions can be forced to go forward when reactant concentrations are increased relative to product concentrations* **(b)**.

releases energy. When water volume is the same in the two reservoirs, water flows from left to right, indicating that the reaction goes forward spontaneously, from reactants to products. However, by adding water to the right-hand reservoir (that is, by increasing the concentration of products), we can make water flow in the opposite direction. Thus, the reaction is forced to go in reverse. In Figure 3.3b, the right-hand reservoir is above the left-hand reservoir, representing a reaction that normally requires energy. When water volume is the same in the two reservoirs, water flows from right to left, indicating that the reaction goes in reverse spontaneously. When enough water is added to the left-hand reservoir, however, the flow goes from left to right. In this case, the reaction is forced to go forward by increasing the concentration of reactants.

Activation Energy

When reactant molecules enter into a reaction, their conversion into product molecules does not occur abruptly. Instead, the reacting molecules go into a high-energy intermediate form called a *transition state,* which then breaks down into the products. The reverse reaction (the conversion of products to reactants) also must go through this transition state. Thus, the transformations that occur in a reaction are continuous and gradual, not sudden, as shown in Figure 3.4a. The "hump" in the middle of the curve, known as an *activation energy barrier,* is due to the fact that the potential energy of the transition state is greater than that of either the reactants or the products.

For reactants to become products or vice versa, molecules must have sufficient potential energy to surmount the activation energy barrier, and they must acquire some "extra" energy, called **activation energy,** which is the

A mixture of molecules dissolved in a liquid is known as a *solution*. The dissolved substance, which is usually a solid or gas in its pure form, is known as the *solute*, whereas the liquid is referred to as the *solvent*. Solute molecules are said to be *dissolved* when they are completely separate from one another and are surrounded by solvent molecules.

The *concentration* of a solution is a measure of the quantity of solute contained in a unit volume of solution. Solute quantity is most commonly expressed in terms of moles (mol), and volume is most often given in liters (L). If 1 mole of solute is present in 1 liter of solution, the concentration is said to be 1 *molar* (1M = 1mol/L). If 1/1000th of a mole (that is, 1 *millimole*) of solute is dissolved in 1 liter of solution, the concentration is 0.001 molar or 1 *millimolar* (1mM = 1×10^{-3} mol/L). For very

dilute solutions, concentrations may be expressed in *micromoles* per liter (μmol/L = 1×10^{-6} mol/L) or *nanomoles* per liter (nmol/L = 1×10^{-9} mol/L). Concentrations of specific substances are often indicated using brackets. For example, [Na^+] represents sodium ion concentration.

Occasionally, concentrations are expressed in terms of the *mass* of solute contained in a unit volume of solution—grams per liter (g/L) or micrograms per liter (μg/L = 1×10^{-6} g/L), for instance. Sometimes the unit of volume is a *deciliter* (dL = 0.1L), which is equivalent to 100 milliliters (100 mL). The concentration of a solution containing 1 gram of solute per deciliter is frequently expressed as 1 *percent* (1%) because physiological solutions are mostly water, 100 mL of which weighs 100 grams.

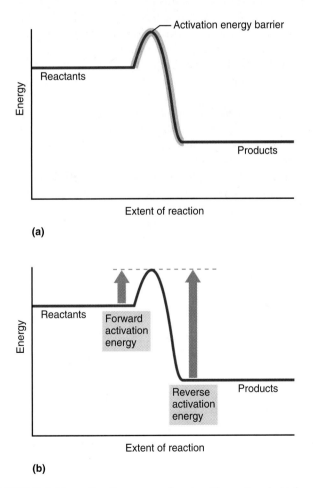

(a)

(b)

FIGURE 3.4 The activation energy barrier. *The reactions depicted are energy-releasing reactions.* **(a)** *An energy diagram similar to that in Figure 3.1a, but with the activation energy barrier included.* **(b)** *An energy diagram in which the forward and reverse activation energies are indicated by the vertical arrows.*

difference between the energy of the transition state and the energy of either the reactants or products. The activation energy is indicated in Figure 3.4b as the vertical distance between the initial or final energies and the peak of the curve. Note that for the reaction shown in the figure, which is an energy-releasing reaction, the activation energy for the forward reaction is less than the activation energy for the reverse reaction. For an energy-requiring reaction, the reverse would be true.

If molecules must acquire extra energy to surmount the activation energy barrier and react, where does this "extra" energy come from? Molecules acquire this energy by colliding with one another, which happens all the time because they are in constant thermal motion. Consider, for example, the reaction A + B → C + D. For molecules of A and B to react, they first must collide. When this happens, some of the molecules' thermal kinetic energy is converted to potential energy. If this gain in potential energy is equal to or greater than the activation energy, then the two molecules will enter the transition state and be converted to the products C and D (Figure 3.5a).

The significance of the activation energy barrier is that it limits how fast a reaction can go, a topic that is explored in detail in the next section. Here it is important to note that not every collision between reactant molecules produces a reaction, because some collisions do not generate enough potential energy to surmount the activation energy barrier. The more quickly two molecules about to collide are moving, the more potential energy they gain in the collision. If the energy gained is less than the activation energy, then the colliding molecules will not enter the transition state and will not react (Figure 3.5b); they will simply emerge from the collision unaltered.

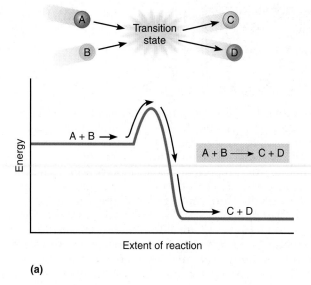

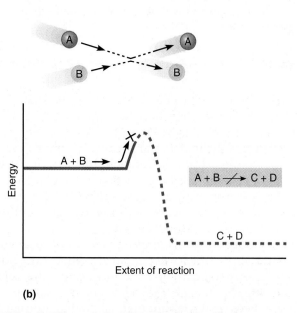

FIGURE 3.5 Effect of the energy activation barrier.
(a) *Collision of two reactant molecules, A and B, with enough energy to enter the transition state. The energy of the collision surmounts the activation energy barrier, resulting in the formation of products C and D.* **(b)** *Collision of A and B with insufficient energy to enter the transition state. The activation energy barrier is not surmounted, so no reaction occurs.*

As the activation energy barrier is lowered, does the energy change in the reaction increase, decrease, or stay the same?

Quick Test 3.2

1. Are energy-releasing reactions usually catabolic or anabolic?

2. In an energy-releasing reaction, which has more energy—the reactants or the products?

Stays the same

3. What factor determines the direction of a reaction? Are reactions that go forward spontaneously energy-releasing or energy-requiring?

4. Give a brief description of the law of mass action.

5. What is activation energy? How does it affect a reaction?

REACTION RATES

The **rate** of a chemical reaction is a measure of how fast it consumes reactants and generates products; it is usually expressed as a change in concentration per unit time (moles/liter-second, or some equivalent). For example, as the reaction A + B → C + D proceeds, the concentrations of A and B diminish while the concentrations of C and D increase. The rate of this reaction can be expressed as the change in the concentrations of C or D per unit time.

The rate of a metabolic reaction is of great physiological significance because proper body function demands that reactions proceed at a rate that matches the body's needs at the moment. Rates higher or lower than required ultimately lead to severe impairment of cellular function. Perhaps the most vivid demonstration of this danger occurs in *hypothermia*, when body temperature falls below normal (see Chapter 1). Any decline in body temperature causes metabolic reactions to slow down, and a decline in core temperature of only a few degrees can cause a person to become weak and disoriented, and possibly even to lose consciousness. A further drop in temperature can precipitate cardiac arrest (stoppage of the heartbeat) and death.

Later in this chapter we will see that metabolic reactions are accelerated or *catalyzed* by special molecules called *enzymes*, and that the rates of metabolic reactions are normally regulated through changes in enzyme activity. First, however, we begin our exploration of reaction rates with a look at the factors that affect the rates of chemical reactions in general.

Factors Affecting the Rates of Chemical Reactions

We know from experience that certain chemical reactions progress more quickly than others. For example, one oxidation reaction—the combustion of gasoline—can occur with explosive speed, whereas the rusting of iron (another oxidation reaction) occurs so slowly that it may take weeks to see the effects. The rate of a reaction is determined by a variety of factors, including (1) reactant and product concentrations, (2) temperature, and (3) the height of the reaction's activation energy barrier.

Reactant and Product Concentrations

When we speak of the rate of a reaction, we are usually referring to its **net rate:** the difference between the rate of the forward reaction (reactants → products) and the rate of the reverse reaction (products → reactants). Generally speaking, any increase in the concentration of reactants will bring about an increase in the forward rate without affecting the reverse rate. Likewise, any increase in the concentration of products will cause the reverse rate to increase without affecting the forward rate. Therefore, an increase in the concentration of reactants relative to products will increase the *net rate* in the forward direction. Conversely, an increase in the concentration of products relative to reactants will decrease the net forward rate, and can even make the reaction go in reverse if the change in concentration is large enough. The effect of reactant and product concentrations on reaction rates is due to the fact that concentrations affect the frequency of collisions between molecules: As the concentration of molecules increases, the number of collisions occurring in a given time period also increases.

Temperature

In general, the rate of a reaction increases with increasing temperature and decreases as the temperature falls. We refrigerate food, for example, because cooling slows down the rate of decomposition reactions, which helps the food "keep" longer.

Temperature influences reaction rates because it affects the frequency and energy of molecular collisions. As temperature increases, the average kinetic energy of molecules increases, which increases the energy of collisions. The result is an increase in the proportion of collisions having enough energy to surmount the activation energy barrier, which increases the rate of a reaction.

The Height of the Activation Energy Barrier

The height of the activation energy barrier differs among reactions. All else being equal, the rate of a reaction increases as the height of the barrier decreases. The reason for this is that at any given temperature, only a fraction of the collisions between molecules has sufficient energy to surmount the barrier and produce a reaction. As the height of the barrier decreases, the proportion of collisions having the requisite energy increases—not because the collisions themselves are any more energetic, but because the minimum energy requirement for a "successful" collision has been reduced. (This situation is analogous to a high-jump competition: If the bar is lowered, the number of contestants making a successful jump will increase.)

Note that as the height of the activation energy barrier decreases, rates of both the reverse reaction and forward reaction increase because the activation energy of both reactions becomes lower. However, for reasons beyond the scope of this book, the net rate of the reaction still increases despite the increase in the rate of the reverse reaction.

The Role of Enzymes in Chemical Reactions

If you were to set up a typical metabolic reaction by mixing reactants together in a beaker, it would proceed very slowly. In fact, most metabolic reactions would proceed too slowly to be compatible with life. These reactions are able to proceed at considerably faster rates in cells because of the presence of **enzymes,** biomolecules (almost always proteins) specialized to act as *catalysts,* a general term for substances that increase the rates of chemical reactions. In this section we examine how enzymes accomplish this task, and how the regulation of enzyme activity controls reaction rates according to the body's needs.

Mechanisms of Enzyme Action

Cells contain a wide variety of enzymes, each specialized to catalyze a particular reaction or group of related reactions. To catalyze a reaction, an enzyme molecule must first bind to a reactant molecule, which in the context of enzyme-catalyzed reactions is called a **substrate.** The enzyme then acts on the substrate to generate a product molecule, which is subsequently released. Thus, the action of the enzyme can be written as the following two-step reaction:

$$\underset{\text{enzyme}}{E} + \underset{\text{substrate}}{S} \rightleftharpoons \underset{\substack{\text{enzyme-}\\\text{substrate}\\\text{complex}}}{E{\cdot}S} \rightarrow \underset{\text{product}}{P} + \underset{\text{enzyme}}{E}$$

The double arrow in the first step (the *binding step*) signifies that it is possible for the substrate to bind to the enzyme and then be released before the enzyme has had a chance to act on it (see Figure 3.6a). Such binding is said to be *reversible* because the substrate can readily dissociate from the enzyme after binding to it. If the substrate molecule stays bound to the enzyme long enough, it will eventually be converted to product in the second step (the *catalytic step*). After the catalytic step, the enzyme molecule emerges from the reaction in the same form as it entered, indicating that it is neither consumed nor otherwise altered in the reaction. The enzyme is then free to act on a new substrate molecule and generate another product. In fact, a single enzyme molecule is capable of catalyzing a reaction repeatedly and thus theoretically could generate an unlimited quantity of product. (Although not shown in the previous reaction, the enzyme can also catalyze the reaction in reverse, such that the product is converted back to substrate.)

Whereas this mechanism illustrates the essentials of enzyme action, it pertains to only the simplest type of enzymatic reaction, in which an enzyme acts on just one

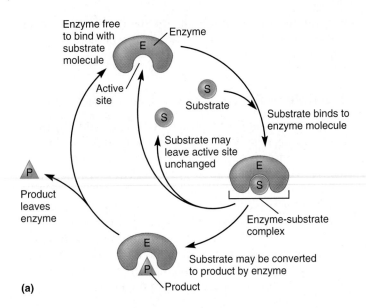

(a)

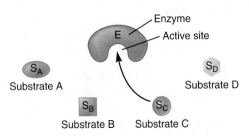

(b)

FIGURE 3.6 The role of substrate specificity in the mechanism of enzyme action. (a) *The mechanism of enzyme action. The closeness of fit between substrate molecule S and the enzyme's active site enables binding and the formation of an enzyme-substrate complex. After binding, the substrate may leave the active site unchanged, or it may be converted to product P by the enzyme. The product then exits the active site, leaving the enzyme free to bind another substrate molecule and catalyze the reaction again.* **(b)** *Substrate specificity. Because only the shape of candidate substrate S_C is complementary to the enzyme's active site, only it can bind with the enzyme.*

substrate molecule at a time to generate a single product, as expressed in the following shorthand notation:

$$S \xrightarrow{\text{E}} P$$

In reality, most enzymes act on two or more different substrates and generate more than one product.

Substrate Specificity

Enzymes are generally able to catalyze one particular reaction because they have the ability to "recognize" and bind to only one particular type of substrate, a phenomenon referred to as **substrate specificity.** This specificity is not necessarily absolute, however, because some enzymes are able to act on a wide variety of different substrate molecules as long as they have certain characteristics.

One example is *pepsin*, an enzyme that is secreted by cells in the stomach lining and breaks proteins in food into smaller fragments. Pepsin can act on almost any protein, so long as the protein contains certain amino acids, but it cannot act on other constituents of food, such as fats or carbohydrates.

The basis of substrate specificity relates to the *complementary* shapes of the enzyme and substrate molecules: Only if a substrate molecule closely fits and binds to a particular site on the enzyme molecule, called the **active site,** can the enzyme act on the substrate and catalyze the reaction. As illustrated in Figure 3.6b, uncomplementary "candidate" substrate molecules cannot fit into the active site very well and thus are not acted upon by the enzyme. This concept of complementary fit is known as the *lock-and-key model* because the substrate matches the active site just as a key matches a lock. A variant of this idea, and probably a more accurate one, is the *induced-fit model,* in which the binding of the substrate alters the shape of the active site by inducing in the enzyme molecule a *conformational change,* a change in the molecule's shape. In this model, the substrate fits the active site more like a foot fits a sock than a lock fits a key. An empty sock is roughly shaped like a foot, but it does not take on the precise shape of a foot until the foot is inserted into it.

Cofactors and Coenzymes

Although enzymes are proteins, many possess additional nonprotein components called **cofactors,** which are necessary for the enzymes to function properly. Many enzymes contain ions of metals such as iron, copper, or zinc, which function as cofactors by binding tightly to the side chains of certain amino acids, thereby holding the enzyme in its normal conformation. Without these ions, the enzymes' shapes would be altered, causing them to lose their activity. Other enzymes contain **vitamins** (various organic molecules that are necessary in only trace amounts but must be obtained through the diet) or vitamin derivatives as cofactors.

Some cofactors function as **coenzymes,** molecules that do not themselves have catalytic activity but nonetheless participate directly in the reactions catalyzed by their enzyme partners. A given coenzyme may be used by more than one type of enzyme; often a coenzyme can dissociate from one enzyme and bind to another, enabling it to participate in more than one reaction. In a few cases, coenzymes are permanently bound to their partner enzymes. Usually a coenzyme carries particular chemical groups from one reaction to another. Like enzymes, coenzymes are not permanently altered by the reactions in which they participate and therefore can be used over and over again.

Two coenzymes that are very important in energy metabolism are *flavin adenine dinucleotide* (or FAD), a derivative of vitamin B_{12} (or riboflavin), and *nicotinamide adenine dinucleotide* (or NAD$^+$), a derivative of the vitamin niacin. Both of these participate as hydrogen (electron) carriers in certain oxidation-reduction reactions and shuttle electrons from one place to another inside cells. FAD acquires electrons by picking up pairs of hydrogen atoms and in the process becomes reduced to FADH$_2$:

$$FAD + 2\,H \rightarrow FADH_2$$

Subsequently, FADH$_2$ releases these electrons to other electron acceptors and in so doing reverts to its oxidized form, FAD, which is then free to pick up another pair of electrons. NAD$^+$ also carries pairs of electrons, but in a different manner: It picks up one electron in a hydrogen atom and the other as a free electron, which it takes from a hydrogen atom, leaving a hydrogen ion (H$^+$) in solution. Thus, the reduction of NAD$^+$ occurs as follows:

$$NAD^+ + 2\,H \rightarrow NADH + H^+$$

Note that in this process NAD$^+$, which carries a single positive charge, is reduced to NADH, which carries no charge. (The positive charge is canceled by the negative charge of the acquired electron.) Subsequently, NADH releases its pair of electrons to other electron acceptors through the reverse of the previous reaction and goes to the oxidized form NAD$^+$, which is free to pick up another pair of electrons. We will see several examples of NAD$^+$ and FAD at work later in the chapter.

Another important coenzyme involved in glucose oxidation is *coenzyme A (CoA)*, a derivative of the vitamin pantothenic acid. This coenzyme picks up chemical groups called *acetyl groups* (—CH$_2$COOH) in certain metabolic reactions and carries them to other reactions. In the process, each CoA becomes covalently bound to an acetyl group, forming a compound called *acetylcoenzyme A (acetyl CoA)*.

Factors Affecting the Rates of Enzyme-catalyzed Reactions

Enzymes accelerate metabolic reactions by reducing the height of the activation energy barrier. Note that the height of the activation energy barrier does not affect the overall energy change of a reaction (ΔE), which depends only on the difference in energy between reactants and products. Therefore, enzymes cannot affect the direction of a reaction or the amount of energy it releases or requires; they can *only* affect the *rate* at which a reaction occurs.

The rate at which an enzyme can catalyze a reaction is, in turn, affected by several factors, including (1) the enzyme's catalytic rate, (2) the affinity of the enzyme for the substrate, (3) the enzyme concentration, and (4) the substrate concentration.

Catalytic Rate   The **catalytic rate** of an enzyme is a measure of how many product molecules it can generate per unit time, assuming that the active site is always oc-

cupied by a substrate molecule. As such, the catalytic rate reflects how fast an enzyme can carry out the catalytic step of the two-step sequence previously described. Some enzymes are inherently faster than others and can convert thousands of substrate molecules into products each second; other enzymes might take more than a minute to act on a single substrate molecule. Other things being equal, the rate of an enzymatic reaction increases as the enzyme's catalytic rate increases.

Affinity   The **affinity** of an enzyme is a measure of how tightly substrate molecules bind to its active site. (Affinity is a general term referring to the tightness of binding between any two molecules.) Generally speaking, higher affinities translate into higher rates for enzyme-catalyzed reactions (except when enzymes are working at their maximum rates, as just described). This makes intuitive sense when you consider that an enzyme's active site must be occupied by a substrate molecule before the enzyme can catalyze any reaction. When substrate molecules are present at a given concentration, an enzyme's active site will be occupied a greater percentage of the time if the enzyme has a high affinity for the substrate. If the affinity is lower, the active site will be occupied for less of the time, all else being equal. Therefore, a high-affinity enzyme is able to generate more product molecules in a given length of time.

A high affinity implies a close fit between an enzyme's active site and a substrate molecule. Such a fit maximizes the area of contact between substrate and enzyme and thus maximizes any attractive forces that might exist between the two. Affinity is also influenced by other factors that increase this force of attraction, such as the presence of opposite electrical charges on the enzyme and the substrate.

Enzyme Concentration   The affinity or catalytic rate of an enzyme influences the rate of an enzymatic reaction because it affects how fast individual enzyme molecules can convert substrates to products. But the rate of an enzymatic reaction also increases in direct proportion to the enzyme concentration: The more enzyme molecules there are, the more product molecules they can generate in a given length of time.

Substrate Concentration   The rate of an enzymatic reaction also increases as the substrate concentration increases. This fact relates to the reversible nature of enzyme-substrate binding, which enables a substrate molecule to exit the active site without being acted upon by the enzyme. Whenever this occurs, another substrate molecule will eventually come along to be acted on. But when the substrate concentration is low, more time will pass before the next substrate molecule binds to the enzyme, and as a result its active site will remain unoccupied for a greater percentage of the time. Under these conditions, the reaction rate will be low because for much

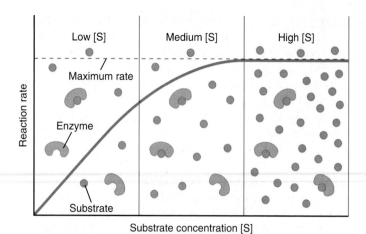

FIGURE 3.7 Influence of substrate concentration on the rate of an enzyme-catalyzed reaction. *A fixed concentration of enzyme is assumed for this plot of reaction rate versus substrate concentration [S]. Binding of substrate to the enzyme increases with increasing [S] until high substrate concentrations are reached, in which case all enzyme molecules are bound (100% saturation).*

As the reaction rate is increasing, is the concentration of free (unbound) enzyme molecules increasing or decreasing?

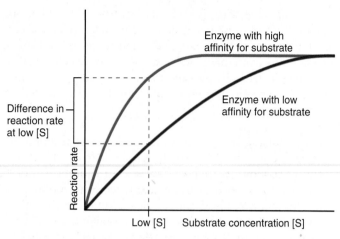

FIGURE 3.8 Influence of enzyme-substrate affinity on the rate of an enzyme-catalyzed reaction. *Plots of reaction rate versus substrate concentration are shown for two enzymes having different affinities for the substrate. For purposes of comparison, it is assumed that the two enzymes have the same catalytic rate and are present at the same concentration. Note that at lower substrate concentrations, the enzyme with higher affinity is able to catalyze the reaction at a faster rate.*

of the time the enzyme is idle. At higher substrate concentrations, less time will pass before the next substrate molecule comes along. This means that the enzyme's active site will be occupied more of the time, which gives the enzyme more opportunity to catalyze the reaction. Consequently, the reaction rate will be higher.

Figure 3.7 shows how the rate of an enzyme-catalyzed reaction varies with the substrate concentration, [S], given a fixed enzyme concentration. At lower and moderate concentrations, the rate of the reaction increases as [S] increases, as we would expect. At higher concentrations, however, the curve levels off, indicating that further increases in [S] fail to raise the rate appreciably. Given a fixed enzyme concentration, when [S] is very high, the active site of every enzyme molecule is occupied by substrate virtually 100% of the time, and the enzyme is said to be *100% saturated*. (The **percent saturation** of an enzyme indicates the proportion of time that the active site is occupied; thus, at 25% saturation, the active site is occupied 25% of the time.) Raising [S] beyond this point does not cause any further increase in the rate because all the enzyme molecules are already bound to substrates and are already catalyzing the formation of products. Under these conditions, the rate of the reaction is limited only by the enzyme's catalytic rate and its concentration.

The degree of saturation of an enzyme and, hence, the rate at which it can catalyze a reaction are affected

not only by the substrate concentration, but also by the affinity of the enzyme for the substrate. At a given substrate concentration, an enzyme with higher affinity will exhibit a higher degree of saturation than will an enzyme with lower affinity, and therefore it reaches its maximum rate at lower substrate concentrations (Figure 3.8).

Other Factors Affecting Enzyme Activity The rates of enzymatic reactions can be affected by still other conditions, such as the temperature of the reaction mixture or its pH (Toolbox: Acids, Bases, and pH). Changes in temperature or pH can alter the shape of an enzyme molecule, which is so crucial to its function. In general, enzyme activities decline if the temperature or pH becomes significantly higher or lower than normal. Within the body, changes in temperature are rarely significant because body temperature is regulated to stay constant. pH is also regulated to stay relatively constant in intracellular and extracellular fluid (within the range 7.2 to 7.4), but it varies appreciably in certain locations within the gastrointestinal tract. The lumen of the stomach, for instance, contains fluid that is highly acidic (the pH can be as low as 2); in the intestine, the luminal fluid is more basic, with a pH over 8. Significantly, the activity of pepsin, which works in the stomach, is highest when the pH is in the acidic range, whereas that of *trypsin*, a similar protein-digesting enzyme that works in the intestine, is highest when the pH is in the basic range.

Decreasing

Certain substances release hydrogen ions or *protons* (H^+) when dissolved in water and are known as *acids*. A familiar example is *hydrochloric acid* (HCl), which dissociates into hydrogen and chloride ions, as follows:

$$HCl \rightarrow H^+ + Cl^-$$

HCl is an example of a *strong acid,* an acid that dissociates completely. Certain other acids are *weak acids,* meaning that they dissociate incompletely. For example, *carboxyl groups* (—COOH), which are found on amino acids and other biomolecules, act as weak acids and dissociate as follows:

$$R—COOH \leftrightharpoons R—COO^- + H^+$$

Here, the double arrow signifies that protons can not only dissociate from the anion ($R—COO^-$), but can also combine with it. Substances that combine with protons are called *bases* and are classified as strong or weak depending on whether they combine completely or incompletely. Examples of weak bases are *amino groups* ($—NH_2$), which are found on amino acids and other compounds and react with protons as follows:

$$R—NH_2 + H^+ \leftrightharpoons R—NH_3^+$$

The *acidity* of a solution is determined by its hydrogen ion concentration, which can be expressed in terms of molarity or denoted by a number called the *pH*, which is defined according to the following expression:

$$pH = \log\{1/[H^+]\} = -\log[H^+]$$

where $[H^+]$ is the hydrogen ion concentration. Note that as $[H^+]$ increases, the pH decreases. Note also that because the pH scale is logarithmic, a change of one pH unit represents a tenfold change in the hydrogen ion concentration. The hydrogen ion concentration in pure water is 10^{-7} M, giving a pH of 7. Solutions with a pH of 7 are said to be *neutral*. A solution whose pH is below 7 is said to be *acidic;* if the pH is above 7, the solution is *basic*.

Regulation of Enzyme Activity

We have just seen how enzymes catalyze reactions, and how an enzyme's ability to catalyze a reaction is influenced by various factors. Now we look at how the body regulates the activity of enzymes to adjust the rates of reactions.

As conditions in the body change, the rates of metabolic reactions are continually adjusted to meet the body's needs. This adjustment is accomplished by means of mechanisms that regulate the activities of certain enzymes. In some cases, cellular enzyme concentrations are adjusted to speed up or slow down reactions. More enzyme molecules may be synthesized, for example, which raises the enzyme concentration and increases reaction rates. Alternatively, enzymes may be broken down or inactivated by other means, which effectively lowers the enzyme concentration and decreases reaction rates. More often, reaction rates are adjusted through changes in the activities of existing enzyme molecules.

Allosteric and Covalent Regulation   Certain enzyme molecules possess a binding site called the **regulatory site** that is specific for molecules known generally as **modulators.** (The regulatory site is distinct from the active site.) A modulator induces a change in an enzyme's conformation that alters the shape of the active site, causing a change in the enzyme's activity by altering its catalytic rate, its affinity for substrate, or both. This type of regulatory mechanism is known as **allosteric regulation** (Figure 3.9).

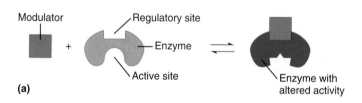

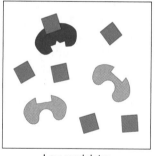 

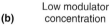

(a)

(b) Low modulator concentration   High modulator concentration

FIGURE 3.9 Allosteric regulation of enzyme activity. (a) *Reversible binding of a modulator molecule to a regulatory site on an enzyme. When bound, the modulator alters the active site and thus the activity of the enzyme.* **(b)** *The effect of modulator concentration on enzyme activity. When the modulator is present at low concentration, relatively few enzyme molecules are bound. At higher modulator concentrations, more enzyme molecules are bound, increasing the modulator's influence on overall enzyme activity.*

Why does the number of bound enzyme molecules increase as the modulator concentration increases?

Because of the law of mass action

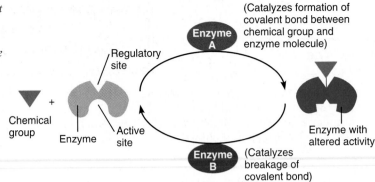

FIGURE 3.10 Covalent regulation of enzyme activity. *Covalent bonding of a chemical group to an enzyme molecule alters the enzyme's activity. Formation of these covalent bonds is catalyzed by another enzyme (A); removal of the chemical group requires the action of yet another enzyme (B), which breaks the covalent bonds.*

In allosteric regulation, the binding of the modulator to the regulatory site is reversible, so the modulator can readily leave the regulatory site after binding to it. Normally, an equilibrium exists between the enzyme and the modulator. When the modulator concentration increases, the equilibrium shifts according to the law of mass action, with more enzymes becoming bound by the modulator (see Figure 3.9b). When the modulator concentration decreases, the equilibrium shifts in the opposite direction, so fewer enzymes are bound by the modulator.

Binding of a modulator to an enzyme can bring about either an increase or a decrease in the enzyme's activity, depending on the enzyme in question. Some enzyme molecules have more than one regulatory site and can respond to more than one type of modulator. Sometimes these different modulators affect the enzyme in different ways—some are inhibitory while others are stimulatory.

Another type of enzyme regulation is **covalent regulation,** in which changes in an enzyme's activity are brought about by the covalent bonding of a specific chemical group to a site on the enzyme molecule (Figure 3.10). In this process, formation of the covalent bond between the chemical group and the enzyme protein is catalyzed by another enzyme (enzyme A in the figure). Because covalent bonds are relatively strong, the chemical group remains attached to the enzyme unless these bonds are broken, which requires the action of other enzymes (enzyme B in the figure).

A common form of covalent regulation involves the phosphorylation and dephosphorylation of enzyme molecules. Phosphorylation of a target enzyme molecule (E) is catalyzed by a type of enzyme called a **protein kinase,** as follows:

$$E + P_i \xrightarrow{\text{protein kinase}} E{-}P$$

Dephosphorylation is catalyzed by an enzyme called a **phosphatase** and can be written as follows:

$$E{-}P \xrightarrow{\text{phosphatase}} E + P_i$$

These kinases and phosphatases are usually specific for particular target enzymes and are themselves the targets of other allosteric or covalent regulatory mechanisms.

Feedback Inhibition Through allosteric or covalent regulation, the activity of an enzyme may be increased or decreased, but the reactions in a metabolic pathway are generally catalyzed by different enzymes, with only certain enzymes being regulated. Thus through control of certain regulated enzymes, cells can exert control over entire pathways. Frequently, this control involves a mechanism known as **feedback inhibition.** Consider the following example of a metabolic pathway consisting of three steps catalyzed by three different enzymes (E_1, E_2, and E_3):

$$A \xrightarrow{E_1} B \xrightarrow{E_2} C \xrightarrow{E_3} D$$

Here, substance C acts through allosteric regulation to inhibit E_2, the enzyme catalyzing the second step, as is indicated by the arrow with the negative sign.

Normally, feedback inhibition regulates enzyme activities so as to hold the rates of metabolic reactions steady. Under these conditions (referred to as a *steady state*), all three steps in the pathway just described proceed at the same rate, so that C is consumed in the third step as fast as it is generated in the second, which keeps [C] steady. If the rate of the second step increased suddenly for some reason, [C] would rise, which would then suppress the activity of E_2 and counteract the change in concentration.

Feedback inhibition can also speed up or slow down metabolic reactions according to changes in the body's needs. Suppose, for instance, that in the previous example the third step speeds up as an appropriate response to an increased cellular demand for the final product, D. The speed-up of the third step will cause C to be consumed at a greater rate, which should cause [C] to begin falling. This fall in [C] will reduce the inhibitory action of C on E_2, thereby allowing the second step to speed up.

In feedback inhibition, an enzyme is usually inhibited by the product of a reaction occurring downstream:

$$A \xrightarrow{E_1} B \xrightarrow{E_2} C \xrightarrow{E_3} D$$

In this case, E_1 is inhibited by D, the end-product of the pathway. This special case of feedback inhibition is called *end-product inhibition.*

Feedforward Activation   Another, less common, mechanism of enzyme regulation is **feedforward activation,** which involves the activation of an enzyme by an intermediate appearing upstream in a metabolic pathway, as follows:

$$A \xrightarrow{E_1} B \xrightarrow{E_2} C \xrightarrow{E_3} D$$

In this example, E_3, which catalyzes the third step, is activated by substance B, which is produced in the first step. As with end-product inhibition, feedforward activation helps to keep reaction rates steady under normal conditions, but it also allows reactions to speed up or slow down when conditions change.

Quick Test 3.3

1. To speed up the net rate of a reaction proceeding in the forward direction, which of the following should increase? temperature, reactant concentrations, product concentrations, the height of the activation energy barrier (choose all that apply)

2. Define the following terms: *enzyme, substrate, substrate specificity, active site, affinity, cofactor, coenzyme, percent saturation, regulatory site, modulator.*

3. How is the rate of an enzymatic reaction affected by the catalytic rate of an enzyme? By its affinity for the substrate?

4. What is the primary distinction between allosteric regulation and covalent regulation? Between feedback inhibition and feedforward activation?

GLUCOSE OXIDATION: THE CENTRAL REACTION OF ENERGY METABOLISM

We have seen that energy is released in certain chemical reactions and consumed in others, and that released energy can be used to perform cellular work. We have also learned about how enzymes catalyze these reactions, so that they can occur quickly enough to meet the body's needs. In this section we bring all this information together to understand how a cell utilizes the breakdown of nutrient molecules to produce and store enough energy to maintain all bodily processes. In this discussion we focus on glucose oxidation, the reaction that virtually all cells rely on to provide their energy needs.

Have you ever wondered why we breathe? The simple answer is that breathing enables us to obtain oxygen from the air when we inhale and get rid of carbon dioxide when we exhale, an exchange of gases that is crucial to our survival because we derive most of our energy from the reaction of oxygen with glucose and other nu-

trient molecules that serve as fuels. When oxygen reacts with these fuels, energy is released and carbon dioxide is generated as a waste product. This is apparent in the reaction of oxygen with glucose ($C_6H_{12}O_6$), which proceeds as follows:

$$C_6H_{12}O_6 + 6\ O_2 \rightarrow 6\ CO_2 + 6\ H_2O + \text{energy}$$

Here we can see that complete oxidation of one molecule of glucose requires six oxygen molecules and generates six molecules each of carbon dioxide and water. We also see that this reaction releases energy, which is why this reaction is of central importance to cells. The energy released amounts to 686 kcal for every mole of glucose that enters the reaction ($\Delta E = -686$ kcal/mole).

ATP: The Medium of Energy Exchange

When energy is released in a reaction, it must be "captured" in certain forms before it can be used to do work; otherwise it is simply dissipated (released into the environment) as heat. When coal burns in a furnace, for example, it releases energy in the form of heat. If this energy is not captured somehow, it just dissipates into the atmosphere, without any work being performed. A power plant, in contrast, captures energy released from the burning of coal by using the heat to generate steam under pressure. This steam is then used to drive an electrical generator, and in doing so it performs work. Thus, the steam acts as a temporary energy store or "middleman" that facilitates the transfer of energy from the burning coal to the electric generator.

Cells harness the energy released in a reaction such as glucose oxidation to synthesize a compound called **adenosine triphosphate (ATP),** which serves as a temporary energy store. ATP is synthesized from a nucleotide called **adenosine diphosphate (ADP)** and inorganic phosphate (P_i) according to the following reaction:

$$ADP + P_i + \text{energy} \rightarrow ATP$$

As Figure 3.11a shows, ATP synthesis is an example of a condensation reaction because water is generated as a product. (This water is often omitted from the reaction notation for simplicity's sake.) The quantity of energy required to make one mole of ATP under normal cellular conditions is about 7 kcal ($\Delta E = +7$ kcal/mole).

We noted that steam produced in a power plant is a temporary energy store because it is eventually used for generating electricity; during this process it loses the energy that it gained from the burning coal. Likewise, ATP is also a temporary energy store because it is eventually broken down to ADP and P_i, in the process losing the energy that was put into it when it was made:

$$ATP \rightarrow ADP + P_i + \text{energy}$$

This reaction is often referred to as *ATP hydrolysis* because water is a reactant (Figure 3.11b), even though it is usually omitted from the reaction notation. (Because ATP

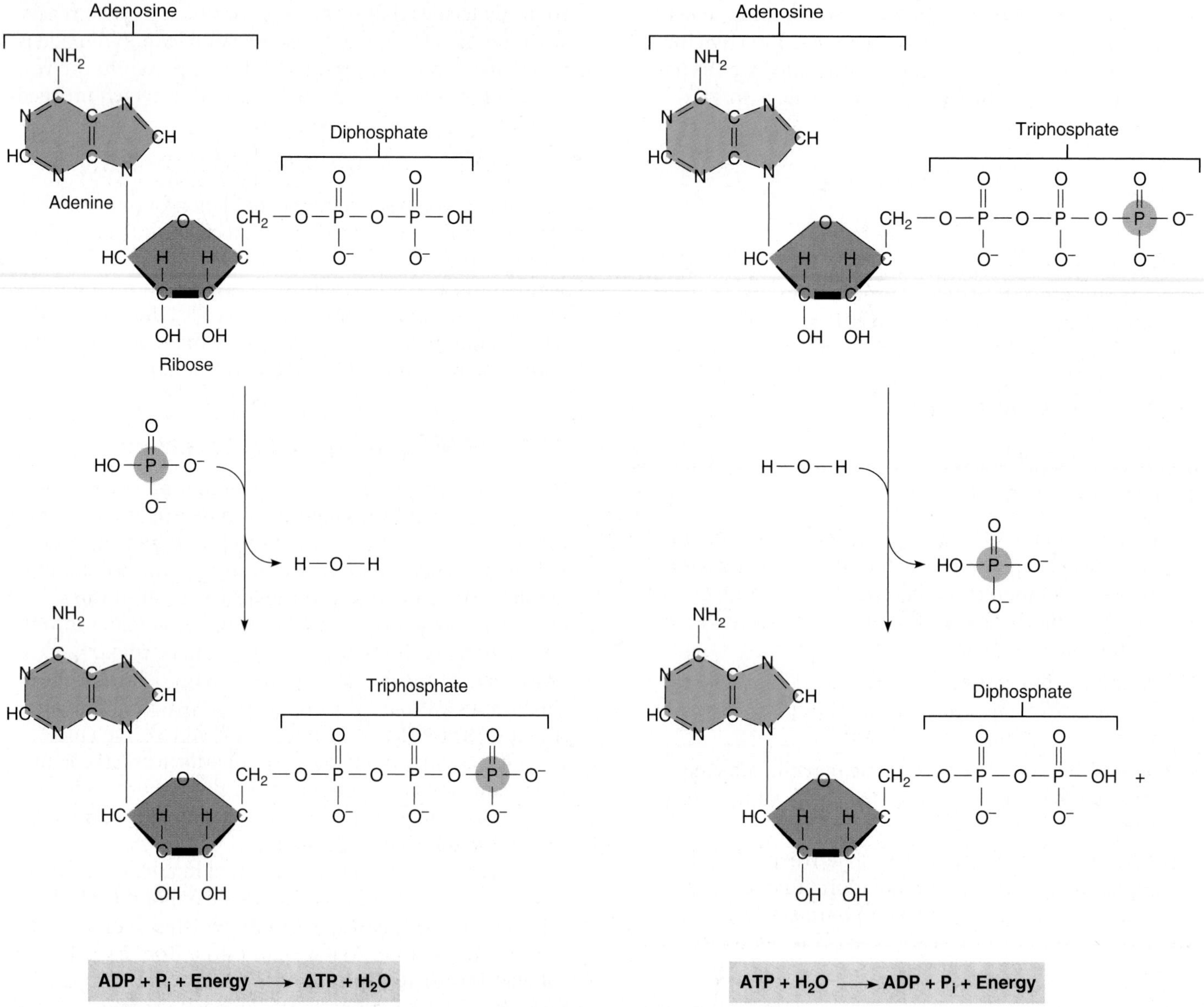

(a) ATP synthesis (condensation)

(b) ATP breakdown (hydrolysis)

FIGURE 3.11 Synthesis and hydrolysis of ATP. (a) *The synthesis of ATP from ADP and* P_i*, which requires energy. ADP consists of adenosine joined to diphosphate, a chain containing two phosphate groups; addition of a third phosphate converts ADP to ATP. This reaction generates water and is therefore a condensation reaction.* **(b)** *The hydrolysis of ATP into ADP and* P_i*, which releases energy. This reaction is a hydrolysis reaction because it consumes water.*

hydrolysis releases energy and involves the splitting of a single bond—the bond between ATP and one of the attached phosphate groups—that bond is commonly termed a *high-energy phosphate bond.*)

When cells need energy to perform work or to run energy-requiring reactions, they obtain it by hydrolyzing previously synthesized ATP. Since glucose oxidation and other energy-releasing reactions supply the energy for making ATP, these reactions are the ultimate sources of cellular energy.

ATP in Action

Cells are able to synthesize ATP using energy derived from glucose oxidation because the reaction serving as the energy source (glucose oxidation) has a negative energy change and therefore occurs spontaneously. In contrast, the synthesis of ATP has a positive energy change and does not occur spontaneously. When cells use glucose oxidation (or any other energy-releasing reaction) to make ATP, some of the energy released in this reaction is used to drive the energy-requiring process of ATP synthe-

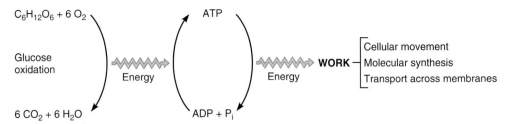

FIGURE 3.12 The coupling of reactions in energy transfer in cells. *Reactions such as glucose oxidation release energy that is used to synthesize ATP. The subsequent breakdown of ATP releases energy that is used to perform various types of cellular work.*

As ATP is broken down to ADP and P_i, does the molecules' potential energy increase or decrease?

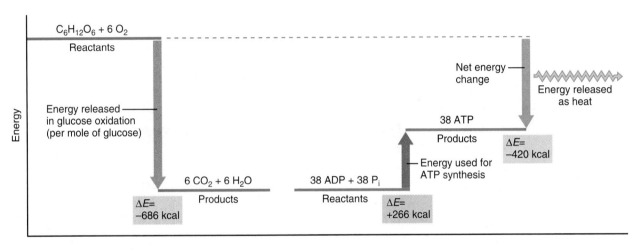

FIGURE 3.13 Energy exchange in ATP synthesis and the release of heat. *The energy changes occurring in glucose oxidation and ATP synthesis, assuming that 38 moles of ATP are synthesized for each mole of glucose oxidized. Note that the net energy change for the combined reaction (glucose oxidation + ATP synthesis) is negative, so energy is released as heat.*

sis. This is accomplished by mechanisms that link or *couple* the energy-releasing reaction to the energy-requiring reaction, so that they must occur together (Figure 3.12). When ATP is subsequently broken down, that energy-releasing reaction is coupled to mechanisms that harness the energy to perform work.

We have seen that the oxidation of one mole of glucose releases 686 kcal of energy. Because only 7 kcal are required to synthesize one mole of ATP, the oxidation of one mole of glucose releases enough energy to make 98 moles of ATP. When the actual amount of ATP is measured, however, it turns out that the quantity of ATP synthesized is only about one-third this amount—about 38 moles of ATP per mole of glucose. Why so few? To understand we must consider the energy change for the entire process.

When a mole of glucose is oxidized and 38 moles of ATP are produced, the entire process can be written in the form of a single reaction, as follows:

$$C_6H_{12}O_6 + 6\ O_2 + 38\ ADP + 38\ P_i \rightarrow$$
$$6\ CO_2 + 6\ H_2O + 38\ ATP$$

The energy change for this reaction is the sum of the energy change for glucose oxidation (-686 kcal) and the energy change for the synthesis of 38 moles of ATP (38 moles $\times$ 7 kcal/mole = 266 kcal). Therefore, as Figure 3.13 shows, glucose oxidation releases more energy than is used in ATP synthesis, giving a net energy change that is negative ($\Delta E = -686$ kcal + 266 kcal = -420 kcal). Because the net energy change is negative, the reaction can proceed in the forward direction, yielding ATP as a product.

This example illustrates the general principle that when an energy-releasing reaction is used to drive an energy-requiring process, the released energy cannot be used 100 percent efficiently; some of the released energy always goes to waste! Because cells normally use about 266 kcal of energy for ATP synthesis for each 686 kcal of energy released from the oxidation of one mole of glucose, only about 40% of the released energy is put to use

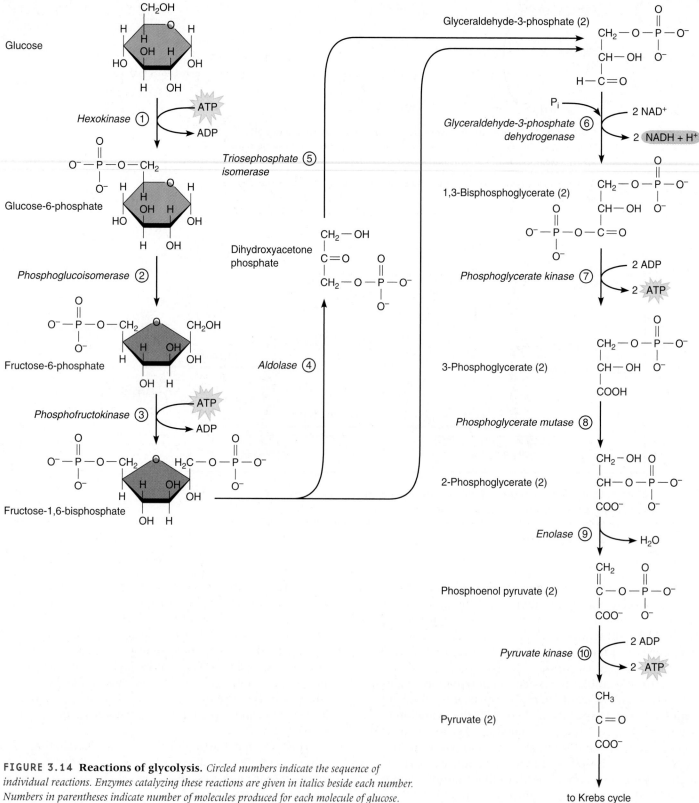

FIGURE 3.14 Reactions of glycolysis. *Circled numbers indicate the sequence of individual reactions. Enzymes catalyzing these reactions are given in italics beside each number. Numbers in parentheses indicate number of molecules produced for each molecule of glucose.*

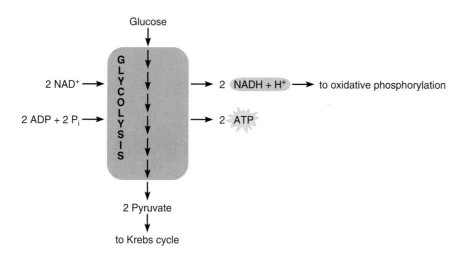

Glucose

2 NAD⁺ →

2 ADP + 2 Pᵢ →

→ 2 NADH + H⁺ → to oxidative phosphorylation

→ 2 ATP

2 Pyruvate

to Krebs cycle

(a)

FIGURE 3.15 Summary of glycolysis. (a) *A diagrammatic summary of the major reactants and products of the glycolysis pathway. Numbers indicate quantities produced or consumed for each molecule of glucose.* **(b)** *Net changes occurring in glycolysis. Numbers with minus signs indicate quantities of reactants consumed; those with plus signs indicate quantities of products produced.* **(c)** *The overall reaction of glycolysis.*

	Glucose	Pyruvate	O_2	CO_2	H_2O	NAD^+	NADH	H^+	FAD	$FADH_2$	ADP	P_i	ATP
(b) Net changes:	−1	+2	0	0	0	−2	+2	+2	0	0	−2	−2	+2

(c) Overall reaction: Glucose + 2 NAD^+ + 2 ADP + 2 P_i ⟶ 2 Pyruvate + 2 NADH + 2 H^+ + 2 ATP

(266 kcal/686 kcal = 0.388 ≈ 0.40). As for the remaining 60% of the released energy, experience tells us at least some of this appears as heat. In fact, "body heat" is actually heat generated as a by-product of metabolic reactions, including glucose oxidation (see Figure 3.13).

Quick Test 3.4

1. Is the oxidation of glucose catabolic or anabolic? Does it release energy or require it?

2. Where does the energy for ATP synthesis come from? When ATP is broken down, what is the released energy used for?

3. When energy from glucose oxidation is used to make ATP, only a certain fraction of the released energy is used for this purpose. What happens to the rest of the released energy?

STAGES OF GLUCOSE OXIDATION: GLYCOLYSIS, THE KREBS CYCLE, AND OXIDATIVE PHOSPHORYLATION

We have seen that the oxidation of glucose releases energy, and that cells utilize some of this released energy to synthesize ATP. Exactly how cells perform this feat is the topic of this section.

In a cell, the oxidation of glucose is not carried out in a single reaction; instead, it occurs in three distinct stages,

which are carried out in different parts of the cell. The stages are (1) *glycolysis,* which takes place in the cytosol; (2) the *Krebs cycle* (also called the *citric acid cycle, tricarboxylic acid cycle,* or *TCA cycle*), which occurs in the mitochondrial matrix; and (3) *oxidative phosphorylation,* which occurs within the inner mitochondrial membrane. Each of these stages involves a series of chemical reactions that are summarized next, beginning with a look at glycolysis.

Glycolysis

Glycolysis, which means *splitting of sugar,* is a metabolic pathway comprising ten reactions, each catalyzed by a different cytosolic enzyme (Figure 3.14). In glycolysis, each glucose molecule (which contains six carbons) is broken down into two molecules of *pyruvate* (the ionized form of *pyruvic acid*) containing three carbons apiece. These molecules of pyruvate are normally broken down further in the subsequent stages of glucose oxidation.

The major results of glycolysis are shown in Figure 3.15 and can be summarized as follows:

1. By the end of glycolysis, each glucose molecule has been split into two molecules of pyruvate.

2. During this process two ATPs are consumed (one each in reactions 1 and 3), but four more are produced (two in reaction 7 and two in reaction 10). This gives a net synthesis of 2 molecules of ATP for each molecule of glucose consumed.

3. Two molecules of NAD^+ are reduced in step 6, yielding 2 molecules of NADH for every molecule of glucose.

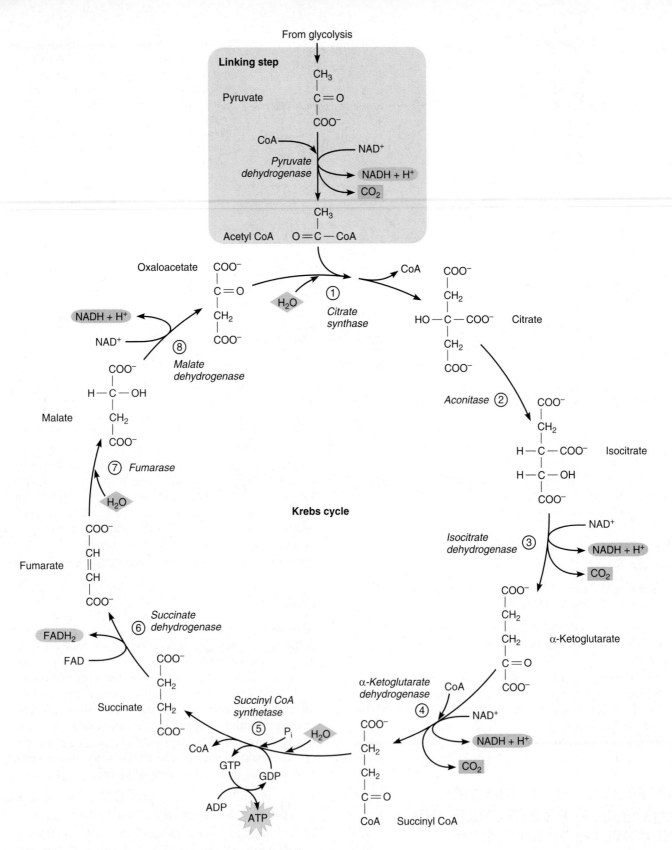

FIGURE 3.16 Reactions of the Krebs cycle and linking step. *Circled numbers indicate the sequence of individual reactions. Enzymes catalyzing these reactions are given in italics beside each number.*

Step 8 of the Krebs cycle is an example of what type of reaction?

An oxidation-reduction reaction

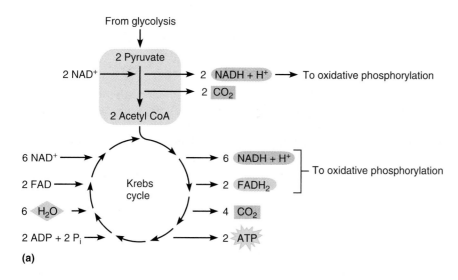

From glycolysis

2 Pyruvate

2 NAD$^+$ → → 2 NADH + H$^+$ → To oxidative phosphorylation

→ 2 CO$_2$

2 Acetyl CoA

6 NAD$^+$ → → 6 NADH + H$^+$ ⎤
⎥ To oxidative phosphorylation
2 FAD → → 2 FADH$_2$ ⎦

Krebs cycle

6 H$_2$O → → 4 CO$_2$

2 ADP + 2 P$_i$ → → 2 ATP

(a)

FIGURE 3.17 Summary of the Krebs cycle. (a) *A diagrammatic summary of the major reactants and products of the Krebs cycle, including the linking step. Numbers indicate quantities produced or consumed for each molecule of glucose that enters glycolysis.* **(b)** *Net changes occurring in the Krebs cycle and linking step. Plus and minus signs indicate products and reactants, respectively.* **(c)** *The overall reaction of the Krebs cycle and the linking step.*

	Glucose	Pyruvate	O$_2$	CO$_2$	H$_2$O	NAD$^+$	NADH	H$^+$	FAD	FADH$_2$	ADP	P$_i$	ATP
(b) Net changes:	0	−2	0	+6	−6	−8	+8	+8	−2	+2	−2	−2	+2

(c) Overall reaction: 2 Pyruvate + 8 NAD$^+$ + 2 FAD + 2 ADP + 2 P$_i$ + 6 H$_2$O ⟶ 6 CO$_2$ + 8 NADH + 8 H$^+$ + 2 FADH$_2$ + 2 ATP

When all reactants and products are accounted for, the overall reaction of glycolysis is as follows:

glucose + 2 NAD$^+$ + 2 ADP + 2 P$_i$ → 2 pyruvate + 2 NADH + 2 H$^+$ + 2 ATP

(The water produced in step 9 is ignored for the sake of simplicity.) Note that during glycolysis, no oxygen is consumed and no carbon dioxide is produced. Although oxygen is indeed consumed in glucose oxidation and carbon dioxide is produced, these events do not occur in glycolysis, but instead in the later stages of glucose oxidation.

Glycolysis is useful to cells because it produces some ATP, but its primary importance is that it sets the stage for subsequent events that yield even more ATP. The NADH that is produced in glycolysis will eventually give up its electrons, thereby releasing energy that will be used to synthesize more ATP by oxidative phosphorylation. Also, the pyruvate that is generated will eventually be catabolized in the next stage of glucose oxidation, the Krebs cycle.

As we see later in the chapter, pyruvate can proceed through further stages of glucose oxidation only when oxygen is readily available inside the cell. Should the availability of oxygen become limited, pyruvate is instead converted to a compound called *lactic acid* (or *lactate*), which is not broken down further. As we continue our discussion of glucose metabolism, assume that oxygen is readily available unless otherwise noted.

The Krebs Cycle

Glycolysis resembles most metabolic pathways in that it is a sequence of steps with distinct starting and ending points. In contrast, the **Krebs cycle,** the next stage of glucose oxidation, has no starting or ending points because it is circular or *cyclic,* as its name implies.

Glucose oxidation does not proceed beyond glycolysis until pyruvate, which is produced in the cytosol, enters the mitochondrial matrix. Although the Krebs cycle is the next major stage of glucose oxidation, pyruvate does not enter the Krebs cycle directly. Instead, it undergoes a reaction that converts it to acetyl CoA, which then enters the cycle. (As we will see, acetyl CoA is also generated in other reactions.) We refer to this reaction as the *linking step* because it links glycolysis to the Krebs cycle. Because glycolysis generates two pyruvate molecules for every molecule of glucose, one glucose molecule ultimately yields two molecules of acetyl CoA in the linking step. Each acetyl CoA molecule participates in one complete "turn" of the Krebs cycle.

Reactions of the Krebs cycle are given in detail in Figure 3.16. This diagram begins with the linking step, which is shaded to distinguish it from the Krebs cycle reactions, which are numbered in sequence. Major results of the Krebs cycle (plus the linking step) are shown in Figure 3.17 and can be summarized as follows:

1. By the end of each turn of the Krebs cycle a total of three carbon dioxide molecules have been generated as end-products (one each in the linking step and in steps 3 and 4), the first time that any carbon dioxide has appeared in glucose oxidation. Because two pyruvate molecules are generated for every molecule of glucose that enters glycolysis, the Krebs cycle must "turn" twice for each glucose molecule, producing

two carbon dioxides during each turn plus two generated in the linking step. Thus for each molecule of glucose that enters glycolysis, the Krebs cycle and the linking step generate a total of six carbon dioxides.

2. Only one ATP is generated directly during the Krebs cycle (in step 5), which translates to two ATP per glucose molecule.

3. In the course of the linking step and the subsequent single turn of the Krebs cycle, a total of five reduced coenzymes—4 NADH and 1 FADH$_2$—are produced (in the linking steps and in steps 3, 4, 6, and 8), which is equivalent to ten reduced coenzymes per glucose (8 NADH and 2 FADH$_2$).

The overall reaction for the linking step plus one turn of the Krebs cycle is as follows:

$$\text{pyruvate} + 4\,NAD^+ + FAD + ADP + P_i + 3\,H_2O \rightarrow$$
$$3\,CO_2 + 4\,NADH + 4\,H^+ + FADH_2 + ATP$$

which translates into the following for each molecule of glucose that enters glycolysis:

$$2\,\text{pyruvate} + 8\,NAD^+ + 2\,FAD + 2\,ADP + 2\,P_i +$$
$$6\,H_2O \rightarrow 6\,CO_2 + 8\,NADH + 8\,H^+ + 2\,FADH_2 + 2\,ATP$$

When combined with the two ATPs per glucose that are netted in glycolysis, the additional two ATPs that are generated in the Krebs cycle makes a running total of four molecules of ATP for every molecule of glucose. As we will see, however, almost all the ATP that is formed is generated during the final stage of glucose oxidation, oxidative phosphorylation. The Krebs cycle and linking step play an important role in this process because these pathways supply 10 of the 12 reduced coenzyme molecules that eventually go to oxidative phosphorylation to give up their electrons and release energy for making ATP.

The Krebs cycle and linking step produce 100% of the six carbon dioxide molecules that result from the complete oxidation of glucose. Note that absolutely no oxygen has been consumed so far, meaning that 100% of the expected oxygen consumption (six molecules of oxygen per molecule of glucose) must occur in oxidative phosphorylation, the final stage of glucose oxidation, if it is to occur at all. As we will see next, oxygen plays a vital role in oxidative phosphorylation, because oxygen is the ultimate acceptor of all the electrons that are given up by NADH or FADH$_2$. Without it, the electrons would have nowhere to go, and oxidative phosphorylation would stop.

Oxidative Phosphorylation

The formation of ATP is a *phosphorylation* reaction because it involves the addition of a phosphate group (P) to another compound, ADP. The formation of ATP that occurs in steps 7 and 10 of glycolysis (see Figure 3.14) or in step 5 of the Krebs cycle (see Figure 3.16) is referred to as **substrate-level phosphorylation** because in each case, an enzyme transfers a phosphate group from one substrate to another. The general form of such a reaction is:

$$X-P + ADP \xrightarrow{E} X + ATP$$

where E represents the enzyme. When the donor substrate (X—P) gives up its phosphate, energy is released, and some of this energy is then harnessed to make ATP.

Most ATP that is made in cells is not made by substrate-level phosphorylation, but instead by an altogether different mechanism called **oxidative phosphorylation,** which involves two simultaneous processes: (1) the transport in the inner mitochondrial membrane of hydrogen atoms or electrons through a series of compounds, known as the *electron transport chain*, which releases energy, and (2) the harnessing of this energy to make ATP, which is carried out by a mechanism called *chemiosmotic coupling.*

In oxidative phosphorylation, the reduced coenzymes (NADH and FADH$_2$) generated in glycolysis and the Krebs cycle serve as the energy source for making ATP. NADH and FADH$_2$ release electrons to the electron transport chain, and as these electrons travel through the chain, energy is released. Much of that energy is captured and harnessed to drive the synthesis of ATP, which is catalyzed by an enzyme called **ATP synthase.** To do this, the enzyme uses free inorganic phosphate (P$_i$) that it obtains from solution, not phosphate donated by another compound:

$$ADP + P_i \xrightarrow{\text{ATP synthase}} ATP$$

This is the essential difference between substrate-level phosphorylation and oxidative phosphorylation. It is called *oxidative* phosphorylation because oxygen is needed to accept all the electrons that pass through the electron transport chain.

The Electron Transport Chain

The **electron transport chain** comprises a set of diverse compounds (including several enzymes) located in the inner mitochondrial membrane. Most of these compounds are proteins specialized to function as electron carriers; that is, they bind electrons reversibly and thus have the ability to pick up electrons and subsequently give them up. Among these electron carriers are a number of compounds called *cytochromes*, which possess special iron-containing chemical groups known as *hemes*. (Heme groups are also found in hemoglobin, as its name suggests.) Also present are various *iron-sulfur proteins*, which contain iron atoms bound to sulfur. Most of these proteins are aggregated into large complexes that are firmly imbedded in the inner mitochondrial membrane. A few electron carriers are individual molecules that move freely in the lipid bilayer and shuttle back and forth, carrying electrons from one complex to another. One,

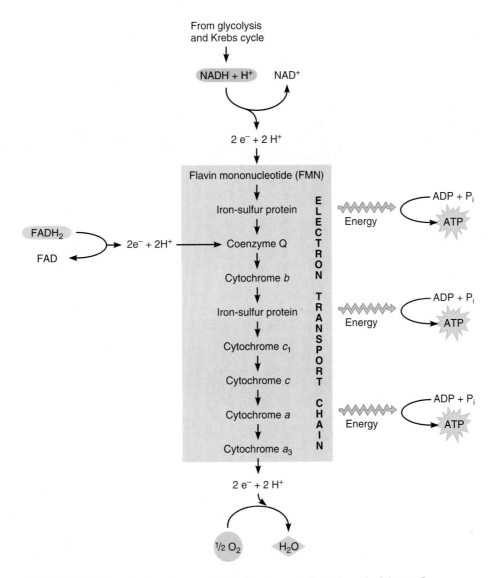

FIGURE 3.18 The electron transport chain. *Arrows indicate the path of electron flow from initial electron donors (NADH or FADH₂) to oxygen, the final electron acceptor. Wavy arrows at right indicate that the movement of electrons through the chain releases energy, which is harnessed to drive ATP synthesis by oxidative phosphorylation.*

called *coenzyme Q*, is not a protein at all, but instead a small molecule composed mainly of hydrocarbon.

Electrons are carried to the electron transport chain by reduced coenzymes (NADH and FADH₂) generated in glycolysis and the Krebs cycle (and in the linking step). These reduced coenzymes then release their electrons to certain components of the chain that function as electron acceptors. These electron acceptors then donate their electrons to other electron acceptors, which pass them on to still other acceptors, and so on. Each time electrons move from one component to the next, they lose some energy, and it is this energy that ultimately is used in making ATP.

As electrons pass through the chain, they go from one electron acceptor to the next in the specific sequence shown in Figure 3.18. NADH donates its pair of electrons

to the first electron acceptor in the chain, called *flavin mononucleotide* (FMN). In the process, NADH becomes oxidized to NAD^+ (which is then free to pick up more electrons in glycolysis or the Krebs cycle), and FMN becomes reduced. FMN then passes its electrons to the next component of the chain, an iron-sulfur protein, and becomes oxidized and ready to pick up another pair of electrons from NADH.

Although electrons leave NADH in the form of hydrogens, most components of the electron transport chain do not carry actual hydrogen atoms, only free electrons. Consequently, at some point electrons are stripped from the original hydrogen atoms, which are left as hydrogen ions (H^+) in solution, and the electrons continue to move through the chain. However, certain components of the chain *do* carry hydrogen atoms, and when

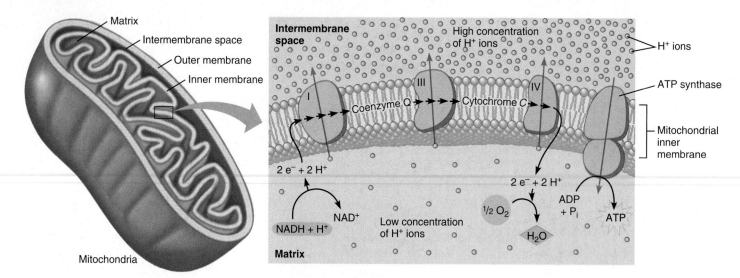

FIGURE 3.19 Chemiosmotic coupling. *The oxidation of NADH yields a pair of electrons that pass through three large inner mitochondrial membrane complexes (I, III, and IV) that contain components of the electron transport chain. (Electrons are carried from complex I to complex III by coenzyme Q, and from complex III to complex IV by cytochrome c.) As these electrons move through the complexes, energy is released, some of which is used to transport hydrogen ions (H⁺) from the mitochondrial matrix to the intermembrane space. Hydrogen ions flow in the opposite direction (down their concentration gradient) through the enzyme ATP synthase, in the process releasing energy that is used to synthesize ATP. (For simplicity, the release of electrons by FADH₂ is omitted from the diagram.)*

the electrons reach these components, they are "reconstituted" into hydrogen atoms by combining with hydrogen ions from solution. In essence, then, we can think of the electron transport chain as something that carries either hydrogen atoms or an equivalent combination of hydrogen ions and electrons. The release of electrons (denoted by the symbol e⁻) to the electron transport chain by NADH is represented at the top of Figure 3.18 as follows:

$$NADH + H^+ \rightarrow NAD^+ + 2 H^+ + 2 e^-$$

Note that the release of electrons by FADH₂ is written in Figure 3.18 in a similar fashion.

After moving from FMN to the iron-sulfur protein, electrons are passed on to coenzyme Q, which passes them on to a cytochrome called cytochrome *b*, and so forth. When eventually the electrons reach the final component of the chain, cytochrome a_3, they recombine with hydrogen ions to form hydrogens, which react immediately with oxygen to form water:

$$2 e^- + 2 H^+ + \frac{1}{2} O_2 \rightarrow H_2O$$

(The single oxygen atom is written as $\frac{1}{2} O_2$ because oxygen enters the reaction in its molecular form, O_2.) By combining this reaction with the previous one, we see that the net result of electron transport is the same as if NADH had simply given up its electrons directly to oxygen to form water, which is an energy-releasing reaction:

$$NADH + H^+ + \frac{1}{2} O_2 \rightarrow NAD^+ + H_2O + energy$$

If this is the net result, then what use is the electron transport chain? The answer is that electron transport provides a way of harnessing this energy so that it can be used to do something useful—namely, to make ATP. (We will see how this is accomplished shortly.) The precise quantity of ATP made depends on conditions in the cell but averages about three ATPs for every pair of electrons released by NADH.

FADH₂ also donates electrons to the electron transport chain, but it does not donate its electrons to FMN for the following reasons: FAD is permanently bound to its partner enzyme succinate dehydrogenase (see step 6 of the Krebs cycle, Figure 3.16), which unlike other Krebs cycle enzymes is imbedded in the inner mitochondrial membrane. For this reason, FADH₂ is not free to move about, as is NADH. Because of its location, FADH₂ is physically removed from FMN and therefore cannot donate electrons to it; instead, it donates its electrons to coenzyme Q (see Figure 3.18). These electrons, like those that come from NADH, eventually combine with oxygen to form water, but because they enter the chain at a point downstream, they release less energy as they move through the chain. As a consequence, only about two ATPs are produced for each pair of electrons released by FADH₂, compared to the three ATPs produced when electrons are released by NADH.

We know that energy is released when electrons move through the electron transport chain, and that this energy is used to make ATP, but what is the mechanism that couples energy release to ATP synthesis? The answer is chemiosmotic coupling, which we discuss next.

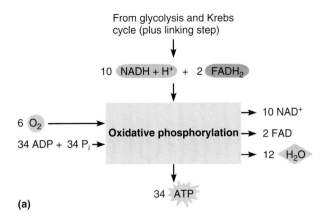

(a)

FIGURE 3.20 **Summary of oxidative phosphorylation.** **(a)** *A diagrammatic summary of the major reactants and products of oxidative phosphorylation. Numbers indicate quantities produced or consumed for each molecule of glucose that enters glycolysis.* **(b)** *Net changes occurring in oxidative phosphorylation. Plus and minus signs indicate products and reactants, respectively.* **(c)** *The overall reaction for oxidative phosphorylation.*

	Glucose	Pyruvate	O_2	CO_2	H_2O	NAD^+	NADH	H^+	FAD	$FADH_2$	ADP	P_i	ATP
(b) Net changes:	0	0	−6	0	+12	+10	−10	−10	+2	−2	−34	−34	+34

(c) Overall reaction: $10\ NADH + 10\ H^+ + 2\ FADH_2 + 34\ ADP + 34\ P_i + 6\ O_2 \longrightarrow 10\ NAD^+ + 2\ FAD + 12\ H_2O + 34\ ATP$

Chemiosmotic Coupling

The process that couples electron transport to ATP synthesis, known as **chemiosmotic coupling,** first uses energy released in the electron transport chain to transport hydrogen ions across the inner mitochondrial membrane against their *concentration gradient*; then it utilizes the energy stored in this gradient to make ATP (Figure 3.19).

The inner mitochondrial membrane contains four distinct complexes containing components of the electron transport chain (see Figure 3.19). Three of these complexes not only function as electron carriers, but also are able to transport hydrogen ions (H^+) across the membrane. As electrons pass through these complexes (designated by Roman numerals I, III, and IV), the complexes utilize the released energy to transport hydrogen ions from the mitochondrial matrix into the intermembrane space, which communicates with the cytosol through large pores in the outer mitochondrial membrane. This movement of hydrogen ions creates a difference in hydrogen ion concentration (a concentration gradient) across the membrane such that the concentration outside is higher than in the matrix (see Figure 3.19). This concentration gradient represents a store of potential energy. (A difference in electrical potential across the membrane also adds to this stored energy.)

The enzyme ATP synthase, which catalyzes the formation of ATP, resides in the inner mitochondrial membrane along with components of the electron transport chain and utilizes energy stored in the hydrogen ion gradient to perform its function. Like the complexes just described, it is also able to transport hydrogen ions across the membrane. In this case, however, the hydrogen ions move down their concentration gradient, not up it (see Figure 3.19). This flow of hydrogen ions releases energy, which ATP synthase harnesses to make ATP.

Summary of Oxidative Phosphorylation

We have seen that in oxidative phosphorylation, the final stage of glucose metabolism, the following events occur:

1. NADH and $FADH_2$, which are generated in the earlier stages of glucose metabolism, release their electrons to the electron transport chain. These electrons flow through the chain and then emerge from it to combine with oxygen, forming water as an end-product.

2. The movement of electrons through the chain releases energy, which is utilized to transport hydrogen ions across the inner mitochondrial membrane. This transport creates a concentration gradient of hydrogen ions across the membrane that stores some of the energy released during electron transport.

3. This stored energy is released when hydrogen ions flow through ATP synthase, which uses the energy to make ATP. On average, three ATPs are made for every pair of electrons released from NADH, while two ATPs are made for every pair of electrons released from $FADH_2$.

Given that 10 molecules of NADH and 2 molecules of $FADH_2$ are generated for every molecule of glucose that is oxidized, we expect a total of 34 molecules of ATP to be made during oxidative phosphorylation [(10 NADH × 3 ATP/NADH) + (2 $FADH_2$ × 2 ATP/$FADH_2$) = 34 ATP]. Because these 12 molecules of reduced coenzymes carry a total of 12 pairs of hydrogens, we also expect 12 molecules of water to be generated, which requires that 6 molecules of oxygen be consumed. Therefore, the overall reaction for oxidative phosphorylation is as follows (Figure 3.20):

$$10\ NADH + 10\ H^+ + 2\ FADH_2 + 34\ ADP + 34\ P_i +$$
$$6\ O_2 \rightarrow 10\ NAD^+ + 2\ FAD + 12\ H_2O + 34\ ATP$$

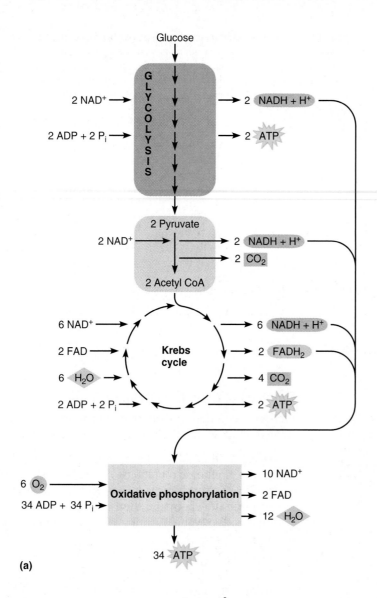

(a)

(b) Net changes

	Glucose	Pyruvate	O_2	CO_2	H_2O	NAD^+	NADH	H^+	FAD	$FADH_2$	ADP	P_i	ATP
Glycolysis:	−1	+2	0	0	0	−2	+2	+2	0	0	−2	−2	+2
Krebs cycle:	0	−2	0	+6	−6	−8	+8	+8	−2	+2	−2	−2	+2
Oxidative phosphorylation:	0	0	−6	0	+12	+10	−10	−10	+2	−2	−34	−34	+34
Total:	−1	0	−6	+6	+6	0	0	0	0	0	−38	−38	+38

(c) Overall reaction: Glucose + 6 O_2 + 38 ADP + 38 P_i ⟶ 6 CO_2 + 6 H_2O + 38 ATP

FIGURE 3.21 **Summary of glucose oxidation. (a)** *A compilation of Figures 3.15a, 3.17a, and 3.20a, showing all three stages of glucose oxidation and their interconnections.* **(b)** *Net changes occurring in glucose oxidation. The first three rows indicate changes occurring during the individual stages; the last row indicates overall changes in the three stages combined.* **(c)** *The overall reaction of glucose oxidation. On average, 38 ATP molecules are produced for every molecule of glucose that is oxidized.*

Note that this accounts for 100% of the oxygen consumed in the oxidation of one molecule of glucose. Perhaps surprisingly, the reaction also shows that more water molecules are produced—12 instead of 6. However, when the 6 water molecules consumed in the Krebs cycle are taken into account, the net result is 6 molecules of water produced, which is in accord with our expectations.

When compared to the quantity of ATP generated by substrate-level phosphorylation in glycolysis and the Krebs cycle, it is clear that oxidative phosphorylation is responsible for most of the ATP that cells produce. Given that substrate-level phosphorylation and oxidative phosphorylation generate 4 and 34 molecules of ATP per glucose, respectively, the latter accounts for nearly 90% of the total of 38 molecules.

Summary of Glucose Oxidation

Now that we have examined the details of all three stages of glucose metabolism, we can "put it all together" to see the "big picture." To help us accomplish this, the whole of glucose metabolism is summarized in Figure 3.21, a compilation of Figures 3.15, 3.17, and 3.20. Note that when all the reactants and products in all three stages are tallied up in Figure 3.21b, the result is that for every molecule of glucose consumed, six molecules of oxygen are also consumed. In addition, six molecules each of carbon dioxide and water are produced. These figures are in exact agreement with the generalized reaction for glucose oxidation ($C_6H_{12}O_6 + 6\ O_2 \rightarrow 6\ CO_2 + 6\ H_2O$). Because 38 molecules of ATP are generated, the overall reaction for the three stages of glucose metabolism is as follows:

$$\text{glucose} + 6\ O_2 + 38\ ADP + 38\ P_i \rightarrow$$
$$6\ CO_2 + 6\ H_2O + 38\ ATP$$

Neither NAD nor FAD appears anywhere in this reaction because these coenzymes are reduced in glycolysis and the Krebs cycle but then oxidized in oxidative phosphorylation and returned to their original forms.

Glucose is not the only sugar that can be broken down by cells and used for energy. Cells can also metabolize other monosaccharides, such as fructose, a component of table sugar; galactose, a component of milk sugar; and another monosaccharide called *mannose* (When It Goes Wrong: Lactose Intolerance, p. 90). To utilize these sugars, cells first convert them to certain intermediates of glycolysis, such as glucose-6-phosphate and fructose-6-phosphate. These intermediates then are simply "fed into" the glycolysis pathway at the appropriate steps and oxidized in the usual manner.

Conversion of Pyruvate to Lactic Acid

Because oxygen is the ultimate electron acceptor in oxidative phosphorylation, it must be supplied to tissues on a continual basis by the blood if glucose oxidation is to go to completion. The rate at which oxygen must be supplied depends on how fast the tissues are consuming it; in other words, it depends on the tissues' *metabolic demand*. If oxygen can be delivered to tissues at a rate high enough to match demand, oxygen will always be available to accept electrons, and glucose oxidation will proceed to completion. In the event that oxygen delivery is no longer able to keep up with demand (due to a reduction in blood flow to a particular organ or tissue, for instance), oxygen concentration in the affected tissue will fall to very low levels, perhaps even approaching zero.

Under these conditions, fewer oxygen molecules will be available to accept electrons as they reach the end of the electron transport chain, so it becomes difficult to "offload" them from the chain's electron carriers. As a result, the flow of electrons through the chain slows, as does the rate of ATP synthesis by oxidative phosphorylation. Under these conditions, an electron "traffic jam" develops in the electron transport chain, which eventually decreases the number of FMN molecules available to accept electrons from NADH. When this happens, coenzyme molecules become "trapped" in their reduced forms, and levels of the coenzyme NAD$^+$ decrease.

Any drop in NAD$^+$ levels in a cell is a potential threat to all ATP production because a supply of NAD$^+$ is necessary for proper operation of both the glycolysis pathway and the Krebs cycle. Of particular interest here is step 6 of glycolysis, which occurs before any of the steps in which ATP is produced by substrate-level phosphorylation. This means that if oxygen becomes depleted and the supply of NAD$^+$ is decreased, not only does oxidative phosphorylation slow, but substrate-level phosphorylation slows as well. Thus, a reduction in oxygen availability has the potential to slow (or perhaps even stop) all ATP-producing steps, which would be disastrous for a cell, for its supply of ATP would eventually become exhausted.

Fortunately, this disaster can be averted. Most cells contain an enzyme called *lactate dehydrogenase*, which can convert pyruvate (the end-product of glucose breakdown in glycolysis) to *lactate* according to the following reaction:

$$\text{pyruvate} + NADH + H^+ \xrightarrow{\text{lactate dehydrogenase}} \text{lactate} + NAD^+$$

The significance of this reaction is as follows: In the event that NADH cannot unload its electrons to the electron transport chain due to a limitation in the availability of oxygen, this reaction provides an alternate pathway that allows electrons to be unloaded in a reaction that generates free NAD$^+$, which can then pick up electrons in step 6 of glycolysis. Therefore, when the electron transport chain is not available to accept electrons, NADH and NAD$^+$ can still shuttle back and forth between the conversion of pyruvate to lactate and step 6 of glycolysis

Most nutrient molecules in our diet must be broken down by enzymes in the digestive system before they can be absorbed. This is certainly true for relatively large molecules such as proteins, but also for some smaller molecules as well. For instance, disaccharides such as sucrose (table sugar) and lactose (milk sugar) must be broken down to monosaccharides, which are then absorbed by the small intestine. The enzymes that perform this task are bound to the plasma membranes of intestinal epithelial cells and are thus positioned to come into contact with nutrients in the lumen of the intestine. After a disaccharide is reduced to monosaccharides, these monosaccharides are absorbed and transported into the bloodstream.

In the epithelium of the small intestine, the lactose present in milk is broken down by the membrane-bound enzyme *lactase* to glucose and galactose, which are then absorbed (see the figure to the right). Normally, the processes of enzymatic breakdown and absorption are so efficient that virtually 100% of the lactose molecules are cleared from the lumen in this manner. In certain individuals with a condition known as *lactose intolerance*,

however, this is not the case. Lactose intolerance develops when cells begin to stop producing lactase, such that the enzyme concentration falls. When this occurs, a person begins to have trouble digesting lactose, with the result that any lactose that may be present in the intestine remains there because it cannot be absorbed. The problem with this is that any lactose remaining in the lumen of the intestine serves as a nutrient for bacteria that normally live there, stimulating bacterial growth and the resultant production of gas and other waste products. These substances irritate the lining of the intestine and cause bloating, discomfort, and diarrhea. Lactose-intolerant individuals

experience these symptoms after drinking milk or eating other dairy products and therefore eventually learn to avoid them.

Lactose intolerance has a genetic basis and is more prevalent among certain peoples, most notably those of African and Asian descent. In most instances it appears in childhood after about age 6, when milk drinking declines. It has been estimated that this condition affects over 50% of adults worldwide. Although there is no cure, new commercial products that allow afflicted individuals to indulge their taste for dairy products include pills containing enzymes that can break down lactose, and milk with reduced levels of lactose.

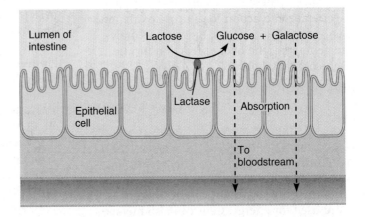

(Figure 3.22). This reaction ensures a steady supply of NAD^+, which enables step 6 to run even when oxygen availability is limited. More importantly, the fact that step 6 can run means that the remaining steps in glycolysis can also run, including steps 7 and 10, which synthesize ATP.

The ability to synthesize ATP in the manner just described allows muscles and other tissues to continue working even when the availability of oxygen is low, but

it is important to realize that this does not mean the body's tissues can live indefinitely in the absence of oxygen. For one thing, only two ATP molecules are produced in glycolysis for every molecule of glucose consumed, which is only about 5 percent of the ATP that is normally produced as a result of complete glucose oxidation. Unless a cell's demand for ATP is very low, it is unlikely that it will be able to supply enough ATP through glycolysis alone. (Interestingly, red blood cells must obtain all their

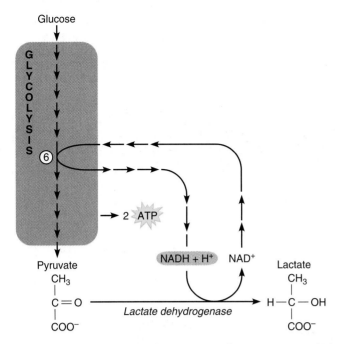

FIGURE 3.22 Conversion of pyruvate to lactate. *NADH, which is generated in step 6 of glycolysis, can be converted back to NAD⁺ in a reaction that generates lactate from pyruvate. This reaction assures a steady supply of NAD⁺, which allows for the continued operation of step 6 and the continued production of ATP even under conditions of reduced oxygen availability.*

Quick Test 3.5

1. Name the three stages of glucose oxidation in the order in which they occur. In which stage(s) is (are) oxygen consumed? In which stage(s) is (are) carbon dioxide produced? Which stage normally produces the most ATP?

2. Distinguish between substrate-level phosphorylation and oxidative phosphorylation. In which stage(s) does (do) substrate-level phosphorylation occur? In which stage(s) does (do) oxidative phosphorylation occur?

3. What conditions are likely to cause an increase in the rate of lactate production by cells? Why?

ENERGY STORAGE AND USE: METABOLISM OF CARBOHYDRATES, FATS, AND PROTEINS

So far, our examination of energy metabolism has been restricted to one energy source—glucose. Nevertheless, the metabolism of fats and proteins (as well as that of carbohydrates other than glucose) also makes a significant contribution to our energy needs. Although a detailed description of fat and protein metabolism is beyond the scope of this book, a general understanding of the connection between these metabolic pathways and those we have just studied is crucial. Eventually we will need this knowledge to understand the role of the endocrine system in regulating energy metabolism, a topic covered in Chapter 20.

The body is able to draw upon fats and proteins as alternate sources of energy whenever glucose is scarce and needs to be conserved. When glucose supplies are limited, the body can break down fats and proteins into smaller molecules, a catabolic process that releases energy. When the supply of glucose is plentiful, the body can run these reactions in reverse to synthesize fats, proteins, and a large glucose storage molecule called *glycogen*. In this manner the body stores energy for future use.

Figure 3.23 shows the relationships among pathways for glucose, fat, and protein metabolism. The three stages of glucose oxidation—glycolysis, the Krebs cycle, and oxidative phosphorylation—are the centerpiece of the figure. Note that the depiction of these three stages is essentially the same as that shown in Figure 3.21, except that only the major products of each stage are shown. In addition, NADH and FADH₂ are simply given the abbreviation *coenzyme-2H*. (For simplicity, reduced coenzymes generated in glycolysis or in the linking step are not shown.) Double arrows indicate reactions that can go in both directions; single arrows denote reactions that essentially go one way only. These reactions are described as being *irreversible*, whereas the two-way reactions are said to be reversible.

ATP from glycolysis because they lack mitochondria.) Furthermore, because the conversion of pyruvate to lactate is a "dead end" reaction, lactate tends to accumulate in cells when produced at a rapid rate, which can cause acidification of the intracellular fluid. Lactate can also leak into extracellular fluid and "spill over" into the bloodstream, causing acidification of the blood. Acidification of tissues will eventually begin to interfere with proper cellular function and therefore cannot continue indefinitely.

If lactate is potentially harmful to cells, how do they get rid of it? When oxygen availability returns to normal, pyruvate begins to proceed to the Krebs cycle as it usually does, and the concentration of pyruvate in cells decreases. As a result of mass action, the lactate dehydrogenase reaction begins to run in reverse, so that lactate is converted back to pyruvate and NAD⁺ is converted to NADH:

$$\text{pyruvate} + \text{NADH} + \text{H}^+ \xleftarrow{\text{lactate dehydrogenase}} \text{lactate} + \text{NAD}^+$$

The NADH that is generated in this reaction proceeds to oxidative phosphorylation, and the pyruvate proceeds to the linking step and the Krebs cycle. Thus, the conversion of pyruvate to lactate represents but a momentary "side trip" from the normal pathway of glucose oxidation, a diversion that ends when lactate is converted back to pyruvate.

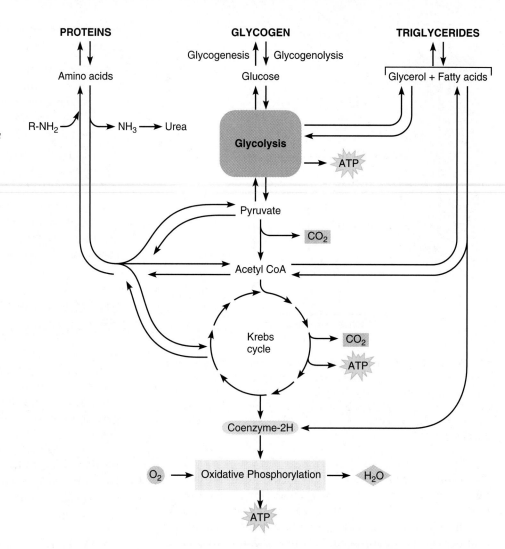

FIGURE 3.23 Metabolic pathways involved in protein, glycogen, and fat metabolism. *The three stages of glucose metabolism are shown in the center. One-way arrows indicate irreversible reactions; double arrows indicate reversible reactions. Arrows leading into or out of this central pathway indicate points of entry or exit for the breakdown products of protein or triglyceride metabolism.*

One-way reactions are not "irreversible" in the absolute sense, but only in the sense that they require large quantities of energy in order to run in reverse. In many cases, however, one-way reactions do run in reverse under certain circumstances. When this occurs, a reaction does not follow exactly the same course while running in reverse as it does when running in the forward direction, and it is catalyzed by a different set of enzymes. For this reason, the reverse reaction is sometimes referred to as a *bypass reaction* because it follows a "route" that bypasses the enzymes that would normally catalyze the reaction when it is running in the forward direction.

In contrast, reversible reactions have smaller energy changes and can be made to go in reverse without large amounts of energy. These reactions readily reverse direction in response to changes in reactant or product concentrations in accordance with the law of mass action. Furthermore, these reactions are catalyzed by the same enzymes, whether they are going forward or in reverse.

Whenever a metabolic pathway consists of an irreversible reaction in combination with a bypass reaction, the reactions are generally regulated so that they do not run simultaneously, which would be counterproductive. When the irreversible (forward) reaction is running,

enzymes catalyzing the bypass reaction are turned off; when the bypass reaction is running, enzymes catalyzing the forward reaction are turned off. In this manner, cells can force the reaction to go in the direction that best suits the body's needs. Thus, when cells need to break down fats for energy, for example, they can turn on the enzymes that perform this function. When they need to synthesize fats to store energy, they can turn these enzymes off and turn on other enzymes that catalyze the necessary bypass reactions. As we will see in Chapter 20, these metabolic adjustments are largely coordinated by hormonal signals that depend on our eating and fasting patterns.

Glycogen Metabolism

Glycogen is a branched-chain molecule found in animal cells; it is similar to starch, a carbohydrate found only in plants. Glycogen is composed of glucose molecules joined together to form a polymer (see Chapter 2). When glucose is in abundant supply, as after a meal, most of the glucose molecules are not metabolized immediately; they are stored as glycogen. This process, which is sometimes called **glycogenesis,** is represented in Figure 3.23 by the vertical arrow running from glucose to glycogen.

PROTEINS

Amino acids

R-NH$_2$ → NH$_3$ → Urea

GLYCOGEN

Glycogenesis ↑↓ Glycogenolysis

Glucose

TRIGLYCERIDES

Glycerol + Fatty acids

Glycolysis → ATP

Pyruvate → Lactate

CO$_2$

Acetyl CoA

Krebs cycle → CO$_2$ → ATP

Coenzyme-2H

O$_2$ → Oxidative Phosphorylation → H$_2$O

ATP

FIGURE 3.24 Metabolic pathways involved in gluconeogenesis. *Pathways relevant to gluconeogenesis are highlighted.*

Tissues differ in their capacity to synthesize and store glycogen; the liver and skeletal muscle are particularly adept at this process. During fasting or when glucose is being used up quickly, the glucose supply is replenished by the breakdown of glycogen into individual glucose molecules, a process known as **glycogenolysis.** The fate of glucose released during glycogenolysis depends on the location where it occurs. When glycogen is broken down in skeletal muscle cells, for example, the glucose is used by those cells. In contrast, when glycogen is broken down in liver cells, the liver releases most of the glucose into the bloodstream for uptake and utilization by other tissues.

Exercise Link

Recall that Bill and Jane "carbo-loaded"—ate lots of starchy and sugary foods—prior to running the marathon. This practice is common among endurance athletes because it increases the glycogen content of skeletal muscles. Obviously, these muscles will have an especially high need for glucose during the race.

Gluconeogenesis: Formation of New Glucose

Although fats and proteins can be used by most tissues as energy substitutes for glucose, an adequate supply of glucose must be maintained in the bloodstream at all times because the nervous system, particularly brain tissue, has a lower capacity than other tissues for switching to alternate energy sources. The nervous system requires an uninterrupted glucose supply; otherwise, loss of consciousness and possibly death may result. Nervous tissue is never entirely free of this glucose requirement, although a limited capacity for utilizing other energy sources does exist.

Under normal conditions, the body's glycogen reserves are sufficient to provide energy for a few hours. If a person fasts for a longer period of time, glycogen stores become depleted, which could potentially cause a dangerous fall in blood glucose concentrations. Fortunately, new glucose molecules can be synthesized from noncarbohydrate precursors via a process called **gluconeogenesis** (Figure 3.24), which is carried out mainly by the liver.

Glucose can be made via gluconeogenesis from three sources: (1) glycerol, which is produced by breaking down triglycerides; (2) lactate; and (3) amino acids, which can be produced in the breakdown of proteins. In gluconeogenesis, glycerol is first converted to glycerol phosphate, an intermediate of the glycolysis pathway,

FIGURE 3.25 Metabolic pathways involved in lipolysis, the breakdown of fats (triglycerides) for energy. *Relevant pathways are highlighted. Note that fatty acid breakdown leads to the production of ketones, which are generated from acetyl CoA through a series of reversible reactions.*

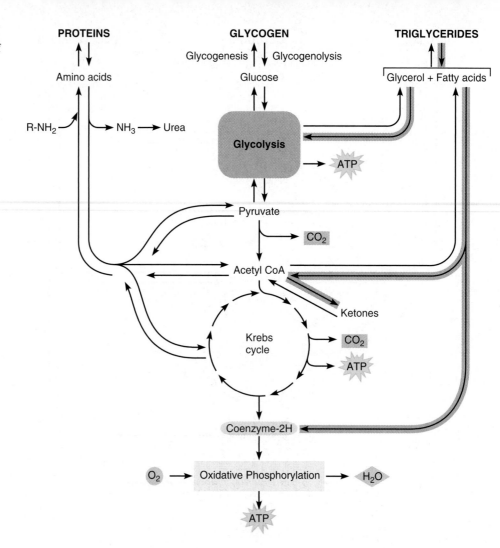

and lactate is first converted to pyruvate, the normal end-product of glycolysis (see Figure 3.24). These molecules then proceed through the glycolysis pathway *in reverse,* such that glucose molecules are generated.

Certain amino acids can be converted to glucose after first being converted to pyruvate, which then enters the glycolysis pathway in reverse. Certain other amino acids, however, are converted to glucose via a more indirect route, which is indicated by the dashed line in Figure 3.24. These amino acids are first converted into a Krebs cycle intermediate (oxaloacetate), which is then converted into a glycolytic intermediate (phosphoenolpyruvate); this molecule then participates in the glycolysis pathway in the reverse direction. Note, however, that some amino acids cannot be converted to glucose; even though these amino acids can be converted to acetyl CoA, there is no pathway permitting the synthesis of glucose from acetyl CoA. (Acetyl CoA can be converted to oxaloacetate in the Krebs cycle, but in the process an equal amount of oxaloacetate is consumed due to the cyclic nature of this series of reactions. Consequently, acetyl CoA cannot be converted to glucose via the dashed line pathway shown in Figure 3.24.) Fatty acids, which like glycerol are generated from triglyceride breakdown, cannot be converted to glucose either because they too are converted to acetyl CoA.

Fat Metabolism

Like glycogen, fats (lipids) can be broken down and used for energy when other energy supplies are running low. Fats can also be synthesized to store energy when they are in abundant supply in the diet. *Adipose tissue*, which contains fat cells, is the primary storage depot for fats.

Lipids are stored predominantly in the form of *triglycerides*, which consist of three fatty acids joined to a glycerol backbone. The first stage of fat breakdown **(lipolysis)** is the separation of fatty acids from the glycerol molecule (Figure 3.25). Glycerol then enters the glycolysis pathway (as dihydroxyacetone phosphate) and proceeds from there to the Krebs cycle and oxidative phosphorylation. Fatty acids are converted to acetyl CoA, which also enters the Krebs cycle and oxidative phosphorylation. Because fatty acids obtained in the diet are usually 12–18 carbons in length, several molecules of acetyl CoA can be made from a single fatty acid molecule. (Recall that glucose yields only two acetyl CoAs.) Thus, many more ATPs can be made from a molecule of fatty acid than from a glucose molecule. The catabolism of fatty acids also generates a large number of reduced coenzyme molecules, which go to the electron transport chain and eventually yield even more ATPs by oxidative phosphorylation. Because a gram of fat yields more energy

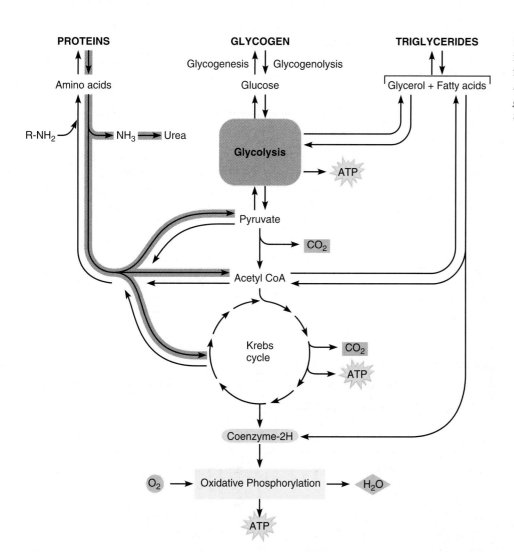

than a gram of glucose or other carbohydrates, fats are known as high-calorie foods.

Note that when fats are broken down for energy, compounds called *ketones* are generated as a by-product (see Figure 3.25). These compounds are synthesized from acetyl CoA in a set of reversible reactions and are synthesized at higher rates when the concentration of acetyl CoA rises. Because many acetyl CoA molecules can be generated from a single fatty acid, the breakdown of fats tends to generate large quantities of ketones, which is important because ketones can be utilized by the nervous system as a partial alternative to glucose. This helps conserve the body's supply of glucose if energy intake becomes limited such that the body must turn to stored fats for energy.

Note as well that the pathways from glycerol to glycolysis, and from fatty acids to acetyl CoA, are double-arrowed in Figure 3.25, which means that it is possible to synthesize fats from other nutrients, a process called **lipogenesis.** This explains why people can become obese from eating too much of nonlipid foods such as carbohydrates or proteins.

Protein Metabolism

In the metabolic breakdown of proteins for energy (Figure 3.26), proteins are first broken down to amino acids, a process known as **proteolysis.** The amino acids are then *deaminated;* that is, an amino group ($—NH_2$) is removed. This deamination process produces ammonia (NH_3), which is toxic. Fortunately, ammonia is converted to another compound that is carried by the bloodstream to the liver, where it is converted to *urea,* which is relatively innocuous and is eventually eliminated by the kidneys.

Exercise Link

In the later stages of the marathon, Bill and Jane's bodies needed to rely more heavily on deaminated proteins as sources of energy because their bodies' glucose stores were almost completely consumed. As a result, their sweat (which is produced from plasma) began to smell of ammonia.

After the amino acid is deaminated, the remainder of the molecule is either converted to pyruvate or acetyl CoA (which then goes into the Krebs cycle), or it is converted to an intermediate that can enter the Krebs cycle directly. Note in Figure 3.26 that the pathways from amino acids to pyruvate, acetyl CoA, and the Krebs cycle are double-arrowed, meaning that nonprotein nutrient molecules such as carbohydrates or fats can serve as raw materials for the synthesis of amino acids, which can then be used to make proteins. (As the figure shows, this requires the addition of an amino group, which is obtained from another molecule, indicated as R—NH_2.) Consequently, certain amino acids can be present in cells even when those amino acids are not present in the diet. Other amino acids, however, cannot be synthesized in the body and must be obtained from the diet; these amino acids are referred to as *essential* amino acids. An **essential** **nutrient** is any biomolecule necessary for proper body function that cannot be synthesized in cells and therefore must be obtained from dietary sources.

Quick Test 3.6

1. Define the following terms: *glycogen, glycogenesis, glycogenolysis, lipogenesis, lipolysis, proteolysis.*

2. What is gluconeogenesis, and why is it important? What three substances can serve as raw materials for this process?

3. What happens to the ammonia that is produced when amino acids are broken down for energy?

4. What is the distinguishing characteristic of an essential nutrient?

CHAPTER SUMMARY

Types of Metabolic Reactions, p. 63

The sum total of chemical reactions occurring in the body is metabolism; reactions of energy metabolism are specifically involved in energy exchange. Catabolic reactions generate smaller products from larger reactants; anabolic reactions generate larger products from smaller reactants. Three common types of metabolic reactions are: (a) hydrolysis and condensation reactions, (b) phosphorylation and dephosphorylation reactions, and (c) oxidation-reduction reactions.

Metabolic Reactions and Energy, p. 65

Metabolic reactions enable cells to transform raw materials from the environment into structural and functional components and also provide cells with energy. The energy change of a reaction determines its direction. Catabolic reactions are generally energy-releasing and proceed spontaneously; anabolic reactions require energy to proceed. A reaction can be made to go in either direction by changing reactant or product concentrations according to the law of mass action. The rate of a reaction is limited by its activation energy, reflecting the fact that molecules must go through a high-energy transition state before they can react.

Reaction Rates, p. 70

Factors affecting the rate of a reaction include reactant and product concentrations, temperature, and the height of the activation energy barrier. Metabolic reactions are catalyzed by special proteins called enzymes, which act on specific substrates that fit the enzymes' active site. Many enzymes require nonprotein cofactors for proper function. Important cofactors are NAD^+ and FAD, which carry electrons between reactions. Rates of enzyme-catalyzed reactions are influenced by the enzyme's catalytic rate, enzyme-substrate affinity, enzyme concentration, and substrate concentration. Certain enzymes possess a regulatory site specific for a certain modulator molecule, whose binding alters the enzyme's activity. In allosteric regulation, binding of the modulator is reversible. In covalent regulation, an enzyme's activity is altered by the covalent binding of a chemical group (often a phosphate group) to a specific site on the enzyme. Metabolic pathways are often regulated by feedback inhibition, in which an enzyme is allosterically inhibited by the product of a reaction occurring downstream. Sometimes enzymes are stimulated by intermediates appearing upstream (feedforward activation).

Glucose Oxidation: The Central Reaction of Energy Metabolism, p. 77

Cells obtain much of their energy from the oxidation of glucose: $C_6H_{12}O_6 + 6O_2 \rightarrow 6CO_2 + 6H_2O$. Cells harness energy released in this and other energy-releasing reactions to synthesize adenosine triphosphate (ATP). Later on, cells use energy released from the breakdown (hydrolysis) of ATP to perform useful work.

Stages of Glucose Oxidation: Glycolysis, the Krebs Cycle, and Oxidative Phosphorylation, p. 81

Glucose oxidation occurs in three stages: glycolysis (in the cytosol), the Krebs cycle (in the mitochondrial matrix), and oxidative phosphorylation (in the inner mitochondrial membrane). Oxidative phosphorylation involves two processes that occur simultaneously: (a) the movement of electrons through the electron transport chain, and (b) chemiosmotic coupling. For each glucose molecule, 4 ATPs are synthesized by substrate-level phosphorylation in glycolysis and the Krebs cycle. 6 CO_2, 10 NADH, and 2 $FADH_2$ are also produced. Later on, NADH and $FADH_2$ release their electrons (hydrogens) to the electron transport chain, and these ultimately react with oxygen to form water. Energy released in this process is used to make ATP by oxidative phosphorylation, which accounts for most of the ATPs that are generated from the complete oxidation of one glucose molecule (34 of 38, on average).

Energy Storage and Use: Metabolism of Carbohydrates, Fats, and Proteins, p. 91

The body is able to draw upon stored fats and proteins as alternate sources of energy when glucose is scarce. In this process, fats and proteins are broken down to smaller molecules (by lipolysis and proteolysis, respectively), which enter the pathway for glucose oxidation at various points. The body can also store energy by synthesizing fats, which are stored primarily in adipose tissue, and proteins. In certain tissues, energy can be stored by converting glucose to glycogen (by glycogenesis), which can be later broken down to yield glucose (glycogenolysis). Glucose can also be synthesized from noncarbohydrate precursors (gluconeogenesis), insuring a steady supply of glucose in the bloodstream, which is necessary for proper nervous system function.

EXERCISES

Multiple-Choice Questions

1. When glucose is oxidized in cells, oxygen reacts with
   a) carbon to form CO_2.
   b) hydrogen to form H_2O.
   c) components of the electron transport chain.
   d) inorganic phosphate to form ATP.

2. Which of the following illustrates *substrate-level* phosphorylation? (P_i = inorganic phosphate, E = enzyme, S = substrate)

   a) $ATP \xrightarrow{E} ADP + P_i$
   b) $ADP + P_i \xrightarrow{E} ATP$
   c) $S-P_i \xrightarrow{E} S + P_i$
   d) $S-P_i + ADP \xrightarrow{E} S + ATP$

3. The enzyme pyruvate dehydrogenase, which converts pyruvate to acetyl CoA, can be activated or inactivated by phosphorylation or dephosphorylation. This is an example of
   a) allosteric regulation.
   b) substrate-level phosphorylation.
   c) saturation.
   d) covalent regulation.

4. The following reactions occur in conjunction with step 5 of the Krebs cycle:

   $$GDP + P_i \rightarrow GTP$$
   $$GTP + ADP \rightarrow GDP + ATP$$

   The net reaction in this process is
   a) $P_i + GTP \rightarrow GDP$
   b) $P_i + GDP \rightarrow GTP$
   c) $GTP + ADP \rightarrow GDP + ATP$
   d) $ADP + P_i \rightarrow ATP$

Questions 5–8: As you know, fatty acids can be oxidized to provide energy for making ATP. This process begins with a set of reactions known as the fatty acid oxidation cycle. Before entering the cycle, a fatty acid reacts with coenzyme A to form a molecule called fatty acyl CoA. This molecule then enters the cycle and continues to go through it until all its carbons have been eliminated, as shown above.

5. From your examination of this diagram, how many turns of the cycle will be necessary for the complete oxidation of a fatty acid that is ten carbons in length?
   a) 1   b) 2   c) 5   d) 10

6. In the fatty acid oxidation cycle, a number of reduced coenzyme molecules are generated. Assuming oxygen is readily available, these coenzymes will release their electrons to the

electron transport chain, which will result in the production of a certain number of ATP molecules. How many ATPs will be produced in this manner from each turn of the cycle?

a) 2
b) 3
c) 4
d) 5

7. From your understanding of energy metabolism, you can surmise that
   a) some of the products generated in this cycle will enter the glycolysis pathway and be oxidized further.
   b) some of the products generated in this cycle will enter the Krebs cycle and be oxidized further.
   c) products generated in this cycle will enter the electron transport chain only.
   d) a fatty acid is completely oxidized in this cycle, and that no additional steps are necessary.

8. Assume that oxygen is not available and oxidative phosphorylation stops completely. Under these conditions, fatty acid oxidation
   a) should slow down due to a decrease in the concentrations of FAD and NAD^+.
   b) should speed up due to an increase in the concentrations of FAD and NAD^+.
   c) should speed up due to an increase in the concentrations of $FADH_2$ and NADH.
   d) should not be affected at all because oxygen is not involved in the fatty acid oxidation cycle.

9. Activity of the enzyme phosphofructokinase, which catalyzes step 3 of glycolysis, is affected by several substances that act as allosteric modulators. Effects of some of these modulators are shown in the following graphs:

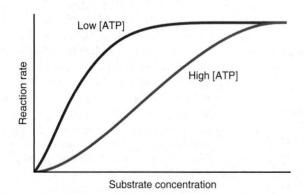

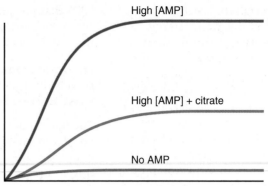

Which of these substances apparently acts to decrease the affinity of the enzyme?

a) ATP
b) AMP (adenosine monophosphate)
c) citrate

10. In the following reaction, NAD^+ receives electrons from the substance AH_2:

$$NAD^+ + AH_2 \xrightarrow{\text{E}}$$
$$NADH + \quad H^+ + A$$

where E is the enzyme that catalyzes the reaction. In this reaction, which substance undergoes oxidation?

a) NAD^+
b) NADH
c) AH_2
d) A

11. Which of the following should decrease the rate of the reaction in Question 10?
    a) an increase in the affinity of the enzyme
    b) an increase in the enzyme concentration
    c) a decrease in the concentration of NADH
    d) a decrease in the concentration of NAD^+

12. Assume that the energy change of the reaction in Question 10 is negative. From this, you know that
    a) the reaction will not go forward unless energy is put into it.
    b) the reaction will go forward spontaneously and will release energy.
    c) the energy change will become more negative when the enzyme concentration is increased.
    d) the products have more energy than the reactants.

13. Under normal circumstances, the energy change of the reaction $NAD^+ + 2H \rightarrow NADH + H^+$ is positive. This means that
    a) energy is released when NADH undergoes oxidation.
    b) energy is released when NAD^+ undergoes reduction.
    c) the reaction requires more energy when the concentration of NAD^+ is higher.
    d) the reaction releases more energy when the concentration of NADH is higher.

14. Which of the following is carried out by enzymes in the cytosol?
    a) glycolysis
    b) the Krebs cycle
    c) oxidative phosphorylation
    d) chemiosmotic coupling

15. In which of the following is carbon dioxide generated?
    a) glycolysis
    b) the Krebs cycle
    c) oxidative phosphorylation
    d) all of the above

Objective Questions

1. The removal of hydrogen atoms from a molecule is an example of (oxidation/reduction).

2. The forward rate of a reaction can be increased by increasing the concentrations of the (reactants/products).

3. For a reaction to proceed spontaneously in the forward direction, the energy change must be (positive/negative).

4. Liver and muscle cells are able to store glucose in the form of a branched-chain molecule called _____.

5. Allosteric regulation involves the attachment of a modulator molecule to an enzyme by means of a covalent bond. (true/false)

6. Glycogenolysis is an example of a catabolic reaction. (true/false)

7. The conversion of noncarbohydrate precursors into glucose is called _____.

8. When a coenzyme molecule such as NAD^+ or FAD is reduced, it (picks up/releases) hydrogens.

9. The mechanism that couples energy release in electron transport to ATP synthesis is called (substrate-level phosphorylation/chemiosmotic coupling).

10. When ATP is synthesized using energy released in glucose oxidation, most of the ATP is synthesized by oxidative phosphorylation. (true/false)

11. Enzymes speed up the rates of reactions by (raising/lowering) the activation energy barrier.

12. When a substrate molecule binds to an enzyme, it binds to a particular location known as the (active/regulatory) site.

13. Unless a substance is an essential nutrient, it is not necessary for proper cellular function. (true/false)

14. An enzyme catalyzes a reaction at the (lowest/highest) possible rate when it is 100% saturated.

15. When an enzyme is 100% saturated, the rate of the reaction is determined by the enzyme's catalytic rate and its (affinity/concentration).

Essay Questions

1. Explain the significance of the energy change in a reaction (ΔE) with regard to (a) the direction of the reaction and (b) whether the reaction releases energy or requires it.

2. Compare and contrast the mechanisms of substrate-level phosphorylation and oxidative phosphorylation, and describe the role of each in glucose oxidation.

3. Explain how the conversion of pyruvate to lactate enables ATP production to continue even when oxygen is not readily available.

4. Compare and contrast allosteric regulation and covalent regulation of enzyme activity.

5. Explain the concept of activation energy and how it influences reaction rates. Include a discussion of how reaction rates are affected by temperature and reactant and product concentrations.

Find the answers to all of these exercises, and additional study tools, at the Physiology Place (www.physiologyplace.com)

Cell Membrane Transport

OBJECTIVES

- Explain the role of chemical, electrical, and electrochemical driving forces in the passive transport of substances across a membrane, and distinguish between passive transport and active transport.

- Identify the three general factors that influence the rate at which a substance can be passively transported across a membrane, and identify the two general factors that influence the rate of active transport.

- Identify four factors affecting the permeability of membranes to molecules that cross by simple diffusion, and explain how and why each affects permeability.

- Identify two factors affecting the permeability of membranes to molecules that cross by facilitated diffusion; compare and contrast the properties of carriers and channels.

- Explain the distinction between primary and secondary active transport, and give examples of each.

- Explain how a difference in solute concentration across a membrane can cause the movement of water, and explain the distinction between the osmolarity and tonicity of a solution.

- Explain in general terms how the polarity of epithelial cells enables them to absorb or secrete materials.

CHAPTER OUTLINE

Factors Affecting the Direction of Transport   101

Factors Affecting the Rate of Transport   108

Simple Diffusion: Passive Transport Through the Lipid Bilayer   111

Facilitated Diffusion: Passive Transport Through Membrane Proteins   113

Active Transport   115

Osmosis: Passive Transport of Water Across Membranes   119

Epithelial Transport: Movement of Molecules Across Two Membranes   124

Above: Scanning electron micrograph of phogocytosis

TABLE 4.1 MILLIMOLAR CONCENTRATIONS OF SELECTED SOLUTES IN INTRACELLULAR FLUID (ICF) AND EXTRACELLULAR FLUID (ECF)		
SOLUTE	**ICF (mM)**	**ECF (mM)**
K^+	140.0	4.0
Na^+	15.0	145.0
Mg^{2+}	0.8	1.5
Ca^{2+}	<0.001*	1.8
Cl^-	4.0	115.0
HCO_3^-	10.0	25.0
P_i	40.0	2.0
Amino acids	8.0	2.0
Glucose	1.0	5.6
ATP	4.0	0.0
Protein	4.0	0.2

*Refers to calcium ions free in the cytoplasm. A significant quantity of intracellular calcium is sequestered in membrane-bounded organelles and/or bound to proteins.

In order to live, the cells in your body need to exchange materials with their immediate environment, the fluid that surrounds them. This occurs, for example, when cells obtain oxygen and dispose of carbon dioxide. Likewise, the body needs to exchange materials with *its* environment, as when you absorb oxygen and eliminate carbon dioxide through your lungs. In either case, molecules must move across cell membranes. We saw in Chapter 1 that a primary function of cell membranes is to prevent intracellular and extracellular fluid from mixing, in the process maintaining necessary differences in composition between the two compartments (Table 4.1). In this function, membranes act as barriers that restrict the passage of molecules between intracellular and extracellular fluid. This function contrasts sharply with another important function of cell membranes, which is to *permit* exchange of certain materials between cells and their environment. How do membranes permit certain molecules to move into or out of cells while restricting the passage of others? How does a molecule move through a membrane? Why do certain molecules enter cells, whereas others leave? The answers to such questions are the subject of this chapter.

We will spend much time in this chapter focusing on basic cellular and molecular theories, for these ideas are crucial to our current understanding of organ system physiology. As you progress in your study of physiology, you will find yourself returning to the subject of membrane transport time and time again.

A typical cell is able to transport a wide range of substances across its plasma membrane, from gases (oxygen and carbon dioxide) to small inorganic ions (such as sodium and potassium) to larger organic molecules (such as glucose, amino acids, fatty acids, and vitamins). To move this great variety of molecules across its plasma membrane, each cell must utilize a number of different transport mechanisms, and different cell types may possess different mechanisms. A relatively few molecules, such as oxygen and carbon dioxide, are able to cross membranes by penetrating through the lipid bilayer, moving into or out of the cell simply as a result of their own thermal motion. This process, known as **simple diffusion,** is described in greater detail later in this chapter. Most molecules, however, can enter or leave a cell only

by moving through specialized transport proteins imbedded in the plasma membrane. This type of transport is referred to as **mediated transport.**

As we progress through this chapter, we will learn about the characteristics of different mechanisms of membrane transport. However, we must first understand the principles that govern all transport processes. Accordingly, in the next two sections we examine the various factors that influence the *direction* and *rate* of transport.

FACTORS AFFECTING THE DIRECTION OF TRANSPORT

In Chapter 3, we saw that the reactant and product molecules in a metabolic reaction have different energies. We also saw that the energy change of a reaction, the difference between the reactant and product energies, determines the direction of the reaction, and whether it proceeds spontaneously or requires energy. In the following section we see that similar principles govern the transport of molecules across membranes.

Passive Transport versus Active Transport

When oxygen enters a cell or carbon dioxide leaves, molecules move across the plasma membrane spontaneously. In other words, the cell need not expend energy to make this happen. In contrast, energy is expended when glucose molecules are transported into epithelial cells in the

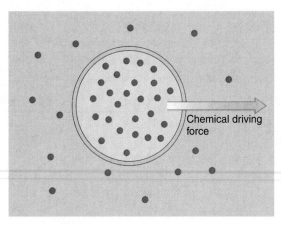

(a)

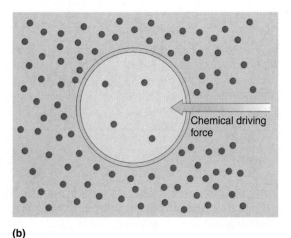

(b)

FIGURE 4.1 Chemical driving forces. (a) *When the concentration of molecules (dots) is higher inside a cell than in extracellular fluid, the direction of the chemical driving force (arrow) is outward.* **(b)** *When the concentration of molecules is higher in extracellular fluid, the direction of the chemical driving force is inward. In both cases, molecules will move passively in the direction of the driving force, or down the concentration gradient.*

intestine. Transport of molecules across a membrane is called **active transport** if it requires energy, and **passive transport** if it does not. Simple diffusion is one form of passive transport, but certain forms of mediated transport are also passive. In contrast, active transport is always mediated by transport proteins referred to as **pumps.**

The distinction between passive transport and active transport is based on the fact that molecules inside and outside cells have different energies, usually because of differences in the molecules' concentrations on either side of the membrane. (Recall that the energy of a molecule increases as its concentration increases.) Thus, when a molecule is transported across a cell membrane, its energy changes. For example, the concentration of glucose is usually about 1 mM in intracellular fluid but nearly 6 mM in extracellular fluid (see Table 4.1), and because of this glucose molecules possess more energy when they are outside a cell than when they are inside a cell.

When molecules are transported across a membrane, the direction of transport is governed by the same tendency that governs the direction of metabolic reactions—namely, the tendency of systems to go spontaneously from higher energy to lower energy. If a difference in energy exists between molecules on either side of a membrane, the molecules will tend to move spontaneously from the side of higher energy to the side of lower energy. Thus, if glucose is allowed to move spontaneously across the membrane of a typical cell, it will move from higher to lower concentration (higher to lower energy) and therefore will move into the cell. Such transport is passive because it requires no energy. In contrast, transport of glucose in the opposite direction (from lower to higher concentration) would be active because it does not occur spontaneously and therefore requires energy.

Driving Forces Acting on Molecules

Any difference in energy existing across a membrane acts as a *driving force* that tends to push molecules in one direction or another. The direction of this force is always from higher to lower energy, which indicates the direction in which molecules will go if they are allowed to move spontaneously. These driving forces can arise as a result of concentration differences or other factors that affect molecular energies. In the following sections we see that molecules are generally influenced by three types of driving forces: chemical, electrical, and electrochemical driving forces.

Chemical Driving Forces

We have seen that when a substance is present in different concentrations on either side of a membrane, a **concentration gradient** is said to exist across the membrane (Chapter 3). We consider the term *concentration gradient* to be synonymous with the difference in concentration and give it the symbol ΔC. (Strictly speaking, the term *concentration gradient* is used in reference to any difference in concentration between one location and another—not just differences across membranes—and is defined as the rate at which the concentration changes with distance.) When molecules are moving from higher to lower concentration, we can say that they are moving *down a concentration gradient;* movement in the opposite direction is *up a concentration gradient.*

Because molecules will move down a concentration gradient spontaneously, we can think of a concentration gradient as a kind of force that "pushes" molecules in that particular direction. Thus, we refer to a concentration gradient as a **chemical driving force,** the direction of which is always down the concentration gradient (Figure 4.1). As we will see, the rate at which a substance is transported varies with the size of the concentration gradient and generally increases as the size of the gradient increases. Therefore, we can say that the magnitude of the chemical driving force increases as ΔC increases. When more than one substance is present, as is the case with real cells, more than one concentration gradient exists. Any chemical driving force that might be acting on a given substance depends only on the concentration gradient of *that particular substance.*

Note that a chemical driving force is fundamentally different from the more familiar types of forces because it does not really act on molecules in the same way that the force of gravity pulls on a rock or a magnetic force pulls on a piece of iron. When molecules move across a membrane down a concentration gradient, they do so simply because there are more molecules on one side of the membrane than on the other. *Individual* molecules are not pushed down the gradient but in fact are equally likely to move in either direction. This issue is explored further later in this chapter.

Quick Test 4.1

1. What is the difference between passive transport and active transport? Between simple diffusion and mediated transport? Is simple diffusion active or passive?

2. Why is a concentration gradient referred to as a driving force?

3. In what direction (into or out of a cell) is the chemical driving force for glucose (see Table 4.1)?

Electrical Driving Forces

In general, molecules moving passively across membranes can be affected by factors other than the chemical driving force. This is particularly true of ions, which are influenced by *electrical driving forces* in addition to chemical driving forces. Electrical driving forces arise due to the **membrane potential,** a difference in *electrical potential* or *voltage* that exists across the membranes of most cells. (Toolbox: Electrical Potential, www.physiologyplace.com, Challenge Yourself.) In the next section we see that the existence of a membrane potential reflects an unequal distribution of positively charged ions and negatively charged ions across the plasma membrane. (We will not worry about what causes this unequal charge distribution until Chapter 6, when we explore the origin of membrane potentials in neurons.) Once we have a firm understanding of what a membrane potential is, we then consider its influence on the forces that affect ion transport.

The Membrane Potential The fluids in the body contain a wide variety of solutes, and many of these are ions, which possess electrical charge. Some ions are *cations,* which possess positive charge; others are *anions,* which possess negative charge. Ions are also present in salt solu-

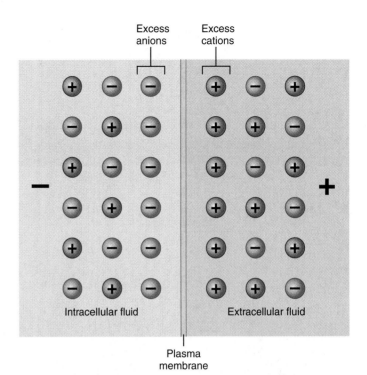

FIGURE 4.2 Separation of charge across a cell membrane. *Under normal conditions, the fluid inside a cell has a slight excess of anions (negative charges), while the fluid outside the cell has a slight excess of cations (positive charges). These excess charges are clustered in the region near the membrane. Net charges inside and outside the cell are indicated by (+) and (−) signs on either side of the membrane.*

tions, such as sea water, but one normally cannot detect the presence of the ions' electrical charges because the number of positive and negative charges are equal. Such a solution is said to be electrically *neutral* because the positive and negative charges cancel one another, giving a net (total) electrical charge of zero. Likewise, the total electrical charge of your body is zero because the number of cations in your body equals the number of anions.

In intracellular or extracellular fluid, cations and anions are present in unequal numbers; consequently, these fluids are not electrically neutral. Intracellular fluid contains a slight excess of anions over cations, giving it a net negative charge. Extracellular fluid contains a slight excess of cations over anions, giving it a net positive charge. Because positive and negative charges are distributed unequally between the inside and outside of a cell, a *separation of charge* is said to exist across the membrane (Figure 4.2). The excess negative and positive charges of intracellular and extracellular fluid tend to be clustered close to the membrane, because the excess negative charges on one side of the membrane are attracted to the excess positive charges on the other side.

A cell's membrane potential reflects this separation of charge and is given in units of electrical potential—in *millivolts* (mV), which are 1/1000 of a volt. The magnitude of the membrane potential (number of millivolts) depends on the degree of charge separation; the greater

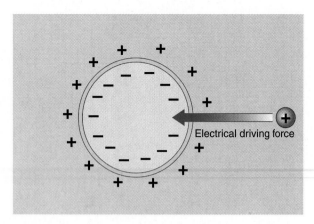

(a)

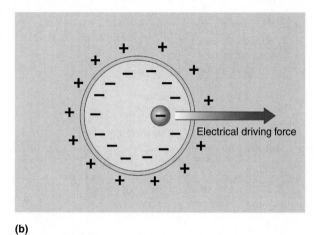

(b)

FIGURE 4.3 Electrical driving forces. (a) *When a positively charged ion crosses a cell membrane, the direction of the electrical driving force (arrow) is inward.* **(b)** *When a negatively charged ion crosses the membrane, the direction of the electrical driving force is outward. In both cases, it is assumed that the membrane potential is negative, as indicated by (+) and (−) signs on either side of the membrane.*

the difference in charge between the two sides of a membrane, the larger the membrane potential. By convention, the *sign* of the membrane potential (positive or negative) is taken to be the sign of the net charge *inside* the cell relative to outside. Since the inside of a cell is typically more negatively charged than the outside, the membrane potential is usually negative. (Under certain conditions, however, the membrane potential can be positive, as we will see in Chapter 6.) For most cells, the membrane potential, which is given the symbol V_m, is around negative 70 millivolts ($V_m = -70$ mV).

How the Membrane Potential Creates an Electrical Driving Force that Acts on Ions As an ion crosses a membrane in the presence of a membrane potential, its electrical charge will be attracted by the net electrical charge existing in the fluid on one side and will be repelled by that existing on the other side. Assuming that no other force (for example, a chemical driving force) is acting on the

ion, it will experience a driving force and move spontaneously from the side that repels it toward the side that attracts it. In this case, the force that is causing it to move is an **electrical driving force,** meaning that it is simply a consequence of the attractive and repulsive forces that inevitably arise between electrically charged objects. Note that this electrical driving force adds to or subtracts from any other driving forces that might also be present, such as a chemical force due to a concentration gradient.

Factors Affecting the Direction and Magnitude of the Electrical Driving Force When an ion is crossing a membrane, the *direction* of the electrical driving force that is acting on it depends on only two things: the sign of the membrane potential and the sign of the ion's *valence* or charge. Determining the direction of the force is easy; just remember that like charges repel and opposite charges attract. If the membrane potential is negative, the electrical force acting on a positively charged ion draws it into the cell because the ion is repelled by the positive charges outside and attracted by the negative charges inside (Figure 4.3). If the ion is negatively charged, a negative membrane potential exerts an electrical driving force that pushes it out of the cell because the ion is repelled by the negative charges inside and attracted by the positive charges outside. Because electrical charges do not act on uncharged molecules such as glucose, substances lacking charges are not affected by the membrane potential.

The *magnitude* of the electrical driving force on an ion depends on the size of the membrane potential and the quantity of charge carried by the ion, and it increases as either of these factors gets larger (Figure 4.4). A larger negative membrane potential, for instance, means a greater number of negative charges inside and positive charges outside, which increases the attractive and repulsive forces acting on an ion. If an ion carries more charge, the attractive and repulsive forces are also increased, which makes the electrical driving force stronger.

Electrochemical Driving Forces

To determine whether ions are being transported passively or actively, a physiologist must identify all the driving forces that might be acting on them. In general, when ions are transported across membranes, two driving forces are influential: (1) a chemical force reflecting the ions' tendency to move down their concentration gradient (from higher to lower concentration), and (2) an electrical force reflecting the ions' tendency to be pushed in one direction or the other by the membrane potential. The total force acting on the ions is the combination of these chemical and electrical driving forces, referred to as the **electrochemical driving force.** (The term *electrochemical driving force* is also used in reference to uncharged molecules, which are not influenced by electrical driving forces; in this case, the electrochemical driving force is synonymous with the chemical driving force.)

The direction of the electrochemical driving force acting on an ion depends on the net direction of the electrical and chemical driving forces. If both forces go in the same direction, then the electrochemical driving force also acts in that direction. If the electrical and chemical forces go in opposite directions, then the electrochemical force acts in the direction of the larger force.

To determine whether the electrical or chemical force is larger, a physiologist must know an ion's **equilibrium potential,** a hypothetical value for the membrane potential at which the electrical driving force is equal and opposite to the chemical driving force, producing an electrochemical driving force of zero. If the membrane potential equals the equilibrium potential for an ion, that ion will not move spontaneously in either direction because the total driving force acting on it is zero. In other words, the ion will be at *equilibrium.* (Recall that a chemical reaction is at equilibrium when its net movement is neither forward nor backward.)

The magnitude and sign of an ion's equilibrium potential depends on the size and direction of the ion's concentration gradient, and on the ion's valence. Larger concentration gradients mean larger equilibrium potentials because a greater electrical force is required to equal or "balance" a larger chemical force. The sign of the equilibrium potential is such that the electrical force goes in the direction opposite to the chemical force. In the case of sodium ions (Na^+), for instance, which are in higher concentration outside a cell, the chemical force is directed inward. Thus, an *outwardly* directed electrical force is required to balance the chemical force. Because Na^+ is positively charged, a *positive* membrane potential will exert an outward electrical force that balances the inward chemical force. This means that the Na^+ equilibrium potential must be positive. If an ion's concentration on either side of a membrane is known, its equilibrium potential can be easily calculated using the Nernst equation (Toolbox: Equilibrium Potentials and the Nernst Equation, p. 107).

Determining the Direction of the Electrochemical Driving Force When an ion is crossing a cell membrane, the direction of the electrochemical driving force acting upon it can be determined using the following simple procedure, which requires knowledge of the ion's equilibrium potential (which is given the symbol E_X for ion X) and the cell's membrane potential:

1. Using the principles described in the preceeding sections, identify the directions of the chemical and electrical driving forces acting on the ion. If the two driving forces are going in the same direction, the electrochemical force also acts in that direction, and no further analysis is necessary.

2. If the chemical and electrical forces act in opposite directions, compare the sizes of the equilibrium potential and membrane potential. If they are equal in

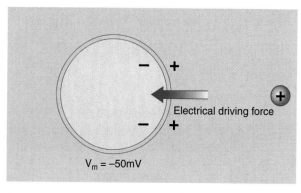

(a)

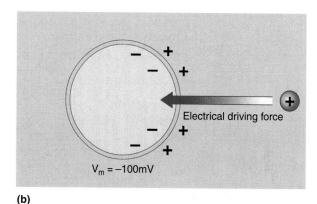

(b)

(c)

FIGURE 4.4 Effects of membrane potential and ion valence (charge) on electrical driving force. (a) *In the presence of a given negative membrane potential ($V_m = -50$ mV), an inward electrical driving force (arrow) acts on a positively charged ion crossing the membrane.* **(b)** *When the magnitude of the membrane potential increases to -100 mV, the magnitude of the electrical driving force also increases (as represented by longer arrow).* **(c)** *For ions of higher valence, the magnitude of the electrical driving force also increases.*

magnitude, then the electrochemical force is zero and the ion is at equilibrium. If not, proceed to step 3.

3. If the equilibrium potential is larger in magnitude than the membrane potential, then the chemical force is larger than the electrical force; the electrochemical force therefore acts in the same direction as the chemical force. If the membrane potential is

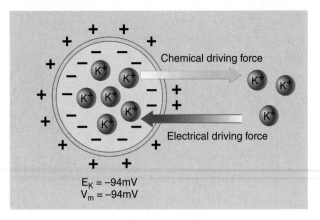

(a)

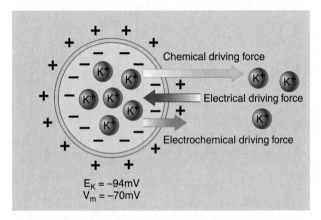

(b)

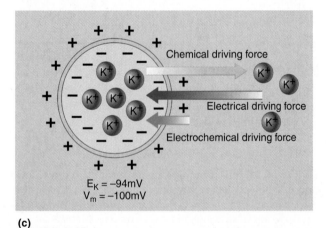

(c)

FIGURE 4.5 Electrical, chemical, and electrochemical driving forces acting on potassium ions crossing a cell membrane. *The equilibrium potential for potassium (K^+) is −94 mV, and the membrane potential is assumed to be negative. (a) When the membrane potential equals the equilibrium potential (V_m = −94 mV), the electrochemical force is zero. (b) When the membrane potential is −70 mV, the electrical force is smaller than the chemical force (compare arrow lengths), and the electrochemical force is outward. (c) When the membrane potential is −100 mV, the chemical force is smaller than the electrical force, and the electrochemical force is inward.*

larger in magnitude than the equilibrium potential, then the electrical force is larger than the chemical force; the electrochemical force therefore acts in the same direction as the electrical force.

Application of this procedure is illustrated in Figure 4.5 for the case of potassium ion (K^+), whose intracellular and extracellular concentrations are 140 mM and 4 mM, respectively (see Table 4.1), giving an outwardly directed chemical force and an equilibrium potential of −94 mV (E_K = −94 mV). In all three panels in Figure 4.5, the membrane potential is negative, producing an inwardly directed electrical force that opposes the chemical force. In Figure 4.5a, the membrane potential and equilibrium potential are equal in size, giving an electrochemical force of zero. In Figure 4.5b, the equilibrium potential is larger than the membrane potential (−70 mV), giving an electrochemical force that is directed out of the cell. In Figure 4.5c, the membrane potential (−100 mV) is larger than the equilibrium potential, so the electrochemical force is directed into the cell.

Significance of the Electrochemical Driving Force Because the electrochemical force is the total driving force acting on transported molecules, it determines the direction in which the molecules will move if they are allowed to cross the membrane spontaneously. When molecules are being transported passively, they always move in the direction of the electrochemical driving force. Put another way, when ions are transported passively, they move *down* their **electrochemical gradient.** When they are being transported actively, they move in the direction opposite to the electrochemical force, or *up* their electrochemical gradient. Table 4.2 summarizes the directions of driving forces for ions and uncharged solutes, and indicates the directions of active and passive transport of these substances relative to the electrochemical gradient.

Quick Test 4.2

1. Define the following terms: *membrane potential, electrical driving force, equilibrium potential, electrochemical driving force.*

2. Determine the direction of the electrical driving force for each of the following ions, assuming the cell membrane potential is negative: Na^+, K^+, Cl^-, HCO_3^-, Ca^{2+}.

3. Assuming that sodium and calcium ions are being transported through a given cell membrane, for which ion—Na^+ or Ca^{2+}—is the electrical driving force larger? Explain. (*Hint:* Look at the valence.)

4. Refer to Table 4.1 for intracellular and extracellular ion concentrations, and determine the direction of the chemical driving force on each of the ions in Question 2. Determine the direction of the electrochemical driving force on Na^+, assuming that E_{Na} = +60 mV and V_m = −70 mV.

	UNCHARGED SOLUTES	CATIONS(+)	ANIONS(−)
Direction of chemical driving force	Down concentration gradient	Down concentration gradient	Down concentration gradient
Direction of electrical driving force	(Not applicable)	From positive to negative	From negative to positive
Direction of electrochemical driving force	Same direction as chemical driving force	Depends on directions and relative magnitudes of chemical and electrical forces	Depends on directions and relative magnitudes of chemical and electrical forces
Direction of passive transport	Down electrochemical gradient	Down electrochemical gradient	Down electrochemical gradient
Direction of active transport	Up electrochemical gradient	Up electrochemical gradient	Up electrochemical gradient

TOOLBOX

**EQUILIBRIUM POTENTIALS AND
THE NERNST EQUATION**

With knowledge of an ion's charge and its intracellular and extracellular concentrations, we can find the equilibrium potential of any ion using the Nernst equation:

$$E = \frac{61}{z} \log \frac{C_o}{C_i}$$

where E is the equilibrium potential, z is the charge (valence) of the ion, and C_o and C_i are the concentrations outside and inside the cell, respectively. In this form, the equation gives the value of the equilibrium potential in millivolts and assumes that the temperature is at or near the normal body temperature of 37°C.

Using typical intracellular and extracellular concentrations, we can find the equilibrium potential for sodium, E_{Na}, by making the appropriate substitutions, as follows:

$$E_{Na} = \frac{61}{1} \log \frac{145 \text{ mM}}{15 \text{ mM}} = 60.1 \text{ mV} \approx 60 \text{ mV}$$

In similar fashion, we can find the equilibrium potential for potassium, E_K:

$$E_K = \frac{61}{1} \log \frac{4 \text{ mM}}{140 \text{ mM}} = -94.2 \text{ mV} \approx -94 \text{ mV}$$

Note that in both cases, the valence is +1, so that the sign of the equilibrium potential depends solely on the direction of the concentration gradient. Note also that E_K is larger in magnitude than E_{Na}. This makes sense because a larger concentration gradient requires a larger membrane potential to balance it, and the K^+ gradient is larger than the Na^+ gradient. If a concentration gradient is small, intracellular and extracellular ion concentrations will be more nearly equal. If the concentrations are identical, the ratio C_o/C_i equals 1, making the equilibrium potential equal to zero. (Recall that the log of 1 is zero.) As the concentrations become more and more dissimilar, the ratio becomes either much larger or much smaller than 1, making the log term larger and more positive, or larger and more negative, respectively.

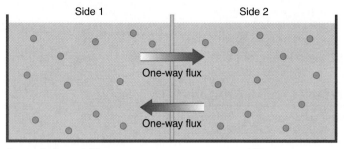

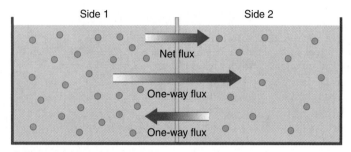

Concentration: 2 M .... 1 M

(b)

FIGURE 4.6 Directional flow of molecules and net flux.
(a) Diffusional equilibrium. When solute molecules (dots) are at the same concentration on both sides of a membrane, the one-way flux from side 1 to side 2 equals that from side 2 to side 1 (arrows). Under these conditions, the net flux across the membrane is zero. (b) Net flux. When the concentration on side 1 is increased by a factor of two, the one-way flux from side 1 to side 2 is twice as large as before, but the one-way flux in the opposite direction is unaffected. The one-way fluxes are now unequal (as indicated by arrow lengths), producing a net flux down the concentration gradient from side 1 to side 2.

FACTORS AFFECTING THE RATE OF TRANSPORT

In Chapter 3, we saw that the rate of a metabolic reaction is important because reactions must proceed at a rate fast enough to meet the body's metabolic demands. But for this to occur, molecules must also be transported across membranes at sufficient rates. When you are resting, for instance, oxygen is transported across the epithelium lining the lungs (the *pulmonary epithelium*) at a rate that matches the rate of oxygen consumption in your body. When you begin exercising, metabolic reactions consume oxygen at a higher rate; to keep up with this increased demand, the rate of transport across the pulmonary epithelium must also increase, or the cells in your body will eventually run out of oxygen. In certain pathological conditions, such as *pulmonary edema* (excess fluid in the lungs), diffusion of oxygen (and of carbon dioxide) is impeded. Under these conditions, a person has difficulty exercising because oxygen cannot move across the pulmonary epithelium fast enough. If the condition is severe, oxygen delivery may be impaired so much that the individual will feel breathless even at rest.

The *rate* at which a substance is transported across a membrane refers to the number of molecules that cross the membrane in a given length of time, which is called the **flux.** Flux is usually expressed in units of moles per second or some equivalent. (Strictly speaking, flux is the number of molecules crossing a membrane per unit time per unit membrane surface area. For simplicity, we omit area from our definition.)

Note that when we say that molecules are transported in one direction or the other (into or out of a cell, for instance), we are referring to a *net* movement of molecules across the membrane. In virtually every case, individual molecules actually move across a membrane in *both* directions, but more of them may move in one direction than in the other. This bidirectional flow of molecules is illustrated in Figure 4.6, which depicts a membrane separating two solutions in a chamber. In this example, molecules are crossing the membrane by simple diffusion under two conditions: when their concentration is the same on both sides, and when the concentration on the left side is twice that on the right side.

In the first condition (Figure 4.6a), molecules cross both from side 1 to side 2 and from side 2 to side 1. The number of molecules crossing the membrane per unit time in just one direction is called a *one-way flux* or *unidirectional flux*. In this case, the one-way flux from side 1 to side 2 equals that from side 2 to side 1, because the number of molecules near the membrane is the same on either side, and each molecule has the same likelihood of crossing to the other side in any given length of time. Because there is no net transfer of molecules from one side to the other, the concentrations do not change, and the system is in *diffusional equilibrium*.

In the second condition (Figure 4.6b), the concentration on side 1 is twice what it originally was, and twice that of side 2. Under these conditions, the one-way flux from side 1 to side 2 has *doubled* (because the number of molecules near the membrane on that side has doubled). Now the one-way flux from side 1 to side 2 is larger than the one-way flux in the opposite direction. As a result, there is net movement of molecules from side 1 to side 2. The rate of this movement is the **net flux,** the difference between the two unidirectional fluxes. Whenever physiologists say that molecules are being transported passively or actively in a certain direction, they are always referring to the direction of the net flux.

As we will see later in this chapter, transport rates are influenced by many variables, some of which affect certain transport mechanisms but not others. In the next two sections we concentrate on those factors that affect the rate of transport regardless of the particular mechanism.

Factors Affecting Rates of Passive Transport

When a substance is transported passively across a membrane, the rate at which it is transported depends on three factors: the magnitude of the driving force, the

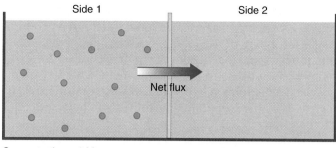

Side 1 | Side 2

Net flux

Concentration: 1 M   0 M (Pure water)

(a)

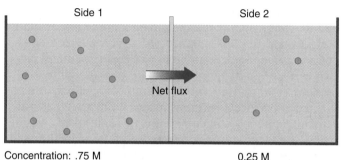

Side 1 | Side 2

Net flux

Concentration: .75 M   0.25 M

(b)

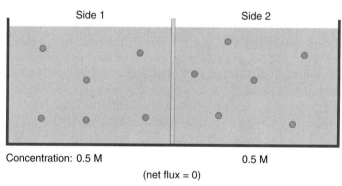

Side 1 | Side 2

Concentration: 0.5 M   0.5 M

(net flux = 0)

(c)

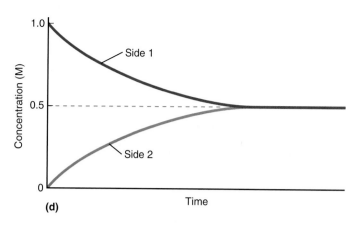

(d)

FIGURE 4.7 Changes in net flux of molecules across a membrane as the concentration gradient changes. (a) *A large concentration gradient creates the net flux indicated by the arrow.* **(b)** *Due to the net flux from side 1 to side 2, the concentration on side 1 has fallen to 0.75 M, while that on side 2 has risen to 0.25 M. The net flux has also decreased, as is indicated by the shorter arrow.* **(c)** *Concentrations on both sides of the membrane have become equal, and the net flux is now zero.* **(d)** *Changes in concentration on the two sides over time.*

As time passes, does the one-way flux from side 2 to side 1 increase, decrease, or stay the same?

membrane surface area, and the **permeability** of the membrane, a measure of the ease with which molecules are able to move through it. Each of these factors is discussed next.

The Magnitude of the Driving Force

When a driving force acts on molecules crossing a membrane, it influences not only the direction in which they move, but also the rate at which they are transported. In most cases the net flux increases as the magnitude of the driving force increases, but not always.

In simple diffusion, the rate of transport is directly related to the size of the driving force. Consider a situation in which a membrane separates two solutions in a chamber (Figure 4.7). Initially (Figure 4.7a), side 1 contains solute at a concentration of 1 M, while side 2 contains an equal volume of pure water. Due to the presence of this concentration gradient, a net flux of solute molecules

from side 1 to side 2 occurs. As a result, the concentration on side 1 falls while the concentration on side 2 rises. In Figure 4.7b, the concentration on side 1 has fallen to 0.75 M while that on side 2 has risen to 0.25 M. Because the size of the concentration gradient has decreased, the net flux has also decreased, as indicated by the shorter arrow. These changes continue until the two concentrations become equal, at which point the net flux equals zero (Figure 4.7c). The graph in Figure 4.7d shows that the concentrations change less rapidly as they become more nearly equal; that is, the net flux diminishes as the size of the concentration gradient decreases. Once the concentrations become equal, they no longer change, because the net flux is zero. Under these conditions, molecules still move back and forth across the membrane, but at equal rates in both directions.

It increases.

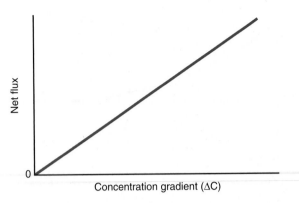

FIGURE 4.8 Direct proportionality between net flux and the magnitude of the concentration gradient for a substance crossing a membrane by simple diffusion.

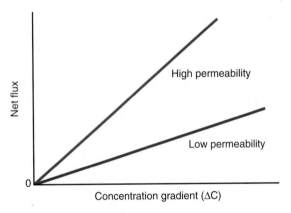

FIGURE 4.9 Effect of permeability on the net flux of a substance crossing a membrane by simple diffusion.

In simple diffusion, the net flux of a substance is directly proportional to the size of the concentration gradient—that is, to the magnitude of the chemical driving force (Figure 4.8). As the concentration gradient increases or decreases by a given factor, the flux increases or decreases by that same factor.

Membrane Surface Area

As a general rule, the rate at which molecules are transported across a membrane varies in direct proportion to the membrane's surface area. To understand why, imagine molecules crossing a membrane in a chamber at a given rate. If two identical chambers were placed side-by-side and combined into one, then the surface area of the membrane would double, and the number of molecules crossing the membrane per unit time would also double. Note that this would be true regardless of the mechanism of transport.

Because the effect of membrane area on transport is so straightforward, it will be omitted from our discussions. Note, however, that various tissues that are specialized for transport—the pulmonary epithelium, the

intestinal epithelium, and the walls of capillaries, for instance—have large membrane surface areas, which enhance the ability of these tissues to transport large quantities of material quickly. The surface areas of the pulmonary and intestinal epithelia are increased by their high degree of folding, whereas capillary surface areas are increased by the extensive branching that characterizes these vessels.

Membrane Permeability

The permeability of a membrane to a particular substance depends on both the nature of the transported substance and the various properties of the membrane (other than surface area) that influence the ease with which molecules are able to penetrate it. For any mechanism of passive transport, a higher permeability translates into a higher rate of transport, other things being equal. The influence of permeability on transport is illustrated graphically in Figure 4.9, which pertains to a substance crossing a membrane by simple diffusion. When permeability is higher, the net flux is larger for a given concentration gradient (Toolbox: Fick's Law and Permeability, p. 112). We will discuss the various factors that affect membrane permeability shortly.

The permeability of a membrane to a substance X is given the symbol P_X and is often referred to as the "permeability of X." For example, P_{Na} would be the "permeability of sodium" (or alternatively, the "sodium permeability"). Although this usage is very common, note that the term *permeability* applies to membranes, not to transported molecules—even though a membrane's permeability is not the same for all substances and therefore depends on the nature of the transported molecules. Strictly speaking, if a substance crosses a membrane, the substance is *permeant* while the membrane itself is *permeable* (able to be permeated).

Factors Affecting Rates of Active Transport

Because active transport is carried out by specific transport proteins called *pumps,* two factors are the sole determinants of the rate at which molecules are actively transported across any membrane: the rate of transport by individual pump proteins, and the number of pumps that are present in the membrane. As either of these variables increases, the rate of transport increases.

The transport rate of an individual pump depends on the nature of the pump; it is also influenced by other factors, such as the concentrations of the transported substance on either side of the membrane, the size of the electrochemical driving force that the pumps must work against, and other variables that affect the conditions under which a pump must operate. These variables affect different pumps in different ways, and a more thorough discussion of this topic is beyond the scope of this book.

1. Define the following terms: *flux, unidirectional flux, net flux.*

2. What does permeability measure?

3. Does a membrane's permeability depend on its surface area? Why or why not?

4. Which of the following (choose all that apply) cause an increase in the net flux of a substance being passively transported across a membrane? a decrease in permeability, an increase in membrane surface area, an increase in the size of the concentration gradient of an uncharged substance.

5. When a substance is being actively transported, each pump protein transports molecules at a certain rate. What other factor influences the rate of transport?

SIMPLE DIFFUSION: PASSIVE TRANSPORT THROUGH THE LIPID BILAYER

Simple diffusion is the least complicated of all transport mechanisms, but in a way this makes it all the more interesting. Although we have already learned quite a bit about some of the characteristics of simple diffusion, here we focus on what makes it work. We will also learn about some of the factors that affect the permeability of a membrane to substances moving by simple diffusion.

The Basis for Simple Diffusion

We use the term *simple diffusion* to describe the passive transport of molecules through a biological membrane's lipid bilayer, but in fact the mechanism of simple diffusion is not strictly biological. When someone opens a bottle of perfume nearby, for example, you will eventually smell the fragrance even if there are no air currents to carry it to your nose. This occurs because certain molecules from the perfume travel from the bottle to your nose by simple diffusion through the air. The molecules travel away from the bottle because they are more concentrated near the bottle and less concentrated farther away, and because these molecules, like any others, move spontaneously down their concentration gradient due to the fact that they are in a state of constant thermal motion.

The movement of molecules from one location to another simply as a result of their own thermal motion is called **diffusion,** a general term that applies to molecules moving through any medium. Thermal motion is often called *random* thermal motion because individual molecules move helter-skelter in various directions due to collisions with other molecules. As a consequence, molecules do not go in any particular direction for long before

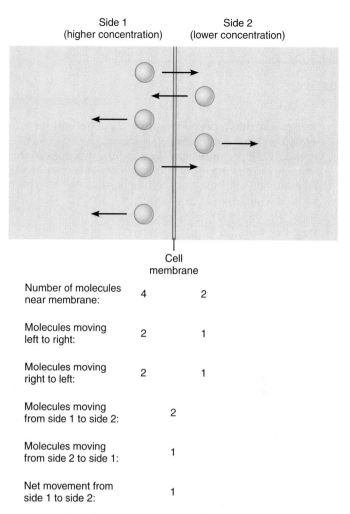

Side 1 (higher concentration) Side 2 (lower concentration)

Cell membrane

	Side 1	Side 2
Number of molecules near membrane:	4	2
Molecules moving left to right:	2	1
Molecules moving right to left:	2	1
Molecules moving from side 1 to side 2:	2	
Molecules moving from side 2 to side 1:	1	
Net movement from side 1 to side 2:	1	

FIGURE 4.10 A simplified model showing how random thermal molecular motion gives rise to movement down a concentration gradient. *Initially, four molecules are present on the left side of a membrane, while two are present on the right, indicating that the concentration is higher on the left side. Because movement due to thermal motion is random, half of the molecules move to the left, and half move to the right. Considering only those molecules that cross the membrane, the general direction of movement is from left to right, or down the concentration gradient.*

changing course, and when this change occurs the direction is unpredictable.

But if thermal motion is random, then how is it that diffusing molecules always move *down* their concentration gradient, which is clearly not a random event? The answer lies in the distinction between *individual molecules,* which move randomly, and a *population of molecules,* which always moves down its concentration gradient (Figure 4.10). Because the motion of each of the molecules in Figure 4.10 is random, each is equally likely to move left or right (and thus to cross the membrane); the net flux from left to right in the figure is simply the result of the greater number of molecules near the left side of the membrane.

Permeability has a precise mathematical definition. We have seen that when a substance crosses a membrane by simple diffusion, the net flux is proportional to the size of the concentration gradient. This relationship is expressed mathematically by *Fick's Law,* which can be written as follows:

$$\text{net flux} = PA(\Delta C)$$

In this expression, A is the membrane surface area, ΔC is the concentration gradient (the difference in concen-tration across the membrane), and the coefficient P represents permeability. The numerical value of P depends on the particular membrane in question and the identity of the transported substance. Note that for a concentration gradient of a given size, the flux increases as P gets larger.

Factors Affecting Membrane Permeability in Simple Diffusion

In simple diffusion, membrane permeability is influenced by properties of the membrane and of the diffusing substance, but it varies in ways that are quite predictable from one membrane to another or one substance to another. Among the factors influencing the permeability of cell membranes are the following:

1. *The lipid solubility of the diffusing substance.* We have seen that molecules vary in the degree to which they can dissolve in lipids. Hydrophobic (nonpolar) substances are the most lipid soluble, while hydrophilic (polar or ionized) substances are the least lipid soluble. Given that molecules must move through a lipid bilayer in simple diffusion, we would expect lipid-soluble molecules to be able to enter the bilayer more readily. Other things being equal, the more lipid soluble a substance is, the greater a membrane's permeability to that substance. Lipid solubility is also influenced by the lipid composition of the membrane, which varies somewhat from one membrane to another.

2. *The size and shape of diffusing molecules.* Molecules vary considerably in their sizes (physical dimensions and molecular weights) and shapes. Generally speaking, larger molecules and those with more irregular shapes move through the bilayer more slowly, making membrane permeability lower.

3. *Temperature.* Molecules move faster at higher temperatures, which increases the permeability. This effect is rarely important in human physiology, however, because body temperature is relatively constant.

4. *Membrane thickness.* Theoretically, a thicker membrane would have a lower permeability because molecules must travel farther to penetrate it. As a reflection of this, tissues specialized for transport, such as capillary walls or the pulmonary epithelium, tend to have relatively thin walls; this thinness increases the permeability and enhances the rate of transport through these tissues.

Of all the factors just mentioned, lipid solubility has the strongest influence on permeability. Differences in lipid solubility are largely responsible for the fact that some substances can cross cell membranes readily by simple diffusion, whereas others cannot. Because most substances in the body are hydrophilic and thus do not penetrate lipid bilayers easily, the list of substances known to be transported by simple diffusion is fairly limited and includes fatty acids, steroid hormones, oxygen, carbon dioxide, and the fat-soluble vitamins (A, D, E, and K).

A final note: Even though our focus here has been solely on simple diffusion across *membranes,* which applies to relatively few substances, cells ultimately depend on simple diffusion for delivery or removal of *all* solutes because molecules must diffuse through *water* in order to move toward and away from membrane surfaces. This is not merely of academic interest. Recall, for instance, that the accumulation of excess fluid in the lungs impedes oxygen transport and thus decreases the lungs' capacity for delivering oxygen to the blood. The reason is that the flux of oxygen across the pulmonary epithelium is affected by the thickness of the fluid layer covering the epithelium. When fluid accumulates, the effect is similar to an increase in tissue thickness, and the diffusion rate slows down.

Quick Test 4.4

1. In simple diffusion, do individual molecules always go down their concentration gradient? Why or why not?

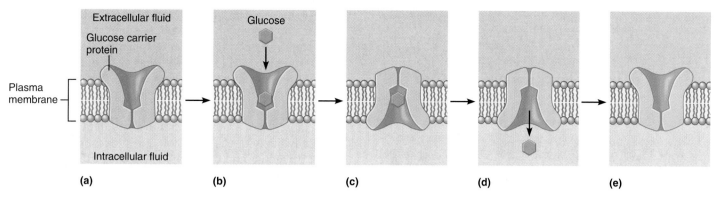

FIGURE 4.11 Transport of a glucose molecule across a cell membrane by a carrier protein. (a) *A carrier protein with an empty binding site.* **(b)** *Binding of a glucose molecule to the protein's binding site, which faces the outside of the cell.* **(c)** *Conformational change in the carrier protein, such that the binding site now faces the inside of the cell.* **(d)** *Release of the glucose molecule. The binding site is once again empty.* **(e)** *Return of the carrier to its original conformation. The carrier is now ready to bind another glucose molecule.*

2. Which properties of transported molecules affect membrane permeability in simple diffusion? Which properties of a membrane affect the permeability?

3. For a substance crossing a tissue by simple diffusion, which of the following (choose all that apply) cause the permeability to increase? increased tissue thickness, decreased size of diffusing molecules, decreased lipid solubility of diffusing molecules, increased temperature.

FACILITATED DIFFUSION: PASSIVE TRANSPORT THROUGH MEMBRANE PROTEINS

We noted earlier that some substances that are transported passively do not cross membranes by simple diffusion but instead cross by way of transport proteins in the membrane—a process known as mediated transport. Physiologists use the term **facilitated diffusion** to distinguish passive mediated transport from active transport, which is also mediated. This term is apt because the function of transport proteins in facilitated diffusion is to aid or *facilitate* the passage of molecules through the membrane. In fact, some physiologists use the term *simple diffusion* (instead of just *diffusion*) to distinguish it from facilitated diffusion.

Transport Proteins in Facilitated Diffusion

Facilitated diffusion is important to cells because many of the substances that cells need—sugars, amino acids, and inorganic ions, for instance—are hydrophilic and have low solubility in lipids. These substances therefore cannot cross membranes by simple diffusion through the bilayer;

to enter or leave cells, these substances require the help of transport proteins. These proteins include two basic types, called *carriers* and *channels,* which are described next.

Carriers

A **carrier** is a transmembrane protein that binds molecules on one side of a membrane and transports them to the other side by means of a *conformational change,* or a change in shape. A carrier possesses one or more binding sites that are usually specific for molecules of certain substances or classes of substances. Examples of substances transported by carriers are sugars and amino acids.

To be transported by a carrier, a molecule must first enter a binding site. Once the molecule is in the binding site, the carrier undergoes a conformational change that exposes the binding site to the fluid on the other side of the membrane. At this point, the molecule is free to dissociate from the carrier and be released into the fluid. Figure 4.11 schematically depicts how a carrier transports a glucose molecule across a cell membrane.

Even though the nature of the conformational changes exhibited by carriers is unknown, they are usually drawn with a kind of hingelike structure, as in Figure 4.11. Regardless of the exact mechanism, conformational changes occur randomly (due to thermal agitation of the carrier protein) and expose the binding site(s) to the fluid first on one side of the membrane and then on the other. (Conformational changes need not be triggered by solute binding and may occur even when binding sites are empty.) When a molecule binds to a carrier on one side of a membrane and is released on the opposite side, the binding site becomes available to a molecule on that side. If a molecule binds, the carrier can then transport it in the opposite direction. If a molecule does not bind, the carrier can revert to its original conformation with the binding site empty, as shown in the figure.

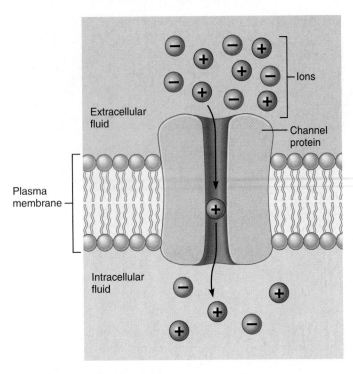

FIGURE 4.12 Transport of ions across a cell membrane through a channel protein.

When a solute is present at the same concentration on both sides of a membrane, a carrier transports molecules in both directions at the same rate, so no net flux occurs. Under these conditions, all individual molecules have the same chance of being transported to the opposite side during a given period of time, because molecules on each side have equal access to the binding site. In other words, the carrier does not show a preference for transporting molecules in one direction as opposed to the other. This means that a net flux exists only in the presence of a concentration gradient (or electrochemical gradient, if ions are being transported). Then, as always, if a concentration gradient exists, the direction of the net flux is down the gradient. (This is because molecules are more likely to bind to the carrier when its binding site faces the side where the concentration is higher, as opposed to the side where the concentration is lower.) In this regard, carrier-mediated transport is like simple diffusion, even though the mechanism is different.

Channels

A **channel** is a transmembrane protein that transports molecules via a passageway or *pore* that extends from one side of the membrane to the other (Figure 4.12). Like carriers, channels are usually specific for certain substances or classes of substances. Most channels transport inorganic ions such as sodium (Na^+), potassium (K^+), or chloride (Cl^-) exclusively, although some are capable of transporting large organic molecules as well.

The mechanism of transport through a channel depends on the type of channel. In some channels the pore appears to act simply as a water-filled passageway through which ions move by diffusion. In other channels, one or more binding sites for ions exist within the pore. When a channel has numerous binding sites, ions move through the pore by "jumping" from one site to the next. In the absence of an electrochemical gradient, a channel is equally likely to transport ions (or other molecules) in either direction, and no net flux will occur. If there is an electrochemical gradient, however, the net flux through the channel will be down the gradient.

Note that the distinction between a channel with binding sites and a carrier is that an empty channel's binding sites are accessible from both sides of the membrane *at the same time*. For a carrier, the sites are accessible from only one side or the other at any given time.

Factors Affecting Membrane Permeability in Facilitated Diffusion

In facilitated diffusion, membrane permeability is determined entirely by two factors: the transport rate of individual carriers or channels, and the number of carriers or channels in the membrane. An increase in either of these factors translates into increased permeability.

Individual carriers or channels transport molecules at different rates, depending on their type, but generally channels are significantly faster than carriers. A glucose carrier typically can transport molecules at a maximum rate of 10,000 per second, whereas a sodium channel might transport 10 million ions per second. As you might expect, the rate of transport depends on other factors, including the magnitude of the concentration (or electrochemical) gradient of the transported substance.

Figure 4.13 shows how the transport rate of a population of carriers varies as the size of the concentration gradient changes. As the concentration on one side of the membrane is raised while that on the other side is kept constant, the net flux increases rapidly at first but then levels off. (Compare with Figure 4.8, which pertains to a substance crossing a membrane by simple diffusion.) This leveling off occurs because the carrier proteins become 100% saturated at high concentrations.

Saturation of carriers is similar to saturation of enzymes (see Chapter 3) and occurs for the same reason: A carrier, like an enzyme, has a fixed number of binding sites. When the concentration of molecules on one side of a membrane is high, the likelihood that at any given time the binding sites will be occupied by molecules is also high. If the concentration is very high, binding sites will be occupied virtually 100% of the time. Under these conditions, the carriers are transporting molecules as fast as they can; increasing the solute concentration further will have no effect on the transport rate.

For channels, the relationship between the flux and the concentration gradient (or electrochemical gradient)

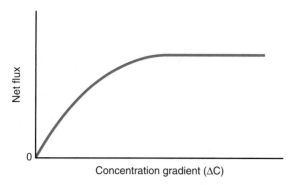

FIGURE 4.13 Relation of net flux and concentration gradient for molecules crossing a membrane via carriers.

depends on the type of channel. Some channels show saturating behavior, whereas others do not.

When a substance crosses a cell membrane by simple diffusion, the cell cannot control the rate of transport because it cannot regulate the permeability of the membrane, which depends on properties of the bilayer. For substances crossing a membrane by facilitated diffusion, however, cells can regulate the activity of individual carriers or channels and/or the number of carriers or channels that are present and functionally active in the membrane. In Chapter 6, for example, we see that the permeability of the neuron membrane to Na^+ and K^+ is regulated by the membrane potential, which affects the number of Na^+ and K^+ channels that are available to transport these ions across the membrane.

Many drugs and hormones exert their actions by altering the function of carriers or channels. One example is a *calcium channel blocker,* a medication commonly used for treating high blood pressure. This medication works by interfering with the ability of certain types of channels to transport calcium ions into heart muscle cells. By reducing the permeability of the plasma membrane to calcium, these drugs decrease the flux of calcium into cells, which reduces the force of heart contractions and thereby lowers blood pressure. Another chemical agent that targets transport proteins is the hormone *insulin,* which acts on certain cells to trigger an increase in the number of glucose carriers in the plasma membrane. This action raises the permeability of cell membranes to glucose, which helps to regulate glucose levels in the blood.

Quick Test 4.5

1. Facilitated diffusion is one form of mediated transport. What is the other?

2. What is the difference between a carrier and a channel? How are they similar?

3. As the concentration of a substance increases, the flux of molecules transported by carriers increases at first but eventually levels off. Why?

4. When an uncharged substance such as glucose enters a cell by facilitated diffusion, does it go up or down its concentration gradient?

ACTIVE TRANSPORT

Even though we have focused nearly all our attention on passive transport, cells perform active transport as well. The importance of active transport to the life of a cell is underscored by the fact that some cells expend large amounts of energy (in some cases, up to 40% of their total ATP production) to actively transport certain molecules across their membranes. Indeed, many vital physiological processes—including the generation of electrical signals in neurons and other excitable cells, the regulation of muscle contraction, the absorption of nutrients and water by the digestive system, and body fluid regulation by the kidneys—depend on active transport either directly or indirectly.

We have seen that active transport requires energy, and that passive transport does not. The reason for this difference lies in the concept of electrochemical gradients: When an electrochemical driving force is acting on molecules, it effectively exerts a "push" in a certain direction. Molecules will move spontaneously in that direction unless something prevents them from doing so. In other words, transport of a substance *down an electrochemical gradient* requires no energy. In contrast, transport of a substance *up an electrochemical gradient* requires an input of energy because molecules are moving against the electrochemical force that is pushing them. Whether a substance is being transported actively or passively relates to the following rules:

1. If the direction of the net flux is down an electrochemical gradient, the transport is passive.

2. If the direction of the net flux is up an electrochemical gradient, the transport is active.

The distinction between the two basic forms of active transport—primary and secondary active transport—relates to the nature of the energy source expended. **Primary active transport** uses ATP or some other chemical energy source *directly* to transport substances. **Secondary active transport** is active transport that is powered by a concentration gradient or an electrochemical gradient that was previously created by primary active transport. In secondary active transport, some of the energy released during ATP hydrolysis is stored in the gradients that are created; this stored energy is then harnessed to drive the active transport of other molecules.

Pumps, the transport proteins that carry out active transport, are similar to carriers in many respects but possess an ability that carriers do not have: Pumps can harness energy to drive the transport of molecules in a preferred direction across a membrane. In contrast, carriers show no such preference; in the absence of an

FIGURE 4.14 Primary active transport by the Na⁺/K⁺ pump. *The pump, which possesses three sodium binding sites and two potassium binding sites, uses ATP directly to transport sodium ions out of the cell and potassium ions into the cell up their electrochemical gradients.* **(a)** *Intracellular Na⁺ bind to the pump protein.* **(b)** *The binding of three Na⁺ triggers phosphorylation of the pump by ATP.* **(c)** *Phosphorylation induces a conformational change in the protein that allows the release of Na⁺ into the extracellular fluid.* **(d)** *Extracellular K⁺ binds to the pump protein and triggers release of the phosphate group.* **(e)** *Loss of the phosphate group allows the protein to return to its original conformation.* **(f)** *K⁺ is released to the inside of the cell, and the Na⁺ sites become available for binding once again.*

In Figure 4.14a, is the affinity of potassium binding sites high or low compared to Figure 4.14d?

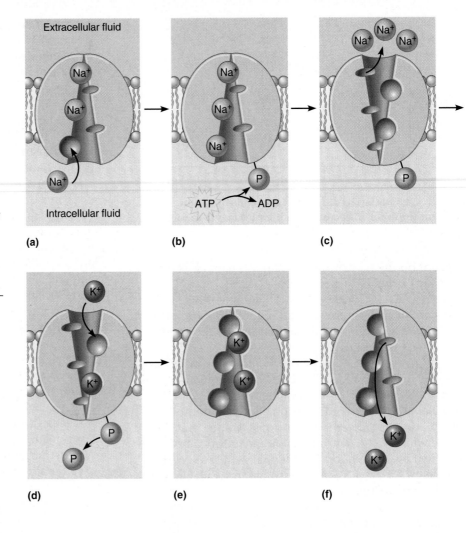

electrochemical gradient, a carrier is equally likely to transport molecules in either direction. Like carriers, pumps are specific for certain molecules and possess fixed numbers of binding sites. Consequently, pumps, like carriers, become saturated as the concentration of transported molecules increases.

In the following sections we explore the mechanisms of primary and secondary active transport in greater detail.

Primary Active Transport

The membrane proteins that perform primary active transport function both as transport proteins and as enzymes. In their capacities as enzymes, most of these proteins harness energy from ATP by catalyzing ATP hydrolysis (see Chapter 3). For this reason, these proteins are frequently referred to as *ATPases*. To understand how primary transport works, we concentrate on the **sodium-potassium pump** (also *Na⁺/K⁺ pump* or *Na⁺/K⁺ ATPase*), which is present in nearly every cell and is crucial to several important physiological processes, including electrical signaling in neurons and the absorption of glucose

by intestinal epithelial cells. A model explaining the pump's action is depicted in Figure 4.14.

Figure 4.14 shows that the Na⁺/K⁺ pump transports Na⁺ and K⁺ ions in opposite directions across the plasma membrane. For each cycle of the pump, three Na⁺ ions are transported out of the cell, and two K⁺ ions are transported in. Transport is active in each case because both types of ion move up their electrochemical gradients. In each cycle of the pump, one ATP is hydrolyzed to provide the energy required for this process.

We can see in Figure 4.14 that the Na⁺/K⁺ pump is like an ordinary carrier protein in two respects: It has specific binding sites for Na⁺ and K⁺ and transports them across the membrane by means of a conformational change. Careful study of the cycle, however, reveals an important difference: An ordinary carrier's binding sites can face either side of the membrane, so that molecules can be transported in both directions (see Figure 4.11); in the Na⁺/K⁺ pump, affinities of the binding sites for Na⁺ or K⁺ change as the pump goes through its cycle, such that a binding site's affinity is higher when it faces one side of the membrane and lower when it faces the opposite side. As a consequence, Na⁺ or K⁺ preferentially bind to the pump protein on one side and are released on the other, which means that Na⁺ or K⁺ are transported in one direction only. As previously mentioned, transport of molecules in a preferred direction is central to the action of any pump.

The details of how the Na^+/K^+ pump transports ions in preferred directions are depicted in Figure 4.14. In the first step of the pump cycle, binding sites face inward and are available for Na^+ in the cell to bind to them (see Figure 4.14a). After three Na^+ bind, ATP hydrolysis occurs (see Figure 4.14b), resulting in phosphorylation of the pump protein. Phosphorylation induces a conformational change in the protein that turns the binding sites outward, so that Na^+ can exit into the extracellular fluid (see Figure 4.14c). During this conformational change, the binding sites change their shape and decrease their affinity for Na^+, making it very unlikely that Na^+ in the extracellular fluid will bind to the pump protein. For these extracellular Na^+ to bind and be carried into the cell would defeat the purpose of the pump. Because the Na^+ binding sites are available on the inside of the membrane but not the outside, the preferred direction of Na^+ transport is out of the cell. In the remaining steps of the pump cycle (see Figure 4.14d–f), a similar process preferentially transports K^+ into the cell.

The Na^+/K^+ pump is responsible for creating the concentration gradients of Na^+ and K^+ listed in Table 4.1. Due to continual outward pumping of Na^+ and inward pumping of K^+, intracellular fluid is rich in K^+ but poor in Na^+ relative to extracellular fluid. As we see in Chapter 6, the concentration gradients of Na^+, K^+, and other ions play a crucial role in generating the membrane potentials essential to transmission of electrical signals in neurons.

The Na^+/K^+ pump is not the only primary active transport system in the cells of your body. For example, pumps in the lining of the stomach transport H^+ out of cells while transporting K^+ in. These pumps are responsible for the secretion of acid in the stomach that is so important in digestion, as described in Chapter 19.

Secondary Active Transport

In Chapter 3 we saw that metabolic reactions can be coupled together, so that an energy-releasing reaction (glucose oxidation, for example) can be used to drive an energy-requiring reaction (such as ATP synthesis). In secondary active transport, something similar happens: A transport protein couples the flow of one substance to that of another. One substance moves passively down its electrochemical gradient, in the process releasing energy that is then used to drive the movement of the other substance up its electrochemical gradient.

Figure 4.15 shows two types of secondary active transport systems in cells. Note that in Figure 4.15a the two transported substances are moving in the same direction (in this case, inward across the cell membrane); in Figure 4.15b they are moving in opposite directions. The transport of two substances in the same direction is called **cotransport** (or sometimes *symport*). The example of co-

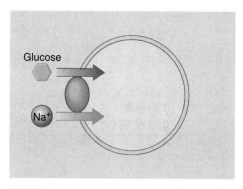

(a) Cotransport

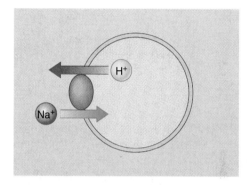

(b) Counter transport

FIGURE 4.15 Secondary active transport. *Circles spanning the cell membrane represent transport proteins.* **(a)** *Sodium-linked glucose transport, an example of cotransport.* **(b)** *Sodium-proton exchange, an example of countertransport.*

transport shown in Figure 4.15a is *sodium-linked glucose transport*, which couples the inward flow of Na^+ with the inward flow of glucose molecules. In this process, Na^+ moves down its electrochemical gradient and releases energy that drives the flow of glucose against its concentration gradient. The transport of two substances in opposite directions is **countertransport** (or sometimes *antiport* or *exchange*). The example of countertransport shown in Figure 4.15b is *sodium-proton exchange*, in which the inward flow of Na^+ is coupled to the outward flow of protons (H^+). Here, energy released from the flow of Na^+ down its electrochemical gradient is harnessed to drive the flow of H^+ up its electrochemical gradient.

These examples illustrate that secondary active transport may involve either cotransport or countertransport, and that the actively transported substance may move into cells in some cases and out of cells in others. Note that the direction of transport *alone* does not tell us anything about which substance is being transported actively or which is flowing passively; this is determined entirely by the directions of the transported substances' electrochemical gradients. The substance that is moving up its electrochemical gradient is the one that is being actively transported, and thus the other substance is moving down its electrochemical gradient and is being passively transported. Energy released from the passive movement

TABLE 4.3 CHARACTERISTICS OF
TRANSPORT PROCESSES

		MEDIATED TRANSPORT			
	SIMPLE DIFFUSION	**Facilitated diffusion**		**Active transport**	
		Channel	**Carrier**	**Primary**	**Secondary**
Direction of net flux	Down electrochemical gradient	Down electrochemical gradient	Down electrochemical gradient	Up electrochemical gradient	Up electrochemical gradient
Transport protein required?	No	Yes	Yes	Yes	Yes
Requires energy?	No	No	No	Yes	Yes
Energy source	(Not applicable)	(Not applicable)	(Not applicable)	ATP or other chemical energy source	Electrochemical gradient of another solute
Saturation?	No	Sometimes	Yes	Yes	Yes
Specificity?	No	Yes	Yes	Yes	Yes
Character of transported substance	Hydrophobic (nonpolar)*	Hydrophilic (ionized or polar)	Hydrophilic (ionized or polar)	Hydrophilic (ionized or polar)	Hydrophilic (ionized or polar)
Examples	Fatty acids, O_2, CO_2	Inorganic ions (Na^+, K^+, Cl^-, Ca^{2+})	Organic molecules (glucose)	Inorganic ions (Na^+, K^+, H^+, Ca^{2+})	Organic molecules and inorganic ions (glucose, amino acids, H^+, Ca^{2+})

*A notable exception to this is water, which can readily penetrate most lipid bilayers due to its small size.

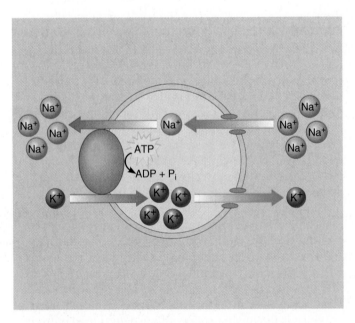

FIGURE 4.16 Pumps and leaks in a cell. *Active transport of Na^+ and K^+ by the Na^+/K^+ pump (oval at left) is counteracted by passive leaks of these ions through channels (at right).*

of this substance down its gradient is used to drive the flow of the other substance up its gradient.

Now that we have discussed active transport, we have examined all the different types of processes that operate in the transport of substances across membranes. The major characteristics of these processes are summarized in Table 4.3.

Pumps and Leaks: How Active and Passive Transport Mechanisms Coexist in Cells

We have seen that intracellular and extracellular fluids differ substantially in composition (see Table 4.1). Now it should be apparent that these differences in composition are created when certain substances are actively transported into or out of cells, which raises or lowers (respectively) the concentration of these substances in intracellular fluid with respect to extracellular fluid.

Normally, the composition of intracellular fluid remains fairly steady, which raises a question: If substances are actively "pumped" into or out of cells, why don't intracellular concentrations change? The answer lies in the fact that these substances are simultaneously transported passively (are "leaked") across the membrane in the opposite direction but at the same rate, such that the net flux across the membrane (active and passive transport combined) is zero (Figure 4.16). Thus even though Na^+, for example, are actively transported out of cells by the Na^+/K^+ pump, they leak passively into cells through channels and other proteins. For every Na^+ that is transported out under normal conditions, another leaks in passively, such that the intracellular Na^+ concentration does not change. Likewise, for every K^+ that is actively pumped in, another passively leaks out, such that the intracellular K^+ concentration does not change.

Quick Test 4.6

1. Define the following terms: *primary active transport, secondary active transport, pump, leak.*

2. When a substance moves up its electrochemical gradient, does it require energy or release energy?

3. Suppose that a transport protein couples inward flow of Na^+ with outward flow of Ca^{2+}. Is cotransport or countertransport occurring?

4. Suppose that glucose moves into a cell by means of sodium-linked cotransport, and that glucose concentrations are 2.5 mM inside the cell and 1.0 mM outside the cell. Is sodium being transported actively or passively?

OSMOSIS: PASSIVE TRANSPORT OF WATER ACROSS MEMBRANES

Normally, the amount of water contained in cells does not change significantly, but it can change in certain extreme or pathological conditions. In some types of kidney dysfunction, for instance, excess water is retained in the body, and water begins to move into cells, causing them to swell. The swelling of brain cells disrupts nervous system function and typically produces headache, nausea, and vomiting, indicating that a condition known as *water intoxication* has occurred. Severe cases can progress to seizures, coma, and death. In contrast, severe dehydration can cause water to move out of cells, causing them to shrink. This, too, is detrimental to brain function and can be fatal.

Under normal conditions, most cells in the body neither swell nor shrink because water does not move across their membranes; forces that might cause water to cross membranes are absent. In some locations in the body, however, movement of water across cell membranes is a normal occurrence. In order for glands to secrete fluids such as sweat, tears, or saliva, for instance, water must be transported across specialized epithelial cells. Also, when you drink water, it moves across epithelial cells lining the intestine to be absorbed into the bloodstream. Such water movement is important because it ultimately affects the volume and composition of all body fluids.

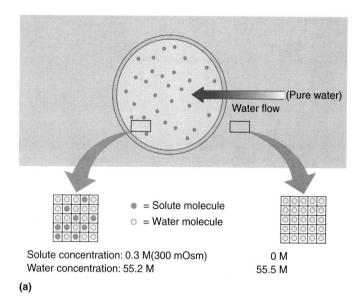

Solute concentration: 0.3 M(300 mOsm)   0 M
Water concentration: 55.2 M   55.5 M

(a)

● = Solute molecule
○ = Water molecule

Solute concentration: 300 mOsm   1000 mOsm
Water concentration: 55.2 M   54.5 M

Water flow → (1 molar sucrose)

(b)

FIGURE 4.17 Osmosis. (a) *Flow of water into a red blood cell placed in pure water. Water molecules flow passively into the cell down their concentration gradient, which is also to an area of higher solute concentration. Boxes in the magnified samples of intracellular and extracellular fluid represent the presence of water or solute molecules.* **(b)** *Flow of water out of a red blood cell placed in 1 M sucrose. Note that water again moves from higher to lower water concentration, and from lower to higher solute concentration.*

Water transport has always had a "special" status in physiology because of its influence on cell volume, and also because its description has its own unique terminology. However, water transport is no different from some other types of transport we've already studied; in fact, it is simpler in many respects. Water transport is simple because water flow across membranes is *always* passive, is unaffected by membrane potentials, and is always driven by its own concentration gradient. The flow of water across a membrane down its concentration gradient is called **osmosis.**

The effects of osmosis are familiar to nearly everyone who has studied high school biology. When you place red blood cells in pure water, water flows into the cells, which then swell up, burst, and release hemoglobin into the water, a process called *hemolysis.* Figure 4.17 depicts the conditions that drive osmosis. Note in Figure 4.17a that when a red blood cell is placed in pure water, water

It might seem strange that physiologists would use the term *osmotic pressure* to describe a solution's total solute concentration, but this is because the definition of osmotic pressure is purely *operational;* that is, it can be properly defined only in reference to an experimental measurement. The relevant experimental set-up for defining osmotic pressure is shown in the box figure, in which two compartments in a chamber are separated by a *semipermeable* membrane, which is permeable to water but not to any solute. One compartment contains pure water, while the other compartment contains the *test solution* having an unknown osmotic pressure. The compartment containing the test solution is equipped with a piston that allows us to apply pressure to that side. When no pressure is applied, water flows down its concentra-

tion gradient into the compartment with the test solution (panel a). However, we can slow down this flow of water by applying pressure to the test solution (panel b). If we increase the pressure to a sufficient degree, water flow stops (panel c). At this point, the applied pressure is equal to the osmotic pressure of the test solution.

As a result of measurements of the type just described, scientists were able to demonstrate that the following three rules apply to the osmotic pressure of a solution: (1) Osmotic pressure increases as the solute concentration rises. (2) As a general rule, osmotic pressure depends on the total solute concentration rather than the molecular identities of the solutes. (3) If a solution contains solute molecules that ionize into separate particles (such as sodium chloride, NaCl, which dissociates

into Na^+ and Cl^- ions), the osmotic pressure depends on the total concentration of particles. Thus, a 1 molar solution of NaCl, which contains 1 mole per liter of Na^+ ions and 1 mole per liter of Cl^- ions, has twice the osmotic pressure of a 1 molar solution of glucose, which does not ionize.

The osmotic pressure of a solution is given by the following equation:

$$\pi = CRT$$

where C is the total solute concentration, R is the universal gas constant (0.082 liter-atm/mole-°K), and T is the absolute temperature (°K). Using this equation, it can be shown that at body temperature, the osmotic pressure of a 1 osmolar solution is surprisingly large—a full 24.6 atmospheres, which is equivalent to over 18,000 mm Hg!

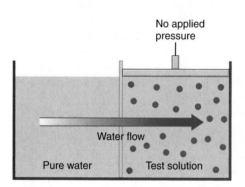

(a)

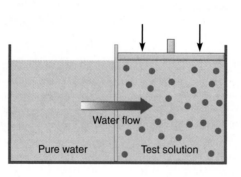

(b)

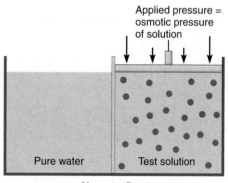

(c)

molecules flow into the cell because in doing so they are flowing down their concentration gradient. This gradient results from the presence within the cell of various solutes (Na^+, K^+, and proteins, among others) at a combined concentration of about 300 millimolar. The presence of these solute molecules reduces the water concentration inside the cell by taking up a certain amount of space that would have been occupied by water molecules, which means that fewer water molecules are present in a given volume of solution. In other words, the *concentration of water within the cell is lower than outside the cell.*

At body temperature the concentration of pure water is 55.5 molar, and the normal total solute concentration in intracellular fluid is 300 millimolar (0.3 molar). Thus the in-

tracellular water concentration is roughly 55.2 molar (55.2 molar − 0.3 molar; see Figure 4.17a). Therefore, when a cell is placed in pure water there is a 0.3 molar smaller water concentration inside it than outside it, so water molecules move passively into the cell down their concentration gradient, causing the cell to swell and eventually burst.

The direction of passive water flow into or out of a cell depends on the direction of the water concentration gradient across the plasma membrane. Thus, for instance, when a cell is placed in a solution of sucrose at a concentration of 1 molar (1000 millimolar), the cell shrinks because water moves out (Figure 4.16b). Outward water movement occurs because the total solute concentration is lower inside the cell than it is outside, which makes the water concentration *higher* inside the cell. Therefore, water moves passively out of the cell.

When discussing osmosis, physiologists prefer to speak in terms of *solute concentration* rather than *water concentration* because the concentration of water in body fluids is only about 0.2% lower than that in pure water. Most differences in water concentration that appear across cell membranes are even smaller than that. However, note that any difference in the total solute concentration across a membrane is always accompanied by a difference in water concentration; it is impossible to have one without the other.

Osmolarity

The total solute concentration of a solution is known as its **osmolarity.** A solution containing 1 mole of any solute per liter is said to be at a concentration of 1 *osmolar* (1 Osm). One mole of solute is referred to as one *osmole.* Thus, 1 liter of a solution containing 0.1 molar glucose plus 0.1 molar sucrose contains 0.2 osmoles of solute and is therefore a 0.2 osmolar (0.2 Osm) solution. In physiology, the terms *milliosmole* and *milliosmolar* (mOsm) are more frequently used because the solute concentration of body fluids is only a fraction of one osmolar. A 1 milliosmolar (1 mOsm) solution contains one milliosmole (1/1000 of an osmole) of solute per liter. The normal osmolarity of intracellular and extracellular fluid is approximately 300 mOsm, which means that its total solute concentration is 300 milliosmoles per liter.

Two solutions having the same osmolarity are said to be **iso-osmotic.** Thus, a 300 millimolar glucose solution is iso-osmotic with intracellular fluid because both solutions are 300 milliosmolar. A solution whose osmolarity is higher than another is said to be **hyperosmotic;** a solution with lower osmolarity is **hypo-osmotic.** When two solutions are iso-osmotic they have not only the same solute concentration, but also the same water concentration. In a hyperosmotic solution, the water concentration is lower because the solute concentration is higher. In a hypo-osmotic solution, the water concentration is higher because the solute concentration is lower.

Note that a solution's osmolarity is determined by the *total* concentration of dissolved particles, which depends not only on the concentrations of particular solutes, but also whether the solutes ionize and to what degree. A solution containing sodium chloride (NaCl) at a concentration of 150 millimolar, for example, contains ions at a concentration of 300 millimolar (twice the NaCl concentration) and therefore has an osmolarity of 300 mOsm. This is because NaCl dissociates completely into Na^+ and Cl^- ions when dissolved in water.

Osmotic Pressure

Sometimes physiologists use the term *osmotic pressure* in place of *osmolarity,* because both terms refer to the same thing—a solution's total solute concentration (Discovery:

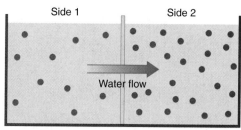

Solute concentration:	300 mOsm (0.3 M)	500 mOsm (0.5 M)
Water concentration:	55.2 M	55.0 M
Osmotic pressure:	7.4 atm	12.3 atm
Osmotic pressure gradient ($\Delta\pi$):	12.3 atm − 7.4 atm = 4.9 atm	

FIGURE 4.18 Movement of water up an osmotic pressure gradient. *When a semipermeable membrane—which is permeable to water but not to solute—separates two solutions in a chamber, water flows down its concentration gradient toward the side with higher solute concentration. Because osmotic pressure increases as solute concentration increases, water also flows from lower to higher osmotic pressure, or up the osmotic pressure gradient ($\Delta\pi$).*

Determining the Osmotic Pressure of a Solution). The **osmotic pressure** of a solution (which is given the symbol π) is an indirect measure of its solute concentration and is expressed in ordinary units of pressure, such as atmospheres or millimeters of mercury (mm Hg). As the total solute concentration (osmolarity) increases, the osmotic pressure increases. Because the osmolarity of a solution is inversely related to its water concentration, as water concentration decreases, osmotic pressure increases.

Due to the relationship between osmotic pressure and water concentration, we can say that whenever there is a water concentration gradient across a membrane, there is also an osmotic pressure gradient. Furthermore, we can say that when water flows down its concentration gradient, it is also flowing *up* an osmotic pressure gradient (Figure 4.18). In the figure the total solute concentration on the left side of a semipermeable membrane (side 1) is less than that on the right (side 2), making the water concentration higher on side 1 than side 2. Water therefore flows from side 1 to side 2 down its concentration gradient. However, the higher solute concentration on side 2 also means a higher osmotic pressure on that side. Thus, water is flowing from lower to higher osmotic pressure, or up the osmotic pressure gradient ($\Delta\pi$).

Note that the tendency of water to flow *up* an osmotic pressure does not violate the rules concerning gradients, which state that substances tend to move passively *down* chemical and electrochemical gradients, not up. When water moves up a gradient of osmotic pressure, it is merely moving down a chemical gradient—its own concentration gradient.

Tonicity

When someone suffers massive hemorrhage, it is quite common for emergency medical personnel to administer

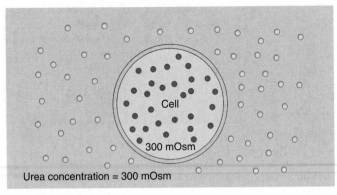

Urea concentration = 300 mOsm

● = Impermeant solutes
○ = Urea

(a)

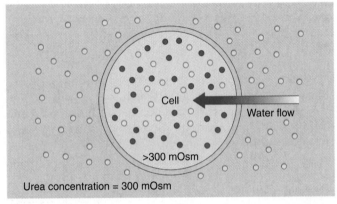

Urea concentration = 300 mOsm

● = Impermeant solutes
○ = Urea

(b)

FIGURE 4.19 The distinction between osmolarity and tonicity. (a) *A cell containing impermeant solutes (purple dots) at a concentration of 300 mOsm placed in an iso-osmotic solution containing the permeant solute urea (yellow dots) at a concentration of 300 mOsm.* **(b)** *Swelling that occurs over time as urea moves into the cell down its concentration gradient. Because of the resultant increase in the solute concentration inside the cell, water then flows into the cell down its concentration gradient, indicating that the extracellular solution is hypotonic.*

cell membranes. A solution is said to be **isotonic** when it does not alter cell volume; when a cell comes into contact with an isotonic solution, it neither shrinks nor swells. (Here, we are referring to the cell's final volume. Under certain conditions, a cell may shrink or swell initially, even if the solution is isotonic.) In contrast, a solution that causes cells to shrink is **hypertonic,** whereas a solution that causes cells to swell is **hypotonic.**

Exercise Link

As Bill and Jane ran the marathon, they began to sweat. Sweat is a fluid, composed of water and solutes (especially sodium and chloride), that is derived from blood plasma, but it is less concentrated than plasma. As a result, Bill and Jane's plasma became more concentrated, causing water to move out of their cells, making them shrink about 2%. This shrinkage during exercise is not usually a serious issue, but cellular volume changes that occurred after Bill drank a large amount of pure water following the marathon caused his severe headache. Based on what you know about osmotic pressure and its influence on cell volume, can you determine what might have caused Bill's headache?

The distinction between osmolarity and tonicity is best illustrated in the situation depicted in Figure 4.19. To start, a cell that has just been placed in a solution of *urea,* a substance that permeates most cell membranes readily (Figure 4.19a). The osmolarity of this solution equals the initial osmolarity inside the cell (300 mOsm), making it iso-osmotic with the intracellular fluid. Under these conditions, water does not move into or out of the cell because the water concentration is the same on both sides of the cell membrane. Over time, urea moves into the cell due to the initial presence of an inwardly directed concentration gradient (Figure 4.19b). Intracellular solutes, in contrast, are retained inside the cell because they are relatively *impermeant;* that is, they cannot permeate the membrane. As a result of urea movement, the cell gains solute, and the osmolarity of the intracellular fluid goes above its initial value of 300 mOsm. (The volume of extracellular solution is taken to be very large, so extracellular osmolarity is not affected by solute movement into the cell.) Since the total solute concentration is now higher inside the cell than outside, water moves into the cell, and the cell swells. The cell swells even though the extracellular solution is originally iso-osmotic to the intracellular fluid. Because the cell swells, the extracellular solution is hypotonic.

Whether a given solution causes cells to swell, shrink, or stay the same is determined solely by the concentration of impermeant solutes contained in it. A solution is

a saline (sodium chloride) solution intravenously to replace the lost blood volume, which is done to keep the person alive while awaiting a blood transfusion. The administered solution is referred to as *isotonic saline* because its total solute concentration has been carefully formulated to match that of extracellular fluid. This formulation is important because such a solution will not alter cell volumes when injected into the bloodstream. A more concentrated saline solution would cause water to flow out of cells, making them shrink; a less concentrated solution would cause water to flow into cells, making them swell.

Whereas a solution's osmolarity is based solely on its total solute concentration, its *tonicity* is determined by how it affects cell volume, which depends not only on the solute concentration, but also the *solute permeability* of

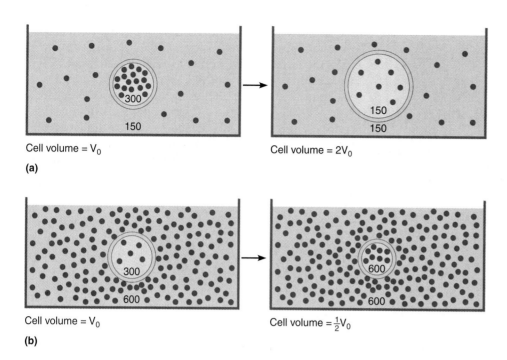

(a)

Cell volume = V_0 → Cell volume = $2V_0$

(b)

Cell volume = V_0 → Cell volume = $\frac{1}{2}V_0$

FIGURE 4.20 Changes in the volumes of cells placed in hypotonic and hypertonic solutions. *All solutes (dots) are assumed to be impermeant; numbers inside and outside cells indicate solute concentrations (mOsm).* **(a)** *A cell placed in a 150 mOsm solution, which is hypotonic, swells until its volume reaches twice the original volume (V_o).* **(b)** *A cell placed in a 600 mOsm solution, which is hypertonic, shrinks until its volume is reduced to one-half the original volume.*

If a cell were placed in a solution containing 100 mOsm impermeant solutes, how would its final volume compare to its initial volume?

isotonic if it contains impermeant solutes at a concentration of 300 mOsm, the normal concentration of impermeant solutes in intracellular fluid. If the concentration of impermeant solutes is greater or less than 300 mOsm, the solution will be hypertonic or hypotonic, respectively. A solution's tonicity is not affected by the concentration of any permeant solutes that may or may not be present. The characteristics that determine a solution's osmolarity or tonicity are summarized in Table 4.4.

When a cell comes into contact with solutions that are hypotonic or hypertonic, the degree to which it swells or shrinks is determined by the initial concentrations of impermeant solutes in intracellular and extracellular fluid. When, for example, a cell containing 300 mOsm impermeant solutes is placed in a large volume of solution containing impermeant solutes at half that concentration (150 mOsm), the cell will swell due to water flowing in (Figure 4.20a). Because the amount of solute inside the cell is fixed, the increase in cell volume causes the intracellular solute concentration to decrease, but the cell continues to swell until the concentration reaches 150 mM, at which point the cell has expanded to twice its original volume (designated V_o). Swelling stops at this point because there is no longer a concentration gradient of water, and hence no driving force to cause water to flow across the membrane. (In Figure 4.17a, the cell continues to swell until it bursts because the solute concentration is always higher inside the cell than outside.) By contrast, when an identical cell is placed in a solution containing 600 mOsm impermeant solutes (Figure 4.20b), it shrinks, which causes the intracellular solute concentration to rise. The cell continues to shrink until the concentration inside the cell reaches 600 mOsm, at which point it has shrunk to one-half its original volume.

TERMS	DEFINITIONS AND SOLUTE CONCENTRATIONS
Osmolarity	Total concentration of permeant and impermeant solutes
Iso-osmotic*	300 mOsm (permeant + impermeant)
Hypo-osmotic*	Less than 300 mOsm (permeant + impermeant)
Hyperosmotic*	Greater than 300 mOsm (permeant + impermeant)
Tonicity	Concentration of impermeant solutes relative to intracellular fluid
Isotonic*	300 mOsm (impermeant)[†]
Hypotonic*	Less than 300 mOsm (impermeant)[†]
Hypertonic*	Greater than 300 mOsm (impermeant)[†]

TABLE 4.4 DISTINCTIONS BETWEEN OSMOLARITY AND TONICITY

*These designations are relative to a cell containing 300 mOsm solutes, which are assumed to be impermeant.
[†]Permeant solutes may or may not be present.

It would be three times larger.

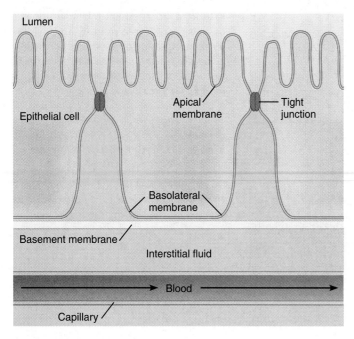

Lumen

Epithelial cell

Apical membrane

Tight junction

Basolateral membrane

Basement membrane

Interstitial fluid

Blood

Capillary

FIGURE 4.21 Diagram of the general structure of an epithelium.

Quick Test 4.7

1. Suppose that a cell containing only impermeant solutes at a concentration of 300 mOsm is placed in a solution whose total solute concentration is 400 mOsm. Will the cell swell or shrink?

2. Is the extracellular solution in Question 1 isotonic, hypertonic, or hypotonic? Iso-osmotic, hyperosmotic, or hypo-osmotic?

3. Why does water move in the situation described in Question 1?

4. Is the osmotic pressure of the solution in Question 1 higher or lower than intracellular fluid?

EPITHELIAL TRANSPORT: MOVEMENT OF MOLECULES ACROSS TWO MEMBRANES

Up to this point we have been focusing on cell membrane function as it pertains to the transport of materials into or out of cells. But in many epithelial tissues, cell membranes function to transport materials *across* cells. This occurs, for instance, when the intestine delivers nutrients to the bloodstream, or when sweat glands produce sweat.

Recall from Chapter 1 that one function of epithelial tissues is to form barriers between the body's internal environment and the external environment, and also between different fluid compartments within the body. But in addition to acting as barriers, certain epithelia (such as those lining the stomach, intestine, and secretory glands)

are also able to transport materials into the internal environment from outside (a process called **absorption**) or from the internal environment to the outside **(secretion).**

In order for an epithelium to absorb or secrete materials, the cells that make up the epithelial tissue must transport substances inward across the membrane on one side of the cell and outward across the membrane on the opposite side of the cell. For this to occur, the membranes on either side must possess different transport systems. Because the membranes on the two sides are distinctly different (in both structure and function), epithelial cells are said to be *polarized.* In this section we see how the polarity of epithelial cells gives them the ability to absorb or secrete materials. We start first with an overview of epithelial structure.

Epithelial Structure

In an epithelial cell layer of the type that is specialized for absorption or secretion (Figure 4.21), one side of an epithelial cell faces the lumen of a body cavity. The membrane on this side is called the *lumen-facing* membrane or **apical membrane.** The membrane on the opposite side faces the internal environment and is in contact with interstitial fluid, which exchanges materials with the blood. This *blood-facing* membrane or **basolateral membrane** rests on a **basement membrane** consisting of noncellular material that is relatively permeable to most substances. The basement membrane anchors the basolateral membrane and provides physical support for the epithelial layer.

An important feature of epithelial tissues is that adjacent cells are joined by tight junctions that limit the passage of material through the spaces between cells (called *paracellular* spaces). These junctions permit the fluids on either side of the cell layer to differ in composition. Tight junctions are important in maintaining homeostasis because even though the composition of fluid in the lumen of an organ may vary widely (consider the variety of foodstuffs that enter the lumen of the stomach), the composition of interstitial fluid should not because it is part of the body's highly regulated internal environment. The "tightness" of these junctions varies from location to location; *tight epithelia* have junctions with extremely low permeabilities, whereas *leaky epithelia* have junctions that are more permeable.

Epithelial Solute Transport

The mechanisms whereby epithelial cells transport solute molecules across an epithelial layer are illustrated in Figure 4.22, which depicts epithelial cells of the kinds that absorb nutrients in the intestines, or that function in body fluid regulation in kidney tubules. Note first that a comparison of the transport systems in the apical and ba-

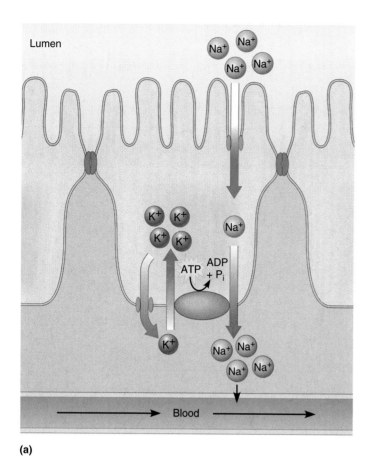

(a)

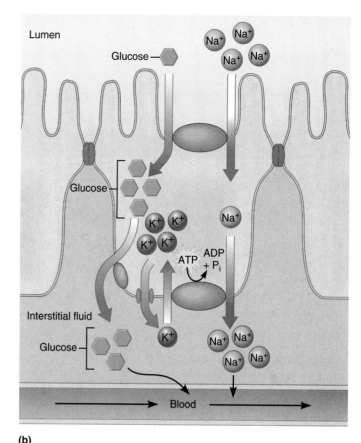

(b)

FIGURE 4.22 Mechanisms of epithelial solute transport. (a) *Absorption of sodium ions.* **(b)** *Absorption of glucose and sodium ions.*

If glucose is being transported across the basolateral membrane by a carrier (see Figure 4.22b), is the glucose concentration in interstitial fluid higher or lower than that inside the cell?

solateral membranes of both cells reveals the previously mentioned polarity: Only the basolateral membranes of these cells have Na^+/K^+ pumps, which transport Na^+ out of the cell and K^+ into the cell. In addition, only the basolateral membranes possess K^+ channels, which allow K^+ to leak out of the cells down its electrochemical gradient. This pump-leak system maintains a nearly constant concentration of K^+ inside the cells, a concentration higher than that outside the cells. The Na^+/K^+ pump also maintains a low Na^+ concentration inside the cells, which creates an inwardly directed Na^+ gradient. Note as well that in both cells the apical membrane possesses transport systems that the basolateral membrane lacks.

In the cell in Figure 4.22a, which absorbs Na^+, Na^+ leaks into the cell through Na^+ channels, which are present in the apical membrane only. This leak is counteracted by the active transport of Na^+ out of the cell across the basolateral membrane. Sodium ions enter the cell on one side and exit it on the other side—they are transported completely across the cell from the lumen to the interstitial fluid.

Note additionally that although Na^+ flows passively across the apical membrane, Na^+ transport across the cell as a whole is *active*, because entry of Na^+ into the cell across the apical membrane depends on the presence of an inwardly directed electrochemical gradient, which in turn depends on the ability of the Na^+/K^+ pump to remove Na^+ from the cell. If the pump were to stop working, the concentration of Na^+ inside the cell would rise; eventually, the Na^+ gradient across the apical membrane would disappear, bringing Na^+ entry to a halt. Because Na^+ transport across the cell is active, Na^+ can be absorbed from lumen to interstitial fluid up an electrochemical gradient.

In the cell in Figure 4.22b, which absorbs both Na^+ and glucose, Na^+ again leaks passively inward across the apical membrane, but it enters the cell by way of a sodium-linked cotransport system, which couples the inward movement of Na^+ with the inward movement of

Lower

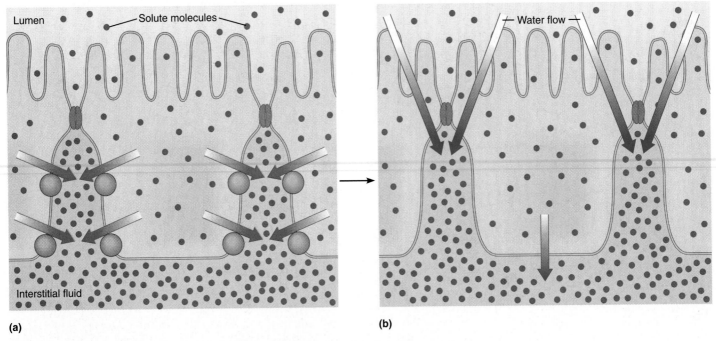

Lumen

Solute molecules

Water flow

Interstitial fluid

(a)

(b)

FIGURE 4.23 Epithelial water transport in an epithelium that absorbs water and solutes. (a) *Pumps (circles) in the basolateral membrane actively transport solute molecules (dots) into interstitial fluid, raising the solute concentration in interstitial fluid, particularly in the paracellular spaces.* **(b)** *Active solute transport creates a gradient of osmotic pressure across the epithelium that drives passive water flow from the lumen to interstitial fluid (large arrows).*

In this situation, does solute transport tend to increase or decrease the water concentration in interstitial fluid?

glucose. The Na^+ electrochemical gradient provides the energy that drives the secondary active transport of glucose up its concentration gradient into the cell. Glucose then passively exits the cell by moving down its concentration gradient via a carrier protein in the basolateral membrane. Although glucose exit across the basolateral membrane is passive, glucose transport across the epithelium as a whole is active because it depends on the presence of the inwardly directed Na^+ electrochemical gradient, which in turn depends on the action of the basolateral Na^+/K^+ pump. Glucose can therefore be transported from the lumen to interstitial fluid up a concentration gradient.

Epithelial Water Transport

Whenever epithelial cells secrete or absorb fluid, water transport occurs by osmosis. This happens, for instance, when you sweat, when the pancreas secretes pancreatic juice, or when you replenish your body fluids by drinking a glass of water.

Epithelia absorb or secrete water by first using the active transport of solutes to create a difference in osmotic pressure—an osmotic pressure gradient—between the solutions on either side of the cell layer. Water then flows across the epithelium passively by osmosis. Because water flow occurs in response to transport of solutes, water transport is said to be *secondary to* solute transport.

In epithelial water transport (Figure 4.23), an epithelium creates an osmotic pressure gradient (or simply an *osmotic gradient*) to absorb water. First, epithelial cells actively transport solute molecules (dots) across the basolateral membrane into the interstitial fluid (Figure 4.23a). As a result, the solute concentration is slightly higher outside the cells than inside—particularly in the paracellular spaces, where the diffusion of solute molecules is hampered by narrowness of the passages. (This is indicated in the figure by the higher density of dots between the cells.) Because this solute transport creates a difference in osmotic pressure between the solutions on either side of the epithelium, water then flows across the cell layer by osmosis (Figure 4.23b). Note that the direction of water flow is toward interstitial fluid, or up the osmotic pressure gradient; for epithelia that secrete fluids, the same idea pertains, except that solute and water transport go in the opposite direction—away from interstitial fluid.

Decrease

Of all lethal hereditary diseases, cystic fibrosis is the most common among Caucasians, affecting about 1 in every 2500 individuals. This condition is notorious for its effects on the respiratory tract, although it affects other systems as well. In afflicted individuals, respiratory passages become clogged with thick, viscous mucus that is difficult to dislodge, even with vigorous coughing. Breathing is difficult, and victims run the risk of choking to death on their own secretions unless strenuous effort is made to clear the lungs several times a day. Victims frequently die in their late 20s of pneumonia, as the mucus-clogged airways offer a fertile environment for bacteria.

In all individuals, mucus is secreted by certain cells in the epithelium that lines respiratory passages. Under normal conditions, the epithelium also secretes a watery fluid that dilutes the mucus, making it thinner and easier to clear from airways. In cystic fibrosis the secretion of the watery fluid is impaired. The undiluted mucus is thicker and quite difficult to clear from the respiratory passages.

In the respiratory epithelium, as in all epithelia that transport fluid, water transport is secondary to solute transport. To drive water secretion, cells of the respiratory epithelium actively transport chloride (Cl^-) from interstitial fluid to the airway lumen, creating a negative electrical potential in the lumen that drives the passive flow of sodium (Na^+) in the same direction. Movement of Na^+ and Cl^- raises the osmotic pressure of fluid bathing the luminal side of the epithelium. As a consequence, water moves passively down the osmotic gradient from interstitial fluid to lumen.

Scientists recently discovered that cystic fibrosis is caused by a defect in a type of chloride channel protein found in the apical membranes of epithelial cells in the respiratory tract and elsewhere. The defect directly impedes Cl^- transport, which indirectly interferes with transport of Na^+ (see the box figure). As a result, the epithelium cannot create the osmotic gradient necessary for water secretion.

It has long been known that cystic fibrosis is caused by a single abnormal gene, and recent biochemical and electrophysiological studies have shown that this gene codes for a portion of the chloride channel protein. Efforts are under way to develop a therapy that will permit normal genes to be inserted into the DNA of diseased epithelial cells. If this can be accomplished, these cells will then manufacture normal channel proteins, which should allow epithelia to secrete fluid in the normal manner.

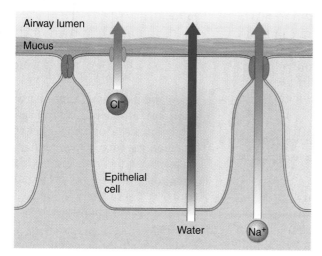

Normal solute and water transport

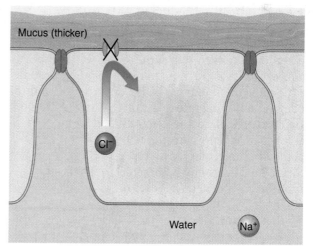

Defective solute and water transport in cystic fibrosis

In certain pathological conditions, epithelial water transport is either excessive or insufficient. In *cholera,* for example, bacterial toxins stimulate certain epithelial cells in the small intestine to oversecrete solutes and water, resulting in massive fluid loss through diarrhea. In *cystic fibrosis,* the opposite problem occurs: The epithelium lining the respiratory airways does not transport enough solute to attain adequate fluid production. As a result the lungs become clogged with mucus, which hampers breathing and can lead to serious respiratory infection (When It Goes Wrong: Cystic Fibrosis, p. 127).

Quick Test 4.8

1. Define the following terms: *absorption, secretion, apical membrane, basolateral membrane, basement membrane.*
2. Why are epithelial cells said to be polarized? How is polarity important to the function of certain epithelia?
3. What does it mean to say that epithelial water transport is secondary to solute transport?
4. In epithelial water transport, is water flow passive or active?

CHAPTER SUMMARY

Factors Affecting the Direction of Transport, p. 101

Molecules that cross cell membranes move either by simple diffusion through the lipid bilayer or by mediated transport, which involves specialized transport proteins. Active transport requires energy and is carried out by proteins called pumps. Passive transport does not require energy and includes simple diffusion and some forms of mediated transport. Transported molecules are generally influenced by three types of driving forces: (a) chemical driving forces, which are due to the presence of concentration gradients; (b) electrical driving forces, which reflect the influence of a cell's membrane potential on the movement of ions; and (c) electrochemical driving forces, a combination of the chemical and electrical driving forces that represents the net force acting on molecules. Passively transported molecules move in the direction of the electrochemical force or down their electrochemical gradient. Actively transported molecules go against the electrochemical force or up their electrochemical gradient.

> IP Nervous 1, The Membrane Potential, pages 1–17
>
> IP Fluids and Electrolytes, Introduction to Body Fluids, pages 12–15

Factors Affecting the Rate of Transport, p. 108

The rate at which a substance moves across a membrane is measured as a flux. The flow of molecules in one direction or the other is the unidirectional flux; the difference between the unidirectional fluxes is the net flux. For a passively-transported substance, the rate of transport depends on three factors: (a) the magnitude of the driving force, (b) membrane surface area, and (c) the permeability of the membrane to that particular substance. For an actively-transported substance, the rate of transport is determined by two factors: (a) the rate at which individual pump proteins transport the substance, and (b) the number of pumps in the membrane.

Simple Diffusion: Passive Transport Through the Lipid Bilayer, p. 111

Diffusion is movement of molecules from one location to another as a result of their random thermal motion. In simple diffusion, membrane permeability is determined by the following factors: (a) the lipid solubility of the diffusing substance, (b) the size and shape of diffusing molecules, (c) temperature, and (d) membrane thickness.

Facilitated Diffusion: Passive Transport Through Membrane Proteins, p. 113

Facilitated diffusion involves two types of transport proteins: (a) carriers, which bind molecules on one side of a membrane and transport them to the other side by means of a conformational change; and (b) channels, which possess pores that extend from one side of the membrane to the other. In facilitated diffusion, membrane permeability is determined by two factors: (a) the transport rate of individual carriers or channels, and (b) the number of carriers or channels in the membrane.

Active Transport, p. 115

There are two basic forms of active transport: primary active transport, which uses ATP or some other chemical energy source, and secondary active transport, which uses the electrochemical gradient of one substance as a source of energy to drive the active transport of another substance.

> IP Nervous 1, The Membrane Potential, pages 13–15
>
> IP Urinary, Early Filtrate Processing, page 11

Osmosis: Passive Transport of Water Across Membranes, p. 119

Water flow across membranes affects cell volumes and is involved in the secretion and absorption of fluids. Water movement (osmosis) is always passive and driven by a concentration gradient of water. Because the water concentration of a solution decreases as its solute concentration increases, a difference in total solute concentration (osmolarity) across a membrane implies the existence of a water concentration gradient. Because the osmotic pressure of a solution increases with increasing solute concentration, the flow of water down its concentration gradient is equivalent to flow up an osmotic pressure gradient. The volume of a cell is determined by the tonicity of the solution surrounding it, which depends on the solute concentration and the permeability of the membrane to the solutes that are present.

IP Urinary, Early Filtrate Processing, pages 5–8

Epithelial Transport: Movement of Molecules Across Two Membranes, p. 124

Certain epithelia are specialized to transport materials into or out of the body's internal environment, termed absorption and secretion, respectively.

Epithelial cells can do this because the membranes on either side differ structurally and functionally, a phenomenon known as polarity. Epithelia form barriers between body fluid compartments or between the internal and external environments. The basolateral membrane, which rests on a noncellular basement membrane faces the internal environment; the opposite membrane is the apical membrane.

IP Urinary, Early Filtrate Processing, pages 5–8

EXERCISES

Multiple-Choice Questions

1. Passive transport of K^+ through channels is an example of
   a) simple diffusion.
   b) facilitated diffusion.
   c) primary active transport.
   d) secondary active transport.

2. For a substance crossing a cell membrane, the chemical driving force
   a) depends only on the concentration gradient, regardless of whether or not the substance is an ion.
   b) depends only on the concentration gradient if the substance is uncharged, but also depends on the electrical force if the substance is an ion.
   c) is the total driving force on the substance, even if it is an ion.
   d) is the force that pushes molecules across the membrane, but only if the substance is actively transported.

3. A factor that does *not* affect membrane permeability in simple diffusion is
   a) the size of diffusing molecules.
   b) membrane thickness.
   c) the size of the concentration gradient.
   d) temperature.

4. One example of primary active transport is the

   a) transport of Ca^{2+} up an electrochemical gradient by a protein that hydrolyzes ATP.
   b) transport of Ca^{2+} up an electrochemical gradient by a protein that couples Ca^{2+} flow to the flow of Na^+ down an electrochemical gradient.
   c) movement of Ca^{2+} down an electrochemical gradient through channels.
   d) transport of glucose molecules down a concentration gradient by carriers.

5. The electrical driving force on Na^+ crossing a cell membrane is
   a) outward if the membrane potential is negative.
   b) inward if the membrane potential is negative.
   c) inward if the membrane potential is positive.
   d) always inward, regardless of the sign of the membrane potential.

6. When the membrane potential is equal to the equilibrium potential of Na^+ (E_{Na} = +60 mV)
   a) Na^+ move into a cell down their electrochemical gradient.
   b) Na^+ move into a cell as a result of the electrical driving force.
   c) Na^+ move out of a cell down their electrochemical gradient.

   d) the net flux of Na^+ is zero because they are at equilibrium.

7. The osmotic pressure of a solution depends on
   a) the concentrations of all solutes contained in it.
   b) the concentrations of all permeant solutes contained in it.
   c) the concentrations of all impermeant solutes contained in it.
   d) pressure exerted on the solution by the atmosphere.

8. Assuming that only impermeant solutes are present, which of the following will occur when a cell is placed in a solution whose osmolarity differs from that of intracellular fluid?
   a) Water will move across the membrane from higher to lower osmotic pressure.
   b) Water will move across the membrane from lower to higher osmotic pressure.
   c) Water will move across the membrane from higher to lower osmolarity.
   d) Water will not move unless it is actively transported across the membrane.

9. A solution is hypotonic if
   a) the concentration of all solutes contained in it is less than 300 mOsm.

b) the concentration of all permeant solutes contained in it is less than 300 mOsm.

c) the concentration of all impermeant solutes contained in it is less than 300 mOsm.

d) its osmolarity is less than 300 mOsm.

10. Movement of Na^+ in sodium-linked glucose transport, in sodium-proton exchange, and via the sodium-potassium pump are all examples of
    a) active transport.
    b) passive transport.
    c) mediated transport.
    d) simple diffusion.

11. The direction of the electrochemical driving force acting on Cl^- crossing a membrane depends on the size and sign of the Cl^- equilibrium potential, the sign of the membrane potential, and
    a) the size of the membrane potential.
    b) the concentrations of other ions that may be present.
    c) whether the Cl^- are crossing the membrane through channels or through carriers.
    d) whether the Cl^- are being actively transported or passively transported.

12. Assuming that a substance is uncharged and is transported across a membrane by carriers, the net flux of that substance will tend to increase as
    a) the membrane surface area decreases.
    b) the magnitude of the concentration gradient decreases.
    c) the membrane potential becomes more positive.
    d) the number of carriers in the membrane increases.

13. In the example in Question 12, the permeability of the membrane increases as
    a) the membrane surface area increases.
    b) the membrane surface area decreases.
    c) the number of carriers in the membrane increases.
    d) the lipid solubility of the transported molecules decreases.

14. What do pumps and carriers have in common?
    a) They both transport molecules up electrochemical gradients.
    b) They both transport lipid-soluble substances preferentially.
    c) They both utilize ATP to transport molecules.
    d) They both are specific for certain molecules.

15. When ions are actively transported across a membrane, the direction of the net flux is
    a) down the electrochemical gradient.
    b) down the concentration gradient.
    c) up the electrochemical gradient.
    d) against the direction of the electrical driving force.

Objective Questions

1. Substances that cross cell membranes by simple diffusion are mostly (hydrophilic/hydrophobic).

2. A channel carries out (active/passive) solute transport.

3. In simple diffusion, an uncharged solute always flows from a region of higher concentration to a region of lower concentration. (true/false)

4. In facilitated diffusion, passive flow of an uncharged solute always goes from higher to lower concentration. (true/false)

5. Passive ion flow always goes from higher to lower concentration, regardless of the membrane potential. (true/false)

6. When an uncharged solute such as glucose flows from a region of lower concentration to a region of higher concentration, its energy (increases/decreases).

7. If the membrane potential is not equal to the equilibrium potential of an ion, that ion will not be at equilibrium. (true/false)

8. A concentration gradient is also referred to as a(n) _____ driving force.

9. When a membrane potential is positive, there is an excess of cations over anions inside the cell. (true/false)

10. Transport of water from the bloodstream to the lumen of the intestine is an example of (secretion/absorption).

11. A cell will shrink if it is placed in a hypertonic solution. (true/false)

12. When water diffuses across a membrane, it normally flows from a region of higher osmotic pressure to a region of lower osmotic pressure. (true/false)

13. The osmotic pressure of a solution depends on the concentrations of all solutes present. (true/false)

14. Junctions connecting adjacent epithelial cells are _____ junctions.

15. When a substance moves by simple diffusion down a concentration gradient, individual molecules all move in the same direction. (true/false)

Essay Questions

1. To determine whether a substance is being transported actively or passively requires knowledge of only two factors: the direction of the electrochemical gradient and the direction of the net flux. Explain.

2. Assuming that $E_{Na} = +60$ mV, $E_{Cl} = -90$ mV, and $V_m = -70$ mV, find the direction of the electrochemical driving forces acting on Na^+ and Cl^- ions. In which direction will the ions move if they are transported passively? If they are transported actively? (Extra challenge: From the sign of E_{Cl}, determine the direction of the concentration gradient of Cl^- ions.)

3. Repeat Question 2 for the case in which $V_m = -100$ mV.

4. Describe the various factors that determine membrane permeability in simple diffusion.

5. Explain the mechanism of glucose absorption by intestinal epithelial cells. Include a discussion of the significance of cellular polarity.

Find the answers to all of these exercises, and additional study tools, at the Physiology Place (www.physiologyplace.com).

5

Chemical Messengers and the Endocrine System

OBJECTIVES

- Compare the six major classes of chemical messengers (paracrines, autocrines, cytokines, neurotransmitters, hormones, and neurohormones) with respect to the cell that releases them and the ways the messenger gets to the target cell.

- Describe the basic structure and function of each class of chemical messenger (amines, catecholamines, peptides, steroids, and eicosanoids) with regard to mechanisms of synthesis, release, transport, and signal transduction.

- Compare the following functional classes of receptors: channel-linked receptors, enzyme-linked receptors, and G-protein-linked receptors.

- Name the primary and secondary endocrine glands, and the hormones associated with each.

- Describe the links of the hypothalamus with the anterior and posterior pituitaries.

- Describe the role of tropic hormones in regulating the release of other hormones. Include feedback loops in your description.

- Describe the types of interactions between hormones acting on the same target cell, including additive, synergistic, and permissive interactions.

CHAPTER OUTLINE

Intercellular Communication 132

Chemical Messengers 133

Signal Transduction Mechanisms 140

Long-Distance Communication Via the Nervous and Endocrine Systems 151

Endocrine Glands 152

Hormone Actions at the Target Cell 161

Above: Photomicrograph of epinephrine

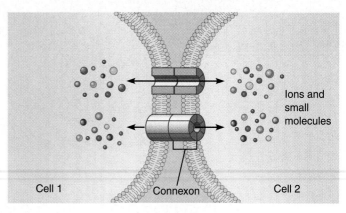

(a) Direct communication through gap junctions

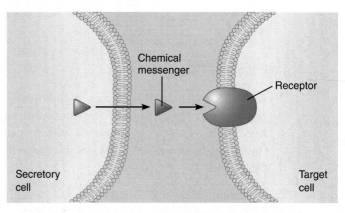

(b) Communication via chemical messengers

FIGURE 5.1 Types of intercellular communication. (a) *Direct communication through gap junctions. Gap junctions are composed of membrane protein structures called connexons that link the cytosols of two adjacent cells, allowing ions and small molecules to move between cells.* **(b)** *Communication via chemical messengers. After a secretory cell releases a messenger into the extracellular fluid, the messenger binds to receptors on target cells, triggering a response in the target cell.*

In previous chapters we studied cells and their functions. However, no cell exists alone in the body. Our bodies are made up of approximately 100 trillion (10^{14}) cells and each depends on the others for survival because they must work together to maintain homeostasis. To do so, cells must be able to communicate with each other to carry out coordinated activities, such as the maintenance of body temperature, as described in Chapter 1, or the beating of the heart. Communication occurs not only between neighboring cells, but also between cells at different locations within the body. This chapter first describes the general mechanisms by which cells communicate, and then it describes in detail one of the body's primary communication systems: the endocrine system.

INTERCELLULAR COMMUNICATION

Virtually all body functions require communication between cells. Seeing an apple requires communication between the cells in the eyes and the cells in the brain. Fighting an infection requires communication among several cell types in the blood. Growing and developing from infants to adults requires communication among cells throughout the body. There are hundreds of other possible examples. Remarkably, however, all the cells in the body use only a few mechanisms to communicate with one another. In relatively few instances cells are physically linked by gap junctions; in most instances cells communicate via chemical messengers.

Direct Communication Through Gap Junctions

Recall from Chapter 2 that gap junctions link adjacent cells and are formed by plasma membrane proteins, called *connexins*, that form structures called *connexons* (Figure 5.1a). These connexons form channels that allow ions and small molecules to pass directly from one cell to another. The movement of ions electrically couples the cells, such that electrical signals in one cell are directly transmitted to the neighboring cells. Gap junctions are found in heart muscle and the smooth muscle of other

internal organs, such as the intestines and blood vessels, where they cause the muscle cells to contract as a unit; that is, the cells contract at the same time. Gap junctions are also found in some glands and between some neurons in the brain and retina.

Indirect Communication Via Chemical Messengers

Most often, cells communicate via chemical messengers (Figure 5.1b), which are all *ligands,* molecules that bind to proteins reversibly. The body has hundreds of chemical messengers with a multitude of functions.

Communication via chemical messengers occurs when one cell releases a chemical into the interstitial fluid, usually by a process called secretion, and another cell, called the **target cell,** responds to the chemical messenger. The target cell, in essence, is the cell at which a message is "aimed." A target cell responds to the chemical

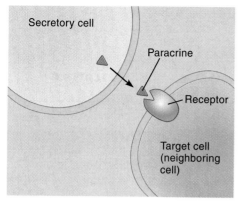

(a) Paracrines

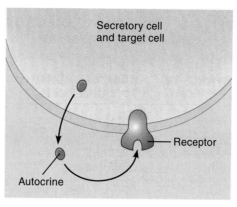

(b) Autocrines

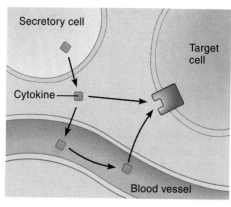

(c) Cytokines

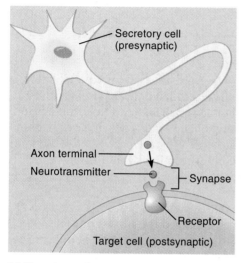

(d) Neurotransmitters

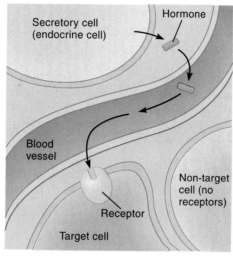

(e) Hormones

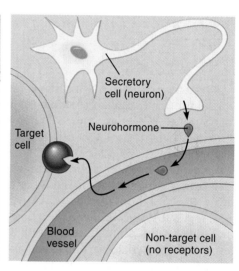

(f) Neurohormones

FIGURE 5.2 Functional classes of chemical messengers.
(a) *Paracrines are secreted by one cell and diffuse to a nearby target cell.*
(b) *Autocrines bind to receptors on the cell that secreted them.* **(c)** *Cytokines are secreted by a cell (often a white blood cell) and then either diffuse to a neighboring target cell or are transported by blood to distant target cells.*
(d) *Neurotransmitters are secreted from neurons at functionally specialized structures called synapses. The axon terminal of a presynaptic cell releases the neurotransmitter, which then diffuses a very short distance to bind to receptors on a very specific target cell, called the postsynaptic cell.* **(e)** *Hormones are secreted by endocrine cells into the interstitial fluid. Hormones then diffuse into the bloodstream for transport to target cells in the body. Target cells are identified by the presence of receptors for the specific hormone. Cells without receptors for the hormone cannot respond to the hormone's signal.*
(f) *Neurohormones are a special class of hormones secreted by neurons. Like hormones, they are secreted into the interstitial fluid and then diffuse into the blood for transport to target cells throughout the body.*

messenger because it has certain proteins, called **receptors,** that specifically recognize and bind the messenger.

The binding of messengers to receptors produces a response in the target cell through a variety of mechanisms referred to as **signal transduction mechanisms.** Generally speaking, the strength of the target cell response increases as the number of bound receptors increases. The number of bound receptors depends on both the concentration of messenger in the interstitial fluid and the concentration of receptors on the target cell.

CHEMICAL MESSENGERS

Chemical messengers can be classified on the basis of their function and chemical structure. First we consider the functional classes of chemical messengers.

Functional Classification of Chemical Messengers

Although there are hundreds of chemical messengers, most can be classified into six main categories:

(1) paracrines, (2) autocrines, (3) cytokines, (4) neurotransmitters, (5) hormones, and (6) neurohormones (Figure 5.2). When released into the interstitial fluid, each of these messenger categories transmits a signal by binding to receptors on a target cell, as described next.

TABLE 5.1   FUNCTIONAL CLASSIFICATION
OF CHEMICAL MESSENGERS

CLASS	SECRETORY CELL TYPE	DISTANCE TO TARGET CELL	MODE OF TRANSPORT TO TARGET CELL	CHEMICAL CLASSIFICATION OF MESSENGER
Paracrine	(Several)	Short	Diffusion	Amines, Peptides, Eicosanoids
Autocrine	(Several)	Short	Diffusion	Amines, Peptides
Cytokine	(Several)	Short*	Diffusion*	Peptides
Neurotransmitter	Neuron	Short[†]	Diffusion	Amino acids, Amines, Peptides
Hormone	Endocrine	Long	Blood	Amines, Peptides, Steroids
Neurohormone	Neuron	Long	Blood	Peptides

*Some cytokines are transported in blood to distant targets.
[†]Even though neurotransmitters diffuse only a short distance to the postsynaptic cell, some neurons are involved in long-distance communication because the neuron that releases the neurotransmitter is often long (up to 1 meter).

Paracrines are chemicals that communicate with neighboring cells. The target cell must be close enough that once the paracrine is secreted into the extracellular fluid, it can reach the target cell by simple diffusion (Figure 5.2a). An example of a paracrine messenger is *histamine,* a chemical that is important in allergic reactions and inflammation and is secreted by *mast cells* scattered throughout the body (Discovery: Antihistamines, p. 136). During allergic reactions, histamine is responsible for the runny nose and red, watery eyes. In response to bacterial infections and various forms of tissue damage, the release of histamine by mast cells is part of a complex response called *inflammation,* which is characterized in part by redness and swelling. In inflammation, the vasodilation produced by histamine increases blood flow to affected tissues (producing redness) and causes fluid to leak out of the blood vessels and into the tissue (producing swelling).

Autocrines are similar to paracrines, except that autocrines act on the same cell that secreted them (Figure 5.2b). Thus, the secretory cell is also the target cell. Often an autocrine also functions as a paracrine or other messenger and regulates its own secretion.

Cytokines (Figure 5.2c) include a wide range of chemical messengers released from a variety of cells, especially white blood cells. Cytokines released from white blood cells function in the body's immune system and help to defend the body against pathogens, including bacteria and viruses. Cytokines usually function as paracrines in that following release into the extracellular fluid they diffuse a short distance to their target cells. However, some cytokines function as autocrines, while others may travel via the bloodstream to more distant target cells, thus functioning more like hormones. Examples of cytokines are *interleukins* and *interferons,* groups of small proteins released from white blood cells as part of the immune response.

Neurotransmitters are released into interstitial fluid from nervous system cells called neurons. Neurotransmitters are released from a specialized portion of the neuron called the *axon terminal* (Figure 5.2d), which is very close to the target cell. Because the juncture between the two cells is called a *synapse,* communication by neurotransmitters is often called **synaptic signaling.** The cell that releases the neurotransmitter is called the **presynaptic cell,** whereas the target cell (which can be another neuron or a gland or muscle cell) is called the **postsynaptic cell.** Upon release from the presynaptic cell, the neurotransmitter quickly diffuses the short distance from the axon terminal and binds to receptors on the postsynaptic cell, triggering a response. Communication between a neuron and its target cell(s) is very specific because it is directed only to cells that are connected to it by synapses. An example of a neurotransmitter is *acetylcholine,* which is released by the neurons that trigger contraction of skeletal muscles.

Hormones are released from *endocrine glands* (or occasionally other types of tissue) into the interstitial fluid, where they can then diffuse into the blood (Figure 5.2e). The hormone then travels in the blood to its target cells, which can be distant from the site of hormone release. The bloodstream distributes a hormone to virtually all cells of the body, but only cells possessing receptors specific for the hormone are able to respond and thus serve as target cells. An example of a hormone is *insulin,* which is secreted by the pancreas and acts on target cells throughout the body to regulate energy metabolism.

A special class of hormones, called **neurohormones,** is released by a special class of neurons called *neurosecretory cells* through a mechanism similar to that of neurotransmitter release (see Chapter 7). Just like the "classical" hormones secreted by endocrine glands, neurohormones are released into the interstitial fluid and then diffuse into the blood, which distributes them to target cells through-

TABLE 5.2 CHEMICAL CLASSIFICATION OF MESSENGERS

CLASS	CHEMICAL PROPERTY	LOCATION OF RECEPTORS ON TARGET CELL	FUNCTIONAL CLASSIFICATION
Amino acids	Lipophobic	Plasma membrane	Neurotransmitters
Amines*	Lipophobic	Plasma membrane	Paracrines, autocrines, neurotransmitters, hormones
Peptides	Lipophobic	Plasma membrane	Paracrines, autocrines, cytokines, neurotransmitters, hormones
Steroids	Lipophilic	Cytosol†	Hormones
Eicosanoids	Lipophilic	Cytosol	Paracrines

*One exception is the thyroid hormones, which, although amines, are lipophilic and have receptors in the nucleus of target cells.
†A few steroid hormones have receptors on the plasma membrane.

out the body (Figure 5.2f). An example of a neurohormone is *vasopressin,* or *ADH* (antidiuretic hormone), which is synthesized by neurosecretory cells originating in an area of the brain called the hypothalamus. Once vasopressin is released from the axon terminals of these neurosecretory cells, which are located in the posterior pituitary gland, it travels in the blood to its target cells. The primary target cells are located in the kidneys, where vasopressin affects the volume of urine that is excreted.

Some characteristics of the functional classes of chemical messengers are summarized in Table 5.1.

Chemical Classification of Messengers

A messenger's chemical structure determines its mechanisms of synthesis, release, transport, and signal transduction. The most important chemical characteristic is whether or not the messenger can dissolve in water or cross the lipid bilayer in the plasma membrane of cells. **Lipophilic** (hydrophobic) molecules are lipid soluble and therefore readily cross the plasma membrane, but they do not dissolve in water. **Lipophobic** (hydrophilic) molecules are water soluble and therefore do not cross the plasma membrane.

In the following sections we discuss the five major classes of chemical messengers: (1) amino acids, (2) amines, (3) peptides, (4) steroids, and (5) eicosanoids (Table 5.2). Other chemical messengers such as acetylcholine and nitric oxide, do not fit into any of these classes and are discussed in later chapters.

Amino Acid Messengers

Four amino acids are classified as chemical messengers because they function as neurotransmitters in the brain and spinal cord: *glutamate, aspartate, glycine,* and *gamma-amino butyric acid (GABA).* Whereas glutamate, aspartate, and glycine are among the 20 amino acids (alpha amino acids) that are used in protein synthesis, GABA belongs to a different class of amino acids (gamma amino acids). Amino acids are lipophobic; therefore, they dissolve in water but do not cross plasma membranes. Because amino acid messengers function only as neurotransmitters, they are described in detail in Chapter 7.

Amine Messengers

Amines, which are chemical messengers derived from amino acids, are so named because they all possess an amine group ($-NH_2$). The amines include a group of compounds called **catecholamines,** which contain a *catechol* group (a six-carbon ring) and are derived from the amino acid tyrosine. Catecholamines include *dopamine, norepinephrine,* and *epinephrine.* Dopamine and norepinephrine are primarily neurotransmitters, whereas epinephrine is primarily a hormone. Other amines include the neurotransmitter *serotonin,* which is derived from tryptophan; the *thyroid hormones,* which are derived from tyrosine; and the paracrine *histamine,* which is derived from histidine. Most of the amines are lipophobic; therefore, they dissolve in water and do not cross plasma membranes. The thyroid hormones are an exception: They are lipophilic and thus do not dissolve in water but readily cross plasma membranes.

Peptide Messengers

Most chemical messengers, including many neurotransmitters and hormones and all cytokines, are polypeptides, chains of amino acids linked together by peptide bonds.

Histamine is a biogenic amine synthesized from the amino acid histidine. Histamine is found in several tissue types, including the nasal cavity, lungs, blood vessels, stomach, small intestine, and skin. Histamine has several physiological functions as a paracrine, an autocrine, or a neurotransmitter. Histamine functions as a paracrine in most tissues, and as a neurotransmitter and an autocrine in the brain, where histamine is associated with increasing mental alertness. As a paracrine, histamine is best known for its role in the immune response and inflammation. In that process, histamine causes dilation of blood vessels, which increases blood flow to an area, and increases the leakiness of blood vessels, which causes tissue swelling as fluid moves out of the blood vessel and into the interstitial fluid. Histamine also acts as a paracrine in the stomach, where it stimulates acid secretion. Treatments of allergic reactions and some ulcers often include administration of *antihistamines*.

In most of these tissues, histamine is stored in mast cells; in the blood, histamine is primarily localized in a type of white blood cell called basophils.

Once released, histamine acts on target cells by binding to one of three types of histamine receptors. H_1 and H_3 receptors have been identified in the brain; H_1 receptors function in neurotransmission associated with mental alertness, whereas H_3 receptors function as *autoreceptors*, with histamine functioning as an autocrine. H_1 and H_2 receptors are found in peripheral tissues, where both receptor types are associated with G proteins and second messenger systems. H_1 receptors are linked to the production of inositol triphosphate and diacylglycerol; H_2 receptors are linked to the production of cAMP. H_1 receptors predominate in the lungs and intestines; H_2 receptors predominate in the stomach. Both H_1 and H_2 receptors are found in the blood vessels.

Histamine binding to H_1 receptors produces several symptoms of an allergic response, including constriction of the airways, increased fluid secretion in the airways, and dilation of the blood vessels (which increases blood flow to the skin but can also decrease blood pressure). During severe allergic reactions, blood pressure drops to levels that can no longer move blood through the circulation adequately, resulting in anaphylactic shock. Also associated with anaphylactic shock is a swelling of the airways, which causes respiratory distress.

Treatment of allergic reactions often includes administration of classic H_1 receptor antagonists, such as diphenhydramine hydrochloride (Benadryl^{TM}) and dimenhydrinate (Dramamine^{TM}). When receptor antagonists are used to treat an ailment, however, they block receptors throughout the body, including those not involved in the ailment, thereby producing *side effects*. In the case of H_1 receptor antagonists, the predominant side effect is drowsiness caused by blockage of the H_1 receptors in the brain that are involved in mental alertness. Recently, H_1 receptor antagonists that do not access the brain have been developed to decrease the sedative effects; examples include loratidine (Claritin^{TM}) and fexofenadine (Allegra^{TM}).

Histamine binding to H_2 receptors stimulates acid secretion by the stomach, which may contribute to the production of *heartburn,* a burning sensation caused by acid irritation of the esophagus, or may produce gastric and intestinal ulcers, perforations of the linings of these organs. As a result, heartburn and some ulcers are treated with administration of H_2 receptor antagonists such as Cimetidine^{TM} and ranitidine (Zantac^{TM}). Side effects associated with H_2 receptor antagonists include headache and drowsiness.

As evident in the actions of drugs such as Benadryl and Cimetidine, the more specific a drug is for a certain receptor type, the more specific the therapeutic actions of the drug and the fewer the side effects. Pharmaceutical companies are continuously designing new drugs to bind specifically to receptor subclasses for a certain messenger. In addition, drugs can be modified chemically to limit access to certain structures, again to eliminate unwanted side effects, as in the case of Claritin. Meanwhile, scientists are discovering new receptor types annually. The combined result is better treatments of ailments, although much work still lies ahead.

These messengers are classified as peptides or proteins based on their size, which varies considerably, from just two amino acids to over a hundred amino acids. The term *peptide* generally refers to chains of fewer than 50 amino acids, whereas proteins are longer chains of amino acids. Peptides are lipophobic; therefore, they dissolve in water but cannot cross plasma membranes.

Steroid Messengers

Steroids are a class of compounds derived from *cholesterol.* All the body's steroid messengers function as hormones. Recall from Chapter 2 that cholesterol is a lipid with a distinctive four-ring structure. Because steroids are derived from cholesterol, which is lipophilic, they too are lipophilic and readily cross plasma membranes and are insoluble in water.

Eicosanoid Messengers

Eicosanoids include a variety of paracrines that are produced by almost every cell in the body. Most eicosanoids are derivatives of arachidonic acid, a 20-carbon fatty acid that is found in various plasma membrane phospholipids. Because eicosanoids are lipids, they readily cross the plasma membrane and are insoluble in water. Eicosanoids include the following families of chemically related compounds: prostaglandins, leukotrienes, and thromboxanes.

Quick Test 5.1

1. Name the six functional classes of messengers. Which messenger(s) is (are) transported in the blood to its (their) target cells? Which are secreted by neurons?
2. Name the five chemical classes of messengers. Which are lipophobic and which are lipophilic?
3. Which chemical class of messengers is derived from cholesterol? To which functional class do these messengers belong?

Synthesis and Release of Chemical Messengers

The general synthetic pathways and mechanisms of release for chemical messengers are similar within a chemical class. In this section we examine the synthesis and release of each class of messenger.

Amino Acids

Of the 20 alpha amino acids, nine are *essential amino acids,* meaning they must be consumed in the diet because the body cannot synthesize them. The other 11 alpha amino acids, including the three that function as neurotransmitters, are both consumed in the diet and synthesized in cells. However, the four amino acids that function as neurotransmitters must be synthesized within the neuron that will secrete them.

Glutamate and aspartate are synthesized from glucose through a three-step series of reactions (see Chapter 3). First, glucose is catabolized to pyruvic acid by glycolysis; pyruvic acid is then converted to acetyl CoA, which then enters the Krebs cycle; and finally the amine groups are added to certain Krebs cycle intermediates to form glutamate or aspartate. Glycine is synthesized from a glycolytic intermediate, 3-phosphoglycerate, in a series of four reactions. GABA is synthesized from glutamate in a single reaction catalyzed by the enzyme *glutamic acid decarboxylase.*

Following their synthesis, amino acid neurotransmitters are transported into vesicles where they are stored until they are released by exocytosis.

Amines

All amines are derived from amino acids, and all except thyroid hormones are synthesized in the cytosol by a series of enzyme-catalyzed reactions. (The synthesis and release of thyroid hormones are described in Chapter 20.) Which amine is produced depends on the enzymes pres-

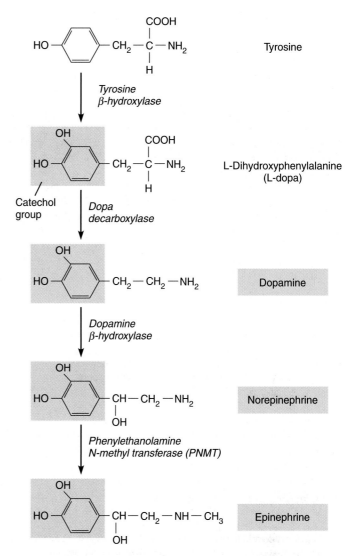

FIGURE 5.3 Catecholamine synthesis. *Catecholamines are synthesized from the amino acid tyrosine by a sequence of enzyme-catalyzed reactions in which one catecholamine functions as the precursor for the next. The names of catecholamines that function as messengers are highlighted.*

What chemical group is removed from L-dopa to form dopamine?

ent in a given cell. Figure 5.3 shows the pathway for synthesis of the catecholamines, which are derived from the amino acid tyrosine. Note that in this pathway, dopamine is the precursor for norepinephrine, which in turn serves as the precursor for epinephrine. Because dopamine is a precursor for the other catecholamines, all catecholamine-secreting cells possess the two enzymes that catalyze its synthesis: *tyrosine β-hydroxylase* and *dopa decarboxylase.* Depending on which catecholamine it secretes, a cell may or may not possess enzymes for the remaining steps. Thus cells that secrete dopamine lack the enzymes necessary for the final two steps. Those that

Carboxyl

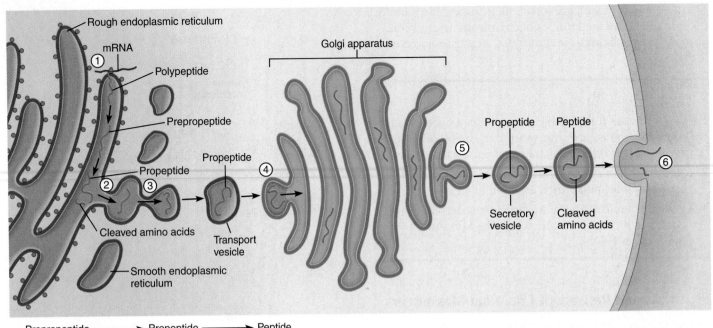

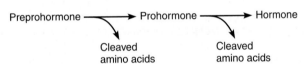

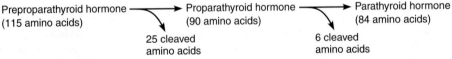

(a) Peptide synthesis

Preprohormone ——→ Prohormone ——→ Hormone
         ↘ Cleaved   ↘ Cleaved
         amino acids   amino acids

(b) Generalized scheme of hormone synthesis

Preproparathyroid hormone ——→ Proparathyroid hormone ——→ Parathyroid hormone
(115 amino acids)   (90 amino acids)   (84 amino acids)
         ↘ 25 cleaved   ↘ 6 cleaved
         amino acids   amino acids

(c) Parathyroid hormone synthesis

FIGURE 5.4 Peptide synthesis and release. (a) *The steps of peptide synthesis. For a detailed description, see the text.* **(b)** *The general steps of hormone synthesis.* **(c)** *Specific example of modification of preproparathyroid hormone, which contains 115 amino acids, to the active parathyroid hormone, which contains 84 amino acids.*

secrete norepinephrine contain the enzyme dopamine β-hydroxylase, whereas those that secrete novepinephrine and epinephrine contain an additional enzyme, *phenylethanolamine N-methyl transferase* (PNMT). Following synthesis, amines are packaged into cytosolic vesicles, where they are stored until their release is triggered. The synthesis of norepinephrine and epinephrine is actually completed within the vesicles, where dopamine β-hydroxylase and PNMT are located. Release occurs by exocytosis.

Peptides and Proteins

Peptides and proteins are synthesized in the same way as other proteins destined for secretion, as described in Chapter 2. Briefly, cytosolic mRNA serves as the template that codes for the amino acid sequence in the peptide or protein. Translation of this mRNA begins on ribosomes free in the cytosol. The subsequent steps, shown in Figure 5.4a using a peptide for this example, are the following:

1. Once translation starts, the ribosomes attach to the rough endoplasmic reticulum, where the rest of translation occurs. The polypeptide is formed inside the lumen of the rough endoplasmic reticulum, first as a *prepropeptide*.

2. In the lumen of the endoplasmic reticulum, *proteolytic enzymes* cleave off some amino acids from the prepropeptide, yielding the *propeptide*.

3. In the smooth endoplasmic reticulum, the propeptide is packaged into transport vesicles.

4. The vesicles transport the propeptide to the Golgi apparatus.

(5) In the Golgi apparatus the propeptide is packaged into a secretory vesicle for storage until release is triggered. More amino acids are cleaved off by proteolytic enzymes in the Golgi apparatus or in the secretory vesicle to give the final product, a peptide.

(6) Release occurs by exocytosis.

Typically, peptide fragments generated by proteolysis are released along with the primary messenger, and they may or may not exert their own biological effects. Figure 5.4c contains a schematic example of the synthesis of a protein hormone, in this case parathyroid hormone.

Steroids

In tissues that secrete steroids, these messengers are synthesized from cholesterol in a series of reactions (Figure 5.5) catalyzed by enzymes located in the smooth endoplasmic reticulum or mitochondria. As a result, the cholesterol molecule is modified slightly, but its basic ring structure remains intact, as does its lipophilic character. Consequently, all steroids are capable of crossing the plasma membrane. Because they are membrane-permeant, steroids cannot be stored prior to release and instead diffuse out of the cell into the interstitial fluid as soon as they are synthesized. Therefore, while cells that secrete peptides or amines can synthesize messengers in advance and store them to be released on demand, steroid hormones are synthesized on demand and released immediately.

Eicosanoids

Like steroids, eicosanoids are synthesized on demand and released immediately because they are lipophilic and able to pass through plasma membranes easily. The first step in eicosanoid synthesis (Figure 5.6) involves an enzyme called *phospholipase A₂*, which is activated in response to chemical signals of various kinds (paracrines, hormones, neurotransmitters, and even foreign chemicals). When active, this enzyme catalyzes the release of arachidonic acid from membrane phospholipids. Once arachidonic acid is released from the membrane, the final product depends on the complement of enzymes present in the particular cell.

To become an eicosanoid, a molecule of arachidonic acid first reacts with one of two enzymes: either *cyclooxygenase* or *lipoxygenase*. Cyclooxygenase is the first enzyme in a series of reactions, called the *cyclooxygenase pathway*, that leads to the synthesis of *prostacyclins, prostaglandins,* or *thromboxanes.* Prostacyclins and thromboxanes are important in blood clotting; prostaglandins are involved in several systems, including the inflammatory response (described in Chapter 22). Lipoxygenase is the first enzyme of a series of reactions, called the *lipoxygenase pathway,* that leads to the synthesis of *leukotrienes,* which also contribute to the inflammatory response.

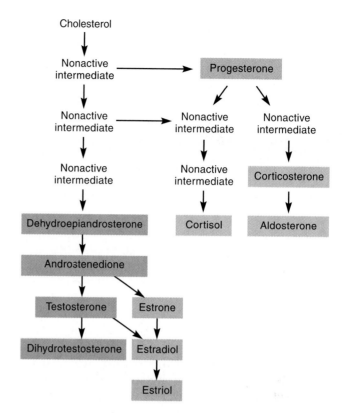

FIGURE 5.5 Synthetic pathway for steroids. *Each arrow indicates an enzyme-catalyzed reaction. Green boxes indicate hormones produced in the adrenal cortex; blue boxes indicate male sex hormones; red boxes indicate female sex hormones.*

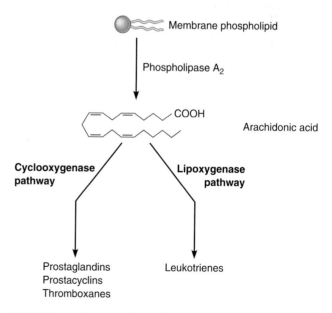

FIGURE 5.6 Eicosanoid synthesis. *Phospholipase A₂ catalyzes the conversion of a membrane phospholipid to arachidonic acid, the precursor for all eicosanoids. Arachidonic acid is converted into eicosanoids via two pathways: The cyclooxygenase-dependent pathway leads to the production of prostaglandins, prostacyclins, and thromboxanes, whereas the lipoxygenase-dependent pathway leads to the production of leukotrienes.*

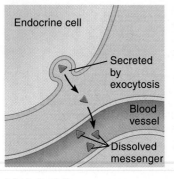
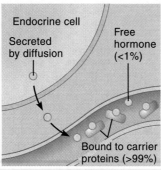

(a) Hydrophilic messenger   **(b)** Hydrophobic messenger

FIGURE 5.7 Transport of messengers in blood. (a) *Hydrophilic messengers are secreted by exocytosis, enter the bloodstream, and dissolve in the plasma.* **(b)** *Hydrophobic messengers are secreted by simple diffusion and then enter the bloodstream. Most of the messenger molecules are transported bound to carrier proteins. Only the small amount of free hormone in the plasma is immediately available for binding with target cell receptors.*

Because of the eicosanoids' role in inflammation, many anti-inflammatory drugs act by targeting enzymes involved in eicosanoid synthesis. (Discovery: Cyclooxygenase Inhibitors, www. physiologyplace.com, Challenge Yourself).

Transport of Messengers

Once released, a messenger must first reach and then bind to receptors on the target cell for the signal to be transmitted. In many instances, the messenger is released from a cell that is near the target cell, such that the messenger reaches the receptor by simple diffusion. This is true of paracrines, autocrines, most cytokines, and neurotransmitters. Typically these messengers are quickly degraded in the interstitial fluid and become inactive, minimizing the spread of their signaling. However, hormones and neurohormones (and a few cytokines) are transported in the blood and thus have access to most cells in the body.

Messengers can be transported in the blood either in dissolved form or bound to *carrier proteins.* To be transported in dissolved form, the messenger must be a hydrophilic messenger (Figure 5.7a). Peptides and amines (except thyroid hormones) are transported in this manner. Because steroids and the thyroid hormones are hydrophobic and thus do not dissolve well in blood, these hormones are largely transported bound to carrier proteins (Figure 5.7b). Although most of the catecholamines that function as neurohormones are hydrophilic and thus are transported in dissolved form, some are bound to carrier proteins. Some carrier proteins are very specific for a particular hormone; one example is *corticosteroid-binding globulin*, which transports the steroid hormone *cortisol.* Other carrier proteins—for example, *albumin*—are not specific and can transport many different hormones.

Even though hydrophobic hormones are transported primarily in bound form, a certain fraction of the hormone molecules dissolve in plasma (generally, <1%). For each such hormone, an equilibrium develops in the bloodstream between the amount of hormone that is bound to a carrier protein (Pr) in the form of a complex (H-Pr)and the amount of free hormone (H) that is dissolved in the plasma:

$$\text{H-Pr} \rightleftharpoons \text{H} + \text{Pr}$$

Only free hormone is available to bind to receptors on target cells. However, once the hormone binds, it is removed from the blood and the equilibrium between bound and free hormone shifts to the right, causing more hormone to be released from the carrier proteins. Likewise, the secretion of hormones into the blood causes the equilibrium to be shifted to the left, such that more hormone binds to carrier proteins.

Blood-borne messengers are generally degraded by the liver and excreted by the kidneys. How long a hormone persists in blood is measured in terms of **half-life,** the time it takes for half of the hormone in the blood to be degraded. Hormones that are present in dissolved form have relatively short half-lives, usually minutes. However, hormones that are bound to carrier proteins are protected from degradation and have longer half-lives, generally hours.

Quick Test 5.2

1. Name the three catecholamines. What amino acid is the precursor for catecholamines? Which catecholamine generally functions as a hormone?

2. Phospholipase A_2 causes the release of what fatty acid from membrane phospholipids? What chemical class of messengers is produced from this fatty acid?

3. What chemical class(es) of hormones is (are) transported in blood bound to carrier proteins? Does such binding generally *increase* or *decrease* the half-life of a hormone?

SIGNAL TRANSDUCTION MECHANISMS

Chemical messengers transmit their signals by binding to target cell receptors located either on the plasma membrane, in the cytosol, or in the nucleus. The location of the receptor depends on whether the messenger is lipophilic or lipophobic. In either case, binding of messenger to receptor either changes the activity of proteins (for example, enzymes) already present in the cell or stimulates the synthesis of new proteins. This section describes properties of receptors and the different signal transduction mechanisms that are set into motion by them.

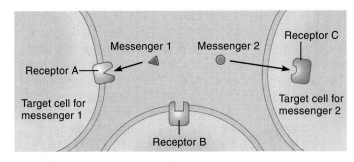

FIGURE 5.8 **Receptor specificity.** *Receptor A is specific for messenger 1, receptor C is specific for messenger 2, and neither messenger can bind to receptor B. Note that receptors can be located either on the plasma membrane (receptors A and B) or inside the cell (receptor C).*

If messenger 2 is to bind receptor C (located inside the target cell), it must cross the plasma membrane of the target cell. What chemical property must messenger 2 possess to enable entry into the cell?

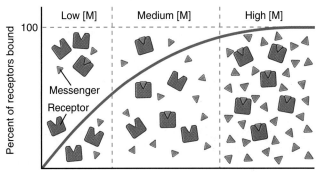

FIGURE 5.9 **Effect of messenger concentration on messenger-receptor binding.** *The proportion of receptors bound increases as the concentration of messenger increases. Because the amount of bound receptor determines the magnitude of target cell response, the y-axis could also have been labeled "target cell response."*

Properties of Receptors

Receptors show **specificity** for the messenger; that is, they generally bind only one messenger or a class of messengers. Observe that in Figure 5.8, messenger 1 can bind to receptor A but not to receptor B or C. Therefore, only cell A is a target cell for messenger 1. Likewise, cell C is the target cell for messenger 2. The binding between a messenger and receptor is a brief, reversible chemical interaction, and the binding is similar to the enzyme-substrate interactions described in Chapter 3. The strength of the binding between a messenger and its receptor is termed affinity.

A single messenger can often bind to more than one type of receptor, and these receptors may have different affinities for it. The affinity of the different receptors for a single messenger often varies. For example, the catecholamine chemical messengers epinephrine and norepinephrine can both bind to *adrenergic* receptors (epinephrine is also called *adrenaline,* and norepinephrine is also called *noradrenaline*). Several different types of adrenergic receptors exist, including α_1, β_1, and β_2. Alpha$_1$ receptors have a greater affinity for norepinephrine than for epinephrine, which means that if norepinephrine and epinephrine are present in equal concentrations, a given receptor will be more likely to bind norepinephrine than epinephrine. Beta$_2$ receptors, on the other hand, have a greater affinity for epinephrine than for norepinephrine. Beta$_1$ receptors have approximately equal affinities for norepinephrine and epinephrine.

A single target cell may have receptors for more than one type of messenger. For example, skeletal muscle cells have receptors for both the neurotransmitter acetylcholine and the hormone insulin. Acetylcholine receptors are directly involved in stimulating muscle contraction, whereas insulin receptors are involved in stimulating glucose uptake and metabolism in the muscle cell.

The Relationship Between Receptor Binding and the Magnitude of the Target Cell Response

As a general rule, the magnitude of a target cell's response to a chemical messenger depends on three factors: (1) the messenger's concentration, (2) the number of receptors present, and (3) the affinity of the receptor for the messenger.

The response of a target cell generally increases as the concentration of messenger increases. This is a consequence of the fact that messengers usually exert their effects by binding reversibly to target cell receptors, as shown in the following reaction:

$$M + R \rightleftharpoons M\text{-}R \rightarrow \text{Response}$$

where M is the messenger, R is the receptor, and M-R is the messenger-receptor complex. As messenger concentration increases, the reaction is driven to the right. The relationship between the concentration of messenger and the number of bound receptors is shown in Figure 5.9. As the concentration of messenger increases, the proportion of bound receptors increases until all receptors have messengers bound to them (100% in Figure 5.9), under which conditions the system is said to be 100% *saturated.*

The target cell's response also depends on the number of receptors it possesses. The more receptors there are (the higher their density), the more likely it is that a messenger will bind to a receptor (Figure 5.10a). This means that at any given concentration of messenger, the

Messenger 2 must be lipophilic (or hydrophobic).

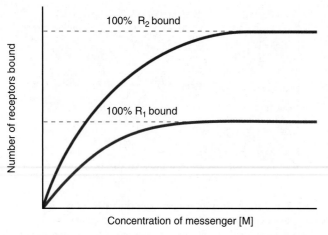

(a) Effects of receptor concentration

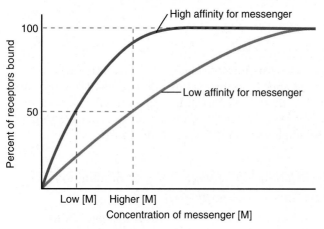

(b) Effects of receptor affinity

FIGURE 5.10 Effects of receptor concentration and affinity on messenger-receptor binding. (a) *Effects of receptor density. The two curves illustrate the effects of doubling the concentration of a given receptor on a target cell.* R_1 *refers to a given concentration of the receptor, whereas* R_2 *refers to twice the concentration of that receptor. When the receptor concentration doubles, the maximum number of receptors that can bind with messenger also doubles.* **(b)** *Effects of receptor affinity. The two curves illustrate the effects of receptor affinity on the proportion of receptors with messenger bound to them. At any concentration of messenger below saturation, a higher proportion of high-affinity receptors have bound messenger compared to low-affinity receptors. The high-affinity receptors reach saturation at a lower messenger concentration than do the low-affinity receptors. Note that the maximum number of sites that can be bound is independent of receptor affinity.*

number of bound receptors will be greater when more receptors are present, and the response will be greater.

The number of receptors that a target cell possesses can vary under different circumstances as a result of the synthesis of new receptors or turnover of old receptors. *Up-regulation,* an increase in the number of receptors compared to "normal" conditions, occurs when cells are exposed to low messenger concentrations for a prolonged period of time. By producing more receptors, target cells

adapt to the relative lack of messenger by becoming more responsive to it. *Down-regulation,* a decrease in the number of receptors, occurs when messenger concentrations are higher than normal for a prolonged period of time. In this case, target cells adapt by producing fewer receptors and becoming less responsive to the messenger.

The target cell response also depends on the affinity of its receptors for the messenger. When a messenger is present at a given concentration, receptors with higher affinity are more likely to become bound than are receptors with lower affinity, as shown in Figure 5.10b. Consequently, target cells possessing high affinity receptors will respond more strongly to a given messenger, all else being equal.

Receptor Agonists and Antagonists

Although target cell responses are always triggered by receptor binding, it is not true that receptor binding always triggers a response. Among other things, it depends on the nature of the ligand that is binding to the receptors. Ligands that bind to receptors and produce a biological response are called **agonists,** whereas **antagonists** are ligands that bind to receptors but do not produce a response. Instead, antagonists may actually compete with agonists for the receptor, decreasing the likelihood that the binding of agonist to receptor will occur and bring about a response. Some therapeutic and experimental drugs developed by pharmaceutical companies are artificial receptor agonists or antagonists (Discovery: Antihistamines, p. 136). For example, under resting conditions, norepinephrine that is released from certain neurons binds to alpha receptors, causing constriction of blood vessels and an increase in blood pressure. The drug *phenylephrine* is an alpha agonist, and exerts the same effects. However, the drug *phenoxybenzamine* is an alpha antagonist, and prevents norepinephrine from binding to alpha receptors. Phenoxybenzamine has no effect itself, but by blocking the effect of norepinephrine, it causes blood pressure to decrease.

Signal Transduction Mechanisms for Intracellular Receptor-Mediated Responses

Receptors for lipophilic messengers are usually located in the cytosol or nucleus of target cells and are readily accessible because these messengers easily permeate the plasma membrane. The binding of the messenger to the receptor alters the synthesis of a specific protein (or proteins) by the mechanism depicted in Figure 5.11, which shows the action of a lipophilic hormone. If a receptor is located in the nucleus, then the hormone diffuses into the nucleus and binds to it, forming a hormone-receptor complex (step 1a). If a receptor is located in the cytosol, then the hormone binds to it there, forming a hormone-receptor complex that then enters the nucleus (step 1b).

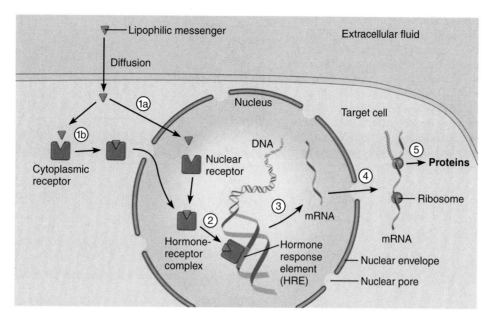

FIGURE 5.11 Actions of lipophilic hormones on the target cell. *Receptors for lipophilic hormones are usually in the cytosol or nucleus. The steps of signal transduction are as follows:* (1a) *When receptors are in the nucleus, the hormone diffuses into the nucleus and binds to the receptor, forming a hormone-receptor complex in the nucleus.* (1b) *When receptors are in the cytosol, the hormone binds to the receptor there, forming a hormone-receptor complex that then moves into the nucleus.* (2) *The hormone-receptor complex binds to the hormone response element, activating (or deactivating) a gene.* (3) *mRNA is transcribed in the nucleus.* (4) *mRNA moves from the nucleus into the cytosol through nuclear pores.* (5) *mRNA is translated to form proteins by ribosomes (either free in the cytosol or attached to the rough endoplasmic reticulum).*

Inside the nucleus, the complex binds to a certain region of DNA called the *hormone response element* (HRE), which is located at the beginning of a specific gene (step 2). Binding of the complex to the HRE activates or deactivates the gene, which affects transcription of mRNA and ultimately increases or decreases synthesis of the protein. In the example shown in Figure 5.11, the gene is activated and mRNA synthesis is increased (step 3). The mRNA moves into the cytosol (step 4), where it is translated by ribosomes to yield proteins (step 5).

Because changes in protein synthesis can take hours or even days, effects of lipophilic messengers are generally slow to develop. In addition, because these newly synthesized proteins often remain in the target cells long after the messenger is gone, the effects can persist a long time. For the few lipophilic messengers that have receptors in the plasma membrane the effects are more rapid, for reasons discussed below.

Signal Transduction Mechanisms for Membrane-Bound Receptor-Mediated Responses

Lipophobic messengers cannot permeate the plasma membrane to any significant degree and thus their receptors are located on the plasma membrane and face the extracellular fluid. The receptors for these messengers fall into three general categories: channel-linked receptors, enzyme-linked receptors, and G-protein-linked receptors. Responses produced at each of these receptor types are described next.

Channel-Linked Receptors

Because the lipid bilayer has virtually no permeability to ions, the ion permeability of the plasma membrane is determined by the presence of ion channels in it (see Chapter 4). These ion channels are generally specific, allowing only one type of ion or class of ions to pass through them. In addition, ion channels are proteins, most of which can be regulated between open and closed states. Ion channels that open or close in response to the binding of a messenger to a receptor are called **ligand-gated channels.** These channels fall into two categories: fast channels, in which the receptor and channel are the same protein (as described next), and slow channels, in which the receptor and channel are separate proteins but are coupled together by a third type of protein, called a *G protein* (described shortly).

Fast ligand-gated channels are proteins that function as both receptors and ion channels. The binding of a messenger to the receptor/ion channel causes the channel to

FIGURE 5.12 Fast ligand-gated channels. *In both situations depicted, binding of a messenger to the receptor/channel opens the ion channel.* **(a)** *The opening of most ion channels results in movement of ions into or out of the cell, which changes the electrical properties of the cell.* **(b)** *The opening of calcium channels enables calcium ions to enter the cell, triggering a response such as secretion of some product by exocytosis, muscle contraction, or change in activity of a protein. In the last instance, calcium acts as a second messenger, binding to the protein calmodulin to form a calcium-calmodulin complex. The calcium-calmodulin complex activates a protein kinase, which phosphorylates a protein that produces a response in the cell.*

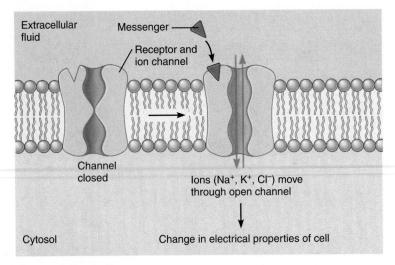

(a) Changing the electrical properties of a cell through ligand-gated channels.

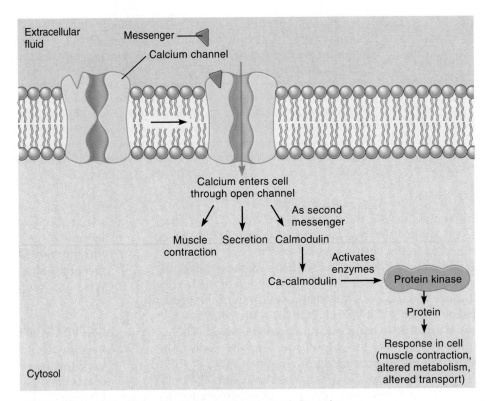

(b) Changing cystolic calcium levels through ligand-gated channels.

open, increasing the membrane's permeability for that specific ion. Open ion channels allow a specific ion or class of ions to move across the plasma membrane. Ion movement into or out of the cell can have two different effects on the target cell: (1) Ions entering and leaving can change the electrical properties of the cell, and (2) entering ions can act as a second messenger inside the cell.

The opening of most ion channels causes effects by changing the electrical properties of the target cell (Figure 5.12a). Recall from Chapter 4 that a difference in potential exists across the plasma membrane at rest, such that the inside of the cell is negative relative to the outside. Ions that move through an open channel carry with them a charge that changes the membrane potential. For example, the neurotransmitter acetylcholine stimulates skeletal muscle contraction by binding to nicotinic cholinergic receptors on skeletal muscle cells, causing the opening of ion channels. Sodium ions move into the cell (potassium also moves out, but to a lesser degree), carrying positive charge into the muscle cell. Given that the receptor and ion channel are the same protein molecule, these changes in membrane potential begin very

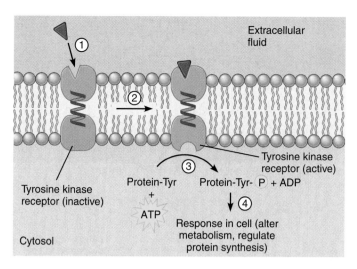

FIGURE 5.13 An enzyme-linked receptor. *The enzyme-linked receptor depicted here is tyrosine kinase. ① Binding of a messenger to its receptor activates the enzyme portion of the receptor. ② Tyrosine kinase is activated and ③ catalyzes phosphorylation of a protein, producing a cellular response ④.*

rapidly, within a millisecond. Because the binding of messenger to receptor is very brief, and the channel is open only while the messenger is bound, the change in membrane potential does not usually last long and terminates in a few milliseconds. In Chapters 6 and 7 we consider in detail how changes in membrane potential are crucial to the function of neurons and muscle cells.

In other cases, fast ligand-gated channels exert their effects by increasing the intracellular concentration of calcium (Figure 5.12b). When these channels open, calcium ions enter the cell. Depending on the target cell, this calcium can then trigger a variety of responses, including muscle contraction, secretion of a product by exocytosis, and a change in the activity of intracellular proteins. In the third case, calcium acts as a **second messenger**—an intracellular messenger produced by the binding of an extracellular messenger (the *first messenger*) to a receptor. As a second messenger, calcium changes the activity of intracellular proteins by binding to a cytosolic protein called **calmodulin**. The resultant calcium-calmodulin complex usually activates a protein kinase, an enzyme that catalyzes the phosphorylation of a protein, thereby altering its structure and function.

Calcium is well suited for its role in intracellular signaling because it is normally present in very low concentrations in the cytosol (10^{-7} to 10^{-6} molar), as compared to 10^{-3} molar in extracellular fluid. The significance of this low cytosolic concentration of calcium is that entry into a cell of even a small quantity of calcium causes a relatively large percentage change in the concentration, which means that the system is very sensitive. Intracellular calcium levels are maintained at their normal low levels by three processes that remove calcium ions from

the cytosol: (1) active transport of calcium across the plasma membrane, (2) sequestration of calcium by binding with proteins in the cytosol, and (3) active transport of calcium into certain organelles, such as the smooth endoplasmic reticulum and mitochondria.

Enzyme-Linked Receptors

Certain receptor proteins, known as **enzyme-linked receptors,** function both as enzymes and as receptors. These are transmembrane proteins, with the receptor side facing the interstitial fluid and the enzyme side facing the cytosol. These enzymes are normally inactive but are activated when a messenger binds to the receptor, which allows them to catalyze intracellular reactions.

Most enzyme-linked receptors are *tyrosine kinases*, which catalyze the addition of a phosphate group to the side chains of the amino acid tyrosine in certain locations in target proteins. The events occurring at tyrosine kinase receptors are as follows (Figure 5.13): First, a messenger binds to the receptor, changing its conformation, which activates the tyrosine kinase. The tyrosine kinase then catalyzes phosphorylation of an intracellular protein. As previously described for calcium-calmodulin, phosphorylation of a protein changes its activity, bringing about a response in the target cell. Other enzyme-linked receptors are *guanylate cyclases*, which catalyze the conversion of GTP *(guanosine triphosphate)* to the second messenger cGMP *(cyclic guanosine monophosphate)*. cGMP then activates a protein kinase, which catalyzes phosphorylation of a protein.

G-Protein-Linked Receptors

G-protein-linked receptors work by activating special membrane proteins called **G proteins.** See When it Goes Wrong: Bacterial Infections that Attack G Proteins, pp. 152–153, as an example of the significance of G proteins. G proteins are located on the intracellular side of the plasma membrane, where they function as links between the G-protein-linked receptor and other proteins in the plasma membrane, such as ion channels or enzymes (Figure 5.14). G proteins, which get their name from their ability to bind guanosine nucleotides, have three subunits: alpha (α), beta (β), and gamma (γ). The guanosine binding site is in the alpha subunit. In its inactive state, a G protein binds GDP (guanosine diphosphate). When a messenger binds to the G-protein-linked receptor (step 1 in Figure 5.14), the G protein releases the GDP, binds a molecule of GTP (guanosine triphosphate), and becomes active (step 2). In its active state, the alpha subunit separates from the other subunits and moves to the target protein (the protein that is to be regulated, either an ion channel or an enzyme), causing the protein to change its activity (step 3). The G protein does not stay active very long because it also functions as

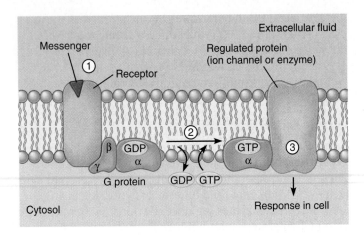

FIGURE 5.14 Actions of G proteins. *G proteins have three subunits: alpha, beta, and gamma. The alpha subunit has binding sites for guanosine nucleotides. In the inactive state, GDP is bound to the alpha subunit.* ① *Binding of a messenger to a G-protein-linked receptor activates the G protein.* ② *The GDP is released as the alpha subunit moves laterally within the membrane and binds a GTP.* ③ *The alpha unit then activates another membrane protein, producing a response in the cell.*

Is the receptor for the messenger an integral membrane protein or a peripheral membrane protein?

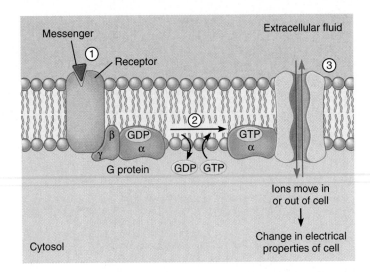

FIGURE 5.15 Action of a G protein on a slow ligand-gated ion channel. ① *Binding of the messenger to the receptor activates the G protein.* ② *The alpha subunit moves to an ion channel in the membrane.* ③ *The ion channel opens or closes, changing the permeability of the membrane to a specific ion. Ions move across the plasma membrane, changing the electrical properties of the cell.*

an enzyme that hydrolyzes the GTP, in the process returning itself to the inactive state.

G proteins are of three basic types: (1) those that affect ion channels, (2) *stimulatory G proteins,* and (3) *inhibitory G proteins.* Stimulatory G proteins (G$_s$ proteins) and inhibitory G proteins (G$_i$ proteins) are associated with the activation and inhibition, respectively, of enzymes that catalyze the production of second messengers in the intracellular fluid. The functions of some different types of G proteins are discussed next.

Slow ligand-gated ion channels are regulated by G proteins, which cause the channels to open or close in response to a messenger binding to its receptor (Figure 5.15). When the G protein is activated, the alpha subunit moves to the ion channel, causing a conformational change that causes it to open or close. These channels exert effects similar to those exerted by fast ligand-gated channels, but there are two important differences between these two classes of channels: (1) Messenger binding to channel-linked receptors only *opens* the channel and therefore increases the permeability of the target cell for the specific ion. By contrast, G-protein-linked ion channels can either be *opened* or *closed* by messenger binding to the receptor. (2) Binding of a messenger to channel-linked receptors produces an immediate and brief (only a few milliseconds) response in the target cell. In contrast, G-protein-linked ion channels are slow to open or close in response to receptor binding and stay open or closed for longer periods of time, often minutes.

G-protein-regulated enzymes are associated with the production of second messengers in the cytosol (Figure 5.16). When the first messenger binds to the receptor, the activated G protein releases the alpha subunit, which binds to the enzyme to either stimulate (activate) or inhibit (inactivate) the enzyme. The enzyme catalyzes the synthesis of a second messenger that ultimately activates a protein kinase.

Five major second messengers account for most of the communication through G-protein-regulated enzymes: (1) *cAMP* (cyclic adenosine monophosphate), (2) cGMP, (3) *inositol triphosphate,* (4) *diacylglycerol,* and (5) calcium. The second messenger systems described next are summarized in Table 5.3.

The mechanisms of action of cAMP, the most common second messenger, are as follows (Figure 5.17a):

① The first messenger binds to the receptor, activating a G$_s$ protein. (Some messengers inhibit the cAMP second messenger system by activating a G$_i$ protein, not shown in the figure.)

② The G protein releases the alpha subunit, which binds to and activates the enzyme **adenylate cyclase.**

③ Adenylate cyclase catalyzes the conversion of ATP to cAMP.

④ cAMP activates *protein kinase A,* also called cAMP-dependent protein kinase.

⑤ The protein kinase catalyzes the transfer of a phosphate group from ATP to a protein, thereby altering the protein's activity.

⑥ Altered protein activity causes a response in the cell.

TABLE 5.3 SECOND MESSENGER SYSTEMS

SECOND MESSENGER	PRECURSOR	ENZYME CATALYZING SYNTHESIS	USUAL ACTION	EXAMPLES OF FIRST MESSENGERS IN THE SYSTEM
Cyclic adenosine monophosphate (cAMP)	ATP	Adenylate cyclase	Activates protein kinase A	Epinephrine, vasopressin
Cyclic guanosine monophosphate (cGMP)	GTP	Guanylate cyclase	Activates protein kinase G	Atrial natriuretic peptide
Diacylglycerol (DAG)	Inositol 4,5-biphosphate (PIP$_2$)	Phospholipase C	Activates protein kinase C	Angiotensin II, histamine, vasopressin
Inositol triphosphate (IP$_3$)	Inositol 4,5-biphosphate (PIP$_2$)	Phospholipase C	Stimulates release of calcium from intracellular stores	Angiotensin II, histamine, vasopressin
Calcium*	None	None	Binds to calmodulin, then activates protein kinase	Angiotensin II, histamine, vasopressin

*Calcium increases in the cytosol in response to opening of ion channels either in the plasma membrane or in certain organelles.

Termination of the actions of cAMP requires its degradation by the enzyme *cAMP phosphodiesterase.* For the actions of the phosphorylated protein to be terminated, the phosphate group must be removed from the protein by a chemical reaction. The enzymes that dephosphorylate a protein are called *phosphoprotein phosphatases.*

The concentration of cAMP in a cell is determined by the relative rates of synthesis and breakdown. When synthesis is faster than breakdown, the concentration rises. When breakdown exceeds synthesis, the concentration falls. Because intracellular levels of cAMP are determined by the rates of two competing enzymes (one that synthesizes it and one that breaks it down), the effects of stimulating one enzyme can be mimicked by inhibiting the other. For example, caffeine, a stimulant found in coffee and in foods and other beverages, inhibits cAMP phosphodiesterase, causing levels of cAMP to rise in many cells. This triggers many of the changes associated with caffeine, including increased heart rate and blood pressure, wakefulness, and heightened alertness.

We previously discussed cGMP as a product of an enzyme-linked receptor, guanylate cyclase. However, guanylate cyclase is more commonly associated with G proteins, in which case the cGMP second messenger system is similar to cAMP but activates *protein kinase G,* also called cGMP-dependent protein kinase.

In the **phosphatidyl inositol system,** a membrane phospholipid undergoes an enzyme-catalyzed reaction that liberates two second messengers, **diacylglycerol** (DAG) and **inositol triphosphate** (IP$_3$); the latter stimulates release of the third second messenger,

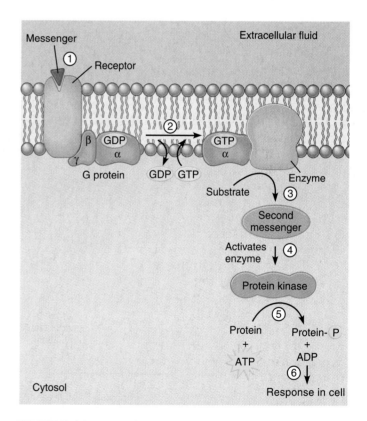

FIGURE 5.16 G-protein-regulated enzymes and second messengers. *In this example the G protein is coupled to an enzyme in the plasma membrane.* ① *Binding of the messenger to its receptor activates the G protein.* ② *The alpha subunit moves to and activates an enzyme in the membrane.* ③ *The activated enzyme catalyzes formation of a second messenger in the cytosol.* ④ *The second messenger activates a protein kinase, which* ⑤ *catalyzes phosphorylation of a protein, which* ⑥ *initiates a response in the cell.*

FIGURE 5.17 Second messenger systems. (a) *The cAMP second messenger system. The steps are identical to those described in the caption for Figure 5.16, except that specific components are identified.* **(b)** *The phosphatidyl inositol second messenger system.* ① *Binding of the messenger to the receptor activates a G protein.* ② *The alpha subunit moves to the enzyme phospholipase C, activating it.* ③ *Phospholipase C catalyzes the conversion of a membrane phospholipid, phosphatidyl inositol biphosphate (PIP₂), to diacylglycerol (DAG) and inositol triphosphate (IP₃).* ④ₐ *DAG activates protein kinase C, which* ⑤ₐ *catalyzes the phosphorylation of a protein in the cytosol,* ⑥ₐ *bringing about a response in the cell.* ④ᵦ *IP₃ enters the cytosol, where it* ⑤ᵦ *stimulates the opening of calcium channels in the endoplasmic reticulum, increasing cytosolic calcium concentration. Calcium can then either* ⑥ᵦ *stimulate secretion or contraction or* ⑥ᵪ *bind to calmodulin to modulate the activity of intracellular proteins.*

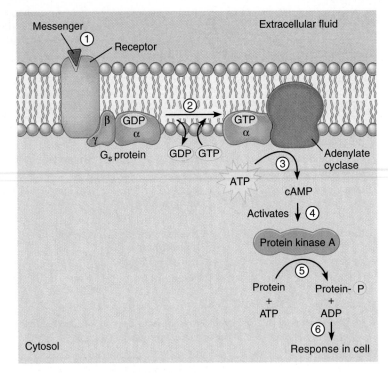

(a) cAMP second messenger system

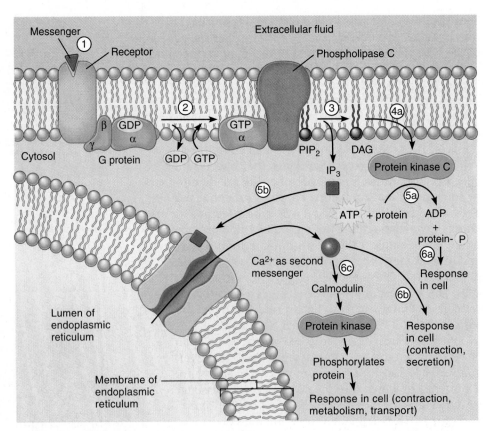

(b) Phosphatidyl inositol second messenger system

TABLE 5.4 SIGNAL TRANSDUCTION
MECHANISMS FOR SELECTED
CHEMICAL MESSENGERS

MESSENGER	FUNCTIONAL CLASS	CHEMICAL CLASS	SIGNAL TRANSDUCTION MECHANISM
Epinephrine	Hormone, neurotransmitter	Amine	G-protein-coupled receptors
Thyroid hormones	Hormone	Amine (lipophilic)	Altered transcription of mRNA
Vasopressin	Hormone, neurotransmitter	Peptide	G-protein-coupled receptors
Insulin	Hormone, neurotransmitter	Peptide	Enzyme-linked receptors
Estrogen	Hormone	Steroid	Altered transcription of mRNA
Glutamate	Neurotransmitter	Amino acid	Channel-linked receptor, G-protein-coupled receptors
Serotonin	Neurotransmitter, paracrine, autocrine	Amine	Channel-linked receptor, G-protein-coupled receptors
Prostaglandins	Paracrine	Eicosanoid	G proteins, unknown for many
Interleukins	Cytokine	Peptide	Enzyme-linked receptors
GABA	Neurotransmitter	Amino acid	Channel-linked receptor

calcium. The action of this system proceeds as follows (Figure 5.17b):

① The messenger binds to its receptor, activating a G protein.

② The G protein releases the alpha subunit, which binds to and activates the enzyme *phospholipase C.*

③ Phospholipase C catalyzes the conversion of a membrane phospholipid called *phosphatidylinositol 4,5-biphosphate* (PIP_2) to DAG and IP_3, each of which functions as a second messenger.

④a DAG remains in the membrane and activates the enzyme *protein kinase C,* which

⑤a catalyzes the phosphorylation of a protein,

⑥a bringing about a response in the cell. At the same time,

④b IP_3 moves into the cytosol, where

⑤b it triggers the release of calcium from the endoplasmic reticulum. Calcium then either

⑥b acts on proteins to stimulate contraction or secretion, or

⑥c acts as a second messenger by binding to calmodulin, activating a protein kinase that phosphorylates a protein that produces a response in the cell.

The signal transduction mechanisms for some selected chemical messengers in various functional and chemical classes are listed in Table 5.4.

Signal Amplification in Chemical Messenger Systems

Given that some cells have simple signal transduction mechanisms such as the one-step activation of tyrosine kinase, one might wonder why second messenger systems are needed, when they ultimately have the same basic effect as tyrosine kinase—phosphorylation of a protein. The reason relates to one of the striking features of second messenger systems—the ability of small changes in the concentration of a chemical messenger to elicit marked responses in target cells, a phenomenon known as *signal amplification.* Figure 5.18 depicts how amplification works, using the cAMP system as an example. The diagram illustrates how a single activated receptor can activate several G proteins, each of which can in turn activate an adenylate cyclase. While active, each adenylate cyclase can generate hundreds of molecules of cAMP, each of which then activates a molecule of protein kinase A. Each protein kinase A molecule can then phosphorylate hundreds of target proteins. The net result is that a very large number of end-product molecules can be regulated in response to the binding of a single ligand molecule to its receptor. In the example shown, one first messenger led to the phosphorylation of approximately 2,500,000 proteins.

The sequence of reactions shown in Figure 5.18 is an example of a **cascade,** a series of sequential steps that progressively increase in magnitude, much as numerous tiny brooks, added together, can eventually become a major river. Cascades of one type or another are common in chemical messenger systems and account for much of the signal amplification that occurs.

Now that our discussion of the various aspects of chemical messengers is complete, see Table 5.5 for a summary of the properties of chemical messengers. We turn next to a discussion of long-distance communication in the body.

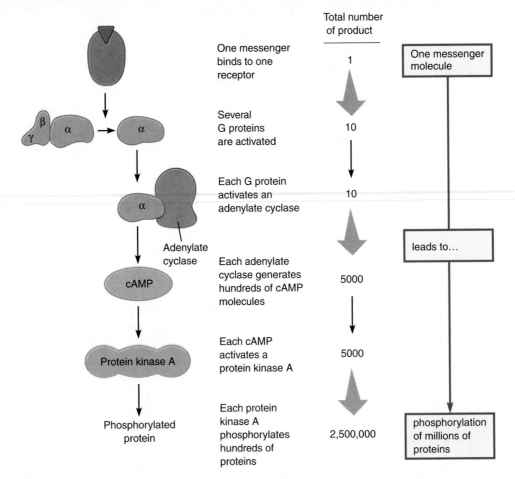

FIGURE 5.18 **Signal amplification, in this case by the second messenger cAMP.**

	TABLE 5.5	PROPERTIES OF CHEMICAL MESSENGERS

PROPERTY	LIPOPHOBIC MESSENGER (HYDROPHILIC)	LIPOPHILIC MESSENGER (HYDROPHOBIC)
Chemical classes	Amino acids, amines, peptides	Steroids, eicosanoids, thyroid hormones*
Storage in secretory cell	Secretory vesicles	None
Mechanism of secretion	Exocytosis	Diffusion
Transport in blood†	Dissolved	Bound to carrier protein
Location of receptor	Plasma membrane	Cytosol or nucleus
Signal transduction mechanism	Open/close ion channels‡ Activate membrane-bound enzymes G proteins and second messenger systems	Alter transcription of mRNA (alter protein synthesis)
Relative time to onset of response	Fast	Slow
Relative duration of response	Short	Long
Relative half-life	Short	Long

*Thyroid hormones are amines but are lipophilic.
†Refers to hormones and certain cytokines only.
‡Some of these effects are mediated by G proteins.

1. Define the following terms: *agonist, antagonist, up-regulation,* and *down-regulation.*

2. Where in the target cell are the receptors for lipophilic messengers located? What is the name of the regulatory region of DNA to which the hormone-receptor complex binds?

3. Name the three categories of membrane-bound receptors.

4. Name five substances that act as second messengers.

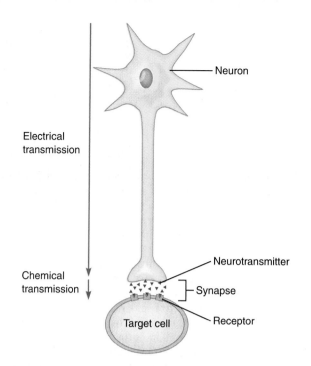

FIGURE 5.19 Signal transmission in neurons. *Neurons transmit both electrical signals within the cell and chemical signals between cells.*

LONG-DISTANCE COMMUNICATION VIA THE NERVOUS AND ENDOCRINE SYSTEMS

In order to maintain homeostasis, it is crucial that cells in one region of the body are able to communicate with cells in distant regions. The body has two organ systems specialized for long distance communication: the nervous system and the endocrine system.

The nervous system consists of neurons and supporting cells called *glial cells.* Neurons are capable of communicating long distances, first by transmitting electrical signals along the length of the cell (some of which are up to a meter long), and then by transmitting chemical signals through the release of a neurotransmitter at a synapse (Figure 5.19). Because of the direct cell-to-cell communication of chemical signals at a synapse, the nervous system is often considered a "wired" system. (One exception is the neurohormones, which travel via the bloodstream.) Communication in the nervous system generally involves the opening and closing of ion channels, which is very fast and generally of short duration. Signals transmitted by the nervous system travel quickly and are generally of short duration, making the system ideal for controlling movements and monitoring the world around us. The nervous system is covered in detail in Chapters 6 through 10.

In contrast, the endocrine system lacks any direct anatomical link between secretory cell and target cell. Instead, the endocrine system communicates through chemical messengers called hormones, which travel via the bloodstream to virtually all cells in the body. Although strictly speaking hormones are secreted into the interstitial fluid and then diffuse into blood, we often speak of the secretion of hormones into blood for the sake of simplicity. Hormones generally communicate by altering protein synthesis or activating G proteins, processes that are considerably slower than the electrical and chemical signals used by the nervous system. The relative slowness of the endocrine system and its ability to broadcast signals over wide areas are important in coordinating metabolic activities among organ systems. The endocrine system is covered in detail in the remainder of this chapter and in Chapter 20. Various aspects of the nervous and endocrine systems are compared in Table 5.6.

TABLE 5.6	CHARACTERISTICS OF THE NERVOUS AND ENDOCRINE SYSTEMS	
	NERVOUS SYSTEM	**ENDOCRINE SYSTEM**
Secretory cell	Neuron	Endocrine cell
Target cell	Neuron, muscle, or gland	Most cell types in body
Messenger	Neurotransmitter	Hormone
Pathway for communication	Across synapse	Via bloodstream
Basis of specificity	Receptors on postsynaptic target cell	Receptors on target cells throughout body
Time to onset of effect	Immediate	Delayed
Duration of effect	Brief	Long

Among the several bacteria that have toxins that attack G proteins and second messenger systems in the body are those that cause cholera and pertussis, severe and often fatal communicable diseases that are rare in the United States but common in developing countries. Although the two diseases affect different regions of the body, they have several things in common: Both are highly communicable, and their signs and symptoms are caused by structurally similar protein toxins secreted by the bacterial cells. More to the point, cholera toxin and pertussis toxin both exert their actions by interfering with the function of G proteins that link receptors to adenylate cyclase, the enzyme that catalyzes formation of the second messenger cAMP.

Cholera, caused by infection with the bacterium *Vibrio cholera*, is the leading cause of death among small children in developing countries. It is most often acquired by ingesting sewage-contaminated water or food. The primary sign of the disease is massive diarrhea, which can result in the loss of 15–20 liters of fluid from the body per day in severe cases. Associated with this problem is the loss of sodium and chloride ions (electrolytes) that accompanies the loss of water, which by altering the electrolyte composition of the plasma can interfere with neural and muscular function. In untreated cases, the mortality rate exceeds 50%; death sometimes occurs within hours after signs first appear. Treatment includes replacement of fluids and electrolytes and antibiotic therapy.

In healthy individuals, certain epithelial cells secrete an electrolyte-rich fluid into the lumen of the small intestine to aid in the digestion and absorption of food. This fluid is subsequently reabsorbed (along with ingested fluid and nutrients) by other epithelial cells lining the intestine, so that little fluid remains in the lumen under normal conditions. But in cholera, the action of cholera toxin on secretory cells increases the rate of secretion so much that absorption cannot keep up with it; the signficant volume of fluid left behind in the lumen is voided as diarrhea.

The effects of cholera toxin result from its actions on a G protein that stimulates adenylate cyclase (see the box figure). The steps of this process are as follows: ① Cholera toxin binds to a membrane *ganglioside* (phospholipid with carbohydrate residues attached) on secretory cells in the small intestine. ② A subunit of the toxin enters the cell and causes the sustained activation of a G protein. ③ This G protein activates adenylate cyclase, which ④ catalyzes the formation of cAMP. ⑤ cAMP activates protein kinases, which ⑥ enhance the secretion of chloride ions. ⑦ The flow of negatively charged chloride ions out of the cell causes positively charged sodium ions to follow them. ⑧ Water follows the electrolytes into the lumen of the small intestine by osmosis. With an excess of cAMP, excess fluid and electrolytes are secreted into the lumen of the small intestine, resulting in severe diarrhea.

Pertussis, also known as *whooping cough*, is a disease caused by infection of the respiratory epithelial cells with the bacterium *Bordatella pertussis*. The disease begins with coldlike symptoms but quickly progresses to a more severe phase characterized by coughing spells that can last up to a minute. The disease's common name refers to

ENDOCRINE GLANDS

Glands are derived from epithelial tissue and function in the secretion of chemical products. Glands are of two main types: *exocrine glands,* which secrete their product (such as sweat or digestive fluids) into ducts that lead into the external environment (for example, the skin surface or the lumen of the digestive tract), and **endocrine glands,** which secrete hormones into the interstitial fluid, from which the hormones then diffuse into the bloodstream. Endocrine glands are found in many organs of the body (Figure 5.20). There are two types of endocrine organs: **primary endocrine organs,** whose primary function is the secretion of hormones, and **secondary endocrine organs,** for which the secretion of hormones is secondary to some other function. Some primary endocrine organs are located within the brain, including the hypothalamus, pituitary gland, and pineal gland; other primary endocrine organs are located outside the nervous system, including the thyroid gland, parathyroid glands, thymus, adrenal glands, pancreas, and gonads (testes in the male and

the high-pitched "whooping" sound made by patients as they gasp for air following these spells. In the United States, mass immunization programs for preschool children almost eradicated the disease; only hundreds of cases are reported per year, with few deaths. However, in recent years, the number of reported cases has been increasing.

Pertussis toxin, like cholera toxin, exerts its effects on a G protein coupled to adenylate cyclase, and it ultimately leads to an increase in cAMP levels. However, there is a crucial difference: Pertussis toxin *inactivates* an inhibitory G protein that *inhibits* the production of cAMP. Because these G proteins normally inhibit adenylate cyclase, their inactivation by pertussis toxin effectively *stimulates* adenylate cyclase, resulting in an inappropriate rise in intracellular cAMP levels. The resultant damage to cells leads to erosion of the respiratory epithelium and discharge of large quantities of mucus-containing fluid, which triggers the coughing spells.

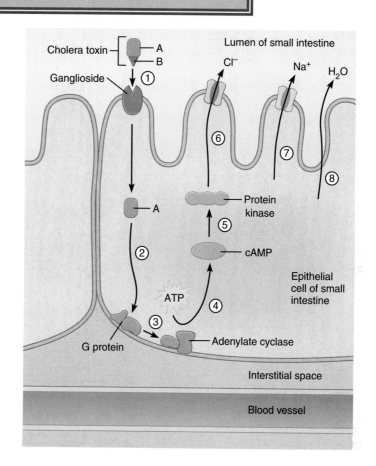

ovaries in the female). The placenta also functions as an endocrine gland in pregnant females. Secondary endocrine glands include organs such as the heart, liver, stomach, small intestine, kidney, and skin.

Primary Endocrine Organs

In this section, we examine the functions of the various primary endocrine organs, beginning with the hypothalamus and pituitary gland.

Hypothalamus and Pituitary Gland

Together, the hypothalamus and pituitary gland (Figure 5.21) function to regulate virtually every body system. The **hypothalamus** is a part of the brain with many functions in addition to its role as an endocrine gland. It is considered a primary endocrine gland because it secretes several hormones, most of which affect the **pituitary gland** (also called the *hypophysis*), a pea-sized structure that is connected to the hypothalamus by a thin stalk of tissue called the *infundibulum*. The pituitary gland

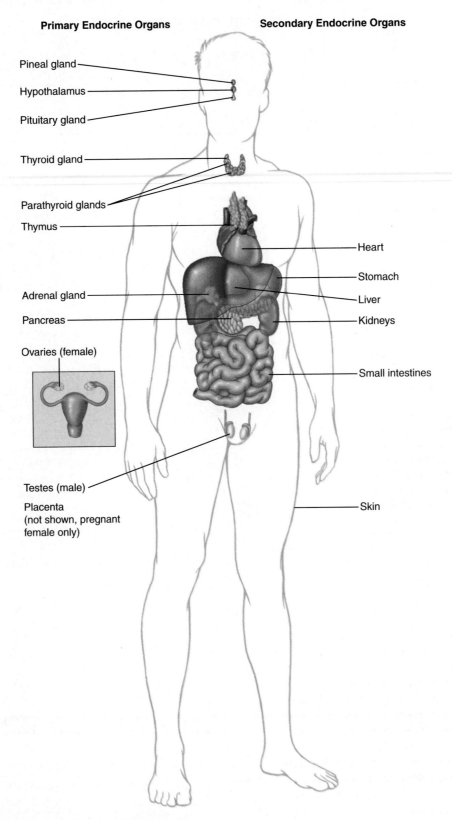

Primary Endocrine Organs

Pineal gland

Hypothalamus

Pituitary gland

Thyroid gland

Parathyroid glands

Thymus

Adrenal gland

Pancreas

Ovaries (female)

Testes (male)

Placenta
(not shown, pregnant
female only)

Secondary Endocrine Organs

Heart

Stomach

Liver

Kidneys

Small intestines

Skin

FIGURE 5.20 Endocrine organs.

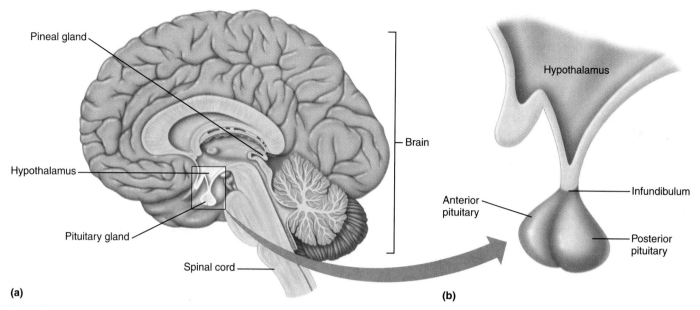

(a)

(b)

FIGURE 5.21 Hypothalamus and pituitary gland. (a) *Locations of the hypothalamus and pituitary gland in the brain.* **(b)** *Connection of the hypothalamus and the pituitary gland by the infundibulum. The pituitary gland is divided into an anterior lobe and a posterior lobe.*

is divided into two structurally and functionally distinct sections called the **anterior lobe** (or *adenohypophysis*) and the **posterior lobe** (or *neurohypophysis*). As we will see shortly, the connections between the hypothalamus and the two lobes of the pituitary gland differ and are critical to the function of both endocrine organs.

The posterior lobe contains neural tissue consisting of the endings of neurons originating in the hypothalamus (Figure 5.22). These neural endings in the posterior pituitary gland secrete two peptide hormones: *antidiuretic hormone* (ADH; also called *vasopressin*) and *oxytocin*. These hormones are generally synthesized in neurons originating in different regions of the hypothalamus; ADH is synthesized in the *paraventricular nucleus,* and oxytocin is synthesized in the *supraoptic nucleus.* Following synthesis, the peptides are packaged into secretory vesicles, which are transported to the neural endings in the posterior pituitary. The hormone is released by exocytosis when these neurons receive a signal, usually from other neurons. The hormones are released into the blood, as are other hormones. Because these hormones are secreted by a neuron instead of an endocrine gland, they are also called *neurohormones.* Antidiuretic hormone regulates water reabsorption by the kidneys and is discussed in Chapter 18; oxytocin stimulates uterine contractions and milk letdown in the breasts and is discussed in Chapter 21.

The anterior lobe and the cells of the hypothalamus that control it secrete primarily **tropic hormones** (also called trophic hormones), which are hormones that regulate the secretion of other hormones. A tropic hormone can be a *stimulating hormone,* which increases the secretion of another hormone, or an *inhibiting hormone,* which decreases the secretion of another hormone. The

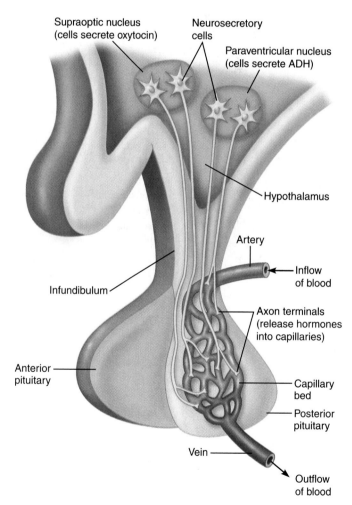

FIGURE 5.22 Connection between hypothalamus and posterior pituitary gland. *Neurons originating in the hypothalamus send projections to the posterior pituitary gland. Neurons originating in the paraventricular nucleus synthesize and secrete ADH; neurons originating in the supraoptic nucleus synthesize and secrete oxytocin. ADH and oxytocin are secreted from nerve terminals into the bloodstream in the posterior pituitary.*

FIGURE 5.23 Connection between hypothalamus and anterior pituitary gland. (a) *The hypothalamic-pituitary portal system. Neurosecretory cells of the hypothalamus secrete tropic hormones into the hypothalamic-pituitary portal system. The tropic hormones travel to the anterior pituitary, where they effect release of anterior pituitary hormones into the blood.* **(b)** *Regulation of anterior pituitary hormone secretion. The hypothalamus secretes into the hypothalamic-pituitary portal vein seven tropic hormones that are either releasing hormones (−RH) or inhibiting hormones (−IH). These tropic hormones act on endocrine cells in the anterior pituitary to stimulate the release of hormones that (except for prolactin) are also tropic hormones. Anterior pituitary tropic hormones act on other endocrine glands to stimulate the release of still other hormones.*

Which hypothalamic tropic hormone indirectly stimulates the release of cortisol?

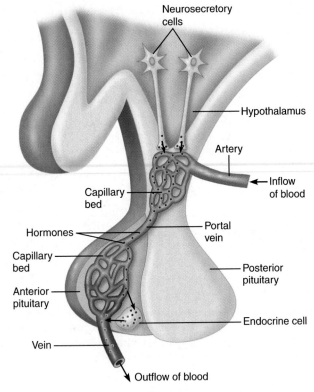

(a) Hypothalamic-pituitary portal system

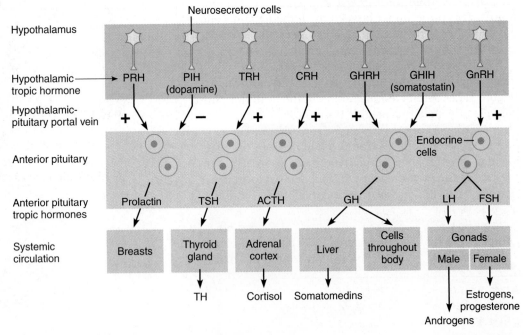

(b) Regulation of anterior pituitary hormone secretion

Corticotropin releasing hormone

156   CHAPTER 5 ■ *Chemical Messengers and the Endocrine System*

general signaling pathway is as follows: the hypothalamus releases a tropic hormone that effects the release of another tropic hormone from the anterior pituitary; this tropic hormone effects the release of a third hormone from another endocrine gland and this hormone exerts effects on target cells throughout the body.

The hormone-secreting cells in the anterior pituitary are the target cells for tropic hormones from the hypothalamus. The hypothalamus and anterior lobe are connected by the **hypothalamic-pituitary portal system** (Figure 5.23a). A *portal system* is a specialized arrangement of blood vessels in which two capillary beds are located in series, one after the other. (Capillaries are the smallest of the blood vessels and are the sites where compounds can be exchanged between blood and tissue.) After the hypothalamus secretes tropic hormones into capillary beds located there (see Figure 5.23), these tropic hormones travel down the infundibulum to the pituitary gland via a portal vein, from which they enter a second capillary bed. The hypothalamic tropic hormones then stimulate or inhibit the release of hormones from the anterior pituitary. The portal system enables the hypothalamic tropic hormones to be delivered directly to their target cells in the anterior pituitary, ensuring that these hormones are not diluted and degraded by enzymes in the general circulation. As a result, the hypothalamic tropic hormones are more concentrated in the portal blood delivered to the anterior pituitary and have a greater effect on hormone release.

Compared to the posterior lobe, the anterior lobe is more like a "typical" endocrine gland in that hormones are synthesized and secreted by cells located entirely within the pituitary gland. Within the anterior lobe, each of several intermingled cell types is responsible for synthesizing and secreting a particular hormone. Secretion of these hormones is regulated by tropic hormones secreted by neurons in the hypothalamus. Secretion of these tropic hormones (five stimulating hormones and two inhibiting hormones) is regulated by neural input to the hypothalamic neurons. The hypothalamic tropic hormones, the anterior pituitary hormones they regulate, and the functions of those anterior pituitary hormones are listed below and shown in Figure 5.23b. With one exception, all these hormones are peptides.

1. *Prolactin releasing hormone* (PRH) stimulates the anterior pituitary to release *prolactin,* which stimulates mammary gland development and milk secretion in females.

2. *Prolactin inhibiting hormone* (PIH), or dopamine (a catecholamine), inhibits the release of prolactin.

3. *Thyrotropin releasing hormone* (TRH) stimulates the release of *thyroid stimulating hormone* (TSH) from the anterior pituitary. TSH then stimulates the secretion of thyroid hormones by the thyroid gland.

4. *Corticotropin releasing hormone* (CRH) stimulates the release of *adrenocorticotropic hormone* (ACTH) by the anterior pituitary. ACTH then stimulates the secretion of other hormones by the adrenal cortex, the outer layer of the adrenal gland.

5. *Growth hormone releasing hormone* (GHRH) stimulates the secretion of *growth hormone* (GH) by the anterior pituitary. GH regulates growth and energy metabolism but also functions as a tropic hormone by stimulating the secretion of *somatomedins* by the liver.

6. *Growth hormone inhibiting hormone* (GHIH), or *somatostatin,* inhibits the secretion of growth hormone by the anterior pituitary.

7. *Gonadotropin releasing hormone* (GnRH) stimulates the release of both *follicle stimulating hormone* (FSH) and *luteinizing hormone* (LH) by the anterior pituitary. LH stimulates ovulation in females, but also stimulates the secretion of sex hormones (*estrogens* and *progesterone* in females and *androgens* in males) by the gonads. FSH promotes the development of egg cells in females and sperm cells in males, but also stimulates the secretion of estrogens in females and *inhibin* in both sexes.

The multistep pathways by which hypothalamic and anterior pituitary tropic hormones are produced are regulated by *feedback loops* (Figure 5.24). The tropic hormone from the anterior pituitary may act through negative feedback on the hypothalamus to decrease its own release. The inhibition of hypothalamic tropic hormones by the anterior pituitary tropic hormone, called *short loop negative feedback,* prevents the build-up of excess anterior pituitary tropic hormone. Additionally, the hormone whose secretion is stimulated by the tropic hormone generally feeds back on the hypothalamus (and often the anterior pituitary as well) to inhibit secretion of the tropic hormone, thereby limiting its own secretion. This phenomenon is called *long loop negative feedback.*

Figure 5.25 depicts the regulation of thyroid hormone release by the thyroid gland as a specific example of negative feedback regulation. Thyrotropin-releasing hormone (TRH) stimulates the release of the anterior pituitary tropic hormone thyroid-stimulating hormone (TSH). Thyroid-stimulating hormone in turn stimulates the release of thyroid hormones (TH) from the thyroid gland. If thyroid hormone levels in the blood increase above a certain level, then negative feedback loops cause a decrease in the release of both TSH and TRH. In both cases, the net result is that the thyroid hormones act to inhibit their own secretion. However, thyroid hormones provide negative feedback only to their own tropic hormones; they have no effect, for example, on the cells that secrete GnRH or LH and FSH.

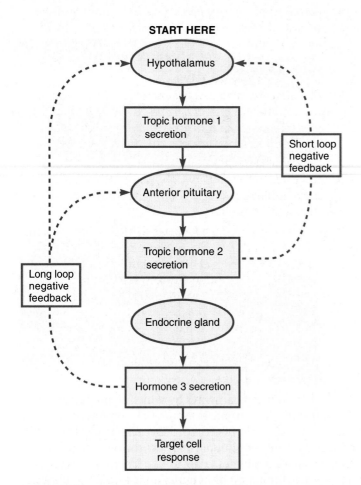

FIGURE 5.24 Negative feedback loops affecting hypothalamic and anterior pituitary tropic hormones.

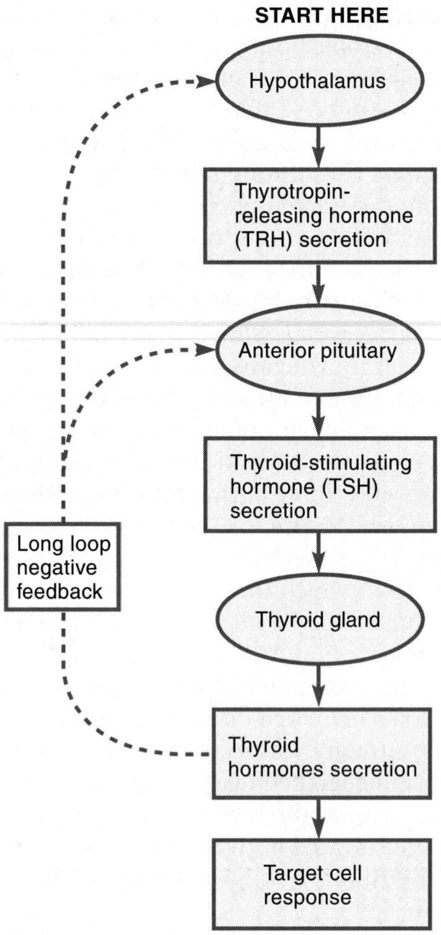

FIGURE 5.25 Regulation of thyroid hormone release.

Pineal Gland

The **pineal gland,** which is located within the brain (see Figure 5.21), is composed of glandular tissue that secretes the hormone *melatonin.* The function of melatonin is still a topic of debate, but most recent studies suggest that it is important in establishing the **circadian rhythm** (a daily rhythm coordinating body activities to the day-night cycle). It is known, for instance, that melatonin secretion rises at night and falls during the day, and that melatonin is a potent sleep-inducing agent when administered therapeutically. Melatonin also enhances immune function and exerts a suppressive effect on reproductive function by interfering with the activity of certain hormones.

Thyroid Gland and Parathyroid Glands

The **thyroid gland** is a butterfly-shaped structure located on the ventral surface of the trachea (Figure 5.26a). The thyroid gland secretes the two thyroid hormones, *tetraiodothyronine* (T_4) and *triiodothyronine* (T_3), as well as *calcitonin.* The thyroid hormones regulate the

body's metabolic rate and are necessary for normal growth and development; calcitonin regulates calcium levels in the blood. The four **parathyroid glands** are smaller structures located on the posterior surface of the thyroid gland (Figure 5.26b). The parathyroid glands secrete parathyroid hormone (PTH), an important regulator of calcium levels in the blood. The thyroid hormones, calcitonin, and parathyroid hormone are discussed in greater detail in Chapter 20.

Thymus

The **thymus** lies close to the heart (see Figure 5.20) and secretes the hormone *thymosin.* In addition, the thymus is critical for normal immune function because immune cells called *T lymphocytes,* which are essential for effective immune responses against invading microorganisms, mature there. Thymosin regulates T cell function.

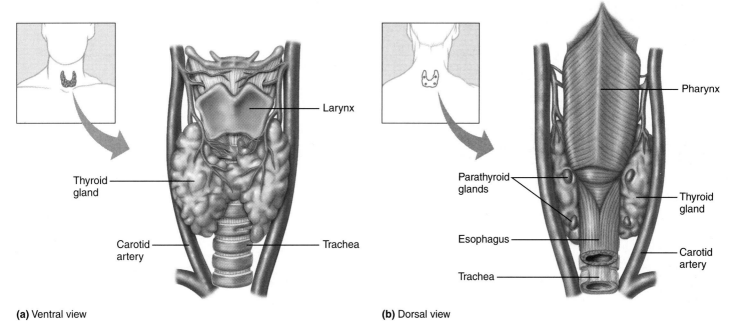

(a) Ventral view

(b) Dorsal view

FIGURE 5.26 **Locations of the thyroid and parathyroid glands.**

Adrenal Glands

The **adrenal glands** are located above the kidneys (see Figure 5.20); for this reason, they are also called the suprarenal glands. Each adrenal gland consists of an outer layer called the *cortex,* which accounts for about 80% of the gland's total mass, and an inner core called the *medulla* (Figure 5.27a). Like the pituitary, each adrenal gland is essentially two glands in one because the cortex and medulla are structurally and functionally distinct.

The **adrenal cortex** is stratified into three distinct layers (Figure 5.27b): (1) an outer layer called the *zona glomerulosa,* (2) a middle layer called the *zona fasciculata,* and (3) an inner layer called the *zona reticularis.* Because cells in these layers possess different complements of the enzymes involved in the synthesis of adrenal cortex hormones, these cells manufacture and secrete various hormones in different proportions.

The adrenal cortex secretes a number of hormones collectively called **adrenocorticoids,** a term that refers to their site of origin (*adren*al *cortex*) and the chemical class of compounds to which they belong (ster*oids*). The adrenocorticoids include hormones of three types:

1. **Mineralocorticoids** (primarily *aldosterone*), which are secreted exclusively by cells in the zona glomerulosa, and regulate sodium reabsorption and potassium excretion by the kidneys.

2. **Glucocorticoids** (primarily *cortisol*), which are secreted mainly by cells in the zona fasciculata and zona reticularis, and regulate the body's response to stress; protein, carbohydrate, and lipid metabolism in a variety of tissues; and blood glucose levels.

3. **Sex hormones** (primarily *androgens*), which are secreted by cells in the zona fasciculata and zona reticularis (and by the gonads), and regulate reproductive function and a variety of other processes. Because androgens are secreted in much larger amounts by the gonads in males, secretion of these hormones by the adrenal cortex is normally of little physiological significance; in females, however, adrenal androgens may stimulate the sex drive.

The **adrenal medulla** contains *chromaffin cells* and secretes catecholamines; about 80% of the secreted hormones is epinephrine, about 20% is norepinephrine, and less than 1% is dopamine. The primary stimulus for secretion of these hormones is neural, as described in greater detail in Chapter 10.

Pancreas

The **pancreas** functions as both an endocrine gland and an exocrine gland. The *exocrine pancreas* includes acinar cells and duct cells that secrete enzymes and fluid into the digestive tract (see Chapter 19); the *endocrine pancreas* consists of cell clusters called *islets of Langerhans,* which are scattered throughout the pancreas in spaces between the ducts (Figure 5.28).

The islets of Langerhans are the source of two major hormones, each of which is secreted by a different cell type: *insulin,* which is secreted by *B cells (beta cells),* and *glucagon,* which is secreted by *A cells (alpha cells).* Both of these hormones are important in the regulation of energy metabolism and blood glucose levels (see Chapter 20). Two other cell types are also located in the islets of Langerhans; *D cells* or *delta cells* secrete *somatostatin,* which helps to regulate digestion and absorption of nutrients and may regulate secretion of other pancreatic hormones (see Chapter 19). (Recall that somatostatin is also a hypothalamic tropic hormone that inhibits secretion of

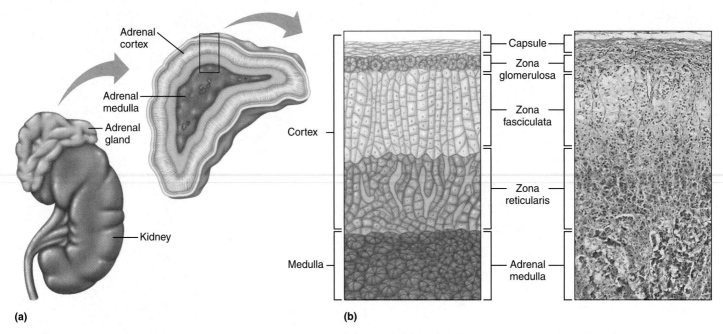

FIGURE 5.27 **Anatomy of the adrenal gland. (a)** *The adrenal glands, located just above the kidneys, are divided into an inner medulla and outer cortex.* **(b)** *The adrenal cortex contains three distinct layers, or zonae.*

What hormones are secreted from the zona reticularis?

growth hormone from the anterior pituitary.) F cells secrete pancreatic polypeptide; the function of this hormone is unknown.

Gonads

The gonads (ovaries and testes) have both endocrine and nonendocrine functions. In both sexes they produce the *gametes*—sperm in males and oocytes in females—and secrete sex hormones. In males, the predominant sex hormones are the androgens, testosterone and androstenedione; in females the major sex hormones are estradiol (which belongs to a group of related hormones known as estrogens) and progesterone.

The placenta also functions as an endocrine gland in pregnant females, secreting primarily estrogens and progesterone. The hormones of reproduction are described in detail in Chapter 21.

Secondary Endocrine Organs

In addition to their primary function, many organs of the body also secrete hormones. These secondary endocrine organs include the heart, kidneys, digestive organs, liver, and skin (see Figure 5.20). The heart secretes *atrial natriuretic peptide* (ANP), which regulates sodium reabsorption

by the kidneys (see Chapter 18). The kidneys secrete *erythropoietin,* which stimulates production of red blood cells by the bone marrow (see Chapter 13). Organs in the gastrointestinal tract secrete several hormones that are important in regulating the digestion and absorption of food (see Chapter 19). The liver secretes *somatomedins,* also called *insulin-like growth factors* (IGF), which promote tissue growth (see Chapter 20). Finally, the skin and kidneys are involved in the production of *calcitrol* (or *vitamin D_3*), which regulates blood calcium levels (see Chapter 18).

The hormones secreted by the primary and secondary endocrine organs, as well as their functions, are summarized in Table 5.7 on pages 162–163.

Quick Test 5.4

1. Which communication system is faster, the nervous system or the endocrine system?

2. What hypothalamic and anterior pituitary tropic hormones regulate the secretion of cortisol from the adrenal cortex?

3. Which two hormones are secreted by the posterior pituitary?

4. Which endocrine gland secretes epinephrine? Aldosterone? Insulin?

Glucocorticoids and sex hormones

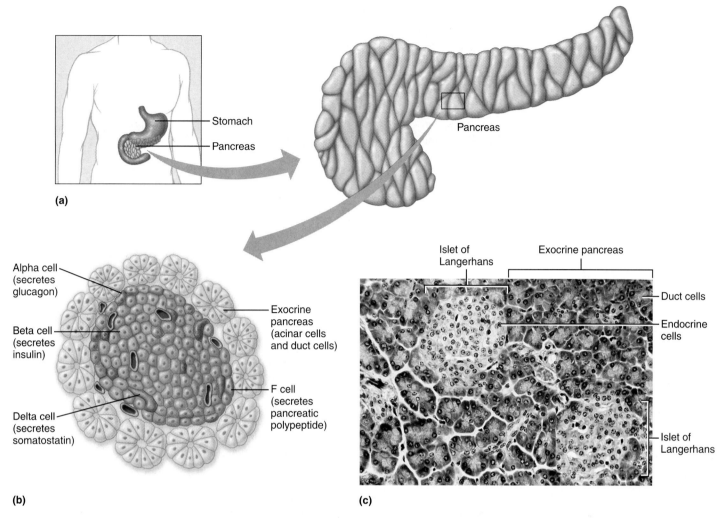

FIGURE 5.28 Anatomy of the pancreas. (a) *Located in the abdominal cavity below the stomach.* **(b)** *Histology of the pancreas. The pancreas consists of acinar cells and duct cells that secrete exocrine products into ducts, and islets of Langerhans, which contain cells that secrete endocrine hormones into the interstitial fluid. Four types of endocrine cells are located in the islets; each type secretes a different hormone.* **(c)** *Photomicrograph of the islets of Langerhans.*

HORMONE ACTIONS AT THE TARGET CELL

The first part of this chapter discussed chemical signaling across all messenger types. In this section, we discuss in detail the actions of hormones at the target cell. Factors influencing the magnitude of a target cell's response to a hormone include the concentration of free hormone in the blood and the types of receptors in the target cell.

Control of Hormone Levels in Blood

The concentration of free hormone in the blood depends on three factors: (1) the rate of hormone secretion, (2) the amount of hormone transported bound to carrier proteins, and (3) the rate at which the hormone is metabolized.

Rate of Hormone Secretion

With few exceptions, endocrine cells and other secretory cells release chemical messengers at variable rates. When these cells receive certain signals, the rate of secretion rises or falls. Faster rates of secretion translate into greater concentration of hormone in blood and more hormone molecules bound to receptors on target cells. Therefore, when hormone levels in blood increase, they "trigger" changes in the target cells. In a few cases, however, hormones are secreted at a relatively steady rate. Because the concentrations of these messengers do not change appreciably, they do not "trigger" changes in the target cell in the same way that other hormones do. Instead, they facilitate processes that are normally ongoing. For example, thyroid hormones are usually secreted at nearly constant rates in adults and are necessary for maintenance of normal metabolism and nervous system function.

TABLE 5.7   ENDOCRINE ORGANS AND THE
HORMONES THEY SECRETE

	HORMONES	FUNCTIONS
PRIMARY ENDOCRINE ORGANS		
Hypothalamus	Releasing and release-inhibiting hormones	Regulate secretion of anterior pituitary hormones
Anterior pituitary gland	Growth hormone (GH)	Essential for growth; stimulates bone and soft tissue growth; regulates protein, lipid, and carbohydrate metabolism
	Adrenocorticotropic hormone (ACTH)	Stimulates glucocorticoid secretion by the adrenal cortex
	Thyroid stimulating hormone (TSH)	Stimulates secretion of thyroid hormones by the thyroid gland
	Prolactin	Stimulates development of breasts and milk secretion by the mammary glands
	Follicle stimulating hormone (FSH)	Females: stimulates growth and development of ovarian follicles, estrogen secretion; males: stimulates sperm production by the testis
	Luteinizing hormone (LH)	Females: stimulates ovulation, transformation of ovarian follicle into corpus luteum, and secretion of estrogen and progesterone; males: stimulates testosterone secretion by the testis
Posterior pituitary gland	Antidiuretic hormone (ADH, vasopressin)	Decreases urine output by the kidneys; promotes constriction of blood vessels (arterioles)
	Oxytocin	Females: stimulates uterine contractions and milk ejection by mammary glands; males: function unknown
Pineal gland	Melatonin	Regulates biological rhythms according to day-night cycles
Thymus	Thymosin	Stimulates proliferation and function of T lymphocytes
Thyroid gland	Thyroid hormones (triiodothyronine and tetraiodothyronine)	Increases metabolic rate of many tissues; necessary for normal development
	Calcitonin	Promotes calcium deposition in bone; lowers blood calcium levels
Parathyroid glands	Parathyroid hormone (PTH)	Promotes calcium release from bone, calcium absorption by intestine, and calcium reabsorption by kidney tubules; raises blood calcium levels; stimulates vitamin D_3 synthesis
Adrenal cortex	Mineralocorticoids (aldosterone)	Stimulates sodium reabsorption and potassium secretion by kidney tubules
	Glucocorticoids (cortisol, corticosterone)	Promotes catabolism of proteins and fats; raises blood glucose levels; adapts the body to stress

In general, endocrine cells alter hormone secretion in response to two types of input, neural signals and *humoral* (blood-borne) signals, each of which may be either stimulatory or inhibitory. Neural signals directly regulate hormone secretion by the hypothalamus, posterior pituitary gland, and adrenal medulla, which can affect the secretion of other hormones. For example, stress activates neural signals that stimulate the hypothalamus to release corticotropin-releasing hormone, which stimulates the release of ACTH from the anterior pituitary (Figure 5.29). ACTH stimulates the adrenal cortex to secrete cortisol, a hormone that helps the body cope with stress. The effects of cortisol on stress are described in detail in Chapter 20.

Humoral signals include three basic categories: (1) hormones, (2) ions, and (3) metabolites. We have already discussed hormonal control of hormone release with respect to tropic hormones of the hypothalamus and

TABLE 5.7
(CONTINUED)

HORMONES		FUNCTIONS
Adrenal cortex (continued)	Androgens (dehydro-epiandrostenedione, androstenedione)	Promote sex drive
Adrenal medulla	Epinephrine	Stimulates "fight-or-flight" response
Pancreas	Insulin	Lowers blood glucose levels; promotes protein, lipid, and glycogen synthesis
	Glucagon	Raises blood glucose levels; promotes glycogenolysis, gluconeogenesis
	Somatostatin	Inhibits secretion of pancreatic hormones; regulates digestion and absorption of nutrients by gastrointestinal system
Gonads		
Testes	Androgens (testosterone, androstenedione)	Necessary for sperm production by testis; promotes sex drive and development of secondary sex characteristics (facial hair, deep voice, etc.)
Ovaries	Estrogens (estradiol)	Necessary for follicular development; promotes development of secondary sex characteristics (breasts, body fat distribution, etc.)
	Progestins (progesterone)	Promotes endometrial growth to prepare uterus for pregnancy
Placenta (during pregnancy)	Chorionic gonadotropin, estrogens, progesterone	Maintain corpus luteum; reinforces actions of hormones secreted by corpus luteum
SECONDARY ENDOCRINE ORGANS		
Heart	Atrial natriuretic peptides (ANP)	Inhibits sodium reabsorption by kidney tubules
Kidneys	Renin	Stimulates aldosterone secretion indirectly via the renin-angiotensin system
	Erythropoietin	Stimulates production of red blood cells in bone marrow
Gastrointestinal tract		
Stomach	Gastrin	Stimulates acid secretion by stomach and intestinal motility
Small intestine	Secretin, cholecystokinin (CCK), glucose-dependent insulinotropic peptide (GIP)	Regulate gastrointestinal motility and secretion; regulate exocrine secretion by liver and pancreas
Liver	Somatomedins (IDGFs)	Promote bone and soft tissue growth
Skin, liver, kidney*	Calcitrol	Promotes absorption of calcium by intestine

*The skin, liver, and kidney are all necessary for the activation of calcitrol.

anterior pituitary and the negative feedback loops. Many hormones regulate the blood concentrations of ions or metabolites, which operate through negative feedback to regulate secretion of the hormones. For example, Figure 5.30a depicts the control of insulin secretion by blood glucose levels. High blood glucose levels stimulate the release of insulin from beta cells of the pancreas.

Insulin stimulates uptake of glucose by most cells of the body, causing a decrease in blood glucose. When glucose levels decrease to normal, the stimulus for insulin secretion is reduced and release slows. Figure 5.30b depicts another example of negative feedback, the regulation of potassium levels in the blood. High levels of potassium ions (K^+) in the blood stimulate cells in the zona glomerulosa of the adrenal cortex to secrete aldosterone. Aldosterone

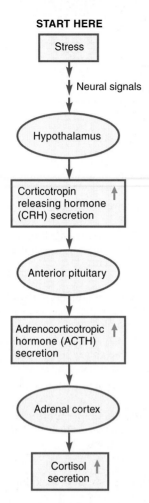

Stress

↓ Neural signals

Hypothalamus

↓

Corticotropin releasing hormone (CRH) secretion ↑

↓

Anterior pituitary

↓

Adrenocorticotropic hormone (ACTH) secretion ↑

↓

Adrenal cortex

↓

Cortisol secretion ↑

FIGURE 5.29 Neural control of hormone release from the hypothalamus. *Neural input resulting from stressful stimuli stimulates the release of CRH from the hypothalamus, which eventually leads to the release of cortisol from the adrenal cortex.*

stimulates the kidneys to excrete potassium ions into the urine, decreasing blood levels of potassium.

Many hormones are secreted according to a circadian rhythm. The mechanisms behind the circadian rhythm are not completely understood, but the rhythm depends on endocrine input and neural input (for example, via the eye). Hormonal control comes, at least partially, from melatonin. The neural input seems to come from an area of the hypothalamus called the *suprachiasmatic nucleus,* which provides neural input (directly or indirectly) to the neurosecretory cells of the hypothalamus that secrete tropic hormones. All of the hypothalamic tropic hormones are affected by circadian rhythm, thereby imparting a rhythm to the release of the anterior pituitary hormones, and, in turn, to the release of hormones affected by them.

Transport of Hormones Bound to Carrier Proteins

Recall that hydrophobic messengers, the steroid and thyroid hormones, are transported in blood bound to carrier

proteins. When hormones are transported in this manner, only the concentration of *unbound* hormones affects the binding of hormone to receptor. Carrier proteins increase the half-life of hormones, so that they are present in the blood for a longer period of time, by decreasing the rate of their metabolism, as described next.

Rate of Hormone Metabolism

Hormones exist in the blood for a relatively short period of time before they are metabolized. Hormones that bind to receptors on target cells are often metabolized by the target cell itself. Even those hormone molecules that bind to membrane receptors often get "internalized" by endocytosis of the hormone-receptor complex. Once inside the cell, the hormone is degraded by enzymes located in lysosomes. Hormones that are free in the blood can be broken down as well. Peptide hormones can be metabolized by proteolytic enzymes that are present in the blood. Hormones can also be metabolized by enzymes in the liver. Breakdown products of these hormones are excreted in the urine, along with certain hormones that are excreted in their original forms.

Steroids and thyroid hormones are metabolized more slowly (have a longer half-life) than are peptides and amines for two reasons: (1) Steroids and thyroid hormones are transported in blood bound to carrier proteins, and (2) the fat-soluble steroids and thyroid hormones can be stored temporarily in fatty tissue. Because both of these processes are reversible, hormones can be released from these pools when the free hormone concentration in the plasma begins to fall. The existence of this releasable pool of hormones thus tends to keep the plasma concentration elevated long after the rate of secretion has returned to its resting level.

Abnormal Secretion of Hormones

Abnormal secretion of hormones can have serious consequences. Some disease conditions are caused by an excess in the secretion of a hormone, termed *hypersecretion;* others are caused by too little secretion of a hormone, termed *hyposecretion.* Hypersecretion occurs for example in *acromegaly,* a disease caused by an excess of growth hormone secretion in adults, which causes the bones to thicken and organs to grow excessively. Hyposecretion occurs for example in *insulin-dependent diabetes mellitus,* a disease caused by insufficient secretion of insulin from the beta cells of the pancreas. With too little insulin in the blood, cells cannot use glucose adequately for energy. In cases of hypersecretion or hyposecretion, the disease process can be primary (acting directly in the endocrine gland) or secondary (involving a problem with the tropic hormone).

In a primary secretion disorder, the abnormality originates in the endocrine gland that secretes the hormone. For example, in primary hypersecretion of thyroid

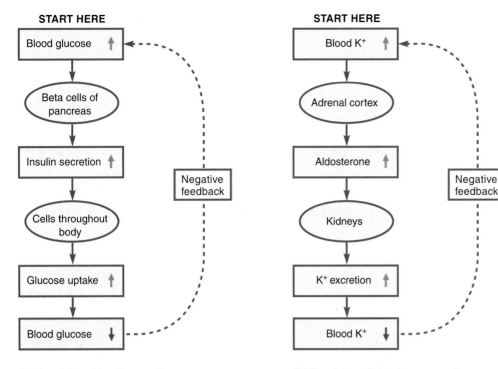

(a) Regulation of insulin secretion
(b) Regulation of aldosterone secretion

FIGURE 5.30 **Examples of humoral control of hormone release. (a)** *Control of insulin release by blood glucose levels.* **(b)** *Control of aldosterone release by blood potassium levels. Check out the flowchart activity that corresponds to this figure under Activities at www. physiologyplace.com.*

hormones, the thyroid gland secretes too much thyroid hormones (Figure 5.31a). One possible cause of hypersecretion is tumors of the endocrine cells. In primary hypersecretion, the blood levels of tropic hormones tend to be lower than normal due to the increased negative feedback from the hormone being regulated by the tropic hormones. In the case of thyroid hormones, levels of TRH and TSH in the blood are low because the excess thyroid hormones inhibit their release through negative feedback. In primary hyposecretion of thyroid hormones, the opposite pattern of hormone levels occurs: Thyroid hormone levels are decreased, but TRH and TSH levels increase due to reduced negative feedback.

In a secondary secretion disorder, the abnormality originates in the endocrine cells of either the anterior pituitary or the hypothalamus, which secrete the tropic hormone. In a secondary hypersecretion of thyroid hormones, for example, blood levels of thyroid hormones increase due to either an excess of TSH secretion by an abnormal anterior pituitary (Figure 5.31b) or an excess of TRH secretion by an abnormal hypothalamus. If excess TSH is secreted from the anterior pituitary, thyroid hormone levels in the blood increase, but levels of TRH fall due to increased negative feedback from the thyroid hormones. If excess TRH is secreted from the hypothalamus, then blood levels of both TSH and thyroid hormones also increase.

Hormone Interactions

As we saw earlier, almost all cells are exposed to hormones, so the target cells for a chemical messenger must have receptors that are specific for that messenger. Because a single hormone may have receptors on different types of cells, that hormone can produce more than one effect in the body. For example, there are ADH receptors both on certain epithelial cells in the kidneys, where ADH increases water reabsorption, and on smooth muscle cells of certain blood vessels, where ADH causes the muscle cells to contract, decreasing the diameter of the blood vessels.

As is evident in Table 5.7, often more than one hormone affects a given body function. For example, blood calcium levels are regulated by calcitonin, parathyroid hormone, and vitamin D_3. Similarly, blood glucose levels are regulated by insulin, glucagon, epinephrine, cortisol, and growth hormone. In some cases the effects of the hormones oppose each other, a process called **antagonism.** For example, parathyroid hormone increases blood calcium levels, whereas calcitonin decreases blood calcium levels. Likewise, glucagon increases blood glucose levels, whereas insulin decreases blood glucose levels.

Exercise Link

As Bill and Jane ran the marathon, glucagon and epinephrine concentrations in their blood increased while insulin levels decreased. All these hormonal changes worked together to release stored glucose into the blood for use as a fuel by their

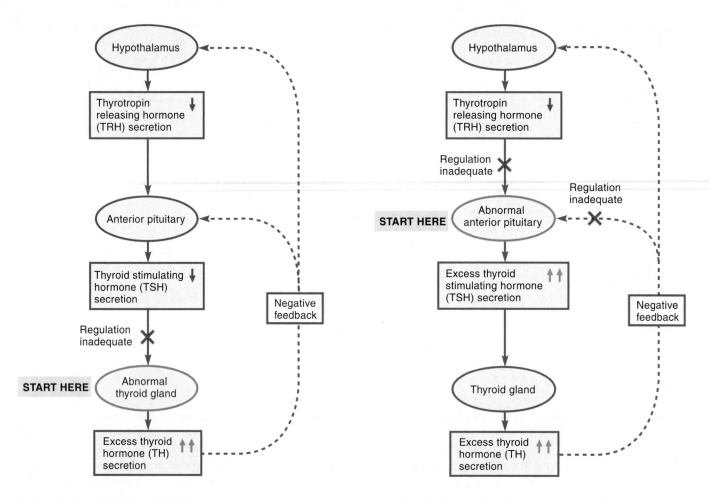

(a) Primary hypersecretion of thyroid hormones

(b) Secondary hypersecretion of thyroid hormones

FIGURE 5.31 Abnormal secretion of thyroid hormones.
(a) *Primary hypersecretion of thyroid hormones. An abnormality of the thyroid gland causes it to secrete too much thyroid hormone. Excess thyroid hormones in the blood induce strong negative feedback, decreasing TRH and TSH release into the blood.* **(b)** *An example of secondary hypersecretion of thyroid hormones. An abnormality of the anterior pituitary causes it to secrete too much TSH. Excess TSH levels in the blood stimulate the thyroid gland to secrete excess thyroid hormones. Negative feedback to the hypothalamus increases, so TRH levels in the blood decrease.*

exercising muscles. This process kept blood glucose concentrations fairly constant, at least for the first 20 miles or so. The fatigue Bill and Jane felt late in the race was due, in part, to falling blood glucose levels. The delivery of glucose to tissues that need it requires adequate blood perfusion through those tissues, which is closely related to the total amount of water within the body. As Bill and Jane lost water (and salt) through sweating, their antidiuretic hormone concentrations increased (which reduced water loss through the kidneys). Likewise, their aldosterone concentrations increased (which reduced salt loss through the kidneys).

In other cases, the hormones produce the same direction of effect, generally by different means. When two or more hormones produce the same type of response in the body, the effect can be either **additive,** in which case the net effect equals the sum of the individual effects, or **synergistic,** in which case the net effect is greater than the sum of the individual effects. Assume, for example, that hormones A and B produce equal effects when present at a concentration of 1 nanogram per deciliter (ng/dL) of blood. If both hormones were present at 1 ng/dL blood, then the response in an additive interaction would be twice that of the individual hormones, but the response in a synergistic interaction would be greater than twice that of the individual hormone responses.

In some cases the presence of one hormone is needed for another hormone to exert its actions, a process called **permissiveness.** One example involves epinephrine, which by binding to beta-adrenergic receptors on smooth muscle cells of bronchioles (small airways of the lungs) causes these airways to increase in diameter (dilate). Thyroid hormones are essential for the synthesis of the beta-adrenergic receptors in these target cells. Therefore, even though thyroid hormones by themselves have no direct effect on diameter of bronchioles, epinephrine cannot trigger dilation of the bronchioles in the absence of thyroid hormones, because there are no receptors to which epinephrine can bind.

Quick Test 5.5

1. Name the three factors affecting the concentration of a hormone in the blood.

2. Define the terms *primary hypersecretion, secondary hypersecretion, primary hyposecretion,* and *secondary hyposecretion.*

3. Define the terms *antagonism, additive effect, synergistic effect,* and *permissiveness* as they relate to hormone interactions.

CHAPTER SUMMARY

Intercellular Communication, p. 132

Virtually all body processes require that cells are able to communicate with each other, which can occur in two basic ways: (1) via gap junctions, which allow electrical signals and small molecules to move directly from one cell to adjacent cells, and (2) via the secretion of chemical messengers, which allow signals to be transmitted from one cell to others that may be at distant locations. Chemical messengers produce responses in target cells by binding to specific receptors.

> IP Cardiovascular, Anatomy Review: The Heart, page 7
>
> IP Nervous II, Anatomy Review, pages 6–7
>
> IP Nervous II, Synaptic Transmission, page 4

Chemical Messengers, p. 133

Chemical messengers fall into six major functional categories: (1) paracrines, (2) autocrines, (3) cytokines, (4) neurotransmitters, (5) hormones, and (6) neurohormones. Many messengers exert their effects only on cells that are close to the cells that secrete them, but hormones and neurohormones can act at distant sites because they are carried in the bloodstream to their target cells.

On the basis of their chemical structure, messengers are divided into five major classes: (1) amino acids, (2) amines (amino acid derivatives), (3) peptides, (4) steroids (cholesterol derivatives), and (5) eicosanoids (derivatives of arachidonic acid). Steroids, eicosanoids, and some amines (the thyroid hormones) are lipophilic and pass through cell membranes easily; other messengers are hydrophilic (lipophobic) and do not.

> IP Nervous I, Ion Channels, pages 4, 7
>
> IP Nervous II, Synaptic Transmission, pages 1–6, 9, 12
>
> IP Nervous II, Anatomy Review, pages 7–8

Signal Transduction Mechanisms, p.140

The magnitude of a target cell's response to a chemical messenger generally increases with increases in the number of bound receptors, which depends on the affinity of the receptors for the messenger, the messenger's concentration, and the number of receptors present. When exposed for long periods of time to messenger concentrations that are very low or high, target cells can alter the number of receptors, bringing about a change in their responsiveness to the messenger. A decrease in the number of receptors is called down-regulation; an increase is called up-regulation.

Lipophilic messengers bind to receptors in the cytosol or nuclei of target cells, with the resulting complex binding to DNA to regulate gene transcription and protein synthesis. Hydrophilic messengers bind to cell surface receptors, which are of three types: (1) channel-linked receptors, which affect the opening and closing of fast ligand-gated channels, (2) enzyme-linked receptors, which catalyze reactions inside cells, and (3) G protein-linked receptors, which activate specific membrane proteins called G proteins. Activated G proteins can themselves activate (or inhibit) a variety of intracellular proteins, including enzymes or channels. Many of these enzymes catalyze the formation of second messengers inside the cells. Among the substances known to act as second messengers are cyclic AMP (cAMP), cyclic GMP (cGMP), inositol triphosphate (IP_3), and calcium ions (which often work by binding to calmodulin, forming a complex that activates protein kinases).

> IP Nervous II, Synaptic Transmission: pages 4, 13
>
> IP Nervous I, Ion Channels, pages 4, 7

Long-Distance Communication via the Nervous and Endocrine Systems, p. 151

In the nervous system, neurons send signals to specific groups of target cells, to which they are connected by synapses. Responses triggered by neural signals are generally fast and brief. The endocrine system broadcasts signals to target cells throughout the body, exerting effects that are generally slow and long lasting. Hormones are secreted by specialized endocrine cells, which may be found in primary or secondary endocrine organs. Primary endocrine organs include the pituitary gland (which is divided into anterior and posterior lobes), pineal gland, thyroid gland, parathyroid glands, thymus gland, pancreas, and gonads. Secretion by the anterior pituitary is regulated by tropic hormones secreted by neurosecretory cells in the hypothalamus. Secretion of these and other hormones is regulated by negative feedback.

IP Nervous II, Orientation, page 1
IP Nervous II, Synaptic Transmission, pages 1–6
IP Urinary, Anatomy Review, page 4

Hormone Actions at the Target Cell, p. 161

The magnitude of a target cell's response to a hormone varies with the hormone's concentration in the plasma, which depends on the rate of hormone secretion; the amount of hormone that is bound to carrier proteins in the blood; and the rate at which the hormone is metabolized. Many hormones are degraded by the liver, with the breakdown products eventually being excreted in the urine. A single hormone may regulate more than one body function, and a given function may be regulated by two or more hormones, which may exert effects that are additive, antagonistic, synergistic, or permissive.

EXERCISES

Multiple-Choice Questions

1. Arachidonic acid is the raw material for the synthesis of
   a) amines.
   b) thyroid hormones.
   c) eicosanoids.
   d) steroids.

2. Epinephrine is a(n)
   a) amino acid.
   b) catecholamine.
   c) eicosanoid.
   d) adrenocorticoid.

3. Most chemical messengers fall into which of the following chemical classes?
   a) amines
   b) amino acids
   c) peptides
   d) steroids

4. All amino acid chemical messengers function as
   a) paracrines.
   b) cytokines.
   c) neurotransmitters.
   d) hormones.

5. All steroid chemical messengers function as
   a) paracrines.
   b) cytokines.
   c) neurotransmitters.
   d) hormones.

6. Which of the following is likely to cause a rise in intracellular cAMP levels?
   a) stimulation of phosphodiesterase activity
   b) activation of an inhibitory G protein targeting adenylate cyclase
   c) binding of chemical messengers to enzyme-linked receptors
   d) stimulation of adenylate cyclase activity

7. Which of the following messenger classes bind to intracellular receptors?
   a) catecholamines
   b) peptides
   c) steroids
   d) both a and c

8. Which of the following is an accurate statement regarding regulation of pituitary hormone secretion by the hypothalamus?
   a) All pituitary hormones are regulated by tropic hormones from the hypothalamus.
   b) All anterior pituitary hormones are regulated by a releasing hormone and a release-inhibiting hormone from the hypothalamus.
   c) All posterior pituitary hormones are regulated by a releasing hormone from the hypothalamus.
   d) All anterior pituitary hormones are tropic hormones.

9. G proteins are involved whenever
   a) binding of messenger molecules to cell surface receptors triggers a target cell response.
   b) binding of ligand molecules to cell surface receptors triggers activation or inhibition of intracellular enzymes.
   c) binding of ligand molecules to cell surface receptors triggers synthesis of second messengers.
   d) binding of ligand molecules to cell surface receptors triggers a change in membrane permeability to ions.

10. Most hypothalamic and pituitary hormones are
    a) peptides.
    b) steroids.
    c) amines.
    d) amino acids.

11. Gonadotropin releasing hormone stimulates release of which of the following from the anterior pituitary?
    a) sex hormones
    b) follicle-stimulating hormone

c) luteinizing hormone

d) both follicle-stimulating hormone and luteinizing hormone

12. Which of the following adrenal hormones is secreted by chromaffin cells?
    a) cortisol
    b) aldosterone
    c) epinephrine
    d) androgens

13. In primary hyposecretion of thyroid hormones,
    a) levels of thyroid hormones in the blood decrease.
    b) levels of TRH in the blood increase.
    c) levels of TSH in the blood increase.
    d) all of the above

14. Which of the following organs secretes glucagon?
    a) liver
    b) pancreas
    c) anterior pituitary
    d) posterior pituitary

15. Which of the following is an example of permissiveness?
    a) Glucagon increases blood glucose levels, and insulin decreases blood glucose levels.
    b) Glucagon, epinephrine, and cortisol all increase blood glucose levels.
    c) Estrogen stimulates synthesis of progesterone receptors in the endometrium.
    d) all of the above

Objective Questions

1. Cells that secrete a messenger are called _____.

2. A (paracrine/autocrine) agent acts on the same cell that secretes it.

3. Endocrine glands release (neurotransmitter/hormone) into the _____ , where they travel to the target cell.

4. An active G protein releases the _____ subunit, which interacts with another membrane protein, altering its activity.

5. The enzyme that catalyzes conversion of ATP to cAMP is called _____.

6. Following activation of the phosphatidyl inositol system, (IP$_3$/DAG) liberates calcium from intracellular stores.

7. (Lipophilic/Lipophobic) messengers exert their effect on target cells by activating or inactivating specific genes.

8. Examples of locally acting chemical messengers are (steroids/eicosanoids).

9. Cytosolic calcium often exerts its effect by binding to cytosolic (protein kinase C/calmodulin).

10. Neural input to the hypothalamus is involved in regulating secretion of hormones by both lobes of the pituitary. (true/false)

11. Epinephrine is secreted by the adrenal (medulla/cortex).

12. Thyroid hormones are classified as (amines/steroids).

13. Lipophobic messengers are secreted by (exocytosis/diffusion across the cell membrane).

14. Calcitonin is secreted by the (thyroid gland/parathyroid gland).

15. Amino acids are (cytokines/neurotransmitters).

Essay Questions

1. Describe the different types of chemical signals (paracrine, autocrine, cytokine, neurotransmitter, hormone, and neurohormone).

2. Describe the role of the hypothalamus in the regulation of anterior pituitary hormone secretion.

3. Compare lipophilic and lipophobic chemical messengers with respect to location of receptors, transport in blood, and general signal transduction mechanisms.

4. Describe the effects of activating the enzyme phospholipase C with regard to second messengers produced and their actions in the cell.

5. Describe the cAMP second messenger system, including all the steps from messenger binding to its receptor to a response in the target cell.

6. Compare long-distance communication by the nervous and endocrine systems with respect to anatomy, speed, and mechanisms of action.

7. Describe the two different types of receptor-mediated channels and their mechanisms of action. Which mechanism is faster?

8. Describe the various factors that affect the concentration of a hormone in plasma.

9. Describe the anatomy of the adrenal gland. What are its two main subdivisions? What are the minor subdivisions of its outer region? Which hormones are secreted from the different subdivisions?

10. Describe the role of protein kinases in signal transduction.

Find the answers to these exercises, and additional study tools, at the Physiology Place (www.physiologyplace.com).

Nerve Cells and Electrical Signaling

OBJECTIVES

- Describe the major components of the nervous system and the direction of information flow within and among them.

- Describe the basic anatomy of a neuron. Compare the functions of each part of a neuron and describe the types of ion channels located in each part. Describe the grouping of neurons within the central nervous system and the peripheral nervous system.

- Describe the structure and function of myelin.

- Explain the ionic basis of the resting membrane potential.

- Describe the various properties of graded potentials, including direction of change in potential, magnitude of change, and temporal and spatial summation. Explain how graded potentials in neurons can trigger an action potential.

- Explain the ionic basis of an action potential. Describe the gating mechanisms for voltage-gated sodium and potassium channels.

- Describe the propagation of action potentials from axon hillock to axon terminal, and compare propagation in myelinated and unmyelinated axons.

- Describe refractory periods, including what causes the absolute and relative refractory periods, and explain their physiological significance.

CHAPTER OUTLINE

Overview of the Nervous System   171

Cells of the Nervous System   172

Electrical Signals in Neurons   177

Action Potentials   187

The Basis of Neural Stability   197

Above: Scanning electron micrograph of juvenile neurons after transplant

Imagine that you are stopped at a traffic light in your car. The light turns green, and you start to enter the intersection. Out of the corner of your eye you notice a truck speeding through the red light, coming at you from the left. Nerve cells in your brain immediately send electrical impulses down your spinal cord, where these impulses are relayed to nerve cells that control the muscles in your legs. In less than a second after seeing the truck, you hit the brake pedal just in time to avoid a serious collision.

Every minute of every day, vast networks of nerve cells throughout the body are firing off messages. In the example just described, the messages were transmitted by way of electrical signals smaller than 1/15 of the voltage of a typical flashlight battery! Some nerve cells send messages from the nervous system to organs, while others send messages from the organs to the nervous system. Such back-and-forth signaling is crucial to short-term maintenance of homeostasis because it provides second-by-second feedback, which is necessary for regulation of the body's internal environment.

In this chapter we begin with an overview of the nervous system and then look in detail at how nerve cells communicate.

OVERVIEW OF THE NERVOUS SYSTEM

The nervous system can be divided into two main anatomical parts: the central nervous system and the peripheral nervous system (Figure 6.1). The **central nervous system (CNS)** consists of the brain and spinal cord; it receives and processes information from the organs and then sends information back to the organs, instructing them to perform various tasks. The central nervous system is also the site of learning, memory, emotions, and other complex functions.

The **peripheral nervous system** consists of nerve cells that provide communication between the central nervous system and organs throughout the body. The peripheral nervous system can be subdivided into two divisions: afferent and efferent. Nerve cells of the **afferent** division transmit information from the organs to the central nervous system. Information transmitted to the central nervous system includes **sensory information** pertaining to the external environment (such as touch, temperature, vision, and sound) and **visceral information** pertaining to the internal environment (such as fullness of the stomach, blood pressure, and blood pH). Nerve cells of the **efferent** division transmit information from the central nervous system to organs in the periphery, called **effector organs,** that perform functions in response to commands from neurons. Effector organs are usually muscles and glands. A neuron capable of transmitting messages to an effector organ is said to **innervate** that effector organ.

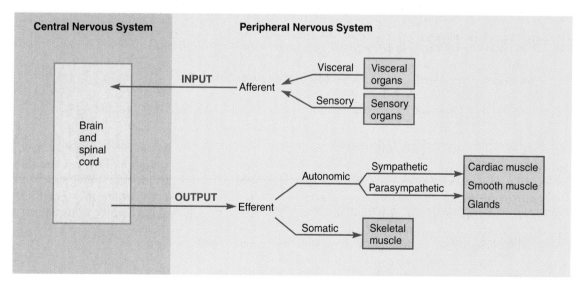

FIGURE 6.1 Organization of the nervous system. *The nervous system has two main parts: the central nervous system and the peripheral nervous system. The peripheral nervous system is functionally divided into afferent and efferent divisions. Arrows indicate the direction of information flow.*

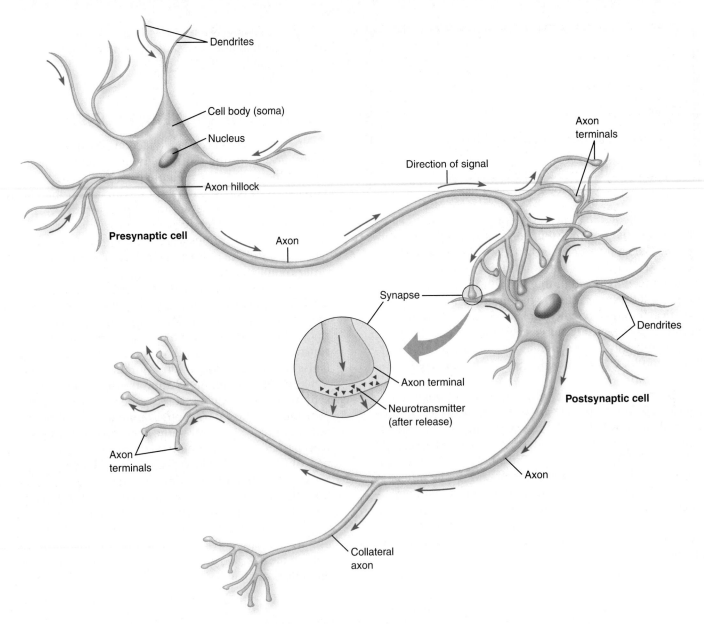

FIGURE 6.2 Structure of a typical neuron. *Two neurons are shown; the upper neuron communicates with the lower neuron, as indicated by the arrows representing information flow. The main parts of a neuron include the cell body (also known as the soma); dendrites, which receive communication from other neurons; and an axon, which is specialized for transmitting electrical impulses. The axon terminals of one neuron release a chemical messenger (neurotransmitter) that communicates with another neuron.*

The efferent division can be further subdivided into two main branches: the somatic and autonomic nervous systems. The **somatic nervous system** consists of the nerve cells (called **motor neurons**) that regulate skeletal muscle contractions. The **autonomic nervous system** consists of nerve cells that regulate the function of internal organs and other structures (such as sweat glands and blood vessels) that are not under voluntary control. The autonomic nervous system can be divided into two branches: the *parasympathetic* and the *sympathetic* nervous systems, which tend to have opposite effects on organs.

The body also has an *enteric nervous system,* which consists of an intricate network of nerve cells in the gastrointestinal tract that can function independently of the rest of the nervous system but communicates with the autonomic nervous system. Chapter 19 describes the enteric nervous system in detail.

CELLS OF THE NERVOUS SYSTEM

The nervous system contains two main classes of cells: neurons and glial cells. Neurons, or nerve cells, are "excitable cells" that communicate by transmitting electrical

For decades neuroscientists believed that the adult human brain could not produce new neurons. Then, in the 1970s, Pasquale Graziadei and colleagues discovered that new olfactory receptor cells (sensory receptor cells for smell) were produced from basal cells in the olfactory epithelium of the nasal cavity. In the 1990s, other researchers have found evidence of *neurogenesis,* the production of new neurons, in the central nervous system.

Neurogenesis in lower animal species has been evident for over 30 years. Elizabeth Gould and her colleagues from Princeton have been studying neurogenesis in primates during the 1990s. In some of these studies, bromodeoxyuridine (BrdU) was administered to the animals as a marker for cell proliferation. Before cells can divide, they must replicate new DNA. BrdU is incorporated into newly synthesized DNA and therefore marks any DNA formed after its administration. Cells that contain BrdU in their DNA are identified postmortem by histological analysis. Using this technique and others, Gould and her colleagues discovered that neurogenesis occurs in several areas of the brain, including the hippocampus, prefrontal cortex, inferior temporal cortex, and parietal cortex of nonhuman primates.

Fred Gage and his colleagues from the Salk Institute for Biological Studies have studied neurogenesis in the hippocampus in lower animal species, and have recently provided evidence that neurogenesis occurs in the human hippocampus as well. Their studies in humans used BrdU in the same manner as Gould's study in lower primates. The human subjects were cancer patients of Peter Eriksson from Sweden, who was treating the patients with BrdU to measure tumor cell proliferation. These patients agreed to donate their brains to Gage's research group upon their deaths. Postmortem analyses of the brains showed that BrdU was incorporated into the DNA of neurons in the hippocampus. These findings are the first evidence of neurogenesis in human brains.

The discovery of neurogenesis in human brains is of profound clinical significance. Many neurological diseases, including Parkinson's disease, Alzheimer's disease, and stroke, involve the loss of functioning neurons. If scientists can learn the mysteries of neurogenesis, they may be able to develop techniques by which the central nervous system can replace lost neurons.

Eriksson, et al., Neurogenesis in the adult human hippocampus. Nat. Med., 4 (1998), 1313–1317.

Gould, et al., Neurogenesis in the Neocortex of Adult Primates. Science, 286 (1999), 548–552.

Graziadei, P.P.C. and Graziadei, G.A.M., The Olfactory System: A Model for the Study of Neurogenesis and Axon Regeneration in Mammals. Neuronal Plasticity, Ed. Carl W. Cotman, Raven Press, New York, 1978.

impulses. **Excitable cells** are defined as cells capable of producing large, rapid electrical signals called *action potentials.* In the nervous system, the neuron is the *functional unit,* the smallest unit of a tissue that can carry out the function of that tissue. *Glial cells,* which constitute 90% of the cells in the nervous system, provide various types of support to the neurons, including structural and metabolic support.

Neurons

Figure 6.2 illustrates the anatomy of typical neurons. Most neurons contain three main components: a cell body and two types of *neural processes* that extend from the cell body: the dendrite(s) and an axon. The **cell body** (or *soma*) contains the cell nucleus and most of the cell's organelles; it carries out most of the functions that other cells perform, such as protein synthesis and cellular metabolism. Although they have a nucleus, mature neurons lose the ability to undergo cell division. Thus in most areas of the nervous system, adults have all the neurons they will ever have. However, we now know that in a few areas of the adult human brain new neurons can develop from undifferentiated cells (see Discovery: Neurogenesis).

Dendrites branch from the cell body and receive input from other neurons at specialized junctions called *synapses.* (Cell bodies themselves can also receive synaptic input.) At a synapse, a presynaptic cell releases a chemical messenger called a neurotransmitter that usually communicates with the dendrite or cell body of a postsynaptic cell.

Neurons have another branch that comes off the cell body called an **axon,** or *nerve fiber.* Unlike a dendrite, whose function is to *receive* information, an axon's job is to *send* information. Generally, a neuron has only one axon, but axons can branch and thus send signals to more than one destination. The branches of an axon are called **collaterals,** and like dendrites, the extent of branching varies among neurons and is indicative of the amount of communication with other cells.

The axon functions in the transmission of information over relatively long distances in the form of electrical signals called **action potentials,** rapid large changes in membrane potential during which the inside of the cell becomes positively charged relative to the outside. (Properties of action potentials are described in detail later in this chapter.) The beginning and end of an axon are specialized structures called the axon hillock and the axon

FIGURE 6.3 **Structural classes of**
neurons. (a) *A bipolar neuron. Afferent*
neurons associated with vision and olfaction are
bipolar neurons. (b) *A pseudo-unipolar neuron.*
The vast majority of afferent neurons are pseudo-
unipolar. (c) *A multipolar neuron. Most neurons*
are multipolar neurons.

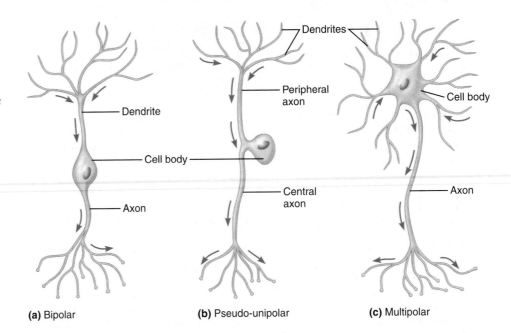

(a) Bipolar   (b) Pseudo-unipolar   (c) Multipolar

terminal, respectively. The **axon hillock,** the site where the axon originates from the cell body, is specialized in most neurons for the *initiation* of action potentials. Once initiated, action potentials are transmitted to the axon terminal, by mechanisms we examine shortly. The **axon terminal** is specialized to release a neurotransmitter upon arrival of an action potential. The released neurotransmitter molecules carry a signal to a postsynaptic cell, usually to a dendrite or the cell body of another neuron or to the cells of an effector organ.

Because different regions of a neuron generally have specialized functions (although there are exceptions), each region has specific types of ion channels that can open or close. The opening or closing of ion channels changes the permeability of the plasma membrane for a specific ion, resulting in ion movement across the membrane that causes a change in the electrical properties of the cell or triggers the release of a neurotransmitter. **Leak channels,** which are found in the plasma membrane throughout a neuron, are always open and are responsible for the resting membrane potential.

Ligand-gated channels open or close in response to the binding of a chemical messenger to a specific receptor in the plasma membrane. In neurons, ligand-gated channels are located in the dendrites and cell body, areas that receive communication from presynaptic neurons in the form of neurotransmitters.

Voltage-gated channels open or close in response to changes in membrane potential. Voltage-gated sodium channels and voltage-gated potassium channels are located throughout the neuron, but most densely in the axon and in greatest density in the axon hillock, and are necessary for the initiation and propagation of action potentials. Voltage-gated calcium channels are located in the axon terminal; these channels open in response to the ar-

rival of an action potential at the axon terminal. When these channels are open, calcium enters the cytosol of the axon terminal and triggers the release of neurotransmitter. Details of signal initiation and propagation are described later in this chapter.

Structural Classification of Neurons

Neurons can be classified structurally according to the number of processes (axons and dendrites) that project from the cell body (Figure 6.3). *Unipolar neurons* have a single projection from the cell body. Unipolar neurons are rare in humans, and are not shown in the figure. *Bipolar neurons* have two projections, an axon and a dendrite, projecting from the cell body. Typical bipolar cells include sensory neurons for olfaction (smell) and vision. Other sensory neurons, including those for touch and pain, are *pseudo-unipolar neurons*, a subclass of bipolar neurons. Although only one process seems to extend in two directions from the cell body, there are actually two processes, an axon and a modified dendrite, that project from the cell body in opposite directions. The modified dendritic process is called the *peripheral axon* because it originates in the periphery with sensory receptors and functions as an axon in that it transmits action potentials. The axon process is called the *central axon* because it ends in the central nervous system, where it forms synapses with other neurons. *Multipolar neurons*, the most common neurons, have multiple projections from the cell body; one projection is an axon, all the others are dendrites.

Functional Classification of Neurons

There are three functional classes of neurons: efferent neurons, afferent neurons, and interneurons (Figure 6.4). **Efferent neurons** transmit information from the central

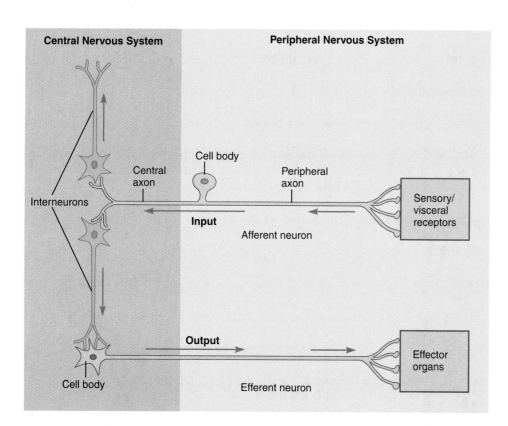

Central Nervous System

Peripheral Nervous System

Interneurons

Central axon

Cell body

Peripheral axon

Input

Afferent neuron

Sensory/ visceral receptors

Output

Cell body

Efferent neuron

Effector organs

FIGURE 6.4 Functional classes of neurons. *Afferent neurons originate in the periphery with sensory or visceral receptors. The peripheral axons of afferent neurons are part of the peripheral nervous system, but the axon terminals are located in the central nervous system, where they communicate with other neurons. Efferent neurons originate in the central nervous system, where the cell body and dendrites receive synaptic communication from other neurons. Efferent axons, however, are part of the peripheral nervous system and terminate at a synapse with an effector organ. Interneurons lie entirely in the central nervous system and can communicate with afferent neurons, efferent neurons, or other interneurons.*

Name two types of effector organs.

nervous system to effector organs. Recall that efferent neurons include the motor neurons extending to skeletal muscle and neurons of the autonomic nervous system (see Figure 6.1). Notice in Figure 6.4 that the cell body and dendrites of efferent neurons are located in the central nervous system (autonomic postganglionic neurons, described in Chapter 10, are an exception). However, the axon leaves the central nervous system and becomes part of the peripheral nervous system as it travels to the effector organ it innervates.

The function of **afferent neurons** is to transmit either sensory information from *sensory receptors* (which detect information pertaining to the outside environment) or visceral information from *visceral receptors* (which detect information pertaining to conditions in the interior of the body) to the central nervous system for further processing. Most afferent neurons are pseudo-unipolar neurons, with the cell body located outside the central nervous system in a *ganglion* (the general term for a cluster of neural cell bodies located outside the CNS). The endings of the peripheral axon are located in the peripheral organ (sensory organ or visceral organ), where they are either modified into sensory receptors or receive communication from separate sensory receptor cells. The central axon terminates in the central nervous system, where it releases a neurotransmitter to communicate with other neurons.

The third functional class of neurons is **interneurons,** which account for 99% of all neurons in the body. They are located entirely in the central nervous system. In-

terneurons perform all the functions of the central nervous system, including processing sensory information from afferent neurons, sending out commands to effector organs through efferent neurons, and carrying out complex functions of the brain such as thought, memory, and emotions.

Structural Organization of Neurons in the Nervous System

Neurons are arranged within the nervous system in an orderly fashion, with those having similar functions tending to be grouped together. In addition, neurons are aligned in such a way that cell bodies and dendrites of adjacent cells tend to be grouped together, and axons of adjacent cells tend to be grouped together. In the central nervous system, cell bodies of neurons are often grouped into **nuclei** (singular = nucleus), and the axons travel together in bundles called **pathways, tracts,** or **commissures.** In the peripheral nervous system, cell bodies of neurons are clustered together in **ganglia** (singular = ganglion), and the axons travel together in bundles called **nerves.**

Muscle and glands

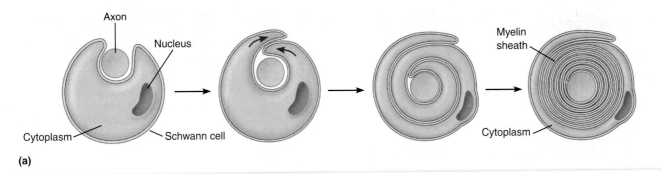

(a)

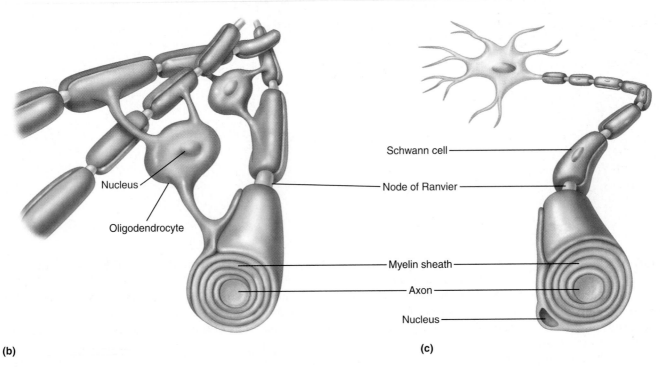

(b)

(c)

FIGURE 6.5 Formation and origins of myelin sheaths. (a) *Formation of a myelin sheath by a Schwann cell. Myelin, which consists of concentric layers of plasma membrane provided by either a Schwann cell or an oligodendrocyte, forms a layer of insulation around an axon.* **(b)** *Arrangement of myelin sheaths formed by oligodendrocytes in the central nervous system. A single oligodendrocyte sends out cytoplasmic processes that form myelin sheaths around several axons. Note the Nodes of Ranvier, gaps in the myelin sheaths.* **(c)** *Arrangement of myelin sheaths formed by Schwann cells in the peripheral nervous system. A given Schwann cell sheathes only a single axon.*

Glial Cells

Glial cells, the second class of cell found in the nervous system, account for 90% of all cells in the nervous system. They do not function directly in signal transmission; instead, they provide structural integrity to the nervous system (*glia* is Latin for "glue"), and they are necessary for neurons to carry out their functions.

There are five types of glial cells: *astrocytes, ependymal cells, microglia, oligodendrocytes,* and *Schwann cells.* Of these glial cells, only Schwann cells are located in the peripheral nervous system; the rest are in the central nervous system. The functions of astrocytes, ependymal cells, and microglia

are covered in Chapter 8. Because the functions of **oligo-dendrocytes** and **Schwann cells** are crucial to electrical transmission in neurons, we discuss these cells here.

The primary function of oligodendrocytes and Schwann cells is to form an insulating wrap of **myelin** around the axons of neurons. Such insulation enables neurons to transmit action potentials more efficiently and rapidly. Figure 6.5a shows the formation of a myelin sheath by a Schwann cell. Myelin consists of concentric layers of the cell membranes of either oligodendrocytes or Schwann cells. Oligodendrocytes form myelin around axons in the central nervous system; one oligodendro-cyte sends out projections providing the myelin segments for many axons (Figure 6.5b). Schwann cells form myelin around axons in the peripheral nervous system, but each Schwann cell provides myelin for only one axon (Figure 6.5c). Many oligodendrocytes or Schwann cells are needed to provide the myelin for a single axon. Because the lipid bilayer of a cell membrane has low permeability to ions, the several layers of membrane that make up a myelin sheath substantially reduce leakage of ions across the cell membrane. However, in gaps in the myelin, called **nodes of Ranvier,** the axonal membrane contains voltage-gated sodium and potassium channels that function in the transmission of action potentials by allowing ion movement across the membrane. We discuss the nature of these electrical signals and how they originate in the next few sections of this chapter.

Quick Test 6.1

1. Name the different parts, divisions, and branches of the nervous system and give the basic functions of each.

2. Draw a neuron and label the following structures: cell body, dendrite, axon, and axon terminal. Briefly state the function of these structures, and the type of ion channels (ligand-gated or voltage-gated) that can be found in each.

3. Which glial cell forms myelin in the central nervous system? Which forms myelin in the peripheral nervous system?

ELECTRICAL SIGNALS IN NEURONS

Neurons communicate by generating electrical signals in the form of changes in membrane potential. To understand these signals, we must first understand the relationship between an electrical potential, current, and resistance. Table 6.1 defines the different types of electrical potentials that are described in this chapter.

Electrical potentials are produced in biological systems by separating oppositely charged ions, which are attracted to each other and will move toward each other if possible. When ions move, they carry their charge with them. The movement of electrical charges is called **current** (I). In biological systems, currents are typically expressed in units of microamps (10^{-6} amperes). The greater the electrical potential, the greater the *force* (voltage) for ion movement, or current. However, force is not the sole determinant of whether ions indeed move.

How easily ions can move depends on the properties of the substance through which they must move. **Resistance** (R) is a measurement of the hindrance to charge movement. The greater a substance's resistance, the more difficult it will be for ions to move through it, and the weaker the current will be. A neuron's plasma membrane

TABLE 6.1 TYPES OF ELECTRICAL POTENTIALS IN BIOLOGICAL SYSTEMS	
POTENTIAL	**DEFINITION**
Potential difference = E	Difference in voltage between two points
Membrane potential = V_m	Difference in voltage across the plasma membrane; always given in terms of voltage inside the cell relative to voltage outside the cell
Resting V_m	Difference in voltage across the plasma membrane when a cell is at rest (not receiving or sending signals)
Graded potential	A relatively small change in membrane potential produced by some type of stimulus that triggers the opening or closing of ion channels; strength of graded potential is relative to strength of stimulus
Synaptic potential	Graded potential produced in the postsynaptic cell in response to neurotransmitters binding to receptors
Action potential	A large, rapid change in membrane potential produced by depolarization of an excitable cell's plasma membrane to threshold
Equilibrium potential	The membrane potential that counters the chemical forces acting to move an ion across the membrane, thereby putting the ion at equilibrium

has high resistance to current flow because its permeability to ions is low. The intracellular and extracellular fluids, by contrast, have low resistance to current flow because these fluids are rich in ions.

The inverse of resistance is **conductance** (g):

$$g = 1/R$$

Because the ability of an ion to cross a plasma membrane depends on the permeability of the plasma membrane to that ion, conductance of a particular ion increases as the membrane's permeability to that ion increases.

The relationship between potential difference, current, and resistance is defined by **Ohm's Law:**

$$I = E/R,$$

where E is the potential difference or voltage. Understanding Ohm's Law is crucial to understanding neural

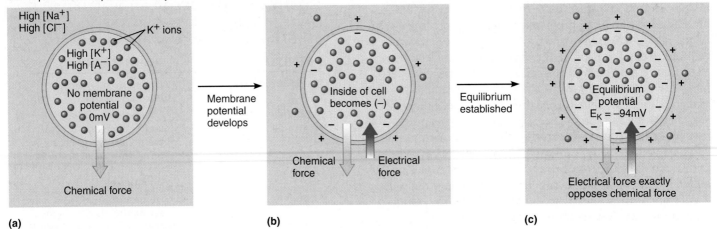

(a)   **(b)**   **(c)**

FIGURE 6.6 **Cell 1: Permeable to potassium only.** *Potassium and organic anions are located in greater concentration inside the cell. Sodium and chloride ions are located in greater concentration outside the cell. The length of an arrow is relative to the strength of ion movement in the direction of the arrow.* **(a)** *Potassium ions move out of the cell due to a chemical force.* **(b)** *As some potassium ions leave the cell taking with them a positive charge, the inside of the cell becomes negative relative to the outside. This change in charge distribution creates an electrical force to move potassium ions into the cell, opposing the chemical force.* **(c)** *Eventually, enough potassium leaves the cell and the electrical force becomes strong enough to oppose further movement of potassium ions out of the cell due to the chemical force, resulting in no net movement of potassium ions. At this membrane potential, potassium is at equilibrium. This potential is thus the potassium equilibrium potential and is approximately −94 mV in neurons.*

physiology because plasma membranes have an electrical potential across them, ions present inside and outside the cell are available to carry charge across the plasma membrane, and resistance to charge movement can be changed by the opening or closing of ion channels.

Resting Membrane Potential

Recall from Figure 4.2 that a cell at rest has a potential difference across its membrane such that the inside of the cell is negatively charged relative to the outside. This difference is called the **resting membrane potential** (resting V_m) because the cell is at rest—is not receiving or transmitting any signals. The resting membrane potential of neurons is approximately −70 mV. Membrane potentials are *always* described as the potential inside the cell relative to outside. Therefore, the inside of a typical neuron at rest is 70 mV more negative compared to the outside. (Although we are discussing the resting membrane potential of neurons, all cells in the body have a negative membrane potential, ranging from −5 mV to −100 mV.) Changes in the membrane potential are necessary for neurons to communicate with other cells by triggering the release of neurotransmitter. In the following sections we explore (1) what is responsible for the existence of the resting membrane potential and (2) what causes the membrane potential to change.

How Chemical and Electrical Forces Acting on Potassium and Sodium Ions Produce the Resting Membrane Potential

The resting membrane potential depends on two critical factors: (1) the concentration gradients of ions (particularly sodium ions and potassium ions) across the plasma membrane and (2) the presence of ion channels in the plasma membrane. Recall from Chapter 4 that the Na^+/K^+ pump creates concentration gradients for sodium and potassium ions. Sodium ions are more highly concentrated outside the cell, and thus there is a chemical driving force tending to push sodium ions into the cell. Potassium ions are more highly concentrated inside the cell, and thus there is a chemical driving force tending to push potassium ions out of the cell. The chemical forces for moving sodium and potassium ions across the plasma membrane, and the differences in the permeability of the plasma membrane to these two ions, establish the resting membrane potential.

To understand what causes the resting membrane potential, consider two hypothetical cells that are identical in all respects, expect for the permeability of their membranes to ions. Cell 1 (Figure 6.6) is permeable to potassium ions only (that is, it has open potassium channels in its plasma membrane), whereas Cell 2 (Figure 6.7) is permeable to sodium ions only (that is, it has open sodium channels in its plasma membrane).

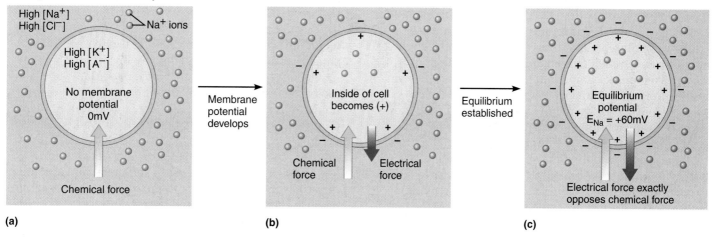

Cell 2: permeable to sodium only

(a)
High [Na⁺]
High [Cl⁻]
Na⁺ ions
High [K⁺]
High [A⁻]
No membrane potential 0mV
Chemical force

Membrane potential develops

(b)
Inside of cell becomes (+)
Chemical force
Electrical force

Equilibrium established

(c)
Equilibrium potential E_{Na} = +60mV
Electrical force exactly opposes chemical force

FIGURE 6.7 Cell 2: Freely permeable to sodium only. *Potassium and organic anions are located in greater concentration inside the cell. Sodium and chloride ions are located in greater concentration outside the cell. The length of an arrow is relative to the strength of ion movement in the direction of the arrow.* **(a)** *Sodium ions move into the cell due to a chemical force.* **(b)** *As some sodium ions enter the cell, taking with them a positive charge, the inside of the cell becomes positive relative to the outside. The change in charge distribution creates an electrical force to move sodium ions out of the cell, opposing the chemical force.* **(c)** *Eventually, enough sodium enters the cell and the electrical force becomes strong enough to oppose further movement of sodium ions into the cell due to the chemical force, resulting in no net movement of sodium ions. At this membrane potential, sodium is at equilibrium. This potential is thus the sodium equilibrium potential and is approximately +60 mV in neurons.*

Consider Cell 1 in Figure 6.6 first. For simplicity, ions are shown in Figure 6.6 only on the side of the plasma membrane where they are found in greater concentration. Sodium ions (Na⁺) are at a higher concentration outside the cell and are balanced electrically by the presence of chloride ions (Cl⁻) outside the cell. Potassium ions (K⁺), by contrast, are at a higher concentration inside the cell and are balanced electrically by the presence of organic anions (A⁻, primarily proteins) inside the cell. We assume at first that no potential difference exists across the cell membrane; that is, the membrane potential is 0 mV. Because Cell 1 is permeable only to potassium ions, potassium will diffuse down its concentration gradient, or out of the cell (Figure 6.6a). As potassium ions move, they carry their positive charge out of the cell, which leaves the inside of the cell negatively charged relative to the outside. As a consequence, a negative membrane potential develops (Figure 6.6b).

Once the membrane potential has developed, two forces are acting on the potassium ions: a *chemical force* due to the concentration gradient and an *electrical force* due to the membrane potential. (Recall from Chapter 4 that the net force acting on an ion is called the *electrochemical force*, the sum of the electrical and chemical forces.) The direction of the chemical force now present in Cell 1 is such that it pushes potassium ions out of the cell; the direction of the electrical force is such that it pulls potassium ions back into the cell because of the attraction of the positively charged potassium ions for the negative charge inside of the cell. Initially the chemical force is greater than the electrical force because the membrane potential is small, and thus potassium ions continue to move out of the cell (see Figure 6.6b). However, the more potassium ions that leave the cell, the greater the membrane potential becomes, and consequently the greater the electrical force to pull potassium back into the cell. Once the electrical force has become just strong enough to exactly balance the opposing chemical force, no net movement of potassium occurs across the membrane (Figure 6.6c). Under these conditions, potassium is said to be at equilibrium because the electrochemical force is zero. The membrane potential under these conditions is equal to the equilibrium potential for potassium (E_K), which is approximately −94 mV. (The equilibrium potential will vary in different neurons based on the concentration gradient for potassium ions.) Recall from Chapter 4 that the equilibrium potential of any ion (E_x) depends only on that ion's charge and the size of its concentration gradient.

The number of potassium ions that move out of Cell 1 to create the potential difference is very small compared to the concentration of potassium ions present in the intracellular and interstitial fluids (less than 0.01%). Therefore, the concentration gradient for potassium does not change to any significant degree.

Now consider Cell 2, which differs from Cell 1 only in that it is permeable to sodium ions, rather than potassium ions (see Figure 6.7). (Figure 6.7 also shows ions only on the side where they are found in greater concentration.) As before, we assume that initially no potential difference exists across the cell membrane. Because sodium can cross the membrane, it diffuses down its concentration gradient into the cell (see Figure 6.7a). As sodium moves, it carries a positive charge into the cell, which makes the inside of the cell positively charged relative to the outside, creating a positive membrane potential. In the presence of this membrane potential, sodium ions are now acted upon by an electrical force in addition to the chemical force. The direction of the chemical force tends to push sodium into the cell, while the electrical force tends to push sodium out of the cell because of the repulsion between the positively charged sodium ions and the net positive charge inside the cell (see Figure 6.7b). Sodium continues to flow into the cell, making the membrane potential more positive, until the electrical force becomes just large enough to exactly balance the chemical force (see Figure 6.7c). At this point, sodium comes to equilibrium, with the membrane potential being equal to the equilibrium potential for sodium (E_{Na}), which is approximately +60 mV. (The actual equilibrium potential for sodium will vary based on the concentration gradient for sodium ions across a given neuron.)

Because the number of sodium ions that must cross the plasma membrane to cause a potential of +60 mV is very small relative to the concentration of ions in the intracellular and interstitial fluids, the concentration gradient for sodium does not change to any significant degree.

Determinants of Resting Membrane Potential

Cells 1 and 2 are clearly hypothetical cells. Now let's consider a more realistic cell, Cell 3 (Figure 6.8). Cell 3 has the same ion gradients across its cell membrane, but because Cell 3 has potassium and sodium channels, the membrane is permeable to both ions. However, the number of open potassium channels far exceeds the number of open sodium channels, and therefore the membrane is approximately 25 times more permeable to potassium than to sodium. Assuming that no membrane potential exists initially, let's consider what happens when potassium and sodium ions are both permeant.

Because potassium and sodium are able to cross the cell membrane, both ions move in the direction of their chemical driving forces: Potassium ions move out of the cell while sodium ions move into the cell (Figure 6.8a). However, the outward movement of potassium at this point exceeds the inward movement of sodium because the permeability of the membrane to potassium is greater than it is to sodium. Under these conditions, a net outward movement of positive charge occurs, which gives rise to a negative membrane potential (Figure 6.8b). As the unequal flows of potassium and sodium continue, the membrane potential becomes more negative, but it does not increase indefinitely, because the negative membrane

potential exerts electrical driving forces on potassium and sodium ions that oppose potassium movement and enhance sodium movement (Figure 6.8c). Therefore, as the membrane potential becomes more negative, outward potassium movement slows down while inward sodium movement speeds up (Figure 6.8d). Eventually, flows of the two ions become equal and opposite, so that there is no net movement of positive charge into or out of the cell (Figure 6.8e). At this point, the membrane potential holds steady at about −70 mV, which is a typical value for the resting membrane potential of a neuron.

In Cell 3, then, both sodium and potassium are moving across the membrane, and the movement of each ion tends to bring the membrane potential toward its respective equilibrium potential (as depicted in Figures 6.6 and 6.7). However, neither ion can ever come to equilibrium because the movement of each opposes the other. Given these circumstances, the final resting membrane potential is a weighted sum of the equilibrium potentials of sodium and potassium. The weight given to each equilibrium potential is based on the relative permeabilities. Because the membrane is much more permeable to potassium, the resting membrane potential, −70 mV, is closer to the potassium equilibrium potential than to the sodium equilibrium potential. The actual resting membrane potential varies among cells, because the types and numbers of ion channels in the plasma membranes of those cells vary, but the resting membrane potential is always negative.

Note that in Cell 3 neither sodium nor potassium is at equilibrium, as was the case with Cell 1 or Cell 2. This is because the membrane potential is not equal to the equilibrium potential of either ion. Therefore, electrochemical forces are acting on both ions, causing sodium to continually *leak into* the cell and potassium to continually *leak out* of the cell. Although these leakages are responsible for creating the membrane potential, they also tend to slowly alter ion concentrations inside the cell, raising the sodium concentration while lowering the potassium concentration. This is a potential problem if left unchecked, because these changes would eventually abolish the concentration gradients of both ions, and the membrane potential would go to zero. However, Na^+/K^+ pumps in the cell membrane avert this problem by actively transporting sodium out of the cell and potassium into the cell using ATP for energy. Normally, sodium is pumped out as fast as it leaks in, and potassium is pumped in as fast as it leaks out. Therefore, the Na^+/K^+ pump not only establishes the concentration gradients, but maintains them as well. Furthermore, because the Na^+/K^+ pump is *electrogenic*—that is, it transports a net positive charge out of the cell—it contributes directly to the resting membrane potential, but this effect is minimal and accounts for only a few millivolts of charge separation. Because energy is required to sustain the resting state of Cell 3, the cell is not at equilibrium; instead it is in a *steady state*.

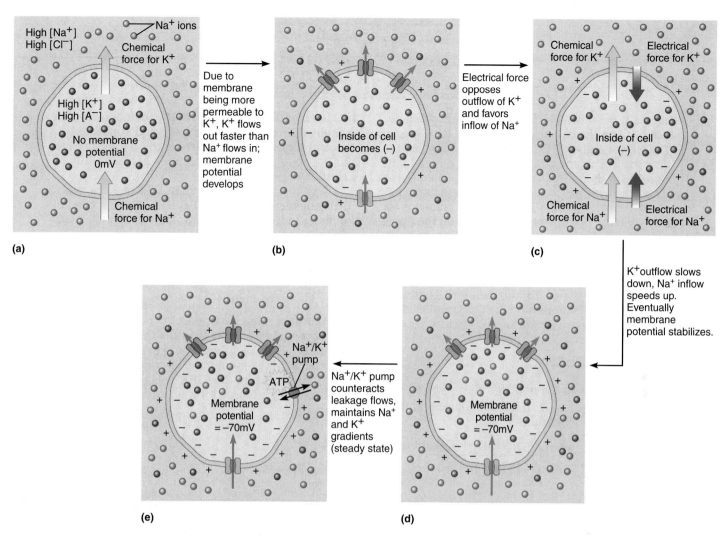

FIGURE 6.8 Cell 3: How a typical animal cell achieves a steady state resting membrane potential. *Potassium and organic anions are located in greater concentration inside the cell. Sodium and chloride ions are located in greater concentration outside the cell. The length of an arrow is relative to the strength of ion movement in the direction of the arrow. The cell is permeable to both sodium and potassium ions, but more permeable to potassium.* **(a)** *Chemical forces act on potassium ions to leave the cell and sodium ions to enter the cell.* **(b)** *More potassium leaves the cell than sodium enters because of the greater permeability for potassium. With more positive charge leaving the cell, a negative membrane potential develops.* **(c)** *Electrical forces now act on the ions, drawing both sodium and potassium ions into the cell, creating a stronger force for sodium to enter the cell and a weaker force for potassium to leave the cell.* **(d)** *Eventually, a steady state is established whereby the movement of sodium into the cell is balanced by the movement of potassium out of the cell, and no net charge movement occurs. This potential is called the resting membrane potential and is approximately −70 mV in neurons.* **(e)** *To prevent the sodium and potassium concentration gradients from dissipating, the Na/K pump moves sodium out of the cell and potassium into the cell, establishing a steady state at −70 mV.*

Neurons at Rest

Although Cell 3 is similar to a real neuron, there are important differences. The sodium and potassium channels that are responsible for the resting membrane potential are leak channels, which are always open. In addition to these leak channels, neurons also have gated ion channels. Furthermore, some cells are permeable to ions other than sodium and potassium, and the flow of these other ions contributes to the resting membrane potential. For example, in some neurons chloride ions make a significant contribution to the membrane potential. Generally speaking, the membrane potential can be affected by two, three, or more ions. Nevertheless, the following rule generally holds true: *A cell's membrane potential is a weighted*

In Chapter 4, we saw how the Nernst equation can be used to calculate the equilibrium potential for a specific ion. The Nernst equation cannot be used, however, to calculate the membrane potential because a membrane is permeable to more than one ion, and the membrane's permeability to various ions differs. The membrane potential thus depends on the concentration gradients for all ions across the plasma membrane, and on the permeability of the membrane to those ions. If the permeability is zero, an ion will not contribute to the membrane potential.

For situations in which only sodium (Na^+) and potassium (K^+) are permeant, the membrane potential (V_m) can be approximated using the Goldman, Hodgkin, Katz (GHK) equation, which is named after its developers, and is given below:

$$V_m = 61 \log \frac{P_{Na}[Na^+]_o + P_K[K^+]_o}{P_{Na}[Na^+]_i + P_K[K^+]_i}$$

If we divide the numerator and denominator both by P_K, then this equation becomes:

$$V_m = 61 \log \frac{(P_{Na}/P_K)[Na^+]_o + [K^+]_o}{(P_{Na}/P_K)[Na^+]_i + [K^+]_i}$$

This form of the equation gives the membrane potential in millivolts. Here, the subscripts o and i indicate concentrations outside and inside the cell, respectively, and P_{Na} and P_K are the membrane's permeabilities to sodium and potassium.

Using concentrations given in Table 4.1, we can calculate the membrane potential under resting conditions, assuming that P_K is 25 times as large as P_{Na} ($P_{Na}/P_K = 1/25 = 0.04$):

$$V_m = 61 \log \frac{(0.04)(145 \text{ mM}) + 4 \text{ mM}}{(0.04)(15 \text{ mM}) + 140 \text{ mM}}$$

$$= 61 \log(0.47) = -70.6 \text{ mV}$$

Note that this value is closer to the potassium equilibrium potential (-94 mV) than the sodium equilibrium potential ($+60$ mV), as we would expect. Should there be a change in the membrane's permeability to sodium relative to potassium, the GHK will give you the new membrane potential, so long as the ratio P_{Na}/P_K is known.

---

sum of the equilibrium potentials of permeant ions, with each ion's contribution depending on its relative permeability. (Those ions at concentration magnitudes less than other ions do not contribute significantly to the resting membrane potential.) *As the cell's permeability to a particular ion increases, the membrane potential moves closer to that ion's equilibrium potential.* (This relationship can be explored further in Toolbox: Resting Membrane Potential and the GHK Equation.)

Another simplification in Figure 6.8 shows sodium, potassium, and chloride ions localized on only one side of the membrane to illustrate the directions of the ions' concentration gradients. In reality, each ion is present both inside and outside the cell, with potassium at a greater concentration inside the cell, and sodium and chloride ions at greater concentrations in the extracellular fluid. Table 4.1 (p. 101) lists the actual ion distribution across a typical cell membrane.

Because neither sodium nor potassium is at equilibrium at the resting membrane potential, a net electrochemical force acts on each ion. The following rule describes electrochemical forces on ions: *The net electrochemical force on an ion tends to move that ion across the membrane in the direction that will move the membrane potential toward that ion's equilibrium potential—that is, bring the ion closer to equilibrium.* Therefore, sodium tends to move into the cell to bring the membrane potential toward +60 mV, and potassium tends to move out of the cell to bring the membrane potential toward −94 mV. The strength of the electrochemical force acting on a specific ion is proportional to the difference between the membrane potential and the equilibrium potential for that ion. Thus, because sodium is 130 mV away from equilibrium whereas potassium is only 24 mV away from equilibrium, the electrochemical force moving sodium into the cell is much greater than the electrochemical force moving potassium out of the cell (indicated by the length of the arrows in Figure 6.8e). In the next section we see that ion movement across the membrane can be altered by changing the permeability of the membrane for that ion—that is, by opening or closing specific ion channels in the plasma membrane.

1. Describe the concentration gradients for sodium and potassium across the plasma membrane. What establishes these concentration gradients?

2. If a neuron had equal permeability to sodium and potassium ions, would the resting membrane potential of that cell be more negative or less negative than −70 mV?

3. When a neuron is at rest, what are the directions of the electrochemical forces that drive sodium and potassium movement?

4. If channels that permitted both sodium ions and potassium ions to move through them suddenly opened in the plasma membrane, in which direction would each ion move, into or out of the neuron? Which ion would move more? Why? What change in membrane potential would occur?

Gated Channels

The sodium and potassium channels responsible for the membrane permeabilities to these ions in cells at rest are primarily leak channels—that is, channels that are always open. In subsequent sections we discuss how neurons communicate through changes in membrane potential that occur when certain ion channels, called *gated channels*, open or close. Gated ion channels open or close in response to particular stimuli, in the process causing changes in membrane permeability. There are three types of gated ion channels: voltage-gated channels, ligand-gated channels, and mechanically gated channels. When closed channels open, permeability to an ion increases and ion conductance increases; when open channels close, the permeability to an ion decreases and ion conductance decreases. (To learn more about ion channels, see Discovery: Thinking About Ion Channel Gating at www.physiologyplace.com, Challenge Yourself.) Gated channels are crucial for normal function of the nervous system; in fact, many toxins exert their poisonous effects by interfering with the normal function of ion channels (When It Goes Wrong: Neurotoxins, p. 185).

Electrical Signaling Through Changes in Membrane Potential

Neurons communicate via changes in membrane potential. Changes in membrane potential are described based on the direction of change relative to the resting membrane potential, as illustrated in Figure 6.9. Because the membrane potential is a difference in potential across the membrane, the membrane is *polarized*. The sign of the membrane potential (positive or negative) always refers to the potential inside the cell relative to the potential outside. Because the resting membrane potential is a

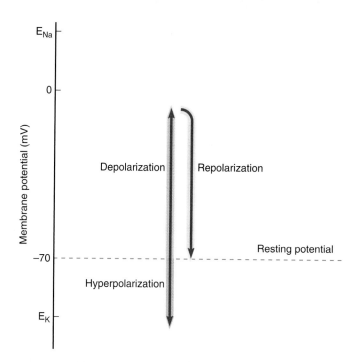

FIGURE 6.9 **Changes in membrane potential.** *The membrane potential can change with the opening or closing of ion channels. A change in membrane potential to a less negative value is called depolarization. The return of a depolarized membrane to the resting membrane potential is called repolarization. A change in membrane potential to a more negative value is called a hyperpolarization. If gated potassium channels open, then potassium moves out of the cell, bringing membrane potential toward the potassium equilibrium potential (E_K), or hyperpolarizing the cell. If gated sodium channels open, then sodium moves into the cell, bringing membrane potential toward the sodium equilibrium potential (E_{Na}), or depolarizing the cell.*

negative value (approximately −70 mV in neurons), a change to a more negative value is a **hyperpolarization** because the membrane becomes *more polarized*. In contrast, a change to a less negative or to a positive potential is a **depolarization** because the membrane becomes *less polarized*. **Repolarization** occurs when the membrane potential returns to the resting membrane potential following a depolarization.

Neurons communicate via two different types of electrical signals that result from the opening or closing of gated ion channels: (1) *graded potentials*, which are small electrical signals that act over short ranges only because they diminish in size with distance, and (2) *action potentials*, which are large signals capable of traveling long distances without decreasing in size. We examine graded potentials in the next section.

1. Define depolarization and hyperpolarization. If the permeability of a membrane to sodium increases, would the membrane depolarize or hyperpolarize? Why?

2. Name the three different gating mechanisms for ion channels.

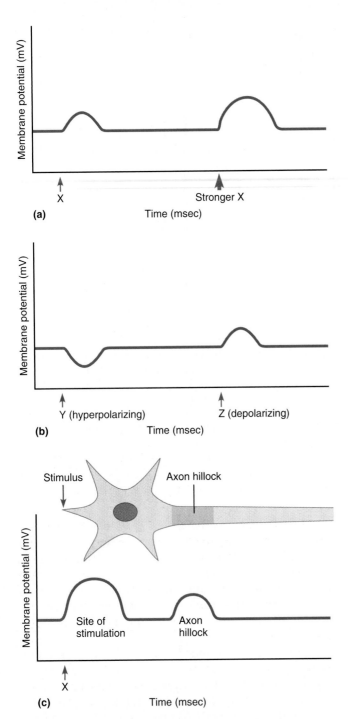

FIGURE 6.10 Properties of graded potentials. (a) *Effect of stimulus strength on size of graded potential. Graded potentials are small changes in the membrane potential (V_m) in response to a stimulus; a stronger stimulus produces a larger change in (V_m).* **(b)** *Effect of stimulus type on graded potentials. Depending on the stimulus and the neuron, graded potentials can cause either depolarization or hyperpolarization.* **(c)** *Effect of distance from site of stimulation on size of graded potentials. The amount of depolarization in this graded potential dissipates as it travels from the site of stimulation to the axon hillock.*

Graded Potentials

Graded potentials are small changes in membrane potential that occur when ion channels open or close in response to a stimulus acting on the cell (Figure 6.10). Graded potentials can be produced as a result of neurotransmitter molecules binding to receptors on a dendrite or the cell body of a neuron; graded potentials can also result from a sensory stimulus, such as touch or light, acting on a sensory receptor at the peripheral ending of an afferent neuron. The magnitude of the change in membrane potential varies with—is *graded*—according to the strength of the stimulus: A weak stimulus produces a small change in membrane potential, whereas a stronger stimulus produces a greater change in membrane potential (Figure 6.10a).

Some graded potentials are depolarizations, whereas others are hyperpolarizations (Figure 6.10b). The direction of change depends on the particular neuron in question, the stimulus applied to it, and the specific ion channels that open or close in response to the stimulus.

The primary significance of graded potentials is that they determine whether or not a cell will generate an action potential. Graded potentials generate action potentials if they depolarize a neuron to a certain level of membrane potential called the **threshold,** a critical value of membrane potential that must be met or exceeded if an action potential is to be generated. Therefore, graded potentials that are depolarizations are described as **excitatory,** because they bring the membrane potential closer to the threshold to generate an action potential. On the other hand, graded potentials that are hyperpolarizations are described as **inhibitory,** because they take the membrane potential away from the threshold to elicit an action potential.

A graded potential can travel away from the site of stimulation, but only for a short distance because it is *decremental*; that is, the change in membrane potential decreases in size as it moves along the membrane away from the site of stimulation. To understand why a graded potential is decremental, consider the following. When a change in potential occurs across a cell membrane at a particular site, this change generates differences in potential within the intracellular and extracellular fluids. Because a separation of charge creates a force for charge to move (current), the graded potential creates charge separation within the intracellular fluid and within the extracellular fluid, which generates currents in these fluids. These currents travel to adjacent areas of the cell membrane, causing voltage changes in these areas. This spread of voltage by passive charge movement is called **electrotonic conduction.** As the graded potential spreads from the site of the stimulation (Figure 6.10c), the current is spread over a larger area, and some current leaks across the plasma membrane. As a result, the size of the membrane potential change decreases as it moves from the site of initial stimulation.

As testimony to their fundamental importance in neural signal transmission, voltage-gated sodium channels are present in distantly related representatives of the animal kingdom, from mammals to mollusks to worms. In further testimony, voltage-gated sodium channels are the targets for some of the most potent neurotoxins known to humanity. A neurotoxin is a toxin that attacks some aspect of nervous system function.

The Chinese proverb "To throw away life, eat blowfish" arose because blowfish (also known as puffer fish) contain the neurotoxin tetrodotoxin (TTX). TTX is very potent, toxic at nanomolar (10^{-9} moles/L) concentrations. Effects of TTX were first described nearly 5000 years ago by the Chinese emperor Shun Nung, who had a dangerous hobby: tasting and cataloguing the effects of various drugs. Apparently, he knew the meaning of "just a taste," because he lived to describe the effects! TTX works by blocking voltage-gated sodium channels necessary for producing an action potential.

Blowfish is one of the most prized delicacies in the restaurants of Japan. This fish is prized not only for its taste, but for the tingling sensation one gets around the lips when eating it. In blowfish, TTX is concentrated in certain organs, including the liver and gonads. Its preparation takes great skill and can only be done by licensed chefs who are skilled at removing the poison-containing organs without crushing them, which can lead to contamination of normally edible parts. The toxin cannot be destroyed by cooking. Lore has it that the most skilled chefs intentionally leave a bit of the poison in, so that diners can enjoy the tingling sensation caused by blockage of nerve signals from the sense receptors on the lips. TTX is also present in some kinds of salamanders, octopus, and goby. Species that carry the toxin have sodium channels that differ slightly, rendering them either less susceptible or completely resistant to the toxin.

A closely related toxin that blocks voltage-gated sodium channels is saxitoxin (STX), which is produced by some marine dinoflagellates and by a freshwater cyanobacterium. These microorganisms become a major health threat during certain periods of the year when they multiply rapidly, or "bloom." At high concentrations they may even impart a reddish color to the water, producing the phenomenon known as "red tide." The primary danger to humans during red tide comes from eating shellfish, which are filter feeders that can accumulate the toxin in their bodies. (Shellfish are resistant to the toxin.) Eating even a single contaminated shellfish can be fatal. Like TTX, STX is not destroyed by cooking. Unfortunately, one cannot always tell whether conditions are hazardous by simple observation; seawater can contain dangerous concentrations of dinoflagellates without appearing red. Fortunately, shellfish are routinely monitored by public health officials for the presence of this toxin, and harvesting of shellfish is banned whenever there is cause for concern.

Sodium channels are not the only ion channels affected by neurotoxins. The table in this box lists a variety of neurotoxins, their source, and their effects on ion channels.

SOURCES AND EFFECTS OF SELECTED NEUROTOXINS

TOXIN	SOURCE	EFFECTS
Tetrodotoxin	Blowfish (puffer fish)	Blocks voltage-gated sodium channels
Saxitoxin	Marine dinoflagellates, freshwater cyanobacterium	Blocks voltage-gated sodium channels
Apamin	Honey bee	Blocks potassium channels
Batrachotoxin	Poison arrow frog	Keeps sodium channels from closing
Calciseptin	Black mamba	Blocks calcium channels
Iberiotoxin	Indian red scorpion	Blocks potassium channels
Phoneutriatoxin	Banana spider	Slows closing of sodium channels
Stichodactyla toxin	Sea anemone	Blocks voltage-gated potassium channels

FIGURE 6.11 **Types of**
neural circuitry. (a) *Conver-*
gence, in which many neurons
communicate with a single neuron.
(b) *Divergence, in which a single*
neuron communicates to many
other neurons.

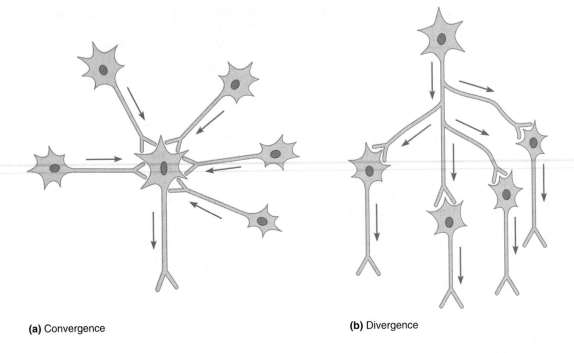

(a) Convergence

(b) Divergence

Neural Integration

Most graded potentials are produced in response to the opening or closing of either ligand-gated ion channels (a neurotransmitter is the ligand) or ion channels associated with sensory receptors. For simplicity, we describe integration of graded potentials in response to ligand-gated ion channels here, although the same principles apply to sensory receptors. Graded potentials produced in response to ligand-gated ion channels are usually generated on the dendrites and cell body of a neuron by neurotransmitter molecules released from other cells. Communication between neurons, however, is not a one-to-one relationship; instead, in some cases many neurons communicate with a given neuron, an arrangement called **convergence** (Figure 6.11a). In convergence, a neuron must function as an *integrator*, taking in information from several different sources in the form of graded potentials, summing it, and then on the basis of the results, either generating an action potential or not. In other cases the axon of one neuron branches into many terminals that communicate with many other cells, an arrangement called **divergence** (Figure 6.11b).

A single graded potential is generally not of sufficient strength to elicit an action potential. However, if graded potentials in a neuron overlap in time, then they can sum, both temporally and spatially. In **temporal summation,** stimuli are applied in such rapid succession that the graded potential from one stimulus does not dissipate before the next graded potential occurs. Thus, the effects

of the potentials sum. The greater the overlap in time, the greater the summation. In **spatial summation,** the effects of stimuli from different sources occurring close together in time sum. Summation of a hyperpolarizing graded potential and a depolarizing graded potential tend to cancel each other out. Figure 6.12 graphically depicts the events in temporal and spatial summation.

Figure 6.12 is a simplified depiction of summation because it takes into account only four synapses. A single neuron may receive communication from hundreds, thousands, or even hundreds of thousands of other neurons. At any given time, some of the neurons will be actively communicating with the cell by releasing neurotransmitter. At most active synapses, graded potentials, also called *synaptic potentials*, are initiated by neurotransmitter molecules binding to receptors. These synaptic potentials spread toward the portion of the cell body where the axon comes off, called the axon hillock, decreasing in size along the way. Those graded potentials arriving at approximately the same time will sum at the axon hillock, and if they depolarize the axon hillock to the threshold, then an action potential will be generated. The axon hillock is also called the **trigger zone** of a neuron because it contains the first voltage-gated sodium and potassium ion channels necessary for the development of an action potential. Summation of synaptic potentials is described in more detail in Chapter 7, and summation of graded potentials produced at sensory receptors is described in Chapter 9.

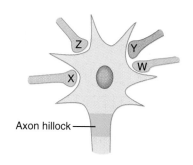

(a)

FIGURE 6.12 **Temporal and spatial summation in a neuron.**
(a) *Convergence of input from neurons W, X, Y, and Z. Neurotransmitters from neurons W, X, and Z produce depolarization, and neurotransmitter from neuron Y produces hyperpolarization, of the postsynaptic plasma membrane.* (b) *Temporal summation of stimulus W resulting in depolarization to above threshold and generation of an action potential.* (c) *Spatial summation of stimuli W and X resulting in depolarization to above threshold and generation of an action potential.* (d) *Spatial summation of stimuli W and Y resulting in no change in membrane potential and therefore, no action potential.* (e) *Spatial summation of stimuli W and Z that fails to reach threshold because the depolarizing graded potential induced by stimulus Z is dissipated over the great distance between the synapse and the axon hillock.*

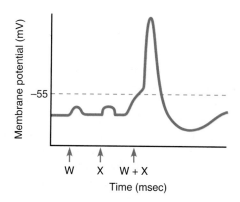

(b)

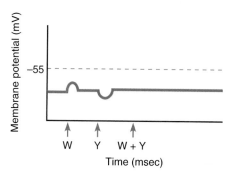

(c)

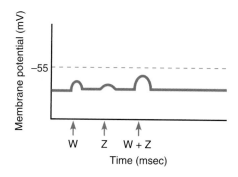

(d)

(e)

Quick Test 6.4

1. What properties of graded potentials result in their being "graded"?

2. What is the distinction between an excitatory graded potential and an inhibitory graded potential?

ACTION POTENTIALS

In most of the remainder of this chapter we discuss action potentials. Action potentials occur in the membranes of excitable tissue (nerve or muscle) in response to graded potentials that reach threshold. Table 6.2 compares the properties of graded and action potentials. During an action potential a large, rapid depolarization occurs in which the polarity of the membrane potential actually reverses; that is, the membrane potential becomes positive for a brief period of time. In fact, the membrane potential changes very quickly (in about 1 msec) from a resting level of approximately -70 mV to $+30$ mV (a change of 100 mV). Once initiated, an action potential, unlike a graded potential, is capable of being propagated long distances along the length of an axon without any decrease in strength.

Ionic Basis of an Action Potential

The generation of an action potential is based on the selective permeability of the plasma membrane and the Na^+ and K^+ electrochemical gradients that exist across the membrane. Recall that at rest the plasma membrane is approximately 25 times more permeable to potassium ions than to sodium ions because of the presence of many more potassium leak channels than sodium leak channels. In excitable cells, changes in the permeability of the plasma membrane resulting from the opening and closing of gated ion channels can produce action potentials.

TABLE 6.2 COMPARISON OF GRADED POTENTIALS AND ACTION POTENTIALS

PROPERTY	GRADED POTENTIAL	ACTION POTENTIAL
Location	Dendrites, cell body, sensory receptors	Axon
Strength	Relatively weak, proportional to strength of stimulus; dissipates with distance from stimulus	100 mV All-or-none
Direction of change in membrane potential	Can be depolarizing or hyperpolarizing depending on stimulus	Depolarizing
Summation	Spatial and temporal	None
Refractory periods	None	Absolute and relative
Channel types involved in producing change in potential	Ligand-gated, mechanically gated	Voltage-gated
Ions involved	Usually Na^+, Cl^-, or K^+	Na^+ and K^+
Duration	Few msec to seconds	1–2 msec (after-hyperpolarization may last 15 msec)

An action potential in a neuron consists of three distinct phases (Figure 6.13a):

1. Depolarization. The first phase of an action potential is a *rapid depolarization* during which the membrane potential changes from −70 mV (rest) to +30 mV. This depolarization is caused by a sudden and dramatic increase in permeability to sodium (Figure 6.13b) followed by an increase in the movement of sodium ions into the cell, down sodium's electrochemical gradient. With permeability to sodium now greater than permeability to potassium, the membrane potential approaches the sodium equilibrium potential of +60 mV. Although the change in membrane potential produced by sodium movement is large (100 mV), the number of sodium ions that actually cross the membrane to produce this change in potential is relatively small compared to the concentration of sodium ions in the intracellular and extracellular fluids. Therefore, the concentrations of sodium in those fluids does not change appreciably.

2. Repolarization. The second phase of an action potential is a *repolarization* of the membrane potential during which the membrane potential returns from +30 mV back to resting levels (−70 mV). Within 1 msec after the increase in sodium permeability, sodium permeability decreases rapidly, reducing the inflow of sodium. At approximately the same time, potassium permeability increases. Potassium then moves down its electrochemical gradient out of the cell, repolarizing the membrane potential to bring it back to resting levels.

3. After-hyperpolarization. The third phase of an action potential is termed **after-hyperpolarization.** Potassium permeability remains elevated for a brief time (5–15 msec) after the membrane potential reaches the resting membrane potential, resulting in an *overshoot* or after-hyperpolarization. During this time the membrane potential is even more negative than at rest as it approaches the potassium equilibrium potential (−94 mV). As with sodium movement, the movement of potassium ions during repolarization and after-hyperpolarization is small, so the concentrations of potassium inside and outside the cell do not change appreciably.

Next we discuss the ion channels responsible for the changes in permeability to sodium and potassium ions during an action potential.

The Role of Voltage-Gated Ion Channels in Action Potentials

The changes in permeability associated with the phases of an action potential are due to the time-dependent opening and closing of voltage-gated sodium and potassium channels located primarily in the plasma membrane of the axon hillock and axon. (These voltage-gated sodium and potassium channels are also found in the plasma membrane of some muscle cells.) In myelinated axons these channels are at a greater concentration at the nodes of Ranvier; in unmyelinated axons these channels are evenly distributed along the entire axon. Because the exact mechanisms of gating in the voltage-gated sodium and potassium channels are not known, we use models to describe their function.

The model for explaining the actions of voltage-gated sodium channels involves two types of gates: activation gates and inactivation gates. **Activation gates** are responsible for the opening of sodium channels during the depolarization phase of an action potential, whereas **inactivation gates** are responsible for the closing of sodium channels during the repolarization phase of an action potential. For a sodium channel to be open, both gates must be open. Both types of gates open and close in response to changes in the membrane potential.

Based on the position of these two gates, a sodium channel can exist in three conformations (Figure 6.14). At rest the inactivation gate is open, but the activation gate is closed. In this state, the channel is *closed but capable of being opened* by a depolarizing stimulus that causes the activation gate to open. With both gates in their open position, the channel is open and sodium ions move through the channel into the cell; this is what occurs during the depolarization phase of an action potential. However, within approximately 1 msec after the initial stimulus to open the activation gate, the inactivation gate closes. The closing is a delayed response initiated by the same depolarization that caused the activation gate to immediately open. With the inactivation gate closed and the activation gate open, the channel is *closed and incapable of opening* in response to another depolarizing stimulus, because the inactivation gate does not open until the membrane potential returns to near its resting value. Once repolarization has occurred, the inactivation gate opens and the activation gate closes, returning the channel to its resting state.

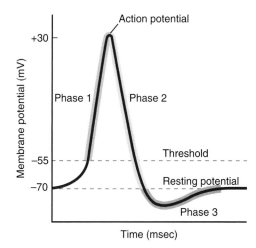

(a) Three phases of an action potential

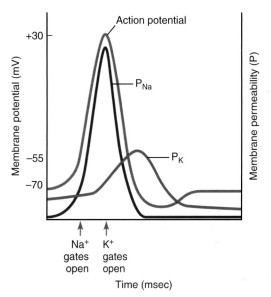

(b) Permeability changes for Na⁺ and K⁺ during an action potential

FIGURE 6.13 The phases and ionic basis of an action potential. (a) *The three distinct phases of an action potential: (1) depolarization, (2) repolarization, and (3) after-hyperpolarization.* (b) *The permeability changes for sodium ions (P_{Na}) and potassium ions (P_K) that occur during an action potential. The rapid depolarization of phase 1 is caused by a rapid sodium permeability increase that enables sodium to move into the cell. The repolarization of phase 2 is caused by a slower increase in potassium permeability, enabling greater movement of potassium out of the cell compared to resting conditions. The after-hyperpolarization of phase 3 is caused by the continuing movement of potassium out of the cell.*

Why does permeability to sodium increase faster than permeability to potassium?

Exercise Link

Although this chapter focuses on nerve cells, it is important to remember that all cells have a resting membrane potential. Furthermore, skeletal muscle cells depolarize to initiate contraction. During exercise, the repetitive depolarization and repolarization of muscle cells causes a rise in extracellular potassium concentration that is proportional to active muscle mass and to exercise intensity and duration. While running the marathon, Bill and Jane's plasma potassium concentrations rose about 10%, which is enough to influence resting membrane potential and may be a factor contributing to muscle fatigue. Exercise training increases the number of Na⁺/K⁺ pumps in skeletal muscle, an adaptation that helps maintain proper potassium (and sodium) concentrations across the cell membranes during exercise.

Potassium channels are slower to open.

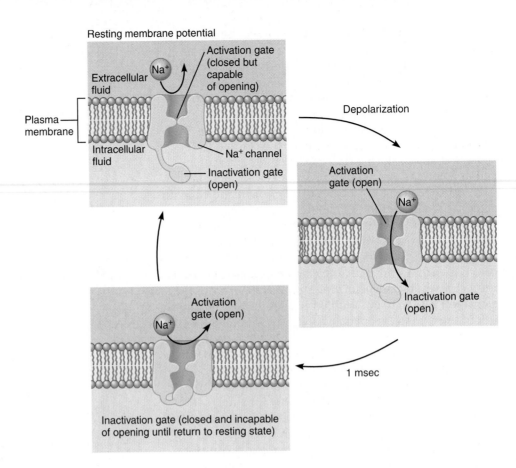

FIGURE 6.14 A model for the operation of voltage-gated sodium channels. *Voltage-gated sodium channels are schematically represented here by two gates. At rest, the sodium inactivation gate is open and the activation gate is closed but can open in response to a depolarizing stimulus. Following a depolarizing stimulus to threshold, both the activation and inactivation gates are open, and sodium can move through the channel. Approximately 1 msec after a depolarizing stimulus, the inactivation gate closes and remains closed until the cell has repolarized to the resting state. Before repolarization the channel cannot open in response to a new depolarizing stimulus.*

The opening of sodium activation gates is a **regenerative** mechanism; that is, the opening of some sodium activation gates causes more sodium activation gates to open by *regenerating* the stimulus to open the gates (depolarization). This regenerative mechanism works as follows: When the membrane is depolarized to threshold, the depolarization initiates a positive feedback loop that triggers the opening of sodium channels and a rapid rise in sodium permeability (Figure 6.15). At first, depolarization triggers the opening of a few sodium channels, which allows some movement of sodium ions into the cell, which depolarizes the cell further. The increased depolarization causes still more sodium channels to open, leading to a larger inflow of sodium ions and more depolarization, and so on. This positive feedback loop causes the very rapid (less than 1 msec) depolarization phase of the action potential. The positive feedback loop terminates when the sodium inactivation gates close.

In contrast to the sodium channels just described, the model for voltage-gated potassium channels describes only a single gate that opens slowly in response to depolarization. At approximately the same time that sodium inactivation gates are closing (1–2 msec after depolarization to threshold), the potassium channels begin to open slowly, and potassium ions begin to leave the cell. This movement of positive charge out of the cell repolarizes the cell. Because the effect of opening potassium chan-

nels (repolarization) is an action opposite to the initial stimulus that opened the potassium channels (depolarization), voltage-dependent potassium channels are part of a negative feedback loop during an action potential (see Figure 6.15). Therefore, as the cell repolarizes, the depolarizing stimulus weakens, and potassium channels slowly close.

Table 6.3 summarizes the conditions surrounding voltage-gated sodium and potassium channels at rest and during the three phases of an action potential.

The threshold for generating an action potential corresponds to the level of depolarization necessary to induce the sodium positive feedback loop. A depolarization that is below threshold (a **subthreshold** stimulus) may open some sodium channels, but not enough of them to produce an inward flow of sodium large enough to overcome the outward flow of potassium through leak channels. Figure 6.16 illustrates the concept of threshold. Note that a subthreshold stimulus produces no action potential, whereas a threshold stimulus elicits an action potential. A stimulus greater than threshold—that is, a **suprathreshold** stimulus—also elicits an action potential; note, however, that the action potential does not increase in size as the strength of a suprathreshold stimulus increases.

Initiation of action potentials follows the **all-or-none principle:** When a membrane is depolarized to threshold or above, an action potential is elicited that is always the

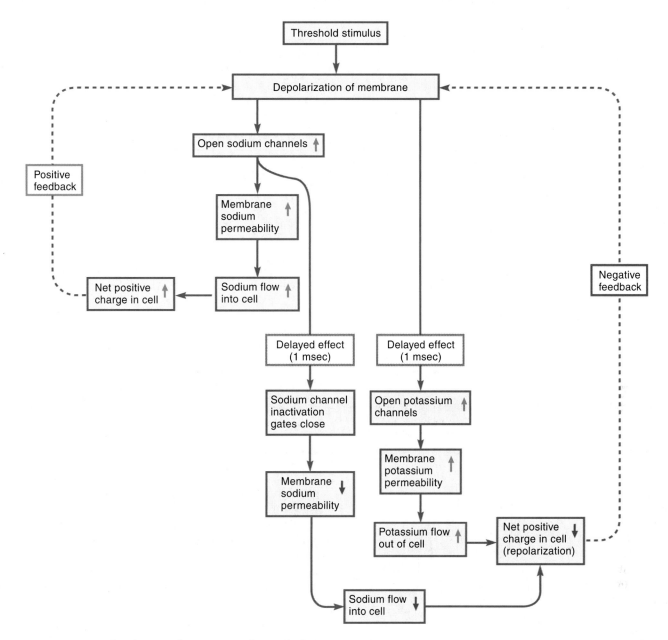

FIGURE 6.15 Gating of sodium and potassium channels during an action potential. *Sodium channel opening is part of a positive feedback loop that allows for the rapid depolarization of the cell. When the cell is depolarized to threshold, sodium channels open. Opening allows sodium to move into the cell, causing further depolarization and opening more sodium channels. The feedback loop continues until the sodium inactivation gates close, approximately 1 msec after the depolarization to threshold. Potassium channel opening and closing is part of a negative feedback loop. Depolarization stimulates the slow opening of potassium channels. This allows potassium to move out of the cell repolarizing it. Since repolarization opposes the depolarizing stimulus for opening potassium channels, the potassium channels close.*

What stops the positive feedback loop for sodium channels to open?

same strength; if the membrane is not depolarized to threshold, no action potential occurs. The level of depolarization reached at the peak of an action potential is caused by the tendency of sodium moving into the cell to depolarize the neuron to its equilibrium potential (+60 mV). However, the neuron can never meet or exceed the sodium equilibrium potential for the same reason that the resting membrane potential can never equal or exceed the potassium equilibrium potential: Sodium movement into the cell is countered by potassium movement out of the cell, primarily through potassium leak channels and later

The closing of sodium inactivation gates and the opening of potassium channels

TABLE 6.3 CHARACTERISTICS OF A NEURON AT REST AND DURING THE DIFFERENT PHASES OF AN ACTION POTENTIAL

	RESTING	DEPOLARIZATION	REPOLARIZATION	AFTER-HYPERPOLARIZATION
Membrane potential	−70 mV	−70 mV to +30 mV	+30 mV to −70 mV	−70 mV to −85 mV
Voltage-gated sodium channel	Closed	Open	Closed	Closed
Activation gate	Closed	Open	Open	Closed
Inactivation gate	Open	Open	Closed	Open
Sodium flow	Low inward, through leak channels	High inward, through voltage-gated channels*	Low inward, through leak channels	Low inward, through leak channels
Voltage-gated potassium channel	Closed	Closed	Open	Closing
Potassium flow	Low outward, through leak channels	Low outward, through leak channels	High outward, through voltage-gated channels*	High outward, through voltage-gated channels, but decreasing*

*Even though at any given time ions move through both voltage-gated channels and leak channels, the conductance through the leak channels is negligible compared to that through voltage-gated channels.

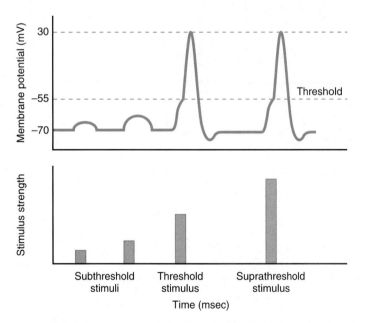

FIGURE 6.16 **The concept of a threshold stimulus.** *A stimulus must reach a critical level of depolarization—threshold—before an action potential is generated. A stimulus less than threshold (a subthreshold stimulus) cannot generate an action potential. Any stimulus greater than threshold (a suprathreshold stimulus) generates an action potential of the same magnitude and duration as a threshold stimulus.*

through the opened voltage-gated potassium channels. Therefore, the constant peak of an action potential is determined primarily by two factors: (1) the concentration gradients for sodium and potassium ions across the plasma membrane, which do not change under normal circumstances, and (2) the number of potassium leak channels in the plasma membrane, which also does not change.

Quick Test 6.5

1. During the depolarization phase of an action potential, is the membrane more permeable to sodium or to potassium? What about during the repolarization phase?

2. Is the action potential produced by a suprathreshold stimulus smaller, larger, or the same size as an action potential produced by a threshold stimulus?

Refractory Periods

During and immediately after an action potential, the membrane is less excitable than it is at rest. This period of reduced excitability is called the **refractory period.** The refractory period can be divided into two phases, the absolute refractory period and the relative refractory period (Figure 6.17). The **absolute refractory period** spans all

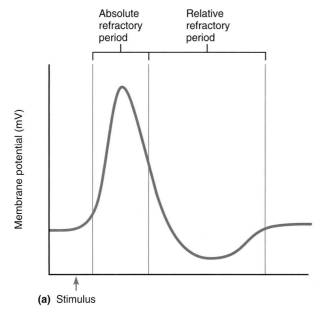

(a) Stimulus

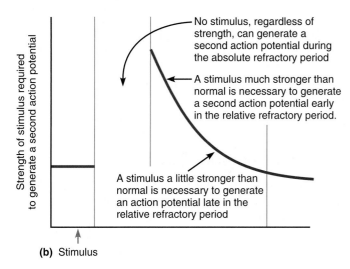

No stimulus, regardless of strength, can generate a second action potential during the absolute refractory period

A stimulus much stronger than normal is necessary to generate a second action potential early in the relative refractory period.

A stimulus a little stronger than normal is necessary to generate an action potential late in the relative refractory period

(b) Stimulus

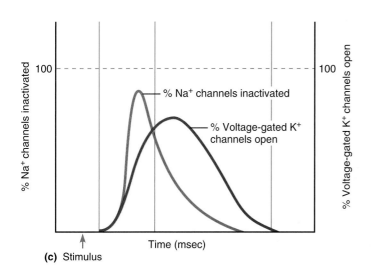

% Na⁺ channels inactivated

% Voltage-gated K⁺ channels open

Time (msec)

(c) Stimulus

of the depolarization phase plus most of the repolarization phase of an action potential (1–2 msec). During this time, a second action potential cannot be generated in response to a second stimulus, regardless of the strength of that stimulus. There are two reasons for the absolute refractory period: (1) During the rapid depolarization phase of an action potential, the regenerative opening of sodium channels that has been set into motion will proceed to its conclusion and will not be affected by a second stimulus. (2) During the beginning of the repolarization phase, most of the sodium inactivation gates are closed and cannot be opened by a second stimulus. A second action potential cannot be generated until the majority of the sodium channels have returned to their resting state, a situation that occurs near the end of the repolarization phase. At that time the activation gates have closed and the inactivation gates have opened, so the sodium channels are closed but capable of opening in response to depolarization.

The **relative refractory period** occurs immediately after the absolute refractory period and lasts 5–15 msec. During this period, it is possible to generate a second action potential, but only in response to a stimulus stronger than that needed to reach threshold under resting conditions. The relative refractory period is primarily due to the increased permeability to potassium that continues beyond the repolarization phase (during after-hyperpolarization). In addition, some sodium inactivation gates may still be closed, especially early in the relative refractory period. Just how much stronger the stimulus must be to elicit a second action potential is a matter of timing. Early in the relative refractory period it takes a stronger stimulus to generate an action potential than is needed late in the relative refractory period, because more sodium inactivation gates are closed and more potassium channels are open early in the refractory period.

The refractory periods establish several properties of an action potential, including the all-or-none property of action potentials, the frequency with which a single neuron can generate an action potential, and the unidirectional propagation of action potentials along an axon.

FIGURE 6.17 Refractory periods associated with an action potential. *The graphs represent* **(a)** *membrane potential changes occurring during the action potential;* **(b)** *the stimulus strength needed to elicit a second action potential; and* **(c)** *the percentages of sodium channels that are inactivated (inactivation gates are closed) and of voltage-gated potassium channels that are open.*

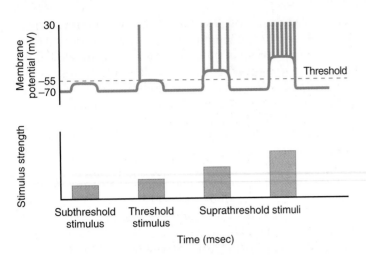

FIGURE 6.18 Frequency coding: how action potentials convey intensity of stimuli. *The subthreshold stimulus does not generate an action potential, and the threshold stimulus generates a single action potential. The weaker of the two suprathreshold stimuli generates a burst of action potentials, whereas the stronger suprathreshold stimulus generates an even higher frequency of action potentials.*

Refractory periods contribute to the all-or-none principle of action potentials described previously. Unlike graded potentials, action potentials are not additive, because the absolute refractory period prevents an overlap of action potentials.

Refractory periods are also important in the coding of information that arrives in the form of action potentials. That action potentials are all-or-none would seem to create a problem: Because graded potentials vary in magnitude based on the strength of a stimulus, the size of a graded potential encodes information about the intensity of the stimulus; but how can action potentials relay information about the intensity of a stimulus (say, the loudness of a sound)? Information pertaining to stimulus intensity is encoded by changes in the *frequency* of action potentials—that is, changes in the number of action potentials that occur in a given period of time. Because graded potentials typically last much longer than action potentials, stronger, longer-lasting graded potentials may generate a burst of action potentials. Depending on the size of the graded potential, these action potentials may occur farther apart or closer together. Thus a loud sound generates action potentials at a higher frequency (more action potentials per unit time) than a soft sound.

Consider Figure 6.18, which compares the effects of subthreshold, threshold, and suprathreshold stimuli on the frequency of action potentials generated in an axon. A subthreshold stimulus does not generate an action potential; a stimulus that just reaches the threshold generates an action potential. In fact, a threshold stimulus could generate more than one action potential *if* the graded potential the stimulus induces lasts longer than the entire refractory period. A suprathreshold stimulus is more likely to generate more than one action potential because it can generate a second action potential during the relative refractory period of the first action potential. To generate a second action potential during the relative refractory period, a stimulus must be strong enough to open enough sodium channels such that sodium inflow overcomes the elevated potassium outflow that occurs during the relative refractory period, and the stimulus may have to overcome some sodium inactivation gates that are still closed. Also, a stronger stimulus can produce a second action potential closer in time to the first action potential (see Figure 6.18). However, because a second stimulus cannot generate a second action potential during the absolute refractory period, the absolute refractory period limits the maximum frequency of action potentials in a neuron to approximately 500–1000 per second.

Quick Test 6.6

1. Can an action potential be generated during the absolute refractory period? During the relative refractory period?

2. During the absolute refractory period, are the majority of sodium channels open, closed but capable of opening, or closed and incapable of opening?

3. How do action potentials encode the intensity of a stimulus?

Propagation of Action Potentials

Once an action potential is initiated in an axon, it is propagated down the length of the axon from the trigger zone to the axon terminal without decrement. An action potential does not actually travel down the axon; instead, an action potential sets up electrochemical gradients in the extracellular and intracellular fluids. Because the extracellular and intracellular fluids have low resistance to current flow, positive charges move from the area where the membrane has been depolarized to the adjacent area of membrane, depolarizing it as well. Propagation of an action potential down an axon is thus analogous to the sequential falling of a row of dominoes set close together on end: When the first domino falls over, it tips the adjacent one over, and so on all the way to the end of the row of dominoes. The first action potential produced at the trigger zone produces current that causes a second action potential in the adjacent membrane, which produces current that causes a third action potential, and so on until an action potential is produced at the axon terminal. Current flows to adjacent areas of the axon's plasma membrane by electrotonic conduction. The propagation mechanisms differ, however, depending on whether the axon is unmyelinated or myelinated.

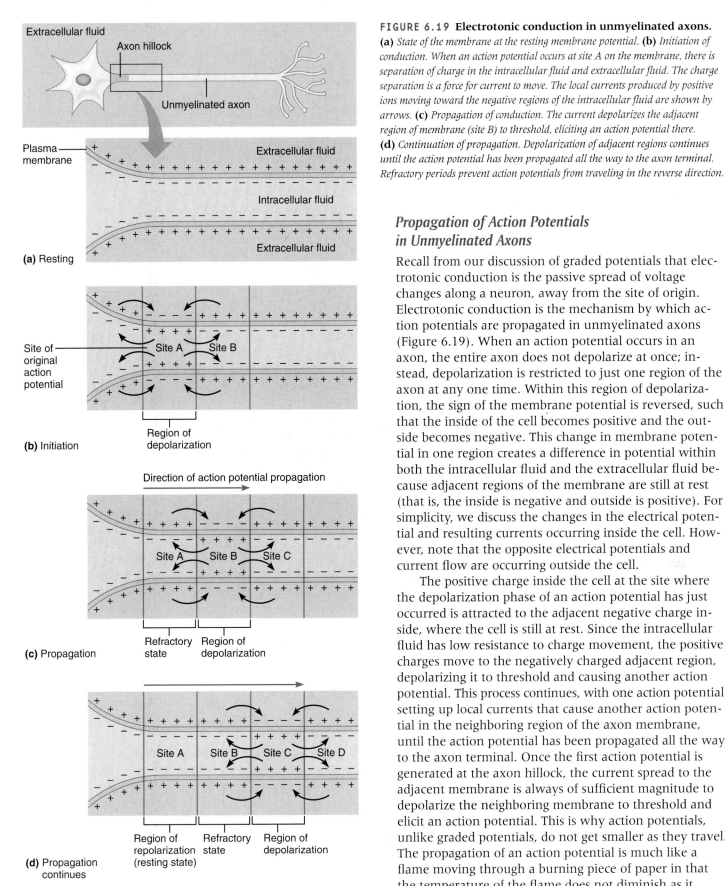

Extracellular fluid

Axon hillock

Unmyelinated axon

Plasma membrane

Extracellular fluid

Intracellular fluid

Extracellular fluid

(a) Resting

Site of original action potential

Site A   Site B

Region of depolarization

(b) Initiation

Direction of action potential propagation

Site A   Site B   Site C

Refractory state   Region of depolarization

(c) Propagation

Site A   Site B   Site C   Site D

Region of repolarization (resting state)   Refractory state   Region of depolarization

(d) Propagation continues

FIGURE 6.19 Electrotonic conduction in unmyelinated axons. *(a) State of the membrane at the resting membrane potential. (b) Initiation of conduction. When an action potential occurs at site A on the membrane, there is separation of charge in the intracellular fluid and extracellular fluid. The charge separation is a force for current to move. The local currents produced by positive ions moving toward the negative regions of the intracellular fluid are shown by arrows. (c) Propagation of conduction. The current depolarizes the adjacent region of membrane (site B) to threshold, eliciting an action potential there. (d) Continuation of propagation. Depolarization of adjacent regions continues until the action potential has been propagated all the way to the axon terminal. Refractory periods prevent action potentials from traveling in the reverse direction.*

Propagation of Action Potentials in Unmyelinated Axons

Recall from our discussion of graded potentials that electrotonic conduction is the passive spread of voltage changes along a neuron, away from the site of origin. Electrotonic conduction is the mechanism by which action potentials are propagated in unmyelinated axons (Figure 6.19). When an action potential occurs in an axon, the entire axon does not depolarize at once; instead, depolarization is restricted to just one region of the axon at any one time. Within this region of depolarization, the sign of the membrane potential is reversed, such that the inside of the cell becomes positive and the outside becomes negative. This change in membrane potential in one region creates a difference in potential within both the intracellular fluid and the extracellular fluid because adjacent regions of the membrane are still at rest (that is, the inside is negative and outside is positive). For simplicity, we discuss the changes in the electrical potential and resulting currents occurring inside the cell. However, note that the opposite electrical potentials and current flow are occurring outside the cell.

The positive charge inside the cell at the site where the depolarization phase of an action potential has just occurred is attracted to the adjacent negative charge inside, where the cell is still at rest. Since the intracellular fluid has low resistance to charge movement, the positive charges move to the negatively charged adjacent region, depolarizing it to threshold and causing another action potential. This process continues, with one action potential setting up local currents that cause another action potential in the neighboring region of the axon membrane, until the action potential has been propagated all the way to the axon terminal. Once the first action potential is generated at the axon hillock, the current spread to the adjacent membrane is always of sufficient magnitude to depolarize the neighboring membrane to threshold and elicit an action potential. This is why action potentials, unlike graded potentials, do not get smaller as they travel. The propagation of an action potential is much like a flame moving through a burning piece of paper in that the temperature of the flame does not diminish as it moves away from the point of origin.

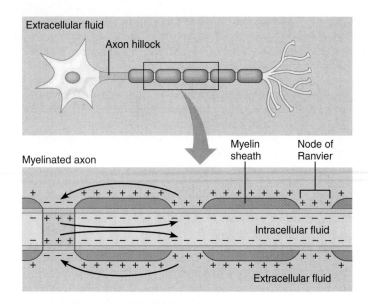

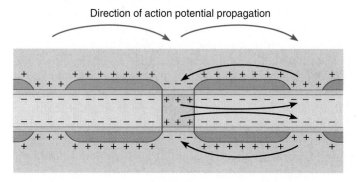

FIGURE 6.20 Saltatory conduction in myelinated axons.
An action potential in a myelinated axon produces electrical gradients in the intracellular and extracellular fluids that are similar to those observed in unmyelinated axons (see Figure 6.19). However, because very little current flows across the membrane where myelin insulates it, the current must flow all the way to the next node of Ranvier, where it depolarizes this area of the membrane to threshold and initiates an action potential.

The diameter of an axon determines how quickly current spreads and, therefore, the conduction velocity of action potentials. The larger the diameter, the less resistance to longitudinal current flow (that is, current flow down the axon). Therefore, action potentials are propagated more quickly from axon hillock to axon terminal in large-diameter axons.

When a particular region of an axon is depolarized during an action potential, the resulting currents travel both downstream and upstream to adjacent regions of the membrane. What prevents an action potential from traveling upstream is that the portion of the membrane nearer to the side closest to the axon hillock has just recently experienced an action potential of its own; therefore, it is in an absolute refractory state. The refractory period thus prevents action potentials from traveling backward, ensuring unidirectional propagation of action potentials (see Figure 6.19c and d).

Action Potential Propagation in Myelinated Axons

In axons that are sheathed in myelin, action potentials are propagated by a specialized type of electrotonic conduction called **saltatory conduction** (Figure 6.20). Myelin provides high resistance to ion movement across the plasma membrane, but longitudinal resistance is low. As noted previously, the nodes of Ranvier are gaps in the myelin where the axon membrane lacks insulation, is exposed to the interstitial fluid, and has the greatest concentration of voltage-gated sodium and potassium channels. In myelinated fibers action potentials are produced at the nodes of Ranvier. The concept is similar to that described for electrotonic conduction, except that action potentials cannot be produced where myelin is present. Therefore, the separation of charge in the intracellular fluid causes current to flow from one node of Ranvier to the next. These currents move rapidly under the myelin sheath but diminish in amplitude because some current leaks across the axon membrane. However, because the distance between nodes of Ranvier is short, the current remains strong enough to depolarize the membrane at the node of Ranvier to threshold and to generate an action potential. In this way, action potentials are generated at each node along the axon until an action potential is generated at the axon terminal. The jumping of action potentials from node to node is the basis of the term *saltatory conduction* (saltatory comes from *saltare*, Latin for "to leap").

We saw in the previous section that conduction velocity is greater in large-diameter axons. Because action potentials can move in large jumps in myelinated axons, conduction velocities in myelinated axons are greater than those in unmyelinated axons. The fastest conducting axons, therefore, are both large-diameter and myelinated (Table 6.4). The fastest conducting axons are generally found in pathways requiring quick action; thus the axons that control skeletal muscle contractions are myelinated and of the largest diameter, and exhibit the fastest conduction velocities.

Quick Test 6.7

1. What is meant by "all-or-none" in reference to action potentials?

2. In which types of axons does saltatory conduction occur?

3. Which of the following axons exhibits the greatest conduction velocity?
   a) an unmyelinated axon with diameter 5 μm
   b) a myelinated axon with diameter 5 μm
   c) an unmyelinated axon with diameter 20 μm
   d) a myelinated axon with diameter 20 μm

TABLE 6.4 CONDUCTION VELOCITIES IN
AXONS OF VARIOUS NERVE FIBER TYPES

FIBER TYPE	MYELIN PRESENT?	EXAMPLE OF FUNCTION	FIBER DIAMETER (μm)	CONDUCTION VELOCITY (m/sec)
A alpha	Yes	Stimulation of skeletal muscle contraction	12–20	70–120
A beta	Yes	Touch, pressure sensation	5–12	30–70
A gamma	Yes	Stimulation of muscle spindle contractile fibers	3–6	15–30
A delta	Yes, but little	Pain, temperature sensation	2–5	12–30
B	Yes	Visceral afferents, autonomic preganglionics	1–3	3–15
C	No	Pain, temperature sensation, autonomic postganglionics	0.3–1.3	0.7–2.3

THE BASIS OF NEURAL STABILITY

In the beginning of this chapter we saw the importance of ion concentration gradients and equilibrium potentials to the establishment of the resting membrane potential; subsequent sections discussed the role of ion movement into or out of the neuron to produce the resting potential, graded potentials, and action potentials. But this raises questions: If ions move across the cell membrane, would not the gradient dissipate, and the equilibrium potentials change? If the gradients dissipate, what happens to the resting membrane potential?

The key to answering these questions lies in the Na^+/K^+ pumps in the neuron membrane. Early in this chapter we discussed the Na^+/K^+ pump in the context of its importance in developing concentration gradients and sustaining the gradients when the cell is at rest, when sodium and potassium ions are moving in and out of the cell through leak channels. To compensate for the movement of sodium and potassium ions in an active cell (a cell propagating action potentials), the Na^+/K^+ pump pumps even more sodium and potassium ions to compensate for the greater movement of these ions through opened voltage-gated channels. Although an action potential is a relatively large change in the membrane potential, we have seen that compared to the total number of ions present in the intracellular and extracellular fluids, very few ions need cross the membrane to produce this change. In fact, even without the Na^+/K^+ pump, no measurable change in the resting membrane potential would occur until a neuron has produced more than a thousand action potentials. Therefore, the action of the Na^+/K^+ pump is sufficient to prevent changes in concentration gradients for sodium and potassium in a cell performing at normal levels of activity.

SUMMARY OF FACTORS THAT INFLUENCE MEMBRANE POTENTIAL

The main focus of this chapter was the generation of electrical signals within a neuron in the form of graded and action potentials. Figure 6.21 summarizes the factors that influence the membrane potential. The three primary factors are the Na^+/K^+ pump, the number of open sodium channels, and the number of open potassium channels. The Na^+/K^+ pump establishes the concentration gradients for sodium and potassium ions across the plasma membrane. These concentration gradients, in turn, determine the equilibrium potentials for sodium and potassium ions. The number of open ion channels for a specific ion determines the membrane permeability to that ion. If the membrane is highly permeable to the ion, then the ion moves to drive the membrane potential toward its equilibrium potential. Under resting conditions, the membrane potential is close to the equilibrium potential for potassium because the membrane is highly permeable to potassium. However, we learned that the membrane potential can change by changing membrane permeability to an ion. These changes in membrane potential occur in the form of graded potentials and action potentials. Action potentials in a single neuron transmit information from one end of the neuron (axon hillock) to the other (axon terminal). The role of this action potential propagation in communication between neurons is the focus of the next chapter.

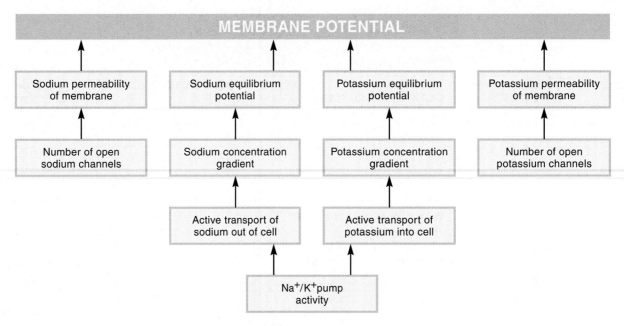

FIGURE 6.21 **Summary of factors that influence the membrane potential.**

CHAPTER SUMMARY

Overview of the Nervous System, p. 171

The nervous system can be divided into the central nervous system and peripheral nervous system. The central nervous system consists of the brain and spinal cord. The peripheral nervous system includes the afferent and efferent divisions. The afferent division consists of neurons that transmit information from the periphery to the central nervous system, whereas the efferent division consists of neurons that transmit information from the central nervous system to the periphery. The efferent division is divided into two main branches: the somatic nervous system, which communicates to skeletal muscle, and the autonomic nervous system, which communicates to smooth muscle, cardiac muscle, glands, and adipose tissue. The autonomic nervous system is divided into the sympathetic and parasympathetic nervous systems.

Cells of the Nervous System, p. 172

The nervous system contains neurons, which are cells specialized for transmitting electrical impulses, and glial cells, which provide metabolic and structural support to the neurons. Parts of a neuron include the cell body, dendrites, and the axon. The dendrites, and to a lesser extent the cell body, receive information from other neurons at synapses. The axon includes an axon hillock, where electrical impulses (action potentials) are initiated, and an axon terminal. The axon terminal transmits information via neurotransmitters to other neurons at synapses.

Neurons are classified functionally into three classes: Efferent neurons transmit information from the central nervous system to effector organs, afferent neurons transmit information from sensory or visceral organs to the central nervous system, and interneurons communicate within the central nervous system.

Two types of glial cells function in forming myelin around axons: oligodendrocytes in the central nervous system, and Schwann cells in the peripheral nervous system. Myelin enhances the propagation of electrical impulses by providing insulation to the axon.

IP Nervous I, Anatomy Review, pages 4–6

IP Nervous I, Ion Channels, pages 3–11

Electrical Signals in Neurons, p. 177

At rest, cells have a membrane potential across them such that the inside of the cell is negative relative to the outside. Changes in the membrane potential can be produced by changing the permeability of the plasma membrane to ions. Graded potentials are small changes in membrane potential in response to a stimulus that opens or closes ion channels. If graded potentials result in a depolarization of the neuron to threshold, an action potential is produced. A single graded potential is usually not of sufficient magnitude to depolarize a neuron to threshold, but graded potentials can be temporally and/or spatially additive.

IP Nervous I, The Membrane Potential, pages 1–13

IP Nervous II, Synaptic Potentials and Cellular Integration, pages 1–10

Action Potentials, p. 187

Action potentials are rapid depolarizations of the plasma membrane that are propagated along axons from the trigger zone to the axon terminal. The rapid depolarization phase of an action potential is caused by the opening of sodium channels and sodium ion movement into the cell. The repolarization phase is caused by the closing of sodium channels and the opening of potassium channels, followed by potassium movement out of the cell. After-hyperpolarization occurs because potassium channels are slow in closing, allowing continued movement of potassium out of the cell for a brief period of time.

Action potentials are all-or-none phenomena, meaning that their size does not vary with the strength of the stimulus eliciting them. The strength of a stimulus is coded by the frequency of action potentials; stronger stimuli produce more action potentials per unit of time. Refractory periods ensure the unidirectional flow of action potentials and limit the frequency of action potentials. The absolute refractory period occurs during all of the depolarization phase and most of the repolarization phase of an action potential and is caused by closed sodium inactivation gates. During the absolute refractory period, a second action potential cannot be produced, regardless of stimulus strength. The relative refractory period immediately follows the absolute refractory period and lasts until the end of the after-hyperpolarization phase. During the relative refractory period, more potassium channels are open than when a neuron is at rest; thus a second action potential can be elicited, but the stimulus must be a suprathreshold stimulus to overcome the increased potassium movement.

IP Nervous I, The Action Potential, pages 4–17*

The Basis of Neuronal Stability, p. 187

The Na^+/K^+ pump is crucial to the normal function of neurons because it establishes the concentration gradients for sodium and potassium ions, thereby generating the chemical gradients that establish the resting membrane potential. The pump also prevents dissipation of the concentration gradients by returning sodium and potassium ions that have crossed the membrane (either through leak channels at rest or through gated channels during activity) to their original sides.

---

* This topic is available on the *InterActive Physiology*® *Sampler CD* that came with the purchase of a new copy of this book.

EXERCISES

Multiple-Choice Questions

1. Depolarization of a neuron to threshold stimulates
   a) opening of sodium channels.
   b) closing of sodium channels.
   c) opening of potassium channels.
   d) all of the above

2. Neurotransmitters are released most commonly from the
   a) cell body.
   b) dendrites.
   c) axon terminals.
   d) axon hillock.

3. If a cation is equally distributed across the cell membrane (that is, its concentration inside the cell equals its concentration outside the cell), then which of the following statements is *false*?
   a) At −70 mV, the chemical force on the ion is zero.
   b) At +30 mV, the chemical force on the ion is zero.
   c) The equilibrium potential for the ion would be 0.
   d) At −70 mV, the electrochemical force on the ion acts to move it out of the cell.

4. The depolarization phase of an action potential is caused by the
   a) opening of potassium channels.
   b) closing of potassium channels.
   c) opening of sodium channels.
   d) closing of sodium channels.

5. During the relative refractory period, a second action potential
   a) cannot be elicited.
   b) can be elicited by a threshold stimulus.
   c) can be elicited by a subthreshold stimulus.
   d) can be elicited by a suprathreshold stimulus.

6. Nerves are found
   a) in the central nervous system.
   b) in the peripheral nervous system.
   c) both a and b
   d) neither a nor b

7. If the membrane potential of a neuron becomes more negative than it was at rest, then the neuron is _____. In this state, the neuron is _____ excitable.
   a) depolarized/more
   b) hyperpolarized/more
   c) depolarized/less
   d) hyperpolarized/less

8. Oubain is a poison that blocks the Na^+/K^+ pump. If this pump is blocked, then the concentration of potassium inside the cell would
   a) increase.
   b) decrease.
   c) not change.

9. If potassium concentrations in the extracellular fluid of the brain increased, activity in the brain would
   a) increase.
   b) decrease.
   c) not change.

10. Which of the following neurons is part of the peripheral nervous system?
    a) motor neuron innervating skeletal muscle
    b) parasympathetic neuron
    c) sympathetic neuron
    d) all of the above

Objective Questions

1. What are the subdivisions of the peripheral nervous system?

2. Information from the periphery is brought to the central nervous system by (afferent/efferent) pathways.

3. Which cell type is more abundant in the nervous system, glial cells or neurons?

4. Voltage-gated calcium channels are located at what region(s) of a neuron?

5. (Schwann cells/Oligodendrocytes) form myelin in the peripheral nervous system, and (Schwann cells/ oligodendrocytes) form myelin in the central nervous system.

6. Myelin (increases/decreases) conduction velocity in axons.

7. If an anion (Z^-) is located in greater concentration outside the cell compared to inside, would the equilibrium potential for that anion be positive, negative, or zero?

8. Which ion is closer to equilibrium at the resting membrane potential of -70 mV, sodium or potassium?

9. In the peripheral nervous system, cell bodies of afferent neurons are located in _____.

10. When one neuron makes synapses with many other neurons, this is called _____. When one neuron receives synaptic communication from many other neurons, this is called _____.

11. Both sodium and potassium channels have inactivation gates that close shortly after the activation gates open. (true or false)

12. When sodium inactivation gates are closed, a second action potential is impossible. (true or false)

13. In myelinated axons, action potentials are propagated by _____ conduction.

14. The Na^+/K^+ pump causes the repolarization phase of an action potential. (true or false)

15. When a neuron is at the peak of an action potential ($+30$ mV), the direction of the electrical force for potassium ions is (into/out of) the cell.

Essay Questions

1. Draw a typical neuron and label the main structures. Then list the functions of each structure.

2. Compare the chemical and electrical forces acting on sodium and potassium ions when a cell is at rest.

3. List some similarities and some differences between graded potentials and action potentials.

4. Explain the ionic basis of an action potential.

5. Predict what would happen to the resting membrane potential and the ability of a neuron to elicit action potentials if the concentration of potassium in the extracellular fluid was decreased to 50% of its normal value. What would happen if the concentration of sodium in the extracellular fluid was decreased to 50% of its normal value?

Find the answers to these exercises, and additional study tools, at the Physiology Place (www.physiologyplace.com).

7

Synaptic Transmission and Neural Integration

OBJECTIVES

- Describe the communication across chemical synapses. Explain how neurotransmitters are released and describe their actions after release.

- Compare fast and slow responses at synapses.

- Describe the process of neural integration and the role of the axon hillock in this process.

- Describe the major classes of neurotransmitters, including chemical structure, synthesis, degradation, and signal transduction mechanisms.

CHAPTER OUTLINE

Types of Synapses   202

Chemical Synapses   202

Neural Integration   210

Presynaptic Modulation   212

Neurotransmitters: Structure, Synthesis, and Degradation   214

Above: Transmission electron micrograph of an axodendritic junction synapse

Joe is the running back on his college football team. During one game he takes a handoff from the quarterback but is immediately hit low from the side by a defensive lineman. The impact breaks his leg, a serious injury that puts Joe in a lot of pain. He is taken off the field and then to the hospital, where a doctor gives him morphine. Within minutes, his pain starts to subside. Why does morphine, a product of the opium poppy, relieve Joe's pain? The answer is related to the ways neurons communicate via neurotransmitters. Morphine alleviates pain because it can bind to certain neurons in the central nervous system at receptors for naturally occurring morphine-like messenger molecules called enkephalins and endorphins. As a result of this binding, morphine suppresses the transmission of pain signals.

TYPES OF SYNAPSES

We saw in Chapter 6 that neurons communicate at specialized structures called *synapses*. There are two types of synapses in the nervous system: electrical synapses and chemical synapses. **Electrical synapses** operate by allowing electrical signals to be transmitted directly from one neuron to another; neurotransmitters are not involved. At these synapses the plasma membranes of adjacent cells are linked together by gap junctions (see Chapter 2, p. 45) such that when an electrical signal, such as an action potential, is generated in one cell, it is directly transferred to the adjacent cell by means of ions flowing through the gap junctions. Because the function of electrical synapses in the nervous system is not well understood, they will not be discussed any further here. However, we will encounter gap junctions again in Chapter 11 in regard to smooth and cardiac muscle, in which gap junctions have an important role in enabling these

muscles to contract as a unit. We now examine the anatomical and functional characteristics of chemical synapses.

CHEMICAL SYNAPSES

Almost all neurons transmit messages to other cells at **chemical synapses.** In a chemical synapse, one neuron secretes a neurotransmitter into the extracellular fluid in response to an action potential arriving at its axon terminal. The neurotransmitter then binds to receptors on the plasma membrane of a second cell, triggering in that cell an electrical signal that may or may not initiate an action potential, depending on a number of circumstances.

A neuron can form synapses with other neurons—the situation that is the topic of this chapter—or with effector cells such as muscle or gland cells. Muscles and glands are generally referred to as *effector organs,* and a synapse between a neuron and an effector cell is called a *neuroeffector junction.* Even though neuroeffector junctions are examined in Chapter 10, they operate according to the same basic principles presented here concerning neuron-to-neuron synapses. In a few instances, non-neuronal cells can form synapses with neurons, as occurs, for example, with taste receptor cells in taste buds on the tongue (see Chapter 9).

Functional Anatomy of Chemical Synapses

Figure 7.1 depicts the possible arrangements in typical neuron-to-neuron synapses, close junctions between the axon terminal of one neuron and the plasma membrane of another neuron. The first neuron, which transmits signals to the second, is designated the **presynaptic neuron;** the second neuron, which receives signals from the first, is referred to as the **postsynaptic neuron.** Most often the presynaptic neuron's axon terminal forms a synapse with either a dendrite or the cell body of the postsynaptic neuron, in which case the synapses are referred to as **axodendritic** or **axosomatic synapses,** respectively. In some cases the presynaptic neuron's axon terminal forms a synapse with the postsynaptic neuron's axon terminal, in which case the synapse is called an **axoaxonic synapse.** In all cases, the axon terminal of the presynaptic neuron releases neurotransmitters into the narrow (30–50 nm) space between the two cells, called the **synaptic cleft** (Figure 7.2a). Once released into the synaptic cleft, the neurotransmitters diffuse rapidly across the cleft and bind to receptors on the postsynaptic neuron. The binding of the neurotransmitter to the receptors produces a response in the postsynaptic neuron by a mechanism called *signal transduction.*

Note that the axon terminal of the presynaptic neuron contains numerous small, membrane-bound compartments called **synaptic vesicles,** which store neurotransmitter molecules. (For more information about synaptic vesicles, see Discovery: Synaptic Vesicles, p. 206.) Most neurotransmitters are synthesized in the cytosol of the axon terminal, where the enzymes for their synthesis are located. After synthesis, neurotransmitters are actively transported into synaptic vesicles, where they are stored until their eventual release by exocytosis. But what triggers this exocytosis?

Recall from Chapter 6 that the membrane of a neuron contains ion channels of various types that depend on their location in the neuron and that location's specific functions. We saw that voltage-gated sodium and potassium channels are abundant in the membrane of the axon to support the propagation of action potentials down the length of the axon. But most abundant in the membrane of the axon terminal are voltage-gated calcium channels. These calcium channels open when the axon terminal is depolarized, which occurs upon arrival of an action potential at the axon terminal (see ① in Figure 7.2b). When the calcium channels open ②, they allow calcium to flow down its electrochemical gradient into the axon terminal, increasing the concentration of cytosolic calcium in the axon terminal. Calcium then causes the membranes of synaptic vesicles to fuse with vesicle attachment sites on the inner surface of the axon terminal membrane and undergo exocytosis, thereby releasing the neurotransmitters into the synaptic cleft ③.

Unless another action potential arrives, neurotransmitter release stops within a few milliseconds, because the voltage-gated calcium channels close soon after opening, and because calcium ions are actively transported out of the axon terminal on a continual basis, bringing the cytosolic calcium concentration back to its resting level. A second action potential will elicit a quick burst of neurotransmitter release similar to the first. When a series of action potentials arrives at an axon terminal, the amount of neurotransmitter released from the presynaptic cell is greater than the amount released in response to a single action potential, and as a result the concentration of neurotransmitter in the synaptic cleft increases as the frequency of action potentials increases.

Once in the synaptic cleft, neurotransmitter molecules diffuse away from the axon terminal and toward the postsynaptic neuron, where they bind to receptors (see ④ in Figure 7.2b), thereby inducing a response in the postsynaptic neuron ⑤. The binding of a neurotransmitter molecule to a receptor is a brief and reversible process. The neurotransmitter is a ligand and the receptor is a protein, and the two bind through weak chemical interactions. If neurotransmitter molecules were to remain indefinitely in the synaptic cleft following their release, they would bind to receptors over and over again, inducing a continual response in the postsynaptic neuron. However, continual binding does not occur

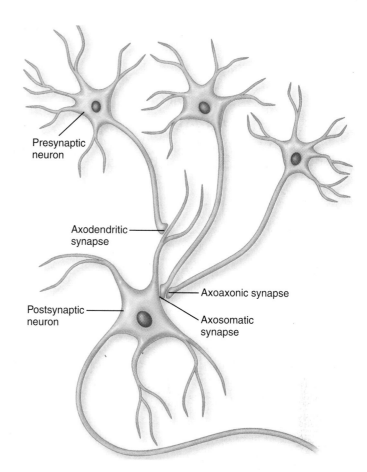

FIGURE 7.1 Neuron-to-neuron chemical synapses. *Synapses can occur at dendrites (axodendritic synapses), at the cell body (axosomatic synapses), or with another axon (axoaxonic synapses).*

because a number of processes quickly clear the neurotransmitter from the cleft, thereby terminating the signal.

Some neurotransmitter molecules are degraded by enzymes, which may be located either on the postsynaptic neuron's plasma membrane (see ⑥ in Figure 7.2b), on the presynaptic neuron's plasma membrane, on the plasma membranes of nearby glial cells, or even in the cytoplasm of the presynaptic neuron or glial cells. Other neurotransmitter molecules can be actively transported back into the presynaptic neuron that released them ⑦, a process known as **reuptake.** Once inside the neuron, these neurotransmitter molecules are usually degraded and the breakdown products recycled to form new neurotransmitter molecules. Still other neurotransmitter molecules in the synaptic cleft simply diffuse out of the cleft ⑧. So long as neurotransmitter molecules are not bound to receptors, they may be subject to any of the fates just described. As a result, neurotransmitter is usually present in the synaptic cleft for only a few milliseconds after its release from the presynaptic neuron.

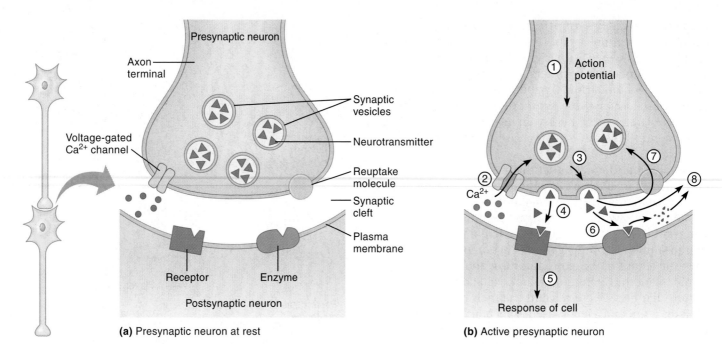

(a) Presynaptic neuron at rest

(b) Active presynaptic neuron

FIGURE 7.2 Functional anatomy of a synapse. (a) *The presynaptic neuron at rest. Calcium channels are closed, and no neurotransmitter is being released.* **(b)** *An active presynaptic neuron. Communication occurs as follows:* ① *The membrane is depolarized by the arrival of an action potential.* ② *Calcium channels open.* ③ *Calcium triggers the release of neurotransmitter by exocytosis.* ④ *Neurotransmitter diffuses across the synaptic cleft; some of it binds to receptors on the postsynaptic cell membrane.* ⑤ *A response is produced in the postsynaptic cell.* ⑥ *Some neurotransmitter is degraded by enzymes.* ⑦ *Some neurotransmitter is taken up by the presynaptic cell.* ⑧ *Some neurotransmitter (or products of its degradation) diffuses away from the synaptic cleft.*

It takes approximately 1–5 msec from the time an action potential arrives at the axon terminal before a response occurs in the postsynaptic cell. This time lag, called the **synaptic delay,** is mostly due to the time required for calcium to trigger the exocytosis of neurotransmitter. Once in the synaptic cleft, diffusion of neurotransmitter to the receptor is so rapid as to be negligible.

Quick Test 7.1

1. What are the essential differences between a chemical synapse and an electrical synapse?

2. List the steps involved in the transmission of signals at a chemical synapse.

3. What is the meaning of the term *synaptic delay*? What events are occurring during this time?

4. Following its release, a neurotransmitter can be cleared from a synaptic cleft by which three processes?

Signal Transduction Mechanisms at Chemical Synapses

The neurotransmitter released by a presynaptic neuron is a messenger; when it binds to receptors on the postsynaptic neuron, a message is conveyed. Neurotransmitters can induce in a postsynaptic neuron either a fast or a slow response. The **fast response** (Figure 7.3a) occurs whenever a neurotransmitter binds to a **channel-linked receptor,** also called an **ionotropic receptor.** All channel-linked receptors are ligand-gated channels (see Chapter 6). The binding of the neurotransmitter opens the ion channel, allowing a specific ion or ions to permeate the plasma membrane and change the electrical properties of the postsynaptic neuron. The typical response is a change in the membrane potential, called a **postsynaptic potential,** or **PSP,** which occurs very rapidly and turns off rapidly (normally within a few milliseconds) because the channel closes as soon as the neurotransmitter leaves the receptor.

Slow responses, on the other hand, are mediated through G-protein-linked receptors called **metabotropic receptors,** which include any receptor that works by

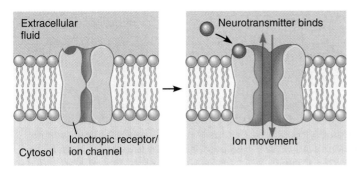

(a) Fast response

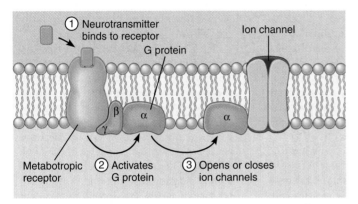

(b) Slow response, direct coupling

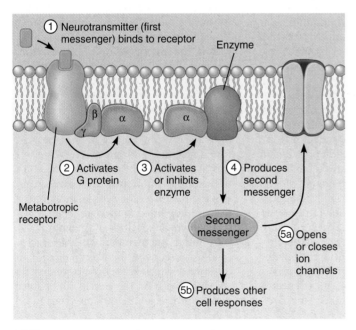

(c) Slow response, second messenger system

FIGURE 7.3 Signal transduction mechanisms at chemical synapses. (a) *Fast responses at a channel-linked receptor. The panel at left shows the closed state of the channel in the absence of neurotransmitter; the panel at right shows the opening of the ion channel by the binding of neurotransmitter to receptor.* **(b, c)** *Slow responses, in which the neurotransmitter receptor is coupled to a G protein. In both slow responses, binding of neurotransmitter to receptor activates a G protein. In direct coupling, the G protein opens or closes an ion channel, thereby changing the electrical properties of the cell. When the G protein functions as part of a second messenger system, it either activates or inhibits an enzyme that produces the second messenger, which then either opens or closes an ion channel or produces some other response in the cell.*

triggering biochemical changes rather than simply inducing a change in permeability. In the nervous system, G proteins can trigger either the opening or the closing of ion channels, depending on the specific synapse. These G-protein-regulated ion channels respond to the binding of neurotransmitter more slowly than the channels that mediate the fast response, with durations ranging from milliseconds to hours, depending on the synapse, but with the same ultimate effect: The opening or closing of ion channels alters the permeability of the postsynaptic neuron to a specific ion or ions, resulting in changes in the electrical properties of the cell. The G protein can serve as a direct coupling between the receptor and the ion channel (Figure 7.3b), or it can trigger the activation or inhibition of a second messenger system (Figure 7.3c). Neurotransmitters that act through G proteins are also called **neuromodulators** because some of the slow responses modulate the fast responses occurring at ionotropic receptors.

We have seen that, in general, a neurotransmitter exerts its effects on a postsynaptic neuron by triggering the opening or closing of ion channels, which typically causes a change in membrane potential. Depending on the type of channel that is opened or closed, the resulting change in membrane potential may be positive or negative; that is, it may cause either a depolarization or a hyperpolarization, respectively. At any given synapse, the direction of the response is always the same, and the synapse is classified as being either excitatory or inhibitory.

Excitatory Synapses

An **excitatory synapse** is one that brings the membrane potential of the postsynaptic neuron closer to the threshold for generating an action potential; that is, excitatory synapses depolarize the postsynaptic neuron. This depolarization is called an **excitatory postsynaptic potential,** or **EPSP,** and can occur as either a fast response or a slow response. EPSPs are graded potentials, with the

Neurotransmitters are packaged and stored in synaptic vesicles in the axon terminals until calcium triggers their release by exocytosis. Let's take a closer look at synaptic vesicles and the neurotransmitters within them.

Synaptic vesicles are initially synthesized as empty (no neurotransmitter) membrane vesicles in the cell body. Following synthesis, the synaptic vesicles must move to the axon terminal, which can be up to a meter away! Because diffusion is much too slow a process, synaptic vesicles are actively transported from cell body to axon terminal.

Neurons have two basic mechanisms for moving products either from the cell body to the axon terminal *(anterograde transport)* or from the axon terminal to the cell body *(retrograde transport)*: (1) *slow axonal transport* and (2) *fast axonal transport.* Both slow and fast axonal transport involve proteins, including microtubules and a variety of neurofilaments. Slow axonal transport (0.5–40 mm/day) is generally associated with movement of soluble components of the cytosol, whereas fast axonal transport (100–400 mm/day) is associated with movement of vesicles, including synaptic vesicles.

Fast axonal transport of synaptic vesicles is shown in **(a).** Microtubules extend the length of the axon and function as "tracks" for transport molecules. One type of molecule that runs on these tracks are *kinesins,* proteins that essentially "walk" down the microtubules, carrying with them a vesicle or organelle. The vesicle can be a synaptic vesicle, or the vesicle can be used to transport other products from the site of synthesis in the cell body to the axon terminal. The process requires ATP.

Once in the axon terminal, neurotransmitter must be concentrated in the synaptic vesicle. The exact mechanism of transport into the vesicle depends on the neurotransmitter. Usually, neurotransmitters are synthesized in the cytosol. Following synthesis, the neurotransmitter is actively transported into the synaptic vesicle by specific carrier molecules. The amount of neurotransmitter packed into a vesicle, called a *quantum,* is generally the same for all vesicles in the cell. (A **quantum** means a specific or definite amount of something, in this case, a specific number of neurotransmitter molecules.)

Evidence of quantal release comes from studies on the spontaneous release of neurotransmitter (acetylcholine) from the motor neurons that innervate skeletal muscles. The spontaneous release of neurotransmitter involves exocytosis in the absence of an action potential in the presynaptic cell. With spontaneous release, only one or very few vesicles will release their neurotransmitter at a given time. Recordings of the skeletal muscle cells show small *quantal* changes in the membrane potential. The minimum change in membrane potential is 0.4 mV. Other spontaneous changes in membrane potential are multiples of 0.4 mV; 0.8, 1.2, 1.6, and so on. Therefore, one synaptic vesicle in motor neurons contains enough acetylcholine to produce a depolarization of 0.4 mV in the skeletal muscle cell.

This analysis can be extended. Estimates suggest that one action potential in the motor neuron stimulates the release of approximately 100–300 quanta of acetylcholine; that is, 100–300 synaptic vesicles release acetylcholine in response to a single action potential. Scientists have also estimated how many acetylcholine molecules are in one quantum. Estimates vary considerably, but there are approximately 5,000–10,000 molecules of acetylcholine in each vesicle. These values are specifically for the synapse between a motor neuron and a muscle cell. In the central nervous system, where synaptic potentials are much smaller, only 1–20 quanta of neurotransmitter are released with each action potential.

An action potential triggers release of neurotransmitter by opening voltage-gated calcium channels in the axon terminal, allowing calcium to enter the cell. Calcium then triggers exocytosis (see **(b)**) by a mechanism not yet fully understood. The site of neurotransmitter release on the membrane of the axon terminal is called the *active zone.* There are various types of proteins located at the active zone, including *docking proteins.* There are two pools of synaptic vesicles in the axon terminal. One pool is docked to the active zone, ready to release the neurotransmitter by exocytosis. The second pool is clustered together in the axon terminal to function as a reserve. The reserve synaptic vesicles will move to the active zone after those already at the active zone have released their neurotransmitter.

Calcium is believed to trigger both the exocytosis of neurotransmitter and the mobilization of reserve synaptic vesicles to the active zone. During exocytosis, calcium causes interactions between membrane proteins and synaptic vesicle proteins, causing the two membranes to fuse. Calcium then triggers reserve vesicles to move into the active zone and bind to the docking proteins. This process involves proteins of the cytoskeleton, but little is known of the mechanism.

Following exocytosis of the neurotransmitter, the membrane of the vesicle that just fused with the plasma membrane during exocytosis is recycled by endocytosis. Once recycled, the vesicle can be filled with neurotransmitter and used again.

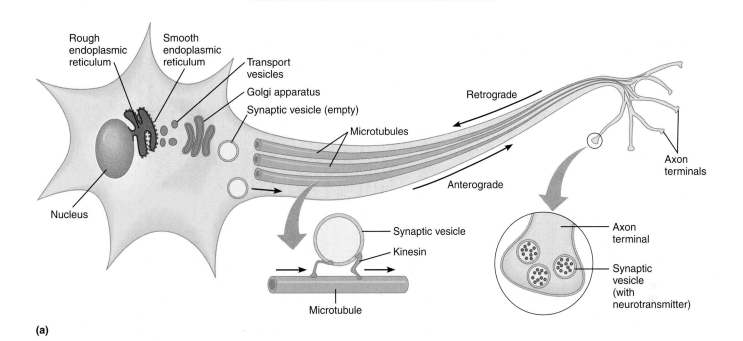

Rough endoplasmic reticulum

Smooth endoplasmic reticulum

Transport vesicles

Golgi apparatus

Synaptic vesicle (empty)

Microtubules

Retrograde

Anterograde

Axon terminals

Nucleus

Synaptic vesicle

Kinesin

Microtubule

Axon terminal

Synaptic vesicle (with neurotransmitter)

(a)

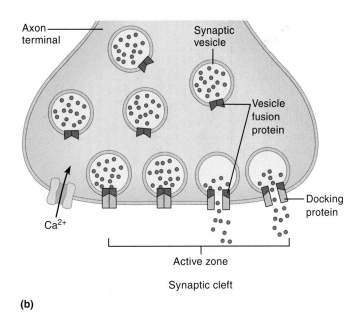

Axon terminal

Synaptic vesicle

Vesicle fusion protein

Ca²⁺

Docking protein

Active zone

Synaptic cleft

(b)

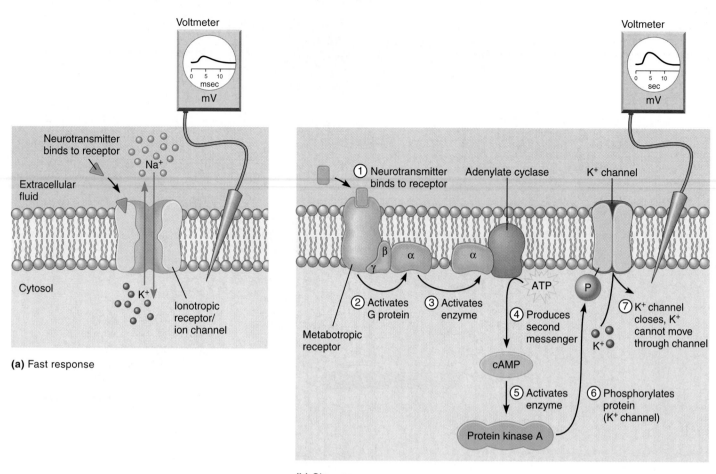

FIGURE 7.4 Excitatory synapses. *An electrode inserted into a neuron and connected to a voltmeter can measure electrical activity in the cell. The type of voltmeter depicted in this and subsequent figures is an oscilloscope.* **(a)** *A fast excitatory synapse. The neurotransmitter opens ion channels, allowing Na^+ to enter the cell and K^+ to leave it. Sodium movement is greater, so the net effect is a depolarization (an EPSP) lasting only 1–5 msec.* **(b)** *A slow excitatory synapse. Activation of a G protein by neurotransmitter leads to the series of events depicted. Phosphorylation of the potassium channel in the final step decreases the leakage of potassium out of the cell, producing a depolarization. Note that this response takes seconds to occur.*

amplitude of depolarization increasing as more neurotransmitter molecules bind to receptors. As with all graded potentials, the depolarization is greatest at the site of origin (usually a dendrite or cell body) and decreases as it moves toward the axon hillock.

Fast EPSPs are generally caused by the binding of neurotransmitter molecules to their receptors on the postsynaptic cell, in the process opening channels that allow small cations (sodium and potassium ions) to move through them (Figure 7.4a). Recall that at rest, sodium is far from equilibrium, and that there is a strong electrochemical force for moving sodium ions into the cell, whereas because potassium is near equilibrium only a weak electrochemical force exists for moving potassium ions out of the cell. Therefore, when channels open, allowing both sodium and potassium ions to move through them, more sodium ions move in than potassium ions move out, producing a net depolarization. Note on

the voltmeter (oscilloscope) in Figure 7.4a that fast EPSPs last only milliseconds.

Slow EPSPs can be caused by various mechanisms, but a typical mechanism is the closing of potassium channels involving the most common second messenger, cAMP (Figure 7.4b). When the neurotransmitter binds to a receptor ① and activates a G protein ②, the G protein then activates the enzyme adenylate cyclase ③, which then catalyzes the reaction converting ATP to cAMP ④. cAMP acts as a second messenger that activates protein kinase A ⑤, which catalyzes the addition of a phosphate group to the potassium channel ⑥. Phosphorylation of the potassium channel closes it ⑦.

Before the potassium channel was phosphorylated, the channel was open, and potassium ions were leaking out of the cell while sodium ions were leaking into the cell. No change in membrane potential was occurring because the sodium and potassium leaks were balanced.

When the potassium channel is closed by phosphorylation, fewer potassium ions leak out of the cell. The net effect is a depolarization, because if permeability to potassium decreases while permeability to sodium remains the same, the membrane potential will move away from the equilibrium potential for potassium and toward the equilibrium potential for sodium (see Chapter 6). Slow EPSPs take longer to develop and last longer (seconds to hours) than fast EPSPs. In the case of cAMP-dependent closure of potassium channels, the membrane potential will not return to resting conditions until the potassium channels are dephosphorylated, which will not occur until enough cAMP has been degraded by phosphodiesterase to return the cAMP concentration to its resting level.

Inhibitory Synapses

An **inhibitory synapse** is one that either takes the membrane potential of the postsynaptic neuron away from the action potential threshold by hyperpolarizing the neuron, or stabilizes the membrane potential at the resting value. In either case, activity at the synapse decreases the likelihood that an action potential will be generated in the postsynaptic neuron—which is why the synapse is said to be *inhibitory*.

At inhibitory synapses, the binding of a neurotransmitter to its receptors opens channels for either potassium ions or chloride ions. When a neurotransmitter causes potassium channels to open (Figure 7.5), potassium will move out of the cell, hyperpolarizing it. This hyperpolarization is called an **inhibitory post-synaptic potential,** or **IPSP.** An IPSP, like an EPSP, is a graded potential: When more neurotransmitter molecules bind to receptors, more potassium channels open and the hyperpolarization is greater. IPSPs (like EPSPs and all other graded potentials) decrease in size as they move toward the axon hillock, where the hyperpolarization moves the membrane potential away from threshold, thereby decreasing the likelihood that an action potential will be generated in the postsynaptic neuron.

What happens when a neurotransmitter causes chloride (Cl^-) channels to open depends on the size and direction of the electrochemical driving force acting to move chloride ions across the postsynaptic neuron membrane. Because the chloride ion has a negative charge (is an anion), the membrane potential at rest is an electrical force that acts to move chloride out of the cell. However, because chloride ions are found in greater concentration outside the cell than inside, there is a chemical force for moving chloride into the cell. The net force depends on the neuron in question. In some neurons, chloride ions are actively transported out of the cell, making the concentration outside much higher than inside the cell. Under these conditions, the equilibrium potential for chloride is negative and larger than the resting membrane potential. In other neurons no active transport of chloride occurs, but there are some ion channels through which

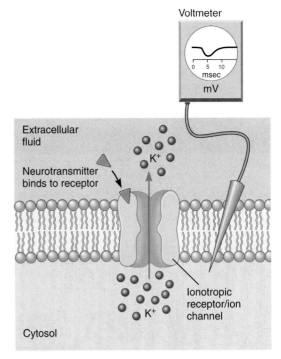

FIGURE 7.5 An inhibitory synapse involving potassium channels. *At this type of inhibitory synapse, potassium channels are opened by the binding of neurotransmitter to the receptor. Potassium flows out of the cell, hyperpolarizing it and producing an IPSP.*

chloride can diffuse. Under these conditions, the passive movement of chloride coupled with the lack of active transport allows chloride ions to be at equilibrium.

In a neuron in which chloride ions are actively transported out of the cell, chloride is concentrated in the extracellular fluid (Figure 7.6a). Therefore, chloride is not at equilibrium; instead, a net electrochemical driving force for moving chloride into the cell exists. When chloride channels are opened by neurotransmitters binding to receptors, chloride moves into the cell, producing a hyperpolarization, or an IPSP. Thus we see the same type of effect that occurs when potassium channels open.

In a neuron that lacks active transport of chloride either into or out of the cell (Figure 7.6b), chloride will simply diffuse through chloride leak channels to distribute itself across the membrane at concentrations that will put it at equilibrium. (Note that the membrane potential determines the concentration gradient for chloride ions, which are not actively transported; this is in contrast to sodium and potassium ions, whose concentration gradients, established by active transport, determine the resting membrane potential.) If chloride is at equilibrium, then the opening of chloride channels will not cause a change in the membrane potential (Figure 7.6b). In this case, perhaps surprisingly, the synapse is still considered to be *inhibitory*. To see why, we must consider how this synapse affects signals that might be generated at other synapses at the same time.

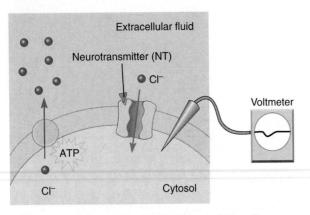

(a) Neuron actively transports chloride out of the cell

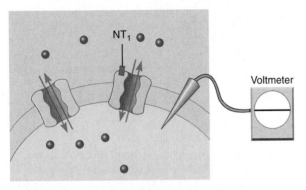

(b) No active transport of chloride, chloride at equilibrium

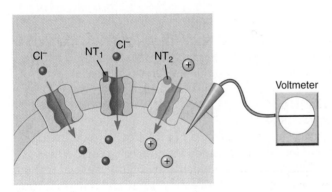

(c) No active transport of chloride, excitatory synapse active

FIGURE 7.6 The roles of chloride channels in inhibitory synapses. (a) *In neurons that actively transport chloride out of the cell, binding of a neurotransmitter that causes chloride channels to open allows chloride to move into the cell, hperpolarizing it (inducing an IPSP).* **(b)** *Cells that lack chloride transporters have chloride leak channels, and chloride is at equilibrium across their membranes. If binding of neurotransmitter 1 (NT$_1$) causes more chloride channels to open, no net movement of chloride occurs because chloride is already at equilibrium.* **(c)** *This figure shows the same cell as in panel b except it includes a second synapse. If NT$_1$ is present at an inhibitory synapse the same time that NT$_2$ is present at an excitatory synapse, chloride will move into the cell at the same time that positive charges enter the channel opened by NT$_2$, opposing any change in membrane potential.*

In part (a), what is the equilibrium potential for chloride ions?

Consider a postsynaptic neuron in which an excitatory synapse and an inhibitory synapse are active at the same time. While the excitatory synapse tends to induce an EPSP (a positive change in membrane potential) in the postsynaptic neuron, the inhibitory synapse also causes chloride channels to open (Figure 7.6c). The result is that the EPSP will be diminished, or may even be absent. The reason for this is that when the membrane becomes depolarized, chloride is no longer at equilibrium. The more positive potential inside the cell pulls the negative chloride ions into the cell, stabilizing the membrane potential and countering the influence of positive charge moving into the cell. This is considered an inhibitory action because it decreases the likelihood that the neuron will reach threshold for an action potential.

As this example illustrates, the activity of a single synapse may influence the likelihood that an action potential will occur, but it usually is not the sole determining factor. Instead, many synapses are normally active at the same time, and it is the particular *combination* of synaptic inputs arriving at any given time that determines whether or not a postsynaptic cell will produce an action potential. The process that produces this determination, known as *neural integration*, is our next topic of study.

Quick Test 7.2

1. Activation of which type of receptor—ionotropic or metabotropic—produces the faster response? Which type of receptor is more likely to be involved in motor control (that is, the ability to move skeletal muscles)?

2. What is the role of G proteins in the operation of metabotropic receptors?

3. If a G protein activates a second messenger that closes sodium channels, will this activation produce an EPSP or an IPSP? Will the response be fast or slow?

NEURAL INTEGRATION

Because a given neuron receives communication from many neurons, one synapse alone generally does not determine whether or not an action potential is generated in that cell. Instead, the axon hillock of the postsynaptic neuron acts as an integrator that in effect adds up all the signals arriving from all active synapses. This summation process is called **neural integration,** and it operates according to one simple rule: *An action potential is triggered if the membrane potential at the axon hillock is depolarized to threshold; if the potential is below threshold, no action potential will occur.*

The resting membrane potential, which is approximately −70 mV

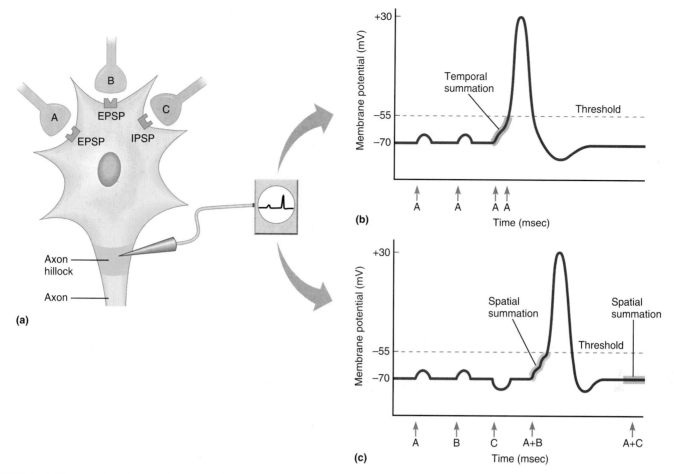

FIGURE 7.7 Temporal and spatial summation. (a) *A postsynaptic neuron receiving excitatory input from neurons A and B, and inhibitory input from neuron C; three different neurotransmitters are involved. If the sum of all synaptic potentials at the axon hillock results in depolarization to threshold, then an action potential is generated.* **(b)** *Temporal summation occurs when action potentials arriving from a presynaptic axon terminal of neuron A occur close enough together in time such that the EPSPs produced in response to the binding of neurotransmitter overlap and sum.* **(c)** *Spatial summation occurs when different synapses are simultaneously active. If EPSPs from synapses with neurons A and B occur at the same time, the EPSPs sum and reach threshold, triggering an action potential. Note that if synapses with neurons A and C were simultaneously active, the resulting IPSP and EPSP would tend to cancel each other out, producing little change in membrane potential.*

The synapse of neuron A with the postsynaptic cell is closer to the axon hillock than is the synapse of neuron B. If a single action potential in each presynaptic neuron results in an 8-mV depolarization at the axon hillock, which presynaptic neuron had to produce the larger depolarization at the site of the synapse?

Summation

Because EPSPs and IPSPs are graded potentials, they are capable of being added together (see Chapter 6). This summation can be either *spatial* or *temporal*, depending on whether the postsynaptic potentials being summed arise at the same synapse or at different synapses.

Temporal Summation

In temporal summation, two or more postsynaptic potentials are generated in rapid succession at the same synapse, such that a postsynaptic potential does not have time to fully dissipate before the next one is generated. When this occurs, the postsynaptic potentials overlap such that the resulting depolarization or hyperpolarization is larger than for a single postsynaptic potential. Figure 7.7a depicts measurement of the membrane potential at the axon hillock of a postsynaptic neuron that receives input from three presynaptic neurons (A, B, and C). The graph

Neuron B, because the resulting graded potential would dissipate more by the time it reached the axon hillock

in Figure 7.7b shows the events in temporal summation. In response to an action potential in neuron A, the postsynaptic cell has an EPSP of about 8 mV, which is not enough to depolarize the membrane to threshold. Once this EPSP has terminated and the membrane potential has returned to the resting value, a second action potential in neuron A triggers another EPSP of the same size. However, when two action potentials in neuron A occur closer together in time (within a few msec) such that the second EPSP is produced before the first has had a chance to dissipate, the two EPSPs sum. In this case, the sum of the two EPSPs is large enough to depolarize the postsynaptic neuron's membrane to threshold and generate an action potential.

Note that IPSPs can also sum temporally in a similar manner. When two IPSPs generated at an inhibitory synapse occur close together in time, they sum such that the resulting hyperpolarization is larger than that produced by a single IPSP. However, the degree of hyperpolarization that can be reached is limited by the equilibrium potential for the ion causing the hyperpolarization. When, for example, the opening of potassium channels produces a hyperpolarization, the more potassium channels that are open, the greater the hyperpolarization will be, until the membrane potential approaches −94 mV (the potassium equilibrium potential). The membrane potential can never exceed and in fact will never reach −94 mV because even when all potassium channels in a membrane are open, open sodium leak channels will allow sodium to leak into the cell, counteracting the movement of potassium.

Temporal summation can occur because postsynaptic potentials last considerably longer than action potentials. This makes it possible for a presynaptic neuron to fire a second action potential, and thus release a second burst of neurotransmitter molecules, while neurotransmitter molecules from the first action potential are still in the synaptic cleft. The higher concentration of neurotransmitter in the synaptic cleft opens more ion channels, causing a greater postsynaptic potential. In contrast, when a presynaptic neuron fires two action potentials farther apart in time, the neurotransmitter released as a result of the first action potential is completely cleared from the synaptic cleft before the next action potential arrives, and therefore the two postsynaptic potentials cannot sum. (Note, however, that when metabotropic (slow response) receptors are involved, a postsynaptic potential may persist even after the neurotransmitter that initiated it has been cleared from the synaptic cleft.)

Spatial Summation

In spatial summation (Figure 7.7c), two or more postsynaptic potentials originating from different synapses are generated at approximately the same time such that they overlap and sum. When neurons A and B have action potentials at different times, each neuron induces an EPSP in the postsynaptic neuron, neither of which is large enough to trigger an action potential. When neurons A and B are activated at the same time, however, the resulting EPSPs sum to produce a depolarization that *is* large enough to trigger an action potential. Note as well that if neurons A and C were active at the same time, no change in the membrane potential of the postsynaptic neuron would result, because the EPSP that originated from synapse A and the IPSP originating from synapse C would cancel each other out. Spatial summation occurs as postsynaptic potentials originating at different synapses spread to the axon hillock, overlapping along the way.

Figure 7.7 illustrates a highly simplified example of the concept of what is involved in neural integration. In reality, the dendrites and cell body of a postsynaptic neuron can receive input from hundreds to hundreds of thousands of different neurons, an arrangement termed convergence (Figure 7.8). Because the number of synapses is so large, and because some are excitatory while others are inhibitory, the number of possible combinations of active synaptic inputs is astronomical. Temporal and spatial summation of postsynaptic potentials occurs continually, and the magnitude of the summed potential at the axon hillock determines if and when action potentials occur.

Frequency Coding

In Chapter 6 we saw that once a depolarizing stimulus exceeds threshold, the degree of depolarization does not affect the size of action potentials, but rather their frequency. We saw that increases in the strength of such suprathreshold stimuli cause the frequency of action potentials to increase, an effect called **frequency coding.** Because the summation of postsynaptic potentials affects the degree of depolarization at the axon hillock, it follows that summation influences the frequency of action potentials in the postsynaptic neuron. When action potentials occur at a higher frequency, more neurotransmitter is released from the neuron. Therefore, a higher frequency of action potentials corresponds to stronger communication to the next neuron in a neural pathway.

PRESYNAPTIC MODULATION

Most synapses occur between the axon terminal of a presynaptic neuron and either a dendrite or the cell body of a postsynaptic neuron. However, we have seen that axoaxonic synapses can form between the axon terminal of one neuron and the axon terminal of another neuron (see Figure 7.1). In axoaxonic synapses, neurotransmitter from the presynaptic neuron (also

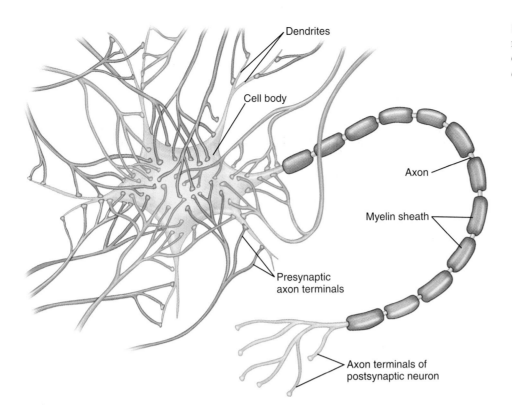

Dendrites

Cell body

Axon

Myelin sheath

Presynaptic
axon terminals

Axon terminals of
postsynaptic neuron

FIGURE 7.8 Convergence, in which many presynaptic cells synapse on one postsynaptic cell. *Most synapses occur on the cell body and dendrites.*

called a *modulating neuron*) does not generate electrical signals in the postsynaptic neuron. Instead, by binding to receptors on the membrane of the axon terminal of the postsynaptic neuron, the neurotransmitter induces a change in the amount of calcium that enters the axon terminal in response to an action potential, in turn altering the amount of neurotransmitter released from the postsynaptic neuron. In some cases the release of neurotransmitter is enhanced, a phenomenon called **presynaptic facilitation;** in other cases the release of neurotransmitter is decreased, a phenomenon called **presynaptic inhibition.** We consider these phenomena next.

Presynaptic Facilitation

Presynaptic facilitation is illustrated in Figure 7.9a. Neurons C and D are presynaptic neurons that form excitatory synapses with postsynaptic neuron X; neuron E is a modulating neuron that forms a synapse on the axon terminal of neuron C. In this example, an action potential in neuron C triggers the release of enough neurotransmitter to produce a 10-mV EPSP in neuron X when E is not active. However, when neuron E is active, a single action potential in neuron C will release enough neurotransmitter to produce a 15-mV EPSP and trigger an action potential. Thus the activity of neuron E increases the amount of neurotransmitter released from

neuron C such that neuron C triggers a larger EPSP in neuron X, compared to when neuron E is inactive. Note that activity in neuron E alone has no direct effect on neuron X because neuron E does not form a synapse with neuron X. Note as well that activity in neuron E has no effect on any EPSPs triggered by input from neuron D.

This example illustrates an important characteristic of presynaptic modulation at axoaxonic synapses that differentiates it from axodendritic and axosomatic synapses: Presynaptic modulation affects transmission to the postsynaptic neuron at one specific synapse, thereby altering the ability of that one synapse to excite or inhibit the postsynaptic neuron. In contrast, at axodendritic and axosomatic synapses, presynaptic neurons nonselectively excite or inhibit the postsynaptic neuron by directly producing changes in membrane potential of the postsynaptic neuron in the form of EPSPs and IPSPs.

Presynaptic Inhibition

Figure 7.9b shows an example of presynaptic inhibition. In this example, neurons F and G make direct excitatory synaptic connections with neuron Y; neuron H is a modulating neuron that forms an axoaxonic synapse with neuron F. When either neuron F alone or neuron G alone is active, an EPSP is produced in neuron Y. When neuron H is active by itself, no change in the membrane potential of neuron Y occurs. However, when neurons F and H are

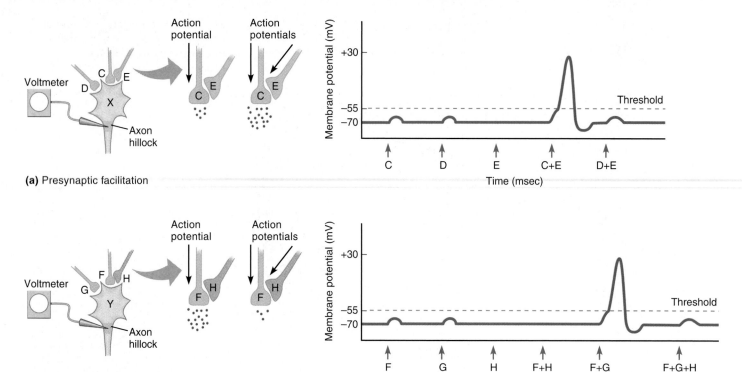

(a) Presynaptic facilitation

(b) Presynaptic inhibition

FIGURE 7.9 Presynaptic modulation at axoaxonic synapses. (a) *Presynaptic facilitation. When neurons C and E are both active, neurotransmitter from neuron E enhances the release of neurotransmitter from neuron C, increasing the strength of the resulting EPSP to threshold and generating an action potential.* **(b)** *Presynaptic inhibition. When neurons F and H are both active, neurotransmitter from neuron H decreases the release of neurotransmitter from neuron F, decreasing the strength of the EPSP. When neurons F and G are both active, their EPSPs sum spatially, and neuron Y is depolarized to threshold, generating an action potential. However, when neurons F, G, and H are all activated, the EPSP from neuron F is diminished due to presynaptic inhibition by neuron H, such that the sum of the EPSPs from neurons F and G is subthreshold, and an action potential is not generated.*

active at the same time, the response of neuron Y to input from neuron F is decreased; that is, neuron H suppresses the release of neurotransmitter from neuron F. In contrast, activity in neuron H does not affect the response of neuron Y to input from neuron G.

Note that although Figure 7.9 illustrates presynaptic facilitation and presynaptic inhibition at excitatory synapses, these phenomena can occur at inhibitory synapses as well.

Quick Test 7.3

1. Define the following terms: *spatial summation, temporal summation, presynaptic facilitation,* and *presynaptic inhibition.*

2. During presynaptic inhibition, does the amount of neurotransmitter released increase or decrease?

NEUROTRANSMITTERS: STRUCTURE, SYNTHESIS, AND DEGRADATION

Neurotransmitters fall into a variety of chemical classes, including acetylcholine, biogenic amines, amino acids, and neuroactive peptides (or neuropeptides). Most are small molecules, with the notable exception of the neuropeptides, which are considerably larger than the others. A few other neurotransmitters, such as *nitric oxide* and ATP, are unlike any others and are quite literally in a class by themselves. Selected members of the different classes of neurotransmitters discussed in the following sections are listed in Table 7.1.

Acetylcholine

Acetylcholine (ACh) is released from neurons in both the central and peripheral nervous systems. It is the most abundant neurotransmitter in the peripheral nervous

TABLE 7.1   SELECTED MEMBERS OF THE CLASSES
OF NEUROTRANSMITTERS

CHOLINE DERIVATIVE	BIOGENIC AMINES	AMINO ACIDS	NEUROPEPTIDES	OTHERS
Acetylcholine	Catecholamines	Glutamate*	TRH	Nitric oxide
	Dopamine	Aspartate*	Vasopressin	ATP
	Epinephrine	Glycine†	Oxytocin	
	Norepinephrine	GABA†	Substance P	
	Serotonin		Endogenous opioids	
	Histamine		Enkephalins	
			Endorphins	

*Excitatory neurotransmitters.
†Inhibitory neurotransmitters.

system, where it is found in efferent neurons of both the somatic and autonomic branches.

Acetylcholine is synthesized in the cytoplasm of the axon terminals of neurons (Figure 7.10). The synthesis of acetylcholine from two substrates, *acetyl CoA* and *choline*, is catalyzed by **choline acetyl transferase (CAT)**, as follows:

$$\text{acetyl CoA} + \text{choline} \xrightarrow{\text{CAT}} \text{acetylcholine} + \text{CoA}$$

We saw in Chapter 3 that one of these substrates, acetyl CoA, is a two-carbon molecule produced during the catabolism of lipids, carbohydrates, and proteins, and the initial substrate for the Krebs cycle. Therefore, it is found in almost all cells of the body, including those neurons that synthesize and release acetylcholine (called *cholinergic* neurons). In contrast, choline cannot be synthesized by neurons (although some is synthesized in the liver); instead, most choline is obtained from the diet and delivered by the bloodstream to cholinergic neurons, where it is taken in via an active transport system.

Once it is synthesized, acetylcholine is stored in synaptic vesicles (see Figure 7.10) until an action potential triggers its release by exocytosis. After release, acetylcholine can bind to receptors on the postsynaptic cell (called **cholinergic receptors**) and/or be degraded by an enzyme called **acetylcholinesterase (AChE),** which is found on either the presynaptic neuron membrane, the postsynaptic neuron membrane, or both. Acetylcholinesterase catalyzes the breakdown of acetylcholine into acetate (an acetyl CoA molecule without the CoA attached) and choline, as follows:

$$\text{acetylcholine} \xrightarrow{\text{AChE}} \text{acetate} + \text{choline}$$

Neither of the breakdown products of this reaction has any transmitter activity, but the choline that is produced is taken

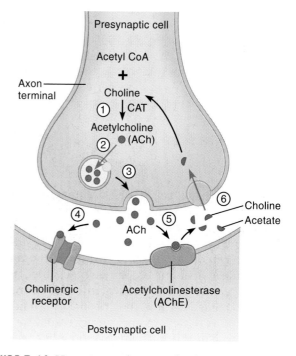

FIGURE 7.10 Neurotransmitter synthesis, action, and degradation at cholinergic synapses. *The sequence of events at a cholinergic synapse. ① The enzyme choline acetyl transferase (CAT) catalyzes the reaction combining acetyl CoA with choline to form ACh (and CoA). ② The ACh is then actively packaged into synaptic vesicles. ③ ACh is released by exocytosis. Once released, ACh may ④ bind to cholinergic receptors on the postsynaptic cell or ⑤ be degraded by the enzyme acetylcholinesterase (AChE) into choline and acetate. ⑥ Choline is actively transported back into the presynaptic neuron, where it can be used to synthesize more acetylcholine.*

What triggers the release of acetylcholine by exocytosis?

Calcium ions entering the cell through voltage-gated calcium channels

FIGURE 7.11 Signal transduction mechanisms at cholinergic receptors. (a) *Nicotinic cholinergic receptors are receptor-operated channels that permit both sodium and potassium ions to move through. When acetylcholine binds to these receptors, the channels open. Because more sodium moves in than potassium moves out, the result is an EPSP.* **(b)** *Muscarinic cholinergic receptors are coupled to G proteins that can either directly open/close ion channels or activate/inhibit an enzyme that catalyzes the production of a second messenger. The second messengers produced can have a variety of effects on the postsynaptic cell, including opening or closing ion channels. The effects on postsynaptic cells can be either excitatory (EPSPs) or inhibitory (IPSPs).*

Which cholinergic receptor type (nicotinic or muscarinic) produces the faster response?

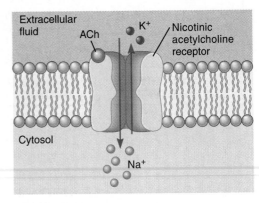

(a) Nicotinic cholinergic receptors

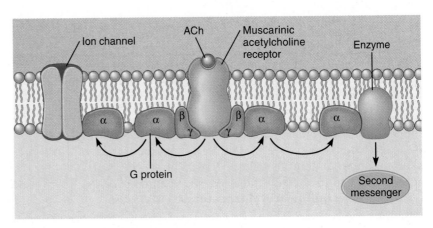

(b) Muscarinic cholinergic receptor

back into the presynaptic cell by an active transport mechanism and can be used to synthesize more acetylcholine. The acetate diffuses away from the synapse and enters the blood.

Cells in the nervous system commonly have different types of receptors for a specific neurotransmitter. Receptors for acetylcholine are of two types: **nicotinic cholinergic receptors** and **muscarinic cholinergic receptors,** so named because the drug nicotine binds to nicotinic receptors but not to muscarinic receptors, whereas the drug muscarine binds to muscarinic receptors but not nicotinic receptors. Even though these receptors share a common messenger—acetylcholine— the effects on the postsynaptic cell are very different (Figure 7.11). Nicotinic cholinergic receptors are ionotropic, and the binding of acetylcoline to them triggers the opening of channels that allow both sodium and potassium to move through, causing an EPSP in the postsynaptic cell (Figure 7.11a). Nicotinic cholinergic receptors are located in several areas of the peripheral nervous system—for example, in certain autonomic neurons and on skeletal muscle cells, which are the effector cells for somatic motor neurons. Nicotinic cholinergic receptors

are also located in some regions of the central nervous system. In contrast, muscarinic cholinergic receptors are metabotropic receptors and operate through the action of a G protein. The effect resulting from the binding of acetylcholine to muscarinic cholinergic receptors depends on the postsynaptic cell in question; these receptors can open or close ion channels or activate enzymes (Figure 7.11b). Muscarinic cholinergic receptors, which are found on some effector organs for the autonomic nervous system, are the dominant cholinergic receptor type found in the central nervous system.

The different effects of acetylcholine at nicotinic and muscarinic cholinergic receptors illustrate an important physiological concept: The action of any chemical messenger ultimately depends, not on the nature of the messenger, but on the type of receptor to which the messenger binds on the target cell.

Biogenic Amines

The **biogenic amines** are a class of neurotransmitters derived from amino acids. They are called *amines* because they all possess an amine group ($-NH_2$). (Note that they do not include amino acids, which also possess amine groups.) The biogenic amines include the catecholamines,

Nicotinic

serotonin, and histamine. The catecholamines contain a *catechol group*, a six-carbon ring with two hydroxyl groups, and include the neurotransmitters dopamine, norepinephrine, and epinephrine. (The synthetic pathway for catecholamines is described in Chapter 5.) Like acetylcholine, the biogenic amines are synthesized in the cytosol of the axon terminal.

Dopamine and norepinephrine are both released primarily by neurons in the central nervous system, but norepinephrine is also released from neurons in the peripheral nervous system. Even though epinephrine is released from some neurons of the central nervous system, it is better known as a hormone (also called adrenaline) that is released from the adrenal medulla in response to commands from the sympathetic nervous system.

The different catecholamines bind to specific receptors. Receptors for epinephrine and norepinephrine are called **adrenergic receptors,** of which there are two main classes, *alpha adrenergic* and *beta adrenergic* receptors. Each of these has subclasses designated by numerical subscripts, with the primary receptor types being $alpha_1$, $alpha_2$, $beta_1$, $beta_2$, and $beta_3$. Epinephrine has the highest affinity for $beta_2$ receptors, whereas norepinephrine has higher affinity for alpha and $beta_1$ receptors, but both compounds bind to all these receptors. Adrenergic receptors are found in the central nervous system and in the effector organs for the sympathetic branch of the autonomic nervous system. Adrenergic receptors are referred to as being *adrenergic* because alternate terms for epinephrine and norepinephrine are *adrenaline* and *noradrenaline*, respectively. In similar fashion, receptors that bind dopamine are called *dopaminergic*.

Catecholamines generally produce slow responses mediated through G proteins and changes in second messenger systems. In addition, they often function as autocrines, binding to receptors (called **autoreceptors**) on the axon terminal of the cell that released them. These receptors enable a neuron to modulate its own release of neurotransmitter, generally by altering the amount of calcium that enters the cell in response to an action potential. Neurons possess autoreceptors for other neurotransmitters as well.

Following their release, catecholamines can be degraded by two enzymes, **monoamine oxidase (MAO)** and **catechol-O-methyltransferase (COMT).** MAO is located in the synaptic cleft, in the axon terminal of the neurons that release the catecholamines, and in some glial cells. COMT is located in the synaptic cleft. To understand the importance of the enzymes that degrade neurotransmitters, see When It Goes Wrong: Treating Depression, p. 218.

Serotonin and histamine are biogenic amines, but they are not catecholamines. Serotonin is found in the central nervous system, particularly the lower portion of the brain called the *brain stem*. Some functions of serotonin include regulating sleep and emotions. Histamine,

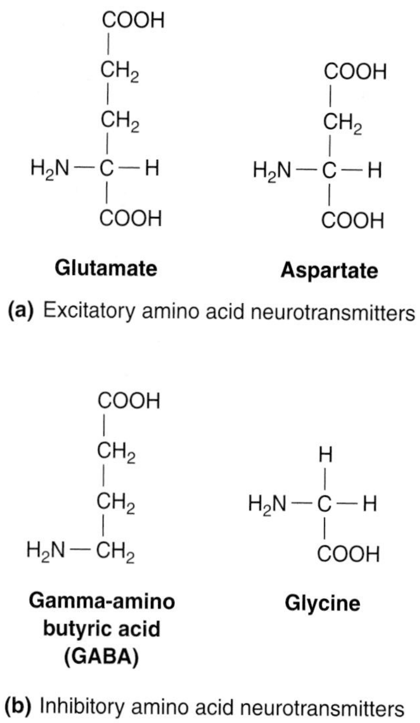

FIGURE 7.12 **The amino acid neurotransmitters. (a)** *The excitatory amino acid neurotransmitters. Note their similar chemical structures.* **(b)** *The inhibitory amino acid neurotransmitters. The similarity of GABA to the excitatory amino acid neurotransmitters results from the fact that GABA is a derivative of glutamate.*

better known for its release from non-neuronal cells during allergic and other types of reactions, can also function as a neurotransmitter. Histamine is found in the central nervous system, primarily in an area of the brain called the *hypothalamus*.

Amino Acid Neurotransmitters

The **amino acid neurotransmitters** are the most abundant class of neurotransmitters in the central nervous system, where they are widely distributed throughout virtually all areas. **Aspartate** and **glutamate** are excitatory neurotransmitters, whereas **glycine** and *gamma-aminobutyric acid*, more commonly known as **GABA,** are inhibitory neurotransmitters (Figure 7.12). Because GABA is derived from one of the other amino acid neurotransmitters, glutamate, it has a structure similar to an excitatory amino acid but produces inhibitory effects. The reason for this apparent contradiction is that the receptors for these compounds are very specific; thus even though GABA and glutamate are similar in structure, they bind to completely separate populations of receptors and produce opposite results.

Neuroactive Peptides

The **neuroactive peptides** (or **neuropeptides**) are short chains of amino acids that are synthesized in the same manner as proteins. More than 50 of these

Most people feel a little blue now and then, and most people have had experiences that made them extremely sad, such as the death of a loved one. At such a time, people may describe themselves as being depressed. But clinical depression is much more than feeling sad, and generally it is not induced by an event.

Depression has many symptoms, including lack of energy, abnormal eating habits (too much or too little), and/or difficulty sleeping or sleeping too much. Often the person feels worthless and may be preoccupied with thoughts of suicide. A depressed person has difficulty functioning in society. The cause(s) of depression and its symptoms are not well understood, but depression is an illness associated with biochemical changes in the brain.

Much evidence suggests that depression is associated with deficiencies in the biogenic amines serotonin and norepinephrine. Indeed, one of the side effects associated with medications that decrease biogenic amines—such as the drug reserpine, used to treat high blood pressure—is depression. Therefore, pharmacological treatment strategies often seek to increase biogenic amine concentrations in the brain.

One class of antidepressants is *monoamine oxidase inhibitors.* Monoamine oxidase is the enzyme that breaks down biogenic amines, including norepinephrine and serotonin. Because these antidepressants inhibit their degradation, these neurotransmitters remain in the synaptic cleft for a longer period of time; the effect is similar to having increased the release of these neurotransmitters. Monoamine oxidase inhibitors, including phenelzine (Nardil™) and isocarboxazid (Marplan™), have been used successfully to treat many cases of clinical depression.

In many neural circuits, monoamine oxidase is located inside the presynaptic neuron, where it degrades neurotransmitters that have been actively transported back into the cell that released them (reuptake). In these cells, inhibition of monoamine oxidase is believed to increase the amount of neurotransmitter that is packaged into synaptic vesicles. The result is that more neurotransmitter is released in response to a given stimulus.

A relatively new class of commonly used antidepressants (first approved by the FDA in 1987) is *selective serotonin reuptake inhibitors (SSRIs).* By decreasing the reuptake of serotonin into the cell that released it, SSRIs selectively increase the amount of serotonin present at the synaptic cleft. SSRIs are more specific than monoamine oxidase inhibitors because they only affect serotonergic synapses. SSRIs include fluoxetine (Prozac™) and paroxetine (Paxil™).

Another class of commonly used antidepressants is the *tricyclics*, named for the presence of three carbon rings. Although tricyclics are the oldest of the antidepressants (first used in the 1950s), their mechanisms of action are the least understood. Several hypotheses on the mechanisms have been proposed, most of which involve alterations of activity at adrenergic or serotonergic synapses. Tricyclics include imipramine (Tofranil™), amitriptyline (Elavil™), and desipramine (Norpramin™).

Does it surprise you that even though a drug has been around for about 50 years, the mechanism of its action is unknown? This is actually a fairly common occurrence, as drugs are often used with little knowledge of their mechanism of action. Even the therapeutic benefits of monoamine oxidase inhibitors and SSRIs for depression are not fully understood, for even though these drugs immediately decrease monoamine oxidase activity or serotonin reuptake, respectively, they do not affect the depression until they have been taken for several weeks. Much research remains to be done on the mechanisms of action of antidepressants and other medications.

compounds have been found in neurons, where their function as neurotransmitters is either well established or probable. Like other peptides or proteins that are to be secreted from cells, neuropeptides are synthesized in the rough endoplasmic reticulum and packaged into secretory vesicles by the Golgi apparatus. In neurons these events take place in the cell body, the only part of the cell that contains the necessary machinery. After this synthesis and packaging, secretory vesicles are slowly transported down the axon to the axon terminal, where they are stored. Many of the neuropeptides are more classically known as hormones, including **TRH,** which regulates release of another hormone called TSH; **vasopressin,** which regulates urine output by the kidney; **oxytocin,** which regulates contractions of the uterus and the flow of milk from the breasts; and **substance P,** which decreases gastrointestinal motility. Also included among the neuropeptides are the **endogenous opioids,** which exert effects similar to the drug morphine. The endogenous opioids include the **enkephalins** and **endorphins,** which were mentioned in the introduction to this chapter.

Exercise Link

As you might imagine, running 26 miles is associated with pain due to impacts, abrasions, heat, and internal changes in pH. Performance depends in part on an ability to tolerate such pain. As Bill and Jane ran, their pituitaries released beta-endorphin resulting in five-fold increases in blood concentrations. Endogenous opioids produced in the central nervous system reduce pain sensations by binding to receptors located where pain-sensitive neurons synapse with spinal cord neurons, causing presynaptic inhibition of neurotransmitter release (due to decreased Ca^{2+} influx); and postsynaptic inhibition (due to increased K^+ conductance). However, *circulating* endorphins do not affect these synapses because the central nervous system is very selective about what it lets in from the blood (the "blood–brain barrier" you will learn about in Chapter 8). Blood–borne endorphins (and those released locally by inflammatory cells) appear to bind receptors on pain-sensitive nerve endings in the skin, joints, and muscles.

Most neuropeptides are colocalized with other neurotransmitters in the same neurons; that is, they are released from the same axon terminal. Neuropeptides often act on metabotropic receptors and modulate the response of the postsynaptic neuron to the colocalized neurotransmitter.

Other Neurotransmitters

Many other chemical substances identified in neurons appear to function as neurotransmitters. These include ATP, more widely known for its role in energy metabolism, and a gas called **nitric oxide.** Nitric oxide has a particularly simple structure, one nitrogen atom and one oxygen atom, and it does not fit the general pattern of neurotransmitter function in that it is not synthesized and stored in synaptic vesicles in advance of its release. Instead it is released as soon as it is synthesized because it easily crosses the plasma membrane. Thus nitric oxide release is controlled via regulation of the rate of synthesis in a reaction catalyzed by the enzyme **nitric oxide synthetase.** Following its release, nitric oxide enters the postsynaptic cell, where it alters the activity of proteins. Because it degrades quickly on its own, without the assistance of enzymes, its lifespan is only a few seconds.

Nitric oxide may not be the only gas that functions as a neurotransmitter. Recent evidence suggests that carbon monoxide may also be a neurotransmitter.

Quick Test 7.4

1. Explain how it is possible for a given neurotransmitter to inhibit one cell and excite another.

2. Name the different classes of neurotransmitters. Which neurotransmitters are dominant in the peripheral nervous system?

3. Most small neurotransmitters are synthesized and stored in the axon terminal before release. How do neuropeptides deviate from this pattern? How does nitric oxide deviate from it?

4. Unlike other neurotransmitters, nitric oxide does not bind to receptors on the surface of a postsynaptic cell. How does it exert its effects?

Types of Synapses, p. 202

Neurons communicate with other neurons or with effector organs at synapses. Neurons communicate via both electrical synapses and chemical synapses. Electrical synapses exist where gap junctions connect two neurons; chemical synapses are more common and involve the release of a neurotransmitter from one neuron to communicate with a second neuron.

IP Nervous II, Anatomy Review, page 6

Chemical Synapses, p. 202

At a chemical synapse, the transmission of a message from a presynaptic neuron to a postsynaptic neuron occurs in response to an action potential in the presynaptic neuron. The action potential will be propagated from the trigger zone to the axon terminal, where it stimulates the opening of voltage-gated calcium channels. Calcium enters the cell, triggering the release of neurotransmitter by exocytosis. The neurotransmitter binds to receptors on the postsynaptic neuron, triggering a response in the postsynaptic neuron, usually a change in the electrical properties of the cell.

Signal transduction refers to the mechanism by which a messenger produces a response in a cell. At ionotropic (fast response) receptors, the neurotransmitter causes a rapid opening of ion channels, producing an immediate change in the electrical properties of the cell. At metabotropic (slow response) receptors, the neurotransmitter activates a G protein, which either opens or closes ion channels to change the electrical properties of the cell, or activates an enzyme that produces a second messenger. The second messenger may then open or close ion channels or produce some other response in the cell.

Synapses are either excitatory or inhibitory. At excitatory synapses, the membrane potential of the postsynaptic neuron is depolarized, bringing it closer to threshold to generate an action potential. This depolarization is called an EPSP. At inhibitory synapses, the membrane potential of the postsynaptic neuron is either hyperpolarized or stabilized, making the neuron less likely to fire an action potential. The hyperpolarization is called an IPSP.

IP Nervous II, Anatomy Review, pages 7–8

IP Nervous II, Synaptic Transmission, pages 2–14

IP Nervous II, Synaptic Potentials and Cellular Integration, pages 4–9

Neural Integration, p. 210

Neural integration is the spatial and temporal summation of synaptic potentials at the axon hillock of a postsynaptic neuron. If the axon hillock is depolarized to threshold, an action potential will be generated. Once depolarization reaches threshold, greater depolarizations will elicit a higher frequency of action potentials.

IP Nervous II, Synaptic Potentials and Cellular Integration, pages 1–10

Presynaptic Modulation, p. 212

Synaptic communication can be modulated. The most common type of modulation occurs at axoaxonic synapses and is called presynaptic modulation. At these synapses, the presynaptic cell modulates the release of neurotransmitter from the postsynaptic cell. In presynaptic facilitation, communication at a specific synapse is enhanced, whereas in presynaptic inhibition, communication at a specific synapse is attenuated.

Neurotransmitters: Structure, Synthesis, and Degradation, p. 214

A variety of neurotransmitters can be released from neurons. Acetylcholine is the most abundant neurotransmitter in the peripheral nervous system, although it is also found in the central nervous system. The biogenic amines include the catecholamines, serotonin, and histamine. The catecholamines include norepinephrine, epinephrine, and dopamine. Norepinephrine is another common neurotransmitter in the peripheral nervous system. Other classes of neurotransmitters include the amino acid neurotransmitters and peptides. Recently discovered neurotransmitters include nitric oxide and ATP.

Neurotransmitters communicate by binding to receptors specific for them. Generally, more than one type of receptor exists for each neurotransmitter, and the response produced in the postsynaptic cell can vary considerably based on the receptor type activated and the coupling mechanism involved.

IP Nervous II, Synaptic Transmission, pages 1–10

Multiple-Choice Questions

1. Suppose that the electrochemical force for anion X (X^-) acts to move the anion out of the cell. If a neurotransmitter binding to its receptor opened channels for X^- on the postsynaptic cell, then the response would
   a) be an EPSP.
   b) be an IPSP.
   c) be stabilization of the membrane.
   d) not occur.

2. Suppose that all the calcium could be removed from the extracellular fluid surrounding a neuron. Such removal would inhibit the ability of a neuron to
   a) produce action potentials.
   b) release neurotransmitter.
   c) respond to the binding of a neuro-transmitter to its receptor.
   d) degrade neurotransmitters.

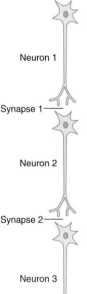

(Questions 3–5): The accompanying diagram depicts three neurons in a neural pathway in which neuron 1 synapses on neuron 2, and neuron 2 synapses on neuron 3. Neuron 1 can be artificially activated through a stimulating electrode that passes current across the membrane. Membrane potentials of all three neurons are recorded by an intracellular electrode connected to a volt-meter. In the following voltage tracing of membrane potential recorded from cell 1, action po-tentials are shown as vertical spikes; the horizontal bar over the voltage tracing indicates the time period during which the stimulating electrode was turned on.

Neuron 1
Synapse 1
Neuron 2
Synapse 2
Neuron 3

3. From this voltage tracing, one can conclude that

V_{m1} ⎯⎯|‖|‖|‖⎯‖‖‖‖‖‖‖‖⎯|‖|‖|‖⎯

   a) the stimulation depolarizes the cell membrane.
   b) the stimulation hyperpolarizes the cell membrane.
   c) the stimulation has no effect on the cell membrane potential.
   d) the stimulation mimics the effect of an inhibitory neurotransmitter.

4. If you were to observe an increase in action potential frequency in cell 2 during stimulation of cell 1, you could conclude that
   a) synapse 1 is excitatory.
   b) synapse 1 is inhibitory.
   c) the findings are inconclusive.

5. Given the following results, what would your conclusion be?

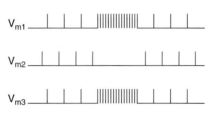

V_{m1}
V_{m2}
V_{m3}

   a) Synapse 1 is excitatory, whereas synapse 2 is inhibitory.
   b) Synapse 1 is inhibitory, whereas synapse 2 is excitatory.
   c) Synapses 1 and 2 are both inhibitory.
   d) Synapses 1 and 2 are both excitatory.
   e) No definite conclusion is possible.

6. Synaptic vesicles
   a) store calcium.
   b) release neurotransmitters by exocytosis.
   c) degrade neurotransmitters.
   d) form gap junctions.

7. If sodium channels closed in response to a stimulus, then
   a) the cell would be depolarized.
   b) the cell would be hyperpolarized.
   c) the membrane potential would be stabilized.
   d) a second messenger would be pro-duced.

8. A fast EPSP is produced by
   a) the opening of sodium-selective channels.
   b) the opening of potassium-selective channels.
   c) the opening of channels selective for both sodium and potassium.
   d) the opening of calcium-selective channels.

9. The enzyme that catalyzes the synthesis of acetylcholine is
   a) adenylate cyclase.
   b) choline acetyl transferase.
   c) monoamine oxidase.
   d) acetylcholinesterase.

10. Which of the following neurotrans-mitters is a biogenic amine but not a catecholamine?
   a) norepinephrine
   b) serotonin
   c) dopamine
   d) epinephrine

Objective Questions

1. At electrical synapses, what type of junction exists between the two cells?

2. The chemical messenger released by a neuron is called a (hormone/neurotransmitter).

3. When the opening of ion channels allows both sodium and potassium ions to move through, no change in membrane potential occurs because sodium moves into the cell and potassium moves out of the cell. (true/false)

4. Neurotransmitter receptors are found at (chemical/electrical) synapses.

5. Whether a synapse is excitatory or inhibitory is determined by the mechanism of coupling between the neurotransmitter receptor and ion channels in the postsynaptic cell. (true/false)

6. The synaptic delay includes the time it takes for an action potential to travel from the trigger zone of a presynaptic cell to the axon terminal. (true/false)

7. A given neurotransmitter might be excitatory at one synapse and in-hibitory at another. (true/false)

8. Given that release of an inhibitory neurotransmitter is altered by presynaptic facilitation, the response in the postsynaptic cell will be a (larger/smaller) degree of hyperpolarization.

9. Neurotransmitter receptors at all inhibitory synapses are coupled to ion channels through activation of a G protein. (true/false)

10. The response to a neurotransmitter is faster at (ionotropic/metabotropic) receptors.

11. When neuron 1 synapses at the axon terminal of neuron 2 and decreases the amount of neurotransmitter released from neuron 2, (presynaptic facilitation/presynaptic inhibition) is occurring.

12. The enzyme that catalyzes the synthesis of acetylcholine is called _____.

13. The enzymes that catalyze the degradation of catecholamines are _____ and _____.

14. Adenylate cyclase catalyzes the formation of _____.

15. The excitatory amino acid neurotransmitters are _____ and _____.

Essay Questions

1. Describe the sequence of events occurring at a chemical synapse, starting with an action potential in the presynaptic cell and ending with a response in the postsynaptic cell.

2. Compare and contrast the events caused by the binding of neurotransmitters to ionotropic and metabotropic receptors.

3. Explain the ionic basis of a fast EPSP.

4. Explain the role of axoaxonic synapses.

5. Describe the steps of the cAMP second messenger system. Explain how cAMP can produce different responses in the many different types of cells that use cAMP as a second messenger.

Find the answers to these exercises, and additional study tools, at the Physiology Place (www.physiologyplace.com).

8

The Nervous System:
Central Nervous System

OBJECTIVES

- Describe the anatomy of the brain and spinal cord, and relate structure to function. Indicate which structures protect the central nervous system, and which are involved in neural signaling.

- Describe the anatomy, physiology, and consequences of the blood-brain barrier.

- Describe the energy supplies of the brain, and explain why blood flow is so critical.

- Define a reflex arc. Describe the following reflex pathways: muscle spindle stretch reflex, withdrawal reflex, and crossed-extensor reflex.

- Describe the areas of the brain that contribute to voluntary control of skeletal muscles, and the basic roles these areas play.

- Describe the different functions of the two language centers: Wernicke's area and Broca's area.

- Describe the different stages of sleep and how the brain shifts from the sleep state to the conscious state.

- Describe the different types of learning and memory. Define neural plasticity and explain how it contributes to learning and memory.

CHAPTER OUTLINE

General Anatomy of the Central Nervous System   224

The Spinal Cord   231

The Brain   237

Integrated CNS Function: Reflexes   244

Integrated CNS Function: Voluntary Motor Control   247

Integrated CNS Function: Language   251

Integrated CNS Function: Sleep   251

Integrated CNS Function: Emotions and Motivation   254

Integrated CNS Function: Learning and Memory   255

Above: Microtubules and nuclei in astrocytes.

STUDY HINTS

1. *Tight junctions, p. 43*

2. *Glycolysis, p. 81*

3. *Organization of the nervous system, p. 171*

4. *Neurotransmitters, p. 214*

5. *Neuron structure, p. 173*

6. *Communication at synapses, p. 202*

7. *Neural integration, p. 210*

As you read this, your central nervous system is performing many tasks, including perceiving words on the page, comprehending them, and making judgments about their importance and whether or not they should be stored in memory. As you read further, you may start daydreaming, or perhaps you may come to feel sleepy and doze off before you finish. In any case, what happens as you read is a consequence of central nervous system processing.

The central nervous system is ultimately responsible for everything we perceive, do, feel, and think. It gives each of us our unique personality and sense of self-identity. It also performs many critical functions that typically escape our notice. For example, it coordinates the activities of all our organ systems, a function that is necessary for the maintenance of homeostasis. One of the things that makes the nervous system so fascinating is that all of its functions are carried out by neurons, about which we learned in the two previous chapters. Given that all neurons operate according to a small set of well-understood principles, how can neurons be responsible for all the nervous system's complexities and mysteries? How are neurons involved when someone feels angry or daydreams? How can neurons enable us to remember what somebody's name is or how the heart works? One neuron alone can perform none of these functions, but when billions of neurons (and the glial cells associated with them) are organized to form the nervous system, neurons perform these—and thousands of other—functions. It is estimated that the central nervous system contains about a hundred billion (10^{11}) neurons and a hundred trillion (10^{14}) synapses, all contained within two remarkable structures, the brain and the spinal cord.

Scientists are unraveling the mysteries of the brain, even as you read this. Because the brain is so complex and its functions so difficult to fathom using current scientific methodology, much of what you will read in this chapter are the commonly accepted portions of a constantly developing theory of central nervous system function. Perhaps someday we will understand the brain as well as we understand, say, the lungs or the kidneys. For now, we begin by examining the anatomy of the central nervous system.

GENERAL ANATOMY OF THE CENTRAL NERVOUS SYSTEM

Protective Structures

The central nervous system (CNS) consists of the **brain** and the **spinal cord.** Because it is made up of soft tissue, with a consistency much like Jello, it is particularly vulnerable to damage by physical trauma. Because the CNS is vital to the operation of the body and to everything we do, feel, or think, it must be protected. Fortunately, structures composed of three different substances—bone, connective tissue, and fluid—protect the brain and spinal cord.

The outermost structures that protect the soft tissues of the central nervous system are the bony skull, or **cranium,** which surrounds the brain, and the bony **vertebral column,** which surrounds the spinal cord (Figure 8.1). Even though these bony structures are clearly beneficial in that they act as rigid armor surrounding delicate nervous tissue, their hardness also poses a potential hazard: What would prevent the soft brain from crashing into the hard inner surface of the skull when, for example, you make a sudden stop in a car moving at freeway speeds? Between the bone and the nervous tissue are a series of three membranes called the meninges and a layer of fluid called cerebrospinal fluid.

The **meninges** are three connective tissue membranes that separate the soft tissue of the CNS from the surrounding bone (see Figure 8.1). The three meningeal membranes are the **dura mater,** the **arachnoid mater,** and the **pia mater.** *Mater* is Latin for mother, indicating the protective nature of the meninges. The dura mater is the outermost layer, closest to the bone. *Dura* is Latin for hard or durable, and the dura mater is a very tough, fibrous tissue. The arachnoid mater is the middle layer. *Arachnoid* is Greek for spider, which appropriately describes the arachnoid mater's weblike structure. Normally, no space exists between the dura and the arachnoid. The innermost layer, the pia mater, is immediately adjacent to the nervous tissue. *Pia* is Latin for tender or kind. The space between the pia mater and arachnoid mater, called the **subarachnoid space,** is filled with cerebrospinal fluid.

Cerebrospinal fluid (CSF) is a clear, watery fluid that bathes the CNS; it is similar (but not identical) in composition to plasma (Table 8.1). Its total volume is only about 125–150 ml. Because the CNS is completely surrounded by CSF and essentially floats in it, the CSF

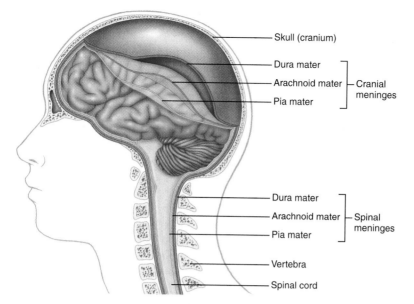

- Skull (cranium)
- Dura mater ⎫
- Arachnoid mater ⎬ Cranial meninges
- Pia mater ⎭

- Dura mater ⎫
- Arachnoid mater ⎬ Spinal meninges
- Pia mater ⎭
- Vertebra
- Spinal cord

(a)

FIGURE 8.1 Protective structures of the CNS. (a, b)
*Sections of CNS protective structures. Bony outer structures include the cranium and vertebrae. The meninges, located between the bony structures and the soft nervous tissue, are composed of three layers: dura mater, arachnoid mater, and pia mater. The cushioning presence of cerebrospinal fluid within the subarachnoid space provides yet another level of protection. The sinus shown in part **b** is a major cerebral vein.*

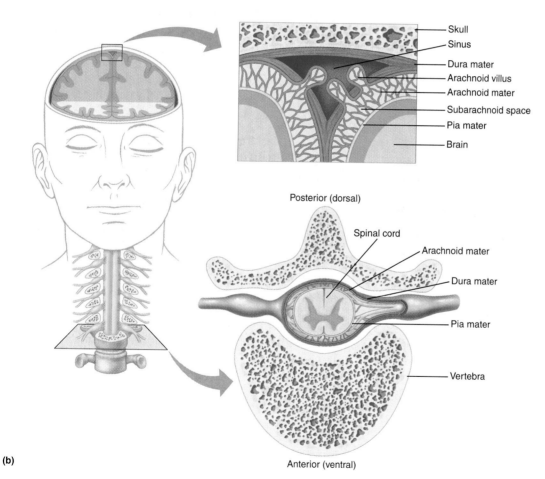

- Skull
- Sinus
- Dura mater
- Arachnoid villus
- Arachnoid mater
- Subarachnoid space
- Pia mater
- Brain

Posterior (dorsal)

- Spinal cord
- Arachnoid mater
- Dura mater
- Pia mater
- Vertebra

Anterior (ventral)

(b)

TABLE 8.1 COMPOSITIONS OF PLASMA AND CSF

	PLASMA	CSF
Glucose (mg/dl)	80–100	45–80
Proteins (mg/dl)	7000	15–45
Sodium (mM)	145	149
Potassium (mM)	4	3.1
Chloride (mM)	103	129
Calcium (mM)	2.5	2.1

acts as a shock absorber that prevents the soft nervous tissue from colliding with the hard bone.

CSF not only surrounds the CNS, but is inside it as well. CSF bathes all the neurons and glial cells of the CNS, and it also fills a number of cavities located within the brain and spinal cord. The brain contains four such cavities, called **ventricles,** which are continuous with the **central canal,** a long thin cylindrical cavity that runs the length of the spinal cord (Figure 8.2). The lining of the ventricles and central canal is composed of glial cells called **ependymal cells,** which are a type of epithelial cell. In some ventricles this lining is vascularized and forms a tissue called the **choroid plexus,** which consists of pia mater, capillaries, and ependymal cells. The choroid plexus produces CSF continuously at a rate of about 400–500 ml/day. As CSF is produced, it circulates through the ventricular system and enters the subarachnoid space through openings of the fourth ventricle. The CSF in the subarachnoid space eventually gets reabsorbed into venous blood through special structures in the arachnoid mater called *arachnoid villi* (singular = villus; see Figure 8.1b) located at the top of the brain.

Blood Supply to the Central Nervous System

Although the CNS accounts for only about 2% of body weight (the adult brain and spinal cord weigh approximately 3–4 pounds), it receives about 15% of the blood that the heart pumps to all the body's organs and tissues under resting conditions. This large blood supply is necessary because CNS tissue has a high rate of metabolic activity compared to most other body tissues, and thus has a high demand for fuel and oxygen to meet its energy needs. Under resting conditions, for example, the brain accounts for about 20% of all the oxygen that the body consumes, and about 50% of all the glucose consumed. To ensure delivery of these needed materials, adequate blood flow to the CNS must be maintained at all times. In fact, the CNS is so dependent on this blood supply that disruption of blood flow for even a few minutes can result in irreversible damage to CNS tissue. Sometimes, pathological conditions can reduce blood flow to a partic-

ular area of the CNS and cause deficits in certain functions, such as the ability to speak or move an arm. These changes in function occur in a *stroke*, an event in which blood flow is interrupted due to a blocked or ruptured blood vessel in the brain (see When It Goes Wrong: Stroke, p. 230).

The CNS is particularly sensitive to interruptions in blood flow because, unlike cells in most other tissues, cells in the CNS cannot store glycogen and therefore must obtain glucose directly from the blood. Furthermore, most cells in the CNS do not have access to fatty acids for energy, which increases their demand for glucose. (Recall that oxidation of a single fatty acid molecule can yield as much energy as the oxidation of several molecules of glucose.) Finally, unlike many other tissues, which can obtain energy from anaerobic metabolism during periods of reduced oxygen availability, nervous tissue cannot do this and therefore requires an uninterrupted supply of oxygen and glucose in order to stay alive.

Although glucose is the primary energy source for cells in the CNS, under certain conditions (such as in starvation or diabetes mellitus) CNS tissues can use ketones to supply up to two-thirds of their energy needs. Recall that ketones are produced as a by-product of lipid catabolism (primarily by the liver) when glucose supplies are limited. However, the CNS cannot rely on these compounds entirely, and thus must always rely on a steady supply of glucose.

The Blood-Brain Barrier

As in the other tissues of the body, the exchange of oxygen, glucose, and other materials between blood and cells in the CNS occurs across the walls of *capillaries*, the smallest blood vessels. Capillary walls are composed of little more than a single layer of endothelial cells (a type of epithelial cell), providing a short diffusion distance for exchange.

In most tissues, small molecules such as gases, inorganic ions, monosaccharides, and amino acids move freely across capillary walls. Hydrophobic molecules diffuse across the membranes of the endothelial cells, whereas hydrophilic molecules diffuse through relatively large gaps (pores) between the endothelial cells. Cells and large molecules such as proteins, however, are too large to move through these gaps; instead, large molecules can be actively transported across the endothelial cells by transcytosis—the movement of a molecule across an endothelial cell by endocytosis into the cell followed by exocytosis out of the cell.

In the CNS, most hydrophobic molecules can diffuse across the endothelial cells of capillaries, as in other tissues. However, transcytosis does not occur across capillary endothelial cells in the CNS, and the movement of hydrophilic molecules across capillary walls is restricted by the **blood-brain barrier** (Figure 8.3), a physical barrier that exists between the blood and CSF, which is the

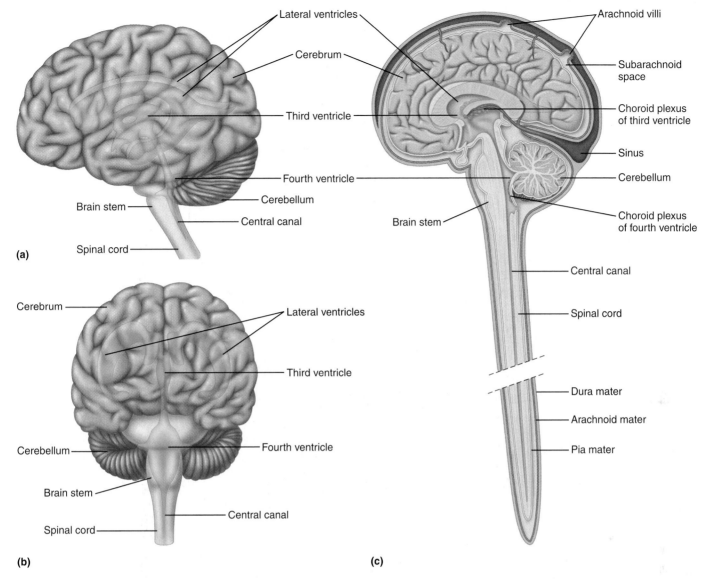

FIGURE 8.2 Ventricular system of the CNS. (a) *Lateral section.* **(b)** *Frontal section. The brain's four ventricles—two lateral ventricles, the third ventricle, and the fourth ventricle—are continuous with the central canal of the spinal cord.* **(c)** *Relationships of the ventricles to other CNS structures. Note the presence of the choroid plexus within the ventricles.*

interstitial fluid in the CNS. The existence of this barrier is due to the presence of tight junctions between the capillary endothelial cells, which eliminate capillary pores and restrict the diffusion of hydrophilic molecules between the cells. Glial cells called **astrocytes** are critical to the formation of the blood-brain barrier in that they stimulate endothelial cells to develop tight junctions by as yet unknown mechanisms.

The blood-brain barrier protects the CNS from harmful substances that may be present in the blood because it restricts the movement of molecules across capillary endothelial cells. In order to enter or leave capillaries, molecules must cross the endothelial cells themselves. Gases and other hydrophobic molecules penetrate through these cells relatively easily because they are able to move across cell membranes by simple diffusion through the lipid bilayer. As a consequence, these molecules are

able to move freely between blood and brain tissue. One example is ethanol (grain alcohol), which depresses CNS function by mechanisms not fully understood. (See Discovery: Alcohol Toxicity, www.physiologyplace.com, Challenge Yourself.)

In contrast, hydrophilic substances such as ions, sugars, and amino acids cannot cross the plasma membranes of any cells by simple diffusion and must therefore rely on *facilitated diffusion* to cross capillary walls in the CNS. Because facilitated diffusion utilizes transport proteins, which are selective to certain molecules, the blood-brain barrier is selectively permeable, allowing only certain compounds to move across. Compounds such as glucose and amino acids can penetrate the blood-brain barrier readily because transport proteins for these compounds are present in cell membranes. Other hydrophilic substances, such as inorganic ions (including H^+) and certain

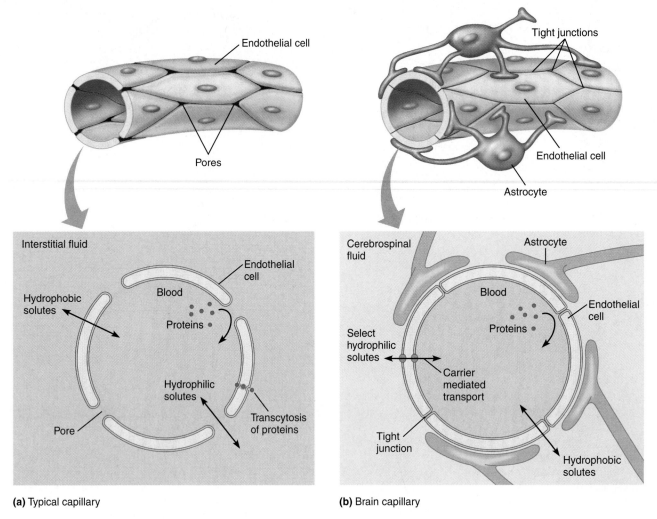

(a) Typical capillary

(b) Brain capillary

FIGURE 8.3 Blood-brain barrier. (a) *Typical capillaries (found in most regions of the body). Whereas exchange of small hydrophilic molecules occurs by simple diffusion between blood and interstitial fluid through pores, proteins are too large to cross through pores; some proteins are transported across capillary walls by transcytosis.* **(b)** *Brain capillaries. Because endothelial cells in these capillaries are connected by tight junctions, hydrophilic molecules must be transported across the wall by mediated transport systems. Proteins cannot cross the blood-brain barrier because transcytosis does not occur in brain capillaries. Even though astrocytes are found in close association with brain capillaries, they do not constitute a functional barrier.*

drugs, cannot penetrate this barrier to any great degree because transport proteins for these compounds do not exist in the endothelial cells of CNS capillaries. It is because of the selective permeability of the blood-brain barrier that the composition of cerebrospinal fluid differs from that of the plasma (see Table 8.1). Therefore, certain CNS diseases can be adequately diagnosed only by sampling cerebrospinal fluid, usually by a procedure called a *spinal tap.*

Gray Matter and White Matter

Although the central nervous system consists only of neurons and glial cells, the CNS is not entirely homogeneous. Instead, it is organized into distinct regions referred to as *gray matter* and *white matter*, based on visible differences in color. Approximately 40% of the CNS is

gray matter, which contains the cell bodies and dendrites of neurons, as well as the axon terminals that form synapses with them (Figure 8.4a). It is in the gray matter that synaptic transmission and neural integration occur. The other 60% of the CNS consists of **white matter,** which contains axons, most of which are myelinated (see Figure 8.4a). The presence of myelin, which has a high fat content, is responsible for the white matter's white appearance. Myelinated axons are specialized for the rapid transmission of information in the form of action potentials over relatively long distances. Glial cells are located throughout the CNS, in both gray matter and white matter.

When one looks at the external surface of the brain, only gray matter is visible, because the prominent globe-like structure (called the *cerebrum*) that makes up the bulk of the brain is entirely covered by a thin layer of gray matter called the **cerebral cortex.** The white matter

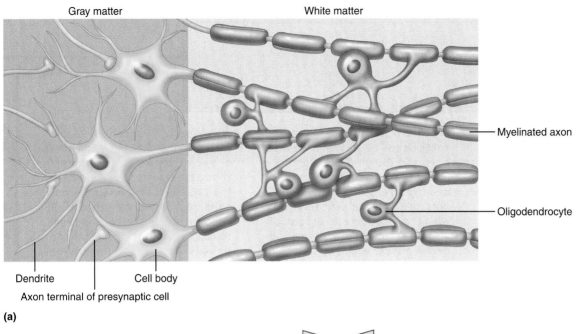

Gray matter   White matter

Myelinated axon

Oligodendrocyte

Dendrite   Cell body

Axon terminal of presynaptic cell

(a)

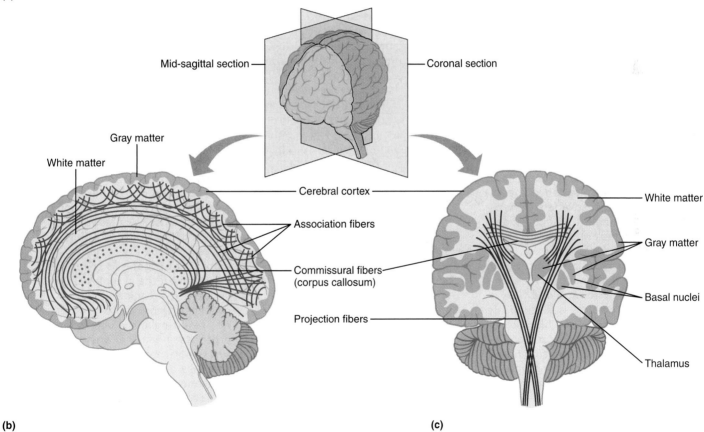

Mid-sagittal section   Coronal section

Gray matter

White matter

Cerebral cortex

Association fibers

Commissural fibers (corpus callosum)

Projection fibers

White matter

Gray matter

Basal nuclei

Thalamus

(b)   **(c)**

FIGURE 8.4 Makeup and arrangement of gray matter and white matter in the CNS. (a) *Histology of gray matter and white matter. Whereas gray matter consists primarily of cell bodies and dendrites and is the site of neural integration, white matter consists primarily of myelinated axons.* **(b, c)** *Midsagittal and coronal sections of the brain, showing association fibers, commissural fibers, and projection fibers, tracts of white matter that connect different areas of the CNS.*

On what part of the neuron are action potentials propagated?

Axons

Cerebrovascular diseases—those that affect blood vessels in the brain—are the third leading cause of death in the United States today, behind only heart disease and cancer. Cerebrovascular diseases are responsible for impairment of brain function in approximately 2 million Americans.

Stroke (also known as *cerebrovascular accident* or CVA) is a cerebrovascular disorder characterized by a sudden decrease or stoppage of blood flow to a part of the brain. Decreased blood flow (ischemia) is dangerous to any tissue, but brain tissue is especially vulnerable, in part because of the high rate of its metabolic reactions, which consume fuel and oxygen quickly. In fact, stoppage of blood flow for no more than three or four minutes may be sufficient to cause death of most brain cells. For this reason, a stroke can kill people within minutes, or leave them with permanent brain damage.

Strokes are classified as either *occlusive* or *hemorrhagic* and may occur in the interior of the brain or on its surface. In occlusive stroke, blood flow through a vessel is blocked, usually as a result of a blood clot. In hemorrhagic stroke, a blood vessel ruptures, causing *hemorrhage* (bleeding). Conditions that increase the likelihood of stroke include *arteriosclerosis* (narrowing of arteries due to buildup of fatty deposits on the inner wall), *aneurysm* (bulging of an artery due to weakening of the wall), and *hypertension* (high blood pressure). Stroke symptoms vary with the location and extent of the affected area, and are sudden in onset, which distinguishes stroke from other neurological disorders. There may be muscle weakness or paralysis, loss of sight (usually in just a part of the visual field), loss of other senses, tingling or other sensory disturbances, disturbances of language or spatial perception, changes in personality, disorientation, or memory loss.

Effects of a stroke may be confined to the initial symptoms, or they may intensify or become more widespread over the course of a few hours or days. Thus a person who first experiences numbness in a hand may find it spreading through the arm and shoulder. There are a number of possible reasons for this. If the stroke is hemorrhagic, blood could pool in the brain tissue and exert pressure on nearby vessels, thereby occluding blood flow over a wider area. Stroke sometimes induces spasms of smooth muscle in blood vessel walls, which also reduces flow.

Another possible cause of progressively worsening symptoms following a stroke is *glutamate excitotoxicity*, a phenomenon discovered only recently. Recall from Chapter 7 that glutamate is an excitatory neurotransmitter that affects virtually every cell in the CNS. Following a stroke, damaged neurons become chronically depolarized, which causes them to release neurotransmitter continually. As a result, the neurotransmitter excites nearby cells, which stimulates them to release more glutamate, which excites more cells, and so on. The result of the release of so much glutamate is that it accumulates in the extracellular fluid at high concentrations, such that the wave of excitatory activity cannot be stemmed. Unfortunately, the story does not end there. Among the receptors activated by glutamate are NMDA receptors in receptor-operated calcium channels. When glutamate binds to NMDA receptors, calcium permeability increases. When glutamate levels are chronically elevated following a stroke, calcium ions diffuse across the plasma membranes of cells at a rate that exceeds the ability of active transport mechanisms to remove calcium from the cytoplasm. Consequently, cytoplasmic calcium concentrations rise to toxic levels, resulting in the activation of proteases and other enzymes whose activities are normally carefully regulated. The result is a derangement of metabolism, and eventually cell death.

of the cerebrum is located beneath this layer. Embedded within this white matter are smaller areas of gray matter known as nuclei. Because they are beneath the cortex, they are often referred to as *subcortical nuclei.* In the spinal cord, the arrangement is the reverse; The white matter is on the outside, whereas the gray matter is on the inside.

In the white matter of the brain and spinal cord, axons (also referred to as **nerve fibers**) are organized into bundles or tracts that connect one region of gray matter with another. Different tracts are classified according to the locations of the particular regions they connect. **Projection fibers** or tracts connect the cerebral cortex with lower levels of the brain or the spinal cord (see Figure 8.4c). Thus, for example, *corticospinal tracts* connect regions of the cortex with gray matter in the spinal cord. **Association fibers** connect one area of cerebral cortex to another area of the cortex on the same side of the brain (see Figure 8.4b). For example, the *arcuate fasciculus* connects two regions known as *Broca's area* and *Wernicke's area*, which are important in language function and are located in different regions of the cerebral cortex on the same side. **Commissural fibers** connect cortical regions on one side of the brain with corresponding cortical regions on the other side. Most commissural fibers are located in a band of tissue called the **corpus callosum** (see Figure 8.4b, c), which connects the two halves of the cerebrum (called the **cerebral hemispheres**) together.

Quick Test 8.1

1. Name the three meninges in order, starting with the outermost.

2. What is cerebrospinal fluid, and where is it found?

3. What is the function of the blood-brain barrier? What anatomical structure between endothelial cells limits the diffusion of water-soluble molecules?

4. What substance is used by the central nervous system as its primary energy source?

THE SPINAL CORD

The spinal cord is a cylinder of nervous tissue that is continuous with the lower end of the brain and is surrounded by the vertebral column (Figure 8.5). The spinal cord is approximately 44 cm long in adults and ranges in diameter from 1 cm to 1.4 cm. Branching off the spinal cord at regular intervals are 31 pairs of **spinal nerves.** Each pair of spinal nerves exits the vertebral column between two adjacent vertebrae. The spinal nerves and the region of spinal cord from which they originate are given designations according to where the nerves emerge from the vertebral column. There are eight pairs of **cervical nerves** (identified as C1–C8), which emerge from the vertebral column in the neck region; 12 pairs of **thoracic nerves** (identified as T1–T12), which emerge in the chest region; five pairs of **lumbar nerves** (identified as L1–L5), which emerge in the region of the lower back; five pairs of **sacral nerves** (identified as S1–S5), which emerge from the region of the tailbone or coccyx; and a single **coccygeal nerve** (identified as C_o), which emerges from the tip of the coccyx. Although spinal nerves emerge along the whole length of the vertebral column, the spinal cord itself extends only about two-thirds of the column's length. Therefore, some of the spinal nerves actually travel downward within the vertebral column before exiting it. In fact, the bottom third of the vertebral column contains individual nerves, but no spinal cord

proper. Because the nerve bundle in this region resembles a horse's tail, it is called the *cauda equina* (Latin for "horse tail"). When physicians administer drugs spinally (give an *epidural*) or take a sample of cerebrospinal fluid (perform a *spinal tap*), they do so in this region to avoid potential damage to the spinal cord itself.

The numerous nerve fibers (or axons) that travel within a single spinal nerve generally travel to adjacent regions of the body. Thus it is possible to map out the body's surface into different sensory regions called *dermatomes*, each of which is served by a particular spinal nerve (Figure 8.6). (The face is not mapped because it is innervated by *cranial nerves*, which emerge from the brain rather than from the spinal cord.) Body maps such as this are useful because they enable clinicians to determine the location of damage to the spinal cord or spinal nerves. For example, numbness in the T1 dermatome on the right side would indicate damage to the first thoracic spinal nerve on that side, whereas numbness in the T1 dermatome and all lower dermatomes would indicate damage to the spinal cord between the level of C8 and T1.

The gray matter of the spinal cord is concentrated in a butterfly-shaped region in the interior of the cord, whereas the white matter is in the surrounding outer region (Figure 8.7). The gray matter contains interneurons, cell bodies, and the dendrites of efferent neurons, and the axon terminals of afferent neurons. The efferent neurons travel in spinal nerves to effector organs; the afferent neurons travel in spinal nerves from sensory receptors in the body's periphery to the spinal cord.

The gray matter of the spinal cord is organized in such a way that different types of neurons are located in different regions. The gray matter includes one *dorsal horn* and one *ventral horn* on each side (see Figure 8.7). The **dorsal horn** encompasses the dorsal (posterior) half of the gray matter on either side; the **ventral horn** encompasses the ventral (anterior) half. Afferent fibers originate in the periphery as sensory receptors and terminate in the dorsal horn, where they synapse on interneurons or efferent neurons. Note that the cell bodies of these afferent fibers are not located in the spinal cord itself; they are located outside the spinal cord in clusters called **dorsal root ganglia.** (Recall from Chapter 6 that *ganglion* is a general term for any cluster of neuron cell bodies outside the CNS.) In contrast, the cell bodies of efferent neurons are located in the spinal cord. Efferent neurons originate in the ventral horn and travel to the periphery, where they form synapses with skeletal muscles. In the thoracic and upper lumbar regions of the spinal cord, there is another area of gray matter called the *lateral horn*, or *intermediolateral cell column*, which is located between the dorsal and ventral horns on either side. The lateral horns are the origins of efferent neurons of the autonomic nervous system (see Chapter 10).

Afferent and efferent axons travel together in spinal nerves, but they separate into different bundles as they enter or leave the spinal cord. The bundles containing afferent axons are called **dorsal roots,** whereas the bundles

FIGURE 8.5 The spinal cord. *(Left) Location of the spinal cord, as seen in a posterior view. (Right) Lateral view of the spinal cord, showing its position within the bony vertebral column. Also shown are the 31 pairs of spinal nerves, which leave the spinal cord between adjacent vertebrae.*

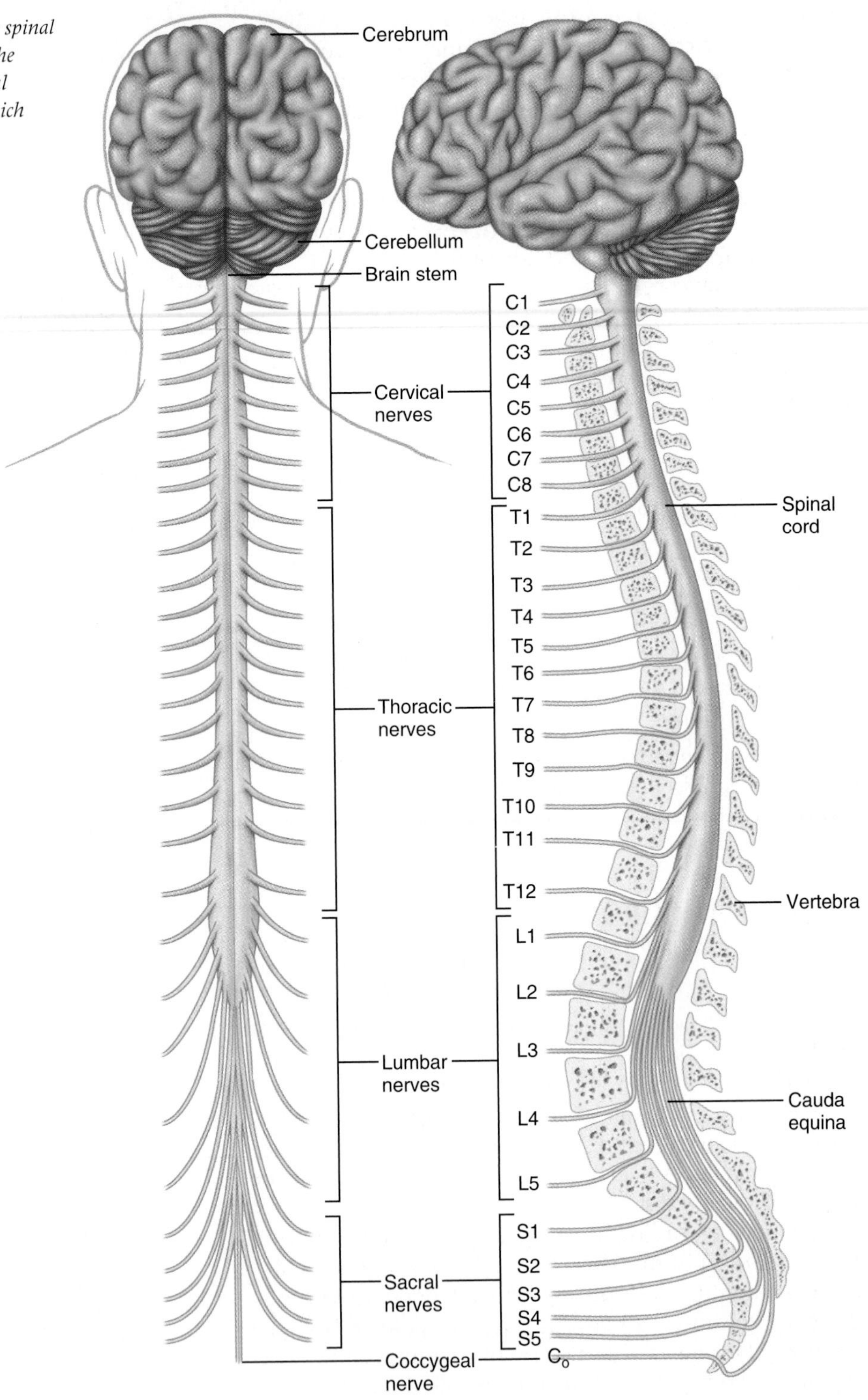

Cerebrum

Cerebellum

Brain stem

Cervical nerves

Thoracic nerves

Lumbar nerves

Sacral nerves

Coccygeal nerve

C1
C2
C3
C4
C5
C6
C7
C8
T1
T2
T3
T4
T5
T6
T7
T8
T9
T10
T11
T12
L1
L2
L3
L4
L5
S1
S2
S3
S4
S5
Co

Spinal cord

Vertebra

Cauda equina

containing efferent axons are called **ventral roots.** A short distance from the spinal cord these dorsal and ventral roots come together to form the spinal nerves. All spinal nerves, therefore, are described as **mixed nerves** because they contain both afferent and efferent axons.

The white matter of the spinal cord consists of tracts that provide communication either between the different levels of the spinal cord or between the brain and various levels of the spinal cord (Figure 8.8). **Ascending tracts** transmit information from spinal cord to brain, whereas **descending tracts** transmit information from brain to spinal cord. Although the tracts in Figure 8.8 are labeled on only one side of the spinal cord for clarity, all tracts are *bilateral;* that is, any given tract is found on both sides of the spinal cord. For example, the *pyramidal tracts* are descending pathways that transmit motor commands to efferent neurons on both sides of the spinal cord. The *dorsal columns* are ascending tracts that transmit sensory information from the periphery to the brain.

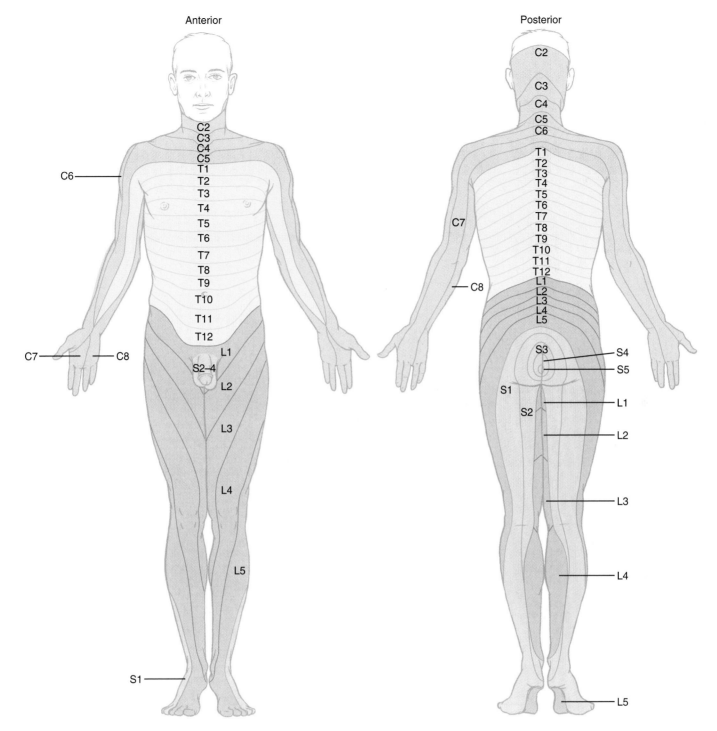

FIGURE 8.6 Dermatomes. *Each dermatome is a sensory region on the surface of the body that is served by the spinal nerve indicated by the abbreviations.*

Numbness in the left thumb indicates damage to which spinal nerve?

The ascending and descending tracts effectively link the peripheral nerves to the brain. When afferent neurons are activated by a stimulus acting on a sensory receptor (for example, when a finger touches a rose petal or a thorn), action potentials travel along the nerve fiber from the sensory receptor to the axon terminal, usually in the dorsal horn of the spinal cord (Figure 8.9a). The axon terminal releases a neurotransmitter that transmits the signal to an interneuron (or, in rare cases, directly to

an efferent neuron). Some of these interneurons form the ascending tracts, which transmit the information to the brain so that perception of the stimulus can occur. Signals from the brain travel along descending tracts to efferent neurons in the ventral horn (Figure 8.9b). For

C6 on the left side

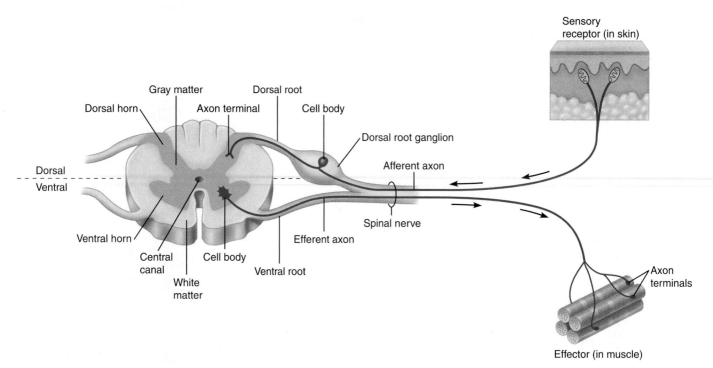

FIGURE 8.7 Spinal cord gray matter and spinal nerves. *This cross section of the spinal cord at the lumbar level reveals the two functional halves of spinal cord gray matter: dorsal and ventral. Axons of afferent neurons enter the spinal cord through the dorsal root and terminate in the dorsal horn; their cell bodies are located in dorsal root ganglia. Axons of efferent neurons originate in the ventral horn and exit through the ventral root. Because they contain axons of both afferent and efferent neurons, spinal nerves are considered mixed nerves.*

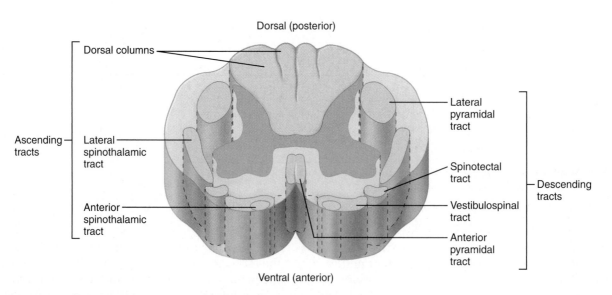

FIGURE 8.8 Cross section of white matter tracts in the spinal cord. *Spinal cord white matter consists of longitudinal tracts that run between the brain and the spinal cord or between different spinal cord segments. Ascending tracts transmit information from spinal cord to brain, whereas descending tracts transmit information from brain to spinal cord. Only selected tracts are illustrated here.*

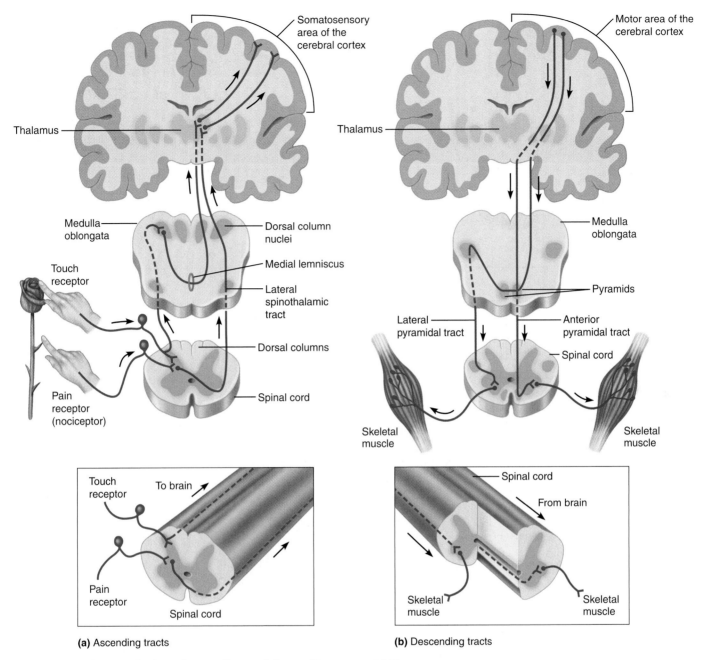

(a) Ascending tracts   **(b)** Descending tracts

FIGURE 8.9 Pathways of selected ascending and descending tracts. (a) *The dorsal column and lateral spinothalamic tracts. Both of these ascending pathways (see Figure 8.8) originate with sensory receptors in the periphery and travel up the spinal cord, eventually communicating sensory information to the thalamus and then to the cerebral cortex. The dorsal column pathway crosses to the contralateral side in the brainstem (medial lemniscus), whereas the spinothalamic tract crosses to the contralateral side in the spinal cord. **(b)** The pyramidal tracts (see Figure 8.8). Both pyramidal tracts originate in the primary motor cortex. The lateral pyramidal tract crosses over in the medullary pyramids, whereas the anterior pyramidal tract does not cross. Both tracts terminate in the ventral horn of the spinal cord, where they communicate to motor neurons innervating skeletal muscle.*

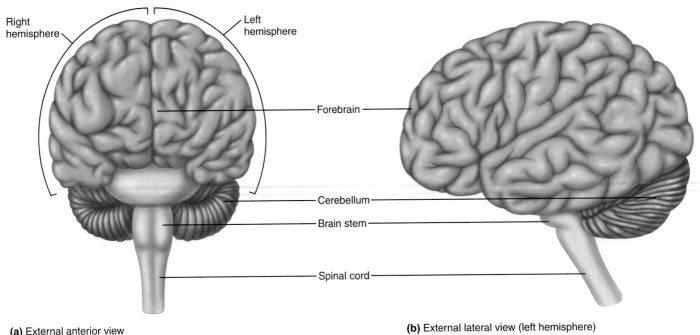

(a) External anterior view

(b) External lateral view (left hemisphere)

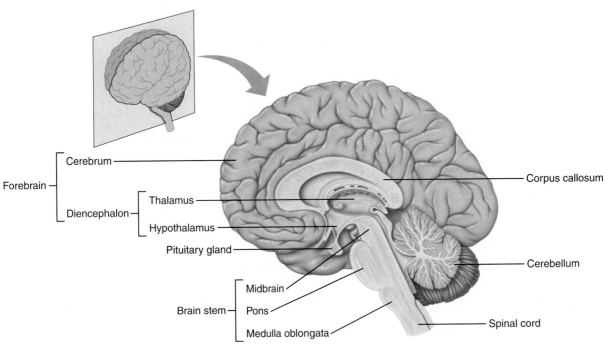

(c) Sagittal section

FIGURE 8.10 The brain. *The brain is composed of three main parts: forebrain, cerebellum, and brainstem. (a, b) External views of the brain, showing its three main parts and their relation to the spinal cord. (c) A mid-sagittal section of the brain, showing the structures of the three main brain parts. Note the corpus callosum, the major fiber tract connecting the left and right cerebral hemispheres.*

example, when you wish to wiggle your finger, the brain transmits commands through descending fibers to the efferent neurons that control the skeletal muscles that cause the finger to move. Controlling efferent neurons is not the only function that descending fibers perform, however. Some descending tracts modulate sensory information. For example, the body has pain-relieving *(analgesic)* systems that block the perception of pain during periods of stress. These systems work by blocking synaptic transmission between a pain-transmitting afferent neuron and interneurons in the spinal cord. By blocking the transmission of pain information to the brain, perception of a painful stimulus is prevented. (This topic is discussed further in Chapter 9.)

Notice that the ascending and descending tracts illustrated in Figure 8.9 generally cross over to the side opposite the side of origin. This is the norm, but there are exceptions.

TABLE 8.2 CRANIAL NERVES

NUMBER	NAME	NERVE CLASS	FUNCTION
I	Olfactory	Sensory	Olfaction (smell)
II	Optic	Sensory	Vision
III	Oculomotor	Mixed	Eye movements; pupillary reflex; accommodation reflex; proprioception (position of muscles and joints)
IV	Trochlear	Mixed	Eye movements; proprioception
V	Trigeminal	Mixed	Motor control of chewing; somatic sensations of face, nose, mouth
VI	Abducens	Mixed	Eye movements
VII	Facial	Mixed	Motor control of facial muscles, salivary glands, tear glands; somatic sensations of face; taste
VIII	Vestibulocochlear	Sensory	Hearing; equilibrium
IX	Glossopharyngeal	Mixed	Motor control of swallowing and salivary glands; taste; visceral afferent from pharyngeal region; visceral afferent from baroreceptors
X	Vagus	Mixed	Motor and visceral afferent of thoracic and abdominal viscera; motor control of larynx and pharynx
XI	Accessory	Motor	Motor control of larynx and pharynx
XII	Hypoglossal	Mixed	Motor and somatic sensations of tongue

When a pathway remains on the same side as its origin, it is called **ipsilateral.** When a pathway crosses to the side opposite its origin, it is called **contralateral.** Because most sensory and motor pathways cross to the opposite side within the CNS, sensory input to the right side of the body is transmitted to the left side of the brain for perception, and motor control of the right side of the body comes from the left side of the brain.

Quick Test 8.2

1. What is the functional difference between the dorsal and ventral horns of the spinal cord?

2. Define dorsal root, ventral root, and spinal nerve. Why are spinal nerves called mixed nerves?

3. What is the functional difference between ascending and descending tracts in the spinal cord?

4. Define the terms *ipsilateral* and *contralateral* in relation to information traveling through the brain and spinal cord.

THE BRAIN

The brain consists of three main parts: the forebrain, the cerebellum, and the brainstem (Figure 8.10). The **forebrain,** the largest and most superior part of the brain, is divided into left and right halves, or *hemispheres*, and consists of the cerebrum and diencephalon (Figure 8.10c).

The **cerebrum** is a large, roughly C-shaped structure containing both gray and white matter. The gray matter areas include the *cerebral cortex* at the surface and deep *subcortical nuclei*. The **diencephalon** consists of the *thalamus* and *hypothalamus*, two mid-line structures located near the base of the forebrain, each of which contains multiple small nuclei.

The **cerebellum** (derived from the Latin word for "little brain") is a bilaterally symmetrical structure, with an outer cortex and inner nuclei, similar to the forebrain. It is located inferior to the forebrain and dorsal or posterior to the brainstem. The cerebellum functions in motor coordination and balance, providing feedback to motor systems to ensure smooth movements of the eyes and body (described in detail later in this chapter).

The **brainstem** is the caudal-most part of the brain and connects the forebrain and cerebellum to the spinal cord. The brainstem consists of three main regions: (1) the **midbrain,** the most rostral portion, which connects to the forebrain; (2) the **pons,** the middle portion, which connects to the cerebellum; and (3) the **medulla oblongata,** the most caudal portion, which connects to the spinal cord. The brainstem contains many nuclei that perform a wide variety of functions. Within the brainstem are processing centers for ten of the 12 pairs of **cranial nerves,** peripheral nerves that emanate directly from the brain rather than the spinal cord. The 12 pairs of cranial nerves and their basic functions are listed in Table 8.2. Also located within the brainstem is the **reticular formation,** a diffuse network of nuclei important in sleep-wake cycles, arousal of the cerebral cortex, and consciousness. In addition, the brainstem is important in

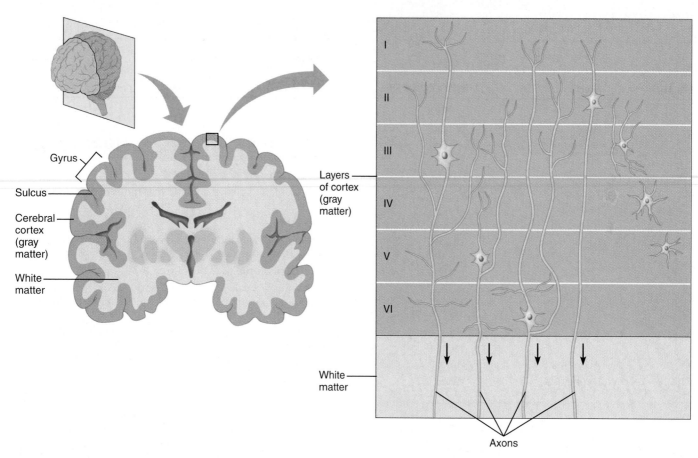

FIGURE 8.11 Organization of the cerebral cortex. *The cerebral cortex, the convoluted outer layer of the cerebrum, is arranged in six distinct layers. The cell bodies are confined to a single layer, but axons and dendrites extend across layers. The actual arrangement of the layers depends on the area of cortex examined.*

What is the purpose of sulci and gyri?

the regulation of many involuntary functions controlled by the autonomic nervous system, such as cardiovascular function, breathing, and vomiting.

Cerebral Cortex

The cerebral cortex is the outermost portion of the cerebrum, consisting of a thin, highly convoluted layer of gray matter (Figure 8.11). The convolutions consist of grooves called **sulci** (singular: **sulcus**) and ridges called **gyri** (singular: **gyrus**). The convolutions allow for a greater volume of cerebral cortex to be accommodated within a given cranial volume. The cerebral cortex is the most advanced area of the brain and, in terms of human evolution, the most recent area to develop. The cerebral cortex ranges from 1.5 mm to 4 mm thick, depending on location, and it contains an estimated billion (10^9) neurons and trillion (10^{12}) synapses. Although the cortex is thin, it is generally arranged in six functionally distinct layers (see Figure 8.11). The specific arrange-

ment of layers varies, depending on the function of the particular cortical region.

The cerebral cortex carries out the highest level of neural processing. It is in the cortex that we perceive our environment, formulate ideas, experience emotions, recall past events, and command our bodies to move. To carry out all these complex functions, the cortex acts as an integrating center; that is, it receives many types of sensory input from different origins, consolidates this information, and uses it to formulate thoughts and actions.

Functional Organization of the Cerebral Cortex

Each of the cerebral hemispheres is divided into four regions known as *lobes* (Figure 8.12). The **frontal lobe** is the anterior part of the cerebrum. Immediately posterior to it is the **parietal lobe.** These two lobes are separated by the *central sulcus*, a prominent groove on each hemisphere of the brain. Located posterior and inferior to the parietal lobe is the **occipital lobe.** The **temporal lobe** is located inferior to the frontal and parietal lobes on the side of the cerebrum; it is separated from the frontal lobe

To increase the volume of cerebral cortex

by a very deep groove called the *lateral sulcus* (or *Sylvian fissure*).

Within each lobe, the cerebral cortex can be subdivided into areas specialized for different functions (Figure 8.13). The occipital lobe is also called the *visual cortex* because the processing of visual information occurs there. Another example is the *auditory cortex*, an area located in the superior temporal lobe that functions in hearing. In the parietal lobe is another specialized area called the **primary somatosensory cortex,** which is involved in the processing of *somatic* sensory information associated with both surface sensations such as touch, itch, temperature, and pain (called *somesthetic sensations*) and awareness of muscle tensions, joint, and limb positions (called *proprioception*). The frontal lobe contains the **primary motor cortex,** which initiates voluntary movement, and other areas involved in motor control. The frontal lobe also contains areas involved in language and planning, and it is important to the establishment of personality (see Discovery: The Story of Phineas Gage, p. 243).

Several functional areas of the cerebral cortex are *topographically* organized, meaning that areas of the cerebral cortex can be mapped according to their function. The best examples of this topographical organization occur in the primary motor cortex in the frontal lobe, and in the primary somatosensory cortex in the parietal lobe. Maps of the *somatotopic organization* of these two cortical areas, in which body parts are mapped onto the cortical surface with which they correspond, are called **motor** and **sensory homunculi** (homunculus = "little man"; Figure 8.14). In homunculi, each body part is shown next to the area of cerebral cortex devoted to it, and the relative size of the body part in the drawing represents the relative size of the cortical area devoted to that body part. Consider the fingers and thumb: Because they are capable of both delicate movements and fine tactile discrimination, the digits are greatly overrepresented in the cortex relative to their actual size. Thus homunculi show that the area of the motor cortex containing neural circuits devoted to the movement of the digits (Figure 8.14a), and the area of the sensory cortex containing neural circuits that process sensory signals coming from the digits (Figure 8.14b), are extensive. Other areas of cortex are also topographically organized. For example, one can construct a map of the visual field (what the eyes see) in the primary visual cortex, or a frequency (tone or pitch of sound) map in the primary auditory cortex.

We have so far discussed the cortex as if each area handles one specific function. However, that is not entirely true. Certain areas of cortex, called **association areas,** are involved in more complex processing that requires integrating different types of information. For example, if you are awakened in the middle of the night by a loud bang outside your bedroom window, you must make a decision as to whether or not action is needed. You would try to determine what made the loud bang, whether it poses potential danger to you, and if so, what

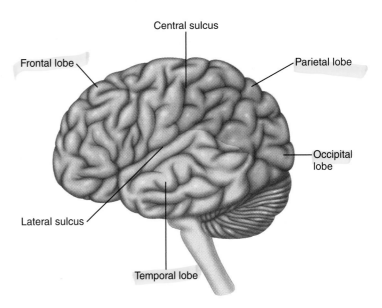

FIGURE 8.12 Lobes of the cerebrum. *This lateral view of the left cerebrum shows its four distinct lobes: frontal, parietal, occipital, and temporal. The central sulcus separates the frontal and parietal lobes; the lateral sulcus separates the temporal lobe from the frontal and parietal lobes.*

you can do to avoid it. You might look outside and listen closely for any other unusual sounds. To make a decision, your brain must consolidate information from sensory systems (such as the eyes and ears) and from memory (What things make noises like that?). Much of this type of processing occurs in association areas.

Functional Specializations of Left and Right Cerebral Cortex

The division of the cerebrum into two hemispheres is functional as well as anatomical. That is, specialization of function occurs in the left and right sides of the brain. The right side of the brain is generally more involved in creative and artistic endeavors, such as art and music, and is better at spatial perception; it also controls movement on the left side of the body, and processes sensory information originating from the left side. The left side of the brain, in contrast, tends to excel in logic, analytical abilities, and the comprehension and expression of language (described in detail later), and it controls sensory and motor function on the right side of the body. Note, however, that this specialization does not mean that no creativity occurs on the left side of the brain, or that language capabilities are absent on the right side—only that each side dominates for certain functions.

Subcortical Nuclei

The subcortical nuclei are regions of gray matter located within the cerebrum. Among the more prominent subcortical nuclei are the basal nuclei, the thalamus, the hypothalamus, and those in the limbic system (Figure

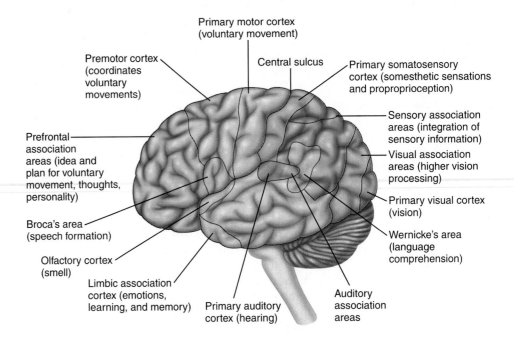

FIGURE 8.13 Functional areas of the cerebral cortex. *Some selected areas of the cerebral cortex and the specific functions associated with them are illustrated.*

Primary motor cortex (voluntary movement)

Premotor cortex (coordinates voluntary movements)

Central sulcus

Primary somatosensory cortex (somesthetic sensations and proprioception)

Sensory association areas (integration of sensory information)

Prefrontal association areas (idea and plan for voluntary movement, thoughts, personality)

Visual association areas (higher vision processing)

Primary visual cortex (vision)

Broca's area (speech formation)

Wernicke's area (language comprehension)

Olfactory cortex (smell)

Limbic association cortex (emotions, learning, and memory)

Primary auditory cortex (hearing)

Auditory association areas

8.15). The **basal nuclei** (also called basal ganglia), which include the *caudate nucleus*, *globus pallidus*, and *putamen*, are notable for their role in modifying movement and are described later in this chapter.

Thalamus

The thalamus is a cluster of subcortical nuclei located in the diencephalon. All sensory information follows a pathway that includes a direct relay through the thalamus to the cerebral cortex (see Figure 8.9 for two examples), with the possible exception of the sense of smell. Much sensory input is filtered by and refined in the thalamus before being transmitted to the cortex. In this manner, the thalamus seems to be important in directing attention, as when a mother is more attentive to her crying baby than to an airplane flying overhead, even though the latter may be louder. The thalamus also plays a role in controlling movement (described later in this chapter).

Hypothalamus

The hypothalamus, located inferior to the thalamus, has many roles in regulating homeostasis. The hypothalamus is the major link between the two communication systems of the body, the endocrine and nervous systems. In response to neural and hormonal input, the hypothalamus releases tropic hormones that regulate the release of anterior pituitary hormones. In addition, the hypothalamus controls the release of hormones from the posterior pituitary, including antidiuretic hormone, which regulates plasma volume and osmolarity, and oxytocin, which regulates uterine contractions and milk ejection.

The hypothalamus also affects many behaviors. The hypothalamus contains satiety and hunger centers, which regulate eating behavior, and the thirst center, which reg-

ulates drinking behaviors. In addition, because the hypothalamus is part of the limbic system (discussed shortly), it affects emotions and behaviors in response to emotions.

Many of the responses produced by the hypothalamus are exerted through the action of the autonomic nervous system. The hypothalamus has both direct and indirect inputs to the autonomic nervous system. For example, emotions can affect cardiovascular, respiratory, and digestive functions through hypothalamic input to autonomic control centers in the brainstem. The hypothalamus also regulates body temperature, which involves several integrated autonomic responses.

Exercise Link

The hypothalamus receives inputs from thermosensitive sensors throughout the body and is itself temperature sensitive. As they ran, Bill and Jane were generating heat as a by-product of their muscle contractions; they also absorbed radiant heat from the sun as it rose higher in the sky. Signals from the anterior hypothalamus to cardiovascular control centers in the medulla oblongata triggered increases in blood flow to the skin and increases in sweating, both of which help dissipate heat by convection and evaporation. Information from the primary motor cortex regarding Bill and Jane's exercise intensity was transmitted to the posterior hypothalamus and then to cardiovascular control centers of the medulla oblongata. This input helped regulate arterial blood pressure, helping to ensure that all tissues and organs received the blood flow they needed.

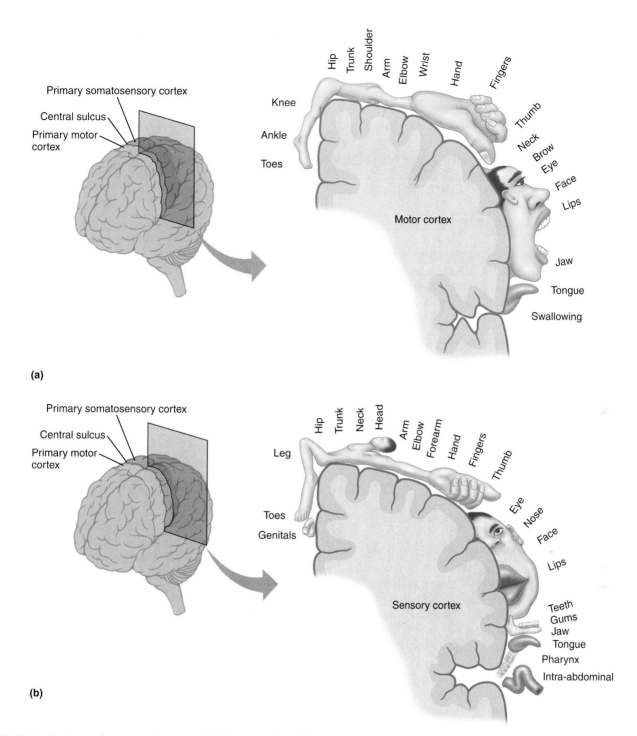

(a)

(b)

FIGURE 8.14 Motor and sensory homunculi. (a) *Cross section of the primary motor cortex, located just anterior to the central sulcus, and the corresponding somatotopic map of body parts.* **(b)** *Cross section of the primary somatosensory cortex, located just posterior to the central sulcus, and the corresponding somatotopic map of body parts.*

The suprachiasmatic nucleus of the hypothalamus generates and regulates the circadian rhythm, endogenous fluctuations in body functions that occur on a 24-hour cycle. Extending from the diencephalon is an endocrine organ called the *pineal gland*, which secretes the hormone *melatonin*, which also plays a role in establishing circadian rhythms.

Limbic System

The **limbic system** is a diverse collection of closely associated cortical regions, subcortical nuclei, and tracts in the forebrain that function in learning and emotions. It includes the *amygdala, hippocampus, fornix,* and *cingulate gyrus* of the cerebral cortex (Figure 8.16), as well as portions of the thalamus and hypothalamus. The limbic system, one

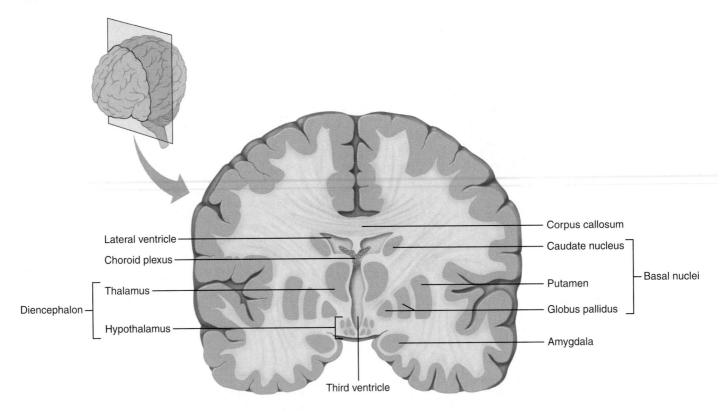

FIGURE 8.15 The subcortical nuclei. *A coronal section of the cerebrum at the level indicated reveals the gray matter areas known as the subcortical nuclei: the basal nuclei (caudate nucleus, putamen, and globus pallidus), thalamus, hypothalamus, and the amygdala (part of the limbic system).*

What is the function of the choroid plexus?

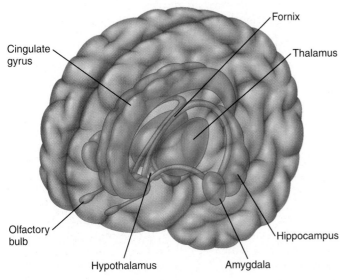

FIGURE 8.16 The limbic system. *The major structures of the limbic system are depicted in this three-dimensional view.*

of the more "primitive" areas of our brain, is involved in basic drives. For example, one of the oldest regions of the brain (in terms of human evolution) is the amygdala, which is involved in memory and emotions, especially fear. The hippocampus, a major component of the limbic system, is involved in learning and memory. The olfactory system provides important sensory input to the limbic system, especially in lower mammalian species.

Quick Test 8.3

1. Name the two structures that make up the diencephalon, and list their major functions.

2. What three structures make up the brainstem?

3. Name the four lobes of the cerebral cortex and describe at least one function associated with each.

4. Name five functions of the hypothalamus.

5. For each of the following functions, indicate whether the left or right cerebral hemisphere plays the predominant role: (a) creative activity, (b) logic, (c) motor control on the left side of the body, (d) sensory perception on the left side of the body, (e) language comprehension and expression.

To synthesize CSF

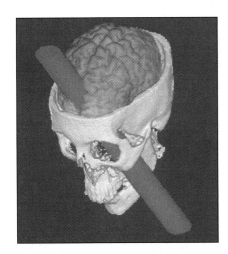

In 1848, while working as a railroad construction foreman, a 25-year-old man named Phineas Gage became the victim of a bizarre accident that yielded early insights into the brain's role in personality. As a result, his story will forever be a part of the history of science. According to eyewitnesses, here is what happened:

On the day of the accident, Gage was supervising workers who were laying a new section of track through the rocky hills of Vermont. To make way for the track, they were preparing to remove a large piece of rock by blasting. To do this, workers would bore a hole into the rock, fill it with gunpowder, cover the powder with a layer of sand to shield it from sparks, tamp down the powder and sand with a heavy metal rod (to make the explosion more powerful), and then ignite the powder with a fuse. Gage was in the middle of this procedure when he was distracted by a sound. After looking around to investigate, he returned to his task, unaware that he had forgotten to add the sand. When he dropped the tamping rod down the hole, it apparently created a spark that ignited the powder. The resulting explosion drove the rod upward into his face underneath the left cheekbone. The missile continued on its relentless course, tearing through bone, muscle, and brain tissue, until it shot from the top of his head with enough force to sail about 50 feet into the air. Gage flew backward, lay unconscious on the ground, and then began to convulse. Within minutes, however, he had regained consciousness and was able to move about (with assistance), despite the fact that a substantial mass of brain and skull tissue had been blown out of his head!

Amazed by what they were witnessing, Gage's coworkers transported him by carriage to a physician in a nearby town. For the next few days, Gage's life hung in the balance; the injury had caused the brain to swell, and infection had set in. Nevertheless, he passed through the crisis and within a few months was fully recovered and able to lead a nearly normal life.

Unfortunately, however, Gage did not emerge unchanged. Although the injury had no apparent effect on his ability to reason, speak, or move, it profoundly affected his personality. Prior to the accident he had been a calm, likeable individual, but afterward he became quarrelsome and would fight with any man at the slightest provocation. His language became coarse and vulgar, and he could not restrain himself from using obscenities. Once a model citizen, even-tempered and sound in judgment, he was now impulsive, hot-headed, and irresponsible. In the words of his physician, he was "no longer Gage." Because he could not control his behavior, he lost his job and was forced to roam the countryside in search of employment. At one point, he took a job with the Barnum and Bailey Circus, displaying himself and his tamping rod as a sideshow attraction.

Phineas Gage died at age 37, 12 years after the accident that robbed him of his personality. His skull and the tamping rod that pierced it are currently kept at Harvard Medical School. By examining a sketch made by Gage's physician indicating the relative sizes of the tamping iron and Gage's skull, and a computer reconstruction of the probable path taken by the rod as it passed through his head we can see that the accident likely obliterated a large portion of his frontal lobes, leaving the rest of the brain intact. Because the frontal lobes are important in goal-directed behavior, they exert a restraining influence on the more emotional, impulsive parts of the brain, such as the limbic system. After the accident, this restraining influence was no longer present, which allowed Gage's emotions to rule his behavior.

Thus far we have focused on identifying specific areas of the CNS and describing their major functions. Although it is true that different areas specialize in different functions, it is important to realize that most tasks carried out by the CNS require the coordinated action of many different areas working together. For example, we can identify an object as an apple because we can see it, touch it, smell it, taste it, put those sensations together into a coherent "picture" of the object, and compare that picture with our concept of what an apple is, which derives from our remembered experiences with apples. Thus, the seemingly simple task of identifying an apple involves the coordinated activity of several sensory areas, memory storage areas, and association areas. In the next section we examine several CNS functions that involve such coordinated actions, starting with the simplest functions: *reflexes*. We then examine more complex tasks, including voluntary motor control, language, sleep and consciousness, emotions and motivation, and learning and memory.

TABLE 8.3   CLASSES OF REFLEXES

BASIS OF CLASSIFICATION	CLASSES	EXAMPLE
Level of neural processing	Spinal	Muscle spindle stretch reflex
	Cranial	Pupillary reflex
Efferent division controlling effector	Somatic	Muscle spindle stretch reflex
	Autonomic	Baroreceptor reflex to control blood pressure
Developmental pattern	Innate	Muscle spindle stretch reflex
	Conditioned	Salivation reflex of Pavlov's dogs
Number of synapses in the pathway	Monosynaptic	Muscle spindle stretch reflex
	Polysynaptic	All other reflexes

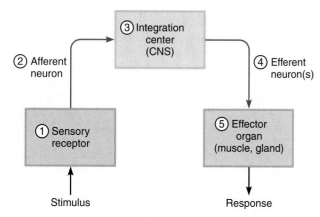

FIGURE 8.17 Schematic representation of a reflex arc. *The five components of a reflex arc are a sensory receptor that detects a stimulus, an afferent neuron that transmits information from the receptor to the CNS, an integration center (which is generally the CNS), an efferent neuron that transmits information from the integration center to the periphery, and an effector organ, which produces a response to the stimulus.*

INTEGRATED CNS FUNCTION: REFLEXES

We have seen that neural pathways are a series of neurons connected by synapses that form a line of communication required for a specific task. One example is a visual pathway that starts in the eye with light-sensitive photoreceptor cells and ends in the area of visual cortex responsible for perception of vision. Pathways in the nervous system vary in their complexity. The simplest pathways in the nervous system are *reflex arcs*.

When we react to some stimuli, we often stop to consider the situation before taking action. Sometimes, however, our responses are automatic, involving no conscious intervention on our part, as when we jump in response to a loud sound. Such an automatic, patterned response to a sensory stimulus is called a **reflex.**

Reflexes can be categorized into the following four groups, each of which contains two classes (Table 8.3):

1. Reflexes can be either *spinal* or *cranial* according to the level of neural processing involved. In spinal reflexes, the highest level of integration occurs in the spinal cord; cranial reflexes require participation by the brain.

2. Reflexes can be either *somatic* or *autonomic,* depending on which efferent division controls the pathway. Somatic reflexes involve signals sent via somatic neurons to skeletal muscle; autonomic reflexes (also called *visceral reflexes*) involve signals sent via autonomic neurons to smooth muscle, cardiac muscle, or glands.

3. Reflexes can be either *innate* (inborn) or *conditioned* (learned).

4. Reflexes can be either *monosynaptic* or *polysynaptic*. In monosynaptic reflexes, the neural pathway consists of only two neurons and a single synapse; polysynaptic reflexes contain more than two neurons and multiple synapses.

Note that a given reflex can belong in more than one class. The reflex involved when the patellar tendon below the knee is tapped and the leg kicks forward is both a spinal reflex (the brain is not involved) and a somatic reflex (because efferent signals are sent to skeletal muscles via somatic neurons). Similarly, the reflex involved when the pupil constricts as a light is shined in the eye is both a cranial reflex (because it involves the brain) and an autonomic reflex (because efferent signals are sent to the smooth muscle surrounding the pupil via autonomic neurons).

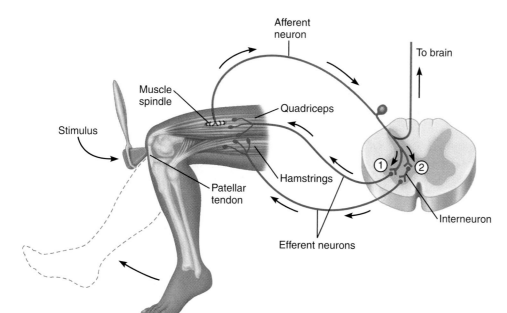

FIGURE 8.18 **The muscle spindle stretch reflex.** *The knee jerk reflex, an example of the monosynaptic muscle spindle stretch reflex, by which a tap on the patellar tendon causes contraction of the quadriceps muscle. Muscle spindle afferent neurons make two synaptic communications in the spinal cord: (1) excitatory synapses with efferent neurons to the quadriceps muscle, and (2) inhibitory synapses with interneurons that communicate with efferent neurons to the hamstring muscles in the same leg. The afferent neurons also have collaterals that travel in the white matter of the spinal cord to the brainstem, where they form synapses with interneurons that transmit information about muscle length to various areas of the brain.*

Note as well that these reflexes are innate, as they need not be learned. The classic example of a learned reflex is the *salivation reflex,* first described by Ivan Pavlov, a Russian physiologist who won the Nobel Prize in 1904 for his study on digestion in dogs. In his classic demonstration of the salivation reflex, over time Pavlov rang a bell just before he fed the dogs. The dogs learned to associate the sound of the bell with the arrival of food, so that eventually they began to salivate upon hearing the bell, even if no food was forthcoming. Thus salivation in response to the sound of a bell is a conditioned (learned) reflex.

The simplest reflex pathways are known as **reflex arcs,** which consist of five components (Figure 8.17): (1) a sensory receptor, (2) an afferent neuron, (3) an integration center, (4) an efferent neuron, and (5) an effector organ. To set the reflex into motion, the receptor first detects a stimulus. Information is then transmitted from the receptor to the CNS via the afferent neuron. The CNS, which functions as the integrator, sends signals to the efferent neuron, which transmits signals to the effector organ, stimulating it to produce a specific response.

Stretch Reflex

The simplest example of a reflex is the **muscle spindle stretch reflex,** of which the knee-jerk reflex (when the leg kicks in response to a tap just below the knee, as previously described) is an example. Figure 8.18 depicts this reflex in detail. The stretch reflex is the only known monosynaptic reflex in the human body. The receptor in this reflex is a *muscle spindle*, a specialized structure found in skeletal muscles that responds when muscles are stretched. In the knee-jerk reflex, tapping the patellar tendon below the knee cap stretches the quadriceps muscle in the upper thigh. This stretch excites muscle spindles

in that muscle, thereby triggering action potentials that travel in afferent neurons to the spinal cord (integration center). In the spinal cord, the afferent neurons make direct excitatory synaptic connections (synapse 1 in Figure 8.18) with efferent neurons that innervate the quadriceps muscle, the same muscle that contains the muscle spindles, stimulating the quadriceps to contract. In addition, for the leg to "kick" effectively, the muscles opposing the kicking motion, the hamstrings, must relax. Afferent neurons from muscle spindles have collaterals that synapse with inhibitory interneurons innervating the motor neurons going to the hamstrings (synapse 2 in Figure 8.18). Simultaneous excitation of the quadriceps and inhibition of the hamstrings by muscle spindle afferents causes leg extension during the knee-jerk reflex.

Neural processing does not end here. Collaterals from muscle spindle afferent neurons also ascend to the brain, forming synapses with various interneurons. The brain uses information from muscle spindles (and other receptors from the muscles, joints, and skin) to monitor the contractile states of the various muscles and the positions of the limbs they control. Because the brain controls skeletal muscle contractions, this information provides feedback that enables the brain to adjust its commands to the muscles, a function that is necessary for the execution of smooth and accurate movements.

Withdrawal and Crossed-Extensor Reflexes

When a portion of the body is subjected to a painful stimulus, it withdraws from the stimulus automatically via a response called the **withdrawal reflex,** as occurs when a person steps on a tack (Figure 8.19). Stepping on a tack is perceived as painful because it activates special sensory receptors called *nociceptors,* which respond to intense stimuli that are damaging (or potentially damaging)

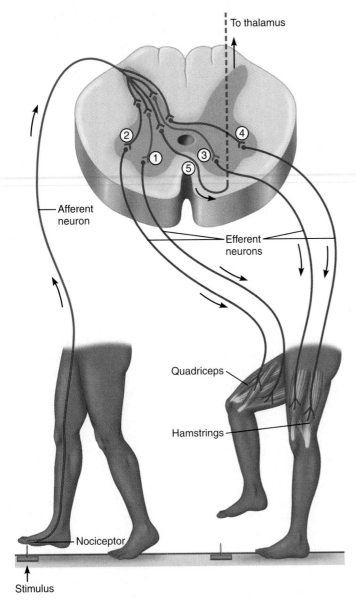

To thalamus

Afferent
neuron

Efferent
neurons

Quadriceps

Hamstrings

Nociceptor

Stimulus

FIGURE 8.19 Withdrawal and crossed-extensor reflexes. *In response to the activation of a nociceptor, an afferent neuron synapses on an excitatory interneuron (1) and an inhibitory interneuron (2), ultimately producing contraction of the hamstrings and relaxation of the quadriceps in the affected leg, and effecting its withdrawal. Simultaneously, the afferent neuron also synapses with an excitatory interneuron (3) and an inhibitory interneuron (4), which ultimately produces contraction of the quadriceps and relaxation of the hamstrings in the other leg. The afferent neuron also synapses with an interneuron (5) that crosses to the opposite site of the spinal cord and travels to the thalamus to convey information about the painful stimulus to the brain.*

to tissue. Afferent neurons from nociceptors transmit this information to the spinal cord, where they have excitatory synapses on interneurons. The interneurons then excite efferent neurons (synapse 1 in Figure 8.19) that innervate the skeletal muscles that cause withdrawal of the limb or other body part. In the example in Figure 8.19, the hamstring muscles contract to withdraw the leg.

For the withdrawal reflex to work effectively, the muscles that cause withdrawal of the affected body part should not be opposed by the actions of other muscles. If during the withdrawal reflex (which causes the hamstrings to contract) the quadriceps muscles were also being stimulated to contract, the contracting quadriceps muscles could slow down or even prevent withdrawal of the leg, which could cause prolonged exposure of the limb to a painful, or perhaps even dangerous, stimulus. Fortunately, this potential problem is avoided because the withdrawal reflex, once set into motion, not only triggers contraction of the muscles that cause withdrawal, but also *inhibition* of muscles whose actions might oppose withdrawal. The reflex does this because branches of the afferents activate inhibitory interneurons to the efferents innervating the quadriceps (synapse 2 in Figure 8.19). Therefore, the withdrawal reflex simultaneously triggers contraction of the hamstrings and relaxation of the quadriceps, thereby allowing flexion of the leg to proceed unimpeded.

But another problem still remains: If you withdraw your leg while it is supporting your weight, you may well lose your balance and fall over. Fortunately, when a painful stimulus triggers the withdrawal reflex, another reflex—the **crossed-extensor reflex**—is initiated simultaneously. The afferent neurons from nociceptors have branches that send signals (via interneurons) to efferent neurons controlling muscles on the opposite leg (synapses 3 and 4 in Figure 8.19). These signals trigger contraction of the extensor muscles and relaxation of the flexor muscles in that leg, so that when the first leg is withdrawn in response to a painful stimulus, the other leg is extended to support the body.

Would the person be aware that he or she stepped on a tack? Of course, because the same afferent neurons that activate both the withdrawal and crossed-extensor reflexes also communicate with ascending pathways that send information to the brain that a painful stimulus has been applied to the body (synapse 5 in Figure 8.19). Given the existence of the withdrawal reflex, does this mean that you will withdraw from every painful stimulus you encounter in life? Probably not, for there are times when reflexes are detrimental. Suppose you were to pick up a cup of tea served in your mother's favorite china and found the cup to be painfully hot. Knowing how much your mother likes her china, you might attempt to override the reflex and hold on to the cup, even if you experience pain by doing so. In this process, the brain activates inhibitory interneurons in the spinal cord that suppress activity in the neurons that control the withdrawal reflex. Remember, postsynaptic cells (in this case, the efferent neurons that control the muscles) sum information that comes from different synaptic inputs. If the inhibitory influence that descends from the brain is greater than the excitatory influence coming from the nociceptor afferents, then the withdrawal reflex will not occur.

Pupillary Light Reflex

The stretch reflex and withdrawal reflex are examples of somatic spinal reflexes. The pupillary light reflex is an example of an autonomic cranial reflex in which the stimulus is light entering one eye and activating photoreceptors. The photoreceptors activate afferent neurons that transmit signals to areas in the midbrain of the brainstem that function as the integration center. These midbrain areas then activate, through a multisynaptic pathway, autonomic efferents that innervate the smooth muscle surrounding the pupils of both eyes. The response is pupillary constriction (reduction in diameter of the pupils) in both eyes.

Quick Test 8.4

1. What is the difference between a spinal reflex and a cranial reflex? Between a somatic reflex and an autonomic reflex? Between an innate reflex and a conditioned reflex? Between a monosynaptic reflex and a polysynaptic reflex?

2. Is salivation in response to a bell an example of an innate reflex or a conditioned reflex? Why?

3. What are the five components of a reflex arc?

4. Briefly describe how a muscle spindle stretch reflex and a withdrawal reflex occur.

INTEGRATED CNS FUNCTION: VOLUNTARY MOTOR CONTROL

Imagine that you are in an operating room, about to undergo surgery for appendicitis. The surgeon explains that before the procedure begins, you will be given two drugs intravenously, an anesthetic and a muscle relaxant. Someone injects a syringe full of liquid into your arm, and soon you feel your arms and legs getting heavy. Within minutes, you find yourself unable to move at all—not even to breathe. (Fortunately, you have been hooked up to a respirator.) Although you expected to become sleepy, you are wide awake and can see, hear, and feel everything that is going on. To your dismay, you notice that the surgeon has picked up a scalpel. Apparently, someone has made a mistake and forgotten to give you the anesthetic. As the blade approaches your skin, you try desperately to yell, to move your arms, to do anything that would let the staff know you are awake, but you can't. All your skeletal muscles are paralyzed, and you can't even blink an eye!

This horrifying scenario illustrates that in order for our thoughts and intentions to become actions, the nervous system must be able to communicate with muscles and trigger muscle contraction. The purpose of this section is to describe what happens when we control our muscles voluntarily; in other words, what happens when our intentions are translated into actions.

Neural Components for Smooth Voluntary Movements

The successful completion of a voluntary motor task involves four components (Figure 8.20):

1. The idea or intention to move must first be formulated. The formulation of intentions, and their translation into plans for movement, is exerted by the cerebral cortex and is the highest level of motor control. Several areas of the cortex are thought to be involved in these processes, including the supplementary motor area (in the frontal lobe), association areas, and the limbic system.

2. A program of motor commands must be developed to ultimately direct the appropriate muscles to contract at the right time to produce the intended motion. Such development involves many areas, including the primary motor cortex, somatosensory areas, the supplementary motor area, and the premotor area (located in the frontal lobe between the primary motor cortex and the supplementary motor area).

3. The program of motor commands must be executed by sending to the muscles commands telling them to contract.

4. As the motor task is being carried out, continual feedback to the CNS enables it to ascertain whether the task is being accomplished according to the intention and the program. Such feedback enables the CNS to make any needed adjustments in the program, and possibly even change the original intention, to ensure smooth and reliable completion of the task. The cerebellum, thalamus, basal nuclei, certain brainstem nuclei, and various cortical association areas provide this feedback and assist with development of motor programming.

The execution of a motor program requires activation of the efferent neurons that innervate skeletal muscle. These efferent neurons originate in the ventral horn of the spinal cord and are called **lower motor neurons,** or simply motor neurons. (For those muscles that are controlled by cranial nerves, the motor neurons originate in the cranial nerve nuclei of the brainstem.) Motor neurons are the only neurons that control contraction of skeletal muscle, and the influence of these neurons is always excitatory; that is, signals in a motor neuron always stimulate a muscle cell to contract. Therefore, initiation of a muscle cell contraction requires excitation of the motor neuron innervating that muscle cell. Relaxation of a skeletal muscle cell simply requires that the motor neuron innervating it not be active.

Input to motor neurons originates from many different levels in the nervous system. In the previous section

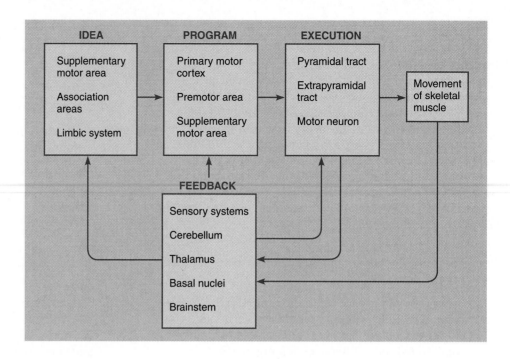

FIGURE 8.20 Four processes in voluntary movement. *Voluntary movement requires four processes to ensure smooth skeletal muscle contraction.*

we saw that motor neurons can be excited or inhibited by input from afferent neurons, as is the case for spinal reflexes. We also saw that activity in motor neurons can be influenced by input descending from the brain, as when one overrides a withdrawal reflex through conscious effort. In this section we are interested primarily in descending input from the brain to motor neurons. In particular, two descending pathways are important in voluntary motor control that provide input to motor neurons: the *pyramidal tracts* (also known as the *corticospinal tracts*) and the *extrapyramidal tracts.*

Cortical Control of Voluntary Movement

The **pyramidal tracts** (Figure 8.21a) are direct pathways from the primary motor cortex to the spinal cord. The axons of neurons in these tracts terminate in the ventral horn of the spinal cord and are called **upper motor neurons;** some synapse directly on motor neurons, whereas others synapse on interneurons. Most axons in the pyramidal tracts cross over to the opposite side of the CNS in an area of the medulla called the *medullary pyramids.* In addition, the axons have collaterals that travel to other motor areas, including brainstem nuclei. The pyramidal tracts are believed to primarily control the fine discrete movements of the distal extremities (portions of limbs farthest from the trunk), especially the forearms, hands, and fingers.

The **extrapyramidal tracts** (Figure 8.21b) include all motor control pathways outside the pyramidal system. Unlike the pyramidal pathways, these pathways are indirect connections between the brain and spinal cord; that is, neurons of the extrapyramidal tracts do not form synapses on motor neurons. The primary influence of the

extrapyramidal tracts are on muscles of the trunk, neck, and proximal portions of the limbs (portions of the limb closest to the trunk). Generally speaking, the extrapyramidal system is involved in controlling large groups of muscles that contract together to maintain posture and balance, whereas the pyramidal system is more involved in the control of smaller groups of muscles that can contract independently. For example, when a seamstress threads a needle, the pyramidal system controls the fine movements in her arms, hands, and fingers while the extrapyramidal system controls the gross movement of her trunk, neck, and legs to maintain appropriate posture.

The pyramidal and extrapyramidal systems illustrate the concept of **parallel processing,** which is the simultaneous transmission of the same general type of information along separate neural pathways. Parallel processing is common in the nervous system in general, not just in motor systems. In parallel processing, the information that is transmitted in the two pathways is processed differently in each. Thus, parallel processing is not quite the same as *redundancy,* in which the two pathways would be duplicates of each other and transmit the same information in basically the same way. However, parallel processing is redundant in the sense that it provides some backup circuitry should one of the pathways fail or malfunction. Thus a person suffering damage to the pyramidal system might recover at least some function, presumably because the extrapyramidal system can partially compensate. Such recovery illustrates another fact about parallel pathways: They are rarely, if ever, truly separate. Thus even though the pyramidal system predominates in the control of distal limb movements, the extrapyramidal system also controls the same muscles to some degree.

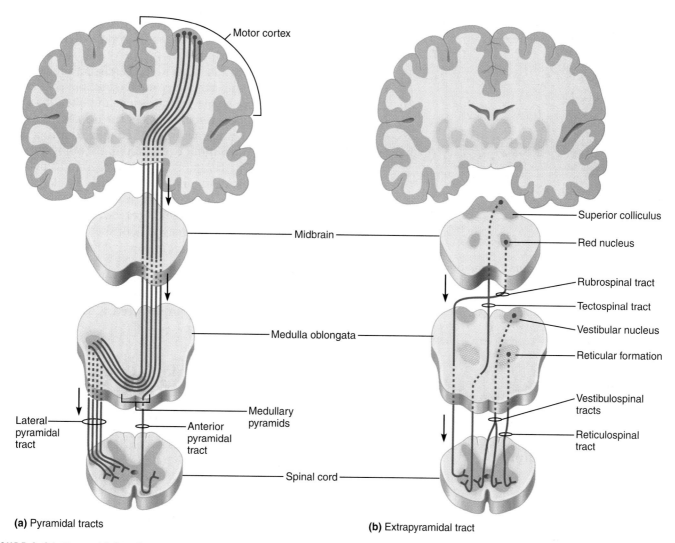

(a) Pyramidal tracts

(b) Extrapyramidal tract

FIGURE 8.21 **Pyramidal and extrapyramidal tracts.**
(a) *Pyramidal tracts. The lateral pyramidal tract is primarily a crossed pathway, whereas the anterior pyramidal tract is primarily uncrossed.* **(b)** *Extrapyramidal tracts. These tracts include all motor tracts except the pyramidal tracts; some cross, some do not, and some are bilateral.*

Where do the pyramidal tracts originate?

The Control of Posture by the Brainstem

The brainstem contains a number of nuclei that project to the spinal cord by way of extrapyramidal tracts. Prominent among these nuclei are the *reticular formation,* the *vestibular nuclei,* and the *red nuclei.* Even though the separate roles of these nuclei are not fully understood, it appears that they are generally involved in the involuntary control of posture. In addition, these nuclei cannot initiate voluntary movements.

Our muscles constantly adjust to changes in posture, whether we are moving or standing still. Even the simple act of reaching for a doorknob requires alteration of muscular tension in the legs and trunk, because we change the position of our body's center of gravity when we extend an arm. To exert postural control, the brainstem uses information from sense receptors in the skin, proprioceptors in muscles and joints, and receptors in the vestibular

system, a set of sense organs associated with the inner ear that detect motion of the head (see Chapter 9). These inputs inform the brainstem about the body's position and the forces acting upon it. The brainstem also receives sensory information from the eyes and ears, so that it can make postural adjustments in response to visual or auditory stimuli. A good example is your reaction when you hear an unexpected sound. When you hear the snap of a twig in the forest, you may automatically stop walking and turn your head toward the source of the sound. This example also illustrates how feedback to motor control systems can alter not only the programming of movement, but also the intention or plan. The original plan may have been to take a long walk through the woods;

In the primary motor cortex

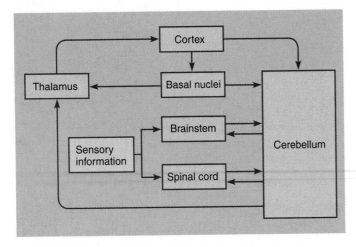

FIGURE 8.22 Major pathways for information flow to and from the cerebellum. *The cerebellum receives information about the status of movement from sensory areas of the cortex, the basal nuclei, the brainstem, and the spinal cord. The cerebellum then transmits information to the cortex via the thalamus, enabling the cortex to alter its output to modify the movement (as needed) to accomplish the task smoothly.*

depending on what snapped the twig, the new task may be to sprint out of the woods.

The Role of the Cerebellum in Motor Coordination

Before launching a spacecraft, mission controllers formulate a trajectory and then program the ship's computer with information regarding the planned route. After the launch, the ship never follows the plan exactly. Instead, an onboard guidance system continually monitors the ship's position and compares it with the planned route, so that it can make course corrections when necessary.

In many respects the cerebellum acts as the "guidance system" for movement. After the cortex formulates plans of action and commands the muscles to execute them, the cerebellum compares the actual movements as they are occurring with the plans, making corrections in the force and direction of movement whenever it detects a discrepancy. In addition, the cerebellum appears to play a role in maintaining muscle tone (a constant low level of tension in muscles at rest). The cerebellum may also store memories of motor activities, so that previously executed movements can be refined and executed more smoothly when they are attempted again.

To do its job, the cerebellum must both receive information from the cortex regarding planned movements and be continually updated regarding how motions are actually being carried out. It receives input from sensorimotor areas of the cortex (areas involved in sensory and/or motor processing), the basal nuclei, the brainstem, the spinal cord, and sensory information originating from all areas of the body (Figure 8.22). It then sends signals

back to the cortex (via a relay in the thalamus), so that the cortex can adjust its output to complete the planned movement successfully.

A person with damage to the cerebellum retains the ability to carry out voluntary movement, but the resulting motions are clumsy, misdirected, and often made with inappropriate force. Consider the numerous finger, wrist, and arm movements required to pick up a pencil, a task you can do without even thinking about it. What would ordinarily be a smooth sequence of movements, one leading seamlessly into the next, may be executed as a series of independent steps, each requiring conscious control. A characteristic sign of cerebellar damage is an *intention tremor,* a sometimes-violent shaking motion that occurs only when a person makes a voluntary movement. When a person with cerebellar damage reaches for a light switch, for example, his or her hand does not head straight toward it but instead begins to make erratic motions that become more pronounced as the hand nears the intended destination.

The Basal Nuclei in Motor Control

The basal nuclei are thought to have functions similar to the cerebellum in that they provide feedback to the cortex for the development of motor strategies and smoothing out movements. Some evidence also suggests that the basal nuclei are necessary for automatic performance of learned repetitive motions. The basal nuclei receive input from the cortex and send output back to the cortex via a relay in the thalamus. One of the functions of this "loop" is to assist the cortex in the selection and initiation of purposeful movement while inhibiting unwanted movements.

By studying two well-known diseases that affect motor control, *Huntington's chorea* and *Parkinson's disease,* we have learned that the effects of the basal nuclei on motor activity are generally inhibitory. Huntington's chorea is a fatal genetic disease characterized in its early stages by the loss of motor coordination and the appearance of exaggerated involuntary jerking or twitching motions involving the limbs and face. Loss of motor control is progressive, so patients eventually become bedridden. At the same time, cognitive function (the ability to think) also declines, progressing ultimately to a complete deficit in mental function, or *dementia.* The initial damage in Huntington's chorea appears to disrupt the transmission of information via pathways projecting from the basal nuclei to the thalamus. Because disruption of these inputs causes certain motor circuits to become overactive, these pathways are believed to exert primarily inhibitory influences over motor control.

Parkinson's disease, like Huntington's chorea, is a progressive neurodegenerative disorder affecting motor control. (See When It Goes Wrong: Parkinson's Disease, www.physiologyplace.com, Challenge Yourself.) Signs include involuntary tremors at rest; a stooped, shuffling gait; movements that are generally slow and stiff; and difficulty in initiating movement, which often requires careful planning and intense effort. Some disruption of cognitive function may occur as well. The initial dysfunction in Parkinson's disease appears to disrupt communication among the various basal nuclei. In particular, the disease is known to target cell bodies of dopaminergic neurons in an area of the midbrain called the *substantia nigra,* with a resultant decline in the release of dopamine in the nuclei to which these cells project.

Quick Test 8.5

1. What are the functional differences between the pyramidal and extrapyramidal systems for motor control?

2. Define parallel processing and give an example of it in voluntary motor control.

3. Briefly describe the role of each of the following structures in motor control: the cerebellum, the basal nuclei, the brainstem nuclei, and the thalamus.

INTEGRATED CNS FUNCTION: LANGUAGE

Compared to the brains of any other species, the human brain excels in its ability to manipulate symbols that it uses to encode thoughts and ideas. Among these symbols are letters, words, and sounds, which we use to formulate *language,* the means by which we communicate our thoughts and ideas to others. Given the importance of language to humans, it should come as no surprise that our brains contain areas specialized for this ability. In fact, two areas of association cortex in each hemisphere are completely devoted to language: **Wernicke's area,** located in the posterior and superior portion of the temporal lobe and the inferior parietal lobe, and **Broca's area,** located in the frontal lobe (see Figure 8.13). Wernicke's area is involved in language comprehension—that is, the ability to understand language. Broca's area, in contrast, is involved in language expression—that is, our ability to speak or write words. Recent studies using *positron emission tomography* (PET), which estimates brain activity by measuring blood flow, suggest that several other areas of the brain also contribute to language. (See Discovery: PET and Language, www.physiologyplace.com, Challenge Yourself.)

Damage to either Wernicke's or Broca's areas produces *aphasia,* or language dysfunction, but the characteristics of the aphasia differ depending on which area is damaged. Damage to Wernicke's area produces *receptive aphasia,* in which a person has difficulty understanding language, regardless of whether the words are spoken or written. Although a person suffering from receptive aphasia can speak words, the words make no sense. Dam-

age to Broca's area, by contrast, produces *expressive aphasia.* In this case, affected individuals comprehend language and know what they want to say but cannot make the correct sounds or write the correct words. Thus if you were to ask someone with expressive aphasia, "How are you doing?" he or she might try to say "Fine" but it would come out garbled and unintelligible. In contrast, a person with receptive aphasia might respond with a meaningless phrase such as "Sun school you see."

INTEGRATED CNS FUNCTION: SLEEP

Everybody knows what sleep is, and we spend approximately one-third of our lives doing it. However, why we do it is not clearly understood, and although it is extensively studied, little is known about how we do it. In this section we examine what is known about this common but mysterious phenomenon.

Functions of Sleep

Sleep is an active process in that it requires energy; it is also a necessary process that all mammals require. Sleep was once considered necessary to let the brain rest. However, we now know that some regions of the brain are more active when a person is asleep than when the person is awake. Sleep is defined as a cyclically occurring state of decreased motor activity and perception. Note that activity is decreased compared to the awake state, but not altogether absent. Occasional bursts of motor activity occur, for example, when a sleeping person shifts position in bed. Perception also occurs because a person can be awakened by sensory stimuli such as sound or touch.

Researchers have proposed many theories to explain the biological usefulness of sleep. Some believe that its function is mainly restorative, that it provides a resting period the body needs to fully recover from its daily activity. Others suggest that sleep allows us to conserve energy. Still others think that sleep may have evolved because dreaming is useful by providing, for example, an opportunity for mentally practicing and refining adaptive behaviors (escaping from predators, for instance) without actually having to perform them, which could place the body in danger. Some evidence suggests that sleep and dreaming may facilitate the storage of long-term memories. Another possibility—that sleep may be necessary for maintaining adequate immune system function—is supported by the observation that animals deprived of sleep for long periods often get infections.

Much of what we know about sleep and its effects on brain function comes from *electroencephalography,* a technique that uses electrodes affixed to the scalp to record electrical activity in the brain. Recordings of such activity, commonly known as *brain waves,* are called **electroencephalograms** (EEGs). Using this method, researchers

TABLE 8.4 CHARACTERISTICS OF SLOW WAVE SLEEP AND REM SLEEP

	SLOW WAVE SLEEP	REM SLEEP
EEG	Slow; medium to high amplitude	Fast; low amplitude
Movement	Moderate muscle tone; some movement, as in shifting position	Little tone; no movement (paralysis) of postural muscles, twitches of distal muscles
Heart and respiration rates	Decreased relative to rest while awake	Increased relative to rest while awake
Dreams	Logical, not detailed	Vivid, illogical
Rapid eye movements	Rare	Frequent
CNS site of induction	Forebrain	Pons

have identified two different kinds of sleep: **slow-wave sleep** (SWS), which is characterized by multiple stages of low-frequency waves in the EEG, and **REM sleep,** which is characterized by high-frequency waves in the EEG and periodic episodes of rapid eye movement (REM).

In SWS, muscle tone is decreased compared to the awake state, but occasional bursts of involuntary activity occur every 10–20 minutes. Activity decreases in most regions of the brain, but activity in the parasympathetic nervous system increases. During SWS, a person may have recurring thoughts and may dream. Thoughts during SWS tend to be more logical and less emotional, compared to those occurring in REM sleep. Dreams during SWS tend to be scant in detail and often include feelings or apparitions as opposed to being narrative. Snoring may occur during SWS.

In REM sleep, postural muscles lose their tone and become paralyzed. However, muscles that control the face, eyeballs, and distal limbs frequently twitch. Jaw muscles relax, frequently causing the mouth to open. Snoring (if it occurs during SWS) stops. Compared to SWS, an overall increase in brain activity occurs, except in the limbic system, where the activity level decreases. Breathing is periodically stimulated, and activity in the sympathetic nervous system increases. Heart rate and blood pressure may rise, and a loss of body temperature control occurs. Dreams in REM sleep are more elaborate and intense than in SWS and are usually narrative (that is, they tell a story). In REM sleep, thoughts are more illogical and bizarre than in SWS. Differences between SWS and REM sleep are summarized in Table 8.4.

Sleep-Wake Cycles

The human body alternates between periods of wakefulness and periods of sleep, normally completing one complete sleep-wake cycle in a 24-hour period. Typically, an average adult spends 16 hours a day awake and eight hours a day sleeping. What causes our body to shift between these different levels of consciousness?

Although the answer to this question is not completely known, certain areas of the brain appear to be involved in the regulation of sleeping and wakefulness. One of these areas is the reticular formation of the brainstem, a diffuse network of nuclei that form the *ascending reticular activating system* (ARAS; Figure 8.23). This area is critical in maintaining alert wakefulness. Input from the ARAS projects to the cortex through relays in the thalamus, hypothalamus, and forebrain and "awakens" the cortex, making it more receptive to incoming signals from other pathways.

Other areas of the brain are involved in inducing sleep. SWS is induced by the forebrain. Recent research suggests that adenosine is a critical neurotransmitter for the induction of SWS. One of the current theories on the stimulatory effects of caffeine suggests that caffeine blocks the release of adenosine, thereby inhibiting sleep. REM sleep, in contrast, is induced by activity originating in the pons. The precise mechanisms of sleep induction or maintenance of the awake state are not yet understood.

Electrical Activity During Wakefulness and Sleep

When you are awake and alert, an EEG shows a pattern of high-frequency, low-amplitude oscillations known as *beta waves,* which reflect electrical signals being generated by large numbers of neurons at different times (Figure 8.24a). When you are awake but resting, the EEG shifts to a new pattern of lower-frequency, higher-amplitude waves known as *alpha waves* (Figure 8.24). Compared to beta waves, alpha waves reflect a greater degree of synchronization of electrical activity among neurons; that is, electrical signals are being generated in large groups of neurons at more-or-less the same time. This change in EEG pattern might be compared to changes in crowd noise in a stadium. When large numbers of people are talking at different times (desynchronized activity, similar to beta waves), the overall sound is a nearly constant din, and it is difficult to make out distinct patterns or words.

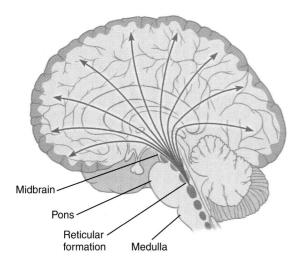

Midbrain

Pons

Reticular
formation   Medulla

FIGURE 8.23 Ascending reticular activating system. *The ascending reticular activating system is part of the reticular formation, a diffuse network of neurons spread throughout the highlighted area of the brainstem. The arrows indicate the spread of excitation that "arouses" the cortex.*

When the crowd begins to chant in unison (synchronized activity, similar to alpha waves), the sound occurs in a series of discernible waves that can be recognized as words.

SWS occurs in four successive stages that are distinguishable by changes in both the EEG and the arousal threshold (the strength of a stimulus required to awaken a sleeper). Stage 1 SWS has the lowest arousal threshold, whereas stage 4 has the highest. In other words, sleep is "lightest" in stage 1 and "deepest" in stage 4. In addition, stages 1–4 show successive increases in the degree of synchronization of the EEG, with stage 4 waves being the highest in amplitude and lowest in frequency, indicating a large degree of synchronization (see Figure 8.24c–e).

During REM sleep, the EEG shows a completely different pattern (Figure 8.24f). Instead of slow, high-amplitude waves, it contains fast, low-amplitude waves resembling those of the awake, alert state. Arousal threshold is higher during REM sleep than at any other time, but a person is more likely to awaken spontaneously during REM sleep than at any other time. Because of this surprising fact, REM sleep is often referred to as *paradoxical* sleep.

During a typical night, a person progresses through the various stages of sleep in a fairly orderly, predictable way (Figure 8.25). When you first fall asleep you go into stage 1 SWS. From there, you proceed through stages 2–4 in order. About an hour after sleep onset you go in reverse order through these stages and then enter into your first period of REM sleep. Following this the pattern repeats a few times, except that you tend to spend less time in the deeper stages of SWS and more time in REM sleep as sleeping continues.

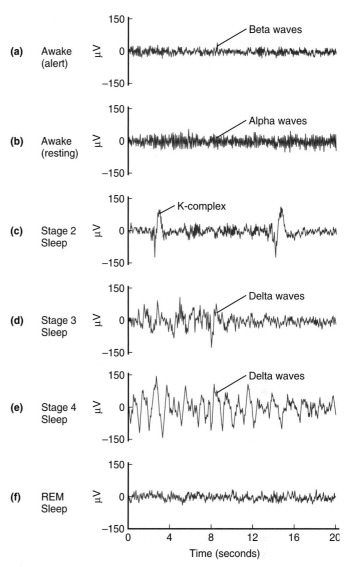

FIGURE 8.24 EEG recordings during wakefulness and sleep. *These graphs are 20-second EEG samples recorded from a 21-year-old female subject. Amplitude of EEG recordings are given in microvolts (μv), and frequencies are given in Hertz (Hz).* **(a)** *Waking EEG is characterized by high-frequency low-amplitude waves.* **(b)** *During relaxed waking, 8–12 Hertz alpha waves are present in the EEG.* **(c)** *The EEG during stage 2 sleep is characterized by high-amplitude k-complexes (seen here at 3 and 14 seconds) and sleep spindles (12–15 Hertz waves seen here at 6, 10, and 12 seconds).* **(d, e)** *High-amplitude delta (0.3–3 Hertz) waves are present in stage 3 sleep and occupy the entire 20-second period in this sample of stage 4 sleep.* **(f)** *EEG amplitude decreases and frequency increases in REM sleep. (Data provided courtesy of I. G. Campball, University of California, Davis.)*

During which phase of sleep is electrical activity of the brain most synchronized?

Notice in Figure 8.25 that REM sleep periods last longer and occur more closely together as sleep continues. In keeping with this, when spontaneous awakenings occur, they occur much more frequently in the last few hours of sleep than in the beginning. As a rule, we re-

Stage 4

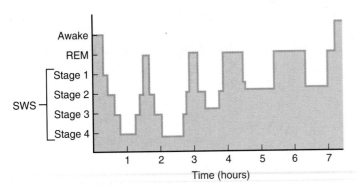

FIGURE 8.25 Stages of sleep. *During eight hours of sleep, a person moves among the different sleep stages. Initially, the person progresses from stage 1 SWS to stage 4 SWS, then returns to REM sleep. As the period of sleep continues, REM sleep becomes longer and more frequent.*

member only those dreams that we experience immediately before waking up. Because this is more likely to happen during REM sleep, we are most likely to recall REM-type dreams.

Quick Test 8.6

1. What is electroencephalography, and how is it useful in the study of sleep?

2. What is the definition of sleep, and what are some of its possible functions?

3. What type of EEG changes are observed as a person progresses from stage 1 to stage 4 of SWS?

4. What is the ascending reticular activating system, and how is it thought to function in regulation of the sleep-wake cycle?

INTEGRATED CNS FUNCTION: EMOTIONS AND MOTIVATION

You know how you feel at this moment. You may be angry because your roommate is making too much noise while you are trying to study, or you may be excited over all the new physiology you are learning. Emotions are highly subjective; they differ among people confronted with identical circumstances, and they also vary among people confronted with the identical set of events if they happen under different circumstances. For example, how you feel when your roommate gets an A on an exam depends on your circumstances. If you are also doing well in school, you will probably feel happy; if you just did poorly on an exam, you might feel jealous. Because emotions are so varied, they are difficult to define and understand.

Emotions involve many diverse areas of the brain, including the cerebral cortex, limbic system, and hypothalamus (Figure 8.26). Emotions are generally triggered by sensory input or memories. The "emotional response" includes autonomic (changes in heart rate, respiration, and so on), motor, and hormonal changes. Currently, many theories pertain to the brain's role in emotions, but little strong evidence supports a single theory. It does appear, however, that different neural pathways come into play for different emotions. For example, the amygdala is critical in feeling fear and anxiety; by contrast, the hypothalamus is associated with anger and aggression, although other structures (including the amygdala and midbrain) are also involved in these emotions.

Closely associated with emotions is *motivation*, the impulse that drives our actions. Motivations can be purely physiological; that is, they can arise simply because some need of the body is not being satisfied. For example, we eat because we feel hunger, which usually arises because the body is lacking certain nutrients. Motivations can also be driven by emotions acting in the absence of any obvious physiological drives. For example, the reason some people eat ice cream when they feel sad, even when they are not hungry, is that eating the ice cream brings pleasure that may override the sadness.

Pleasure is a highly motivating emotion. The brain contains "pleasure centers" that can be acted on by many different stimuli. For example, the euphoria that many perceive when they consume alcohol, opioids, amphetamines, or nicotine may be due to the actions of these substances on pleasure centers. All these substances are thought to produce euphoria by activating dopaminergic systems. The motivation that is elicited by these substances is so strong that animals, including people, can develop addictions to them. Sometimes these addictions override other motivations, even those closely connected to the body's needs. Addicted animals, for example, have been known to die of starvation because they habitually ignored food in favor of other pleasurable stimuli.

Quick Test 8.7

1. Recall the single event in your life that scared you the most. What happened to your heart rate? What is happening to your heart rate now? Are you feeling similar emotions simply through memory? What area of the brain is driving this autonomic response to emotion?

2. What emotion leads to addiction? What neurotransmitter is critical in the development of drug addiction?

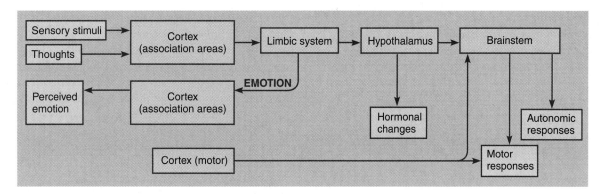

FIGURE 8.26 The CNS structures involved in emotions. *Various structures of the brain are involved in producing emotions, both the "feeling" and the responses associated with them. Cortical association areas integrate thoughts and sensory information and communicate to the limbic system. The limbic system "creates" the emotion, but we are not aware of the emotion until it is transmitted back to the cortex for perception. Meanwhile, the limbic system also communicates the emotion to the hypothalamus, which is responsible for bodily responses coupled with the emotion, including hormonal changes (for example, adrenaline release), motor responses (for example, frowning), and autonomic responses (changes in heart rate and respiration).*

INTEGRATED CNS FUNCTION: LEARNING AND MEMORY

Whereas **learning** is the acquisition of new information or skills, **memory** is the retention of information, skills, or thoughts. Learning is such an important part of life; we do it many times every day. In fact, every new experience is an opportunity for learning. However, only a portion of what we experience is actually put into memory so that it can be recalled at some later time. If we were able to store and retrieve everything we ever experienced, then life would be much easier—we would never need to study for exams or look at a road map a second time! Learning and memory are closely linked, as are the brain areas involved.

Learning

Although learning has a simple definition, psychologists have long recognized two major types of learning: associative learning and nonassociative learning.

Associative learning is the type of learning that requires making connections between two or more stimuli. The salivation reflex in Pavlov's dogs is an example of associative learning in which the dogs learned to "associate" the ringing of the bell with the arrival of food. If you were to try to teach a younger sister not to touch something that is hot, she might not listen when you tell her it will hurt; however, when she touches a hot stove, the pain she feels will teach her not to do it again. In this case she may learn two associative lessons: Do not touch something that is hot because it will hurt, and pay heed when your older sibling tells you to do something.

Nonassociative learning occurs in response to repetition of a single stimulus (or similar stimuli) and includes the processes of habituation and sensitization. **Habituation** is a decrease in response to a repeated stimulus. For example, if construction workers were to start hammering outside your room while you are studying, you would first be distracted and likely unable to study. However, as the noise persists, you would likely find that you are no longer as aware of it. Studying would become possible, even though circumstances might not be ideal. In this case your brain has become *habituated* to a repetitive stimulus. In contrast, **sensitization** is an increase in response to repeated stimulus. If your mother ever forced you to eat something you did not like when you were a small child, the first bite may have seemed bad enough, but each additional bite seemed to taste worse. In this case, you became *sensitized* to the taste. Whether we become sensitized or habituated to repetitive stimuli depends on the circumstances. If we perceive a stimulus as important (such as a signal of impending danger), we tend to become sensitized to it; if we deem the stimulus to be unimportant, we tend to become habituated to it.

Memory

We saw how a younger sister learned to associate hot things with pain, but now we want her to remember that lesson so that she does not repeatedly burn herself. To do so, the information she learned must be put into memory. Just as there are two types of learning, there are also two types of memory: procedural memory and declarative memory.

Procedural memory is the memory of learned motor skills and behaviors. Think back to when you first tried to play a musical instrument or a sport. At first, learning was difficult; each movement had to be thought through. But now you are more proficient, because you have learned and memorized each skill. The type of memory involved when someone says, "It's like riding a bike—you never forget," is procedural memory. This type of memory is thought to reside, at least partially, within the cerebellum.

Declarative memory is the memory of learned experiences, such as facts and events and other things that can be stated verbally. You can state that the first three letters of the alphabet are A, B, and C (a declarative memory), but you cannot state the memories you use while riding a bicycle (procedural memory). The ability to remember and explain what generates action potentials is an example of declarative memory. Remembering your birthdate and your parents' anniversary are other examples. Declarative memory is more closely associated with the everyday use of the word "memory." This type of memory involves the hippocampus.

Memory occurs at two levels, *short-term memory* and *long-term memory*. In the current model of how information is processed and stored into memory, incoming information first enters the CNS and is stored as *short-term memory*, which is temporary storage lasting only a few seconds or up to a few hours. Limited space is available for short-term memory, and information placed in short-term memory will be lost if it is not *consolidated* into **long-term memory**, which can last years or a lifetime. The mechanisms of consolidation are not well understood, but repetition appears to help (your little sister may not believe that something hot will always hurt her until she touches hot objects a few times). By contrast, repetition is not necessary for certain memories. That most brides and grooms remember their wedding day, even though it only occurred once, suggests that the importance of an experience also plays a role in memory.

Plasticity in the Nervous System

Learning and memory are able to occur because the nervous system is endowed with **plasticity,** the limited ability to alter its anatomy and function in response to changes in its activity patterns. This plasticity derives from both the fact that the function of existing synapses can be altered for long periods of time (hours to days), and the fact that new synaptic connections can develop. Recent studies suggest that our brains may even grow new neurons in areas involved in memory.

One example of plasticity in the nervous system is *long-term potentiation* (LTP), which in mammals was first discovered in the hippocampus and occurs at preexisting synapses. In LTP, repetitive stimulation of a particular synapse eventually leads to an increase in the *strength* of that synaptic connection; that is, repetition increases the

likelihood that synaptic input will be able to trigger an action potential in the postsynaptic cell (assuming the synapse is excitatory). Such increases in synaptic strength occur not because of a decrease in the threshold for action potential generation in the postsynaptic cell, but because of an increase in the size of the EPSPs that are being generated in the postsynaptic cell. Generally speaking, an increase in synaptic strength can be due to an increase in the postsynaptic cell's sensitivity to the neurotransmitter released at that synapse, to an increase in the quantity of neurotransmitter released by the presynaptic cell with each action potential, or to both. LTP is thought to be important in the consolidation of long-term memory because it provides a mechanism whereby repetitive activity in particular neural pathways, such as might occur during repetition of a learned fact, can leave a more-or-less permanent "record" of itself once the activity has ceased.

One mechanism of long-term potentiation is illustrated in Figure 8.27. In this example, the presynaptic cell releases the excitatory neurotransmitter glutamate. The postsynaptic cell has two different types of glutamate receptors: AMPA (α-amino-3-hydroxy-5-methylisoxazolepreprionic acid) receptors and NMDA (N-methyl D-aspartate) receptors. NMDA and AMPA are glutamate agonists that bind to glutamate receptors. When glutamate binds to NMDA receptors, calcium channels open, but when it binds to AMPA receptors, sodium channels open. When the presynaptic cell releases glutamate, it binds to both receptor types.

The concentration of glutamate in the synaptic cleft depends on the frequency of action potentials in the presynaptic cell. When low to moderate levels of glutamate are present in the synaptic cleft (Figure 8.27a), some AMPA receptor-dependent sodium channels open, and sodium diffuses into the cell, causing a depolarization. Some NMDA receptor-dependent calcium channels also open, but calcium cannot move through these channels because they are blocked by magnesium ions. The end result is an EPSP that may or may not reach threshold for an action potential.

When the action potential frequency increases in the presynaptic cell, large amounts of glutamate are released (Figure 8.27b), which produces a greater depolarization of the postsynaptic cell. The strong depolarization exerts on the magnesium ions a repulsive force that forces them out of the NMDA receptor channels. With magnesium gone and the channels open, calcium is now able to flow into the postsynaptic cell, raising the cytosolic calcium concentration and triggering activation of other second messengers that activate protein kinases. One protein kinase phosphorylates the AMPA receptor channels, thereby increasing their sensitivity to glutamate; another protein kinase causes the postsynaptic cell to release a paracrine that reaches the presynaptic cell and stimulates it to release more glutamate. The net result of increased glutamate release from the presynaptic cell and height-

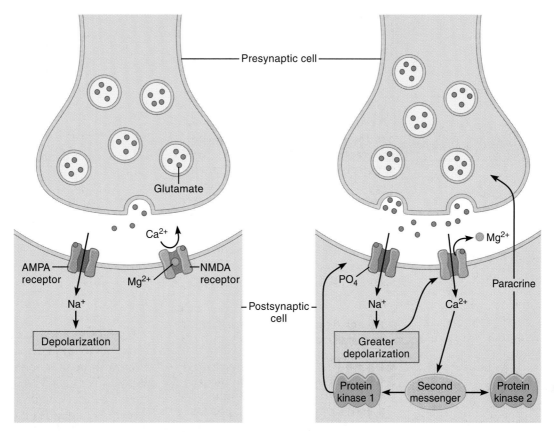

(a) Low levels of activity in presynaptic cell

(b) High levels of activity in presynaptic cell

FIGURE 8.27 A mechanism of long-term potentiation. *The figure depicts the events whereby repetitive stimulation of a synapse increases the likelihood that synaptic input will produce an action potential.* **(a)** *At low levels of activity in the presynaptic cell, glutamate released from that cell binds to both receptor types, but the presence of magnesium in the NMDA receptor prevents calcium influx; the net effect is a depolarization of the postsynaptic cell.* **(b)** *At high levels of activity in the presynaptic cell, increased glutamate release allows more sodium to enter the postsynaptic cell, producing a greater depolarization that forces magnesium out of the NMDA receptor channel. The resulting increased calcium inflow activates protein kinases. One protein kinase phosphorylates the sodium channel, making it more sensitive to glutamate; a second protein kinase triggers production of a paracrine that causes the presynaptic cell to produce more glutamate. The net effect of more glutamate acting on a more sensitive postsynaptic cell is a prolonged depolarization.*

What triggers exocytosis of glutamate?

ened responsiveness to glutamate by the postsynaptic cell is larger and more prolonged EPSPs, which increase the likelihood that action potentials will be generated in response to signals being transmitted at that synapse.

Quick Test 8.8

1. When a dog is trained to sit by rewarding him with food each time he obeys a command, is associative learning or nonassociative learning involved?

2. What is the difference between habituation and sensitization?

3. What is the distinction between procedural memories and declarative memories?

4. Define long-term potentiation, and explain how it relates to the concept of neural plasticity.

Calcium entering the axon terminal after an action potential has opened voltage-regulated calcium channels.

General Anatomy of the Central Nervous System, p.224

The central nervous system (CNS) consists of the brain and spinal cord. CNS tissue is organized into gray and white matter. Gray matter consists primarily of cell bodies, dendrites, and axon terminals; white matter consists mostly of myelinated axons. Glial cells are also located throughout the CNS. The CNS is protected by various structures, including the cranium and vertebral column, the meninges (dura mater, arachnoid mater, and pia mater), and the blood-brain barrier. Also providing cushioning protection is cerebrospinal fluid.

The Spinal Cord, p.231

Thirty-one pairs of spinal nerves come off the spinal cord, one pair at each vertebral level. All spinal nerves are mixed nerves because they contain both afferent and efferent axons. The white matter of the spinal cord contains ascending and descending tracts that carry information to and from the brain, respectively. The gray matter is divided into ventral and dorsal horns; cell bodies and dendrites of efferents are located in the ventral horn, axon terminals of afferents are located in the dorsal horn, and small interneurons are scattered throughout.

The Brain, p.237

The brain has three major parts: forebrain, brainstem, and cerebellum. The forebrain includes the cerebrum (cerebral cortex and subcortical nuclei) and diencephalon (thalamus and hypothalamus). The brainstem includes the midbrain, pons, and medulla oblongata. The cerebrum is responsible for higher brain functions including motor control, sensory perception, language, emotions, learn-ing, and memory. The subcortical nuclei include the basal nuclei and nuclei of the limbic system. The basal nuclei are important in motor control, whereas the limbic system is involved with emotions, learning, and memory. The diencephalon, located beneath the cerebrum at the core of the forebrain, includes the thalamus, which serves as a relay station for information traveling to the cortex, and the hypothalamus, which regulates a number of body functions to maintain homeostasis.

Integrated CNS Function: Reflexes, p.244

The simplest actions of the nervous system are reflexes, or automatic responses to a particular stimulus. Reflexes can be mediated spinally or cranially, they may be either somatic or autonomic, and they may be either conditioned or innate. Reflexes can be described as a reflex arc. Two spinal reflexes are the muscle spindle stretch reflex and the withdrawal-crossed extensor reflex.

Integrated CNS Function: Voluntary Motor Control, p.247

Voluntary control of muscles involves many regions of the CNS. The cerebral cortex formulates plans for movement and sets up a command. The pyramidal and extrapyramidal tracts transmit information about the plan to the motor neurons in the spinal cord, which are the neurons that stimulate the skeletal muscle to contract. Continuing modulation through sensory systems, the cerebellum, and basal nuclei help make the movement smooth. Involuntary contractions of skeletal muscles also occur. The extrapyramidal tracts are more important for supportive movements such as maintaining posture than for precise voluntary movement.

Integrated CNS Function: Language, p.251

The areas of the brain important for language include Wernicke's area in the temporal lobe and Broca's area in the frontal lobe. An abnormality in language skills is called aphasia. Damage to Wernicke's area causes receptive aphasia, whereas damage to Broca's area causes expressive aphasia.

Integrated CNS Function: Sleep, p.251

The purpose and mechanisms of sleep are not well understood. The stages of sleep can be monitored using an EEG. Slow wave sleep is characterized by low-frequency, high-amplitude EEG waves, whereas REM sleep is characterized by high-frequency, low-amplitude EEG waves. The body moves through various stages throughout periods of sleep.

Integrated CNS Function: Emotions and Motivation, p.254

Emotions are generated in the limbic system based on sensory input. The limbic system then transmits information to the cortex, where the emotion is actually perceived. Emotion-related responses also include changes in the autonomic nervous system. Motivation directs actions and is closely associated with pleasure.

Integrated CNS Function: Learning and Memory, p.255

Learning is the acquisition of new skills or information, whereas memory is the retention of knowledge. Learning and memory require plasticity. One type of plasticity is long-term potentiation, in which communication across a synapse is prolonged.

Multiple-Choice Questions

1. To perceive touch, sensory neurons from the skin transmit information to
   a) the basal nuclei.
   b) the somatosensory cortex.
   c) the hypothalamus.
   d) the limbic system.

2. Which of the meninges is closest to the neural tissue?
   a) dura mater
   b) arachnoid mater
   c) pia mater

3. At rest, the brain accounts for how much of the oxygen consumption by the body?
   a) 2%
   b) 5%
   c) 15%
   d) 50%

4. Most of the fibers connecting the left and right cerebral hemispheres are located in the
   a) hippocampus.
   b) corpus callosum.
   c) corticospinal tract.
   d) central canal.

5. All spinal nerves contain
   a) afferent fibers.
   b) efferent fibers.
   c) both afferent and efferent fibers.
   d) neither afferent nor efferent fibers.

6. The origin of neurons of the corticospinal tract is the
   a) ventral horn of the spinal cord.
   b) somatosensory cortex.
   c) primary motor cortex.
   d) cerebellum.

7. Which of the following is *not* a component of the limbic system?
   a) hippocampus
   b) cerebral cortex
   c) amygdala
   d) midbrain

8. The circadian rhythm is established by what brain area?
   a) suprachiasmatic nucleus
   b) amygdala
   c) thalamus
   d) occipital lobe of cerebral cortex

9. Parkinson's disease is due to loss of dopaminergic innervation of the
   a) cerebellum.
   b) basal nuclei.
   c) brainstem.
   d) primary motor cortex.

10. Which of the following reflexes is monosynaptic?
    a) withdrawal reflex
    b) crossed-extensor reflex
    c) muscle spindle stretch reflex
    d) all of the above

11. The area of the brain important for the maintenance of posture is the
    a) primary motor cortex.
    b) basal nuclei.
    c) brain stem.
    d) cerebellum.

12. Expressive aphasia results from damage to
    a) the corpus callosum.
    b) the cerebellum.
    c) Broca's area.
    d) Wernicke's area.

13. Long-term potentiation is
    a) a prolonged increase in synaptic activity at a given synapse.
    b) the formation of new synapses.
    c) memory from greater than 10 years ago.
    d) activated through the ascending reticular activating system.

Objective Questions

1. Where is cerebrospinal fluid synthesized?

2. What type of junction between the endothelial cells of brain capillaries produces the blood-brain barrier?

3. Myelinated axons are found in (gray/white) matter.

4. Somatic efferents originate in the (dorsal/ventral) horn of the spinal cord.

5. The major function of the (cerebrum/cerebellum) is to coordinate body movements.

6. What three structures make up the brainstem?

7. Motor aphasia is due to damage to (Broca's/Wernicke's) area.

8. Which side of the brain is generally associated with creativity?

9. What are the two main structures of the diencephalon?

10. What is the major sensory relay nucleus to the cortex?

11. Parkinson's disease is associated with a decrease in the activity of what neurotransmitter?

12. The area of the brain most closely associated with fear is the _____.

13. The _____ system is associated with emotions, learning, and memory.

14. The ability to recall information when taking physiology exams is an example of _____ memory.

15. The afferents that activate reflex motor actions are different from those that activate ascending tracts to communicate a sensory input to the brain. (true/false)

Essay Questions

1. Describe the different structures that protect central nervous system tissues.

2. Describe the energy supply to the brain.

3. Identify the four lobes of the cerebral cortex, and name at least one function associated with each.

4. Name the different components involved in accomplishing a voluntary motor task. Identify the structure(s) associated with each component.

5. Name the different stages of sleep. During which sleep stage is a person most likely to roll over?

6. Describe the roles of NMDA and AMPA receptors in long-term potentiation.

Find the answers to these exercises, and additional study tools, at the Physiology Place (www.physiologyplace.com).

9

The Nervous System: Sensory Systems

OBJECTIVES

- Describe the function of sensory receptors, and explain how they perform that function.

- Describe the sensory transduction mechanism for each of the special senses.

- Describe the neural pathways of the different sensory systems, from sensory receptor to cerebral cortex.

- Describe how light is focused on the retina.

- Describe how the iris regulates the amount of light that enters the eye.

- Explain the differences between rods and cones.

- Describe the two sensory systems of the ear.

CHAPTER OUTLINE

General Principles of Sensory Physiology   261

The Somatosensory System   270

Vision   277

The Ear and Hearing   291

The Ear and Equilibrium   299

Taste   303

Olfaction   306

Above: Scanning electron micrograph of cilia in the organ of Corti.

STUDY HINTS

1. Ion channels, p. 114

2. Signal transduction mechanisms, p. 140

3. Cerebral cortex, p. 238

4. Spinal cord dorsal horn, p. 231

5. Ascending tracts, p. 232

S it back for a moment and reflect on the world around you. Are you sitting or lying down? If you are indoors, what color are the walls of the room in which you are currently sitting? Are you warm or cold? Are the people near you now noisy? Maybe someone has the stereo on so loud that you can feel the vibrations. Your perception of the world around you is achieved via the functioning of your sensory systems. But you cannot perceive everything in your environment. For example, if you have ever had an X ray taken, you know you could neither see nor hear the rays as they passed through your body, and when you put some food in the microwave, you cannot see the waves that heat your food.

Why can we perceive some things in the world and not others? Humans have special sensory receptors that can detect specific forms of energy in the environment, such as light waves, sound waves, or vibrations, but we do not have receptors for X rays or microwaves. Therefore, we cannot perceive these energy forms, even when they are present in our immediate environment. Furthermore, our perceptions of the stimuli that we *can* perceive are not absolute. When you jump into a lake, for example, you immediately perceive the cold water, but over time your body adapts, and within minutes the water does not feel as cold—even though the temperature of the water has not changed.

This chapter focuses on the sensory systems involved in **perception,** the conscious interpretation of the world based on the sensory systems, memory, and other neural processes. We first describe general properties of sensory systems. Then we discuss those senses that allow us to perceive touch and other stimuli associated with the skin surface, and that also allow us to perceive such things as the position of our limbs and bodies. Finally, we discuss the *special senses.*

GENERAL PRINCIPLES OF SENSORY PHYSIOLOGY

The afferent division of the peripheral nervous system transmits information from the periphery to the central nervous system. The information is detected by *sensory receptors* that respond to specific types of stimuli. Whereas some of these receptors detect stimuli in the external environment, other receptors detect stimuli that arise within the body. Information relating to the latter stimuli is transmitted to the CNS by a class of afferent neurons known as *visceral afferents;* the sensory receptors that detect these stimuli are called *visceral receptors.*

Examples of the many types of visceral receptors include *chemoreceptors* in major blood vessels that monitor O_2 and H^+ levels in the blood; *baroreceptors* in certain blood vessels that monitor blood pressure; and *mechanoreceptors* in the intestinal tract that monitor the degree of stretch or distension. Although the brain receives information from these receptors and utilizes this information for regulatory purposes, we are not consciously aware of these stimuli. For example, we cannot sense what the pH of our own blood is, unless we take a blood sample and analyze it. However, pH receptors in certain arteries and in the brainstem tell the brain whether or not the pH is normal, information that enables the brain to make regulatory decisions. The specifics of these and other visceral receptors are covered in chapters that discuss particular organ systems.

Exercise Link

Although a person does not consciously sense blood pH or the other individual inputs from visceral receptors, there may be times when we are able to sense the overall message being sent by these receptors. In the later stages of the marathon, Jane and especially Bill experienced two sensations that we call fatigue. One was a reduction in the responsiveness of their leg muscles because metabolic fuels were running low and waste products were building up in ways that hampered normal function. The other sensation is variously called "central" or "psychological" fatigue. Although physiologists do not fully understand this second type of fatigue, it is suspected that visceral receptor inputs, along with influences of body temperature and blood glucose levels, feed back on the motor cortex, reducing the drive to exercise. In addition, these inputs touch on the edge of our consciousness to produce those hard-to-describe feelings we associate with fatigue.

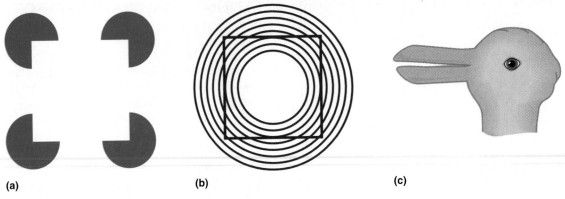

FIGURE 9.1 The fallibility of perception. *These optical illusions demonstrate that our perception is not always accurate.* **(a)** *Where parts of the four green disks are missing, the brain perceives a square.* **(b)** *The effect of the concentric circles is to make straight lines appear curved.* **(c)** *This shape can resemble a duck or a rabbit—or both or neither.*

This chapter covers the sensory systems that detect information about the external environment. This information is transmitted to the cerebral cortex, reaches consciousness, and is perceived, making us aware of the world around us. But note that our perception of the world is not always fully accurate. That this is so is demonstrated in Figure 9.1.

In Figure 9.1a you perceive something that is not present—a square. Part of your perception is based on previous experiences; in this example, your brain perceives what it expects to see. In Figure 9.1b you perceive curved lines when in actuality the lines are straight, which shows that our perceptions may deceive us. In Figure 9.1c some people perceive a rabbit, others a duck, others both, and still others neither. Individuals may perceive things differently, based in part on their individual past experiences.

The sensory systems that enable us to perceive the external environment include the somatosensory system and the special sensory systems. The **somatosensory system** is necessary for perception of sensations associated with receptors in the skin **(somesthetic sensations)** and for **proprioception,** the perception of the position of the limbs and the body. Proprioception depends on specific proprioceptors in muscles and joints and on more generalized receptors in the skin. The **special sensory systems** are necessary for senses of vision, hearing, balance and equilibrium, taste, and smell.

Receptor Physiology

Sensory receptors are specialized neuronal structures that detect a specific form of energy in either the internal or external environment. The energy form of a stimulus is called its **modality.** Some examples of modalities include light waves, sound waves, pressure, temperature, and chemicals. The **law of specific nerve energies** states that a given sensory receptor is specific for a particular modality. Thus special cells in the eye called *photo-*

receptors detect light waves, but not sound waves. The modality to which a receptor responds best is called the **adequate stimulus.** Table 9.1 lists the characteristics of several classes of sensory receptors. Modalities other than the adequate stimulus may activate receptors, but only if at relatively high energy levels. Moreover, a sufficiently strong "inadequate" stimulus will induce the perception of the adequate stimulus, as when a blow to the eye (pressure) activates photoreceptors, causing the person to perceive light.

Sensory Transduction

The function of sensory receptors is **transduction**—that is, the conversion of one form of energy into another. In **sensory transduction,** receptors convert the energy of a sensory stimulus into changes in membrane potential called **receptor potentials** or **generator potentials.** Receptor potentials resemble postsynaptic potentials (see Chapter 7) in that they are graded potentials caused by the opening or closing of ion channels. The greater the strength of the stimulus, the greater the change in membrane potential. However, unlike postsynaptic potentials, which are triggered by the binding of neurotransmitter to receptors, receptor potentials are triggered by sensory stimuli.

In some cases, a sensory receptor is a specialized structure at the peripheral end of an afferent neuron (Figure 9.2a); in other cases it is a separate cell that communicates via a chemical synapse with an associated afferent neuron (Figure 9.2b). When the sensory receptor is a specialized structure, and if the receptor potential depolarizes the cell to threshold, then an action potential is generated in the afferent neuron and propagated to the CNS, thereby transmitting information about the stimulus. When the sensory receptor is a separate cell, changes in the receptor cell's membrane potential cause the release of a chemical messenger or transmitter. (In some cases a stimulus triggers a change in membrane potential that *inhibits* the release of a transmitter.) The greater the

TABLE 9.1 CHARACTERISTICS OF SENSORY RECEPTORS

RECEPTOR CLASS	SENSATION/VISCERAL INFORMATION	MODALITY
Photoreceptors	Vision	Photons of light
Chemoreceptors	Taste	Chemicals dissolved in saliva
	Smell	Chemicals dissolved in mucus
	Pain	Chemicals in extracellular fluid
	Blood oxygen	Oxygen dissolved in plasma
	Blood pH	Free hydrogen ions in plasma
Thermoreceptors		
Warm receptors	Warmth	Increase in temperatures between 37°C and 45°C
Cold receptors	Cold	Decrease in temperatures between 37°C and 20°C
Mechanoreceptors		
Baroreceptors	Blood pressure	Stretch of specific blood vessel wall
Osmoreceptors	Osmolarity of extracellular fluid	Swelling (stretch) of receptor cells
Hair cells	Sound	Sound waves
	Balance and equilibrium	Acceleration

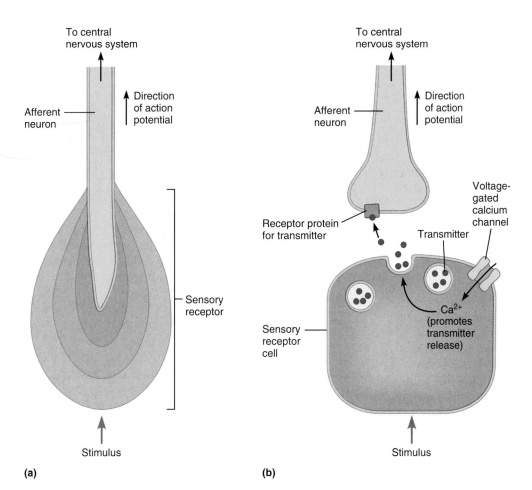

FIGURE 9.2 Structure and function of sensory receptors.
(a) *A sensory receptor that is a specialized ending of an afferent neuron. The stimulus acts on the sensory receptor by opening or closing ion channels, thus producing a receptor potential.* **(b)** *A sensory receptor that is a separate cell from the afferent neuron. The stimulus changes the membrane potential of the receptor cell, which opens or closes a calcium channel, and cytosolic calcium concentration increases or decreases. Changes in calcium concentration trigger or inhibit the release of a chemical transmitter by exocytosis. The transmitter communicates to the afferent neuron by binding to receptors on the afferent ending.*

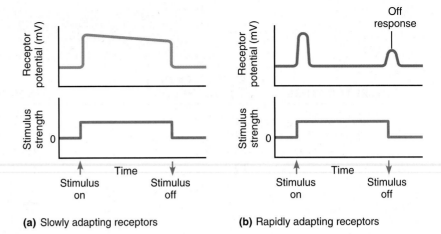

FIGURE 9.3 Responses of slowly adapting receptors and rapidly adapting receptors. (a) *Slowly adapting receptors respond with a change in receptor potential that persists for the duration of the stimulus.* **(b)** *Rapidly adapting receptors respond with a change in receptor potential at the onset of a stimulus, but then adapt. The "off response" is a second, smaller response that occurs upon termination of a stimulus.*

(a) Slowly adapting receptors

(b) Rapidly adapting receptors

excitatory stimulus, the greater the amount of transmitter released. This transmitter then binds to receptors on the afferent neuron and causes changes in the membrane potential of that cell. If the afferent neuron is depolarized to threshold, then an action potential is generated and transmitted by that cell to the CNS. The mechanism of transduction varies for different types of receptors and is covered with the individual sensory systems later in this chapter.

Receptor Adaptation

Some receptors continue to respond to a stimulus for as long as the stimulus is applied. However, most receptors *adapt* to the stimulus; that is, their response declines with the passage of time. **Adaptation** is a decrease over time in the magnitude of the receptor potential in the presence of a constant stimulus. **Slowly adapting** or **tonic receptors** show little adaptation and therefore can function in signaling the intensity of a prolonged stimulus (Figure 9.3a). Examples of slowly adapting receptors are muscle stretch receptors, which detect the degree to which muscles are stretched, and proprioceptors, which detect the position of the body in three-dimensional space. For proper movement and balance, it is critical that these receptors continually inform the CNS of the position of muscles and joints. **Rapidly adapting** or **phasic receptors** adapt quickly, and thus function best in detecting changes in stimulus intensity (Figure 9.3b). Rapidly adapting receptors respond at the onset of a stimulus, then adapt. Some rapidly adapting receptors also show a second, smaller response upon the termination of a stimulus, called the "off response." Examples of rapidly adapting receptors are *olfactory receptors*, which detect odors, and *Pacinian corpuscles*, which detect vibration in the skin.

Quick Test 9.1

1. Define the following terms: *modality, adequate stimulus, transduction, adaptation,* and the *law of specific nerve energies.*

2. How can a sensory receptor cell that is not part of an afferent neuron cause action potentials in that neuron?

3. Which type of receptor is better for communicating changes in the intensity of a sensory input, rapidly adapting receptors or slowly adapting receptors?

Sensory Pathways

The sensory information communicated to the CNS by afferent neurons can trigger reflexes, as described in Chapter 8, but it can also be relayed to areas of the brain for further processing, including perception of the stimulus. Here we focus on the sensory pathways involved in perception.

A **sensory unit** is a single afferent neuron and all the receptors associated with it (Figure 9.4). All the receptors associated with a given afferent neuron are of the same type. Activation of any of the associated receptors may trigger action potentials in the afferent neuron. The area over which an adequate stimulus can produce a response (which can be either excitatory or inhibitory) in the afferent neuron is called the **receptive field** of that neuron; it corresponds to the region containing receptors for that afferent neuron. In the skin, for example, a single afferent neuron possessing Pacinian corpuscles branches in the periphery, with each branch terminating with a Pacinian corpuscle. Vibration on any area of the skin where these Pacinian corpuscles are located can depolarize the afferent neuron and, if strong enough, can generate action potentials. If pressure applied to any given area of the skin causes a response in a neuron, that area is included within that neuron's receptive field. Note, however, that receptive fields in the visual system are associated with areas in the visual field (the area of the external environment from which light rays can enter the eyes and can be seen). Afferent neurons in the visual pathway respond to light emanating from specific areas of the visual field.

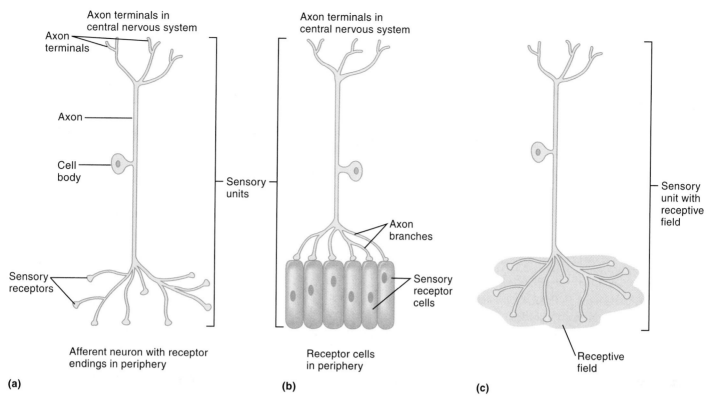

Axon terminals in central nervous system

Axon terminals

Axon

Cell body

Sensory receptors

Sensory units

Afferent neuron with receptor endings in periphery

(a)

Axon terminals in central nervous system

Axon branches

Sensory receptor cells

Receptor cells in periphery

(b)

Sensory unit with receptive field

Receptive field

(c)

FIGURE 9.4 Sensory units and receptive fields. (a) *A sensory unit in which the receptors are specialized endings of the afferent neuron.* **(b)** *A sensory unit in which the receptors are separate cells, each of which communicates to a single afferent neuron.* **(c)** *The receptive field for the sensory unit depicted in* **a.**

The specific neural pathways that transmit information pertaining to a particular modality are referred to as **labeled lines,** and each sensory modality follows its own labeled line or pathway. Activation of a specific pathway causes perception of the associated modality, regardless of the actual stimulus that activated the pathway. Thus, as we have seen, pressure on the eye may trigger light flashes because the pressure generates signals in the pathway for perception of light. The pathways for different modalities terminate in different sensory areas of the cerebral cortex (Figure 9.5). Although the pathways are distinct, we can make some generalizations concerning most pathways.

A generalized pathway for the transmission of sensory information is illustrated in Figure 9.6. The afferent neuron that transmits information from the periphery to the CNS is called the **first-order neuron.** A single first-order neuron may diverge within the CNS and communicate with several interneurons. In addition, interneurons may receive converging input from several first-order neurons. Some of these interneurons transmit the information to the thalamus, the major relay nucleus for sensory input; such interneurons are examples of **second-order neurons.** In the thalamus, these second-order neurons form synapses with **third-order neurons** that transmit information to the cerebral cortex, where sensory perception occurs. Different sensory pathways travel through different areas of the thalamus and cortex. The concept of receptive fields previously described for first-order neurons also applies to second- and third-order neurons, and even to higher-level neurons in sensory pathways. The receptive field of any neuron in a sensory pathway is the area in which an adequate stimulus can induce a response in that neuron.

Quick Test 9.2

1. Define a sensory unit. How many different types of receptors are associated with a single sensory unit?

2. Describe the general pathway by which information is transmitted from a sensory receptor to the cerebral cortex for perception.

3. What is another name for a first-order neuron?

Sensory Coding

In the previous section we saw that specific neural pathways transmit information from sensory receptor to the central nervous system. But when a stimulus acts on a sensory receptor, how does the nervous system identify

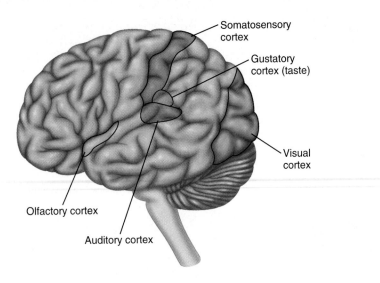

FIGURE 9.5 **Sensory areas of the cerebral cortex.**

Name the lobe that corresponds to each of the following: visual cortex, auditory cortex, and somatosensory cortex.

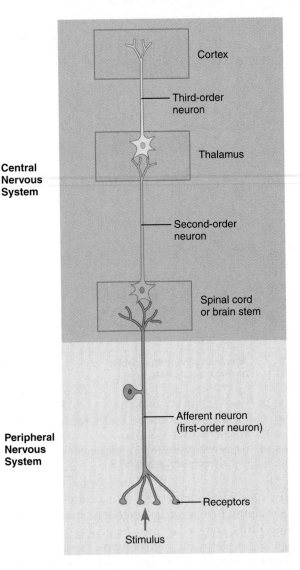

FIGURE 9.6 **Generalized pathway for sensory systems.**

the type, strength, and location of that stimulus? These tasks are made possible by *sensory coding.*

Coding for Stimulus Type

Stimulus type is coded by the receptor and pathway activated when the stimulus is applied. For example, light waves activate photoreceptors, which communicate via a specific pathway to the visual cortex. In fact, even if the visual pathway is inappropriately activated, as in the example of the blow to the eye, light will still be perceived because the visual pathway was activated.

Our perceptions of stimuli are not necessarily based on a single sensory pathway; often the brain must integrate information from different sensory systems. For example, we perceive when our skin is wet, even though we do not have wetness receptors, because thermoreceptors and touch receptors, when activated appropriately, transmit a combination of signals interpreted by the brain as wetness. To demonstrate that there are no wetness receptors, do the following: Put a thin latex glove on your hand and then hold it under running water. You will be convinced that the glove is leaking water, even when it is not! If wet receptors existed, they would not have been activated by the inadequate stimulus in this example.

Coding for Stimulus Intensity and Duration

Stimulus intensity is coded by the frequency of action potentials *(frequency coding)* and the number of receptors ac-

tivated *(population coding)*; by contrast, stimulus duration is coded by *changes* in the frequency of action potentials and *changes* in the number of receptors activated. In frequency coding (Figure 9.7), a stronger stimulus results in a larger receptor potential. Recall from Chapter 6 that as long as the graded potential (in this case, the receptor potential) exceeds threshold, stronger depolarizations can overcome the relative refractory period of an action potential and generate a second action potential more quickly than can weaker depolarizations. Therefore, stronger stimuli produce a higher frequency of action potentials.

Slowly adapting receptors are better able to code the intensity of a stimulus for its entire duration (Figure 9.8), whereas rapidly adapting receptors are better able to code changes in the intensity of a stimulus. For example, when a stimulus is applied to a slowly adapting receptor (Figure 9.8a), action potentials are produced in the afferent neuron for the entire time the stimulus is present, even though the frequency of action potentials may slightly

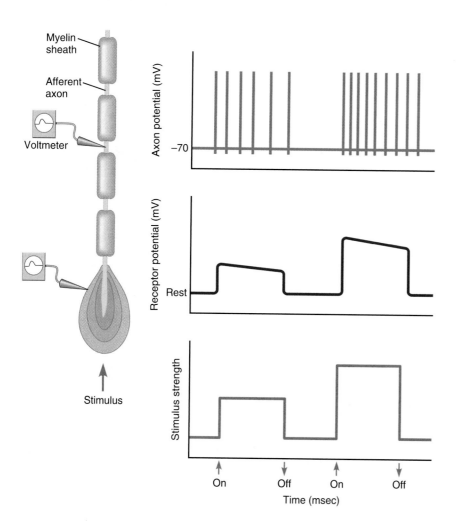

FIGURE 9.7 Coding of stimulus intensity.
Changes in membrane potential along the axon (top) and at the receptor (middle) are plotted for two different stimulus intensities (bottom). Detection of a stronger stimulus by the receptor leads to more frequent action potentials along the axon.

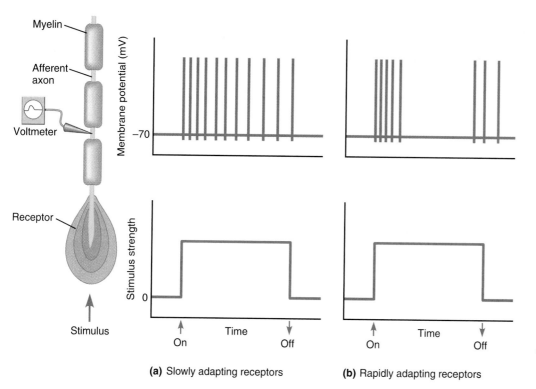

(a) Slowly adapting receptors

(b) Rapidly adapting receptors

FIGURE 9.8 Coding of stimulus duration. (a) *In axons associated with slowly adapting receptors, action potentials are propagated for the duration of the stimulus and generally decline slightly in frequency while the stimulus persists.* **(b)** *Rapidly adapting receptors do not code for stimulus duration; they generate action potentials when the stimulus is first applied and when it ceases.*

Can a stimulus of very strong intensity produce action potentials at such a fast rate that they sum? Explain why or why not.

No, because action potentials have refractory periods.

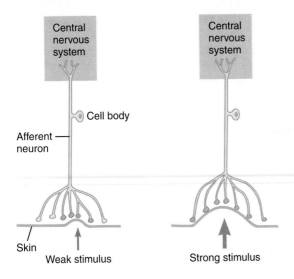

(a) Single sensory unit stimulated

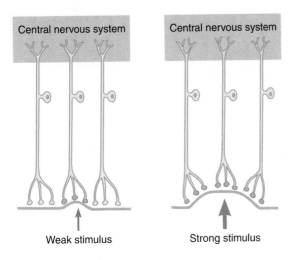

(b) Multiple sensory units stimulated

FIGURE 9.9 Coding of stimulus intensity by recruitment (population coding). (a) *Receptor recruitment within a single sensory unit.* **(b)** *Receptor recruitment of additional sensory units.*

decrease over time (in which case the intensity would be interpreted as being weaker than it really is). When the stimulus is removed, action potentials are no longer produced. However, in rapidly adapting receptors (Figure 9.8b), action potentials are produced at a high frequency at the onset of the stimulus, but not for the entire time the stimulus is present. In some but not all phasic receptors, action potentials are produced again when the stimulus is terminated, but at a lower frequency than at the onset. Therefore, rapidly adapting receptors detect changes in the intensity of a stimulus but not the duration of the stimulus.

In population coding, a stronger stimulus activates, or recruits, a greater number of receptors (Figure 9.9). These receptors may be associated with a single afferent neuron (Figure 9.9a), in which case the receptor potentials that

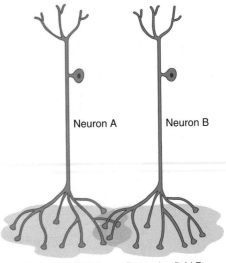

FIGURE 9.10 Overlapping receptive fields of two afferent neurons.

are generated at the individual receptors sum and produce a greater frequency of action potentials in that neuron. A stimulus may also recruit receptors associated with different afferent neurons (Figure 9.9b), in which case more afferent neurons transmit signals to the CNS concerning the presence of the stimulus. In either case, a greater frequency of action potentials is transmitted to the CNS in response to the stimulus, indicating that the stimulus is stronger.

Coding for Stimulus Location

The basis for coding for the locations of most stimuli is receptive fields. Recall that when an adequate stimulus is applied within a particular receptive field, it activates receptors associated with a specific afferent neuron. This concept is best illustrated with sensory receptors in the skin.

The precision with which the location of a stimulus is perceived is called **acuity.** In sensations associated with the skin, acuity depends on the size and number of receptive fields, the amount of overlap between the receptive fields, and a phenomenon called *lateral inhibition* (discussed shortly). If a specific afferent neuron is activated, then the stimulus must be located within that neuron's receptive field. However, the sizes of receptive fields vary considerably over the body. Localization of a stimulus is better in areas served by neurons with smaller receptive fields. Furthermore, information from one afferent neuron alone does not provide precise localization of a stimulus because the stimulus could be anywhere within its receptive field. Localization is improved by the overlapping of the receptive fields of different afferent neurons (Figure 9.10). Such overlapping improves localization via two phenomena: (1) any stimulus that occurs within the region of overlap between the receptive fields of two

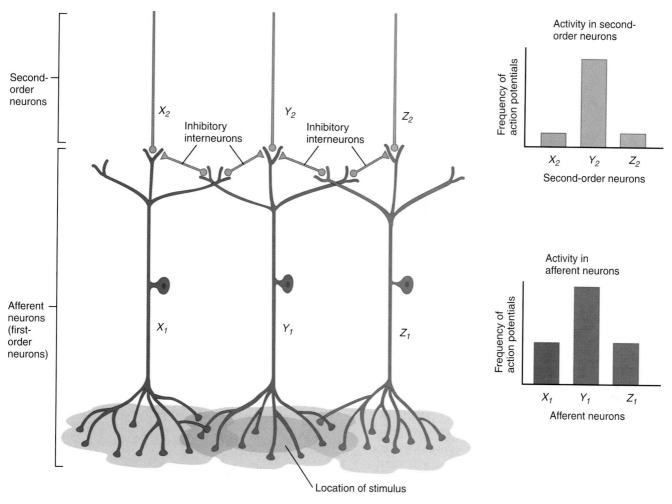

FIGURE 9.11 Lateral inhibition. *Lateral inhibition enhances contrast between sites of strong and weak stimulation. In this example, the stimulus is applied in the center of the receptive field for afferent neuron Y_1, and in the periphery of the receptive fields for afferent neurons X_1 and Z_1. Lateral inhibition occurs when collaterals of one afferent neuron (here, Y_1) activate interneurons that inhibit communication between neighboring afferent neurons and their second-order neurons. The graphs show the frequencies of action potentials in the afferent and second-order neurons.*

afferent neurons activates both neurons, and (2) lateral inhibition.

In **lateral inhibition,** a stimulus that strongly excites receptors in a given location inhibits activity in the afferent pathways of other nearby receptors. In the simplified pathways depicted in Figure 9.11, each afferent neuron synapses with only one second-order neuron. Note that the afferent neurons have collaterals that communicate with inhibitory interneurons in the CNS. In this case, collaterals of afferent neuron Y_1 activate inhibitory interneurons that decrease the communication between afferent neurons with neighboring receptive fields (neurons X_1 and Z_1) and their second-order neurons (X_2 and Z_2, respectively).

Lateral inhibition increases acuity because it increases the contrast of signals in the nervous system—it allows the transmission of strong signals in some neurons while suppressing transmission of weaker signals arising in nearby neurons. When a stimulus is applied in the center of the receptive field for neuron Y_1, that neuron is more

strongly activated than are neurons X_1 or Z_1, as evidenced by a higher frequency of action potentials (see lower graph in Figure 9.11). Because neuron Y_1 is the most highly activated, its activation of the inhibitory interneurons to neurons X_1 and Z_1 exceeds the inhibitory effect of the other interneurons, producing a greater inhibition of neurons X_2 and Z_2. Decreased activity occurs in all the second-order neurons, but the activities of X_2 and Z_2 are decreased more than that of neuron Y_2 (see upper graph in Figure 9.11). The result is a greater difference in the frequency of action potentials among second-order neurons than in afferent neurons, but more importantly the frequency of action potentials in Y_2 is *much greater* than the frequencies in X_2 and Z_2. The resulting enhanced contrast between more important and less important neural signals allows better localization of the stimulus, thereby increasing tactile acuity.

One measure of tactile acuity is **two-point discrimination,** the ability of a person to perceive two fine points

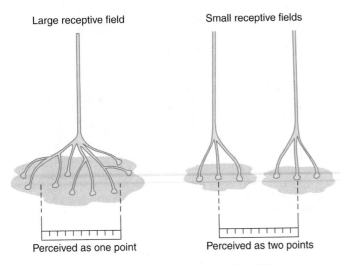

Large receptive field   Small receptive fields

Perceived as one point   Perceived as two points

FIGURE 9.12 Two-point discrimination. *The ability to discriminate between two separate points depends on the activation of separate receptive fields. The smaller the receptive field, the greater the ability for two-point discrimination, and the greater the tactile acuity.*

pressed against the skin as two distinct points. The minimum distance that must exist between two points for them to be perceived as separate is termed the *two-point discrimination threshold;* points that are closer together are perceived as a single point. Two-point discrimination occurs only if the two points are applied to the receptive fields of two different afferent neurons (Figure 9.12). The smaller the receptive fields, the greater the ability to discriminate two points, and the greater the tactile acuity.

Tactile acuity varies over different regions of the body. The skin in some areas of the body is innervated by afferent neurons with few branches and, therefore, small receptive fields, whereas other areas are innervated by afferent neurons with extensive branching and large receptive fields. In addition, greater overlap occurs for the smaller receptive fields, which also contributes to greater tactile acuity. Table 9.2 lists two-point discrimination thresholds in several selected areas of the body. In the lips, the most sensitive areas, points as close together as 1 mm can be distinguished as two distinct points. The fingertips are also quite sensitive. Areas on the back, thigh, and forearm, by contrast, are not very sensitive. In fact, two points as far apart as 40–50 mm (almost 2 inches!) may be indistinguishable as separate points in these areas.

Localization in some sensory systems is unrelated to coding by receptive fields. Even though a person can determine the direction from which an odor or a sound comes, neither olfactory nor auditory afferent neurons have receptive fields. Instead, these systems code for the quality and intensity of smell, and the pitch and loudness of sound, respectively. Localization in olfaction and in hearing is based on the arrival of stimuli at the two nostrils or at the two ears at slightly different times (Figure 9.13). The brain uses the difference in the time of arrival of action potentials at the olfactory or auditory cortex to determine where the stimulus originated.

TABLE 9.2 TWO-POINT DISCRIMINATION THRESHOLDS FOR SELECTED AREAS OF THE BODY

BODY REGION	TWO-POINT DISCRIMINATION THRESHOLD (mm)*
Lips (greatest acuity)	1
Index finger	2
Thumb	3
Palm of hand	10
Big toe	10
Forehead	18
Sole of foot	22
Breast	31
Abdomen	36
Shoulder	38
Back	42
Thigh	46
Upper arm	47
Calf (least acuity)	48

*Smaller distances indicate greater tactile acuity.

Source: Weinstein and Kenshalo, editors, *The Skin Senses*, copyright 1968. Courtesy of Charles C. Thomas Publisher, Ltd., Springfield, Illinois.

Quick Test 9.3

1. What is the difference between population coding and frequency coding? Do these processes code a stimulus's type or its intensity?

2. What is the difference between rapidly adapting receptors and slowly adapting receptors?

3. Explain the concept of two-point discrimination. Is two-point discrimination better on the palm of the hand or on the thigh? Why? Where is the ability to locate a stimulus better, on the palm of the hand or on the thigh? Why?

4. In lateral inhibition, the nervous system enhances signal contrast by increasing the relative strength of a strong stimulus compared to a weaker stimulus. How does this improve our ability to locate a stimulus?

THE SOMATOSENSORY SYSTEM

The somatosensory system is involved with body sensations such as pressure, temperature, pain, and body position (recall from Chapter 6 that *soma* means "body"). We first consider the types of receptors in this system.

TABLE 9.3 SENSORY RECEPTORS
IN THE SKIN

RECEPTOR CLASS	TYPE	ASSOCIATED AFFERENT TYPE	LOCATION	RECEPTIVE FIELD SIZE	ADAPTATION	MODALITY
Mechano-receptors	Free nerve ending	A-delta, C	Superficial, all skin	Small	Slow	Light touch
	Merkel's disk	A-beta	Superficial, all skin	Small	Slow	Pressure
	Pacinian corpuscle	A-beta	Deep, all skin	Large	Rapid	Vibration
	Meissner's corpuscle	A-beta	Superficial, glabrous skin	Small	Rapid	Vibration
	Hair follicle receptor	A-beta	Superficial, hairy skin	Small	Rapid	Bending of hair
	Ruffini's ending	A-beta	Deep, hairy skin	Large	Slow	Pressure
Thermo-receptors	Warm receptors (free nerve endings)	C	Superficial, all skin	Small	Rapid	Increase in skin temperature
	Cold receptors	A-delta	Superficial, all skin	Small	Rapid	Decrease in skin temperature
Nociceptors	Mechanical (free nerve endings)	A-delta	Superficial, all skin	Large	Slow	Intense mechanical stimulus
	Thermal (free nerve endings)	A-delta	Superficial, all skin	Small	Rapid	Intense hot or cold stimulus
	Polymodal (free nerve endings)	C	Superficial, all skin	Large	Slow	Intense mechanical or thermal stimulus; specific chemicals

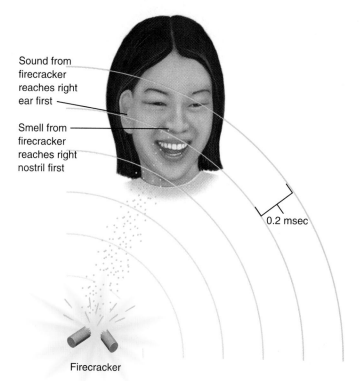

Sound from firecracker reaches right ear first

Smell from firecracker reaches right nostril first

0.2 msec

Firecracker

FIGURE 9.13 Localization in hearing and olfaction. *Localization of a sound or an odor depends on the difference between the time the stimulus reaches the left and right ears or nostrils, respectively.*

Somatosensory Receptors

The somatosensory system responds to a variety of stimuli arising in many areas of the body, and thus it utilizes many receptor types. For example, proprioception of our body's position requires receptors in the muscles, tendons, ligaments, and joints, as well as receptors in the skin. Such receptors are called proprioceptors, one example of which is the muscle spindle, discussed in Chapter 8. Somesthetic sensations of stimuli associated with the surface of the body require **mechanoreceptors** to detect pressure, force, or vibration; **thermoreceptors** to detect temperature; and **nociceptors** to detect tissue-damaging stimuli. Of all the sensory systems, the somatosensory system has the widest variety of receptor types; Table 9.3 lists and characterizes the various types of receptors in the skin.

FIGURE 9.14 **Sensory receptors in the skin.**

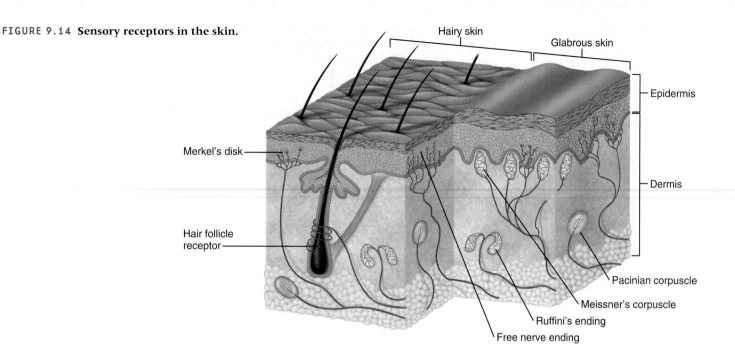

Mechanoreceptors in the Skin

Most somatosensory receptors in the skin are specialized structures at nerve endings, which are easily identified under the light microscope. However, a few somatosensory receptor types lack identifiable specialized structures and are therefore called **free nerve endings.** The different structures are designed to respond to particular modalities of stimuli impinging on the skin.

Mechanoreceptors in the Skin

The various types of mechanoreceptors in the skin are depicted in Figure 9.14. Some receptors are found in superficial layers of the skin close to the *epidermis,* or outer layer of the skin. Superficial receptors include *Merkel's disks* and *Meissner's corpuscles.* Other receptors are located deeper, in the *dermis* or inner layer of the skin, including *hair follicle receptors, Pacinian corpuscles,* and *Ruffini's endings.* Meissner's corpuscles are found only in *glabrous* skin (hairless skin), whereas hair follicle receptors are only found in hairy skin. Note that slowly adapting mechanoreceptors respond to pressure (a sustained stimulus), whereas rapidly adapting receptors respond best to vibration (a constantly changing stimulus). Note as well that the sizes of the receptive fields for the various mechanoreceptors vary greatly, and that the smaller receptive fields provide better tactile acuity.

Thermoreceptors in the Skin

Thermoreceptors respond to the temperature of the receptor endings themselves and the surrounding tissue, not the temperature of the surrounding air. There are two types of thermoreceptors: warm receptors and cold receptors. **Warm receptors** are free nerve endings that respond to temperatures between 30°C and 43°C; the frequency of action potentials increases as the temperature increases. **Cold receptors** respond to temperatures between 35°C and 20°C; the frequency of action potentials increases as the temperature falls and the receptors get colder. The structure of cold receptors is not known, although some evidence suggests that they too are free nerve endings.

Nociceptors in the Skin

Nociceptors are the sensory receptors responsible for the transduction of noxious stimuli that we perceive in the brain as pain. Nociceptors are free nerve endings that respond to tissue-damaging (or potentially damaging) stimuli. There are three types of nociceptors: **mechanical nociceptors,** which respond to intense mechanical stimuli, such as when you stub your toe; **thermal nociceptors,** which respond to intense heat (greater than 44°C), such as when you touch a hot stove; and **polymodal nociceptors,** which respond to a variety of stimuli, including intense mechanical stimuli, intense heat, intense cold, and chemicals released from damaged tissue. Chemicals that are released from damaged tissue and are capable of activating polymodal nociceptors include *histamine, bradykinin,* and *prostaglandins.*

1. Name the three classes of receptors for somesthetic sensations and the adequate stimulus (or stimuli) for each.

2. Name the two types of thermoreceptors and describe the types of stimuli that excite them.

3. Name the three types of nociceptors and the stimuli that activate them.

The Somatosensory Cortex

The perception of somatic sensations from all parts of the body begins in the primary somatosensory cortex (although recent studies suggest that some "crude" perception may occur in the thalamus). Recall from Chapter 8 that the somatosensory cortex is topographically oriented; that is, sensory information arising in neighboring areas of the body generally projects to neighboring areas of the cortex. Recall as well that the size of the area of the cortex devoted to somatic sensations from a specific area of the body is not necessarily proportional to the size of the body region, but rather to the sensitivity of the body region. The lips and fingertips, for example, are very sensitive areas because they have small two-point discrimination thresholds (see Table 9.2), and these body regions also have large areas of the primary somatosensory cortex devoted to them (see Figure 8.14, p. 241).

Recall also that the cerebral cortex has columnar organization. In the primary somatosensory cortex, vertical columns are organized according to sensory modality. For example, in the area of the primary somatosensory cortex for the thumb, one column is associated with pressure to the thumb, another column with vibration to the thumb, another column with cold, and so on. The next section describes the pathways along which information travels from the receptors to the primary somatosensory cortex.

Somatosensory Pathways

Two main pathways transmit information from peripheral somatosensory receptors to the central nervous system: the *dorsal column–medial lemniscal pathway* and the *spinothalamic tract.* These pathways transmit different types of sensory information to the thalamus, and then to the primary somatosensory thalamus. In both cases the pathways enter the spinal cord on one side and cross to the other side before reaching the thalamus. Therefore, somatosensory information from the right side of the body is perceived in the left somatosensory cortex, and vice versa.

The Dorsal Column–Medial Lemniscal Pathway

The **dorsal column–medial lemniscal pathway** transmits information from mechanoreceptors and proprioceptors to the thalamus; it crosses to the other side of the CNS in the medulla oblongata (Figure 9.15a). In this pathway, first-order neurons originate in the periphery and enter the dorsal horn of the spinal cord. Collaterals from the main axon may terminate in the spinal cord, communicating with interneurons involved in spinal reflexes. However, the main branch of the axon ascends from the spinal cord to the ipsilateral (same side as the stimulus) brainstem in the *dorsal columns,* which are tracts of white matter that run dorsal and medial to the dorsal horn. The first-order neurons terminate in *dorsal column nuclei,* which are located in the medulla, where they form synapses with second-order neurons. The second-order neurons then cross over to the contralateral side of the medulla in a tract called the *medial lemniscus,* and then ascend to the thalamus. In the thalamus the second-order neurons synapse with third-order neurons, which transmit information from the thalamus to the somatosensory cortex.

The Spinothalamic Tract

The spinothalamic tract transmits information from thermoreceptors and nociceptors to the thalamus; it crosses to the other side of the CNS within the spinal cord before it reaches the brain (Figure 9.15b). In this pathway, first-order neurons originate in the periphery at either thermoreceptors or nociceptors and enter the dorsal horn of the spinal cord. Here, first-order neurons may ascend or descend for a short distance (a few spinal segments) along *Lissauer's tract,* but eventually they form synapses with second-order neurons in the dorsal horn. The second-order neurons cross over to the contralateral side of the spinal cord, ascend in the anterolateral quadrant of the spinal cord through the brainstem, and then terminate in the thalamus. In the thalamus, second-order neurons form synapses with third-order neurons that ascend to the somatosensory cortex.

Quick Test 9.5

1. If a person suffers damage to the left side of the spinal cord, which somatic sensations would be lost from which side of the body?

2. How do the dorsal column–medial lemniscal and the spinothalamic pathways differ with regard to the types of sensory information they transmit? Where does each pathway cross over to the other side of the CNS?

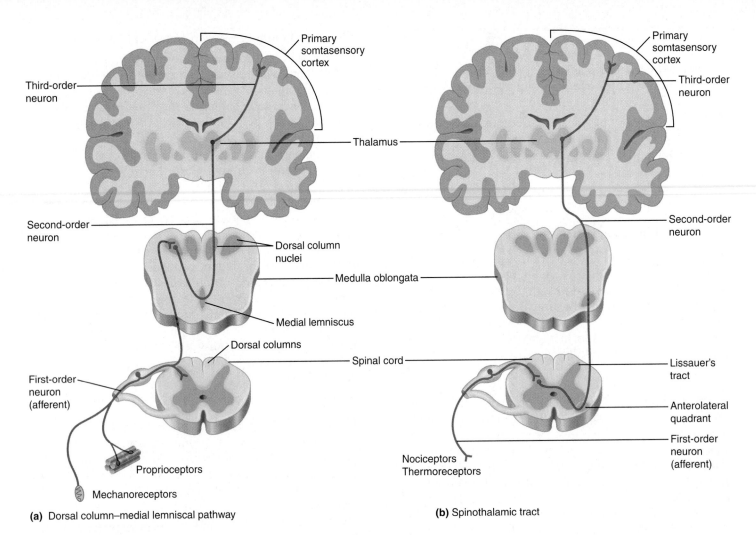

FIGURE 9.15 **The two somatosensory pathways. (a)** *Dorsal column–medial lemniscal pathway, which transmits information from touch, pressure, vibration, and proprioceptors to the CNS.* **(b)** *Spinothalamic tract, which transmits information from temperature and pain to the CNS.*

If a person received damage to the right dorsal columns, would they lose perception of touch on the right or left side?

Pain Perception

Pain, one of the somesthetic sensations, is important because it teaches us to avoid subsequent encounters with potentially damaging stimuli (see When It Goes Wrong: Insensitivity to Pain). Pain is also important clinically, because it is a sign that tissue damage may have occurred. Thus the mechanisms of pain perception warrant special consideration.

The Pain Response

The activation of nociceptors leads not only to the perception of pain, but also to a variety of other body responses, including one or more of the following:

(1) autonomic responses, such as increases in blood pressure and heart rate, increases in blood epinephrine levels, increases in blood glucose, dilation of the pupils of the eye, or sweating; (2) emotional responses such as fear or anxiety; and (3) a reflexive withdrawal from the stimulus. The level of perceived pain varies considerably among individuals based on their past experiences and the circumstances under which the stimulus is applied. Thus a toothache that is barely noticeable during a busy day, for example, may be excruciating when one tries to fall asleep at night. To better understand how pain perception can vary, let's take a closer look at the mechanisms of pain perception.

Each of two types of pain, fast pain and slow pain, is perceived differently and is transmitted by a different class of afferent neurons. **Fast pain** is perceived as a

The right side

Have you ever thought what it would be like to not feel pain? Sounds good, doesn't it? No more pain because you mishandled a knife or stubbed your toe. No more worrying about bullies threatening to hurt you, because they cannot hurt you. No more time lost from your favorite athletic activity because you sprained an ankle. You could do things others fear doing because you know it would not hurt. You would be a celebrity, the idol of all your friends. Or would you?

Pain affects many aspects of our lives, and no one likes it, but that we are lucky to have it is best illustrated by those who cannot perceive pain. Although it is rare, some people are born with a *congenital insensitivity to pain*; due to a genetic abnormality, such people lack some component of the sensory systems for pain. The exact causes are unknown and seem to vary among patients, but the outcome is often deadly: People who do not perceive pain generally do not survive childhood.

Injuries are common in those afflicted with congenital insensitivity to pain. These people often lose digits from handling sharp objects. They may also suffer severe burns. Their knees often show sores from kneeling too long. People with normal pain perception are less likely to suffer from such injuries because they would withdraw from sharp or hot objects, and pain from kneeling would cause them to change position long before any tissue damage occurred.

Clearly, pain has a purpose: It is a warning signal that enables us to protect ourselves from bodily harm. Pain also enhances our ability to learn what types of stimuli can damage our bodies. So the next time something hurts, be glad that you feel it and are able to protect yourself from further harm. With care you may not have to feel it again.

sharp pricking sensation that can be easily localized; it is transmitted by *Aδ* (A delta) *fibers,* thin, lightly myelinated axons with a conduction velocity of approximately 12–30 m/sec. **Slow pain** is perceived as a poorly localized, dull aching sensation; it is transmitted by *C fibers,* thin, unmyelinated axons with a conduction velocity of approximately 0.2–1.3 m/sec. Think about the last time you stubbed your toe. An initial sharp pain (fast pain), was followed by a prolonged aching pain (slow pain).

The primary afferents, whether Aδ or C fibers, form synapses with second-order neurons in the dorsal horn of the spinal cord. Communication between these first- and second-order neurons involves different neurotransmitters, one of which is *substance P.* Substance P is released from primary afferent neurons and binds to receptors on second-order neurons. The second-order neurons ascend to the thalamus via the spinothalamic tract, the pathway involved in the perception and discrimination of pain. Nociceptive afferents also activate different ascending pathways that are necessary for interpreting the affective components of pain. The affective pathways ascend to the reticular formation of the brainstem, the hypothalamus, and the limbic system.

Visceral Pain

Pain is not limited to the body surface. For example, most people have suffered pain in muscles after overexercising, or experienced a stomach ache. The viscera are subject to tissue damage, and nociceptors in the organs detect this damage. But what does a person perceive when her appendix becomes inflamed or when he suffers a heart attack? Generally, activation of nociceptors in the viscera produces pain that is called **referred pain** (because it has been "referred" to the body surface). For example, a person having a heart attack generally complains about pain in the left chest, upper arm, and shoulder—not in the heart itself. Referred pain is due to the fact that the second-order neurons that receive input from visceral afferents also receive input from somatic afferents (Figure 9.16a). According to one theory, the brain interprets information based on past experiences. Throughout a person's life, these second-order neurons are activated primarily by the somatic afferents, and thus the brain has learned that signals from these neurons are somatic in origin. When a person has a heart attack, the brain interprets the signals as a somatic disturbance, because that is what such signals have meant in the past. Physicians use maps of surface locations of referred pain (Figure 9.16b) to ascertain which internal organ(s) may be causing a patient's pain.

Modulation of Pain Signals

Signals about sensory information can be modulated as they are transmitted along sensory pathways; that is,

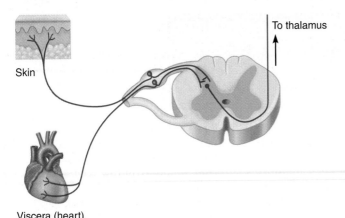

Skin

To thalamus

Viscera (heart)

(a) Mechanism of referred pain

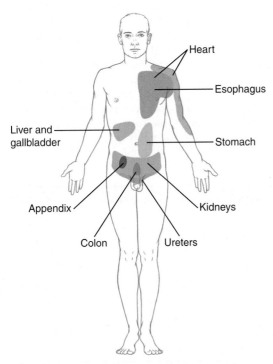

Heart

Esophagus

Liver and gallbladder

Stomach

Appendix

Kidneys

Colon

Ureters

(b) Clinical map of referred pain

FIGURE 9.16 Mechanisms and sites of referred pain. (a) *Referred pain occurs when visceral and somesthetic afferents converge on the same second-order neurons in the spinal cord.* **(b)** *Pain in specific visceral organs is generally referred to the areas of the body surface indicated in this map.*

facilitation or inhibition of signals can result in changes in the final perception of that information. Sensory signals can be modulated wherever there is a synapse in the pathway. The mechanisms involved in the modulation of pain have profound clinical significance. The **gate-control theory** of pain modulation, which states that somatic signals of nonpainful sources can inhibit signals of pain, was developed in 1967 and revolutionized pain research and the clinical management of pain.

The gate-control theory describes modulation of pain at the spinal level (Figure 9.17). Among the various interneurons within the spinal cord are interneurons that inhibit the second-order neurons that transmit pain information. When these interneurons are active, the transmission of pain signals is suppressed, and the perception of pain is lessened. When information about a painful stimulus is being transmitted to the spinal cord by C fibers (Figure 9.17a), the collaterals of these C fibers *inhibit* the activity of the inhibitory interneuron, which allows transmission to proceed to the second-order neuron. But the same inhibitory interneuron is *stimulated* by collaterals from large-diameter afferents (Aβ fibers) associated with mechanical stimuli such as touch, pressure, and vibration (Figure 9.17b). If a nonpainful mechanical stimulus is applied simultaneously with a painful stimulus, the collaterals from the Aβ fibers stimulate the inhibitory interneuron, thereby decreasing the transmission of pain signals.

The gate-control theory describes why rubbing a painful area relieves the pain. It is also the basis for using TENS, or *transcutaneous electrical nerve stimulation,* to treat pain. In TENS, a small current applied through the skin overlying a nerve activates large-diameter afferents, which relieves the pain.

The gate-control theory describes how afferent signals to the spinal cord can influence the perception of pain. The brain also has the ability to block pain, or produce **analgesia,** through descending pathways that are part of the pain-blocking **endogenous analgesia systems.** Pain can be debilitating, and there are times when perceiving pain is disadvantageous. For example, the chances that seriously wounded soldiers will survive are reduced if they are immobilized by pain while their lives are at stake. Such wounded soldiers can often function without perceiving pain because the endogenous analgesia systems become active and block the pain, allowing the body to cope with a more pressing need—survival.

Many brain areas are involved in the endogenous analgesia systems. One of the best defined pathways is illustrated in Figure 9.18. Stressful situations can activate an area in the midbrain called the *periaqueductal gray matter.* This area communicates to an area in the medulla called the *nucleus raphe magnus,* and to the *lateral reticular formation,* which extends the length of the brainstem. Neurons from these areas descend to the spinal cord, where they block the communication between nociceptive afferent neurons and second-order neurons, as follows.

Recall that substance P is one neurotransmitter released from nociceptive afferents that communicate with second-order neurons. Inhibitory interneurons in the spinal cord form synapses on the cell body and dendrites of the second-order neurons and also on the axon terminal of the nociceptive afferent neuron. These inhibitory interneurons release the endogenous opiate neurotransmitter *enkephalin* (discussed in Chapter 7),

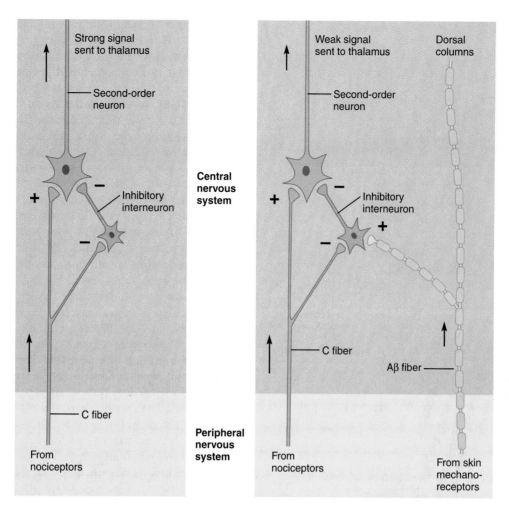

(a) Unmodulated pain

(b) Modulation of pain

FIGURE 9.17 Gate-control theory of pain. (a) *In unmodulated perception of pain, collaterals of the nociceptor afferents (C fibers) inhibit inhibitory interneurons, allowing transmission of pain signals to second-order neurons in the dorsal horn of the spinal cord and then to the thalamus.* **(b)** *In the modulation of pain, collaterals of large-diameter afferents (Aβ fibers) extending from touch and pressure receptors excite the inhibitory interneuron, thereby decreasing the transmission of pain signals.*

which binds to opioid receptors on the second-order neuron and induces inhibitory postsynaptic potentials. Enkephalin also binds to opioid receptors on the axon terminal of the nociceptive afferent neuron, which inhibits the release of substance P. Both these actions suppress signal transmission from the afferent neuron to the second-order neuron, thereby decreasing the transmission of pain signals to the brain. These inhibitory interneurons are activated by descending neurons of the nucleus raphe magnus.

Quick Test 9.6

1. Distinguish between fast pain and slow pain. Which class of afferent neuron is responsible for fast pain, and which is responsible for slow pain?

2. What kind of pain is experienced in the left shoulder by a person having a heart attack?

3. What are endogenous analgesia systems?

VISION

Most of what we learn about the world we learn through seeing. In this section we examine the visual system, which endows us with this important capability. We begin with the anatomy of the eye, the sensory organ of the visual system.

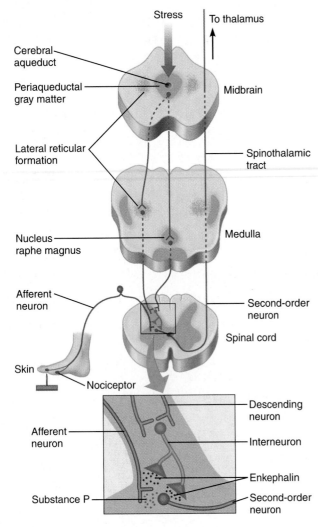

FIGURE 9.18 Endogenous analgesia systems. *When a painful stimulus activates nociceptors, information is transmitted to the CNS via the spinothalamic tract. In the presence of stress, endogenous analgesia systems can block pain transmission at the level of the synapse between the nociceptive afferent neuron and the second-order neuron, as follows: The periaqueductal gray matter in the midbrain communicates to the lateral reticular formation and to the nucleus raphe magnus of the medulla. These regions have neurons that descend to the dorsal horn of the spinal cord. Neurons from the nucleus raphe magnus activate inhibitory interneurons that release the neurotransmitter enkephalin, which then blocks communication between the nociceptive afferent and the second-order neuron via two mechanisms: presynaptic inhibition of substance P release from the nociceptive afferent, and production of IPSPs on the second-order neuron.*

Anatomy of the Eye

The important structures of the eye are shown in Figure 9.19. The eye can be divided into three concentric layers. The outermost layer consists of the sclera and cornea. The **sclera,** a tough connective tissue, makes up the "white" of the eye. In the front of the eye, however, the sclera gives way to the **cornea,** a transparent structure that allows light to enter the eye.

The middle layer of the eye consists of the choroid, ciliary body, and iris. The **choroid** is a highly pigmented

layer of tissue beneath the sclera. Its pigment absorbs light that reaches the back of the eye so that it is not reflected, which would cause distortion of the visual image. The choroid also contains blood vessels that nourish the inner layer of the eye. The **ciliary body** contains the **ciliary muscles,** which are attached to the lens by strands of connective tissue called **zonular fibers.** The **lens** focuses the light on the retina, where the visual information is transduced. The ciliary muscles change the shape of the lens to focus light waves. The **iris,** which consists of two layers of pigmented smooth muscle, is located in front of the lens. The pigmentation of the iris determines eye color. The **pupil** is a hole in the center of the iris that allows light to enter the posterior part of the eye; it is not a structure. The iris regulates the diameter of the pupil, thereby regulating the amount of light that reaches the back of the eye.

The innermost layer of the eye is the **retina.** The retina consists of neural tissue and contains the **photoreceptors,** cells that detect the light waves. Photoreceptors are of two types, *rods* and *cones,* which detect dim light and bright light, respectively. The retina, therefore, functions in *phototransduction,* the conversion of light energy to electrical energy. Two areas of the retina are worth noting. One is the **fovea,** the central point on the retina, where light from the center of the visual field strikes. (The transparent lens focuses light waves on the retina, much as a camera lens focuses light waves on film.) For reasons described later, the fovea is the area of the retina with the greatest visual acuity. The other area of note is the **optic disk,** the portion of the retina where the optic nerve and blood vessels supplying the eye pass through the retina. Because there are no photoreceptors in the optic disk, this area is a **blind spot,** a region where light striking the retina cannot be transduced into neural impulses and thus cannot be perceived.

The lens and ciliary body separate the eye into two fluid-filled chambers. In front of the lens and ciliary body is the **anterior cavity** containing a clear, watery fluid called **aqueous humor,** which supplies nutrients to the cornea and lens. The cornea and lens are transparent so that light can pass through them easily. If these structures relied on blood to supply their nutrients, the presence of blood vessels would obstruct the light. Behind the lens and ciliary body is the **vitreous chamber** containing a firmer, jellylike material called **vitreous humor,** which maintains the spherical structure of the eye.

The Nature and Behavior of Light Waves

Light is a form of energy. Specifically, light exists as *electromagnetic waves.* Along with other forms of electromagnetic energy, including radio waves, television waves, X rays, and gamma rays, light is a part of the electromagnetic spectrum (Figure 9.20). Visible light includes those electromagnetic waves having wavelengths between

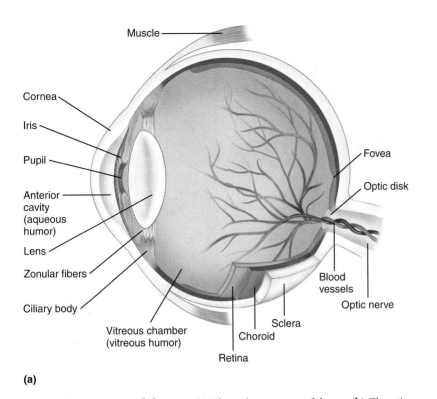

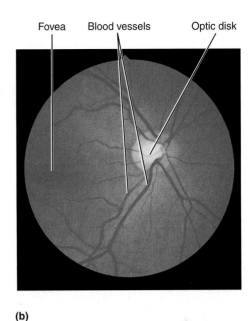

(a)

(b)

FIGURE 9.19 **Anatomy of the eye. (a)** *The major structures of the eye.* **(b)** *The retina as viewed through an ophthalmoscope.*

about 350 nm and 750 nm; different colors correspond to different wavelengths within this range.

Because light is a wave, it exhibits the usual properties of waves, including reflection and refraction. **Reflection** is a phenomenon in which light waves strike and bounce off a surface. Reflection is important in vision because much of the light we perceive has reflected off the objects we are observing. (We also see emitted light, such as the light coming directly to the eye from the sun or a light bulb.) Light that is absorbed by objects is not perceived; thus we perceive an object as green because it reflects to the eyes light of the wavelength corresponding to green (approximately 530 nm) while absorbing all other wavelengths. **Refraction** refers to the bending of light waves as they pass through transparent materials of different densities. This property is important in vision because in its path from objects to the photoreceptors in the retina, light must pass through several different materials, including air, the cornea, the lens, and the vitreous and aqueous humors. Because refraction is important in the focusing of light waves on the retina, we consider it here in greater detail.

Try this at home: Place a straw in a clear glass half full of water and let it rest at an angle. So oriented, the straw looks broken at the air-water interface (Figure 9.21a). Now hold the straw straight up so that it is perpendicular to the air-water interface. The straw now appears to be in one piece again (Figure 9.21b). Light waves are refracted as they travel from one medium to another if they strike the second medium at an angle other than perpendicular.

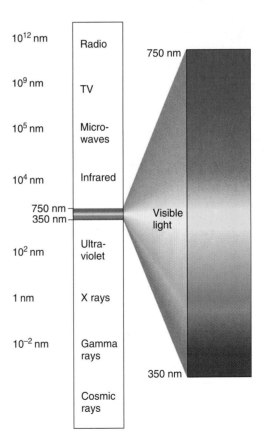

FIGURE 9.20 **The electromagnetic spectrum.** *The numbers indicate wavelength in nanometers (nm = 10^{-9} meter). The band of visible light is highlighted.*

FIGURE 9.21 **Refraction of light waves passing through different media. (a)** *Straw at an angle to the air-water interface.* **(b)** *Straw perpendicular to the air-water interface.*

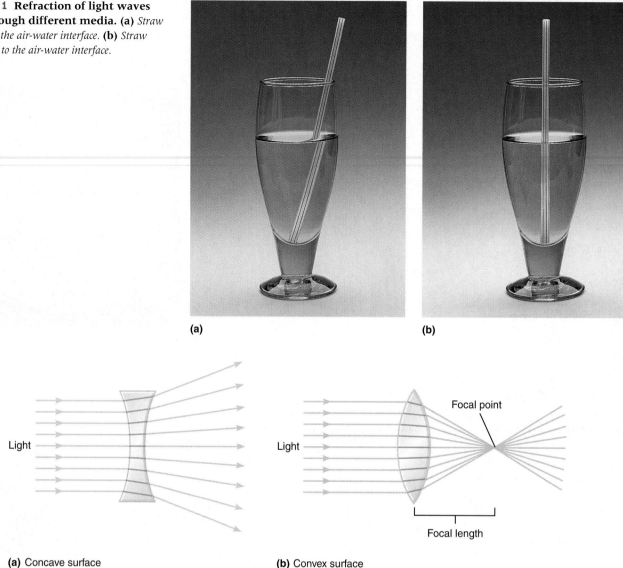

(a)   (b)

(a) Concave surface   **(b)** Convex surface

FIGURE 9.22 **Refraction of light waves passing through curved surfaces.**
(a) *Concave surfaces cause divergence of light waves.* **(b)** *Convex surfaces cause convergence of light waves to a focal point. The distance from the long axis of a convex lens to the focal point is called the focal length.*

How much the light waves refract depends on the differences in the densities of the two media and the angle at which the light strikes them. Try holding the straw at different angles.

Figure 9.22 illustrates the refraction of parallel light waves as they pass through *concave* and *convex* surfaces often used in lenses, such as those in eyeglasses or telescopes. Note that with either type of surface, light waves striking the surface perpendicular to it pass straight through. However, as the light strikes the surfaces at other angles, the concave lens causes the once-parallel light waves to diverge (move farther apart), whereas the convex lens causes the light waves to converge at a single point called the *focal point*. The distance from the long axis

of the convex lens to the focal point is called the *focal length.*

Both the cornea and the lens have convex surfaces that function to converge the light waves entering the eye onto the retina, a process that is necessary if visual images are to be in focus. Figure 9.23 shows how light waves from a viewed object are projected onto the retina. To see items in focus, light from a given point in the *visual field*—what we are looking at—must converge to a single point on the retina. Although the cornea has greater refractive power than the lens due to its greater curvature, the refractive power of the lens can be varied as needed to focus light on the retina. For example, to view very near objects, the lens becomes rounder, increasing its re-

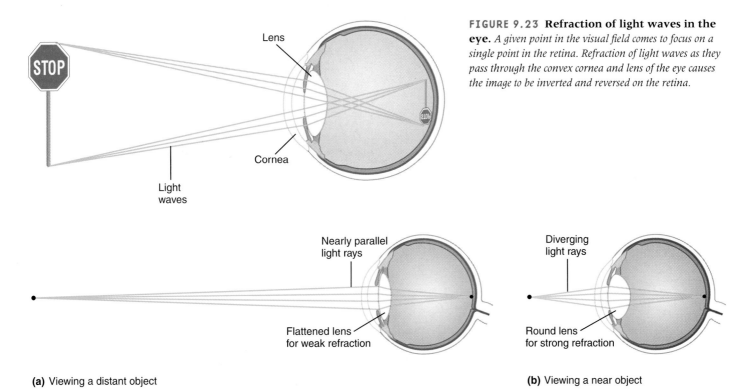

FIGURE 9.23 Refraction of light waves in the eye. *A given point in the visual field comes to focus on a single point in the retina. Refraction of light waves as they pass through the convex cornea and lens of the eye causes the image to be inverted and reversed on the retina.*

Lens

Light waves

Cornea

Nearly parallel light rays

Flattened lens for weak refraction

(a) Viewing a distant object

Diverging light rays

Round lens for strong refraction

(b) Viewing a near object

FIGURE 9.24 Focusing light from distant and near sources. (a) *Light waves reflected from a distant object approach the lens parallel to one another. A relatively flat (weak) lens is sufficient to converge the light waves on the retina.* **(b)** *Light waves reflected from a near object diverge as they approach the lens. A rounder (strong) lens is needed to converge the light waves on the retina.*

fractive power to focus the image on the retina. The ability of the lens to adjust its refractive power for near and distant objects is part of a process called **accommodation,** which we explore next.

Accommodation

If an object is to be seen clearly, the light reflected from any given point on the object must converge at a single point on the retina. When viewing something far away, the light waves enter the eye almost parallel to each other (Figure 9.24a), so little refractive power is needed to focus the light on the retina. However, light waves from close-up objects diverge as they enter the eye (Figure 9.24b), so the greater refractive power of a rounder lens is necessary to overcome this divergence and focus the light on the retina.

The shape of the lens is controlled by the circularly arranged ciliary muscle through the tension it applies to the zonular fibers, which attach the ciliary muscle to the lens. The more a circular muscle contracts, the smaller the diameter of the circle becomes. For viewing distant objects, the ciliary muscle is relaxed, which increases the diameter of the circle and tightens the zonular fibers, pulling the lens into a flattened shape (Figure 9.25a). To achieve accommodation for viewing close-up objects

(Figure 9.25b), the ciliary muscle contracts, reducing the diameter of the circle and lessening the tension on the zonular fibers. Due to its inherent elasticity, the lens becomes rounder when the tension on the zonular fibers is reduced. Accommodation is under control of the parasympathetic nervous system, which triggers contraction of the ciliary muscle for near vision. In the absence of parasympathetic activity, the ciliary muscle relaxes.

Clinical Defects in Vision

If light waves are not adequately focused on the retina, then vision is blurred. There are many causes of blurred vision, and different types of corrective lenses to improve vision.

Common visual defects include near-sightedness or **myopia,** and far-sightedness or **hyperopia.** In **emmetropia** or normal vision (Figure 9.26a), a person can see both distant and close-up objects clearly because the eye can focus light from far sources without accommodation, and from near sources with accommodation. In myopia or hyperopia, a mismatch exists between lens or cornea strength and eyeball length. The defect could be related to the lens, the cornea, or the eyeball length.

In myopia (Figure 9.26b), a person can see near objects clearly, but not distant objects because the lens or

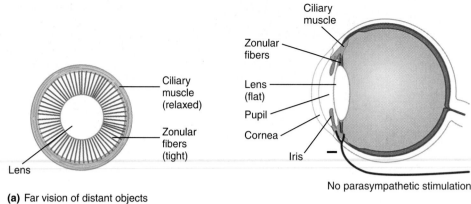

(a) Far vision of distant objects

No parasympathetic stimulation

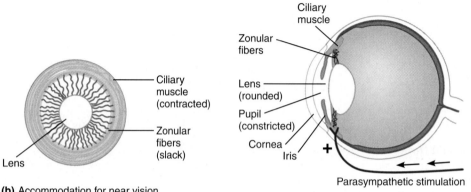

(b) Accommodation for near vision

Parasympathetic stimulation

FIGURE 9.25 **Mechanism of accommodation. (a)** *Vision of distant objects. In the absence of parasympathetic stimulation, the ciliary muscle relaxes, putting tension on the zonular fibers. The zonular fibers pull on the lens, flattening the lens.* **(b)** *Accommodation for near vision. Under parasympathetic stimulation, the ciliary muscle contracts, reducing the tension on the zonular fibers and enabling the elastic lens to become rounder.*

cornea is too strong for the length of the eyeball and therefore bends light rays too much. In this situation, close-up objects can be focused without accommodation, but light from distant objects is focused in front of the retina, resulting in a blurred image. To correct for myopia, a concave lens is placed in front of the eye. The lens causes light waves to diverge before reaching the eye. Under these conditions, the eye must accommodate to view close-up objects, and distant objects will be in focus without accommodation.

In hyperopia (Figure 9.26c), the lens or cornea is too weak for the length of the eyeball. Therefore, distant objects can be focused only with accommodation, which means that the lens cannot increase accommodation enough to adjust for near vision. Light from close-up objects comes to focus behind the retina, resulting in a blurred image. To correct for hyperopia, a convex lens is placed in front of the eye. The lens causes light waves to converge before reaching the eye. Now the eye can see distant objects without accommodation, which gives the lens enough leeway to enable it to accommodate for near objects.

Many other clinical defects affect the ability to focus light on the retina. In *astigmatism,* irregularities on the surface of the cornea or lens cause erratic bending of light waves. *Presbyopia* is a hardening of the lens that occurs with aging; as the lens hardens, the loss of elasticity decreases its ability to become spherical, making accommodation for near vision more difficult. A *cataract* is an age-related discoloration of the lens that decreases its transparency. In *glaucoma,* an increase in the volume of aqueous humor raises pressure in the anterior cavity of the eyeball, which can distort the shape of the cornea and shift the position of the lens. Shifting of the lens can transmit the increased pressure to the vitreous chamber, where it can compress the blood vessels that supply the retina. Excessive pressure can reduce the blood supply significantly, leading to permanent blindness.

Regulating the Amount of Light Entering the Eye

The eyes are capable of regulating the amount of light that enters them by varying the size of the pupils. In

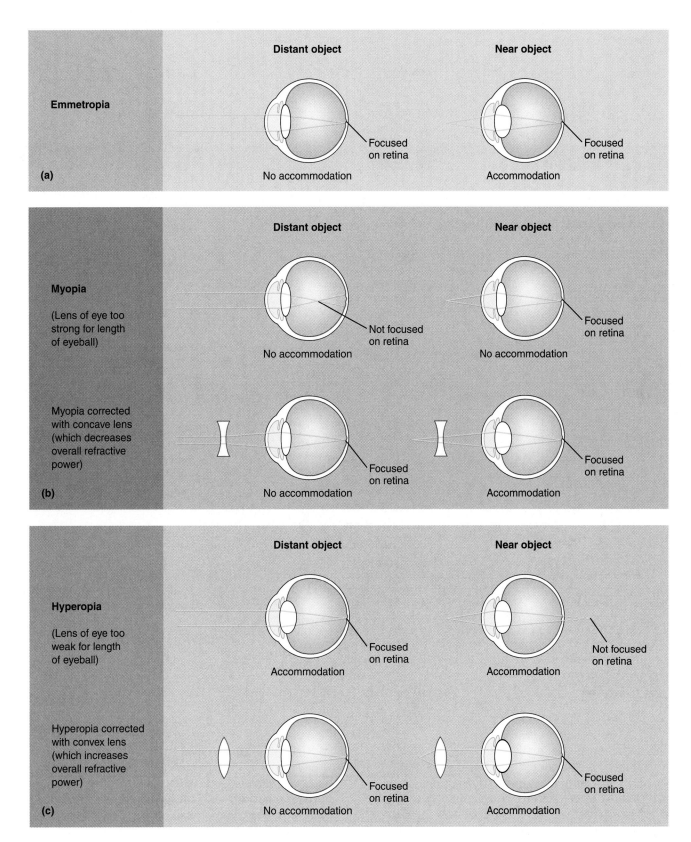

FIGURE 9.26 Normal, near-sighted, and far-sighted vision.
(a) *In emmetropia, or normal vision, distant objects are focused on the retina without accommodation, and near objects are focused with accommodation.* (b) *In myopia, or near-sightedness, the lens (or cornea) is too strong for the length of the eyeball. Near objects are focused without accommodation, and distant objects come into focus in front of the retina even without accommodation. Myopia can be corrected by using a concave lens to produce divergence of light waves before they enter the eye.* (c) *In hyperopia, or far-sightedness, the lens (or cornea) is too weak for the length of the eyeball. Distant objects are focused with accommodation, and near objects come into focus behind the retina, even with accommodation. Hyperopia can be corrected by using a convex lens to produce convergence of light waves that supplements the convergence produced in the eye.*

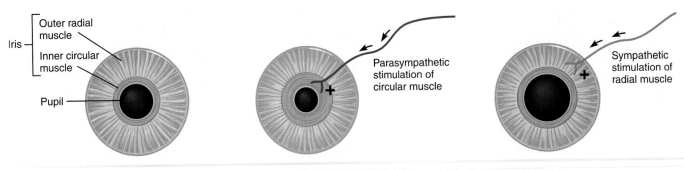

Iris
— Outer radial muscle
— Inner circular muscle
— Pupil

Parasympathetic stimulation of circular muscle

Sympathetic stimulation of radial muscle

(a) Anatomy of iris and pupil   **(b)** Pupillary constriction   **(c)** Pupillary dilation

FIGURE 9.27 Regulation of the amount of light entering the eye. (a) *The iris, which consists of two layers of smooth muscle—an inner circular layer and an outer radial layer—controls pupil size. The size of the pupil determines the amount of light that enters the eye.* **(b)** *Pupillary constriction, which is caused by parasympathetic stimulation of the circular muscle layer of the iris.* **(c)** *Pupillary dilation, which is caused by sympathetic stimulation of the radial muscle layer of the iris.*

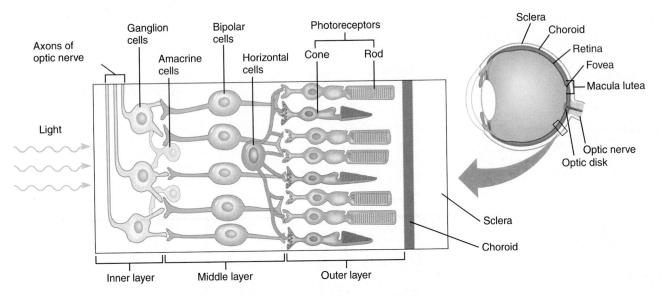

Axons of optic nerve
Ganglion cells
Amacrine cells
Bipolar cells
Horizontal cells
Photoreceptors
Cone   Rod
Light

Inner layer   Middle layer   Outer layer

Sclera
Choroid
Retina
Fovea
Macula lutea
Optic nerve
Optic disk
Sclera
Choroid

FIGURE 9.28 Anatomy of the retina. *Located on the inner surface of the eye, the retina consists of three layers of neural tissue composed of the various types of cells depicted. Note that light must pass through the inner and middle layers of the retina before striking the photoreceptors in the outer layer. Deep to the retina is the choroid, which absorbs light.*

bright light, the pupils are small, or *constricted,* so that the photoreceptors do not become "bleached out" by too much light. In dim light, by contrast, the pupils are large or *dilated* to allow more light in, which enhances the ability to see. The size of the pupil is controlled by the iris.

Recall that the iris consists of two layers of smooth muscle around the pupil. These two layers of smooth muscle are an inner **circular muscle** layer, also called the *constrictor muscle,* and an outer **radial muscle** layer, also called the *dilator muscle* (Figure 9.27a). The circular muscles form concentric rings around the pupil; when they contract, the diameter of the pupil decreases. Thus

contraction of the circular muscles causes *pupillary constriction* (Figure 9.27b). The radial muscles are arranged like spokes in a wheel; when they contract, the diameter of the pupil increases. Thus contraction of the radial muscles causes *pupillary dilation* (Figure 9.27c). The iris is under control of the autonomic nervous system. Parasympathetic neurons innervate the circular muscles; activity in these neurons causes the circular muscles to contract, producing pupillary constriction. Sympathetic neurons innervate the radial muscles; activity in these neurons causes the radial muscles to contract, producing pupillary dilation.

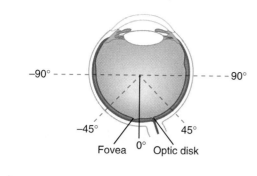

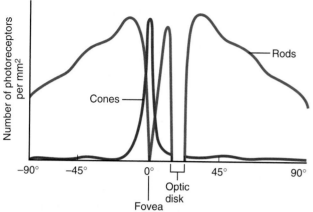

FIGURE 9.29 Distribution of rods and cones in the retina.
The abundance of cones is greatest at the fovea and declines rapidly with distance from it. Rods are absent from the fovea but very abundant near it; they slowly decrease in abundance with distance from the fovea.

The Retina

The retina, which is composed of neural tissue, is the location of **photoreceptors,** the rods and cones. **Rods** provide the ability to see in black and white during relatively low light conditions (that is, up to the brightness of a full moon). **Cones** provide us with color vision, but they are active only in relatively bright light (that is, illumination levels greater than that provided by a full moon).

The retina consists of three distinct layers (Figure 9.28): (1) an inner layer containing neurons called ganglion cells, (2) a middle layer containing other neurons called bipolar cells, and (3) an outer layer containing rods and cones. Also present are *amacrine cells* and *horizontal cells,* which modulate (for example, by lateral inhibition) communication between the cells in the retina.

Note in Figure 9.28 that because the photoreceptors are in the outer layer of the retina, light must pass through the inner and middle layers before striking them. In addition, blood vessels are in the light's path to the photoreceptors. However, to provide light a clear path to the fovea, the bipolar and ganglion cells are laterally displaced, creating a depression in the center of the retina called the **macula lutea.** The fovea contains cones only; the ratio of rods to cones increases with distance from the fovea, until at the periphery of the retina only rods are

present (Figure 9.29). This is why in dim light we see objects better if we do not look directly at them, and why when we do so we see things in black and white only.

Quick Test 9.7

1. What is the effect of contraction of the ciliary muscles on the lens? Does this help vision of near objects or of distant objects?
2. What effects do the two branches of the autonomic nervous system have on the diameter of the pupil?
3. What are the locations of the vitreous humor and the aqueous humor? What is the function of each?
4. What cells are in each of the three layers of the retina?

Phototransduction

Phototransduction, the conversion of light energy into electrical signals, is carried out by the rods and cones. The basic morphology of the two types of photoreceptors is the same; each consists of two major portions referred to as *outer* and *inner segments* (Figure 9.30). The outer segment contains invaginations with membranous disks that

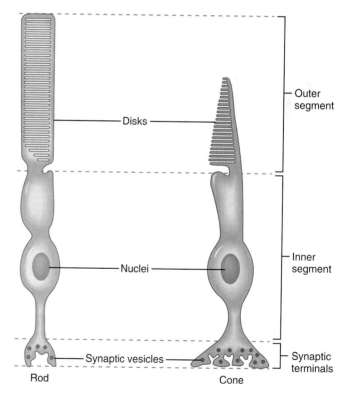

FIGURE 9.30 Morphology of the photoreceptors. *Rods and cones have the same basic structural components: The outer segment consists of disks that contain the photopigment; the inner segment contains the nucleus and most of the organelles. The synaptic terminal contains the synaptic vesicles, which store a chemical transmitter used for communication.*

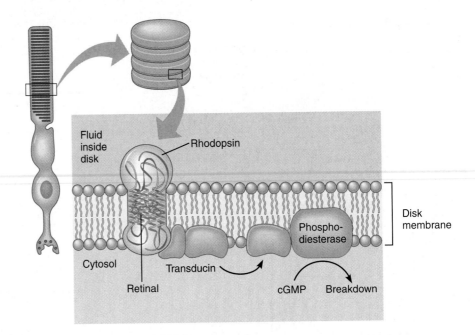

Fluid inside disk

Rhodopsin

Cytosol

Retinal

Transducin

Phospho-diesterase

cGMP   Breakdown

Disk membrane

contain the molecules that absorb light waves, giving the photoreceptors the ability to respond to light. The inner segment contains the cell nucleus and various organelles and ends at the receptor's synaptic terminal, which is analogous to an ordinary neuron's axon terminal and is where a chemical messenger is stored in synaptic vesicles.

The absorption of light is the first step in phototransduction, and the molecule in the photoreceptors that absorbs light is a **photopigment.** Each of the four different types of photoreceptors contains a different photopigment. One type of photopigment is found in rods, and the other three types are found in three types of cones, each of which contains a photopigment that best absorbs light of a particular range of wavelengths and is thus most responsive to certain colors. Each photopigment molecule contains a light-absorbing portion called *retinal* and a protein called an *opsin.* The retinal portion is the same in all photopigments, but the kind of opsin present determines which light wavelengths are absorbed by a given photopigment by altering the electromagnetic energies to which the retinal is sensitive.

The components of photoreceptors involved in phototransduction are shown in Figure 9.31 using rods as an example. The photopigment of rods, *rhodopsin,* is located in the membrane of the disks. Also within the disk membrane is a G protein called *transducin* and the enzyme *phosphodiesterase,* which catalyzes the degradation of cGMP (which if present in the cytosol, opens sodium channels located in the plasma membrane of the photoreceptor).

Let's examine the process of phototransduction as it occurs in rods. (The same process occurs in the three types of cones, but different photopigments are involved.) We begin by considering the state of the photoreceptor in the dark (Figure 9.32a). In the dark, levels of the second

messenger cGMP levels are high inside the outer segment ①, so cGMP opens sodium channels in the plasma membrane of the outer segment ②. Therefore, sodium ions are moving into the cell, and the photoreceptor is *depolarized* ③. This depolarization spreads to the inner segment and ④ opens calcium channels that are also present in the plasma membrane. Calcium enters the cell ⑤, triggering the release of transmitter by exocytosis *in the dark.* The transmitter communicates to bipolar cells ⑥.

When the photoreceptor is exposed to light (Figure 9.32b), light is absorbed by the rhodopsin ①. The retinal component changes its conformation and dissociates from the opsin ②, leaving what is called "bleached opsin." (When opsin is bleached, the photoreceptors become less sensitive to light, a phenomenon called *light adaptation.*) The bleached opsin activates transducin ③, which activates the enzyme phosphodiesterase ④, which then catalyzes the breakdown of cGMP.

With cGMP levels in the outer segment decreased ⑤, sodium channels close ⑥. Potassium leaking out of the cell causes a hyperpolarization that is no longer opposed by sodium movement into the cell ⑦. The hyperpolarization causes closing of calcium channels on the inner segment ⑧. With less calcium entering the cell, release of transmitter decreases ⑨. Therefore, in the light, less transmitter is released from the photoreceptor terminal. Information about the presence of light is relayed, therefore, by a decrease in signaling to the next cells in the visual pathway, the bipolar cells ⑩.

Rods are very sensitive to light of a wide range of wavelengths, but most sensitive to light in the blue-green range (Figure 9.33). Rods are so sensitive to light that they can respond to a single photon (the unit of electromagnetic energy) of light. In bright light, however, rods become saturated (completely bleached); that is, they be-

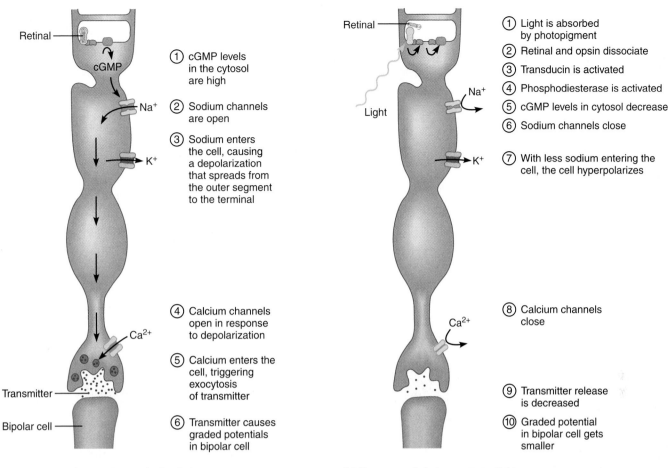

| (a) Response of photoreceptor in the dark | (b) Response of photoreceptor to light |

① cGMP levels in the cytosol are high
② Sodium channels are open
③ Sodium enters the cell, causing a depolarization that spreads from the outer segment to the terminal
④ Calcium channels open in response to depolarization
⑤ Calcium enters the cell, triggering exocytosis of transmitter
⑥ Transmitter causes graded potentials in bipolar cell

① Light is absorbed by photopigment
② Retinal and opsin dissociate
③ Transducin is activated
④ Phosphodiesterase is activated
⑤ cGMP levels in cytosol decrease
⑥ Sodium channels close
⑦ With less sodium entering the cell, the cell hyperpolarizes
⑧ Calcium channels close
⑨ Transmitter release is decreased
⑩ Graded potential in bipolar cell gets smaller

FIGURE 9.32 Phototransduction of light. (a) *In the dark, photoreceptors release their chemical transmitter.* **(b)** *When light is present, it is absorbed by the photopigment, initiating a sequence of events that decreases release of the transmitter.*

come as hyperpolarized as possible and thus cannot code for any additional brightness.

The process of phototransduction in cones is similar to that in rods. However, cones are not as sensitive to light as rods and therefore do not function well in dim light. In addition, cones respond best to light within a narrower range of wavelengths than rods. Blue cones are most sensitive to light at a wavelength near 430 nm, green cones are most sensitive to light at 530 nm, and red cones are most sensitive to light at 560 nm (see Figure 9.33). However, because the absorbance spectra of the three types of cones overlap, many colors can be perceived based on the patterns of activation of the different cones. A comparison of the characteristics of rods and cones can be found in Table 9.4.

Adaptation to Light and Dark

To a certain extent the eyes adapt to changes in brightness via pupillary dilation and constriction. But dilation and constriction are not enough with more drastic changes in light intensity. Recall the last time you went to a matinee in a theater, where it was very dark inside.

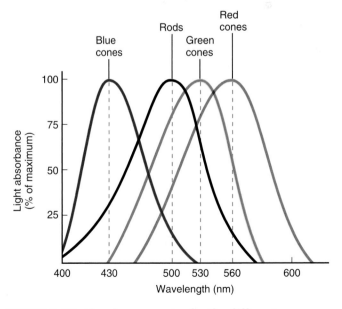

FIGURE 9.33 Absorbence spectra for the different photoreceptors. *Rods can absorb light over the widest range of wavelengths. The absorbence spectra for of the three types of cones overlap.*

TABLE 9.4	CHARACTERISTICS OF RODS AND CONES	
	RODS	**CONES**
Types of vision	Black and white; night (dim light)	Color; day (bright light)
Sensitivity to light	High	Low
Abundance	100 million per retina	3 million per retina
Visual acuity	Low	High
Site of greatest concentration	Periphery of retina	Fovea
Degree of convergence with bipolar cells	High	Low

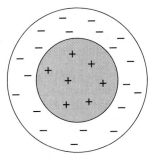

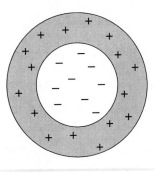

(a) ON-center, OFF-surround ganglion cell

(a) OFF-center, ON-surround ganglion cell

FIGURE 9.34 Receptive fields of ganglion cells. *Ganglion cells respond with both excitation (+) and inhibition (−) to light applied to the visual field.* **(a)** *ON-center, OFF-surround ganglion cells respond with excitation to light in the center of their receptive field and with inhibition to light in the surrounding area of their receptive field.* **(b)** *OFF-center, ON-surround ganglion cells respond with inhibition to light in the center of their receptive field, and with excitation to light in the surrounding area of their receptive field.*

When you walked outside into the bright sunlight, your eyes were overwhelmed by the intensity of the light. Similarly, when a person moves from the sunny outdoors into a dark room, the ability to see is greatly impeded. In either case, however, the eyes adapt in a matter of minutes to enable clear vision. What allows our eyes to adjust to these varying intensities of light?

When exposed to bright light, the rods become "bleached"; that is, most of the rhodopsin has absorbed light, and the opsin is in its active form. As a result, no more light can be absorbed until the rhodopsin has been returned to its original or "unbleached" state. Under these conditions, which correspond to when you first enter a dark room, rods are much less sensitive to light. Unbleaching of rods occurs in dim light, when the opsin returns to its inactive state. The retinal and opsin reassociate, and retinal becomes sensitive to light again. By contrast, unbleached rods are extremely sensitive to light. Therefore, when you have been in the dark for a period of time and then emerge into the daylight, the bright light overwhelms the rods until they become bleached.

Neural Processing in the Retina

The transmitter released from rods and cones communicates light and dark signals to bipolar cells in the retina. Some degree of convergence generally exists between photoreceptors and bipolar cells; that is, more than one photoreceptor communicates to a single bipolar cell. However, the extent of this convergence is greater with rods than with cones. Thus in the fovea and macula, where cones predominate, little convergence occurs; only a few photoreceptors converge onto one bipolar cell. In contrast, in the periphery of the retina, where there are only rods, thousands of rods converge on one bipolar cell. Recall that in the somatosensory system, less convergence results in greater tactile acuity and two-point discrimination. Similarly, less convergence in the visual system

provides greater visual acuity, because two separate light sources can be discriminated as distinct sources only if they trigger responses in separate cells along the visual pathway. Greater convergence, on the other hand, provides greater sensitivity to light, because of spatial summation of inputs from several photoreceptors onto one bipolar cell.

Bipolar cells are capable of transmitting graded potentials, but not action potentials. In some synapses between photoreceptor and bipolar cells, the transmitter is excitatory (depolarizes the bipolar cell), whereas in others the transmitter is inhibitory (hyperpolarizes the bipolar cell). Therefore, light excites some bipolar cells and inhibits others. In addition, the synapses between photoreceptors and bipolar cells are subject to lateral modulation (excitation or inhibition) by horizontal cells.

When depolarized, bipolar cells release a transmitter that communicates to ganglion cells. Again, at some synapses the transmitter is inhibitory, whereas at other synapses the transmitter is excitatory. In addition, the synapses between bipolar cells and ganglion cells are subject to lateral modulation by amacrine cells. The ganglion cells are the first neurons in the visual pathway that are capable of transmitting action potentials. The axons of the ganglion cells make up the **optic nerve** and are the output neurons of the visual pathway.

Based on this description, one would expect the receptive field properties of ganglion cells to be complex. That is indeed the case. The receptive field for a ganglion cell is the area of the visual field in which a light stimulus either increases or decreases the frequency of action potentials in the cell. Figure 9.34 shows the receptive fields of two types of ganglion cells. The first type of ganglion cell is called an *ON-center, OFF-surround cell* (Figure 9.34a). In these ganglion cells, light in the center of the receptive field excites the cell (turns the cell "on"), whereas light in the surrounding area of the receptive field (the "surround") inhibits the cell (turns the cell

The ability to perceive colors is based on the presence of three types of cones that respond best to light of different wavelengths. That activation of a single cone cannot distinguish between colors is illustrated in figure **(a),** which shows the absorbance spectrum for one type of photoreceptor only. This hypothetical photoreceptor absorbs light at 450 nm and at 550 nm equally and thus is unable to distinguish between the two wavelengths. But how do three types of cones allow us to perceive the full spectrum of colors?

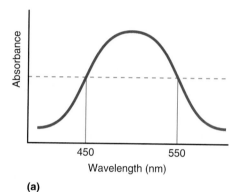

(a)

Each type of cone responds best to a specific wavelength, but each cone also responds over a range of wavelengths, though to different degrees. Whereas a given wavelength of light might elicit a response in more than one type of cone, the different cones generally respond to different degrees. Thus, our brains discern different colors by *comparing* the responses of the different cone types to each wavelength.

However, the relative responses of cones alone does not fully explain color perception. For example, why is there no such color as reddish green or yellowish blue? Properties like these can be explained by the *opponent-process theory,* which states that red and green, or blue and yellow, or black and white are opponent colors such that stimulation of one color inhibits the other. Therefore, we cannot see reddish green because the presence of green inhibits the perception of red. The opponent-process theory per-

tains to the level of ganglion cells, where some ganglion cells are excited by red in their visual fields and inhibited by green in the same regions.

The opponent-process theory also explains the concept of afterimages. To observe afterimages, perform the following: Stare at figure **(b)** for approximately 30 seconds, and then stare at a blank sheet of white paper for approximately 30 seconds. What do you see?

(b)

When you looked at the white area, you should have seen the opponent colors because of adaptation to the original colors. Recall that most of the light we perceive is light waves reflected off objects. White is observed when the full spectrum of light waves is reflected to the eye and all three cones are activated. To understand your perception, let's concentrate on what happened in the area of the visual field where you were looking at green. While looking at the picture, the green cones were activated or bleached. When vision was shifted to the white area of paper, all wavelengths of light were reflected to the eye. However, in the visual

field that originally detected green, the green cones were bleached and thus did not respond as strongly to the green wavelengths that were present in the white light. With less inhibition from green, red signals were transmitted to the CNS more strongly, and red was observed.

People who lack a specific type of cone lose their ability to distinguish between certain colors; this is the basis of color blindness. The most common type of color blindness is red-green color blindness, in which red and green colors cannot be distinguished from each other. This type of color blindness is generally caused by a genetic defect in the photo pigments of red or green cones. The genes that code for the red and green photopigments are recessive and are located on the X chromosome; because males only have one X chromosome, they are more likely to inherit this recessive trait. Blue color blindness also exists, but it is rarer and is not linked to the X chromosome.

A common test for color blindness employs what are known as *Ishihara charts.* In these charts, numbers are hidden within a pattern of colored dots. An example is shown in figure **(c).** A person with normal vision can identify the number, but a person who is color blind is not able to see the number.

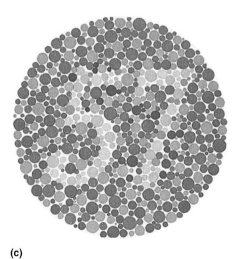

(c)

FIGURE 9.35 Neural pathways for vision.

Damage to the optic radiations on the left side would produce what type of visual deficit?

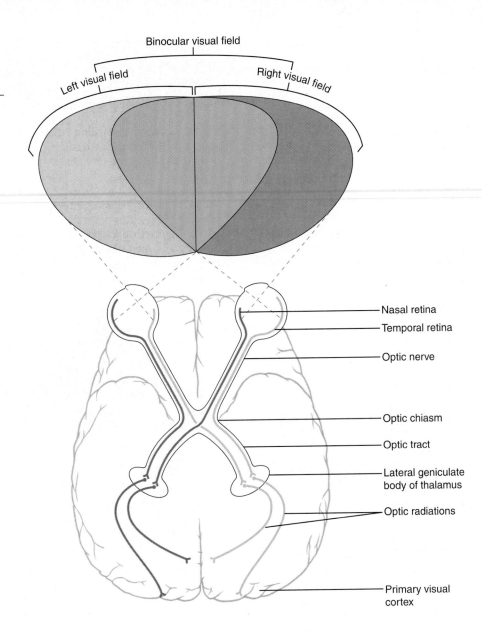

Binocular visual field

Left visual field

Right visual field

Nasal retina

Temporal retina

Optic nerve

Optic chiasm

Optic tract

Lateral geniculate body of thalamus

Optic radiations

Primary visual cortex

"off"). Diffuse light over the entire receptive field produces a small excitation (relative to diffuse dark). The other type of ganglion cell is called an *OFF-center, ON-surround cell* (Figure 9.34b). In these ganglion cells, light in the center of the receptive field inhibits the cell, whereas light in the surround excites the cell. Diffuse light over the entire receptive field produces a small inhibition (relative to diffuse dark).

Receptive fields of ganglion cells become even more complex when color is considered. For example, some ganglion cells are excited by red in the center of the visual field and inhibited by green in the surround; others are excited by blue in the center of the visual field and inhibited by yellow in the surround. (For more on color vision, see Discovery: Color Vision, p. 289.)

Loss or impaired perception of the right visual field

Neural Pathways for Vision

As previously described, the ganglion cells are the output neurons from the retina because they generate action potentials that are transmitted to the CNS. The axons of ganglion cells form the optic nerve (cranial nerve II). The two optic nerves exit each eye at the optic disk and combine at the base of the brain just in front of the brainstem to form the **optic chiasm.** In the optic chiasm, half the axons from each eye cross over to the other side of the brain (Figure 9.35). Note in Figure 9.35 that input from the left visual field strikes the nasal retina (side closest to the nose) of the left eye and the temporal retina (side closest to the side of the head) of the right eye. Likewise, input from the right visual field strikes the nasal retina of the right eye and the temporal retina of the left eye. Therefore, both eyes receive information from both visual fields.

In the optic chiasm, axons originating from nasal ganglion cells cross to the opposite side, whereas axons originating from temporal ganglion cells stay on the side of origin. The result is that after the optic chiasm, all input from the right visual field travels in axons in the left side of the brain, and all input from the left visual field travels in axons in the right side of the brain. Although the axons are still those of ganglion cells, after the optic chiasm the axons travel in what is called the **optic tract.** The ganglion cells terminate in a nucleus in the thalamus called the **lateral geniculate body,** where they form synapses with neurons that ascend to the primary visual cortex in the occipital lobe. Pathways from the lateral geniculate body to the visual cortex on either side are called the **optic radiations.**

In the somatosensory cortex, the topographic organization of the cortex is such that adjacent areas of the body are usually represented on adjacent areas of cortex. The visual cortex also has topographic organization, but in this case the *visual field* is mapped onto the cortex. Due to the crossing of axons in the optic chiasm, the right visual field is mapped onto the left visual cortex, and the left visual field is mapped onto the right visual cortex.

Parallel Processing in the Visual System

The visual system clearly possesses **parallel processing,** in which parallel pathways transmit different qualities of a stimulus. For example, information about the color of an observed object is transmitted by some neurons in the visual pathway, whereas other neurons transmit information about shape or movement. The coding stays distinct all the way to the primary visual cortex. However, higher cortical areas integrate the different qualities of a stimulus so that we can perceive, for example, a red fire truck moving quickly away from us.

Depth Perception

Depth perception requires that the brain receives input from both eyes. Figure 9.35 shows that images in most (but not all) areas of the left and right visual fields are detected by both eyes. The portion of the visual fields that is detected by both eyes, called the *binocular visual field,* is the portion of the visual fields where we are capable of depth perception. Depth perception depends on the fact that the left and right eyes see images from slightly different angles due to their different positions in the head. The cortex uses these differences to construct a three-dimensional image of the world (that is, one that includes depth) rather than a flat, two-dimensional image. To test your depth perception, try the following: Hold two pencils at arm's length, one in each hand. With one eye open, try to touch the point of one pencil to the point of the other. Repeat with both eyes open.

Quick Test 9.8

1. Put the following components of the visual pathway in order such that they correctly reflect the path of transmission of visual information: optic tract, ganglion cell, photoreceptor, optic radiation, optic chiasm, bipolar cell, optic nerve, lateral geniculate nucleus, and visual cortex.

2. When are cGMP levels greatest in photoreceptors: when they are exposed to light or when they are exposed to dark? What effect does cGMP have on sodium channels in the outer segment of the photoreceptors?

3. What are the two components of photopigment molecules? Which of these components absorbs light?

We now turn our attention to the sensory systems of the ear: first to the *auditory system,* which is responsible for hearing, and then to the *vestibular system,* which is responsible for balance or equilibrium. Although the stimuli involved vary, both of these sensory systems rely on hair cells to detect within the cavities of the ear the movement of fluid, which causes a receptor potential and release of a transmitter that communicates to an afferent neuron.

We begin our examination of the auditory system by considering the anatomy of the ear.

THE EAR AND HEARING

Anatomy of the Ear

The ear can be divided into three parts: the *external ear, middle ear,* and *inner ear* (Figure 9.36). The external and middle ears are air-filled cavities, whereas the inner ear is fluid filled.

The external ear includes the *pinna* and the **external auditory meatus,** or ear canal. The primary function of the external ear is to gather sound waves and conduct them to the *tympanic membrane* (or eardrum), which separates the external and middle ears. Within the middle ear are three **ossicles,** or small bones, called the *malleus, incus,* and *stapes.* The three ossicles extend from the tympanic membrane to a thin membrane called the **oval window,** which is one connection between the middle and inner ears. The **round window** also connects the middle and inner ears. The function of the middle ear is amplification of sound waves in preparation for the transmission of those waves from air to a fluid environment.

The **eustachian tube,** which connects the middle ear with the *pharynx,* or throat, helps maintain normal pressure in the middle ear. Pressure changes, which may occur in the middle ear while flying or scuba diving, dur-

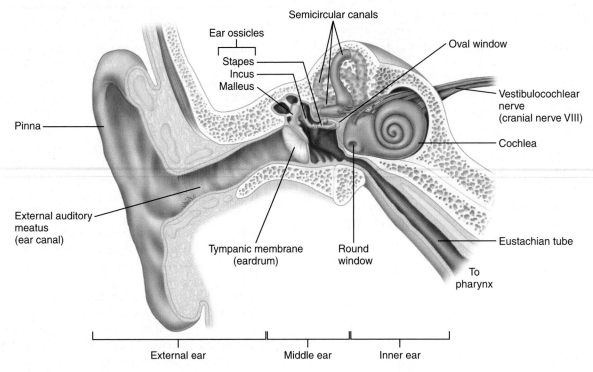

FIGURE 9.36 Anatomy of the ear. *The major structures of the external ear, middle ear, and inner ear are labeled.*

ing ear infections, or even while going up or down in an elevator, can be painful and could cause rupture of the tympanic membrane if they become large enough. Even mild pressure changes can change the ability to hear, as the pressure changes dampen sound vibrations, just as occurs when a tympanist puts a hand on the drum head to stop a note. Opening the eustachian tube allows the pressure in the middle ear to equilibrate with the pressure in the pharynx, which alleviates any pressure difference across the eardrum. Swallowing or yawning can facilitate the opening of the eustachian tube.

The inner ear contains structures associated with both hearing and equilibrium. The **cochlea** is a spiral-shaped structure that contains the receptor cells for hearing. (The structures of the *vestibular apparatus,* including the *semicircular canals,* are discussed in the section on equilibrium.) The nerve that contains the afferents for both hearing and equilibrium, cranial nerve VIII, is also called the **vestibulocochlear nerve.**

The Nature of Sound Waves

Sound waves are mechanical waves caused by air molecules put into motion. Figure 9.37a shows the production of sound waves by a vibrating tuning fork. When the tines of a tuning fork vibrate, they generate waves in the air, similar to those produced when you touch your hand to the surface of a still pond. The waves consist of areas where the air molecules are closer together or *compressed,* and areas where the air molecules are further apart or *rarefied.*

Figure 9.37b illustrates the properties of sound waves: loudness and pitch. The loudness (amplitude) of a sound is proportional to the difference in the densities of air molecules between the areas of compression and the areas of rarefaction. The amplitude of a sound is most conveniently expressed in units called *decibels* (dB) on a logarithmic scale, (see Toolbox: Decibels, p. 294).

The pitch of the sound is determined by the frequency of sound waves. Low-frequency sound waves produce low-pitch sounds, such as those produced by a tuba. High-frequency sound waves produce high-pitch sounds, such as the squealing of car brakes. The frequency of sound waves is measured as the number of waves per second, or *Hertz* (Hz). The average person can hear sound waves with frequencies ranging from 20 to 20,000 Hz; the greatest auditory sensitivity occurs in the range between 1000 Hz and 4000 Hz.

The coding of amplitude and pitch of sound must be maintained both while sound waves move from the air-filled middle ear to the fluid-filled inner ear, and during the transduction process in the inner ear. The anatomy of the ear is exquisitely designed to sustain the coding.

Sound Amplification in the Middle Ear

When sound waves enter the ear, they strike the tympanic membrane, causing it to oscillate (vibrate) back and forth. The oscillations occur at the same frequency as that of the sound waves; and with an amplitude that increases and decreases as the amplitude of the sound waves increases or decreases. Because the malleus, the first of the

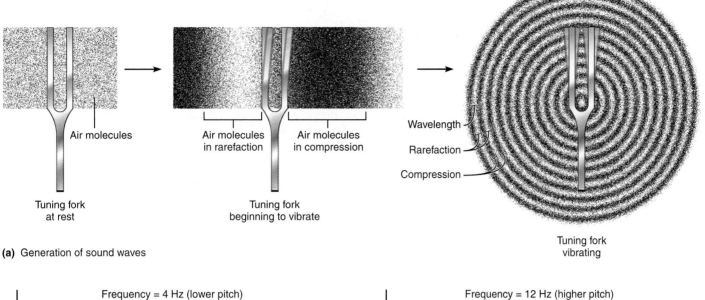

Air molecules

Tuning fork
at rest

Air molecules
in rarefaction

Air molecules
in compression

Tuning fork
beginning to vibrate

Wavelength
Rarefaction
Compression

Tuning fork
vibrating

(a) Generation of sound waves

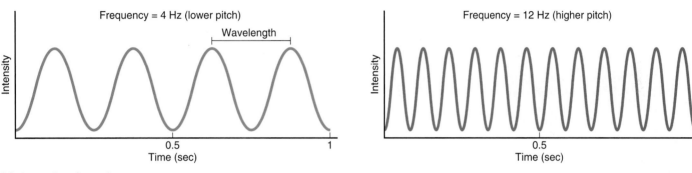

Frequency = 4 Hz (lower pitch)

Wavelength

Intensity

0.5   1
Time (sec)

Frequency = 12 Hz (higher pitch)

Intensity

0.5   1
Time (sec)

(b) Properties of sound waves

FIGURE 9.37 **The nature of sound waves. (a)** *Vibrations produce sound waves, which consist of areas of high-density air molecules (compression) separated by areas of low-density air molecules (rarefaction).* **(b)** *Sound waves have an intensity (loudness) and a frequency (pitch).*

small ossicles, is connected to the tympanic membrane (Figure 9.38), oscillations of the tympanic membrane cause the malleus to oscillate at the same frequency and with an amplitude reflecting that of the tympanic membrane's vibrations. The three ossicles are arranged in such a manner that they function as a series of levers, with movement of the malleus causing a greater movement of the incus, which causes an even greater movement of the stapes. The net effect is amplification of the motion produced by the initial sound waves. The stapes overlies the oval window to the fluid-filled cochlea, such that oscillations of the stapes generates waves in the fluid of the cochlea. However, because it takes a greater pressure to produce waves in fluid than in air, amplification is required.

To generate enough pressure to produce waves in the fluid of the cochlea, the sound waves are amplified as they travel from middle to inner ear by two primary means: First, the ossicles act as a lever system so that small amplitude vibrations of the tympanic membrane are translated into larger amplitude vibrations at the stapes; second, the much larger diameter of the tympanic membrane than that of the oval window produces amplification, because a given force acting on a smaller surface

produces a greater pressure. As an analogy, consider hammering a nail into wood. The head of a nail has a larger area relative to the point. When you place the point of the nail against a block of wood and hit the head of the nail with a hammer, the force of the blow is transmitted through the nail to the wood. Because the area of the point is so small, however, the point exerts much more pressure on the wood than the hammer exerts on the nail, and therefore the nail penetrates the wood. In essence, the nail amplifies the pressure exerted on it by the hammer.

Signal Transduction for Sound

The cochlea is the organ in which sound transduction occurs. Next we explore the anatomy of the cochlea, and how it relates to the mechanism of sound transduction.

Functional Anatomy of the Cochlea

Understanding the mechanisms of sound transduction requires in-depth knowledge of the anatomy of the cochlea (Figure 9.39). From the outside (Figure 9.39a, b), the cochlea looks like a spiral sea shell; the point at the end

The loudness (amplitude) of sound is based on the difference in density of air molecules in the areas of rarefaction and compression. Because the ear functions over a wide range of amplitudes, loudness is generally expressed in logarithmic units known as decibels (dB).

Measurements in decibels always compare the ratio of two intensities according to the following equation:

$$dB = 20 \log \frac{\text{sound amplitude}}{\text{reference amplitude}}$$

In hearing, the reference amplitude is the threshold for normal hearing, that is, the minimum amplitude that can be perceived by humans, and it is given a value of 1. Thus, with the denominator set at 1, the equation becomes:

$$dB = 20 \log (\text{sound amplitude})$$

At the threshold amplitude, the number of decibels is 0 ($20 \log (1) = 0$). A sound that is ten times the threshold for hearing is 20 dB ($20 \log (10) = 20$), whereas a sound that is 100,000 times the threshold is 100 dB ($20 \log (100,000) = 100$).

The intensities of various familiar sounds are listed in the table. Sounds approaching 100 dB create the potential for hearing loss; sounds at 130–140 dB approach the pain threshold.

INTENSITIES OF SOME FAMILIAR SOUNDS

SOUND	INTENSITY (DECIBELS)
Ticking watch	20
Elevator music	40
Conversational speech	50–60
Alarm clock	80
Live rock band	100
Jackhammer	110
Propeller airplane	120
Jet airplane	130

FIGURE 9.38 **Structures that transmit sound waves in the middle ear.** *Sound waves hit the tympanic membrane, initiating vibrations in the ossicles. The stapes pushes the oval window, and vibrations are passed on to fluid in the cochlea.*

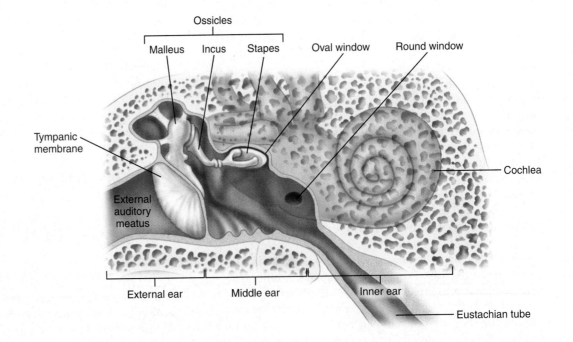

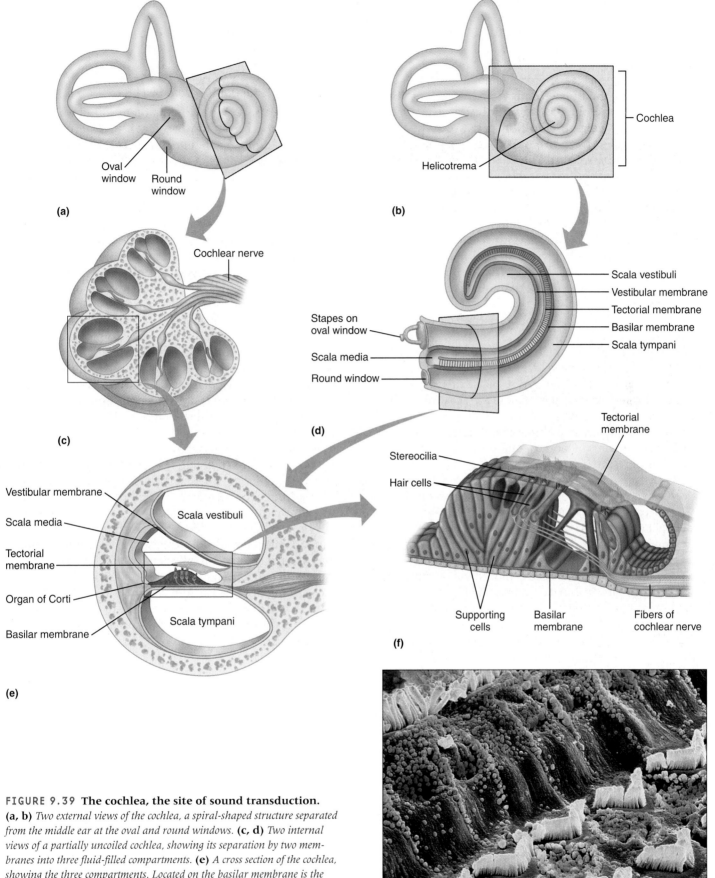

(a)

Oval window   Round window

(b)

Helicotrema   Cochlea

Cochlear nerve

(c)

Scala vestibuli
Vestibular membrane
Tectorial membrane
Basilar membrane
Scala tympani

Stapes on oval window
Scala media
Round window

(d)

Vestibular membrane
Scala media
Tectorial membrane
Organ of Corti
Basilar membrane

Scala vestibuli
Scala media
Scala tympani

(e)

Tectorial membrane

Stereocilia
Hair cells

Supporting cells   Basilar membrane   Fibers of cochlear nerve

(f)

(g)

FIGURE 9.39 The cochlea, the site of sound transduction.
(a, b) Two external views of the cochlea, a spiral-shaped structure separated from the middle ear at the oval and round windows. (c, d) Two internal views of a partially uncoiled cochlea, showing its separation by two membranes into three fluid-filled compartments. (e) A cross section of the cochlea, showing the three compartments. Located on the basilar membrane is the organ of Corti. (f) An enlargement of the organ of Corti, showing the hair cells with stereocilia that are embedded in the tectorial membrane. (g) An electron micrograph of hair cells in the cochlea.

of the spiral is called the **helicotrema.** To view the inside of the cochlea, the spiral in Figure 9.39c, d is partially uncoiled. A cross section of the cochlea is shown in Figure 9.39e. Inside the cochlea are two membranes that separate it into three fluid-filled compartments: The **vestibular membrane** and the **basilar membrane** separate the cochlea into the **scala vestibuli** (or **vestibular duct),** the **scala tympani** (or **tympanic duct),** and the **scala media** (or **cochlear duct).** The vestibular and basilar membranes join at the helicotrema; thus there is an opening between the scala vestibuli and scala tympani at the helicotrema. The fluid in the scala vestibuli and scala tympani is called **perilymph;** it differs from the fluid in the scala media, which is called **endolymph.** Perilymph is similar in composition to cerebrospinal fluid, but endolymph is closer in composition to intracellular fluid, with a high concentration of potassium ions and low concentration of sodium ions.

At the oval window, a membrane separates the middle ear from the scala vestibuli; at the round window, a membrane separates the middle ear from the scala tympani. Therefore, the cochlea is a closed fluid-filled structure. When the stapes vibrates in response to sound waves, waves are produced in fluid of the scala vestibuli and travel to the scala tympani. These waves cause movement of the vestibular and basilar membranes, which ultimately leads to transduction of sound waves. The round window is necessary for movement of the oval window, and for the production of waves in the cochlea, because it allows movement of fluid, the volume of which cannot change because fluid is incompressible. For the oval window to vibrate by itself would require compressing the perilymph. However, when the oval window vibrates, the energy of the resulting waves causes movement of the round window so that the fluid is moved, not compressed. Consider the following analogy: If you fill a syringe with water and cap off the opening, then you cannot move the plunger because the water can neither expand nor compress. However, if you connect two water-filled syringes with a tube, you can push one plunger down because the resulting force rushes the other plunger out. This arrangement allows you to move the plunger (the stapes) back and forth despite the fact that the water remains at a constant volume.

Functional Anatomy of the Organ of Corti

The **organ of Corti,** the sensory organ for sound, is located on top of the basilar membrane (Figure 9.39f). The organ of Corti contains **hair cells,** supporting cells, and an overlying membrane called the *tectorial membrane.* The receptor cells are called hair cells because they end in hairlike projections called **stereocilia,** the tips of which are embedded in the **tectorial membrane.** The anatomy of the organ of Corti is such that sound waves cause me-

chanical bending of the stereocilia, which causes receptor potentials in the hair cells.

Sound Transduction by Hair Cells

Figure 9.40a shows the conduction of sound waves into the internal ear. Sound waves enter the external ear and cause vibrations of the tympanic membrane, which in turn causes vibrations of the ossicles of the middle ear. Vibrations of the stapes cause movement of the oval window, which sets up waves in the perilymph in the scala vestibuli of the inner ear. The energy from waves in the perilymph causes the vestibular and basilar membranes to move relative to one another (Figure 9.40b), which causes the stereocilia to bend back and forth. Depending on the direction the stereocilia bend, potassium channels in the hair cells either open or close. Because endolymph has a higher concentration of potassium than that inside the hair cell (which is opposite to the conditions in most of the body, where potassium is in greater concentration inside the cell compared to outside), the opening of potassium channels results in potassium diffusing into the hair cell, causing depolarization; by contrast, closing of the potassium channels prevents potassium from diffusing into the hair cell, causing hyperpolarization.

Figure 9.41 shows how the bending of stereocilia causes potassium channels to open or close. Note that the stereocilia are linked together by elastic protein filaments and are of different sizes, such that one side of the hair cell has the tallest stereocilia. When the stereocilia extend straight up (no sound waves present), some tension on the elastic filaments holds the potassium channels in a partially opened state, allowing some potassium to diffuse into the hair cell and partially depolarize it (Figure 9.41a). This partial depolarization of the hair cell at rest causes the opening of some calcium channels in the hair cell, so calcium enters the cell and causes the release of transmitter by exocytosis. The transmitter communicates to an afferent neuron that is part of the cochlear nerve (see Figure 9.40b).

When the stereocilia are bent in the direction of the tallest stereocilium (Figure 9.41b), then more tension is put on the elastic filaments and the potassium channels are opened farther. More potassium diffuses into the hair cell, producing a more depolarized state. More calcium channels open, resulting in the release of more transmitter.

When the stereocilia are bent away from the tallest stereocilium, the elastic filaments go slack and the potassium channels close (Figure 9.41c). With the potassium channels closed, potassium cannot diffuse into the cell, and it becomes hyperpolarized compared to the resting state. Under these conditions, calcium channels close and no transmitter is released.

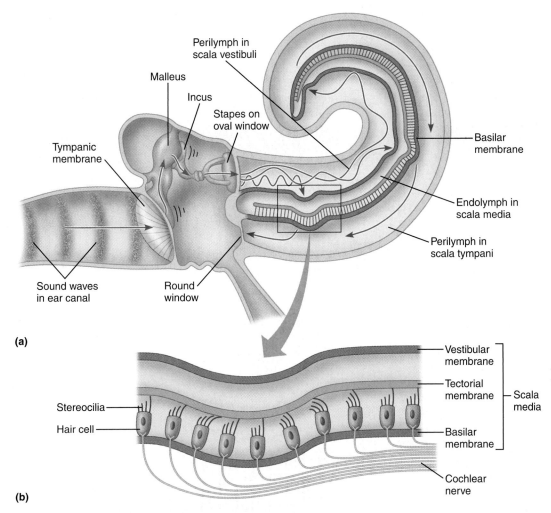

(a)

(b)

FIGURE 9.40 Conduction of sound waves in the ear. (a) *Sound waves that enter the ear through the pinna and external auditory meatus strike the tympanic membrane, causing it to vibrate. The ossicles vibrate in response to the tympanic membrane and transmit the vibrations to the oval window. The vibrating oval window causes waves in the fluid (perilymph) of the cochlea.* **(b)** *Waves in the perilymph cause deflection of the membranes in the cochlea. When the membranes oscillate, the stereocilia of the hair cells bend, causing the opening or the closing of potassium channels.*

Coding of Sound Intensity and Pitch in the Cochlea

Given that receptor potentials are produced by bending of stereocilia, how is the intensity and pitch of sound coded? Louder sounds cause the stereocilia to bend farther in either direction, causing larger changes in the number of open potassium channels. This results in larger receptor potentials and larger variations in transmitter release.

The coding of sound frequency is based on the fact that sound waves of different frequencies cause deflection of the basilar membrane at different regions, causing hair cells located in different regions to be stimulated to different degrees (Figure 9.42). The structure of the basilar membrane varies over its length: The basilar membrane is stiff and narrow near the oval and round windows, and wide and flexible near the helicotrema. Therefore, high-frequency sound waves associated with high pitch cause greatest deflection of the basilar membrane in the region closer to the oval and round windows, which activates hair cells located in this region. Low-frequency sound waves, by contrast, cause greatest deflection of the basilar membrane in the region closer to the helicotrema, which activates hair cells located in this region of the basilar membrane. The frequency of sound, therefore, is coded by which hair cells are activated most strongly.

Neural Pathways for Sound

The transmitter released from hair cells binds to receptors on afferent neurons of the *cochlear nerve*, part of cranial nerve VIII. The hair cell transmitter depolarizes the affer-

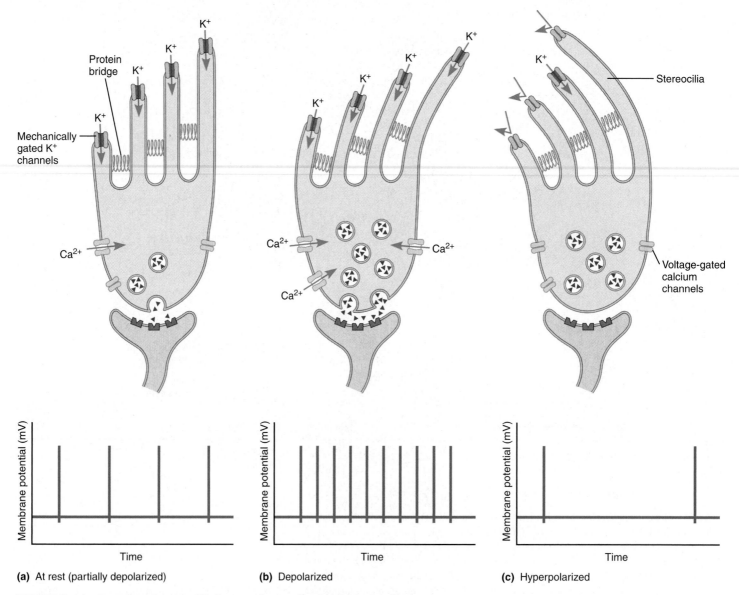

(a) At rest (partially depolarized) **(b)** Depolarized **(c)** Hyperpolarized

FIGURE 9.41 The roles of stereocilia in sound transduction by hair cells. (a)
The mechanically gated potassium channels in stereocilia are partially opened when the cell
is at rest (stereocilia stand upright), and potassium ions enter the cell, producing a small
depolarization that is sufficient to release transmitter that communicates to the afferent
neuron; the result is a low frequency of action potentials. (b) When the stereocilia bend
toward the taller stereocilium, the potassium channels open more, and more potassium ions
enter the cell, producing a greater depolarization and a higher frequency of action potentials in
the afferent neuron. (c) When the stereocilia bend away from the taller stereocilium, the
potassium channels close, and little potassium can enter the cell. Less transmitter is released,
and the frequency of action potentials in the afferent neuron decreases.

ent neuron; the greater the degree of depolarization, the greater frequency of action potentials in the afferent neuron, which therefore codes for the intensity of the sound. The afferent neurons terminate in the cochlear nuclei in the brainstem, where they indirectly communicate with second-order neurons that travel to a nucleus of the thal-amus called the **medial geniculate body.** In the medial geniculate body, the second-order neurons form synapses with third-order neurons that transmit information to the **auditory cortex** in the temporal lobe. Like other sensory systems, the auditory cortex has topographical organiza-tion. Its organization is **tonotopic;** that is, it maps the frequency of sound.

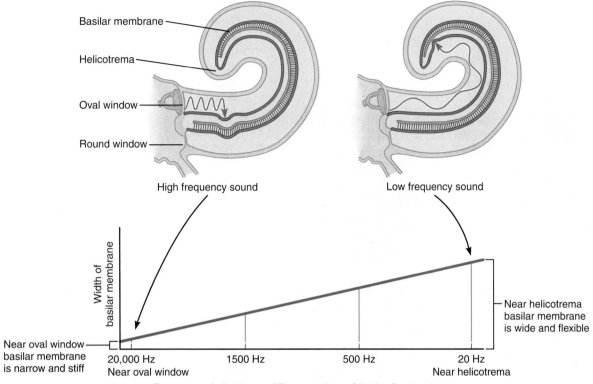

FIGURE 9.42 Coding for the frequency of sound. *The frequency of sound is coded based on the location along the basilar membrane of activated hair cells. Variations in width and flexibility of the basilar membrane along its length dictate that high-frequency sounds cause deflections near the oval window, whereas low-frequency sounds cause deflections near the helicotrema.*

Quick Test 9.9

1. Define the terms *outer ear, middle ear,* and *inner ear,* and describe the major structures contained in each.

2. Explain how the pitch of a sound is coded. How is the loudness of sound coded?

3. When stereocilia of hair cells are bent in a certain direction, what ion moves into the hair cell to depolarize it?

THE EAR AND EQUILIBRIUM

The previous section discussed the anatomy of the ear as it relates to transduction of sound waves. This section presents ear anatomy as it pertains to the ability to detect acceleration of the body and position of the head, information required if balance and equilibrium are to be maintained. Acceleration is a change in an object's velocity, which depends on its speed and direction. A change in speed without a change in direction is called *linear acceleration;* it occurs, for example, in a car that is gaining speed while moving in a straight line. However, acceleration also occurs when the car turns, even if its speed remains constant; this type of acceleration is called *rotational acceleration.*

Anatomy of the Vestibular Apparatus

The vestibular apparatus is located in cavities of the temporal bones called the *bony labyrinth.* Because the vestibular apparatus consists of membrane-bound structures within the bony labyrinth, these structures are also called the *membranous labyrinth* (the cochlea is also part of the membranous labyrinth). The membranous labyrinth is filled with endolymph, whereas the space between the membranous labyrinth and the bony labyrinth contains perilymph; note that these fluids are the same as those found in the cochlea.

Figure 9.43 provides a close-up view of the vestibular apparatus of the inner ear. The **vestibular apparatus** consists of the semicircular canals and the utricle and the saccule. Notice in Figure 9.43 that the three semicircular

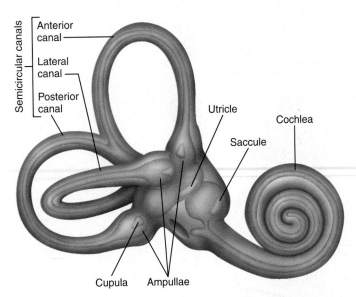

Semicircular canals

Anterior canal

Lateral canal

Posterior canal

Utricle

Cochlea

Saccule

Cupula   Ampullae

FIGURE 9.43 Anatomy of the vestibular apparatus in the inner ear. *The vestibular apparatus includes the three semicircular canals, the utricle, and the saccule. The hair cells for detecting rotational acceleration are located in the ampullae of the semicircular canals.*

canals in each ear, which detect rotational acceleration, are oriented in planes that are perpendicular to each other, which enables them to detect rotational movements of the head in three planes. The anterior canal detects rotation of the head up and down, as when nodding "yes." The posterior canal detects rotation of the head up and down to the side, as in moving the ear toward the shoulder. The lateral canal detects rotation of the head from side to side, as when shaking the head to indicate "no." The utricle and saccule detect linear acceleration; the utricle detects acceleration forward and backward, whereas the saccule detects acceleration up or down.

The Semicircular Canals and the Transduction of Rotation

At the base of each semicircular canal is an enlarged area called an **ampulla** (Figure 9.44a), which contains the receptor cells for equilibrium, another type of hair cell. Within each ampulla is the **cupula,** a gelatinous area separated from the endolymph by a membrane. Among the support cells at the base of the cupula are hair cells that are similar to those in the cochlea, including stereocilia that project upward into the cupula (Figure 9.44b). However, one of the stereocilia is much larger than the others and is called a **kinocilium.** As in hearing, mechanical bending of the stereocilia causes ion channels to open or close, resulting in a change in membrane potential in the hair cells. Next we examine how rotation can cause the stereocilia to bend.

When the head is at rest (Figure 9.44c), no force is acting on the cupula, so the stereocilia are upright. In this

case the hair cells are partially depolarized, which leads to a low frequency of action potentials in the associated afferent neuron. When the head begins to rotate, the bony labyrinth rotates with it, but the endolymph lags behind the motion of the head, which exerts a force on the cupula and causes the stereocilia to bend in the direction that is opposite that of rotation. When the direction of rotation is such that the stereocilia bend away from the kinocilium (Figure 9.44d), the hair cell is hyperpolarized, and the frequency of action potentials in the afferent neuron declines. When the head rotates such that the stereocilia bend toward the kinocilium (Figure 9.44e), the hair cells are depolarized, and the frequency of action potentials in the afferent neuron increases.

Note that when the head continues to rotate at a constant speed (see Figure 9.44c), the movement of fluid eventually catches up to the movement of the bony labyrinth. In that case, no force is acting on the cupula, so the stereocilia are no longer bent, and the frequency of action potentials returns to that occurring when the head is at rest. Thus the semicircular canals detect only changes in the rate of rotation, not constant rotation.

What happened when as a child you spun around to make yourself dizzy and then suddenly stopped spinning? When the constant rotation suddenly stops, the head and bony labyrinth stop as well, but the endolymph keeps moving for a time. Thus for awhile rotation is detected, even though you are standing still, but it feels like it is occurring in the direction opposite that of the original rotation.

The Utricle and Saccule and the Transduction of Linear Acceleration

The utricle and saccule function in a manner similar to the semicircular canals in that they contain hair cells with stereocilia that bend, but their anatomy is somewhat different. The utricle and saccule, which are bulges located between the semicircular canal and cochlea of the inner ear (see Figure 9.43), have hair cells with stereocilia extending up into a gelatinous material (Figure 9.45). However, within the upper edges of the gelatinous material are **otoliths,** small calcium carbonate crystals, which add mass to the gelatinous material. The hair cells of the utricle are oriented in horizontal rows in the head, with the stereocilia extending up vertically; the hair cells of the saccule are oriented in vertical rows in the head, such that the stereocilia are oriented horizontally. Because of these orientations, the utricle detects forward and backward linear acceleration, whereas the saccule detects up and down linear acceleration.

Figure 9.45b–d illustrates the bending of the stereocilia within the utricle during forward and backward linear acceleration. Although all kinocilia are shown oriented in the same direction (posterior) in this example, some kinocilia are located anteriorly and others posteri-

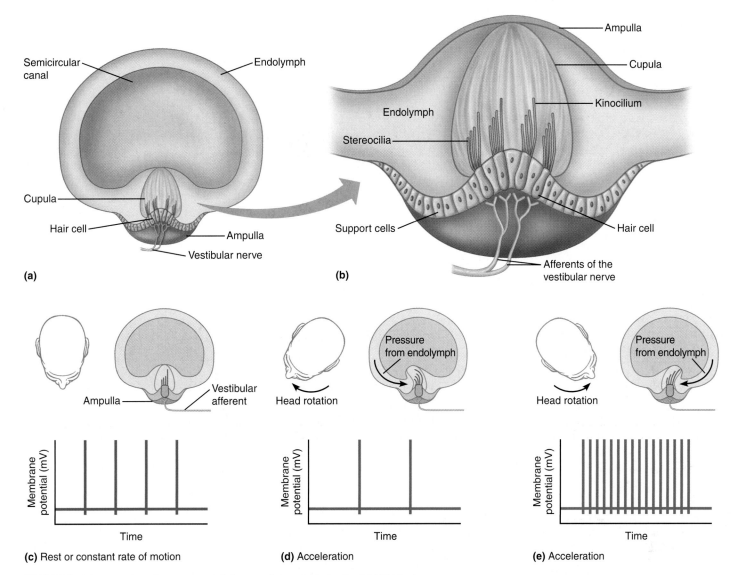

(a)

(b)

(c) Rest or constant rate of motion

(d) Acceleration

(e) Acceleration

FIGURE 9.44 Functional anatomy of the semicircular canals. (a) *Within the ampulla, a swelling located at the base of each semicircular canal, is the cupula, a gelatinous area separated from the endolymph by a membrane.* **(b)** *Hair cells, the receptor cells for acceleration, have stereocilia that extend into the gelatinous cupula; the longest stereocilium is the kinocilium. Hair cells communicate to afferent neurons.* **(c)** *When the head is still or moving at a constant rate of motion, no net force is acting on the cupula, and the stereocilia are upright. The hair cells are partially depolarized, releasing some chemical transmitter and causing a low frequency of action potentials in the afferent neuron.* **(d)** *During acceleration of the head, inertia of the fluid causes it to lag behind the motion of the head. As it does so it pushes against the cupula in the direction opposite that of rotation, causing the stereocilia to bend. When the stereocilia bend away from the kinocilium, the hair cell is hyperpolarized, and the frequency of action potentials in the afferent neuron decreases.* **(e)** *During acceleration of the head in the opposite direction, the stereocilia bend toward the kinocilium, the hair cells are depolarized, and the frequency of action potentials in the afferent neuron increases.*

Stereocilia are composed of what type of protein filament?

Microtubules

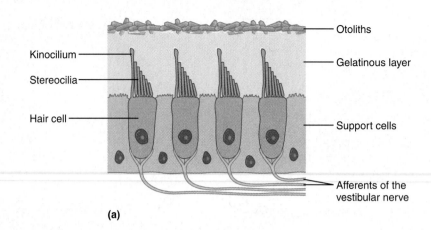

(a)

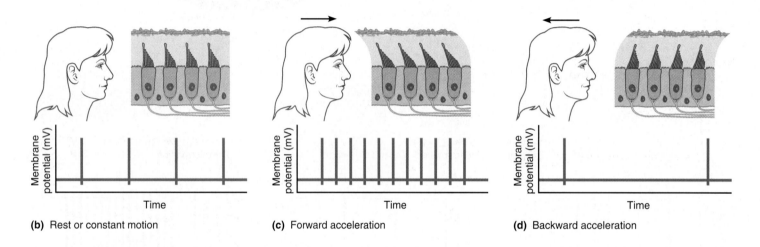

(b) Rest or constant motion

(c) Forward acceleration

(d) Backward acceleration

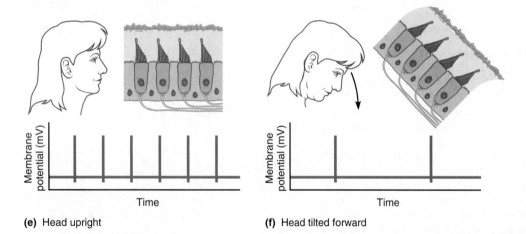

(e) Head upright

(f) Head tilted forward

FIGURE 9.45 Functional anatomy of the utricle and saccule.
(a) *The receptor cells for linear acceleration in both the utricle and the saccule are hair cells with stereocilia extending into a gelatinous layer containing otolith crystals. Parts b–f depict events associated with the utricle only.* (b) *At rest or in constant motion, the stereocilia in the utricle stand erect, and the hair cell is partially depolarized. The afferent neuron has a low frequency of action potentials.* (c) *During forward acceleration, the otoliths cause a drag on the gelatinous mass of the utricle, causing the stereocilia to bend. In this example, the stereocilia are bending toward the kinocilium. The hair cell is depolarized, and the frequency of action*

potentials in the afferent neuron is increased. (d) *During backward acceleration, the otoliths in the utricle cause a drag on the gelatinous mass in the opposite direction as in c. The stereocilia bend away from the kinocilium, causing a hyperpolarization of the hair cell and a decrease in the frequency of action potentials in the afferent neuron.* (e, f) *Effect of gravity on the utricle. In the example shown, when the head is tilted forward, gravity causes the otoliths to fall, pulling on the gelatinous mass. The stereocilia bend, causing a hyperpolarization and a decrease in the frequency of action potentials in the afferent neuron.*

orly in the utricle. Hair cells with anterior kinocilia will behave in a manner opposite that described here. When a person is at rest or moving forward or backward at a constant speed (Figure 9.45b), the hair cells are vertical, and the resulting depolarization results in moderately frequent action potentials in the afferent neuron. When the person begins to walk forward (Figure 9.45c), the otoliths cause the gelatinous mass to drag behind, bending the stereocilia toward the kinocilium and causing greater depolarization and more frequent action potentials. Backward linear acceleration (Figure 9.45d) causes bending of stereocilia away from the kinocilium, producing hyperpolarization and less frequent action potentials. When linear motion continues at a constant rate (see Figure 9.45b), the gelatinous material catches up with the movement of the body, and the stereocilia return to the at-rest position.

The utricle also plays a role in detecting the position of the head relative to gravity, a force that can cause acceleration. When the head is upright (Figure 9.45e), the stereocilia in the utricle are vertical, but when the head is tilted forward (Figure 9.45f), gravity pulls the weighted otoliths downward, bending the stereocilia downward and producing hyperpolarization.

The saccule functions in the same way as the utricle, but because of its orientation when the head is held upright it detects up and down linear acceleration, as occurs when one rides in an elevator.

Neural Pathways for Equilibrium

At rest, hair cells of the vestibular apparatus are partly depolarized and release some transmitter that communicates to afferents in the vestibular nerve. Therefore, at rest, a low frequency of action potentials is traveling along the vestibular afferents. When a hair cell becomes depolarized due to bending of the stereocilia toward the kinocilium, more transmitter is released, and the frequency of action potentials in the afferent neuron increases. Similarly, when a hair cell becomes hyperpolarized due to bending of the stereocilia away from the kinocilium, less transmitter is released, and the frequency of action potentials in the afferent neuron decreases.

The vestibular afferents enter the brainstem as part of the vestibular nerve. Most terminate in the *vestibular nuclei,* although some travel directly to the cerebellum to provide immediate feedback for equilibrium and balance, which is important in motor coordination. The vestibular nuclei have projections to the cortex that allow perception of acceleration, but the pathways are not well understood. The vestibular nuclei are important in feedback control of movement for balance and in controlling eye movements. To carry out these functions, the vestibular nuclei also receive input from different sensory systems, including those involved in vision, somesthetic sensations, and proprioception. This sensory information is used to provide output to motor neurons involved in bal-

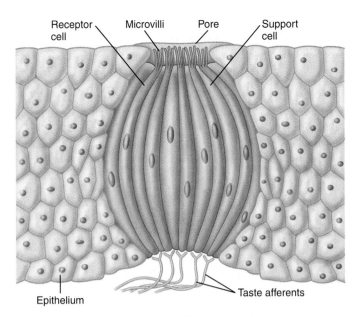

FIGURE 9.46 Taste receptor cells in a taste bud. *Taste receptors contain microvilli that extend into a pore at the top of the taste bud, where they are exposed to food particles dissolved in saliva.*

ance and to match eye movements with body movements. If the eyes and head do not move together, vision will be blurred and motion sickness can occur.

Quick Test 9.10

1. Of the organs of the vestibular apparatus, which detect linear acceleration?

2. Why do the semicircular canals detect rotation when it starts or stops, but not constant rotation?

3. What two nerves combine to form the vestibulocochlear nerve? What types of sensory information are carried by each?

TASTE

Although many of us might think of chocolate as nectar of the gods, on another level chocolate is simply a bunch of chemicals mixed in just the right way to stimulate chemoreceptors in the mouth in a way that is pleasurable to most of us. This section describes how the sense of taste detects and interprets the chemicals in food.

Anatomy of Taste Buds

We can taste food because chemoreceptors in the mouth respond to certain chemicals in food. The chemoreceptors for taste are located in structures called **taste buds,** each of which contains 50–150 receptor cells and numerous support cells (Figure 9.46). At the top of each bud is a pore that allows receptor cells to be exposed to saliva and dissolved food molecules. Each person has over 10,000

FIGURE 9.47 Taste transduction
mechanisms for the four primary
tastes. (a) *Sour is produced by the presence of
hydrogen ions.* (b) *Salty is produced by the
presence of sodium ions.* (c) *Sweet is produced
by molecules that can bind to a receptor on the
plasma membrane of receptor cells.* (d) *Bitter
taste can be produced by two different
mechanisms shown here.*

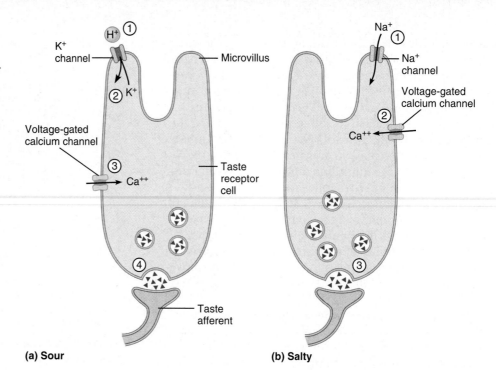

(a) Sour

(b) Salty

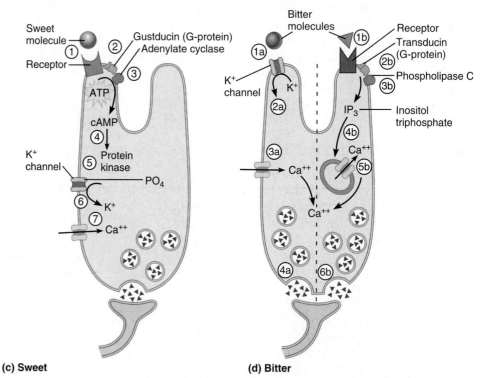

(c) Sweet

(d) Bitter

taste buds, located primarily on the tongue and the roof
of the mouth, but also located in the pharynx.

Taste receptor cells are modified epithelial cells with
microvilli that extend into the pore. Located on the
plasma membrane of the microvilli are taste receptors
that bind selectively to the different chemicals, called **tastants,** that give food its flavor. For a chemical to bind to
the taste receptor, the chemical must dissolve in saliva,
which is why a person does not taste food as well when
the mouth is dry. The binding of the tastant to a taste receptor opens or closes ion channels, causing potential
changes in the taste receptor cell.

Signal Transduction in Taste

There are four primary tastes—sour, salty, sweet, and bitter—each requiring its own transduction mechanism
(Figure 9.47). Sour tastes are caused by the presence of
hydrogen ions (acid) in the food (Figure 9.47a). These
hydrogen ions bind to potassium channels ①, blocking
them, and preventing potassium from leaving the cell ②.
The result is that the taste receptor cell depolarizes, which
opens voltage-gated calcium channels and allows calcium
to enter the cell ③, where it triggers the release of a
transmitter by exocytosis ④.

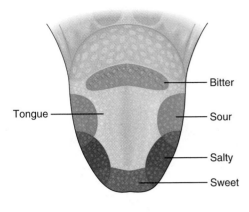

(a) Map of taste sensitivity

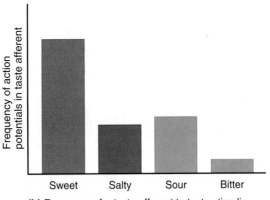

(b) Response of a taste afferent to taste stimuli

FIGURE 9.48 Neural coding of taste. (a) *Different areas of the tongue are more sensitive to a particular taste.* **(b)** *Responses of a single taste afferent to the four primary tastes. This particular taste afferent responds best to sweet.*

Salty tastes are caused by the presence of sodium ions in the food (Figure 9.47b). When sodium levels outside the cell are increased by the consumption of salty foods, the inward electrochemical driving force on sodium ions increases, causing an increased flow of sodium into the cell ①, which depolarizes it. Voltage-gated calcium channels then open ②, and calcium triggers the release of a transmitter ③.

Sweet tastes are caused by the presence of organic molecules with structures similar to sucrose (Figure 9.47c). The binding of the organic molecule to a plasma membrane receptor ① activates a G protein called *gustducin* ②, which stimulates the production of cAMP ③. cAMP activates a protein kinase ④ that catalyzes the phosphorylation of potassium channels such that they close ⑤, which decreases the leakage of potassium out of the cell ⑥ and causes a depolarization. Voltage-gated calcium channels then open ⑦, and calcium triggers the release of a transmitter.

Bitter tastes are associated with a wide variety of nitrogen-containing compounds, including some that are toxic; that animals tend to avoid eating things that taste extremely bitter is thus a protective mechanism. There are several different types of bitter substances, and two known mechanisms for transduction of bitter taste (Figure 9.47d). Some bitter foods have molecules, such as quinine, that can block potassium channels ①ₐ and thereby decrease potassium diffusion out of the cell ②ₐ, depolarizing the cell and opening voltage-gated calcium channels ③ₐ. Calcium enters the cell and triggers release of a transmitter ④ₐ. Other bitter molecules act as ligands that bind to receptors on the plasma membrane ①ᵦ activating a G-protein called transducin ②ᵦ. Transducin activates the enzyme phospholipase C ③ᵦ, which catalyzes the formation of inositol triphosphate ④ᵦ. Inositol triphosphate opens calcium channels on the membrane of organelles ⑤ᵦ, causing calcium release into the cytosol. Calcium triggers the release of a transmitter ⑥ᵦ.

Researchers have identified a fifth taste that seems to differ from the primary tastes in that it produces no taste sensation of its own, but instead enhances other tastes: *umami,* which is Japanese for "delicious." Chemicals associated with umami seem to be flavor enhancers. Umami is associated with amino acids—most commonly, glutamate—and thus monosodium glutamate (MSG) is often added to food to enhance its flavor. Glutamate binds to receptors on taste receptor cells, opening channels that allow both sodium and potassium to move through. The net effect is a depolarization, which ultimately triggers the release of a transmitter.

Each taste receptor cell has all four transduction mechanisms, and thus each responds to all four primary tastes. However, one receptor cell tends to respond *stronger* to one primary taste than to the others. Receptor cells that respond most strongly to a given taste are differentially distributed in the tongue, causing different regions of the tongue to be more sensitive to certain tastes (Figure 9.48a). Several receptor cells communicate with a single afferent neuron. This extensive convergence of information creates complex properties of taste afferent neurons, as shown in Figure 9.48b, which contains a plot of the responses of an afferent neuron that responds most strongly to sweet. Therefore, whereas most sensory systems code the quality of the stimulus based on receptor specificity and a labeled-line pathway, taste does not have that specificity. The perception of taste is complex and seems to depend on the pattern of activity in multiple afferent neurons, and it is further complicated by the fact that it is dependent on the sense of smell.

Neural Pathway for Taste

Taste receptor cells communicate to afferent neurons that travel in three cranial nerves: VII, IX, and X. These taste afferents terminate in the *gustatory nucleus* of the medulla, where they form synapses with second-order neurons. The second-order neurons travel to the contralateral thalamus, where they form synapses with third-order neurons that terminate in the *gustatory cortex,* which is near the region of the somatosensory cortex that corresponds to the mouth.

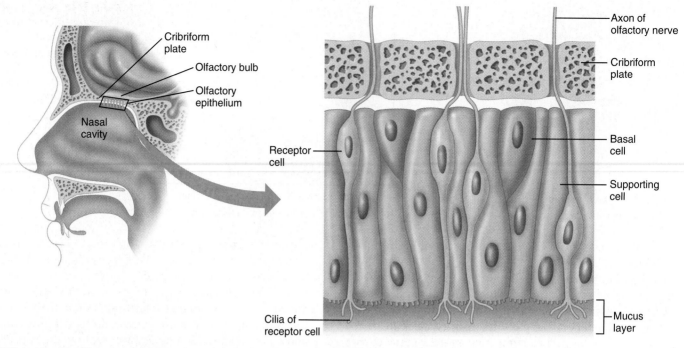

FIGURE 9.49 The olfactory epithelium. *The olfactory epithelium, located in the nasal cavity, contains three types of cells: receptor cells, supporting cells, and basal cells. Cilia of the receptor cells extend into a layer of mucus in the nasal cavity, whereas axons of the receptor cells ascend to the olfactory bulb of the brain through pores in the cribriform plate of the skull.*

Quick Test 9.11

1. Name the four primary tastes.
2. What is the difference between a taste bud and a taste receptor?

OLFACTION

The sensation of *olfaction,* or smell, depends on many of the same mechanisms as the sensation of taste. In olfaction, however, specific chemical substances called **odorants** must dissolve in *mucus* if they are to bind to specific chemoreceptors and be smelled.

Anatomy of the Olfactory System

Within the nasal cavity is the organ for smell, the **olfactory epithelium** (Figure 9.49). The epithelium is composed of three main classes of cells: supporting cells, basal cells, and olfactory receptor cells. The **supporting cells** provide structural support to the epithelium and secrete mucus; the **olfactory receptor cells,** the neurons that respond to olfactory stimuli, are replaced continuously throughout life by development of the **basal cells.** Olfactory receptor cells are the only neurons in the body currently known to be replaced regularly. Axons of olfactory receptor cells ascend to the olfactory bulb, and thus enter the CNS, through pores in the cribriform plate of the skull.

The olfactory receptor cells have cilia that project down into the mucus lining the nasal cavity. The cilia contain receptors that bind with specific odorant molecules. Within the mucus are **olfactory binding proteins,** which carry an odorant molecule to the receptor on the cilia, in a way similar to the way that carrier proteins in the blood transport steroid hormones. Binding of the odorant molecule to the receptor initiates signal transduction.

Olfactory Signal Transduction

Although there are probably primary odors, a single basic transduction process seems to be universal in olfactory receptor cells. In addition, olfactory transduction is quite different from transduction mechanisms for other sensory systems. Binding of an odorant molecule to a membrane receptor activates a G protein called G_{olf}, which in turn activates the enzyme adenylate cyclase, which catalyzes the formation of cAMP. Even though cAMP typically functions as a second messenger by activating a protein kinase, in olfactory receptor cells cAMP instead binds to cation channels, opening them and allowing both sodium and calcium ions to enter the cell. The primary effect of sodium and calcium entry is depolarization. However, the entry of calcium into the cell also causes chloride channels to open, allowing chloride to move out of the cell and increasing depolarization of the receptor cell. If the

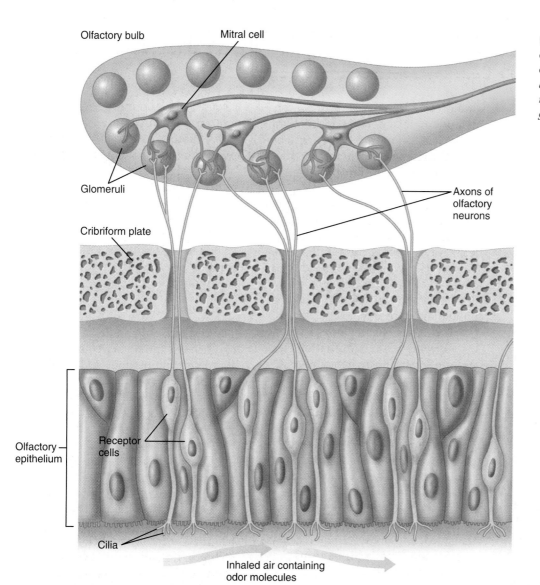

Olfactory bulb

Mitral cell

Glomeruli

Cribriform plate

Axons of olfactory neurons

Olfactory epithelium

Receptor cells

Cilia

Inhaled air containing odor molecules

FIGURE 9.50 The initial portion of the neural pathway for olfaction. *Communication between olfactory afferents and second-order neurons called mitral cells occurs in glomeruli of the olfactory bulb.*

depolarization is great enough, action potentials are generated on the axon of the receptor cell.

Neural Pathway for Olfaction

Unlike other special sense receptors, olfactory receptors are specialized endings of afferent neurons, not separate cells. The axons of the afferent neurons make up the **olfactory nerve** (cranial nerve I). If a receptor cell is depolarized to threshold, action potentials are produced and transmitted to the axon terminal in the brain. Figure 9.50 illustrates a portion of the neural pathway for olfaction. The axons terminate in an area of the brain called the olfactory bulb, where they communicate with second-order neurons called *mitral cells.* The synapses between the afferents and second-order neurons occur in clusters called *glomeruli,* with much convergence of input onto the second-order neurons.

The second-order neurons form the *olfactory tract,* which is the neural pathway to an area of the brain called the *olfactory tubercle.* From the tubercle, transmission of ol-

factory information can take two routes. In the first route, which transmits information pertaining to the perception and discrimination of smells, some olfactory afferent neurons form synapses with neurons in the olfactory tubercle that travel to the thalamus. In the thalamus, these neurons form synapses with other neurons that travel to the olfactory cortex. The other route involves other olfactory afferent neurons that do not synapse in the olfactory tubercle, but instead travel directly to the limbic system. This route is important in the triggering of olfaction-driven behaviors such as sexual behavior.

Quick Test 9.12

1. Name the three types of cells found in the olfactory epithelium, and describe their functions.

2. Give three similarities between the systems responsible for taste and smell.

3. What is unique about the neurons in cranial nerve I?

General Principles of Sensory Physiology, p. 261

Our ability to perceive the world around us depends on the presence of sensory receptors and specific neural pathways to communicate information to the cerebral cortex. Sensory systems must code for different qualities of the stimulus. Stimulus type is coded by the receptor and the pathway activated. Stimulus intensity is coded by frequency coding and population coding. The ability to locate a stimulus depends on the size of receptive fields, the degree of overlap of receptive fields, and lateral inhibition.

The Somatosensory System, p. 270

The somatosensory system enables perception of stimuli associated with the body surface (somesthetic sensations) or body position (proprioception). Some receptors in the somatosensory system are specialized nerve endings, whereas others are free nerve endings. Information about touch, pressure, vibration, and proprioception is transmitted to the thalamus via the dorsal column–medial lemniscal pathway. Information about pain and temperature is transmitted to the thalamus via the spinothalamic tract. Information from the thalamus is transmitted to the primary somatosensory cortex.

Vision, p. 277

The eyes function in transduction of light energy. To focus light on the retina, the eye can change lens shape and thus its refractive power. Several clinical defects affect the ability of the eyes to focus light, including myopia, hyperopia, astigmatism, presbyopia, cataracts, and glaucoma. The pupils of the eye are capable of constricting or dilating to regulate the amount of light that enters the eye.

Phototransduction occurs in the retina, which consists of neural tissue forming three layers: photoreceptors, bipolar cells, and ganglion cells. The photoreceptors, rods and cones, contain photopigments that absorb light. In the dark, photoreceptors are depolarized and release a chemical transmitter. In the light, the photopigments absorb the light and dissociate, causing a chain of reactions leading to hyperpolarization of the photoreceptor and decreased release of the transmitter. Photoreceptors communicate to bipolar cells, which communicate to ganglion cells. If the ganglion cell is depolarized to threshold, then an action potential results.

The axons of ganglion cells make up the optic nerve. Information is transmitted from the optic nerve to the optic chiasm, where half of the axons from each eye cross to the opposite side of the CNS such that all input from the right visual field is now on the left side, and all information from the left visual field is on the right side. The axons of ganglion cells after the optic chiasm make up the optic tract. The optic tract terminates in the lateral geniculate body of the thalamus, where the ganglion cell axons communicate with neurons that transmit information to the visual cortex.

The Ear and Hearing, p. 291

The ear contains the receptor cells for two sensory systems, hearing and equilibrium. For hearing, sound waves must enter the external ear, be amplified in the middle ear, and then be transduced into neural impulses in the cochlea of the inner ear. Sound transduction occurs in hair cells of the organ of Corti, located in the cochlea. When sound waves reach the fluid-filled cochlea, they set up waves that cause movement of the basilar membrane, which causes stereocilia to bend. This bending causes the opening or closing of potassium channels, which changes the electrical properties of the hair cells. When a hair cell is depolarized, a chemical transmitter is released and communicates with the afferent neurons in the cochlear nerve. Sound information is transmitted to the medial geniculate body of the thalamus and then to the auditory cortex. In the auditory cortex is a tonotopic map. The loudness of a sound is coded by the degree of bending of the stereocilia, whereas the pitch of sound is coded for by the location of hair cells on the basilar membrane.

The Ear and Equilibrium, p. 299

The vestibular apparatus of the inner ear includes the semicircular canals (for detecting rotation) and the utricle and saccule (for detecting linear acceleration). The vestibular apparatus contains hair cells with stereocilia that bend with acceleration of the head. In the semicircular canal, the hair cells are located in the ampulla. Hair cells in the utricle and saccule have stereocilia that extend into a gelatinous mass containing otoliths. The bending of the stereocilia opens or closes ion channels, affecting the release of a chemical transmitter that communicates with afferent neurons in the vestibular nerve. Vestibular information is transmitted to the vestibular nuclei of the brainstem, which communicate to the thalamus and then the cortex for perception of equilibrium. These nuclei also communicate to the cerebellum for maintaining balance, and to brainstem nuclei that regulate eye movement.

Taste, p. 303

Both chemical senses—taste (gustation) and smell (olfaction)—depend on the binding of specific chemicals in food or the air to chemoreceptors on receptor cells. Taste receptor cells are located within taste buds. Molecules that bind to taste receptors are called tastants. Each of the four primary tastes requires a different transduction mechanism. A single taste receptor cell responds to all four primary tastes, but most strongly to only one. Information is first transmitted via cranial nerves to the gustatory nucleus of the medulla, then relayed through the thalamus to the gustatory cortex.

Olfaction, p. 306

Olfactory receptors, located in the olfactory epithelium of the nasal cavity, respond to odorants dissolved in the mucus found there. Also present in the olfactory epithelium are basal cells, which are precursors for receptor cells. To bind to olfactory receptors, an odorant must bind to olfactory-binding proteins in the mucus. Upon binding to receptors, odorants trigger the production of cAMP in the cytosol of the receptor cell, which through a series of steps depolarizes the cell. If depolarized to threshold, an action potential is transmitted along the axon of the receptor cell. Axons of the receptor cells, which together constitute the olfactory nerve, form synapses with mitral cells in glomeruli of the olfactory bulb. Mitral cells transmit information along two pathways, one terminating in the olfactory cortex and the other in the limbic system.

EXERCISES

Multiple-Choice Questions

1. The strength of a stimulus is coded by
   a) the size of the receptor potential.
   b) the size of the action potentials.
   c) the frequency of action potentials.
   d) a and c
   e) all of the above

2. The mechanism by which a receptor converts a stimulus into an electrical signal is called
   a) conduction.
   b) convection.
   c) transduction.
   d) modulation.
   e) propagation.

3. In lateral inhibition,
   a) the nervous system produces contrast to emphasize more-important information over less-important information.
   b) afferent neurons with neighboring receptive fields inhibit each other's communication to second-order neurons.
   c) the ability to locate the site of a stimulus is enhanced.
   d) a and c
   e) all of the above

4. Which of the following observations best illustrates the concept of the labeled line?
   a) When a boxer gets punched in the eye, he perceives light.
   b) Rotation of the head stimulates certain receptors in the vestibular system but not those in the visual system.
   c) Information from different photoreceptors converges on a single ganglion cell that projects to the lateral geniculate nucleus.
   d) Hair cells in the cochlea are stimulated by sound vibrations over a wide range of frequencies.

5. Which of the following best illustrates the concept of an adequate stimulus?
   a) When a boxer gets punched in the eye, he perceives light.
   b) Rotation of the head stimulates certain receptors in the vestibular system but not those in the visual system.
   c) Information from different photoreceptors converges on a single ganglion cell that projects to the lateral geniculate nucleus.
   d) Hair cells in the cochlea are stimulated by sound vibrations over a wide range of frequencies.

6. Rubbing a sore area can decrease the sensation of pain by
   a) activating the endogenous analgesia systems.
   b) referring the pain to another area of the body.
   c) activating larger-diameter afferents, which activate an inhibitory interneuron, which inhibits the second-order neurons for pain.
   d) decreasing the number of action potentials in nociceptor afferents.
   e) presynaptic inhibition of substance P release.

7. In the dorsal column–medial lemniscal pathway,
   a) proprioception information is transmitted to the brain.
   b) the first-order neuron communicates to the second-order neuron in the dorsal horn of the spinal cord.
   c) the pathway crosses to the contralateral side in the spinal cord.
   d) all of the above

8. Which of the following is the correct name of the pathway from the retina to the optic chiasm?
   a) optic tract
   b) optic radiations
   c) optic nerve
   d) optic disk

9. Which of the following is the correct name of the pathway from the lateral geniculate nucleus of the thalamus to the visual cortex?
   a) optic tract
   b) optic radiations
   c) optic nerve
   d) optic chiasm
   e) optic disk

10. Where would you expect to find the ascending tracts for somatosensory information?
    a) in the white matter of the spinal cord
    b) in a spinal nerve
    c) in the gray matter of the spinal cord
    d) none of the above

11. The ability to perceive different frequencies in sound vibrations is based on the fact that
    a) the stereocilia of any given hair cell respond to one frequency only.
    b) different areas of the basilar membrane resonate at different frequencies, such that sound of a particular frequency causes only a

certain region of the membrane to vibrate.

c) the frequency of action potentials in the cochlear nerve varies in proportion to the frequency of a sound stimulus.

12. When you walk straight ahead at a constant speed,
a) no forces are acting on any of the vestibular organs.
b) the output of all vestibular organs should be constant.
c) fluid circulates in all semicircular canals at the same speed.
d) none of the above

13. The stereocilia for hearing are exposed to
a) endolymph in the vestibular duct.
b) perilymph in the vestibular duct.
c) endolymph in the cochlear duct.
d) perilymph in the cochlear duct.
e) endolymph in the tympanic duct.

14. The parasympathetic nervous system causes
a) contraction of the radial muscle of the iris.
b) contraction of the ciliary muscle.
c) pupillary dilation.
d) a and c
e) all of the above

Objective Questions

1. The two types of thermoreceptors are _____ and _____.

2. Receptors are most sensitive to energy from the _____ stimulus.

3. A phasic receptor adapts (quickly/slowly) to a constant stimulus.

4. The three types of nociceptors are _____ , _____ , and _____ .

5. Information about touch detected on the left side of the body is transmitted to the brain in the dorsal columns on the _____ side of the spinal cord.

6. When a photopigment absorbs light, cGMP levels (increase/decrease).

7. The first neurons that support production of action potentials in the visual pathway are (photoreceptors/bipolar cells/ganglion cells).

8. The pitch of sound vibration reflects its (amplitude/frequency).

9. A hair cell in the cochlea can be excited by sounds of different frequencies. (true/false)

10. The process by which the lens becomes stronger for close-up vision is called _____ .

11. Rods and cones differ with regard to the type of (retinal/opsin) they contain.

12. A single ganglion cell will either be excited or inhibited by light applied to its visual field. (true/false)

13. The visual cortex on the left side of the brain receives information from the right eye only. (true/false)

14. Odorant molecules must be dissolved in mucus if they are to bind to olfactory receptors. (true/false)

15. A given taste receptor cell responds to only one of the four primary tastes. (true/false)

Essay Questions

1. Compare the response of rapidly and slowly adapting touch receptors when you place your hand on a vibrating speaker.

2. Explain the concepts of topographic organization of the cerebral cortex. Compare the topographic organization of the somatosensory cortex, the visual cortex, and the auditory cortex.

3. Explain how it is possible for one person's perception to differ from another person's perception.

4. Diagram the general sensory pathway for transmitting information from a receptor to the cortex. Try fitting each sensory system into the general pathway; which sensory systems do not exactly fit this scheme?

5. Describe the sequence of events that occurs when photoreceptors are exposed to light.

6. Make a list of similarities between the olfactory and gustatory systems.

7. Make a list of the different types of sensory receptors in your body that are being activated at this moment.

Find the answers to these exercises, and additional study tools, at the Physiology Place (www.physiologyplace.com).

The Nervous System: Autonomic and Motor Systems

OBJECTIVES

- Describe the anatomy of the somatic nervous system and of the two branches of the autonomic nervous system.

- Describe the chemical messengers and receptor types associated with the peripheral nervous system.

- Explain the basic function of the two branches of the autonomic nervous system and the concept of dual innervation.

- Describe how the central nervous system regulates or controls the autonomic branch of the peripheral nervous system.

CHAPTER OUTLINE

The Autonomic Nervous System   312

The Somatic Nervous System   323

Above: Photomicrograph of motor neurons from a spinal smear

STUDY HINTS

1. G proteins, p. 145

2. cAMP second messenger system, p. 146

3. Phosphatidyl inositol second messenger system, p. 147

4. Acetylcholine, p. 214

5. Catecholamines, p. 216

6. Cholinergic receptors, p. 215

7. Adrenergic receptors, p. 217

8. Synapses, p. 202

Imagine it's a Sunday afternoon, and you've just finished a big lunch and are sitting in a recliner watching a movie. Your heart is beating slowly while your gastrointestinal organs are actively digesting and absorbing nutrients. All of a sudden, the fire alarm goes off and you see smoke. Your heart starts pounding. Blood flow is diverted from your gastrointestinal organs to your skeletal muscles, and you flee the house as flames emerge. Shortly after you've reached safety, your body starts to return to its resting state.

How does your body shift so rapidly from a relaxed, resting state to one prepared for rapid action? Your central nervous system (CNS) controls your muscles and other organs by way of signals sent through the efferent branch of the peripheral nervous system. Recall from Chapter 6 that the efferent nervous system has two branches, the *autonomic* and *somatic* nervous systems, and that the autonomic nervous system is further divided into two subdivisions, the *sympathetic* and *parasympathetic*. Activity in the parasympathetic nervous system is responsible for the resting state; when the body was put in danger, however, the characteristic "fight-or-flight" response of the sympathetic nervous system quickly took over. The somatic nervous system was also important in the response to the fire, as skeletal muscles enabled the body to move from the burning house to safety.

THE AUTONOMIC NERVOUS SYSTEM

The autonomic nervous system innervates all effector organs and tissues in the body except the skeletal muscles, including cardiac muscle, the smooth muscle found in blood vessels and various visceral organs (for example, the stomach and the respiratory airways), glands (for example, sweat glands, salivary glands, and some endocrine

glands), and adipose tissue. It is called "autonomic" because its functions occur at a subconscious level. For example, a person does not consciously decide to increase his or her heart rate; instead, heart rate increases through the operation of subconscious neural mechanisms when the need arises, such as during exercise. For this reason, the autonomic nervous system is also sometimes referred to as the *involuntary nervous system*.

Dual Innervation in the Autonomic Nervous System

Innervation of organs by the two branches of the autonomic nervous system is depicted in Figure 10.1; note that *both* branches of the autonomic nervous system innervate most organs, an arrangement called *dual innervation*. Although both branches innervate most organs, their functions are generally opposite in nature, and thus their effects are opposite. This poses no conflict, however, because the two autonomic divisions are typically most active under different conditions. For instance, the parasympathetic nervous system is most active during resting conditions, when it both stimulates the digestive organs (enhancing the digestion and absorption of nutrients) and inhibits the cardiovascular system (decreasing heart rate). In contrast, the sympathetic nervous system is most active during periods of excitation or physical activity, when it coordinates a group of physiological changes known as the **"fight-or-flight" response** that prepares the body to cope with threatening situations. During the fight-or-flight response, the rate and force of the heart's contractions increase, blood flow shifts from the gastrointestinal organs to skeletal and cardiac muscles, and energy stores are mobilized. These and other changes, all mediated by the sympathetic nervous system, prepare the body for intense physical exertion and otherwise adapt it for a possible response to a threatening situation.

The primary function of the autonomic nervous system is to regulate the function of effector organs in order to maintain homeostasis. At rest, both the sympathetic and parasympathetic branches are active, but the parasympathetic nervous system dominates. When the body gets excited or stressed, then the pattern shifts to less parasympathetic activity and more sympathetic activity. The existence of both systems working at cross purposes provides a backup mechanism that permits greater control over the effector organs. Control of heart rate, for example, is exerted by both sympathetic and parasympathetic neurons; to bring about an increase in heart rate, sympathetic activity increases (that is, the frequency of action potentials in sympathetic neurons increases) while parasympathetic activity decreases (that is, the frequency of action potentials in parasympathetic neurons decreases). Additional coverage of the autonomic nervous system appears in the chapters that discuss the individual organ systems.

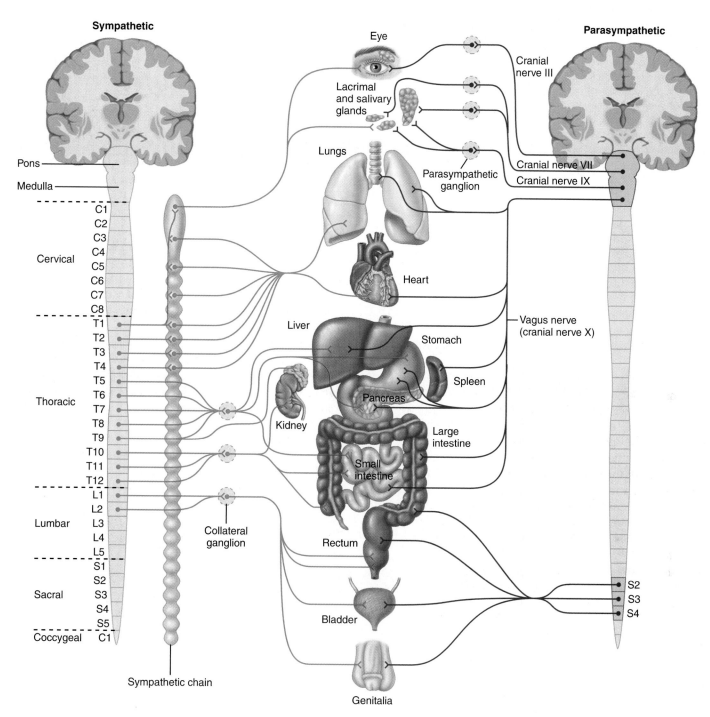

Sympathetic

Pons

Medulla

Cervical
- C1
- C2
- C3
- C4
- C5
- C6
- C7
- C8

Thoracic
- T1
- T2
- T3
- T4
- T5
- T6
- T7
- T8
- T9
- T10
- T11
- T12

Lumbar
- L1
- L2
- L3
- L4
- L5

Sacral
- S1
- S2
- S3
- S4
- S5

Coccygeal C1

Collateral ganglion

Sympathetic chain

Eye

Lacrimal and salivary glands

Lungs

Parasympathetic ganglion

Heart

Liver

Stomach

Spleen

Pancreas

Kidney

Large intestine

Small intestine

Rectum

Bladder

Genitalia

Parasympathetic

Cranial nerve III

Cranial nerve VII

Cranial nerve IX

Vagus nerve (cranial nerve X)

- S2
- S3
- S4

FIGURE 10.1 Dual innervation in the autonomic nervous system. *In this detailed schematic diagram of the autonomic nervous system, parasympathetic pathways are shown in purple, and sympathetic pathways in green. Both branches of the autonomic nervous system generally innervate the same organs. (Only one side of the body is shown for each system; thus there are actually two sympathetic chains, two of each cranial nerve, and so on.)*

Anatomy of the Autonomic Nervous System

The autonomic nervous system consists of efferent pathways containing two neurons arranged in series that communicate between the CNS and the effector organ (Figure 10.2). The neurons communicate with each other through synapses located in peripheral structures called **autonomic ganglia.** The neurons that travel from the CNS to the ganglia are called **preganglionic neurons;** the neurons that travel from the ganglia to the effector organs are called **postganglionic neurons.** Within each ganglion, therefore, are the axon terminals of preganglionic neurons and the cell bodies and dendrites of

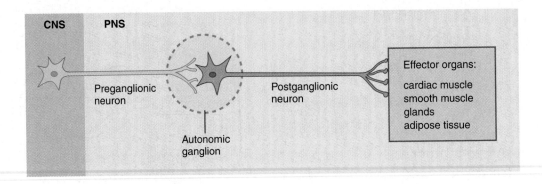

postganglionic neurons. A single preganglionic neuron generally synapses with several postganglionic neurons. In addition, other neurons located entirely within each ganglion, called *intrinsic neurons,* modulate the flow of information to the target organs. Although the two branches of the autonomic nervous system are similar in this regard, they show characteristic anatomical differences that are discussed in the following two sections.

Anatomy of the Sympathetic Nervous System

Because the preganglionic neurons in the sympathetic nervous system emerge from the thoracic and lumbar portions of the spinal cord (see Figure 10.1), the sympathetic nervous system is sometimes called the *thoracolumbar* division of the autonomic nervous system. The preganglionic neurons originate in a region of gray matter called the **lateral horn** or the *intermediolateral cell column.* Preganglionic and postganglionic sympathetic neurons are anatomically arranged in three patterns (Figure 10.3).

In the most common of these arrangements, the preganglionic neurons have short axons that originate in the lateral horn of the spinal cord and exit the spinal cord in the ventral root (Figures 10.3 and 10.4). Immediately after the dorsal and ventral roots merge to form a spinal nerve, the axon of the preganglionic neuron leaves the spinal nerve via a branch called a *white ramus* to enter one of several sympathetic ganglia located just outside the spinal cord. Here the preganglionic neuron synapses with several postganglionic neurons that have long axons that travel to the effector organ. Most of these postganglionic axons return to the spinal nerve via a branch called the *gray ramus.* The various sympathetic ganglia are linked together to form structures that run parallel to the spinal column on either side of it, called the **sympathetic chains** or **sympathetic trunks** (see Figure 10.4). Because a preganglionic neuron that enters a particular ganglion may have collaterals that travel up or down these chains to make synapses with postganglionic neurons of other ganglia, activation of the sympathetic nervous system produces a widespread action that affects many different target organs at once, as is evident in the fight-or-flight response.

The second anatomical arrangement of sympathetic fibers, and a very important exception to the arrangement just described, is a group of long preganglionic neurons that innervate endocrine tissue—the *adrenal medulla* of the adrenal gland—instead of synapsing on postganglionic neurons (see Figure 10.3). Each of the two adrenal glands, located in fat pads on top of each kidney (Figure 10.5a), is divided into an outer cortex and inner medulla. The medulla consists of modified sympathetic postganglionic cells, called *chromaffin cells,* that developed into endocrine cells instead of neurons (Figure 10.5b). Upon stimulation by the sympathetic nervous system, the adrenal medulla releases catecholamines. Of the three catecholamines released by the adrenal gland, approximately 80% is epinephrine (also known as adrenaline), 20% is norepinephrine (also known as noradrenaline), and a very small amount is dopamine. Like other endocrine glands, the adrenal medulla releases its products

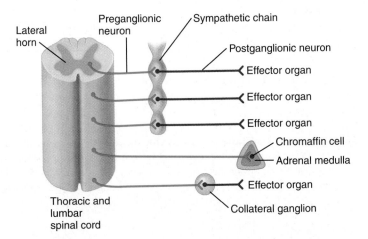

FIGURE 10.3 Anatomical pathways of preganglionic and postganglionic neurons in the sympathetic nervous system. *Most sympathetic preganglionic neurons synapse with postganglionic neurons in ganglia in the sympathetic chain; some sympathetic preganglionic neurons innervate secretory cells of the adrenal medulla; still other sympathetic preganglionic neurons synapse in collateral ganglia that are independent of the sympathetic chain.*

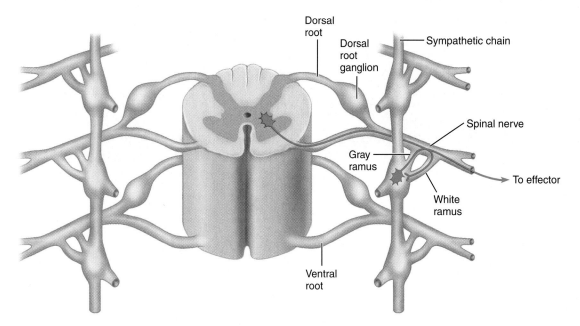

Dorsal root

Dorsal root ganglion

Sympathetic chain

Spinal nerve

Gray ramus

To effector

White ramus

Ventral root

FIGURE 10.4 The most common pathway of sympathetic fibers. *The sympathetic preganglionic neuron exits the spinal cord as part of the ventral root and travels to the spinal nerve. Shortly after entering the spinal nerve the preganglionic neuron exits the nerve and travels via a white ramus to a ganglion in the sympathetic chain, where the neuron synapses with a postganglionic neuron. The postganglionic neuron then returns to the spinal nerve via a gray ramus.*

In addition to sympathetic postganglionic axons, what types of axons travel in a spinal nerve?

directly into the bloodstream, and thus these products function as hormones. Because these catecholamines travel throughout the body in the blood, this endocrine component of the sympathetic nervous system contributes to the widespread effects of sympathetic activation.

The third anatomical arrangement of sympathetic fibers includes preganglionic neurons that synapse with postganglionic neurons in structures called **collateral ganglia** situated somewhere between the CNS and the effector organ (see Figure 10.3). An example of a collateral ganglion is the celiac ganglion, which contains the cell bodies of postganglionic neurons innervating some of the digestive organs. In this case, a preganglionic neuron exits the spinal cord in the ventral root and then enters the sympathetic chain via a white ramus, as previously described. However, the axon of the preganglionic neuron continues through this ganglion without forming a synapse, and travels to a collateral ganglion via a sympathetic nerve. Within the collateral ganglion, the preganglionic neuron forms synapses with several postganglionic neurons that travel to target tissues. Because these ganglia are not interconnected, as are those of the sympathetic chain, they provide routes that enable the sympathetic nervous system to target select groups of target tissues and thus exert more discrete effects.

Anatomy of the Parasympathetic Nervous System

The preganglionic neurons of the parasympathetic nervous system originate in either the brainstem or the sacral spinal cord (see Figure 10.1), and thus the parasympathetic nervous system is sometimes called the *craniosacral division* of the autonomic nervous system. Generally speaking, parasympathetic preganglionic neurons are relatively long and terminate in ganglia located near the effector organ (Figure 10.6); in the ganglia they form synapses with short postganglionic neurons that travel to the effector organ.

In the cranial portion of the parasympathetic system, axons of the preganglionic neurons originate from cranial nerve nuclei located in the brainstem, and travel with axons in cranial nerves. One important cranial nerve is the **vagus nerve** (cranial nerve X), which originates in the medulla oblongata and innervates much of the viscera, including the lungs, heart, stomach, small intestines, and liver (see Figure 10.1). Other cranial nerves containing parasympathetic preganglionic axons are the *oculomotor nerve* (cranial nerve III), the *facial nerve* (cranial nerve VII), and the *glossopharyngeal nerve* (cranial nerve IX), which innervate structures of the head and neck region.

Axons of afferent neurons and motor neurons

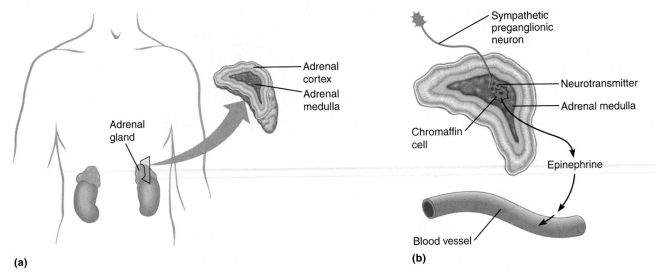

(a)

(b)

FIGURE 10.5 Sympathetic innervation of the adrenal gland. (a) *The adrenal glands, located within fat pads on top of each kidney, consist of an outer cortex and inner medulla.* **(b)** *Sympathetic preganglionic neurons innervate chromaffin cells of the adrenal medulla, stimulating the release of epinephrine (and a small amount of norepinephrine) into the blood.*

What three hormones are secreted by chromaffin cells, and what are the proportions of the total amount secreted?

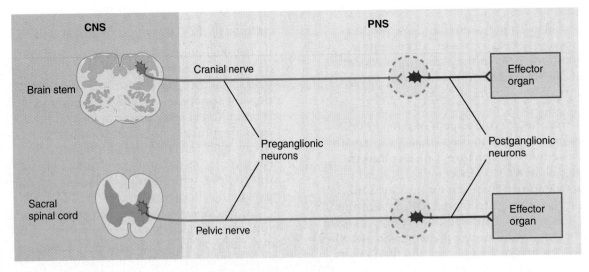

FIGURE 10.6 Parasympathetic nervous system pathways. *Parasympathetic preganglionic neurons originate in either the brainstem or the sacral spinal cord, and their axons are found in cranial and pelvic nerves, respectively. These axons form synapses with postganglionic neurons in ganglia close to or within the effector organs.*

Epinephrine (80%), norepinephrine (20%), dopamine (less than 1%)

Unlike sympathetic preganglionic neurons, parasympathetic preganglionic neurons that originate in the spinal cord do not join with the spinal nerve. Instead, they join with other parasympathetic preganglionic neurons to form distinct pelvic nerves (see Figures 10.1 and 10.6).

The Mixed Composition of Autonomic Nerves

Although it was long thought that nerves of the autonomic nervous system contained only efferent fibers, recent studies have shown that autonomic nerves are generally mixed nerves, with some (for example, the vagus) actually containing more afferent fibers than efferent fibers. Although in this chapter we focus on the efferents, it is important to remember that afferent fibers also travel in autonomic nerves and transmit signals from the viscera that inform the CNS about the status of the various organs. For example, some afferent fibers inform the CNS about whether or not the stomach is full. These and other afferent fibers have an important role in the reflex control of visceral function and homeostasis.

Quick Test 10.1

1. Define the following terms: *preganglionic neuron, postganglionic neuron, autonomic ganglia, sympathetic chain,* and *collateral ganglia.*

2. Which autonomic division regulates the release of epinephrine from the adrenal medulla?

3. Which system, sympathetic or parasympathetic, produces the more diffuse response? Why?

Autonomic Neurotransmitters and Receptors

The two neurotransmitters in the peripheral nervous system are acetylcholine and norepinephrine. Neurons that release the more common of the two, acetylcholine, are referred to as **cholinergic.** Acetylcholine is released by preganglionic neurons of both the sympathetic and parasympathetic branches of the autonomic nervous system, and by parasympathetic postganglionic neurons. The sympathetic preganglionic neurons that innervate the chromaffin cells of the adrenal medulla release acetylcholine, just like all other preganglionic neurons. However, in this case, acetylcholine acts on endocrine cells in the adrenal medulla to stimulate the release of epinephrine. Acetylcholine is also the sole neurotransmitter of the somatic branch of the efferent nervous system, a topic that is discussed later in this chapter.

Norepinephrine is released by almost all sympathetic postganglionic neurons. Neurons that release norepinephrine are referred to as **adrenergic.** (There are some sympathetic postganglionic neurons that release acetylcholine, specifically those that innervate the sweat glands).

Recall from Chapter 7 that the effects of a neurotransmitter depend on the type of postsynaptic receptors activated when the neurotransmitter binds to them. Acetylcholine and norepinephrine can each bind to different classes and subclasses of cholinergic and adrenergic receptors, respectively. We also consider receptors for the hormone epinephrine in this section because it is released from the adrenal medulla during activation of the sympathetic nervous system.

Types of Cholinergic Receptors

The two classes of cholinergic receptors—*nicotinic receptors* and *muscarinic receptors*—are distinguished on the basis of pharmacological studies using two acetylcholine agonists (chemicals that bind to a receptor and produce the biological effect): *nicotine* (found in tobacco) and *muscarine* (a toxin found in certain mushrooms). Nicotinic cholinergic receptors are located on the cell bodies and dendrites of sympathetic and parasympathetic postganglionic neurons (Figure 10.7), on chromaffin cells of the adrenal medulla, and on skeletal muscle cells. Muscarinic cholinergic receptors are found on effector organs of the parasympathetic nervous system, such as the heart, smooth muscles controlling the diameter of the pupil of the eye, and smooth muscle in the digestive tract (Figure 10.7).

The signal transduction mechanisms of cholinergic receptors are summarized in Table 10.1. Recall from Chapter 7 that nicotinic cholinergic receptors are associated with channels that allow both sodium and potassium ions to move through them. When acetylcholine binds to these receptors, these cation channels open, allowing sodium to diffuse into the cell and potassium to diffuse out of the cell. Because sodium is farther from equilibrium, the flow of sodium into the postsynaptic cell exceeds the flow of potassium out of the cell and leads to depolarization. Thus nicotinic cholinergic receptors are associated with depolarization, or excitation, of the postsynaptic cell.

In contrast, muscarinic cholinergic receptors are coupled to G proteins. The responses triggered by the binding of acetylcholine can be either excitatory or inhibitory, depending on the target cell in question and the nature of the signal transduction pathway. The binding of acetylcholine at muscarinic receptors can induce either changes in second-messenger systems or the opening or closing of specific ion channels.

Types of Adrenergic Receptors

There are two major classes of adrenergic receptors located in effector organs of the sympathetic nervous system: **alpha (α) receptors** and **beta (β) receptors.** Each of these is further divided into subclasses: α_1 and α_2, and β_1, β_2, and β_3.

Adrenergic receptors are coupled to G proteins that either activate or inhibit second-messenger systems (Figure 10.8). Binding of norepinephrine or

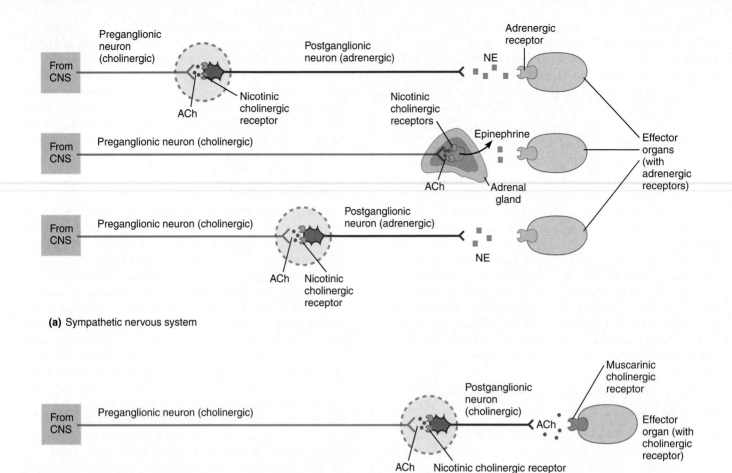

(a) Sympathetic nervous system

(b) Parasympathetic nervous system

FIGURE 10.7 Neurotransmitters and receptors in the autonomic nervous system. (a) *Neurotransmitters and receptors for the three distinct anatomical pathways of the sympathetic nervous system. In all cases, the preganglionic neuron releases acetylcholine, which then binds to nicotinic cholinergic receptors on either postganglionic neurons or endocrine cells in the adrenal medulla. The postganglionic neurons release norepinephrine, which binds to adrenergic receptors on the effector organs.* **(b)** *Neurotransmitters and receptors in the parasympathetic pathway. Acetylcholine is released from both preganglionic and postganglionic neurons; it binds to nicotinic cholinergic receptors on the postganglionic neuron, and to muscarinic cholinergic receptors at the effector organ.*

epinephrine to an α_1 receptor activates a G protein that in turn activates the enzyme phospholipase C, which catalyzes the conversion of phosphatidyl inositol biphosphate (PIP_2) to inositol triphosphate (IP_3) and diacylglycerol (DAG) (Figure 10.8a). IP_3 enters the cell and triggers the release of calcium from intracellular stores. Calcium then causes a response in the cell. DAG activates the enzyme protein kinase C, which catalyzes the phosphorylation of a protein causing a response in the cell.

Binding of norepinephrine or epinephrine to an α_2 receptor activates an inhibitory G protein (G_i) that decreases the activity of the enzyme adenylate cyclase, thereby suppressing the synthesis of cAMP (Figure 10.8b). Binding of norepinephrine or epinephrine to a β receptor, on the other hand, activates a stimulatory G protein (G_s) that increases the activity of the enzyme adenylate cyclase, thereby enhancing the synthesis of cAMP (Figure 10.8b).

Whereas acetylcholine is the only endogenous messenger that binds to cholinergic receptors, both norepinephrine and epinephrine bind to adrenergic receptors. However, the subclasses of adrenergic receptors have different affinities for these two messengers (Table 10.2). The α receptors have a greater affinity for norepinephrine than epinephrine and are generally excitatory; that is, they stimulate muscle cells to contract or other cells to secrete a product. β_1 and β_3 receptors have approximately equal affinities for norepinephrine and epinephrine, and tend to be excitatory. β_2 receptors have a much greater affinity for epinephrine than norepinephrine, and as a consequence norepinephrine does not normally function as a chemical messenger at these receptors. Activation of β_2 receptors generally produces an inhibitory response.

Autonomic receptors are of great clinical significance. For example, people suffering from asthma or nasal congestion often take β_2 adrenergic agonists, such as

TABLE 10.1 CHOLINERGIC RECEPTORS

RECEPTOR TYPE	SIGNAL TRANSDUCTION MECHANISM	TARGET CELL	EFFECT ON TARGET CELL
Nicotinic	Opens channels for sodium and potassium ions	Postganglionic cell body, chromaffin cells, skeletal muscle cells	Excitatory
Muscarinic	G-protein-coupled; opens or closes specific ion channel	Effector organs of parasympathetic nervous system	Excitatory or inhibitory

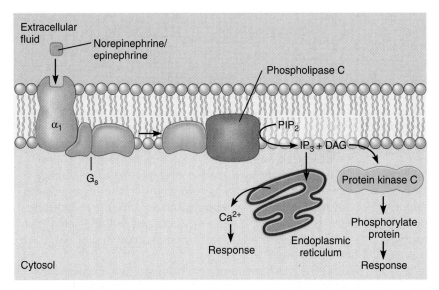

(a) Phosphatidyl inositol biphosphate (PIP$_2$) and α_1 receptors

FIGURE 10.8 Signal transduction mechanism at effector organs of the sympathetic nervous system. (a) *Binding of norepinephrine or epinephrine to an α_1 adrenergic receptor triggers activation of the phosphatidyl inositol second messenger system.* **(b)** *Binding of norepinephrine or epinephrine to an α_2 adrenergic receptor triggers inhibition of the cAMP second messenger system, whereas binding to a β adrenergic receptor triggers excitation of the cAMP second messenger system.*

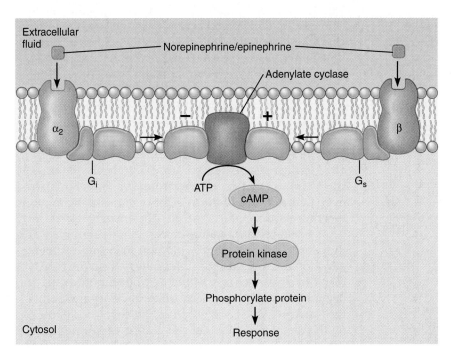

(b) cAMP and α_2 and β receptors

TABLE 10.2 ADRENERGIC RECEPTORS

RECEPTOR TYPE	EFFECTOR ORGAN WITH RECEPTOR TYPE	RELATIVE AFFINITIES*	SIGNAL TRANSDUCTION MECHANISM	EFFECT ON EFFECTOR ORGAN†
α_1	Most vascular smooth muscle, pupils	NE > Epi	Activates IP$_3$	Excitatory
α_2	CNS, platelets, adrenergic nerve terminals (autoreceptors), some vascular smooth muscle, adipose tissue	NE > Epi	Inhibits cAMP	Excitatory
β_1	CNS, cardiac muscle, kidney	NE = Epi	Activates cAMP	Excitatory
β_2	Some blood vessels, respiratory tract, uterus	Epi >>> NE	Activates cAMP	Inhibitory
β_3	Adipose tissue	NE = Epi	Activates cAMP	Excitatory

*NE = norepinephrine; Epi = epinephrine; > = greater than; >>> = much greater than
†Effects are generalizations and not absolute.

epinephrine or ephedrine, to produce dilation of the respiratory airways; α_1 agonists can also be used to treat congestion. Hypertension can be treated with drugs such as propranolol, which are called beta-blockers because they block the effects of norepinephrine or epinephrine at beta adrenergic receptors. Atropine, a muscarinic cholinergic antagonist, comes from the plant *Atropa belladonna*, so named because extracts from the plant were once used to dilate the pupils of women's eyes for cosmetic purposes (*belladonna* means "beautiful woman"). Atropine is sometimes used to dilate the pupils of the eye before an eye exam and is used to treat intestinal spasms and nausea.

Quick Test 10.2

1. What are the two neurotransmitters of the efferent branch of the peripheral nervous system, and which is the more common of the two?

2. What peripheral neurons release the less common of the two neurotransmitters in the efferent branch of the PNS?

3. What branch of the autonomic nervous system would be affected by administration of a nicotinic cholinergic antagonist, and what action would such a drug have?

Autonomic Neuroeffector Junctions

The synapse between an efferent neuron and its effector organ is called a *neuroeffector junction*. Synapses between autonomic postganglionic neurons and their effector organs differ from ordinary neuron-to-neuron synapses in that the postganglionic neurons do not have discrete axon terminals. Instead, neurotransmitters are released from numerous swellings located at intervals along the axons of these neurons, called **varicosities** (Figure 10.9a). Within these varicosities, neurotransmitters are synthesized and then stored in vesicles (Figure 10.9b). The membrane of the axon contains the usual voltage-gated sodium and potassium channels that support propagation of action potentials. In addition, the membrane in the region of each varicosity contains voltage-gated calcium channels that open when an action potential reaches them.

The mechanism of neurotransmitter release from these varicosities is similar to the mechanism of transmitter release from an ordinary axon terminal (Figure 10.10). An action potential arriving at a varicosity opens voltage-gated calcium channels, which allows calcium to enter the cytoplasm and stimulates the release of neurotransmitter by exocytosis. However, unlike ordinary synapses, no discrete anatomical arrangement exists between the varicosity and the effector organ. First, because a single postganglionic axon has several varicosities, an action potential propagated along the axon triggers the release of neurotransmitter from all the varicosities. Second, because the distance between the varicosity and effector organ is greater than the width of a synaptic cleft, the neurotransmitter released from a varicosity diffuses over a greater area of the effector organ, binding to receptors on cells throughout the effector organ.

The effects of the neurotransmitter are terminated as they are in a typical neuron-to-neuron synapse: either diffusion of the neurotransmitter away from the receptors, or active re-uptake of the neurotransmitter, or degradation of the neurotransmitter by enzymes occurs. One such enzyme is *acetylcholinesterase*, which is located on the membrane of either the postganglionic neuron or the effector organ at cholinergic neuroeffector junctions. Following degradation of acetylcholine to acetate and

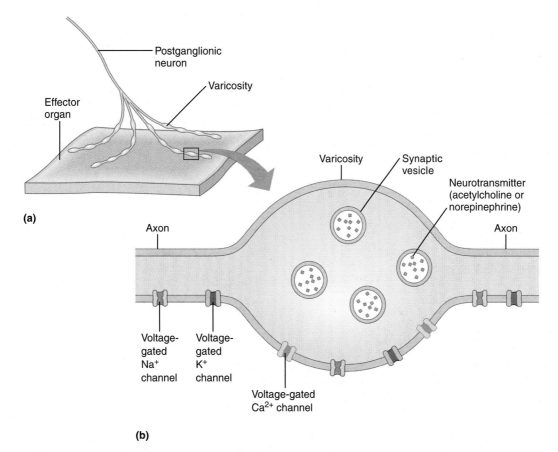

(a)

(b)

FIGURE 10.9 Neuroeffector junctions of the autonomic nervous system.
(a) *Near the effector organ the axon of a postganglionic neuron splits into branches bearing swellings called varicosities.* **(b)** *An enlarged view of a varicosity. Whereas voltage-gated sodium and potassium channels are present throughout the axon to support propagation of action potentials, voltage-gated calcium channels are concentrated at the varicosity. Neurotransmitter is stored in synaptic vesicles in the varicosity.*

Where with respect to the varicosity depicted in Figure 10.9b are action potentials propagated: at the left, at the right, or on both sides?

choline by acetylcholinesterase, choline is actively transported back into the postganglionic varicosity and used to synthesize more acetylcholine. Monoamine oxidase, another degradative enzyme, is located at adrenergic neuroeffector junctions within mitochondria in postganglionic neurons, where it degrades catecholamines that have been actively transported back into the cells.

Regulation of Autonomic Function

Balance of activity levels between the two branches of the autonomic nervous system is necessary for the maintenance of homeostasis. Under resting conditions, the body expends less energy than at other times, with the primary energy expenditure going into the digestion and absorption of nutrients from recently ingested food. With low energy demands, the heart does not need to work as hard, because organs' demand for blood is low. Appropri-

ately, the parasympathetic nervous system exerts dominant control over the body's organs under these conditions. However, when the body is active, energy demands of the tissues increases. These demands must be met, even at the expense of suspending operation of the gastrointestinal organs. Under these conditions, parasympathetic activity decreases and sympathetic activity increases. The result is increased activity of the heart, which helps increase blood flow to the skeletal muscles. But how does the brain regulate the balance between parasympathetic and sympathetic activity to meet the body's changing demands?

Most changes in the activity of the autonomic nervous system are accomplished through the operation of **visceral reflexes,** automatic changes in the functions of

On both sides

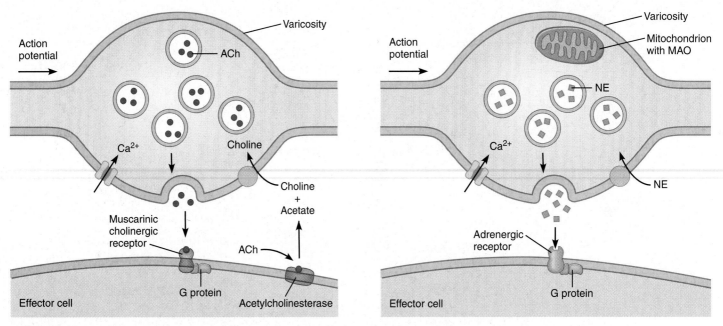

(a) Acetylcholine release from postganglionic neuron

(b) Norepinephrine release from postganglionic neuron

FIGURE 10.10 Neurotransmitter release from varicosities. (a) *Release of acetylcholine from a varicosity. When an action potential depolarizes a varicosity, voltage-gated calcium channels open, allowing calcium ions to enter the cell and triggering the release of acetylcholine by exocytosis. Acetylcholine binds to muscarinic cholinergic receptors on the effector organ membrane, thereby activating a G protein. Once acetylcholine is degraded by acetylcholinesterase located on the effector organ, choline is actively transported back into the postganglionic varicosity and used to synthesize more acetylcholine.* **(b)** *Release of norepinephrine from a varicosity. When an action potential depolarizes the varicosity, voltage-gated calcium channels open, allowing calcium ions to enter the cell and trigger the release of norepinephrine by exocytosis. Norepinephrine binds to adrenergic receptors on the effector organ membrane, thereby activating a G protein. Some norepinephrine is actively transported back into the postganglionic neuron's varicosity, where it is degraded by monoamine oxidase (MAO) in mitochondria.*

organs that occur in response to changing conditions within the body. Consider, for example, the autonomic response that operates when a person stands up rapidly (Figure 10.11). A drop in blood pressure that occurs because blood pools in the lower limbs due to the force of gravity is detected by receptors in some of the major arteries (aorta and carotid arteries), and that information is then relayed to cardiovascular regulatory areas in the medulla oblongata through an afferent pathway. In response, these areas adjust their output to the sympathetic and parasympathetic nervous systems. Sympathetic output to the heart and blood vessels is enhanced, which brings blood pressure back to normal. At the same time, parasympathetic output to the heart is decreased, which allows the blood pressure to increase unopposed. (Recall that parasympathetic activity tends to slow the heart. This action tends to lower blood pressure.) Blood pressure typically returns to normal in just a few seconds, so the person is unaware of any changes that might have occurred. This reflex arc acts to prevent reduced blood flow to the brain and loss of consciousness when a person arises from lying down.

The major areas of the brain that regulate autonomic function include the hypothalamus, pons, and medulla oblongata (Figure 10.12). The hypothalamus initiates the "fight-or-flight" response to elicit widespread activation of the sympathetic nervous system when a person is in danger or is otherwise excited. The hypothalamus also contains the regulatory centers for body temperature, food intake, and water balance, all of which are regulated in some way by autonomic efferent neurons. The medulla oblongata and pons contain cardiovascular and respiratory regulatory centers that control the heart, blood vessels, and smooth muscle in the respiratory airways and regulate the automatic breathing patterns that do not require conscious thought. These areas of the brain receive input from other brain regions, including the hypothalamus, cerebral cortex, and limbic system. They also receive afferent information needed for reflex control of visceral function, such as that just described concerning blood pressure. Other autonomic reflexes involving the brainstem include the pupillary light reflex, the accommodation reflex, the vomiting reflex, and the swallowing

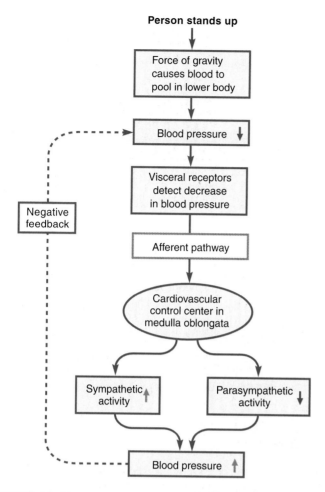

FIGURE 10.11 **Autonomic reflex response that controls blood pressure when a person stands up.**

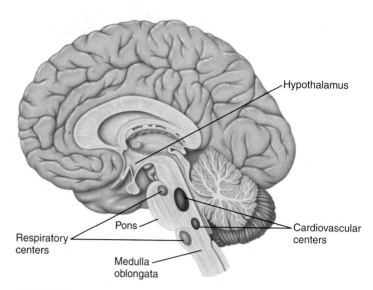

FIGURE 10.12 **Areas of the brain that regulate autonomic function.** *This midsagittal section of the brain reveals the pons and medulla oblongata of the brainstem and the hypothalamus, all of which regulate autonomic function.*

reflex, which are discussed in other chapters. Some visceral reflexes are coordinated by neural circuits in the spinal cord and are thus called spinal reflexes. Examples include the reflexes that control urination (Chapter 17), defecation (Chapter 19), erection (Chapter 21), and ejaculation (Chapter 21).

Recall from Chapter 8 that one area of the brain that influences activity in autonomic control centers is the limbic system, which is involved in the development of emotions. Emotions have a strong effect on activity of the autonomic nervous system and thus affect the functions of effector organs controlled by it. We are all aware of some of the more common autonomic responses to emotions, including a racing heartbeat, upset stomach ("butterflies"), blushing, fainting, and sweating.

The autonomic nervous system is also responsible for many of the symptoms associated with motion sickness, which is generally caused by a mismatch of sensory inputs, including those from the vestibular apparatus (which detects motion of the body), the visual system, and proprioceptors throughout the body (which detect the body's position in three-dimensional space). Common symptoms include nausea and sweating. The neural

mechanisms mediating motion sickness are not fully understood, but communication between the vestibular system and autonomic control centers in the brainstem may be involved, with both branches of the autonomic nervous system being excited. A common treatment for motion sickness is the drug scopolamine, a muscarinic cholinergic antagonist that can be administered as a tablet or via a patch.

Table 10.3 summarizes the effects of the two branches of the autonomic nervous system. We turn our discussion to the somatic nervous system in the next section.

Quick Test 10.3

1. What is a neuroeffector junction? In what respects are autonomic neuroeffector junctions different from ordinary neuron-to-neuron synapses? In what respects are they the same?

2. What structures in postganglionic neurons store and release neurotransmitters? Where are these structures located?

3. What are visceral reflexes, and how are they involved in the control of autonomic functions?

4. What areas of the CNS exert primary control over the autonomic nervous system?

THE SOMATIC NERVOUS SYSTEM

Unlike the autonomic nervous system, which controls the functions of many different types of effector organs, the somatic nervous system controls only one type of effector

TABLE 10.3 EFFECTS OF INNERVATION BY THE
AUTONOMIC NERVOUS SYSTEM

ORGAN SYSTEM	PARASYMPATHETIC NERVOUS SYSTEM* Effect	SYMPATHETIC NERVOUS SYSTEM Effect	Adrenergic receptor class
Heart			
SA Node	Decreases heart rate	Increases heart rate	β_1
AV Node	Decreases conduction velocity	Increases conduction velocity	β_1
Force of contraction	Decreases (small effect)	Increases	β_1
Blood Vessels			
Arterioles to most of body	None	Vasoconstriction	α_1
Arterioles to skeletal muscle	None	Vasoconstriction Vasodilation (epinephrine)	α_1 β_2
Arterioles to brain	None	None	
Veins	None	Vasoconstriction Vasodilation (epinephrine)	α_1 β_2
Lungs			
Bronchial muscle	Contraction	Relaxation	β_2
Bronchial glands	Stimulates secretion	Inhibits secretion	α
Digestive Tract			
Motility	Increased	Decreased	$\alpha_1, \alpha_2, \beta_2$
Secretions	Stimulated	Inhibited	α_2
Sphincters	Relaxation	Contraction	α_1
Pancreas			
Exocrine glands	Stimulates secretion	Inhibits secretion	α
Endocrine glands	Stimulates secretion	Inhibits secretion	α_2
Salivary glands	Stimulates watery secretion	Stimulates mucus secretion	α_1
Kidneys			
Renin release	None	Stimulated	β_1

organ—skeletal muscle. Most skeletal muscles are connected to bones and thus function in the support and movement of the body. In addition, the somatic nervous system has only a single type of efferent neuron: motor neurons, the neurons that innervate skeletal muscle. Most skeletal muscle is under voluntary control; that is, one can consciously decide to contract a muscle. Therefore, the somatic nervous system is also referred to as the *voluntary nervous system.*

Anatomy of the Somatic Nervous System

In the somatic nervous system, a single motor neuron travels from the central nervous system to a skeletal muscle (Figure 10.13); recall that two neurons are present in the autonomic nervous system pathway to an effector organ. Motor neurons originate in the ventral horn of the spinal cord (or the analogous brainstem nuclei) and receive input from multiple sources, including affer-

ents (for spinal reflexes), the brainstem, and the cerebral cortex. A single motor neuron innervates many muscle cells (called **muscle fibers**), but each muscle fiber is innervated by only one motor neuron. A motor neuron plus all the muscle fibers it innervates is a **motor unit** (Figure 10.14). When a motor neuron is activated, it stimulates all the muscle fibers in its unit to contract. The functional importance of motor units is discussed in Chapter 11.

The Neuromuscular Junction

Each branch of a motor neuron synapses with a skeletal muscle fiber at a single highly specialized central region of the fiber called the **neuromuscular junction** (Figure 10.15a). The axon terminals of the motor neuron, called **terminal boutons,** store and release acetylcholine, the only neurotransmitter in the somatic nervous

TABLE 10.3 EFFECTS OF INNERVATION BY THE
AUTONOMIC NERVOUS SYSTEM

ORGAN SYSTEM	PARASYMPATHETIC NERVOUS SYSTEM* Effect	SYMPATHETIC NERVOUS SYSTEM Effect	Adrenergic receptor class
Urinary Bladder			
Bladder wall	Contraction	Relaxation (small effect)	β_2
Sphincter	Relaxation	Contraction	α_1
Male Reproductive Tract			
Blood vessels (erection)	Vasodilation	None	
Vas deferens and seminal vesicles (ejaculation)	None	Ejaculation	α_1
Female Reproductive Tract			
Uterus, nonpregnant	Unknown	Relaxation	β_2
Uterus, pregnant	Unknown	Contraction	α_1
Skin			
Sweat glands	Stimulates secretion	Stimulates secretion	α_1, muscarinic[†]
Piloerector muscles	None	Contraction (hairs stand up)	α_1
Eye			
Iris muscles (pupil size)	Contraction of circular muscle (pupillary constriction)	Contraction of radial muscle (pupillary dilation)	α_1
Ciliary muscles (accommodation)	Contraction for near vision	Relaxation for far vision (small effect)	β_2
Metabolism			
Liver	None	Stimulates glycogenolysis and gluconeogenesis	α_1, β_2
Adipose tissue	None	Stimulates lipolysis	β_3

*Receptor types for the parasympathetic nervous system are not given because *all* effector organs have muscarinic cholinergic receptors.
†Sympathetic postganglionic neurons to the sweat glands release acetylcholine as the neurotransmitter.

system. Opposite these terminal boutons is a specialized region of the muscle fiber's plasma membrane called the **motor end plate,** which has invaginations containing large numbers of acetylcholine receptors. These receptors are of the nicotinic cholinergic variety, although pharmacological studies indicate that these nicotinic cholinergic receptors differ somewhat from those found on postganglionic neurons in autonomic ganglia. Acetylcholinesterase, which is found between the invaginations of the motor end plate, terminates the excitatory signal, which allows the muscle fiber to relax. The action of this enzyme serves to terminate the excitatory signal, which allows the muscle fiber to relax.

Exercise Link

One of the most obvious ways that exercise training improves muscle function is an increase in the speed of contraction. But the speed of relaxation is just as important. Rapid clearance of acetylcholine from the neuromuscular junction is essential in preparing the muscle fiber to receive the next neuronal signal. One of the benefits Jane and Bill gained in their premarathon training was an increase in acetylcholinesterase at the motor end plates of the muscle fibers used in running.

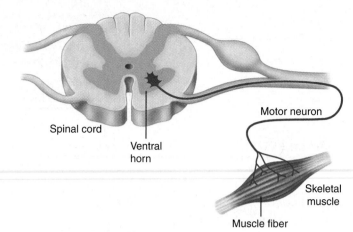

FIGURE 10.13 Anatomy of somatic nervous system pathways.
The somatic nervous system consists of motor neurons, which originate in the ventral horns of the spinal cord and terminate on skeletal muscle fibers throughout the body.

Spinal cord

Ventral horn

Motor neuron

Skeletal muscle

Muscle fiber

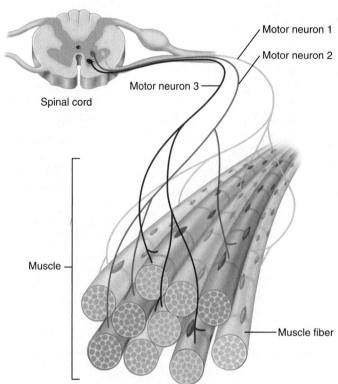

Motor neuron 1

Motor neuron 2

Motor neuron 3

Spinal cord

Muscle

Muscle fiber

FIGURE 10.14 Motor units. *A motor unit consists of a motor neuron and all the muscle fibers it innervates. Whereas a single neuron innervates many muscle fibers, a given muscle fiber is innervated by one motor neuron only. Note that the muscle fibers within a given motor unit are scattered throughout the muscle.*

The signal transmission mechanism at the neuromuscular junction is similar to that at excitatory neuron-to-neuron synapses (Figure 10.15b). When a motor neuron is activated by converging synaptic input, action potentials are propagated to the terminal boutons at the neuromuscular junctions of all muscle fibers in the motor unit. The resulting depolarization causes voltage-gated calcium channels in the boutons to open, allowing calcium to enter the cytosol and triggering the release of acetylcholine by exocytosis. Acetylcholine diffuses across the synaptic cleft and binds to nicotinic cholinergic receptors at the motor end plate, causing cation channels to open. This allows sodium to flow into the muscle fiber and produces a depolarization that is similar in many ways to the excitatory postsynaptic potentials (EPSPs) generated in neurons; in this instance it is called an **end-plate potential (EPP).** The one major difference is that end-plate potentials are normally of sufficient magnitude to depolarize the muscle fiber to threshold, which generates an action potential and triggers contraction of the muscle fiber by mechanisms described in Chapter 11.

Whereas hormonal regulation of skeletal muscle *metabolism* occurs, only innervation by a motor neuron controls skeletal muscle *contraction*. Therefore, all neural

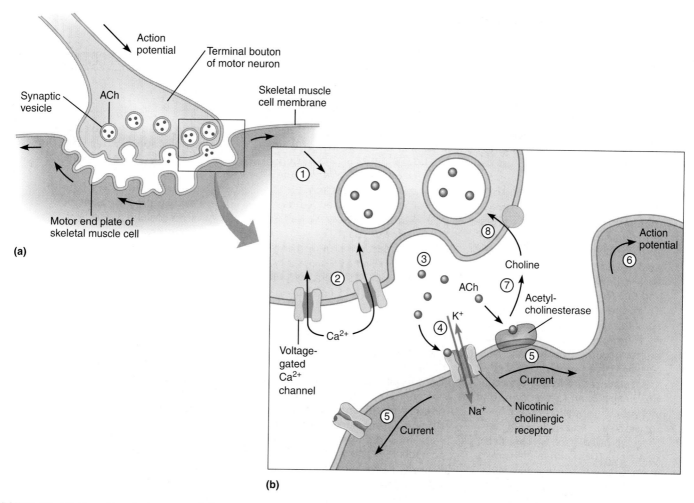

FIGURE 10.15 Functional anatomy of the neuromuscular junction. (a) *Both the axon terminal of the motor neuron and the portion of the plasma membrane of the skeletal muscle called the motor end plate are specialized at the neuromuscular junction.* **(b)** *Communication at the neuromuscular junction. When an action potential arrives at the axon terminal of a motor neuron* ①, *voltage-gated calcium channels open and calcium enters the cytoplasm* ②. *The entry of calcium triggers the release by exocytosis of acetylcholine* ③, *which diffuses to and binds to nicotinic cholinergic receptors at the motor end plate, opening small cation channels. Sodium enters the cell* ④, *producing an end-plate potential that generates currents throughout the plasma membrane of the skeletal muscle cell* ⑤, *depolarizing the membrane to threshold to generate an action potential* ⑥. *The action potential spreads along the skeletal muscle cell membrane, ultimately stimulating contraction. Acetylcholinesterase degrades acetylcholine to produce acetate and choline* ⑦, *which is actively transported into the terminal bouton* ⑧, *where it can be used to synthesize more acetylcholine.*

communication to skeletal muscle is excitatory; that is, it stimulates the muscle to contract. For skeletal muscle to relax, the neural stimulation must cease.

Clearly, normal signal transmission at the neuromuscular junction is necessary for normal control of skeletal muscle contraction. If transmission at the neuromuscular junction is altered, as occurs in the disease *myasthenia gravis,* then normal function of skeletal muscle is lost (see When It Goes Wrong: Myasthenia Gravis, p. 328).

The neuromuscular junction is also targeted by many of the toxins present in animal venoms. The venom of

the black widow spider, for example, contains the toxin *latroxin,* which by stimulating the release of acetylcholine at the neuromuscular junction induces muscle spasms and rigidity. Because the respiratory muscles (which are skeletal muscles) are affected, this venom can cause respiratory failure and death by inducing spastic contractions of these muscles. The venom of the rattlesnake, by contrast, contains the toxin *crotoxin,* which has the opposite effect of latroxin. Crotoxin inhibits the release of acetylcholine, which induces flaccid paralysis of skeletal muscles. Another toxin with paralyzing effects on skeletal

All of us have experienced muscle fatigue, but for persons afflicted with myasthenia gravis, muscle fatigue is more than just a nuisance—it can be debilitating and sometimes fatal. *Myasthenia gravis* is a disease affecting transmission at neuromuscular junctions. Its victims are mostly women, and it usually strikes between the ages of 20 and 50. The defining characteristic of the disease is fatigue of unusually rapid onset and severity following the use of certain muscle groups. Because the muscles most frequently affected are those of the head, difficulties in speaking (*dysarthria*) and in swallowing (*dysphagia*) are common symptoms; drooping of the eyelids (*ptosis*) is also a common sign. The disease also frequently targets limb muscles, with resultant weakness in the arms and legs. In certain persons, muscles used in breathing can be affected, sometimes necessitating the use of a mechanical ventilator. Before the development of modern treatments, the mortality rate for those afflicted with myasthenia gravis was over 30%, mostly as a result of respiratory problems.

Compared with other neuromuscular diseases, the time course of myasthenia gravis is unusual. Following onset of the disease, the severity of symptoms waxes and wanes from day to day or month to month, or even within the course of a single day. During certain periods the disease might even appear to be in complete remission.

Although the disease has been known to medical practitioners for at least the past few hundred years, the underlying mechanism remained a mystery until the mid-20th century, when techniques permitting the recording of electrical signals in nerves and muscles were developed. These techniques revealed that symptoms of myasthenia gravis were due to failure of motor neurons to excite muscle cells to contract. Even though motor neurons were capable of transmitting action potentials that were followed by end-plate potentials (EPPs) in muscle cells, these EPPs were often smaller than normal, particularly if neurons were repetitively stimulated. When this happened, muscle cell membrane potentials would often fail to reach threshold, resulting in a "dropped" contraction.

These observations suggested a number of possible mechanisms for the development of the disease. Perhaps a reduction in the amount of acetylcholine released by motor neurons was responsible, or maybe the problem was some deficiency in acetylcholine receptor function. We now know that myasthenia gravis is an *autoimmune* disease—a disease in which the immune system attacks proteins that are normal components of body tissues.

In myasthenia gravis, the immune system produces antibodies against acetylcholine receptors at the neuromuscular junctions. These antibodies bind to nicotinic cholinergic receptors, triggering their removal from the plasma membrane and subsequent destruction by immune cells. The resulting decrease in the number of functional receptors on the cell surface impairs the ability of these muscle cells to respond to acetylcholine.

Current therapies attempt to blunt the immune system's action by reducing antibody levels in the blood. One approach is *thymectomy,* removal of the thymus gland, which plays an important role in immune function. This treatment reverses symptoms in about 50% of patients. *Plasmapheresis* is another approach. In this procedure, blood is removed from a patient (not all of it at once!), the *plasma* (liquid component of blood, which contains the antibodies) is separated from the cells, and the cells are then returned to the patient. The availability of these and other treatments has reduced the mortality rate for myasthenia gravis to near zero and has enabled most persons stricken with this disease to live normal lives.

Curare is an extract of a plant *(Chondrodendron tomentosum)* found in South America. Indians of that region crush and cook the roots and stems of the plant to produce a poison for the tips of their arrows and darts. (The word *curare* comes from the Indian word for poison.) Curare was valued by Indians because they used the poison in hunting game. When an animal was struck by a curare-laced arrow or dart, it would become paralyzed and eventually die from respiratory failure.

The effective component of curare is a compound called *tubocurarine,* which blocks communication at the neuromuscular junction. Tubocurarine binds to nicotinic cholinergic receptors, thereby preventing acetylcholine from binding. When this occurs, skeletal muscles are unable to contract even when action potentials are transmitted by the motor neurons that innervate them.

In the late 19th and early 20th centuries, curare was studied for its possible

pharmaceutical benefits. Curare was first used as a skeletal-muscle relaxant to supplement general anesthesia in the 1930s. Today, curare has numerous clinical uses, including dilating hollow organs such as the rectum and relaxing the throat, which enables easier examination. Curare is also used for relief of *spastic paralysis,* a type of paralysis that can result from excessive skeletal muscle activity.

TABLE 10.4   PROPERTIES OF THE AUTONOMIC AND SOMATIC NERVOUS SYSTEMS

	AUTONOMIC: PARASYMPATHETIC	AUTONOMIC: SYMPATHETIC	SOMATIC
Origin	Brainstem or lateral horns of sacral spinal cord	Lateral horns of thoracic and lumbar spinal cord	Ventral horns of spinal cord
Neurons in pathway	Two (preganglionic and postganglionic)	Two (preganglionic and postganglionic)	One (motor neuron)
Effector organs	Cardiac muscle, smooth muscle, glands	Cardiac muscle, smooth muscle, glands, adipose tissue	Skeletal muscle
Neurotransmitters at neuroeffector junction	Acetylcholine	Norepinephrine	Acetylcholine
Receptor type at effector organ	Muscarinic cholinergic	Adrenergic (all classes)	Nicotinic cholinergic
Effects on effector organ	Either excitation or inhibition	Either excitation or inhibition	Excitation
Control	Primarily involuntary	Primarily involuntary	Primarily voluntary

muscle is *curare,* a poison used by native South Americans on the tips of the darts used in blow guns (see Discovery: Curare).

Quick Test 10.4

1. Describe two differences between the somatic nervous system and the two branches of the autonomic nervous system with regard to the anatomical arrangement of their efferent neurons.

What are the efferent neurons of the somatic nervous system called?

2. What neurotransmitter is released by motor neurons at the neuromuscular junction? To what type of receptor does this neurotransmitter bind?

3. Explain how respiratory failure can result from either abnormal inhibition or abnormal excitation of the neuromuscular junction.

4. How many motor neurons innervate a single skeletal muscle fiber?

Urinary System

Autonomic and somatic neurons regulate urination

Autonomic neurons regulate glomerular filtration rate

Digestive System

Autonomic neurons regulate smooth muscle and glands of gastrointestinal tract

Autonomic neurons regulate enteric nervous system

Somatic neurons regulate skeletal muscles of esophagus and sphincters

Respiratory System

Respiratory centers in brainstem regulate rate and depth of ventilation

Nervous System

Reproductive System

Autonomic neurons control arousal, erection and ejaculation

Afferent neurons provide input to endocrine cells controlling parturition

Afferent neurons provide input to endocrine cells controlling milk production and ejection from breasts

Endocrine System

Neural input affects secretion of hormones by hypothalamic neurosecretory cells

Neural input determines secretion of vasopressin and oxytocin from posterior pituitary

Autonomic neurons regulate hormone secretion from adrenal medulla

Immune System

Hypothalamus induces fever in response to chemical mediators of immune response

Autonomic neurons regulate lymphoid organs

Stress stimulates CRH secretion (and thus, cortisol secretion) through neural pathways

Cardiovascular System

Cardiovascular centers in brainstem regulate heart beat and blood vessel radius through autonomic neurons

Muscles

Somatic neurons stimulate skeletal muscle contraction

Autonomic neurons stimulate/ inhibit smooth muscle contraction

Autonomic neurons increase/ decrease rate or strength of cardiac muscle contraction

The Autonomic Nervous System, p.312

There are two main branches of the efferent nervous system: the autonomic nervous system and the somatic nervous system. Table 10.4 (p. 329) compares the properties of the two branches of the autonomic nervous system with those of the somatic nervous system. The autonomic nervous system includes the parasympathetic and sympathetic nervous systems, which innervate cardiac muscle, smooth muscle, glands, and adipose tissue. Effector organs are generally innervated by both the parasympathetic and sympathetic divisions, an arrangement termed dual innervation. The parasympathetic nervous system is most active during rest, whereas the sympathetic nervous system is most active during periods of activity or excitation and is responsible for the fight-or-flight response.

Pathways in the autonomic nervous system consist of two neurons that communicate between the CNS and the effector organ: preganglionic neurons and postganglionic neurons. Postganglionic neurons innervate the effector organ. The sympathetic nervous system also has an endocrine component because one set of preganglionic neurons innervates the adrenal medulla, stimulating the release of the hormone epinephrine. All preganglionic neurons contain the neurotransmitter acetylcholine. The parasympathetic postganglionic neurons also contain the neurotransmitter acetylcholine, but most sympathetic postganglionic neurons contain the neurotransmitter norepinephrine. The receptors for acetylcholine on postganglionic neurons are nicotinic cholinergic receptors, whereas the receptors for acetylcholine on effector organs in the parasympathetic nervous system are muscarinic cholinergic receptors. The receptors for norepinephrine and epinephrine on the effector organs in the sympathetic nervous system are adrenergic receptors.

The synapse between an efferent neuron and its effector organ is called a neuroeffector junction. At the neuroeffector junctions between autonomic postganglionic neurons and their effector organs, neurotransmitter is released diffusely from varicosities and then binds to receptors on the effector organ. The mechanism of release of neurotransmitter is similar to that for neuron-to-neuron synapses.

The autonomic nervous system is under involuntary control. Areas of the brain that influence autonomic activity include the brainstem, hypothalamus, and limbic system.

> **IP** Nervous II, Synaptic Transmission, pages 3–11
>
> **IP** Nervous II, Ion Channels: pages 4–5, 7*
>
> **IP** Nervous I, The Membrane Potential, pages 3–6; 11

The Somatic Nervous System, p.323

The somatic division of the efferent nervous system consists of pathways composed of single motor neurons. Motor neurons originate in the ventral horn of the spinal cord and innervate skeletal muscle cells. A single motor neuron and the muscle cells it innervates is called a motor unit. The synapse between a motor neuron and a skeletal muscle fiber is called a neuromuscular junction. The motor neuron contains the neurotransmitter acetylcholine. The receptors in skeletal muscle are nicotinic cholinergic. Binding of acetylcholine to nicotinic cholinergic receptors at the motor end plate produces an end-plate potential, which ultimately causes the skeletal muscle fiber to contract.

> **IP** Nervous II, Synaptic Transmission, page 11
>
> **IP** Muscular, The Neuromuscular Junction, pages 1–10

---

*This topic is available on the *InterActive Physiology*® Sampler CD that comes with the purchase of a new copy of this book.

Multiple-Choice Questions

1. Effector organs of the autonomic nervous system include all of the following *except*
   a) the heart muscle.
   b) smooth muscle in the pupils of the eye.
   c) respiratory muscles.
   d) sweat glands.

2. According to the concept of dual innervation by the autonomic nervous system, if sympathetic activity inhibits pancreatic secretions, then the parasympathetic nervous system should
   a) inhibit pancreatic secretions as well.
   b) stimulate pancreatic secretions.

c) have no effect on pancreatic secretions.

3. The adrenal medulla
   a) contains sympathetic postganglionic neurons.
   b) is part of the brainstem.
   c) releases epinephrine into the blood.
   d) is part of the parasympathetic nervous system.

4. Which of the following cranial nerves does *not* contain parasympathetic preganglionic neurons?
   a) oculomotor (cranial nerve III)
   b) facial (cranial nerve VII)
   c) glossopharyngeal (cranial nerve IX)
   d) vagus (cranial nerve X)
   e) hypoglossal (cranial nerve XII)

5. Which of the following receptor types does *not* activate G proteins?
   a) nicotinic cholinergic
   b) muscarinic cholinergic
   c) α_1 adrenergic
   d) β_3 adrenergic

6. The origin of spinal preganglionic neurons is the
   a) ventral horn of the spinal cord.
   b) dorsal horn of the spinal cord.
   c) lateral horn of the spinal cord.

7. The origin of motor neurons is the
   a) ventral horn of the spinal cord.
   b) dorsal horn of the spinal cord.
   c) lateral horn of the spinal cord.

8. Which of the following second messengers stimulates the release of calcium from intracellular stores?
   a) cAMP
   b) inositol triphosphate
   c) diacylglycerol

9. Which of the following is the location of the cardiovascular regulatory centers?
   a) hypothalamus
   b) limbic system
   c) pons
   d) medulla oblongata

10. How many motor neurons innervate a single skeletal muscle cell?
    a) zero
    b) one
    c) several
    d) millions

11. The motor end plate is
    a) the specialized synaptic terminal of the motor neuron.
    b) the specialized synaptic terminal of autonomic postganglionic neurons.
    c) the specialized region of skeletal muscle innervated by a motor neuron.
    d) the specialized region of an effector organ innerved by an autonomic postganglionic neuron.

12. Neurotransmitter is released from which portion of a postganglionic neuron?
    a) terminal bouton
    b) axon terminal
    c) varicosity
    d) cell body

Objective Questions

1. Which branch of the autonomic nervous system has longer preganglionic neurons?

2. Which branch of the autonomic nervous system is most active when the body is at rest?

3. The communication between preganglionic neurons and postganglionic neurons in the autonomic nervous system is one-to-one. (true/false)

4. What part of the adrenal gland secretes epinephrine?

5. Name the four cranial nerves that contain parasympathetic preganglionic neurons.

6. Autonomic nerves contain only efferent neurons. (true/false)

7. Which neurons in the peripheral nervous system are cholinergic?

8. Which neurons in the peripheral nervous system are adrenergic?

9. What enzyme catalyzes the formation of diacylglycerol and inositol triphosphate?

10. A decrease in cAMP is associated with what class of adrenergic receptor?

11. Activation of α adrenergic receptors usually produces (excitation/inhibition).

12. β_2 adrenergic receptors have a greater affinity for (epinephrine/norepinephrine).

13. The motor end plate has (nicotinic/muscarinic) cholinergic receptors.

14. Skeletal muscle can be excited to contract only; that is, it cannot be inhibited to relax. (true/false)

15. The enyzme that degrades acetylcholine in the synaptic cleft is called _____.

Essay Questions

1. Describe the different anatomical arrangements found in the two branches of the autonomic nervous system and in the somatic nervous system.

2. Explain the concept of dual innervation.

3. Explain why sympathetic activation produces a more diffuse effect compared to activation of the parasympathetic nervous system.

4. Compare the signal transduction mechanisms for the different types of adrenergic receptors.

5. What areas of the brain regulate autonomic function?

Find the answers to these exercises, and additional study tools, at the Physiology Place (www.physiologyplace.com).

11

Muscle Physiology

OBJECTIVES

- Name the major structural features of a skeletal muscle cell, and briefly describe each feature's roles in muscle contraction.

- Describe the sequence of events that occurs in the crossbridge cycle, and relate this sequence to the sliding-filament model of muscle contraction.

- Identify the various factors that affect the force of muscle contraction, and define the following terms: *twitch, summation, tetanus, isotonic contraction, isometric contraction, motor unit, recruitment,* and *size principle.*

- Name the three types of skeletal muscle fibers, and describe the major differences among them.

- Explain how antagonistic muscle groups work, and how the lever action of bones affects the force of muscle contraction.

- Describe the major characteristics of smooth and cardiac muscle, and compare these muscle types to skeletal muscle.

CHAPTER OUTLINE

Skeletal Muscle Structure   334

The Mechanism of Force Generation in Muscle   338

The Mechanics of Skeletal Muscle Contraction   344

Types of Skeletal Muscle Fibers   353

Skeletal Muscles at Work   357

Other Muscle Types   360

Above: Electron micrograph of skeletal muscle

Without a doubt, muscles are among the most intricate machines on the planet. Muscle cells possess many of the special components of neurons (neurotransmitter receptors, voltage-sensitive ion channels, and so on) and can generate electrical signals in response to neural input, just as neurons do. In addition, muscle cells have an important defining characteristic—the ability to generate force and movement. They have this capability because they possess an elegant array of *contractile proteins* that can convert chemical energy derived from the hydrolysis of ATP into mechanical energy.

Muscles are remarkable not only for their ability to contract, but also for the precision of the mechanisms that regulate these contractions. To perform their jobs properly, the muscles that move the body must be able to respond faithfully and quickly to commands from the nervous system. Indeed, a typical muscle can activate its contractile machinery within milliseconds of receiving a neural signal, and can turn it off nearly as quickly. It is this quickness of response that enables us to perform complicated motions such as running, jumping, or throwing, in which the precise timing of muscle contractions is critical. In addition, muscles can vary the amount of force they generate to accommodate various activities. For instance, you can use a given set of muscles in your arm to lift a paper clip, a book, or a chair.

We begin our study of muscle physiology with an examination of muscle anatomy, moving from the gross anatomical level to the molecular level, with special emphasis on the structures that generate and regulate contractile force. Then we examine contractile and regulatory mechanisms to see how they work. Although we concentrate on *skeletal muscle,* most of the basic principles apply to each of three muscle types found in the body. At the end of the chapter we examine the special properties of the other two muscle types—*cardiac muscle* and *smooth muscle.*

SKELETAL MUSCLE STRUCTURE

With few exceptions, **skeletal muscles,** such as the *biceps* of the arm, are connected to at least two bones. In contrast, the other muscles of the body—those found in internal organs, blood vessels, and certain other structures—are not connected to bones. The exceptions to this rule include certain skeletal muscles that are connected to the skin (as is the case with some facial muscles), to cartilage (for example, muscles of the larynx), or to other muscles (the external anal sphincter, for instance). Muscles are connected to bones by **tendons,** cords of elastic connective tissue that transmit force from the muscle to the bone.

Structure at the Cellular Level

The part of the muscle that generates force is called the **body,** the "meaty" part of the muscle (Figure 11.1). The body contains many bundles (called *fascicles*) of individual muscle cells, as well as connective tissue, blood vessels, and nerves. Each fascicle contains hundreds or thousands of muscle cells, which are called **muscle fibers** because of their elongated shape. Each muscle fiber runs the full length of the muscle (often running at a diagonal) and is encased in a sheath of connective tissue. Unlike most cells, which have a single nucleus, muscle fibers have many because each muscle fiber is formed during embryonic life from the fusion of several cells. These nuclei lie immediately below the muscle fiber's plasma membrane, which is called the **sarcolemma.**

A muscle fiber's semifluid cytoplasm is packed with mitochondria and hundreds of banded, rodlike elements called **myofibrils,** which contain the fiber's contractile machinery (Figure 11.2). Each myofibril is a bundle of overlapping thick and thin filaments made of the proteins *myosin* and *actin*, respectively. A saclike membranous network called the **sarcoplasmic reticulum** surrounds each of the myofibrils and is closely associated with other structures called **transverse tubules** (or *T tubules*), which are connected to the sarcolemma and penetrate into the cell's interior. The sarcoplasmic reticulum and T tubules play an important role in the activation of muscle contractions because they help transmit signals from the sarcolemma to the myofibrils, enabling a muscle cell to respond to neural input. The functions of the sarcoplasmic reticulum are to store calcium ions (Ca^{2+}) and to release them into the cytosol when the muscle cell is stimulated to contract. As we will see, these calcium ions are released in response to electrical signals that travel from the sarcolemma to the T tubules, and they serve as chemical messengers that carry these signals to the cell's interior, where the myofibrils are located.

Structure at the Molecular Level

When viewed under a microscope, skeletal muscle cells have a striped appearance, and for this reason this muscle (and also cardiac muscle) is often referred to as **striated muscle.** A close-up view shows that these striations are due to the orderly arrangement of protein fibers in the

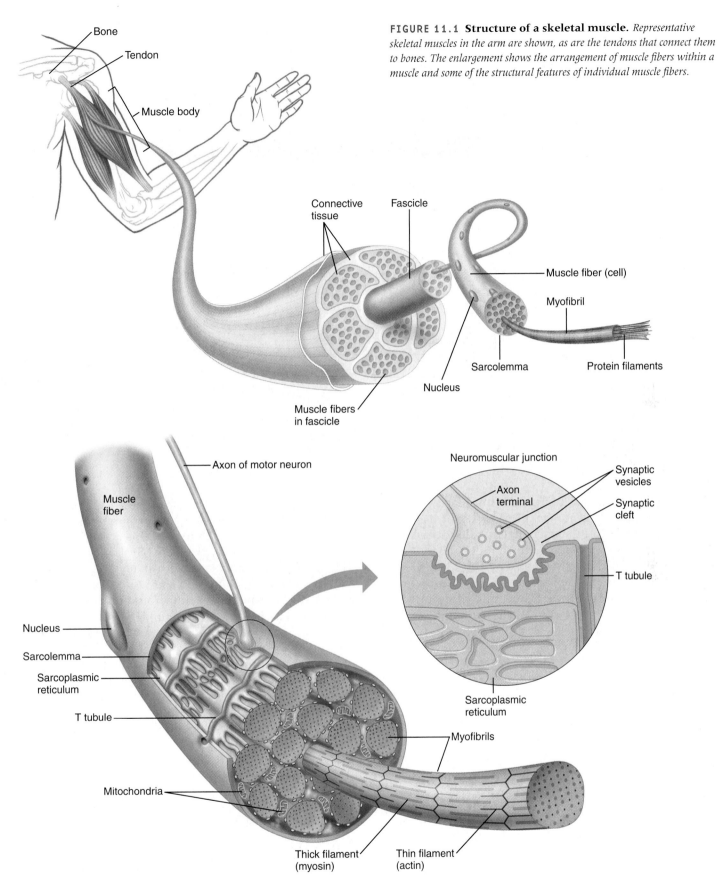

FIGURE 11.1 Structure of a skeletal muscle. *Representative skeletal muscles in the arm are shown, as are the tendons that connect them to bones. The enlargement shows the arrangement of muscle fibers within a muscle and some of the structural features of individual muscle fibers.*

Bone

Tendon

Muscle body

Connective tissue

Fascicle

Muscle fiber (cell)

Myofibril

Sarcolemma

Nucleus

Protein filaments

Muscle fibers in fascicle

Axon of motor neuron

Neuromuscular junction

Axon terminal

Synaptic vesicles

Synaptic cleft

T tubule

Muscle fiber

Nucleus

Sarcolemma

Sarcoplasmic reticulum

T tubule

Sarcoplasmic reticulum

Myofibrils

Mitochondria

Thick filament (myosin)

Thin filament (actin)

FIGURE 11.2 Structure of a skeletal muscle fiber. *Major internal components of a muscle fiber are shown. A single myofibril in the muscle fiber has been extended and slightly enlarged to reveal the arrangement of thick and thin filaments within it. The enlarged view shows a magnified image of a neuromuscular junction.*

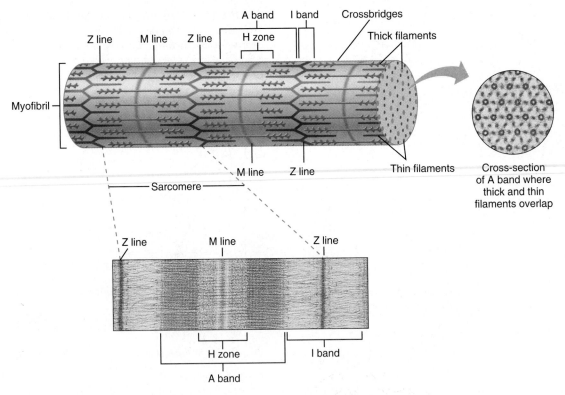

FIGURE 11.3 Sarcomere structure. *A drawing of a myofibril, showing the regular arrangement of protein filaments within sarcomeres. The lower photomicrograph shows the banding pattern typical of striated muscle; the photomicrograph at right shows a cross section through the A band of a myofibril, in which the three-dimensional arrangement of thick and thin filaments can be clearly seen.*

myofibrils called **thick filaments** and **thin filaments,** which run parallel to the muscle cell's long axis. Myofibrils are composed of a fundamental unit called a **sarcomere,** that repeats over and over (Figure 11.3). Each sarcomere is bordered on either end by **Z lines,** which run perpendicular to the long axis and anchor the thin filaments at one end. The thick filaments in a sarcomere are connected together by **M lines,** which also run perpendicular to the long axis.

Before the structure of the sarcomere was elucidated and its protein filaments identified, early investigators used the terms *A band, I band,* and *H zone* to designate certain regions in the muscle fiber's banding pattern (see Figure 11.3). The **A band,** the region that appears darkest under the microscope, spans the length of the thick filaments. The **H zone** is a region in the center of the A band that appears noticeably lighter. In all parts of the A band except the H zone, thin filaments are present along with the thick filaments and overlap them. (A cross section of the fiber reveals that six thin filaments surround each thick filament.) It is the absence of thin filaments in the H zone that gives it its lighter appearance. The **I band,** the brightest region of the muscle fiber, occupies the space between the A bands of adjacent sarcomeres. It contains only thin filaments and the Z line that connects them, and the absence of thick filaments accounts for the brightness of the I band.

The thin and thick filaments of the sarcomere are made up of two proteins called **actin** and **myosin,** which are referred to as **contractile proteins** because they constitute the machinery that generates contractile force. The thin filaments are made of actin; myosin makes up the thick filaments. Notice in Figure 11.3 that both ends of each thick filament have numerous protrusions that are called **crossbridges** because under certain conditions (discussed shortly) they bridge the gap between the thick and thin filaments.

Just as myofibrils are made up of orderly, repeating structures, thick and thin filaments are also made up of structures arranged in an orderly, repeating fashion. The basic components of each thin filament are actin monomers called *G* actin (because they are *g*lobular proteins) linked together end to end, like links in a chain, to form two helically arranged fibrous polymer strands (Figure 11.4a). Each actin monomer possesses a myosin-binding site; as we will see, the ability of actin and myosin to bind together under certain conditions is crucial to a muscle's ability to generate force.

Also present in thin filaments are two special proteins called *regulatory proteins* that enable muscle fibers to start or stop contracting: tropomyosin and troponin (Figure 11.4b). **Tropomyosin** is a long fibrous molecule that extends over numerous actin molecules in such a way that it blocks the myosin-binding sites in muscles at rest. **Troponin** is a complex of three proteins—one that attaches

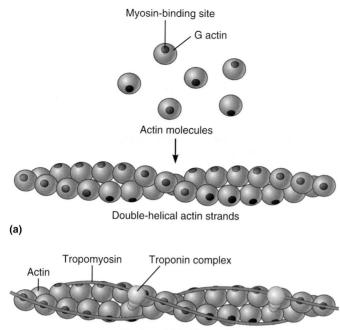

(a)

(b)

Portion of thin filament

FIGURE 11.4 Structure of a thin filament. (a) *The backbone of a thin filament consists of two strands of polymerized actin molecules wound together to form a double helix. Myosin-binding sites on individual actin molecules (G-actin) are represented by dark dots.* **(b)** *A portion of a thin filament showing troponin and tropomyosin in their normal resting positions on the actin strands. Notice that actin's myosin binding sites are covered by tropomyosin when a muscle cell is at rest.*

to the actin strand, another that binds to tropomyosin, and a third containing a site to which calcium ions can bind reversibly. As we will see, the binding of calcium to this site triggers muscle contraction by causing troponin to move tropomyosin aside, thereby exposing the myosin-binding sites on the actin molecules.

Each thick filament is made of hundreds of myosin molecules, each of which looks a bit like two golf clubs wrapped around each other (Figure 11.5a). Each myosin molecule is a dimer consisting of two intertwined subunits, each having a long tail and a fat, protruding head. (These heads are the crossbridges shown in Figure 11.3.) Within a thick filament, the myosin molecules bind to each other at their tail ends so that their heads extend in opposite directions away from the center (Figure 11.5b). The tails of adjacent myosin molecules are also arranged in a staggered fashion so that their heads protrude from the thick filament in an orderly helical pattern (Figure 11.5c). Because the middle of the thick filament is devoid of crossbridges, this region is appropriately called the *bare zone.*

The head is the "business end" of the myosin molecule because it is the part that actively generates a muscle's mechanical force. Each head possesses two sites that are critical to its force-generating ability: an *actin-binding site,* which is capable of binding to the actin monomers in the thin filaments, and an *ATPase site,* which has enzymatic activity and hydrolyzes ATP (see Figure 11.5a).

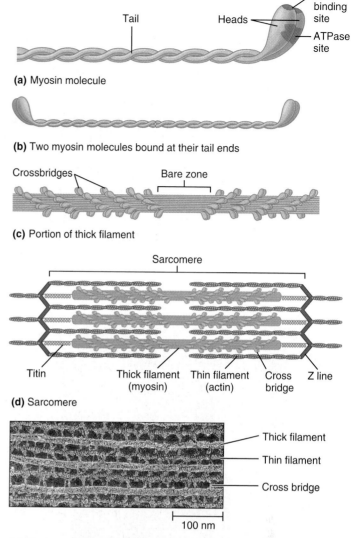

(a) Myosin molecule

(b) Two myosin molecules bound at their tail ends

(c) Portion of thick filament

(d) Sarcomere

(e) Electron micrograph of portion of sarcomere

FIGURE 11.5 Structure of a thick filament. (a) *A myosin molecule, a dimer composed of two subunits wound together. Note the actin-binding and ATPase sites in the head region.* **(b)** *Two myosin molecules joined tail to tail.* **(c)** *A portion of a thick filament showing myosin heads (crossbridges) protruding at either end but not in the middle region (the bare zone).* **(d)** *A detailed view of a sarcomere showing the relative positions of thick and thin filaments and the protein titin, which anchors the thick filaments in place.* **(e)** *A photomicrograph of a sarcomere showing thick and thin filaments and crossbridges.*

Like thin filaments, thick filaments have additional proteins associated with them, most notably *titin,* an extraordinarily elastic protein that can be stretched to more than three times its unstressed length (Figure 11.5d). Strands of titin extend along each thick filament from the M line to each Z line, thereby anchoring the thick filaments in their proper positions relative to the thin filaments. When a relaxed muscle is stretched passively (that is, in response to an external force applied to it), its muscle fibers elongate, and the thick and thin filaments slide past each other so that the sarcomeres also lengthen.

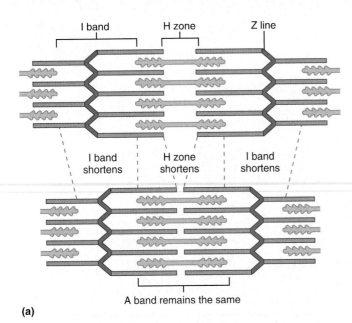

(a)

I band | H zone | Z line

I band shortens | H zone shortens | I band shortens

A band remains the same

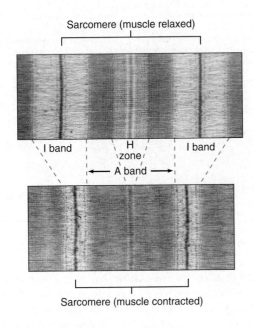

Sarcomere (muscle relaxed)

I band | H zone | I band

A band

(b)

Sarcomere (muscle contracted)

FIGURE 11.6 How changes in striation pattern are explained by the sliding-filament model of muscle contraction. *A schematic drawing* **(a)** *and photomicrographs* **(b)** *showing the relative positions of the thick and thin filaments in sarcomeres in relaxed muscle (top) and contracted muscle (bottom).*

When the force is removed, the muscle fibers and sarcomeres spring back to their original resting lengths automatically due to the actions of titin. When an external stretching force is applied to a muscle, titin strands (see Figure 11.5d) elongate as the sarcomeres lengthen, and these strands begin to exert an opposing force, just as a spring resists stretching. When the external force is removed, this opposing force pulls the Z lines and thick filaments closer together and causes the sarcomeres to shorten, allowing the titin strands to spring back to their original length. As this occurs, individual muscle fibers shorten, as does the whole muscle.

Quick Test 11.1

1. Define the following terms: *muscle fiber, myofibril, Z line, sarcomere, crossbridge.*

2. Name and describe the locations and general functions of the two contractile proteins and the two regulatory proteins present in sarcomeres.

3. What two functions are performed by the heads of myosin molecules?

THE MECHANISM OF FORCE GENERATION IN MUSCLE

If you knew nothing about cells, it is doubtful that you could figure out how their parts work just by looking at them through a microscope. Could you, for instance, conclude that a mitochondrion carries out oxidative phosphorylation simply by noting its two membranes and its oblong shape? With a muscle cell, though, things are different. As you examine the intricate structure of a myofibril, with its elegant array of slender filaments, you cannot escape the thought that within this structure lies the key to understanding how muscle contraction works—and in thinking this, you would be correct. In muscle physiology there is but one rule: Form follows function.

The Sliding-Filament Model

When physiologists first discovered the presence of actin and myosin in myofibrils, they thought that muscle contractions were caused by shortening of the proteins themselves. As advances in microscopy occurred, researchers discovered that during muscle-cell contraction the A band does not change in length, but the I bands and the H zone shorten (Figure 11.6). Given that the A band spans the length of the thick filaments in a sarcomere, this means the thick filaments do not change length when the muscle cell contracts. Researchers realized that the shortening of the I bands (which contain only thin filaments) occurred not because thin filaments contract, but because they slide past the thick filaments, moving deeper into the H zone and decreasing its width. As this occurs, adjacent A bands move closer together, which decreases the width of the I bands. The end result is that the Z lines at either end of a sarcomere move closer together, and thus the sarcomere shortens. As sarcomeres shorten, myofibrils also shorten, as do muscle fibers and ultimately whole muscles. In other words, muscles contract because the thick and thin filaments of the myofibrils slide past each other. Appropriately, this is called the **sliding-filament model** of muscle contraction.

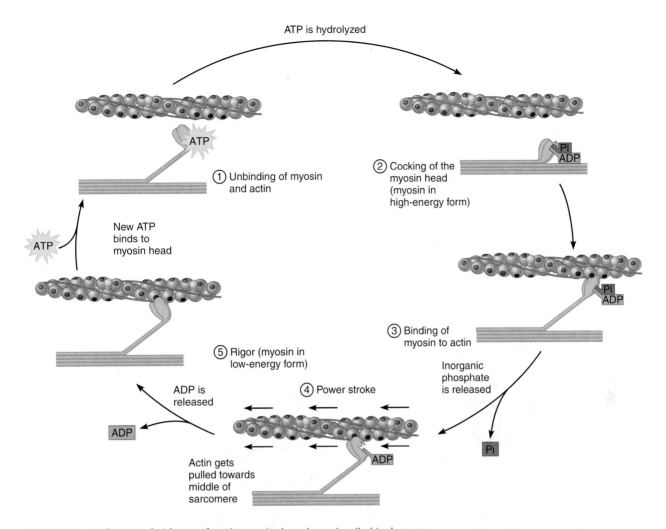

ATP is hydrolyzed

① Unbinding of myosin and actin

New ATP binds to myosin head

ATP

② Cocking of the myosin head (myosin in high-energy form)

Pi ADP

Pi ADP

③ Binding of myosin to actin

⑤ Rigor (myosin in low-energy form)

Inorganic phosphate is released

ADP is released

④ Power stroke

ADP

Actin gets pulled towards middle of sarcomere

ADP

Pi

FIGURE 11.7 The crossbridge cycle. *The steps in the cycle are described in the text.*

The Crossbridge Cycle: How Muscles Generate Force

Knowledge of the fact that thick and thin filaments slide past one another during muscle contraction does not tell us anything about the mechanism that drives the sliding motion and powers the contraction, which is referred to as the **crossbridge cycle** (Figure 11.7). At the heart of this mechanism is an oscillating, back-and-forth motion of myosin crossbridges that is powered by ATP hydrolysis. Coupled with this is cyclic binding and unbinding of crossbridges to the thin filaments, which occurs in such a way that the motion of the cross-bridges pulls the thin filaments toward the center of the sarcomere.

The back-and-forth movement of crossbridges is due to changes in the conformation (shape) of the myosin molecules. These conformational changes not only cause the heads to change position, but also alter both their ability to bind to actin monomers in the thin filaments and the *energy content* of the myosin molecules. One con-formation of myosin is referred to as the *high-energy form*, which is indicated in step 2 of Figure 11.7. Myosin heads

go into this conformation after they hydrolyze ATP. It is called the *high-energy form* because the myosin molecule stores energy that is released in the hydrolytic splitting of ATP. Myosin heads go into the other conformation, called the *low-energy form*, after the stored energy is released to drive the movement of the thin filaments.

Each crossbridge cycle involves the following five steps (see Figure 11.7):

① *Unbinding of myosin and actin.* We start with the myosin head in its low-energy form and bound to an actin monomer. ATP enters the ATPase site on the myosin head, triggering a conformational change in the head that causes it to detach from the thin fila-ment. Although bound to ATP, myosin remains in the low-energy form.

② *Cocking of the myosin head.* Soon after it binds to myosin's ATPase site, ATP is split by hydrolysis into ADP and P_i (inorganic phosphate), which releases en-ergy. Some of this energy is captured by the myosin molecule, which then goes into its high-energy con-formation. Although ATP has been hydrolyzed at this point, the end-products of the reaction (ADP and P_i) remain bound to the ATPase site.

③ *Binding of myosin to actin.* The myosin head binds to an actin monomer in the adjacent thin filament, which triggers the release of P_i from the ATPase site.

④ *Power stroke.* The release of P_i allows the myosin molecule to return to its low-energy form. As it does so, the myosin head pivots toward the middle of the sarcomere, pulling the thin filament along with it. During this process, the myosin molecule releases the ADP from its ATPase site, making the site available for the future binding of another ATP.

⑤ *Rigor.* In releasing its stored energy, the myosin head returns to its low-energy form. At this point, myosin and actin are tightly bound together and unable to separate, a condition called *rigor* (after *rigor mortis,* the stiffening of the body that occurs after death because the crossbridge cycle gets stuck at this step due to a lack of ATP; rigor mortis continues until enzymes leaked by disintegrating cellular components begin to break down the myofibrils). In the crossbridge cycle in living muscle cells, the binding of a "new" ATP molecule to the myosin head allows it to detach from the thin filament, thereby breaking the rigor.

The crossbridge cycle starts again when another "new" ATP molecule binds to the myosin head, triggering unbinding. The entire sequence of steps then repeats: The myosin binds to another actin monomer in the thin filament, pulls it toward the middle of the sarcomere, and so on.

Although a given crossbridge generates force only part of the time while it is active (during the power stroke), a muscle cell generates force continually during a contraction, because many crossbridges go through the cycle simultaneously but out of phase ("out of step") with each other. Thus at any given time, some crossbridges are starting the cycle, others are finishing it, and still others are at various stages in between. To see the significance of this, consider what happens when you walk: Your legs move back and forth (as crossbridges do during a muscle contraction), but your body moves forward in a smooth, continuous fashion. As one leg moves backward, it pushes you forward; this is the "power stroke." When you lift that leg up and move it forward, it does not propel you but instead simply returns to its original position; let us call this the "return stroke." When you walk, you move forward smoothly because one leg goes through the power stroke while the other is going through the return stroke.

Because the crossbridges at opposite ends of a thick filament are oriented in opposite directions from each other (see Figures 11.3 and 11.5), the power strokes of crossbridges at opposite ends of the thick filament move in opposite directions, pulling the thin filaments on either side of the A band in toward the center and causing the sarcomere to shorten. When the crossbridge cycle stops and the contraction ends, the thin filaments passively slide back to their original position. During a contraction, each myosin head completes only about five crossbridge cycles per second, but because each thick filament has several hundred heads, thousands of power strokes can occur each second. For this reason, sarcomeres—and entire muscle fibers—can shorten very rapidly, in many cases taking less than a tenth of a second to contract fully.

In theory the crossbridge cycle could go on indefinitely, so long as there is a continual supply of ATP and actin sites are available for binding with myosin. But what initiates the cycle, and what keeps the cycle from occurring when a muscle is relaxed? In the next section we see that the answer to both questions lies in the actions of the regulatory proteins troponin and tropomyosin.

Quick Test 11.2

1. Which of the following shortens during a muscle contraction: thick filaments, thin filaments, A bands, I bands, H zones, sarcomeres? (choose all that apply)

2. To what does the sliding-filament model refer?

3. When ATP is hydrolyzed, myosin crossbridges change conformation. Do they go into the high-energy form or the low-energy form?

4. What triggers the power stroke of the crossbridge cycle?

Excitation-Contraction Coupling: How Muscle Contractions Are Turned On and Off

We saw in Chapter 10 that the central nervous system ultimately controls skeletal muscle contractions, with *motor neurons* delivering to the muscles commands telling them when and when not to contract. We saw that input from motor neurons always has an excitatory effect on muscle cells and serves to trigger contraction of those cells. Like neurons, muscle cells are *excitable,* meaning that they are capable of generating action potentials if their plasma membranes are depolarized to a sufficient degree. When a muscle cell receives input from a motor neuron, the cell depolarizes and fires an action potential that then stimulates contraction. The sequence of events that links the action potential to the contraction is referred to as **excitation-contraction coupling** (Figure 11.8).

The Role of the Neuromuscular Junction in Excitation-Contraction Coupling

We have seen that the connection between a motor neuron and a muscle cell, referred to specifically as a *neuromuscular junction,* is fundamentally no different from an "ordinary" synapse between two neurons in the nervous system. The motor neuron (the presynaptic cell) transmits an action potential and secretes the neurotransmitter *acetylcholine* upon its arrival at the axon terminal (step 1, Figure 11.8). After release, acetyl-

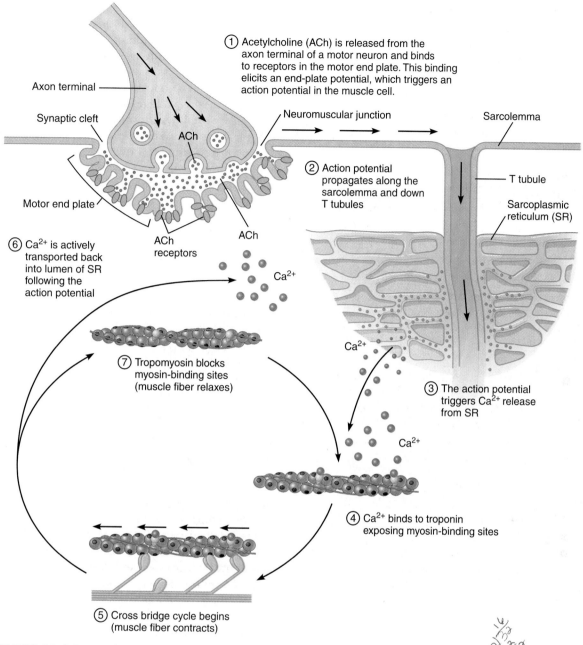

① Acetylcholine (ACh) is released from the axon terminal of a motor neuron and binds to receptors in the motor end plate. This binding elicits an end-plate potential, which triggers an action potential in the muscle cell.

Axon terminal

Synaptic cleft

ACh

Neuromuscular junction

Sarcolemma

Motor end plate

② Action potential propagates along the sarcolemma and down T tubules

T tubule

Sarcoplasmic reticulum (SR)

ACh receptors

ACh

⑥ Ca^{2+} is actively transported back into lumen of SR following the action potential

Ca^{2+}

⑦ Tropomyosin blocks myosin-binding sites (muscle fiber relaxes)

Ca^{2+}

③ The action potential triggers Ca^{2+} release from SR

Ca^{2+}

④ Ca^{2+} binds to troponin exposing myosin-binding sites

⑤ Cross bridge cycle begins (muscle fiber contracts)

FIGURE 11.8 Events in excitation-contraction coupling. *Contraction of a skeletal muscle fiber is initiated and maintained by the arrival of action potentials at the axon terminal of a motor neuron. Upon the cessation of action potentials, the contraction stops following the transport of calcium back into the SR, and the muscle fiber relaxes.*

What enzyme breaks down acetylcholine after it is released?

choline diffuses to the plasma membrane of the muscle cell (the postsynaptic cell), where it binds to specific receptors, triggering a change in ion permeability that results in a change in membrane potential—specifically, depolarization.

Despite its similarity to an ordinary synapse, recall that a neuromuscular junction has many characteristics that make it special (see Chapter 10). Although a motor neuron typically branches and innervates more than one

muscle cell, each muscle fiber receives input from only one motor neuron. At the neuromuscular junction, the motor neuron's terminal boutons fan out over a wide area of the sarcolemma. Opposite these boutons is a specialized region of the sarcolemma called the *motor end plate,* which is highly folded and contains a high density

Acetylcholinesterase

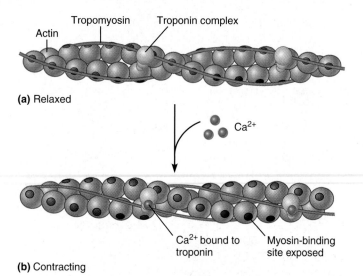

(a) Relaxed

Tropomyosin   Troponin complex
Actin

Ca²⁺

Ca²⁺ bound to   Myosin-binding
troponin   site exposed

(b) Contracting

FIGURE 11.9 Actions of troponin and tropomyosin in excitation-contraction coupling. (a) *In relaxed muscle, tropomyosin covers up actin's myosin-binding sites, which prevents the crossbridge cycle from occurring.* **(b)** *Following their release from the sarcoplasmic reticulum, calcium ions bind to troponin, causing a conformational change in the troponin complex that shifts tropomyosin's position on the actin filament and exposes the myosin-binding sites.*

of acetylcholine receptors (see Figure 11.8). An action potential in the motor neuron triggers the release of acetylcholine from each of its many terminal boutons, which causes many acetylcholine receptors to become activated. As a consequence, the resulting depolarization (called an *end-plate potential*) is much larger than an ordinary postsynaptic potential—so large, in fact, that it is always above threshold and triggers an action potential in the muscle cell. Thus, if an action potential occurs in a motor neuron, it is always followed by an action potential in the muscle cells it innervates (step 2, Figure 11.8). Once an action potential is initiated in a muscle cell, it propagates through the entire sarcolemma and also travels through the T tubules, which are essentially extensions of the sarcolemma. As the action potential travels through the T tubules, it triggers the release of calcium from the nearby sarcoplasmic reticulum (step 3, Figure 11.8). This calcium then serves as the signal that initiates the crossbridge cycle and, hence, contraction of the muscle cell. (When It Goes Wrong: Muscular Atrophies and Dystrophies, www.physiologyplace.com, Challenge Yourself)

The Roles of Calcium, Troponin, and Tropomyosin in Excitation-Contraction Coupling

When a muscle cell is relaxed, the concentration of calcium in the cytosol is very low, and little binding occurs between calcium and troponin. Troponin is in its normal (resting) conformation and because of this, tropomyosin is positioned on the thin filaments in such a way that it blocks actin's myosin-binding sites, so the crossbridge cycle cannot occur (Figure 11.9a). The cytosolic calcium level is normally low because the membrane of the sar-

coplasmic reticulum (SR) is equipped with pumps that actively transport calcium ions into the lumen from the cytosol. Due to the action of these pumps, the SR is able to accumulate calcium against a concentration gradient and therefore function as a calcium storehouse.

In addition to calcium pumps, the membrane of the SR also contains voltage-gated calcium channels that are normally closed, which prevents calcium inside the SR from leaking out. When an action potential travels through the T tubules, however, it causes these channels to open briefly, allowing calcium to flow out into the cytosol. The end result is a rise in the cytosolic calcium concentration.

Whereas the calcium channels that allow Ca²⁺ out of the SR are voltage-gated, it is an unusual kind of voltage-gating because the electrical signal that triggers gating occurs in the membrane of the T tubule—not the membrane of the SR itself. An action potential in the T tubule can trigger the release of Ca²⁺ from the SR because adjacent T tubules and SR membranes are physically linked by proteins called *foot structures* (or *ryanodine receptors*) that bridge the gap between them and also function as calcium channels. Where the foot structures come into contact with the T-tubule membranes are other proteins (called *dihydropyridine receptors*) that function as voltage sensors. When an action potential travels through the T tubules, these voltage sensors react (presumably by undergoing a conformational change) and in so doing transmit a signal directly to the foot structures with which they are in contact. This signal triggers the opening of calcium channel pores within the foot structures, which allows Ca²⁺ to flow out of the SR. As the ions enter the cytosol, some of them bind to specific sites on other SR calcium channels and cause them to open. In this manner the initial release of calcium triggers the release of even more calcium from the SR.

As the cytosolic calcium concentration rises, some of it binds to one of the three proteins making up each troponin complex (step 4 in Figure 11.8, and Figure 11.9b), which then undergoes a conformational change that causes tropomyosin to shift out of its normal resting position, thereby exposing the myosin-binding sites on the actin monomers. With the myosin heads of the thick filament now able to bind to actin, the crossbridge cycle can begin (step 5, Figure 11.8), and the sarcomere contracts. The design of the SR and T-tubule network surrounding the myofibrils permits nearly simultaneous delivery of calcium to all sarcomeres of a muscle fiber, so the sarcomeres contract in unison, as does the entire muscle fiber.

A muscle cell stops contracting when it no longer receives input from its motor neuron, and action potentials no longer occur in the sarcolemma. When an action potential triggers the release of calcium from the SR, this release does not continue indefinitely because as the cytosolic calcium concentration rises, calcium ions begin to bind to certain sites on the SR calcium channels, causing them to close. (These sites are distinct from those that trigger channel opening and have a lower affinity for calcium, so they do not come into play until cytosolic calcium has risen to a sufficiently high level). The closure of these channels turns off the release of calcium and enables the active transport of calcium back into the SR (an ongoing process) to clear calcium from the cytosol (step 6 in Figure 11.8), which causes the calcium concentration to fall. Because the binding of calcium to troponin is reversible, this concentration change causes calcium to dissociate from troponin, which allows both troponin and tropomyosin to revert to their original positions (step 7 in Figure 11.8). The number of exposed sites on the actin filament therefore decreases, leading to a decline in the number of active crossbridges. Eventually, as calcium concentration returns to normal, the number of active crossbridges reaches zero, and the muscle contraction ends.

Muscle Cell Metabolism: How Muscle Cells Provide ATP to Drive the Crossbridge Cycle

To see that muscles must be able to begin contracting within a fraction of a second after receiving neural input and must sometimes be capable of sustaining contractile activity for a time, we need only think of a competitive sprinter, whose muscles begin to work near maximum capacity just after the starting gun goes off. For muscles to perform in this manner, it is necessary that muscle cells always have a readily available supply of ATP, even when the demand for ATP increases suddenly and rapidly.

The Role of the Creatine/Creatine Phosphate System

When a muscle fiber is resting, its demand for ATP is small, but when it is signaled to contract the demand for ATP soars. At rest, a muscle cell contains a small store of ATP, but it cannot rely on this ATP for long once it begins contracting. To keep from depleting its ATP supply, it must "gear up" ATP production to keep pace with the increased rate of utilization.

The ATP that powers muscle contraction is produced in muscle cells, as in other cells, by substrate-level phosphorylation and oxidative phosphorylation. When a cell's rate of ATP utilization increases, the concentration of ATP inside the cell falls and the concentration of ADP rises. These and other changes then stimulate the enzymes that control ATP-producing reactions, such that ATP is generated at a higher rate. But even though this occurs once a muscle cell begins to contract, these reactions need a few seconds to "come up to speed." To ensure a steady supply of ATP in the meantime, muscles rely on an immediately available store of high-energy phosphate that is present in the form of a compound called **creatine phosphate,** which donates its phosphate to ADP (which is always present) to form ATP. A resting cell contains enough creatine phosphate to supply four to five times the quantity of preformed ATP, which is sufficient to "tide the cell over" until the other ATP-producing reactions can take over.

The reaction of creatine phosphate with ADP is catalyzed by the enzyme *creatine kinase* and is reversible:

$$\text{creatine phosphate} + \text{ADP} \xrightleftharpoons[]{\text{creatine kinase}} \text{creatine} + \text{ATP}$$

Thus this reaction generates ATP and **creatine** as endproducts when it proceeds to the right, and generates ADP and creatine phosphate when it proceeds to the left. When a muscle cell is at rest, the reaction is at equilibrium, and for every molecule of creatine phosphate formed, another is broken down to creatine. When muscle activity begins, however, ATP levels fall and ADP levels rise, which drives the reaction to the right by mass action. As a result, some ADP is converted to ATP that can be used in the crossbridge cycle, but creatine phosphate is also consumed. Because the supply of creatine phosphate is limited, this reaction can produce ATP for only a short while, but this is sufficient to give ATP-producing metabolic reactions time to gear up. When the muscle cell relaxes, the supply of creatine phosphate is replenished because the reduced demand for ATP causes the ATP concentration to rise and the ADP concentration to fall, which pushes the reaction to the left such that creatine is converted back to creatine phosphate. Thus creatine phosphate remains "on tap" for the next burst of contractile activity.

How Muscle Cell Metabolism Changes with Exercise Intensity

When a muscle is exercised continuously at a moderate rate (such that the oxygen supply is sufficient to keep up with demand), most of the needed ATP is supplied by oxidative phosphorylation. For the first few seconds of exercise, muscles rely on their own stored glycogen to supply glucose as fuel for ATP production. As exercise continues, they come to rely on glucose and fatty acids delivered to them by the bloodstream. After about 30 minutes, glucose utilization decreases, and fatty acids become the dominant energy source.

In heavy exercise, oxidative phosphorylation becomes less important as a source of ATP, and substrate-level phosphorylation (particularly glycolysis) becomes more important, for reasons that will be explained later in the chapter. Even though glycolysis is capable of producing ATP all by itself, one result is the production of lactic acid, which accumulates in muscle tissue and can spill over into the bloodstream. (Recall that in glycolysis, glucose is converted to pyruvate, which normally undergoes further oxidation in the Krebs cycle and oxidative phosphorylation; when pyruvate is produced at a rate exceeding the rate at which it is oxidized, it builds up and is converted to lactic acid.) This buildup of lactic acid is believed to be the reason for the burning sensation one feels in the muscles after heavy exercise.

So far, we have seen how muscle cells respond to neural signals, how they use ATP to power contractions, and how they ensure an adequate supply of ATP. But what makes one muscle stronger than another? How is it possible that we can use a given group of muscles, such as those of the arm and hand, to lift a chair one moment and a paper clip the next? We address these and related issues in the following section, which focuses on the *mechanics* of skeletal muscle contraction—the various factors that influence contractile speed and force.

Quick Test 11.3

1. Define the following terms: *neuromuscular junction, motor end plate, end-plate potential.*

2. What is the function of the sarcoplasmic reticulum in skeletal muscle?

3. How does an increase or a decrease in the cytosolic calcium level cause the crossbridge cycle to turn on or off?

4. How does creatine phosphate help to maintain an adequate supply of ATP at the start of a muscle contraction?

THE MECHANICS OF SKELETAL MUSCLE CONTRACTION

When you use a given set of muscles in different activities, the contractions of those muscles vary in force and duration. As you lift a chair, for example, your arm muscles must exert more force than when you lift a paper clip. When you throw a baseball, muscle contractions are brief and explosive, but when you carry a suitcase, the contractions are smooth and sustained. In view of this great variability, you may be surprised to learn that when a muscle cell contracts in response to a single action potential, the result is always the same: Within a fraction of a second the force rises to a maximum and then falls to zero nearly as quickly. This event, called a *twitch,* is like an action potential in that it is a reproducible, all-or-nothing event; if the cell is stimulated again, it gives the identical response. Surprising as it may seem, all skeletal muscle contractions—whether strong, weak, short, or long—are built upon this simple muscle twitch, which we discuss next.

The Twitch

We saw in Chapter 10 that the axons of motor neurons typically branch before reaching their target cells, so that one neuron innervates several muscle fibers. Thus an action potential in a motor neuron triggers contraction of *all* the muscle cells that are connected to that neuron, and it is not possible to stimulate one cell without stimulating the others. As we also saw, a motor neuron and all

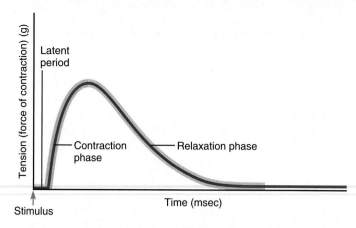

FIGURE 11.10 Phases of a twitch. *The tension that is actively generated by a muscle in response to a single stimulus is plotted as a function of time.*

the muscle fibers it innervates is referred to as a **motor unit.** A **twitch** is the mechanical response of an individual motor unit (or muscle cell) to a single action potential.

Phases of the Twitch

With the possible exception of an eye blink, a single isolated twitch exists only under artificial conditions produced in the laboratory. To observe twitches, a muscle can be surgically removed from an animal and anchored in an apparatus that permits contractile force to be measured and displayed on an oscilloscope, strip chart recorder, or other recording device. The muscle is then stimulated electrically, which causes individual muscle cells to have action potentials at the same time and contract together. When each end of a muscle is anchored in place, a single brief stimulus applied, and the contractile force recorded, the force first rises, then falls. When the force tracing is observed, it is seen to consist of a latent period followed by a contraction phase and a relaxation phase (Figure 11.10). The **latent period** is the delay of a few milliseconds that occurs between the action potential in a muscle cell and the start of contraction, when the cell first begins to generate force. This time lag exists because the events of excitation-contraction coupling must occur before crossbridge cycling—and hence force generation—can begin. The **contraction phase,** which can range from 10 milliseconds to over 100 milliseconds for different muscles, starts at the end of the latent period and ends when muscular tension peaks. (*Tension* is synonymous with *force* and is commonly expressed in units of mass, such as grams; muscle tension of 1 gram is equivalent to the force generated by a 1-gram weight hanging on a string.) The **relaxation phase,** typically the longest of the three phases, is the time between peak tension and the end of the contraction, when tension returns to zero.

As mentioned, a characteristic feature of a muscle twitch is its reproducibility; that is, repetitive stimulation of a muscle produces several twitches in a row, each having the same magnitude and shape. (Later we will see that this is not true when the frequency of stimulation is

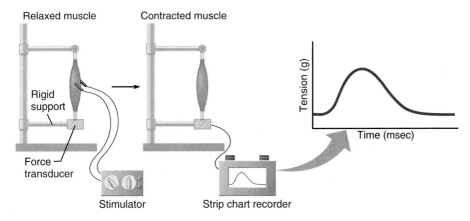

(a) Isometric muscle contraction

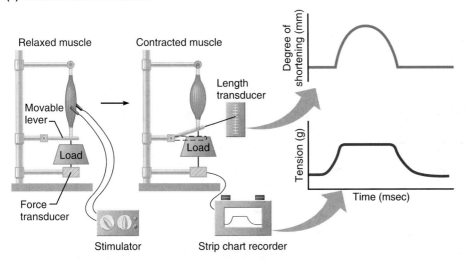

(b) Isotonic muscle contraction

FIGURE 11.11 **Isometric and isotonic muscle contractions. (a)** *An isometric contraction, in which a muscle mounted with both ends firmly anchored cannot shorten upon contraction. The result of stimulation is an isometric twitch, in which tension (force) rises to a peak and then declines.* **(b)** *An isotonic contraction, in which contraction of the muscle is allowed to lift a load. The result of stimulation is an isotonic twitch, in which tension increases and then plateaus as the muscle begins to shorten. Note that the plateau phase of the tension curve corresponds to the period during which muscle shortening occurs.*

high enough such that the twitches follow one another closely.) The reason for this reproducibility relates to the all-or-nothing character of the muscle cell's action potentials: Assuming that over time no changes occur in the cell's properties (such as occurs with fatigue, discussed later), an action potential always triggers the same degree of calcium release from the SR, which causes the same rise in cytosolic calcium levels, activates the same number of crossbridges, and produces the same force.

Note that although twitches are reproducible for any given muscle or muscle fiber, twitches vary considerably from one muscle or muscle cell to another. One reason is that certain muscle fibers are inherently stronger than others—that is, are capable of generating more force; larger-diameter fibers can exert more force than smaller-diameter fibers (discussed later in this chapter). Another reason is that muscle fibers differ in the speed with which they can achieve peak force; differences in these so-called *fast-twitch fibers* and *slow-twitch fibers* are also discussed later in the chapter.

Isometric and Isotonic Twitches

There are two variations on the twitch, known as *isometric* and *isotonic* twitches, which differ not in the basic mecha-

nism of force generation, but in whether or not the muscle is allowed to shorten as it contracts. When a muscle contracts isotonically, it generates a tension at least equal to any forces opposing it (called *loads;* an example is the weight of a barbell), and so the muscle shortens. When a muscle contracts isometrically, it creates tension but does not shorten because the load is greater than the force generated by the muscle. This occurs, for example, when you try to lift an object that is too heavy for you to move, or when you stand still and your postural muscles hold your body upright. In the latter instance, the muscles do not shorten because they are held in place by bones, which are stationary under those conditions.

Figure 11.11 shows how isometric and isotonic contractions are measured in a laboratory. In both cases, twitches are measured as the response of a whole muscle to a single stimulus; one end of the muscle is anchored while the other end is attached to a device (a force transducer) that detects mechanical force and displays it on a strip chart recorder. To measure an **isometric twitch,** both ends of the muscle are rendered immobile. When the muscle is electrically stimulated, it develops tension but cannot shorten (Figure 11.11a). The force tracing shows that tension rises to a peak and then declines to the resting level. To measure an **isotonic twitch,** the lower end

FIGURE 11.12 Effect of load on peak tension in an isotonic twitch. *The responses to stimulation of a muscle subjected to four different loads are plotted. The muscle contracts isotonically when the loads are 5, 10, or 15 grams, and in each of these cases muscle tension plateaus at a level equal to the load. When the load is 20 grams, however, the muscle is unable to shorten because the load exceeds the maximum tension it can generate in a single twitch (17 grams), and thus the muscle contracts isometrically. Note that as load increases, the duration of muscle shortening decreases, and that the latent period before the onset of muscle shortening increases.*

If a 16-gram load were placed on the muscle, would the contraction be isotonic or isometric?

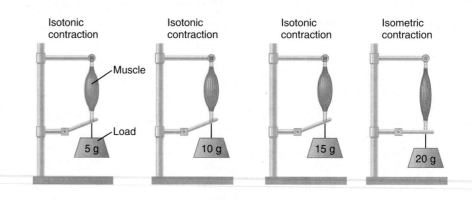

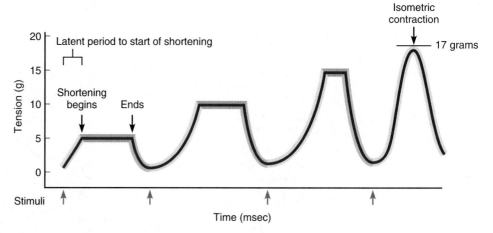

of the muscle is not anchored in place but is instead attached to a moveable load, such that the muscle is able to shorten when it contracts (Figure 11.11b). Note that the curve for the isotonic twitch shows a distinct plateau, indicating that the force is constant for a period of time (hence the name *isotonic,* which means "same tension"). It is during this plateau phase that the muscle shortens and the load moves. Before the plateau phase the force increases but the muscle does not shorten because it is not yet generating enough force to lift the load. Only when the force becomes equal to the load does the muscle begin to shorten. This force remains constant so long as the load is moving, but eventually the muscle starts to relax and the load starts to fall. When the load comes to rest, the plateau phase ends and the force begins to decline.

Unlike an isometric twitch, an isotonic twitch is not an all-or-nothing event—its size and shape depend on the size of the load that is placed on the muscle. When the load is increased, for example, greater tension is needed to overcome it, and for this reason the force tracings show plateaus at higher tension levels (Figure 11.12). At the same time, the time lag (latent period) between the stimulus and the beginning of muscle shortening (the

start of the plateau) also increases, because it takes the muscle longer to develop the force required to move the load. When the load exceeds the amount of force the muscle can generate, the muscle cannot move it and therefore contracts isometrically. Under these conditions, the force tracing has the rounded peak characteristic of an isometric twitch (Figure 11.12, far right).

When a muscle contracts isometrically, its sarcomeres shorten even though the whole muscle does not. This is possible because the sarcomeres (collectively referred to as the muscle's **contractile component,** CC) do not extend the entire length of each muscle fiber and therefore do not transmit force directly to the ends of the cells. Instead, the force is transmitted through certain cellular components that connect the myofibrils to the ends of the cells and then through connective tissue that anchors the ends of the cells in place and extends through the tendons. Those parts of a muscle (or muscle cell) that do not actively generate force but only serve to passively transmit force to the ends of the muscle are collectively referred to as the muscle's **series elastic component** (SEC). When a muscle contracts isometrically, the CC shortens and stretches the SEC, causing it to pull on the ends of the muscle. In so doing, the SEC lengthens as the CC shortens, giving an overall length change of zero.

Note that although muscles do contract isometrically in the body, they rarely (if ever) contract in a strictly isotonic fashion—that is, with a constant force as they shorten. When you walk, run, lift objects, or otherwise move such that contracting muscles shorten, your nervous system continually adjusts its input to the muscles to ensure that the muscular force is appropriate for the intended activity. This is quite different from the laboratory situation depicted in Figures 11.11 and 11.12, in which a single invariant stimulus is delivered to a muscle. Furthermore, when a muscle contracts in the body, the load placed on the muscle is rarely constant (as it is in Figures 11.11 and 11.12), even when a constant weight is being lifted. This is the case because the positions of the joints change as you move, and this alters the loads placed on muscles. You can easily demonstrate this by attempting to hold a 10-pound weight steady in your hand under two conditions: with your elbow close to your body and bent at a 90-degree angle, and with your arm extended straight out in front of you. Without a doubt you will find it more difficult to hold the weight steady in the second instance because the extended position of the arm puts greatly increased stress (that is, *load*) on certain muscles of the arm and shoulder.

Factors Affecting the Force Generated by Individual Muscle Fibers

In this section we examine the factors that influence the force generated by the contraction of *individual muscle fibers:* frequency of stimulation, fiber diameter, and changes in fiber length. Note, however, that these are not the only factors that determine the force of muscular contraction, because not all fibers are active at all times; we examine the factors that affect the force generated by whole muscles in the next section.

Frequency of Stimulation

Isometric muscle twitches are in fact reproducible, all-or-nothing events only if a muscle is stimulated at a frequency low enough to ensure that twitches are well separated in time. When a muscle is stimulated at a sufficiently high frequency, such that twitches follow one another closely, peak tension rises in a step-wise fashion with each twitch, until eventually it reaches a constant level. This step-wise increase in peak tension is called *treppe* (Figure 11.13). The cause of treppe is not known, but it may be the result of a higher-than-normal calcium concentration in the cytosol between twitches, perhaps due to incomplete removal by the pumps of the sarcoplasmic reticulum.

Compared to an action potential, which takes at most a few milliseconds to complete, a muscle twitch is fairly slow, taking anywhere from tens to hundreds of milliseconds to complete (Figure 11.14). Because of this, a muscle fiber can have several action potentials in the time it takes to complete one twitch. When a muscle is stimulated repetitively such that additional action potentials

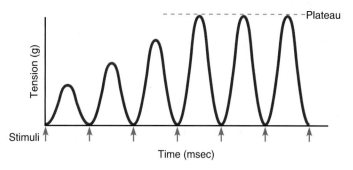

FIGURE 11.13 Treppe. *In response to sufficiently frequent repetitive stimuli (arrows), peak tension in a muscle contracting isometrically rises with each of the first few twitches but eventually reaches a plateau.*

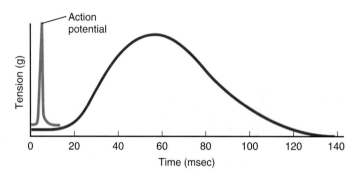

FIGURE 11.14 Duration of an isometric twitch. *The twitch of a muscle fiber lasts significantly longer than the action potential that stimulated it.*

What term is given to the series of events occurring in the time interval between an action potential and the rise in muscle tension?

arrive before twitches can be completed, the twitches superimpose on one another, yielding a force greater than that of a single twitch; this process is called **summation** (Figure 11.15). Summation happens whenever twitches occur so frequently that muscle fibers cannot relax completely between twitches. However, when stimuli are delivered over a prolonged time period, the force does not continue to mount but instead reaches a plateau, during which it oscillates about a constant average level (curve 3 in Figure 11.15). Under these conditions, the muscle is said to be in a state of incomplete (unfused) **tetanus.** The term *tetanus* also refers to a disease in which toxins produced in a bacterial infection cause motor neurons to stimulate muscle contraction inappropriately (see When It Goes Wrong: Tetanus, p. 349).

If the muscle is stimulated with still greater frequency, the individual twitches become indistinguishable because they follow each other so closely. In this case the

Excitation-contraction coupling

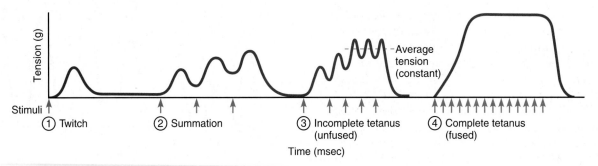

FIGURE 11.15 Effects of high stimulus frequency: summation and tetanus.
In response to repetitive stimuli (arrows) delivered close together in time, muscle twitches superimpose in summation. A train of more frequent stimuli causes tension to rise with each twitch until the muscle reaches incomplete tetanus, characterized by a plateau composed of individually distinguishable twitches. Still greater stimulus frequency produces complete tetanus, in which force increases swiftly and smoothly to a plateau in which individual twitches are no longer distinguishable.

tension rises swiftly and smoothly at the beginning of stimulation and eventually reaches a higher plateau; this condition, shown in curve 4 of Figure 11.15, is called *complete (fused) tetanus*. If the stimulus frequency is increased still further, tetanic tension (that is, the tension generated during tetanus) increases, but only up to a point; further increases in frequency beyond this point yield no further increases in force. Under these conditions the muscle is generating all the force it can, which is referred to as **maximum tetanic tension.**

Fiber Diameter

We accept as a fact of everyday life that some muscles have an inherent ability to generate more force than others. Why else would we equate a weightlifter's bulging muscles with superior strength? The inherent ability of a muscle to generate force is referred to as the muscle's *force-generating capacity*, which is usually assessed by measuring maximum tetanic tension or peak tension in an isometric twitch.

The force-generating capacity of a muscle fiber depends on both the number of crossbridges in each sarcomere and the geometrical arrangement of the sarcomeres. Other things being equal, a muscle whose sarcomeres contain more crossbridges can generate more force, just as the force exerted on the rope in tug-of-war increases when more participants are added. In addition, a muscle that has more sarcomeres—and hence more thick and thin filaments—arranged in parallel can generate more force than a muscle with fewer sarcomeres arranged in parallel. Because the number of thick and thin filaments per unit of cross-sectional area does not vary significantly from one muscle to another, a fiber's *diameter* is the crucial variable that determines its force-generating capacity. The greater the fiber diameter, the greater its cross-sectional area, and the more force it can generate. This is why a weightlifter's

bulging muscles are stronger than the average person's slimmer muscles.

Note that although the number of parallel sarcomeres strongly affects a muscle's force-generating capacity, the force-generating capacity does *not* depend on the number of sarcomeres that are joined in series (end to end). This means that two muscles, one longer than the other but otherwise identical, have the same force-generating capacity. Consider a chain with a weight dangling from one end: Assuming that the weight of the chain itself is negligible, each of its links exerts on its neighbors a force that is equal to the force exerted on the chain by the weight. This will be true regardless of the number of links in the chain.

Changes in Fiber Length

Although a muscle fiber's length does not affect its force-generating capacity insofar as it reflects the number of sarcomeres in series, *changes* in a fiber's length do influence its ability to generate force. For each fiber, maximum force-generating capacity occurs over a certain range of lengths. When a fiber either shortens beyond or is stretched beyond this optimum range, its force-generating capacity decreases, because such changes in length alter the length of individual sarcomeres and reduce their ability to generate force.

Figure 11.16 shows how the force of contraction, measured as percent of maximum tetanic tension or peak tension in an isometric twitch, varies with a muscle's length. The graph is an example of a **length-tension curve,** and its shape is a consequence of the sliding-filament model and the nature of the crossbridge cycle. We have seen that the force of a muscle contraction is related to the number of active crossbridges—the greater this number, the greater the force—and that crossbridge activity requires that myosin crossbridges be able to bind to

Just a few decades ago, U.S. schoolchildren faced the very real possibility of contracting any one of several severe, often-fatal diseases that have since largely disappeared from public consciousness thanks to widespread immunization programs. One of these diseases—*tetanus*, characterized by severe muscle spasms and convulsions—now affects fewer than 100 Americans each year.

Tetanus results from infection with *Clostridium tetani*, an anaerobic bacterium whose spores are found in soil and animal feces. The disease most commonly results from contamination of deep puncture wounds or burns, although it can occur after relatively minor wounds. Deep wounds are most typically involved because the anaerobic bacteria grow well in the oxygen-poor conditions found in such wounds. Symptoms of the disease appear after an incubation period of between 2 days and 50 days, although 5–10 days is most common.

Symptoms of the disease result from the action of a toxin (known as *tetanospasmin*) secreted by the bacteria at the wound site. The toxin exerts its effects after reaching the CNS, where it binds to synaptic terminals and blocks the transmission of signals that normally inhibit the activity of motor neurons. As a consequence, motor neurons become hyperexcitable, leading to inappropriate stimulation of skeletal muscles. Depending on the extent of the toxin's spread, effects may be localized to muscles in the vicinity of the wound or may involve muscles all over the body. Local effects include soreness and increased muscle tone in the affected area, whereas systemic effects include spastic movements and intermittent convulsions.

The most frequent early systemic symptom of tetanus is stiffness of the jaw. Commonly, stiffness spreads to other areas, including the neck, arms, or legs and may be accompanied by sore throat, headache, fever, restlessness, irritability, or difficulty in swallowing. As the disease progresses, muscle spasms may cause a patient to experience difficulty opening the jaw (*trismus*), which accounts for the disease's common name—*lockjaw*. Other facial muscles may become affected, causing the patient to take on a bizarre facial expression characterized by a fixed smile and raised eyebrows (*risus sardonicus*). Death, if it comes, normally results from asphyxiation due to spasm of laryngeal or thoracic muscles.

The low incidence of tetanus in the United States is largely a result of mass immunization programs involving preschool children. Immunity is conferred by injection of inactivated tetanus toxin, toxoid, which stimulates the body to produce antibodies against the toxin. The tetanus vaccine is usually combined with vaccines against diphtheria and pertussis (whooping cough) to form a *DPT shot*. To maintain immunity, booster shots must be given every ten years. For unimmunized individuals who sustain a puncture wound, some protection can be conferred by injection of antitoxins (antibodies produced against the toxin in vaccinated humans or animals). The mortality rate for unimmunized persons who contract the disease is estimated to be about 60%.

Barring effective prophylaxis, measures that can manage the symptoms include administration of drugs that interfere with neuromuscular transmission and thus reduce or eliminate muscle spasms. In all cases, adequate air flow to the lungs must be ensured, usually by the insertion of a breathing tube into a surgically created opening in the trachea. If neuromuscular block is strong enough to interfere with normal breathing, the patient must be placed on an artificial ventilator.

actin. If we start with a muscle fiber at optimum length and then stretch it, the tension generated by the fiber decreases linearly as its length increases (segment b–c in Figure 11.16). The reason is that the degree of overlap between thick and thin filaments decreases as sarcomeres lengthen, and crossbridges that are not overlapped by thin filaments cannot bind to actin and therefore cannot generate force.

If the muscle is significantly shorter than its optimum length, tension declines in two stages as the length decreases. At first tension decreases relatively slowly, but once the length decreases beyond a certain point (point a in Figure 11.16), tension decreases more rapidly. The decrease in tension that occurs with decreasing length is not the result of any change in the force generated by crossbridges,

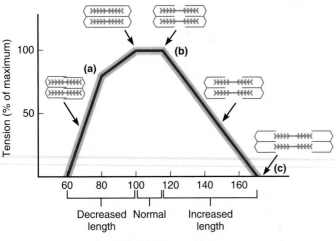

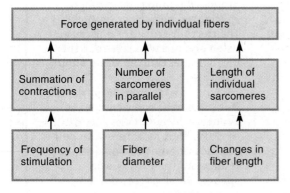

FIGURE 11.16 **A length-tension curve.** *The curve plots the tension developed by a muscle as a function of the muscle's resting length. The center bracket indicates the normal range of muscle lengths in situ. Drawings of sarcomeres indicate changes in the relative positions of thick and thin filaments as muscle length changes.*

FIGURE 11.17 **Factors affecting the force generated by individual muscle fibers.**

because the crossbridges are completely overlapped by thin filaments, and all are active. However, as sarcomeres shorten beyond their optimum length, the thin filaments at opposite ends of the sarcomere begin to overlap each other, which interferes with their movement. Then, as the sarcomeres shorten beyond point a in Figure 11.16, the Z lines eventually come into contact with thick filaments, so most of the force generated by crossbridges is exerted on the sarcomere itself instead of being transmitted to the ends of the muscle fiber.

The constant force-generating capacity of a muscle fiber at the peak of the length-tension curve in Figure 11.16 occurs for two reasons: All crossbridges are overlapped by thin filaments and therefore are capable of generating force, and sarcomeres are not short enough to allow thin filaments to come into contact, so interference between them does not occur.

The fact that muscles shorten when they contract does not mean that muscles in the body compromise their force-generating capacity when they contract. Note the bracket in Figure 11.16 indicating the normal operating range of muscles in situ (in the body), which shows that muscles in situ are always close to their optimum lengths, even when they have been maximally shortened or stretched. The reason is that the movement of a muscle is constrained by the bones to which it is attached; muscles can lengthen or shorten only so far because the possible range of joint angles is limited by the architecture of the skeleton. As a result, muscles usually operate within the range of lengths in which they can generate maximum force.

Thus far we have seen that the force of muscle contraction is influenced by three factors acting on individual fibers (Figure 11.17): the frequency of stimulation, which alters force through the summation of twitches; fiber diameter, which affects force by virtue of its relationship to the number of sarcomeres working in parallel; and changes in fiber length, which affect force by altering the tension generated by individual sarcomeres. When you use a particular set of muscles for different activities—for example, using your arm to pick up a chair, as opposed to picking up a paper clip—the muscles' contractile force varies because the nervous system alters the pattern of commands it sends to them. One way it does this is by varying the action potential frequency in individual motor units; another way is by varying the number of motor units stimulated at any given time. In the next section we explore these issues further.

Regulation of the Force Generated by Whole Muscles

Based on what we have seen concerning summation of contractions, we know that the action potential frequency in a muscle's motor units has a direct bearing on the force the muscle generates. The most force a single muscle fiber can develop contracting isometrically is the maximum tetanic tension, which for most muscle fibers is only about five times greater than peak tension in a single twitch. Given that muscular tension can vary over several orders of magnitude (for example, the forces required to hold a paper clip compared to that required to hold a chair), it is clear that variation in action potential frequency can account for only a small fraction of the range of forces a muscle can generate.

When a muscle contracts, only rarely do *all* of its fibers actively generate force. Some motor units are active, but the fibers in other motor units simply "go along for the ride," passively shortening in response to forces generated by actively contracting fibers. When larger forces are needed, the nervous system can activate some of these "extra" fibers, thereby increasing the total number of active fibers. Indeed, the nervous system exerts most of its control over muscular force by varying the number of active motor units; variation in the frequency of stimulation of individual fibers plays a secondary role. An increase in the number of active motor units is called **recruitment.**

Recruitment

We have seen that within a muscle, fibers belonging to a given motor unit are intermixed with fibers from other motor units. But not all motor units are created equal; they often differ in size, with some having relatively more fibers and others having relatively few. Figure 11.18b shows two motor units (X and Y) in a muscle, the fibers of which have identical force-generating properties and have been stimulated to give maximum tetanic tension; motor unit X contains five fibers, whereas motor unit Y contains seven fibers. When motor unit X is stimulated to contract, it generates five times the force of a single fiber because its fibers are working together in parallel; likewise, stimulation of motor unit Y results in a force seven times that of a single fiber, and stimulation of both motor units produces a force 12 times greater than that of a single fiber (Figure 11.18c). Because a muscle may contain hundreds of motor units, muscular tension can be varied over a wide range merely by varying the number of active motor units.

Muscles differ in regard to the numbers of motor units they contain, from a handful in the muscles that control movements of the eyes to hundreds in larger muscles such as the biceps. Within a given muscle, the various motor units differ in both the numbers of fibers they possess and in the diameter and strength of those fibers. The fibers within any given motor unit tend to fall within a narrow range of sizes, with some consisting mostly of small fibers and others consisting mostly of large fibers. Furthermore, motor units that have larger fibers also tend to have *more* fibers.

The Size Principle

When a muscle is called upon to generate small forces, generally only the smaller motor units come into play; when larger forces are needed, larger motor units are recruited. This correspondence between the size of motor units and the order of recruitment is known as the **size principle.** In addition, when contractions are sustained over a long time, motor units are activated asynchronously—as one becomes active, another ceases its activity. In this manner the total force of the muscle is maintained at a constant level without overworking any of the individual motor units.

The fact that motor units differ in size has practical implications for precise control of muscular force. As a general rule, fine control is easier when muscular forces are small, because only the smaller motor units are re-

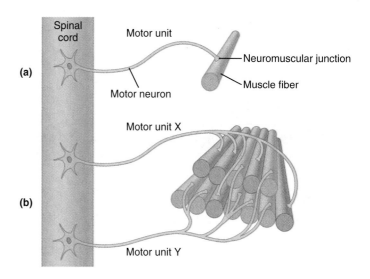

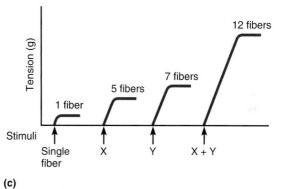

FIGURE 11.18 Increases in force generation with recruitment of motor units. (a) *A hypothetical motor unit consisting of a motor neuron and a single muscle fiber.* **(b)** *Motor units X and Y, which possess five fibers and seven fibers, respectively.* **(c)** *Tension developed by the single fiber, by motor unit X, by motor unit Y, and by motor units X and Y together.*

cruited. Smaller variations in force are possible under these conditions because the recruitment of additional motor units causes only small increases in the total number of active fibers. In contrast, when large forces are involved, only larger increments in force are possible because larger motor units are recruited.

The basis for the size principle is not only that motor units vary in size, but that the *motor neurons* that control them also vary in size. Larger motor units are controlled by motor neurons with larger-than-average cell bodies and axon diameters, whereas smaller motor units are controlled by neurons with smaller-than-average cell bodies and axon diameters. This distribution of neuron sizes has important consequences for muscular control. For complicated reasons, larger cells are harder to depolarize to threshold; more excitatory synaptic input (a higher action potential frequency in the presynaptic cell) is required to induce a larger neuron to fire. Thus, when gradually increasing synaptic input is delivered to a set of motor neurons, the small neurons will fire first and the large ones last.

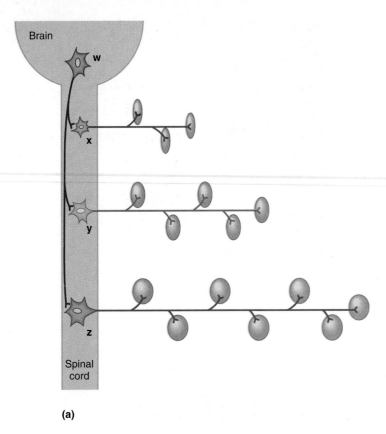

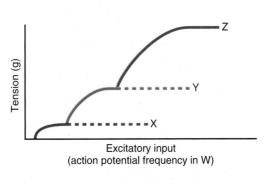

(a)

(b)

FIGURE 11.19 The size principle. (a) *The anatomical relationship of three motor units (X, Y, and Z) of increasing size to an excitatory neuron W within the CNS.* **(b)** *As the frequency of action potentials in neuron W increases, the order of motor units activated proceeds from smallest (X) to largest (Z).*

This idea is illustrated schematically in Figure 11.19, in which small, medium, and large motor units (X, Y, and Z, respectively) are controlled by excitatory input from a neuron (W) located in the CNS. Note that as the action potential frequency in neuron W increases, the motor units become active in order of increasing size. Note as well that as each motor unit is recruited, the force increases in stepwise increments that reflect both the increased number of fibers in the larger units and the larger size of those fibers.

Exercise Link

Bill experienced a consequence of the size principle in the late stages of the marathon. For the first 20 miles or so, recruitment of small- and medium-size motor units was sufficient to produce the force in his leg muscles needed to propel him onward. During this time, if fibers in these motor units became temporarily fatigued, other small- or medium-size motor units could be recruited to "share the burden." But after about 2 hours of constant activity, *all* of the small- and medium-size motor units were essentially exhausted, so more and more larger motor units had to be called upon. As a result, Bill experienced a greater perception of effort (more excitatory synaptic input was needed), and he ran with jerky, stiff-legged strides because the stepwise increments in force were too large to allow controlled, fluid movement.

Now we can see that the force generated by whole muscles is affected by a combination of the factors acting on individual fibers and the number of fibers that are active (Figure 11.20).

But more than just the force of muscle contraction is important in movement; the speed with which muscles contract is important as well.

Velocity of Shortening

To determine a muscle's velocity of shortening, the muscle is stimulated to contract isotonically, and while it does so the distance it shortens is plotted over time, usually while different loads are placed on it. The results of these measurements, shown in Figure 11.21, reveal three effects: (1) The latent period of shortening (the time between the stimulus and the beginning of shortening) increases with increasing load; (2) the duration of shortening (the time during which the muscle is shorter than its resting length) decreases with increasing load; and (3) the velocity of shortening decreases with increasing load.

The velocity of shortening is defined as the *rate of change* of the distance shortened, which is the slope of each curve in Figure 11.21. (Because the slope changes continually throughout the period of shortening, it is customary to use the initial slopes of the curves as the measures of shortening velocity.) When velocity of shortening is plotted as a function of the load, the result is a load-velocity curve like that in Figure 11.22. Note that as the load increases, the velocity of shortening gradually de-

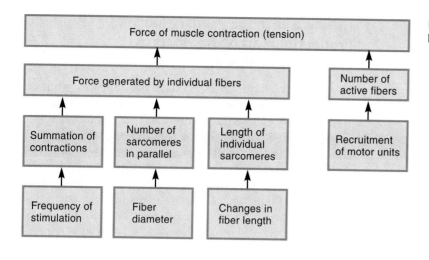

FIGURE 11.20 **Factors affecting the force developed by whole muscles.**

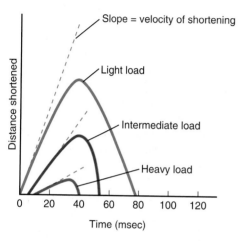

FIGURE 11.21 **The effect of load on muscle shortening.** *The distance a muscle shortens while contracting isotonically is plotted over time for three different loads. Dashed lines represent the initial slopes of the curves, a measure of the muscle's velocity of shortening.*

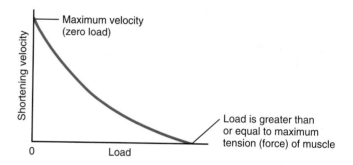

FIGURE 11.22 **A load-velocity curve.** *The curve plots the velocities of shortening (as determined in Figure 11.21) over a range of loads.*

creases, eventually reaching zero when the load is equal to or greater than the maximum tension that can be generated by the muscle, and that velocity of shortening is greatest when no load is placed on the muscle. Far from being surprising, these observations are in accord with our everyday experience: We know that one cannot lift a box full of books as quickly as one can lift an empty box.

Measurements of shortening velocity have proven to be informative because they have revealed that muscles differ in ways other than force-generating capacity. Researchers have found, for example, that certain types of muscle fibers can shorten faster than others. The bases for such differences are discussed in the next section.

Quick Test 11.4

1. What is a twitch? How does an isometric twitch differ from an isotonic twitch?

2. Define the following terms: *summation* (of contractions), *contractile component*, *series elastic component*, *tetanus*, *recruitment*, *length-tension curve*.

3. What is the size principle?

4. What is the meaning of the term *velocity of shortening*?

TYPES OF SKELETAL MUSCLE FIBERS

Although all skeletal muscle fibers are fundamentally alike with respect to the mechanisms of excitation-contraction coupling and force generation, there are significant differences among them in terms of how quickly they can contract and how they produce most of their ATP. In this section we examine these differences and their effects on muscle performance in everyday life.

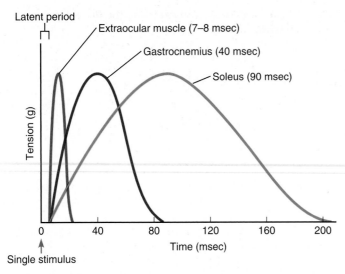

FIGURE 11.23 Differences in speed of contraction for three selected muscles. *The curves represent isometric twitches; times in parentheses are those required for each muscle to develop maximum tension following a single stimulus (arrow). To facilitate comparison, the curves have been adjusted to make their peaks the same height, even though these muscles develop different maximum tensions.*

Differences in Speed of Contraction: Fast-Twitch Fibers and Slow-Twitch Fibers

When different muscles are stimulated to contract isometrically, some take longer to reach peak tension than others (Figure 11.23). The reason is these muscles contain different populations of fibers. Some muscles (such as the soleus muscle of the leg) contain mostly **slow-twitch fibers,** which contract relatively slowly. In other muscles (such as the extraocular muscles, which control eye movements) the predominant fibers are **fast-twitch fibers,** which contract relatively quickly. In still other muscles (such as the gastrocnemius of the leg) the proportion of slow-twitch and fast-twitch fibers is intermediate. Differences between the two fiber types are not seen in isometric contractions only, but in isotonic contractions as well. Fast-twitch fibers attain peak isometric tension sooner than slow-twitch fibers and also have higher maximum shortening velocities when they contract isotonically (compared to slow-twitch fibers of similar length).

The difference between fast-twitch and slow-twitch fibers is based not on their size or shape, but instead on the type of myosin present in their thick filaments. So-called *fast myosin* has the inherent ability to hydrolyze ATP at a faster rate than *slow myosin,* and this ATPase rate has been found to correlate strongly with a fiber's speed of contraction. The higher ATPase rate of fast myosin implies that this form of myosin can complete more cross-bridge cycles per second, which means that sarcomeres shorten faster, all else being equal.

Differences in the Primary Mode of ATP Production: Glycolytic Fibers and Oxidative Fibers

Even though all muscle fibers have the ability to produce ATP by both oxidative phosphorylation and substrate-level phosphorylation, they differ in their capacities for doing so and are grouped into two general categories on this basis. **Glycolytic fibers** have high cytosolic concentrations of glycolytic enzymes and therefore can generate ATP rapidly via glycolysis (substrate-level phosphorylation); these fibers have a relatively low capacity for generating ATP via oxidative phosphorylation because they contain relatively few mitochondria, where oxidative phosphorylation occurs. By contrast, **oxidative fibers** are rich in mitochondria and have a high capacity for producing ATP via oxidative phosphorylation. However, these fibers contain relatively low concentrations of glycolytic enzymes and therefore have a low glycolytic capacity. Both fiber types are found in all muscles of the body, but their proportions vary among muscles.

The distinction between glycolytic and oxidative fibers goes beyond differences in glycolytic enzyme content or numbers of mitochondria. Oxidative fibers are generally of smaller diameter and well-supplied with capillaries, whereas glycolytic fibers are of larger diameter and are surrounded by fewer capillaries (Figure 11.24). This makes sense given that oxidative fibers have a higher capacity for utilizing oxygen and therefore rely more heavily on rapid oxygen delivery for proper function. A rich capillary supply ensures rapid oxygen delivery to the interstitial fluid surrounding the fibers, whereas the fibers' small diameter minimizes the distance oxygen must diffuse to reach the mitochondria.

Another difference between the fiber types is that oxidative fibers contain an oxygen-binding protein known as **myoglobin,** whereas glycolytic fibers lack it. Myoglobin, like hemoglobin (the oxygen-carrying protein in red blood cells), is a reddish molecule that binds oxygen reversibly. Its function is to serve as an oxygen buffer—to store a supply of oxygen that can be released whenever the oxygen concentration inside cells declines (as can happen when a muscle contracts strongly and compresses nearby blood vessels, thereby interrupting the blood supply). Because this oxygen store is limited, however, it can supply adequate amounts of oxygen for only a short time before it must be replenished, which occurs when blood flow is restored and the oxygen concentration rises. Because myoglobin imparts a reddish-brown color to oxidative fibers, these fibers are often referred to as *red muscle.* In contrast, glycolytic fibers, which lack myoglobin and this reddish color, are referred to as *white muscle.* Familiar examples of red and white muscle are the "dark" and "white" meat found in chicken.

Glycolytic fibers produce ATP less efficiently than oxidative fibers because fewer ATP molecules are synthe-

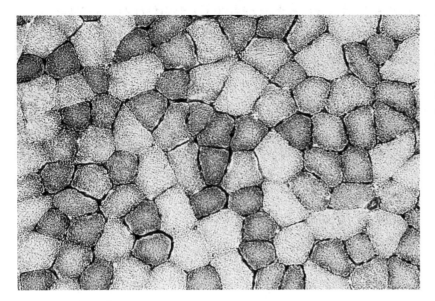

FIGURE 11.24 Oxidative and glycolytic muscle fibers. *A micrograph of a cross-section through a skeletal muscle showing numerous muscle fibers. The muscle has been stained to reveal mitochondria and also capillaries surrounding the fibers. Smaller fibers stain more deeply, indicating a higher density of mitochondria, which is typical of oxidative fibers. Note that capillary density is highest around these fibers.*

sized per unit of fuel consumed. However, glycolytic fibers are better able to produce ATP when oxygen availability is low because glycolysis does not require oxygen. When glycolytic fibers are active and producing ATP at a high rate, lactic acid is produced as a by-product, because these cells have a low oxidative capacity in addition to their high glycolytic capacity. One consequence of these differing capacities is that pyruvate, the end-product of glycolysis, is generated faster than it can be consumed and therefore accumulates in these cells. As it accumulates, it is converted to lactic acid, which has been implicated as a cause of muscle fatigue (discussed shortly). For this reason, glycolytic fibers fatigue more rapidly than oxidative fibers. In contrast, oxidative fibers produce little lactic acid so long as they are supplied with adequate amounts of oxygen, and as a consequence they are more resistant to fatigue. Lactic acid does not normally accumulate in these cells because their high oxidative capacity enables them to oxidize pyruvate as fast as it is produced.

Slow Oxidative, Fast Oxidative, and Fast Glycolytic Fibers

We have seen that skeletal muscle fibers can be classified as fast-twitch fibers or slow-twitch fibers on the basis of their contractile speeds, and as glycolytic fibers or oxidative fibers on the basis of their metabolic capacities. Not surprisingly, various combinations of contractile speed and oxidative or glycolytic capacity are possible. Indeed three major classes of skeletal muscle fibers have been identified: *slow oxidative fibers,* the relatively rare *fast oxidative fibers,* and *fast glycolytic fibers.* Muscles generally contain all three fiber types, but in different proportions.

As their name implies, **slow oxidative fibers** contain slow myosin and have a high oxidative capacity, producing most of their ATP by oxidative phosphorylation. **Fast oxidative fibers** also have a high oxidative capac-

ity, but they contain fast myosin. (Actually, the myosin ATPase activity in these fibers is intermediate between the slowest and fastest myosin.) **Fast glycolytic fibers** contain fast myosin and have a high glycolytic capacity, producing most of their ATP through glycolysis.

Although there is no direct connection between a muscle fiber's type and its ability to generate force, there is an indirect connection because the three fiber types also differ in diameter. Slow oxidative fibers are the smallest in diameter and are therefore capable of generating only small forces. Fast glycolytic fibers have the largest diameter and generate the highest forces, whereas fast oxidative fibers are intermediate in terms of diameter and force-generating capacity. Properties of the three types of fibers are summarized in Table 11.1.

Response of the Three Fiber Types to Exercise

Even though the different types of fibers are intermixed in skeletal muscles, all three types are not typically utilized every time a muscle contracts. For one thing, fiber types are segregated into different motor units, such that a given motor unit usually contains fibers of only one type. Furthermore, because a correlation exists between the size of a motor unit and the type of fiber it contains, the different fiber types are recruited in a specific order (in accordance with the size principle) as muscle tension increases.

Slow oxidative fibers are recruited first because these fibers are located in the smaller motor units. Fast oxidative fibers are recruited next because these fibers are found in intermediate motor units. Fast glycolytic fibers are the last to be recruited because they are found in the larger motor units. These fibers are not usually recruited unless a muscle is generating a large amount of force, as occurs in high-intensity exercises such as weight-lifting or sprinting.

TABLE 11.1 PROPERTIES OF SKELETAL MUSCLE
FIBER TYPES

	SLOW OXIDATIVE (RED)	FAST OXIDATIVE (RED)	FAST GLYCOLYTIC (WHITE)
Oxidative capacity	High	High	Low
Glycolytic capacity	Low	Intermediate	High
Speed of contraction	Slow	Fast	Fast
Myosin ATPase activity	Low	High	High
Mitochondrial density	High	High	Low
Capillary density	High	High	Low
Myoglobin content	High	High	Low
Resistance to fatigue	High	Intermediate	Low
Fiber diameter	Small	Intermediate	Large
Force-generating capacity	Low	Intermediate	High

Resistance to Fatigue

It is not uncommon for competitive cyclists to ride all day, covering well over 100 miles in the process. By comparison, even Olympic-class weightlifters cannot lift at or near their maximum capacity for more than a few seconds. This difference is due to the fact that muscles differ in their ability to resist **fatigue,** a decline in a muscle's ability to maintain a constant force of contraction in the face of long-term, repetitive stimulation. Although fatigue eventually sets in after any kind of muscular activity, it generally occurs more quickly when a muscle is stimulated at higher frequencies and when larger forces are generated (Figure 11.25).

Although the precise causes of muscle fatigue are not fully understood, different types of exercise are known to induce fatigue for different reasons. In high-intensity exercise, glycolytic muscle fibers are recruited and, as previously mentioned, have a tendency to generate lactic acid because of their low oxidative capacity. As a consequence, a rapid buildup of lactic acid in muscles is common in high-intensity exercise. When contractions are strong and sustained, an additional factor can come into play: Strong contractions can compress vessels that supply blood to the muscles, which interrupts or reduces the muscles' blood supply; the resulting decrease in oxygen delivery causes muscle cells to produce larger amounts of lactic acid, which by lowering intracellular pH, can alter enzyme activities and interfere in a variety of metabolic processes. In low-intensity exercise, by contrast, lactic acid accumulation is not generally a problem because few glycolytic fibers are recruited; the active fibers are mostly oxidative fibers, which have little tendency to produce lactic acid. The cause of fatigue in low-intensity exercise, which takes a longer time to develop, is thought to be linked to the depletion of energy reserves, glycogen in particular. Full recovery from this type of fatigue requires about 24 hours, whereas recovery from the more rapid-onset type of fatigue requires only minutes or a few hours.

Very-high-intensity exercise can induce *neuromuscular fatigue,* which occurs when motor units are stimulated to contract at high frequencies, as occurs when large forces are generated. When cells are stimulated to contract strongly over long periods, repeated firing of motor neurons can deplete synaptic terminals of acetylcholine, which ultimately causes failure of neuromuscular transmission.

In addition to these mechanisms of fatigue, which affect either a muscle cell's ability to generate force or the nervous system's ability to trigger muscle contractions, fatigue also has a psychological component that resists any purely physiological explanation. It is widely accepted among athletes that performance is influenced by mental state as well as by physical condition. Often, the athletes who run the fastest or the longest are the ones with the strongest "will to win." Competitors with less desire are more likely to succumb to discouragement when fatigue sets in and muscles begin to ache; as a consequence, they may well put forth less effort.

Long-Term Responses of Muscles to Exercise

Athletes go into training not simply to hone the skills needed in their particular sport, but also to change the body's physical condition so that it is fitter to perform those skills. Such changes involve, among other things, changes in muscles' cellular architecture that result from regular exercise over time. As a result of these changes, the muscles' capacity to generate force and resist fatigue is altered.

As every athlete knows, different types of exercise affect muscles in different ways, and thus it is important to choose an exercise regimen that can achieve the desired results. For example, an aspiring marathon runner, who requires great endurance for her sport, should train using long-duration, low-intensity exercises ("aerobic" exercise) such as jogging. In contrast, an aspiring boxer, who desires bigger and stronger muscles, would do well to include short-duration, high-intensity exercise in his regimen.

Aerobic exercise increases the oxidative capacity of muscle fibers, thereby increasing their resistance to fatigue. As a result of such training, some fast glycolytic fibers are effectively converted to fast oxidative fibers. (However, because exercise does not alter the type of myosin present in muscle fibers, slow-twitch fibers remain slow and fast-twitch fibers remain fast.) Changes include increases in the size and number of mitochondria within the fibers, and an increase in the number of capillaries surrounding the fibers. In addition, the average diameter of the fibers decreases, which facilitates the movement of oxygen into the cells but also decreases the cells' force-generating capacity. In contrast, high-intensity exercise decreases the oxidative capacity of muscle fibers and increases their glycolytic capacity, thereby converting a portion of the fast oxidative fibers into fast glycolytic fibers. Changes include decreases in the size and number of mitochondria, increases in the concentration of glycolytic enzymes, and increases in average fiber diameter. However, the declining oxidative capacity of these fibers reduces their resistance to fatigue. As fibers grow, new myofibrils are synthesized, which enables the fibers to generate more force; reflecting the increase in average fiber diameter, entire muscles also become bulkier and more massive. Note that muscle growth is not due to the addition of new fibers because muscle fibers are *post-mitotic*—that is, they cannot divide to form new cells. Even though new fibers can be generated from immature precursors called *satellite cells*, this normally happens only when fibers die and must be replaced.

Quick Test 11.5

1. What is the major difference between slow-twitch fibers and fast-twitch fibers?

2. How do glycolytic fibers and oxidative fibers differ in regard to their primary mode of ATP production? What are some other differences between these types of fibers?

3. What three types of fibers are found in skeletal muscles?

4. How does aerobic exercise affect resistance to fatigue in muscles? How is resistance to fatigue affected by high-intensity exercise? Why?

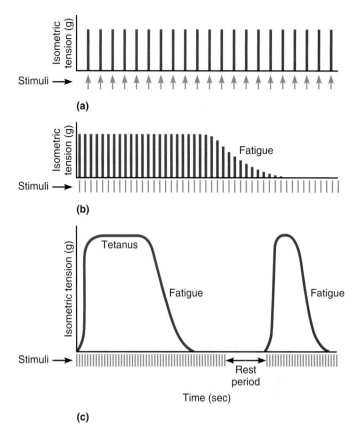

FIGURE 11.25 Muscle fatigue. *In these records of tension developed by a muscle in response to repetitive stimuli (arrows) the time scale is compressed, so individual twitches appear as spikes.* **(a)** *Stimulation at a relatively low frequency produces little, if any, fatigue. Note that peak tension is constant over time.* **(b)** *Stimulation at a higher frequency produces fatigue, apparent in the decline in peak tension near the end of the record.* **(c)** *A muscle stimulated at a frequency high enough to yield maximum tetanic tension fatigues even more rapidly. After a recovery period, tension returns to previous levels upon resumption of stimulation, but fatigue sets in more rapidly than before.*

SKELETAL MUSCLES AT WORK

Even though we have concentrated our attention almost entirely on how skeletal muscles operate, we must not forget that the ability of these muscles to cause body movement derives from their attachment to bones.

Attachment of Muscles to Bones

As mentioned previously, almost all skeletal muscles are connected to at least two bones by tendons. When a muscle contracts, one of these bones typically moves while the other remains relatively stationary. A muscle's point of attachment to the stationary bone is called the **origin,** whereas its point of attachment to the movable bone is the **insertion** (Figure 11.26). Depending on a muscle's function and location, its tendons may be short or long. The biceps muscle, for instance, is connected to the bones of the arm by short tendons, but the calf muscle called the *gastrocnemius* is connected to the heel bone by a long tendon (the *Achilles tendon*), an arrangement that frees

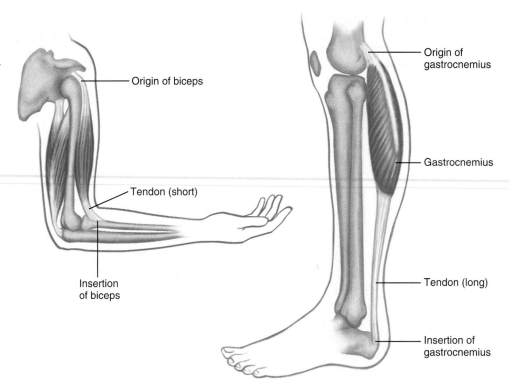

Origin of biceps

Tendon (short)

Insertion of biceps

Origin of gastrocnemius

Gastrocnemius

Tendon (long)

Insertion of gastrocnemius

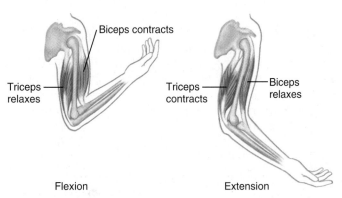

Biceps contracts

Triceps relaxes

Triceps contracts

Biceps relaxes

Flexion

Extension

FIGURE 11.27 **An antagonistic muscle group, the biceps and triceps of the arm.** *In flexion of the forearm (left), the biceps contracts while the triceps relaxes. In extension of the forearm (right), the triceps contracts while the biceps relaxes.*

the ankle of both bulk and weight. Similarly, long tendons permit the muscles that control the fingers to reside in the forearm, which is useful because a muscle-bound hand would be detrimental to finger dexterity.

A muscle can actively exert force only by contracting, which in the case of skeletal muscle means *pulling*, not pushing, on a bone. Nevertheless, it is clear that you can use muscular force to move the elbow joint, for example, in opposite directions: You can *flex* the forearm (that is, reduce the angle of the elbow joint) by contracting the biceps muscle, and you can *extend* the forearm (increase the angle of the elbow joint) by contracting the triceps muscle, which is located opposite the biceps in the arm (Figure 11.27). The biceps and triceps are examples of muscles that are **antagonistic**—each exerts force in a di-

rection that opposes the action of the other. (Most other skeletal muscles are arranged in antagonistic groups as well.) To flex the forearm, the biceps generates force actively while the triceps relaxes and stretches passively—that is, in response to forces exerted on it by the action of the biceps. To extend the forearm, the triceps contracts actively, and the biceps stretches passively. At times, simultaneous contraction of antagonistic muscle groups is useful. Thus when the biceps and triceps are stimulated to contract at the same time, as when you "brace yourself" to receive a package whose weight you are unsure of, the elbow joint stiffens to resist motion, such that the forearm remains stationary.

Bones as Levers: Implications for Muscle Contraction

When you use your muscles to lift a weight or move a part of your body, the force your muscles must generate is actually much greater than you might think. For instance, to hold a 15-kg weight (about 33 pounds) in the palm of your hand, the biceps in your arm generates an impressive 105 kg (about 232 pounds) of force! The reason relates to the fact that the insertion of the biceps is located close to the elbow joint (about 5 cm away), whereas the hand is located farther from the joint (about 35 cm away) (Figure 11.28a). To lift the weight, the biceps pulls upward on the bones of the lower arm, causing the hand to exert an upward force against the weight. (When the weight is being held steady or is being lifted at a constant speed, the magnitude of this force equals the downward

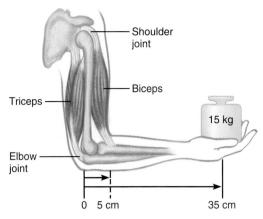

(a) Distance from elbow

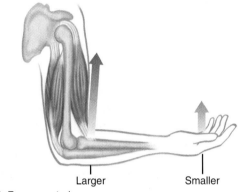

(b) Force exerted

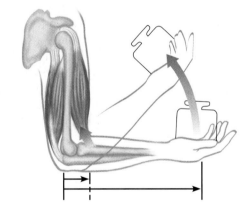

(c) Radius of arc (distance from elbow)

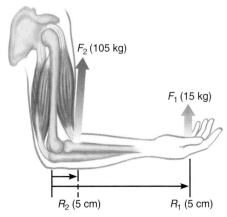

(d) Radius of arc (distance from elbow)

FIGURE 11.28 Force that must be generated by the biceps to hold or lift a weight held in the hand. (a) *Contraction of the biceps to hold a 15-kg weight steady in the palm of the hand. Distances from the elbow joint are indicated for the biceps' insertion and the weight.* **(b)** *Forces exerted by the biceps on the bones of the forearm and by the hand on the weight.* **(c)** *Thick arrows show distances (arcs) traveled by the insertion of the biceps and the hand as the forearm pivots at the elbow. Thin arrows show the radii of these arcs—the distances from the elbow joint.* **(d)** *Relevant forces and distances for calculating work performed: F_1 = force exerted by hand; F_2 = force exerted by biceps; R_1 = distance of hand from elbow joint; R_2 = distance of insertion of biceps from elbow joint. Note that these forces are valid only when the elbow joint is bent at a 90-degree angle, as shown.*

If the weight were increased to 30 kg, how much force would the biceps need to generate to hold it steady?

force of gravity acting on the weight.) Because the insertion of the biceps is close to the elbow joint, which acts as a pivot, the muscle must exert a force significantly greater than the force exerted by the hand (Figure 11.28b).

To see why the biceps must exert more force, it is instructive to consider what happens when the weight is being lifted, as shown in Figure 11.28c. Whenever a force is applied to an object, causing it to move a certain distance, a quantity of *work* is performed:

$$\text{work} = \text{force} \times \text{distance}$$

In this case, the biceps performs a certain amount of work to move the arm, and the hand performs an equal quantity of work to move the weight. As the forearm pivots at the elbow, the hand and the biceps' insertion both move through an arc and travel a certain distance, but the hand travels farther because its arc has a greater radius, as shown in the figure. Because the hand and biceps perform the same amount of work, the product of force exerted and distance traveled is the same for both. Given that the insertion of the biceps travels a shorter distance than the hand, it follows from the previous equation that the biceps must exert a greater force than the hand.

Using these principles, we can determine how much force the biceps must generate in order to lift a given weight. Since the length of an arc is proportional to its radius, the distance the hand or the insertion of the biceps travels is proportional to its distance from the elbow joint. Therefore, if the hand exerts a force F_1 and is located a distance R_1 from the elbow, it performs a quantity of work proportional to the product $F_1 \times R_1$ (Figure 11.28d). Likewise, if the biceps exerts a force F_2 and its insertion is located a distance R_2 from the elbow, the work it performs is proportional to the product $F_2 \times R_2$.

Because the hand and biceps perform equal amounts of work, the force of the hand and biceps are related by the following expression:

$$F_1 \times R_1 = F_2 \times R_2$$

or

$$F_2 = F_1 \times (R_1/R_2)$$

Using values given previously, we find that when the hand is lifting a 15-kg weight, the biceps exerts a force given by:

$$F_2 = 15 \text{ kg} \times (35 \text{ cm}/5 \text{ cm}) = 105 \text{ kg}$$

This is also true when the hand is holding the weight steady.

In the example above, the bones of the forearm act as a lever that pivots at the elbow. The biceps pulls upward on this lever, while the weight in the hand pushes down on it. To hold the weight steady (or lift it at constant speed), the biceps must exert a force greater than the downward force of the weight because its *lever arm*—the distance between the lever's pivot point and the point at which the force is applied—is shorter than that of the weight. As a reflection of this, the shorter lever arm is said to put the biceps at a *mechanical disadvantage*. Consequences of the lever arm effect may be familiar to you: If you have ever had to change an automobile tire, for instance, you know that is easier to loosen the lug nuts with a longer wrench than with a shorter wrench. The shorter wrench puts you at a mechanical disadvantage, such that you have to exert more force to loosen the nuts.

Even though the lever arm effect puts muscles at a mechanical disadvantage with respect to the force they must generate, the situation is actually advantageous in another way: As a muscle contracts it shortens with a certain velocity, but the limb (or other body part) to which it is attached moves significantly faster. If, for example, the biceps shortens at a rate of 2 cm/sec, the hand moves at a rate seven times as great—14 cm/sec. Thus the lever arm effect makes it possible for the biceps and other muscles to move body parts more quickly than the muscles themselves can shorten. This is important because it not only makes activities such as running and throwing possible; it also enables us to move out of danger quickly in times of emergency.

Quick Test 11.6

1. What is a muscle's insertion, and what is its origin?

2. Why are some muscle groups described as being antagonistic? Why is the existence of such muscle groups useful?

3. If you attempt to lift a weight and are unable to move it, is any work performed on the weight? Why or why not?

OTHER MUSCLE TYPES

When people hear the word *muscle*, most automatically think of *skeletal muscle*, which has been our sole focus so far in this chapter. However, two other types of muscle in the body—*smooth muscle* and *cardiac muscle*—perform their functions without attracting much notice. We now shift our focus to these muscles, and to the special properties that make them different from skeletal muscle. For easy comparison, the three types of muscle are shown side-by-side in Figure 11.29. The specific features of smooth muscle and cardiac muscle that are apparent in this diagram will be discussed in the next two sections.

Smooth Muscle

Smooth muscle, which gets its name from the fact that it lacks the striations characteristic of skeletal and cardiac muscle and therefore appears uniformly bright under the light microscope (see Figure 11.29), is the type of muscle found in internal organs, blood vessels, and other structures that are not under voluntary control. Functions performed by these muscles are many and varied and depend on the organ in which they are located. In the gastrointestinal tract, for example, smooth muscle contractions mix the ingested food with digestive secretions and propel it from one location to another. In blood vessels, smooth muscle regulates blood flow to organs and tissues by causing the vessels to constrict or dilate.

Like skeletal muscle, smooth muscle has thick and thin filaments and generates force through the crossbridge cycle. However, the filaments are not arranged in sarcomeres, which accounts for the lack of striations. Even though thick and thin filaments are arranged in parallel with each other, as in skeletal muscle, they tend to run obliquely in various directions, which means that contraction occurs along several axes (Figure 11.30). *Dense bodies,* points of attachment between these filaments and connective tissue inside the cells, serve to transmit contractile force to the cell's exterior.

The Mechanism of Excitation-Contraction Coupling

Smooth muscle contractions are regulated by intracellular calcium, but the sarcoplasmic reticulum is not as extensive as in skeletal muscle. Moreover, much of the calcium that triggers contractions comes from outside the cells, because when the cell is depolarized, voltage-gated calcium channels in the plasma membrane open and allow calcium to flow in. (Depolarization also triggers the release of calcium from the sarcoplasmic reticulum.)

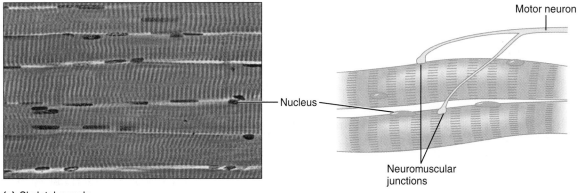

(a) Skeletal muscle

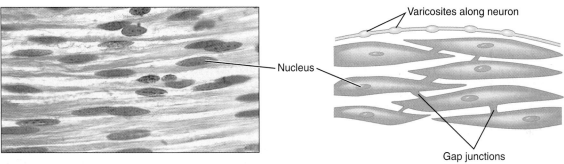

(b) Smooth muscle

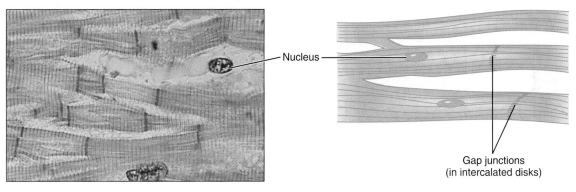

(c) Cardiac muscle

FIGURE 11.29 **Photomicrographs (left) and schematic diagrams (right) of the three types of muscle in the body.**

The neurons that control skeletal muscles and those that control smooth muscle belong to different branches of the efferent nervous system. What are these branches called?

Cytosolic calcium activates the crossbridge cycle in smooth muscle, but it does so differently than in skeletal muscle because contractions are not regulated by the troponin-tropomyosin system (Figure 11.31). Instead, contractions in smooth muscle are triggered when calcium binds reversibly to *calmodulin,* a cytosolic protein that regulates many processes in almost all cells of the body (see p. 145). This binding triggers a conformational change that enables

the calcium-calmodulin complex to bind to and activate an enzyme called **myosin kinase** (or *myosin light-chain kinase*). The activated kinase then catalyzes the phosphorylation of myosin crossbridges, which activates them and initiates crossbridge activity. Crossbridge cycling then proceeds

The somatic and autonomic nervous systems, respectively

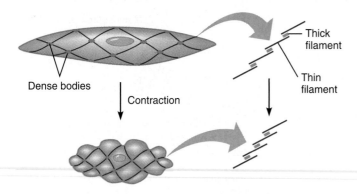

Dense bodies

Contraction

Thick filament

Thin filament

FIGURE 11.30 Oblique arrangement of thick filaments and thin filaments in a smooth muscle cell. *The relative positions of the filaments are shown for a relaxed cell (top) and a contracting cell (bottom).*

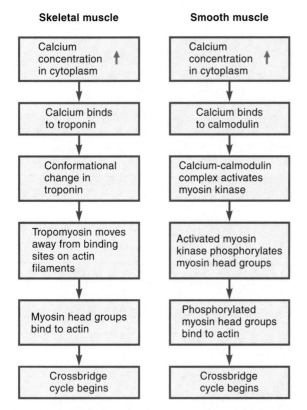

Skeletal muscle	Smooth muscle
Calcium concentration in cytoplasm ↑	Calcium concentration in cytoplasm ↑
Calcium binds to troponin	Calcium binds to calmodulin
Conformational change in troponin	Calcium-calmodulin complex activates myosin kinase
Tropomyosin moves away from binding sites on actin filaments	Activated myosin kinase phosphorylates myosin head groups
Myosin head groups bind to actin	Phosphorylated myosin head groups bind to actin
Crossbridge cycle begins	Crossbridge cycle begins

FIGURE 11.31 Comparison of the events leading up to activation of the crossbridge cycle in skeletal and smooth muscle.

essentially as shown in Figure 11.7 on page 339. Note that in smooth muscle the calcium signal that triggers crossbridge activity targets *myosin* filaments, whereas in skeletal muscle it targets *actin* filaments because this is where troponin and tropomyosin are located.

Termination of the crossbridge cycle in smooth muscle requires more than just removal of calcium from the cytosol, because the phosphate groups attached to myosin are covalently bound and therefore do not dissociate readily. For this reason, termination of the crossbridge cy-

cle in smooth muscle requires the action of phosphatase enzymes, which by removing phosphate groups inactivates myosin. Because these phosphatases (which are continually active) compete with myosin kinase, activation of myosin occurs only when enough calcium is present to activate myosin kinase to a degree sufficient to overcome the action of the phosphatases.

Because the mechanism of excitation-contraction coupling is entirely different from that in skeletal muscle, it takes a lot longer to initiate and also to terminate smooth muscle contractions. This is no real handicap because smooth muscle isn't exactly "built for speed" anyway; myosin ATPase activity in smooth muscle is anywhere from 10 to 100 times lower than in skeletal muscle, which means that smooth muscle contraction is an inherently slow process.

Neural Regulation of Contraction

Unlike skeletal muscle, which is regulated by motor neurons, smooth muscle is regulated by *autonomic neurons.* A given smooth muscle cell may be regulated by sympathetic neurons, parasympathetic neurons, or frequently both. Whereas motor neurons always exert excitatory effects on skeletal muscle cells, the effect of autonomic input to smooth muscle cells may be excitatory or inhibitory, depending on whether the neurons in question are sympathetic or parasympathetic; these two types of neuron almost always affect a given smooth muscle cell in opposite ways. For instance, intestinal smooth muscle contracts in response to parasympathetic input and relaxes in response to sympathetic input. Furthermore, whether a muscle cell contracts or relaxes depends on where it is located. Thus, for example, sympathetic input induces relaxation of smooth muscle in the intestine but contraction of smooth muscle in most blood vessels. These opposite responses are due to differences in the type of neurotransmitter receptors found on the muscle cells, not to differences in the type of neurotransmitter released, because sympathetic or parasympathetic neurons release the same neurotransmitters (norepinephrine and acetylcholine, respectively) at virtually all their target tissues.

Another difference between skeletal and smooth muscle concerns the specificity of neural connections. In skeletal muscle, neural input is delivered to each cell individually because motor neurons are connected to specific cells via neuromuscular junctions. In contrast, smooth muscle cells do not receive neural input by "personal telegram" but instead by "bulk mailing," because an autonomic neuron does not make synaptic connections to specific cells. Instead, neurotransmitter is released from varicosities (swellings) located at intervals along the axon and diffuses over a relatively long distance to large groups of cells (see Figure 11.29b). Consequently, neighboring muscle cells tend to contract or relax together. This synchronized activity is also promoted by the presence in most smooth muscle tissue of *gap junctions* that allow ions (and other small molecules) to move from one cell to another, so an electrical signal initiated in one cell spreads to neighboring cells.

Skeletal and smooth muscles also differ in regard to the electrical signal generated in response to neural input. Whereas skeletal muscle cells always respond to neural input with an action potential that triggers a reproducible twitch, in smooth muscle cells slow but twitchlike contractions may be elicited in response to action potentials, but this is not necessarily so. The membrane potential of most smooth muscle cells varies in a graded fashion in response to neural input, depolarizing if the input is excitatory and hyperpolarizing if it is inhibitory, which causes the contractile force to increase or decrease in a graded fashion. Thus contractions in smooth muscle need not be elicited by action potentials. In fact, some smooth muscle cells do not have action potentials at all. When action potentials do occur, they are not usually followed by real twitches (that is, individual contractions, one in response to each action potential), just greater tension.

Some types of smooth muscle cells are able to actively exert tension even in the absence of external stimulation, because resting calcium levels are high enough to maintain a constant low level of crossbridge activity. This resting tension is referred to as *tone*. (Skeletal muscles also exhibit some degree of basal tone, but in their case it is due to a constant low level of neural stimulation.) Many smooth muscle cells have the ability to contract in response to hormones and other chemical agents, independent of neural input. Furthermore, some smooth muscle cells exert active tension in response to mechanical stretch.

Single-Unit and Multi-Unit Smooth Muscle Smooth muscle tissue varies in both the degree to which muscle cells are connected by gap junctions and the pattern of innervation. In some places, most smooth muscle cells are not connected by gap junctions but instead are largely separate and richly supplied with neurons; smooth muscle of this type is referred to as **multi-unit smooth muscle** (Figure 11.32a). In other locations, smooth muscle cells are extensively linked by gap junctions, such that electrical signals originating in a few cells are transmitted to the rest of the cells; such smooth muscle is innervated by relatively few neurons and is referred to as **single-unit smooth muscle** (Figure 11.32b). Example of organs containing single-unit smooth muscle are the gastrointestinal tract and the uterus, in which large groups of cells contract synchronously. Multi-unit smooth muscle occurs in the large respiratory airways and large arteries, where the number of active smooth muscle cells may be large or small depending on the circumstances.

Pacemaker Activity In some cases, smooth muscle cells exhibit spontaneous depolarizations that occur on a regular basis and that may or may not be accompanied by ac-

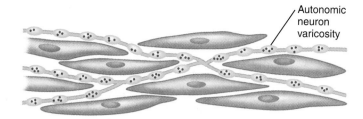

(a) Multi-unit smooth muscle

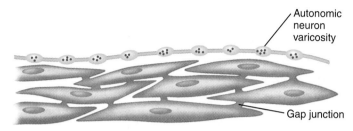

(b) Single-unit smooth muscle

FIGURE 11.32 Multi-unit and single-unit smooth muscle.

tion potentials. These depolarizations are called **pacemaker potentials,** and the cells that generate them are called **pacemakers.** In single-unit smooth muscle, pacemaker activity causes cells to contract and relax in unison. Although the frequency or amplitude of electrical signals in pacemakers can be influenced by neural activity, pacemaker signals occur even in the absence of any neural influence.

Cardiac Muscle

Cardiac muscle is similar to skeletal muscle in that it is striated (see Figure 11.29c), has the same sarcomere structure, and has contractions that are regulated by the troponin/tropomyosin system. Cardiac muscle cells are similar to smooth muscle cells in that they are extensively connected by gap junctions, such that an action potential, once initiated, travels throughout the entire cell network.

Cardiac action potentials are broad and last for hundreds of milliseconds, making them quite different from the "spikey" action potentials that occur in skeletal muscle and most neurons and last for 1–2 milliseconds. The relatively long duration of cardiac action potentials is significant: Because they last nearly as long as it takes for cardiac muscle cells to contract and relax (Figure 11.33), summation of cardiac muscle contractions cannot occur, even when the action potential frequency is high and the heart is beating rapidly. This is good, because summation would be detrimental to the heart's pumping action in that the heart would not be able to relax completely and fill with blood between contractions.

Certain heart muscle cells that are concentrated in two regions known as the *sinoatrial* and *atrioventricular nodes* exhibit pacemaker activity. The heartbeat is triggered by action potentials originating in pacemaker cells and does not depend on neural input; this is why a heart removed from the body continues to beat, even though all neural connections to the heart have been severed. (Because the signals that trigger the heartbeat originate within the heart muscle itself, the heart's contractile activity is said to be *myogenic*, whereas the contractile activity of skeletal muscle is said to be *neurogenic*.) However, the autonomic nervous system does regulate the heart muscle by modulating the frequency and force of heart muscle contractions. Properties of cardiac muscle and their importance to the heart's function are discussed further in Chapter 12.

Quick Test 11.7

1. How does smooth muscle differ from skeletal muscle with respect to the arrangement of thick and thin filaments? With respect to the mechanism of excitation-contraction coupling?

2. What is the difference between multi-unit smooth muscle and single-unit smooth muscle?

3. What does a pacemaker cell do, and in what types of muscle are they found?

4. What prevents summation of cardiac muscle contractions, and what is the significance of this phenomenon?

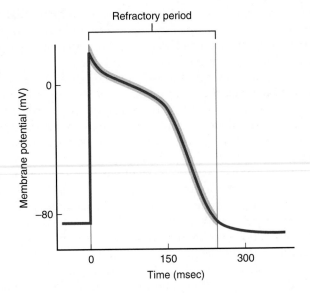

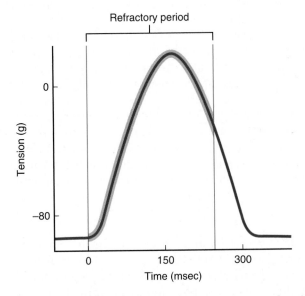

FIGURE 11.33 Relative durations of contractions and action potentials in cardiac muscle. *The common time scale for the two graphs allows comparison of an action potential (above) with the tension developed in response to it (below) in a cardiac muscle cell.*

Urinary System

Smooth muscle in afferent and efferent arterioles regulates glomerular filtration

Smooth muscle in wall of urinary bladder contracts during micturition to drive urine flow

Digestive System

Smooth muscle in the wall of the gastrointestinal tract mixes and propels chyme

Muscular sphincters regulate the flow of chyme at various places in the gastrointestinal tract

Respiratory System

Action of diaphragm and other skeletal muscles works to expand lungs during inspiration

Smooth muscles in small airways and pulmonary blood vessels regulate ventilation and perfusion of alveoli

Muscles

Reproductive System

Smooth muscle contractions propel sperm or eggs through the reproductive tract

Smooth muscle contractions in the uterus propel the fetus through the birth canal during parturition

Endocrine System

Contraction of cardiac muscle drives blood flow, which delivers hormones to target tissues

Immune System

Smooth muscle in the walls of larger lymphatic ducts propels lymph flow

Skeletal muscle contractions aid in the flow of lymph

Cardiovascular System

Cardiac muscle generates arterial pressure, which drives blood flow throughout the body

Vascular smooth muscle regulates resistance of blood vessels, which is important in the control of blood pressure

Nervous System

Smooth muscle in cerebral vasculature adjusts the distribution of blood flow to different regions, according to changes in brain activity

Muscles in the eyes and ears help adapt them to changing conditions

Skeletal Muscle Structure, p. 334

Most skeletal muscles are connected to bones by tendons and contain numerous elongated cells (muscle fibers) that generate contractile force using energy from ATP hydrolysis. Within muscle fibers are rodlike elements (myofibrils) that contain the contractile machinery. The sarcoplasmic reticulum surrounds the myofibrils, stores calcium ions, and is closely associated with transverse (T) tubules, which penetrate into the cell interior from the sarcolemma. Skeletal and cardiac muscle are striated, reflecting the orderly arrangement of thick and thin filaments in the myofibrils, which are made up of fundamental force-generating units (sarcomeres) joined end-to-end. Thick and thin filaments contain the contractile proteins myosin and actin, respectively. The heads (crossbridges) of myosin molecules are responsible for generating the motion that drives contraction and possess two important sites: an actin-binding site and an ATPase site. Two regulatory proteins (troponin and tropomyosin) present on the thin filaments serve to initiate and terminate contractions.

> **IP** Muscular, Anatomy Review: Skeletal Muscle Tissue, pages 4–11
>
> **IP** Muscular, Sliding Filament Theory, pages 3–15

The Mechanism of Force Generation in Muscle, p. 338

When a muscle contracts, thick and thin filaments slide past one another. This sliding is driven by the crossbridge cycle, in which the motion of crossbridges is coupled to their cyclic binding and unbinding to actin molecules in adjacent thin filaments. In skeletal muscle, each fiber receives input from one motor neuron, which typically branches and innervates more than one fiber. An action po-

tential in a motor neuron triggers the release of acetylcholine, which binds to receptors in the muscle fiber's motor end-plate. The result is an electrical signal (end-plate potential) that triggers an action potential in the sarcolemma. This is followed by propagation of the action potential through the T tubules, release of Ca^{2+} from the sarcoplasmic reticulum, binding of Ca^{2+} to troponin, movement of tropomyosin away from actin's myosin-binding sites, and initiation of the crossbridge cycle.

The Mechanics of Skeletal Muscle Contraction, p. 344

A motor neuron plus the muscle fibers it innervates constitutes a motor unit. When a motor neuron fires an action potential, all fibers in the motor unit contract together. The mechanical response of a motor unit to a single action potential is a twitch, which is reproducible in size. Twitches can be isometric, in which case the muscle generates force but does not shorten, or isotonic, in which case the muscle shortens. The force generated by an entire muscle is determined by both the force generated by individual fibers (which depends on the frequency of stimulation, fiber diameter, and changes in fiber length) and the number of fibers that are active. Stimulation at high frequencies causes summation of twitches, such that the force eventually reaches a plateau (tetanus).

The central nervous system regulates muscular force by varying both the action potential frequency in motor neurons and the number of active motor units (recruitment). As muscular force increases, motor units are recruited in order of increasing size, a phenomenon referred to as the size principle.

> **IP** Muscular, Sliding Filament Theory, pages 1; 17–29*
>
> **IP** Muscular, Contraction of Whole Muscle, pages 1–16
>
> **IP** Muscular, The Neuromuscular Junction, pages 1–16

Types of Skeletal Muscle Fibers, p. 353

Skeletal muscles contain different types of fibers in various proportions. Fast-twitch fibers and slow-twitch fibers differ in their speed of contraction, which is related to the type of myosin they contain. Glycolytic fibers synthesize most of their ATP via glycolysis and generate lactic acid, which makes them fatigue rapidly. Oxidative fibers synthesize ATP mostly via oxidative phosphorylation and are more resistant to fatigue.

> **IP** Muscular, Muscle Metabolism, pages 3–23

Skeletal Muscles at Work, p. 357

When a muscle shortens in the body, one of the bones to which it is connected moves while the other usually remains stationary. The point of attachment of a muscle's tendon to the stationary bone is the origin; the point of attachment to the movable bone is the insertion. Most muscles are arranged in antagonistic groups in which one muscle (or set of muscles) exerts force in a direction that opposes the action of the other muscle(s). The lever action of bones places many muscles at a mechanical disadvantage, which increases the force that they must generate to move a given load.

> **IP** Muscular, Sliding Filament Theory, pages 22–25; 29*

Other Muscle Types, p. 360

Smooth muscle is found in internal organs and other structures that are not under voluntary control and is

---

*This topic is available on the *InterActive Physiology*® *Sampler CD* that comes with the purchase of a new copy of this book.

regulated by autonomic neurons. Contractions are triggered by the binding of Ca^{2+} to calmodulin, which activates myosin kinase, resulting in the phosphorylation of myosin cross-bridges.

In cardiac muscle, contractions are triggered by action potentials initiated in pacemaker cells. Action potentials travel from cell to cell through gap junctions, so the entire network of cells contracts as a unit.

IP Muscular, Anatomy Review: Skeletal Muscle Tissue, page 3

IP Cardiovascular, Anatomy Review: The Heart, pages 6–7

EXERCISES

Multiple-choice Questions

1. A physiologist measures peak twitch tension for a series of muscle contractions performed with different loads. If peak tension is plotted as a function of load, the results should look something like

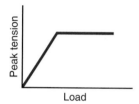

(a)

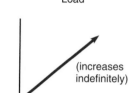

(b) (increases indefinitely)

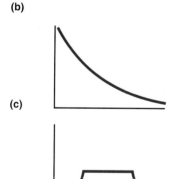

(c)

(d)

2. Following the contraction phase of a muscle twitch, cytoplasmic calcium concentration declines, and calcium ions dissociate from troponin. When the calcium concentration returns to its resting level, ATP hydrolysis ceases because
   a) ATP concentration goes to zero.
   b) low calcium levels cause the energy of ATP hydrolysis to become positive, favoring ATP synthesis.

   c) ATP binding sites on myosin molecules lose their ability to catalyze ATP hydrolysis.
   d) the number of available (unoccupied) myosin ATPase sites declines to virtually zero.

3. Two muscle fibers (X and Y) generate the same isometric peak twitch tension, but fiber X contains myosin that hydrolyzes ATP at a high rate, whereas fiber Y has myosin that hydrolyzes ATP more slowly. Properties of these fibers are consistent with which of the following sets of curves?

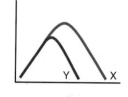

(a)

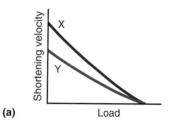

(b)

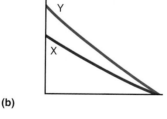

(c)

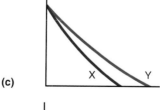

(d)

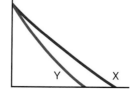

4. For the muscle fibers described in Question 3, distance of shortening was measured under conditions of zero load. Which of the following sets of curves is consistent with the results of Question 3?

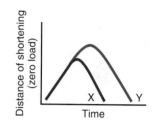

(a)

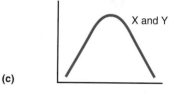

(b)

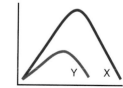

(c)

(d)

5. When a muscle cell is relaxed and intracellular ATP levels are normal, a crossbridge will remain in which of the following states?
   a) bound to actin and in the low-energy form
   b) bound to actin and in the high-energy form
   c) in the high-energy form, with ADP and P_i bound to it
   d) in the high-energy form, with ATP bound to it

6. During a muscle contraction, which of the following does *not* change length?
   a) the distance between Z lines
   b) the width of I bands
   c) the width of A bands
   d) none of the above

7. Suppose you have a beaker containing a suspension of purified myosin molecules and ATP in a saline solution. Based on your understanding of events occurring during the crossbridge cycle,
   a) ATP hydrolysis should occur.
   b) ATP hydrolysis will occur if actin is added to the solution.
   c) ATP hydrolysis will occur if calcium is added to the solution.
   d) ATP hydrolysis will occur if calcium and troponin are added to the solution.

8. Which of the following would tend to *reduce* the concentration of lactic acid that accumulates in a muscle cell as a result of contractile activity?
   a) increasing the concentration of glycolytic enzymes
   b) increasing the diameter of the cell
   c) increasing the number of mitochondria in the cell
   d) all of the above

9. Which of the following statements is a valid generalization regarding the properties of smooth muscle?
   a) Neurotransmitters can either excite or inhibit smooth muscle contraction, but any given neurotransmitter is always excitatory or inhibitory, regardless of where the muscle is located.
   b) A given smooth muscle cell can respond to more than one type of neurotransmitter.
   c) Smooth muscle cells are generally unresponsive to neurotransmitters of all types.
   d) none of the above

10. In which of the following sets of curves do the labels for X, Y, and Z illustrate that varying the load on a muscle not only affects shortening velocity but also changes the distance the muscle shortens during a single twitch?

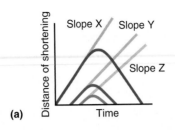

(a)

(b)

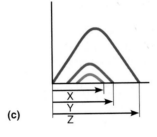

(c)

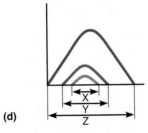

(d)

Objective Questions

1. When calcium is released from the sarcoplasmic reticulum, it binds to (troponin/tropomyosin) to initiate the crossbridge cycle.

2. When a muscle fiber contracts, the I bands shorten. (true/false)

3. Glycolytic fibers generate more force than oxidative fibers because they are larger in diameter. (true/false)

4. (Glycolytic/Oxidative) fibers contain high concentrations of the oxygen-binding protein myoglobin.

5. A crossbridge exerts force on a thin filament when it goes from its high-energy form to its low-energy form. (true/false)

6. The plasma membrane of a muscle cell is also known as the _____.

7. During muscle contraction, ATP hydrolysis is catalyzed by (myosin head groups/actin monomers).

8. During an (isometric/isotonic) muscle contraction, a muscle develops contractile force but does not change in length.

9. The velocity of contraction of a muscle fiber is directly related to its (diameter/myosin ATPase activity).

10. A reduction in the number of active crossbridges is responsible for a decrease in force-generating capacity of a muscle fiber that is significantly (longer/shorter) than its optimum length.

11. (Oxidative/Glycolytic) muscle fibers are more resistant to fatigue.

12. According to the size principle, the force-generating capacity of a muscle fiber increases in direct proportion to its length. (true/false)

Essay Questions

1. Compare and contrast mechanisms of excitation-contraction coupling in striated muscle and smooth muscle. Be sure to include a description of the mechanisms responsible for termination of muscle contractions.

2. Compare and contrast skeletal muscle and smooth muscle with respect to regulation of contractile activity by the nervous system.

3. Describe the relationship between contractile force and action potential frequency in skeletal muscle, and explain how the situation differs in cardiac muscle.

4. Explain the size principle and how it relates to the amount of tension developed by a skeletal muscle.

5. Discuss how oxidative muscle fibers differ from glycolytic fibers, and explain how each of these factors relates to the ability of a fiber to resist fatigue.

6. Describe the role of creatine kinase in muscle cell metabolism.

Find the answers to these exercises, and additional study tools, at the Physiology Place (www.physiologyplace.com).

12

The Cardiovascular System: Cardiac Function

OBJECTIVES

- Identify the major components of the cardiovascular system, and briefly describe their functions.

- Identify the major structures of the heart, and describe the path of blood flow through the heart and vasculature.

- Explain the following events in the cardiac cycle: changes in ventricular, aortic, and atrial pressure; changes in ventricular volume; and heart sounds.

- Trace the path of action potentials through the conduction system of the heart, and relate the heart's electrical activity to its pumping action.

- Describe how the phases of the electrocardiogram relate to the events of the cardiac cycle.

- Explain the difference between extrinsic control and intrinsic control, and explain how these terms relate to the regulation of the heart.

- Explain how each of the following variables affects cardiac output: sympathetic and parasympathetic nervous activity, circulating epinephrine, afterload, preload, end-diastolic volume, ventricular contractility, and filling time.

CHAPTER OUTLINE

An Overview of the Major Components of the Cardiovascular System and Their Functions   370

The Path of Blood Flow Through the Heart and Vasculature   374

Electrical Activity of the Heart   377

The Cardiac Cycle   386

Cardiac Output and Its Control   392

Above: Arteriography of the hand

The ability of cells to exchange materials with their immediate environment is an absolute requirement of life. When exchanging materials with interstitial fluid, each cell relies on diffusion to bring needed materials, such as oxygen and nutrients, to it and to carry unwanted materials, such as carbon dioxide and other wastes, away. Because these materials ultimately come from or go into the *external* environment, which is at a considerable distance from most body cells, diffusion alone cannot provide all cells with what they need as quickly as they need it; as a mechanism of transport, it is too slow. Providing much more efficient transport is the purpose of the cardiovascular system, the subject of this and the next two chapters.

The **cardiovascular system** consists of three components: (1) **blood**—a fluid that circulates around the body, carrying materials to and from the cells; (2) **blood vessels**—conduits through which the blood flows; and (3) the **heart**—a muscular pump that drives the flow of blood through blood vessels. The cardiovascular system can deliver materials throughout the body faster than diffusion could because molecules in the blood move by *bulk flow.* In bulk flow, molecules move in a stream, as do molecules of water in a river. Bulk flow is faster than diffusion because the predominant molecular motion in bulk flow is nonrandom—neighboring molecules are carried in the same general direction rather than zigzagging back and forth, as in diffusion.

The circulation of blood is so crucial to your existence that if blood flow were suddenly to stop, you would experience no more than a few seconds of consciousness before blacking out—and then you would have but a few minutes to live. Clearly, the heart must perform its functions continuously and nearly flawlessly for every minute of every day that you live. In this chapter we focus on how the heart performs its vital pumping action and look at the various factors that influence cardiac function. We begin with an overview of the cardiovascular system's major components and their functions, and then we trace the path of blood flow through the cardiovascular system. Next, we focus on the internal workings of the heart to see how it pumps blood and how this pumping action is set into motion by electrical signals originating within the heart muscle. We finish with a look at various factors operating both inside and outside the heart that affect its ability to pump blood.

AN OVERVIEW OF THE MAJOR COMPONENTS OF THE CARDIOVASCULAR SYSTEM AND THEIR FUNCTIONS

At first glance, the function of the cardiovascular system is simple: The heart pumps blood through blood vessels to various organs, while the blood carries oxygen and nutrients to tissues and removes carbon dioxide and other wastes. However, the heart is more than a pump because it performs sensory and endocrine functions that help regulate cardiovascular variables such as blood volume and pressure. The blood vessels are not just conduits for blood but are also important sensory and effector organs that regulate blood pressure and the distribution of blood to various parts of the body. The blood not only carries nutrients and wastes but also transports hormones from one part of the body to another and thus serves as a communications link acting in conjunction with the nervous system. As we will see, it is not possible to fully understand how one part of the system works—the heart, for instance—without understanding how the rest of the system works. Furthermore, regulation of the cardiovascular system involves interactions with several other organ systems, including the nervous system, the endocrine system, and the kidneys.

As you study the individual mechanisms of cardiovascular function in this and the next chapter—the action of the autonomic nervous system on the heart, or the effect of blood pressure on blood flow, for example—it is important to bear in mind that these are just small parts of a bigger picture. In Chapter 14 we will see how all these things interact to enable the cardiovascular system to accomplish its ultimate function: *To provide adequate blood flow to all the organs and tissues of the body.* Before we examine the workings of the heart, the central focus of this chapter, we discuss the major components of the cardiovascular system and their functions.

Blood

Although blood is a fluid, nearly half its volume is composed of cells. The most numerous cells are **erythrocytes,** also known as *red blood cells.* These cells contain *hemoglobin,* a protein that carries oxygen and carbon dioxide. The presence of hemoglobin in these cells gives them their characteristic red color. The remainder of the cells are **leukocytes** or *white blood cells,* which come in a variety of types and help the body defend itself against invading microorganisms. Also present are **platelets,** which are not cells but instead cell fragments that play an important role in blood clotting. The liquid portion of the blood, called plasma, is made up of water containing dissolved proteins, electrolytes, and other solutes. The composition of blood is discussed in greater detail in Chapter 13.

Blood Vessels

When blood moves through the body, it travels in a circular pattern through a system of blood vessels that carry it from the heart to the various organs and then back to the heart again. This system of blood vessels is generally referred to as the **vasculature.** As blood flows away from the heart, blood vessels branch repeatedly, becoming more numerous and smaller in diameter, just as the limbs of a tree become smaller and more numerous as you move from the trunk to the outer branches. As blood flows back to the heart, the vessels conveying it converge, becoming less numerous and larger in diameter, just as the limbs of a tree do as one moves from the outer branches to the trunk. When blood leaves the heart, it is transported to the body's organs and tissues in relatively large vessels called **arteries,** which branch repeatedly within the organs and tissues. The smallest arteries branch into still-smaller vessels called **arterioles,** which carry blood to the smallest vessels, called **capillaries.** From the capillaries, blood moves to larger vessels called **venules,** which lead to still-larger vessels called **veins,** which then carry blood back to the heart.

The various types of blood vessels differ not only in diameter, but also in the thickness and composition of their walls. Capillaries have very thin walls that permit the ready exchange of oxygen, carbon dioxide, and other materials between blood and interstitial fluid. Arteries, by contrast, have thick elastic walls that enable them to withstand the pressure of the blood, which is higher in arteries than anywhere else in the vasculature. Arterioles have walls containing rings of smooth muscle that contract or relax in response to various signals, thereby regulating blood flow downstream. The structure and function of blood vessels are discussed further in Chapter 13.

The Heart

The heart generates the force that propels blood through the blood vessels. It is a muscular organ enclosed within a membranous sac called the *pericardium* and is centrally located in the *thoracic cavity* (chest cavity) just above the *diaphragm,* a muscular partition that separates the thoracic cavity from the *abdominal cavity* (Figure 12.1). It is about the size of a fist, weighing approximately 300–350 grams in males and 250–300 grams in females. Figure 12.2 shows that the heart has four chambers. The two upper chambers, called **atria** (singular, *atrium*), receive the blood that comes back to the heart from the vasculature and transfer it to the two lower chambers, called ventricles, which are significantly larger than the atria and make up the bulk of the heart. The ventricles generate the pressure that pushes blood away from the heart and through the vasculature.

The atrium and ventricle on the left side of the heart constitute the **left heart;** the atrium and ventricle on the right side constitute the **right heart.** The atria and ventricles on either side of the heart are separated by a wall called the **septum** that prevents blood in the left heart from mixing with blood in the right heart. The portion separating the left and right atria is referred to as the *interatrial septum;* the portion separating the left and right ventricles is the *interventricular septum.* Just as the heart has left and right sides, it also has a "top" and "bottom." The wider upper pole (end) of the heart is known as the *base;* the narrower lower pole is the *apex.*

The heart's pumping action is conferred by the rhythmic contraction and relaxation of *cardiac muscle,* which makes up the bulk of the heart's mass and which is found in the outer wall and septum. The whole cardiac muscle mass is called the **myocardium.** When muscle in the wall of an atrium or ventricle contracts, the wall moves inward and squeezes the blood in the chamber. This squeezing increases the pressure within the chamber and forces the blood out. When the muscle relaxes, the chamber expands and fills with blood.

Note in Figure 12.2 that ventricular muscle is substantially thicker than atrial muscle. This reflects the fact that because the ventricles pump blood over relatively long distances through the vasculature (and not just into the next chamber, as the atria do), they must work harder to pump a given volume of blood. Note as well that the ventricular muscle is much thicker on the left side than it is on the right (Figure 12.3). The thicker muscle enables the left ventricle to develop greater pressure than the right ventricle. It is important for the left ventricle to develop greater pressure because it pumps blood to all the organs in the body except the lungs, whereas the right ventricle pumps blood only to the lungs, as we will see. The pressure required to pump blood at a given rate through the whole body is greater than that required to pump blood at the same rate through the lungs.

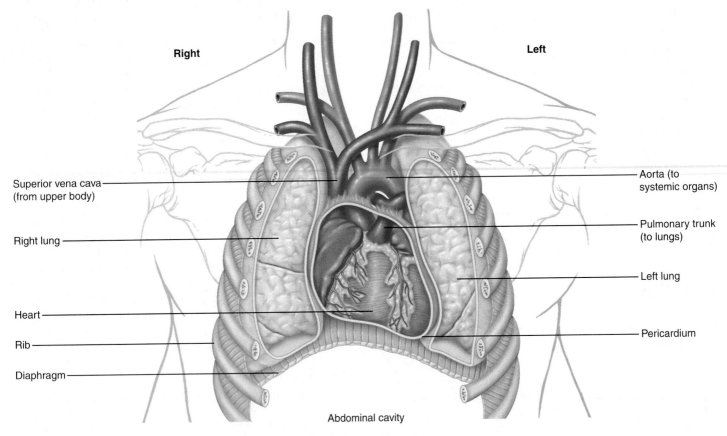

Right   **Left**

Superior vena cava
(from upper body)

Right lung

Heart

Rib

Diaphragm

Aorta (to
systemic organs)

Pulmonary trunk
(to lungs)

Left lung

Pericardium

Abdominal cavity

FIGURE 12.1 Location of the heart in the thoracic cavity. *Relative positions of the heart, rib cage, and diaphragm. Also shown are the major blood vessels connecting to the heart, and the lungs.*

What we call the *heartbeat* is actually a wave of contraction that sweeps through heart muscle fibers (cells) in an orderly, coordinated fashion. The atria contract first, driving blood into the ventricles; then the ventricles contract, driving blood to the organs. Although the entire heart muscle functions as a unit, atrial muscle (the *atrial myocardium*) and ventricular muscle (the *ventricular myocardium*) are physically anchored to and separated by a layer of fibrous connective tissue called the *fibrous skeleton* of the heart.

The heart has four valves that keep blood flowing in the proper direction within the heart itself and between the heart and the arteries connected directly to it (the *aorta* and the *pulmonary trunk*) (see Figure 12.2). The atrium and ventricle on each side are separated by **atrioventricular valves** *(AV valves)*, which permit blood to flow from the atrium to the ventricle but not in the opposite direction. AV valves open or close in response to cyclic changes in pressure that occur with every heartbeat (Figure 12.4). When atrial pressure is higher than ventricular pressure, the valves open; when ventricular pressure becomes higher than atrial pressure, the valves close. The AV valve on the left consists of two flaps or *cusps* of connective tissue and is thus called the **bicuspid valve**

(mitral valve). The right AV valve has three cusps and is called the **tricuspid valve.**

When a ventricle contracts, the increased ventricular pressure exerts an upward force against the AV valve. Because of this force, there is a potential danger that one or more valve cusps could be pushed into the atria, a condition called *prolapse*. If this were to happen, the edges of the cusps would no longer meet properly when the valve closes, and the valve would not be able to seal completely. Prolapse of the AV valves is normally prevented because the valve cusps are held in place by strands of connective tissue (known as the *chordae tendineae*) that extend from the edges of the cusps to *papillary muscles*, which protrude from the ventricular wall. During ventricular contraction, the papillary muscles also contract, which exerts tension on the chordae tendineae. The chordae tendineae pull downward on the valve cusps, thereby enabling the AV valves to seal properly while resisting the upward force of ventricular pressure. The cusps of these valves (and also those of the *semilunar valves,* discussed next) are anchored at their bases to rings of connective tissue formed by the fibrous skeleton.

In addition to the AV valves, other valves, called **semilunar valves,** are located between the ventricles

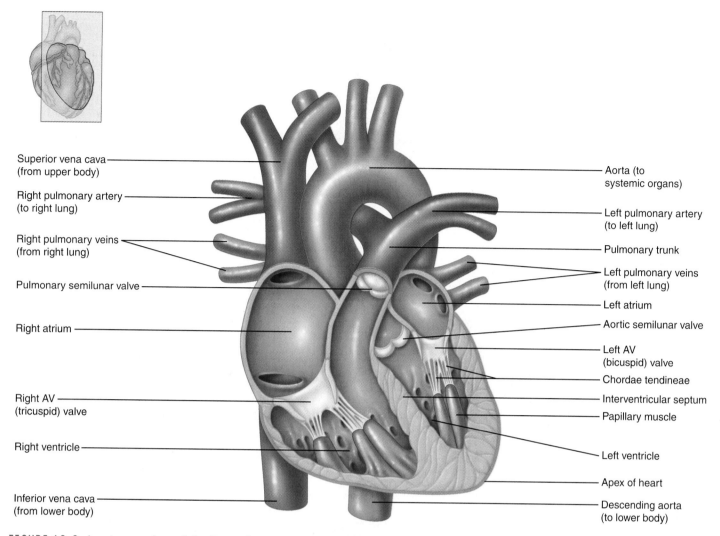

Superior vena cava
(from upper body)

Right pulmonary artery
(to right lung)

Right pulmonary veins
(from right lung)

Pulmonary semilunar valve

Right atrium

Right AV
(tricuspid) valve

Right ventricle

Inferior vena cava
(from lower body)

Aorta (to
systemic organs)

Left pulmonary artery
(to left lung)

Pulmonary trunk

Left pulmonary veins
(from left lung)

Left atrium

Aortic semilunar valve

Left AV
(bicuspid) valve

Chordae tendineae

Interventricular septum

Papillary muscle

Left ventricle

Apex of heart

Descending aorta
(to lower body)

FIGURE 12.2 A cutaway view of the heart showing the atria, ventricles, atrioventricular valves, and connections to major blood vessels.

and arteries. The **aortic semilunar valve** (or *aortic valve*) is located between the left ventricle and the aorta, and the **pulmonary semilunar valve** *(pulmonary valve)* is located between the right ventricle and the pulmonary trunk. The function of these valves is similar to that of the AV valves—to permit blood to flow forward while preventing it from flowing backward (Figure 12.5). The aortic and pulmonary valves open when ventricular pressure is greater than arterial pressure (when the ventricles contract). This allows blood to leave the ventricles and enter the arteries. When the ventricles relax and ventricular pressure becomes lower than arterial pressure, the valves close, thus preventing blood from flowing back into the ventricles from the arteries.

The cardiovascular system's major components and their functions are summarized in Table 12.1 on page 376.

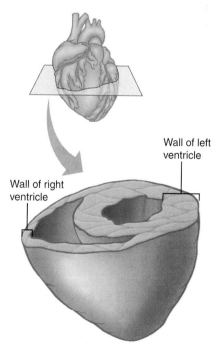

Wall of left
ventricle

Wall of right
ventricle

FIGURE 12.3 **Right and left ventricle muscle thickness.** *The greater thickness of the muscle of the left ventricle allows it to generate the force necessary to pump blood throughout the body.*

FIGURE 12.4 Action of the atrioventricular valves.

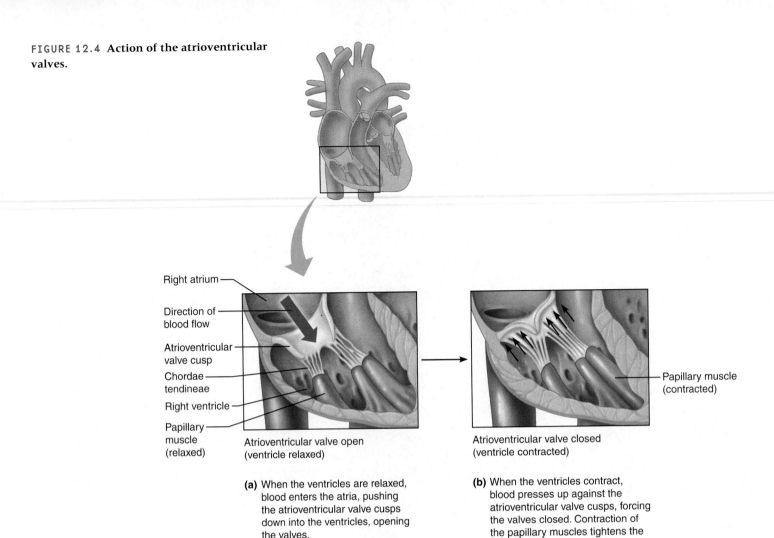

Right atrium

Direction of blood flow

Atrioventricular valve cusp

Chordae tendineae

Right ventricle

Papillary muscle (relaxed)

Atrioventricular valve open (ventricle relaxed)

Papillary muscle (contracted)

Atrioventricular valve closed (ventricle contracted)

(a) When the ventricles are relaxed, blood enters the atria, pushing the atrioventricular valve cusps down into the ventricles, opening the valves.

(b) When the ventricles contract, blood presses up against the atrioventricular valve cusps, forcing the valves closed. Contraction of the papillary muscles tightens the chordae tendineae, preventing the valve cusps from being pushed into the atria.

Quick Test 12.1

1. What is the liquid portion of the blood called? What two types of cells are found in the blood?

2. What are the five types of blood vessels found in the vasculature?

3. What are the "receiving chambers" of the heart called? the "pumping chambers"?

4. Where are the atrioventricular valves located? What is their function? What is the location and function of the semilunar valves?

THE PATH OF BLOOD FLOW THROUGH THE HEART AND VASCULATURE

The cardiovascular system is often referred to as the *circulatory system* because blood follows an essentially circular path as it travels through the body. Although the sheer number of blood vessels makes the structure of the circulatory system complex, the layout of the system is simple in concept. In this section we look at the path of blood flow through the circulatory system.

The general pattern of blood flow through the circulatory system is shown in Figure 12.6. In the diagram you can see that the circulatory system consists of two divisions: the **pulmonary circuit,** which consists of all blood vessels within the lungs and also those connecting the lungs with the heart, and the **systemic circuit,** which encompasses the rest of the blood vessels in the body. Note that these two divisions are supplied with blood by different sides of the heart. The *right* heart supplies blood to the pulmonary circuit, whereas the *left* heart supplies blood to the systemic circuit. Notice that blood on one side of the heart never mixes with blood on the other side. Thus the heart is actually two separate pumps housed within a single organ.

The pulmonary and systemic circuits both possess dense networks of capillaries called *capillary beds*, where exchange of nutrients and gases (oxygen and carbon

FIGURE 12.5 **Semilunar valve operation.**

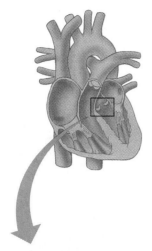

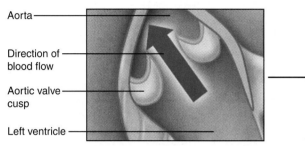
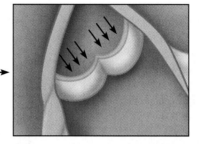

Aorta

Direction of
blood flow

Aortic valve
cusp

Left ventricle

Aortic semilunar valve open
(ventricle contracted)

Aortic semilunar valve closed
(ventricle relaxed)

(a) When the ventricles contract, blood
presses up against the semilunar
valve cusps, forcing the valves
open and allowing blood to flow
into the aorta and pulmonary
artery.

(b) When the ventricles relax, blood in
the aorta and pulmonary artery
presses down against the valve
cusps, forcing them to close.

dioxide) takes place. In pulmonary capillaries, oxygen (O_2) moves into the blood from air in the lungs while carbon dioxide (CO_2) leaves the blood. When it leaves pulmonary capillaries, the blood is relatively rich in oxygen and is thus called *oxygenated* blood. Capillary beds in the systemic circuit are located in all other organs and tissues in the body besides the lungs. In these organs and tissues, cells consume oxygen and generate carbon dioxide, so as blood travels through systemic capillaries, oxygen leaves the blood and carbon dioxide enters. Blood leaving these capillaries is called *deoxygenated* blood because it is relatively oxygen poor.

When oxygenated blood becomes deoxygenated or vice versa, its color actually changes. Oxygenated blood is bright red, whereas deoxygenated blood is a darker red. Despite the fact that both forms of blood are red, the terms *red blood* and *blue blood* are frequently used to denote oxygenated and deoxygenated blood, respectively. Deoxygenated blood is called *blue* because it imparts a bluish color to veins in view beneath the skin. Oxygenated and deoxygenated blood are indicated by red and

blue colors in Figure 12.6.

As blood flows through the circulatory system, it travels through the pulmonary and systemic circuits in an alternating fashion, returning to the heart each time (see Figure 12.6). Let us follow the path of blood flow step-by-step, starting in the left ventricle:

1. The left ventricle pumps oxygenated blood into the **aorta,** a major artery whose branches carry blood to capillary beds of all organs and tissues in the systemic circuit.

2. Blood becomes deoxygenated in systemic tissues and then travels back to the heart in the **venae cavae** (singular: *vena cava*), two large veins that carry blood into the right atrium. (The *superior* vena cava carries blood from parts of the body above the diaphragm, whereas the *inferior* vena cava carries blood from parts below the diaphragm.)

3. From the right atrium, blood passes through the tricuspid valve into the right ventricle.

TABLE 12.1 COMPONENTS OF THE CARDIOVASCULAR SYSTEM

COMPONENT	FUNCTION
BLOOD	
Plasma	Fluid in which cells and platelets are suspended; contains various solutes including electrolytes and proteins
Erythrocytes (red blood cells)	Contain hemoglobin, which carries oxygen and carbon dioxide
Leukocytes (white blood cells)	Participate in immune and defense-related functions
Platelets	Participate in blood clotting
BLOOD VESSELS	
Arteries	Carry blood away from heart
Arterioles	Carry blood toward capillaries; regulate blood flow
Capillaries	Allow exchange of materials between blood and interstitial fluid
Venules	Carry blood away from capillaries; participate in some exchange of materials
Veins	Carry blood toward heart
HEART	
Right atrium	Receives blood returning from systemic circuit
Right ventricle	Pumps blood to pulmonary circuit
Left atrium	Receives blood returning from pulmonary circuit
Left ventricle	Pumps blood to systemic circuit

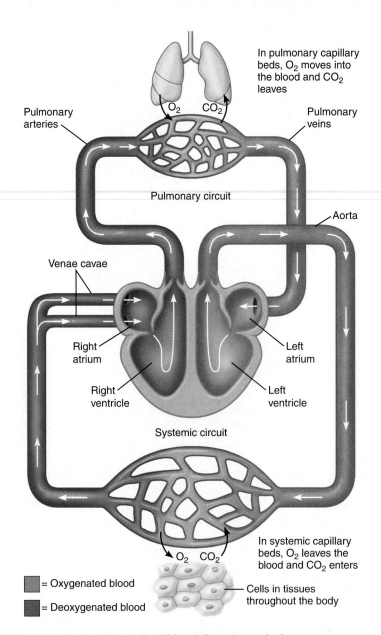

FIGURE 12.6 **The path of blood flow through the cardiovascular system.** *The pulmonary and systemic circuits and major blood vessels connecting with the heart are shown. Arrows indicate direction of blood flow.*

4. The right ventricle pumps blood into the pulmonary trunk, which almost immediately branches into the **pulmonary arteries,** which carry deoxygenated blood to the lungs. Note that the pulmonary arteries are the only arteries in the body carrying deoxygenated blood. They are called *arteries* because they carry blood away from the heart.

5. Blood becomes oxygenated in the lungs and then travels to the left atrium in the **pulmonary veins.** These are the only veins in the body carrying oxygenated blood and are called *veins* because they carry blood toward the heart.

6. From the left atrium, blood passes through the bicuspid valve into the left ventricle, which is where we started. The whole cycle then repeats.

Quick Test 12.2

1. Which organs are supplied by blood flowing through the *pulmonary* circuit? the *systemic* circuit?

2. Arrange the order of the following terms so that they correctly describe the path of blood flow through the body: *left ventricle, pulmonary arteries, bicuspid valve, aortic semilunar valve, pulmonary semilunar valve, right ventricle, aorta, pulmonary veins, vena cava, tricuspid valve, right atrium, left atrium.*

ELECTRICAL ACTIVITY OF THE HEART

For the heart to adequately pump blood through the circulatory system, the cardiac muscle must contract in a highly synchronized manner, with the contraction of both atria first followed by contraction of both ventricles. Cardiac contractions are coordinated by an elaborate conduction system that determines the sequence of excitation of cardiac muscle cells.

The Conduction System of the Heart

In Chapter 11 we saw that cardiac muscle, unlike skeletal muscle, does not require commands from the central nervous system to contract. Instead, cardiac muscle contractions are triggered by signals originating from within the muscle itself. For this reason, the contractile activity of cardiac muscle is said to be *myogenic*. (The contractile activity of skeletal muscle is said to be *neurogenic* because the initial signal for contraction originates in neurons.) The ability of the heart to generate signals that trigger its contractions on a periodic basis—that is, to generate its own rhythm—is called **autorhythmicity.** The heart's autorhythmicity is due to the action of a small percentage of muscle cells, called autorhythmic cells, that generate little or no contractile force but that are critical to the heart's pumping action because they are specialized to initiate and/or conduct the action potentials that trigger heart muscle contractions. These cells make up the **conduction system** of the heart. The cells that generate the contractile force are called *contractile cells.*

Contractions of the heart are initiated by specialized muscle cells called **pacemaker cells,** which spontaneously generate action potentials (see p. 363). As their name suggests, *pacemaker* cells determine the rate or *pace* of the heartbeat by firing action potentials on a regular basis. Pacemaker cells are concentrated primarily in two specific regions of the myocardium: the **sinoatrial node** (SA node), located in the wall of the upper right atrium near where it joins with the superior vena cava, and the **atrioventricular node** (AV node), which is located near the tricuspid valve in the interatrial septum. Although cells exhibiting spontaneous pacemaker activity are located in nearly all parts of the heart, contractions of the heart muscle are normally driven only by cells in the SA node. (The mechanism of action potential generation in the heart is discussed in detail later in this chapter.)

Pacemaker cells are closely associated with other muscle fibers, called *conduction fibers,* which are specialized to quickly conduct the action potentials generated by the pacemaker cells from place to place through the myocardium, thus triggering heart muscle contractions. Although all cardiac muscle fibers are capable of transmitting action potentials, conduction fibers differ in that they are larger in diameter and can therefore conduct action potentials more rapidly than "ordinary" fibers (see p.

195). Action potentials can travel up to 4 meters per second in certain parts of the conduction system, as opposed to 0.3–0.5 meter per second in most cardiac muscle fibers.

The association of pacemaker cells and conduction fibers is important because proper pumping action of the heart requires not only that action potentials be generated on a rhythmic basis (the job of the pacemaker cells), but that they be rapidly conducted to the right locations in the heart muscle at the right times (the job of the conduction fibers). Together, the pacemaker cells and conduction fibers make up the conduction system of the heart.

Once an action potential is initiated in pacemaker cells, action potentials move rapidly through the conduction cells to coordinate the spread of excitation. The conduction system causes a wave of excitation to move first through the atria, causing them to depolarize and then contract as a unit. Next, the wave of excitation moves through the ventricles, causing them to contract as a unit. Rapid transmission of action potentials from pacemakers to conduction fibers to the contractile cells is possible because all cardiac muscle cells are connected to their neighbors by gap junctions, which permit electrical current to pass from one cell to another across their plasma membranes. In the heart, gap junctions are concentrated in structures called *intercalated disks,* which form the junctions between adjacent muscle fibers (Figure 12.7). Intercalated disks also contain large numbers of *desmosomes,* areas in which protein fibers link adjacent cells together, forming a physical bond between them that resists mechanical stress. This is important because it enables the myocardium to resist stretching, which occurs every time the heart fills with blood, and also because it enables the myocardium to withstand the tension that is generated every time the muscle cells contract.

Initiation and Conduction of an Impulse During a Heartbeat

The sequence of electrical events that normally triggers the heartbeat occurs as follows (Figure 12.8):

1. An action potential is initiated in the SA node. From the SA node, impulses travel to the AV node by way of internodal pathways—systems of conduction fibers that run through the walls of the atria. As these signals move through the internodal pathways, they also spread through the bulk of the atrial muscle.

2. The impulse is conducted to cells of the AV node, which transmit action potentials less rapidly than other cells of the conduction system. As a result, the impulse is momentarily delayed (by about 0.1 second) before moving onward.

3. From the AV node, the impulse travels through the atrioventricular bundle, also known as the *bundle of His* (pronounced "hiss"), a compact bundle of muscle

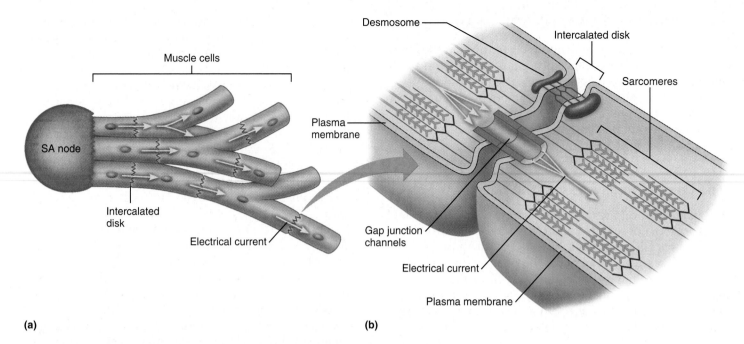

FIGURE 12.7 Electrical connections between cardiac muscle cells. (a) *An action potential generated spontaneously in cells of the SA node spreads to adjacent muscle cells by means of electrical current passing through gap junctions in intercalated disks.* **(b)** *A schematic view of the junction between two adjacent muscle cells showing a gap junction and a desmosome.*

Besides cardiac muscle, what other type of muscle possesses gap junctions?

fibers located in the interventricular septum. This is the only electrical connection between the atria and the ventricles, which are otherwise separated by the fibrous skeleton.

④ The signal travels only a short distance through the atrioventricular bundle before it splits into left and right bundle branches, which conduct impulses to the left and right ventricles, respectively.

⑤ From the bundle branches, impulses travel through an extensive network of branches referred to as *Purkinje fibers,* which spread through the ventricular myocardium. From these fibers, impulses travel through the rest of the myocardial cells.

Control of the Heartbeat by Pacemakers

Although the SA node and the AV node are both capable of generating spontaneous action potentials, the heartbeat is almost always triggered by impulses originating from the SA node. The AV node rarely initiates contractions for two reasons. The first is that action potentials originating in the SA node travel through the AV node on their way to the ventricles. When this happens, cells in the AV node go into a *refractory period,* during which they cannot generate their own action potentials. The second reason is that the SA node has a higher "beat frequency" than the AV node. This means that if we were to measure the action potential frequency in isolated cells of either node under normal resting conditions, we would see that SA node cells would fire more frequently than AV node cells—about 70 impulses/minute for the SA node, as opposed to 50 impulses/minute for the AV node. Thus the AV node rarely has a chance to fire an action potential because the SA node always "beats it to the punch."

However, if the SA node fails to fire an action potential or if it slows down dramatically—that is, if it "misses a beat," so to speak—the AV node *will* initiate action potentials, which travel through the conducting system and trigger ventricular contraction in the normal manner. The AV node can also take over control of the heartbeat if conduction between the nodes is blocked or slowed down for some reason. In these circumstances, the AV node functions as an "emergency backup system" that keeps the ventricles beating. If for some reason the AV node is unable to drive ventricular contraction, the heart has yet another backup system: Certain cells in the Purkinje fibers (sometimes referred to as *idioventricular pacemakers*) can take over. However, the beat frequency of these cells is only 30–40 impulses per minute.

Smooth muscle

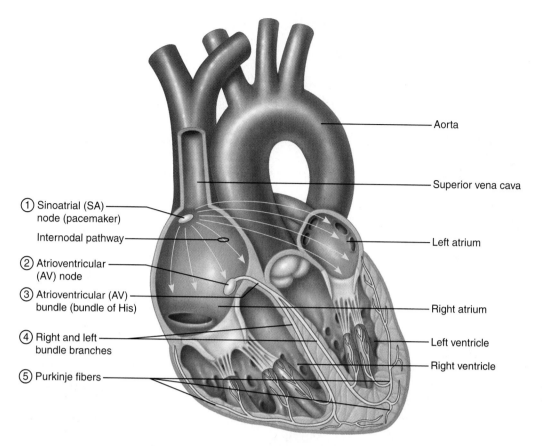

Sinoatrial (SA) node (pacemaker)
Internodal pathway
Atrioventricular (AV) node
Atrioventricular (AV) bundle (bundle of His)
Right and left bundle branches
Purkinje fibers

Aorta
Superior vena cava
Left atrium
Right atrium
Left ventricle
Right ventricle

FIGURE 12.8 The conduction system of the heart. *The pathway of impulse conduction through the heart is indicated in this longitudinal section.*

Spread of Excitation Through the Heart Muscle

As impulses propagate through the heart muscle, they travel in an orderly pattern as a kind of wavefront—a "wave of excitation." As this wave of excitation spreads, contraction of the muscle follows. The pattern of excitation is shown in Figure 12.9.

The wave of excitation starts at the SA node and then spreads outward through the atria. The wave then "funnels" through the atrioventricular bundle by way of the AV node, which acts as a kind of bottleneck due to the relative slowness of impulse conduction in this region. This delay is essential for efficient cardiac function; it allows the wave of excitation to spread completely through the atria before it reaches the ventricles, thus ensuring that atrial contraction is complete before ventricular contraction starts. Given that the function of atrial contraction is to drive blood into the ventricles, if no such delay occurred, ventricular contraction would work against the pumping action of the atria.

Once impulses reach the bundle branches and the Purkinje fibers, they are carried relatively quickly to the lower portion of the ventricles. From there, the wave of excitation fans out through the entire ventricular muscle. Thus ventricular contraction begins at the apex and spreads upward. This makes sense when you consider that blood exits the ventricles from the top (see Figure 12.2, p. 373). Ventricular contraction is thus reminiscent of how one should squeeze a tube of toothpaste—from the bottom up.

The Ionic Basis of Electrical Activity in the Heart

We now know that the heartbeat is triggered by action potentials that originate in pacemaker cells and propagate through the heart muscle in an orderly, predictable fashion. Here we examine the cellular mechanisms responsible for generating these electrical signals, beginning with events occurring in the membrane of pacemaker cells.

Electrical Activity in Pacemaker Cells

A cardiac contractile cell fires an action potential only when it is depolarized to threshold by a stimulus. Normally this stimulus is a circulating electrical current that

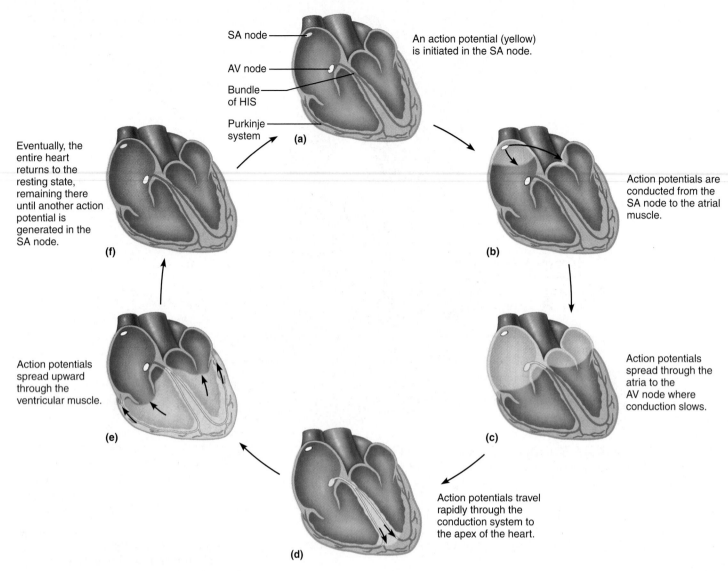

SA node

AV node

Bundle of HIS

Purkinje system **(a)**

An action potential (yellow) is initiated in the SA node.

Eventually, the entire heart returns to the resting state, remaining there until another action potential is generated in the SA node.

(f)

Action potentials are conducted from the SA node to the atrial muscle.

(b)

Action potentials spread upward through the ventricular muscle.

(e)

Action potentials spread through the atria to the AV node where conduction slows.

(c)

Action potentials travel rapidly through the conduction system to the apex of the heart.

(d)

FIGURE 12.9 The spread of action potentials through the heart. *The sequence of electrical excitation during a single heartbeat, starting with depolarization of the SA node* **(a)** *and ending with the return of the heart to the resting state* **(f).**

originates in neighboring cells that are firing action potentials. We have seen that this current enters the cell through gap junctions that connect it with its neighbors. After entering, the current exits the cell by passing through the plasma membrane, and in doing so it triggers depolarization.

Recall that pacemaker cells are different because they can fire action potentials in the absence of any external stimulus, and do so in a regular, periodic fashion. A pacemaker cell is able to fire action potentials spontaneously because it does not have a steady resting potential. After an action potential, a pacemaker cell immediately begins to depolarize slowly and continues to do so until its membrane potential reaches threshold, which triggers another action potential (Figure 12.10a). Following this, the membrane potential returns to about −60 to −70 mV

and then begins another round of slow depolarization until another action potential is triggered. The slow depolarizations or "ramps" that lead up to each action potential are referred to as **pacemaker potentials.**

In pacemaker cells and other cardiac muscle cells, electrical signals are caused by changes in plasma membrane ion permeability brought about by the opening and closing of specific types of ion channels, just as in any other type of cell. To understand how these permeability changes affect the membrane potential, recall the following rule, which we first encountered in Chapter 6: *As a membrane's permeability to a particular ion increases relative to that of other ions, the membrane potential moves toward the equilibrium potential of that ion.* In cardiac muscle cells, the most important permeability changes involve sodium, potassium, and calcium ions (Na^+, K^+, and Ca^{2+}, respec-

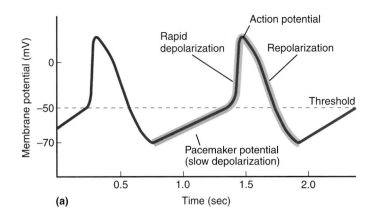

(a)

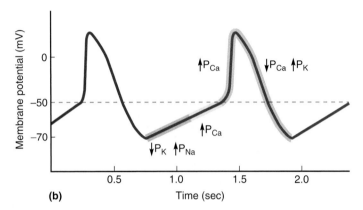

(b)

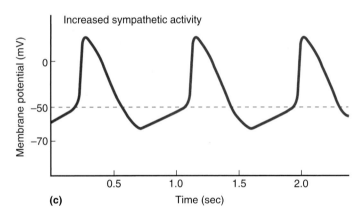

(c)

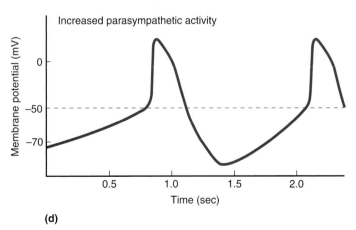

(d)

FIGURE 12.10 Electrical activity in a pacemaker cell. (a) *A recording of the membrane potential showing action potentials and pacemaker potentials.* **(b)** *Changes in membrane permeability to potassium, sodium, and calcium (P_K, P_{Na}, and P_{Ca}, respectively) occurring during a pacemaker potential and the subsequent action potential.* **(c)** *Effect of increased sympathetic activity to pacemaker cells.* **(d)** *Effect of increased parasympathetic activity to pacemaker cells.*

tively). Because ion concentrations in cardiac muscle cells are similar to those in other cells—with the intracellular fluid being rich in potassium but poor in sodium and calcium with respect to extracellular fluid—the equilibrium potential of potassium is negative, whereas the equilibrium potentials of sodium and calcium are both positive. (Approximate values are $E_K = -94$ mV, $E_{Na} = +60$mV, and $E_{Ca} = +130$mV.) Therefore, increased sodium or calcium permeability (P_{Na} or P_{Ca}) tends to make the membrane potential become more positive, whereas increased potassium permeability (P_K) tends to make it become more negative.

In pacemaker cells, electrical signals are triggered by changes in P_K, P_{Na}, and P_{Ca}, as shown in Figure 12.10b. The slow depolarization that occurs in the early stages of the pacemaker potential is due to closing of potassium channels and opening of so-called *funny channels*. Potassium channels open during repolarization of the action potential, and then close when the membrane returns to its polarized state. Funny channels, so named because investigators noticed that they had some unusual characteristics, open after the cell repolarizes and allow sodium and potassium ions to cross the plasma membrane. With potassium channels closed and funny channels open during the early stages of the pacemaker potential, potassium movement out of the cell decreases, whereas sodium movement into the cell increases, causing the initial depolarization.

The funny channels, however, are open for only a brief period of time, closing when the membrane potential approaches -55 mV, approximately 5 mV short of the threshold needed to generate an action potential. However, the initial depolarization triggers the opening of voltage-gated calcium channels. This raises P_{Ca}, which depolarizes the cell even further. Although these channels (called *T-type channels*) stay open for only a short time before inactivating, the resulting depolarization triggers the opening of a second population of voltage-gated calcium channels *(L-type channels)*, which stay open longer and inactivate only slowly. The result is a large increase in P_{Ca} that produces the rapid depolarization characteristic of the upswing of the action potential. (As these calcium channels open, they also allow some sodium to flow into the cell, which increases P_{Na} and adds to the depolarizing effect.) This depolarization triggers the opening of potassium channels and, consequently, a rise in P_K that occurs shortly after the increase in P_{Ca} and acts to pull the

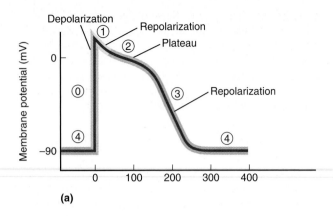

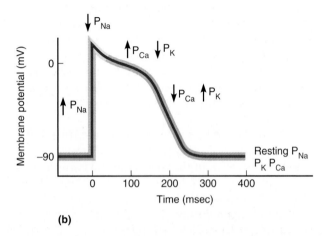

FIGURE 12.11 The cardiac action potential. (a) *An action potential recorded from a ventricular muscle cell.* **(b)** *Changes in membrane permeability to sodium, potassium, and calcium (P_{Na}, P_K, and P_{Ca}, respectively) occurring during the various phases of the action potential.*

membrane potential back down. The resulting fall in potential removes the stimulus for calcium channel opening, allowing these channels to begin closing. This reduces P_{Ca} and decreases the flow of calcium into the cell, which works along with the increase in P_K to repolarize the membrane and terminate the action potential.

Because the heart initiates its own action potentials, it does not require neural input to trigger its contractions. However, as discussed later in this chapter, autonomic neurons do exert control over the rate and force of these contractions. The autonomic influence on rate is exerted by sympathetic and parasympathetic neurons that provide direct input to pacemaker cells in the SA node. Increased activity in sympathetic neurons augments the opening of funny channels and T-type calcium channels, causing an increase in the slope of the spontaneous depolarization such that threshold for an action potential is reached more quickly (Figure 12.10c). Thus sympathetic nerve activity increases the frequency of action potentials in the SA node, which increases the rate of contraction (the heart rate). Increased activity in parasympathetic

neurons augments the opening of potassium channels and suppresses the opening of funny channels and T-type calcium channels, causing a decrease in the slope of the spontaneous depolarization and a hyperpolarization of the membrane potential (Figure 12.10d). Both of these effects decrease the frequency of action potentials in the SA node, thereby decreasing the heart rate.

Electrical Activity in Cardiac Contractile Cells

Not all action potentials in cardiac contractile cells are alike; instead, they differ from place to place in the heart in regard to their shapes (time courses) and speeds of propagation, because contractile cells differ not only in their physical dimensions, but also in the type and number of ion channels they possess. Despite these differences, two important events characterize most cardiac action potentials: (1) During a typical cardiac action potential, P_K *decreases* due to the action of a certain type of voltage-gated potassium channel that *closes* in response to depolarization. (Recall that in pacemaker cells and most other excitable tissues, P_K increases during an action potential because these tissues contain voltage-gated potassium channels that open in response to depolarization.) (2) During a cardiac action potential, depolarization causes the opening of voltage-gated *calcium* channels, which not only affects the membrane potential but also is instrumental in triggering muscle cell contractions.

The majority of ventricular muscle cells, which make up the bulk of the myocardium, are unlike pacemaker cells in that they have stable resting potentials. They also have longer-lasting action potentials with a distinctive shape that can be divided into several phases (designated 0–4), as shown in Figure 12.11a. Permeability changes occurring during each of these phases are described next and summarized in Figure 12.11b:

⓪ Phase 0: Phase 0 of the cardiac action potential is similar to the upswing of a neuronal action potential and is caused by similar events: Depolarization of the membrane triggers the opening of voltage-gated sodium channels, which raises P_{Na} and increases the flow of sodium ions into the cell. Consequently, the membrane potential becomes more positive, which triggers the opening of more sodium channels, additional increases in P_{Na}, more depolarization, and so on. The result is a rapid rise in membrane potential that peaks at between +30 and +40 mV.

① Phase 1: The sodium channels that were opened in phase 0 start to inactivate, which reduces P_{Na}. This decreases the flow of sodium into the cell and causes the membrane potential to fall toward more negative values. The membrane potential drops only a small amount, however, because the depolarization of the membrane that began in phase 0 has set into motion two additional events that are also occurring at this time: (1) the closing of voltage-gated potassium channels (known as inward rectifier channels), which reduces P_K and decreases the flow of potassium out of the cell; and (2) the opening of voltage-gated calcium channels (L-type channels), which raises P_{Ca} and increases the flow of calcium into the cell. Both of these changes act to depolarize the membrane, thereby counteracting the effect of sodium channel inactivation.

② Phase 2: During this phase, which is also referred to as the plateau phase, most of the potassium channels that were closed in phase 1 stay closed, so that P_K remains lower than its resting value. At the same time, most of the calcium channels that opened in phase 1 remain open, and P_{Ca} remains elevated. The lowered P_K and elevated P_{Ca} both act to keep the membrane in its depolarized state.

③ Phase 3: During this phase P_K increases, partly because of the action of a second population of potassium channels similar to those in neurons (called *delayed rectifier channels*), which open in response to depolarization. These channels begin to open during phases 1 and 2 but do not exert a significant influence on the membrane potential until phase 3 because they open slowly. As P_K rises, the flow of potassium out of the cell increases, which pulls the membrane potential down toward more negative values. Furthermore, this fall in potential removes the stimulus that kept the inward rectifier channels closed in phase 2, and as a consequence these channels begin to open, which raises P_K even further. The fall in potential also removes the stimulus that kept the calcium channels open during phase 2 and allows them to begin closing, which lowers P_{Ca} and reduces the flow of calcium into the cell. This works hand in hand with the increase in P_K to repolarize the membrane, thereby terminating the action potential.

④ Phase 4: During this phase, which corresponds to the *resting potential*, P_K, P_{Na}, and P_{Ca} are at their resting values. Because P_K is much greater than P_{Na} or P_{Ca} under these conditions, the membrane potential is around -90 mV, which is close to the equilibrium potential of potassium.

Excitation-Contraction Coupling in Cardiac Contractile Cells

The mechanism by which a cardiac action potential stimulates contraction is similar to that for skeletal muscle in many respects. An action potential travels down the T tubules, causing voltage-sensitive calcium channels on the sarcoplasmic reticulum to open, releasing calcium into the cytosol. Calcium binds to troponin, shifting tropomyosin off of the binding sites on actin for myosin, and crossbridge cycling occurs. Some of the calcium that enters the contractile cell during the plateau phase of the

action potential also binds to troponin, thereby contributing to crossbridge cycling. More important, calcium that enters during the plateau phase acts on the voltage-sensitive channels that trigger calcium release from the sarcoplasmic reticulum and stimulates them to stay open longer. As a result, the sarcoplasmic reticulum releases more calcium with each action potential. This phenomenon is known as *calcium-induced calcium release*.

Relaxation of cardiac muscle requires removal of calcium from the cytosol, which occurs by three mechanisms: (1) As in skeletal muscle, a Ca^{2+}-ATPase located in the membrane of the sarcoplasmic reticulum actively transports calcium from the cytosol into the lumen of the sarcoplasmic reticulum. (2) Cardiac muscle also has a Ca^{2+}-ATPase located in the plasma membrane that actively transports calcium from the cytosol into the interstitial fluid. (3) Cardiac muscle also has a Na^+-Ca^{2+} exchanger in the plasma membrane that actively transports calcium out of the cell by countertransport with sodium.

Recording the Electrical Activity of the Heart with an Electrocardiogram

The electrocardiogram (ECG or EKG; the *K* is for the German form of the word, *elektrokardiograph*) is a noninvasive means for monitoring the electrical activity of the heart. To understand the interpretation of ECGs, it is important to remember that the recorded electrical events cause the contraction of cardiac muscle. Physicians use ECG recordings to determine whether problems exist in the electrical activity of the heart (Discovery: Myocardial Infarction, www.physiologyplace.com, Challenge Yourself); ECG recordings do not give information about mechanical problems of the heart unless those problems result from an electrical problem. An ECG, for example, would not be useful for identifying a mechanical malfunction of a valve.

The ECG is a record of the overall spread of electrical current through the heart as a function of time during the cardiac cycle. The ECG is usually recorded by means of electrodes placed on the skin. The concept of recording ECGs is similar to that for recording EEGs described in Chapter 8; namely, electrical activity generated in nervous or muscle tissue spreads through the body because body fluids function as conductors. The more synchronized the activity, the larger the amplitude of signals that are recorded at a distance from the source. Because the electrical activity of the heart is highly synchronized, relatively large amplitude electrical potentials that correspond to distinct electrical phases of the heart can be detected at the surface of the skin.

A Dutch physiologist, William Einthoven, developed the technique of ECG recordings. The procedure for standard ECG recording is based on an imaginary equilateral triangle surrounding the heart. The triangle is expanded until its corners fall on the right arm, left arm, and left

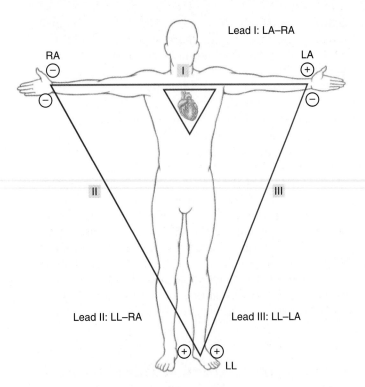

Lead I: LA–RA

RA (−)
(−)

LA (+)
(−)

I

II   III

Lead II: LL–RA   Lead III: LL–LA

(+) (+)
LL

FIGURE 12.12 Einthoven's triangle. *Three electrodes are placed on limbs to form an equilateral triangle around the heart.*

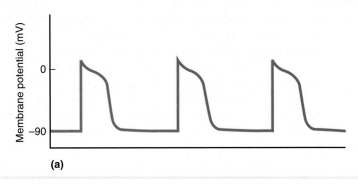

(a)

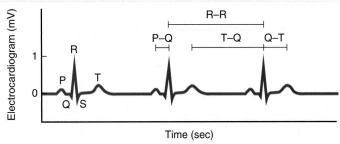

Time (sec)

Component	Amplitude (mV)	Duration (sec)
P wave	0.2	0.10
QRS complex	1.0	0.08–0.12
T wave	0.2–0.3	0.16–0.27
P–Q interval	N/A	0.12–0.21
Q–T interval	N/A	0.30–0.43
T–Q segment	N/A	0.55–0.70
R–R interval	N/A	0.85–1.00

(b)

FIGURE 12.13 Electrical activity of the heart. (a) *Recording of the membrane potential in a ventricle contractile cell.* **(b)** *Recording of a lead II ECG. The table gives normal values for ECG waves, intervals, and segments.*

leg, a pattern known as Einthoven's triangle (Figure 12.12). Electrodes placed on the skin at the corners of the triangle are connected in pairs to a voltage-measuring device such as an oscilloscope or chart recorder. Certain pairs of electrodes are referred to as *leads* and are designated by Roman numerals. One electrode in each lead is designated as the positive electrode, the other as the negative.

Each specific lead detects the difference in the surface electrical potential between the positive and negative electrodes. Lead I detects the potential at the left arm minus that at the right arm; lead II detects the potential at the left leg minus that at the right arm; lead III detects the potential at the left leg minus the left arm. The direction of the recorded waveforms (up or down) depends on whether the difference in potential between the two electrodes is positive (which gives an upward deflection) or negative (which gives a downward deflection).

The waveforms recorded with a standard lead II ECG and the action potential of a ventricular contractile cell are shown in Figure 12.13. ECGs are recorded on chart paper at a rate of 25 mm/sec with an amplitude of 1 mV/cm. The ECG normally shows three characteristic waveforms: (1) The **P wave** is an upward deflection occurring as a result of atrial depolarization. (2) The **QRS complex** is a series of sharp upward and downward deflections as a result of ventricular depolarization; it is correlated with phase 0 of the ventricular contractile cell action potential. (3) The **T wave** is an upward deflection

caused by ventricular repolarization; it is correlated with phase 3 of the ventricular contractile cell action potential. Between the waves, a normal ECG trace consists of a horizontal line, called the *isoelectric line,* because no changes in electrical activity are occurring.

It is important to note that although the phases of the ECG are due to action potentials traveling through the heart muscle, the ECG is *not* simply a recording of an action potential. During the heartbeat, cells fire action potentials at different times, and the ECG reflects *patterns* of action potential firing in the entire population of cells that make up the heart muscle.

In addition to waves, certain intervals and segments can provide important information about the function of the heart. The *P-Q* or *P-R interval* occurs between the onset of the P wave and the onset of the QRS complex and is an estimate of the time of conduction through the AV node. The *Q-T interval* is the time from the onset of the QRS complex to the end of the T wave and is an estimate of the time the ventricles are contracting, called ventricular systole. The *T-Q segment* is the time from the end of the T wave to the beginning of the QRS complex and is an estimate of the time the ventricles are relaxing, called ventricular diastole. The *R-R interval* is the time between the peaks of two successive QRS complexes; it represents the time between heartbeats. Heart rate can be determined by dividing 60 seconds by the R-R interval. If the R-R interval is 1 second, for example, then the heart rate is 60 beats per minute.

Figure 12.14 compares normal ECGs to those recorded during examples of abnormal electrical activity of the heart, called *cardiac arrhythmias*. Abnormal SA nodal firing can cause either a sinus tachycardia, which is an abnormally fast resting heart rate (greater than 100 beats/min.), or a sinus bradycardia, which is an abnormally slow resting heart rate (less than 50 beats/min.). Altered conduction through the AV node can cause various degrees of heart block: During first degree heart block, conduction through the AV node is slowed, causing a longer delay in AV nodal conduction (an increased P-Q interval). During second degree heart block, conduction through the AV node does not always occur. If conduction does not occur, the ventricles do not depolarize (as shown by the absence of a QRS complex and T wave) and thus do not contract. The 1:1 relationship between atrial contractions and ventricular contractions, therefore, is lost. During third degree heart block, conduction through the AV node does not occur at all, causing complete dissociation of atrial and ventricular contractions. The atria contract at the rate of SA nodal discharge, but the ventricles contract at the rate of Purkinje fiber discharge, which is only about 30–40 times per minute. This slow rate of ventricular contraction is insufficient to supply the body with the oxygen and nutrients required, and thus third degree block is deadly if not immediately corrected.

Sometimes the heart is depolarized by an electrical stimulus arising outside the normal conduction pathway. Because cardiac muscle cells are connected by gap junctions, an abnormal depolarization will spread throughout the heart, causing an extra contraction called an *extrasystole*. If the depolarization occurs in an atrium, then a premature atrial contraction (PAC) occurs, and conduction follows through the AV node, causing contraction of the ventricle. If the depolarization occurs in a ventricle, then a premature ventricular contraction (PVC) occurs, and no atrial contraction precedes it. PACs and PVCs are generally of little clinical significance unless they occur at very high frequencies.

More serious arrhythmias are fibrillations, which occur when the heart muscle no longer has synchronized depolarization. In atrial fibrillation, atrial muscle fibers depolarize independently, so atrial contraction is inefficient

Normal

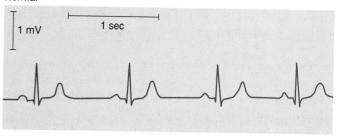

Sinus tachycardia (with inverted T wave)

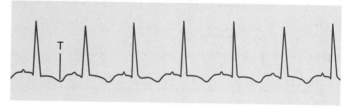

Sinus bradycardia

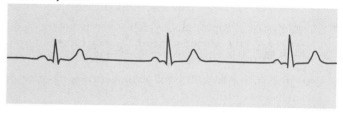

Heart block, 3rd degree

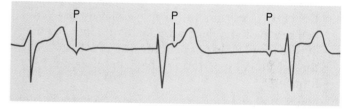

Pre-mature atrial contraction (PAC)

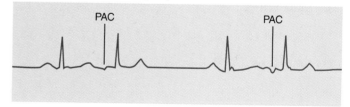

Ventricular fibrillation

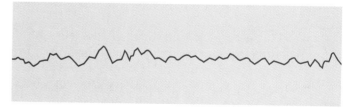

FIGURE 12.14 Lead II ECGs showing various arrhythmias.
25 mm on the horizontal = 1 sec; 1 cm on the vertical = 1 mV. Some data courtesy of C.V. Massey, University of South Alabama.

in pumping blood to the ventricle. Atrial fibrillation results in weakness and light-headedness due to decreased blood flow, but as long as the ventricles still contract at a sufficient rate, atrial fibrillation is generally not deadly because contraction of the atria contributes little to ventricular filling (most ventricular filling is passive). Ventricular fibrillation, by contrast, can cause death within minutes. When ventricular muscle cells depolarize independently, the ventricles can no longer efficiently pump the blood out to the tissues, including the brain. Clinicians must quickly *defibrillate* the ventricular muscle to keep the person alive. Defibrillation is often done by passing a large current through the chest wall to the heart such that the externally applied current depolarizes all the muscle cells at the same time, returning synchronous electrical activity to the heart.

Quick Test 12.3

1. Define the following terms: *autorhythmicity, pacemaker cell, conduction fiber, pacemaker potential.*

2. Under normal conditions, which controls the heartbeat—the SA node or the AV node? Explain why.

3. Arrange the order of the following terms so that they describe the normal path of electrical impulses through the heart: *Purkinje fibers, atrioventricular bundle, AV node, bundle branches, SA node, ventricular muscle, atrial muscle.*

4. The entry of calcium into a ventricular muscle cell helps to maintain depolarization of the membrane during the plateau phase of the action potential, but this calcium also performs what other important function?

5. Match the following terms—*P wave, QRS complex,* and *T wave*—with the following events: ventricular depolarization, ventricular repolarization, atrial depolarization.

THE CARDIAC CYCLE

The **cardiac cycle** includes all the events associated with the flow of blood through the heart during a single complete heartbeat. In our discussion we concentrate on the following aspects of the cardiac cycle: (1) the various phases in the pumping action of the heart, often called the *pump cycle;* (2) periods of valve opening and closure; (3) changes in aortic, ventricular, and atrial pressure, which reflect contraction and relaxation of the heart muscle; (4) changes in ventricular volume, which reflect the amount of blood entering and leaving the ventricle during each heartbeat; and (5) the two major heart sounds.

The relationships among the various aspects of the cardiac cycle are depicted in Figure 12.15. (The pressure graphs pertain to the left heart only; the graphs for pressures in the right heart are similar, except that the peak pressures are lower.)

The Pump Cycle

Because the cardiac cycle involves the events of one heartbeat, a complete cycle involves both ventricular contraction and ventricular relaxation. As a result, the cycle can be divided into two major stages: **systole,** the period of ventricular contraction, and **diastole,** the period of ventricular relaxation. (Even though the atria also undergo periods of contraction and relaxation—termed atrial systole and diastole, respectively—we use the terms *systole* and *diastole* to refer to ventricular events.)

We begin our examination of the cardiac cycle in the middle of diastole, a time at which the atria and ventricles are completely relaxed:

1. During mid-to-late diastole (phase 1 in Figure 12.15), blood returning to the heart via the systemic and pulmonary veins enters the relaxed atria and passes through the AV valves and into the ventricles under its own pressure; that is, the pressure in the veins is sufficiently high to drive the flow of blood into the heart (called **venous return**). As the ventricles fill, the pulmonary and aortic (semilunar) valves are closed because ventricular pressure is lower than that in the aorta and pulmonary arteries.

    Late in diastole (at the end of phase 1), the atria contract, driving more blood into the ventricles. Shortly thereafter, the atria relax and systole begins.

2. At the beginning of systole (phase 2), the ventricles contract, which raises the pressure within them. When ventricular pressure exceeds atrial pressure (which occurs very early in systole), the AV valves close; the semilunar valves remain closed because ventricular pressure is not yet high enough to force them open. At this point, no blood is flowing into or out of the ventricles because all the valves are closed. Thus even though the ventricles are contracting, the volume of blood within them remains constant, so phase 2 is termed **isovolumetric contraction** (*iso* = same). Phase 2 ends when the ventricular pressure is great enough to force open the semilunar valves so that blood can leave the ventricles.

3. In the remainder of systole (phase 3), blood is ejected into the aorta and pulmonary arteries through the open semilunar valves, and ventricular volume falls. During the exit of blood from the ventricles, referred to as **ventricular ejection,** ventricular pressure rises to a peak and then begins to decline. When it falls below aortic pressure, the semilunar valves close, ending ejection (and systole) and marking the beginning of diastole.

4. At the onset of early diastole (phase 4), the ventricular myocardium is relaxing. Some blood is present in the ventricles, and it is still under pressure, because it

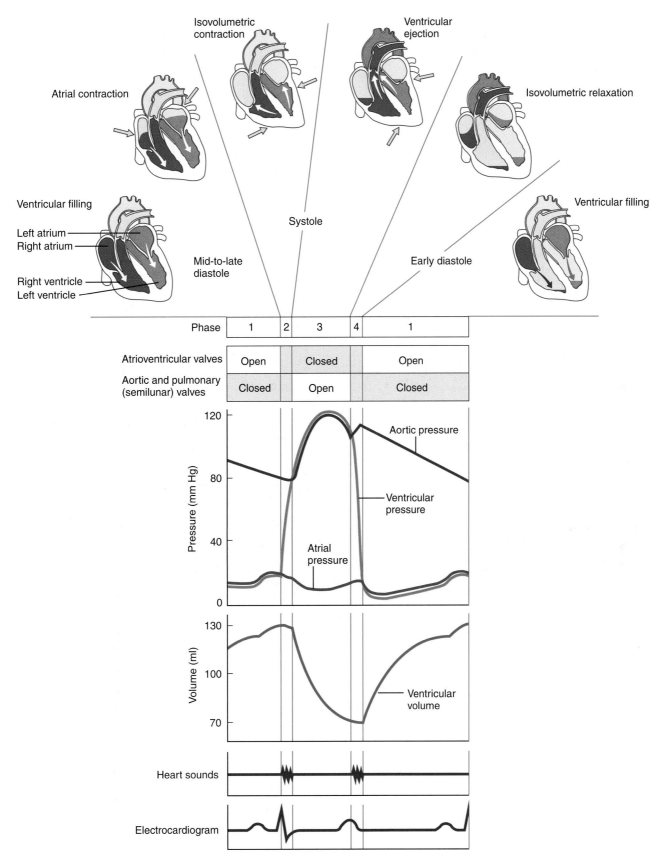

Atrial contraction

Isovolumetric contraction

Ventricular ejection

Isovolumetric relaxation

Ventricular filling

Left atrium
Right atrium

Right ventricle
Left ventricle

Ventricular filling

Systole

Mid-to-late diastole

Early diastole

Phase	1	2	3	4	1

| Atrioventricular valves | Open | Closed | | Open | |
| Aortic and pulmonary (semilunar) valves | Closed | | Open | Closed | |

Aortic pressure

Ventricular pressure

Atrial pressure

Pressure (mm Hg)
120
80
40
0

Volume (ml)
130
100
70

Ventricular volume

Heart sounds

Electrocardiogram

FIGURE 12.15 Cardiac cycle.

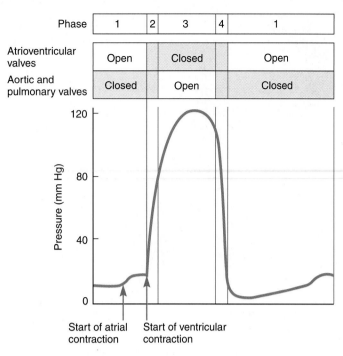

FIGURE 12.16 Ventricular pressure during the cardiac cycle.

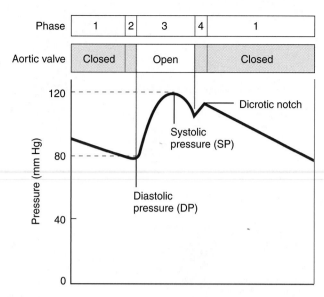

FIGURE 12.17 Aortic pressure during the cardiac cycle. *An aortic pressure wave is shown along with periods of opening and closing of the aortic valve.*

takes time for the tension in the ventricular muscle to wane. Ventricular pressure is simultaneously too low to keep the semilunar valves open and too high to allow the AV valves to open. Because all valves are closed and the volume of blood remains constant within the relaxing ventricles, phase 4 is referred to as **isovolumetric relaxation.**

Once ventricular pressure becomes low enough to permit the AV valves to open again, blood enters the ventricles from the atria. This marks the beginning of phase 1, termed **ventricular filling,** and the pump cycle begins once again.

The durations of systole and diastole are not equal. For a heart beating at the normal resting rate of 72 beats per minute (one beat every 0.8 second), most of the cardiac cycle (about 65%, or 0.5 second) is spent in diastole; systole lasts only about 0.3 second. This longer diastole gives the heart adequate time to fill with blood, which is essential for efficient pumping, and it also gives the heart muscle more time to relax, which helps prevent fatigue.

Now that our examination of the pump cycle has provided an overview of the cardiac cycle, we examine its other aspects in detail.

Atrial and Ventricular Pressure

By convention, cardiovascular pressures (the pressure of blood in the chambers of the heart or in the vasculature) are given in millimeters of mercury (mm Hg). Recall that atmospheric pressure is also measured in millimeters of mercury, and that normal atmospheric pressure at sea level is 760 mm Hg. In cardiovascular physiology, all pressures are given relative to atmospheric pressure, which is taken to be *zero.* Thus, when physiologists say that blood is at a pressure of 100 mm Hg, they mean that it is 100 mm Hg *above* atmospheric pressure.

The rise in atrial pressure that occurs in late diastole indicates the beginning of atrial contraction (see Figure 12.15). This rise in pressure is small and short-lived, however, and is followed by a series of similar small increases at various times throughout the cardiac cycle. Because these changes in pressure are of little significance to overall cardiac function, we will not discuss them further.

In mid-diastole (phase 1), ventricular pressure stays very low until the end of the phase, when an abrupt but small rise occurs (Figure 12.16). This rise is due to atrial contraction, which adds a small volume of blood to the ventricle; the increase in volume causes ventricular pressure to rise. Shortly thereafter, a much larger increase in pressure that corresponds to ventricular systole occurs. In early ventricular diastole, the pressure falls to near zero, reflecting relaxation of the myocardium. Through the remainder of diastole, ventricular pressure slowly creeps upward as the ventricle passively fills with blood returning from the pulmonary circulation.

Aortic Pressure

During diastole, no blood enters the aorta because the aortic valve is closed. However, blood is able to leave the aorta by flowing through blood vessels downstream in the systemic circuit. This continual exit of blood causes the volume of blood in the aorta to decrease during diastole, which causes a slow decline in aortic pressure (Figure 12.17). (Imagine the pressure falling as air escapes from a leaky tire.) At the end of this decline, aortic pres-

The following diagram is a record of aortic pressure during a single cardiac cycle.

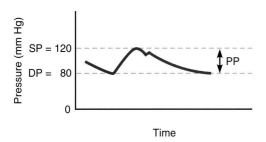

The pulse pressure (PP) is the difference between systolic pressure (SP) and diastolic pressure (DP):

$$PP = SP - DP$$

Using average numbers for a normal healthy adult, the pulse pressure is

$$PP = 120 \text{ mm Hg} - 80 \text{ mm Hg} = 40 \text{ mm Hg}$$

In older persons, a pulse pressure that is abnormally high may indicate *hardening of the arteries*, a condition in which the arteries become thickened and more rigid, which decreases their ability to stretch. To understand the connection between pulse pressure and hardening of the arteries, consider that arteries normally expand and contract during systole and diastole, respectively, reflecting the fact that the pressure of blood inside them is rising and falling. Because the expansion of blood vessels tends to relieve the pressure inside, whereas contraction of blood vessels raises it, the *lack* of expansion and contraction in hardened arteries tends to raise systolic pressure while lowering diastolic pressure. As a result, the difference between systolic and diastolic pressure—the pulse pressure—increases.

Mean arterial pressure (MAP), the average pressure occurring in the aorta during one cardiac cycle, is given by the following expression:

$$MAP = DP + 1/3 \ PP$$

For a normal healthy adult, this becomes

$$MAP = 80 \text{ mm Hg} + 1/3 \ (40) = 93.3 \text{ mm Hg}$$

which is indicated by the dashed line in the following diagram:

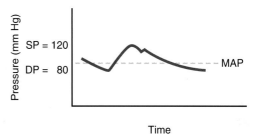

Notice that the mean arterial pressure is *not* obtained simply by taking the average of the systolic and diastolic pressures, which is (120 mm + 80 mm) ÷ 2 = 100 mm. Instead, MAP is lower than this. The reason has to do with the changes in aortic pressure as seen in the shape of the pressure wave: During a single cardiac cycle, aortic pressure is near its maximum for a relatively short period and is closer to the minimum most of the time. Any change away from this pattern will affect the mean arterial pressure—even if systolic and diastolic pressures do not change. For this reason, the equation for mean arterial pressure should be regarded as only a rule of thumb.

sure reaches a minimum called the **diastolic pressure** (DP). As the next systolic period begins (phase 2), the aortic pressure continues to fall because the aortic valve opens only when the ventricular pressure becomes high enough to force it open.

When the aortic valve opens and the ejection phase begins (phase 3), the aortic pressure rises quickly. This increased pressure reflects an increase in the volume of blood contained in the aorta, which occurs because blood is flowing into the aorta faster than it can flow out. (Imagine the pressure rising as you pump air into a leaky tire, even though air continues to escape from it.) Within a short period of time, however, the flow of blood from the heart begins to slow down. As a reflection of this, the

aortic pressure does not continue to rise but, instead, reaches a maximum, called the **systolic pressure** (SP), and then starts to fall. At the end of systole, the aortic valve closes, terminating the flow of blood into the aorta from the heart (beginning of phase 4). This event is marked by a "wiggle" in the aortic pressure curve called the *dicrotic notch.*

The average aortic pressure occurring during the cardiac cycle is known as the **mean arterial pressure** (MAP) (Toolbox: Pulse Pressure and Mean Arterial Pressure). The mean arterial pressure is an extremely important variable in cardiovascular physiology because, as we see in Chapter 13, it represents the driving force that pushes blood through the systemic circuit. As we see

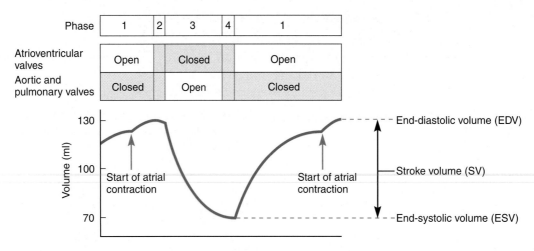

FIGURE 12.18 Changes in ventricular volume during the cardiac cycle.
The volume of blood ejected from the ventricle in one beat is the stroke volume.

This figure shows stroke volume for the left ventricle. Will stroke volume for the right ventricle be *higher*, *lower*, or the *same*?

in Chapter 14, mean arterial pressure is regulated by a negative feedback system involving pressure-sensitive receptors in the walls of some of the major arteries.

When you go to a physician to have your blood pressure taken, what is actually measured is an *estimate* of the aortic pressure. Because there is no convenient way to measure aortic pressure directly, pressure is usually measured in the *brachial artery,* which runs through the upper arm. Pressure measured in this manner is close to aortic pressure because the brachial artery is not far from the heart and is also at about the same height as the aorta. (Blood pressure tends to be lower in upper regions of the body and higher in lower regions due to the force of gravity acting on blood.) By convention, blood pressure is usually recorded as systolic pressure over diastolic pressure (in mm Hg). For a healthy resting adult, the systolic pressure is about 120 mm Hg and the diastolic pressure is around 80 mm Hg. Therefore, the blood pressure is recorded as 120/80 (Discovery: The Role of Turbulence in Blood Pressure Measurement, p. 425).

It is apparent in Figure 12.15 that the overall variation in aortic pressure that occurs during the cardiac cycle is much less than the variation in left ventricular pressure. During the ejection phase (phase 3), aortic and ventricular pressures are virtually identical because the aorta and ventricle are in communication through the aortic valve. Throughout the remainder of the cycle, however, the difference between aortic and ventricular pressure is substantial—ventricular pressure drops abruptly while aortic pressure remains elevated and decreases only slowly. This difference in pressure reflects the fact that the aorta and ventricle are no longer in communication because the aortic valve is closed.

Aortic pressure is higher than ventricular pressure during diastole because the aorta is able to store pressure during systole—pressure that is subsequently released during diastole. When the volume of blood contained in the aorta rises during ejection, the vessel expands and its wall stretches. By stretching, the aorta stores some of the energy that is generated by the heart; this energy becomes evident as a rise in *pressure.* (Because the aorta and other arteries have this ability to store pressure, they are said to function as *pressure reservoirs,* as is explained in Chapter 13.) During diastole, the heart is no longer actively generating pressure, but the pressure in the aorta remains elevated due to the inward force exerted on the blood by the stretched elastic tissue in its wall. As blood leaves the aorta and the pressure falls, the wall recoils. In doing so, the aorta releases the energy that it stored during systole. This energy drives the flow of blood through downstream vessels during diastole. Thus blood moves through the vasculature during diastole *even though no blood is being ejected from the heart at this time.* As a consequence, blood flows through the vasculature in a more-or-less continual fashion throughout the cardiac cycle, despite the fact that it exits the heart in spurts.

Ventricular Volume

Figure 12.18 shows a curve that traces changes in *ventricular volume,* in this case the volume of blood in the left ventricle, during the cardiac cycle. During mid-to-late diastole (phase 1), the volume of blood contained in the ventricle rises during ventricular filling. Volume increases quickly at first and then rises less rapidly as time goes on. Toward the end of diastole, a small but abrupt rise in volume reflects blood being pumped into the ventricle as a

The same

Stroke volume (SV) can be calculated as the difference between end-diastolic volume (EDV) and end-systolic volume (ESV), as follows:

$$SV = EDV - ESV$$

For an adult at rest, normal end-diastolic volume is about 135 ml, whereas end-systolic volume is 65 ml. This gives a stroke volume of

$$SV = 135 \text{ ml} - 65 \text{ ml} = 70 \text{ ml}$$

These volumes are indicated in the following diagram, which is a record of ventricular volume during one cardiac cycle:

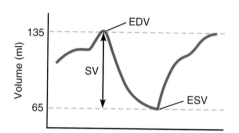

The ejection fraction (EF) is the ratio of the volume ejected in one beat (stroke volume) to the volume contained in the ventricle immediately prior to ejection (end-diastolic volume):

$$EF = SV/EDV$$

Using normal values just given, we obtain an ejection fraction of around 52%:

$$EF = 70 \text{ ml}/135 \text{ ml} = 0.52$$

The ejection fraction is important because it tells us whether or not the heart is pumping blood efficiently. An ejection fraction that is abnormally low may indicate that the ventricular muscle is weak.

result of atrial contraction. The volume of blood in the ventricle at the end of diastole, referred to as the **end-diastolic volume** (EDV), represents the maximum ventricular volume attained during the cardiac cycle, which is reached just before the start of ejection. The volume does not change during the period of isovolumetric contraction (phase 2) but begins to fall as soon as the aortic valve opens at the start of ejection (beginning of phase 3). It continues to fall until the aortic valve closes at the end of the ejection period, at which time it stays constant until the AV valve opens at the start of ventricular filling (beginning of phase 1). The volume of blood in the ventricle at the end of systole, called the **end-systolic volume** (ESV), represents the minimum ventricular volume, which is attained just after ejection.

The difference between end-diastolic volume and end-systolic volume represents the volume of blood ejected from the heart during one beat, which is the **stroke volume** (SV) (Toolbox: Stroke Volume and Ejection Fraction):

stroke volume =
  volume of blood in ventricle just before ejection
  minus volume of blood in ventricle just after ejection

stroke volume =
  end-diastolic volume − end-systolic volume

$$SV = EDV - ESV$$

Notice in Figure 12.18 that when the ventricle contracts (phase 3), it does not eject all of the blood contained within it; about 65 ml of blood remain in the ventricle at the end of systole under normal resting conditions. End-systolic volume is determined by a number of factors, including the force of ventricular contraction, which can be altered by autonomic neurons and hormones. An increase in the force of contraction, such as would occur during a fight-or-flight response, raises the stroke volume, so that a greater fraction of the end-diastolic volume is ejected, and lowers the end-systolic volume. A decrease in the force of contraction has the opposite effect: It reduces the stroke volume, so that a smaller fraction of the end-diastolic volume is ejected, and raises the end-systolic volume. The fraction of end-diastolic volume ejected during a heartbeat is known as the **ejection fraction** (EF) (see Toolbox, above).

Heart Sounds

The sounds of the beating heart that can be heard through a stethoscope are called *heart sounds*. In most people, each heartbeat comprises two distinct sounds: a soft, low-pitched "lub," designated the "first sound," and a louder, sharper, higher-pitched "dup," designated the "second sound." Thus the sounds of the beating heart are often described as lub-DUP, lub-DUP, lub-DUP. . . .

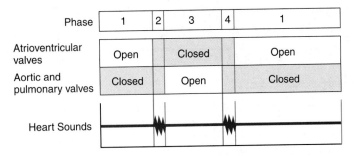

Phase	1	2	3	4	1
Atrioventricular valves	Open		Closed		Open
Aortic and pulmonary valves	Closed		Open		Closed

FIGURE 12.19 Heart sounds. *First and second heart sounds occur as the atrioventricular valves (first) and aortic and pulmonary valves (second) close.*

Comparing the timing of the heart sounds to the events in the pump cycle reveals that the heart sounds occur at the start of systole (phase 2), when the AV valves close, and at the start of diastole (phase 4), when the semilunar valves close (Figure 12.19). But contrary to what seems obvious, heart sounds are not caused by the valve cusps slapping together as they snap shut. Instead, the sounds are caused by the turbulent rushing of blood through the valves as they are narrowing and about to close.

Quick Test 12.4

1. Define the following terms: *cardiac cycle, diastole, systole, systolic pressure, diastolic pressure, mean arterial pressure, isovolumetric contraction, isovolumetric relaxation, end-diastolic volume, end-systolic volume, stroke volume.*

2. What event is associated with the first heart sound? With the second heart sound?

CARDIAC OUTPUT AND ITS CONTROL

The ability of the cardiovascular system to deliver blood to organs ultimately depends on the rate at which the heart's ventricles are able to pump blood. When a person is at rest, the left and right ventricles each pump a little over 5 liters through the vasculature every minute. Because the total volume of blood in the body is around 5 liters, it only takes a minute for a ventricle to pump the equivalent of the entire blood volume. This translates into over 2.6 million liters per year, and remember that this only pertains to *one* ventricle! The rate at which a ventricle pumps blood is called the **cardiac output** (CO), and it is usually expressed in liters per minute.

With each heartbeat, the left and right ventricles contract together. Thus, the number of contractions per minute (called the **heart rate,** HR) is the same for both ventricles. The cardiac output is determined by the heart rate and the volume of blood that is pumped from each ventricle with every beat, which we know as the *stroke volume* (Toolbox: Cardiac Output, p. 394):

$$\text{cardiac output} = \text{heart rate} \times \text{stroke volume}$$
$$\text{CO} = \text{HR} \times \text{SV}$$

The cardiac output of the left ventricle equals the rate of bloodflow through the systemic circuit; cardiac output of the right ventricle equals bloodflow through the pulmonary circuit. Over the long run, the left and right sides of the heart must have the same cardiac output, or else blood volume would shift from the pulmonary circuit to the systemic circuit, or vice versa. Because the heart rate *and* cardiac output are the same for the right and left sides of the heart, both ventricles must also have the same average stroke volume for their output to be equal.

Because we know that cardiac output is determined entirely by two variables, heart rate and stroke volume, our goal in this section is to examine the various factors that influence these two variables, so that we can develop an understanding of how the factors interact to determine cardiac output.

Although contractions of the heart are not *triggered* by input from the central nervous system, the nervous system does regulate various aspects of cardiac function, including the *rate* and *force* of heart muscle contraction. For this reason, neural input to the heart exerts a significant influence on cardiac output. In addition, cardiac output is influenced by hormones that circulate in the bloodstream. Regulation of the heart (or any other organ or tissue) by neural input, circulating hormones, or any other factor originating from outside the organ is referred to as **extrinsic control.** When the function of an organ or tissue is regulated by factors originating from within the organ or tissue itself, the function is said to be under **intrinsic control** (also known as *autoregulation* or *local regulation*). Like most organs, the heart is regulated by extrinsic *and* intrinsic control, and we will see examples of both in the following sections.

Exercise Link

There are times when cardiac output must be increased. Under hot conditions, more blood must flow to the skin in order to dissipate heat to the external environment. During exercise, large amounts of oxygen, nutrients, carbon dioxide, heat, and wastes must be exchanged with skeletal muscles. By augmenting heart rate and stroke volume, cardiac output can increase up to sevenfold. This provides a large reserve capacity that meets almost all the needs of everyday life. Only when faced by multiple stresses simultaneously (such as the combination of exercise AND a hot environment AND fluid [blood volume]

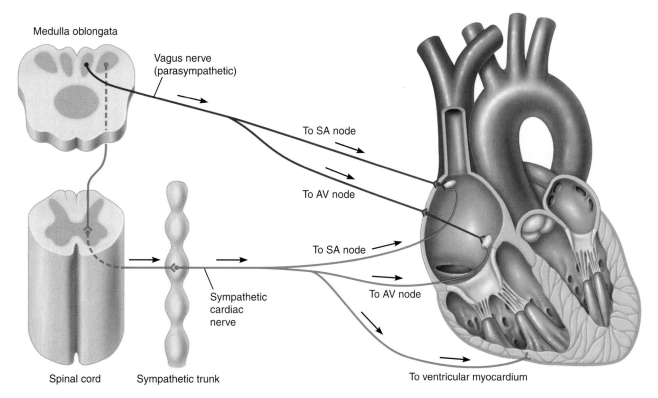

FIGURE 12.20 Major autonomic inputs to the heart. *Sympathetic nerves travel to the SA and AV nodes as well as to the ventricular myocardium; parasympathetic nerves travel mainly to the nodes. Pathways of transmission from the central nervous system are also indicated.*

What neurotransmitter is secreted from axon terminals in the sympathetic trunk?

losses due to sweating—which Bill and Jane faced during the marathon) does the capacity of the cardiovascular system to provide adequate flow to all organs and tissues begin to fail.

Autonomic Input to the Heart

Neural control of the heart is carried out by the autonomic nervous system, which is the branch of the central nervous system that regulates "automatic" functions—functions not under conscious control, such as those of internal organs. Fibers of the autonomic nervous system project to nearly every region of the heart, including the conduction system and the bulk of the myocardium, and regulate both heart rate and stroke volume. These fibers include those belonging to the sympathetic nervous system, which release the neurotransmitter norepinephrine at target cells, and those belonging to the parasympathetic nervous system, which release acetylcholine (see p. 317). At the various sites of action in the heart, sympathetic and parasympathetic neurons exert opposite effects (as is generally true throughout the body). Sympathetic input to the heart is transmitted by preganglionic fibers traveling in several pairs of nerves that emerge from the lower cervical and upper thoracic regions of the spinal cord. (Recall that in the autonomic nervous system, in-

formation is carried away from the brain or spinal cord by *preganglionic* neurons; these neurons synapse with *postganglionic* neurons, which transmit information to the organs; see p. 313.) In contrast, parasympathetic input to the heart is carried by preganglionic fibers traveling in a single pair of cranial nerves—the *vagus* nerves, which emerge from the medulla oblongata. Although most regions of the heart receive sympathetic and parasympathetic input, the distribution of parasympathetic fibers is relatively sparse in the ventricles. As a consequence, the ventricular myocardium is regulated primarily by the sympathetic nervous system. Major autonomic inputs to the heart are shown in Figure 12.20.

Factors Affecting Cardiac Output: Changes in Heart Rate

One person's heart rate is not necessarily the same as another's and is not necessarily the same from day to day or even from minute to minute. This is because a person's heart rate depends on many factors, including his or her age, general health, level of muscular activity, and emotional state. When a person exercises or when he or she is excited, anxious, or frightened, the heart rate can increase

Acetylcholine

Cardiac output (CO) depends on the number of heartbeats per minute (heart rate, HR) and the volume of blood ejected from one ventricle with each beat (stroke volume, SV). It can be calculated as follows:

$$CO = HR \times SV$$

For an adult at rest, the normal resting heart rate is around 72 beats/minute and the stroke volume is 70 ml (0.07 liter); thus a cardiac output is

$$CO = 72 \text{ beats/min.} \times 0.07 \text{ liter/beat} = 5.0 \text{ liter/min.}$$

During periods of exercise, the heart rate and stroke volume both increase. The resulting increase in cardiac output, which can go up to 35 liters/min. in trained athletes, helps to increase the rate at which blood delivers oxygen and nutrients to exercising muscles.

from the normal resting value of around 70 beats per minute to over 100 beats per minute, sometimes going as high as 180 beats per minute. In well-trained athletes, the resting heart rate is lower than average, being typically around 50 beats per minute. In children, the resting heart rate is higher than in adults. In this section we focus on various factors involved in the minute-by-minute regulation of heart rate, which is entirely under extrinsic control.

Sympathetic Control of Heart Rate

As previously mentioned, pacemaker cells of the SA node receive direct input from sympathetic neurons. This is important because input from these neurons alters the frequency of action potentials generated by these cells, which is normally the sole determinant of heart rate. As sympathetic activity increases, action potential frequency increases. Consequently, heart rate increases, which tends to increase cardiac output.

Sympathetic neurons also project to the AV node and other parts of the conduction system, where they influence the speed with which action potentials are conducted. As sympathetic activity increases, action potentials move faster, which decreases the delay of impulse conduction between the atria and ventricles and shortens the time it takes for action potentials to travel through the ventricles. As a result, ventricular contraction starts sooner after atrial contraction and proceeds more quickly, which decreases the duration of systole.

Parasympathetic Control of Heart Rate

Pacemaker cells of the SA node receive input from parasympathetic neurons in addition to sympathetic neurons. As we have seen, parasympathetic input has the opposite effect of sympathetic input and causes the action potential frequency of the SA node to decrease. This lowers the heart rate and therefore tends to reduce cardiac output. Parasympathetic neurons also influence impulse conduction through the AV node and the rest of the conduction system. As activity in these neurons increases, the speed of impulse conduction decreases, which increases the delay of conduction between the atria and ventricles and lengthens the time required for impulses to travel through the ventricles. As a result, the duration of systole increases.

Hormonal Control of Heart Rate

Although the function of the heart can be affected by a number of hormones, one—*epinephrine*—is significant in the minute-to-minute regulation of cardiac function. The effects of epinephrine, which is secreted by the adrenal glands and comes to the heart by way of the bloodstream, are similar to those exerted by sympathetic neural activity: Epinephrine increases action potential frequency at the SA node and therefore increases heart rate. In addition, epinephrine increases the velocity of action potential conduction through heart muscle fibers. Because increased sympathetic nervous activity is usually coupled with enhanced epinephrine secretion, the hormone's actions generally reinforce the effects of sympathetic neural input.

Other hormones that directly affect cardiac function include the *thyroid hormones,* which are secreted by the thyroid gland, and *insulin* and *glucagon,* which are secreted by the pancreas. These hormones primarily increase the force of myocardial contraction, but glucagon also promotes increased heart rate. The importance of these hormones, if any, in the short-term regulation of cardiac function is unclear.

Integration of Heart Rate Control

We have seen that the heart rate is determined entirely by the frequency of action potential firing by the SA node, which is in turn regulated primarily by the following: (1) activity in *sympathetic neurons* projecting to the SA node, which tends to raise the heart rate; (2) activity in *parasympathetic neurons* projecting to the SA node, which tends to lower the heart rate; and (3) levels of circulating *epinephrine,* which acts to raise the heart rate (Table 12.2).

Both divisions of the autonomic nervous system are active at all times. As a result, the heart simultaneously receives signals from the sympathetic and parasympathetic nervous systems that act in opposite directions.

Whether heart rate goes up or down depends on the relative rates of activity in the two branches. Activity varies in a "push-pull" manner, in which increases in sympathetic activity are usually accompanied by decreases in parasympathetic activity, and vice versa. Because the two divisions exert opposing effects, this push-pull arrangement ensures that changes in sympathetic and parasympathetic activity reinforce one another. For example, increases in sympathetic activity *and* decreases in parasympathetic activity both act to increase heart rate.

Laboratory experiments have shown that the SA node fires action potentials at a "natural" frequency of about 100 per minute in the absence of neural or hormonal influences. The fact that the heart rate in a person at rest is significantly lower (about 70 per minute) indicates that under normal resting conditions, the influence of parasympathetic input to the SA node predominates over the combined influence of sympathetic neurons and epinephrine, giving a net *suppressive* effect on the firing frequency. Any increase in firing frequency is usually triggered by an increase in sympathetic input (and a rise in epinephrine levels) in conjunction with a decrease in parasympathetic input. Any decrease in firing frequency is usually brought about by neural and hormonal changes in the opposite direction. As heart rate increases or decreases, cardiac output tends to increase or decrease, respectively.

Quick Test 12.5

1. Define the following terms: *cardiac output, stroke volume, heart rate.*
2. What is the difference between extrinsic control of an organ's function and intrinsic control of that organ's function?
3. What parts of the heart receive input from autonomic neurons? What effect does autonomic input have on each of these parts? (Include both sympathetic and parasympathetic influences.)
4. What hormone reinforces the effects of the sympathetic nervous system on the heart?

Factors Affecting Cardiac Output: Changes in Stroke Volume

The second important determinant of cardiac output is stroke volume. Like heart rate, stroke volume can vary from moment to moment and depends on several factors. In the following sections we examine the primary factors that affect stroke volume: (1) *ventricular contractility*, a measure of the ventricles' capacity for generating force; (2) end-diastolic volume; and (3) *afterload*, the pressure that the ventricles have to work against as they pump

TABLE 12.2 MAJOR FACTORS AFFECTING HEART RATE		
FACTOR	**EFFECT ON SA NODE PACEMAKER CELLS**	**EFFECT ON HEART RATE**
Autonomic nervous system		
Sympathetic neurons to heart	↑Action potential frequency	↑HR
Parasympathetic neurons to heart	↓Action potential frequency	↓HR
Hormones		
Epinephrine	↑Action potential frequency	↑HR

blood out of the heart. As we will see, these variables are themselves influenced by a number of other factors, including neural input to the heart, hormones, and various physical variables affecting bloodflow into or out of the heart.

The Influence of Ventricular Contractility on Stroke Volume

The first two factors just mentioned—ventricular contractility and end-diastolic volume—both affect stroke volume because they affect the force of ventricular contraction. What, then, is the difference between *contractility* and *contractile force*? The answer is that a change in **ventricular contractility** means a change in the force of ventricular contraction at any given end-diastolic volume. (The term *contractility* can be used to describe any muscle's capacity for generating force.) Generally speaking, any factor that causes the ventricles to contract with more force will tend to make stroke volume larger, which will increase cardiac output. This is true, regardless of whether the change in contractile force is due to a change in contractility or a change in end-diastolic volume.

Sympathetic Nervous Control of Ventricular Contractility

Ventricular contractility, like heart rate, is regulated by the autonomic nervous system, but autonomic control of stroke volume is exerted almost entirely by the sympathetic nervous system. There is little or no parasympathetic influence on ventricular contractility because of the sparse distribution of parasympathetic fibers in the ventricular myocardium.

As mentioned previously, sympathetic neurons project not only to the heart's conduction system, but also to the muscle cells that make up the bulk of the myocardium. Some of these neurons project to the atria and influence the force of atrial contraction. Increased

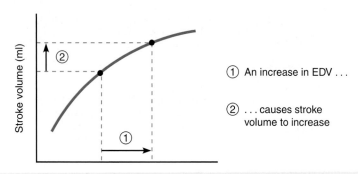

① An increase in EDV . . .

② . . . causes stroke volume to increase

End-diastolic volume (ml)

FIGURE 12.21 A Starling curve showing how stroke volume changes in response to changes in end-diastolic volume.

sympathetic activity causes the atria to contract with more force, which raises atrial pressure and increases the volume of blood the atria pump into the ventricles. More importantly, sympathetic neurons project to the ventricular myocardium (see Figure 12.20), where they exert a direct influence on myocardial contractility. As sympathetic activity increases, ventricular contractility increases, which tends to raise cardiac output.

Sympathetic neurons exert their influence over ventricular contractility in the following manner: Action potentials in these neurons trigger the release of norepinephrine, which binds to beta adrenergic receptors on the surface of heart muscle cells. Binding of neurotransmitter molecules to these receptors causes the activation of adenylate cyclase, the enzyme that catalyzes formation of cyclic AMP from ATP (see p. 146). The resulting increase in cyclic AMP levels in the muscle cell then alters the function of plasma membrane calcium channels, such that the amount of calcium entering cells with each action potential increases. The end result is a general increase in intracellular calcium concentrations, which has a twofold effect: (1) Binding of calcium to troponin increases, which results in more active crossbridges and more contractile force; and (2) calcium ions trigger an increase in the calcium permeability of the sarcoplasmic reticulum (see p. 383), which increases the amount of calcium released during an action potential. This increase augments the first effect.

Hormonal Control of Ventricular Contractility Ventricular contractility is affected by a number of hormones, including insulin, glucagon, and thyroid hormones, but most importantly it is regulated by epinephrine. Like norepinephrine, epinephrine binds to beta receptors on heart muscle cells, and it affects intracellular cAMP levels in the same manner. Therefore, epinephrine increases myocardial contractility, thereby promoting increases in stroke volume and cardiac output.

The Influence of End-diastolic Volume on Stroke Volume: Starling's Law

We have seen that the force of ventricular contraction is influenced by the sympathetic nervous system and circulating epinephrine; thus it is under extrinsic control. In this section we see that the force of ventricular contraction also varies in response to how much the ventricular myocardium is stretched when it fills with blood. Because this effect results from a mechanism operating entirely within the heart and does not depend on the actions of extrinsic factors such as nerves or hormones, it is an example of an intrinsic control.

Intrinsic control of cardiac function is exemplified by **Starling's Law of the Heart,** which can be stated as follows: *When the rate at which blood flows into the heart from the veins (that is, venous return) changes, the heart automatically adjusts its output to match the inflow.* The basis of Starling's Law is the following observation, called the *Starling effect:* If an increase in end-diastolic volume occurs, the force of ventricular contraction rises, producing an increase in stroke volume and cardiac output. If the end-diastolic volume decreases, the force of ventricular contraction declines, producing a decrease in stroke volume and cardiac output.

The physiological basis for the Starling effect is related to the fact that increases in end-diastolic volume cause muscle fibers in the ventricular myocardium to lengthen. Such stretching of the muscle fibers causes an increase in the force of contraction by two mechanisms. First, unlike skeletal muscle, the optimum length for cardiac muscle is much greater than its resting length and is never reached in a healthy heart. Therefore, increasing the length of the muscle by increasing the end-diastolic volume brings the muscle fibers closer to their optimum length for contraction, and thus they contract with greater force. Second, stretching of the muscle fibers induces an increase in the affinity of troponin for calcium. As a consequence, binding between troponin and calcium is increased, which increases the number of crossbridges that are activated with each contraction.

The Starling Curve The Starling effect is illustrated in Figure 12.21, which shows a graph referred to as a *Starling curve* or *cardiac function curve.*

When interpreting this graph and others like it, note that stroke volume depends on other factors besides end-diastolic volume. Earlier in the chapter we saw that stroke volume is influenced by sympathetic input to the ventricular myocardium. The curve in Figure 12.21 is drawn assuming that the degree of sympathetic input, and all other factors that might affect stroke volume, are held constant. Further, note that such a curve pertains only to what happens in normal, healthy hearts. Normally, an increase in venous return causes an increase in end-diastolic volume, which triggers an increase in stroke volume according to the Starling effect. In persons whose hearts are chronically expanded, prolonged stretching of the heart muscle can bring about weakening of the connective tissue, which causes EDV to gradually increase. Eventually, the volume becomes so large that the slope of the cardiac function curve actually becomes *negative*—stroke volume *decreases* as end-diastolic volume gets larger. Under these conditions, the heart is capable of generating only weak contractions and is therefore unable to adjust its volume to normal because it cannot expel the excess blood that it has accumulated.

Because changes in either sympathetic activity or end-diastolic volume affect the force of ventricular contraction, but by different mechanisms, it is possible to alter stroke volume either by changing sympathetic activity without changing end-diastolic volume, or by changing end-diastolic volume without changing sympathetic activity. Cardiac function is therefore described not by a single Starling curve but instead by a *family* of curves, each of which pertains to a given level of sympathetic activity (Figure 12.22). An increase in sympathetic activity shifts the Starling curve upward. As a result, stroke volume *at any given end-diastolic volume* increases, reflecting the fact that there has been an increase in ventricular contractility. A decrease in contractility means that the Starling curve shifts downward, so that stroke volume at any given end-diastolic volume decreases. In actuality, end-diastolic volume and contractility can change simultaneously.

Significance of Starling's Law

On the face of it, Starling's Law may seem a mere curiosity of no special significance. After all, if more blood flows into the heart, one would expect that more blood should come out, even if there were no such thing as the Starling effect. But consider what would happen if venous return were suddenly to increase but stroke volume did not? (For simplicity, assume that heart rate is constant.) If cardiac output and venous return were equal to begin with, an increase in venous return with no change in stroke volume would cause the volume of blood remaining in the heart to increase because the amount flowing in would be greater than the amount flowing out. With each beat, the heart would accumulate more blood, which would eventually cause it to expand far beyond its normal size. From this viewpoint, Starling's Law takes on a new significance: By adjusting stroke volume so that cardiac output matches venous return, the heart regulates its *size.*

Certain pathological conditions can lead to chronic enlargement of the heart, which can be detrimental to cardiac function for the following reasons: As the wall of a ventricle stretches, the muscle must develop more tension and therefore work harder just to maintain the same

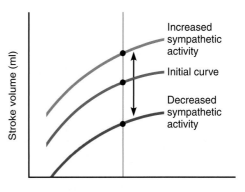

FIGURE 12.22 A family of Starling curves, which shows the influence of sympathetic input on ventricular contractility. *At any given end-diastolic volume, a rise in sympathetic activity induces an increase in the stroke volume, which is reflected as an upward shift in the Starling curve. Under the same conditions, a decrease in sympathetic activity induces a decrease in the stroke volume, which is reflected as a downward shift. Upward or downward shifts in Starling curves reflect increases or decreases, respectively, in ventricular contractility.*

Does a shift from one Starling curve to another reflect intrinsic or extrinsic control of cardiac function?

pressure on the blood contained within it (Toolbox: Laplace's Law, p. 398). If the ventricles get too big, they become unable to generate enough pressure to maintain adequate cardiac output; such an inability is called *heart failure.*

Factors Affecting End-diastolic Volume

According to Starling's Law, the force of ventricular contraction (and hence stroke volume) rises or falls as the end-diastolic volume rises or falls, respectively. In turn, end-diastolic volume is determined by a number of factors, each of which indirectly influences the stroke volume through its effect on end-diastolic volume. Here we examine the various factors that influence end-diastolic volume.

End-diastolic volume is primarily determined by *end-diastolic pressure*, sometimes referred to as **preload.** Ventricular end-diastolic pressure is called preload because it places tension (or *load*) on the myocardium *before* it begins to contract. When a ventricle fills with blood during diastole, the process is similar to what happens when you blow up a balloon with air: As pressure inside rises, the balloon expands. Therefore, the final volume of a given balloon is determined by the final pressure of the air inside it. Likewise, the end-diastolic volume of a ventricle is determined by the pressure of the blood inside it at the end of diastole. As preload (end-diastolic pressure) increases, end-diastolic volume increases, and stroke volume increases according to Starling's Law.

Extrinsic control

The detrimental effect of increased size on the heart's pumping action has its basis in fundamental physical principles: In a vessel containing gas or liquid under constant pressure, such as a balloon or blood vessel, the outward force exerted by the pressure stretches the wall of the vessel, creating tension. This tension exerts an inward force that balances the outward force of the pressure, so that the system comes to equilibrium and the vessel neither expands nor contracts. For a spherical container, wall tension is proportional to the interior pressure (P) and to the diameter of the container (D):

$$\text{tension} \propto P \times D$$

This relationship is known as Laplace's Law.

A consequence of Laplace's Law is that if two vessels of different size contain gas or liquid at the same pressure, wall tension will be greater in the larger vessel. Thus, if two different soap bubbles, one large and one small, contain air at the same pressure, the larger one will be subject to more tension because its diameter is larger. This is why a larger bubble is more apt to break than a smaller one, all else being equal. Because the heart is not spherical, the relationship between wall tension, pressure, and size is more complex than for a sphere, but the same basic idea applies: If a ventricle contains a large volume of blood, the muscle of the wall has to exert greater tension to generate a given pressure.

Preload is determined by a number of factors, including (1) filling time, which depends on heart rate, and (2) atrial pressure, which is determined by venous return and the force of atrial contraction. As heart rate decreases, filling time increases because diastole increases in duration. At a heart rate of 60 beats per minute, diastole is approximately 0.6 second long; when the heart rate increases to 180 beats per minute, diastole decreases to a little over 0.1 second in length. Because more time is allowed for the entry of blood into the ventricles when the heart rate is lower, a decrease in heart rate (increase in filling time) tends to increase preload and end-diastolic volume. Preload and end-diastolic volume also tend to increase when atrial pressure rises, because atrial pressure is virtually identical to ventricular pressure when the AV valves are open (that is, during diastole). Atrial pressure, in turn, rises in response to increases in either venous return or the force of atrial contraction.

As discussed in Chapter 13, the most important influence on venous return is *central venous pressure,* the pressure of blood contained in the large veins that lead into the heart. As we will see, central venous pressure is affected by many variables, including changes in blood volume, muscular activity, and even changes in posture (as when a person stands up or lies down). As central venous pressure rises, venous return increases because the increased pressure forces more blood to flow into the atria. This raises atrial pressure, which leads to an increase in preload. Consequently, end-diastolic volume increases, which produces an increase in stroke volume via the Starling effect.

The Influence of Afterload on Stroke Volume

The previously mentioned factors influence stroke volume by altering the force of ventricular contraction. However, stroke volume depends not only on how much force the ventricular muscle develops, but also on how large a force it has to work against. (Consider a person attempting to push a wagon up a slope: The speed of the wagon depends not only on how much force the person exerts, but also on how much the wagon weighs.) When the heart ejects blood, the ventricular muscle works against arterial pressure in the same way that your muscles of respiration work against the pressure inside a balloon you are blowing up. For this reason, increases in arterial pressure tend to cause stroke volume to decrease. Because arterial pressure places a load on the myocardium *after* contraction starts, it is called **afterload.** For the left ventricle, afterload is determined by the pressure in the aorta *during the ejection period.* Generally speaking, afterload increases as mean arterial pressure rises.

Summary of Factors Affecting Stroke Volume

We have seen that stroke volume is influenced by three major factors: (1) ventricular contractility, which is regulated by sympathetic nervous activity and circulating epinephrine; (2) end-diastolic volume, which is influenced primarily by end-diastolic pressure (preload); and (3) afterload, which depends on arterial pressure. Note that a change in stroke volume (and hence in cardiac output) can result from a change in just one variable (say, an increase in sympathetic activity) or a simultaneous change in several variables (an increase in sympathetic activity coupled with an increase in end-diastolic volume, for instance). Because these variables interact and are them-

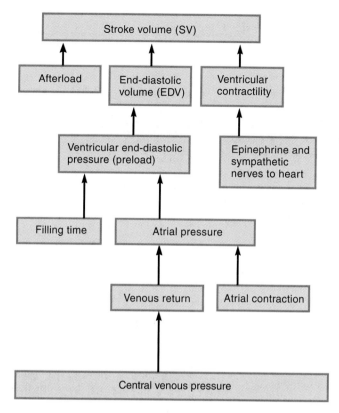

FIGURE 12.23 **Factors influencing stroke volume.**

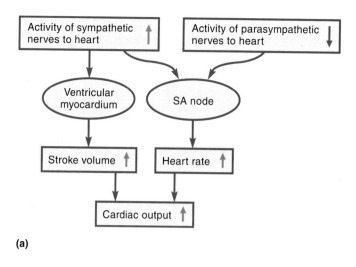

(a)

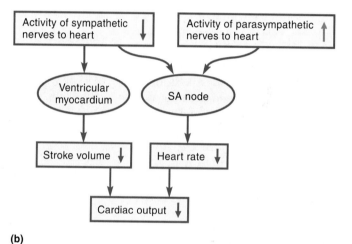

(b)

FIGURE 12.24 **A process diagram summarizing autonomic regulation of cardiac output.** (a) *Increases in sympathetic activity are typically accompanied by decreases in parasympathetic activity. As a result, heart rate and stroke volume tend to increase, which raises cardiac output.* (b) *Decreases in sympathetic activity are typically accompanied by increases in parasympathetic activity. As a result, heart rate and stroke volume tend to decrease, which reduces cardiac output.*

selves affected by several other factors, it may not always be possible to determine how the stroke volume behaves in every conceivable situation. However, you should understand how each variable *tends* to affect the stroke volume. These factors are summarized in Figure 12.23.

Integration of Factors Affecting Cardiac Output

To this point we have focused on factors that affect heart rate and stroke volume individually, but heart rate and stroke volume can change simultaneously, and it is the change in the *combination* of these two variables that determines whether cardiac output increases or decreases or stays the same. An increase in sympathetic activity, for instance, is usually coupled with a decrease in parasympathetic activity, both of which lead to an increase in heart rate (Figure 12.24a); the rise in sympathetic activity also causes the stroke volume to increase. The net result is an increase in cardiac output. In contrast, a decrease in sympathetic activity and an increase in parasympathetic activity combine to produce decreases in heart rate, stroke volume, and cardiac output (Figure 12.24b).

Quick Test 12.6

1. What is the distinction between ventricular contractility and force of contraction?

2. What is the Starling effect? Why is it an example of intrinsic control?

3. Given that all other variables remain constant, explain whether stroke volume increases or decreases following an *increase* in each of the following variables: sympathetic activity, end-diastolic volume, afterload, preload, filling time.

4. What is preload? afterload? How does each affect stroke volume?

An Overview of the Major Components of the Cardiovascular System and Their Functions, p. 370

The cardiovascular system includes the heart, blood, and blood vessels. Blood, which consists of a liquid (plasma) in which the other components (erythrocytes, leukocytes, and platelets) are suspended, acts as a medium that carries oxygen and nutrients to the body's cells while carrying away carbon dioxide and other waste products. The blood vessels (arteries, arterioles, capillaries, venules, and veins) function as conduits for blood flow. The heart, which acts as a pump that drives the flow of blood, is a muscular organ possessing four chambers: the left and right atria, which receive blood as it returns to the heart from the vasculature, and the left and right ventricles, which pump blood away from the heart through the vasculature.

> **IP** Cardiovascular, Anatomy Review: Blood Vessels, p. 6
>
> **IP** Cardiovascular, Anatomy Review: The Heart, p. 4
>
> **IP** Cardiovascular, Cardiac Action Potential, pp. 4–5
>
> **IP** Cardiovascular, Cardiac Cycle, p. 3*

The Path of Blood Flow Through the Heart and Vasculature, p. 374

The vasculature is divided into a pulmonary circuit, which supplies blood to the lungs, and a systemic circuit, which supplies blood to all the other organs and tissues of the body. In the pulmonary circuit, blood becomes oxygenated and gives up carbon dioxide; in the systemic circuit it becomes deoxygenated and picks up carbon dioxide. Blood exiting the right ventricle passes through the pulmonary semilunar valve into the pulmonary trunk, which divides into left and right pulmonary arteries, carrying blood to the lungs. The pulmonary veins carry blood away from the lungs and deliver it to the left atrium. From there, blood moves through the bicuspid valve into the left ventricle. As it exits the left ventricle, it passes through the aortic semilunar valve to enter the aorta, which delivers it to the systemic organs and tissues. Blood returns to the heart by way of the venae cavae, which carry it to the right atrium. From there, the blood passes through the tricuspid valve and enters the right ventricle.

> **IP** Cardiovascular, Anatomy Review: The Heart, pp. 6–7

Electrical Activity of the Heart, p. 377

The heart muscle fibers that make up the heart's conduction system are specialized to initiate action potentials and conduct them rapidly through the myocardium. Contractions of the heart are triggered on a regular basis by action potentials initiated by pacemaker cells concentrated in certain regions of the myocardium. Normally the heartbeat is driven by pacemakers in the sinoatrial (SA) node, located in the upper right atrium. Following each action potential, pacemaker cells exhibit slow, spontaneous depolarizations (pacemaker potentials) that eventually depolarize the membrane to threshold and trigger the next action potential. In most cardiac muscle cells, action potentials are characterized by a broad plateau phase that largely results from an increase in the cell membrane's calcium permeability; the flow of calcium into the cells is important in triggering heart muscle contractions.

The heart's electrical activity can be recorded using electrodes placed on the skin surface, yielding an electrocardiogram (ECG), which consists of three phases: a P wave, corresponding to atrial depolarization; a QRS complex, corresponding to ventricular depolarization; and a T wave, corresponding to ventricular repolarization.

> **IP** Cardiovascular, Intrinsic Conduction System, pp. 3–4
>
> **IP** Cardiovascular, Cardiac Action Potential, p. 3

The Cardiac Cycle, p. 386

The cardiac cycle is divided into two distinct periods: diastole (ventricular relaxation), during which ventricular filling occurs; and systole (ventricular contraction), during which the exit of blood from the ventricles (ejection) occurs. Aortic pressure varies throughout the cardiac cycle; it rises to a maximum (systolic pressure, SP) during systole and falls to a minimum (diastolic pressure, DP) during diastole. The average pressure throughout the cycle, which represents the driving force for blood flow through the systemic circuit, is the mean arterial pressure (MAP). Ventricular volume falls to a minimum at the end of systole (end-systolic volume, ESV) and rises to a maximum at the end of diastole (end-diastolic volume, EDV). The difference between these volumes is the stroke volume (SV), the volume pumped by each ventricle in a single heartbeat.

> **IP** Cardiovascular, Cardiac Cycle, pp. 4–19*
>
> **IP** Cardiovascular, Factors That Affect Blood Pressure, p. 14
>
> **IP** Cardiovascular, Cardiac Output, p. 6
>
> **IP** Cardiovascular, Intrinsic Conduction System, pp. 5–6

Cardiac Output and Its Control, p. 392

The volume of blood pumped by each ventricle per minute is the cardiac output (CO), which depends on the heart rate (HR) and stroke volume: $CO = HR \times SV$. The heart is regulated by sympathetic and parasympathetic neurons and hormones (extrinsic control), and by factors operating entirely within the heart (intrinsic control). Heart rate, determined by the firing frequency of the SA node, is entirely under extrinsic control. Stroke volume is under extrinsic and intrinsic control and is affected by three major factors: ventricular contractility, which is regulated by sympathetic neurons and epinephrine; end-diastolic volume, which depends on preload; and afterload, which depends on arterial pressure. The influence of end-diastolic volume on stroke volume is the basis of Starling's Law of the Heart, an example of intrinsic control of cardiac function.

IP Cardiovascular, Cardiac Output, pp. 4–9

IP Cardiovascular, Blood Pressure Regulation, pp. 4–9

---

* This topic is available on the *InterActive Physiology*® *Sampler CD* that came with the purchase of a new copy of this book.

EXERCISES

Multiple-Choice Questions

1. Minimum aortic pressure during the cardiac cycle is attained
   a) immediately after closure of the aortic semilunar valve.
   b) immediately before opening of the aortic semilunar valve.
   c) immediately before opening of the atrioventricular valves.
   d) in mid-diastole.

2. The first heart sound occurs when the atrioventricular valves close, and thus it marks
   a) the end of the ejection period.
   b) the start of the ejection period.
   c) the start of systole.
   d) the start of isovolumetric contraction.

3. If you know end-diastolic volume, the only other thing you need to know to be able to determine stroke volume is
   a) afterload.
   b) ventricular contractility.
   c) end-systolic volume.
   d) heart rate.

4. As a result of the Starling effect, stroke volume should increase following an increase in
   a) mean arterial pressure.
   b) sympathetic activity.
   c) afterload.
   d) preload.

5. Which of the following blood vessels contains smooth muscle that is capable of regulating blood flow through capillary beds?
   a) arteries
   b) arterioles
   c) capillaries
   d) all of the above

6. Sympathetic and parasympathetic input to the SA node influences
   a) ventricular filling time.
   b) ventricular contractility.
   c) afterload.
   d) atrial contractility.

7. Which of the following contains deoxygenated blood?
   a) the right ventricle
   b) the left ventricle
   c) pulmonary veins
   d) the aorta

8. Which of the following is *not* normally apparent in the ECG?
   a) atrial depolarization
   b) atrial repolarization
   c) ventricular depolarization
   d) ventricular repolarization

9. The second heart sound occurs when the semilunar valves close, and thus it marks
   a) the end of the ejection period.
   b) the start of the ejection period.
   c) the start of systole.
   d) the start of isovolumetric contraction.

10. The QRS complex of the ECG is due to
    a) atrial depolarization.
    b) atrial repolarization.
    c) ventricular depolarization.
    d) ventricular repolarization.

11. As a wave of action potentials travels from the atria to the ventricles, it is momentarily delayed by about 0.1 second as a result of slow conduction through
    a) the AV node.
    b) the atrioventricular bundle.
    c) the left and right bundle branches.
    d) Purkinje fibers.

12. Which of the following is most likely to cause a *decrease* in the stroke volume of the left ventricle?
    a) an increase in mean arterial pressure
    b) an increase in end-diastolic pressure
    c) an increase in end-diastolic volume
    d) an increase in the activity of sympathetic nerves to the heart

13. Left ventricular pressure and aortic pressure are virtually identical during
    a) isovolumetric contraction.
    b) isovolumetric relaxation.
    c) systole.
    d) the ejection period.

Objective Questions

1. Heart rate is normally determined by the action potential frequency in the (SA/AV) node.

2. According to the Starling effect, stroke volume should increase if end-diastolic volume (increases/decreases).

3. Heart rate is determined entirely by the inherent action potential frequency in cells of the SA node, with no external influences. (true/false)

4. Blood flow through the systemic circuit is driven by contractions of the (right/left) ventricle.

5. The valve located at the junction between the left ventricle and the aorta is an example of a(n) (atrioventricular/semilunar) valve.

6. (Isovolumetric contraction/Ejection) comes immediately after diastole.

7. Maximum aortic pressure during the cardiac cycle is called (diastolic/systolic) pressure.

8. Under normal conditions, pressures in the left and right ventricles are equal during systole. (true/false)

9. Stroke volume and _____ completely determine cardiac output.

10. If end-diastolic volume does not change but end-systolic volume decreases, stroke volume (increases/decreases).

11. If end-diastolic volume does not change but end-systolic volume decreases, ejection fraction (increases/decreases).

12. If sympathetic and parasympathetic inputs are constant and end-diastolic volume increases, contractility of the ventricular myocardium increases. (true/false)

13. The period of relaxation of the heart muscle is known as _____.

14. The (P/T) wave of the ECG corresponds to ventricular repolarization.

15. The difference between systolic pressure and diastolic pressure is known as _____ pressure.

16. Action potentials generated by pacemaker cells are called *pacemaker potentials*. (true/false)

Essay Questions

1. Discuss autonomic regulation of cardiac function. Include in your discussion a description of the effects of autonomic activity on the rate and force of ventricular contraction. Feel free to use Starling curves to clarify your discussion.

2. Describe the process of action potential propagation through the heart. Include a description of the role of pacemaker cells and gap junctions in cardiac electrical activity.

3. Discuss the interplay of the various influences on stroke volume. Be sure to include a discussion of the Starling effect and the influence of autonomic neurons. Use graphs or charts in your explanation if you feel that it is appropriate.

4. Clarify the distinction between ventricular *contractility* and *force of contraction*. Use graphs or charts in your explanation if applicable.

Find the answers to these exercises, and additional study tools, at the Physiology Place (www.physiologyplace.com).

The Cardiovascular System: Blood, Blood Flow, and Blood Pressure

OBJECTIVES

- Identify the major components of the blood. List the five major types of blood vessels, and describe the primary function of each.

- Describe the mechanism of blood clot formation.

- Describe the difference between series and parallel flow; explain how the parallel arrangement of organs enables blood flow to each organ to be regulated independently.

- Explain how pressure and resistance affect blood flow through vessels; explain how mean arterial pressure influences blood flow to individual organs and to the entire systemic circuit, and identify the factors that determine mean arterial pressure.

- Identify the factors that influence central venous pressure; explain how central venous pressure affects cardiac output.

- Explain how net filtration pressure governs the movement of fluid across capillary walls, and identify the factors that influence net filtration pressure.

CHAPTER OUTLINE

The Composition of Blood   404

Platelets and Hemostasis   407

The Structure and Function of Blood Vessels   411

Patterns of Blood Flow Within the Cardiovascular System   417

Physical Laws Governing Blood Flow and Blood Pressure   420

Factors Affecting Flow and Distribution of Blood to Organs   427

How Changes in Central Venous Pressure Affect Blood Flow to Organs   430

Movement of Fluid Across Capillary Walls   432

Above: Red blood cells, platelets, and a T-lymphocyte

When people donate blood, they are usually advised to lie down for a few minutes after the blood has been drawn. Those who ignore this advice and attempt to walk away often find themselves feeling dizzy as soon as they stand up, and some even pass out. Why? When someone gives blood, the quantity of blood removed—about one pint—represents roughly 10% of the average total blood volume in an adult. Any reduction in blood volume, but particularly a reduction as large as this, tends to reduce *venous return,* the rate at which blood flows back to the heart from the vasculature. When a person who has lost blood stands up, venous return drops even further because the force of gravity pulls blood toward the legs and away from the upper parts of the body. The decrease in venous return triggers a drop in cardiac output and mean arterial pressure, and blood flow to the brain falls, resulting in dizziness and possibly even a loss of consciousness.

In everyday life we experience a number of things that have the potential to affect us in the same way as a loss of blood. If you get up out of bed suddenly, for instance, you may experience temporary dizziness because gravity pulls blood toward your legs. You may also experience dizziness if you go for a long period without drinking liquids, which can reduce your blood volume by causing you to become dehydrated. In these circumstances, you usually do not lose consciousness unless the problem is severe. How is this so?

Arterial blood pressure and blood flow to the organs (including the brain) are regulated to deliver adequate supplies of oxygen and nutrients to organs and tissues under most circumstances. As we see in Chapter 14, the mechanisms that regulate arterial pressure and organ blood flow involve control of the heart and blood vessels by neural and hormonal signals, and by other factors. In this chapter we establish the groundwork necessary for understanding these regulatory mechanisms by looking at how various aspects of cardiovascular function, such as heart rate, stroke volume, and venous return, interact to determine arterial pressure and the flow of blood to the organs. In this effort we draw upon many of the principles of cardiac function from Chapter 12. We begin with a discussion of the properties of blood.

THE COMPOSITION OF BLOOD

Blood consists of two components: a fluid portion called plasma, and formed elements, which include cells (erythrocytes and leukocytes) and cell fragments called platelets (Figure 13.1). Here we briefly examine the properties of these components, beginning with plasma.

Plasma

Plasma is an aqueous solution in which a great variety of solutes are dissolved. These solutes include proteins, small nutrients (such as glucose, lipids, and amino acids), metabolic waste products (such as urea and lactic acid), gases (oxygen, carbon dioxide, nitrogen, and others), and electrolytes (such as sodium, potassium, and chloride). Although the proteins are the most abundant solutes in the plasma by weight, the smaller solutes are generally present in higher concentrations.

With respect to small solutes—that is, solutes other than proteins—the composition of plasma is very similar to that of interstitial fluid. This similarity occurs because capillary walls (which separate plasma from interstitial fluid) are highly permeable to small solutes, which allows these solutes to move freely between plasma and interstitial fluid. With respect to proteins, however, plasma and interstitial fluid differ greatly in composition; the concentration of proteins in the plasma is significantly greater than that in interstitial fluid. This difference in concentration is maintained by the low permeability of capillary walls to proteins, which limits proteins' ability to move out of the plasma.

Plasma proteins are categorized into three main groups: *albumins, globulins,* and *fibrinogen.* The albumins, which are synthesized by the liver, are the most abundant plasma proteins and make a large contribution to the osmotic pressure of plasma, a point whose significance is discussed when we examine the movement of fluid across capillaries. The globulins encompass a large number of different proteins that transport lipids, steroid hormones, and other substances in the blood; play a critical role in the blood's ability to form clots; and are important in defending the body against foreign substances. Fibrinogen is synthesized by the liver and is a key substance in the formation of blood clots. Major components of the plasma are listed in Table 13.1.

Formed Elements

The formed elements of the blood include cells (erythrocytes and leukocytes) and platelets, which are not cells but cell fragments. All blood cells are produced in **bone marrow,** a soft tissue located inside bones that contains cells dispersed within a porous matrix. Although the marrow in all bones produces blood cells in children, only bones of the skull, chest, pelvis, and the long bones in the

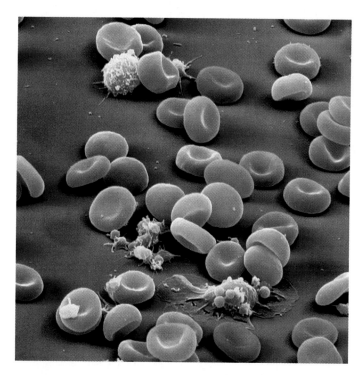

FIGURE 13.1 Formed elements of the blood. (Electron micrograph about 3000×)

TABLE 13.1
COMPONENTS OF PLASMA

COMPONENT	DESCRIPTION AND IMPORTANCE
WATER	Makes up 90% of plasma volume; provides dissolving and suspending medium for solutes and formed elements
SOLUTES	
Proteins	Account for 8% of plasma (by weight); most are synthesized by liver
Albumin	60% of plasma proteins; largely responsible for plasma osmotic pressure
Globulins	36% of plasma proteins; include clotting proteins, antibodies secreted by certain leukocytes during the immune response, and proteins that bind to lipids, fat-soluble hormones, and metal ions to transport these substances in the blood
Fibrinogen	Important in the formation of blood clots
Others	Enzymes, hormones, and antibacterial proteins
Nitrogenous waste products	By-products of metabolism, such as urea, uric acid, and creatinine
Organic nutrients	Materials absorbed from the intestines and used by cells throughout the body; includes glucose and other simple sugars, amino acids, fatty acids, glycerol, triglycerides, cholesterol, and vitamins
Electrolytes	
Cations	Sodium, potassium, calcium, magnesium (important in neuromuscular signaling), and trace metals (important in normal enzyme activity
Anions	Chloride (important in neuromuscular signaling), bicarbonate, and phosphate (important in maintenance of normal plasma pH)
Respiratory gases	Oxygen and carbon dioxide; most oxygen and some carbon dioxide is bound to hemoglobin in erythrocytes; a significant fraction of carbon dioxide is found in the plasma in the form of bicarbonate

upper portion of the limbs are actively involved in blood cell production in adults.

Erythrocytes

Erythrocytes (red blood cells) are the most abundant cells in the blood, numbering about 5 million per cubic millimeter of blood. Among cells of the body, erythrocytes are unique in that they lack nuclei, mitochondria, and other organelles, such as ribosomes, that are necessary for manufacturing proteins. Erythrocytes are shaped like disks and are about 7 μm in diameter and a little more than 2 μm thick (see Figure 13.1). (A μm is one *micrometer,* one millionth of a meter.) Erythrocytes are often described as *biconcave disks* because they are indented on both sides.

The major function of erythrocytes is to transport oxygen and carbon dioxide in the blood. Such transport is essential for the delivery of oxygen from the lungs to respiring cells, and for the delivery of carbon dioxide from respiring cells to the lungs, where it is eliminated from the body. Erythrocytes have a high capacity for carrying these gases because they contain in their cytoplasm large amounts of a protein called hemoglobin. This protein has the ability to bind oxygen and carbon dioxide reversibly, so it can pick up these materials at certain locations and release them at other locations. Each erythrocyte normally contains over 250 million hemoglobin molecules.

Hemoglobin is composed of four subunits of two types, each of which contains a special ring structure known as a *heme group,* which contains an atom of iron in an oxidized state (Figure 13.2). This iron is the site to which a molecule of oxygen binds. Carbon dioxide, by contrast, binds to other sites on the hemoglobin molecule. (Transport of oxygen and carbon dioxide by hemoglobin is discussed in more detail in Chapter 16.) The

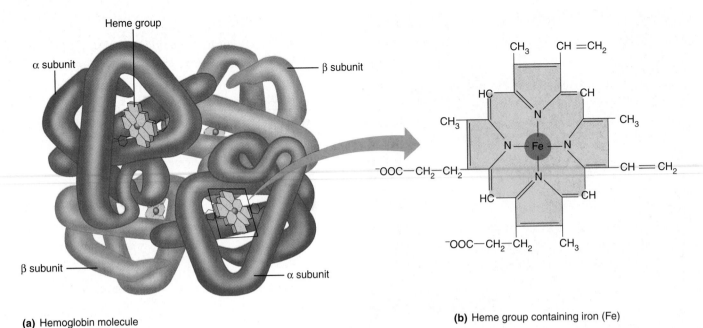

(a) Hemoglobin molecule

(b) Heme group containing iron (Fe)

FIGURE 13.2 Hemoglobin. (a) *The hemoglobin molecule consists of four protein subunits, designated as either α or β, each covalently bound to a heme group containing iron.* **(b)** *Chemical structure of a heme group.*

oxidized iron of the heme group gives hemoglobin a red color, which is responsible for the red color of erythrocytes and of blood.

Production of erythrocytes (erythropoiesis) is regulated primarily by the hormone *erythropoietin,* which is secreted into the bloodstream by certain cells in the kidneys and stimulates erythrocyte precursor cells in the bone marrow to proliferate and mature. Once erythrocytes are released into the bloodstream, they remain there for about 120 days, after which time they are removed from the blood by the liver and spleen and destroyed. Enzymatic degradation of hemoglobin causes the release into the bloodstream of a number of breakdown products, including *bilirubin,* a yellow compound that is responsible for both the characteristic yellowish tinge of plasma and the yellow color of urine.

Because iron is necessary for the production of hemoglobin (and other molecules in the body), sufficient quantities of iron must be obtained from dietary sources if a normal concentration of hemoglobin is to be maintained in the blood (16 grams per 100 ml in men, and 14 grams per 100 ml in women). The dietary requirement for iron is reduced somewhat by the fact that when hemoglobin is broken down, the released iron is carried in the bloodstream from the liver and spleen to the bone marrow, where it is used in making new hemoglobin. In addition to iron, production of erythrocytes also requires *folic acid* and *vitamin B_{12}* in the diet, because these vitamins are necessary for DNA synthesis and cell division.

A shortage of the dietary requirements needed for erythropoiesis can produce either a reduction in the amount of hemoglobin per cell or a reduction in the number of erythrocytes in the blood, either of which reduces the blood's oxygen-carrying capacity. Any reduction in the oxygen-carrying capacity of the blood is referred to as *anemia.* If the condition is due to a lack of iron in the diet, it is called *iron-deficiency anemia;* if it is caused by a lack of vitamin B_{12}, it is *pernicious anemia.* Anemia can also result from either hemorrhage (bleeding) or an abnormally high rate of erythrocyte destruction. Symptoms of anemia include chronic fatigue and shortness of breath upon exertion.

The total volume of blood in a normal healthy adult is about 5 liters, the vast majority of which is erythrocytes and plasma. Typically, erythrocytes make up about 45% of total blood volume, with most of the remainder consisting of plasma. The fractional contribution of erythrocytes to the blood volume is called the **hematocrit** (abbreviation, *hct*), which is determined by centrifuging a sample of blood in a tube (Figure 13.3). Because erythrocytes are denser than other elements of the blood, they are pulled to the bottom of the tube. To determine hematocrit, the heights of the erythrocyte column and the whole blood column are measured: hematocrit (hct) = height of erythrocyte column ÷ height of whole blood column.

The hematocrit is a useful clinical measure because it indicates whether or not a person has a normal complement of erythrocytes. A low hematocrit indicates a lower-than-normal concentration of erythrocytes in the blood. For men, the normal range of the hematocrit is 40–54, meaning that erythrocytes occupy from 40% to 54% of blood volume. For women, the normal range of the hematocrit is 37–47.

Leukocytes

Compared to erythrocytes, leukocytes (white blood cells) are far less numerous in the blood, numbering only about 4000–10,000 per cubic millimeter, which is lower than the red cell count by roughly a thousandfold. White cells are nucleated and possess all the normal cellular machinery, and they are thus the only fully functional cells in the blood. All white cells participate in some way in defending the body against invading microorganisms and other foreign materials, and each type is specialized to perform certain functions. (Details of leukocyte function are presented in Chapter 22, which covers the immune system.)

Unlike red cells, white cells are normally found not only in the bloodstream, but also in other tissues of the body. The presence of white cells outside blood vessels results from their mobility, which allows them to squeeze through pores in capillaries and migrate through tissues. This ability to migrate is important to their defensive function because it enables them to reach infected areas.

Platelets

Platelets are cell fragments that arise when portions of large bone-marrow cells called *megakaryocytes* break off. They are smaller than erythrocytes and contain mitochondria, smooth endoplasmic reticulum, and cytoplasmic granules, but no nucleus. Platelets number from 100,000 to 500,000 per cubic millimeter and are important in triggering the sequence of events that leads to the formation of blood clots.

PLATELETS AND HEMOSTASIS

Blood vessels get damaged frequently, leading to both internal and/or external bleeding. Although bleeding is usually minor, without mechanisms for stopping the bleeding—called **hemostasis**—even a superficial cut could cause a person to bleed to death. The process of hemostasis occurs in three steps: vascular spasm, formation of a platelet plug, and formation of a blood clot, or **thrombus**. Each of these steps is described next.

Vascular Spasm

When a blood vessel is damaged, intrinsic mechanisms trigger a constriction called a *vascular spasm,* which increases resistance to blood flow. Damage also tends to activate the sympathetic nervous system, which causes further vasoconstriction. With less blood flowing to the area of damage, blood loss is minimized. However, decreasing blood loss is not sufficient; blood loss must be stopped altogether.

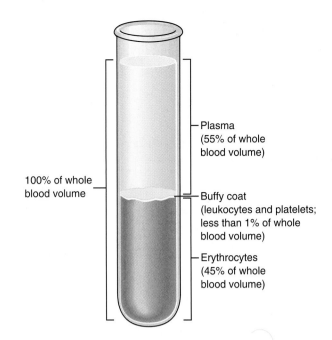

FIGURE 13.3 Determination of hematocrit. *When a blood sample is centrifuged, erythrocytes are pulled to the bottom of the tube because they are denser than other blood elements.*

Platelet Plug

Platelets, also called thrombocytes, are non-nucleated fragments of megakaryocytes. Platelets possess granules containing a variety of substances that can be secreted into the plasma, including ADP, serotonin, epinephrine, and a variety of chemicals that participate in the formation of a blood clot. Platelets also are "sticky" under certain circumstances, allowing them to adhere to surfaces, especially those of damaged blood vessels.

Both the formation of platelet plugs and the subsequent blood clot require the presence of platelets and a fairly large set of specific plasma proteins. In platelet plug formation, the key protein is **von Willebrand factor** (vWf), which is secreted by megakaryocytes, platelets, and endothelial cells lining blood vessels. Although vWf is present in the plasma at all times, it accumulates at the site of vessel damage.

The first step in platelet plug formation is *platelet adhesion*, which occurs when blood vessel damage exposes tissue underlying the endothelium (the layer of endothelial cells that line a vessel), called *subendothelial tissue.* When blood contacts subendothelial tissue, vWf binds to collagen fibers in the subendothelium, triggering the binding of platelets to vWf, which anchors platelets in place. Contact with vWf also activates platelets, changing their metabolism and surface properties (making them "sticky") and stimulating the secretion of certain products.

Two of the secretory products of activated platelets—serotonin and epinephrine—cause vasoconstriction,

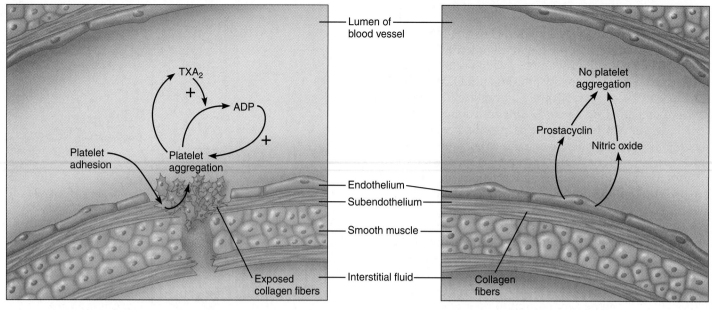

(a) Damaged blood vessel endothelium

(b) Normal blood vessel endothelium

FIGURE 13.4 Formation of a platelet plug. (a) *Platelets adhere to collagen fibers at the site of vessel damage. Adhered platelets secrete ADP to stimulate platelet aggregation at the site of adhesion, causing more platelets to secrete ADP. Aggregated platelets also release thromboxane A₂ (TXA₂) into the lumen of the blood vessel, which stimulates further platelet aggregation.* **(b)** *Healthy endothelial cells secrete nitric oxide (NO) and release prostacyclin, both of which inhibit platelet aggregation.*

which increases resistance to blood flow and minimizes blood loss. A third secretory product of activated platelets, ADP, causes *platelet aggregation*, the second step in platelet plug formation. ADP stimulates morphological changes in the platelets that cause them to adhere to one another such that they form a mass, or aggregate. Aggregated platelets secrete more ADP, which stimulates further platelet aggregation, thereby providing a positive feedback loop that increases the rate of platelet plug formation (Figure 13.4a). ADP also stimulates the production of **thromboxane A₂** (TXA₂), which further supports platelet aggregation.

TXA₂ is formed from a phospholipid, **arachidonic acid,** located in the membrane of platelets. TXA₂ has many roles in hemostasis, including stimulation of platelet aggregation, stimulation of ADP secretion (providing more positive feedback for platelet plug formation), and vasoconstriction (reducing blood flow to the area).

Platelet plug formation is limited to the area of blood vessel damage to prevent unnecessary blood clots that, by occluding blood flow, might deprive tissue of its needed nutrients and allow waste products to accumulate. Platelet plugs do not form on normal endothelium because healthy endothelial cells continuously release **prostacyclin** (also called prostaglandin I₂) and nitric oxide, both of which inhibit platelet aggregation (Figure 13.4b). Whereas aggregated platelets convert arachidonic

acid to TXA₂ to facilitate platelet plug formation, healthy endothelial cells convert arachidonic acid to prostacyclin to inhibit platelet plug formation.

Platelets contain high concentrations of the contractile proteins actin and myosin. As platelets aggregate and form a plug, they contract to increase the "tightness" of the plug.

Formation of a Blood Clot

Integral to the formation of a blood clot is a plasma protein called **fibrin,** which is necessary for the blood to coagulate, or be converted into a gel that traps erythrocytes and plugs the damage to the blood vessel and thus prevents blood loss. For this reason blood clots are also called *fibrin clots* (Figure 13.5). Fibrin clot formation is secondary to platelet plug formation because it requires that a number of steps occur on phospholipids located on the surface of activated platelets, and because secretory products of aggregated platelets are necessary for clot formation.

The formation of a fibrin clot requires a sequence of reactions called the *coagulation cascade*. During this cascade, plasma proteins called *coagulation factors*, which are always present in the plasma in their inactive form, undergo a series of proteolytic activations resulting from hydrolysis of certain peptide bonds. Most coagulation factors are designated by Roman numerals; the number of the factor provides little insight on function or location in the

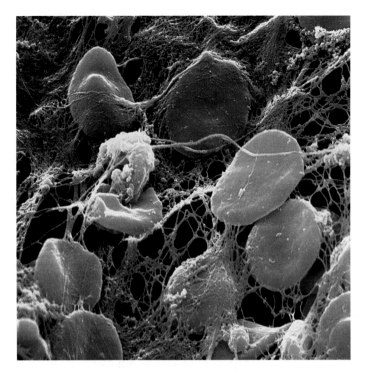

FIGURE 13.5 **A fibrin clot.**

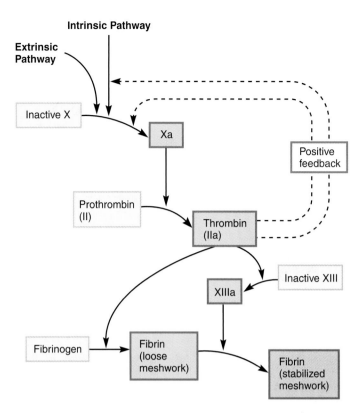

Intrinsic Pathway

Extrinsic Pathway

FIGURE 13.6 **Role of thrombin in forming fibrin clot.**
Thrombin is activated by the action of factor Xa on prothrombin and provides positive feedback for its own synthesis. Once active, thrombin converts fibrinogen to fibrin and activates factor XIII.

coagulation cascade, as these coagulation factors were numbered based on the order of discovery. A lowercase "a" following the Roman numeral indicates activation of the factor. Most of the activated coagulation factors function as proteolytic enzymes for the next step of the cascade, although some serve as cofactors.

The ultimate step of clot formation is conversion of a filamentous plasma protein, *fibrinogen*, into its active form, fibrin. Conversion of fibrinogen to fibrin is a proteolytic reaction catalyzed by **thrombin**, the active form of another coagulation factor, prothrombin. Once formed, fibrin molecules adhere to each other, forming a loose meshwork of strands. The meshwork is then stabilized by formation of covalent linkages between strands, a reaction catalyzed by another coagulation factor, *factor XIII* (also called fibrin-stabilizing factor). Like fibrin and thrombin, factor XIII is present in the plasma in an inactive form and must be activated to XIIIa before participating in clot formation.

Central to formation of the stable fibrin meshwork is activation of thrombin (Figure 13.6). Factor Xa converts prothrombin (factor II) to thrombin. Thrombin has several roles in clot formation, including converting fibrinogen to fibrin, activating factor XIII, and providing positive feedback for its own activation. Although not shown in the figure, thrombin also contributes to additional platelet aggregation and stimulates platelets to secrete several products.

Two pathways lead to the activation of thrombin: an intrinsic pathway involving coagulation factors and other necessary chemicals that are already present in the plasma, and an extrinsic pathway involving some coagulation factors present in damaged tissue adjacent to the

site of vessel damage. Both clotting pathways are generally activated simultaneously because rarely is vessel damage not accompanied by damage to other tissues. The stimuli that initiate the pathways, however, differ.

The intrinsic pathway starts when circulating factor XII (also called Hageman factor) is activated by contact with substances in the subendothelium, including collagen and phospholipids (Figure 13.7). Activation of factor XII starts a cascade of reactions that leads to activation of factor X, to activation of thrombin, and ultimately to fibrin formation. The extrinsic pathway starts when tissue damage allows *tissue factor* (factor III) to contact the plasma and react with inactive factor VII to form a complex, which activates factor VII. The complex of factor VIIa and tissue factor then activates factor X, leading to the activation of thrombin. Even though the intrinsic and extrinsic pathways start from separate places, they eventually merge at activation of factor X to form a common pathway.

Only some of the several other chemicals in blood that participate in clot formation are shown in Figure 13.7. At several steps, Ca^{2+} (factor IV) and **platelet factor 3** (PF3) are needed. PF3, located on the surface of activated platelets, is a phospholipid necessary for activation of thrombin by the intrinsic pathway.

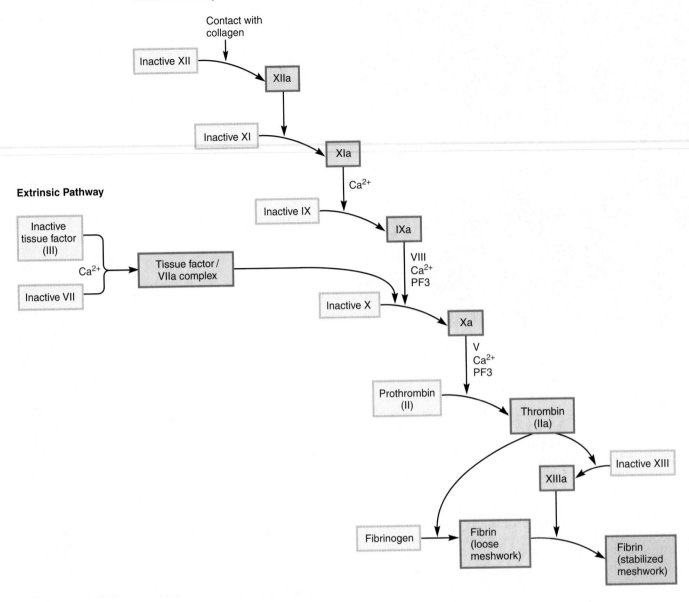

FIGURE 13.7 The intrinsic and extrinsic coagulation pathways. *Formation of a fibrin clot is a sequence of reactions involving several coagulation factors, platelet factors (such as PF3), and calcium. All components of the intrinsic pathway are located in the blood. The extrinsic pathway requires a factor not in the blood: tissue factor (factor III).*

Factors Limiting Clot Formation

Like platelet aggregation, fibrin clot formation is limited to the immediate vicinity of the damaged area, because certain proteins found in plasma and on the surface of endothelial cells act as *anticoagulants*, chemicals that inhibit blood clotting or coagulation. During the initial phase of clotting, **tissue factor pathway inhibitor** is secreted by healthy endothelial cells and inhibits the extrinsic pathway. Additionally, healthy endothelial cells secrete a molecule called *thrombomodulin*, which binds to thrombin, forming a complex. While a part of the thrombomodulin-thrombin complex, thrombin cannot convert fibrinogen to fibrin; instead it activates **protein C,** another plasma protein that continuously circulates in the plasma. Protein C is an anticoagulant that inhibits both the intrinsic and extrinsic pathways. Thus thrombin not only promotes clotting at injury sites; it also indirectly inhibits clotting in healthy tissue.

Once formed, fibrin clots are eventually dissolved by **plasmin,** a protein derived from the plasma protein *plasminogen*. Plasmin dissolves clots by enzymatically breaking down fibrin. Plasminogen is converted to plasmin by *plasminogen activators* secreted by a variety of cell types. One example of a plasminogen activator is *tissue plasminogen activator* (TPA), which is secreted by endothelial cells during clot formation. Fibrin activates TPA, which subsequently converts plasminogen to plasmin.

The Role of Coagulation Factors in Clot Formation Disorders

Most coagulation factors are synthesized in the liver, which releases them into the plasma in their inactive form. **Serum** is plasma from which these coagulation factors have been removed.

Several coagulation factors are essential for the formation of a fibrin clot. The lack of any essential factor impairs clot formation and thus results in excessive bleeding. Hemophilia, a genetic disorder caused by the deficiency of the gene for a specific coagulation factor, is most commonly due to a deficiency of factor VIII in the blood. Another genetic bleeding disorder, von Willebrand's disease, is characterized by reduced levels of vWf, which interferes with platelet plug formation. However, because vWf also serves as a plasma carrier for factor VIII, the absence of vWf causes factor VIII to be less stable and to break down more quickly, leading to lower factor VIII levels in the bloodstream. Excessive bleeding can also occur if vitamin K is lacking in the diet, because the liver requires vitamin K for the synthesis of many of the proteins necessary in blood clotting.

Aspirin as an Anticoagulant

Aspirin, one of the most commonly used analgesics, has received a lot of attention lately for another of its actions: It prevents blood clotting. Aspirin is one of many anti-clotting drugs given to people who are susceptible to stroke or coronary artery disease. Formation of blood clots in cerebral arteries (in the brain) or coronary arteries (in the heart) can have serious deleterious effects, including death. Given at low dosages, aspirin acts as an anticoagulant by inhibiting the formation of TXA_2, decreasing platelet aggregation and platelet plug formation. At high dosages, however, aspirin decreases formation of prostacyclin, which actually increases the likelihood of clot formation. Clinical studies have shown that low doses of aspirin both reduce the incidence of subsequent heart attacks and decrease the severity of damage when given after a heart attack.

Quick Test 13.1

1. What are the three components that make up the formed elements of the blood, and what are the primary functions of each?
2. How do the concentrations of small solutes in the plasma compare to their concentrations in interstitial fluid? How do the two fluids compare in regard to their protein concentrations?
3. What is hemoglobin? Why is it important to the function of red blood cells?
4. What is the hematocrit, and how is it determined?
5. Describe the roles of thrombin in hemostasis.

THE STRUCTURE AND FUNCTION OF BLOOD VESSELS

Blood vessels are classified according to whether they carry blood away from or to the heart, and according to size. Arteries and smaller arterioles carry blood from the heart and to capillaries, which are drained by venules and then larger veins, which return the blood to the heart (Figure 13.8). All blood vessels possess a hollow interior called the *lumen,* through which blood flows; the lumen of all blood vessels (and of the heart as well) is lined by a layer of epithelium called the *endothelium.* Surrounding the lumen is a wall that varies in thickness and composition from one vessel type to another.

The smallest of all blood vessels, capillaries, consist only of a layer of endothelial cells and a basement membrane; the walls of all other blood vessels contain various amounts of *smooth muscle* and fibrous and/or elastic connective tissue (Figure 13.9). Within the fibrous connective tissue are extracellular fibers made of a protein called *collagen,* which lends tensile strength to vessel walls, enabling them to withstand the pressure of blood within them without rupturing. Elastic connective tissue contains fibers of a highly stretchable extracellular protein called *elastin,* which enables vessels to expand or contract as the pressure of blood within them changes.

We next examine the structure and function of the various types of blood vessels, beginning with arteries.

Arteries

Arteries conduct blood away from the heart and toward the body's tissues; they have relatively large diameters and thick walls. The largest artery, the aorta, has an internal diameter of about 12.5 mm and a wall that is 2 mm thick. The smaller arteries that branch off the aorta have internal diameters ranging from 2 mm to 6 mm and a wall thickness of about 1 mm. Arterial walls contain large amounts of elastic and fibrous tissue, enabling arteries to withstand relatively high blood pressures, which are higher in these vessels than anywhere else in the vasculature. The thickness of arterial walls, coupled with the relative abundance of elastic tissue, gives arteries both a certain stiffness and the ability to expand and contract as the blood pressure rises and falls with each heartbeat.

This combination of stiffness and flexibility enables arteries to perform one of their major functions—acting as *pressure reservoirs* (storage sites for pressure) to ensure a

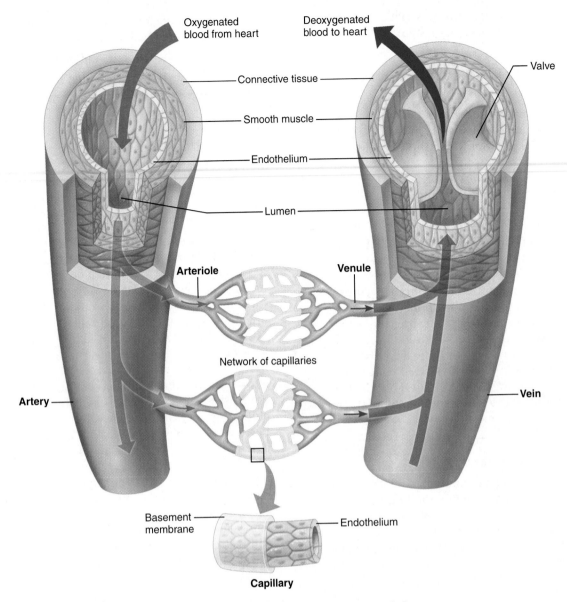

Oxygenated blood from heart

Deoxygenated blood to heart

Valve

Connective tissue

Smooth muscle

Endothelium

Lumen

Arteriole

Venule

Network of capillaries

Artery

Vein

Basement membrane

Endothelium

Capillary

FIGURE 13.8 The relationships of blood vessels according to size and the direction of blood flow. *The arrow at the upper left represents oxygenated blood (red) arriving from the heart, whereas the arrow at the upper right represents deoxygenated blood (blue) returning to the heart. Note the differences in lumen diameter and wall thickness between arteries and veins, and the presence of one-way valves in the vein.*

continual, smooth flow of blood through the vasculature even when the heart is not pumping blood (diastole). Arteries are able to act as pressure reservoirs because their ability to expand when the pressure within them rises—a property called **compliance** (Toolbox: Compliance, p. 415)—is low. In vessels with low compliance, such as arteries, a given increase in blood pressure causes a small degree of vessel expansion. Put another way, a small degree of expansion is accompanied by a large change in pressure. Therefore, when the heart ejects blood into the arteries during systole and causes them to expand, the resulting rise in pressure is greater than it would be if arteries' compliances were higher. This stored pressure is released during diastole and propels the blood forward during this period.

Arterioles

Arterioles are blood vessels that conduct blood from small arteries to capillaries. Arterioles, unlike arteries, are not visible to the unaided eye. (In this regard, arterioles are like capillaries and venules; together, these three types of vessels constitute the body's *microcirculation*.) The inner diameter of arterioles averages 0.03 mm (30 μm), whereas the wall thickness is about 6 μm. The number of arterioles in the body is estimated to be in the hundreds of thousands.

The major function of arterioles is to serve as points of control for regulating the flow of blood through the capillary beds downstream from them (Figure 13.10). Blood flow is regulated in arterioles by the contraction or relaxation of circular smooth muscle, which is present in

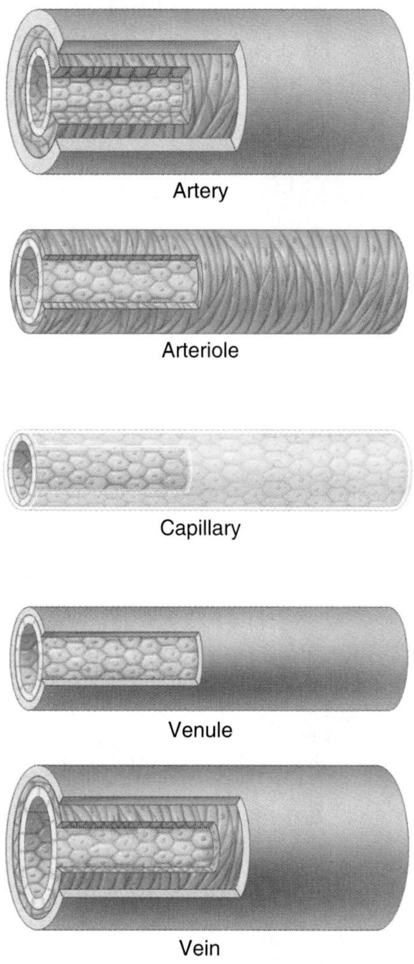

Average internal diameter (mm)	Average wall thickness (mm)		Special features
4.0	1.0	Artery	Muscular, highly elastic
0.03	0.006	Arteriole	Muscular, well innervated
0.008	0.0005	Capillary	Thin-walled, highly permeable
0.02	0.001	Venule	Thin-walled, some smooth muscle
5.0	0.5	Vein	Thin-walled (compared to arteries), fairly muscular, highly distensible

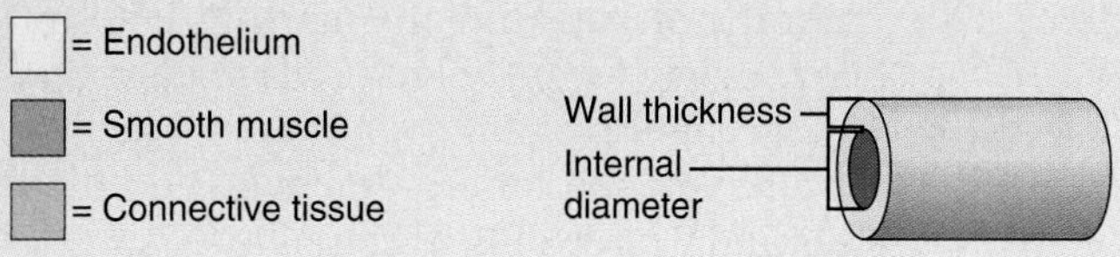

☐ = Endothelium
■ = Smooth muscle
▨ = Connective tissue

Wall thickness ⎯
Internal diameter ⎯

FIGURE 13.9 Structural characteristics of the five blood vessel types.

the walls of these vessels in greater abundance (relative to their wall thickness) than in other vessel types (see Figure 13.9). Contractile activity of arteriolar smooth muscle is regulated in most regions of the body by the autonomic nervous system and by local chemical agents and hormones. Contraction of circular smooth muscle cells causes arterioles to constrict, which decreases the flow of blood through them and through the capillaries downstream; relaxation of these smooth muscle cells allows arterioles to dilate, increasing the flow of blood.

The more systemic effects of the control of blood flow by arterioles are the regulation of blood flow to various organs and tissues throughout the body, and the regulation of mean arterial pressure, both of which are discussed in greater detail later in this chapter.

Most tissues contain blood vessels called *metarterioles,* which are structurally intermediate between arterioles and capillaries; instead of the continuous layer of smooth muscle that surrounds arterioles, metarterioles possess isolated rings of smooth muscle that act as "gatekeepers" at strategic points (see Figure 13.10). Unlike arterioles, which direct blood into the interbranching vessels in a capillary bed, metarterioles serve as bypass channels or *shunts* by directly connecting arterioles to venules. The presence of these shunts allows blood to continue flowing from arterioles to venules even when constriction in arterioles prevents the blood from moving through capillaries.

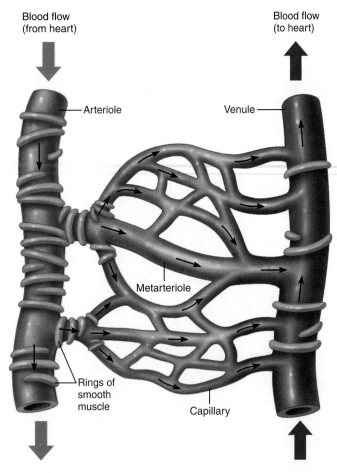

FIGURE 13.10 The role of arterioles in regulating the flow of blood through a capillary bed. *Direction of blood flow is indicated by arrows within vessels. Contraction or relaxation of abundant smooth muscle cells in the arteriolar wall regulates the flow of blood into the capillary bed. Note the metarteriole directly connecting the arteriole and the venule.*

Capillaries

Capillaries, which are the smallest and most numerous blood vessels in the body, ranging from 5 μm to 10 μm in diameter and numbering around 10 billion, are also the vessels with the thinnest walls (about 0.5 μm). The thinness of capillary walls—they are little more than a single cell layer thick, consisting of a layer of endothelial cells and a basement membrane—facilitates the capillary's primary function: to permit the exchange of materials between cells in tissues and the blood. The thinness of capillary walls permits small molecules (for example, oxygen, carbon dioxide, sugars, amino acids, and water) to enter and leave capillaries readily, promoting efficient material exchange. The exchange of materials across capillaries is also facilitated by the large number of them: The body's total capillary surface area available for the exchange of materials likely exceeds 6000 square meters.

In most regions of the body, small solutes readily enter and leave the bloodstream by simple diffusion across capillary walls. (A notable exception is in the brain, where the blood-brain barrier limits the diffusion of solutes across capillary walls; see p. 226.) However, the permeability of capillaries varies from region to region because capillaries differ with regard to the physical properties of their walls. Based on these anatomical differences, capillaries are grouped into two major classes: continuous capillaries and fenestrated capillaries (Figure 13.11).

In *continuous capillaries* (Figure 13.11a), which are the more common, the endothelial cells are joined together such that the spaces between them are relatively narrow. These capillaries are highly permeable to substances having small molecular sizes and/or high lipid solubilities (such as oxygen, carbon dioxide, and steroid hormones) and are somewhat less permeable to small water-soluble substances (such as sodium, potassium, glucose, or amino acids). The permeability of continuous capillaries to proteins and other large molecules is very low because these substances can neither readily cross membranes of endothelial cells nor easily penetrate the gaps between cells.

In *fenestrated capillaries* (Figure 13.11b), the endothelial cells possess relatively large pores (*fenestrations*) that are wide enough to allow proteins and other large molecules to pass through. In some fenestrated capillaries, the gaps between endothelial cells are also wider than usual, enabling large proteins and in some cases even entire cells to pass through easily. For this reason, fenestrated capillaries are not only highly permeable to small molecules (whether lipid-soluble or water-soluble), but to large molecules as well. Fenestrated capillaries are found mostly in organs whose functions depend on the rapid movement of materials across capillary walls, including the kidneys, liver, intestines, and bone marrow. In the liver, the presence of fenestrated capillaries allows newly synthesized proteins such as albumin or clotting factors to enter the plasma. In bone marrow, fenestrated capillaries allow newly formed blood cells to enter the circulation.

Mechanisms of transport across capillary walls differ for different substances, depending on their molecular sizes and degree of lipid solubility (Figure 13.12). The high permeability of continuous capillaries to small lipid-soluble substances is attributable to the fact that these substances readily diffuse through cell membranes and therefore can easily cross endothelial cells. Continuous capillaries show a somewhat lower permeability to small water-soluble solutes because these substances are mostly restricted to moving through the water-filled gaps between the endothelial cells. The permeability to proteins and other large molecules is negligible because these substances are too large to pass through any but the largest water-filled pores and are thus prevented from passing through endothelial cells or around them. This is not true of *all* proteins, however, because certain proteins (referred to as *exchangeable proteins*) are selectively transported across endothelial cells by a slow, energy-requiring process known as *transcytosis*. In this process, endothelial cells engulf proteins in the plasma within capillaries by endocytosis. The proteins are then ferried across the cells by vesicular transport and released by exocytosis into the interstitial fluid on the other side.

A blood vessel (indeed, any hollow structure) tends to expand as the pressure inside it rises, and to contract as the pressure falls. Strictly speaking, it is not just the pressure *inside* that determines whether a vessel expands or contracts, but rather the *difference* between the pressure inside and the pressure outside. This pressure difference is called the *distending pressure* (or *transmural pressure*). When the pressure inside a vessel is greater than the pressure outside, the distending pressure is positive, and a net outward force acts on the wall of the vessel, tending to make it expand. As the distending pressure increases, the volume of the vessel increases. If the pressure inside a vessel is less than the pressure outside, the distending pressure is negative, and a net inward force acts on the wall and tends to compress the vessel.

The *compliance* of a vessel is strictly defined as the change in volume per unit change in distending pressure. Mathematically, this is expressed as

$$\text{compliance} = \Delta V / \Delta (P_{\text{inside}} - P_{\text{outside}})$$

where ΔV is the change in volume, and P_{inside} and P_{outside} are pressures inside and outside, respectively. The term within the parentheses is the distending pressure.

The following diagram, which is a pressure-volume curve, graphically depicts the concept of compliance:

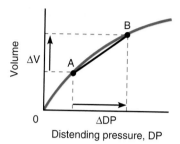

An increase in the distending pressure in a vessel, as when one goes from point A to point B in the diagram, produces a certain increase in the vessel's volume. The increase in volume is given by the vertical distance between the two points. The compliance is therefore the change in vertical distance (volume) divided by the change in horizontal distance (distending pressure). In other words, the compliance of the vessel over the range between points A and B is the *slope* of the line joining the two points.

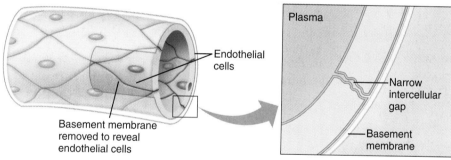

(a) Continuous capillary

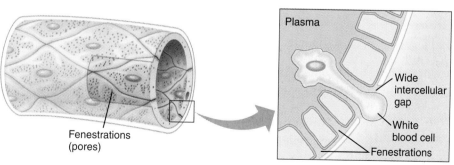

(b) Fenestrated capillary

FIGURE 13.11 Two types of capillaries. (a) *A continuous capillary, featuring narrow, water-filled gaps between endothelial cells.* (b) *A fenestrated capillary, which possesses pores (fenestrations) that penetrate through endothelial cells, in addition to intercellular gaps between endothelial cells. In some fenestrated capillaries, gaps between endothelial cells are large enough to permit blood cells to move through them.*

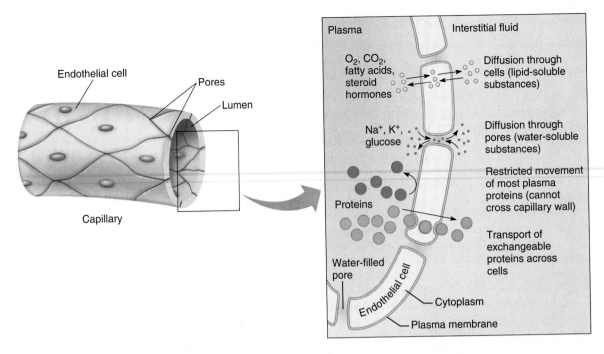

FIGURE 13.12 Exchange of materials across the wall of a continuous capillary. *Lipid-soluble substances are able to diffuse through endothelial cells, whereas passage of small water-soluble substances is restricted to water-filled pores. Whereas most proteins in the plasma are unable to cross the wall, certain proteins cross the wall via transcytosis.*

Is the movement of glucose across capillary walls an *active* or *passive* process?

Venules

Venules are slightly smaller than arterioles, averaging about 20 μm in diameter. The walls of venules, however, contain little or no smooth muscle and are about one-sixth as thick as those of arterioles (see Figure 13.9). These vessels, like capillaries, function in the exchange of materials, and they conduct blood from capillaries to veins.

Veins

Veins have roughly the same diameter as arteries, but have walls about one-half as thick. A typical vein has an internal diameter of 5 mm but a wall thickness of only 0.5 mm. The largest veins, the venae cavae, are even larger in diameter than the aorta (30 mm, as opposed to 12.5 mm) but have a wall thickness of only 1.5 mm, compared to 2 mm for the aorta. The relative thinness of the walls of veins reflects the fact that blood pressure in the veins is significantly lower than in arteries. The walls of veins are similar to those of arteries in that they contain smooth muscle and elastic and fibrous connective tissue (see Figure 13.9).

Unlike any other blood vessels in the body, veins are equipped with one-way valves that permit blood to flow toward the heart but prevent it from flowing back toward organs and tissues (see Figure 13.8). These valves are present in veins located outside the thoracic cavity *(peripheral veins)* but are absent from veins located within the thoracic cavity *(central veins)*. The significance of this difference will become apparent when we discuss the action of the *respiratory pump* later in this chapter.

Unlike arteries, which function as pressure reservoirs, veins function as *volume reservoirs*, a property that is also related to vessel compliance. Because veins are thin-walled and easily stretched, they have high compliances—a relatively small increase in the pressure within them causes a relatively large degree of expansion (increase in volume). Put another way, veins can accommodate a large increase in blood volume in response to a small increase in blood pressure and therefore are good at storing volume. As a result of their high compliance, veins can hold a larger volume of blood than arteries can at a given pressure (Figure 13.13). In fact, the veins in the human body contain a substantially greater volume of blood than do the arteries (Figure 13.14), even though the pressure within veins is much lower than that within arteries.

The volume reservoir function of veins is not merely a curiosity; it has important practical consequences.

Passive

When one donates blood or loses a fraction of total blood volume for any reason, most of the lost blood volume comes from the veins. As the volume of blood in the veins decreases, however, the accompanying drop in venous pressure is relatively small, owing to the fact that veins have high compliances. Therefore, the veins can lose a substantial volume of blood before the drop in venous pressure becomes large enough to cause a significant decrease in ventricular pressure and, hence, cardiac output. (Recall that central venous pressure is an important determinant of end-diastolic volume, which influences the stroke volume and cardiac output by virtue of the Starling effect; see p. 396.) If veins had lower compliances, the loss of a given volume of blood would cause a larger drop in venous pressure and cardiac output. The existence of a volume reservoir in the veins is important for another reason, too: In exercise and other circumstances in which it is desirable to increase the cardiac output or maintain it at an adequate level, a number of mechanisms (discussed later in the chapter) come into play to force much of the blood volume out of the veins and toward the heart, thereby promoting increased ventricular filling and cardiac output.

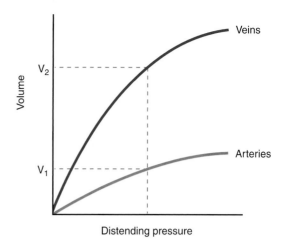

FIGURE 13.13 **Curves showing how the volume of blood contained in arteries and veins varies with the pressure inside them.** *Comparison of the two curves shows that at a given pressure, veins hold more blood (V_2) than arteries (V_1).*

Quick Test 13.2

1. Where is the endothelium of a blood vessel located? What is it?

2. In regard to the regulation of blood flow and blood pressure, what two important functions are performed by arterioles?

3. How do fenestrated capillaries differ from continuous capillaries?

4. When physiologists speak of the compliance of a blood vessel, to what are they referring? Which have higher compliances—arteries or veins?

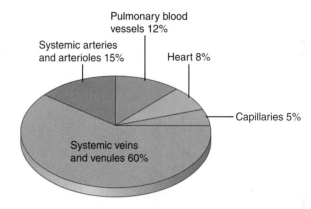

FIGURE 13.14 **Distribution of blood volume in the various portions of the cardiovascular system.** *Percentages indicate proportions of total blood volume.*

PATTERNS OF BLOOD FLOW WITHIN THE CARDIOVASCULAR SYSTEM

In Chapter 12 we saw that the cardiovascular system is divided into a *pulmonary circuit,* which delivers blood to the lungs, and a *systemic circuit,* which delivers blood to the rest of the organs and tissues of the body. We also saw that blood flow through the pulmonary circuit is driven by the right heart, whereas flow through the systemic circuit is driven by the left heart. If we take any location as a starting point and follow the flow of blood back to that point, we find that the pulmonary and systemic circuits are *in series* with each other; that is, for the cardiovascular system as a whole, blood must pass through the two circuits in sequence before it can return to the starting point (Figure 13.15a). However, if we look at blood flow *within*

either the systemic or pulmonary circuits, we see a different pattern—*parallel flow.*

Parallel Flow

Figure 13.15b shows why blood flow in the systemic circuit is called *parallel flow* (and why the systemic organs are said to be *in parallel* with one another). In the systemic circuit, blood does not flow from one organ directly to the next. Instead, blood travels through the aorta and the arteries that branch off it to reach only one organ at a time before flowing through veins that converge to either the superior or inferior vena cava. Moreover, the pattern of blood flow *within* organs, including the lungs in the pulmonary circuit, is also parallel because arteries branch to arterioles, which branch to capillaries, and so on.

Note in Figure 13.15b that the *heart* is in parallel with the other organs in the systemic circuit. Even though the heart pumps a large volume of blood, the blood within

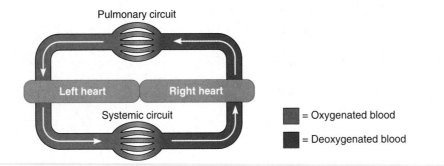

(a) Series flow

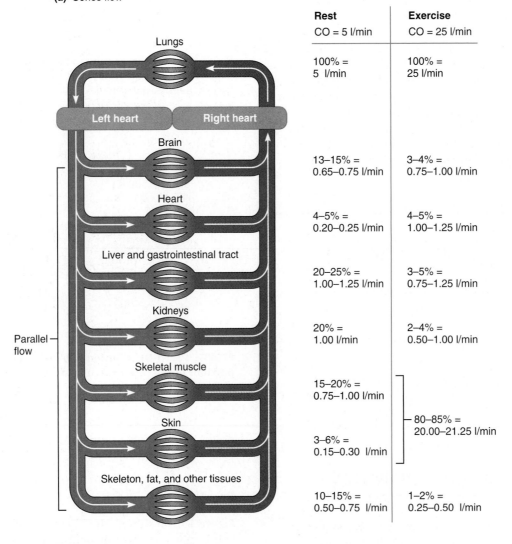

	Rest CO = 5 l/min	Exercise CO = 25 l/min
Lungs	100% = 5 l/min	100% = 25 l/min
Brain	13–15% = 0.65–0.75 l/min	3–4% = 0.75–1.00 l/min
Heart	4–5% = 0.20–0.25 l/min	4–5% = 1.00–1.25 l/min
Liver and gastrointestinal tract	20–25% = 1.00–1.25 l/min	3–5% = 0.75–1.25 l/min
Kidneys	20% = 1.00 l/min	2–4% = 0.50–1.00 l/min
Skeletal muscle	15–20% = 0.75–1.00 l/min	80–85% = 20.00–21.25 l/min
Skin	3–6% = 0.15–0.30 l/min	
Skeleton, fat, and other tissues	10–15% = 0.50–0.75 l/min	1–2% = 0.25–0.50 l/min

(b) Parallel flow

FIGURE 13.15 Blood flow patterns in the cardiovascular system. (a) *Series flow through the pulmonary and systemic circuits.* **(b)** *Parallel flow through organs in the systemic circuit. The table at right lists the blood flow to the respective organs (expressed as both proportion of cardiac output and flow rate) at rest and during exercise.*

the heart's chambers does not supply the heart muscle with significant quantities of oxygen or nutrients. Instead, the heart muscle obtains most of its nourishment from blood via the *coronary arteries,* which branch off the aorta near its base and run through the heart muscle.

Even though Figure 13.15b is helpful for understanding the *general* parallel pattern of blood flow in the systemic circuit, it is not a detailed map. Some exceptions to the parallel flow pattern do not appear in the diagram. For instance, when blood leaves the intestines, it does not travel directly to the vena cava and back to the heart; instead, it goes to the liver via the *hepatic portal vein* and then to the vena cava via the *hepatic vein.* Thus, the liver and intestines, which are in series with each other, are lumped together (with other organs in the gastrointestinal tract) in Figure 13.15b to show that the *combination* of these organs is in parallel with the other organs in the systemic circuit. The path of blood flow from the intestines to the liver is an example of *portal circulation,* in which blood flows from one organ to a second organ before returning to the heart. Portal circulation also occurs within the kidneys, where blood flows from one set of vascular beds to a second set of vascular beds.

The parallel arrangement of organs in the systemic circuit confers two distinct advantages. First, because each organ is fed by a separate artery, each receives fully oxygenated blood—that is, blood that has not been depleted of oxygen as a result of having already flowed through another organ. Thus, for instance, when a muscle contracts and extracts oxygen from the blood, the muscle's increased metabolic rate does not deprive other organs of fully oxygenated blood because blood leaving the muscle returns to the heart and is reoxygenated in the lungs before it returns to the systemic circuit. Second, because blood reaches the organs via parallel paths, blood flow to the organs can be independently regulated, enabling blood flow to be adjusted to match the constantly changing metabolic needs of organs. At any given instant, blood flow can be increased to more active organs and decreased to less active organs.

Independent Regulation of Blood Flow in the Systemic Circuit

Independent regulation of organ blood flow is illustrated in the righthand portion of Figure 13.15b, which lists cardiac output and blood flow to the various organs at rest and during heavy exercise. At rest, cardiac output (CO) is 5 liters per minute; during exercise, CO rises dramatically to 25 liters per minute, a fivefold increase.

We can see that blood flow to organs is independently regulated by comparing the proportions of total blood flow each organ receives at rest and during exercise (see Figure 13.15b). (A comparison of these proportions or relative *shares* of cardiac output enables us to identify what is termed the *distribution* of blood flow to organs.) If blood flow to organs were not regulated independently, the proportion of CO each organ receives would remain constant, and blood flow to every organ would rise as cardiac output rises during exercise.

But whereas skeletal muscle and skin combined receive 18–26% of cardiac output at rest, during exercise the proportion of CO these organs receive increases to 80–85%. At the same time, blood flow to the liver and gastrointestinal tract declines to 3–5% of CO from the resting level of 20–25%. Clearly, when the body makes a transition from rest to exercise, blood flow is diverted away from the liver and GI tract (where metabolic demand is low) and toward the muscles and skin, which have high metabolic demands during exercise. The increased blood supply to muscles supplies the oxygen and nutrients needed to generate contractile force; increased flow to the skin both helps to dissipate excess heat at the body surface and supplies sweat glands with the energy and water they need in order to cool the body via the evaporation of sweat.

Blood flow distribution is adjusted continually on a moment-to-moment basis to accommodate the ever-changing metabolic needs of organs and tissues. The same idea applies to blood flow *within* an organ or tissue, because metabolic activity can vary from place to place. In this case, the distribution of blood flow among the various capillary beds is adjusted according to the metabolic needs of local tissues.

The key players in the regulation of blood flow to organs are *arterioles* (and arteries smaller than 0.2 mm in diameter), which regulate flow by narrowing or widening in response to contraction or relaxation of smooth muscle in their walls. An organ's increased share of the cardiac output results from a general relaxation of vascular smooth muscle and an increase in the average diameter of arterioles within that organ. A decrease in an organ's share of cardiac output results from the contraction of vascular smooth muscle, causing the average diameter of arterioles within that organ to decrease. Arterioles also regulate the distribution of blood flow *within* an organ or tissue because they control flow through individual capillary beds. Factors that influence the contractile activity of vascular smooth muscle are discussed in Chapter 14.

Quick Test 13.3

1. What does it mean when we say that organs are in parallel? in series?

2. What are the advantages of a parallel-flow system?

3. When physiologists talk about the distribution of blood flow to organs, to what are they referring?

FIGURE 13.16 **Pressures in the vasculature.** *Distances along the horizontal axis represent the relative distances blood travels in the various portions of the vasculature.* **(a)** *Representative blood pressures and pressure drops for the various vessel types in the systemic circulation.* **(b)** *Representative pressures and pressure drops in the systemic and pulmonary circuits.*

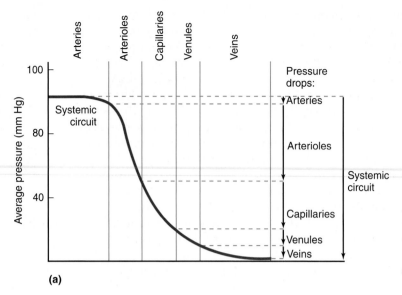

(a)

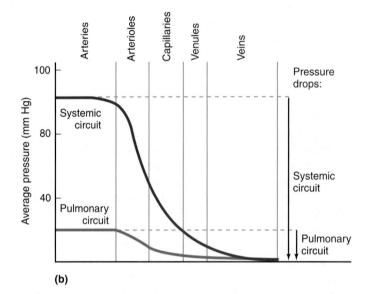

(b)

PHYSICAL LAWS GOVERNING BLOOD FLOW AND BLOOD PRESSURE

In a certain sense the vasculature is an elaborate "system of pipes" that runs through the body, so the fundamental physical laws that describe the flow of any liquid through a system of pipes also pertain to blood flow in the cardiovascular system. The rule that is pertinent to our discussion here states that the flow rate of a liquid (the volume flowing per unit of time) through a pipe is directly proportional to the pressure gradient along the pipe—to the difference between the pressures at the two ends of the pipe—and inversely proportional to the resistance of the pipe:

$$\text{Flow} = \text{pressure gradient/resistance} = \Delta P/R$$

The quantity ΔP, the size of the **pressure gradient,** represents the driving force that *pushes* the flow of liquid

through a pipe; the quantity R, the resistance, is a measure of the various factors that *hinder* the flow of liquid through a pipe.

This rule is so crucial to our understanding of blood flow that it is the starting point for all our discussions pertaining to flow, pressure, and resistance in the cardiovascular system. It is so universally applicable that it applies to liquids flowing in a single pipe or blood vessel, or in a system of pipes or blood vessels, no matter how complicated. As we will see in Chapter 15, it even pertains to the flow of air into and out of the lungs.

In this section we examine the general principles governing how pressure gradients and resistance affect blood flow in individual vessels and networks of vessels; in the following section we apply these principles to understand the various factors that influence the flow of blood through organs and tissues. First we consider pressure gradients.

Pressure Gradients in the Cardiovascular System

When you inflate a balloon, it expands because the pressure the air exerts on the inside of the balloon is greater than the pressure the air exerts on the outside. If you remove your fingers from the nozzle of the balloon, air rushes out for the same reason—air pressure is greater inside the balloon than outside. Whenever there is a difference in pressure between two locations, the pressure gradient drives the flow from a region of higher pressure to one of lower pressure, or *down the pressure gradient.*

Air flowing out of a balloon, like the flow of blood through the cardiovascular system, is an example of *bulk flow.* Regardless of whether the flowing medium is a gas or a liquid, the driving force for bulk flow is always a pressure gradient, and the direction of flow is always down the gradient from a region of greater pressure to a region of lower pressure. This rule applies to blood flow and to all other examples of bulk flow that occur in the body, such as the flow of air into and out of the lungs.

The Role of Pressure Gradients in Driving Blood Flow

As we saw in Chapter 12, the primary function of the heart is to generate the pressure that drives the flow of blood through the vasculature. Strictly speaking, however, it is not absolute pressure that drives blood flow, but rather a pressure gradient. By pumping blood into the arteries, the heart raises mean arterial pressure, which creates a difference in pressure between the arteries and veins that drives the flow of blood.

Figure 13.16a shows that as blood flows from arteries to veins, pressure decreases gradually. A difference in pressure across any portion of the vasculature is also called the *pressure drop* across that part. The largest pressure drop occurs along the arterioles: In the systemic circuit, blood enters them at an average pressure of about 90 mm Hg and leaves them at a pressure of about 40 mm Hg (see Figure 13.16a). In contrast, the pressure drops in the larger vessels (the arteries and veins) are quite small, such that the pressure in these vessels is nearly uniform. Figure 13.16b shows that pressures in the pulmonary circuit are lower than pressures in the systemic circuit, reflecting the fact that the right ventricle generates less pressure than the left ventricle.

Figure 13.17 shows a useful model for explaining the relationship between pressure and flow in blood vessels. In the diagram, a tube or "blood vessel" connects two large reservoirs containing liquid. The pressure at either end of the vessel is determined by the vertical distance from the vessel to the surface of the liquid, the so-called *hydrostatic column.* (The pressure also depends on the *density* of the liquid—its mass per unit volume—which we assume is constant.) The higher the hydrostatic column, the greater the pressure. (You can feel the effect of a hydrostatic column by diving underwater. The deeper you go, the more pressure the water exerts on your body.)

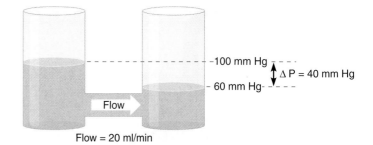

(a)

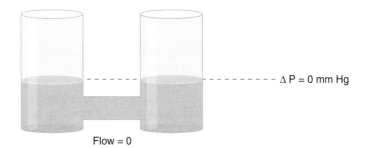

(b)

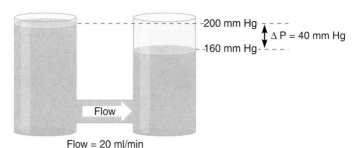

(c)

FIGURE 13.17 A model that relates blood flow to the pressure gradient. *A single blood vessel is represented by a tube connecting two reservoirs, in which the depth of liquid determines the pressure.* **(a)** *The difference in the pressures in the two reservoirs produces a pressure gradient (ΔP) of 40 mm Hg, creating a flow of 20 ml/min., as indicated by the arrow.* **(b)** *When the levels of the liquid are the same in both reservoirs, the pressure gradient is zero, and hence flow is zero.* **(c)** *When the levels in both reservoirs are raised such that ΔP remains at 40 mm Hg, flow remains at 20 ml/min., indicating that the pressure* gradient, *not absolute pressure, determines flow.*

When the liquid level is different on the two sides (Figure 13.17a), a pressure gradient exists, and liquid flows at a rate of 20 ml/min. through the tube from the high-pressure side (100 mm Hg) to the low-pressure side (60 mm Hg), or down the pressure gradient (ΔP = 40 mm Hg). As a result, the level on one side drops while that on the other side rises. Eventually, the levels become equal (Figure 13.17b), and flow stops because a pressure gradient no longer exists. Figure 13.17c illustrates that the rate of flow through the tube depends only on the *difference* between the pressures at either end, not the absolute pressure. When the liquid level is raised on both sides, the

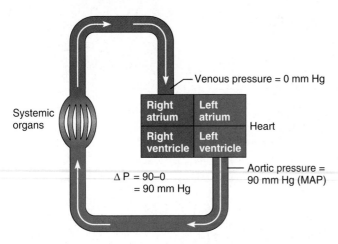

FIGURE 13.18 A pressure gradient is the driving force for blood flow. *Aortic pressure averages about 90 mm Hg, whereas the pressure in the vena cava is close to zero where it joins the heart. This creates a pressure gradient of 90 mm Hg, which represents the overall driving force that pushes the flow of blood through the systemic circuit. Note that this pressure gradient is virtually identical to the mean arterial pressure (MAP).*

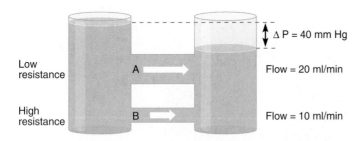

FIGURE 13.19 The effect of resistance on flow. *In this model, two blood vessels are depicted as two tubes (A and B) connecting the reservoirs. Even though the pressure gradient is the same (40 mm Hg) for both vessels, blood flows through vessel B at a lower rate than through vessel A. Because it has a smaller diameter, vessel B has a higher resistance than vessel A.*

If the pressure difference decreased to 20 mm, what will be the flow rate in tube A? in tube B?

pressure on both sides increases. However, if the difference between the levels remains constant (here, ΔP still equals 40 mm Hg), the flow does not change (flow still equals 20 ml/min.).

Pressure Gradients Across the Systemic and Pulmonary Circuits

In the systemic circuit, *mean arterial pressure* (the average pressure in the aorta throughout the cardiac cycle) is about 90 mm Hg (see Figure 13.16a). At the other end of the circuit, in the large veins in the thoracic cavity that

lead to the heart, the pressure—known as the **central venous pressure** (CVP)—is very close to 0 mm Hg. The difference between the mean arterial pressure and central venous pressure is the pressure gradient that drives blood flow through the systemic circuit. Because central venous pressure is so small, we can say that *the pressure gradient (ΔP) driving blood flow through the systemic circuit is virtually identical to the mean arterial pressure* (Figure 13.18).

Blood flow through the pulmonary circuit is also driven by a pressure gradient—the difference between the pressure in the pulmonary arteries and the pressure in the pulmonary veins. However, this pressure gradient is smaller than the one that drives flow through the systemic circuit because pulmonary arterial pressure is lower than aortic pressure (see Figure 13.16b). During the cardiac cycle, pulmonary arterial pressure averages about 15 mm Hg, as opposed to about 90 mm Hg in the aorta. (Pulmonary venous pressure, like central venous pressure, is close to zero.)

Quick Test 13.4

1. What is a pressure gradient? Where are pressure gradients present in the vasculature?

2. Define mean arterial pressure.

3. Define central venous pressure. Is it higher or lower than mean arterial pressure?

4. What is the pressure gradient driving blood flow through the systemic circuit? through the pulmonary circuit? Besides the pressure gradient, what other factor determines the rate at which blood flows through a system of blood vessels?

Resistance in the Cardiovascular System

In Chapter 12 we saw that the blood flow through the pulmonary circuit is identical to that through the systemic circuit (about 5 liters per minute at rest). But if the pressure gradient in the pulmonary circuit is lower than that in the systemic circuit, then how is it possible to have the same blood flow in the two circuits? The answer can be deduced from the rule, Flow $= \Delta P/R$: The pulmonary circuit offers less *resistance* (because of its physical characteristics), so a smaller pressure gradient can achieve the same flow.

Here we examine in turn the factors that determine the resistance of individual blood vessels and of networks of vessels such as the systemic and pulmonary circuits.

Resistance of Individual Blood Vessels

The resistance of any tube (including a blood vessel) is a measure of the degree to which the tube hinders or resists the flow of liquid through it. From the flow rule, it's apparent that for a given pressure gradient, a vessel with

10 ml/min; 5 ml/min

In examining membrane transport in Chapter 4, we encountered Fick's Law, which relates the flux to the concentration gradient for a substance crossing a membrane by simple diffusion. In simple diffusion, a concentration gradient of a given magnitude produces a larger or smaller flux, depending on a variable called *permeability*, which in turn depends on properties of the diffusing substance and of the membrane through which the substance is moving. Permeability therefore relates a flow of something (the flux) to a driving force (the concentration gradient).

In this chapter we explore another relationship between a flow and a driving force. In this case the flow is the bulk flow of a fluid, such as blood, and the driving force is a pressure gradient. The parameter relating them is the *resistance*. Resistance, like permeability, depends on properties of the substance that is flowing and of the structure through which the substance is moving. The following table compares bulk flow and simple diffusion:

	BULK FLOW	**SIMPLE DIFFUSION**
Driving force	Pressure gradient (ΔP)	Concentration gradient (ΔC)
Flow	Volume of fluid moving through a tube or vessel per unit time (flow)	Number of molecules crossing a membrane per unit time (flux)
Variable relating flow to driving force	Resistance of blood vessel (R)	Permeability of membrane (P)
Mathematical relationship	Flow = $\Delta P/R$	Flux = $PA(\Delta C)$ (A = membrane area)
Direction of flow	From higher to lower pressure (down the pressure gradient)	From higher to lower concentration (down the concentration gradient)
Characteristics of moving substance that affect resistance or permeability	Fluid's viscosity	Diffusing molecules' size, shape, and lipid solubility
Characteristics of vessel or membrane affecting resistance or permeability	Blood vessel's radius and length	Membrane's thickness and lipid composition

higher resistance yields a lower flow (Figure 13.19). Put another way, for a given pressure gradient, blood flow is greater when resistance is lower because it is easier for blood to flow.

If you have ever drunk liquid through a drinking straw, you have experienced the effects of resistance. You likely noticed that it is easier to drink through a wide straw than through a narrow one, and that it is easier to drink through a short straw than through a long one. You also know that it's more difficult to drink a milkshake through a straw than it is to drink a soda. Resistance, then, depends on the physical dimensions of the tube and properties of the fluid flowing through it (Discovery: Permeability and Resistance), namely the tube's *radius* and *length*, and the fluid's *viscosity* ("thickness" or "syrupiness"); it also depends on the manner of flow—whether the flow is turbulent or more streamlined. Next we examine how these factors affect resistance in the vasculature.

1. *Vessel radius.* Changes in resistance to blood flow in the cardiovascular system almost always result from changes in the radii of blood vessels: As radius decreases, resistance increases. Vascular resistance is controlled by relaxation and contraction of vascular smooth muscle in the walls of small arteries and arterioles. Relaxation causes an increase in vessel radius (called **vasodilation**) and a decrease in resistance; contraction causes a decrease in vessel radius (called **vasoconstriction**) and an increase in resistance.

2. *Vessel length.* Even though longer vessels have greater resistance than shorter ones (all else being equal), changes in vascular resistance are rarely due to changes in vessel length; vessels do not change length except as a person grows.

3. *Blood viscosity.* Vascular resistance increases as viscosity increases, but blood viscosity does not change appreciably under normal conditions. The major

A fluid flowing through a tube or blood vessel encounters resistance, some of which is due to frictional forces acting between the fluid and the walls of the tube or vessel. Friction within the fluid itself also contributes to the resistance, which is why some fluids, such as molasses, flow more slowly than others, such as water. The speed at which a fluid moves varies from one location to another within the moving liquid. Therefore, layers of fluid moving at different speeds rub against each other, which creates friction and dissipates energy. For a fluid moving by laminar flow through a cylindrical tube, the resistance (R) is given by the following equation, which is called *Poiseuille's Law:*

$$R = 8L\eta/\pi r^4$$

where L is the length of the tube, η is the viscosity of the fluid, and *r* is the tube's internal radius. Note that the resistance is strongly affected by the internal diameter of the tube because it depends on the *fourth power* of the radius. Therefore, if one tube is half as wide as another of the same length, its resistance is 16 times as great!

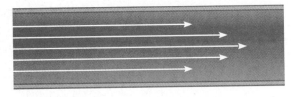

(a) Laminar flow

(b) Turbulent flow

FIGURE 13.20 Laminar versus turbulent blood flow. (a) *When flow is laminar, blood moves in the same direction in all parts of a vessel. Here, the length of each line represents the velocity of the blood, which is greatest in the center of the vessel and declines nearer the sides, due to friction with the blood vessel's wall.* **(b)** *When blood flow is turbulent, the direction of flow varies from place to place within a vessel.*

determinant of blood viscosity is the hematocrit; as the hematocrit increases (whether from an increase in erythrocytes or a decrease in plasma volume), viscosity increases. One example occurs in *polycythemia,* in which the problem is overproduction of erythrocytes in the bone marrow, which can lead to a hematocrit over 70% instead of the normal value of 45%. This condition more than doubles vascular resistance, which can be harmful because it increases the workload of the heart. Severe dehydration can also cause an increase in the hematocrit by reducing plasma volume.

4. *Manner of flow.* The resistance of blood flow through a vessel is also affected by the manner in which blood flows through it. Under most conditions, blood moves through the cardiovascular system via *laminar (streamlined) flow,* similar to water that slowly leaves a faucet in a thin, silent stream. In laminar flow within a blood vessel, all the blood is flowing smoothly along the length of the vessel (Figure 13.20a), and blood flowing in the center of the vessel moves faster than blood nearer the wall because friction between the moving blood and the wall slows the blood down. This friction is one reason why blood encounters resistance as it flows (Toolbox: Poiseuille's Law).

However, when a liquid is forced to move through a tube quickly enough, its flow becomes *tur-*

bulent, just as water leaving a faucet more quickly begins to flow roughly and noisily. In turbulent flow (Figure 13.20b), movement of blood in a vessel is disrupted into eddies or whirls in which the blood flows in different directions and speeds in different places within the vessel. These variations increase the resistance and create vibrations that can be heard as sounds.

Whether fluid flow is laminar or turbulent depends on several factors, including the velocity at which the fluid is moving. Generally speaking, the faster a fluid moves, the greater likelihood the flow will become turbulent. At most locations in the cardiovascular system, blood flows slowly enough that flow is always laminar. However, the velocity in certain locations *can be* high enough to cause turbulence. In the heart, turbulence occurs for short periods immediately before the AV valves and semilunar valves close, and this turbulence is what causes the first and second heart sounds. Turbulence occurs because the narrowing of the valve openings as they are about to close forces the blood moving through them to go faster. Sounds created by turbulence are useful in the measurement of blood pressure (Discovery: The Role of Turbulence in Blood Pressure Measurement).

Resistance of Blood Vessel Networks: Total Peripheral Resistance

Although thus far we have limited our consideration to the resistance of individual blood vessels, a *network* of blood vessels (such as the systemic or pulmonary circuits, the vasculature within an organ, or even a single capillary bed) also has a resistance. For blood vessel networks, the rules governing flow, pressure, and resistance are fundamentally the same as for individual

The most common method for measuring blood pressure is based on the phenomenon of turbulence and the noise that accompanies it. When technicians take your blood pressure, they use a device called a *sphygmomanometer*, which consists of an inflatable cuff and a pressure-measuring device (either a mercury column or a meter) that displays the air pressure inside the cuff.

To measure blood pressure, a technician places the cuff around the upper arm and then inflates it by squeezing a rubber bulb, which raises the cuff pressure. This pressure is transmitted through the tissues of the arm to the *brachial artery*, which runs to the lower arm. The technician increases cuff pressure until it is above systolic arterial pressure, which causes the artery to collapse, which stops blood flow. Next, the technician opens a valve to slowly let air out of the cuff, allowing cuff pressure to fall, while she listens for a pulse by using a stethoscope placed on the inside of the elbow. When cuff pressure drops to where it is just slightly below systolic arterial pressure, the artery opens briefly with each heartbeat, because the pressure inside the artery is higher than that outside it, forcing the vessel open. When this happens, blood flows through the artery, but in a turbulent fashion because it is forced through a narrow opening. This turbulence creates sounds, called *Korotkoff sounds*, that can be heard through the stethoscope. (These sounds are not to be confused with those made by the heart.) When the Korotkoff sounds first appear, the technician notes the cuff pressure and records it as systolic arterial pressure.

The technician continues to let air out of the cuff, which further lowers cuff pressure. As a consequence, the artery remains open for longer periods during each heartbeat. Eventually, cuff pressure falls just below diastolic arterial pressure, from which point the artery stays open throughout the entire cardiac cycle because pressure inside the artery is always higher than that outside it. Under these conditions, blood flows more slowly and without turbulence, because it is no longer forced to go through a narrow passage. Therefore, the Korotkoff sounds disappear. When the technician first notes the absence of sounds, she notes the cuff pressure and records it as diastolic pressure.

The events involved in blood pressure measurement are illustrated in the diagram in this box; the straight line represents cuff pressure, and the wavy line represents arterial pressure.

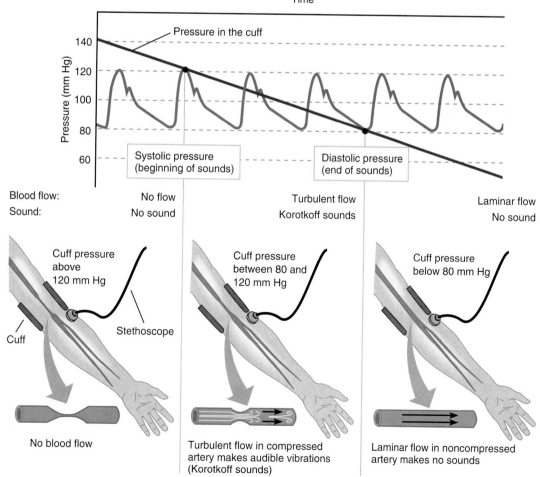

VARIABLE IN FLOW RULE	DEFINITION	EQUIVALENT VARIABLE IN SYSTEMIC CIRCUIT	DEFINITION
Flow	Volume of liquid moving through system per unit time	CO (cardiac ouput)	Volume of blood flowing through systemic circuit per minute (equals volume of blood pumped by left ventricle per minute)
ΔP (pressure gradient)	The force that drives the flow (equals pressure difference across the system)	MAP (mean arterial pressure)	Average pressure in the aorta throughout the cardiac cycle
R (resistance)	A measure of factors hindering flow in the system	TPR (total peripheral resistance)	Combined resistance of the systemic organs plus all blood vessels leading to and from them

Law governing flow in any system:

$$\text{Flow} = \Delta P/R$$

Law governing blood flow in the systemic circuit:

$$CO = MAP/TPR$$

vessels: For any network, the total flow (F) increases in proportion to the pressure gradient (ΔP) along the network and decreases as the resistance (R) of the network increases.

As we might expect, the resistance of a vascular network depends on the resistances of all the individual blood vessels it contains. Any factor that causes the resistance of individual vessels in a network to increase or decrease also tends to cause a respective increase or decrease in the resistance of the network as a whole. From this, it follows that *vasoconstriction anywhere within a network of blood vessels tends to increase the resistance of the network, whereas vasodilation anywhere within a network tends to decrease the resistance of the network.*

In the systemic circuit, the combined resistances of all the organs, including the blood vessels leading toward and away from them, is known as **total peripheral resistance (TPR).** Even though total peripheral resistance is influenced by the resistance of all blood vessels in the systemic circuit, most of it is attributable to the resistance of arterioles and small arteries, which are appropriately referred to as **resistance vessels.** (In fact, over 60% of TPR is attributable to arterioles; whereas individual capillaries have greater resistance than individual arterioles, the greater abundance of capillaries lowers their combined resistance.) Any changes in TPR are almost entirely due to alterations in the internal diameters of resistance vessels.

Most of the time, TPR stays fairly constant; even though the vascular beds in the organs are in a constant state of flux, with vasoconstriction occurring in some beds and vasodilation in others, these changes tend to cancel each other out. Sometimes, however, the resistances of vascular beds in several organs change in a concerted manner to produce an increase or decrease in total resistance. As we discuss in Chapter 14, these concerted changes are usually triggered by neural and hormonal inputs to vascular smooth muscle, which then alters the diameters of resistance vessels (especially arterioles).

Relating Pressure Gradients and Resistance in the Systemic Circulation

We can express the relationships among pressure, resistance and flow in the systemic circuit by making some substitutions in the flow rule, $F = \Delta P/R$. First, because all the blood that flows from the heart goes through the systemic circuit, the flow is equal to the volume of blood flowing through the circuit each minute, or cardiac output (CO). Next, we have already seen that this flow of blood is driven by the pressure gradient represented by the difference between mean arterial pressure (MAP) and central venous pressure, which we also learned is virtually identical to MAP. And finally, we know that the resistance in the systemic circuit is TPR. Accordingly, substituting these variables into the flow rule yields

$$CO = MAP/TPR$$

The relationships among the quantities governing blood flow, and their application to the systemic circulation, are summarized in Table 13.2.

1. Of the following changes, which would cause an *increase* in the resistance to blood flow through a vessel? (choose all that apply): an increase in vessel diameter; an increase in blood viscosity; a decrease in hematocrit; a transition from laminar to turbulent flow

2. Define total peripheral resistance.

3. Assuming that mean arterial pressure remains constant, how will an increase in total peripheral resistance affect the flow of blood through the systemic circuit? Why?

4. What types of blood vessels are classified as resistance vessels? Why? What role do these vessels play in controlling total peripheral resistance?

FACTORS AFFECTING FLOW AND DISTRIBUTION OF BLOOD TO ORGANS

As we see in Chapter 14, the cardiovascular system's primary function—to provide adequate blood flow to the body's organs and tissues—involves the interplay of a number of regulatory mechanisms that control cardiac function and vascular resistance. Our purpose in the following two sections is to lay the groundwork necessary for understanding these regulatory mechanisms.

Determinants of Organ Blood Flow: Mean Arterial Pressure and Organ Resistance

Explaining the factors that determine blood flow to organs is a simple matter of applying the flow rule. We have already seen that the flow rule applies to single vessels and networks of vessels, and it applies to organs as well.

In this case we consider three organs arranged in parallel (Figure 13.21a); blood flow for each organ is driven by the same pressure gradient: the difference between mean arterial pressure and central venous pressure, which effectively equals MAP. Suppose the flows through organs A, B, and C are 1.5 liters/min., 0.5 liter/min., and 1.0 liter/min., respectively (Figure 13.21b). Given that all these flows are driven by the same pressure gradient, the differences in flow must result from differences in resistance, according to the flow rule:

$$\text{organ blood flow} = \text{MAP/organ resistance}$$

On the basis of this relationship, the flows depicted in Figure 13.21b indicate that resistance is lowest in organ A, intermediate in organ C, and highest in organ B.

The vascular resistance of an organ (or tissue), like that of a vessel or system of vessels, is altered by the con-traction or relaxation of smooth muscle in arterioles and small arteries. These changes in vessel caliber are in turn regulated by a number of factors, including neural input, hormones, and local chemical agents.

The effect of organ resistance on the *distribution* of blood flow is illustrated by comparing parts (b) and (c) of Figure 13.21. In both situations MAP is constant; the resistances, and thus the blood flows, of organs A and B are unchanged. But the resistance of organ C is higher in Figure 13.21c, which results in a decrease in blood flow compared to that in Figure 13.21b. Note that because the flow to organ C decreases while the flows to organs A and B remain unchanged, total flow declines, and organ C's *share* of cardiac output decreases while the *shares* of organs A and B increase.

These findings have an important implication: *Changes in the distribution of blood flow to organs—that is, changes in the percentage of cardiac output supplied to each organ—are due to changes in the vascular resistance of individual organs.* Any change in mean arterial pressure in the absence of any change in resistance in individual organs affects blood flows to all organs equally, and thus each organ's share of cardiac output does not change. Therefore, any observed change in an organ's share of cardiac output (as occurs during the transition from rest to exercise) results from a change in that organ's resistance relative to other organs.

The matching of blood flow to each organ's metabolic needs is accomplished mainly via the operation of local factors that alter vascular resistance, as we see in Chapter 14.

Determinants of Mean Arterial Pressure: Heart Rate, Stroke Volume, and Total Peripheral Resistance

In the previous discussion we saw that two factors influence blood flow to an organ: mean arterial pressure and the organ's resistance. Given that mean arterial pressure influences blood flow to all organs in the systemic circuit, it becomes clear why matching blood flow to organs' needs requires that the cardiovascular system maintain adequate MAP: Any decline in MAP tends to compromise blood flow to all the systemic organs. Let's further explore the determinants of mean arterial pressure.

We previously saw that one variation of the flow rule is CO = MAP/TPR. This expression can be rearranged algebraically to give MAP = CO × TPR. Recalling from Chapter 12 that cardiac output is determined by heart rate (HR) and stroke volume (SV), and substituting these terms into the expression, we see that

$$\text{MAP} = \text{HR} \times \text{SV} \times \text{TPR}$$

Thus, mean arterial pressure is completely determined by three factors: (1) heart rate, (2) stroke volume, and (3) total peripheral resistance.

FIGURE 13.21 The effects of pressure gradients and resistance on blood flow to organs. (a) *Blood flow through three parallel organs. The pressure gradient driving flow is the difference between MAP and CVP.* **(b)** *Differences among blood flows to organs are due to differences in resistance in the organs.* **(c)** *Given constant MAP and CVP, an increase in resistance in one organ (organ C) reduces flow to that organ alone. Note that this results in a reduction in total flow and in a change in the distribution of blood flow.*

In part c, how do the resistances of tubes B and C compare?

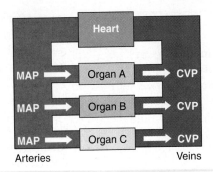

(a)

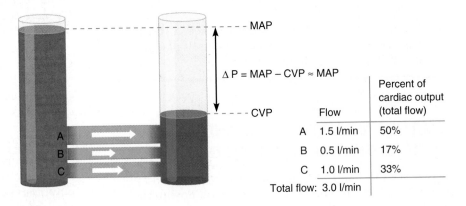

(b)

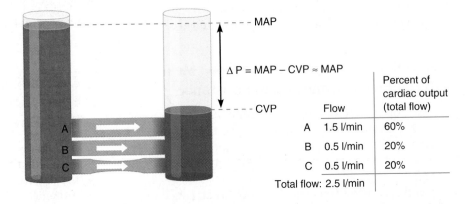

(c)

The previous expression indicates that mean arterial pressure should rise following an increase in heart rate or stroke volume (which tend to increase cardiac output) or total peripheral resistance. We can get an intuitive feeling for why increases in cardiac output or total peripheral resistance should cause MAP to rise by looking at Figure 13.22, which shows blood flows (wide arrows) into and out of the aorta in different circumstances. Total peripheral resistance is shown schematically as a constriction at the distal end of the aorta, where the blood flows out. When mean arterial pressure is steady (Figure 13.18a), blood flows into the aorta at the same rate as it flows out, such that the volume of blood contained within it does not change. Because the flow into the aorta is cardiac output, the flow out of the aorta under these conditions equals cardiac output. (The flows shown in Figure 13.22 represent *average* blood flows over several cardiac cycles. Within a single cardiac cycle, the flows into and out of the aorta differ from one another as aortic pressure cycles up and down; see p. 388.)

Figure 13.22b shows how mean arterial pressure would change given a sudden increase in cardiac output

They are the same.

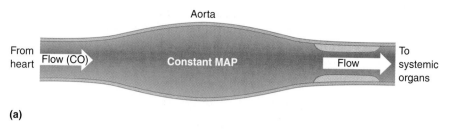

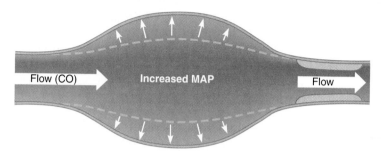

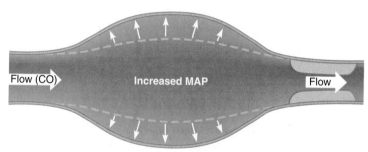

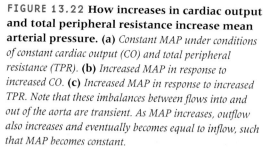

FIGURE 13.22 How increases in cardiac output and total peripheral resistance increase mean arterial pressure. (a) *Constant MAP under conditions of constant cardiac output (CO) and total peripheral resistance (TPR).* (b) *Increased MAP in response to increased CO.* (c) *Increased MAP in response to increased TPR. Note that these imbalances between flows into and out of the aorta are transient. As MAP increases, outflow also increases and eventually becomes equal to inflow, such that MAP becomes constant.*

(due to an increase in heart rate and/or stroke volume), with total peripheral resistance remaining constant. Under these conditions, blood flows into the aorta faster than it flows out, so the blood volume within the aorta increases and the vessel expands. This expansion stretches the wall of the aorta, causing it to exert a large inward force on the blood, so that the pressure of the blood rises. Thus an increase in heart rate or stroke volume causes mean arterial pressure to rise. (If cardiac output remains elevated, the rising aortic pressure causes the flow of blood out of the aorta to increase. Eventually, mean arterial pressure reaches a steady level, with rates of blood flow into and out of the aorta being equal.) When a sudden increase in total peripheral resistance occurs while cardiac output remains constant (Figure 13.18c), what ensues is basically the same as in the previous situation: The increase in total peripheral resistance

reduces the flow of blood out of the aorta, so that blood flows in faster than it flows out, and mean arterial pressure rises.

As we see in Chapter 14, each of the three factors that determine mean arterial pressure (heart rate, stroke volume, and total peripheral resistance) is regulated by reflex mechanisms involving the autonomic nervous system: Sympathetic and parasympathetic inputs to the SA node control *heart rate,* and sympathetic input to the ventricular myocardium controls *stroke volume.* Sympathetic activity also controls the *total peripheral resistance* via *sympathetic vasoconstrictor nerves,* which regulate contraction of smooth muscle in small arteries and arterioles in many organs and tissues. Activity in these nerves induces vasoconstriction, which raises the total peripheral resistance and thus tends to increase the mean arterial pressure.

1. What two factors determine the blood flow to a given organ in the systemic circuit?

2. When an organ's share of the cardiac output changes, what is responsible for the change?

3. What three factors determine mean arterial pressure?

4. Of the following changes, which would tend to cause an *increase* in mean arterial pressure? (choose all that apply): vasodilation in systemic organs; an increase in stroke volume; a decrease in heart rate; a decrease in total peripheral resistance

HOW CHANGES IN CENTRAL VENOUS PRESSURE AFFECT BLOOD FLOW TO ORGANS

Central venous pressure has an important, though indirect, influence on mean arterial pressure, and therefore it affects the flow of blood to all systemic organs. This is illustrated by the events following a loss of blood.

When a person loses blood, it should come as no surprise that arterial pressure falls; we would expect the pressure in the cardiovascular system to decrease with blood loss, just as the pressure in a leaky tire falls as air escapes from it. But along with the fall in arterial pressure is a fall in central venous pressure—the pressure of blood in the large veins in the thoracic cavity. The drop in venous pressure starts a sequence of events that reduces the heart's ability to pump blood, producing a larger decline in arterial pressure than would otherwise occur: Lowered venous pressure causes a fall in venous return, which leads to a reduction in end-diastolic volume. By virtue of the Starling effect, the decline in end-diastolic volume triggers a decrease in stroke volume and cardiac output, which causes mean arterial pressure to fall.

In this section we examine four factors that affect CVP and thus indirectly affect blood flow to organs: the skeletal muscle pump, the respiratory pump, blood volume, and venomotor tone.

Exercise Link

While Bill and Jane were running the marathon, several factors were contributing to diminished venous return and therefore to central venous pressure, including increased blood flow to skeletal muscle and skin, upright posture, and fluid loss through sweat. Fortunately, the compensatory processes that are described next became functional during exercise and played important roles in maintaining central venous pressure.

The Skeletal Muscle Pump

We have seen that peripheral veins contain one-way valves that allow blood to flow forward toward the heart but prevent it from flowing backward. When skeletal muscles contract, they press against veins traveling between them, which raises the pressure of blood within them. This increased pressure forces the more distal valves to close, preventing blood from flowing backward, and forces the more proximal valves to open, allowing blood to flow toward the heart (Figure 13.23a). When the muscles relax and the pressure drops, the reduced pressure allows the distal valves to open, so that blood can flow forward into the previously compressed vein, and also causes the proximal valves to close, thereby preventing blood from flowing away from the heart (Figure 13.23b). By alternately contracting and relaxing, muscles act as "pumps" or "auxiliary hearts" that help drive blood toward the central veins, which raises central venous pressure. For this reason, any exercise that involves rhythmic muscle contractions, such as walking or running, promotes an increase in venous return, increased stroke volume, and increased cardiac output.

The Respiratory Pump

Just as exercise helps to move blood back to the heart through the action of the skeletal muscle blood pump, the vigorous respiratory movements that accompany exercise also help to move blood back to the heart. We call the effect of respiratory movements on venous return the *respiratory pump.*

The respiratory pump works in the following manner: When you inhale, your diaphragm pulls downward and your rib cage expands, which lowers pressure in the thoracic cavity and raises pressure in the abdominal cavity. This action creates a pressure gradient that promotes the movement of blood from abdominal veins to the central veins located in the thoracic cavity. When you exhale, thoracic pressure rises and abdominal pressure falls. This creates a pressure gradient that would tend to favor the backward movement of blood from the central veins to the abdominal veins, but such backward flow is prevented by the closure of valves in the abdominal veins. Instead, the rise in thoracic pressure drives the forward movement of blood from the central veins to the heart, thereby promoting increased end-diastolic volume and cardiac output.

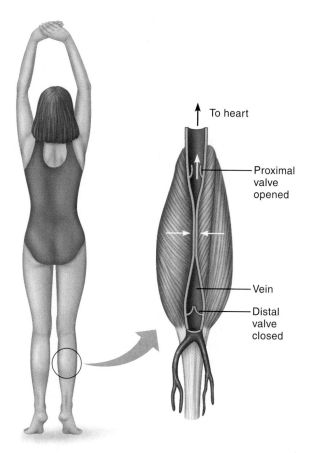

(a) Skeletal muscle contracted

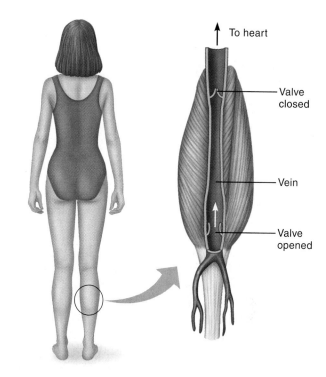

(b) Skeletal muscle relaxed

FIGURE 13.23 The skeletal muscle pump. (a) *When a muscle contracts, it presses against veins, driving blood toward the heart (left).* **(b)** *When the muscle relaxes, backward flow is prevented by closure of one-way valves in the veins.*

Would an increase in ventricular end-diastolic pressure tend to cause an increase or a decrease in stroke volume?

Blood Volume

The body's total blood volume has an important influence on mean arterial pressure through its effect on central venous pressure. The relationship between blood volume and central venous pressure is a simple one: An increase in blood volume produces an increase in venous pressure, and a decrease in blood volume produces a decrease in venous pressure. If blood volume falls as a result of bleeding, dehydration, or for any other reason, central venous pressure falls, as do venous return and end-diastolic volume. The resulting fall in cardiac output then triggers a fall in mean arterial pressure. Conversely, a rise in blood volume has the opposite effect, tending to increase mean arterial pressure. Certain forms of *hypertension* (elevated arterial pressure), for example, are due to failure of the kidneys to excrete adequate amounts of salt and water, which results in the retention of excess amounts of fluid in the body. (Hypertension is discussed in more detail in Chapter 14.) This excess fluid causes blood volume to increase, which raises the mean arterial pressure.

Because arterial pressure is strongly affected by blood volume, control of blood volume is an important part of blood pressure regulation. A fall in blood volume activates reflex mechanisms that act to reduce the kidneys' output of water in the urine, which helps the body to conserve water, thereby maintaining blood volume and venous pressure. At the same time, activation of thirst centers in the brain induces a person to drink fluids. Absorption of these fluids by the gastrointestinal tract serves to restore lost blood volume, thereby maintaining adequate venous pressure. Compared to the mechanisms that regulate mean arterial pressure by controlling heart rate, stroke volume, and total peripheral resistance (which can act within seconds), these reflexes act relatively slowly and thus regulate blood volume and blood

An increase

pressure over the long term (hours or days). (Long-term regulation of blood pressure by the kidneys is discussed in greater detail in Chapter 18.)

Under certain circumstances the high compliance of veins actually works to the detriment of the heart's pumping action by leading to **venous pooling**—accumulation of blood in veins. When a person stands up, for instance, the force of gravity increases the pressure on the blood in the lower veins of the body, causing those veins to expand and enabling the volume of blood within them to increase. This pooling of venous blood is detrimental to the heart's pumping action because it reduces venous return; instead of returning to the heart, much of the blood entering the veins remains there. Thus venous pooling reduces central venous pressure by reducing the volume of blood in the central veins, and it therefore lowers arterial pressure in the same manner as does a reduction in blood volume.

A drop in mean arterial pressure upon standing (referred to as *orthostatic hypotension*) may cause a person to feel dizzy, but reflex mechanisms normally quickly compensate for it. The presence of aggravating factors, such as dehydration or a failing heart, may cause a person to faint upon standing. In such a case, fainting is actually advantageous because once a person has fallen over, blood that had pooled in the veins in the legs moves toward the central veins, just as water flows out of a glass when it is tipped over. This increases central venous pressure, which promotes an increase in venous return and an increase in cardiac output. The rise in cardiac output raises mean arterial pressure, which helps restore blood flow to the brain.

Exercise Link

These compensatory mechanisms were operating when Bill almost fainted just after the marathon. He was hypovolemic in the later stages of the race, but skeletal muscle pumps prevented excessive blood pooling in his legs so long as he kept running. As soon as Bill stopped running, however, the pumping action ceased, blood pooled in his legs, and central venous pressure decreased abruptly, leading to orthostatic hypotension and the necessity for a corrective change in posture.

Venomotor Tone

The smooth muscle in the walls of veins contracts or relaxes in response to input from autonomic nerves and certain chemical agents. In particular, activity of *venoconstrictor* neurons of the sympathetic nervous system trig-

gers increased contractile activity in venous smooth muscle, with a resulting rise in tension referred to as **venomotor tone.**

An increase in venomotor tone has two effects: (1) Constriction of veins raises the pressure of blood within them, which forces blood to return to the heart and briefly increases the stroke volume, and (2) increased wall tension reduces venous compliance, which raises central venous pressure and produces a sustained increase in stroke volume. Therefore, an increase in venomotor tone promotes a rise in cardiac output and mean arterial pressure. Changes in venomotor tone are an important component of the reflexes that regulate arterial pressure. When arterial pressure falls as a result of blood loss, for example, activity in venoconstrictor nerves increases, which acts to raise arterial pressure.

Figure 13.24 summarizes the ways these factors influence central venous pressure. This diagram illustrates that increases in muscle pump activity, respiratory pump activity, blood volume, or venomotor tone all act to raise central venous pressure and therefore tend to raise mean arterial pressure.

Quick Test 13.7

1. Does an increase in central venous pressure tend to increase or decrease cardiac output? Explain.

2. What is the skeletal muscle blood pump? Does it raise or lower central venous pressure?

3. Of the following, which would tend to *increase* central venous pressure? (choose all that apply): standing up; an increase in blood volume; contraction and relaxation of skeletal muscles; decreased venomotor tone

---

MOVEMENT OF FLUID ACROSS CAPILLARY WALLS

In *heart failure*, which is commonly caused by weakening of the ventricular myocardium, the heart becomes unable to maintain an adequate cardiac output (When It Goes Wrong: Heart Failure, p. 434). This condition is often called *congestive* heart failure because it is frequently accompanied by swelling of tissues *(edema)*, particularly in the lower extremities. This swelling is due to the accumulation of excess fluid in the interstitial space (also known as the *interstitium*), which is brought about by the following series of events: When the heart fails, venous pressure rises, which raises the pressure of blood in capillaries. Because capillary walls are freely permeable to water and small solutes, the increased pressure forces fluid out of the capillaries, so that a portion of the plasma volume moves into the interstitial space, causing it to expand.

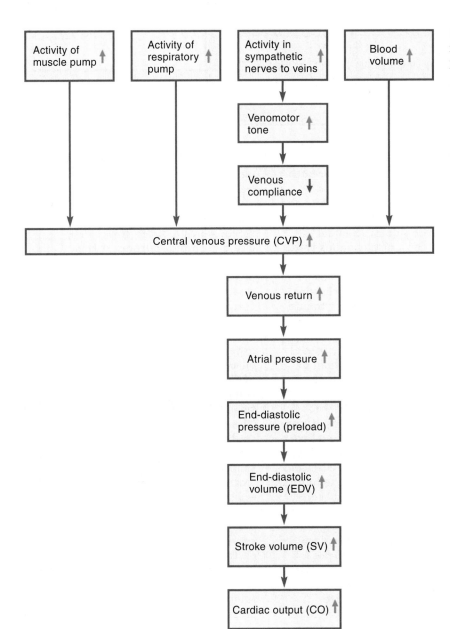

FIGURE 13.24 A process chart illustrating the influence of several factors on central venous pressure.

This raises an interesting question: Given that the blood in capillaries is always under pressure (not just in heart failure), why doesn't fluid leak out of the capillaries all the time? The answer is that it *does;* however, this leakage of fluid is not readily apparent because it is balanced by the movement of fluid in the opposite direction—from the interstitium into the capillaries. Whether fluid moves from the plasma to the interstitium or vice versa is determined by forces called *Starling forces.* In some locations the balance of these forces acts such that fluid moves out of the capillaries and into the interstitium, a process called **filtration.** In other locations the balance of forces acts in the opposite direction, and fluid moves into capillaries from the interstitium, a process called **absorption.**

The processes of capillary filtration and absorption are important to cardiovascular function because they affect the plasma volume, which has a direct effect on blood vol-

ume. Excess filtration of fluid from capillaries is potentially dangerous because it affects arterial pressure the same way hemorrhage does: The loss of fluid reduces the blood volume, which lowers central venous pressure, end-diastolic volume, cardiac output, and ultimately, mean arterial pressure. Although this situation might seem to suggest that capillary filtration is a bad thing, it is actually essential for survival. Capillary filtration is the first step in the formation of urine by the kidneys, for instance, which is necessary for ridding the body of certain wastes and for regulating the volume and composition of body fluids. Capillary filtration is also responsible for the formation of *lymphatic fluid (lymph),* which is important in defending the body against invading microorganisms. In the following section we examine the various factors that influence the movement of fluid across capillary walls—the Starling forces that make filtration and absorption happen.

The term *heart failure* refers to any change in the heart's condition that reduces its ability to maintain an adequate cardiac output. Most commonly this is a result of a chronic decrease in ventricular contractility. (Recall that the contractility of a ventricle is a measure of its ability to generate force at a given end-diastolic volume.) However, heart failure can occur even if ventricular contractility is normal or above normal. Obstruction of an atrioventricular valve, for example, can cause heart failure because it slows ventricular filling, which reduces end-diastolic volume. Alternatively, chronic *hypertension* or *stenosis* (constriction) of the aorta may induce overgrowth and thickening *(hypertrophy)* of the ventricular myocardium due to the increased workload placed on the heart. The increase in ventricular thickness makes the left ventricle harder to stretch (i.e., it reduces ventricular compliance), which tends to reduce the end-diastolic volume. Either way, a decrease in end-diastolic volume causes stroke volume and cardiac output to fall by virtue of the Starling effect. This constitutes heart failure because it interferes with the ability of the

heart to maintain an adequate cardiac output.

In heart failure, one or both sides of the heart may be affected. Signs of heart failure can thus depend on which side is failing. If one side of the heart fails, it can cause changes that place stress on the other side of the heart, causing it to fail as well. For this and other reasons, the initial cause of heart failure can be difficult to unravel. One thing is certain, however: Once the heart begins to fail, it sets into motion a complex chain of events that only worsens the condition.

What might cause a heart to fail in the first place? Sometimes heart failure is caused by a heart attack or *myocardial infarction,* in which the death of muscle cells leads to weakening of the heart. Abnormalities in electrical conduction can also lead to weakening the heart. In third-degree *heart block,* for example, impulse conduction to the ventricles is prevented, which causes a dramatic reduction in heart rate. The decrease in heart rate causes the end-diastolic volume to become chronically elevated because ventricular filling time increases. When this happens, the myocardium becomes

overstretched and weakened. Bacterial and viral infections can also induce heart failure by weakening the myocardium.

Once the heart muscle weakens, it generates less contractile force, which reduces stroke volume and cardiac output. If this happens in the left heart, mean arterial pressure falls. Signs include weakening of the pulse, fatigue, and *cyanosis*—a bluish tinge to the skin indicating inadequacy of the oxygen supply. Another sign of left heart failure is *pulmonary edema* (accumulation of fluid in the lungs) due to a rise in pressure on the venous side of the pulmonary circuit. (The rise in venous pressure causes edema because it causes pressure in pulmonary capillaries to increase, which promotes filtration of fluid across capillary walls.) The resulting accumulation of fluid in the lungs can impair gas exchange, causing shortness of breath or even respiratory distress (gasping for air) in severe cases.

To understand why ventricular failure causes a rise in venous pressure, we must consider how venous pressure is affected by the pumping action of the heart. To begin, imagine that the heart has stopped pumping blood.

Forces Driving the Movement of Fluid Into and Out of Capillaries

The movement of fluid across the wall of a capillary is governed by two forces: (1) a *hydrostatic pressure gradient*—the difference between the hydrostatic pressure of fluid inside the capillary and the hydrostatic pressure of fluid outside the capillary, and (2) an *osmotic pressure gradient*—the difference between the osmotic pressure of fluid inside the capillary and the osmotic pressure of fluid outside

the capillary. These gradients of hydrostatic and osmotic pressure are the *Starling forces* we alluded to earlier.

The Hydrostatic Pressure Gradient

Whenever a hydrostatic pressure gradient exists across a semipermeable barrier such as a capillary wall, water tends to move from the side where hydrostatic pressure is higher to the side where this pressure is lower. The hydrostatic pressure gradient across the wall of a capillary is the difference between the pressure of blood inside the

Under these conditions, only a small pressure (a few mm Hg) would exist in the cardiovascular system due to blood filling up blood vessels and stretching their walls. However, the pressure would be the same in the arteries, veins, and everywhere else, so no blood flow would occur. Now imagine that the heart were to suddenly resume its pumping action. In doing so it would receive blood from veins and push it into the arteries, which would effectively transfer a certain fraction of the total blood volume from the veins to the arteries. This transfer of blood would raise mean arterial pressure and lower venous pressure, as you can see in the following diagram where the resumption of pumping occurs at the arrow.

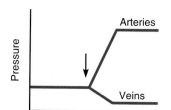

Arterial pressure rises more than venous pressure falls because the arteries have a lower compliance than the veins. In

heart failure, the heart's ability to pump blood is reduced, which causes a portion of the total blood volume to shift back into the veins from the arteries. This reduces arterial pressure and also raises venous pressure, as occurs at the arrow in the following diagram:

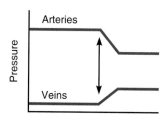

In heart failure, pressure can rise in either the systemic veins or the pulmonary veins, depending on which side of the heart is affected. Failure of the left ventricle (*left-sided* failure) causes pressure to rise in the pulmonary veins because they carry blood to the left side of the heart. In contrast, failure of the right ventricle (*right-sided* failure) causes pressure to rise in the vena cava and other systemic veins. In left-sided failure, however, this is only the beginning of the problem. The fall in mean arterial pressure that accompanies left-sided failure usually triggers a multitude of reflex re-

sponses, including a reduction in salt and water excretion by the kidneys. This leads to accumulation of excess fluid in the body (*fluid retention*), which raises blood volume. Because the veins act as volume reservoirs, most of this excess volume ends up on the venous side of the circulation, which increases venous pressure even further.

Edema, a common sign of heart failure, is a consequence of increased venous pressure. As venous pressure rises, the increase in pressure is transmitted to vessels upstream, including capillaries. When the left side of the heart fails, edema occurs in the lungs because the pressure increases in pulmonary capillaries. When the right side of the heart fails, it tends to cause edema in systemic tissues because pressure rises in systemic capillaries. Edema is most noticeable in parts of the body that hang down, such as the wrists and ankles. When edema occurs, the condition is referred to as *congestive heart failure*.

capillary (P_{CAP}) and the pressure of interstitial fluid outside the capillary (P_{ISF}):

$$\text{hydrostatic pressure gradient } (\Delta P) = P_{CAP} - P_{ISF}$$

Because in capillaries the pressure of blood inside is generally greater than the pressure of the interstitial fluid outside, the hydrostatic pressure gradient is directed *outward*—it tends to drive water out of the capillaries.

The hydrostatic pressure within a capillary varies along its length, because the pressure of blood declines continually as blood flows from the arteriolar end of the

capillary to the venular end (Figure 13.25a). In contrast, little or no variation in hydrostatic pressure exists outside a capillary. Although the actual value of the interstitial hydrostatic pressure is controversial and difficult to measure, it is significantly lower than the pressure inside capillaries.

Suppose that the hydrostatic pressure inside a capillary declines from 38 mm Hg at the arteriolar end to 16 mm Hg at the venular end, and that the pressure outside the capillary is 1 mm Hg. In this case, the hydrostatic pressure gradient drops from 38 − 1 = 37 mm Hg at the arteriolar end to 16 − 1 = 15 mm Hg at the venular end.

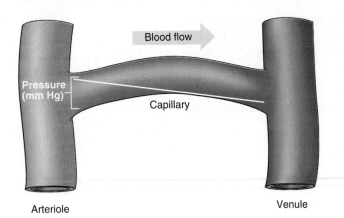

Blood flow

Pressure (mm Hg)

Capillary

Arteriole

Venule

(a)

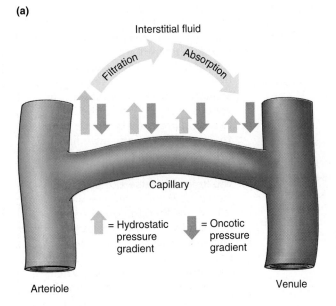

Interstitial fluid

Filtration

Absorption

Capillary

↑ = Hydrostatic pressure gradient

↓ = Oncotic pressure gradient

Arteriole

Venule

(b)

FIGURE 13.25 Forces affecting the flow of fluid across capillary walls. (a) *Blood pressure falls as blood flows from the arteriolar end of a capillary to the venular end. This pressure is the hydrostatic pressure in the capillary. As the capillary hydrostatic pressure decreases, the size of outwardly directed hydrostatic pressure gradient also falls. This pressure gradient tends to force fluid out of the capillary.* **(b)** *How filtration and absorption occur in the same capillary. At the arteriolar end of the capillary, the outwardly directed hydrostatic pressure gradient is larger than the inwardly directed oncotic pressure gradient, as indicated by the relative lengths of the arrows. As a result, fluid flows outward in this region* (filtration). *At the venular end of the capillary, the hydrostatic pressure gradient is smaller than the oncotic pressure gradient, due to the fall in capillary hydrostatic pressure. Consequently, fluid flows inward in this region* (absorption).

Would an increase in the plasma protein concentration tend to favor increased filtration or absorption?

Absorption

The Osmotic Pressure Gradient

When an *osmotic pressure gradient* exists across a capillary wall, water tends to flow from the side where the osmotic pressure is *lower* to the side where the osmotic pressure is *higher*. Recall that when a semipermeable barrier separates two solutions with different solute concentrations, water tends to flow from the side where the solute concentration is lower to the side where the solute concentration is higher. The osmotic pressure of a solution is a measure of its solute concentration, and it increases as the solute concentration increases. Because the smaller solutes such as sodium or glucose are normally present at the same concentration in plasma and interstitial fluid, these solutes do not contribute to the osmotic pressure gradient and therefore do not affect water movement into or out of capillaries. In contrast, the difference in *protein concentration* between plasma and interstitial fluid does affect water movement because it creates a difference in osmotic pressure between the inside and outside of capillaries. Osmotic pressure that is exerted by proteins is referred to as **oncotic pressure.**

The oncotic pressure gradient is the difference between the oncotic pressure of plasma inside the capillary (π_{CAP}) and the oncotic pressure of interstitial fluid outside the capillary (π_{ISF}):

$$\text{oncotic pressure gradient } (\Delta\pi) = \pi_{CAP} - \pi_{ISF}$$

Because the concentration of proteins in the plasma is higher than the concentration of proteins in the interstitial fluid, the *oncotic pressure gradient* is directed *inward;* that is, it tends to drive water into the capillaries.

Under normal conditions, the concentration of proteins in the plasma is 6–8 grams per 100 ml, which is many times the protein concentration in interstitial fluid. The oncotic pressure of plasma is approximately 25 mm Hg, whereas the oncotic pressure of interstitial fluid is negligible. Therefore, the oncotic pressure gradient across the capillary wall is 25 − 0 = 25 mm Hg. Although the capillary oncotic pressure is affected by any movement of water into or out of capillaries that might occur as blood flows through them, for the sake of simplicity we will ignore this and assume that the oncotic pressure of plasma does not change as blood flows from one end of a capillary to the other.

Net Filtration Pressure

The direction of water flow across the wall of a capillary is determined by the **net filtration pressure** at any given location—the difference between the hydrostatic and oncotic pressure gradients:

net filtration pressure (NFP) =
   hydrostatic pressure gradient (ΔP) −
               oncotic pressure gradient ($\Delta\pi$)

When the sign of the net filtration pressure is positive, the hydrostatic gradient is greater than the oncotic gradient, and fluid flows *outward* (filtration); when it is nega-

TABLE 13.3 FORCES AFFECTING THE MOVEMENT OF FLUID ACROSS CAPILLARY WALLS

VARIABLE		DEFINITION	SIGNIFICANCE
ΔP (hydrostatic pressure gradient)	$= P_{CAP} - P_{ISF}$	Difference between hydrostatic pressures inside capillary and in interstitial fluid outside capillary	Force that tends to drive fluid out of capillary
$\Delta \pi$ (oncotic pressure gradient)	$= \pi_{CAP} - \pi_{ISF}$	Difference between oncotic pressures inside capillary and in interstitial fluid outside capillary	Force that tends to drive fluid into capillary
NFP (net filtration pressure)	$= \Delta P - \Delta \pi$	Difference between hydrostatic and oncotic pressure gradients	Net driving force that determines direction of fluid flow

CONDITIONS	SIGN OF NFP	RESULT
$\Delta P > \Delta \pi$	Positive	Fluid moves out of capillary (filtration)
$\Delta \pi > \Delta P$	Negative	Fluid moves into capillary (absorption)

tive, the oncotic gradient is greater than the hydrostatic gradient, and fluid flows *inward* (absorption).

Assuming a hydrostatic pressure gradient of 37 mm Hg at the arteriolar end of a capillary and an oncotic pressure gradient of 25 mm Hg, the net filtration pressure is $37 - 25 = 12$ mm Hg, which favors filtration. Assuming that the hydrostatic pressure falls to 15 mm Hg at the venular end of the capillary, the net filtration pressure at that end is $15 - 25 = -10$ mm Hg, which favors absorption. Therefore, filtration and absorption can occur within the same capillary, as illustrated in Figure 13.25b.

Forces governing the movement of fluid across capillary walls are summarized in Table 13.3.

Factors Affecting Filtration and Absorption Across Capillaries

The rate at which fluid is filtered or absorbed across capillary walls is influenced by any factor that alters the relative sizes of the hydrostatic and oncotic pressure gradients and, hence, the net filtration pressure. Increased filtration is favored by any change that either increases the size of the outwardly directed hydrostatic gradient or decreases the size of the inwardly directed oncotic gradient. Changes in the opposite direction, of course, favor an increase in the rate of fluid absorption.

Under normal conditions, the total volume of fluid that filters out of an individual's capillaries amounts to about 20 liters per day—more than six times total plasma volume! About 17 liters of fluid is absorbed into capillaries in the same period of time, giving a net volume of 3 liters per day that is filtered, which is roughly equal to an individual's entire plasma volume. If this volume of fluid shifts from the plasma to the interstitial space every day, why don't tissues swell from edema? Why doesn't blood

pressure fall because of the loss of blood volume? The answer is that the 3 liters or so of filtered fluid is picked up from the interstitium and returned to the cardiovascular system by the *lymphatic system*, which is described in the next section.

The balance between filtration and absorption can be altered as a result of certain pathological conditions, or even as a result of everyday occurrences. When a person stands up, for example, capillary hydrostatic pressure increases in the lower parts of the body because the column of blood raises the hydrostatic pressure in the lower arteries and veins. Any increase in pressure, whether it be at the arteriolar end of a capillary or at the venular end, tends to raise the pressure of blood in the capillary, which increases capillary hydrostatic pressure and causes an increase in the rate of filtration. When a person stands very still, which minimizes the action of the skeletal muscle pump, the problem is exacerbated because pressure in the lower veins rises even further due to venous pooling, which augments the increase in capillary pressure. (Venous pooling also lowers central venous pressure.) Over a prolonged period, the loss of fluid from capillaries can result in a significant decline in blood volume, which can trigger dizziness or fainting. It is for this reason that soldiers may faint after standing at attention on the parade grounds for long periods of time.

An increase in capillary filtration, with accompanying tissue swelling, also occurs in response to certain types of injuries. When the skin is cut or abraded, for instance, the affected part becomes swollen within a few minutes. In part this results from damage to capillaries, which allows protein-rich fluid to leak out. Consequently, the concentration of proteins in the interstitial fluid rises, which raises its oncotic pressure. In addition, certain cells in the injured area release a chemical called *histamine*, which increases

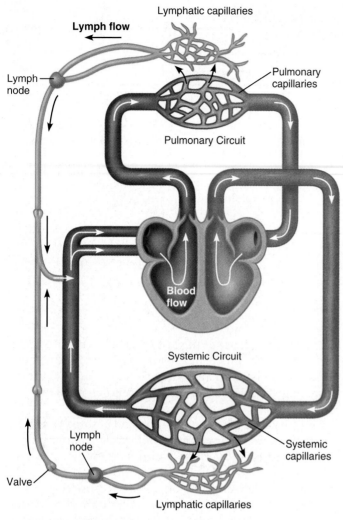

Lymph flow

Lymphatic capillaries

Lymph node

Pulmonary capillaries

Pulmonary Circuit

Blood flow

Systemic Circuit

Lymph node

Valve

Systemic capillaries

Lymphatic capillaries

FIGURE 13.26 The lymphatic system. *This highly schematic view depicts fluid leaving blood capillaries and entering lymphatic capillaries in the lungs and systemic tissues. This fluid, called* lymph, *flows through lymphatic ducts and lymph nodes to veins in the systemic circuit.*

The Lymphatic System

Even though approximately 3 liters of fluid leak out of the capillaries each day, the tissues do not normally swell because this fluid enters the **lymphatic system,** a network of vessels (often referred to as *ducts*) that courses throughout the body (Figure 13.26). Once the fluid gets into the lymphatic system, it is carried through the ducts and is eventually returned to the cardiovascular system. The lymphatic system is a sort of "silent partner" to the cardiovascular system, because even though fluid continually moves through it, this fluid (called **lymphatic fluid** or *lymph*) often goes unnoticed because it is clear, with only a slight yellow tinge to it. If you have ever scraped your knee slightly and noticed clear fluid oozing out of it, this is mostly lymphatic fluid.

Fluid enters the lymphatic system by way of small, blind-ended ducts called *lymphatic capillaries,* the walls of which have large pores that allow water, small solutes, and even proteins and larger particles to pass through. As a consequence, any fluid that routinely leaks out of "ordinary" blood capillaries can easily move into the lymphatic system from the interstitium. From the lymphatic capillaries, fluid moves through a series of successively larger ducts until it eventually reaches one of the two ducts that drain into the bloodstream. In this manner, filtered fluid is returned to the cardiovascular system. The flow of lymphatic fluid is driven by contractions of smooth muscle in the walls of the larger ducts, which are equipped with one-way valves similar to those in veins. The flow of fluid is also aided by body movements because contractions of skeletal muscles compress lymphatic vessels that run between them, producing an effect similar to the skeletal muscle pump.

At certain points in the lymphatic system, lymph passes through structures known as *lymph nodes.* In lymph nodes, any particles that may be present in the lymph, including bacteria or other foreign matter, are filtered out and removed by phagocytic cells called **macrophages.** Lymphocytes and other cells of the immune system also congregate in the lymph nodes, making them important sites for the cellular interactions that are an integral part of the immune response (see Chapter 22).

the permeability of capillary walls to proteins, thereby increasing the leakage of proteins into the interstitium.

Other conditions that promote increased capillary filtration and edema include liver and kidney disease. Because most plasma proteins are manufactured by the liver, damage to the liver can result in a decrease in plasma protein concentration, which lowers plasma oncotic pressure. Certain forms of damage to the kidneys interfere with their ability to eliminate excess water (and solutes) in the urine, which results in the accumulation of excess fluid in the body. As a consequence, blood volume increases and blood pressure rises throughout the cardiovascular system. This increase in pressure raises capillary hydrostatic pressure, which increases filtration. Damage to the kidneys can also cause them to eliminate significant quantities of plasma proteins in the urine. (Normally, only negligible quantities of protein are voided in the urine.) This loss of protein triggers increased capillary filtration because it reduces plasma oncotic pressure.

Quick Test 13.8

1. To what does the term *oncotic pressure* refer?
2. Of the following changes, which would tend to cause an *increase* in the rate at which fluid is filtered from capillaries? (choose all that apply): a decrease in plasma oncotic pressure; a decrease in interstitial oncotic pressure; an increase in venous pressure; an increase in plasma protein concentration
3. What is lymphatic fluid? What are its origins?
4. How do lymphatic capillaries differ from "ordinary" blood capillaries?

The Composition of the Blood, p. 404

Blood consists of plasma, composed of water and a variety of dissolved solutes (including proteins), and formed elements, which include red and white blood cells (erythrocytes and leukocytes) and platelets. Erythrocytes contain large amounts of hemoglobin and function primarily to transport oxygen and carbon dioxide. The hematocrit, which is the fraction of blood volume occupied by erythrocytes, is an indicator of the blood's oxygen-carrying capacity. Leukocytes help to defend the body against invading microorganisms and other foreign material, whereas platelets are important in blood clotting.

Platelets and Hemostasis, p. 407

Mechanisms for stoppage of bleeding (hemostasis) include vascular spasm (blood vessel constriction), platelet plug formation, and blood clot formation, which occur in response to blood vessel damage. In platelet plug formation, platelets aggregate around the site of damage, forming a physical barrier to blood leakage. During this process, platelets also become activated, which sets the stage for clot formation. In clot formation, fibrinogen (a soluble plasma protein) is transformed by proteolytic cleavage into fibrin, which forms a fibrous mesh around the platelet plug. This transformation is set into motion following a cascade of reactions (the coagulation cascade) involving activated platelets and several clotting factors in the plasma. The coagulation cascade can be initiated by an intrinsic pathway, which involves only components present in plasma, or an extrinsic pathway, which involves factors present in tissues outside blood vessels. Spread of the blood clot beyond the site of damage is prevented by substances secreted by undamaged tissues and by other mechanisms.

The Structure and Function of Blood Vessels, p. 411

All blood vessels possess a lumen and are lined by a layer of endothelial cells, and their walls contain varying amounts of smooth muscle and connective tissue. Arteries, which have thick walls that enable them to withstand the high pressure of blood within them, have relatively low compliance and function as pressure reservoirs. Arterioles contain relatively large amounts of smooth muscle, which enables them to expand or contract, thereby regulating blood flow through capillary beds. Arterioles are important in controlling mean arterial pressure and the distribution of cardiac output to tissues. Capillaries have the thinnest walls and are highly permeable to water and small solutes; their primary function is to permit exchange of materials between the blood and the tissues. Venules are also thin walled and participate in material exchange. Veins are large thin-walled vessels; most possess valves that permit blood to flow toward the heart but not away from it. Veins have a high compliance and function as volume reservoirs.

IP Cardiovascular, Anatomy Review: Blood Vessel Structure and Function, pp. 3–19.

Patterns of Blood Flow Within the Cardiovascular System, p. 417

Because blood moves through the pulmonary and systemic circuits alternately, these circuits are in series with one another. Organs in the systemic circuit are in parallel with one another because blood is delivered to all the organs simultaneously. This parallel arrangement both ensures that each organ receives fully oxygenated blood and enables independent regulation of each organ's blood flow. The distribution of blood flow among the organs is regulated by contraction and relaxation of arteriolar smooth muscle.

IP Cardiovascular, Anatomy Review: The Heart, pp. 6–7.

Physical Laws Governing Blood Flow and Blood Pressure, p. 420

The flow of blood through any vessel or network of vessels depends on the pressure gradient (ΔP) and the resistance (R) of the vessel or network: Flow = $\Delta P/R$. The overall pressure gradient driving flow through the systemic circuit is the difference between mean arterial pressure (MAP) and central venous pressure (CVP), which is virtually identical to mean arterial pressure. The most important influence on vascular resistance is vessel radius. Contraction of smooth muscle in arterioles and small arteries (vasoconstriction) decreases vessel radius, increases resistance, and tends to decrease flow; relaxation of this smooth muscle (vasodilation) has the opposite effects. The resistance of the entire systemic circuit (total peripheral resistance, TPR) is controlled by vasoconstriction or vasodilation of these small resistance vessels.

IP Cardiovascular, Factors That Affect Blood Pressure, pp. 3–8.

Factors Affecting Flow and Distribution of Blood to Organs, p. 420

Blood flow through any given systemic organ is determined by mean arterial pressure and that organ's vascular resistance: Organ blood flow = MAP/organ resistance. Mean arterial pressure is determined by three factors: heart rate, stroke volume, and total peripheral resistance: MAP = HR × SV × TPR. Mean arterial pressure is regulated by neural input to the heart (which controls HR and SV) and to resistance vessels (which controls TPR). Hormones also play a role in regulating MAP.

How Changes in Central Venous Pressure Affect Blood Flow to Organs, p. 430

Central venous pressure influences arterial pressure because it affects venous return, end-diastolic volume, stroke volume, and cardiac output. As CVP rises or falls, cardiac output and MAP also tend to rise or fall, respectively. Factors affecting CVP include: activity of the skeletal muscle pump, activity of the respiratory pump, blood volume, and venomotor tone (which is regulated by sympathetic input to veins).

Movement of Fluid Across Capillary Walls, p. 432

The movement of fluid across capillary walls is driven by the net filtration pressure, which has two components: a hydrostatic pressure gradient across the wall (which favors filtration), and an oncotic pressure gradient (which favors absorption). Most of the fluid that is filtered from capillaries is returned to the cardiovascular system by absorption. Excess filtered fluid is returned to the cardiovascular system by the lymphatic system.

IP Cardiovascular, Autoregulation and Capillary Dynamics, pp. 14–37.

EXERCISES

Multiple-Choice Questions

1. Total peripheral resistance is
   a) the combined resistance of all organs in the body.
   b) the resistance of capillaries located in distal body parts.
   c) the combined resistance of all organs in the systemic circuit.
   d) the combined resistance of all the blood vessels within an organ or tissue.

2. Central venous pressure increases
   a) when blood volume decreases.
   b) as a result of venous pooling.
   c) as a result of an increase in venomotor tone.
   d) when a person stands up.

3. If total peripheral resistance is constant, an increase in mean arterial pressure could be the result of
   a) vasoconstriction in the systemic circuit.
   b) an increase in sympathetic input to the ventricular myocardium.
   c) a decrease in blood volume.
   d) none of the above

4. Which of the following tends to promote edema in systemic tissues?
   a) an increase in the concentration of plasma proteins
   b) a decrease in pressure in the vena cava
   c) a decrease in arterial pressure
   d) leakage of proteins from capillaries into the interstitial fluid

5. If mean arterial pressure is constant, which of the following will *not* affect blood flow to a particular organ?
   a) constriction of arterioles within that organ
   b) an increase in blood viscosity
   c) constriction of arterioles in a neighboring organ
   d) a decrease in the vascular resistance of that organ

6. Which of the following tends to cause a decrease in ventricular end-diastolic volume?
   a) an increase in central venous pressure
   b) an increase in skeletal muscle pump activity
   c) a decrease in filling time
   d) an increase in blood volume

7. Knowing stroke volume, heart rate, and mean arterial pressure provides sufficient information to determine
   a) total peripheral resistance only.
   b) cardiac output only.
   c) total peripheral resistance and cardiac output.
   d) total peripheral resistance, cardiac output, and combined blood flows to all systemic organs.

8. Which of the following blood vessels possess valves that prevent blood from flowing backward?
   a) arteries
   b) arterioles
   c) venules
   d) veins

9. Exchange of oxygen and carbon dioxide between blood and respiring tissues occurs across the walls of
   a) arteries.
   b) arterioles.
   c) capillaries.
   d) veins.

10. Most total blood volume is contained in the
    a) heart.
    b) arteries.
    c) capillaries.
    d) veins.

11. Which of the following tends to cause a decrease in mean arterial pressure?
    a) a drop in total peripheral resistance
    b) an increase in stroke volume of the left ventricle
    c) an increase in heart rate
    d) an increase in venous return

12. Lymphatic capillaries differ from blood capillaries in that
    a) lymphatic capillaries have a lower permeability to water.
    b) lymphatic capillaries have a lower permeability to small solutes.
    c) lymphatic capillaries are blind ended.
    d) lymphatic capillaries are not connected to any other vessels.

13. Which of the following tends to cause increased venous pooling?
    a) a decrease in venomotor tone
    b) a decrease in the osmotic pressure of plasma proteins
    c) exercise
    d) dehydration

14. Moment-to-moment changes in total peripheral resistance are normally due to changes in
    a) the lengths of blood vessels in the systemic circuit.
    b) the radius of certain blood vessels in the systemic circuit.
    c) the viscosity of blood.
    d) the mean arterial pressure.

15. Which of the following does *not* tend to increase as a result of the action of the skeletal muscle pump?
    a) venous pooling
    b) venous return
    c) end-diastolic volume
    d) stroke volume

16. Which of the following opposes formation of a platelet plug?
    a) thromboxane A_2
    b) prostacyclin
    c) ADP
    d) von Willebrand factor

Objective Questions

1. An increase in total peripheral resistance tends to lower mean arterial pressure. (true/false)

2. Plasma oncotic pressure is the osmotic pressure due to all solutes dissolved in plasma. (true/false)

3. Total peripheral resistance (increases/decreases) when general vasodilation of smaller arteries and arterioles occurs in systemic tissues.

4. Changes in blood flow distribution among organs are brought about by changes in mean arterial pressure. (true/false)

5. If the resistance of a single organ in the systemic circuit increases, but the resistance of all other organs stays the same, total peripheral resistance (increases/decreases/is not affected).

6. Vascular resistance is affected by the hematocrit because it affects the blood's ———.

7. Veins have a (larger/smaller) compliance than arteries.

8. Action of the skeletal muscle pump tends to (raise/lower) end-diastolic volume.

9. Blood flow to an organ is determined by both its vascular resistance and the ———.

10. A(n) (increase/decrease) in the radius of a blood vessel raises its resistance.

11. Over the course of a day, capillary filtration and absorption occur at the same rate, so that the net volume of fluid filtered from capillaries is zero. (true/false)

12. Venous pooling is (promoted/counteracted) by increases in venomotor tone.

13. The resistance of a blood vessel increases as its radius decreases. (true/false)

14. An increase in the hydrostatic pressure of blood in capillaries tends to increase the rate at which fluid is filtered across capillary walls. (true/false)

15. The hematocrit is the fraction of the blood volume that is occupied by ———.

16. Red cell production is regulated by a hormone called ———, which is secreted by the kidneys.

17. The clotting factor that converts fibrinogen to fibrin is ———.

Essay Questions

1. After working in a field for a whole day without drinking fluids, a man feels faint when he stands up. Using your understanding of the factors influencing mean arterial pressure, explain why.

2. Using your understanding of principles relating blood flow, pressure, and resistance, explain why mean arterial pressure depends only on heart rate, stroke volume, and total peripheral resistance.

3. Discuss the various factors that affect venous return to the heart.

4. Explain how blood flow to an organ can change even if mean arterial pressure does not.

5. When the left side of the heart begins to fail, blood pressure tends to rise in the pulmonary veins and vessels upstream in the pulmonary circuit. Using your understanding of the factors affecting capillary filtration, explain why edema of the lungs is common in left heart failure.

Find the answers to these exercises, and additional study tools,
at the Physiology Place (www.physiologyplace.com).

14

The Cardiovascular System: Regulation of Function

OBJECTIVES

- Describe what the arterial baroreceptor reflex is, and explain how it regulates mean arterial pressure.

- Describe the difference between how a baroreceptor reflex affects blood flow to the brain following hemorrhage, and how it affects the flow to other organs, such as those of the gastrointestinal system; explain the significance of this difference.

- Explain in general terms how intrinsic control of vascular resistance regulates blood flow to organs.

- Give examples of active and reactive hyperemia, and describe the sequences of events that trigger these responses.

CHAPTER OUTLINE

Extrinsic Control of Cardiovascular Function: Regulation of Mean Arterial Pressure   443

Intrinsic Control of Cardiovascular Function: Regulation of Blood Flow Distribution to Organs   455

Other Cardiovascular Regulatory Processes   461

Above: Scanning electron micrograph of blood flow from the aorta to the renal artery

In people who are allergic to bee stings, even a single sting can cause loss of consciousness and even death within a few minutes. The reason is that a toxin contained in bee venom triggers release of massive amounts of *histamine* and other chemicals from certain cells found in tissues throughout the body. (Histamine is normally released in tissues in response to injury, and it acts locally to increase the permeability of capillaries, so that white blood cells can enter the injured area to fight possible infection.) These chemicals enter the bloodstream and induce relaxation of smooth muscle in blood vessels throughout the body. The resulting vasodilation causes a dramatic drop in total peripheral resistance and mean arterial pressure, the latter of which then causes a reduction in the flow of blood to all tissues of the body, including the brain. The reduction in blood flow to the brain leads to loss of consciousness and then death.

To distribute blood to the tissues adequately, the cardiovascular system must perform a delicate balancing act. As the previous example illustrates, some degree of constriction in blood vessels is required to maintain a mean arterial pressure high enough to drive sufficient blood flow to the organs. If all arterioles were to dilate at once, arterial pressure would plummet to dangerously low levels. If they all were to constrict, blood flow to some tissues would be inadequate. What actually happens is a kind of compromise: Arterioles in some capillary beds are open, so that cells in those regions receive blood; at the same time, other arterioles are closed, so the cells and tissues they serve are not receiving blood. At a later time, many of the open capillary beds close and vice versa. In this manner the blood supply is shared among capillary beds, so that no area of tissue is deprived of blood for too long.

Getting the right amount of blood to the right place at the right time requires that the actions of the heart and the blood vessels be coordinated. How, then, do arterioles "know" when to open and when to close? How does the cardiovascular system maintain adequate flow when a person is dehydrated or has lost blood? How does it compensate for changes in the metabolic activity of tissues during exercise? These questions concern cardiovascular *regulation,* the focus of this chapter.

In the first section of this chapter we examine various mechanisms that normally operate to ensure that mean arterial pressure stays constant. As we will see, these mechanisms involve control of the heart and other organs by signals originating from outside the organs themselves; thus, regulation of mean arterial pressure exemplifies the concept of extrinsic control. In the next section we look at mechanisms that regulate the distribution of blood flow to the body's organs and tissues. As we will see, these mechanisms involve control of an organ's blood vessels by signals originating from within the organ itself; thus, regulation of organ blood flow distribution exemplifies the concept of intrinsic control. In the final section we focus on a select set of regulatory mechanisms that also influence cardiovascular function but are secondarily important or operate only under special circumstances.

EXTRINSIC CONTROL OF CARDIOVASCULAR FUNCTION: REGULATION OF MEAN ARTERIAL PRESSURE

Recall from Chapter 13 that mean arterial pressure is the driving force for blood flow to all systemic organs and tissues, and that the blood flow to a particular organ also depends on that organ's vascular resistance. As a consequence, the proper functioning of these organs and tissues hinges on the ability of the cardiovascular system to maintain a mean arterial pressure that is high enough to drive adequate blood flow. How the cardiovascular system regulates mean arterial pressure is the topic of this section.

Control of mean arterial pressure is accomplished by extrinsic regulatory mechanisms—that is, mechanisms involving the control of organs and tissues by the central nervous system and circulating hormones. In this section we concentrate most of our effort on understanding the *short-term* regulation of arterial pressure, which takes place over a time span of seconds to minutes. *Long-term* regulation of arterial pressure, which is discussed briefly in this chapter but explained more fully in Chapter 18, involves control of the blood volume by the kidneys and takes place over minutes to days. To see how short-term and long-term regulation work together, consider this example: When a person loses blood, the resulting decrease in blood volume causes a fall in mean arterial pressure. (Recall that a drop in blood volume causes a fall in central venous pressure, which causes a decrease in venous

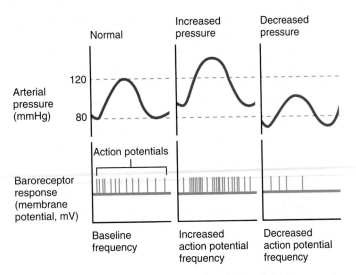

FIGURE 14.1 Response of arterial baroreceptors to changes in arterial pressure. *Top trace in each panel shows an arterial pressure wave; vertical lines in the lower traces represent action potentials recorded from a carotid sinus baroreceptor.*

Here, a change in the intensity of a stimulus (arterial pressure) triggers a change in the frequency of action potentials in afferent neurons. What is the general term for this phenomenon?

return, end-diastolic volume, and cardiac output.) As we will see, this drop in arterial pressure usually triggers a number of neural and hormonal responses that act within seconds to raise the arterial pressure back toward its normal level. At the same time, the drop in arterial pressure also triggers a decrease in the rate of urine output by the kidneys, which helps the body to maintain the blood volume by conserving water. In contrast, when a person drinks excess fluids, blood volume and mean arterial pressure rise above normal. This rise in arterial pressure triggers neural and hormonal responses that act quickly to lower the pressure, returning it to near normal. At the same time, the rise in pressure triggers an increase in the rate of urine output by the kidneys that rids the body of excess water, thereby returning blood volume to normal.

Neural Control of Mean Arterial Pressure

When the body is at rest, extrinsic regulatory mechanisms work to keep mean arterial pressure at a constant level. If for some reason arterial pressure drops, regulatory responses "kick in" to bring it back up to the normal level; if mean arterial pressure rises, regulatory responses work to bring it back down. Thus mean arterial pressure is a *regulated variable* that is regulated by *negative feedback control.* As we saw in Chapter 1, *sensors* monitor the regulated variable in any mechanism involving negative feed-

back control; the sensors monitoring mean arterial pressure are called *arterial baroreceptors,* which we examine next.

Arterial Baroreceptors: Sensors of Mean Arterial Pressure

A **baroreceptor** is a general term for a type of sensory receptor neuron in blood vessels and the heart that responds to changes in pressure within the cardiovascular system. (Recall that a *barometer* is an instrument that monitors atmospheric pressure.) *Arterial* baroreceptors respond specifically to the stretching that occurs during pressure changes in arteries. The sensory endings of arterial baroreceptors are embedded within arterial walls. When arterial pressure (or more precisely *distending pressure*) rises, the arteries expand, stretching the walls of the arteries and the sensory endings of the baroreceptors within them, and inducing depolarization. Depolarization triggers action potentials, which are then conducted to the central nervous system by the baroreceptors' axons. Increased pressure induces greater stretch of the sensory endings, which produces greater depolarization and an increase in action potential frequency (Figure 14.1).

Arterial baroreceptors are found in only two locations: the *aortic arch,* the curved portion of the aorta located close to where it emerges from the heart, and the *carotid sinuses* of the *carotid arteries,* which are located in the neck (see Figure 14.2). These baroreceptors are strategically placed because pressure in the aorta affects blood flow to every organ in the systemic circuit, and because pressure in the carotid arteries affects blood flow to the brain, which is exceedingly sensitive to any reduction in its blood supply.

Arterial baroreceptors are important in the regulation of mean arterial pressure because they relay pressure information to the central nervous system, which exerts control over cardiovascular function via autonomic neurons projecting to the heart and blood vessels (Figure 14.2). When mean arterial pressure changes, activity in these neurons changes, which causes the heart and blood vessels to alter their function in response. Baroreceptor input to the nervous system also triggers changes in the secretion of several hormones that target the heart and blood vessels, (discussed later in the chapter). In the next section we focus on the control of cardiovascular function by autonomic neurons.

Autonomic Control of Cardiovascular Function

Neural control of mean arterial pressure is orchestrated primarily by the medulla oblongata, which possesses a diverse set of neural networks encompassing several nuclei that regulate different aspects of cardiovascular function. These medullary networks receive input from arterial baroreceptors and other receptors found in various loca-

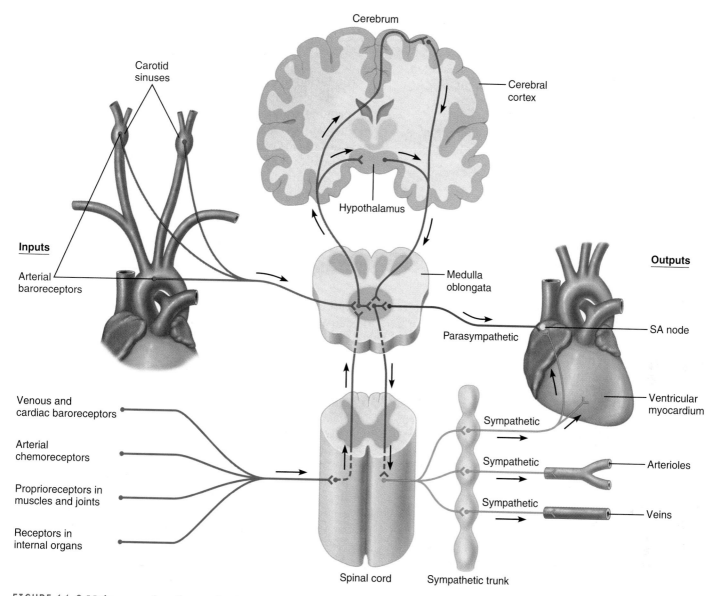

Carotid sinuses

Cerebrum

Cerebral cortex

Hypothalamus

Inputs

Arterial baroreceptors

Outputs

Medulla oblongata

SA node

Parasympathetic

Ventricular myocardium

Venous and cardiac baroreceptors

Arterial chemoreceptors

Proprioreceptors in muscles and joints

Receptors in internal organs

Sympathetic

Sympathetic

Arterioles

Sympathetic

Veins

Spinal cord

Sympathetic trunk

FIGURE 14.2 **Major neural pathways in the control of cardiovascular function.** *Cardiovascular function is regulated primarily by the medulla oblongata, which relays signals to the heart and blood vessels by way of sympathetic and parasympathetic neurons. Higher brain areas also play a role in regulating cardiovascular function. The medulla and other regions of the brain receive input from arterial baroreceptors and other sensory receptors throughout the body.*

Does the sympathetic trunk belong to the central nervous system or to the peripheral nervous system?

tions throughout the body. Using this input, as well as input it receives from other areas of the brain, the medulla is able to assess various indicators of cardiovascular performance (such as arterial pressure) and decide whether or not this performance is sufficient to meet the body's current needs. If it is not, the medulla instructs the cardiovascular system to make appropriate adjustments. The medulla exerts control over cardiovascular function by sending output to effectors via autonomic nerves.

Sensory Input to the Medulla Figure 14.2 shows a variety of sensory receptors that project to the medulla and function in cardiovascular regulation. Foremost among these are the arterial baroreceptors, which inform the medulla about current pressures in the aortic arch and carotid sinuses. Other receptors include baroreceptors in the heart and large systemic veins, which monitor venous pressure, and *chemoreceptors* in the brain and carotid arteries that monitor concentrations of oxygen, carbon dioxide, and hydrogen ions in arterial blood. Functions of these receptors are discussed later in the chapter. Figure 14.2 also includes proprioceptors in skeletal muscles and joints, which sense body movement and joint position, and other receptors of various types in internal organs throughout the body; these receptors are important in certain circumstances and are mentioned later in conjunction with the cardiovascular response to exercise.

Input From Higher Brain Areas As Figure 14.2 shows, the medulla receives input not only from sensory receptors but also from higher brain areas, including the cerebral cortex and hypothalamus, that influence cardiovascular function. The hypothalamus is important in orchestrating fight-or-flight responses of the cardiovascular system, and it also regulates the resistance of blood vessels in the skin in response to changes in body temperature. These changes in resistance control blood flow through the skin, which helps to regulate the rate of heat loss from the body. (The role of skin blood vessels in body temperature regulation is discussed later in the chapter.) The precise nature of cortical influences on cardiovascular function is not fully understood, but these influences are thought to be wide ranging. The cortex is involved in the cardiovascular changes that occur in response to pain and emotional states (such as the rise in blood pressure that frequently accompanies anxiety) and in exercise. The cortex also exerts continual control over cardiovascular function by modulating the medulla's responses to its sensory inputs.

Autonomic Inputs to Cardiovascular Effectors Sensory inputs to the medulla (and inputs from other brain areas) exert their control over cardiovascular function by influencing levels of activity in sympathetic and parasympathetic neurons that project from the medulla to the heart and blood vessels. Baroreceptor input to the medulla is relayed to sympathetic and parasympathetic neurons in such a way that when one set of neurons is activated, the other is inhibited. When arterial baroreceptors detect a fall in arterial pressure, for example, they trigger an increase in sympathetic activity and a decrease in parasympathetic activity. A rise in arterial pressure has the opposite effect.

Parasympathetic input to the heart travels via preganglionic fibers in the vagus nerve (cranial nerve X) that synapse with postganglionic neurons in the heart itself. (Recall that in the autonomic nervous system, signals are transmitted to effector organs by two sets of neurons: *Preganglionic* neurons carry signals from the central nervous system to clusters of nerve cells called *ganglia*, where they synapse with *postganglionic* neurons, which carry signals to the organs.) Sympathetic input to the heart and blood vessels travels via preganglionic fibers that emerge from the spinal cord and synapse with postganglionic neurons

in the sympathetic trunk. Signals are carried from the sympathetic trunk to effectors by these postganglionic neurons, whose axons travel in a number of different nerves. Signals from the various sensory receptors are conducted to the medulla by cranial and spinal nerves.

Major autonomic inputs to the cardiovascular system include the following (see Figure 14.2): (1) sympathetic and parasympathetic nerves to the sinoatrial node, which control heart rate; (2) sympathetic nerves to the ventricular myocardium, which control ventricular contractility; (3) sympathetic nerves to arterioles and other resistance vessels; which control vascular resistance; and (4) sympathetic nerves to veins, which control venomotor tone. Clearly, the primary neural influence on cardiovascular function is *sympathetic;* parasympathetic input comes into play only at the SA node.

When arterial pressure falls, for example, and an increase in sympathetic activity and a decrease in parasympathetic activity occur, both of these changes act at the sinoatrial node to produce an increase in heart rate. However, the change in sympathetic activity alone triggers all the other responses: (1) an increase in ventricular contractility, which tends to raise the stroke volume; (2) constriction of arterioles and other resistance vessels in most tissues of the body, which raises total peripheral resistance; and (3) an increase in venomotor tone, which raises central venous pressure and increases the flow of blood back to the heart (Figure 14.3a). When arterial pressure rises, sympathetic activity decreases and parasympathetic activity increases, both of which lower heart rate. All other changes are brought about by the decrease in sympathetic activity—decreases in ventricular contractility, total peripheral resistance, and venomotor tone (Figure 14.3b).

The Baroreceptor Reflex

Most of us have probably felt the effects of baroreceptors in action. We have seen that when a person rises quickly from a reclining position, for example, venous pooling causes a drop in central venous pressure, which then causes a fall in mean arterial pressure. Because this fall in pressure reduces blood flow to the brain, many people frequently experience a slight dizziness upon standing. Within seconds, however, the dizziness goes away because the pressure has risen to nearly normal. This occurs because the fall in arterial pressure is detected by arterial baroreceptors, which trigger an increase in sympathetic activity and a decrease in parasympathetic activity, which brings about increases in heart rate, myocardial contractility, and vascular resistance. This sequence of events is called the **baroreceptor reflex.**

To appreciate the operation of the baroreceptor reflex we must understand how each response that is triggered by the reflex affects mean arterial pressure (Figure 14.4). We know that a drop in arterial pressure triggers an increase in heart rate (HR) and an increase in myocardial

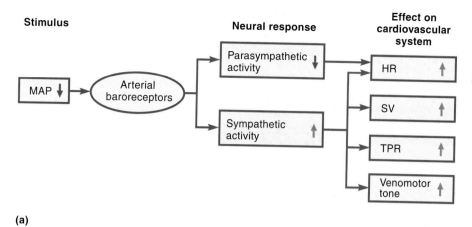

Stimulus

Neural response

Effect on cardiovascular system

FIGURE 14.3 Cardiovascular responses to changes in mean arterial pressure mediated by arterial baroreceptors. *Vertical arrows indicate directions of changes.* **(a)** *Responses to a fall in arterial pressure.* **(b)** *Responses to a rise in arterial pressure.*

(a)

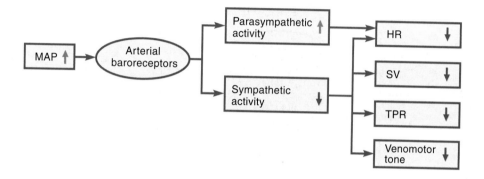

(b)

contractility and venomotor tone, both of which tend to raise stroke volume (SV). Along with this, the fall in arterial pressure triggers vasoconstriction, leading to an increase in total peripheral resistance (TPR). As we saw in Chapter 13, this combination of effects raises mean arterial pressure (MAP):

$$\uparrow HR \times \uparrow SV \times \uparrow TPR = \uparrow MAP$$

Figure 14.5 illustrates in graphical form the baroreceptor response to a drop in blood volume due to hemorrhage. When blood volume falls, the end-diastolic volume decreases, and the stroke volume and cardiac output also fall. At the same time that cardiac output drops, mean arterial pressure falls. (Heart rate and total peripheral resistance do not change until reflex responses are activated and changes in autonomic nervous activity occur.) The drop in arterial pressure then activates the baroreceptor reflex, which drives the pressure back up to near its initial value. The pressure does not return to *exactly* the original level because if it did, baroreceptors would relay signals to the cardiovascular regulatory center, telling it that the pressure was normal; in that case the compensatory response to blood loss would shut down, and the pressure would drop again. Thus a small error signal is necessary for keeping the response activated and compensating for the blood loss.

Because a drop in arterial pressure triggers a subsequent compensatory rise, the baroreceptor reflex acts by negative feedback to keep pressure constant. The same principle holds when arterial pressure rises. In this case the change in pressure triggers responses in the opposite direction, producing a decline in pressure, which sometimes can have unfortunate consequences. In certain persons, even wearing a tight-fitting collar is sufficient to exert enough pressure on the carotid baroreceptors to "trick" them into signaling that arterial pressure is abnormally high when in fact it is normal. This false signal triggers a baroreceptor reflex and a subsequent drop in pressure that can cause loss of consciousness. The same effect can be produced by manually massaging certain areas of the neck. In rare instances, manual pressure on the neck has even been known to cause the heart to stop because such stimulation can trigger inappropriately high levels of activity in parasympathetic nerves to the sinoatrial node, which suppresses pacemaker activity.

The action of the baroreceptor reflex raises an interesting question: If baroreceptor reflexes work to keep mean arterial pressure constant at normal levels, why do certain persons suffer *hypertension*—chronically high arterial blood pressure? The reason is that hypertension is a condition that develops slowly over long periods of time. The accompanying gradual rise in arterial pressure causes baroreceptors to lose their sensitivity such that they reset

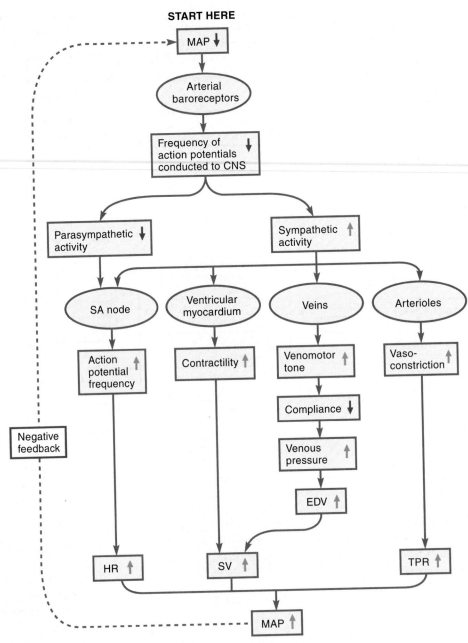

FIGURE 14.4 The events in the baroreceptor reflex in response to a drop in mean arterial pressure.

at a new, higher pressure, which effectively becomes "normal." Under these conditions the baroreceptors still function to regulate arterial pressure, but they act to maintain it at a level higher than it would normally be. Once this reset has occurred, baroreceptors cannot correct the problem (When It Goes Wrong: Hypertension, p. 450).

When all is considered, the *immediate* danger posed by low arterial pressure (*hypotension*) is far greater than that posed by hypertension, because a low mean arterial pressure acts to reduce blood flow to all systemic organs, which can compromise their function and even permanently damage them. This is not to say that hypertension is not dangerous, but hypertension usually takes years to kill, whereas hypotension can kill in minutes. Thus the

most important function of the baroreceptor reflex is to counteract potentially dangerous reductions in organ blood flow. But if a fall in mean arterial pressure triggers an increase in total peripheral resistance, which tends to *reduce* organ blood flow, then how does an increase in peripheral resistance protect *against* reductions in blood flow?

When a baroreceptor reflex triggers an increase in sympathetic nervous activity, the resistance in most, but not *all*, organs increases. In particular, resistance in the brain and heart are affected very little by sympathetic influences and are therefore not altered significantly by baroreceptor reflexes. When a baroreceptor reflex triggers a compensatory rise in mean arterial pressure, it acts to

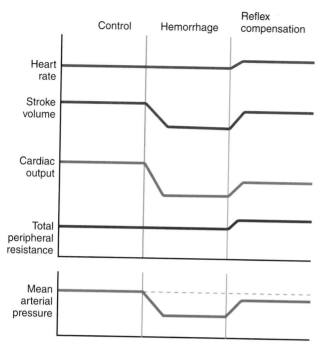

FIGURE 14.5 Baroreceptor-mediated responses to hemorrhage. *Changes in cardiovascular variables are indicated relative to normal values before hemorrhage occurs (the "control" condition in the graph). The delay between hemorrhage and onset of reflex compensation is exaggerated for clarity. Note that reflex compensation raises mean arterial pressure only to near-normal levels.*

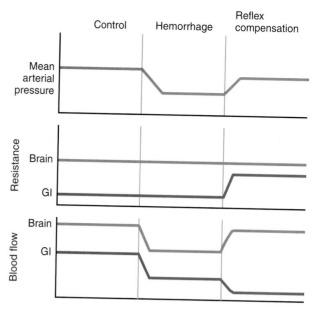

FIGURE 14.6 Effects of a hemorrhage-induced baroreceptor reflex on mean arterial pressure (top trace), and on vascular resistance (middle traces) and blood flow (bottom traces) in the brain and gastrointestinal tract. *In the GI tract, increased vascular resistance results in lower-than-normal blood flow to these organs. By contrast, virtually no change occurs in vascular resistance in the brain; as a result, blood flow to this vital organ returns to near normal as the reflex restores MAP to near normal.*

In this example, which of the following decreases following hemorrhage—preload, afterload, or both?

maintain flow to the heart and brain—but it does so *at the expense of other organs,* whose blood flow actually decreases as a result of vasoconstriction. (Recall from Chapter 13 that if mean arterial pressure is kept constant and an organ's vascular resistance increases, blood flow to that organ decreases.) The end result is that blood flow is shifted *away from* certain organs (such as the skin and gastrointestinal system), and *toward* other organs (such as the brain and heart) that are more vital to the short-term survival of the body (Figure 14.6).

Exercise Link

As Bill and Jane ran the marathon, their cardiovascular systems made similar adjustments. Blood flow increased (1) to their legs, providing oxygen and energy to the working muscles; (2) to their skin, transferring heat out of their bodies; (3) and to their heart muscles, supporting the work of increased cardiac output. As compensation, blood flow was directed away from their livers, kidneys, and gastrointestinal tracts, the functions of which were not directly or immediately necessary to support exercise.

Although this shift in blood flow might seem a bit like "robbing Peter to pay Paul," we can see why it is beneficial by considering what would happen if the baroreceptor reflex promoted vasoconstriction in all organs, such that the resistance of the heart and brain *did* increase: Assume that a drop in mean arterial pressure has triggered a baroreceptor reflex that returns the pressure to normal. Based on what we learned in Chapter 13, if the reflex triggered an increase in resistance in the heart and brain, the normal arterial pressure in conjunction with the increased resistance would lead to lower-than-normal blood flow in these organs. This would be detrimental to survival because the heart and brain require continual delivery of oxygen and nutrients for their function and are therefore exquisitely sensitive to any reduction in blood flow. Thus even though the baroreceptor reflex might be able to maintain a normal mean arterial

Both

Hypertension (high blood pressure), which is generally defined as a chronic elevation of diastolic pressure over 90 mm Hg or systolic pressure over 140 mm Hg, affects at least 60 million persons in the United States. Because it is painless, those affected by hypertension can remain unaware of it for years—until it causes irreversible damage to the cardiovascular system and other organs. Because elevated arterial pressure increases the workload on the heart, it can increase the likelihood of myocardial infarction. It can also lead to cardiac failure, because it increases afterload and can chronically raise end-diastolic volume. Moreover, it also stresses arterial walls, which can trigger a stroke or damage vessels in a way that predisposes them to development of atherosclerosis. Hypertension-induced damage to blood vessels can also lead to kidney failure or loss of vision.

Even though the precise cause is unknown in many persons with hypertension, the disease is known to be associated with certain risk factors, including obesity, stress, smoking, and genetic disposition. Hypertension of unknown origin is termed *primary hypertension* (or *essential hypertension*). If the cause of hypertension is known, it is called *secondary hypertension*. In certain forms of secondary hypertension, the cause is on the "arterial side"—that is, mean arterial pressure is elevated be-

cause cardiac output or total peripheral resistance is inappropriately high. In other cases the problem is on the "venous side"—blood volume is elevated, which raises mean arterial pressure indirectly.

In *renal* hypertension, the primary cause is a disorder of kidney function. Kidney disease may cause failure to excrete normal amounts of salt and water, leading to fluid retention and expansion of blood volume. Sometimes the problem is occlusion of blood flow to a kidney or within a kidney, which can trigger inappropriately high rates of renin release. Abnormally high levels of renin results in overproduction of angiotensin II in the plasma, which stimulates vasoconstriction and increased peripheral resistance and also stimulates the kidneys to retain salt and water. In *endocrine* hypertension, the primary cause is inappropriate secretion of a hormone by an endocrine gland. A tumor of the adrenal medulla, for example, can oversecrete epinephrine, which stimulates increased cardiac output.

Regardless of the primary cause, hypertension cannot be "cured" or compensated for by baroreceptor reflexes or any of the normal mechanisms that regulate blood pressure. The reason is that baroreceptors are part of the problem: Under conditions of chronically elevated pressure, the baroreceptors reset, and regulatory mechanisms then work to maintain the new, higher pressure.

Fortunately, hypertension is easily detectable during routine blood pressure checks. (Repeated measurements over time are necessary for a firm diagnosis, however.) Unfortunately, the condition can be permanently reversed only on rare occasions because the cause is unknown in the vast majority (over 90%) of cases. Instead, the disease is controlled through chronic medication or other means.

Treatments for hypertension include diuretics, which promote increased excretion of salt and water by the kidneys, and specific antihypertensive drugs, such as beta blockers and calcium channel blockers. *Beta blockers* reduce cardiac output by interfering with the ability of epinephrine and norepinephrine to bind to beta receptors, thereby reducing the stimulatory influence of these agents on the heart. *Calcium channel blockers* reduce the flow of calcium into vascular smooth muscle cells across the plasma membrane, which reduces vasomotor tone and lowers peripheral resistance. Other drugs, called *converting enzyme inhibitors*, reduce plasma levels of angiotensin II by blocking the enzyme that catalyzes its formation from angiotensin I. Other treatment options include exercise, which has been shown to lower resting blood pressure, and weight reduction.

pressure, the effort would be in vain because the resulting vasoconstriction in the heart or brain would compromise flow to those organs.

Note that when a drop in arterial pressure is due to a loss of blood volume, the baroreceptor reflex is only a "quick fix." So long as the drop in volume is not so great that the body cannot compensate for it, the baroreceptor reflex will keep arterial pressure high enough to ensure survival until the lost volume can be replaced by drinking fluids (or by blood transfusion, if necessary). Long-term regulation of arterial pressure is accomplished by controlling blood volume, which depends on adjusting the balance between fluid intake and excretion. This is primarily the job of the kidneys, as we see in Chapter 18.

When individuals lose a lot of blood, the baroreceptor "quick fix" can sometimes get them into a "real fix" because the baroreceptor reflex reduces blood flow to most organs. Eventually these organs require restoration of normal flow; otherwise they will be damaged by either the reduced availability of oxygen and nutrients or the accumulation of toxic metabolic by-products. Because arteriolar smooth muscle is sensitive to local levels of oxygen, carbon dioxide, and other factors, these chemical changes have important consequences for cardiovascular function. When tissues are deprived of an adequate blood supply for a prolonged period, local chemical changes cause vascular smooth muscle to relax. Unless lost blood volume is restored quickly (within one or two hours in the case of severe hemorrhage), these local influences will override the influence of sympathetic vasoconstrictor nerves on vascular smooth muscle, and vascular resistance will begin to fall. As a result, arterial pressure will fall, despite the actions of compensatory mechanisms to raise it. When this happens, blood flows to the heart and brain fall. This condition, known as *circulatory shock*, may become irreversible, in which case it progresses inexorably to death.

Quick Test 14.1

1. What is a baroreceptor? Where are arterial baroreceptors located?

2. What area of the brain is primarily responsible for controlling activity in autonomic neurons to the heart and blood vessels?

3. What is a baroreceptor reflex? How is negative feedback involved in this process?

4. Indicate whether each of the following autonomic nervous activities increases or decreases when arterial pressure falls: sympathetic nervous activity, parasympathetic nervous activity, heart rate, myocardial contractility, vascular resistance (in most tissues), venomotor tone.

5. Name two organs whose vascular resistance is generally unaffected by baroreceptor reflexes. Why is this beneficial?

Hormonal Control of Mean Arterial Pressure

Arterial baroreceptors exert control over cardiovascular function not only via the baroreceptor reflex, but also by regulating the secretion of certain hormones, including *epinephrine, vasopressin,* and *angiotensin II,* which work hand in hand with the autonomic nervous system to regulate mean arterial pressure. We discuss these hormones briefly in the following sections.

Epinephrine

The effects of epinephrine, a hormone secreted by the adrenal medulla, are generally similar to those of the sympathetic neurotransmitter norepinephrine. The rate of epinephrine secretion is controlled by sympathetic nerves to the adrenal gland in such a way that increases in sympathetic activity are generally accompanied by increased epinephrine release. Low arterial pressure is a stimulus for epinephrine secretion, although evidence suggests that secretion is enhanced only when the drop in pressure is relatively severe. Here we first discuss this hormone's effects on cardiac function, and then on vascular function.

The effects of epinephrine on cardiac function mirror the actions of sympathetic input. At the sinoatrial node, epinephrine increases the action potential frequency of pacemaker cells, which increases heart rate. In the myocardium, it stimulates increased contractility. In both cases, these effects are brought about as a result of epinephrine binding to *beta receptors,* the predominant type of adrenergic receptor in cardiac tissue (see p. 317). These are the same receptors that bind norepinephrine, which explains the similarity between epinephrine's actions and those of the sympathetic nervous system.

Epinephrine's effects on vascular resistance are more complicated than its effects on the heart because vascular smooth muscle possesses two types of adrenergic receptors—*alpha* and *beta* receptors, which bind both epinephrine and norepinephrine. Alpha receptors are clustered around sympathetic nerve endings and promote vasoconstriction, whereas beta receptors are more widely scattered (not concentrated around nerve endings) and promote vasodilation. In most tissues, sympathetic nervous activity has a predictable effect on vascular resistance because it activates alpha receptors and therefore triggers vasoconstriction. In contrast, the effects of epinephrine are less predictable because it circulates in the blood and thus has access to both types of receptors. When epinephrine is present at lower concentrations, it binds primarily to beta receptors and promotes vasodilation because it has greater affinity for beta receptors than for alpha receptors. At higher concentrations, epinephrine binds to both types of receptors, and for this reason the hormone's effect on vascular resistance depends on which receptor type predominates, which differs from one vascular bed to another.

Because alpha receptors outnumber beta receptors in most locations, high concentrations of epinephrine usually promote vasoconstriction (the same effect as sympathetic input). Because beta receptors exert the predominant effect on blood vessels in cardiac and skeletal muscle, vasodilation occurs in these tissues in response to epinephrine. At the same time that this vasodilation promotes increased blood flow to these tissues, vasoconstriction decreases blood flow elsewhere. The significance of this increased blood flow is clear given that lots of epinephrine is released during the fight-or-flight response, which adapts the body for vigorous physical exercise. In such exercise the workload of the heart and skeletal muscles increases, and greater blood flow is required to meet the increased metabolic demand.

Vasopressin (Antidiuretic Hormone)

Vasopressin is a hormone secreted by the pituitary, a gland connected to the base of the brain by a stalk (see Chapter 5). Because it acts on the kidneys to limit urine output, vasopressin is also known as *antidiuretic hormone.* (Increased urine flow is called *diuresis.*) Along with this effect, and more to the point here, it also promotes vasoconstriction in most tissues—hence, the name *vasopressin.* (Effects that tend to raise blood pressure are referred to as *pressor* effects.) Vasopressin secretion is regulated by a variety of factors, including the level of activity in arterial baroreceptors. When arterial pressure falls, vasopressin release is enhanced, which promotes vasoconstriction, increased total peripheral resistance, and thus increased mean arterial pressure.

In addition to its effects on vascular resistance, vasopressin also acts on the kidneys to promote decreased urine output, which by reducing the rate at which water is lost from the body conserves blood volume. This is important because any drop in blood volume (due to hemorrhage or severe sweating, for example) tends to lower mean arterial pressure by reducing venous return and cardiac output. Because vasopressin reduces the loss of body water, the hormone acts indirectly to maintain mean arterial pressure by counteracting further reductions in blood volume. Actions of vasopressin and regulatory mechanisms affecting its secretion are discussed fully in Chapter 18.

Angiotensin II

Angiotensin II is a protein derived from a precursor called *angiotensinogen,* which is always present in plasma. The generation of angiotensin II from angiotensinogen is a two-step process: Angiotensinogen is first converted to *angiotensin I* by *renin,* an enzyme secreted by the kidney (see Chapter 18). Angiotensin I is then converted to angiotensin II by *angiotensin converting enzyme,* which is present on the inner surface of blood vessels in many parts of the body, particularly in the lungs.

When arterial pressure falls, the release of renin by the kidneys is stimulated both directly by reduced arterial pressure and via activity of sympathetic nerves to the kidneys. As a result of the increase in renin secretion, the plasma concentration of angiotensin I rises, and this is followed by an increase in the concentration of angiotensin II. One of the effects of angiotensin II is to promote vasoconstriction, which causes an increase in total peripheral resistance. Thus angiotensin II works hand in hand with vasopressin and sympathetic vasoconstrictor nerves to increase total peripheral resistance, thereby helping to raise mean arterial pressure back to normal.

Angiotensin II acts to maintain arterial pressure not only by promoting vasoconstriction, but also by reducing urine output by the kidneys. In this regard, the actions of angiotensin II resemble those of vasopressin. Unlike vasopressin, however, angiotensin II acts on the kidneys indirectly. Angiotensin II affects urine flow by acting on the adrenal glands to stimulate the secretion of the hormone *aldosterone.* (Aldosterone is secreted by cells in the outer layer of the adrenal glands, called the *cortex.*) Aldosterone then targets the kidneys, promoting a reduction in urine flow. Actions of angiotensin II and mechanisms controlling its release are discussed further in Chapter 18.

Extrinsic factors involved in the control of mean arterial pressure are summarized in Table 14.1.

Control Initiated by Cardiac and Venous Baroreceptors (Volume Receptors)

In addition to the baroreceptors in the aortic arch and carotid arteries, which monitor systemic arterial pressure, other baroreceptors monitor pressures elsewhere in the cardiovascular system. Particularly important are baroreceptors in the walls of large systemic veins and in the walls of the heart that monitor pressure in the veins and cardiac chambers. These receptors function in the same way as arterial baroreceptors in that they have receptor endings that respond to stretch, but because of their locations they monitor pressure on the *venous* side of the systemic circulation and therefore act directly to detect changes in *blood volume.* (Recall from Chapter 13 that most of the blood volume is contained in veins, and that venous pressure rises and falls in response to changes in blood volume.) Because the venous side of the circulation acts as a volume reservoir, venous and cardiac baroreceptors are frequently referred to as *volume receptors.* Additional baroreceptors located in the pulmonary vasculature (which also functions as a volume reservoir) act indirectly to monitor systemic venous pressure.

TARGET ORGAN OR TISSUE	NEURAL OR HORMONAL FACTOR	FACTOR'S EFFECT ON TARGET	INFLUENCE ON MEAN ARTERIAL PRESSURE
Heart			
Sinoatrial node	Sympathetic nerves	↑HR	↑MAP
	Parasympathetic nerves	↓HR	↓MAP
	Epinephrine	↑HR	↑MAP
Ventricular myocardium	Sympathetic nerves	↑Contractility (↑SV)	↑MAP
	Epinephrine	↑Contractility (↑SV)	↑MAP
Arteriolar smooth muscle (most tissues)	Sympathetic nerves	Vasoconstriction (↑TPR)	↑MAP
	Epinephrine	Vasoconstriction or vasodilation, depending on concentration and location	Variable
	Vasopressin	Vasoconstriction (↑TPR)	↑MAP
	Angiotensin II	Vasoconstriction (↑TPR)	↑MAP
Venous smooth muscle	Sympathetic nerves	↑Venomotor tone	↑MAP
	Epinephrine	↑Venomotor tone	↑MAP

Because arterial pressure is influenced by venous pressure, the functions of venous, cardiac, and arterial baroreceptors are intertwined; levels of activity in these receptors frequently vary in the same direction, because when venous pressure increases or decreases, arterial pressure also tends to increase or decrease, respectively. In addition, venous and cardiac baroreceptors exert many of the same actions as arterial baroreceptors. When venous pressure falls, for example, venous and cardiac baroreceptors trigger an increase in both sympathetic nerve activity and vasopressin secretion.

Integrated Neural and Hormonal Control of Cardiovascular Function: Restoration of Mean Arterial Pressure Following Hemorrhage

So far we have seen that mean arterial pressure is controlled by a fairly large number of neural and hormonal inputs that produce overlapping effects. To help clarify how these numerous inputs act in a coordinated way, we now explore how MAP is regulated in response to hemorrhage (Figure 14.7). (Because knowledge of kidney function is necessary for a full understanding of blood pressure regulation, the explanation presented here is partial; full treatment is given in Chapter 18.)

Hemorrhage triggers a drop in mean arterial pressure through a series of events we previously discussed: The loss of blood produces a decrease in total blood volume that lowers venous pressure; lowered venous pressure reduces the flow of blood to the heart, causing a reduction in end-diastolic volume that, by virtue of the Starling effect, reduces stroke volume. Reduced stroke volume then leads to reduced MAP. Mean arterial pressure is raised toward normal through various negative feedback loops composed of actions triggered by both the decrease in arterial pressure (detected by arterial baroreceptors) and the drop in venous pressure (detected by cardiac and venous baroreceptors).

Responses Triggered by Arterial Baroreceptors

The fall in mean arterial pressure causes a decrease in the frequency of action potentials conducted to the central nervous system by arterial baroreceptors. As a result, the activity of sympathetic neurons increases, and the activity of parasympathetic neurons decreases. By virtue of direct neural inputs to cardiovascular effectors, these changes trigger the following characteristic responses of the baroreceptor reflex (see Figure 14.7):

1. The frequency of action potentials in the sinoatrial node increases, causing an increase in heart rate.

2. The contractility of the ventricular myocardium increases, promoting an increase in stroke volume.

START HERE

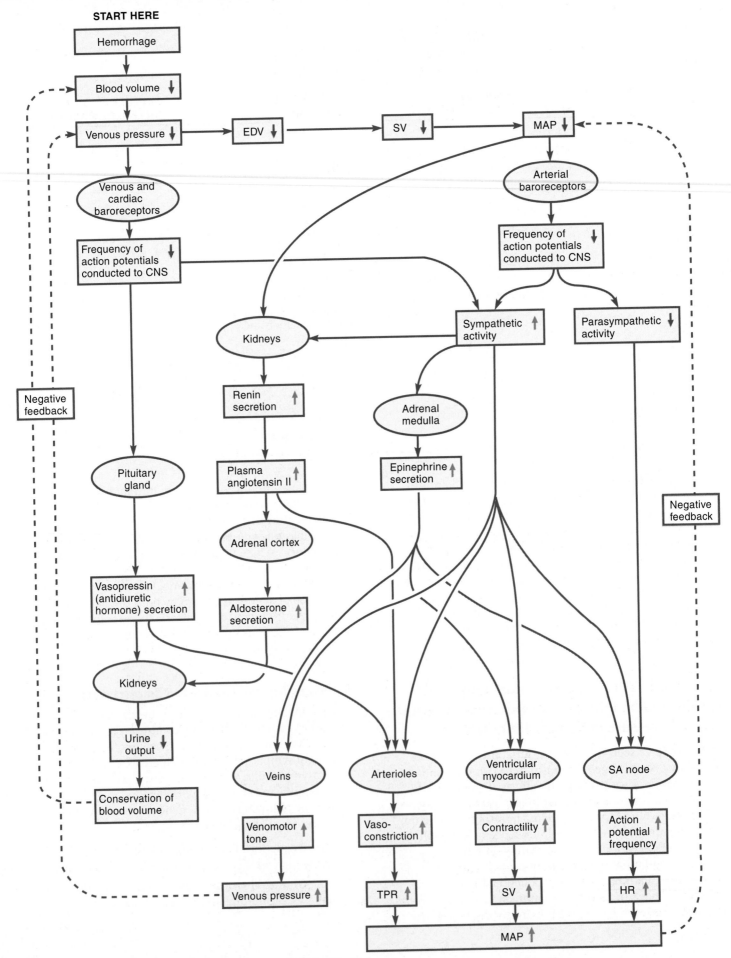

FIGURE 14.7 **Neural and hormonal regulation of cardiovascular function in response to hemorrhage.**

Assuming that MAP is held constant through the operation of reflex mechanisms, will blood flow to the brain increase, decrease, or stay the same as a result of reflex compensation?

3. In many tissues, the degree of constriction in arterioles and other resistance vessels increases, leading to an increase in total peripheral resistance.

4. Venomotor tone increases, causing a rise in venous pressure.

As a consequence of the increase in heart rate, stroke volume, and total peripheral resistance, mean arterial pressure rises. Because this rise in pressure counteracts the initial fall, the sequence of events just described constitutes negative feedback control of arterial pressure. In addition, the rise in venous pressure caused by the increase in venomotor tone counteracts the initial decline in venous pressure, thereby acting indirectly to counteract the fall in arterial pressure.

In addition to these responses, the increase in sympathetic activity triggered by arterial baroreceptors also stimulates increases in the secretion of several hormones that raise arterial pressure, as follows (see Figure 14.7):

1. Sympathetic neurons to the adrenal medulla trigger the secretion of epinephrine, which acts on the sinoatrial node, ventricular myocardium, and veins to increase heart rate, stroke volume, and venomotor tone.

2. Sympathetic nerves to the kidneys trigger the secretion of renin, resulting in an increase in the plasma concentration of angiotensin II. (The drop in MAP also directly stimulates the kidneys to secrete renin.) The rise in angiotensin II acts on arterioles to raise total peripheral resistance. Angiotensin II also stimulates the adrenal cortex to secrete aldosterone, which acts on the kidneys to promote a reduction in urine output, thereby helping to conserve blood volume. By acting to prevent further reductions in blood volume, the reduction in urine output works indirectly to counteract the fall in arterial pressure.

Responses Triggered by Venous and Cardiac Baroreceptors

The fall in venous pressure causes a decrease in the frequency of action potentials conducted to the central nervous system by venous and cardiac baroreceptors (see Figure 14.7). This signal triggers increased sympathetic nervous activity, thereby reinforcing the sympathetic actions described in the previous section. In addition, signals sent to the pituitary gland cause it to increase the secretion of vasopressin. This hormone acts on arterioles to promote increased total peripheral resistance, which

reinforces the effects of sympathetic input and angiotensin II on these vessels. Vasopressin also acts on the kidneys to promote a reduction in urine output, thereby reinforcing the action of aldosterone.

Quick Test 14.2

1. How does epinephrine affect cardiac function? Where is it secreted?

2. Where are vasopressin and angiotensin II secreted?

3. How do vasopressin and angiotensin II affect vascular resistance? Do these hormones increase or decrease arterial pressure?

4. Vasopressin helps to regulate arterial pressure not only because of its action on blood vessels, but also because of its action on the kidneys. What effect does this hormone have on the kidneys, and why is it important in the control of arterial pressure?

INTRINSIC CONTROL OF CARDIOVASCULAR FUNCTION: REGULATION OF BLOOD FLOW DISTRIBUTION TO ORGANS

When extrinsic regulatory mechanisms such as the baroreceptor reflex are working correctly, they can accomplish only one thing: maintaining arterial pressure at a level sufficient to drive adequate blood flow through the systemic circuit. But even when the cardiovascular system is maintaining adequate mean arterial pressure, the job of supplying blood to organs is only half done, because the ultimate goal is to provide blood flow that is matched to *each organ's* metabolic needs. As we saw in Chapter 13, mechanisms adjust organs' blood flow by adjusting vascular resistance; in this section we examine the ways organs and tissues sense their blood flow needs and adjust vascular resistance accordingly.

Why Intrinsic Control?

As we have seen, the *distribution* of blood flow among the organs—which organs get more and which get less—depends entirely on the relative vascular resistances of the

Stay the same

individual organs. It makes sense, then, that the distribution of flow among the organs is regulated by the organs themselves—that is, by *intrinsic control* (also known as *local regulation* or *autoregulation*) of organ resistance. Intrinsic control of organ resistance is accomplished through dilation or constriction of arterioles, which lowers or raises organ resistance, respectively, thereby increasing or decreasing blood flow. Intrinsic control mechanisms regulate not only the distribution of blood flow among the organs, but also the distribution of blood flow *within* organs. These mechanisms are responsible for the sharing of flow among capillary beds mentioned in the introduction to this chapter.

The distribution of blood flow is analogous to the problem of regulating water usage in a neighborhood: Whereas it is the job of the utility company to provide adequate water pressure for all houses in the system, the regulation of the flow of water to any individual house can be accomplished only by the people who live there, because they are the only ones who know of their moment-to-moment need for water. Similarly, extrinsic control mechanisms provide adequate arterial pressure for all organs in the systemic circuit, but only an individual organ or tissue can really "know" how much blood it needs at any given time, and so it regulates its own blood flow through local control.

Intrinsic control mechanisms are especially important in regulating blood flow to the heart, brain, and skeletal muscles, because neural mechanisms play only a minor role in regulating blood flow to the heart and brain, and because metabolic activity in the heart and skeletal muscles can vary widely to meet their widely changing demands for oxygen and nutrients. Even though the metabolic activity of the brain as a whole is relatively constant, activity varies from region to region within the brain. These activity patterns change frequently, depending on what a person is doing at any given moment. Intrinsic mechanisms work on a continual basis to increase blood flow to regions that are becoming more active while reducing flow to those whose activity is decreasing.

Local Factors that Control Vascular Resistance

Given that intrinsic control of organ blood flow is accomplished through relaxation or contraction of smooth muscle in arterioles, which control the flow through individual capillary beds within an organ or tissue, what senses whether blood flow is adequate or not? Asked another way, what is it that tells the smooth muscle whether to contract or relax? The answer to both questions is *vascular smooth muscle* itself. In the following sections we examine the responses of vascular smooth muscle in response to four factors: changes in metabolic activity, changes in blood flow, stretch of arteriolar smooth muscle, and local chemical messengers.

Regulation in Response to Changes in Metabolic Activity

Vascular smooth muscle cells in arterioles are sensitive to conditions in extracellular fluid and respond to changes in the concentrations of a wide variety of chemical substances, including oxygen, carbon dioxide, potassium ions, hydrogen ions, and others. These changes in concentration occur as a result of metabolic activity, and arteriolar smooth muscle either contracts or relaxes depending on whether concentrations of particular substances rise or fall. The general rule of thumb is this: *Changes associated with increased metabolic activity generally cause vasodilation, whereas changes associated with decreased metabolic activity induce vasoconstriction.* (An important exception to this rule occurs in the pulmonary vasculature, as we see in Chapter 16.)

We can demonstrate how this rule works by considering the oxygen concentration in tissue as an example. When blood flow is matched to the tissue's metabolic needs, the concentration of oxygen in the tissue is in a steady state: The rate at which oxygen enters the tissue from the blood equals the rate at which it is consumed by cells. Now suppose that the metabolic rate increases, such that the rate of oxygen consumption rises. The increased rate of oxygen consumption causes a decrease in tissue oxygen concentration, a condition called *hypoxia*, because oxygen is being used up faster than it is delivered by the blood. In other words, blood flow is insufficient to keep up with metabolic demand; such an insufficiency in blood flow is known as **ischemia.** According to our rule, the fall in oxygen concentration causes arteriolar smooth muscle to relax. When the muscle relaxes, vascular resistance in the tissue drops, and blood flow in that region increases, as does the rate of oxygen delivery. As a result of this increase in blood flow, the oxygen concentration eventually reaches a new steady state in which the rate of oxygen delivery and the rate of consumption are both higher than they were originally. The increase in blood flow following an increase in metabolic activity is termed **active hyperemia** (Figure 14.8). (*Hyperemia* is a general term for a higher-than-normal rate of blood flow.) Note that these changes in blood flow are a result of a direct effect of the reduced oxygen concentrations on the arterioles themselves, and that no nerves or hormones are involved. Thus active hyperemia is clearly an example of *intrinsic* control.

The same mechanism decreases blood flow if the local metabolic rate falls: Because the rate at which oxygen enters the tissue from the blood is greater than the rate at which it is consumed, tissue oxygen concentration rises at first. Rising oxygen concentration is normally associated with *falling* metabolic rates, and vascular smooth muscle in that region constricts. Vasoconstriction in the arterioles reduces blood flow to the tissue, thereby reducing the rate of oxygen delivery to keep it in line with the reduced

FIGURE 14.8 Active hyperemia, a response to an increase in metabolic activity.

(a) *The delivery and consumption of oxygen during the steady state.* (b–d) *Changes in local conditions in response to increased metabolic rate.* (e) *The negative feedback loop by which active hyperemia increases blood flow and tissue oxygen delivery in response to an increase in local metabolic demand.*

As the tissue oxygen concentration decreases in this example, what is happening to the concentration of carbon dioxide in the tissue?

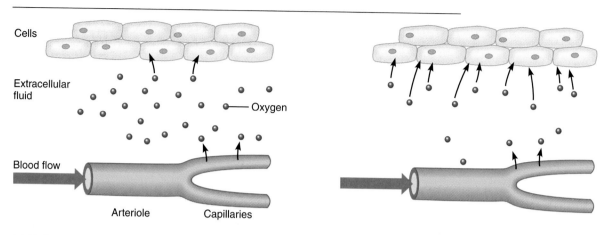

Cells

Extracellular fluid

— Oxygen

Blood flow

Arteriole   Capillaries

(a) Under normal steady state conditions, oxygen (dots) is delivered to tissues by the blood as fast as it is consumed by cells.

(b) An increase in the metabolic rate causes oxygen to be consumed faster than it is delivered. The oxygen concentration in extracellular fluid decreases.

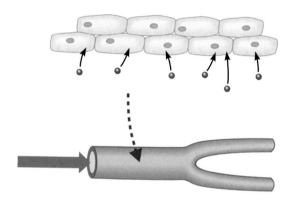

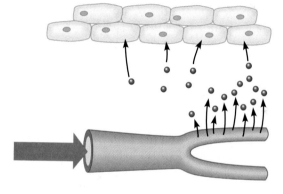

(c) The decreased oxygen concentration acts on arteriolar smooth muscle to promote vasodilation.

(d) Vasodilation promotes increased blood flow and increased oxygen delivery to cells.

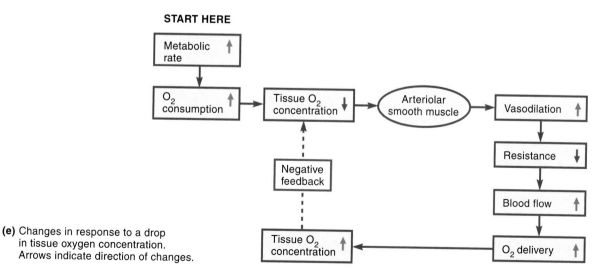

START HERE

(e) Changes in response to a drop in tissue oxygen concentration. Arrows indicate direction of changes.

Metabolic rate ↑ → O₂ consumption ↑ → Tissue O₂ concentration ↓ → Arteriolar smooth muscle → Vasodilation ↑ → Resistance ↓ → Blood flow ↑ → O₂ delivery ↑ → Tissue O₂ concentration ↑ → Negative feedback ⇢ Tissue O₂ concentration

FIGURE 14.9 **Events in active hyperemia in response to increased production and tissue levels of carbon dioxide.**

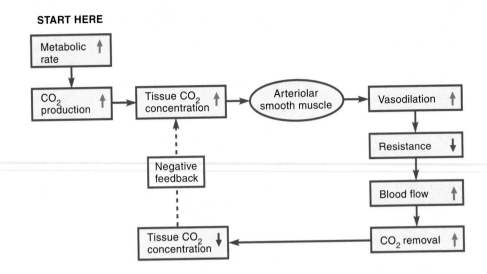

FIGURE 14.9 **Events in active hyperemia in response to increased production and tissue levels of carbon dioxide.**

demand. Eventually, the oxygen concentration reaches a new steady state in which blood flow and the rates of delivery and consumption are equal but lower than before.

If we consider the concentration of carbon dioxide in the tissue rather than oxygen, the line of reasoning is the same. When the metabolic rate of a tissue increases, cells produce more carbon dioxide, and the tissue concentration of carbon dioxide rises. Our rule of thumb tells us that the rise in tissue carbon dioxide concentration induces vasodilation, which increases blood flow, which in turn increases the rate at which carbon dioxide is *removed from* the tissue by the blood (Figure 14.9). Eventually the carbon dioxide concentration reaches a new steady state in which the rates of production and removal are equal. In this new steady state, blood flow and the rates of carbon dioxide production and removal are all higher than before. If a decrease in metabolic rate occurs instead, tissue carbon dioxide levels will fall, causing vasoconstriction and a reduction in blood flow.

Regulation in Response to Changes in Blood Flow

In the previous examples, blood flow increased in response to changes in metabolic activity; tissue oxygen and metabolite concentrations can also change as a result of changes in *blood flow*. For example, if blood flow is blocked or reduced below adequate levels for any reason (such as in circulatory shock), the oxygen concentration falls and the carbon dioxide level rises because rates of oxygen consumption and carbon dioxide production exceed rates of delivery and removal, respectively. By our rule, both of these changes induce vasodilation and a reduction in vascular resistance, which tend to increase blood flow. Once the blockage is removed, the rate of flow will be higher than normal and will remain elevated until excess metabolites are removed and tissue oxygen concentration is restored to normal.

Such an increase in blood flow in response to a previous reduction in blood flow is called **reactive hyperemia** (Figure 14.10). Note that the basic mechanism underlying reactive hyperemia is the same as that for active hyperemia—changes in tissue oxygen and metabolite concentrations induce vasodilation and an increase in blood flow. The only difference is the cause of the changes in concentration (compare Figure 14.8e and Figure 14.10e). The sequence of events occurring in reactive hyperemia also works in reverse: If blood flow rises above what is required for metabolic needs, intrinsic control mechanisms will induce vasoconstriction and a reduction in blood flow.

Regulation in Response to Stretch of Arteriolar Smooth Muscle

In certain tissues, arteriolar smooth muscle is responsive to stretch and responds to changes in the *pressure* of blood *within* the arterioles as well as to changes in the concentrations of chemicals in the fluid surrounding it. When these fibers are stretched, they respond by contracting. (Muscle fibers that respond in this manner are often described as being *stretch sensitive.*) A change in vascular resistance that occurs in response to stretch of blood vessels, and that does not require the action of sympathetic nerves, blood-borne hormones, or other chemical agents, is described as a **myogenic** response.

The pressure gradient that drives blood flow through a given organ or tissue is called the *perfusion pressure.* (In the case of a systemic organ, the perfusion pressure is virtually equivalent to mean arterial pressure.) If perfusion pressure increases in an organ or tissue, blood flow increases and arteriolar pressure rises, which increases the degree of stretch of arteriolar walls because it increases the distending pressure across them. In arterioles containing stretch-sensitive smooth muscle, the muscle fibers

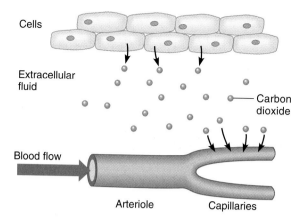

Cells

Extracellular fluid

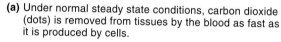

Carbon dioxide

Blood flow

Arteriole   Capillaries

(a) Under normal steady state conditions, carbon dioxide (dots) is removed from tissues by the blood as fast as it is produced by cells.

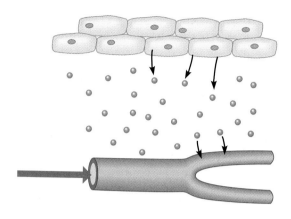

(b) A decrease in blood flow causes the rate of carbon dioxide removal to go below the rate of production. The carbon dioxide concentration in extracellular fluid increases.

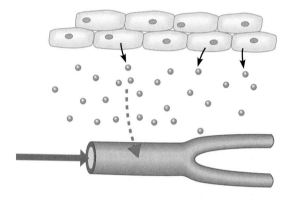

(c) The increased carbon dioxide concentration acts on arteriolar smooth muscle to promote vasodilation.

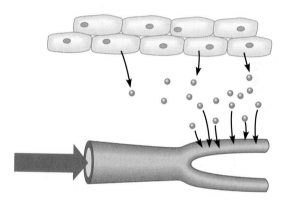

(d) Vasodilation promotes increased blood flow and increased carbon dioxide removal from tissue.

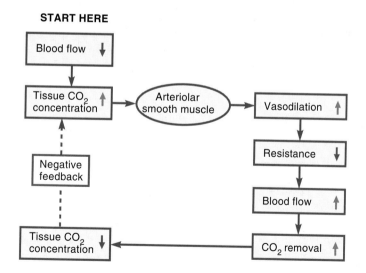

(e) Changes in response to a decrease in blood flow. Note that changes following the drop in flow are the same as those occurring after an increase in metabolic rate (Fig. 14.9).

FIGURE 14.10 **Reactive hyperemia, a response to a reduction in blood flow.**

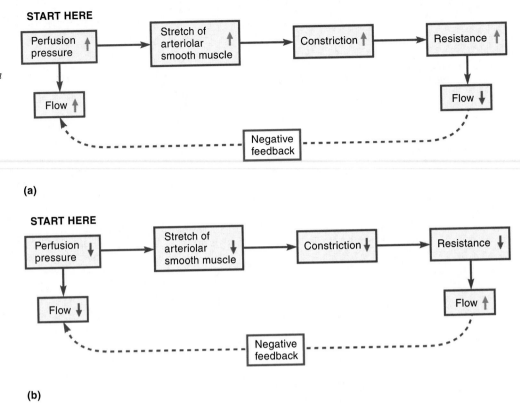

FIGURE 14.11 The myogenic response to changes in perfusion pressure. (a) *Changes occurring in response to a rise in perfusion pressure.* **(b)** *Changes occurring in response to a fall in perfusion pressure.*

then contract, which increases the arteriole's resistance and decreases the flow of blood through it (Figure 14.11a). A decrease in perfusion pressure (which causes blood flow to decrease) brings about the opposite response—vasodilation and an increase in blood flow (Figure 14.11b).

Myogenic control of vascular resistance in response to changes in perfusion pressure is clearly similar to the other intrinsic control mechanisms we have discussed in that it involves a negative feedback loop. Note, however, that unlike either active or reactive hyperemia, the variable that remains constant in myogenic control is blood flow. Local regulation that tends to keep blood flow constant is called *flow autoregulation.*

It can be difficult to ascertain whether flow autoregulation is due to a myogenic response or to changes in metabolite concentrations because the two processes are interconnected and tend to go in the same direction. For example, an increase in perfusion pressure causes a rise in tissue oxygen levels and a fall in metabolite concentrations (assuming no change in metabolic activity) because the rise in pressure causes blood flow to increase. By themselves, these changes in concentration cause vasoconstriction, which reduces blood flow. But the increase in perfusion pressure also increases stretch of arteriolar walls, which also induces vasoconstriction and a reduction in flow.

Regulation by Locally Secreted Chemical Messengers

Contractile activity of vascular smooth muscle is also affected by a variety of chemical substances, most of which are secreted by blood vessel endothelial cells or by cells in surrounding tissues (Table 14.2). One such substance is *nitric oxide,* which is released on a continual basis by endothelial cells in arterioles and acts on smooth muscle to promote vasodilation. Synthesis of nitric oxide is stimulated by other substances, such as *bradykinin* and *histamine,* which are produced by inflamed tissues. The resulting increase in blood flow accounts for the redness of inflamed areas. Another potent vasodilator is *prostacyclin,* a member of a family of locally acting chemical messengers called *eicosanoids* (see Chapter 5). Among the substances that promote vasoconstriction is *endothelin-1,* which is also produced by endothelial cells.

We have seen that an organ's vascular resistance can be affected by a variety of factors, both extrinsic (nerves and hormones) and intrinsic (local metabolites, chemical messengers, and arteriolar stretch). However, the relative importance of these factors differs from organ to organ for a number of reasons. Blood vessels in some organs are richly innervated by sympathetic nerves, for instance, whereas those in other organs are only sparsely innervated, if at all. In addition, the sensitivity of vascular smooth muscle to stretch or to particular chemicals varies from location to location. Table 14.3 summarizes the neural and local control of vascular resistance for selected systemic organs.

SUBSTANCE	SOURCE	EFFECT ON VASCULAR SMOOTH MUSCLE
Oxygen	Delivered to tissues by blood; consumed in aerobic metabolism	Vasoconstriction
Carbon dioxide	Generated in aerobic metabolism	Vasodilation
Potassium ions	Released from cells (particularly in muscle) as a result of repeated depolarization occurring during activity	Vasodilation (vasoconstriction at high concentrations)
Acids (hydrogen ions)	Generated during anaerobic metabolism (lactic acid) and by reaction of carbon dioxide with water (carbonic acid)	Vasodilation
Adenosine	Released by cells in certain tissues in response to hypoxia	Vasodilation
Nitric oxide	Released by endothelial cells on a continuous basis and in response to various chemical signals	Vasodilation
Bradykinin	Generated from a precursor protein (kininogen) by action of an enzyme (kallakrein) secreted by cells in certain tissues in response to various chemical signals	Vasodilation
Endothelin-1	Released by endothelial cells in response to various chemical signals and mechanical stimuli	Vasoconstriction
Prostacyclin	Released by endothelial cells in response to various chemical signals and mechanical stimuli	Vasodilation

Quick Test 14.3

1. Define the following terms: *ischemia, hyperemia, flow autoregulation, perfusion pressure.*

2. What is the difference between active hyperemia and reactive hyperemia?

3. To what does the term *myogenic regulation* refer?

4. If a substance is vasoactive and is produced in metabolic reactions, would you expect it to cause vasoconstriction or vasodilation if its concentration rises?

OTHER CARDIOVASCULAR REGULATORY PROCESSES

In addition to the intrinsic and extrinsic mechanisms described in the previous sections, other processes control cardiovascular function. Some of these, such as the cardiovascular response to exercise, act to alter the pattern of blood flow so that the body can adapt to a particular set of circumstances. Others, such as *sinus arrhythmia,* have no known functional significance. What follows is a sampling of these other regulatory processes.

TABLE 14.3 NEURAL AND LOCAL CONTROL OF VASCULAR RESISTANCE IN SELECTED SYSTEMIC ORGANS

ORGAN	NEURAL INFLUENCE ON VASCULAR RESISTANCE	LOCAL CONTROL OF VASCULAR RESISTANCE
Brain	Minor	Metabolic activity; flow autoregulation
Gastrointestinal system (gastrointestinal tract, liver, pancreas)	Sympathetic nerves (vasoconstriction)	Metabolic activity; neurotransmitters and hormones secreted within the gastrointestinal tract
Heart	Minor	Metabolic activity; flow autoregulation
Kidney	Sympathetic nerves (vasoconstriction)	Flow autoregulation
Skeletal muscle	Sympathetic nerves (vasoconstriction)	Metabolic activity
Skin	Sympathetic nerves (vasoconstriction)	Metabolic activity; chemicals released from sweat glands (bradykinin)

Respiratory Sinus Arrhythmia

Sinus arrhythmia is a rhythmic variation in heart rate in which inspiration is accompanied by increases in sympathetic activity and heart rate, whereas expiration is accompanied by increases in parasympathetic activity and a decrease in heart rate. The incompletely understood mechanism involves sympathetic and parasympathetic influences on the sinoatrial node. It is seen in only some people and appears to be more pronounced in children.

Respiratory sinus arrhythmia has at least one known practical consequence: Competitive archers and target shooters use it to aid their accuracy. By taking a deep breath and then breathing out while aiming, respiratory sinus arrhythmia slows the heart rate, enabling shooters to fire between heartbeats and thus avoid even a slight movement of the body that could throw off their aim.

Chemoreceptor Reflexes

As we see in Chapter 16, control of respiration is governed by *chemoreceptors* located in the carotid sinuses and the brain. These chemoreceptors are neurons specialized to monitor concentrations of oxygen and carbon dioxide in the blood. When oxygen levels fall or carbon dioxide rises, chemoreceptors stimulate breathing, so that more oxygen is brought into the lungs and more carbon dioxide is eliminated. These receptors also influence cardiovascular function, but the directions and sizes of these responses are difficult to predict, because some cardiovascular effects are primary (that is, due to stimulation of chemoreceptors themselves), whereas others are secondary (that is, due to changes in respiration triggered by stimulation of chemoreceptors).

One of the primary cardiovascular effects of chemoreceptor stimulation when arterial oxygen levels fall is to trigger a decrease in heart rate and an increase in peripheral resistance. These responses seem counterproductive, because when the concentration of oxygen in the blood falls, blood flow to tissues should be *increased* in order to maintain the same rate of oxygen delivery. However, when viewed from another perspective, these responses are entirely appropriate. Due to the decrease in heart rate, oxygen consumption by the heart muscle itself is diminished, which tends to conserve oxygen. Although the fall in heart rate also tends to reduce mean arterial pressure, the increase in peripheral resistance offsets this effect. As a result, blood flow to the brain is maintained at normal or near-normal levels.

The chemoreceptor-mediated increase in peripheral resistance may also be adaptive in another way: When arterial oxygen levels fall, the concentration of oxygen falls in tissues all over the body. Given that vascular smooth muscle is sensitive to local oxygen concentrations, such a fall in arterial oxygen levels should trigger vasodilation in many tissues, leading to a potentially dangerous drop in peripheral resistance and mean arterial pressure. By stimulating an increase in peripheral resistance, chemoreceptor reflexes work to protect against this possibility.

Thermoregulatory Responses

The ability to control heat loss through the skin is an essential component of the body's ability to regulate its own temperature. As we saw in Chapter 1, the body's response to changes in temperature is mediated by a *thermoregulatory center* in the hypothalamus that receives input from *thermoreceptors*, heat-sensitive neurons found at various locations throughout the body. The thermoregulatory center receives from these thermoreceptors information concerning skin and core temperatures.

Under normal conditions, there is a significant level of activity in sympathetic nerves that project to blood vessels in the skin. When the body's *heat load* increases (that is, when the body's heat content increases), the resulting rise in temperature decreases the level of sympathetic activity in nerves supplying the skin, which induces relaxation of vascular smooth muscle. (Over time, other factors that promote vasodilation are thought to come into play.) As a result, blood vessels dilate, the skin's vascular resistance decreases, and blood flow to skin increases. In addition, under conditions of heat stress, sweat glands produce bradykinin, whose vasodilatory effects tend to decrease vascular resistance in the skin. As a result of the increase in blood flow, the rate of heat loss through the skin increases, which helps to counteract the rise in body temperature. A decrease in the heat load has the opposite effect: Increased activity in the skin's sympathetic nerves stimulates contraction of smooth muscle in the blood vessels, which increases the resistance of blood vessels in the skin and decreases the blood flow through the skin. As a consequence, blood is diverted away from the skin and toward the deeper structures of the body, so that heat is lost at a lower rate.

Because the influence of body temperature on skin blood vessels takes precedence over other reflex controls in most circumstances, changes in skin resistance can sometimes be dangerously maladaptive. For instance, when a person experiences severe blood loss on a very hot day, baroreceptor reflexes trigger constriction of blood vessels throughout the body, including those in the skin. If that person then becomes overheated, the thermoregulatory center takes control, inducing dilation of skin blood vessels to increase heat loss. The resulting decrease in vascular resistance in skin vessels tends to cause a decrease in the total peripheral resistance. As a consequence, arterial pressure may fall, counteracting the baroreceptor reflex, which should be acting to raise total peripheral resistance in order to maintain arterial pressure.

Responses to Exercise

As we saw in Chapter 13, exercise is accompanied by profound alterations in cardiovascular function. Figure 14.12 shows that jogging increases cardiac output from the resting value of 5 liters per minute to over 11 liters per minute; in highly trained athletes, cardiac output can reach 35 liters per minute. Heart rate also rises from the resting average of 72 beats per minute to about 135 beats per minute. In addition to these changes in cardiac function, dramatic changes occur in blood flow: Flow to skeletal muscles (and to cardiac muscle and skin as well) increases while (as we saw in Chapter 13) flow to the liver and the gastrointestinal tract decreases. Clearly, these

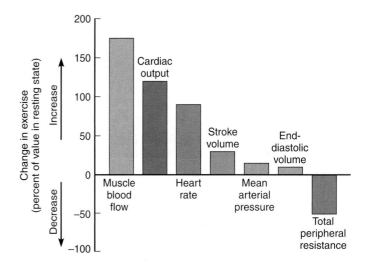

FIGURE 14.12 Cardiovascular responses to light exercise. *Note that the decrease in TPR allows MAP to remain nearly constant despite the huge increase in CO.*

changes are beneficial in several ways: (1) Delivery of oxygen and nutrients to cardiac muscle and active skeletal muscles is increased, which is appropriate, given the increase in metabolic activity in these organs; (2) oxygen and nutrients are conserved by curtailing delivery to tissues for which the need for nutrients is not as acute; and (3) increased blood flow to the skin aids the body in getting rid of excess heat generated during exercise.

In exercise, the drop in vascular resistance in skin and muscle is not quite compensated for by the increase in the resistance of other organs, so total peripheral resistance drops. Blood pressure *rises* slightly, however, due to the increase in cardiac output. Note in Figure 14.12 that stroke volume increases markedly despite the fact that only a modest increase in end-diastolic volume occurs (in light exercise). This indicates that the increase in stroke volume is due not to the Starling effect, but to an increase in ventricular contractility resulting from an increase in sympathetic nervous activity and increases in the levels of circulating epinephrine.

What triggers these responses to exercise? In the central nervous system, cortical and limbic regions of the brain exert a direct influence on the output of sympathetic and parasympathetic neurons, resulting in increased sympathetic activity and decreased parasympathetic activity to the heart that account for the increase in heart rate and ventricular contractility. In addition, there is an increase in sympathetic activity in the digestive system and other organs (resulting in vasoconstriction) and a *decrease* in sympathetic activity in the skin (which promotes vasodilation). This change in sympathetic input to the skin is a thermoregulatory response triggered by the rise in body temperature that accompanies exercise. Vasodilation in the

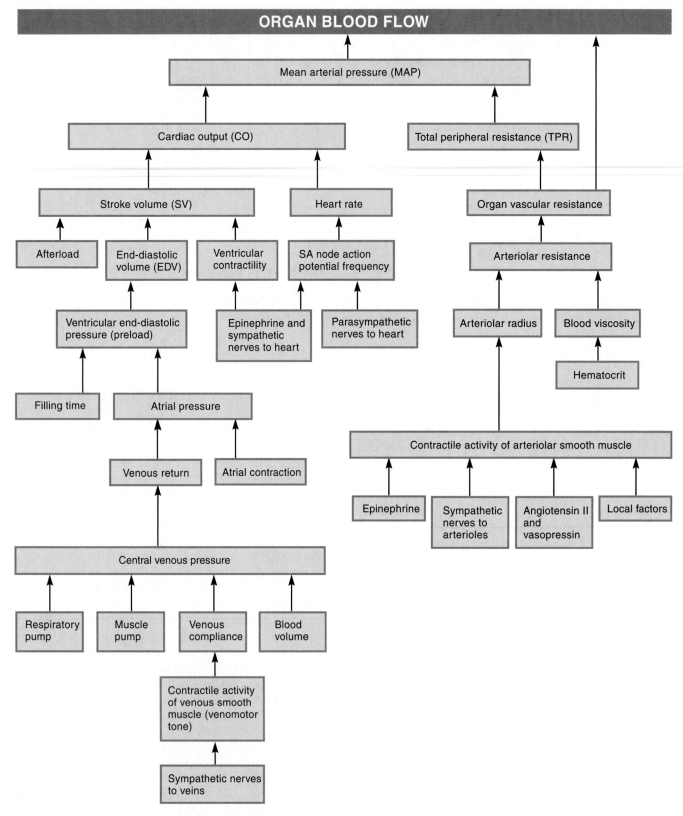

FIGURE 14.13 The relationships of the factors that affect the delivery of blood to systemic organs by the cardiovascular system.

skin is also promoted by bradykinin released from sweat glands, whose activity increases in exercise. The changes in sympathetic activity are facilitated by the medulla oblongata, whose output to cardiovascular targets is also modulated by input from various types of receptors in muscles, including chemoreceptors responsive to local chemical factors and mechanoreceptors sensitive to muscular physical activity. The increase in blood flow to skeletal and cardiac muscles is largely a result of the action of local chemical factors on vascular smooth muscle.

Other mechanisms also promote increased venous return in exercise, thereby facilitating the rise in cardiac output. One such mechanism is the skeletal muscle pump; another is the respiratory pump, whose activity increases in exercise because breathing becomes faster and deeper. A third mechanism is an increase in venomotor tone, which results from an increase in the activity of sympathetic venoconstrictor nerves.

Quick Test 14.4

1. What is respiratory sinus arrhythmia?

2. What is a chemoreceptor? What cardiovascular changes are triggered when chemoreceptors detect a drop in arterial oxygen levels?

3. When body temperature rises, does vascular resistance in the skin increase or decrease? How does this change help the body regulate its temperature?

4. Name two organs in which blood flow increases during exercise.

SUMMARY OF FACTORS INFLUENCING ORGAN BLOOD FLOW

Figure 14.13 is a chart summarizing the many concepts we have encountered in this and the two preceding chapters. This chart can help you recall and understand the many factors that bear on the cardiovascular system's primary function, which is to maintain adequate blood flow to the body's organs. The chart is also an emphatic reminder that all these factors are interconnected, a truth it illustrates in a way that no words ever could. The chart can help you understand relationships between certain variables. For example, the fact that arrows from two boxes labeled *heart rate* and *stroke volume* lead to the box labeled *cardiac output* should remind you that cardiac output is determined by these two variables (CO = HR × SV), a concept discussed in Chapter 12. If you have difficulty understanding any portion of the chart, review the appropriate sections in the three cardiovascular chapters.

SYSTEMS INTEGRATION

CARDIOVASCULAR SYSTEM

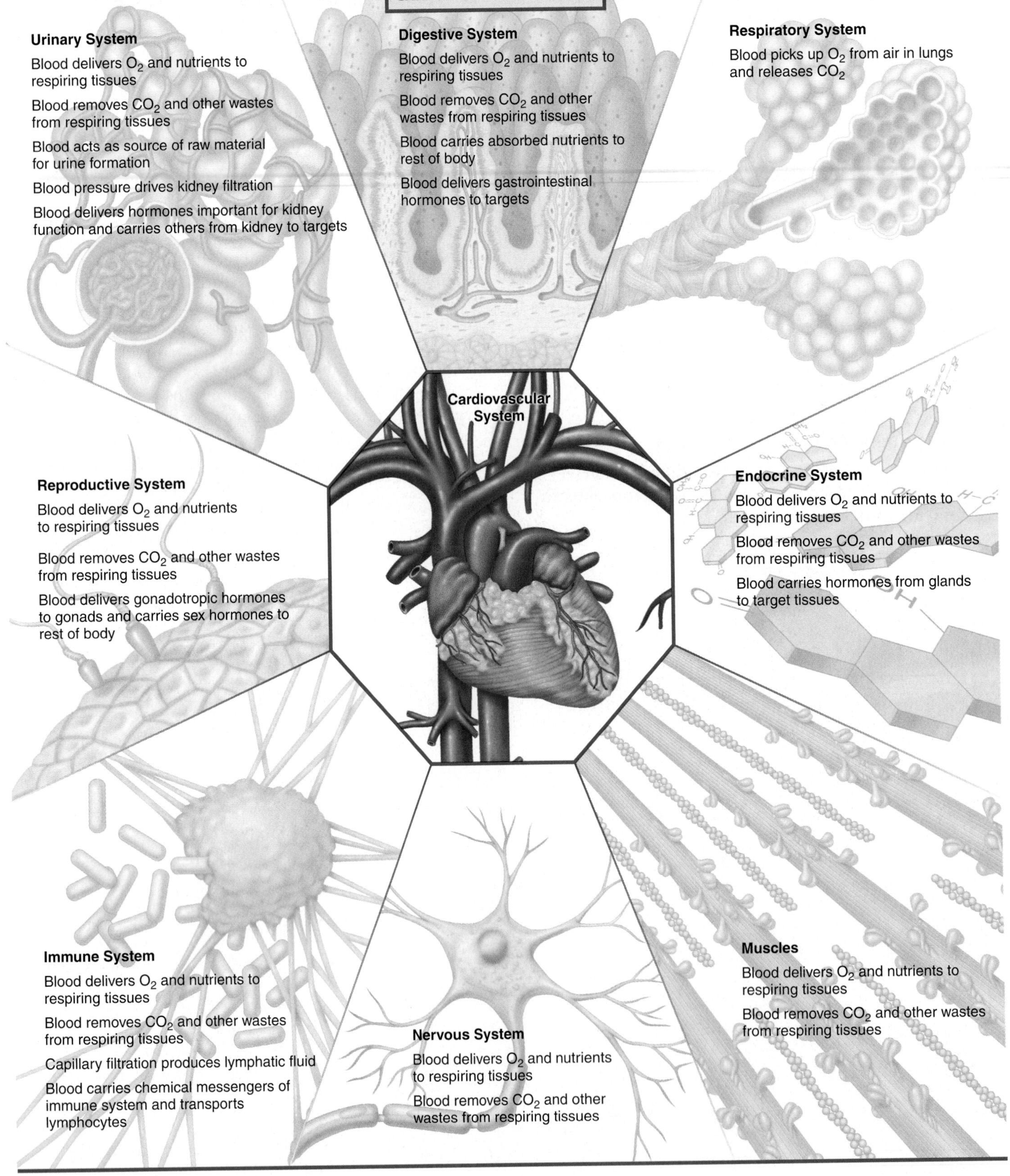

Urinary System
Blood delivers O_2 and nutrients to respiring tissues
Blood removes CO_2 and other wastes from respiring tissues
Blood acts as source of raw material for urine formation
Blood pressure drives kidney filtration
Blood delivers hormones important for kidney function and carries others from kidney to targets
Digestive System
Blood delivers O_2 and nutrients to respiring tissues
Blood removes CO_2 and other wastes from respiring tissues
Blood carries absorbed nutrients to rest of body
Blood delivers gastrointestinal hormones to targets
Respiratory System
Blood picks up O_2 from air in lungs and releases CO_2
Cardiovascular System
Reproductive System
Blood delivers O_2 and nutrients to respiring tissues
Blood removes CO_2 and other wastes from respiring tissues
Blood delivers gonadotropic hormones to gonads and carries sex hormones to rest of body
Endocrine System
Blood delivers O_2 and nutrients to respiring tissues
Blood removes CO_2 and other wastes from respiring tissues
Blood carries hormones from glands to target tissues
Immune System
Blood delivers O_2 and nutrients to respiring tissues
Blood removes CO_2 and other wastes from respiring tissues
Capillary filtration produces lymphatic fluid
Blood carries chemical messengers of immune system and transports lymphocytes
Nervous System
Blood delivers O_2 and nutrients to respiring tissues
Blood removes CO_2 and other wastes from respiring tissues
Muscles
Blood delivers O_2 and nutrients to respiring tissues
Blood removes CO_2 and other wastes from respiring tissues

Extrinsic Control of Cardiovascular Function: Regulation of Mean Arterial Pressure, p. 443

The cardiovascular system's primary function is to provide adequate blood flow to the body's organs and tissues. Mean arterial pressure is controlled by both short-term and long-term extrinsic regulatory mechanisms. Whereas short-term regulation is achieved through neural and hormonal control of cardiovascular function, long-term regulation is achieved through control of blood volume, which involves the kidneys. Important in the regulation of MAP are arterial baroreceptors located in the aortic arch and carotid arteries. These receptors monitor systemic arterial pressure and relay information to the medulla oblongata, which controls autonomic output to the heart and vasculature. Autonomic control of MAP is accomplished through (a) sympathetic and parasympathetic input to the SA node, which controls heart rate; (b) sympathetic input to the myocardium, which controls ventricular contractility and stroke volume; and (c) sympathetic input to arteriolar smooth muscle in most tissues, which regulates total peripheral resistance. Arterial baroreceptors are aided by volume receptors in the heart and large veins that monitor venous pressure. Baroreceptors also influence the secretion of several hormones that influence cardiovascular function, including epinephrine, vasopressin, and angiotensin II.

IP Cardiovascular, Blood Pressure Regulation, pp. 1, 3–7, 8–12, 15–20

Intrinsic Control of Cardiovascular Function: Regulation of Blood Flow Distribution to Organs, p. 455

Regulation of the distribution of blood flow among the various systemic organs is achieved through intrinsic control of organ vascular resistance. The resistance of an organ or tissue can change in response to variations in the metabolic activity of that organ or tissue because arteriolar smooth muscle is sensitive to local concentrations of chemicals produced or consumed in metabolism, including oxygen and carbon dioxide. Chemical changes associated with increased metabolic activity lead to vasodilation, decreased resistance, and increased blood flow (active hyperemia). The resistance of an organ or tissue can also change in response to local variations in blood flow, even in the absence of any change in metabolic activity. If blood flow becomes insufficient to meet metabolic demands (ischemia), local mechanisms induce vasodilation and a resultant increase in blood flow (reactive hyperemia). In those tissues in which vascular smooth muscle is responsive to stretch, an increase in perfusion pressure causes arterioles to stretch, which stimulates vasoconstriction and a reduction in blood flow. A response of this type is termed a myogenic response.

IP Cardiovascular, Autoregulation and Capillary Dynamics, pp. 1–13

Other Cardiovascular Regulatory Processes, p. 461

In addition to the mechanisms just described, other processes regulate cardiovascular function in specific situations. For example, cardiovascular function is influenced by activity in arterial chemoreceptors, which monitor concentrations of oxygen and carbon dioxide in arterial blood. Regulation of blood flow to the skin, which is controlled by sympathetic nerves to skin blood vessels, is important in body temperature regulation, which is controlled by a thermoregulatory center in the hypothalamus. Cardiovascular responses to exercise are largely achieved through changes in the activity of autonomic nerves to the heart and blood vessels, changes that are orchestrated by cortical and limbic brain regions. Blood flow to the heart and to exercising skeletal muscle is also regulated by local factors operating within these tissues.

EXERCISES

Multiple-Choice Questions

1. A person stands up abruptly after lying down. Before any reflex mechanisms are activated, you would expect
   a) a decrease in central venous pressure to occur.
   b) a decrease in venous return to occur.
   c) a decrease in end-diastolic volume to occur.
   d) all of the above

2. The situation in Question 1 sets in motion a reflex that causes the heart rate to speed up. This reflex is triggered by
   a) a fall in heart rate.
   b) a decrease in mean arterial pressure.
   c) an increase in total peripheral resistance.
   d) an increase in preload.

3. In the situation in Question 1, the reflex should trigger
   a) an increase in parasympathetic nervous activity.
   b) an increase in heart rate.

c) a decrease in sympathetic nervous activity.

d) a decrease in venomotor tone.

4. If arterial pressure is elevated, baroreceptor signals trigger which of the following responses?

a) a rise in vasopressin secretion

b) a fall in plasma angiotensin II levels

c) increased activity in sympathetic vasoconstrictor nerves

d) an increase in epinephrine secretion

5. The baroreceptor reflex

a) is an example of intrinsic control of vascular resistance.

b) serves to maintain blood flows to all organs at nearly constant levels.

c) serves to maintain mean arterial pressure at a nearly constant level.

d) triggers a rise in arterial pressure following a sudden increase in cardiac output.

6. During exercise, carbon dioxide produced by muscle cells causes vasodilation in skeletal muscle. This is an example of

a) active hyperemia.

b) reactive hyperemia.

c) flow autoregulation.

d) extrinsic control of vascular resistance.

7. During exercise, reduction in blood flow to gastrointestinal organs results from

a) an increase in the activity of sympathetic nerves to blood vessels in those organs.

b) an increase in the activity of parasympathetic neurons to blood vessels in those organs.

c) a fall in mean arterial pressure.

d) a decrease in cardiac output.

8. Which of the following observations best demonstrates that reflex responses to hemorrhage trigger changes in the vascular resistance of organs?

a) Reflex responses cause mean arterial pressure to return to near normal levels.

b) In moderate hemorrhage, blood flow to the brain may be reduced only slightly.

c) Following hemorrhage, cardiac output decreases, but heart rate increases.

d) Reflex responses trigger a decrease in blood flow to some, but not all, organs.

9. Of the following substances, which has an effect on vascular resistance that is opposite to the effect of the others?

a) vasopressin

b) bradykinin

c) norepinephrine

d) angiotensin II

10. Which of the following observations indicates that myogenic responses of smooth muscle in blood vessels are causing changes in the vascular resistance of a tissue?

a) When the tissue perfusion pressure increases, blood flow decreases.

b) When the tissue perfusion pressure increases, blood flow increases.

c) Blood flow increases following a period of ischemia in that tissue.

d) Blood flow increases following an increase in metabolic activity in that tissue.

Objective Questions

1. A drop in arterial blood pressure triggers (an increase/a decrease) in sympathetic nervous activity.

2. A drop in arterial blood pressure triggers (an increase/a decrease) in vasopressin secretion.

3. When venous baroreceptors detect a fall in pressure, arterial baroreceptors tend to detect a change in the opposite direction. (true/false)

4. An increase in blood flow in response to a rise in metabolic activity is referred to as _____ hyperemia.

5. In certain tissues, a rise in perfusion pressure causes vascular smooth muscle to contract in response to stretch. This is known as a _____ response.

6. If blood flow to a tissue increases in the absence of any change in metabolic activity, the local oxygen concentration (increases/decreases).

7. If blood flow to a tissue increases in the absence of any change in metabolic activity, the local carbon dioxide concentration (increases/decreases).

8. When a fall in blood volume triggers a baroreceptor reflex, the reflex acts to maintain nearly normal levels of blood flow to all systemic tissues. (true/false)

9. Activity in parasympathetic neurons to the heart tends to promote decreased (heart rate/myocardial contractility).

10. Epinephrine always affects vascular smooth muscle in the same way that norepinephrine does. (true/false)

11. Angiotensin II has the same general effect on vascular resistance as vasopressin. (true/false)

Essay Questions

1. Describe how a drop in blood volume resulting from dehydration will affect mean arterial pressure, and then describe how the baroreceptor reflex should work to restore the pressure to normal.

2. Explain the concept of flow autoregulation and how it differs from other forms of intrinsic control of vascular resistance. Explain how both the myogenic response and metabolic chemical factors can produce the phenomenon of flow autoregulation. Use the example of a compensatory decrease in flow following an increase in perfusion pressure.

3. Explain how vascular resistance is influenced by the actions of alpha and beta adrenergic receptors. On the basis of differences between these receptors, explain how it is possible for sympathetic nerve activity and circulating epinephrine to affect vascular resistance in opposite ways.

4. Describe the changes in cardiovascular function that occur in response to exercise. Explain how these changes help to adapt the body to the particular stresses that occur in exercise.

Find the answers to these exercises, and additional study tools, at the Physiology Place (www.physiologyplace.com).

15

The Respiratory System:
Breathing Mechanics

OBJECTIVES

- Compare internal respiration to external respiration, and describe the processes occurring in each.

- Describe the major structures of the respiratory system, and list the functions of each.

- Describe the anatomy of the respiratory membrane, and explain how the structure facilitates the diffusion of gases.

- Describe the anatomy of alveoli. Explain the roles of type I cells, type II cells, and alveolar macrophages in respiratory function.

- Explain the function of pulmonary surfactant.

- Describe the mechanics of breathing. Name the muscles of respiration. List the different pulmonary pressures and explain their roles in ventilation.

- Describe the roles of lung compliance and airway resistance in ventilation.

- List the different lung volumes and capacities. Explain the clinical applications of lung volumes, forced vital capacity, and forced expiratory volume.

CHAPTER OUTLINE

Overview of Respiratory Function   470

Anatomy of the Respiratory System   470

Forces for Pulmonary Ventilation   475

Factors Affecting Pulmonary Ventilation   484

Clinical Significance of Respiratory Volumes and Air Flows   488

Above: Scanning electron micrograph of cilia

We all know that the depth of breathing changes with changing circumstances. When you exercise, for example, you breathe more deeply than while resting, and when something startles you, you might hold your breath for a moment before breathing a deep "sigh of relief." In either case, a deeper breath brings more air into the lungs, where oxygen in inhaled air moves into the blood, and where carbon dioxide moves out of the blood and leaves the body in exhaled air. In this chapter we explore breathing mechanics, the factors that influence *pulmonary ventilation*—the movement of air into and out of the lungs that facilitates the exchange of oxygen and carbon dioxide between the air and the blood.

After a brief overview of respiratory function we examine the anatomy of the respiratory system, and consider how its various structures carry out their functions. Then we explore how the muscles of respiration produce several different pulmonary pressures, and the roles of those pressures in pulmonary ventilation. Next we examine the factors that affect the rates of air flow during pulmonary ventilation. Finally, we conclude by considering the bases for and clinical significances of various lung volumes and capacities.

OVERVIEW OF RESPIRATORY FUNCTION

The respiratory system is so named because its function is **respiration,** the process of gas exchange. This exchange of gases occurs at two levels, termed internal respiration and external respiration (Figure 15.1). Whereas **internal respiration** (or cellular respiration) refers to the use of oxygen within mitochondria to generate ATP by oxidative phosphorylation, and the production of carbon dioxide as a waste product, **external respiration** refers to the exchange of oxygen and carbon dioxide between the atmosphere and body tissues, which involves both the respiratory and circulatory systems. In this chapter we focus on certain aspects of external respiration.

External respiration encompasses four processes:

1. **Pulmonary ventilation,** the movement of air into the lungs (inspiration) and out of the lungs (expiration) by bulk flow

2. Exchange of oxygen and carbon dioxide between lung air spaces and blood by diffusion

3. Transportation of oxygen and carbon dioxide between the lungs and body tissues by the blood

4. Exchange of oxygen and carbon dioxide between the blood and tissues by diffusion

Whereas this chapter and the next address all four aspects of external respiration, our focus in this chapter is on the first process, pulmonary ventilation.

In addition to its main function—respiration—the respiratory system also performs several other functions, including (1) contributing to the regulation of acid-base balance in the blood, (2) enabling vocalization, (3) participating in defense against pathogens and foreign particles in the airways, (4) providing a route for water and heat losses (via the expiration of air that was moistened and warmed during inspiration), (5) enhancing venous return (through the respiratory pump), and (6) activating certain plasma proteins as they pass through the pulmonary circulation.

ANATOMY OF THE RESPIRATORY SYSTEM

The major organs of the respiratory system are the *lungs,* which are located in the thoracic cavity. Each lung is divided into lobes; the right lung has three lobes and the left lung has two lobes. Air gets into and out of the lungs by way of the *upper airways* and a network of tubes forming a system of passageways called the *respiratory tract* (Figure 15.2). The following sections describe how these various components of the respiratory system are specialized to carry out their specific functions.

Upper Airways

The term **upper airways** refers to air passages in the head and neck. Air enters the *nasal cavity* and/or the *oral cavity*, both of which lead to the **pharynx,** a muscular tube that serves as a common passageway for both air and food. After the pharynx, the passageways for food and air diverge; food enters the esophagus, a muscular tube leading to the stomach, whereas air enters the first structure in the respiratory tract, the larynx.

The Respiratory Tract

The **respiratory tract** includes all air passageways leading from the pharynx to the lungs (including those present within the lungs themselves), starting with the

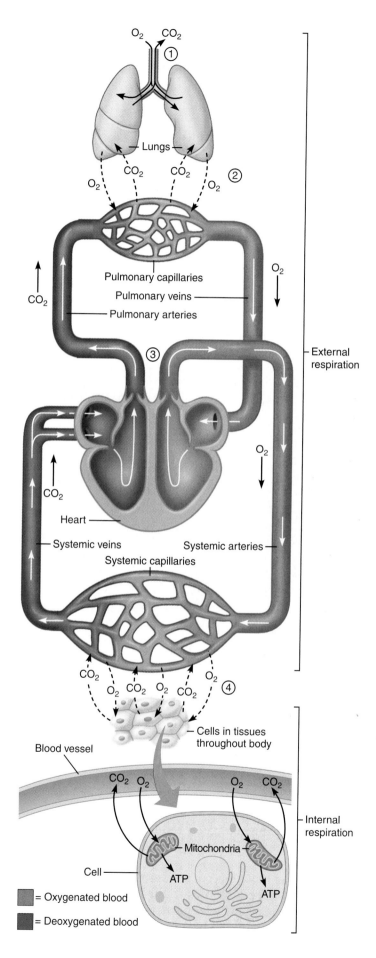

O$_2$ CO$_2$
①

Lungs

CO$_2$ CO$_2$
O$_2$ O$_2$
②

Pulmonary capillaries

O$_2$

CO$_2$

Pulmonary veins

Pulmonary arteries

O$_2$

③

External respiration

O$_2$

Heart

CO$_2$

Systemic veins

Systemic arteries

Systemic capillaries

CO$_2$ O$_2$
O$_2$ CO$_2$ O$_2$ CO$_2$
④

Cells in tissues throughout body

Blood vessel

CO$_2$ O$_2$ O$_2$ CO$_2$

Internal respiration

Mitochondria

Cell

ATP   ATP

= Oxygenated blood

= Deoxygenated blood

FIGURE 15.1 Relationship between external respiration and internal respiration. *In external respiration,* ① *air moves between the atmosphere and the lungs,* ② *oxygen and carbon dioxide are exchanged between lung tissue and the blood,* ③ *oxygen and carbon dioxide are transported in the blood, and* ④ *oxygen and carbon dioxide are exchanged between systemic tissues and the blood. Internal respiration is the use of oxygen and production of carbon dioxide by cells, primarily within the mitochondria.*

What metabolic pathway in the mitochondria utilizes oxygen?

larynx, a tube held open by *cartilage* (a dense connective tissue) in its walls. To keep food from entering the respiratory tract, the opening to the larynx, called the **glottis,** is covered by a flap of tissue called the **epiglottis,** which during swallowing is forced down over the glottis and prevents food or water from entering the larynx. The larynx houses the vocal cords (or vocal folds), which generate sounds by vibrating when air passes over them.

The respiratory tract can be functionally divided into two components: a conducting zone and a respiratory zone (Figure 15.3). The **conducting zone,** the upper part of the respiratory tract, functions in conducting air from the larynx to the lungs, whereas the **respiratory zone,** the lowermost part of the respiratory tract, contains the sites of gas exchange within the lungs. The primary anatomical difference between the conducting and respiratory zones that determines whether or not gas exchange occurs is in the *thickness* of the walls surrounding the air spaces; only air spaces with sufficiently thin walls can participate in gas exchange.

The Conducting Zone

After the larynx, the next component of the respiratory tract is the **trachea,** a tube about 2.5 cm in diameter and 10 cm long that runs parallel with and anterior to the esophagus. Unlike the esophagus, which is collapsed except during swallowing, the trachea stays open because the fronts and sides of its wall contain 15–20 C-shaped bands of cartilage that provide structural rigidity. This rigidity is important because without it, the decline in air pressure that occurs in the trachea during inspiration would collapse it and cut off the flow of air. The absence of cartilage at the posterior side of the trachea allows the esophagus to expand with the passage of food.

After it enters the thoracic cavity, the trachea divides into left and right **bronchi** (singular: *bronchus*) that conduct air to each lung. Like the trachea, the bronchi contain cartilage; however, the cartilage forms rings around the entire circumference of the bronchus. Within each lung, the bronchi divide into smaller tubes called **secondary bronchi;** three secondary bronchi conduct air to

The electron transport chain

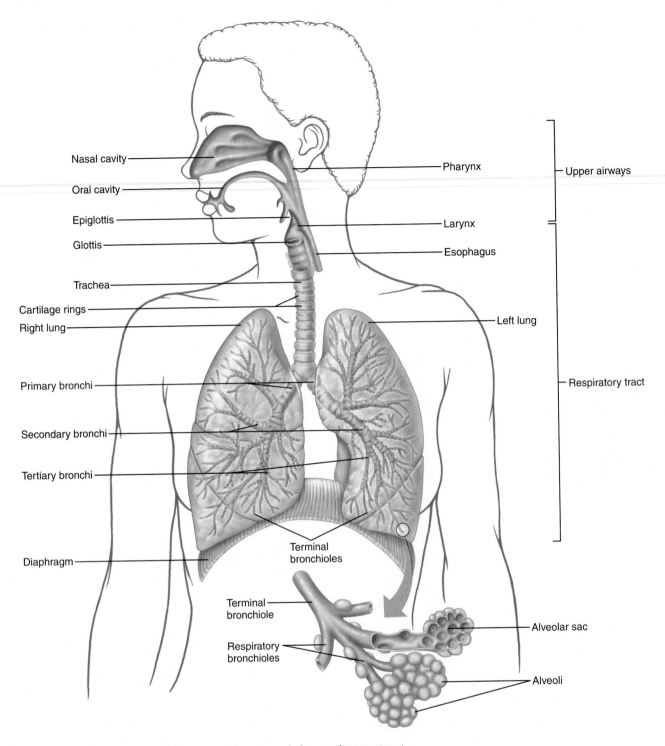

FIGURE 15.2 Anatomy of the upper airways and the respiratory tract.

Labels in figure:
Nasal cavity
Oral cavity
Epiglottis
Glottis
Trachea
Cartilage rings
Right lung
Primary bronchi
Secondary bronchi
Tertiary bronchi
Diaphragm
Pharynx
Larynx
Esophagus
Left lung
Upper airways
Respiratory tract
Terminal bronchioles
Terminal bronchiole
Respiratory bronchioles
Alveolar sac
Alveoli

the lobes of the right lung, and two secondary bronchi conduct air to the lobes of the left lung. The cartilage in secondary bronchi is less abundant than that in the primary bronchi and occurs as plates. Each secondary bronchus divides into smaller *tertiary bronchi,* which in turn divide into successively smaller bronchi such that approximately 20–23 orders of branching occurs. The ex-

tensive branching ultimately results in approximately 8 million tubules, the smallest being less than 0.5 mm in diameter.

Once the tubules become less than 1 mm in diameter, they are called **bronchioles** ("little bronchi"). Unlike the larger bronchi, bronchioles have no cartilage and are thus capable of collapsing. To help prevent collapse, the walls

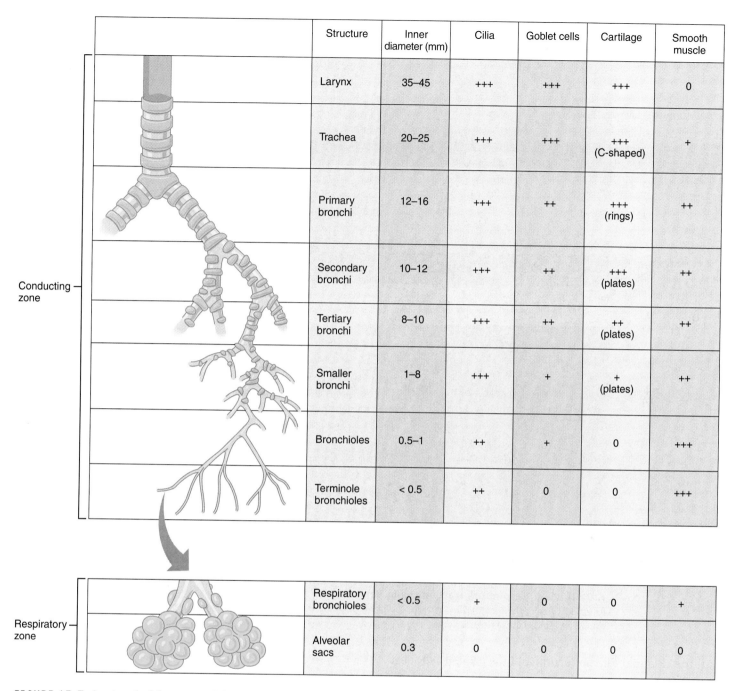

Structure	Inner diameter (mm)	Cilia	Goblet cells	Cartilage	Smooth muscle
Larynx	35–45	+++	+++	+++	0
Trachea	20–25	+++	+++	+++ (C-shaped)	+
Primary bronchi	12–16	+++	++	+++ (rings)	++
Secondary bronchi	10–12	+++	++	+++ (plates)	++
Tertiary bronchi	8–10	+++	++	++ (plates)	++
Smaller bronchi	1–8	+++	+	+ (plates)	++
Bronchioles	0.5–1	++	+	0	+++
Terminole bronchioles	< 0.5	++	0	0	+++
Respiratory bronchioles	< 0.5	+	0	0	+
Alveolar sacs	0.3	0	0	0	0

Conducting zone

Respiratory zone

FIGURE 15.3 Anatomical features of the conducting and respiratory zones of the respiratory tract. *0 indicates not present, + indicates sparse, ++ indicates present, and +++ indicates abundant.*

of bronchioles contain elastic fibers. The bronchioles divide further into **terminal bronchioles,** the smallest component of the conducting zone.

The primary function of the conducting zone is to provide a passageway through which air can enter and exit the respiratory zone, where gas exchange occurs. No gas exchange occurs in the conducting zone. The conducting zone holds approximately 150 ml of air and is considered "dead space," because the air does not participate in gas exchange with blood. The functional significance of the dead space is described later. As air travels

through the conducting zone, its temperature is adjusted to body temperature and humidified to keep the respiratory tract moist.

The conducting zone is lined by an epithelium that changes in composition as the tubules become smaller in diameter. The epithelium lining the larynx and trachea (and to a lesser extent, the bronchi) contains numerous **goblet cells;** also abundant in the epithelium throughout the conducting zone are **ciliated cells** (Figure 15.4). Goblet cells secrete a viscous fluid called *mucus,* which coats the airways and traps foreign particles in inhaled

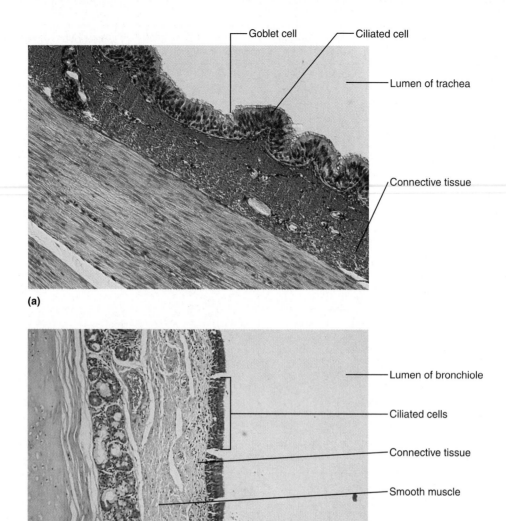

Goblet cell Ciliated cell

Lumen of trachea

Connective tissue

(a)

Lumen of bronchiole

Ciliated cells

Connective tissue

Smooth muscle

(b)

FIGURE 15.4 Respiratory tract epithelia. (a) *A photomicrograph of tracheal epithelium, located in the conducting zone.* **(b)** *A photomicrograph of respiratory bronchiolar epithelium, located in the respiratory zone.*

air; the cilia (hairlike projections) of the ciliated cells beat in a whiplike fashion to propel the mucus containing the trapped particles up toward the glottis and then into the pharynx, where the mucus is then swallowed. This process, called the *mucus escalator*, prevents mucus from accumulating in the airways and clears trapped foreign matter. Accumulation of mucus in the airways increases the likelihood of infections such as bronchitis and pneumonia, because it promotes retention and growth of bacteria. Because cilia are easily paralyzed by tobacco smoke, smoking disables the mucus escalator, so that mucus and trapped debris accumulate in the airways and can only be cleared by coughing; this is one reason for the familiar "smoker's cough." At levels below the bronchioles, phagocytic cells called *macrophages* engulf foreign matter in the interstitial space and on the surface of the epithelium.

In addition to changes in the epithelium, other tissue changes occur as the airways become smaller. As previously noted, cartilage is plentiful in the walls of the trachea and bronchi, but it becomes less abundant as the diameter of the bronchi decreases, until in the bronchioles it is completely absent. Smooth muscle is sparse in the trachea and bronchi but increases in abundance as the airways become smaller. The lack of cartilage and the presence of circular smooth muscle within the bronchioles enable these airways to change their diameter; such changes alter the resistance to air flow, just as the actions of circular smooth muscle in arterioles enables them to alter resistance to blood flow.

The Respiratory Zone

We saw in Chapter 4 that the rate at which substances diffuse across a membrane increases as membrane surface area increases and as membrane thickness decreases. The arrangement of structures in the respiratory zone, the site of gas exchange, maximizes surface area and minimizes thickness, such that the diffusion of oxygen and carbon dioxide between air and blood is facilitated.

Past the site where the terminal bronchioles of the conducting zone branch, the respiratory zone begins (Figure 15.5). The first respiratory zone structures, **respiratory bronchioles,** terminate in **alveolar ducts,** which lead to **alveoli,** the primary structures where gas exchange occurs (Figure 15.5a). Most alveoli occur in clusters called **alveolar sacs,** which resemble clusters of grapes; some alveoli open off of respiratory bronchioles.

Alveolar structure facilitates diffusion of gases between blood and air. The wall of an alveolus consists primarily of a single layer of epithelial cells called **type I cells** overlying a basement membrane (Figure 15.5c). Recall from Chapter 13 that a capillary wall consists of a single layer of endothelial cells and an underlying basement membrane. In many places in the lungs, the alveolar epithelial cells and the endothelial cells of the nearby capillaries are so close together that their basement membranes are fused (Figure 15.5d). Together, the capillary and the alveolar wall form a barrier, called the **respiratory membrane,** that separates air from blood. The thinness of the respiratory membrane—only about 0.2 micron thick (or 0.2 millionth of a meter)—is essential for efficient gas exchange.

In addition to the thinness of the respiratory membrane, the abundance of alveoli and capillaries also facilitates the diffusion of gases. The approximately 300 million alveoli in the lungs have a total surface area of approximately 100 square meters, which is about the size of a tennis court. Because these alveoli have a supply of capillaries that is so rich, many physiologists think of the pulmonary vasculature not in terms of arterioles and capillary beds, but instead liken it to a "sheet of blood" surrounding the alveoli.

Also located in the alveoli are *type II alveolar cells* (described later) and **alveolar macrophages,** which engulf foreign particles and pathogens inhaled into the lungs. These macrophages are cells that are free to roam around in the alveoli by amoeboid movements. Dead macrophages are moved from the alveoli into the conducting zone where the mucus escalator carries the macrophages to the pharynx, so that they can be swallowed with mucus.

Adjacent alveoli are not completely independent structures. Because they are connected by **alveolar pores**, air flows between alveoli, allowing equilibration of pressure within the lungs.

Structures of the Thoracic Cavity

The lungs are located within the thoracic cavity. The **chest wall** is composed of structures that protect the lungs (Figure 15.6): the rib cage (consisting of 12 pairs of ribs), the sternum (breastbone), the thoracic vertebrae, and associated muscles and connective tissue (primarily hyaline cartilage). Muscles of the chest wall, which are responsible for breathing, are the **internal** and **external intercostals,** located between the ribs, and the dome-shaped **diaphragm,** which seals off the lower end of the chest wall and separates the thoracic and abdominal cavities. Muscles and connective tissue in the neck close off the chest wall at its upper end. Because the chest wall forms a continuous barrier around the lungs, the compartment enclosing the lungs is airtight.

The interior surface of the chest wall and the exterior surface of the lungs are lined by a membrane called the **pleura,** which is composed of a layer of epithelial cells and connective tissue; each lung is surrounded by a separate **pleural sac** (see Figure 15.6). The side of the pleural sac attached to the lung tissue is called the *visceral pleura;* the side attached to the chest wall is called the *parietal pleura.* Between the two pleurae is a very thin compartment called the **intrapleural space,** which is filled with a small volume (approximately 15 milliliters) of **intrapleural fluid.** Because intrapleural fluid is mostly water, which is incompressible, its volume remains virtually constant, even when the chest wall expands and contracts.

Quick Test 15.1

1. Arrange the following structures in order, so that they reflect the path of air flow into the lungs: respiratory bronchioles, alveolar sacs, trachea, glottis, terminal bronchioles, bronchi, alveolar ducts, larynx, bronchioles.

2. What is the difference between the conducting zone and the respiratory zone of the respiratory tract?

3. Name two structural features that facilitate gas exchange in the alveoli.

4. What structures make up the respiratory membrane?

5. Name the structures of the chest wall.

FORCES FOR PULMONARY VENTILATION

Air flow into or out of the lungs (*breathing* or *ventilation*) resembles blood flow through the vasculature in that both are examples of bulk flow driven by pressure gradients. Ventilation occurs because of the presence of pressure gradients between the alveoli and the outside air (atmosphere). Air moves down such a pressure gradient, from an area of high pressure to one of low pressure. Inspiration occurs when the pressure in the atmosphere exceeds the pressure in the alveoli, creating a pressure gradient for air to move into the alveoli; expiration occurs

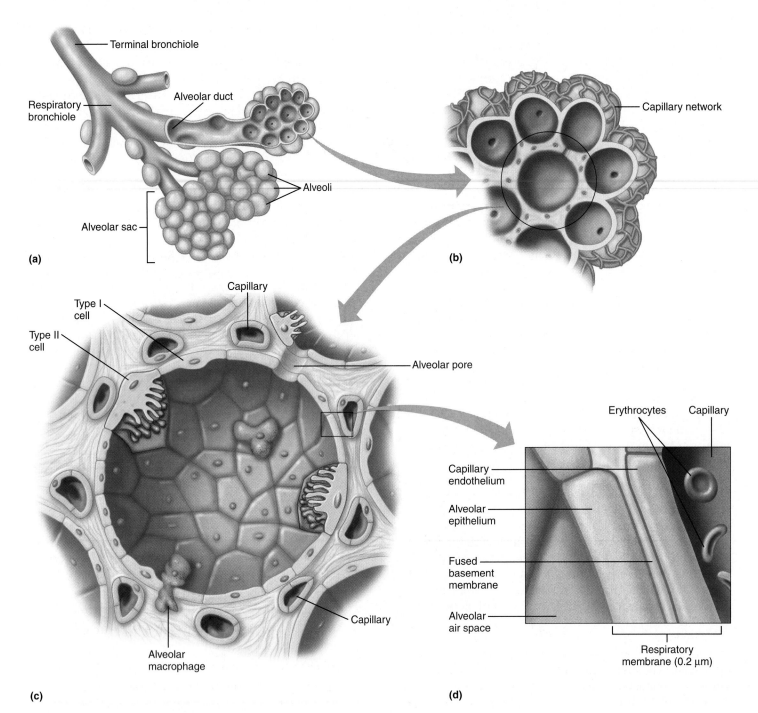

(a)

(b)

(c)

(d)

FIGURE 15.5 Anatomy of the respiratory zone. (a) *Structures in the respiratory zone, which begins where terminal bronchioles branch into respiratory bronchioles. Alveoli are shown both in clusters called alveolar sacs at the ends of alveolar ducts, and associated with alveolar ducts and respiratory bronchioles.* **(b)** *The dense capillary network surrounding alveoli.* **(c)** *Wall of an alveolus in cross section. The alveolar wall contains type I cells, which make up the structure of the wall, and type II cells, which secrete surfactant. Also found in alveoli are macrophages.* **(d)** *Enlargement of the respiratory membrane showing the close association between alveolar and capillary walls.*

Macrophages are derived from what type of white blood cell?

_____

Monocytes

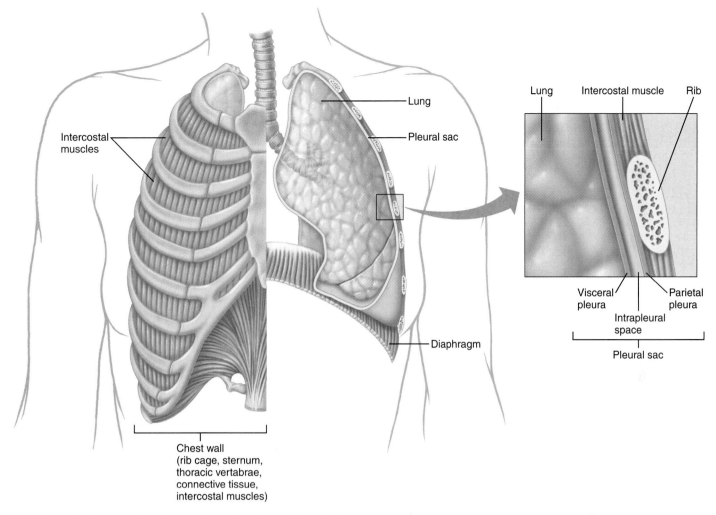

FIGURE 15.6 Chest wall and pleural sac. *The chest wall includes the ribs, sternum, thoracic vertebrae, connective tissue, and intercostal muscles. The side of the pleural sac attached to the lung is called the visceral pleura; the side of the sac attached to the chest wall is called the parietal pleura. The fluid-filled intrapleural space is much thinner than shown here, with a total volume of approximately 15 mL.*

when the pressure in the alveoli exceeds the pressure in the atmosphere, creating a pressure gradient for air to leave the alveoli. First we discuss the pressure gradients; then we discuss how these pressure gradients are created.

Pulmonary Pressures

Three primary pressures are associated with ventilation (we discuss a fourth pressure, transpulmonary pressure, later in this section): atmospheric pressure, intra-alveolar pressure, and intrapleural pressure. Figure 15.7a illustrates these pressures when a lung is at rest (between breaths, at the end of a quiet expiration). **Atmospheric pressure** (P_{atm}) is the pressure of the outside air. At sea level, atmospheric pressure is normally 760 mm Hg, although it varies slightly with weather. For simplicity we

assume that atmospheric pressure is constant. All other lung pressures are expressed relative to atmospheric pressure. **Intra-alveolar pressure** (P_{alv}) is the pressure of air within the alveoli; at rest, it is equal to atmospheric pressure, and thus is 0 mm Hg. However, intra-alveolar pressure varies during the phases of ventilation; in fact, the difference between intra-alveolar pressure and atmospheric pressure is the pressure gradient that drives ventilation. When atmospheric pressure exceeds intra-alveolar pressure (when intra-alveolar pressure is negative), inspiration occurs; when intra-alveolar pressure exceeds atmospheric pressure (when intra-alveolar pressure is positive), expiration occurs. **Intrapleural pressure** (P_{ip}) is the pressure inside the pleural space. Recall that the intrapleural space contains intrapleural fluid, not air. At

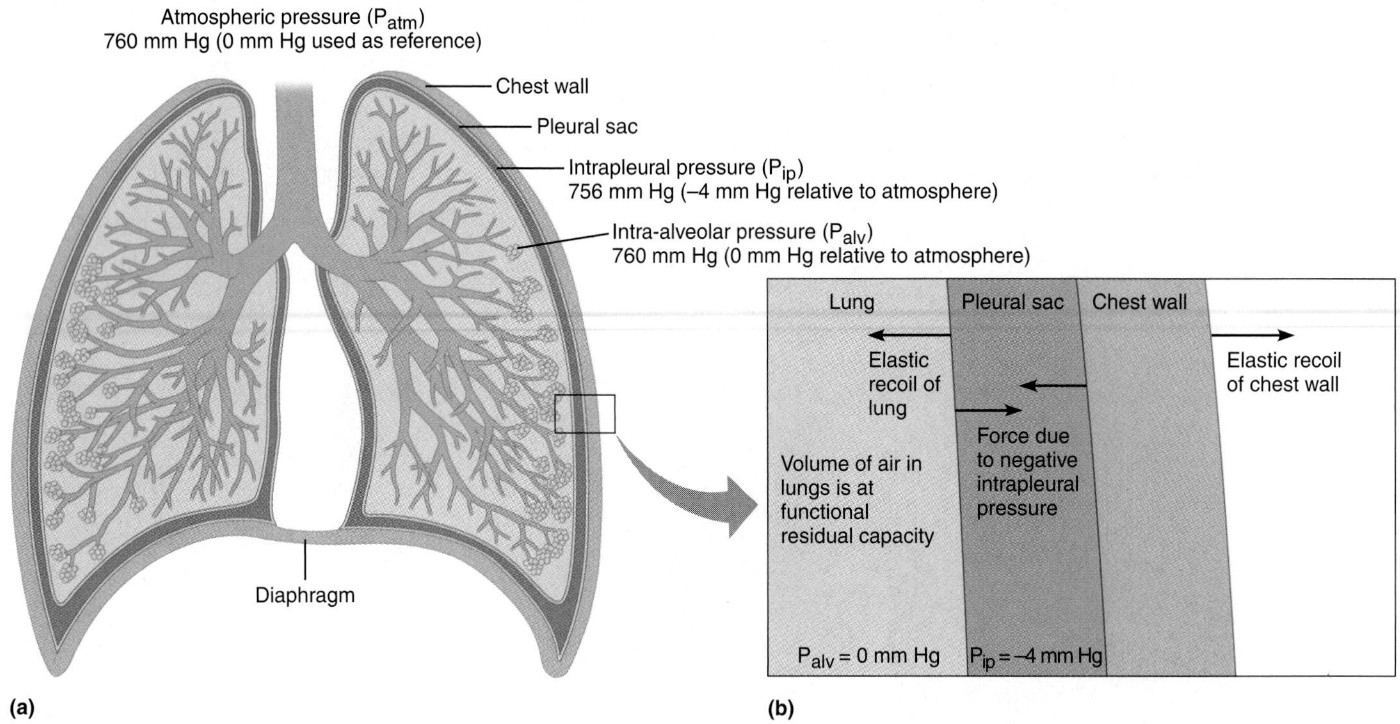

FIGURE 15.7 Lung pressures. (a) *Pulmonary pressures for a lung at rest. Intra-alveolar pressure is the pressure within the alveoli; intrapleural pressure is the pressure in the pleural sac. All pressures are given as absolute pressures and as pressures relative to atmospheric pressure.* **(b)** *Pressures and elastic forces when the lungs are at the functional residual capacity. At the functional residual capacity, intra-alveolar pressure (P_{alv}) = atmospheric pressure = 0 mm Hg. The lung is distended, and an elastic recoil force tends to collapse it inward; the chest wall is compressed, and an elastic recoil force tends to expand it outward. The net force these two opposing forces exert on the two sides of the pleural sac creates a negative intrapleural pressure (P_{ip}). The entire system is stable in that the elastic recoil forces of the lungs and chest wall are in balance, and no net change in size of the lung or chest wall occurs.*

rest, intrapleural pressure is −4 mm Hg. However, the intrapleural pressure, like intra-alveolar pressure, varies with the phase of ventilation. Intrapleural pressure is always less than the intra-alveolar pressure and is always negative during normal breathing because opposing forces exerted by the chest wall and the lungs tend to pull the parietal pleura and visceral pleura apart, as described next.

The lungs and the chest wall are both elastic, meaning that if they are stretched or compressed out of their natural positions, they tend to recoil, or spring back into their natural position. At rest, the chest wall is compressed and tends to recoil outward (as would a compressed spring), whereas the lungs are stretched and tend to recoil inward, like an inflated balloon (Figure 15.7b). These forces tend to move the chest wall and lungs apart, but they do not separate because the surface tension of the intrapleural fluid keeps the parietal pleura and visceral pleura from pulling apart. (This phenomenon is similar to what happens when two wet microscope slides are stacked: The surface tension of the water holds the two slides tightly together, and it takes a lot of force to pull

them apart.) Therefore, the chest wall pulls outward on the intrapleural space while the lungs pull inward (which tends to separate the visceral and parietal plurae), but this in turn creates a negative intrapleural pressure that opposes the separation and thus opposes the outward and inward recoil forces of the chest wall and lungs, respectively.

When the lungs are at rest all breathing muscles are relaxed, and the volume of air in the lungs under these conditions is called the **functional residual capacity** (FRC). When the respiratory system is at rest, no air is moving into or out of the lungs because there is no pressure gradient to drive air movement. To move air into or out of the lungs requires muscular force to create a pressure gradient. The generation of that muscular force is discussed in the following section on the mechanics of breathing—that is, how air moves into and out of the lungs.

To maintain the negative intrapleural pressure, the pleural sac must be airtight. If the pleural sac is broken, such as can occur as a result of a knife or gunshot wound to the chest, then the negative intrapleural pressure is lost as it equilibrates with atmospheric pressure. Without the negative intrapleural pressure, the lungs recoil and collapse while the chest wall recoils and expands (Figure 15.8). This results in a condition called **pneumothorax** (air in the intrapleural space), which is quite dangerous. Fortunately, each lung is isolated in its own pleural cavity, so that if one lung collapses, the other can continue to function.

Trauma is not the only possible cause of a pneumothorax. A *spontaneous pneumothorax* occurs if disease damages the wall of the pleura adjacent to a bronchus or alveolus such that air from inside the lungs enters the intrapleural space. Common diseases that may cause spontaneous pneumothorax include pneumonia and emphysema.

Mechanics of Breathing

Air flow into and out of the lungs is driven by pressure gradients that the muscles of respiration create by changing the volume of the lungs. The relationship between pressure and volume follows **Boyle's Law,** which states that *for a given quantity of any gas (such as air) in a container, the pressure is inversely related to the volume of the container.* In other words, if the volume of the container increases, the pressure exerted by the gas falls, and if the volume decreases, pressure rises (Toolbox: Boyle's Law and the Ideal Gas Law, p. 482). Therefore, the pressure in the lungs changes when their volume changes.

We saw in Chapter 13 that blood flow is determined by a pressure gradient and a resistance, as expressed by the following rule: Flow equals the pressure gradient divided by the resistance. Air flow into and out of the lungs also occurs by bulk flow, with the rate of flow determined by a pressure gradient ($P_{atm} - P_{alv}$) and resistance as follows:

$$\text{Flow} = \frac{P_{atm} - P_{alv}}{R}$$

Thus the driving force for air movement is the difference between atmospheric and intra-alveolar pressure. Because atmospheric pressure is constant, changes in alveolar pressure determine the direction of air movement. (The role of resistance is described later in the chapter.)

Determinants of Intra-alveolar Pressure

Intra-alveolar pressure is determined by two factors: the quantity (moles) of air in the alveoli, and the volume of the alveoli themselves. At rest the alveoli contain a volume of air at atmospheric pressure that is equal to the functional residual capacity previously described. At the start of inspiration, the lungs expand as a result of contraction of the inspiratory muscles (described later) which increases the volume of the alveoli, thereby lowering intra-alveolar pressure according to Boyle's Law. The reduction in intra-alveolar pressure creates a pressure gradient that draws air into the lungs. During expiration,

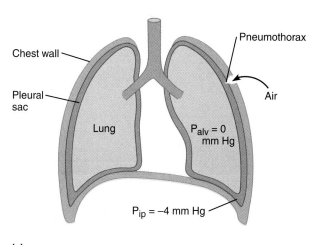

(a)

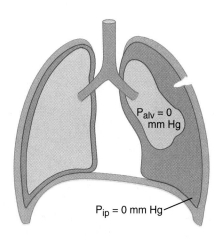

(b)

FIGURE 15.8 Pneumothorax. (a) *Normal intrapleural pressure at rest is −4 mm Hg. Air entering the intrapleural space through a hole in the chest wall creates pneumothorax.* **(b)** *As a consequence of pneumothorax, intrapleural pressure equilibrates with atmospheric pressure. Without the negative force of intrapleural pressure drawing the lung outward, the lung collapses due to elastic recoil forces.*

the reverse occurs: The chest and lungs recoil, decreasing the volume of the alveoli and raising the intra-alveolar pressure. This creates a pressure gradient that drives air out of the lungs.

The pressure changes that occur during inspiration and expiration are shown in Figure 15.9. Note that during inspiration, intra-alveolar pressure falls at first but eventually rises to zero, because intra-alveolar pressure depends on both the quantity of alveolar air (number of molecules) and its volume. Initially, the lungs expand and alveolar pressure decreases. However, as air flows into the alveoli during inspiration, the number of air molecules in the alveoli rises, so alveolar pressure increases (becomes less negative). Air stops flowing in when alveolar pressure rises to zero (which is atmospheric pressure); that is, air stops flowing in when there is no longer a pressure gradient to drive the flow. The same line of reasoning also explains why alveolar pressure rises and then

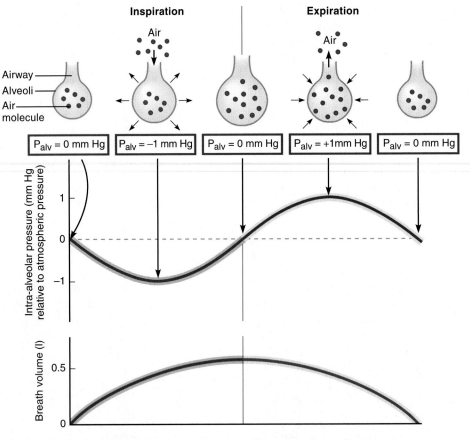

FIGURE 15.9 Changes in alveolar pressure and breath volume during inspiration and expiration. *Before inspiration, intra-alveolar pressure is 0 mm Hg. During inspiration, expansion of the lungs causes intra-alveolar pressure to decrease. Air flow increases the quantity of gas in the lungs, which increases intra-alveolar pressure. At the end of inspiration, intra-alveolar pressure is equal to atmospheric pressure. During expiration, the lungs collapse inward, causing intra-alveolar pressure to increase. Air flows out of the lungs down a pressure gradient. At the end of expiration, intra-alveolar pressure is equal to atmospheric pressure, and air flow is zero.*

If a person inhaled more than 0.5 liters of air in the same length of time as shown above, would intra-alveolar pressure increase (become less negative or positive) or decrease (become more negative)?

falls during expiration. As the lung volume decreases, alveolar pressure increases, causing air to flow out. However, as air flows out, the quantity of air in the alveoli decreases, which lowers the pressure toward zero. Air flow stops when alveolar and atmospheric pressures become equal.

The changes in the volume of the alveoli are produced by changes in the volume of the thoracic cavity, which involve the respiratory muscles, or muscles of ventilation (Figure 15.10a). The diaphragm and the external intercostal muscles are the primary **inspiratory muscles,** whereas the internal intercostals and abdominal muscles are the primary **expiratory muscles,** although

expiration is primarily a passive process not requiring any muscle contraction.

Inspiration

The process of inspiration depicted in Figure 15.10b is initiated by neural stimulation of the inspiratory muscles (Figure 15.11). These skeletal muscles are stimulated to contract by the release of acetylcholine at the neuromuscular junction, as described in Chapter 11. Contraction of the diaphragm causes it to flatten and move downward; meanwhile, contraction of the external intercostals causes the ribs to pivot upward and outward, expanding the chest wall. These combined actions increase the volume of the thoracic cavity. Other muscles of the neck (scalenes and sternocleidomastoids) and chest region (pectoralis minor) play subsidiary roles in inspiration, especially during forceful inspiration.

Decrease

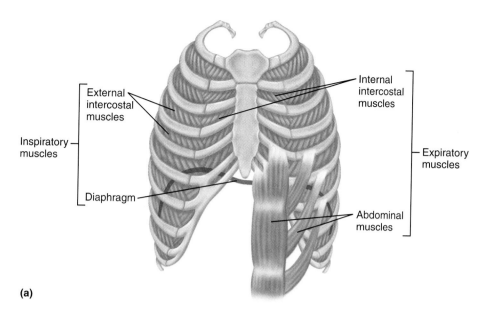

(a)

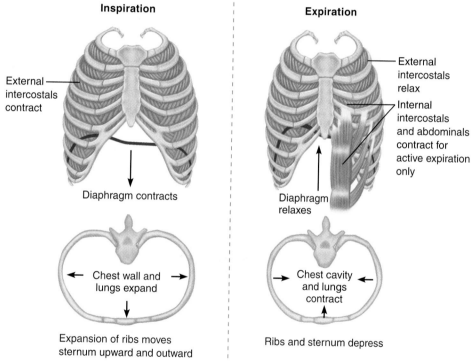

Inspiration

External intercostals contract

Diaphragm contracts

Chest wall and lungs expand

Expansion of ribs moves sternum upward and outward

Expiration

External intercostals relax

Internal intercostals and abdominals contract for active expiration only

Diaphragm relaxes

Chest cavity and lungs contract

Ribs and sternum depress

(b)

FIGURE 15.10 **Muscles of ventilation.** (a) *Locations of inspiratory and expiratory muscles.* (b) *Actions of muscles of ventilation. When inspiratory muscles contract (at left), the chest wall expands, causing the lungs to expand. Quiet expiration (at right) occurs passively by relaxation of the muscles of inspiration, which allows the lungs and chest wall to recoil to their original positions. Active expiration requires contraction of the muscles of expiration, while the muscles of inspiration relax.*

A fundamental relationship governing the behavior of gases is the *ideal gas law,* which is expressed as follows:

$$PV = nRT$$

where P is the gas pressure (atmospheres), V is the volume in which the gas is contained (liters), n is the quantity of gas (moles), R is the universal gas constant (0.083 liter-atm/mole-°K), and T is the absolute temperature (°K). This equation can be rearranged such that pressure is given in terms of the other variables, as follows:

$$P = nRT/V$$

Given a fixed quantity of gas in a container at a given temperature, n and T are both constant, which means that the numerator in the previous equation is constant. Under these conditions, the pressure is inversely proportional to the volume of the container: $P \propto 1/V$. This is Boyle's Law, illustrated in the box figure.

If instead the volume of the container is fixed but the quantity of gas is allowed to vary, the quantity RT/V becomes constant. Under these conditions, the pressure is proportional to the number of molecules in the container: $P \propto n$.

When we consider why gases exert pressure, we can see why the pressure varies with the number of gas molecules in a container. Gas molecules are normally in a state of constant thermal motion. Gas molecules inside a container continually collide with the wall and rebound off it. Every collision exerts a certain amount of force on the wall. The *pressure* is the total force exerted by all such collisions divided by the surface area of the wall (force per unit area). If the number of molecules is held constant but the volume of the container decreases, the density of molecules increases, so any given area of the wall experiences more collisions per second, and the pressure is higher. If more molecules are added to a container with a fixed volume, the same thing happens: Molecules are more densely spaced, which increases the number of collisions per unit area of wall, which raises the pressure.

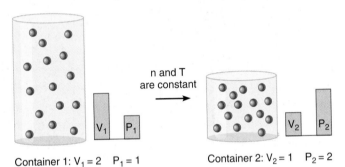

n and T are constant

Container 1: $V_1 = 2$ $P_1 = 1$ Container 2: $V_2 = 1$ $P_2 = 2$

As the chest wall expands, it pulls outward on the intrapleural fluid, causing the intrapleural pressure to decrease. This decrease in intrapleural pressure causes an increase in a fourth pulmonary pressure—the **transpulmonary pressure,** which is the difference between the intrapleural pressure and the intra-alveolar pressure ($P_{alv} - P_{ip}$; Figure 15.12). An increase in transpulmonary pressure due to a decrease in P_{ip} creates a larger distending pressure across the lungs, so the lungs (alveoli) expand with the chest wall. When the lungs expand, pressure in the alveoli decreases to less than atmospheric pressure, so air flows into the alveoli by bulk flow, and continues to flow in until the pressure in the alveoli increases to atmospheric pressure. Stronger contractions of the inspiratory muscles produce a greater expansion of the thoracic cavity, making intrapleural pressure even more negative and creating a greater transpulmonary pressure, resulting in greater lung expansion and a deeper inspiration; that is, a larger volume of air moves into the lungs.

Expiration

During quiet breathing, expiration is a *passive* process in that it does not require muscle contraction. At the end of an inspiration, the chest wall and lungs are expanded by muscle contraction. By simply relaxing these muscles, which occurs when motor neurons to the inspiratory muscles stop firing, the elastic chest wall and lungs recoil to their resting position. As the chest wall and lungs recoil, the volume of the lungs decreases, causing alveolar pressure to increase to values greater than atmospheric pressure. Air flows out (expiration occurs) due to the pressure gradient until the volume in the lungs equals the FRC.

A more forceful expiration can be produced by contraction of the expiratory muscles in a process called *active expiration.* Contraction of the expiratory muscles produces a greater and more rapid decrease in the volume of the thoracic cavity, which increases intra-alveolar pressure and causes a greater pressure gradient for air flow out of the alveoli.

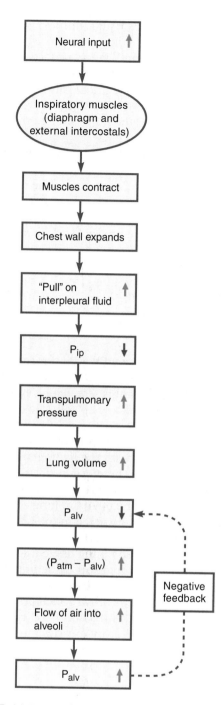

FIGURE 15.11 **Events in the process of inspiration.** *This flowchart is available as an interactive exercise under Activities at www.physiologyplace.com.*

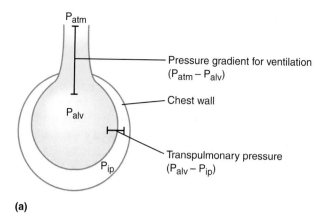

(a)

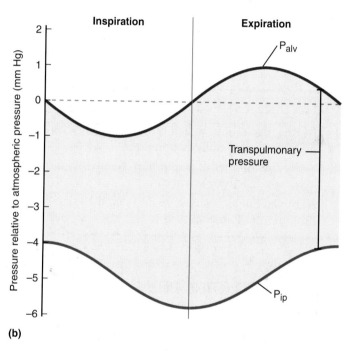

(b)

FIGURE 15.12 **Volume and pressure changes during inspiration and expiration. (a)** *Whereas the difference between atmospheric and intra-alveolar pressures (the pressure gradient for ventilation) provides the force for moving air into or out of the lungs, the transpulmonary pressure provides the force for the expansion of the lungs.* **(b)** *Changes in the intra-alveolar and intrapleural pressures that occur during breathing are such that the transpulmonary pressure increases during inspiration and during the beginning of expiration, and then decreases as expiration continues.*

Quick Test 15.2

1. During inspiration, which is greater, intra-alveolar pressure or atmospheric pressure?

2. Name the muscles of inspiration. Name the muscles involved in active expiration.

3. At the end of an inspiration, in what direction do the lungs naturally recoil?

FACTORS AFFECTING PULMONARY VENTILATION

The rate at which air flows into or out of the lungs is determined by two factors: the pressure gradient between the atmosphere and alveoli, and the airway resistance. In the previous section, we saw how pressure gradients produced by changes in the volume of the lungs drive inspiration and expiration. In this section, we consider the various factors that affect development of those pressure gradients and factors that affect airway resistance.

Lung Compliance

We have already seen that lungs are elastic and recoil after being stretched. A measure of the *ease* with which they can be stretched is called *compliance*. Lung compliance is defined as the change in lung volume (ΔV) that results from a given change in transpulmonary pressure $\Delta(P_{alv} - P_{ip})$ (see the Toolbox on page 415):

$$\text{Lung compliance} = \frac{\Delta V}{\Delta(P_{alv} - P_{ip})}$$

A large lung compliance is advantageous, because a smaller change in transpulmonary pressure is needed to

bring in a given volume of air, and thus less work or muscle contraction is required. Lung compliance depends on the elasticity of the lungs, and on the surface tension of the fluid lining the alveoli. Because lungs are elastic due to the presence of elastic fibers in the connective tissue, forces exerted by these elastic fibers generally oppose lung expansion because as the lungs stretch, the fibers tend to recoil.

The surface tension of a liquid is a measure of the work required to increase its surface area by a certain amount. The greater the surface tension, the more work needed to spread the fluid out. The surface tension of the lungs is caused by the air-liquid interface formed by the thin layer of fluid lining the internal surface of the alveoli (Toolbox: Pulmonary Surfactant and the Law of LaPlace, p. 486). As lung tissue expands, so does the fluid layer in the alveoli. Therefore, as the lungs expand, work is required not only to stretch the elastic tissue but also to increase the surface area of the water layer. Consequently, the surface tension of water acts to *decrease* lung compliance.

The presence of a detergentlike substance called **pulmonary surfactant** decreases the surface tension in alveoli. Pulmonary surfactant is secreted by **type II alveolar cells** located in the walls of alveoli. The surface tension of the fluid lining the alveoli is reduced (but not eliminated) by the action of pulmonary surfactant because surfactant interferes with the hydrogen bonding between water molecules. Therefore, surfactant increases lung compliance and decreases the work of breathing. Figure 15.13 summarizes the effects of lung compliance and pulmonary pressures (discussed in the previous section) on lung volume.

Compliance is decreased if lung tissue thickens, such as occurs with the formation of scar tissue in tuberculosis, or if surfactant production is decreased, such as occurs in *infant respiratory distress syndrome* (in which premature babies do not produce enough surfactant). When lung compliance decreases, the respiratory muscles must do more work to expand the lungs to a given volume.

Airway Resistance

In the context of this discussion, the term **airway resistance** refers to the resistance of the entire system of airways in the respiratory tract; it is analogous to *total peripheral resistance* in cardiovascular physiology (see Chapter 13). Airway resistance is determined primarily by the resistances of individual airways and is affected most strongly by changes in airway *radius:* As radius decreases, airway resistance increases.

In healthy lungs, resistance to air flow into and out of the lungs is low, because the radii of the tubes in the conducting zone are relatively large. Although the tubes decrease in radius as air moves down the conducting zone toward the alveoli, the total cross sectional area of the smaller tubules increases due to the extensive branching.

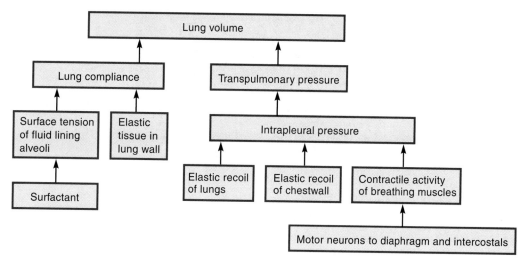

FIGURE 15.13 **Factors affecting lung volume.**

Therefore, overall resistance in the conducting zone is low. (This situation is similar to the low combined resistance of capillaries in the circulatory system.) This low resistance means that alveolar pressure need not differ much from atmospheric pressure to achieve normal rates of air flow under normal conditions. During quiet breathing *(eupnea)*, the difference between alveolar and atmospheric pressure is generally less than 2 mm Hg. The effect of airway resistance on breathing is illustrated in Figure 15.14, which shows intra-alveolar pressure under normal conditions and when airway resistance is increased. (For purposes of illustration, we assume that the rates of air flow are the same in both conditions.) Note that when resistance increases, a larger pressure gradient is required to produce a given rate of air flow.

The resistance to air flow is affected by a number of factors, including passive forces exerted on the airways, contractile activity of smooth muscle in the bronchioles, and secretion of mucus into the airways. The passive forces are responsible for changes in airway resistance that occur in a single breath, whereas changes in smooth muscle contractile activity are responsible for long-term variations in airway resistance.

The passive forces include changes in transpulmonary pressure that occur during inspiration and expiration, and *tractive forces* exerted on the airways by the pulling action of tissues surrounding them. During inspiration, the transpulmonary pressure increases because intrapleural pressure decreases more than intra-alveolar pressure decreases (see Figure 15.12). The increase in transpulmonary pressure that occurs during inspiration pulls outward on the airways, causing them to distend, which decreases resistance as inspiration continues. During expiration, however, the transpulmonary pressure decreases, which reduces the distending force on airways and allows them to contract, thereby increasing the resistance. As the surrounding tissue moves away from the airways during inspiration, tractive forces exerted by the tissues increase, which pulls the airways open and

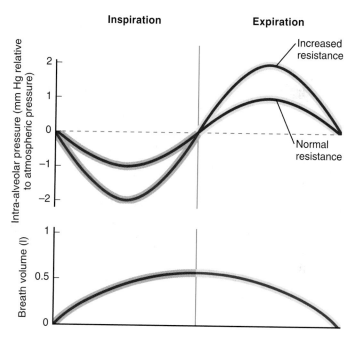

FIGURE 15.14 **Effects of increasing airway resistance on the pressure changes required to move a fixed volume of air.** *When airway resistance is increased over normal, a greater intra-alveolar pressure is required to move a given volume of air (in this case, 0.5 mL) into and out of the lungs in a given period of time.*

reduces resistance. During expiration, removal of these tractive forces reduces airway diameter and increases resistance.

Resistance to air flow in the bronchioles can also change as a result of contraction or relaxation of smooth muscle in the walls of the bronchioles. When this smooth muscle contracts, it decreases the radius of the bronchioles (called *bronchoconstriction*), which increases resistance. The contraction and relaxation of bronchiolar smooth muscle is subject to both extrinsic control (neural and hormonal signals) and intrinsic control (local chemical mediators).

The thin film of water that lines the alveoli increases the muscular work required to inflate the lungs because the water's surface tension decreases the lungs' compliance, making them harder to stretch. This surface tension creates another problem as well—hydrogen bonding between adjacent water molecules tends to pull them into a round droplet, which tends to pull the walls of an alveolus inward and make it collapse. What, then, prevents all the alveoli from collapsing into one large "air bubble"?

To understand why the alveoli do not collapse, consider what happens in a soap bubble: The wall of a soap bubble contains water, which exerts surface tension and thus tends to pull the wall inward. As the bubble shrinks, however, it raises the pressure of the air contained within it, creating a distending pressure that opposes the tendency of the bubble to collapse. The bubble remains at a stable volume, neither collapsing nor expanding, as long as the distending pressure is just large enough to balance the inwardly-directed forces created by the surface tension. When the lungs are not expanding or contracting, the volume of an alveolus remains stable for the same reason—the pressure of the air inside it balances the inward forces that would otherwise cause it to collapse.

According to the *Law of LaPlace*, the air pressure (*P*) necessary to prevent the collapse of an alveolus (which is assumed to be spherical) is directly proportional to the surface tension (*T*) and inversely proportional to the alveolus' radius (*r*):

$$P = \frac{2T}{r}$$

Here, *P* is the air pressure inside relative to the pressure outside, which is taken to be zero. Thus, if two alveoli—one larger and one smaller—are subject to the same surface tension, the smaller one will require a greater pressure inside to keep from collapsing.

At the end of an inspiration or expiration, air pressure is the same inside all the alveoli. (If air is not flowing, the pressure of the air must be the same everywhere.) However, alveoli are not all the same size. This situation poses an interesting problem that is illustrated in the diagram on the next page, which shows two adjacent alveoli of unequal sizes.

Let us assume for the moment that no surfactant is present. If the air pressure is just high enough to prevent the large alveolus from collapsing, then according to the law of LaPlace, it *cannot* be large enough to prevent the smaller alveolus from collapsing. Thus, if the two alveoli were at the same pressure initially, the smaller alveolus should collapse, which will raise the air pressure inside it (P_1), making it higher than the pressure in the larger alveolus (P_2). Air should then flow down the pressure gradient (from P_1 to P_2) and thus flow from the smaller to the larger alveolus, as shown in the left part of the diagram.

The smooth muscle of the bronchioles is influenced by the autonomic nervous system: Sympathetic stimulation causes relaxation of the smooth muscles and increases the radius of bronchioles (called *bronchodilation*), whereas parasympathetic stimulation causes contraction of the smooth muscle and *bronchoconstriction*. Epinephrine released from the adrenal medulla during sympathetic stimulation also causes bronchodilation.

Histamine, a chemical released locally during allergic reactions (see Chapter 5), causes contraction of the smooth muscle resulting in bronchoconstriction. This effect only partially accounts for the difficulty in breathing during an allergic reaction, however, because histamine also stimulates mucus secretion, which builds up in the airways and increases resistance to air flow.

Another important local chemical that affects the radius of bronchioles is carbon dioxide. When carbon dioxide levels are high, bronchioles dilate; when carbon dioxide levels are low, bronchioles constrict. Carbon dioxide levels are an important regulator of bronchiole radius in the process of matching air flow (ventilation) to blood flow (perfusion), a topic explored in Chapter 16.

Airway resistance can be increased in a number of pathological states. One such condition, *asthma*, is associated with an increase in airway resistance caused by spastic contractions of the smooth muscle in bronchioles coupled with increased mucus secretion and inflammation of the walls of the bronchioles. Symptoms include coughing, *dyspnea* (labored breathing), and wheezing. Asthma is of-

In real lungs, however, surfactant is present and is actually more highly concentrated in the smaller alveoli. As a consequence, the surface tension in a small alveolus is lower than that in a large alveolus, which reduces the amount of pressure that must exist inside the small alveolus to prevent it from collapsing. Thus, small and large alveoli can both have stable volumes at the same pressure, as shown in the right part of the diagram.

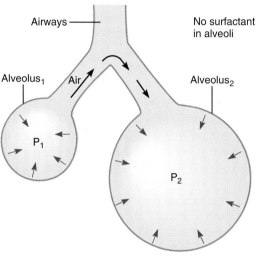

$r_1 < r_2$, $P_1 > P_2$
Smaller alveolus collapses into larger

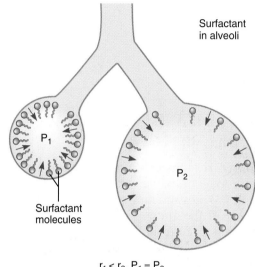

$r_1 < r_2$, $P_1 = P_2$
Alveoli do not collapse

ten the result of hypersensitivity to certain *allergens* (substances that stimulate allergic responses) such as fungi, dust mites, or animal dander, but it can also be induced by stress, exercise, eating certain foods, or breathing cold air. Because of increased airway resistance, significantly larger pressure gradients are required to produce comparable rates of air flow, which greatly increases the work of breathing. The treatment for asthma varies for individuals, but it can include *bronchodilators,* which induce relaxation of airway smooth muscle, and *corticosteroids,* which reduce inflammation. Other pathological states that increase airway resistance are *chronic obstructive pulmonary diseases,* or *COPD.* Whereas asthma involves acute (temporary) increases in airway resistance, COPDs are associated with chronic (long-lasting) increases in airway resistance.

Quick Test 15.3

1. If scarring of the lung resulted from some disease process, what would happen to lung compliance?

2. If surfactant secretion decreased, what would happen to lung compliance?

3. Will contraction of bronchiole smooth muscle cause the resistance to air flow to *increase* or *decrease*?

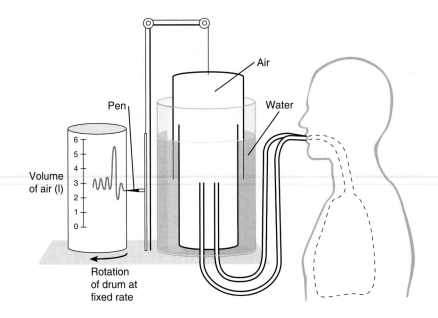

FIGURE 15.15 Spirometry. *While a subject breathes air in and out, a pen attached via a pulley system records changes in the volume of air in the inverted bell.*

CLINICAL SIGNIFICANCE OF RESPIRATORY VOLUMES AND AIR FLOWS

Certain pathological conditions affect the volume of air contained in the lungs or the rate of air flow into or out of the lungs. Clinicians measure lung volumes, calculate lung capacities (which are the sums of two or more measured lung volumes), and measure air flow rates in order to gain information concerning pulmonary function. These values can be measured using a technique called *spirometry.*

Spirometry is a technique for measuring the volumes of inspired and expired air using a device called a **spirometer** (Figure 15.15). One type of spirometer consists of an inverted, air-filled bell in a tub of water, plus a connection from the bell to an outlet via a hose that goes to the patient. The patient breathes air in and out of the bell. When the patient inhales, the volume of air in the bell decreases and the bell descends deeper into the water. When the patient exhales, the volume of air in the bell increases and the bell rises in the water. The bell is connected via a pulley system to a pen that moves up and down when the bell moves up and down. The pen is positioned to write on paper attached to a drum that rotates at a set speed. The movement of the pen up and down leaves marks on the paper that are calibrated to the volume of air that moved into or out of the lungs.

Lung Volumes and Capacities

Using spirometry, clinicians can measure three of the four nonoverlapping **lung volumes** that together make up the *total lung capacity.* The data in Figure 15.16 were obtained from a healthy 70-kg male sitting at rest. (Typically, the values would be smaller for a female and would change with a change in posture.) The volume of air that moves into and out of the lungs during a single, unforced breath is called the **tidal volume** (V_T); the average resting tidal volume is 500 mL. The maximum volume of air that can be inspired from the end of a normal inspiration is called the **inspiratory reserve volume** (IRV) and averages about 3000 mL. The maximum volume of air that can be expired from the end of a normal expiration is called the **expiratory reserve volume** (ERV) and averages about 1000 mL. However, even following a maximum expiration, some air is left in the lungs and airways, because the negative intrapleural pressure prevents complete collapse of the lungs. The volume of air remaining in the lungs after a maximum expiration is called the **residual volume** (RV) and averages about 1200 mL.

The residual volume cannot be measured by spirometry, because spirometry only measures air that actually moves into or out of the lungs. One method of calculating residual volume is the *helium dilution method,* in which the subject breathes a gas mixture containing a known concentration of helium, which is not exchanged with the blood to any significant degree. During inspiration, helium in the inspired air mixes with all the air that is present in the lungs, which dilutes the helium. When the subject exhales, the concentration of helium in the expired air is measured and compared with the original concentration. With some simple calculations, the spirometer operator can then determine the residual volume.

Lung capacities are sums of two or more of the lung volumes described above. The **inspiratory capacity** (IC) is the maximum volume of air that can be inspired at the end of a resting expiration; it is the sum of tidal volume and inspiratory reserve volume ($IC = V_T + IRV$) and is approximately 3500 mL. The **vital capacity** (VC) is the maximum volume of air that can be expired following a maximum inspiration; it is the sum of the tidal volume, inspiratory reserve volume, and expiratory reserve volume ($VC = V_T + IRV + ERV$) and is approximately 4500 mL. The functional residual capacity (FRC) is the volume of air remaining in the lungs at the end of a tidal expiration; it is equal to the expiratory reserve volume and the

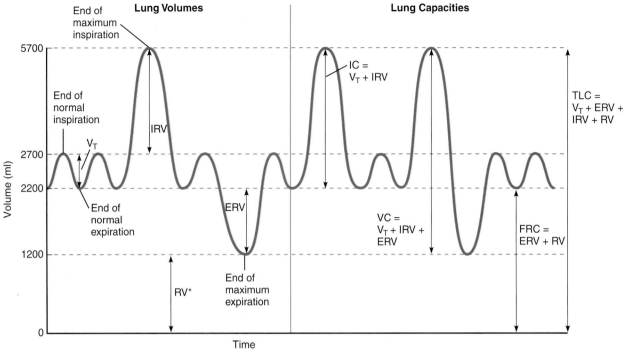

Lung Volumes

V_T = Tidal volume = 500 ml
IRV = Inspiratory reserve volume = 3000 ml
ERV = Expiratory reserve volume = 1000 ml
RV = Residual volume* = 1200 ml

*Cannot be measured by spirometry

Lung Capacities

IC = Inspiratory capacity = V_T + IRV = 3500 ml
VC = Vital capacity = V_T + IRV + ERV = 4500 ml
FRC = Functional residual capacity = ERV + RV = 2200 ml
TLC = Total lung capacity = V_T + ERV + IRV + RV = 5700 ml

FIGURE 15.16 Lung volumes and capacities measured using spirometry. *The curves shown were produced by spirometry (see Figure 15.15) and represent average values for a 70-kg male.*

residual volume ($FRC = ERV + RV$) and is approximately 2200 mL. The functional residual capacity is the volume of air in the lungs when the lungs are between breaths and the respiratory muscles are relaxed. Under these conditions, the elastic recoil of the lungs is balanced by the elastic recoil of the chest wall. The **total lung capacity** (TLC) is the volume of air in the lungs at the end of a maximum inspiration; it is the sum of the tidal volume, inspiratory reserve volume, expiratory reserve volume, and residual volume ($TLC = V_T + IRV + ERV + RV$) and is approximately 5700 ml.

Pulmonary Function Tests

Spirometry is a simple method to test for respiratory disorders. A simple measure of lung volumes and calculations of lung capacities can help distinguish between *obstructive pulmonary diseases,* which involve increases in airway resistance, and *restrictive pulmonary disorders,* in which something interferes with lung expansion. In obstructive disorders, for example, the residual volume often increases because an increase in resistance not only makes it harder to inspire, but it is also harder to expire.

The lungs become overinflated and the functional residual capacity and total lung capacity are often increased. In contrast, restrictive disorders often involve structural damage to the lungs, pleura, or chest wall that decreases the total lung capacity and vital capacity.

Additional useful information can be obtained by testing either forced vital capacity or forced expiratory volume. When testing the **forced vital capacity** (FVC), the patient takes a maximum inspiration and then forcefully exhales as much and as rapidly as possible. A low FVC is indicative of restrictive pulmonary disease. The **forced expiratory volume** (FEV) is a measure of the percentage of the FVC that can be exhaled within a certain length of time, most commonly one second (FEV_1). A normal FEV_1 is 80%, meaning that a person should be able to exhale 80% of the forced vital capacity within 1 second. If a person has a forced vital capacity of 4000 ml, for example, then following a maximum inspiration that person should be able to exhale 3200 ml in 1 second. An FEV_1 that is less than 80% is indicative of increased resistance, which is characteristic of obstructive pulmonary disease (see When It Goes Wrong: Chronic Obstructive Pulmonary Disease, p. 490).

The term *obstructive pulmonary disease* is an umbrella term for a number of disorders, all of which are characterized by increased airway resistance. Three major disorders falling under this classification are *emphysema*, *chronic bronchitis,* and *asthma*. Whereas asthma is acute, the others are chronic. Chronic obstructive pulmonary disease is frequently referred to by the acronym *COPD*. Together, asthma and COPD currently affect about 20 million persons in the United States.

Unlike asthma, COPD is chronic and progressive; it is also largely preventable because it is most often associated with cigarette smoking. Other precipitating causes include exposure to air pollutants and occupational exposure to certain chemicals. Symptoms include dyspnea, coughing, and frequent infections. Although the progression of COPD differs depending on the particular disease, it is invariably accompanied by destruction of lung tissue. When fatal, the most frequent cause of death is respiratory failure leading to *hypoxemia* (reduced oxygen concentration in the blood) and carbon dioxide retention, which may be due to either inadequate air flow to alveoli (due to airway obstruction) or impairment of gas exchange within the alveoli. Alternatively, it may be due to spontaneous pneumothorax following weakening and rupture of the pulmonary wall. Because capillaries are frequently destroyed along with other lung tissues, COPD may also be accompanied by right heart failure brought about by increased resistance in the pulmonary blood vessels.

Emphysema is a permanent enlargement of airspaces in the respiratory zone accompanied by destruction of airway walls. According to a widely accepted theory, tissue destruction is a result of the action of *proteases,* enzymes secreted by macrophages and other white blood cells during chronic inflammation. These proteases destroy tissue by breaking down proteins, including those in elastic connective tissue fibers. As a result, lungs lose elasticity, and compliance increases. Loss of these fibers also reduces tractive forces on airways, which reduces their diameter and makes them more likely to collapse, especially during expiration, when airway distending pressures are lower. The increase in compliance both increases functional residual capacity (that is, it reduces the volume that could be exhaled) and reduces the effort required to inhale. However, the loss of airway traction makes exhaling more difficult. As a consequence, those afflicted with emphysema tend to breathe at higher lung volumes (that is, the residual volume increases), resulting in a decrease in the vital capacity. Destruction of tissue also reduces the combined surface area of alveolar walls, which diminishes the capacity for gas exchange.

Chronic bronchitis is an inflammation of the airways that lasts for at least three months a year for at least two consecutive years. It is characterized by inflammation and thickening of the airway lining, which reduces airway diameter and which can also lead to destruction of the normal tissue and fibrosis (thickening and scarring by the formation of connective tissue). Along with these symptoms comes an abnormally high rate of mucus secretion, which exacerbates the problem of airway obstruction and also predisposes the lungs to infection.

Treatments for COPD include *bronchodilators* (such as β_2 adrenergic receptor agonists) and *anti-inflammatory drugs* (such as corticosteroids). Administration of oxygen is also helpful in combating hypoxemia when it is a problem.

Alveolar Ventilation

Minute ventilation, which is the total amount of air that flows into or out of the respiratory system in a minute, can be calculated as the tidal volume times the number of breaths per minute, or **respiration rate.** At rest, the average respiration rate is 12 breaths per minute. Therefore, the average resting minute ventilation is:

$$\text{Minute ventilation} = V_T \times \text{respiration rate}$$

$$= (500 \text{ ml/breath}) \times (12 \text{ breaths/min})$$

$$= 6000 \text{ ml/min}$$

More important than the minute ventilation is the amount of fresh air that reaches the alveoli. Only a portion of the air breathed in actually participates in gas

TIDAL VOLUME (ML)	RESPIRATION RATE (BREATHS/MIN)	MINUTE VENTILATION (ML/MIN)	MINUTE ALVEOLAR VENTILATION (ML/MIN) *
300	20	6,000	3,000
500	12	6,000	4,200
1000	6	6,000	5,100
500	24	12,000	8,400
1000	12	12,000	10,200

*Assumes dead space volume = 150 mL

exchange, because a significant fraction of the air simply fills up the volume of airways in the conducting zone. For example, of a normal breath of 500 mL of air, only about 350 mL actually reaches the alveoli; the remaining 150 mL (30% of the tidal volume of 500 mL in this example) fills the trachea, the bronchi, and the bronchioles. The combined volume of these nonexchanging airways is referred to as the **anatomical dead space.** The effects of the anatomical dead space on the amount of "fresh air" reaching the lungs is illustrated in Figure 15.17. At the end of an expiration (Figure 15.17a), the conducting zone is full of "old" air—that is, air with less oxygen and more carbon dioxide than atmospheric air because it once occupied the alveoli where oxygen and carbon dioxide were exchanged with blood. On the next inspiration (Figure 15.17b), that 150 mL old air moves into the alveoli along with 350 mL fresh air (assuming a tidal volume of 500 mL). Therefore, from a functional view, only 350 mL fresh air enters the alveoli with each breath. **Alveolar ventilation** is a measure of the volume of fresh air reaching the alveoli each minute. It is similar to minute ventilation, except that the tidal volume has been corrected for the dead space volume (DSV):

$$\text{Alveolar ventilation} = (V_T \times \text{respiration rate}) - (DSV \times \text{respiration rate})$$

$$= (500 \text{ mL/breath} \times 12 \text{ breaths/min}) - (150 \text{ ml/breath} \times 12 \text{ breaths/min})$$

$$= 4200 \text{ mL/min}$$

During times when oxygen demand of the tissues is increased, such as during exercise, alveolar ventilation must increase to meet those demands, either by increasing tidal volume, or by increasing respiration rate, or both. As illustrated in Table 15.1, it is more efficient to increase alveolar ventilation by increasing tidal volume than by increasing respiration rate. The reasoning is as

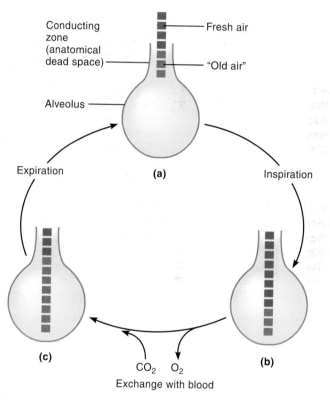

FIGURE 15.17 The effects of anatomical dead space on alveolar ventilation. (a) *At the end of expiration, all the air in the conducting and respiratory zones is stale air.* **(b)** *During inspiration, stale air from the conducting zones (anatomical dead space) enters the respiratory zones first, followed by atmospheric air.* **(c)** *During expiration, atmospheric air in the conducting zones is expired first, followed by stale air.*

follows: When respiration rate is increased, the dead space volume is effectively subtracted out of each additional breath. However, when tidal volume is increased, the total increase in volume in excess of the dead space volume adds to the fresh air reaching the alveoli.

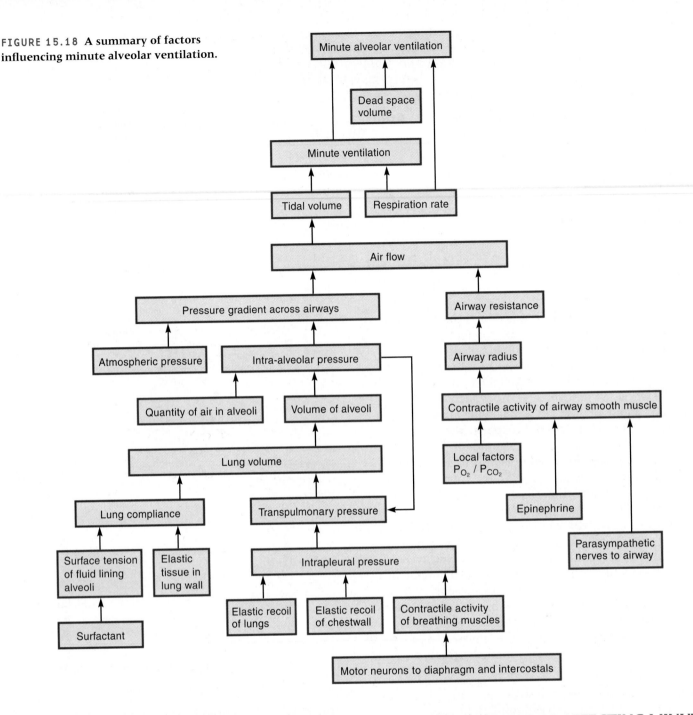

FIGURE 15.18 A summary of factors influencing minute alveolar ventilation.

Minute alveolar ventilation

Dead space volume

Minute ventilation

Tidal volume

Respiration rate

Air flow

Pressure gradient across airways

Airway resistance

Atmospheric pressure

Intra-alveolar pressure

Airway radius

Quantity of air in alveoli

Volume of alveoli

Contractile activity of airway smooth muscle

Lung volume

Local factors P_{O_2} / P_{CO_2}

Lung compliance

Transpulmonary pressure

Epinephrine

Surface tension of fluid lining alveoli

Elastic tissue in lung wall

Intrapleural pressure

Parasympathetic nerves to airway

Surfactant

Elastic recoil of lungs

Elastic recoil of chestwall

Contractile activity of breathing muscles

Motor neurons to diaphragm and intercostals

Quick Test 15.4

1. Define the following terms: *tidal volume, minute ventilation, intra-alveolar pressure, anatomical dead space, alveolar ventilation.*

2. Following a maximum inspiration, a person expires maximally. Which lung capacity corresponds to the volume of air that has been expired? What is the volume of air left in the lungs called?

3. Given a tidal volume of 450 ml, a dead space volume of 150 ml, and a respiration rate of 15 breaths per minute, what is the minute ventilation? What is the minute alveolar ventilation?

SUMMARY OF FACTORS AFFECTING MINUTE ALVEOLAR VENTILATION

Figure 15.18 is a summary of how breathing mechanics determine the minute alveolar ventilation. Minute alveolar ventilation is determined by the minute ventilation minus the dead space volume times the respiration rate. Minute ventilation depends on air flow into the lungs, which depends on pressure gradients and airway resistance. The various factors that contribute to development of the pressure gradients and airway resistance are shown. Each of these factors and their contributions to pressure or resistance has been described in this chapter; if you are unsure of the role of any factor, review the appropriate portion of the chapter.

Overview of Respiratory Function, p. 470

Respiration is the process of gas exchange and includes internal and external respiration. The four processes of external respiration are (1) pulmonary ventilation, (2) exchange of oxygen and carbon dioxide between lung air spaces and blood, (3) transport of oxygen and carbon dioxide in blood, and (4) exchange of oxygen and carbon dioxide between the blood and systemic tissues. Functions of the respiratory system include supplying oxygen to the tissues and eliminating carbon dioxide, acid-base balance of the blood, vocalization, and protection against pathogens and irritants in the air.

Anatomy of the Respiratory System, p. 470

The upper airways include the nasal cavity, oral cavity, and pharynx. After the pharynx, a common passageway for air and food, the passageways for food and air diverge. The respiratory tract is the pathway for air and can be functionally divided into two components: a conducting zone and a respiratory zone. The conducting zone (larynx, trachea, bronchi, and bronchioles) functions in conducting air from the larynx to the lungs. The conducting zone is lined by epithelium that contains goblet cells and ciliated cells. The respiratory zone (respiratory bronchioles, alveolar ducts, alveoli, and alveolar sacs) is the site of gas exchange within the lungs; the alveoli are the primary sites of exchange. The wall of an alveolus contains type I cells and type II cells. The type I cells and endothelial cells of capillaries form the respiratory membrane across which gas exchange occurs. The type II cells secrete pulmonary surfactant. Also located in the alveoli are alveolar macrophages.

The pleurae are membranes that line the chest wall and lung, forming a pleural sac around each lung. The space between the two membranes, called the intrapleural space, is filled with a thin layer of intrapleural fluid. The chest wall consists of the rib cage, the sternum, the thoracic vertebrae, and associated muscles and connective tissue. Muscles of the chest wall include the internal and external intercostals and the diaphragm.

IP Respiratory, Anatomy Review: Respiratory Structures, pp. 4–12

IP Respiratory, Pulmonary Ventilation, pp. 4–6

Forces for Pulmonary Ventilation, p. 475

Atmospheric pressure is the pressure of the air outside the body. Intra-alveolar pressure is the pressure of air within the alveoli. Intrapleural pressure is the pressure of the intrapleural fluid. Because the lungs and the chest wall are elastic, between breaths the chest wall tends to recoil outward, and the lungs tend to recoil inward. These forces tend to separate the chest wall from the lungs, creating a negative intrapleural pressure.

Inspiration and expiration are driven by differences in atmospheric and intra-alveolar pressures. These pressure gradients are created when the volume of the lungs is changed. Inspiration is caused by contraction of the diaphragm and the external intercostal muscles; when these muscles contract, the volume of the thoracic cavity increases. As the thoracic cavity expands, the intrapleural pressure decreases, creating a force that expands the lungs as the chest wall expands. Intra-alveolar pressure decreases below atmospheric pressure, and inspiration occurs. During quiet breathing, expiration occurs when the chest wall and lungs return passively to their original positions. Active expiration involves contraction of the internal intercostals and abdominal muscles.

IP Respiratory, Pulmonary Ventilation, pp. 3–13

Factors Affecting Pulmonary Ventilation, p. 484

The rate of air flow into or out of the lungs is determined by the magnitude of the pressure gradient driving the flow and airway resistance. Lungs have a high compliance; that is, they are easily stretched to increase lung volume for inspiration. Airway resistance depends primarily on the radius of the tubules of the respiratory tract. Airway resistance generally is low but can be affected by breathing mechanics, the autonomic nervous system, chemical factors, and pathological states.

IP Respiratory, Pulmonary Ventilation, pp. 14–18

Clinical Significance of Respiratory Volumes and Air Flows, p. 488

Lung volumes and capacities can be measured using spirometry. Lung volumes include tidal volume, inspiratory reserve volume, expiratory reserve volume, and residual volume. The lung capacities include inspiratory capacity, vital capacity, functional residual capacity, and total lung capacity. Other lung measurements take into account the rate of air flow. The forced vital capacity is the amount of air a person can expire following a maximum inspiration, expiring as forcefully and rapidly as possible. The forced expiratory volume is a measure of the percentage of the forced vital capacity that can be exhaled within a certain time frame.

Minute ventilation is the total amount of air that flows into or out of the respiratory system in a minute. Minute alveolar ventilation is a measure of the volume of fresh air reaching the alveoli each minute, which is minute ventilation corrected for dead space volume. To increase minute alveolar ventilation, it is more efficient to increase tidal volume than respiration rate.

EXERCISES

Multiple-Choice Questions

1. Which of the following is a component of internal respiration?
   a) ventilation
   b) transport of oxygen in the blood
   c) diffusion of carbon dioxide from tissues to blood
   d) oxidative phosphorylation

2. Which of the following is *not* a function of the conducting zone of the respiratory system?
   a) humidifying the air
   b) adjusting the air to body temperature
   c) exchanging gases between respiratory system and blood
   d) secreting mucus

3. The smallest airways in the conducting zone are
   a) terminal bronchioles.
   b) respiratory bronchioles.
   c) alveolar ducts.
   d) alveolar sacs.

4. Surfactant is secreted by
   a) goblet cells.
   b) alveolar macrophages.
   c) type I cells.
   d) type II cells.

5. The product of tidal volume and breathing frequency gives
   a) total lung capacity.
   b) alveolar ventilation.
   c) minute ventilation.
   d) dead space volume.

6. When all muscles of respiration are relaxed and alveolar pressure is zero, lung volume is equal to
   a) residual volume.
   b) vital capacity.
   c) functional residual capacity.

d) tidal volume.

7. Which of the following statements describes the lungs at the functional residual capacity?
   a) Atmospheric, intra-alveolar, and intrapleural pressures are all equal.
   b) The lungs tend to collapse due to their elastic properties.
   c) The chest wall tends to collapse due to its elastic properties.
   d) all of the above

8. Which of the following factors decreases airway resistance?
   a) activation of the parasympathetic nervous system
   b) epinephrine
   c) histamine

9. Pulmonary surfactant
   a) prevents collapse of alveoli.
   b) prevents small alveoli from joining with larger alveoli.
   c) increases lung compliance.
   d) all of the above

10. Which of the following muscles contracts during quiet expiration?
    a) diaphragm
    b) internal intercostals
    c) external intercostals
    d) none of the above

Objective Questions

1. Contraction of the diaphragm increases the rate of air flow during forced expiration. (true/false)

2. During inspiration, transpulmonary pressure (increases/decreases).

3. During inspiration, intrapleural pressure becomes (more/less) negative.

4. If airway resistance increases, a (higher/lower) alveolar pressure is

required to produce a given rate of air flow during expiration.

5. Pulmonary surfactant (increases/decreases) the surface tension of water.

6. Pulmonary surfactant (increases/decreases) lung compliance.

7. Dead space volume is the volume of air in the (conducting zone/respiratory zone).

8. (Obstructive/Restrictive) lung diseases are characterized by increased airway resistance.

Essay Questions

1. Describe the various ways in which the structure of the respiratory system is adapted to facilitate gas exchange.

2. Describe the balance between distending pressures and recoil forces acting on the lungs and chest wall at FRC. Include in your discussion an explanation of why intrapleural pressure is subatmospheric.

3. Explain the difference between minute ventilation and minute alveolar ventilation. Include a discussion of the effect of tidal volume on alveolar ventilation.

4. Describe the role of surfactant in decreasing the surface tension in alveoli. What are some of the consequences of decreasing surface tension?

5. Describe the mechanics of ventilation. What muscles contract for inspiration? What muscles contract for expiration? How does the elasticity of the lungs and chest wall affect ventilation?

Find the answers to these exercises, and additional study tools, at the Physiology Place (www.physiologyplace.com).

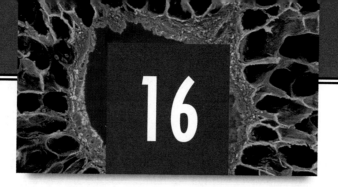

16

The Respiratory System: Gas Exchange and Regulation of Breathing

OBJECTIVES

- Describe the circulatory pathway for oxygenated and deoxygenated blood. Describe the exchange of gas in the lungs and in systemic tissues.

- List the normal partial pressures of oxygen and carbon dioxide in arterial and mixed venous blood, and explain how they contribute to the exchange of gases.

- Describe the mechanisms of transport of oxygen and carbon dioxide in the blood.

- Explain the relationship between the P_{CO_2} of blood and the pH of blood. Describe the actions of carbonic anhydrase in erythrocytes as blood passes through the systemic and pulmonary circulations.

- Describe the neural mechanisms that establish the respiratory rhythm. Distinguish between the respiratory centers that establish the rhythm and those that regulate the rhythm.

- Explain the role of peripheral and central chemoreceptors in the control of ventilation

- Explain how carbon dioxide affects ventilation.

- Describe how changes in P_{O_2} and P_{CO_2} in lung tissues can alter ventilation. Explain the ventilation/perfusion ratio.

- Explain how the respiratory system regulates acid-base balance of the blood by varying the rate of carbon dioxide expiration.

Above: Scanning electron micrograph of human lung airways

CHAPTER OUTLINE

Overview of the Pulmonary Circulation   496

Diffusion of Gases   497

Exchange of Oxygen and Carbon Dioxide   500

Transport of Gases in the Blood   505

Central Regulation of Ventilation   514

Control of Ventilation by Chemoreceptors   518

Local Regulation of Ventilation and Perfusion   521

The Respiratory System in Acid-Base Homeostasis   522

STUDY HINTS

1. *Diffusion, p. 111*

2. *Affinity, p. 73*

3. *Structures of the brainstem, p. 237*

4. *Control of skeletal muscle contraction, p. 340*

5. *Pathway of blood flow, p. 374*

6. *Components of blood, p. 404*

7. *Hemoglobin, p. 405*

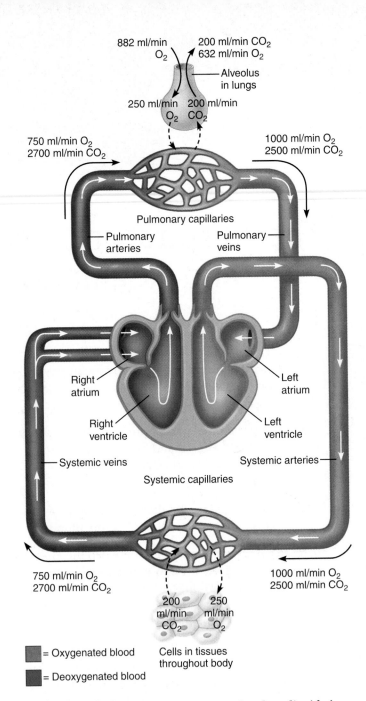

FIGURE 16.1 Movements of oxygen and carbon dioxide in pulmonary and systemic tissues during rest. *In capillary beds in the lungs, oxygen diffuses from alveoli to blood and carbon dioxide diffuses from blood to alveoli. In systemic capillary beds, oxygen diffuses from blood to the cells, whereas carbon dioxide diffuses from cells to the blood.*

Under resting conditions, how much blood is pumped to the lungs each minute?

The body may hunger for food and thirst for water, but it is absolutely voracious in its need for oxygen. The cells of the body consume approximately 250 mL of oxygen and produce approximately 200 mL of carbon dioxide each minute when at rest. To maintain homeostasis, the body must obtain the oxygen from the atmosphere and expel the carbon dioxide into the atmosphere.

To provide the cells with the oxygen they need, an average adult male inhales 6000 mL of air per minute, 4200 mL of which reaches the alveoli. About 882 mL of this air (21%) is oxygen. About 250 ml of this oxygen diffuses from the alveoli to the blood for subsequent consumption by the cells of the body; the remaining 632 mL is exhaled. With a consumption of 250 mL of oxygen per minute, the body consumes 360,000 mL of oxygen per day—nearly 100 gallons.

In Chapter 15 we saw how air is moved into and out of the lungs. In this chapter we learn how oxygen and carbon dioxide are exchanged efficiently between the alveoli and the blood and how the blood transports oxygen and carbon dioxide to and from the tissues, respectively. We also learn how ventilation is regulated to maintain adequate oxygen supply to tissues, even when demands change.

OVERVIEW OF THE PULMONARY CIRCULATION

Under most conditions, concentrations of oxygen and carbon dioxide in systemic arterial blood are maintained at relatively constant levels because oxygen moves from alveolar air into the blood at the same rate it is consumed

by the tissues, and because carbon dioxide moves into alveolar air from the blood at the same rate it is produced in the tissues (Figure 16.1). The ratio of the amount of carbon dioxide produced by the body to the amount of oxygen consumed is called the **respiratory quotient.** On average, cells consume 250 mL of oxygen per minute

5 liters

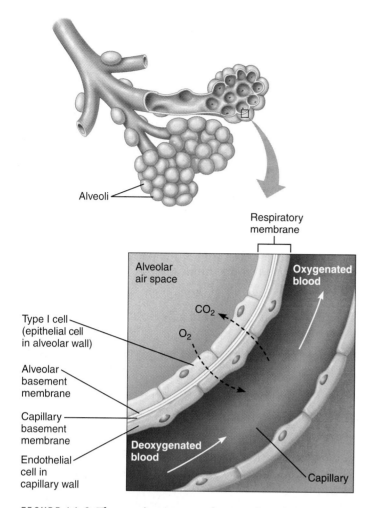

Alveoli

Respiratory membrane

Alveolar air space

Oxygenated blood

Type I cell (epithelial cell in alveolar wall)

CO_2

O_2

Alveolar basement membrane

Capillary basement membrane

Endothelial cell in capillary wall

Deoxygenated blood

Capillary

FIGURE 16.2 The respiratory membrane. *The exchange of oxygen and carbon dioxide between alveoli and blood occurs by simple diffusion across the respiratory membrane, which is composed of a single layer of type I epithelial cells lining the alveolus, a single layer of endothelial cells lining the capillary, and the alveolar and capillary basement membranes.*

while producing 200 mL of carbon dioxide under resting conditions, so the average respiratory quotient at rest is 0.8 (200/250 = 0.8).

Figure 16.1 shows the movements of oxygen and carbon dioxide into and out of the lungs and tissues under resting conditions. Oxygen enters the alveoli and carbon dioxide leaves the alveoli by bulk flow of air during ventilation. Deoxygenated blood returns via systemic veins to the right atrium of the heart. The blood then enters the right ventricle, which pumps blood through pulmonary arteries to the lungs. Each minute the deoxygenated blood in the pulmonary arteries transports approximately 750 mL of oxygen and 2700 mL of carbon dioxide to the lungs. In the pulmonary capillaries, oxygen diffuses from the alveoli to the blood at a rate of 250 mL/min., and carbon dioxide diffuses from the blood to the alveoli at a rate of 200 mL/min. Therefore, the oxygenated blood in the pulmonary veins transports approximately 1000 mL of oxygen per minute (750 mL/min. already in deoxy-

genated blood +250 mL/min. added from alveoli) and 2500 mL of carbon dioxide per minute (2700 mL/min. in deoxygenated blood −200 mL/min. lost to alveoli). The oxygenated blood leaves the lungs and returns to the left atrium via pulmonary veins. Blood then moves into the left ventricle, which pumps the blood through the systemic arteries to the cells of the body. In the systemic capillaries, 250 mL of oxygen diffuses from the blood to the cells each minute, and 200 mL of carbon dioxide diffuses from the cells to the blood each minute. The now deoxygenated blood returns to the right atrium and the cycle begins again.

The movement of oxygen and carbon dioxide between alveolar air and blood occurs by diffusion down concentration gradients. Oxygen is at a higher concentration in the alveoli and diffuses into the blood, whereas carbon dioxide is at a higher concentration in the blood and diffuses into the alveoli. In Chapter 4 we learned that oxygen and carbon dioxide are able to cross cell membranes by simple diffusion; we also learned that the rate of transport in simple diffusion is proportional to the magnitude of the concentration gradient, the surface area of the membrane through which it is moving, and the permeability of the membrane to that particular substance. Figure 16.2 shows the structure of the respiratory membrane, which is composed of three layers: type I epithelial cells in the alveolar wall and endothelial cells in the capillary wall "sandwiched" around their basement membranes. The respiratory membrane provides a large surface area of very thin membrane, favoring fast rates of diffusion for oxygen and carbon dioxide between alveolar air and blood.

Quick Test 16.1

1. Under normal conditions, if respiring tissues consume oxygen at a rate of 250 mL/min, at what rate does oxygen diffuse from the alveoli to blood?

2. If the cells of the body consume 300 mL of oxygen per minute and produce 270 mL of carbon dioxide per minute, then what is the respiratory quotient?

3. Which blood vessel contains blood with the greater concentration of oxygen, a pulmonary artery or a pulmonary vein?

DIFFUSION OF GASES

We have seen that the overall movements of oxygen and carbon dioxide between lungs and systemic tissues occur by diffusion down concentration gradients. In this section we look more closely at how these gradients are established and find that they are affected by the *partial pressures* and *solubilities* of oxygen and carbon dioxide.

According to the Ideal Gas Law, for any pure gas, $PV = nRT$, where P = pressure, V = volume, n = number of moles, R = universal gas constant, and T = temperature. If V and T are constant, then the pressure exerted by a gas is directly proportional to the number of moles of the gas. If 1 mole of helium produces a pressure of 100 mm Hg, for example, then 2 moles of helium at the same volume and temperature would exert a pressure of 200 mm Hg.

What about a mixture of two gases? If we start with 1 mole of helium at 100 mm Hg, then add 1 mole of oxygen, what will happen to the pressure? Note that the Ideal Gas Law does not depend on the kind of gas; all gases behave identically so far as pressure is concerned.

Therefore, the pressure of the mixture would be 200 mm Hg, just as if we had added another mole of helium.

In 1801, the English chemist John Dalton formulated a law regarding mixtures of gases. Dalton's Law states that the pressure exerted by a mixture of gases is equal to the sum of the pressures exerted by the individual gases occupying the same volume alone. The pressure exerted by the individual gases in a mixture is called the *partial pressure* of that gas. Therefore, the total pressure of a mixture of gases is a sum of the partial pressures exerted by each gas in the mixture:

$$P_{\text{total}} = P_1 + P_2 + P_3 + \cdots + P_n$$

where n is the total number of gases in the mixture.

Partial Pressure of Gases

We saw in Chapter 15 that the pressure of a gas depends on its temperature and the number of gas molecules contained in a given volume. (The exact relationship was described by the Toolbox entitled Boyle's Law and the Ideal Gas Law, p. 482.) A gas (air, for instance) is often a mixture of more than one type of molecule. The total pressure of such a gas is the sum of the pressures of the individual gases that make up the mixture:

$$P_{\text{total}} = P_1 + P_2 + P_3 + \cdots + P_n$$

where n is the number of gases. In any gas mixture, the **partial pressure** of an individual gas is the proportion of the pressure of the entire gas that is due to the presence of the individual gas (see Toolbox: Partial Pressures and Dalton's Law). For example, if helium and nitrogen are mixed together in equal proportions and the pressure of the mixture is 500 mm Hg, then half the pressure (250 mm Hg) is exerted by helium, and half by nitrogen. In this example, 500 mm Hg is the *total* pressure of the mixture, and 250 mm Hg is the *partial pressure* of helium or nitrogen.

The partial pressure of a gas mixture is determined by two factors: (1) the *fractional concentration* of that gas, which is the quantity of that gas (moles) relative to the total quantity of gas in the mixture, and (2) the total pressure exerted by the gas mixture. To find the partial pressure of a gas, one multiplies these two factors together.

Air is almost entirely composed of two gases: nitrogen and oxygen. (Other gases such as carbon dioxide, argon, neon, helium, and methane are found in only minute amounts in air). The amount of water vapor in air depends on the humidity. The total pressure of air can be described as the sum of the partial pressures of the gases found in air, plus the pressure of water vapor, as follows:

$$P_{\text{air}} = P_{N_2} + P_{O_2} + P_{H_2O}$$

On a molar basis, nitrogen is the most abundant gas in air, making up 79% of the molecules in air. Oxygen is the next most abundant at 21%. Depending on the humidity, water may become a critical component of air, decreasing the contribution of the other two gases to the total pressure of air. At sea level, air pressure is 760 mm Hg. Assuming zero humidity, the partial pressures of the two primary gases in air are:

$$P_{N_2} = 0.79 \times 760 \text{ mm Hg} = 600 \text{ mm Hg}$$

$$P_{O_2} = 0.21 \times 760 \text{ mm Hg} = 160 \text{ mm Hg}$$

Carbon dioxide accounts for only 0.03% of the air molecules, and thus the partial pressure of carbon dioxide in air is:

$$P_{CO_2} = 0.0003 \times 760 \text{ mm Hg} = 0.23 \text{ mm Hg}$$

Recall from Chapter 15 that as air moves through the conducting zone, it is humidified to saturation (100% humidity). At 100% humidity and 37°C, the partial pressure of water is 47 mm Hg. With water contributing significantly to the pressure of air that enters the lungs, the partial pressures of the other gases in the air decrease to the following: P_{N_2} = 575 mm Hg, P_{O_2} = 152 mm Hg, and P_{CO_2} = 0.21 mm Hg.

Solubility of Gases in Liquids

Gas molecules can either exist in gaseous form or be dissolved in liquids. The ability of a gas to dissolve in water has physiological significance because oxygen and carbon dioxide are exchanged between air in the alveoli and blood (which is primarily water) and are then transported by the blood. Even when gas molecules are dissolved in a liquid, they have a certain partial pressure. When a gas mixture and a liquid are in contact with each other, gas molecules will dissolve in the liquid until the system reaches equilibrium, at which point gas molecules dissolve in the liquid at the same rate that gas molecules move from the liquid to the gaseous state. At equilibrium, the dissolved gas molecules and those in the gaseous phase are said to be at the same partial pressure. Under these conditions, the concentration of gas molecules in the liquid is proportional to the partial pressure of the gas.

At a given partial pressure, the relative concentrations of different dissolved gas molecules differ from one gas to another because some gases are more soluble in a given liquid than other gases; that is, some gases mix with the liquid more easily. For example, carbon dioxide is nearly 30 times more soluble in blood than is oxygen. Therefore, the concentration of dissolved gas molecules depends not only on the partial pressure, but also the solubility of the gas in that particular liquid. The relationship among the concentration, partial pressure, and solubility of a gas is described by Henry's Law, mathematically expressed as:

$$c = kP$$

where c is the molar concentration (moles of gas/liter of liquid) of the dissolved gas, P is the partial pressure of the gas in atmospheres (1 atmosphere = 760 mm Hg), and k is Henry's Law constant, which varies based on the gas and the temperature (see Toolbox: Henry's Law and Solubility of Gases, p. 502).

Let's first look at two common examples of Henry's Law. Think about the last time that you opened a bottle of a carbonated beverage. The beverage was bottled under high pressure to force more gas, primarily carbon dioxide, to dissolve in the liquid. When the bottle was uncapped, the gas pressure in the bottle decreased as it equilibrated with atmospheric pressure; the pressure of carbon dioxide in the bottle decreased to approximately 0.23 mm Hg. When the partial pressure of carbon dioxide in the gas in the bottle decreased, carbon dioxide was no longer at equilibrium in the gas and in the liquid, so gas bubbles emerged from the liquid as they came out of so-

lution, which decreased the partial pressure of carbon dioxide in the liquid until a new equilibrium was established. A similar phenomenon occurs during the *bends*, a condition that occurs when a scuba diver rises too rapidly from deep water to the surface. Under water, pressure increases 760 mm Hg, or 1 atmosphere, for every 30 feet of depth. Gases, especially nitrogen, dissolve more readily in the blood as pressure increases in deep water. If a diver comes to the surface too quickly, nitrogen gas bubbles form in the blood as the pressure decreases. These gas bubbles commonly lodge in joints and in the nervous system, causing problems as minor as discomfort upon the bending of joints or as severe as paralysis. Gas bubbles can also clog blood vessels, producing *air embolisms*. By rising to the surface slowly, divers can protect themselves from the bends by allowing the nitrogen gas to come out of solution slowly.

Figure 16.3a illustrates the solubility of oxygen in water. The system starts off exposing air, with a partial pressure of 100 mm Hg of oxygen, to water. Initially, no oxygen is dissolved in the water. Over time, oxygen dissolves in the water, and the concentration of oxygen in the water increases until equilibrium is reached—that is, until the number of oxygen molecules dissolving in water per unit time equals the number of oxygen molecules leaving the water. When a gas is in equilibrium with a liquid, the partial pressure of the gas in the liquid equals the partial pressure of the gas in the air; at equilibrium, the partial pressure of oxygen in the water is 100 mm Hg.

That the partial pressures of oxygen in air and water are equal at equilibrium does not mean that the *concentrations* of oxygen in the air and in the water are equal. In fact, the concentrations are far from equal because oxygen has low solubility in water. The concentration of oxygen in the air at 100 mm Hg and 37°C (as determined by Boyle's Law) is 5.2 mmoles/liter, whereas the concentration of oxygen in the water at 100 mm Hg and 37°C (as determined by Henry's Law) is 0.15 mmoles/liter.

We can perform the same analysis with carbon dioxide, again starting with a partial pressure of 100 mm Hg (Figure 16.3b). When at equilibrium with water, the concentration of carbon dioxide in air is 5.2 mmole/liter, the same as that of oxygen at 100 mm Hg. However, the concentration of carbon dioxide in water is 3.0 mmoles/liter, which is much greater than the concentration of oxygen in water at the same partial pressure. Thus we can see that carbon dioxide is much more soluble in water than is oxygen.

Quick Test 16.2

1. At sea level, atmospheric pressure is 760 mm Hg. However, at the top of Mount Everest, atmospheric pressure is only 250 mm Hg. Given that air in both places is 21% oxygen, what is the partial pressure of oxygen at the top of Mount Everest?

FIGURE 16.3 Solubilities of oxygen and carbon dioxide in water. *The beakers at left depict the initial conditions in which a gas at 100 mm Hg has just been placed over water; the beakers at right show the conditions after equilibration, when the gas has dissolved in the water until the gas's partial pressures in both media are equal.* **(a)** *After equilibration, the concentration of oxygen in water is much less than its concentration in air, indicating that the solubility of oxygen in water is low.* **(b)** *After equilibration, the concentration of carbon dioxide in water is greater than that of oxygen in water at the same partial pressure, indicating that carbon dioxide is more soluble in water than is oxygen.*

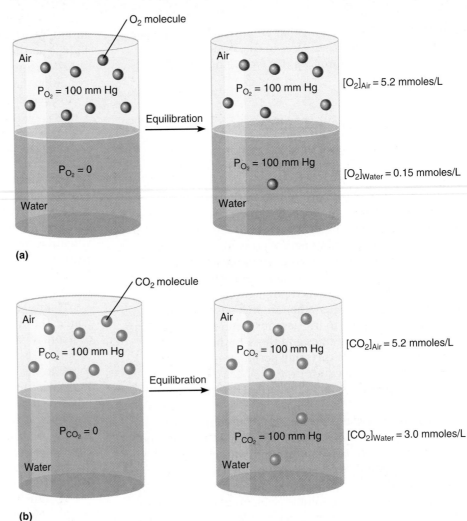

(a)

(b)

2. The partial pressure of helium in a gas mixture is 50 mm Hg. Helium is at equilibrium with water. What is the partial pressure of helium dissolved in water?

EXCHANGE OF OXYGEN AND CARBON DIOXIDE

In mixtures of gases, a particular gas will diffuse down its *partial pressure gradient*. The exchange of oxygen and carbon dioxide between the alveoli and the blood, and between the blood and systemic tissues, occurs by the same mechanism: Each gas diffuses down its own partial pressure gradient. This section describes the partial pressures that exist in the alveoli and blood, and the resulting exchange of gases that occur.

Gas Exchange in the Lungs

Representative partial pressures of oxygen and carbon dioxide in alveolar air, blood, and respiring tissues are shown in Figure 16.4 and tabulated in Table 16.1. Recall that in atmospheric air, the P_{O_2} is 160 mm Hg, and the P_{CO_2} is 0.23 mm Hg. Notice in Figure 16.4, however, that the P_{O_2} in the alveoli is only 100 mm Hg, whereas the P_{CO_2} is 40 mm Hg. Alveolar gas pressures differ from atmospheric pressures for three reasons: (1) Exchange of gases occurs continually between alveolar air and capillary blood, (2) upon inspiration, fresh atmospheric air mixes with air rich in carbon dioxide and relatively poor in oxygen in the dead space of the conducting zone, and (3) air in the alveoli is saturated with water vapor.

The blood entering the pulmonary capillaries is deoxygenated blood, with a P_{O_2} of 40 mm Hg and a P_{CO_2} of 46 mm Hg. As this blood passes by the alveoli, oxygen and carbon dioxide diffuse down their partial pressure gradients: Oxygen diffuses from alveoli to blood (P_{O_2} alveolar = 100 mm Hg; P_{O_2} blood = 40 mm Hg) while carbon dioxide diffuses from blood to alveoli (P_{CO_2} blood = 46 mm Hg; P_{CO_2} alveolar = 40 mm Hg). Eventually, diffusion results in an equilibrium between alveolar air and capillary blood such that the blood leaving the pulmonary capillaries and entering the pulmonary veins has a P_{O_2} of 100 mm Hg and a P_{CO_2} of 40 mm Hg (the same partial pressures as those in the alveoli).

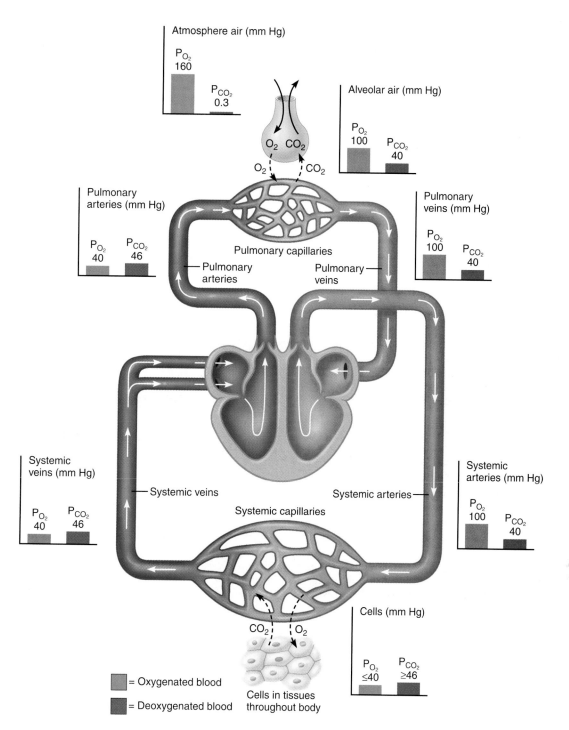

FIGURE 16.4 Partial pressures of oxygen and carbon dioxide in atmospheric air, alveolar air, and at various sites in the body. *Oxygen and carbon dioxide diffuse down their concentration gradients from high partial pressures to low partial pressures. In pulmonary capillary beds, oxygen diffuses from the alveoli to blood, and carbon dioxide diffuses from blood to alveoli. In cells, oxygen diffuses from blood to the cells, and carbon dioxide diffuses from cells to blood.*

At what body sites does the partial pressure of carbon dioxide exceed the partial pressure of oxygen?

Cells, systemic veins, the right atrium and ventricle, and pulmonary arteries

Gases can dissolve in liquids. The solubility of a gas in a liquid depends on temperature, the partial pressure of the gas over the liquid, the chemical properties of the gas, and the chemical properties of the liquid.

If a gas exists over a liquid, then the gas will dissolve in the liquid until equilibrium is reached. At equilibrium, the number of gas molecules dissolving in the liquid is equal to the number of gas molecules leaving the liquid. However, the concentrations of gas (moles/L) in the air and in the liquid are not equal at equilibrium. The concentration of gas in the liquid depends on the specific solubility for that gas in that particular liquid: the more soluble, the greater the concentration of gas in the liquid.

In 1803, William Henry developed a law to describe the solubility of gases in liquids. Basically, Henry's Law states that when the temperature is constant, the concentration of a gas in a liquid is proportional to its partial pressure. In mathematical terms, Henry's Law is

$$c = kP$$

where c is the molar concentration of the dissolved gas (moles/liter), P is the partial pressure of the gas, and k is Henry's Law constant at a specific temperature (determined experimentally; constants can be found in tables). The relationship between concentration and pressure can be written as

$$k = c/P$$

Because k is a constant, if the pressure changes, the relationship between the concentration of dissolved gas at the initial pressure (P_1) and at the subsequent pressure (P_2) can be described by

$$c_1/P_1 = c_2/P_2$$

Therefore, if pressure is doubled, then the concentration of gas dissolved in the liquid also doubles.

In the human body, gases such as nitrogen, oxygen, and carbon dioxide dissolve in plasma (and other body fluids) according to Henry's Law. The partial pressure of the dissolved gas equals the partial pressure of the gas surrounding the liquid. There is a direct relationship

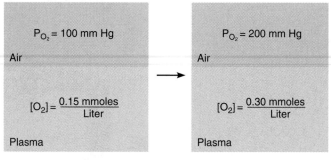

(a) Oxygen solubility in plasma

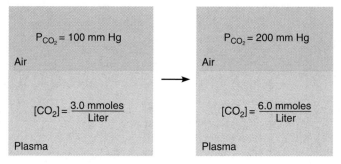

(b) Carbon dioxide solubility in plasma

between the partial pressure of a gas in plasma and the concentration of that gas in plasma because as the pressure increases, the concentration also increases. For example, the box figure shows the concentration of oxygen (a) and carbon dioxide (b) in plasma at two partial pressures, 100 mm Hg and 200 mm Hg. Notice that for either gas, the concentration dissolved in plasma doubles as the partial pressure doubles. Also note that more carbon dioxide dissolves in plasma than oxygen at any given partial pressure because carbon dioxide is more soluble in plasma. The concentration of a gas in plasma is directly related to the partial pressure of the gas, and movement of a gas by diffusion occurs from areas of high partial pressure (or high concentration) to areas of low partial pressure (low concentration).

Diffusion is a very rapid process (taking approximately 0.25 second for the blood to equilibrate with alveolar air) that is nearly complete by the time blood has traveled only one-third the length of a capillary (Figure 16.5). The rapidity of gas exchange provides a margin of safety in that gases can still equilibrate between capillary blood and alveolar air even if blood is flowing at a rate up to three times faster than the resting rate, such as can occur during exercise. The diffusion rate is rapid because of the thinness of the respiratory membrane.

Whenever the respiratory membrane is effectively thickened, gas exchange is hampered. During pulmonary

TABLE 16.1 TYPICAL PARTIAL PRESSURES OF OXYGEN AND CARBON DIOXIDE IN ATMOSPHERIC AIR AND IN VARIOUS SITES IN THE BODY

	OXYGEN	CARBON DIOXIDE
Atmospheric air	160 mm Hg	0.3 mm Hg
Alveolar air	100 mm Hg	40 mm Hg
Pulmonary veins	100 mm Hg	40 mm Hg
Systemic arteries	100 mm Hg	40 mm Hg
Cells	$\leq$40 mm Hg	$\geq$46 mm Hg
Systemic veins	40 mm Hg	46 mm Hg
Pulmonary arteries	40 mm Hg	46 mm Hg

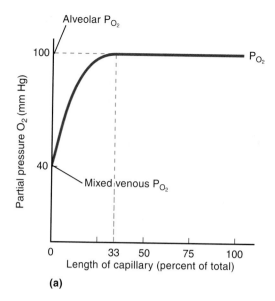

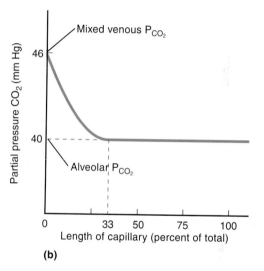

FIGURE 16.5 Gas exchange as a function of pulmonary capillary length. *In the rapid exchange of gases between alveolar air and pulmonary capillary blood, equilibration of the partial pressures of oxygen* **(a)** *and carbon dioxide* **(b)** *occurs within the first 33% of a capillary's length. Mixed venous blood refers to the blood in the pulmonary artery that contains blood returned to the heart from the systemic veins.*

edema, for example, excess fluid builds up in the interstitial spaces of the lungs and in the alveoli, effectively increasing the distance between alveoli and capillaries and thus increasing the thickness of the diffusion barrier between air and blood. This thicker barrier decreases the rate of diffusion of gases, which reduces the exchange of gases between alveolar air and capillary blood, resulting in systemic arterial blood with a lower P_{O_2} and a higher P_{CO_2} than normal. The body will try to compensate for the decreased P_{O_2} and increased P_{CO_2} in arterial blood by increasing ventilation. (See When It Goes Wrong: Pulmonary Edema, p. 504, for more details.)

Gas Exchange in Respiring Tissue

Blood oxygenated in the pulmonary capillaries returns to the left atrium via the pulmonary veins. The blood then flows into the left ventricle, which pumps it out to the systemic capillaries, where exchange between tissue cells and blood occurs. Blood entering the systemic capillaries has a P_{O_2} of 100 mm Hg and a P_{CO_2} of 40 mm Hg (see Figure 16.4). In the interstitial fluid surrounding the capillaries, the P_{O_2} is lower due to oxygen consumption by respiring cells, and the P_{CO_2} is higher due to its production by the cells. When oxygen and carbon dioxide diffuse down their pressure gradients, oxygen moves from blood to the tissue, and carbon dioxide moves from the tissue to blood.

The actual partial pressures of oxygen and carbon dioxide in the venous blood leaving the systemic capillaries vary depending on the metabolic activity of the tissue and the blood flow to the tissue. During intense exercise, for example, the metabolic activity of skeletal muscle is high relative to its blood flow, and thus the P_{O_2} in the interstitial fluid around the muscle tissue will be lower, and the P_{CO_2} will be higher, than they are in less active tissue. With larger pressure gradients between the capillary blood and the interstitial fluid, gas exchange occurs more rapidly, and more gas molecules are exchanged. More

oxygen diffuses from the blood to the interstitial fluid, and more carbon dioxide diffuses from the interstitial fluid to the blood. Therefore, venous blood coming from capillaries that supply active tissue will have a lower P_{O_2} and a higher P_{CO_2} than venous blood coming from capillaries supplying less active tissue.

All venous blood returns to the right atrium and *mixes* together before being pumped by the right ventricle into the pulmonary artery. Therefore, the blood in the pulmonary artery is called **mixed venous blood.** Typically, the P_{O_2} of mixed venous blood is 40 mm Hg for a person at rest, whereas the P_{CO_2} is 46 mm Hg (see Figure 16.4). During strenuous exercise, however, the value for P_{O_2} decreases, and that for P_{CO_2} increases.

Pulmonary edema, the accumulation of excess fluid in the lungs, is a fairly common but dangerous disorder. The condition is marked by signs of respiratory distress (including *tachypnea*, or rapid, shallow breathing) and by low oxygen levels in the tissues (*hypoxia*), which may be evident as a bluish coloration (*cyanosis*) of the skin and mucous membranes.

Pulmonary edema occurs in two stages: *interstitial edema*, in which excess fluid accumulates in the interstitial spaces in lung tissue, and *alveolar edema*, in which fluid accumulates in the alveoli. In severe cases, fluid may even move into the airways,

in which case the affected person may cough up a frothy foam. Pulmonary edema interferes with breathing in two ways: (1) by increasing the distance gases must diffuse to move between alveolar air and capillary blood, which impedes gas exchange, and (2) by interfering with the action of pulmonary surfactant, which causes a decrease in lung compliance and thus an increase in the work of breathing.

Pulmonary edema is similar to systemic edema (described in Chapter 14). The most common cause of pulmonary edema is increased hydrostatic pressure in the pulmonary capillaries. Left heart failure (see Chapter 13,

p. 434) can cause a backup of blood in the pulmonary circulation, which increases the pressure in the pulmonary veins. As pressure builds up in the pulmonary capillaries, fluid is pushed out of the capillaries and into the interstitial space of the lungs.

Treatment of pulmonary edema is critical because it is a life-threatening situation. The symptoms of pulmonary edema are treated by administering oxygen and diuretics (medications that increase fluid output by the kidneys). However, the cause of the pulmonary edema must be determined and treated appropriately once the symptoms are stabilized.

Quick Test 16.3

1. List the normal resting values for P_{O_2} and P_{CO_2} (in mm Hg) for alveolar air, arterial blood, and mixed venous blood.

2. When oxygen diffuses from alveolar air to pulmonary capillary blood, what causes it to move? Why does carbon dioxide leave the blood and enter the alveoli?

3. At a certain time oxygen stops diffusing into the blood as the blood moves through the lungs. Why?

4. Why is blood in the pulmonary artery referred to as mixed venous blood?

Determinants of Alveolar P_{O_2} and P_{CO_2}

Because alveolar P_{O_2} and P_{CO_2} determine systemic arterial P_{O_2} and P_{CO_2}, it is important to understand the factors that affect alveolar partial pressures. Alveolar P_{O_2} and P_{CO_2} are normally determined by only three factors: (1) the P_{O_2} and P_{CO_2} of inspired air, (2) minute alveolar ventilation (the volume of fresh air reaching the alveoli each minute), and (3) the rates at which respiring tissues consume oxygen and produce carbon dioxide.

Under most conditions, partial pressures in inspired air do not change appreciably and can be assumed to be constant (exceptions include changes in altitude and the use of artificial breathing mixtures such as 95% O_2 : 5% CO_2 for therapeutic purposes). Therefore, alveolar partial pressures are normally determined by alveolar ventilation and the rates of oxygen consumption and carbon dioxide production. In particular, the crucial factor is the alveolar ventilation *relative to* the rate of oxygen consumption or carbon dioxide production. When alveolar ventilation (or air flow) is constant but oxygen consumption increases, the P_{O_2} of alveolar air decreases. When oxygen consumption is constant but alveolar ventilation decreases, the P_{O_2} of alveolar air also decreases. However, if oxygen consumption and alveolar ventilation *both* increase proportionally, alveolar P_{O_2} does not change.

The dependence of alveolar partial pressures on alveolar ventilation, oxygen consumption, and carbon dioxide production can be summed up in the following two rules:

1. When alveolar ventilation increases relative to oxygen consumption and carbon dioxide production (that is, when alveolar ventilation exceeds the demands of the tissues), alveolar P_{O_2} increases and P_{CO_2} decreases.

2. When alveolar ventilation decreases relative to oxygen consumption and carbon dioxide production (that is, when alveolar ventilation does not meet the demands of the tissues), alveolar P_{O_2} decreases and P_{CO_2} increases.

Under normal circumstances, alveolar ventilation matches the demands of the tissues; that is, if oxygen consumption and carbon dioxide production increase, then alveolar ventilation will increase as well. This appropriate increase in alveolar ventilation is called **hyperpnea,** and the mechanisms that trigger it are described later in this chapter. In hyperpnea, arterial P_{O_2} and P_{CO_2} do not change because alveolar ventilation increases "in step with" the increased rates of oxygen consumption and carbon dioxide production. In **hypoventilation,** alveolar ventilation is insufficient to meet the demands of the tissues; in this case, arterial P_{CO_2} rises above the normal value of 40 mm Hg, and arterial P_{O_2} decreases below the normal value of 100 mm Hg. In **hyperventilation,** alveolar ventilation exceeds the demands of the tissues, and arterial P_{CO_2} decreases to less than 40 mm Hg and P_{O_2} increases to greater than 100 mm Hg. Both hypoventilation and hyperventilation can alter normal functioning of the body, including the nervous system. (A review of terms used in respiratory physiology is provided in Table 16.2.)

TABLE 16.2 SOME TERMS USED IN RESPIRATORY PHYSIOLOGY

TERM	DEFINITION
Hyperpnea	An increase in ventilation to meet an increase in the metabolic demands of the body
Dyspnea	Labored or difficult breathing
Apnea	Temporary cessation of breathing
Hyperventilation	A condition in which ventilation exceeds the metabolic demands of the body
Hypoventilation	A condition in which ventilation is insufficient to meet the metabolic demands of the body
Hypoxia	A deficiency of oxygen in the tissues
Hypoxemia	A deficiency of oxygen in the blood
Hypercapnia	An excess of carbon dioxide in the blood
Hypocapnia	A deficiency of carbon dioxide in the blood

Quick Test 16.4

1. What three factors determine alveolar P_{O_2} and P_{CO_2}?
2. When alveolar ventilation increases, what happens to alveolar P_{O_2} and P_{CO_2}?
3. What happens to alveolar P_{CO_2} when a person hyperventilates? What happens to alveolar P_{CO_2} when a person hypoventilates?
4. If alveolar P_{O_2} decreases and P_{CO_2} increases, what happens to arterial P_{O_2} and P_{CO_2}?

TRANSPORT OF GASES IN THE BLOOD

The previous section explained how gas is exchanged between air and blood, and between tissue and blood. This section explains how the gas is transported within the blood. We saw that gases are not very soluble in blood. For example, when arterial P_{O_2} is at the normal value of 100 mm Hg, blood contains only about 3 mL of dissolved oxygen per liter of blood. To deliver dissolved oxygen to tissues at the normal resting rate of approximately 250 mL/min, cardiac output would have to be about 83 liters/min if the total oxygen concentration in the blood were this low. Because the normal resting cardiac output is approximately 5 liters/min, oxygen must be trans-

ported in blood by a more efficient means other than merely being dissolved in plasma. The same is true for carbon dioxide. The next sections describe how oxygen and carbon dioxide are transported in blood.

Oxygen Transport in the Blood

Every liter of arterial blood contains about 200 mL of oxygen. Approximately 3 mL of this oxygen (1.5%) is dissolved in the plasma or in the cytosol of erythrocytes, and only this dissolved oxygen contributes to the P_{O_2} in blood. The remaining 197 mL oxygen (98.5%) is transported bound to **hemoglobin,** a protein found in erythrocytes. Although the bound oxygen does not contribute to the P_{O_2}, it is in equilibrium with the dissolved oxygen, and thus the amount of oxygen bound to hemoglobin is a function of the P_{O_2}.

As we saw in Chapter 13, hemoglobin consists of four subunits, each of which contains a *globin* (globular polypeptide chain) and a *heme group* that contains iron. The heme groups are each capable of binding one oxygen molecule, so each hemoglobin molecule can carry four oxygen molecules. When all oxygen-binding sites on a hemoglobin molecule are occupied, the hemoglobin is said to be *saturated.* The complex of hemoglobin and bound oxygen is called *oxyhemoglobin;* a hemoglobin molecule without any oxygen is called *deoxyhemoglobin.*

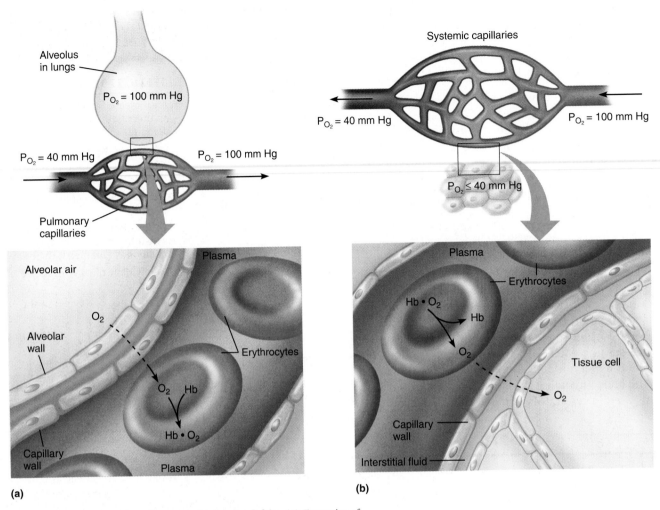

FIGURE 16.6 Transport of oxygen by hemoglobin. (a) *Formation of oxyhemoglobin. Once oxygen diffuses from alveolar air to blood in pulmonary capillaries, it diffuses into erythrocytes and binds to hemoglobin for transport in the blood.* **(b)** *Dissociation of oxygen from hemoglobin. In systemic capillaries, hemoglobin in erythrocytes releases oxygen, which then diffuses from the blood into tissue cells.*

What is the oxygen that moves into respiring tissue used for?

In the lungs, as oxygen molecules move from alveolar air to capillary blood, they bind to hemoglobin (Figure 16.6a); when the blood reaches respiring tissues, oxygen molecules dissociate from the hemoglobin and diffuse to the cells (Figure 16.6b). For hemoglobin to function in oxygen transport, it is critical that it binds the oxygen *reversibly;* that is, tightly enough so that it can pick up large quantities of oxygen in the lungs, but not so tightly that it cannot release the oxygen into the respiring tissues later on.

The binding or release of oxygen depends on the P_{O_2} in the fluid surrounding hemoglobin. High P_{O_2} facilitates the binding of oxygen with hemoglobin, whereas low P_{O_2} facilitates release of oxygen from hemoglobin. The reaction of oxygen with hemoglobin can be written as

$$Hb + O_2 \rightleftharpoons Hb \cdot O_2$$

where Hb is deoxyhemoglobin, O_2 is the dissolved oxygen in blood, and $Hb \cdot O_2$ is oxyhemoglobin. The law of mass action states that an increase in the concentration of the reactants drives the reaction to the right, resulting in the generation of more product. Therefore, as oxygen levels in the pulmonary capillaries increase, more oxyhe-

To generate ATP through the electron transport chain and oxidative phosphorylation

moglobin is formed. Conversely, as oxygen levels in the systemic capillaries decrease, the reaction is driven to the left to release oxygen from the hemoglobin.

Note, however, that because hemoglobin can bind up to four oxygen molecules, the previous equation could be written as follows:

$$Hb \underset{\uparrow O_2}{\rightleftharpoons} Hb{\cdot}O_2 \underset{\uparrow O_2}{\rightleftharpoons} Hb{\cdot}(O_2)_2 \underset{\uparrow O_2}{\rightleftharpoons} Hb{\cdot}(O_2)_3 \underset{\uparrow O_2}{\rightleftharpoons} Hb{\cdot}(O_2)_4$$

The law of mass action is still in effect in that the more oxygen is available, the more oxyhemoglobin is formed. When all oxygen-binding sites on a hemoglobin molecule are occupied, the hemoglobin molecule is said to be 100% saturated.

When hemoglobin is 100% saturated, 1 gram of hemoglobin carries 1.34 mL of oxygen. The normal concentration of hemoglobin in the blood is 12–17 gm/dL, or an average of 150 gm/liter. Therefore, the oxygen-carrying capacity of the hemoglobin in blood is about 200 mL of oxygen per liter of blood (1.34 mL/gram × 150 grams/liter). At the normal arterial P_{O_2} of 100 mm Hg, hemoglobin is at approximately 98% of its oxygen-carrying capacity (is 98% saturated) (Figure 16.7a). When cardiac output is 5 liters per minute, the blood supplies almost 1000 mL of oxygen to respiring tissues each minute (5 liters blood/min. × 200 mL O_2/liter blood = 1000 mL O_2/min.). Because respiring tissues need only about 250 mL of O_2 per minute, only 25% of the oxygen diffuses into respiring cells, which means that 75% of the binding sites on hemoglobin are still occupied when blood leaves the tissue, which occurs at a P_{O_2} of 40 mm Hg (Figure 16.7b). Therefore, in mixed venous blood under resting conditions, hemoglobin is still 75% saturated.

Anemia is a decrease in the oxygen-carrying capacity of blood. There are many causes of anemia, including a deficiency or defect in hemoglobin (see When It Goes Wrong: Anemia, p. 510). With less functioning hemoglobin in the blood, the oxygen-carrying capacity is decreased, and tissues may not be supplied with the oxygen they need, even when the P_{O_2} of blood is normal. Therefore, people suffering from anemia tire more easily.

The Hemoglobin-Oxygen Dissociation Curve

The relationship between P_{O_2} and hemoglobin saturation just described can be summarized in the hemoglobin-oxygen dissociation curve, a plot of the percent saturation of hemoglobin as a function of the P_{O_2} (Figure 16.8). Although the percent saturation of hemoglobin increases as the P_{O_2} increases, the binding of oxygen to hemoglobin is not linear, but instead S-shaped (sigmoidal), because the ability of hemoglobin to bind oxygen depends on how many oxygen molecules are already bound. Specifically, the binding of one oxygen molecule to hemoglobin increases the affinity of the hemoglobin molecule for oxygen and therefore increases the likelihood that another

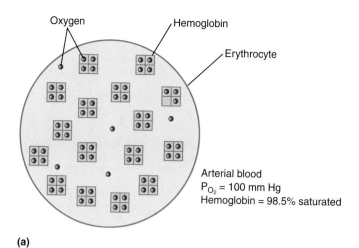

(a)

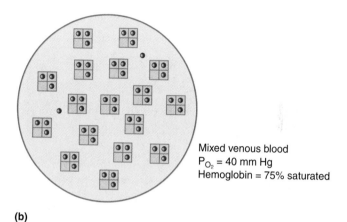

(b)

FIGURE 16.7 Saturation of hemoglobin. (a) *In arterial blood the partial pressure of oxygen is 100 mm Hg and hemoglobin is 98.5% saturated; that is, almost all binding sites are occupied by oxygen. Very little oxygen is dissolved in the cytosol of the erythrocyte.* **(b)** *In mixed venous blood the partial pressure of oxygen is 40 mm Hg and hemoglobin is 75% saturated; that is, three of every four binding sites are occupied by oxygen.*

oxygen will bind with hemoglobin. The binding of oxygen to one of the subunits of a hemoglobin molecule induces a conformational change in the molecule that increases the affinity of the remaining subunits for oxygen (a process called *positive cooperativity*). Once this conformational change occurs, a given increase in P_{O_2} yields a larger increase in percent saturation.

At very low partial pressures (less than 15 mm Hg, which are not usually found in blood), most hemoglobin molecules have no oxygen bound to them. Under these conditions, the affinity of hemoglobin for oxygen is relatively low, and a given increase in P_{O_2} yields a small increase in percent saturation. As the P_{O_2} increases, more hemoglobin molecules have acquired at least one oxygen molecule, causing an increase in the affinity of hemoglobin for other oxygen molecules. This relationship is seen in the steep part of the hemoglobin-oxygen dissociation curve at P_{O_2} values between approximately 15 mm Hg

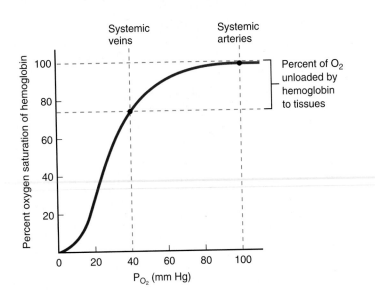

	P_{O_2} (mm Hg)	Percent hemoglobin saturation
Systemic arteries	100	98%
Systemic veins	40	75%

FIGURE 16.8 Hemoglobin-oxygen dissociation curve. *The binding of oxygen to hemoglobin depends on the partial pressure of oxygen in blood. At low partial pressures, little oxygen is bound to hemoglobin. As P_{O_2} increases, binding increases and then levels off as saturation approaches 100%. The table shows that at the normal resting partial pressures of the systemic arteries and veins, the percent saturation of hemoglobin varies by only about 25%.*

According to the graph, at what partial pressure of oxygen is hemoglobin 50% saturated?

and 60 mm Hg. At P_{O_2} values higher than approximately 60 mm Hg, the slope of the curve decreases because fewer binding sites are available as saturation increases. Above a P_{O_2} of approximately 80 mm Hg, the slope of the curve becomes nearly flat.

We can relate the hemoglobin-oxygen dissociation curve to events occurring in the lungs and respiring tissues. Figure 16.8 indicates the P_{O_2} values and percent saturation of hemoglobin in the systemic arteries and veins under resting conditions. The P_{O_2} in the systemic arteries is approximately 100 mm Hg, and at this P_{O_2} hemoglobin is approximately 98% saturated. (To reach 100% saturation of hemoglobin would require a P_{O_2} of approximately 250 mm Hg.) In systemic veins, the P_{O_2} is approximately 40 mm Hg, and hemoglobin is approximately 75% saturated. Therefore, at rest, respiring tissues take up only 25% of the oxygen transported in

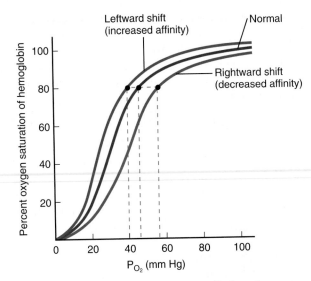

FIGURE 16.9 Effects of changes in the affinity of hemoglobin for oxygen on the hemoglobin-oxygen dissociation curve. *An increase in affinity of hemoglobin for oxygen causes a leftward shift in the curve, whereas a decrease in affinity of hemoglobin for oxygen causes a rightward shift in the curve.*

blood, leaving a large reserve of oxygen should the demands of respiring tissues increase.

Other Factors Affecting the Affinity of Hemoglobin for Oxygen

At least four other factors (discussed following this paragraph) affect the affinity of hemoglobin for oxygen. Changes in the affinity of hemoglobin for oxygen are reflected in rightward or leftward shifts in the hemoglobin-oxygen dissociation curve (Figure 16.9). Decreases in the affinity cause the curve to shift rightward, indicating that a higher P_{O_2} is required to achieve any given level of saturation; a rightward shift also indicates that oxygen is *unloaded* more easily from hemoglobin, making it more available to the tissue. Increases in affinity cause leftward shifts, indicating that a lower P_{O_2} is required to achieve any given level of saturation; a leftward shift also indicates that oxygen is *loaded* more easily onto hemoglobin. Under normal conditions, for example, a P_{O_2} of approximately 45 mm Hg produces 80% saturation of hemoglobin. With a rightward shift, a P_{O_2} of greater than 45 mm Hg is required to produce the same 80% saturation; with a leftward shift, a P_{O_2} of less than 45 mm Hg produces 80% saturation.

Now we consider the factors that affect the affinity of hemoglobin for oxygen. The first three factors—temperature, pH, and P_{CO_2}—all work to promote oxygen unloading from hemoglobin in respiring tissues and oxygen loading in the lungs, both of which increase the efficiency of oxygen exchange at and transport to the tissues that need it.

26 mm Hg

1. *Temperature.* Temperature affects the affinity of hemoglobin for oxygen by altering the structure of the hemoglobin molecule. (This factor is nonspecific in that temperature affects the tertiary structure of all proteins.) This structural change has important functional consequences. As metabolism of tissue increases, temperature increases, thereby decreasing the affinity of hemoglobin for oxygen (Figure 16.10a). As a result, more oxygen is unloaded in tissue that is highly active. Similarly, the decrease in temperature of blood as it travels to the lungs increases hemoglobin's affinity for oxygen, promoting oxygen loading.

2. *pH.* The effect of pH on the hemoglobin-oxygen dissociation curve is known as the **Bohr effect,** which is based on the fact that when oxygen binds to hemoglobin, certain amino acids in the protein release hydrogen ions:

$$Hb + O_2 \rightleftharpoons Hb{\cdot}O_2 + H^+$$

By the law of mass action, an increase in hydrogen ion concentration (a decrease in pH) pushes the reaction to the left, causing some oxygen to dissociate from hemoglobin, even when P_{O_2} is kept constant. The Bohr effect is important because when hydrogen ions bind to hemoglobin, they decrease the affinity of hemoglobin for oxygen (Figure 16.10b), and oxygen is unloaded. The hydrogen ion concentration tends to increase in active tissue, which facilitates oxygen unloading.

3. P_{CO_2}. The P_{CO_2} affects the affinity of hemoglobin for oxygen because carbon dioxide reacts reversibly with certain amino groups in hemoglobin to form **carbaminohemoglobin** ($HbCO_2$):

$$Hb + CO_2 \rightleftharpoons HbCO_2$$

Therefore, an increase of P_{CO_2} in the blood, such as occurs when metabolic activity increases, pushes the reaction to the right, forming more carbaminohemoglobin. When carbon dioxide is bound to hemoglobin, it changes hemoglobin's conformation and decreases its affinity for oxygen; this is known as the **carbamino effect.** The binding of carbon dioxide to hemoglobin is also one of the mechanisms of carbon dioxide transport in the blood (discussed in the next section).

4. *2,3-DPG.* The fourth factor, 2,3-DPG (2,3-diphosphoglycerate), is a chemical compound produced in erythrocytes from an intermediate compound in glycolysis, the anaerobic pathway by which erythrocytes obtain all their energy. When oxyhemoglobin is present in high concentrations, it inhibits the enzyme that forms 2,3-DPG, and thus 2,3-DPG levels are low and have little effect on hemoglobin affinity. However, if oxyhemoglobin levels are low, which occurs when oxygen supply is limited, then 2,3-DPG synthe-

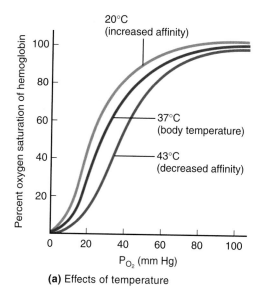

(a) Effects of temperature

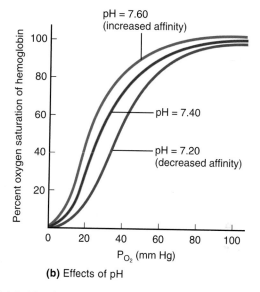

(b) Effects of pH

FIGURE 16.10 **Effects of temperature and pH on the hemoglobin-oxygen dissociation curve. (a)** *Increases or decreases in temperature from normal body temperature (37°C) cause decreases or increases, respectively, in the affinity of hemoglobin for oxygen.* **(b)** *Increases or decreases in pH from normal arterial pH (7.4) cause increases or decreases, respectively, in the affinity of hemoglobin for oxygen.*

sis occurs, and 2,3-DPG decreases the affinity of hemoglobin for oxygen. This enhances the unloading of oxygen in the respiring tissues that need it. Conditions that enhance the production of 2,3-DPG include anemia and high altitudes.

Another factor that affects the binding of oxygen to hemoglobin is carbon monoxide. Carbon monoxide is toxic, because when present it binds to hemoglobin more readily than oxygen, which prevents oxygen from binding and decreases the oxygen-carrying capacity of blood. (When It Goes Wrong: Carbon Monoxide Poisoning, www.physiologyplace.com, Challenge Yourself.)

Anemia, defined as a decrease in the oxygen-carrying capacity of blood, is generally associated with a low hematocrit, which can result from either a decrease in the number of erythrocytes or a decrease in the size of erythrocytes. However, a person can still have anemia if his or her hematocrit is normal but each red cell contains less than the normal concentration of hemoglobin. Because most of the oxygen transported in blood is bound to hemoglobin in erythrocytes, a decrease in erythrocyte abundance or size is associated with low levels of hemoglobin. There are six general categories of anemia:

1. *Nutritional anemias* are caused by a dietary deficiency, most commonly an iron deficiency. Because iron is a component of hemoglobin, an iron deficiency decreases hemoglobin synthesis. Even though the number of erythrocytes is normal, less hemoglobin is present in each erythrocyte, causing the erythrocytes to be smaller. Less hemoglobin means a decrease in the oxygen-carrying capacity of blood.

Another essential nutrient for normal oxygen transport is folic acid, which is required for the synthesis of thymine, one of the bases in DNA. A lack of folic acid affects all cells of the body that undergo rapid cell division, because DNA replication is required for cell division. Given that each second the body produces 2–3 million erythrocytes by the rapid cell division of certain precursor cells, insufficient folic acid in the diet results in fewer erythrocytes, each of which is larger than normal. Large erythrocytes, however, tend to be more fragile and have a shorter life span.

2. *Pernicious anemia* is caused by a deficiency of intrinsic factor, which is required for the absorption of vitamin B_{12} in the intestinal tract. Because vitamin B_{12}, like folic acid, is required for the synthesis of thymine, this anemia has the same characteristics as anemia caused by a folic acid deficiency.

3. *Aplastic anemia* is caused by a defect in the bone marrow, the primary site of erythrocyte and leukocyte production; the result is a deficiency of both classes of blood cells. Fewer erythrocytes causes anemia, whereas fewer leukocytes impairs the body's defense against pathogens such as bacteria and viruses.

4. *Renal anemia* is associated with decreased production of the hormone erythropoietin due to a pathological state in the kidneys. Because erythropoietin stimulates the synthesis of erythrocytes in the bone marrow, a decrease in erythropoietin production results in fewer erythrocytes in the blood.

5. *Hemorrhagic anemia* is caused by the rapid loss of blood. Within a few days following a severe hemorrhage, fluid intake and the shifting of body fluids between compartments increases the

Carbon Dioxide Transport in the Blood

Of all the carbon dioxide in the blood, 5–6% is dissolved, 5–8% is bound to hemoglobin as carbaminohemoglobin (as previously described), and 86–90% is dissolved in the blood as bicarbonate ions (HCO_3^-) (Table 16.3).

Bicarbonate is formed from carbon dioxide within erythrocytes in systemic capillaries, where carbon dioxide levels are highest. Erythrocytes are able to form bicarbonate because they contain the enzyme carbonic anhydrase.

The Role of Carbonic Anhydrase in Carbon Dioxide Transport

Carbonic anhydrase catalyzes the reversible reaction of carbon dioxide with water, which converts carbon dioxide to carbonic acid (H_2CO_3):

$$CO_2 + H_2O \xrightleftharpoons{\text{Carbonic anhydrase}} H_2CO_3$$

The pathway continues with the reversible dissociation of carbonic acid, which yields a hydrogen ion and a bicarbonate ion:

$$H_2CO_3 \rightleftharpoons H^+ + HCO_3^-$$

blood volume back to normal. However, erythrocyte production takes a few weeks, so even as blood volume returns to normal, the hematocrit remains low until new erythrocytes can be produced.

6. *Hemolytic anemia* is caused by the rupture, or *hemolysis,* of an excessive number of erythrocytes. An increase in hemolysis can be caused by infections such as malaria, or by defects in the erythrocytes, such as occurs in *sickle cell anemia.*

Sickle cell anemia is a hereditary disease caused by a defect in the gene that codes for hemoglobin. Individuals who inherit the gene from both parents have *sickle cell disease,* whereas those who inherit the gene from only one parent have *sickle cell trait.* About 1 in 12 African-Americans have sickle cell trait. A single base in the gene coding for the beta chain is wrong, causing it to code for the wrong amino acid. (Specifically, the sixth amino acid in the beta chain is normally glutamate, but is replaced by valine.) This single amino acid substitution causes the hemoglobin molecules in an erythrocyte to polymerize (link to one another) when oxygen levels are low, making the erythrocyte sickle-shaped and fragile (see the figure). Such erythrocytes also can adhere to blood vessels, producing blockages of blood flow.

Symptoms of sickle cell disease are indicative of the anemia caused by excessive hemolysis of the fragile erythrocytes and blockage of blood vessels. The blockage causes the tissue in the area to become *hypoxic* (low P_{O_2}), often resulting in pain. Other symptoms include swelling of the extremities, decreased immune function, gallstones, and kidney failure. Individuals with sickle cell disease are also more prone to strokes. (Sickle cell disease symptoms may appear in individuals with sickle cell trait if they are subjected to conditions of low oxygen.) Therapy for sickle cell disease is limited to treating the symptoms. Pain is treated with *analgesics* (pain blocking

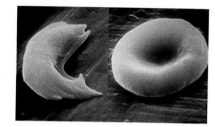

drugs), and the diet is often supplemented with folic acid.

Diseases can often be associated with a particular ethnic group or a geographical location. The predominance of sickle cell disease in African-Americans may have developed as a protection against malaria, which is caused by a parasite spread by mosquitoes. The parasites invade erythrocytes, where they proliferate. Sickle cell disease actually protects against malarial infections because the malaria parasite enhances the "sickling" of erythrocytes in individuals with sickle cell disease. Once "sickled," the erythrocytes are more fragile and are recognized as abnormal cells by the spleen. The spleen destroys these erythrocytes and the malaria parasite within them.

The complete pathway can be written as follows:

$$CO_2 + H_2O \xrightleftharpoons{\text{Carbonic anhydrase}} H_2CO_3 \rightleftharpoons H^+ + HCO_3^-$$

As shown in this final equation, carbon dioxide can be converted into H^+ and HCO_3^-. This reaction follows the law of mass action, meaning that an increase in carbon dioxide concentration (or an increase in P_{CO_2}) drives the reaction to the right, producing more hydrogen and bicarbonate ions; a decrease in carbon dioxide concentration (or a decrease in P_{CO_2}) drives the reaction to the left, producing carbon dioxide from hydrogen and bicarbonate ions. In addition to being important in the transport and exchange of carbon dioxide (as described in the next sec-

tion), this reaction is also important in acid-base balance, because an increase in P_{CO_2} causes an increase in the acidity of the blood (the concentration of hydrogen ions increases), and a decrease in P_{CO_2} causes a decrease in the acidity of the blood (the concentration of hydrogen ions decreases). The role of carbon dioxide in acid-base balance is described later in this chapter.

Carbon Dioxide Exchange and Transport in Systemic Capillaries and Veins

Respiring cells produce carbon dioxide at a rate of approximately 200 mL/min. at rest, and this carbon dioxide must be removed by the circulatory and respiratory

TABLE 16.3 CARBON DIOXIDE TRANSPORT IN BLOOD

FORM	SYSTEMIC ARTERIAL BLOOD		SYSTEMIC VENOUS BLOOD	
	Volume (mL CO_2/Liter of Blood)	% of Total CO_2 in Blood	Volume (mL CO_2/Liter of Blood)	% of Total CO_2 in Blood
Dissolved in blood	27	5.5	31	5.8
Dissolved as bicarbonate	439	89.6	470	87.0
Bound to hemoglobin	24	4.9	39	7.2
Total	490	100	540	100

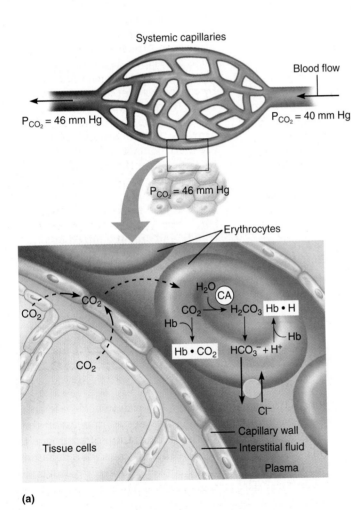

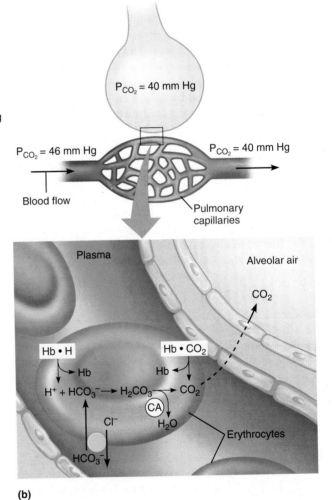

(a)

(b)

FIGURE 16.11 Carbon dioxide exchange and transport in blood. (a) *Carbon dioxide produced in the tissues diffuses into the blood and then into an erythrocyte. The increased P_{CO_2} in the erythrocyte causes most carbon dioxide molecules to be converted to bicarbonate, some to bind to hemoglobin, and some to remain dissolved in the blood. The bicarbonate is transported out of the erythrocyte into the plasma. As bicarbonate is produced, the hydrogen ions produced are buffered by binding to hemoglobin.* **(b)** *In the lungs, carbon dioxide diffuses from blood to alveolar air, decreasing the P_{CO_2} in blood. As P_{CO_2} in the erythrocytes decreases, bicarbonate enters the erythrocytes, and hydrogen ions are released from hemoglobin. The bicarbonate and hydrogen ions are then converted to carbon dioxide, which diffuses into the alveoli.*

systems. The carbon dioxide produced in respiring cells diffuses, based on its partial pressure gradient, first into interstitial fluid and then into the plasma (Figure 16.11a). Once CO_2 is dissolved in the plasma, P_{CO_2} increases, creating a pressure gradient between plasma and erythrocytes such that the pressure in the plasma is greater. Thus carbon dioxide diffuses from plasma into the erythrocytes. Whereas some carbon dioxide remains dissolved in the blood, and some binds to hemoglobin, most of the carbon dioxide is converted to bicarbonate and hydrogen ions by actions of carbonic anhydrase in the erythrocytes. This chemical reaction removes dissolved carbon dioxide from the blood and decreases the P_{CO_2}. The lower P_{CO_2} creates a greater gradient for diffusion of carbon dioxide from tissues to blood.

If bicarbonate and hydrogen ions were allowed to accumulate in erythrocytes, the reaction of carbon dioxide with water would slow down and eventually come to equilibrium. Under these conditions, carbon dioxide would no longer be converted to bicarbonate and hydrogen ions; instead, it would simply accumulate in the cells in its dissolved form, which could cause P_{CO_2} inside the cells to rise. As a result, the ability of the blood to transport carbon dioxide as bicarbonate would be hindered. To prevent this from occurring, bicarbonate and hydrogen ions must be removed from the erythrocytes. As bicarbonate ion levels in the erythrocyte increase, bicarbonate ions are transported into the plasma in exchange for chloride ions via a transport protein in the erythrocyte plasma membrane. The coupled movement of chloride ions into the erythrocyte and bicarbonate into the plasma is called the **chloride shift.** Meanwhile, many of the hydrogen ions produced by the reaction are buffered by binding to hemoglobin. Because of these chemical reactions and ion exchanges, most of the carbon dioxide in blood is transported as bicarbonate ions dissolved in the plasma (although some dissolved bicarbonate ions remain in the erythrocytes).

Carbon Dioxide Exchange and Transport in Pulmonary Capillaries and Veins

In the lungs, carbon dioxide diffuses from the blood in pulmonary capillaries to alveoli to be exhaled, thereby decreasing blood carbon dioxide levels (Figure 16.11b). The loss of carbon dioxide causes bicarbonate and hydrogen ions in erythrocytes to combine and form carbonic acid, which is converted by carbonic anhydrase to carbon dioxide and water. The carbon dioxide then diffuses into alveoli to be exhaled. As bicarbonate ions in the erythrocyte are used in the reaction, the levels inside the cell decrease. Bicarbonate ions are then transported from the plasma into erythrocytes in exchange for chloride ions. Most of the carbon dioxide formed by this reaction diffuses into the alveoli to be exhaled, further driving the reaction in the direction that favors the formation of more carbon dioxide.

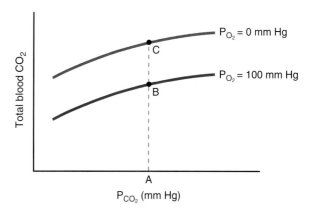

FIGURE 16.12 Effect of P_{O_2} on carbon dioxide transport. *Increased P_{O_2} in blood decreases the affinity of hemoglobin for carbon dioxide, which decreases the ability of the blood to transport carbon dioxide. Thus at a given P_{CO_2} (point A), more carbon dioxide is transported at a $P_{O_2} = 0$ mm Hg (point C) than at a $P_{O_2} = 100$ mm Hg (point B).*

Effect of Oxygen on Carbon Dioxide Transport

Just as the P_{CO_2} of the blood is one of many factors that affect the affinity of hemoglobin for oxygen, and thus the oxygen-carrying capacity of blood, the P_{O_2} of blood also affects the affinity of hemoglobin for carbon dioxide, and the blood's ability to transport carbon dioxide. This effect is illustrated in Figure 16.12, which shows the relation between total blood carbon dioxide and P_{CO_2} both in the absence of oxygen ($P_{O_2} = 0$ mm Hg) and in the presence of oxygen ($P_{O_2} = 100$ mm Hg). Under both conditions, total blood carbon dioxide rises with increasing P_{CO_2}, as expected. At a given P_{CO_2} (point A), however, the carbon dioxide content of the blood falls as the P_{O_2} rises (compare points B and C), a phenomenon known as the **Haldane effect.**

One reason for the Haldane effect is that binding of oxygen to hemoglobin decreases the affinity of hemoglobin for carbon dioxide. Therefore, when P_{O_2} increases, the amount of carbon dioxide that can be transported bound to hemoglobin decreases because the hemoglobin is carrying more oxygen. Conversely, a decrease in P_{O_2} promotes the binding of carbon dioxide to hemoglobin.

In respiring tissues, where P_{O_2} is low and P_{CO_2} is high, the Haldane effect promotes the loading of carbon dioxide onto hemoglobin while both the Bohr effect (effect of pH on hemoglobin's affinity for oxygen) and the carbamino effect (effect of carbon dioxide levels on hemoglobin's affinity for oxygen) work to promote oxygen unloading. In the lungs, where P_{O_2} is high and P_{CO_2} is low, the Haldane effect promotes unloading of carbon dioxide while the Bohr effect and the carbamino effect promote oxygen loading. These interactions are summarized in Figure 16.13.

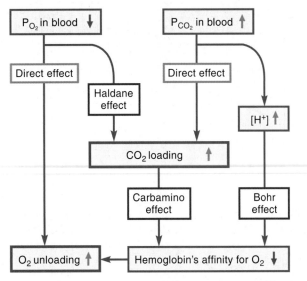

(a) CO$_2$ loading and O$_2$ unloading of hemoglobin in respiring tissue

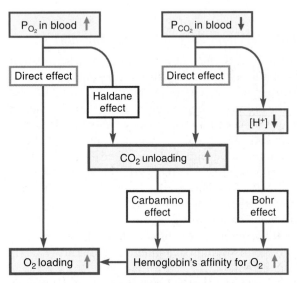

(b) CO$_2$ unloading and O$_2$ loading of hemoglobin in lungs

FIGURE 16.13 Effects of P_{O_2} and P_{CO_2} on carbon dioxide and oxygen loading and unloading. (a) *In systemic tissues, as P_{O_2} decreases and P_{CO_2} increases, carbon dioxide loading and oxygen unloading occur.* **(b)** *In the lungs, as P_{O_2} increases and P_{CO_2} decreases, carbon dioxide unloading and oxygen loading occur.*

Quick Test 16.5

1. Name three factors that affect the affinity of hemoglobin for oxygen.

2. What is the Bohr effect?

3. Carbon dioxide is carried in the blood in what three forms? How is the majority of carbon dioxide transported?

CENTRAL REGULATION OF VENTILATION

In the simplest terms, the job of the respiratory system is to deliver oxygen to and remove carbon dioxide from cells at a rate sufficient to keep up with metabolic demands. The signals that indicate whether the respiratory system is doing this job adequately are the partial pressures of oxygen and carbon dioxide in systemic arterial blood. If the respiratory system can maintain P_{O_2} and P_{CO_2} at or near the levels that are normal for systemic arterial blood (100 mm Hg and 40 mm Hg, respectively), then it is delivering oxygen and removing carbon dioxide at adequate rates.

To maintain normal arterial partial pressures, minute alveolar ventilation (the volume of air reaching the alveoli each minute) must be regulated such that it is neither too high nor too low. Recall from Chapter 15 that alveolar ventilation depends on the frequency and volume of breaths. In this section we will see how the breathing rhythm is generated, and how it is regulated to serve the metabolic demands of the body.

Neural Control of Breathing by Motor Neurons

Breathing is a cycle of inspiration followed by expiration. We saw in Chapter 15 that inspiration is an active process that requires contraction of the inspiratory muscles, including the diaphragm and external intercostals, and that during quiet breathing, expiration is a passive process in which no muscle contraction is required. During quiet breathing, therefore, the cycle consists of contraction of the inspiratory muscles during inspiration, followed by relaxation of the same muscles during expiration. During more active breathing, however, expiration becomes active and requires contraction of the expiratory muscles. Therefore, during active breathing the respiratory cycle consists of contraction of the inspiratory muscles and relaxation of the expiratory muscles during inspiration, followed by relaxation of the inspiratory muscles and contraction of the expiratory muscles during expiration.

Because the muscles of respiration are skeletal muscles, they are stimulated to contract by neural input from somatic motor neurons. The **phrenic nerve** innervates the diaphragm, whereas the internal and external **intercostal nerves** innervate the intercostal muscles. Figure 16.14 compares peripheral events occurring during quiet breathing to those occurring during active breathing. During quiet breathing, action potential bursts occur in inspiratory motor neurons. During active breathing, action potential bursts in inspiratory motor neurons occur asynchronously with bursts in expiratory motor neurons. The neural signals that control the cyclical contractions of the respiratory muscles are generated in respiratory control regions located in the brainstem.

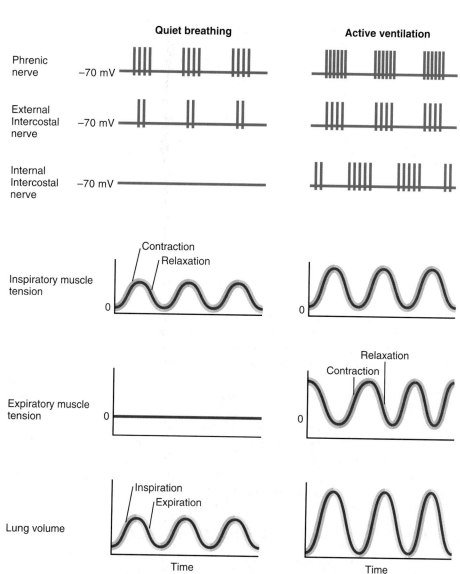

Quiet breathing   **Active ventilation**

Phrenic nerve   −70 mV

External Intercostal nerve   −70 mV

Internal Intercostal nerve   −70 mV

Inspiratory muscle tension — Contraction / Relaxation — 0

Expiratory muscle tension — 0 — Relaxation / Contraction

Lung volume — Inspiration / Expiration — 0 — Time   Time

FIGURE 16.14 A comparison of quiet breathing and active ventilation. *Ventilation involves cyclical changes in neural stimulation of respiratory muscles, which cause cyclical changes in lung volume. The nerve traces indicate the occurrences of action potentials over time. In quiet breathing, expiration is a passive process, and thus no neural or muscle activity of the expiratory muscles is present. During active ventilation, inspiratory neurons and muscles become more active, and expiratory neurons and muscles become active.*

Generation of the Breathing Rhythm in the Brainstem

Breathing is under both voluntary and involuntary control. We alter our breathing rhythm voluntarily when we speak, sing, hold our breath, or sigh in dismay or exasperation. Most of the time, however, our breathing is automatic and requires no conscious effort.

Central control of respiration is not fully understood, but research indicates that respiratory control regions are present in the medulla and pons of the brainstem (Figure 16.15). Two general classes of neurons located in these regions, **inspiratory neurons** and **expiratory neurons,** generate action potentials during inspiration and expiration, respectively. Within these two general classes are subclasses of neurons that differ slightly in their patterns of activity and functions.

Respiratory Control Centers of the Medulla

Two respiratory control centers are located on each side of the medulla: a ventrally located **ventral respiratory group** (VRG) and a more dorsally located **dorsal respiratory group** (DRG). The VRG contains two regions of primarily expiratory neurons and one region of primarily inspiratory neurons (see Figure 16.15). The inspiratory neurons show a ramp increase in activity during inspiration: The action potential frequency is low at the onset of inspiration but gradually increases until it reaches a maximum at the peak of inspiration, when it abruptly terminates and expiration begins (Figure 16.16). The DRG contains primarily inspiratory neurons, although there are some expiratory neurons (see Figure 16.15). The DRG inspiratory neurons have more complex patterns of activity than the VRG inspiratory neurons in that their firing pattern depends on the degree of stretch of the lungs.

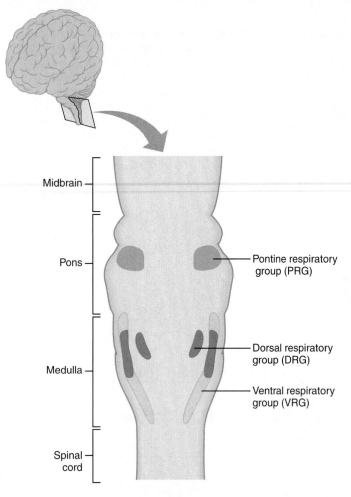

FIGURE 16.15 Brainstem centers of respiratory control.
Areas containing predominantly inspiratory neurons are indicated by blue, areas of predominantly expiratory neurons are indicated by yellow, and areas of scattered inspiratory and expiratory neurons as well as mixed neurons are indicated by green.

The figure is labeled as follows (top to bottom on the left): Midbrain, Pons, Medulla, Spinal cord. On the right: Pontine respiratory group (PRG), Dorsal respiratory group (DRG), Ventral respiratory group (VRG).

Current theories suggest that inspiratory neurons in the VRG and DRG control the motor neurons in the cervical spinal cord that control the inspiratory muscles. Put another way, VRG and DRG inspiratory neurons stimulate motor neurons of the phrenic and external intercostal nerves, which then cause contraction of the inspiratory muscles. Some inspiratory neurons of the VRG provide input to accessory respiratory muscles that do not participate in inspiration or expiration directly but instead provide supportive movements such as expansion of the opening to the larynx during inspiration. The role of the expiratory neurons is not as well understood, but one of their functions may be to suppress the activity of inspiratory neurons during expiration. Some expiratory neurons of the VRG are known to stimulate the motor neurons to expiratory muscles and to be involved in active expiration.

Respiratory Control Centers of the Pons

The respiratory center of the pons, called the **pontine respiratory group** (PRG; formerly called the pneumo-

taxic center), contains both inspiratory and expiratory neurons as well as *mixed neurons,* which have activity associated with both inspiration and expiration (see Figure 16.15). The PRG may facilitate the transition between inspiration and expiration.

Central Pattern Generator

We have seen that inspiratory neurons of the medulla control the motor neurons to the inspiratory muscles, and that these neurons generate action potentials during inspiration but not during expiration. The source of this cycle of activity is called the central pattern generator.

The **central pattern generator** (CPG) is a network of neurons that generates a regular, repeating pattern of neural activity called the respiratory rhythm. The location of the CPG and its mechanism of action are unknown. Most likely, the CPG is located in the medulla, although the pons has not been completely ruled out.

Even though the location of the CPG is not known, two primary hypotheses have been proposed to explain how it functions. One hypothesis suggests that certain neurons in the CPG have pacemaker activity; that is, they spontaneously depolarize and generate action potentials in a cyclical manner similar to pacemaker cells of the heart. However, no studies have yet identified any cells with pacemaker activity in the respiratory centers of the brainstem. The second hypothesis suggests that complex interactions between networks of neurons are responsible for generating the breathing rhythm.

Model of Respiratory Control During Quiet Breathing

Figure 16.17 shows a simplified model for quiet breathing in which the breathing rhythm is produced by the CPG. This rhythm is communicated to inspiratory neurons of the DRG and VRG, causing a ramp increase in the frequency of action potentials in these cells, which communicate with motor neurons of the phrenic and external intercostal nerves. Early in inspiration—when the frequency of action potentials in the inspiratory neurons is relatively low—only a few motor neurons are recruited, and contraction of inspiratory muscles is weak. As inspiration continues, the greater frequency of action potentials in the inspiratory neurons causes recruitment of more motor neurons, causing the recruitment of additional inspiratory muscle fibers and a stronger overall contraction, which causes lung volume to increase. After a couple of seconds, activity in the inspiratory neurons terminates abruptly, causing a similar abrupt termination in motor neuron activity. As a result, the inspiratory muscles stop contracting, inspiration stops, and expiration begins. Following a brief period of quiescence, another burst of action potentials begins, marking the beginning of the next inspiration.

In this model, respiratory control areas of the medulla are primarily responsible for controlling breath-

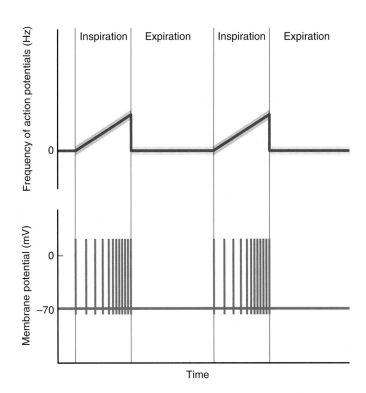

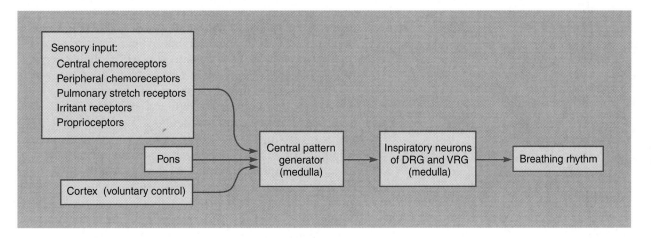

FIGURE 16.16 **Activity of inspiratory neurons.** *Inspiratory neurons show a ramp increase in the frequency of action potentials during inspiration, followed by a sudden termination of all activity at the end of inspiration and beginning of expiration.*

FIGURE 16.17 **Model of respiratory control during quiet breathing.**

ing, but breathing is also affected by activity in other brain regions, including the pons, cerebral cortex, cerebellum, limbic system, hypothalamus, and medullary cardiovascular regulatory areas. This explains, for example, why breathing patterns change when a person experiences feelings of rage or fear.

Peripheral Input to Respiratory Centers

Several types of sensory input can alter respiration, presumably through indirect communication with the central pattern generator (see Figure 16.17). Particularly important in this regard are signals from central and peripheral **chemoreceptors** (chemically sensitive receptor cells) located in the brain and in systemic arteries. These chemoreceptors monitor chemical conditions in cerebrospinal fluid and arterial blood and are primarily responsible for regulating ventilation under resting conditions (as discussed in the next section). Additional sensory inputs that affect breathing come from a variety of receptors located in the respiratory system and elsewhere. These include pulmonary stretch receptors in the smooth muscle of pulmonary airways, irritant receptors in the lining of the respiratory tract, proprioceptors in muscles and joints (which detect movement of the body), arterial baroreceptors (which detect changes in blood pressure), and nociceptors and thermoreceptors located throughout the body.

Pulmonary stretch receptors are excited by inflation of the lungs and do not appear to play a significant role in

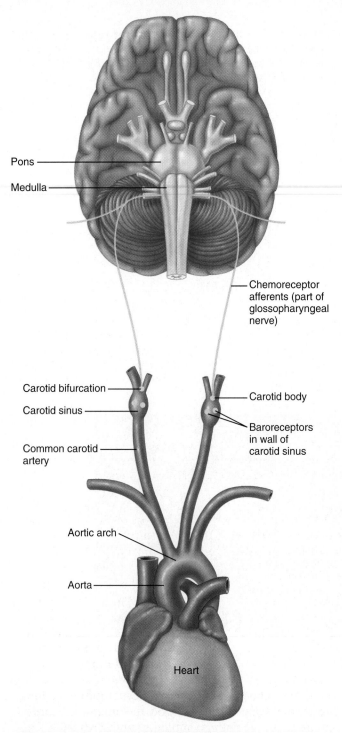

Pons

Medulla

Chemoreceptor afferents (part of glossopharyngeal nerve)

Carotid bifurcation

Carotid sinus

Carotid body

Common carotid artery

Baroreceptors in wall of carotid sinus

Aortic arch

Aorta

Heart

FIGURE 16.18 Location of peripheral chemoreceptors in the carotid bodies. *Afferents from the chemoreceptors ascend to the medulla, but not directly to the respiratory centers.*

substances from the lungs. Coughing is initiated by irritant receptors in the trachea, whereas sneezing is triggered by irritant receptors in the nose and pharynx. Input to respiratory centers from muscle and joint proprioceptors plays a role in stimulating the increase in ventilation that occurs during exercise.

Quick Test 16.6

1. Where are inspiratory neurons located? Where are expiratory neurons located?
2. What is the function of the central pattern generator for breathing?
3. What is the function of irritant receptors?

CONTROL OF VENTILATION BY CHEMORECEPTORS

Anyone who has ever experienced pneumonia, an asthma attack, or similar breathing difficulty can testify that few sensations are quite so unsettling as the feeling of lacking air, known as *air hunger*. You can demonstrate this to yourself by attempting to voluntarily slow down your own breathing while trying to keep the size of each breath the same. If you try this, you will probably begin to feel discomfort within the first seconds. Soon you will be completely unable to keep a steady breathing rhythm, and your breathing will become faster and deeper.

Those who attempt this demonstration inevitably fail, because altering the breathing pattern alters the chemical composition of systemic arterial blood. The reduction in breathing frequency coupled with a constant tidal volume causes alveolar ventilation to fall. This decrease in alveolar ventilation causes changes in the partial pressures of gases in alveolar air, which cause corresponding changes in arterial blood. Changes in chemical concentrations in the blood are detected by chemoreceptors located in major arteries and in the brain, which relay signals to the respiratory control center via afferent neurons. As a result of this neural input, the respiratory control center triggers an increase in the rate and depth of breathing to restore partial pressures in arterial blood to their normal values. The chemoreceptors also relay information to the cerebral cortex, giving rise to the conscious sensation of air hunger.

Chemoreceptors

Chemoreceptors monitor partial pressures of oxygen and carbon dioxide in arterial blood and relay this information to the respiratory control center, so that it can adjust ventilation in response to changes in these variables. Chemoreceptors involved in the control of breathing are classified as either peripheral or central, depending on their location. **Peripheral chemoreceptors** are located

regulating breathing in humans. In other animal species, however, activation of these receptors inhibits inspiration and may act to protect the lungs against overinflation. *Irritant receptors* are stimulated by inhaled particulates such as smoke or dust, and by certain chemicals such as sulfur dioxide. Stimulation of these receptors triggers bronchoconstriction, hyperpnea, and coughing and sneezing, which under certain circumstances help to clear these

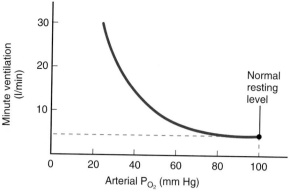

(a) Effects of arterial P_{O_2} on ventilation

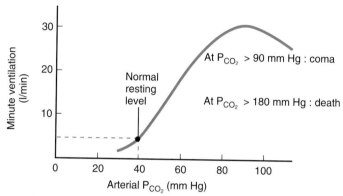

At $P_{CO_2} > 90$ mm Hg : coma

At $P_{CO_2} > 180$ mm Hg : death

(b) Effects of arterial P_{CO_2} on ventilation

FIGURE 16.19 Respiratory control by chemoreceptors.
(a) *Declining arterial P_{O_2} has little effect on minute ventilation until the P_{O_2} drops to less than 60 mm Hg.* **(b)** *Increasing arterial P_{CO_2} has large effects on minute ventilation as P_{CO_2} increases above or decreases below normal. At P_{CO_2} greater than 90 mm Hg, coma and then death can occur.*

Which class of chemoreceptors is responsible for the increased ventilation that occurs when P_{O_2} decreases to less than 60 mm Hg?

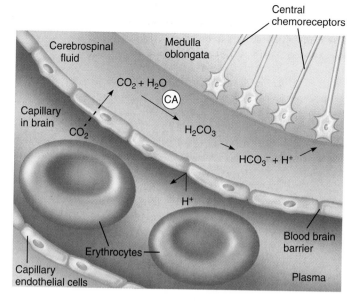

FIGURE 16.20 Activation of central chemoreceptors in the medulla oblongata. *Central chemoreceptors respond best to changes in pH in the cerebrospinal fluid. However, hydrogen ions cannot cross the blood-brain barrier. Instead, carbon dioxide in the blood diffuses into the cerebrospinal fluid, where carbonic anhydrase (CA) catalyzes the conversion of carbon dioxide and water to carbonic acid (H_2CO_3), which dissociates to bicarbonate (HCO_3^-) and hydrogen ions (H^+). The hydrogen ions can then activate the central chemoreceptors.*

What type of junction links the endothelial cells of the capillary wall together to prevent hydrogen ions from passing between plasma and cerebrospinal fluid?

in the *carotid bodies* near the carotid sinus (Figure 16.18). (Other peripheral chemoreceptors called *aortic bodies* are located in the aortic arch and regulate respiration in many animal species, but not in humans.) The **central chemoreceptors** are located in the medulla oblongata.

Peripheral and central chemoreceptors differ not only in their location, but also in their structures and chemical sensitivities. Peripheral chemoreceptors are specialized chemically sensitive cells that are in direct contact with arterial blood and communicate (through secretion of a chemical messenger) with afferent neurons projecting to medullary respiratory control regions. Peripheral chemoreceptors respond to changes in arterial P_{O_2} or pH (which changes when P_{CO_2} changes). Decreases in P_{O_2} can directly activate peripheral chemoreceptors, but only when the arterial P_{O_2} drops to below 60 mm Hg (Figure 16.19a). Because 60 mm Hg is an extreme drop in arterial P_{O_2}, oxygen is usually not a primary factor in peripheral chemoreceptor activation. However, low P_{O_2} also

increases the sensitivity of the peripheral chemoreceptors to carbon dioxide. Peripheral chemoreceptors can also directly respond to changes in the blood's pH (hydrogen ion concentration). In fact, changes in hydrogen ion concentration are the primary stimulus for the peripheral chemoreceptors. The hydrogen ions can come from many sources, but the main source is the reaction of carbon dioxide with water. Therefore, increases in arterial P_{CO_2} can also activate the peripheral chemoreceptors (Figure 16.19b), but only indirectly; first, the carbon dioxide must be converted to hydrogen ions and bicarbonate.

Central chemoreceptors are neurons in the medulla that respond directly to changes in hydrogen ion concentration in the cerebrospinal fluid surrounding this area. Because these chemoreceptors are located in the brain, they are protected by the blood-brain barrier, which separates cerebrospinal fluid from the blood. Hydrogen ions cannot cross this barrier, but carbon dioxide can. This carbon dioxide does not affect the central chemoreceptors directly, but instead is converted to hydrogen ions and bicarbonate by carbonic anhydrase in the cerebrospinal fluid; the chemoreceptors are activated by the hydrogen ions that are generated as a result (Figure 16.20). These chemoreceptors, unlike the arterial chemoreceptors, are not sensitive to changes in P_{O_2}.

Peripheral chemoreceptors

Tight junctions

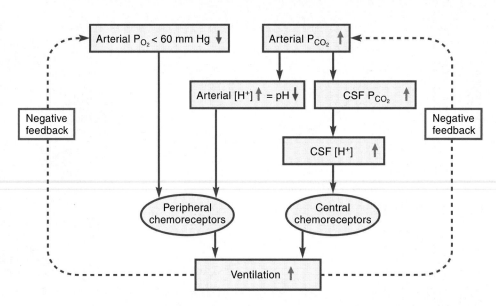

FIGURE 16.21 Chemoreceptor reflexes: The effects of changes in arterial P_{O_2}, P_{CO_2}, and pH on ventilation. *A decrease in arterial P_{O_2} to less than 60 mm Hg activates the peripheral chemoreceptors but has no effect on the central chemoreceptors. An increase in arterial P_{CO_2} activates both the peripheral and central chemoreceptors, but only after carbon dioxide is converted to hydrogen ions (and bicarbonate). A decrease in arterial blood pH activates the peripheral chemoreceptors. When active, chemoreceptors stimulate an increase in ventilation, which provides negative feedback to the initial stimulus. This flowchart is available as an interactive exercise under Activities at www.physiologyplace.com.*

Chemoreceptor Reflexes

Figure 16.21 summarizes the role of chemoreceptor reflex activity in maintaining normal arterial P_{CO_2} and P_{O_2}. Both the central and peripheral chemoreceptors respond indirectly to changes in arterial P_{CO_2}. In fact, changes in P_{CO_2} are the primary stimuli for changes in ventilation under normal conditions. The peripheral chemoreceptors respond to changes in arterial blood pH, which may come from carbon dioxide or other sources (such as lactic acid produced by cellular metabolism). The central chemoreceptors respond to changes in the concentration of hydrogen ions in the CSF but do not respond to changes in the hydrogen ion concentration of the blood because these ions cannot cross the blood-brain barrier. For this reason, hydrogen ions generated by lactic acid or other acids in the blood do not affect the activity of central chemoreceptors. The peripheral chemoreceptors are sensitive to arterial P_{O_2}, but only when it drops to very low levels (below 60 mm Hg). In all cases, activation of the chemoreceptors results in an increase in ventilation; decreased activation of the chemoreceptors results in a decrease in ventilation.

Exercise Link

At the start of the marathon, Bill and Jane's oxygen consumption and carbon dioxide production increased over 12-fold, compared with resting rates. Hyperpnea developed, but with essentially no changes in their arterial P_{O_2} or P_{CO_2}. In other words, their ventilation increased with little or no involvement of their chemoreceptor reflexes. In this situation, the respiratory control center triggered increased ventilation in response to signals from the motor cortex involved in initiating muscle contraction, reflexes from receptors in joints and muscles, increases in body temperature and increased circulating epinephrine concentrations.

When a person at sea level breathes atmospheric air with normal partial pressures (P_{O_2} = 160 mm Hg and P_{CO_2} = 0.2 mm Hg), alveolar and arterial partial pressures are also near normal (P_{O_2} = 100 mm Hg and P_{CO_2} = 40 mm Hg), so long as alveolar ventilation is matched to the body's metabolic needs. During hypoventilation, alveolar ventilation is less than it should be, which causes arterial P_{O_2} to decrease and P_{CO_2} to increase (Figure 16.22a). Under these conditions, chemoreceptors are stimulated by the increased P_{CO_2} (only severe hypoventilation will decrease P_{O_2} to levels that stimulate chemoreceptors) and trigger an increase in the rate and depth of breathing. During hyperventilation, the alveolar ventilation is greater than it should be, which causes arterial P_{CO_2} to decrease (Figure 16.22b). Under these conditions, chemoreceptor stimulation decreases due to the decreased P_{CO_2} and triggers a decrease in the rate and depth of breathing. In both conditions, the changes in ventilation should return the arterial P_{O_2} and P_{CO_2} toward normal values.

Quick Test 16.7

1. Describe the locations of the peripheral chemoreceptors and the central chemoreceptors.

2. What stimuli activate the peripheral chemoreceptors? The central chemoreceptors?

3. How does carbon dioxide activate the peripheral and central chemoreceptors?

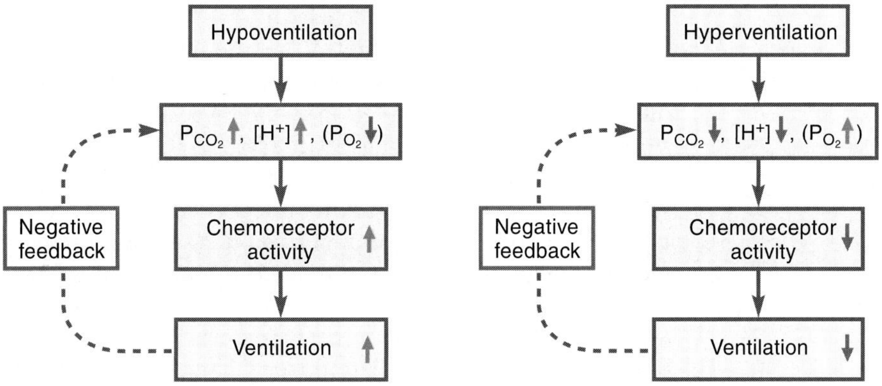

FIGURE 16.22 The effects of hypoventilation and hyperventilation on minute ventilation.

(a) Hypoventilation

(b) Hyperventilation

LOCAL REGULATION OF VENTILATION AND PERFUSION

In cases of obstructive lung diseases such as emphysema or bronchitis, it is not uncommon for airways in some regions of the lungs to be completely obstructed by mucus or other material that prevents air from flowing into alveoli (see Chapter 15). When this occurs, blood flowing to these alveoli is "wasted" in that it cannot participate in gas exchange. In other types of lung diseases, the pulmonary capillaries may be damaged, leading to the opposite problem: Certain alveoli may receive air but not blood. In this case, air flowing to these alveoli is "wasted" in that it does not participate in gas exchange.

Ventilation-Perfusion Ratios

In the normal lung, the rate of air flow to the alveoli (ventilation = $\dot{V}_A$) is matched to the rate of blood flow (perfusion = $\dot{Q}$); the dots over the abbrevations indicate that these are rates. The relationship of ventilation to perfusion is called the **ventilation-perfusion ratio** and is abbreviated $\dot{V}_A/\dot{Q}$. In the normal lung, $\dot{V}_A/\dot{Q}$ is approximately 1 (Figure 16.23a). Under these circumstances, the P_{O_2} and P_{CO_2} of any alveolus are at the normal values of 100 and 40 mm Hg, respectively. The P_{O_2} and P_{CO_2} in the blood emerging from the alveolus are also 100 and 40 mm Hg, respectively, because partial pressures in the capillary blood equilibrate with those in alveolar air.

The conditions mentioned in the introductory paragraph for this section illustrate extreme forms of ventilation-perfusion inequality. When airways are obstructed, for example, $\dot{V}_A$ in certain alveoli decreases, and the blood traveling in the capillaries to those alveoli does not undergo adequate gas exchange (Figure 16.23b). The result is that blood and air in these alveoli will have a lower P_{O_2} and a higher P_{CO_2} than normal. If ventilation of the alveoli is completely prevented, then the blood that emerges from the capillaries will have the same partial pressures of oxygen and carbon dioxide as the blood enter-ing the capillaries. When pulmonary capillaries are damaged, perfusion is obstructed, causing a decrease in $\dot{Q}$ (Figure 16.23c). The result is that blood and air in these alveoli will have a higher P_{O_2} and a lower P_{CO_2} than normal.

Ventilation-perfusion inequalities do not occur in disease states only. Under normal conditions, gravity affects the ventilation-perfusion ratio. Capillaries in the lungs can collapse if the pressure in them is not greater than the pressure of air in the alveoli. Gravity causes a greater capillary hydrostatic pressure, and thus, greater blood flow, in the lower regions of the lungs. For example, in the uppermost regions of the lungs of a standing or sitting person, capillary pressure is greater than alveolar pressure at the arterial end of the capillary, but on the venous end, alveolar pressure is greater than capillary pressure. Blood flow is thus intermittent as the capillary alternately opens and closes. In the lower regions of the lungs, capillary pressure is always greater than alveolar pressure, such that the capillary is always open and blood flow is continuous.

Local Control of Ventilation and Perfusion

Given that differences in the ventilation or perfusion of certain alveoli can occur, whether by diseases or gravity, local control mechanisms operate to match ventilation and perfusion to ensure efficient gas exchange. To obtain a ventilation-perfusion ratio of 1, areas with low ventilation require low perfusion, and areas with high ventilation require high perfusion. Changes in P_{O_2} and P_{CO_2} in alveoli and the interstitial fluid of the lungs affect the contractile activity of smooth muscle, causing changes in bronchiolar diameter and thus in air flow (Figure 16.24), and changes in arteriole diameter and thus in perfusion of individual alveoli (Figure 16.25). The effects of changes in the partial pressures of oxygen and carbon dioxide on the radii of bronchioles and pulmonary arterioles are summarized in Table 16.4.

Oxygen acts primarily on the pulmonary arterioles; a low P_{O_2} causes a vasoconstriction (decrease in $\dot{Q}$). Carbon dioxide acts primarily on the bronchioles; a high P_{CO_2}

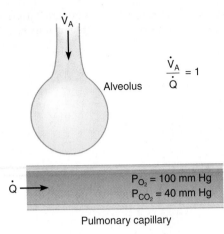

$\dfrac{\dot{V}_A}{\dot{Q}} = 1$

Alveolus

$P_{O_2} = 100$ mm Hg
$P_{CO_2} = 40$ mm Hg

Pulmonary capillary

(a) Normal

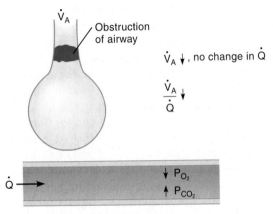

Obstruction of airway

$\dot{V}_A \downarrow$, no change in $\dot{Q}$

$\dfrac{\dot{V}_A}{\dot{Q}} \downarrow$

$\downarrow P_{O_2}$
$\uparrow P_{CO_2}$

(b) Decreased ventilation

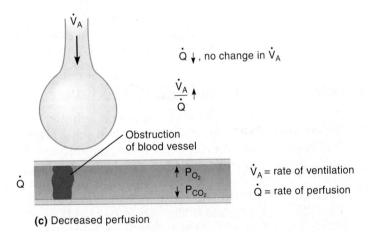

$\dot{Q} \downarrow$, no change in $\dot{V}_A$

$\dfrac{\dot{V}_A}{\dot{Q}} \uparrow$

Obstruction of blood vessel

$\uparrow P_{O_2}$
$\downarrow P_{CO_2}$

$\dot{V}_A$ = rate of ventilation
$\dot{Q}$ = rate of perfusion

(c) Decreased perfusion

FIGURE 16.23 Ventilation-perfusion ratios. *Dots over V_A and Q indicate that these quantities are rates.* **(a)** *Under normal conditions, ventilation and perfusion are matched ($\dot{V}_A/\dot{Q} = 1$).* **(b)** *If ventilation decreases and perfusion is normal, then $\dot{V}_A/\dot{Q} < 1$; arterial P_{O_2} decreases and P_{CO_2} increases.* **(c)** *If perfusion decreases and ventilation is normal, then $\dot{V}_A/\dot{Q} > 1$; arterial P_{O_2} increases and P_{CO_2} decreases.*

causes bronchodilation (increase in $\dot{V}_A$). Therefore, in regions of the lungs with a high $\dot{V}_A/\dot{Q}$, the increase in P_{O_2} and decrease in P_{CO_2} cause bronchoconstriction and vasodilation to decrease ventilation while increasing perfusion. In regions of the lungs with a low $\dot{V}_A/\dot{Q}$, the increase in P_{CO_2} causes bronchodilation and the decrease in P_{O_2} causes vasoconstriction. The net effect is an increase in ventilation and a decrease in perfusion. Note that the effects of oxygen and carbon dioxide on pulmonary arterioles is the opposite of the effects of these gases on systemic arterioles, where oxygen causes vasoconstriction and carbon dioxide causes vasodilation (see Chapter 14). The radii of arterioles and bronchioles can also be affected by other factors, such as epinephrine (see Chapters 14 and 15).

Quick Test 16.8

1. How is the P_{O_2} of alveolar air affected when the ventilation-perfusion ratio increases? How is P_{CO_2} affected?

2. What effect does a decrease in P_{CO_2} in the interstitial fluid around an alveolus have on the radii of arterioles and bronchioles in the region?

3. If blood flow to an alveolus is decreased, what should happen to the radius of the bronchiole leading to the alveolus if a normal ventilation-perfusion ratio is to be maintained?

THE RESPIRATORY SYSTEM IN ACID-BASE HOMEOSTASIS

Although the primary function of the respiratory system is to control the oxygen and carbon dioxide content of arterial blood, it also plays an important role in regulating the blood's hydrogen ion concentration or pH. (Toolbox: Review of pH, www.physiologyplace.com, Challenge Yourself) We saw in the previous section that hydrogen ions can be buffered by hemoglobin, and that carbon dioxide is a source of hydrogen ions. In this section we focus on the role of hemoglobin and, more importantly, carbon dioxide in acid-base balance—that is, in maintaining normal blood pH.

Acid-Base Disturbances in Blood

The normal pH of arterial blood is 7.4, which is slightly more basic than pure water (pH = 7.0). Blood pH is tightly regulated by both the respiratory system and the kidneys (see Chapter 18), such that the pH rarely varies by more than a few hundredths of a unit in either direction; the normal pH range is 7.38–7.42. In fact, changes in blood pH of just a few tenths of a unit can have serious, even life-threatening consequences. If blood pH in-

CHANGE IN GAS COMPOSITION IN LUNGS	RESPONSE OF BRONCHIOLES	RESPONSE OF PULMONARY ARTERIOLES
Increased P_{CO_2}	Dilation (increased $\dot{V}_A$*)	Weak constriction (decreased $\dot{Q}$†)
Decreased P_{CO_2}	Constriction (decreased $\dot{V}_A$)	Weak dilation (increased $\dot{Q}$)
Increased P_{O_2}	Weak constriction (decreased $\dot{V}_A$)	Dilation (increased $\dot{Q}$)
Decreased P_{O_2}	Weak dilation (increased $\dot{V}_A$)	Constriction (decreased $\dot{Q}$)

*$\dot{V}_A$ = ventilation
†$\dot{Q}$ = perfusion

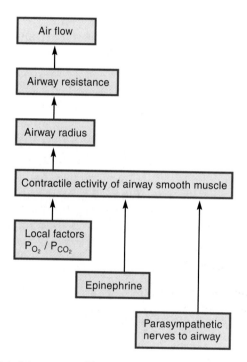

FIGURE 16.24 Factors affecting airway resistance and thus air flow.

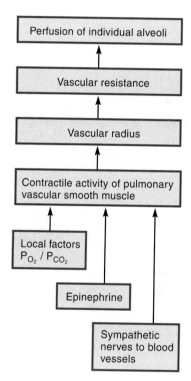

FIGURE 16.25 Factors affecting the perfusion of individual alveoli.

creases to 8.0 or decreases to 6.8 for even just a few seconds, the result is death!

A change in arterial blood pH can have profound effects on body function because it alters the pH of fluids throughout the body. A change in pH alters the distribution of electrical charges in protein molecules, thereby altering their shape and interfering with their normal functions. Given that proteins have such a wide variety of functions in the body (as enzymes and receptors for messenger molecules, to name but two), changes in pH have widespread effects on body functions.

Arterial blood is considered to be excessively acidic if the pH is 7.35 or lower, a condition known as **acidosis.**

(Note that acidosis occurs even when blood pH is in fact alkaline—that is, greater than 7.0—because acidosis refers to a pH that is *more acidic* than the normal blood pH of 7.4.) A pH of 7.45 or greater indicates that the blood is excessively alkaline or basic, a condition known as **alkalosis.** The danger of severe acidosis is that it depresses central nervous system activity. In severe cases, the depression of CNS activity can progress to coma and, ultimately, fatal respiratory failure. When the pH becomes too high (alkalosis), the nervous system becomes overly excitable, which can lead to uncontrollable muscle seizures and convulsions. Frequently, death occurs as a result of spasmodic contraction of respiratory muscles, which makes normal breathing movements impossible.

The relationship between carbon dioxide and acidity can be described using the Henderson–Hasselbalch equation, which is based on the equilibrium for the dissociation of an acid (HA) into a free hydrogen ion (H^+) and a base (A^-):

$$HA \leftrightarrow H^+ + A^-$$

The equilibrium constant (K) for this reaction is given by the following expression:

$$K = \frac{[H^+][A^-]}{[HA]}$$

which can be rearranged to give the hydrogen ion concentration under equilibrium conditions:

$$[H^+] = \frac{K[HA]}{[A^-]}$$

Because the hydrogen ion concentration is usually reported as pH, we can reformulate this equation by taking the negative log of both sides, which yields the Henderson–Hasselbalch equation:

$$pH = -\log K + \log \frac{[A^-]}{[HA]} = pK + \log \frac{[A^-]}{[HA]}$$

Note here that the negative log of the equilibrium constant is designated as *pK*, which is analogous to pH, the negative log of the hydrogen ion concentration.

We can apply the previous equation to the dissociation of carbonic acid (H_2CO_3) into bicarbonate (HCO_3^-) and hydrogen ions:

$$H_2CO_3 \leftrightarrow HCO_3^- + H^+$$

by substituting $[HCO_3^-]$ for $[A^-]$ and $[H_2CO_3]$ for [HA]. However, we can also express the equation in terms of the carbon dioxide concentration because carbonic acid is generated in the reaction of CO_2 with water:

$$CO_2 + H_2O \leftrightarrow H_2CO_3$$

which is described by the following equilibrium constant:

$$K = \frac{[H_2CO_3]}{[CO_2][H_2O]}$$

Because $[H_2O]$ is virtually constant, this equation tells us that the ratio $[H_2CO_3]/[CO_2]$ is also constant. In other words, $[H_2CO_3]$ will always be proportional to $[CO_2]$. As a consequence, we can substitute $[CO_2]$ for $[H_2CO_3]$ to yield the Henderson–Hasselbalch equation in the following form (in which the appropriate real numbers have been substituted):

$$pH = 6.1 + \log \frac{[HCO_3^-]}{[CO_2]}$$

This equation tells us that maintenance of a normal arterial pH of 7.4 requires that the ratio $[HCO_3^-]/[CO_2]$ remain constant at 20:1. If this ratio changes, blood pH must also change. The lungs and kidneys both play a role in helping to keep the ratio constant; the lungs regulate $[CO_2]$ while the kidneys regulate $[HCO_3^-]$.

Because blood carbon dioxide levels are commonly expressed in terms of the partial pressures, the previous equation can also be written in the following form:

$$pH = 6.1 + \log \frac{[HCO_3^-]}{(0.03)P_{CO_2}}$$

where 0.03 is a proportionality constant that converts the partial pressure of carbon dioxide (mm Hg) to a concentration (millimoles/liter).

The Role of the Respiratory System in Acid-Base Balance

The respiratory system and kidneys work together to regulate blood pH; a more thorough discussion of acid-base balance is presented in Chapter 18. In this section we concentrate on aspects of the respiratory system involved in acid-base balance.

Hemoglobin as a Buffer

As previously stated, hemoglobin is a buffer because it can bind or release hydrogen ions. Deoxyhemoglobin's greater affinity for hydrogen ions than oxyhemoglobin's is a component of the Bohr effect (described on p. 509). In the tissues, hemoglobin unloads oxygen and binds hydrogen ions:

$$Hb \cdot O_2 \rightarrow O_2 + Hb \qquad Hb + H^+ \rightarrow HbH$$

This is important because the tissues are producing carbon dioxide, which is quickly converted to bicarbonate and hydrogen ions. Some of these hydrogen ions can be buffered by hemoglobin, which helps to prevent the pH from becoming too acidic:

$$CO_2 + H_2O \rightarrow H_2CO_3 \rightarrow HCO_3^- + H^+ \qquad H^+ + Hb \rightarrow HbH$$

At high altitudes, the atmospheric pressure decreases, so the partial pressures of the gases in air decrease, even though gas composition of the air does not change. For example, in Denver, Colorado (the "Mile High City"), the atmospheric pressure of air is 630 mm Hg, whereas the atmospheric pressure of air at the top of Mount Everest is 250 mm Hg. The partial pressures of oxygen and carbon dioxide at these two locations, therefore, would be:

Denver: $P_{O_2} = (0.21) \times (630 \text{ mm Hg})$
$= 132.3 \text{ mm Hg}$

$P_{CO_2} = (0.0003) \times (630 \text{ mm Hg})$
$= 0.2 \text{ mm Hg}$

Mount Everest: $P_{O_2} = (0.21) \times (250 \text{ mm Hg})$
$= 52.5 \text{ mm Hg}$

$P_{CO_2} = (0.0003) \times (250 \text{ mm Hg})$
$= 0.1 \text{ mm Hg}$

Thus as altitude increases, the air that reaches the alveoli contains less oxygen. Carbon dioxide partial pressures do not change much because this gas is only a minor component of air.

Because the alveolar P_{O_2} is decreased at high altitudes, the arterial P_{O_2} also de-creases, a condition called *hypoxemia*. With low oxygen levels in blood, less oxygen is available for respiring tissues, resulting in *hypoxia* (decreased oxygen in tissue). The degree of hypoxia depends on the altitude and how long a person has been at that altitude.

Initially the hypoxemia triggers compensatory responses in an attempt to reestablish arterial P_{O_2}. If the P_{O_2} drops below 60 mm Hg, peripheral chemoreceptors are activated, causing the respiratory center to increase ventilation. If ventilation increases relative to metabolic demands, however, both arterial P_{CO_2} and the concentration of hydrogen ions in the blood will decrease, which decreases the activation of both peripheral and central chemoreceptors, countering the effects of low oxygen. In addition, a *respiratory alkalosis* occurs. With a decrease in the blood's acidity comes a shift in the hemoglobin-oxygen dissociation curve to the left as the affinity of hemoglobin for oxygen increases. An increase in affinity means that less oxygen is unloaded in the tissues, but it also means that more oxygen is loaded in the lungs.

If exposed to high altitudes for a few days, the body begins to acclimate. The kidneys help maintain acid-base balance by excreting bicarbonate to match the loss of hydrogen ions that accompanies the reduction in arterial P_{CO_2}. If exposed for longer periods of time, other acclimatizations occur. In response to the hypoxia, the kidneys secrete the hormone erythropoietin, which stimulates erythrocyte synthesis, causing a rise in the hematocrit up to 60%, a condition called *polycythemia*. With an increase in erythrocytes comes an increase in the concentration of hemoglobin in blood and thus an increase in the oxygen-carrying capacity of the blood.

Upon exposure to low oxygen levels, oxyhemoglobin levels are decreased, causing an increase in the production of 2,3-DPG by the erythrocytes. 2,3-DPG decreases the affinity of hemoglobin for oxygen, which increases oxygen unloading in the tissues, countering the effects of the alkalosis.

Sometimes, high-altitude conditions cannot be tolerated by the body, and *chronic mountain sickness* develops. Early symptoms include headache, dizziness, fatigue, and shortness of breath. The condition can progress to disorientation and heart failure. The symptoms of mountain sickness are primarily due to the hypoxia and polycythemia. Pulmonary vasoconstriction also occurs, causing the right side of the heart to work harder against a greater resistance.

In the lungs, the reaction goes in reverse: Hemoglobin releases hydrogen ions and loads oxygen. The clearance of carbon dioxide tends to reduce the hydrogen ion concentration, an effect countered somewhat by the release of hydrogen ions from hemoglobin as it binds oxygen:

$$HbH \rightarrow H^+ + Hb \qquad Hb + O_2 \rightarrow Hb \cdot O_2$$

Bicarbonate Ions as a Buffer

Bicarbonate is another major buffering system in blood. If the hydrogen ion concentration increases, hydrogen ions bind to bicarbonate to form carbon dioxide. Likewise, if carbon dioxide in the blood increases, it can be converted to bicarbonate and hydrogen ions. The production of hydrogen ions from carbon dioxide is important in acid-base balance: If carbon dioxide levels in the blood are allowed to increase, acidosis will result.

The relationship between carbon dioxide and acidity can be described using the Henderson–Hasselbalch equation, which can be applied to the equilibrium for the series of reactions that ultimately convert carbon dioxide and water to bicarbonate and hydrogen ions. The Henderson–Hasselbalch Equation for this reaction is:

$$pH = 6.1 + \log \frac{[HCO_3^-]}{[CO_2]}$$

(see Toolbox: The Henderson–Hasselbalch Equation).

This equation tells us that maintaining a normal blood pH of 7.4 requires that the ratio of bicarbonate to carbon dioxide remain constant at 20:1. If this ratio changes, blood pH will change. The lungs regulate the concentration of carbon dioxide, whereas the kidneys regulate the concentration of bicarbonate ions.

Respiratory disturbances in acid-base balance are due to changes in the concentration of carbon dioxide. **Respiratory acidosis** is an increase in the acidity of the blood due to increased carbon dioxide, which occurs, for example, during hypoventilation. **Respiratory alkalosis** is a decrease in the acidity of the blood due to decreased carbon dioxide, which occurs, for example, during hyperventilation or at high altitudes (see Discovery: The Effects of High Altitude, p. 525). Other disturbances in acid-base balance, called metabolic acidosis and metabolic alkalosis, are discussed in Chapter 18.

SUMMARY OF FACTORS AFFECTING OXYGEN DELIVERY TO AND CARBON DIOXIDE REMOVAL FROM TISSUES

Figure 16.26 summarizes the topics discussed in Chapters 15 and 16. The primary function of the respiratory system is oxygen delivery to tissues and carbon dioxide removal from tissues. Accomplishing this task requires concerted activities of the respiratory and cardiovascular systems. Air flow, or ventilation, and blood flow to the lungs must be maintained to ensure adequate exchange of gases between alveolar air and blood. Air flow depends on the pressure gradient across airways and the airway radius. Blood flow depends on the pressure gradient across blood vessels and vessel radius. The mechanics of air flow was the topic of Chapter 15, where we saw that pressure gradients between the alveoli and the atmosphere drive air into or out of the lungs. These pressure gradients are created by cyclical contraction-relaxation of respiratory muscles. In this chapter we saw that diffusion of gases between blood and tissues depends on pressure gradients as well, and that diffusion occurs down partial pressure gradients for specific gases, mainly oxygen and carbon dioxide.

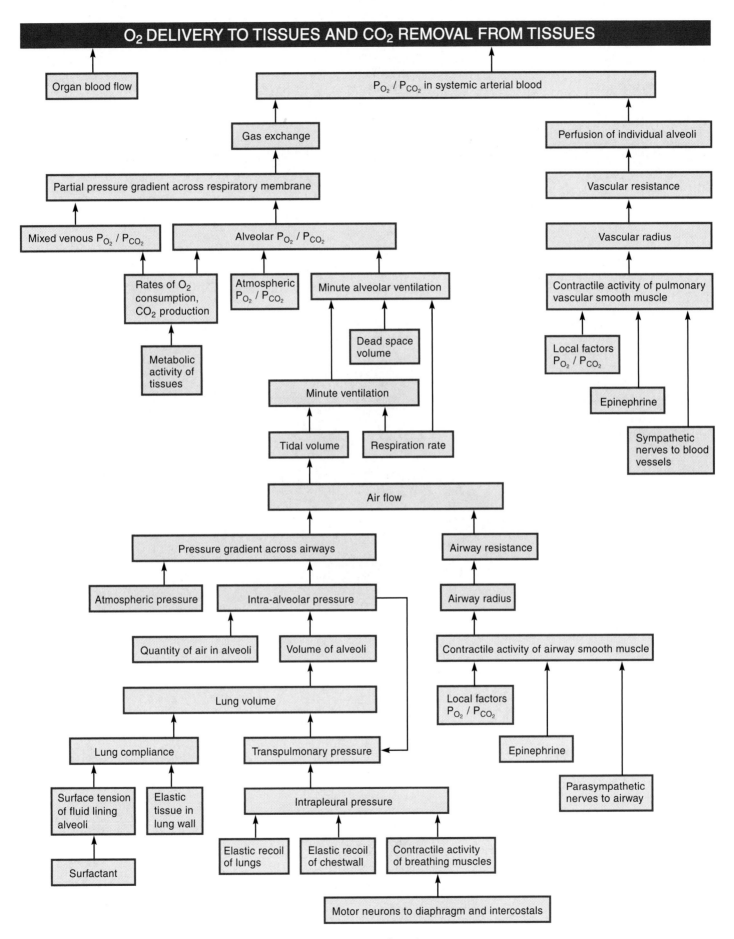

FIGURE 16.26 Summary of the factors affecting respiratory system function.

SYSTEMS INTEGRATION

RESPIRATORY SYSTEM

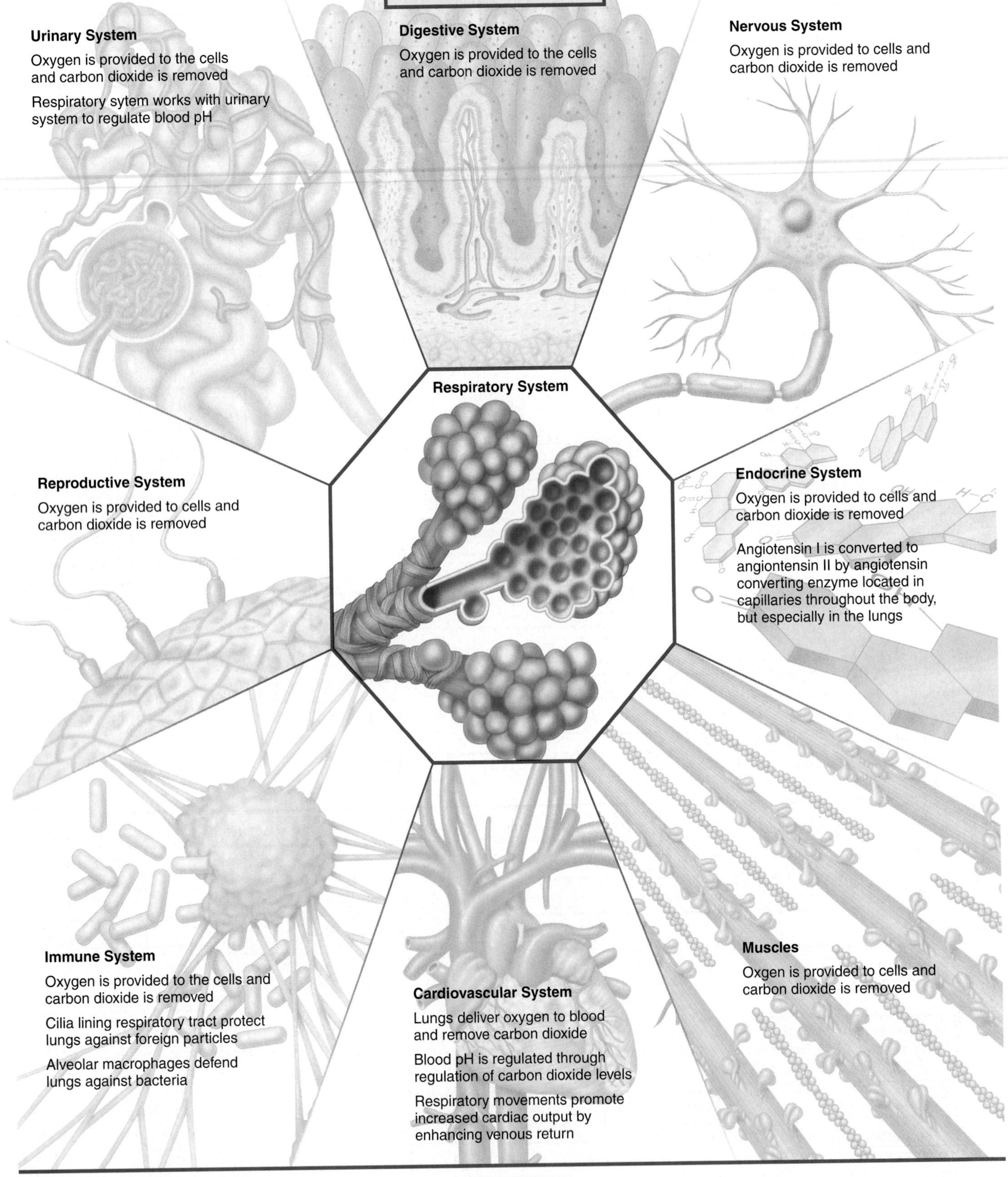
Urinary System
Oxygen is provided to the cells and carbon dioxide is removed
Respiratory sytem works with urinary system to regulate blood pH
Digestive System
Oxygen is provided to the cells and carbon dioxide is removed
Nervous System
Oxygen is provided to cells and carbon dioxide is removed
Respiratory System
Reproductive System
Oxygen is provided to cells and carbon dioxide is removed
Endocrine System
Oxygen is provided to cells and carbon dioxide is removed
Angiotensin I is converted to angiontensin II by angiotensin converting enzyme located in capillaries throughout the body, but especially in the lungs
Immune System
Oxygen is provided to the cells and carbon dioxide is removed
Cilia lining respiratory tract protect lungs against foreign particles
Alveolar macrophages defend lungs against bacteria
Cardiovascular System
Lungs deliver oxygen to blood and remove carbon dioxide
Blood pH is regulated through regulation of carbon dioxide levels
Respiratory movements promote increased cardiac output by enhancing venous return
Muscles
Oxgen is provided to cells and carbon dioxide is removed

Overview of the Pulmonary Circulation, p. 496

The right heart pumps deoxygenated blood to the pulmonary capillaries, where oxygen diffuses from the alveoli to blood, and carbon dioxide diffuses from blood to the alveoli. The respiratory membrane provides a large surface area and short distance for diffusion, so diffusion rate is very rapid. The now oxygenated blood returns to the left side of the heart, where it is pumped to the systemic capillaries in the tissues of the body. Oxygen diffuses from the blood to tissue, and carbon dioxide diffuses from tissue to the blood. The now deoxygenated blood returns to the right side of the heart.

IP Respiratory, Anatomy Review: Respiratory Structures, pp. 7–11

IP Cardiovascular, Anatomy Review: The Heart, p. 5

Diffusion of Gases, p. 497

The pressures of individual gases in a mixture are called partial pressures, and they are equal to the fractional concentration of the gas multiplied by the total pressure. Gases can dissolve in liquids to varying degrees based on their solubility and partial pressure. The greater the solubility and the greater the partial pressure, the more of the gas dissolves in the liquid. Neither oxygen nor carbon dioxide is very soluble in water, but carbon dioxide is approximately 30 times more soluble than oxygen.

IP Respiratory, Gas Exchange, pp. 3–5*

Exchange of Oxygen and Carbon Dioxide, p. 500

Gas exchange occurs by diffusion down partial pressure gradients. In the lungs, oxygen diffuses from alveoli to blood, and carbon dioxide diffuses from blood to alveoli. In respiring tissues, oxygen diffuses from blood to tissue, and carbon dioxide diffuses from tissue to blood. The amount of oxygen and carbon dioxide diffusing across a particular systemic capillary depends on the activity of the tissue; more active tissues have greater partial pressure gradients, leading to greater diffusion rates.

Alveolar P_{O_2} and P_{CO_2} are determined by (1) the P_{O_2} and P_{CO_2} of inspired air, (2) the alveolar ventilation, and (3) the rates of oxygen consumption and carbon dioxide production in respiring tissue. In turn, alveolar P_{O_2} and P_{CO_2} determine arterial P_{O_2} and P_{CO_2}. Normally, alveolar ventilation is matched to oxygen consumption and carbon dioxide production. If metabolic activity increases, ventilation increases to match the demands of the tissue, a condition called hyperpnea.

IP Respiratory, Gas Exchange, pp. 6–16*

Transport of Gases in the Blood, p. 505

Oxygen is transported in blood two ways, dissolved in blood (1.5%) and bound to hemoglobin (98.5%). The relationship between P_{O_2} and the amount of oxygen bound to hemoglobin is shown in the hemoglobin-oxygen dissociation curve. Many factors in blood influence the binding of oxygen to hemoglobin, including temperature, pH, P_{CO_2}, 2,3-DPG, carbon monoxide, and P_{O_2}. The Bohr effect is the decrease in affinity of hemoglobin for oxygen that occurs when hydrogen ions bind to hemoglobin. The carbamino effect is the decrease in affinity of hemoglobin for oxygen that occurs when carbon dioxide binds to hemoglobin. The Haldane effect is the decrease in affinity of hemoglobin for hydrogen ions and carbon dioxide that occurs when oxygen binds to hemoglobin.

Carbon dioxide is transported in blood in three ways: dissolved in blood (5–6%), bound to hemoglobin (5–8%), and as dissolved bicarbonate ions in blood (86–90%). The conversion of carbon dioxide to bicarbonate plays a significant role in maintaining acid-base balance in the blood, and bicarbonate is the main form in which carbon dioxide is transported between tissues and lungs. Carbonic anhydrase, an enzyme found in erythrocytes, catalyzes the reversible reaction that converts carbon dioxide and water to carbonic acid, which then dissociates to hydrogen ions and bicarbonate.

IP Respiratory, Gas Transport, pp. 3–15

Central Regulation of Ventilation, p. 514

Breathing is a rhythmic process caused by cyclical neural excitation of respiratory muscles. The generation of the breathing rhythm requires the action of respiratory centers in the brainstem. The medullary respiratory control center includes the dorsal respiratory group and the ventral respiratory group. Inspiratory neurons in these regions activate the motor neurons that innervate the inspiratory muscles, causing inspiration. The pontine respiratory group may be involved in the transition between inspiration and expiration. Higher brain areas can influence respiration. Various stimuli affect ventilation, including changes in arterial P_{O_2} and P_{CO_2}, stretch of the lungs, irritants in the airways, proprioceptors, arterial baroreceptors, nociceptors, thermoreceptors, emotions, and voluntary controls.

IP Respiratory, Control of Respiration, pp. 1–5, 14–15

Control of Ventilation by Chemoreceptors, p. 518

Peripheral and central chemoreceptors detect changes in P_{O_2}, P_{CO_2}, and pH of arterial blood. P_{CO_2} is the primary stimulus to the peripheral and central chemoreceptors, but its effects are always indirect: CO_2 must first be converted to hydrogen ions (and bicarbonate). The peripheral chemoreceptors located in the carotid bodies respond directly to changes in pH and to decreases in P_{O_2} to less than 60 mm Hg, and indirectly to changes in P_{CO_2}. Central chemoreceptors are located in the medulla oblongata and respond to changes in the pH of cerebrospinal fluid.

> **IP** Respiratory, Control of Respiration, pp. 6–10

Local Regulation of Ventilation and Perfusion, p. 521

The ratio between air flow to the alveoli and blood flow to the capillaries supplying the alveoli is called the ventilation-perfusion ratio. In normal lungs, air flow and ventilation are matched, and the ventilation-perfusion ratio is 1. If ventilation to a particular alveolus is decreased, perfusion will be decreased by vasoconstriction to sustain the normal ventilation-perfusion ratio. Likewise, if perfusion to a particular alveolus is decreased, then airflow will be decreased by bronchoconstriction.

The Respiratory System in Acid-Base Homeostasis, p. 522

The pH of blood is highly regulated between 7.38 and 7.42 to maintain the normal function of proteins necessary for homeostasis. Acidosis is a decrease in pH to 7.35 or less, whereas alkalosis is an increase in pH to 7.45 or greater. The respiratory and urinary systems work together to maintain normal blood pH (to regulate acid-base balance). The primary contribution of the respiratory system to acid-base balance is the regulation of arterial P_{CO_2}. Because carbon dioxide can be converted to carbonic acid, a change in P_{CO_2} can cause either respiratory acidosis or respiratory alkalosis. The respiratory system works in concert with the kidneys to maintain a ratio of bicarbonate to carbon dioxide of 20:1.

> **IP** Respiratory, Control of Respiration, pp. 11–13

---

*This topic is available on the *InterActive Physiology*® *Sampler CD* that comes with the purchase of a new copy of this book.

EXERCISES

Multiple-Choice Questions

1. Under steady-state conditions, the rate at which oxygen enters pulmonary capillaries from alveolar air is equal to
   a) the rate at which oxygen is delivered to alveoli in inspired air.
   b) the rate at which oxygen is carried out of the alveoli in expired air.
   c) the rate at which oxygen is consumed in respiring tissues.
   d) the rate at which carbon dioxide is produced in respiring tissues.

2. At the normal resting P_{O_2} of mixed venous blood, hemoglobin is
   a) nearly 100% saturated.
   b) nearly 97% saturated.
   c) nearly 75% saturated.
   d) nearly 25% saturated.

3. In respiring tissues, a rise in blood P_{CO_2} causes all of the following *except*
   a) an increase in the hydrogen ion concentration.
   b) a rise in bicarbonate concentration.
   c) a rise in the concentration of carbaminohemoglobin.
   d) an increase in the affinity of hemoglobin for oxygen.

4. Which of the following does *not* affect alveolar P_{O_2}?
   a) the rate of oxygen consumption by respiring tissues
   b) alveolar ventilation
   c) the P_{O_2} of inspired air
   d) the volume of air contained in the alveoli

5. During hyperventilation, which of the following would be expected to happen?
   a) an increase in the P_{O_2} of arterial blood
   b) an increase in the P_{CO_2} of arterial blood
   c) an increase in the acidity of arterial blood
   d) an increase in the bicarbonate concentration of arterial blood

6. Under normal conditions, the P_{O_2} and P_{CO_2} of arterial blood are determined by
   a) the metabolic rate of respiring tissues.
   b) alveolar ventilation.
   c) alveolar P_{O_2} and P_{CO_2}.
   d) all of the above

7. Which of the following would be expected to cause a decrease in the percent saturation of hemoglobin?
   a) an increase in P_{O_2}
   b) a decrease in blood pH
   c) a decrease in P_{CO_2}
   d) a decrease in temperature

8. Suppose that alveolar $P_{O_2} = 100$ mm Hg and $P_{CO_2} = 60$ mm Hg. Which of the following is true?
   a) pH will be less than normal.
   b) Percent saturation of hemoglobin by oxygen will be below normal.
   c) Bicarbonate concentration will be above normal.
   d) all of the above

9. Suppose a person's arterial P_{O_2} and P_{CO_2} are normal ($P_{O_2} = 100$ mm Hg; $P_{CO_2} = 40$ mm Hg). Which of the following would most likely stimulate an increase in ventilation?
   a) a decrease in P_{O_2} to 90 mm Hg
   b) a decrease in P_{CO_2} to 35 mm Hg
   c) an increase in P_{O_2} to 110 mm Hg
   d) an increase in P_{CO_2} to 45 mm Hg

10. A rise in arterial P_{CO_2} triggers an increase in ventilation by stimulating both central and peripheral chemoreceptors. The response of central chemoreceptors is due to
    a) diffusion of carbon dioxide into brain extracellular fluid, which stimulates chemoreceptors directly.
    b) diffusion of hydrogen ions into brain extracellular fluid, which stimulates chemoreceptors directly.
    c) diffusion of carbon dioxide into brain extracellular fluid, which reacts with water to form hydrogen ions, which stimulates chemoreceptors directly.
    d) direct stimulation by hydrogen ions in arterial blood.

11. When a person exercises, ventilation increases to meet the demands of more active tissues. This is an example of
    a) hyperventilation.
    b) hypoventilation.
    c) hyperpnea.
    d) hypoxia.

12. The normal ratio of bicarbonate concentration to carbon dioxide concentration in arterial blood is
    a) 1:5.
    b) 5:1.
    c) 20:1.
    d) 1:20.

13. Which of the following can hemoglobin bind and transport in blood?
    a) oxygen
    b) carbon dioxide
    c) hydrogen ions
    d) all of the above

14. Which of the following areas of the brain contain inspiratory neurons?
    a) the dorsal respiratory group only
    b) the ventral respiratory group only
    c) both the dorsal and ventral respiratory groups
    d) neither the dorsal nor ventral respiratory groups

Objective Questions

1. Under normal conditions, the rate at which oxygen is brought into the alveoli in inspired air is (the same as/greater than/less than) the rate at which it is consumed in respiring tissues.

2. Under resting conditions, tissues normally extract (exactly half/more than half/less than half) of the oxygen that is delivered to them in arterial blood.

3. The amount of carbon dioxide in systemic arterial blood is less than 50% of that in mixed venous blood. (true/false)

4. When the P_{CO_2} of the blood increases, the concentration of bicarbonate (increases/decreases), and the concentration of hydrogen ions (increases/decreases).

5. The enzyme that catalyzes the conversion of carbon dioxide to carbonic acid is _____.

6. As the pH of the blood increases, the affinity of hemoglobin for oxygen (increases/decreases).

7. When a person hypoventilates, the P_{CO_2} of arterial blood (increases/decreases).

8. A decrease in alveolar ventilation would be expected to cause a(n) (increase/decrease) in arterial P_{O_2}, and a(n) (increase/decrease) in arterial P_{CO_2}.

9. Hemoglobin with carbon dioxide bound to it is called _____.

10. In gas exchange in both the lungs and respiring tissues, oxygen and carbon dioxide always move down their partial pressure gradients. (true/false)

11. (Central/peripheral) chemoreceptors respond directly to hydrogen ions produced during metabolism.

12. A rise in arterial P_{CO_2} would be expected to cause an increase in the hydrogen ion concentration of arterial blood and cerebrospinal fluid. (true/false)

13. Coughing is triggered by stimulation of pulmonary _____ receptors.

14. Assuming that arterial P_{O_2} and P_{CO_2} are normal, an increase in arterial pH should activate (peripheral/central/both types of) chemoreceptors, triggering a(n) (increase/decrease) in ventilation.

15. In respiratory acidosis, arterial P_{CO_2} is (higher/lower) than normal.

16. An increase in the P_{CO_2} of alveolar air would be expected to trigger local (bronchoconstriction/bronchodilation) in airways.

17. An increase in the P_{O_2} of alveolar air would be expected to trigger local (vasoconstriction/vasodilation).

Essay Questions

1. Explain how changes in blood P_{CO_2} affect loading and unloading of oxygen in the lungs and in respiring tissues.

2. Sketch a hemoglobin-oxygen dissociation curve, and explain how it is affected by pH and P_{CO_2}. Include in your explanation both changes in affinity and shifts in the hemoglobin-oxygen dissociation curve.

3. Describe what happens to oxygen and carbon dioxide as blood travels through the circulatory system. Start in the left ventricle and finish in the left atrium.

4. Suppose that alveolar P_{O_2} and P_{CO_2} are normal. If a sudden increase occurs in tissue metabolic activity and CO_2 production, but no change in minute alveolar ventilation occurs, then what would you expect to happen to arterial P_{CO2}, mixed venous P_{CO2}, and alveolar P_{CO2}?

5. Describe how chemoreceptors work to keep arterial P_{CO_2} constant. Include an explanation of how arterial P_{CO_2} affects both central and peripheral chemoreceptors.

Find the answers to these exercises, and additional study tools, at the Physiology Place (www.physiologyplace.com).

17

The Urinary System: Renal Function

OBJECTIVES

- Identify and describe the functions of the following structures in the urinary system: nephron, glomerulus, renal tubule, collecting duct, ureter, bladder, and urethra.

- Describe how the urinary excretion of solutes and water influences the volume and composition of plasma, and identify other processes that affect plasma volume and composition.

- Explain how the basic renal exchange processes of filtration, secretion, and reabsorption affect the rate at which materials are excreted in the urine.

- Define the following terms: *filtered load, glomerular filtration rate, clearance, transport maximum,* and *renal threshold.*

- Describe the events that occur during micturition.

CHAPTER OUTLINE

Functions of the Urinary System 533

Anatomy of the Urinary System 534

Basic Renal Exchange Processes 538

Regional Specialization of the Renal Tubules 549

Excretion 553

Above: Transfer electron micrograph of lysosomes in a renal cell

Several years ago, Joaquin was in an accident that severely damaged his left kidney. More recently, his right kidney has started to fail, and his doctors predict that he will suffer from complete renal failure within six months. Donor kidneys are scarce, and the chance of his receiving one is unlikely. This leaves Joaquin in a life-threatening situation. Fortunately for him, his sister Elena has agreed to donate one of her kidneys to him.

Unlike most organ transplants, in which donor organs are removed from the deceased, kidneys can be obtained from living donors. Because the kidneys normally function far below their maximum capacity, the loss of one kidney has little, if any, long-term effect on the body's ability to maintain homeostasis. Although this might seem to suggest that the kidneys do not work very hard, the contrary is true: The kidneys filter the entire plasma volume (approximately 3 liters) every 22 minutes, and they are essential to maintaining the normal extracellular fluid environment that bathes the cells of the body. Without medical intervention, the loss of both kidneys would result in death within a couple of weeks.

In this chapter we describe basic renal processes and the formation of urine. In Chapter 18 we will describe the regulation of these processes to maintain homeostasis.

FUNCTIONS OF THE URINARY SYSTEM

Of all the organs of the body, the kidneys are perhaps the most misunderstood and underappreciated. Because the kidneys filter the blood and produce **urine,** a fluid that is eliminated from the body, many people think that the sole function of the kidneys is to clear the blood of waste products. Such a view, however, overlooks the many other vital functions performed by these amazingly versatile organs. Although urine does contain metabolic by-products and other substances properly described as "wastes," it also contains water and solutes (such as sodium and potassium) that must be maintained at certain levels in the plasma and other body fluids. The significance of this is that the rate at which these materials are eliminated from the body, or *excreted*, by the kidneys has a significant impact on the volume and composition of these fluids and therefore is highly regulated according to the body's needs. The kidneys perform the following primary functions:

1. *Regulation of plasma ionic composition.* By increasing or decreasing the excretion of specific ions in the urine, the kidneys regulate the concentration of inorganic ions in the plasma. Ions whose concentrations are regulated by the kidneys include sodium (Na^+), potassium (K^+), calcium (Ca^{2+}), magnesium (Mg^{2+}), chloride (Cl^-), bicarbonate (HCO_3^-), and phosphates (HPO_4^{2-} and $H_2PO_4^-$).

2. *Regulation of plasma osmolarity.* Because the kidneys vary the rate at which they excrete water relative to solutes, they have the ability to regulate the osmolarity (solute concentration) of the plasma.

3. *Regulation of plasma volume.* By controlling the rate at which water is excreted in the urine, the kidneys regulate plasma volume, which has a direct effect on total blood volume, and thus on blood pressure.

4. *Regulation of plasma hydrogen ion concentration (pH).* By regulating the concentration of bicarbonate and hydrogen ions in the plasma, the kidneys partner with the lungs to regulate the pH of the blood.

5. *Removal of metabolic waste products and foreign substances from the plasma.* Because the kidneys excrete wastes and other undesirable substances in the urine, they clear the plasma of waste products and eliminate them from the body. These materials include metabolic by-products such as urea and uric acid that are generated during protein and nucleic acid catabolism, respectively, as well as foreign substances such as food additives, drugs, or pesticides that enter the body from the external environment.

Because a free exchange of water and small solutes occurs between the plasma and interstitial fluid throughout most of the body, as the kidneys regulate the volume and composition of the plasma they also regulate the volume and composition of interstitial fluid. In addition, changes in the interstitial fluid affect the intracellular fluid. Thus the kidneys ultimately control the volume and composition of all the body's fluids. As we will see, the ability of the kidneys to form urine, and thus to perform their primary functions, hinges on their ability to filter and process large quantities of solutes and water.

The kidneys perform several secondary functions as well. The kidneys are endocrine organs because they secrete the hormone erythropoietin (which stimulates erythrocyte production by the bone marrow) and the enzyme renin (which is necessary for the production of angiotensin II, a hormone important in regulating salt and water balance for long-term control of blood pressure). The kidneys are also necessary for the activation of vitamin D_3 (ultimately to $1,25(OH)_2$ vitamin D_3, or calcitrol), an important factor in regulating blood calcium and phosphate levels. Furthermore, the kidneys can function during periods of fasting to maintain a steady supply

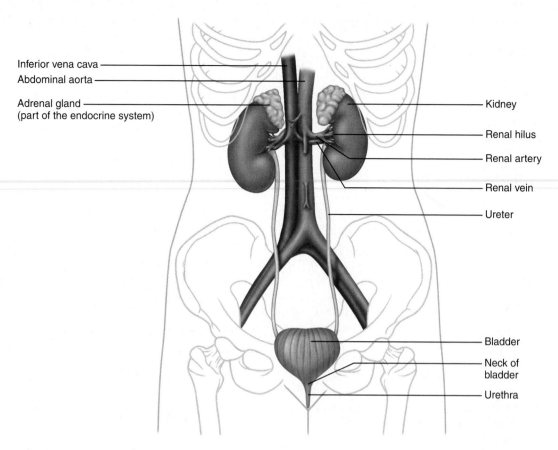

Inferior vena cava
Abdominal aorta
Adrenal gland
(part of the endocrine system)

Kidney
Renal hilus
Renal artery
Renal vein
Ureter

Bladder
Neck of bladder
Urethra

FIGURE 17.1 **Structures of the urinary system.**

of plasma glucose by carrying out gluconeogenesis, the process by which molecules such as glycerol and certain amino acids are used to synthesize glucose (described in detail in Chapter 20). Our goal in this chapter is to gain an understanding of the basic process of urine formation; Chapter 18 describes the basic mechanisms that control urine content and thus regulate plasma levels of water and solutes.

ANATOMY OF THE URINARY SYSTEM

Before we examine the macroscopic and microscopic features of the kidneys that enable them to perform their many crucial functions, we first take a brief look at the various structures that make up the urinary system.

Structures of the Urinary System

The **urinary system** consists of two kidneys, two ureters, the urinary bladder, and the urethra (Figure 17.1). Once formed by the kidneys, the urine is conducted to the bladder by the ureters. The bladder stores the urine until it is time to excrete it; at this time the urine moves through the urethra and into the external environment.

The kidneys are paired organs lying at the rear wall of the abdominal cavity just above the waistline, at about the level of the 12th rib. Each kidney is roughly bean shaped and is about the size of a fist. Even though most abdominal organs are enclosed within the *peritoneum,* a clear membrane that lines the abdominal cavity, the kidneys are located between the peritoneum and the wall of the abdominal cavity.

The kidneys receive their blood supply from the **renal arteries,** which branch off the aorta and enter each kidney at a region called the *renal hilus.* Each kidney weighs only 115–170 grams (less than half a pound); their combined weight is less than 1% of the body weight of an average adult. Despite their small fraction of body weight, the kidneys receive about 20% of the cardiac output under normal resting conditions. This rich blood supply is crucial to the kidneys' function not only because it provides them with oxygen and nutrients (kidneys account for 16% of total body ATP usage), but also because it enables the kidneys to remove (or *clear*) unneeded solutes and water from the blood at a rapid rate and eliminate them as urine. Meanwhile, the blood (minus these cleared materials) returns to the general circulation via the **renal veins,** which run parallel to the renal arteries and drain into the inferior vena cava.

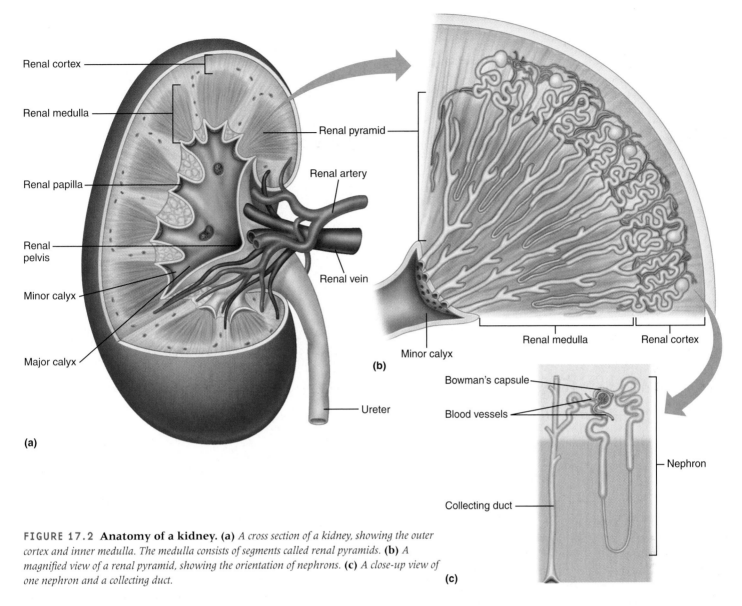

Renal cortex

Renal medulla

Renal papilla

Renal pelvis

Minor calyx

Major calyx

(a)

Renal pyramid

Renal artery

Renal vein

Renal medulla

Renal cortex

Minor calyx

(b)

Bowman's capsule

Blood vessels

Nephron

Collecting duct

(c)

Ureter

FIGURE 17.2 **Anatomy of a kidney.** **(a)** *A cross section of a kidney, showing the outer cortex and inner medulla. The medulla consists of segments called renal pyramids.* **(b)** *A magnified view of a renal pyramid, showing the orientation of nephrons.* **(c)** *A close-up view of one nephron and a collecting duct.*

Macroscopic Anatomy of the Kidney

A cross section of a kidney reveals that it contains two major regions: a reddish-brown outer layer called the **cortex,** and an inner region called the **medulla,** which is darker and has a striped appearance (Figure 17.2a). The medulla is subdivided into a number of conical sections called *renal pyramids* (Figure 17.2b). At the tips of the renal pyramids, called *papillae* (singular: papilla), tubules called *collecting ducts* drain into common passageways called *minor calyces* (singular: calyx; see Figure 17.2a). The minor calyces converge to form two or three larger passageways called *major calyces,* which drain into a single funnel-shaped passage called the **renal pelvis,** the initial portion of the ureter.

Within a kidney's many renal pyramids are over a million microscopic subunits called **nephrons** (Figure 17.2c), which are the functional units of the kidneys; they do the work of filtering the blood and forming the

urine. The most obvious feature of the nephron is a long, coiled tube (called a **renal tubule**) that forms a hairpin loop about midway along its length. During the process of urine formation, fluid flows through the renal tubules, during which time the fluid's composition is modified. Fluid from individual tubules eventually drains into a set of common passageways called **collecting ducts.** The composition of the fluid is further modified as it travels through the collecting ducts. The fluid that exits the collecting ducts is called *urine.*

Microscopic Anatomy of the Kidney

Each nephron is in essence a complete, self-contained "mini-kidney" that filters blood and forms urine. An individual nephron is composed of two parts: a *renal corpuscle* that filters the blood, and a renal tubule through which the filtrate travels and becomes modified in the formation of urine (Figure 17.3).

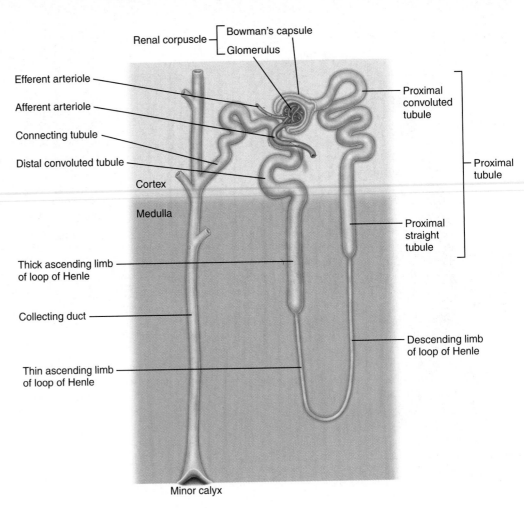

FIGURE 17.3 Anatomy of a nephron. *A nephron has two parts, a renal corpuscle and a renal tubule. Each renal corpuscle consists of a Bowman's capsule and a glomerulus; each renal tubule consists of numerous continuous tubular segments. Also shown are the blood supply to the renal corpuscle and the collecting duct associated with the nephron.*

Renal corpuscle — Bowman's capsule / Glomerulus
Efferent arteriole
Afferent arteriole
Connecting tubule
Distal convoluted tubule
Cortex
Medulla
Proximal convoluted tubule
Proximal tubule
Proximal straight tubule
Thick ascending limb of loop of Henle
Collecting duct
Descending limb of loop of Henle
Thin ascending limb of loop of Henle
Minor calyx

Renal Corpuscle

A renal corpuscle consists of two parts: a spherical structure at the inflow end of the renal tubules called **Bowman's capsule,** and a tuft of capillaries called the **glomerulus.** The renal corpuscle is the site where blood is filtered and where tubular fluid, or *filtrate*, has its origin. Before the blood is filtered, it enters the glomerular capillaries via an *afferent arteriole*. As the blood flows through the glomerular capillaries, protein-free plasma filters across the walls of the capillaries into Bowman's capsule by a process called **glomerular filtration.** The remaining blood leaves the glomerulus via an *efferent arteriole*. This arrangement of two arterioles in series with a capillary bed between them is unique to the renal corpuscle and allows greater regulation of glomerular filtration. The walls of the afferent and efferent arterioles contain smooth muscle that can contract or relax in response to input from the sympathetic nervous system, thereby regulating their diameter (described in greater detail later in this chapter).

Renal Tubule

As the glomerular filtrate is formed, it flows from Bowman's capsule to the initial portion of the renal tubule, called the **proximal convoluted tubule** because of its proximity to the capsule and its highly folded or *convoluted* structure, and then to the **proximal straight tubule.** The two tubules together are called the **proximal tubule.** The proximal tubule empties into the **loop of Henle,** the portion of the tubule that makes up the hairpin loop within the medulla. The loop of Henle is divided into three sections: (1) the descending limb, (2) the thin ascending limb, and (3) the thick ascending limb. The **descending limb** is a thin tubule leading from the proximal tubule and extending into the renal medulla. At the tip of the loop, the tubule reverses direction, becoming the **thin ascending limb,** which extends toward the cortex. As the tubule approaches the cortex, it widens into the **thick ascending limb.** From the ascending limb of the loop of Henle, the fluid flows into the **distal convoluted tubule,** which resembles the proximal tubule in appearance but is considerably shorter. The fluid then enters a short, straight terminal portion of the nephron, called the *connecting tubule*, which joins the nephron with the collecting duct. Several tubules empty their fluid into a single collecting duct. The collecting ducts then empty into the minor calyces, as previously described.

Cortical and Juxtamedullary Nephrons

There are two classes of nephrons based on their location: *cortical nephrons* and *juxtamedullary nephrons* (Figure 17.4).

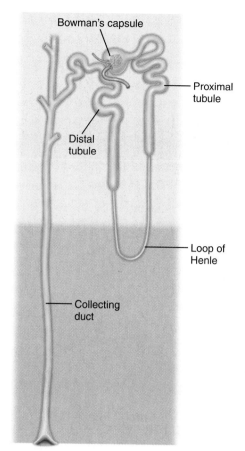

(a) Cortical nephron

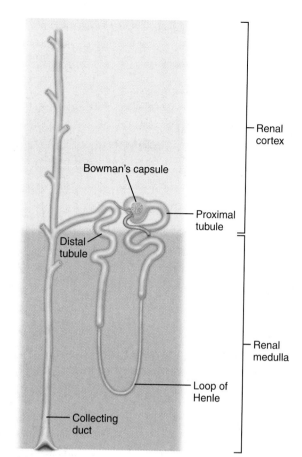

(b) Juxtamedullary nephron

FIGURE 17.4 **Locations of cortical and juxtamedullary nephrons. (a)** *A cortical nephron, which is located primarily in the renal cortex; only the tip of the loop of Henle dips into the renal medulla.* **(b)** *A juxtamedullary nephron, which is located in both the cortex and medulla; Bowman's capsule and part of the proximal and distal tubules are located in the renal cortex, whereas the loop of Henle dips deep into the medulla.*

The vast majority of nephrons in the kidneys are cortical nephrons, which are located almost entirely within the renal cortex; only the tip of the loop of Henle dips into the renal medulla. In juxtamedullary nephrons, which constitute about 15–20% of all nephrons, the renal corpuscle is located near the border between the cortex and medulla. The glomerulus, proximal convoluted tubule, and distal convoluted tubule are located in the cortex, whereas the loop of Henle dips deep into the renal medulla. Although the two types of nephrons are fundamentally similar, there are important functional differences: Whereas both cortical and juxtamedullary nephrons function directly in the processes involved in urine formation, juxtamedullary nephrons also function in maintaining an osmotic gradient in the renal medulla that is crucial to the kidneys' ability to produce highly concentrated urine and thus conserve water under certain conditions (described in Chapter 18).

The Juxtaglomerular Apparatus

At a site where the initial portion of the distal tubule comes into contact with a nephron's afferent and efferent arterioles is a structure called the **juxtaglomerular apparatus** (Figure 17.5). The juxtaglomerular apparatus has two components: (1) a specialized cluster of the tubule's epithelial cells, called the **macula densa,** and (2) **granular cells** (or **juxtaglomerular cells**), specialized

cells in the wall of the afferent and efferent arterioles that have granular cytoplasms due to the presence of numerous secretory granules containing a product called *renin*. The juxtaglomerular apparatus plays an important role in regulating blood volume and blood pressure.

Blood Supply to the Kidney

The blood supply to the kidney is illustrated in Figure 17.6a. Within the kidney, the renal artery branches into *segmental arteries,* which branch into a number of smaller *interlobar arteries* that feed into another set of arteries called *arcuate arteries.* The arcuate arteries then branch into *interlobular arteries,* from which blood is carried to individual nephrons by the afferent arterioles, which lead into the glomerular capillary beds. Coming off each of the glomerular capillary beds is the efferent arteriole, which then gives rise to one of two types of capillary beds (Figure 17.6b): **peritubular capillaries,** which branch from the efferent arterioles of cortical nephrons and are located close to the renal tubules, and **vasa recta,** which branch from the efferent arterioles of juxtamedullary nephrons and are networks of blood vessels forming hairpin loops that run along the loops of Henle and collecting ducts, dipping deep into the renal medulla. Each of these capillary beds has a distinct function in the formation of urine, as described later in this chapter and in Chapter 18.

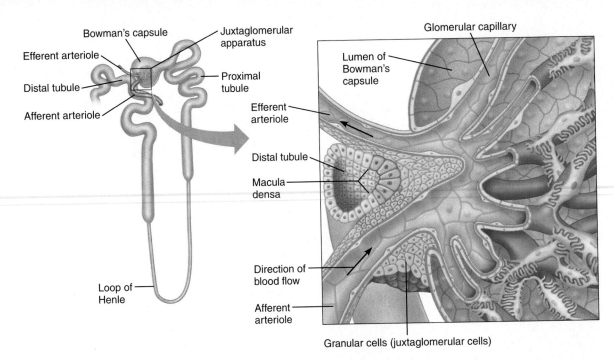

FIGURE 17.5 The juxtaglomerular apparatus. *Located where the initial part of the distal tubule passes through the fork between the afferent and efferent arterioles, the juxtaglomerular apparatus consists of granular cells of the afferent and efferent arterioles and the macula densa of the distal tubule.*

The peritubular capillaries and vasa recta drain into the *interlobular veins.* From here, blood is carried away from nephrons by the *arcuate veins,* and then *interlobar veins,* which run parallel to their respective arterial counterparts, eventually draining into the renal vein.

Quick Test 17.1

1. What structures make up the urinary system? What is the function of each?

2. What is a nephron? What are the two types of nephrons? How do they differ in location?

3. Arrange the following structures in order so that they correctly describe the path of fluid as it flows through the urinary system: renal pelvis, glomerulus, bladder, collecting duct, proximal convoluted tubule, ureter, Bowman's capsule, distal convoluted tubule, urethra, descending limb of the loop of Henle, thick ascending limb of the loop of Henle, thin ascending limb of the loop of Henle.

4. What structures make up the juxtaglomerular apparatus? Where are they located?

BASIC RENAL EXCHANGE PROCESSES

In the kidneys, water and solutes are exchanged between plasma and fluid in the renal tubules to regulate the composition of plasma. Substances ultimately removed from the plasma are excreted in urine. The following three exchange processes occur within the renal nephrons (Figure 17.7):

1. Glomerular filtration is the bulk flow of protein-free plasma from the glomerular capillaries into Bowman's capsule.

2. **Reabsorption** is the selective transport of molecules from the lumen of the renal tubules to the interstitial fluid outside the tubules. Reabsorbed molecules eventually enter the peritubular capillaries by diffusion, and are then returned to the general circulation.

3. Secretion is the selective transport of molecules from the peritubular fluid to the lumen of the renal tubules. These secreted molecules originate from the plasma of the peritubular capillaries.

First we describe these three exchange processes; then we look at how the renal tubules are specialized for reabsorption and secretion. The final part of this chapter describes a fourth renal process: **excretion,** the elimination of materials from the body in the form of urine.

Glomerular Filtration

The renal corpuscle is where blood is filtered and where tubular fluid has its origin. Filtration is driven by Starling forces (hydrostatic and osmotic pressure gradients) existing across the walls of glomerular capillaries. These Starling forces are the same forces that drive the filtration of fluid from capillaries throughout the body (see Chapter 13). The filtrate resembles plasma in composition, except that it lacks most of the proteins present in plasma.

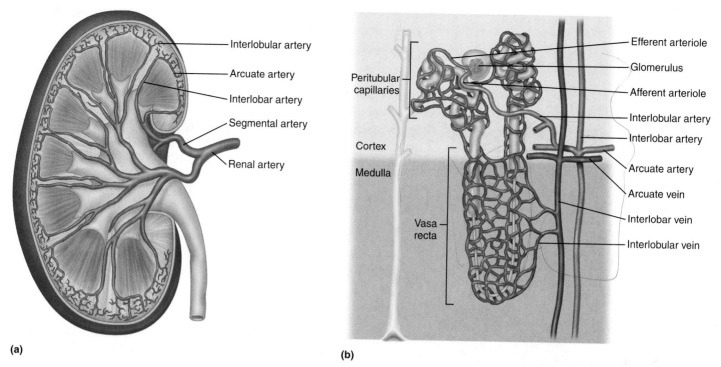

(a)

(b)

FIGURE 17.6 **Blood supply to the kidneys. (a)** *The renal arteries supply blood to the kidneys and branch into the smaller arteries indicated.* **(b)** *The efferent arterioles lead into two different types of capillary beds: the peritubular capillaries located around the renal tubules, and the vasa recta located around the loops of Henle.*

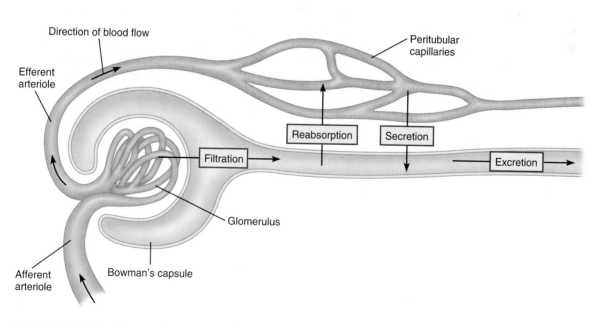

FIGURE 17.7 **The three exchange processes in the renal tubules.** *Filtration, which occurs in the renal corpuscle, is the bulk flow of protein-free plasma from the glomerulus into Bowman's capsule. Reabsorption, which occurs along the tubules, is the movement of water or solute from the lumen of the tubules into the peritubular capillaries. Secretion also occurs along the tubules, but it is the movement of solute from the peritubular capillaries into the lumen of the tubules. A fourth process, excretion, is the bulk flow of urine out of the body.*

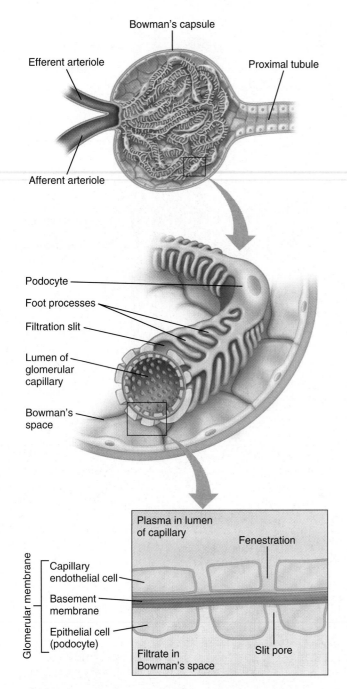

Bowman's capsule
Efferent arteriole
Proximal tubule
Afferent arteriole

Podocyte
Foot processes
Filtration slit
Lumen of glomerular capillary
Bowman's space

Plasma in lumen of capillary
Fenestration

Glomerular membrane
Capillary endothelial cell
Basement membrane
Epithelial cell (podocyte)

Filtrate in Bowman's space
Slit pore

FIGURE 17.8 Anatomy of the renal corpuscle. *The most magnified view shows the glomerular membrane, which includes the capillary endothelial cells, the basement membrane, and the epithelial cells of Bowman's capsule. Fenestrations and slit pores favor the bulk flow of plasma.*

Figure 17.8 shows a magnified view of the renal corpuscle. The wall of Bowman's capsule and the wall of the renal tubule are composed of a continuous layer of epithelial cells. In Bowman's capsule, this epithelium folds upon itself to envelop the glomerular capillaries. Below the epithelium is a basement membrane that acts as the primary filtration barrier for proteins. The glomerular fil-

trate must cross three barriers to enter Bowman's capsule: (1) the capillary endothelial cell layer, (2) the surrounding epithelial cell layer, and (3) the basement membrane that is sandwiched between them. The combination of these three layers makes up what is sometimes called the *glomerular membrane* or *filtration barrier.* The epithelial cells that cover the glomerular capillaries have special extensions or *foot processes,* giving them their name *podocytes.* As fluid moves out of the glomerular capillaries, it passes through gaps between the podocytes, called *slit pores.* The size of slit pores are regulated by *slit diaphragms.*

The presence of fenestrations (pores) in the capillary endothelium, the large number of slit pores in the surrounding capsule epithelium, and the large surface area of the filtration barrier combine to make the renal corpuscle favorable for the bulk flow of protein-free fluid between blood and the lumen of Bowman's capsule (called *Bowman's space*).

The sum of the Starling forces in the renal corpuscle is called the **glomerular filtration pressure,** which is analogous to the *net filtration pressure* discussed in Chapter 13. The four Starling forces are the following (Figure 17.9):

1. ***Glomerular capillary hydrostatic pressure.*** The glomerular capillary hydrostatic pressure (P_{GC}) favors filtration and is equal to the blood pressure in the glomerular capillaries—approximately 60 mm Hg. This pressure is substantially higher than the hydrostatic pressure in most other capillaries because of the high resistance of the efferent arteriole, which is located downstream from the glomerular capillaries. As a general rule, the presence of a high resistance in any network of vessels tends to raise the pressure in vessels located upstream while lowering the pressure downstream, just as tightening a clamp on a water hose causes the pressure downstream to fall (and reduces the flow of water) but increases the pressure upstream due to the "backup" of water. (Discovery: The Physics Behind Glomerular Capillary Hydrostatic Pressure, www.physiologyplace.com, Challenge Yourself)

2. ***Bowman's capsule oncotic pressure.*** The oncotic pressure in Bowman's capsule (π_{BC}) favors filtration because the presence of proteins in the interstitial fluid surrounding the glomerulus would tend to pull fluid out of the capillaries and into the capsule. Because very little protein leaves the capillaries with the filtrate, the protein concentration in Bowman's capsule is very small, and thus the oncotic pressure is negligible under normal conditions. (However, in certain diseases that cause damage to the glomerulus, significant quantities of protein can leak out of glomerular capillaries.) The net pressure favoring filtration at the renal corpuscle under normal conditions is

$$P_{GC} + \pi_{BC} = 60 \text{ mm Hg} + 0 \text{ mm Hg} = 60 \text{ mm Hg}$$

3. ***Bowman's capsule hydrostatic pressure.*** Bowman's capsule hydrostatic pressure (P_{BC}) opposes filtration

and is typically about 15 mm Hg. This pressure is considerably higher than the hydrostatic pressure in the interstitial fluid surrounding most capillary beds because the relatively large volume of fluid that filters out of the glomerular capillaries is "funneled" into the restricted space of Bowman's capsule. However, because the hydrostatic pressure in Bowman's capsule is still considerably lower than the capillary hydrostatic pressure, there is a net hydrostatic pressure gradient for filtration; that is, favoring the movement of fluid from the capillary into the capsule.

4. **Glomerular oncotic pressure.** The glomerular oncotic pressure (π_{GC}) opposes filtration. Recall that oncotic pressure of a fluid is due to the presence of nonpermeant solutes. Because proteins are generally the only solute that cannot move between plasma and Bowman's capsule, oncotic pressure is the osmotic force exerted by the presence of proteins. The glomerular oncotic pressure opposes filtration because the presence of proteins in the plasma tends to draw filtrate back into the glomerulus. The oncotic pressure in the glomerulus is approximately 29 mm Hg, which is higher than the typical 25 mm Hg oncotic pressure found in most systemic capillaries, because the blood that flows through these capillaries loses a substantial fraction of its water as a result of glomerular filtration, and this loss of water causes the concentration of plasma proteins to increase. The net pressure opposing filtration at the renal corpuscle under normal conditions is:

$$P_{BC} + \pi_{GC} = 15 \text{ mm Hg} + 29 \text{ mm Hg} = 44 \text{ mm Hg}$$

Glomerular Filtration Rate

Given that the oncotic pressure in Bowman's capsule is negligible, the average net glomerular filtration pressure is given by

$$\text{glomerular filtration pressure} = P_{GC} - (P_{BC} + \pi_{GC})$$

Substituting the values for the Starling forces into this equation, glomerular filtration pressure is thus

$$60 \text{ mm Hg} - (15 \text{ mm Hg} + 29 \text{ mm Hg}) = 16 \text{ mm Hg}$$

Under normal conditions, approximately 625 mL of plasma flow through the kidneys each minute. The volume of the plasma filtered per unit time is called the glomerular filtration rate or GFR, and is approximately 125 mL/min. Thus over the course of a day, the kidneys filter 180 liters of plasma:

$$\text{GFR} = \frac{125 \text{ mL}}{\text{min}} \times \frac{1 \text{ Liter}}{1000 \text{ mL}} \times \frac{60 \text{ min}}{\text{hour}}$$
$$\times \frac{24 \text{ hour}}{\text{day}} = 180 \text{ liters/day}$$

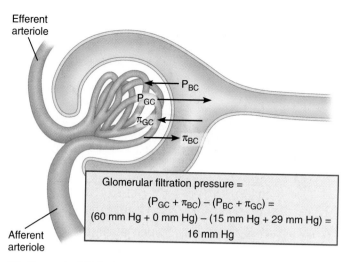

$$\text{Glomerular filtration pressure} =$$
$$(P_{GC} + \pi_{BC}) - (P_{BC} + \pi_{GC}) =$$
$$(60 \text{ mm Hg} + 0 \text{ mm Hg}) - (15 \text{ mm Hg} + 29 \text{ mm Hg}) =$$
$$16 \text{ mm Hg}$$

(a) Glomerular filtration pressure

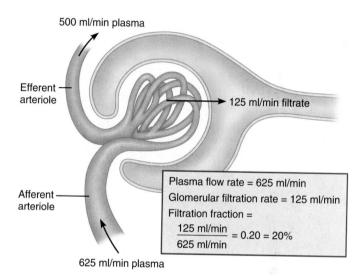

Plasma flow rate = 625 ml/min
Glomerular filtration rate = 125 ml/min
Filtration fraction =
$$\frac{125 \text{ ml/min}}{625 \text{ ml/min}} = 0.20 = 20\%$$

(b) Filtration fraction

FIGURE 17.9 Glomerular filtration. (a) *Glomerular filtration pressure is the result of four Starling forces: (1) hydrostatic pressure in the glomerular capillaries (P_{GC}), (2) hydrostatic pressure in Bowman's capsule (P_{BC}), (3) oncotic pressure in the glomerular capillaries (π_{GC}), and (4) oncotic pressure in Bowman's capsule (π_{BC}). The net filtration pressure is 16 mm Hg.* **(b)** *The filtration fraction is the proportion of renal plasma that is filtered into Bowman's capsule. The normal filtration fraction is 20%.*

If proteins leaked out of the glomerular capillaries (which would decrease π_{GC} and increase π_{BC}), what would happen to the glomerular filtration pressure and to the glomerular filtration rate?

This is an enormous volume of fluid, considering that the body's total plasma volume is only approximately 2.75 liters for an average adult. In fact, the GFR is so high that a volume of fluid equivalent to total plasma volume filters through the glomeruli every 22 minutes.

Both would increase.

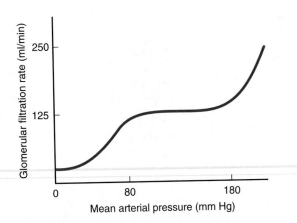

FIGURE 17.10 Effect of mean arterial pressure on glomerular filtration rate. *Between 80 mm Hg and 180 mm Hg, glomerular filtration rate does not vary with mean arterial pressure due to the actions of intrinsic regulatory mechanisms.*

If mean arterial pressure increased to 200 mm Hg, what would happen to the filtered load of any given solute?

Filtration Fraction

The fraction of the renal plasma volume that is filtered is called the *filtration fraction* and is equal to the glomerular filtration rate divided by the renal plasma flow rate (Figure 17.9b):

$$\text{Filtration fraction} = \frac{\text{GFR}}{\text{Renal plasma flow}}$$

$$= \frac{125 \text{ mL/min}}{625 \text{ mL/min}} = 0.20 = 20\%$$

Filtered Load

When molecules of a solute are small enough to move across the glomerular membrane without significant restriction, that solute is said to be *freely filterable*. When such a substance is filtered, its concentration in the glomerular filtrate is virtually identical to its concentration in the plasma. Consequently, the quantity of solute that is filtered in any given period of time is entirely determined by two factors: the solute's plasma concentration and the GFR. The quantity of a particular solute that is filtered per unit time is known as the **filtered load,** usually expressed in mole/min or the equivalent. The filtered load equals the product of the GFR and the solute's plasma concentration (which is given the symbol P):

$$\text{Filtered load} = \text{GFR} \times \text{plasma concentration of X}$$

$$= \text{GFR} \times P_X$$

We already know that the normal GFR is 125 mL/min. Given that the normal plasma glucose concentration is

100 mg/dL = 1 mg/mL, then the filtered load of glucose under normal conditions is

$$\text{Filtered load} = 125 \text{ mL/min} \times 1 \text{ mg/mL} = 125 \text{ mg/min}$$

Note that the filtered load increases if either the plasma concentration of the solute or the GFR increases.

Regulation of Glomerular Filtration Rate

Although 180 liters of fluid filters into the renal tubules every day, only about 1.5 liters of urine are normally excreted during the same period of time. The reason for this low excretion rate is that over 99% of the fluid filtered out of the plasma is normally reabsorbed. Because the GFR is so large, however, even a small percentage change in its value will have an enormous effect on the volume of fluid filtered, and therefore on the quantity of material that must be reabsorbed in order to maintain the same urinary output. Thus a 10% increase in GFR, for example, would translate into an extra 18 liters of fluid entering the kidney tubules per day.

Under most circumstances, changes in GFR do not cause a large increase or decrease in urine output, because the GFR is regulated to stay relatively constant by the intrinsic and extrinsic mechanisms we discuss next.

Intrinsic Control of Glomerular Filtration Changes in mean arterial pressure can potentially alter the GFR because the arterial pressure affects the glomerular capillary pressure, which in turn influences the glomerular filtration pressure. Although mean arterial pressure is regulated to stay constant by baroreceptor reflexes and other mechanisms, it does change in certain situations, such as exercise. When arterial pressure increases, glomerular capillary pressure also tends to rise, which increases the glomerular filtration pressure and hence GFR. Conversely, when mean arterial pressure falls, the glomerular capillary pressure also tends to fall, which decreases the glomerular filtration pressure and GFR. Such changes in GFR are undesirable because they tend to make urine flow increase or decrease, respectively, which interferes with the kidneys' ability to regulate the volume and composition of the plasma.

Although variations in mean arterial pressure pose a potential problem, the kidneys can tolerate a change in mean arterial pressure over a fairly wide range (approximately 80–180 mm Hg) with very little change in GFR (Figure 17.10), because three intrinsic mechanisms regulate the GFR in the face of changes in arterial pressure. Two of these intrinsic mechanisms, *myogenic regulation* of afferent arteriolar smooth muscle and *tubuloglomerular feedback,* operate by changing the resistance of the afferent arteriole; the third factor, *mesangial cell contraction,* acts by changing the permeability of the filtration barrier.

Myogenic regulation of the GFR (Figure 17.11a) is similar to the myogenic regulation of blood flow that occurs in other parts of the body. The smooth muscle of the

The filtered load would increase because GFR increased.

afferent arteriole is sensitive to stretch and responds to stretch by contracting. When mean arterial pressure rises, pressure in the afferent arteriole also rises, which causes its wall to stretch. Pressure also rises in the glomerular capillaries, which raises the glomerular filtration pressure and the GFR. In response to stretching, the afferent arteriole constricts, which increases its resistance to blood flow. As a result, the pressure in blood vessels downstream, including the glomerular capillaries, decreases. This decrease in pressure counteracts (but does not change) the initial rise in pressure that triggered constriction of the afferent arteriole. Thus, through this negative feedback mechanism the glomerular capillary pressure, and hence the GFR, tends to stay nearly constant. As expected, a fall in mean arterial pressure has the opposite effects, triggering relaxation of the afferent arteriole and a subsequent rise in glomerular capillary pressure that counteracts the initial fall.

The smooth muscle of the afferent arteriole is sensitive not only to stretch, but also to chemical agents that are secreted by cells of the macula densa, which are located in the nearby distal tubule. In **tubuloglomerular feedback,** a change in GFR causes a change in the flow of tubular fluid past the macula densa, which alters the secretion of certain (as yet unidentified) paracrines from the macula densa. These paracrines then trigger contraction or relaxation of the afferent arteriole, which causes a change in glomerular capillary pressure and GFR in the direction opposite to that of the original change. As a consequence of this negative feedback control of GFR, the flow of fluid past the macula densa changes such that it opposes the change in flow that triggered the response initially. For example, if an increase in GFR causes the flow of tubular fluid to increase, then the afferent arteriole constricts and the GFR decreases, thereby decreasing the flow. Thus the resistance of the afferent arteriole varies in such a way that the flow of fluid past the macula densa tends to be held constant. Figure 17.11b shows what happens when an increase in blood pressure causes an increase in GFR.

The third mechanism of GFR autoregulation is similar to the first, except it targets mesangial cells instead of blood vessels. Mesangial cells are modified smooth muscle cells located around glomerular capillaries. An increase in blood pressure, which increases GFR, stretches mesangial cells. In response to the stretch, the mesangial cells contract, decreasing the surface area of capillaries available for filtration, which decreases the GFR back to normal.

Extrinsic Control of Glomerular Filtration and Renal Blood Flow

Although intrinsic control mechanisms work to keep the GFR constant when mean arterial pressure changes, these mechanisms work over a limited range of pressures only. When mean arterial pressure goes above or below this range, the GFR rises or falls, respectively, because intrinsic mechanisms are no longer

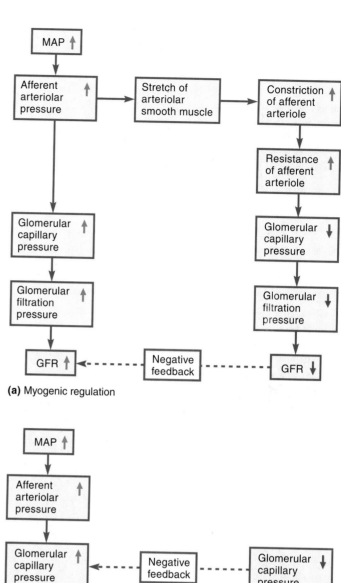

(a) Myogenic regulation

(b) Tubuloglomerular feedback

FIGURE 17.11 **Intrinsic controls of glomerular filtration rate. (a)** *In response to a change in mean arterial pressure, myogenic regulation of afferent arteriolar resistance prevents significant changes in GFR by affecting glomerular capillary pressure.* **(b)** *Glomerulotubular feedback, triggered by chemical signals released from cells in the macula densa in response to increased GFR, prevents significant changes to GFR when mean arterial pressure changes by affecting glomerular capillary pressure.*

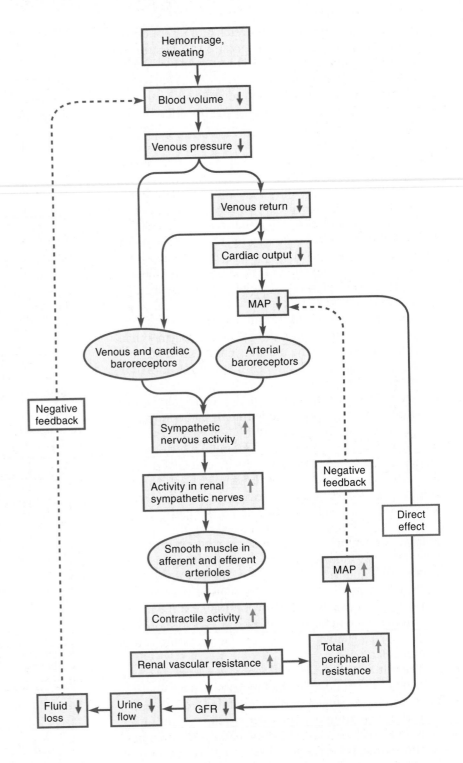

FIGURE 17.12 Extrinsic control of GFR and renal vascular resistance during fluid loss due to hemorrhage or sweating. *Loss of fluid decreases MAP, which directly decreases the filtration pressure and thus the GFR. The decrease in MAP also reflexively increases sympathetic nerve activity, which causes constriction of the afferent and efferent arterioles, which by decreasing blood flow to the glomerulus decreases the filtration pressure and GFR.*

able to prevent the glomerular capillary pressure from changing. Figure 17.12 shows what happens when the mean arterial pressure falls due to hemorrhage or excessive sweating. When MAP falls, GFR decreases directly due to the lowered filtration pressure. In addition, the fall in MAP triggers an increase in sympathetic nervous activity via baroreceptor reflexes. In response to increased input from sympathetic nerves, smooth muscle in the afferent and efferent arterioles contract, and both afferent and efferent arterioles constrict, which increases the overall resistance of the renal vasculature and decreases the GFR. The increase in renal vascular resistance acts to decrease renal blood flow and also raises the total peripheral resistance, which increases MAP. The decrease in GFR also decreases urine output, which helps the body conserve water. This process minimizes reductions in blood volume, which in turn counteracts further decreases in arterial pressure.

SUBSTANCE	FILTRATION RATE	REABSORPTION RATE	PERCENT OF FILTERED LOAD REABSORBED
Water	180 liters/day	178.5 liters/day	99.2%
Glucose	800 millimoles/day	800 millimoles/day	100%
Urea	933 millimoles/day	467 millimoles/day	50%
Na^+	25.20 moles/day	25.05 moles/day	99.4%
K^+	720 millimoles/day	620 millimoles/day	86.1%
Ca^{2+}	540 millimoles/day	530 millimoles/day	98.1%
Cl^-	18.00 moles/day	17.85 moles/day	99.2%
HCO_3^-	4.320 moles/day	4.318 moles/day	>99.9%

Exercise Link

During long-duration exercise in warm conditions, such as Jane and Bill's marathon, water loss due to sweating can exceed a liter per hour. Sympathetic activation decreases GFR and thus limits urine production. But this mechanism of water conservation, and others described later, is insufficient to maintain blood volume in these circumstances. Jane and Bill attempted to replace lost water by drinking during the race but they were still unable to maintain fluid balance. Their urine production rates remained depressed for two days after the end of the marathon.

Quick Test 17.2

1. Of the four renal exchange processes, which two move a substance into the renal tubules, and which two move a substance out of the renal tubules?

2. Indicate whether the GFR will tend to rise or fall in response to an increase in each of the following: the glomerular filtration pressure, the glomerular capillary hydrostatic pressure, the glomerular capillary oncotic pressure, the hydrostatic pressure in Bowman's capsule, the oncotic pressure in Bowman's capsule, the concentration of proteins in the plasma.

3. Name three intrinsic control mechanisms that regulate the GFR, and briefly explain how these mechanisms work to keep the GFR constant in the face of changes in the mean arterial pressure.

4. Is the control of afferent and efferent arteriolar resistance by renal sympathetic neurons an example of intrinsic control or extrinsic control? Why? How does an increase in sympathetic activity tend to affect the GFR? Renal vascular resistance?

Reabsorption

Reabsorption refers to movement of filtered solutes and water from the lumen of the tubules back into the plasma. If reabsorption did not occur, all filtered materials would be excreted, and it would take just eight minutes for a person to lose a liter of fluid in the urine (assuming that the GFR stays constant at its normal rate). Given that 1 liter of fluid is equivalent to 20% of total blood volume, the ability of the kidneys to reabsorb filtered solutes and water is clearly an absolute necessity of life. In fact, 100% of many substances filtered at the glomerulus are reabsorbed by the renal tubules. The reabsorption of other substances is regulated to vary their excretion rate, which in turn regulates the concentration of these substances in the plasma. First we look at general properties of solute reabsorption; then we describe specific examples of solute and water reabsorption across the tubular epithelium.

Solute Reabsorption

Table 17.1 lists the quantities of several freely filtered substances that are filtered and reabsorbed in a single day under normal circumstances. The sheer mass of material that must be transported from tubular fluid back into the plasma by the renal tubule epithelium every day reveals how extensive a process reabsorption is. Many solutes are reabsorbed actively; that is, they are transported against their electrochemical gradients as they move from the tubular lumen to the plasma.

Reabsorption of most solutes occurs in the proximal and distal convoluted tubules. Recall that the efferent arteriole branches into peritubular capillary beds adjacent

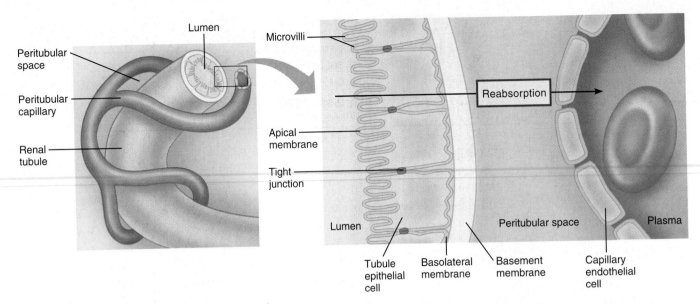

FIGURE 17.13 The barriers to reabsorption. *For reabsorption to occur, a substance must cross the epithelial cells of the renal tubules and the endothelial cells of the capillary. The primary barriers are the apical and basolateral membranes of the tubule epithelial cells because capillary walls are relatively permeable and many substances can move between the endothelial cells.*

to the renal tubules in the renal cortex (see Figure 17.6b). The interstitial fluid between the epithelium of the renal tubules and the peritubular capillaries *(peritubular fluid)* fills the *peritubular space.* When a substance is reabsorbed, it must move across two barriers: the tubule epithelium and the capillary endothelium (Figure 17.13). Because tight junctions connect the epithelial cells lining the renal tubules, movement of molecules between cells is restricted. The plasma membrane of the epithelial cells facing the tubule lumen is called the apical membrane and has microvilli; the plasma membrane facing the interstitial fluid is called the basolateral membrane. Microvilli on the apical membrane are abundant in the more proximal portions of the tubule but are sparse in the distal portions. Beneath the basolateral membrane is a basement membrane that does not contribute significantly as a barrier and thus will not be discussed further.

Some molecules are passively reabsorbed, whereas others require energy. Figure 17.14 depicts the reabsorption of three uncharged solutes (X, Y, and Z) and water. (Note that the movement of charged substances is affected by electrical forces in addition to the forces described next.) Although each substance moves in the same direction, the actual mechanism of reabsorption differs.

Substance X is passively reabsorbed via diffusion (Figure 17.14a). For this to occur, two conditions must be satisfied: the concentration of X must be greater in the tubular fluid than in the plasma, and X must be able to permeate the plasma membranes of the tubule epithelium and the capillary endothelium. The tubular fluid, which came from plasma, has a higher concentration of X than

fluid in the capillary lumen, because most water that is filtered is reabsorbed by mechanisms described later. As water leaves the lumen of the tubules and enters the plasma, the tubular concentration of X increases, whereas the plasma concentration of X decreases. Therefore, X diffuses down its concentration gradient from the tubular fluid to the plasma. Stated another way, reabsorption of X follows water reabsorption. An example of a molecule that is passively reabsorbed in this manner is urea.

In the examples shown in Figure 17.14b, substances Y and Z are both actively transported, but through different mechanisms. Both substances are transported by mechanisms involving active transport across one membrane in conjunction with passive movement across the other membrane. Recall that active transport requires energy to move a molecule against its concentration gradient. Energy can come either directly from ATP (primary active transport) or from an ion electrochemical gradient created at the expense of ATP (secondary active transport).

Active transporters for substance Y located on the basolateral membrane of the tubule epithelial cell transport Y out of the cell and into the peritubular fluid; Y then diffuses into the plasma (Figure 17.14b). This process keeps the concentration of Y inside the tubule epithelial cell low. Carrier proteins for Y are located on the apical membrane. Because the concentration of Y inside the cell is low, Y moves into the cell by facilitated diffusion. Therefore, the net movement of Y is from the tubular fluid to the plasma.

Active transporters for substance Z located on the apical membrane of the tubule epithelial cell transport Z into the cell, creating a high intracellular concentration of Z (Figure 17.14b). Carrier proteins for Z are located on the basolateral membrane. As the concentration of Z inside the cell rises due to active transport, it moves out of the cell into the peritubular fluid by facilitated diffusion; Z then diffuses into the plasma.

Water Reabsorption

We saw that the reabsorption of solute X depends on water movement; likewise, water reabsorption depends on solute movement. Water diffusion is based on differences in osmolarity. As solutes such as Y and Z are actively reabsorbed, they increase the osmolarity of the plasma while decreasing the osmolarity of the tubular fluid. Therefore, water diffuses down its concentration gradient to a region of greater osmolarity (Figure 17.14c). Stated another way, water reabsorption follows the active reabsorption of solute. For water to follow solute, water must be able to permeate the tubule epithelium. Although water can permeate most plasma membranes in the body, we see in Chapter 18 that some plasma membranes in the renal tubules are impermeable to water.

Transport Maximum

When solutes are transported from filtrate to plasma across the tubular epithelium by carrier proteins or pumps, those modes of transport can become saturated (see Chapter 4); that is, when solute concentration is high enough, all carrier proteins and pumps are occupied, and the system is operating at **transport maximum,** or Tm.

The transport maximum is best understood for substances that are normally 100% reabsorbed in the renal tubules, such that none of the substance is excreted in the urine. When the plasma concentration of the solute rises, however, the amount of solute filtered at the glomerulus rises, causing an increase in the concentration of solute in the filtrate. At a certain plasma concentration of the solute, the resulting filtrate concentration saturates the carrier proteins in the tubules, and some of the substance appears in the urine. The plasma concentration of solute at which "spillover" into the urine occurs is called the **renal threshold.** As the plasma concentration rises above the renal threshold, the rate at which the solute is excreted in the urine becomes progressively higher. A well-known example of transport maximum occurs in the active reabsorption of glucose.

Glucose, which is freely filtered from the glomerulus, is normally completely reabsorbed by active transport in the proximal tubule, so no glucose is excreted in the urine. Figure 17.15 shows the mechanism of glucose reabsorption, which is similar to that for solute Z in Figure 17.14b. Glucose is actively transported across the apical membrane by a sodium-linked active transport; that is,

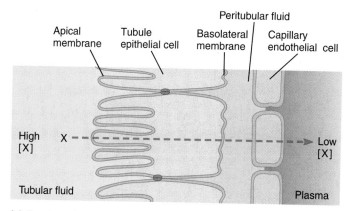

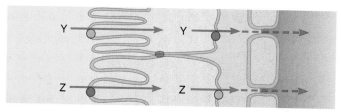

(a) Passive solute reabsorption via diffusion

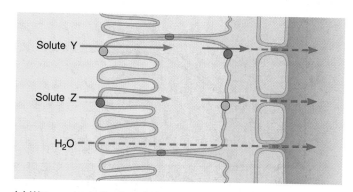

(b) Active solute reabsorption

(c) Water reabsorption (passive)

FIGURE 17.14 Mechanisms of solute and water reabsorption. (a) *Passive reabsorption of a solute. Reabsorption of solute X occurs via diffusion down its concentration gradient.* **(b)** *Active reabsorption of solutes. Solutes can be actively transported against their concentration gradients across either the basolateral membrane (solute Y) or the apical membrane (solute Z).* **(c)** *Passive reabsorption of water. The active transport of solutes Y and Z shown in part b increases the osmolarity of peritubular fluid and plasma, creating conditions that enable reabsorption of water via osmosis.*

glucose is cotransported with sodium ions from the tubular fluid into the epithelial cell. Such cotransport concentrates glucose inside the epithelial cell. A carrier protein for glucose is located on the basolateral membrane. Because glucose is in high concentration inside the epithelial cell, glucose is transported by this carrier down its concentration gradient into the peritubular fluid, where it can diffuse into the plasma.

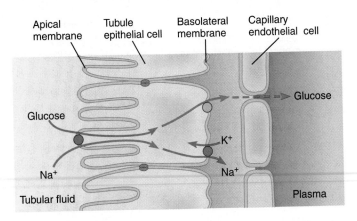

FIGURE 17.15 Mechanism of glucose reabsorption. *Glucose is actively reabsorbed by cotransport with sodium across the apical membrane, followed by facilitated diffusion across the basolateral membrane.*

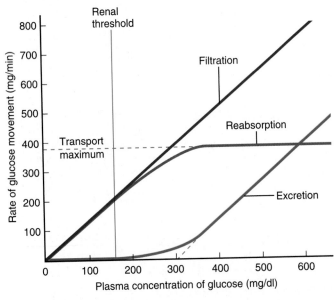

FIGURE 17.16 Glucose filtration, reabsorption, and excretion as a function of plasma glucose concentration. *The excretion rate is the difference between the filtration and reabsorption rates. GFR is assumed to be constant at 125 mL/min. Note that the "theoretical" renal threshold (300 mg/dL) need not be reached before glucose begins appearing in the urine.*

Figure 17.16 shows the relationship between the plasma concentration of glucose and the handling of glucose by the kidneys. The transport maximum for glucose reabsorption is 375 mg/min. The normal plasma glucose level is 80–100 mg/dL. Given that GFR is 125 mL/min (= 1.25 dL/min), then the filtered load of glucose when its plasma concentration is 100mg/dL is

$$\text{GFR} \times \text{P}_{glucose} = 1.25 \text{ dL/min} \times 100 \text{ mg/dL}$$

$$= 125 \text{ mg/min}$$

which is well below the transport maximum of 375 mg/min. Therefore, all the glucose is reabsorbed, and

none is excreted in the urine. However, if glucose levels increase, eventually the amount of glucose in the filtrate will exceed the capacity for reabsorption, and some glucose will be excreted in the urine.

The "theoretical" renal threshold for glucose—the plasma concentration at which the amount of glucose in the filtrate exceeds the transport maximum, and glucose appears in the urine—can be calculated as follows:

$$\text{GFR} \times \text{renal threshold} = \text{transport maximum}$$

$$1.25 \text{ dL/min} \times \text{renal threshold} = 375 \text{ mg/min}$$

$$\text{renal threshold} = \frac{375 \text{ mg/min}}{1.25 \text{ dL/min}}$$

$$= 300 \text{ mg/dL}$$

However, the true renal threshold for glucose is 160–180 mg/dL (see Figure 17.16). At these plasma levels of glucose, the filtered load of glucose is approximately 225 mg/min. Although this value is considerably lower than the transport maximum for glucose, some glucose molecules in the filtrate avoid contact with carrier proteins and are therefore excreted in the urine even though the carrier proteins are not 100% saturated. (Notice the nonlinear relationship between glucose reabsorption and excretion rates and plasma concentration of glucose as the rates approach the maximum in Figure 17.16).

Before modern methods for detecting glucose in urine were invented, a common test for diabetes mellitus was to taste the urine for sweetness. In untreated diabetes mellitus, plasma glucose levels are elevated, and may be several times the normal value. When the concentration of glucose exceeds the renal threshold, glucose appears in the urine—thus the sweet taste.

Secretion

In tubular secretion, molecules move from the plasma into the renal tubules to become part of the filtrate. Secretion does not include filtration at the glomerulus. Secretion follows the same basic processes as reabsorption and involves the same barriers, except that movement is in the reverse direction. Some substances diffuse from plasma into the filtrate, whereas others are actively transported. Secretion by active transport requires either that proteins in the basolateral membrane actively transport the solute from interstitial fluid to inside the epithelial cell, or that proteins in the apical membrane actively transport the solute from inside the epithelial cell into the filtrate. Among the substances actively secreted by the renal tubules are ions such as potassium ions and hydrogen ions, waste products such as choline and creatinine, and foreign substances such as the antibiotic penicillin. The end result of secretion is an increase in the quantity of solute excreted in the urine, which decreases the solute's plasma concentration.

To learn more about the necessity of filtration, reabsorption, and secretion, see Discovery: Dialysis, page 550.

1. Define the term *freely filterable*.

2. Describe the barriers for reabsorption and secretion. Be sure to include all plasma membranes that must be crossed by the substance being reabsorbed or secreted.

3. Describe the general mechanism of water reabsorption.

4. Define the terms *transport maximum* and *renal threshold*.

REGIONAL SPECIALIZATION OF THE RENAL TUBULES

Because the properties of the tubule epithelium vary from region to region along the length of the tubule, both the substances transported and the mechanisms of transport differ in different regions of the tubule. In the following sections we focus on the reabsorption of sodium, for three reasons: (1) Sodium is the single most abundant solute in the plasma (and thus in the glomerular filtrate), (2) the reabsorption or secretion of many other solutes is either directly coupled to or indirectly affected by the reabsorption of sodium, and (3) the reabsorption of water is dependent on sodium reabsorption.

Mechanisms of Sodium Reabsorption in the Renal Tubule

In all tubular segments where sodium is reabsorbed, sodium ions are actively transported. This active reabsorption is driven by Na^+/K^+ pumps located in the basolateral membrane of renal tubule epithelial cells. (Recall that these pumps utilize energy from ATP hydrolysis to transport sodium and potassium ions against their electrochemical gradients.) Because the active transport of sodium out of the epithelial cell keeps its concentration low in the intracellular fluid, sodium passively enters the cell from the tubular lumen across the apical membrane. Even though this latter step is passive, the overall movement of sodium ions across the cell (that is, from tubular fluid to peritubular fluid) is active because it depends on the active transport of sodium across the basolateral membrane. The active transport of sodium affects the movement of water across the renal tubules by osmosis.

In the proximal tubule, the entry of sodium into the tubule epithelial cells is carried out by transport proteins in the apical membrane that couple sodium movement to the flow of other solutes. Figure 17.17a shows two such pathways for sodium entry into cells: (1) cotransport with solutes such as glucose and amino acids (designated X in

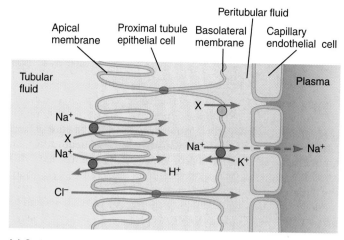

(a) Sodium reabsorption in the proximal tubule

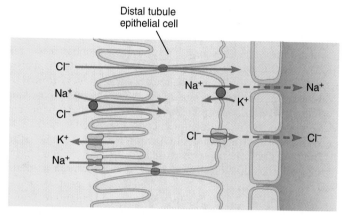

(b) Sodium reabsorption in the distal tubule

FIGURE 17.17 **Mechanisms of sodium reabsorption in the proximal and distal tubules. (a)** *Sodium reabsorption in the proximal tubule. Sodium is actively transported across the basolateral membrane by the Na^+/K^+ pump. Sodium moves across the apical membrane either by cotransport with an organic molecule (X) such as glucose or an amino acid, or by countertransport with another ion such as hydrogen. Note that chloride follows sodium reabsorption.* **(b)** *Sodium reabsorption in the distal tubule. Again, sodium is actively transported across the basolateral membrane by the Na^+/K^+ pump. Sodium moves across the apical membrane either by cotransport with chloride ions or through sodium channels. Potassium secretion from peritubular fluid to the tubule lumen sometimes accompanies sodium reabsorption.*

the figure), and (2) countertransport with hydrogen ions. In the first of these processes, energy released by the passive entry of sodium is harnessed to drive the flow of glucose or amino acids against their electrochemical gradients as they enter the epithelial cell. These solutes then exit the cell passively across the basolateral membrane. Given that the transport of glucose and amino acids from the tubular fluid to peritubular fluid requires an active step, the reabsorption of these solutes is active and requires energy. The ultimate source of this energy is ATP hydrolysis, because the reabsorption of these solutes is

Chronic renal failure is the progressive and irreversible loss of kidney structures that can either result from a variety of kidney diseases or be secondary to another disease such as diabetes or hypertension. Symptoms do not appear until the disease is advanced due to the remarkable ability of the kidneys to compensate for loss in function. However, once they appear, symptoms are widespread and severe.

When the GFR is decreased to less than 5% of normal, the result is *end-stage renal disease* (ESRD), which each year occurs in approximately 180 people per million population. Among the symptoms of ESRD are either dehydration or water retention, edema, electrolyte imbalance, acid-base disturbances, anemia due to the decrease in erythropoietin secretion, weakened bones due to calcium imbalance and impaired activation of vitamin D, and hypertension due to increased plasma volume and increased renin secretion.

Because of the many systems affected by ESRD, it is a fatal disease unless treated by kidney transplant or dialysis. Kidney transplant is the optimum treatment for a patient, but the number of kidneys available for transplant is scarce. Dialysis is the use of a semipermeable membrane to al-

low small solutes to exchange freely between two liquids. Although dialysis is not a cure, it can prolong the life of the patient until a kidney becomes available. Approximately 85,000 people are maintained by dialysis therapy each year.

There are two types of dialysis: hemodialysis and peritoneal dialysis. In *hemodialysis*, a person's blood is pumped through a hemodialysis system, also called an artificial kidney (figure a). The dialysis machine includes a membrane that separates the machine into two compartments, one that contains dialysis fluid and another that contains blood. A catheter from an artery takes blood from the patient to the dialysis machine. As the blood moves through the machine, exchange between blood and dialysis fluid occurs across the membrane. The membrane is semipermeable, allowing all molecules except blood cells and proteins to diffuse down their electrochemical gradients. The composition of the dialysate can be varied to favor movement of molecules in a particular direction. For example, if the patient is suffering from edema, then the solute concentration of the dialysate is made high to favor movement of water from blood to dialysate. After exchange, the blood leaves the machine and returns

to the patient via a catheter connected to a vein.

In *peritoneal dialysis* (figure b), a membrane inside the body, the *peritoneum,* is used in the same manner as the dialysis membrane just described. In peritoneal dialysis, the dialysate is infused through a catheter into the peritoneal cavity, which functions as the dialysate chamber. As blood flows through the capillaries of the peritoneum, materials are exchanged between the plasma and the dialysate. After a prescribed amount of time, generally 4–6 hours, the dialysate is drained out of the peritoneal cavity and discarded.

Dialysis is able to prolong and improve the life of patients with kidney failure, but it is not a cure. A better treatment for renal failure is kidney transplantation. In kidney transplants, a kidney from a deceased or living donor is placed below one of the diseased kidneys of the patient (figure c). An artery and a vein are connected to the donor kidney, which then takes over the function of the diseased kidneys. Not all kidney transplants are successful. After one year, approximately 85% of all transplants are functioning; after five years the number of functional transplants decreases to about 60%.

---

coupled to the flow of sodium, which ultimately depends on the ATP-driven Na^+/K^+ pumps. Other transport proteins couple the passive entry of sodium to the active secretion of hydrogen ions into the tubule lumen, a process that is important in acid-base regulation (see Chapter 18).

Figure 17.17b shows the mechanism of active sodium reabsorption in the distal tubule. The process involves passive movement of sodium across the apical membrane into the tubule epithelial cell and active transport of sodium across the basolateral membrane out of the epithelial cell and into the peritubular fluid. In this regard, sodium reabsorption in the distal tubule resembles that in the proximal tubule. However, the two processes differ with regard to the mechanism of sodium transport across the apical membrane. In the distal tubule, sodium enters the epithelial cell by two means: by cotransport with chloride ions, and by facilitated diffusion through sodium channels.

Sodium reabsorption in the distal tubule is also often coupled to potassium secretion, which, along with the cotransport of sodium with chloride, minimizes changes in the electrical potential that exists across the walls of the tubules. If the electrical potential is to be maintained, the reabsorption of a cation such as sodium must be balanced by reabsorption of anions (and to a lesser extent by secretion of other cations). Because chloride and bicarbonate are the most abundant anions in the tubular fluid, they account for the bulk of the anions that are reabsorbed with sodium. Potassium and hydrogen ions are the most common cations secreted as sodium ions are reabsorbed.

Nonregulated Reabsorption in the Proximal Tubule

In the proximal tubule, mechanisms of reabsorption are so efficient that 70% of the sodium and water that are

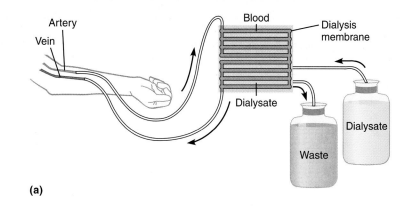

(a)

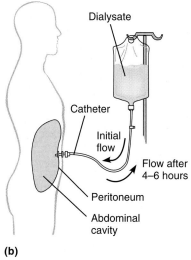

(b)

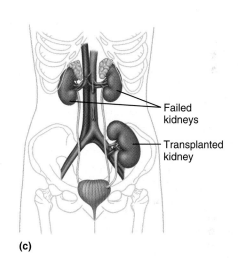

(c)

filtered is reabsorbed by the time tubular fluid reaches the beginning of the loop of Henle. Solute and water reabsorption are coupled such that the process is iso-osmotic; that is, no change in solute concentration occurs in either the plasma or the filtrate. Some solutes (including glucose) are virtually 100% reabsorbed by this time. For this reason, the proximal tubule is said to function as a *mass absorber* whose function is to reabsorb the bulk of filtered solutes and water, thereby preventing their loss from the body. This reabsorption of solutes in the proximal tubule is an ongoing process that generally is *not* regulated.

The epithelium of the proximal tubule has a number of features that facilitate mass absorption (Figure 17.18a). First the apical membrane is highly folded into many microvilli, a conformation known as a *brush border;* such folding increases the total surface area of the apical membrane, which facilitates transport. Second, transporting cells possess large numbers of mitochondria, which sup-

ply the large quantity of ATP necessary to drive active transport. Finally, the tight junctions between epithelial cells have a relatively high permeability to small solutes and water, making this a "leaky" epithelium. This high permeability at the tight junctions facilitates the diffusion of solutes and water between cells (called *paracellular transport*), further enabling the epithelium to transport large quantities of materials.

Regulated Reabsorption and Secretion in the Distal Tubule and Collecting Duct

In contrast to the proximal tubule, the distal tubule and collecting duct are specialized to allow *regulation* of reabsorption and secretion. In the epithelium of these tubules, the brush border is much less prominent than in the proximal tubule, or is even lacking altogether (Figure

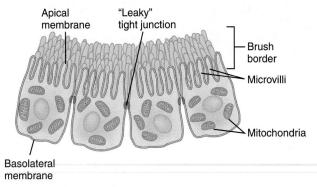

Apical membrane "Leaky" tight junction Brush border Microvilli Mitochondria Basolateral membrane

(a) Proximal tubule epithelium

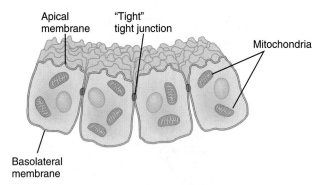

Apical membrane "Tight" tight junction Mitochondria Basolateral membrane

(b) Distal tubule and collecting duct epithelium

FIGURE 17.18 Epithelial cells in selected portions of a renal tubule. (a) *Epithelial cells in the proximal tubule.* **(b)** *Epithelial cells in the distal tubule and collecting duct. Epithelial cells in the proximal tubule have a much more extensive brush border, more mitochondria, and leakier tight junctions than those in more distal portions of the renal tubule.*

Is paracellular transport between epithelial cells more likely in the proximal tubule, or in the distal tubule? Why?

17.18b). In addition, epithelial cells have fewer mitochondria, and tight junctions are far less permeable, making this an example of a "tight" epithelium. Furthermore, tubule epithelial cells have receptors for hormones such that the transport of water and several solutes is under hormonal control, and water reabsorption does not always follow solute reabsorption. In the distal tubule and collecting duct, for example, sodium reabsorption is stimulated by aldosterone, a steroid hormone secreted by the adrenal cortex, and inhibited by atrial natriuretic peptide, a peptide hormone secreted by the atria of the heart. In these tubular segments, water reabsorption is stimulated by antidiuretic hormone (ADH), a peptide hormone secreted by the posterior pituitary. Actions of these hor-

mones and others in fluid and electrolyte balance are described in Chapter 18.

Water Conservation in the Loop of Henle

The loop of Henle of juxtamedullary nephrons is specialized to create an osmotic gradient in the renal medulla, such that the fluid in the outer portion of the medulla (near the cortex) is at a lower osmolarity than the fluid at the inner portion of the medulla (near the renal pelvis). This gradient is critical to the kidneys' ability to conserve water. Both the loops of Henle of juxtamedullary nephrons and the collecting ducts are important in water reabsorption. The functions of the loops of Henle are described in detail in Chapter 18.

The sites at which water and major plasma solutes are reabsorbed and secreted are listed in Table 17.2.

TABLE 17.2 SITES AT WHICH SUBSTANCES ARE REABSORBED AND SECRETED ACROSS RENAL TUBULES

TUBULE SEGMENT	SUBSTANCES REABSORBED		SUBSTANCES SECRETED
Proximal tubule	Na^+	Glucose	H^+
	Cl^-	Amino acids	
	K^+	Vitamins	
	Ca^{2+}	Urea	
	HCO_3^-	Choline	
	Water		
Loop of Henle (descending limb)	Water		
Loop of Henle (ascending limb)	Na^+	Mg^{2+}	
	Cl^-	Ca^{2+}	
	K^+		
Distal tubule	Na^+		K^+
	Ca^{2+}		H^+
	Cl^-		
	Water		
Collecting duct	Na^+	HCO_3^-	K^+
	K^+	H^+	H^+
	Cl^-	Urea	
	Ca^{2+}	Water	

Quick Test 17.4

1. On what membrane of renal tubule epithelial cells is the Na^+/K^+ pump located? In which direction is sodium transported?

2. What solutes are transported with sodium in the proximal tubule? in the distal tubule?

In the proximal tubule, because it has leakier tight junctions.

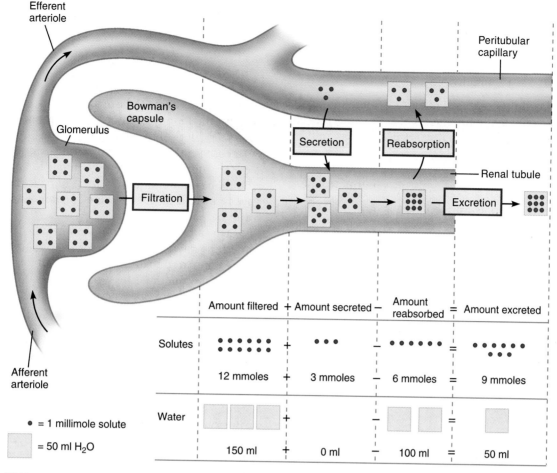

FIGURE 17.19 Schematic representation of the four basic renal processes.
Each dot represents 1 millimole of a solute, and each box represents 50 mL of water.

3. Name three structural differences between the epithelial cells of the proximal tubule and the distal tubule. The structure of which tubule is more favorable for exchange?

EXCRETION

Excretion by the kidneys is the elimination of solute and water from the body in the form of urine. The rate at which a substance is excreted in the urine is an important quantity because it has a direct bearing on the volume and composition of the plasma. For any substance, the quantity that is excreted over a period of time is determined by a simple rule: *Material that enters the lumen of the renal tubules is excreted unless it is reabsorbed.* Because a substance can enter renal tubules either by filtration or by secretion, the quantity of a substance that is excreted in the urine over any given time period is given by the following expression:

amount excreted = amount filtered +

amount secreted − amount reabsorbed

or

$$E = F + S - R$$

Excretion Rate

The equation for calculating the amount of a substance excreted ($E = F + S - R$) indicates that the rate at which a solute is excreted in the urine (moles/min or the equivalent) depends entirely on three factors: (1) the filtered load, (2) the rate at which the solute is secreted, and (3) the rate at which solute is reabsorbed. Figure 17.19 schematically depicts the renal processing of a hypothetical solute and water. In this figure, a single tubule represents all renal tubules combined, each dot represents 1 millimole of the solute, and each box represents 50 mL of water.

In the figure, three boxes are shown entering Bowman's capsule in one minute, so GFR is 150 mL/min. In the plasma, each box contains four dots, so solute concentration is 4 mmole/50 mL, or 0.08 millimole/mL.

Therefore, the filtered load is calculated as follows:

$$\text{Filtered load} = \text{GFR} \times \text{plasma concentration}$$

$$= 150 \text{ mL/min} \times 0.08 \text{ mmole/mL}$$

$$= 12 \text{ mmole/min}$$

Figure 17.19 also shows that in one minute 3 millimoles of solute are secreted and 6 millimoles of solute are reabsorbed. Therefore, the excretion rate is calculated as

$$\text{Excretion rate} = \text{filtered load} + \text{secretion rate}$$
$$- \text{reabsorption rate}$$
$$= 12 \text{ mmole/min} + 3 \text{ mmole/min}$$
$$- 6 \text{ mmole/min}$$
$$= 9 \text{ mmole/min}$$

Accordingly, nine dots are shown exiting the end of the tubule in Figure 17.19.

By calculating the filtered load of a specific solute and comparing this value to the amount of the solute excreted per minute, the net effect of renal processing (reabsorption or secretion) of the solute can be determined by two simple rules:

1. *If the amount of solute excreted per minute is less than the filtered load, then the solute was reabsorbed in the renal tubules.*

2. *If the amount of solute excreted per minute is greater than the filtered load, then the solute was secreted in the renal tubules.*

Note that only the *net* effect can be determined, because some substances are both reabsorbed and secreted. The amount excreted relative to the filtered load depends on which is greater, reabsorption or secretion. In the example depicted in Figure 17.19, both secretion and reabsorption of the solute occurred. Because the filtered load of the solute (12 mmole/min) exceeded the excretion rate (9 mmole/min), net reabsorption occurred.

Clearance

The rate at which a substance is excreted can be described in terms of **clearance,** a virtual measure of the volume of plasma from which a substance is completely removed or "cleared" by the kidneys per unit time (usually expressed in liters/hour). Given that we know how much of a substance is excreted per unit time, the clearance depends on the volume of plasma that contained that amount of substance. The equation to calculate clearance is

$$\text{Clearance} = \frac{\text{excretion rate}}{\text{plasma concentration}}$$

To understand the meaning of clearance, let us return to the example in Figure 17.19. The excretion rate for this solute was 9 mmole/min, or 540 mmole/hour. The plasma concentration of the solute was 0.08 mmole/mL, or 80 mmole/liter. Therefore, the clearance of this solute was

$$\text{Clearance} = \frac{540 \text{ mmole/hr}}{80 \text{ mmole/liter}} = 6.75 \text{ liters/hour}$$

This 6.75 liters of plasma is called a "virtual" volume because the kidneys do not actually remove all the solute from 6.75 liters of plasma. (In fact, the body does not even contain this much plasma.) Instead, the kidneys excrete an amount of solute that equals the amount that would be contained in 6.75 liters of plasma. The clearance is a result of the combined actions of filtration, reabsorption, and secretion.

Clearance is an important concept because it tells us something about how the kidneys are handling one substance compared to another. If the kidneys simply excreted urine that had the same composition as the glomerular filtrate, for example, the clearance would be the same for all solutes despite the fact that the *excretion rate* (moles/min) would be different for each. If instead the kidneys excreted urine containing equal sodium and potassium concentrations, the excretion rate would be the same for the two solutes, but the potassium clearance would be greater than the sodium clearance because the plasma potassium concentration is lower than the plasma sodium concentration. The relative clearances of solutes tells us how urinary excretion affects the plasma concentration of one solute *relative to another.* Clearance is also used clinically to estimate GFR and renal blood flow rate, as described next.

Clinical Uses of Clearance

In renal physiology, the clearance of a substance is most often expressed in terms of the following three measurable variables: the concentration of the substance in the urine (U_X), the concentration of the substance in the plasma (P_X), and the urine flow rate (V). Although the symbol V is frequently used in scientific fields to designate volume, the use of V in this context designates the volume of urine produced *per unit time.* The product of the urinary concentration (in mole/liter) and the urine flow rate (in liter/min) gives the excretion rate (in mole/min):

$$\text{Excretion rate} = U_x \times V$$

Therefore, the equation for calculating clearance can be written as

$$\text{Clearance} = \frac{U_x \times V}{P_x}$$

This form of the clearance equation is useful because it gives the clearance in terms of variables that can be readily measured. U_X and V can be determined by taking urine samples. To estimate V, it is only necessary to measure the volume of urine collected during a given period

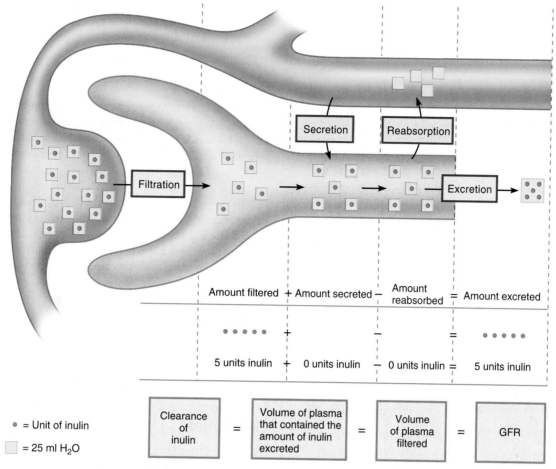

Amount filtered	+ Amount secreted	− Amount reabsorbed	= Amount excreted
● ● ● ● ●	+	−	= ● ● ● ● ●
5 units inulin	+ 0 units inulin	− 0 units inulin	= 5 units inulin

● = Unit of inulin

▢ = 25 ml H₂O

Clearance of inulin	=	Volume of plasma that contained the amount of inulin excreted	=	Volume of plasma filtered	=	GFR

FIGURE 17.20 Clearance of inulin. *Each dot represents an arbitrary unit of inulin, and each box represents 25 mL of water. Because inulin is neither reabsorbed nor secreted, the amount excreted equals the amount filtered. Therefore, clearance of inulin = GFR = 125 mL/min.*

of time; the volume of urine collected divided by the elapsed time gives the urine flow rate. For many substances the urinary concentration U_X can be determined by standard laboratory tests. Similarly, the plasma concentration P_X can be determined by drawing blood samples and running standard tests.

To see how to calculate the clearance, suppose that a person produces 450 mL of urine in an hour. The urine flow rate is thus

$$\text{Flow rate} = 450 \text{ mL}/60 \text{ min} = 7.5 \text{ mL/min}$$

If the sodium concentration in the urine is determined to be 15 mM, and the plasma concentration is found to be the normal value of 145 mM, then the sodium clearance (C_{Na}) is calculated as follows:

$$C_{Na} = \frac{(15 \text{ mmole/liter})(7.5 \text{ mL/min})}{(145 \text{ mmole/liter})} = 0.78 \text{ mL/min}$$

Estimates of Glomerular Filtration Rate Clearance measurements also provide a relatively easy way of determining the GFR. If a substance is freely filtered and is neither reabsorbed nor secreted, then the amount of the substance in the urine must be equal to the amount filtered, or the filtered load. Under these conditions, the substance is entirely cleared from the volume of plasma that was filtered; in other words, the clearance of the substance is equal to the GFR. One substance that meets these requirements is *inulin,* a polysaccharide that is not produced in the body but can be injected into the bloodstream for the purpose of measuring the GFR. That inulin clearance is equal to the GFR is illustrated in Figure 17.20 and can be shown mathematically. Because inulin is neither secreted nor reabsorbed, its excretion rate is equal to the filtered load, which by definition equals GFR times the plasma concentration:

$$\text{excretion rate} = \text{filtered load} = \text{GFR} \times P_{\text{inulin}}$$

herefore, the GFR equals the excretion rate divided by the plasma concentration, which by definition is the clearance:

$$\text{GFR} = \text{excretion rate}/P_{\text{inulin}} = (U_{\text{inulin}})(V)/P_{\text{inulin}} = C_{\text{inulin}}$$

Let's look at an example in which GFR is determined experimentally in a subject. Suppose that the subject completely empties her bladder at time 0, and then inulin is infused into her to maintain a constant plasma concentration level of 4 mmole/liter. After one hour, the subject completely empties her bladder again, and the volume of the urine is determined to be 120 mL. The urine flow rate can be calculated from the volume of this sample:

$$V = 120 \text{ mL/hour} = 2.0 \text{ mL/min}$$

The concentration of inulin in the urine sample is measured to be 250 mmole/liter. Therefore, the clearance of inulin is

$$C_{\text{inulin}} = \frac{(U_{\text{inulin}})(V)}{(P_{\text{inulin}})} = \frac{(250 \text{ mmole/liter})(2.0 \text{ mL/min})}{(4 \text{ mmole/liter})}$$
$$= 125 \text{ mL/min}$$

The clearance of any substance that is freely filtered but neither reabsorbed nor secreted is equal to the GFR. Because the intravenous infusion of inulin is an invasive procedure, it is desirable to determine GFR using molecules that are normally present in the blood, even though there are none that fit the requirements exactly. The chemical in the body that "best" meets the requirements to estimate GFR is creatinine, a waste product of muscle metabolism. Creatinine is freely filtered at the glomerulus and is not reabsorbed, but a small amount of it is secreted. Because the amount of creatinine excreted slightly exceeds the amount of creatinine filtered, the clearance of creatinine is slightly greater than the GFR. Still, creatinine clearance provides a suitable clinical *estimate* of GFR.

Determining the Fates of Solutes in Renal Tubules

Clearance can be used to determine whether there has been a *net* reabsorption or *net* secretion of a certain substance in the renal tubules, but some substances are both reabsorbed and secreted. In such cases, the clearance depends on which was greater (or which direction was the net effect), reabsorption or secretion. The use of clearance in determining whether reabsorption or secretion of a substance occurred can be summarized by two rules:

1. *If the clearance of a substance is greater than the GFR, then that substance was secreted in the renal tubules.*

2. *If the clearance of a substance is less than the GFR, then that substance was reabsorbed in the renal tubules.*

We saw earlier that glucose is completely reabsorbed at normal plasma levels. Therefore, we should predict that the clearance of glucose would be zero: Because no glucose is excreted, no plasma has been cleared of glucose. This is shown in Figure 17.21a and can be expressed mathematically as follows:

$$C_{\text{glucose}} = \frac{(0 \text{ mmole/liter})(V)}{(P_{\text{glucose}})} = 0 \text{ mL/min}$$

Another example is para-aminohippuric acid (PAH), a foreign substance used clinically to measure renal blood flow. PAH is freely filtered at the glomerulus and is not reabsorbed. In addition, any PAH remaining in the plasma of renal capillaries following filtration is secreted into the renal tubules; that is, the plasma that enters the kidneys is completely cleared of PAH. Therefore, the clearance of PAH is equal to the flow of plasma that entered the kidney, or the *renal plasma flow* (Figure 17.21b). This can easily be converted to *renal blood flow* by measuring the hematocrit. For example, if the clearance of PAH is 625 mL/min, then renal plasma flow is 625 mL/min. With a normal hematocrit of 45, then

$$\text{Renal blood flow} = \frac{625 \text{ mL/min}}{1 - 0.45} = 1136 \text{ mL/min}$$

Table 17.3 lists the clearance rates for some of the common substances processed by the kidneys.

Micturition

The fluid that remains in the renal tubules after filtration, reabsorption, and secretion is excreted as urine. The fluid drains from the collecting ducts into the renal pelvis and then into the ureter, a thin tube leading to the bladder (see Figure 17.1). Wavelike contractions of the smooth muscle in the walls of the ureter propel the urine toward the bladder. The bladder stores the urine until it is eliminated in the process called **micturition.**

The anatomy of the bladder is shown in Figure 17.22. The wall of the bladder contains smooth muscle fibers that are connected by gap junctions and thus function as a unit. The smooth muscle fibers of the bladder are known collectively as the **detrusor muscle.** In the neck of the bladder, muscle fibers converge and overlap, forming a thickening of the wall referred to as the **internal urethral sphincter** that acts as a valve to regulate the flow of urine from the bladder. (A *sphincter* is any muscular structure that forms a ring around a hollow organ or tube and regulates the flow of material by contracting or relaxing.) The flow of urine from the bladder is also controlled by the contraction and relaxation of skeletal muscle that surrounds the urethra as it passes through the pelvic floor; this muscle is referred to as the **external urethral sphincter.**

Micturition or urination is under both voluntary and involuntary control. Micturition occurs when the detrusor muscles contract, the smooth muscle of the internal

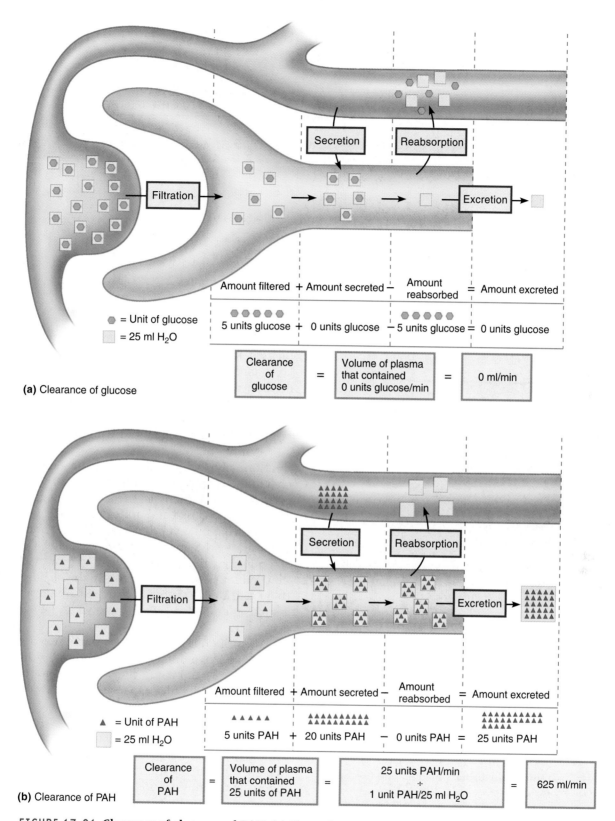

FIGURE 17.21 Clearance of glucose and PAH. (a) *Glucose clearance. Each hexagon represents a unit of glucose; each box represents 25 mL of water. Because normally all of the glucose filtered is reabsorbed, zero glucose is excreted, and glucose clearance = 0 mL/min.* (b) *PAH clearance. Each triangle represents a unit of PAH; each box represents 25 mL of water. Because PAH is completely secreted from the plasma, any PAH in the plasma entering the peritubular capillaries is secreted.*

TABLE 17.3 CLEARANCE OF SOME COMMON
SUBSTANCES PROCESSED BY THE KIDNEYS

SUBSTANCE	CLEARANCE RATE (mL/min)	NET RENAL PROCESSING (REABSORPTION OR SECRETION) *
PAH	650	Secretion
Creatinine	140	Secretion
Inulin	125	None
Glucose	0	Reabsorption
Sodium	0.9	Reabsorption
Potassium	12.0	Reabsorption
Chloride	1.3	Reabsorption

*GFR = 125 mL/min. If clearance is greater than GFR, net secretion has
occurred; if clearance is less than GFR, net reabsorption has occurred.

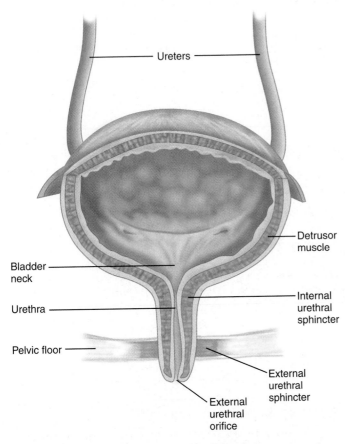

FIGURE 17.22 Anatomy of the urinary bladder and urethra.
*Structures are shown as they occur in females; in males the urethra extends
into the penis. The bladder wall contains smooth muscle called the detrusor
muscle. Two sphincters control the flow of urine out of the bladder: the
internal urethral sphincter, which is composed of smooth muscle, and the
external urethral sphincter, which is composed of skeletal muscle.*

Which urethral sphincter is under voluntary control?

The external urethral sphincter.

urethral sphincter relaxes and this sphincter opens, and
the skeletal muscle of the external urethral sphincter relaxes and this sphincter opens. The smooth muscle of the
detrusor muscle and the internal urethral sphincter are
under autonomic control, whereas the skeletal muscle is
under voluntary control. Let's look at the activity of these
muscles while the bladder fills with urine, and then during micturition.

Normally, the urine formed by the kidneys is delivered to the bladder on a continual basis. The internal and
external sphincter muscles are contracted, and thus the
sphincters are closed and urine does not leave the bladder, so the volume of urine in the bladder increases. The
detrusor muscles are relaxed and the bladder expands to
accommodate the increasing volume. However, as expansion continues, stretch receptors in the bladder wall become activated. Eventually, activation of these receptors
is sufficient to induce what is known as the *micturition reflex.*

Micturition is regulated by a spinal reflex that can be
overridden by voluntary control in a trained individual
(Figure 17.23). In infants, the pathway is purely reflexive. As the bladder fills, expansion of the wall is detected
by stretch receptors and transmitted to the spinal cord. In
the spinal cord, signals are relayed via interneurons to
three sets of neurons that project back to the bladder and
associated structures: parasympathetic neurons, which innervate the detrusor muscle; sympathetic neurons, which
innervate the internal urethral sphincter; and somatic
motor neurons, which innervate muscles of the external
urethral sphincter. Stimulation of stretch receptors excites
parasympathetic neurons that travel in the pelvic nerve
and stimulate the detrusor muscle to contract. Contraction of the detrusor muscle increases the pressure on the
contents of the bladder, and when the contractions of the
detrusor muscle become strong enough they cause the in-

ternal urethral sphincter to open. At the same time, stretch receptor activity inhibits sympathetic neurons projecting to the internal urethral sphincter and also inhibits somatic motor neurons projecting to the external urethral sphincter, allowing the sphincters to relax and open. The opening of both sphincters coupled with the contraction of the detrusor muscle allows the bladder to empty.

In older children and adults, the micturition reflex can be overridden by voluntary control. Signals from the stretch receptors that detect filling of the bladder are transmitted in ascending pathways to the cerebral cortex, giving rise to the conscious sensation of fullness of the bladder and serving as a signal to activate descending pathways that *inhibit* the parasympathetic neurons controlling the detrusor muscle, and to activate descending pathways that *excite* the motor neurons controlling the external urethral sphincter, allowing postponement of the micturition reflex if needed. Such postponement cannot be continued indefinitely, however, because continued filling of the bladder leads to greater excitation of stretch receptors. Eventually the level of activity in these neurons becomes high enough to trigger the micturition reflex despite conscious efforts to the contrary, triggering uncontrollable urination.

In addition to voluntary control to prevent micturition, one can initiate micturition through voluntary relaxation of the external urethral sphincter and the lowering of the pelvic floor. This lowering both causes the bladder to drop, which pulls open the internal urethral sphincter, and stretches the bladder wall, which activates the stretch receptors and induces the micturition reflex. Contraction of the diaphragm and abdominal muscles can increase the volume of urine voided by increasing the pressure in the abdominal cavity, which increases the pressure on the bladder for micturition.

Quick Test 17.5

1. Define the following terms: *filtered load, clearance,* and *micturition.*

2. The clearance of what two substances can be used to estimate the GFR? The clearance of what substance can be used to estimate renal plasma flow?

3. If the clearance of molecule X is greater than the GFR, was X reabsorbed or secreted in the renal tubules?

4. Describe the innervation of the bladder, internal urethral sphincter, and external urethral sphincter. Voluntary inhibition of micturition is mediated through innervation to which of these structures?

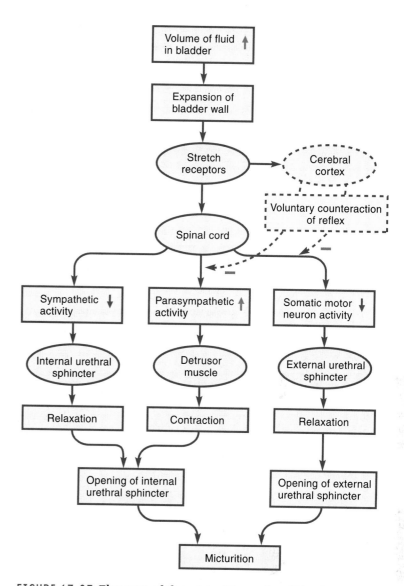

FIGURE 17.23 Elements of the micturition reflex. *Voluntary control over the reflex is indicated by the dashed arrows. This flowchart is available as an interactive exercise under Activities at www. physiologyplace.com.*

Figure 17.24 is a chart summarizing many of the concepts covered in this chapter. Note that regulation of GFR, reabsorption, and secretion determine the plasma volume and solute concentration. Whereas the mechanisms for regulating GFR were described in this chapter, those for regulating reabsorption and secretion of water and solutes are described in Chapter 18.

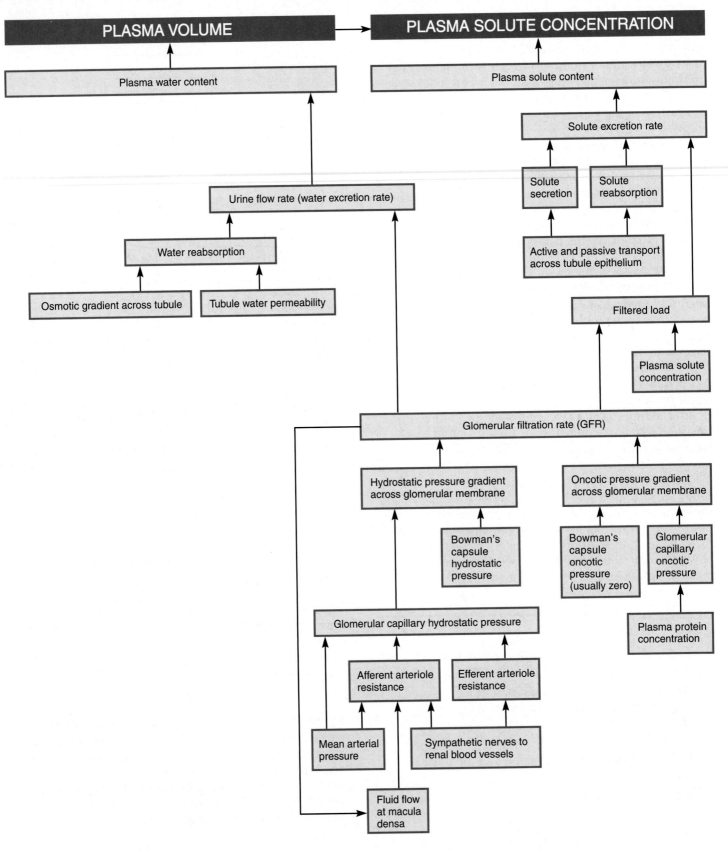

FIGURE 17.24 Role of renal processes in determining plasma volume
and composition.

CHAPTER SUMMARY

Functions of the Urinary System, p. 533

The primary function of the kidneys is to filter the blood to regulate the ionic composition, osmolarity, volume, and pH of plasma, and to remove metabolic waste products and foreign substances from the plasma. Urine is formed in this process.

IP Urinary System, Anatomy Review, pages 1–3

Anatomy of the Urinary System, p. 534

The urinary system consists of the kidneys, ureters, bladder, and urethra. The functional units of the kidneys are nephrons, consisting of Bowman's capsule, proximal tubule, descending loop of Henle, ascending loop of Henle, and distal tubule. The distal tubule drains into a collecting duct. Filtration occurs at the renal corpuscle, which includes Bowman's capsule and the glomerulus. The glomerular filtrate resembles plasma in composition except that it lacks proteins. As the filtrate moves through the nephron, its volume and composition change as a result of the reabsorption and secretion of water and solutes. Reabsorbed materials move from the tubular fluid in the lumen of the tubule to the peritubular fluid that surrounds the tubule, and then into the plasma of peritubular capillaries that surround the tubule. Secretion moves material in the opposite direction, from the plasma into the filtrate.

The kidneys receive a large proportion of the cardiac output via the renal artery. An afferent arteriole leads into each glomerulus, and an efferent arteriole leaves the glomerulus. The efferent arteriole branches into either peritubular capillaries or vasa recta, which drain into veins that eventually lead to the renal vein. The juxtaglomerular apparatus, which consists of the macula densa cells in the distal tubule and granular cells in the walls of the afferent and efferent arterioles, is important in the regulation of glomerular filtration and in salt and water reabsorption.

IP Urinary System, Glomerular Filtration, page 3*

IP Urinary System, Anatomy Review, pages 5, 9–20

Basic Renal Exchange Processes, p. 538

Glomerular filtration is driven by the four Starling forces that contribute to the glomerular filtration pressure: (1) the glomerular capillary hydrostatic pressure, (2) the hydrostatic pressure inside Bowman's capsule, (3) the oncotic pressure of plasma in glomerular capillaries, and (4) the oncotic pressure of fluid in Bowman's capsule. The glomerular filtration pressure and the presence of fenestrations in the glomerular capillaries and slit pores in the epithelium of Bowman's capsule favor the bulk flow of protein-free fluid between blood and the lumen of Bowman's capsule. The normal glomerular filtration rate is approximately 125 mL/min. The filtration fraction is the percentage of renal plasma flow that is filtered; on average it is approximately 20%. The filtered load is the quantity of a certain solute that is filtered at the glomerulus. For a solute that is freely filtered, the filtered load equals the product of the GFR and the plasma concentration of the solute.

Under normal conditions, the glomerular filtration rate is regulated to stay nearly constant by three intrinsic control mechanisms: (1) myogenic regulation of smooth muscle in the afferent arteriole, (2) tubuloglomerular feedback, and (3) mesangial cell contraction. Extrinsic control of GFR includes sympathetic nervous control of smooth muscle in the afferent and efferent arterioles.

When solutes are transported across the tubular epithelium by carrier proteins during reabsorption or secretion, the transport is subject to a transport maximum, which occurs when the concentration of solute is great enough to saturate the carrier proteins. The plasma concentration at which the solute appears in the urine is called the renal threshold.

IP Urinary System, Glomerular Filtration, pages 1–14*

Regional Specialization of the Renal Tubules, p. 549

The proximal tubule is specialized to reabsorb large quantities of solutes and water, and returns most filtered material to the bloodstream. In contrast, the distal tubule and collecting duct are specialized for the regulation of transport, which is important in controlling the volume and composition of the plasma. The transport of water and many solutes in the distal tubule and collecting duct is regulated by hormones.

Sodium reabsorption is important not only in regulating plasma sodium composition, but also affects the reabsorption of other solutes and water and the secretion of some solutes. Sodium is actively reabsorbed across the renal tubules, and this reabsorption is driven by the Na^+/K^+ pumps located in the basolateral membrane of tubule epithelial cells.

IP Urinary System, Early Filtrate Processing, pages 1–13

IP Urinary System, Late Filtrate Processing, pages 1–14

Excretion, p. 553

The rate at which a substance is excreted in the urine is determined by three factors: the rate at which it is filtered at the glomerulus, the rate at which it is reabsorbed, and the rate at which it is secreted. If the amount of solute excreted per minute is less than the filtered load, then the solute was reabsorbed in the renal tubules. If the amount of solute excreted per minute is greater than the filtered load, then the solute was secreted in the renal tubules.

Clearance is a measure of the volume of plasma from which a substance is completely removed or "cleared" by the kidneys per unit time. The clearance of inulin or creatinine can be used to estimate the GFR. The clearance of PAH can be used to estimate renal plasma flow and thus renal blood flow.

The fluid that remains in the renal tubules after filtration, reabsorption, and secretion is excreted as urine. The fluid drains from the collecting ducts into the renal pelvis and then into the ureter. Wavelike contractions of smooth muscle in the wall of the ureter propel the urine toward the bladder. The bladder stores the urine until it is excreted during micturition. Micturition is under both reflex and voluntary control. The micturition reflex is triggered by stretching of the bladder wall.

*This topic is available on the *InterActive Physiology*® *Sampler CD* that comes with the purchase of a new copy of this book.

EXERCISES

Multiple-Choice Questions

1. Which structure of the urinary system stores urine until it is excreted?
   a) kidneys
   b) bladder
   c) ureter
   d) urethra

2. Which of the following is *not* one of the mechanisms by which a solute can be exchanged between the plasma and the renal tubules?
   a) glomerular filtration
   b) secretion
   c) excretion
   d) reabsorption

3. What type of specialized junction connects epithelial cells lining the renal tubules?
   a) gap junctions
   b) tight junctions
   c) desmosomes
   d) intercalated disks

4. Which of the following does *not* favor a large glomerular filtration rate?
   a) slit pores
   b) fenestrations
   c) high resistance in the afferent arteriole
   d) high resistance in the efferent arteriole

5. Most reabsorption of water and solutes occurs in the
   a) proximal tubule.
   b) descending limb of the loop of Henle.
   c) ascending limb of the loop of Henle.
   d) distal tubule.

6. In which of the following are microvilli most abundant?
   a) Bowman's capsule
   b) glomerular capillaries
   c) distal tubule
   d) proximal tubule

7. Which of the following would occur if mean arterial pressure increased from 95 mm Hg to 125 mm Hg?
   a) Glomerular filtration rate would increase due to the increased glomerular capillary hydrostatic pressure.
   b) Glomerular filtration rate would decrease due to increased Bowman's capsule hydrostatic pressure.
   c) Glomerular filtration rate would not change due to autoregulation.
   d) Glomerular filtration rate would not change due to activation of the sympathetic nervous system.

8. The normal fasting plasma glucose concentration is 100 mg/deciliter, and the renal threshold is 300 mg/deciliter. If the plasma concentration doubles to 200 mg/deciliter, then
   a) the rate at which glucose is reabsorbed will double.
   b) the capacity of the renal tubule for transporting glucose will be exceeded.
   c) urinary water excretion will increase.
   d) glucose clearance will increase.

9. A substance S is freely filterable and is excreted at a rate (in moles/min) that is lower than the filtered load.

On the basis of this information alone, which of the following is the *most precise* conclusion that can justifiably be drawn regarding the kidneys' processing of S?
   a) S is neither reabsorbed nor secreted.
   b) S is definitely reabsorbed and may be secreted.
   c) S is definitely secreted and may be reabsorbed.
   d) S is definitely both reabsorbed and secreted.

10. Which of the following observations would enable you to definitely conclude that a substance X is being secreted?
   a) The clearance of X is greater than the GFR.
   b) The concentration of X in the urine is greater than its concentration in the plasma.
   c) The concentration of X in the plasma is decreasing over time.
   d) any of the above

11. Micturition occurs in response to
   a) relaxation of the detrusor muscle.
   b) contraction of the internal and external urethral sphincters.
   c) activation of parasympathetic neurons to the bladder.
   d) activation of somatic motor neurons to the bladder.

Objective Questions

1. The (ureter/urethra) carries urine from the bladder to the outside of the body.

2. Urinary excretion is the elimination of urine from the bladder. (true/false)

3. The (afferent/efferent) arteriole carries blood toward the glomerulus.

4. The combination of a glomerulus and the surrounding Bowman's capsule is called a(n) _____ _____.

5. The hydrostatic pressure in glomerular capillaries is (higher/lower) than that in most capillaries of the body.

6. The glomerular filtration rate tends to (increase/decrease) as the concentration of proteins in the plasma increases.

7. Autonomic neurons regulate contraction of the (internal/external) urethral sphincter.

8. The glomerular filtration pressure is synonymous with the hydrostatic pressure inside glomerular capillaries. (true/false)

9. In the proximal tubule, the Na^+/K^+ pump is located on the (apical/basolateral) membrane.

10. The filtered load of a solute is determined by its plasma concentration and the (glomerular filtration rate/urine flow rate).

11. If the clearance of a substance is greater than the glomerular filtration rate, then that substance must have undergone (reabsorption/secretion) in the renal tubules.

12. The clearance of (PAH/creatinine) is approximately equal to the renal plasma flow rate.

13. Substances that are reabsorbed move into the (peritubular capillaries/tubule lumen).

14. An increase in the flow rate through the macula densa causes a(n) (increase/decrease) in the glomerular filtration rate.

15. Glucose reabsorption occurs primarily in the (proximal tubule/distal tubule).

Essay Questions

1. Describe the pathway of filtrate flow from the renal corpuscle to elimination from the body.

2. Explain why the glomerular filtration rate does not change with a moderate decrease in mean arterial pressure.

3. Compare the mechanisms of active and passive reabsorption of solute.

4. Explain the concept of transport maximum. Why does glucose appear in the urine of diabetic patients?

5. Explain why the proximal tubules are considered mass absorbers.

6. Suppose that sodium is being excreted in the urine at a rate of 0.5 millimoles per minute, and that sodium is present in the plasma at a concentration of 150 mM. If the GFR is 125 mL/min, what is the filtered load of sodium? What is the sodium clearance? Is sodium being reabsorbed or secreted by the renal tubules? In your own words, describe what the term *clearance* means.

Find the answers to these exercises, and additional study tools, at the Physiology Place (www.physiologyplace.com).

18

The Urinary System: Fluid and Electrolyte Balance

OBJECTIVES

- Explain the concept of balance.

- Describe the different sources of body water input and output.

- Explain the control of water balance and osmolarity by antidiuretic hormone.

- Describe how aldosterone and atrial natriuretic peptide regulate plasma sodium levels. Include an explanation of the stimuli that release these hormones.

- Describe the major mechanisms whereby water and sodium balance influence mean arterial pressure.

- Explain the role of aldosterone in potassium balance.

- Describe the major hormone systems that regulate calcium balance.

- Describe various factors that influence acid-base balance.

- Explain how buffers in the blood, actions of the respiratory system, and the kidneys compensate for acid-base disturbances.

CHAPTER OUTLINE

The Concept of Balance   565

Water Balance   567

Sodium Balance   575

Potassium Balance   579

Calcium Balance   580

Interactions Between Fluid and Electrolyte Regulation   581

Acid-Base Balance   583

Above: Intravenous urogram showing kidneys, renal pelvises, and ureters

STUDY HINTS

1. *Transport across epithelia, p. 124*

2. *Osmolarity and osmosis, p. 119*

3. *Movement of ions across cell membranes, p. 114*

4. *Posterior pituitary, p. 155*

5. *Adrenal cortex, p. 159*

The body's ability to regulate fluid and electrolyte composition is remarkable when we consider the vast differences in individuals' diets and levels of physical activity. When you drink a large volume of water, it isn't long before your body eliminates the excess water consumed, and when you consume salty food, such as french fries or potato chips, your body retains fluid and you feel thirsty. We do not regulate what we take into our bodies very well, but our bodies regulate the fluid and electrolyte composition by varying the amounts of water and solutes excreted. This chapter describes how the kidneys maintain normal fluid and electrolyte composition, which is critical to maintaining normal blood pressure and the normal functioning of cells.

THE CONCEPT OF BALANCE

To maintain homeostasis, the human body must be kept in balance. To be in balance in this context means that what comes into the body and what is produced by the body must equal the sum of what is used by the body and what is eliminated from the body, as shown in the following equation:

$$\text{Input} + \text{Production} = \text{Utilization} + \text{Output}$$

The kidneys play a key role in regulating fluid and electrolyte balance and acid-base balance.

Factors Affecting the Plasma Composition

The kidneys exert control over the volume and composition of plasma by regulating its solute and water content. The volume and composition depend on each other and must be maintained within narrow limits. The volume of plasma is determined almost entirely by its water content because solutes make only a negligible direct contribution to the volume. However, the amount of solute in plasma indirectly affects plasma volume because changes in plasma osmolarity (solute concentration) can cause water to shift between the plasma and other body fluid compartments. Plasma volume has an important influence on the body's homeostasis because it affects mean arterial pressure. As we will see later in this chapter, mean arterial pressure is regulated over the long run by changes in plasma volume.

The solute and water content of the plasma is affected both by the movement of materials into and out of the body and by the movement of materials between different compartments within the body. Figure 18.1 shows pathways for the movement of water and solutes small enough to permeate capillary walls and cell membranes and thus to move freely into and out of the plasma. The plasma and interstitial fluid compartments are combined in this diagram to emphasize their similarity in composition, which is due to the free exchange of water and small solutes between them.

As Figure 18.1 shows, the plasma can gain or lose materials by exchange with cells or with extracellular connective tissue, such as the bone matrix. When the bone is resorbed, for example, calcium and phosphates are released into the plasma, which raises the concentrations of these substances; conversely, the deposition of calcium and phosphates into bone lowers their plasma concentrations. Figure 18.1 shows three other routes by which the plasma and the external environment can exchange materials: The plasma can (1) either gain or lose materials by exchange with the lumen of the gastrointestinal tract, (2) either gain or lose materials by exchange with the lumens of the renal tubules, and (3) lose materials through sweating, hemorrhage, or respiration. (The plasma normally gains oxygen through respiration, however.)

Solutes and water are absorbed into the plasma from the gastrointestinal tract and also move from the plasma to the lumen of the tract in the form of saliva, bile, pancreatic juice, and other gastrointestinal secretions. Under normal conditions, the absorptive capacity of the gastrointestinal tract is sufficient to recover virtually 100% of all secreted materials as well as those that enter the tract via ingestion (eating or drinking). The rate at which solutes and water are lost by elimination from the gastrointestinal tract is very small compared to the rate at which these substances are lost through the excretion of urine by the kidneys. Consequently, *the transport of materials across the wall of the gastrointestinal tract normally amounts to a net gain of solutes and water by the body.*

Small solutes and water move from the plasma to the lumen of the renal tubules by glomerular filtration and secretion. Solutes and water are returned to the plasma from the renal tubules by reabsorption. Since not all this material is reabsorbed, *the transport of materials across the walls of the renal tubules amounts to a net loss of water and solutes by the body.* These lost materials are excreted in the urine.

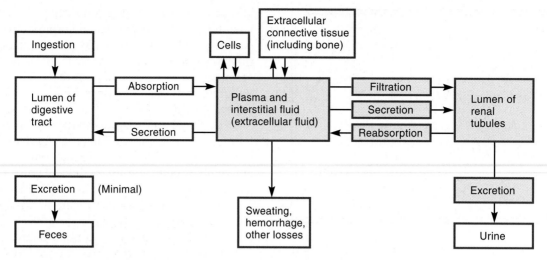

FIGURE 18.1 Material exchanges affecting plasma content. *Materials (water and solutes) enter the body through ingestion into the digestive tract. Once in the body, materials are exchanged among the digestive tract, the plasma (and the interstitial fluid), cells of the body, connective tissue, and the renal tubules. The highlighted areas represent exchanges between plasma and the renal tubules.*

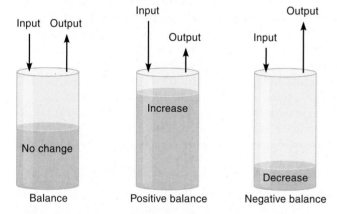

FIGURE 18.2 The concept of balance. *A substance is in balance in the body when input equals output. A substance is in positive balance when input exceeds output, causing a net increase in the amount of that substance. A substance is in negative balance when output exceeds input, causing a net decrease in the amount of that substance.*

Solute and Water Balance

When solutes and water enter and exit the plasma at the same rate, the plasma's volume and composition do not change, and it is said to be in balance. Changes in volume and/or composition occur when materials enter the plasma faster than they exit or vice versa (Figure 18.2). When a substance enters the body faster than it exits, the substance is said to be in a state of *positive balance;* under these conditions the quantity of that substance in the plasma tends to increase, unless the substance enters cells or is metabolized within the body. If a substance leaves the body faster than it enters, the substance is in a state of

negative balance; under these conditions the quantity of that substance in the plasma tends to decrease.

For certain substances whose plasma concentrations are controlled by specific regulatory mechanisms, a state of positive balance or negative balance can exist with little or no change in the plasma concentration. When you eat a meal containing glucose, for example, your body goes into a state of positive glucose balance. The absorption of glucose from the lumen of the gastrointestinal tract causes a rise in the plasma glucose level, but this rise is transient because it triggers insulin secretion and other hormonal changes. Insulin causes an increase in cellular glucose uptake which, combined with other hormonally triggered responses, quickly lowers the blood glucose concentration, eventually restoring it to normal. Once the glucose enters cells, it is catabolized for energy or converted to glycogen or fats for storage.

In Chapter 17 we saw that the kidneys filter 180 liters of plasma per day. Approximately 70% of the filtered water and sodium is reabsorbed in the proximal tubule in the absence of any regulation. However, the body can regulate the remaining filtered water and sodium to vary the amount excreted or retained based on the body's requirements for maintaining balance. The kidneys also function in the regulation of potassium and calcium levels in the blood and of acid-base balance. The regulation of renal excretion occurs primarily in the late distal tubule and collecting duct. Two types of epithelial cells line these tubules: principal cells and intercalated cells. Water and electrolyte balance is regulated through hormonal actions on principal cells, whereas acid-base balance is regulated through processes occurring within the intercalated cells.

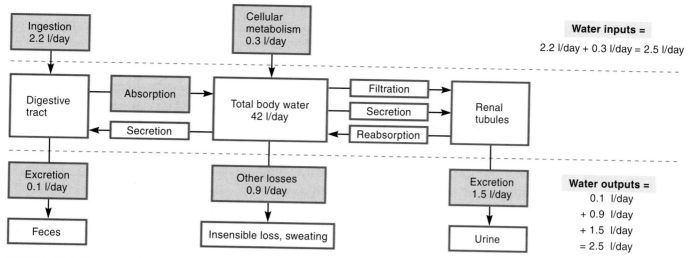

FIGURE 18.3 Factors affecting water balance. *Body water input includes absorption across the digestive tract and production by cellular metabolism totaling approximately 2.5 L/day. Body water output includes excretion in the urine and feces and loss through sweat and respiration totaling approximately 2.5 L/day. Under these circumstances the body is in water balance.*

WATER BALANCE

For the body to be in water balance, water intake plus any metabolically produced water must equal water output plus any water used in chemical reactions (Figure 18.3). Nearly all water enters the body through the digestive tract in food and drinks consumed, but a small amount of water is produced by cellular metabolism (for example, during oxidative phosphorylation or condensation reactions). Water leaves the body through several routes, including *insensible loss* during respiration and through the skin, sweating, elimination of feces by the digestive tract, and excretion of urine by the kidneys. Only the kidneys, however, regulate the rate of water loss for the purpose of maintaining water balance.

When over a given period of time a person drinks a quantity of water that exceeds the quantity of water lost, his or her body is said to be in a state of positive *fluid balance,* and if plasma volume (and total blood volume) is normal to begin with, it will rise above normal, a condition called *hypervolemia.* If over that time the quantity of water lost is greater than the quantity gained, the body is in a state of negative fluid balance, and if plasma volume (and total blood volume) is normal initially, it will go below normal, a condition called *hypovolemia.* A state of normal blood volume is referred to as *normovolemia.*

The control of urinary water excretion is instrumental in regulating plasma volume and osmolarity. Plasma volume is directly related to blood pressure: Increases in plasma volume increase mean arterial pressure, whereas decreases in plasma volume decrease mean arterial pressure. Plasma volume can also affect osmolarity: An increase in plasma water with no increase in solute decreases osmolarity; a decrease in plasma water with no decrease in solute increases osmolarity. Changes in plasma osmolarity affect the movement of water between fluid compartments—in particular, between intracellular fluid and extracellular fluid. In turn, water movement can cause changes in the volume of cells throughout the body, affecting their function.

Osmolarity and the Movement of Water

The kidneys vary the volume of water excreted by regulating water reabsorption in the late distal tubules and collecting ducts. The forces acting on water across these tubules are osmotic gradients between the tubular fluid and the peritubular fluid. However, water can move only if the tubule membranes are permeable to water. This section describes the mechanisms responsible for establishing the osmotic gradients and regulating water permeability, and thus water reabsorption.

Under normal conditions, the total solute concentration of the plasma is approximately 300 mOsm (300 milliosmoles of solute per liter of plasma); this is also the normal osmolarity of interstitial fluid and intracellular fluid. Because the osmolarity is the same in both extracellular fluid (interstitial fluid and plasma) and intracellular fluid, the osmotic pressure is also the same, and there is no net force for water to either leave cells or enter them. Under these conditions, cell volumes do not change, and the intracellular fluid and extracellular fluid are said to be in *osmotic equilibrium.*

If a person's plasma volume is normal and then she drinks a significant quantity of pure water, her plasma volume will increase, and the concentration of solutes in the plasma will decrease. If her plasma osmolarity was normal initially, then the absorption of water into the plasma will cause plasma osmolarity to go below normal. Unless it is corrected, this decrease in osmolarity will

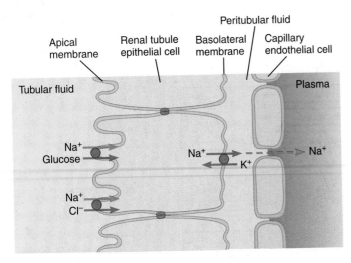

FIGURE 18.4 Mechanism of sodium reabsorption in the proximal tubule. *Sodium is actively transported across the basolateral membrane into the peritubular fluid by the Na⁺/K⁺ pump. Sodium moves from the tubular fluid into the epithelial cell through sodium channels or via cotransport with other molecules such as glucose.*

What happens to the glucose and chloride ions that are cotransported across the apical membrane with glucose?

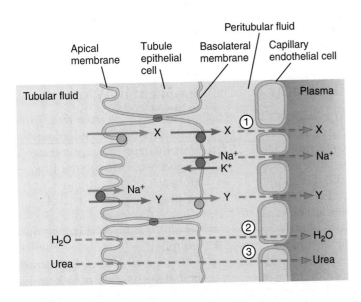

Steps for water and urea reabsorption:

① Solutes (Na⁺, X, Y) are actively reabsorbed. Osmolarity of peritubular fluid and plasma increases.

② Water is reabsorbed by osmosis.

③ Urea (permeating solute) is reabsorbed passively.

FIGURE 18.5 Mechanism of water reabsorption. *When solutes such as sodium ions or molecule X are actively reabsorbed, the osmolarity of the peritubular fluid and plasma increase, so water then moves from the lumen of the renal tubules into the peritubular fluid and then into plasma by osmosis. Permeating solutes such as urea follow water reabsorption.*

They move across the basolateral membrane by facilitated diffusion (glucose) or through channels (chloride) and are reabsorbed.

cause cells to swell, because water will flow spontaneously from extracellular fluid, where the osmolarity is lower (and the concentration of water is higher), to intracellular fluid, where the osmolarity is higher (and water concentration is lower). Such swelling can have deleterious effects on the functioning of cells all over the body, but cells in the brain are particularly sensitive. The swelling of brain cells can alter neurological function, producing headache, nausea, confusion, seizures, or coma. We do not worry about the possible adverse effects of drinking water because when excess water is ingested, the kidneys increase the rate at which they excrete water, which quickly restores the volume and osmolarity of the plasma to normal. Ideally, the kidneys would excrete a volume of pure water equal to the volume of water that was drunk, but the kidneys cannot excrete pure water, so instead they excrete a large volume of urine with a low osmolarity.

Exercise Link

Brain swelling due to low plasma osmolarity was the cause of Bill's headache and nausea after the marathon. In his case, the situation was compounded by two factors. First, because he gradually lost electrolytes in his sweat during the race, the osmolarity of his extracellular fluid became *very* low after he drank the large volume of water. Second, he was also hypovolemic due to sweat losses. This meant that there were conflicting signals influencing renal function: hypo-osmolarity indicated a need for increased water excretion, and hypovolemia indicated a need for decreased water excretion.

Now suppose that a person whose plasma volume and osmolarity are normal eats a large quantity of salty pretzels without drinking any water. In this case, plasma volume will not change, but salt will be absorbed into the plasma, raising its osmolarity. If not corrected, the increased plasma osmolarity will cause cells to shrink because the osmolarity of the extracellular fluid exceeds that of intracellular fluid. Such cell shrinkage induces many of the neurological problems that cell swelling causes. To avert this possibility, the kidneys excrete the excess solutes that were ingested while minimizing water loss by excreting a small volume of water. Ideally, the kidneys would excrete the excess solutes only, but this is not possible. Thus, when little or no water is ingested, the kidneys excrete a small volume of highly concentrated urine.

In these examples, the kidneys compensate for changes in plasma volume and osmolarity by adjusting the rate of water excretion. Because water is reabsorbed

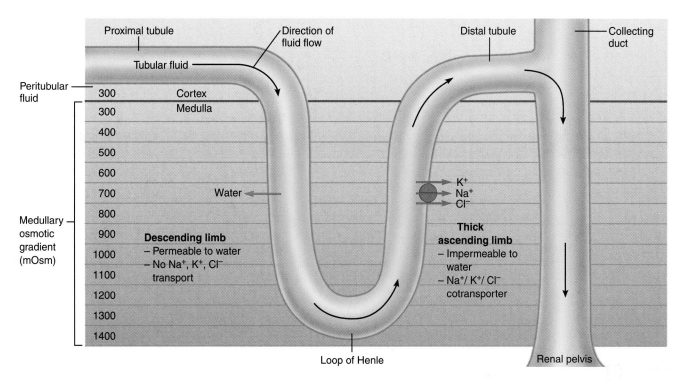

FIGURE 18.6 The medullary osmotic gradient. *Because the descending limb of the loop of Henle is permeable to water whereas the thick ascending limb is impermeable to water and contains transporters for Na⁺, Cl⁻, and K⁺, differences in the transport of materials establish an osmotic gradient in the medullary interstitial fluid. The osmolarity of medullary interstitial fluid is 300 mOsm near the cortex and increases continuously to a maximum of about 1400 mOsm near the renal pelvis.*

Which of the two types of nephrons in the body is depicted in this figure?

but not secreted, these adjustments are achieved through changes in the rate of water reabsorption.

In the renal tubules, water reabsorption is passive and is coupled to the active reabsorption of solutes. Solute transport creates an osmotic gradient across the tubule epithelium. The precise mechanism responsible for creating the osmotic gradient varies in the different segments of the renal tubules. In the next sections we see that in the proximal and distal tubules, reabsorption of solute increases the osmolarity of the peritubular fluid (which drives reabsorption of water by osmosis), and that in the collecting duct the *medullary osmotic gradient* drives reabsorption of water.

Water Reabsorption in the Proximal Tubule

Because the primary solute in extracellular fluid is sodium and most of the filtered sodium is reabsorbed in the proximal tubule, sodium is the solute that is primarily responsible for producing the osmotic gradient that drives water reabsorption. Although the exact mechanism varies in different segments of the renal tubules, sodium reabsorption always involves the active transport of sodium across the basolateral membrane from the epithelial cell of the tubule into the peritubular fluid, where it can diffuse into the plasma of peritubular capillaries (Figure

18.4). Sodium crosses the apical membrane by a variety of mechanisms, especially via mediated transport during which the movement of sodium is coupled to the movement of another molecule (such as glucose) or another ion (such as chloride).

Water reabsorption occurs through osmosis, as illustrated in Figure 18.5. Active reabsorption of sodium and other solutes in the proximal tubule creates an osmotic gradient, and therefore water follows the solutes. Because reabsorption of water creates a concentration gradient for permeating solutes such as urea to move from tubular fluid to the plasma in peritubular capillaries, the movement of permeating solutes follows water reabsorption.

Role of the Medullary Osmotic Gradient in Water Reabsorption

Within the interstitial fluid of the renal medulla is the **medullary osmotic gradient;** the outer regions of the medulla have a lower osmolarity than the inner regions (Figure 18.6). The osmolarity varies from 300 mOsm at the edge of the medulla to approximately 1200–1400 mOsm at the innermost portion of the medulla near the

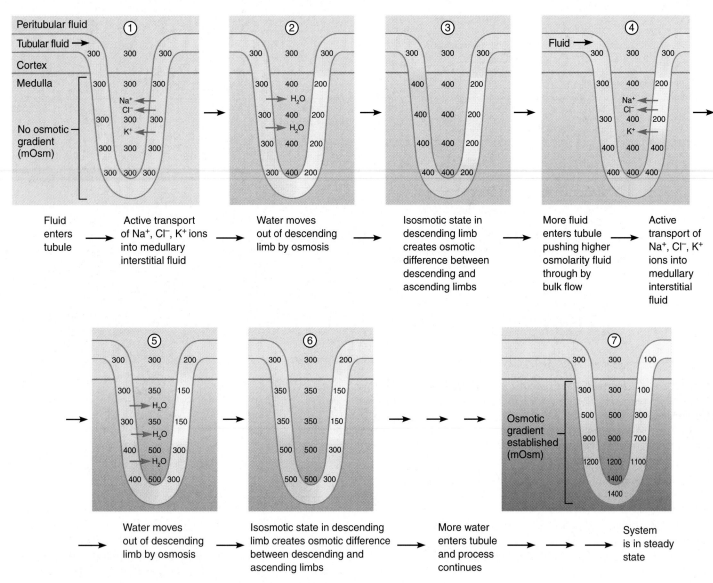

FIGURE 18.7 How the countercurrent multiplier establishes the medullary osmotic gradient. *Initially all fluids are iso-osmotic at 300 mOsm ①. Active transport of solutes (Na⁺, Cl⁻, K⁺) from the ascending limb of the loop of Henle into the medullary interstitial fluid increases the osmolarity of the interstitial fluid and decreases the osmolarity of the tubular fluid in the ascending limb ②. The increased osmolarity of the medullary interstitial fluid draws water from the lumen of the descending limb of the loop of Henle into the interstitial fluid, increasing the osmolarity of the tubular fluid in the descending limb ③. More tubular fluid then enters the loop of Henle ④, pushing the fluid farther into the renal tubules. The process of solute transport from the ascending limb followed by water movement from the descending limb ⑤, followed by more tubular fluid entering the loop of Henle ⑥, repeats until the medullary osmotic gradient is established ⑦.*

renal pelvis. This gradient, which is necessary for water reabsorption from the collecting duct, exists because of a mechanism known as the countercurrent multiplier.

Countercurrent Multiplier

The properties of different portions of the loops of Henle of juxtamedullary nephrons are critical to the countercurrent multiplier and establishment of the osmotic gradient in the renal medulla. The descending limb is permeable to water, so water diffuses out of the tubule under

the influence of an osmotic gradient. The thick ascending limb, by contrast, is impermeable to water and has sodium/potassium/chloride cotransporters that the descending limb lacks. The term *countercurrent* refers to the fact that fluid flowing through the descending and ascending limbs, which parallel one another, moves in opposite directions (see Figure 18.6).

Figure 18.7 illustrates how the countercurrent multiplier creates the medullary osmotic gradient. The figure starts with no osmotic gradient along the tubules or within the medullary interstitial fluid. The fluid that enters the descending limb from the proximal tubule is iso-osmotic with the interstitial fluid, at 300 mOsm ①. The fluid in the proximal tubule is iso-osmotic because water freely crosses the wall of the tubule and is therefore reabsorbed along with solutes. (The leakiness of the tubular epithelium makes it impossible for significant differences in osmolarity to exist between tubular fluid and peritubular fluid.) As fluid moves down the descending limb, there is no net movement of water across the tubule wall because there is no osmotic gradient. As the fluid begins to travel up the ascending limb of the loop of Henle, sodium, chloride, and potassium are actively transported from the tubule into the medullary interstitial fluid, increasing the osmolarity of the interstitial fluid from 300 mOsm to 400 mOsm and lowering the osmolarity of fluid in the ascending limb to 200 mOsm. ②. When the osmolarity of the peritubular fluid increases, water moves out of the descending limb and into the peritubular fluid until the two are iso-osmotic again at 400 mOsm ③. This creates a difference in osmolarity between the fluid in the descending limb and the fluid in the ascending limb, with the latter at a lower osmolarity (200 mOsm versus 400 mOsm).

As more fluid at 300 mOsm enters the loop of Henle from the proximal tubule ④, this fluid pushes the fluid ahead of it through the tubule, pushing the fluid with the higher osmolarity deeper into the medulla. Active transport of sodium, chloride, and potassium in the ascending limb raises the osmolarity of deeper ⑥ medullary interstitial fluid from 400 mOsm to 500 mOsm ⑤, which causes water movement into the medullary interstitial fluid from the descending limb. More fluid at 300 mOsm enters the loop of Henle from the proximal tubule ⑥, and the process continues until the medullary osmotic gradient is created and the system is in a steady state ⑦. Whereas fluid entering the loop of Henle from the proximal tubule is iso-osmotic to extracellular fluid, at 300 mOsm, the osmolarity of the tubular fluid within both limbs of the loop of Henle is greater in the deeper portions of the renal medulla. At the tip of the loop of Henle, the osmolarity of the tubular fluid is approximately 1400 mOsm.

Note that at any given level in the medulla, the osmolarity of the fluid in the ascending limb is always lower than the osmolarity of fluid in the descending limb, because the ascending limb actively transports solutes out of the tubular fluid but prevents water from following them. As the tubular fluid leaves the loop of Henle and enters the distal tubule, it is hypo-osmotic to extracellular fluid at approximately 100–200 mOsm.

Role of Urea in the Medullary Osmotic Gradient

Although urea freely crosses most membranes and therefore tends to distribute itself at equal concentrations across them, in the collecting duct it is actively transported out of the tubules and into the peritubular fluid, where it contributes approximately 40% of the osmolarity of the medullary osmotic gradient.

Role of the Vasa Recta in Preventing Dissipation of the Medullary Osmotic Gradient

As blood flows into the renal medulla to supply it with nutrients and oxygen, water tends to diffuse out of capillaries, and solutes tend to diffuse into them. However, the anatomical arrangement of the vasa recta capillaries, which accompany the loops of Henle as they dip into the medulla and return to the cortex (see Figure 17.6b), prevents the diffusion of water and solutes from dissipating the medullary osmotic gradient (Figure 18.8). As the vasa recta capillaries enter the renal medulla, the plasma has an osmolarity of 300 mOsm. As the capillaries enter regions of the medulla with higher osmolarity, water leaves the capillaries by osmosis, and solutes enter the plasma by diffusion, which would tend to reduce the osmolarity of the interstitial fluid if left unchecked. This process continues to the tip of the vasa recta due to the increasing osmolarity of the medullary interstitial fluid. However, as the blood flows back toward the cortex, the direction of the osmotic gradient across the capillary walls reverses, so water moves into the plasma and solutes move into the interstitial fluid, which tends to raise the osmolarity of the interstitial fluid. As a result, the osmolarity of the interstitial fluid stays relatively constant and the osmolarity of plasma leaving the renal medulla in the vasa recta capillaries is almost equal to that of the plasma entering the renal medulla.

Water Reabsorption in the Distal Tubule and Collecting Duct

Recall that 70% of the water filtered from plasma at the renal corpuscle is reabsorbed in the proximal tubule. Approximately 20% of the filtered water is reabsorbed in the distal tubule, and most of the remaining 10% is reabsorbed in the collecting ducts. In the initial portion of the distal tubule, the lumenal fluid (100–200 mOsm) is hypo-osmotic to the peritubular fluid (300 mOsm). As fluid moves down the collecting duct, the osmolarity of the lumenal fluid is always less than the increasing osmolarity of the medullary interstitial fluid, so water moves from the lumen of the collecting duct to the medullary interstitial fluid, and from there into the plasma (that is, water is reabsorbed) when the wall of the collecting duct is permeable to water.

The epithelial cells lining the late distal tubule and collecting duct are connected by tight junctions such that water cannot pass between cells from peritubular fluid to tubular fluid or vice versa. In addition, the lipid bilayers of these cells' plasma membranes are not permeable to water. The ability of water to cross the plasma membrane (and therefore the epithelial layer) depends on the presence of water channels or pores, called *aquaporins,* in the plasma membrane of principal cells. Aquaporin-3 is present in the basolateral membrane of principal cells at all times, whereas aquaporin-2 is present in the apical membrane only in the presence of the hormone ADH (discussed in the next section).

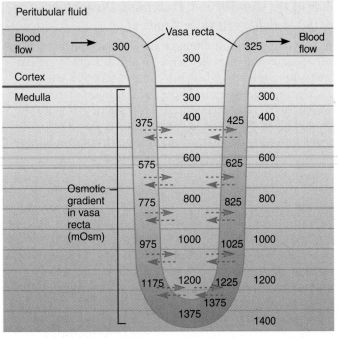

Water movement

Solute flow

FIGURE 18.8 How the vasa recta prevents the dissipation of the medullary osmotic gradient. *As the vasa recta capillaries accompany the loops of Henle through the medulla, plasma water losses and solute gains on the way into the medulla are counteracted by plasma water gains and solute losses on the way out of the medulla.*

Figure 18.9b shows how the medullary osmotic gradient enables the kidneys to excrete highly concentrated urine when the collecting duct is permeable to water. In Figure 18.9a, the walls of the late distal tubule and collecting duct are impermeable to water. As tubular fluid of 100 mOsm enters the late distal tubule, an osmotic gradient for water movement exists, but because of the impermeable membrane, water cannot move. The osmotic gradient gets larger as the tubular fluid at 100 mOsm travels down the collecting duct toward the renal pelvis, but water still cannot cross the impermeable membrane. The final result is excretion of a large volume of urine with low osmolarity.

Figure 18.9b shows how the kidneys can conserve water when the late distal tubule and collecting duct are made highly permeable to water. In the early portion of the collecting duct, the tubular fluid is initially hypo-osmotic to the cortical interstitial fluid, and water is reabsorbed. As the collecting duct leaves the cortex, the tubular fluid is iso-osmotic with the interstitial fluid at 300 mOsm. As the fluid moves down the collecting duct, water continues to be reabsorbed from the collecting duct into the medullary interstitial space such that the fluid in the collecting duct always remains very nearly iso-osmotic with the medullary interstitial fluid; eventually the fluid reaches an osmolarity of 1400 mOsm at the end of the

collecting duct. Tubular fluid osmolarity can never exceed the medullary interstitial fluid osmolarity because water will stop moving across the wall once the osmolarity inside the tubule becomes equal to that outside the tubule. Therefore, the maximum osmolarity of urine is 1400 mOsm. Because those solutes that are not 100% reabsorbed must be excreted in the urine and because there is an upper limit on the osmolarity of urine, a minimum volume of water must be excreted to eliminate the solutes. This volume is the **obligatory water loss,** which is approximately 440 mL of water per day under normal conditions (See Discovery: Don't Drink the Water, p. 576).

The length of the loop of Henle determines the maximum concentration of urine. Longer loops of Henle can form a larger medullary osmolarity gradient by the countercurrent multiplier and thereby allow greater water reabsorption. Camels, for example, have longer loops of Henle than humans and can generate urine with a concentration of 2800 mOsm. Australian hopping mice, which have the longest loops of Henle of any known species, can concentrate urine to 9800 mOsm. Because of their strong ability to conserve water, Australian hopping mice are subjected to very small obligatory water loss and can survive with little water to drink.

Quick Test 18.1

1. Where does most water reabsorption occur in the renal tubules? Where does the regulation of water reabsorption in the renal tubules occur?

2. Define the terms *iso-osmotic, hypo-osmotic,* and *hyperosmotic.* Describe the movement of water across plasma membranes when the osmolarity of nonpermeating solutes in extracellular fluid differs from that in intracellular fluid.

3. Compare the movements of solute and water across the epithelial cells lining the descending limb of the loop of Henle and those lining the ascending limb of the loop of Henle. Explain how this differential movement allows establishment of an osmotic gradient in the renal medulla.

Effects of ADH on Water Permeability in the Late Distal Tubules and Collecting Ducts

For the kidneys to produce urine at a concentration of 1400 mOsm, the walls of the late distal tubules and collecting ducts must be made highly permeable to water. By varying the degree of permeability to water in certain portions of the renal tubule, the kidneys can vary the concentration of the urine and the volume of water excreted. **Antidiuretic hormone,** or **ADH** (also known as vasopressin), regulates the permeability of the late distal tubules and collecting ducts to water.

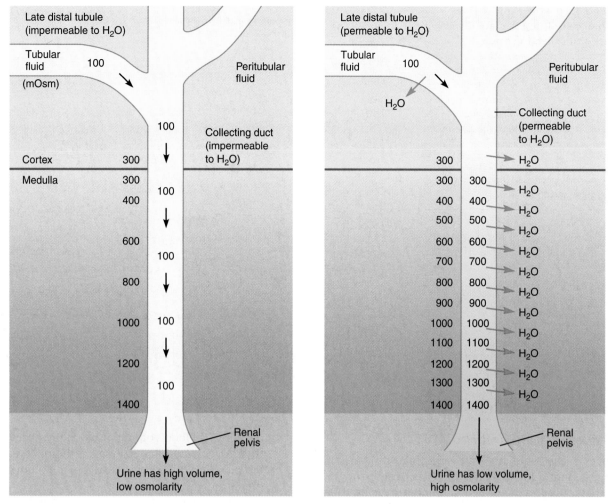

(a) Late distal tubule and collecting duct impermeable to water

(b) Late distal tubule and collecting duct permeable to water

FIGURE 18.9 Water reabsorption across the late distal tubule and collecting duct. *The fluid in the distal tubule and collecting duct is hypo-osmotic to the medullary interstitial fluid, creating an osmotic force for water to leave the tubules.* **(a)** *When the membrane of the late distal tubule and collecting duct is impermeable to water, water cannot leave the tubule and is excreted in the urine, producing a large volume of urine with low osmolarity.* **(b)** *When the membrane is permeable to water, water can leave the tubule. If the membranes are highly permeable, the final urine will be iso-osmotic with the deepest layers of the renal medulla, producing a small volume of urine with high osmolarity.*

ADH is secreted from the posterior pituitary gland by neurosecretory cells that originate in the hypothalamus. ADH stimulates synthesis of aquaporin-2 and its insertion into the apical membrane of principal cells in the late distal tubules and collecting ducts. In the absence of ADH, the apical membrane is impermeable to water, so water reabsorption cannot occur. Water reabsorption and urine volume are thus regulated by variations in the plasma levels of ADH, which by determining the number of aquaporin-2 pores determines the permeability of the membranes to water.

Aquaporin-2 is stored in the membrane of cytoplasmic vesicles of principal cells, awaiting insertion into the apical membrane in response to ADH. ADH acts on renal tubule cells by binding to receptors on the plasma membrane (Figure 18.10). These receptors are coupled to a G protein that activates the enzyme adenylate cyclase, which then catalyzes the production of the second messenger cAMP. cAMP activates protein kinase A, which stimulates the insertion of aquaporin-2 pores into the apical membrane by exocytosis. ADH also stimulates the synthesis of new aquaporin-2 molecules. At high concentrations of ADH, water reabsorption is high and urine output is low. (The word *antidiuretic* refers to something that counters the effects of *diuresis,* or increased urine flow; a diuretic is a drug that promotes urine flow.) At low concentrations of ADH, water reabsorption is low and urine output is high.

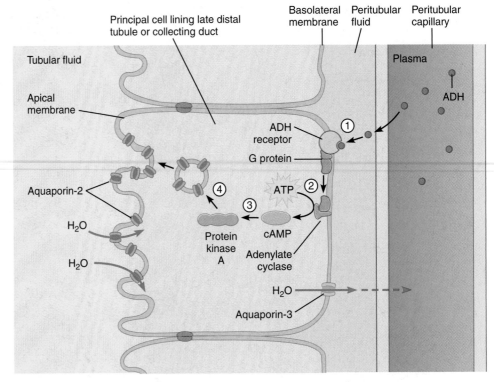

FIGURE 18.10 Effects of ADH on principal cells lining the late distal tubules and collecting ducts. ① *ADH binds to receptors on the membrane of principal cells.* ② *Activation of a G protein activates the enzyme adenylate cyclase, which catalyzes the formation of cAMP, which* ③ *activates protein kinase A.* ④ *The protein kinase A stimulates insertion of new water channels (aquaporin-2) into the apical membrane, which increases its water permeability.*

What type of hormone is ADH, and what endocrine gland secretes it?

Regulation of ADH Secretion

Changes in the osmolarity of extracellular fluid are the strongest stimuli for ADH release. **Osmoreceptors** in the hypothalamus and digestive tract monitor the osmolarity of extracellular fluid. If the osmolarity increases, then ADH secretion is stimulated and increases water reabsorption (Figure 18.11). Conversely, if the osmolarity of extracellular fluid decreases, ADH secretion is inhibited, which decreases water reabsorption and increases water excretion.

Plasma levels of ADH also depend on signals arising in baroreceptors that detect blood volume and blood pressure (Figure 18.12). As we saw in Chapter 14, baroreceptors in the atria respond to changes in blood volume, and those in the aortic arch and carotid sinus (sinoaortic baroreceptors) respond to changes in blood pressure. As blood volume and/or blood pressure decrease, the frequency of action potentials in baroreceptor afferents also decreases. Decreased baroreceptor activity stimulates increases in the secretion of ADH, which increases water reabsorption and minimizes the stimuli for the release of ADH (see Figure 18.12). Note that if plasma volume is below normal, the kidneys cannot raise it back to normal; they can only minimize further fluid loss, which then minimizes any additional reductions in plasma volume. Achieving an increase in body water content requires behavioral changes—specifically, an input of water through drinking. If blood volume or blood pressure increases, then ADH secretion decreases, resulting in increased water excretion, which decreases blood volume and blood pressure.

Inadequate reabsorption of water occurs in the disease *diabetes insipidus*, which is caused by a deficiency in ADH secretion from the posterior pituitary. The deficiency can be caused by a head injury, inflammation of the hypothalamus, or tumors in the hypothalamus or posterior pituitary. Because ADH increases water reabsorption, a deficiency in ADH causes excessive urination *(polyuria)* and excessive fluid intake *(polydipsia)* as a compensatory response. People suffering from diabetes insipidus may lose up to 20 liters of water per day! Diabetes insipidus results in high plasma sodium levels and increased plasma osmolarity. Treatment includes fluid replacement.

How Regulation of GFR Regulates Water Excretion

We saw in Chapter 17 that normally the glomerular filtration rate (GFR) is autoregulated to minimize fluctuations in water and electrolyte loss. However, if blood pressure drops to sufficiently low levels (mean arterial pressure less than 80 mm Hg), then the GFR decreases. With less water filtered, less will be excreted. Likewise, if blood pressure increases to sufficiently high levels (mean arterial pressure greater than 180 mm Hg), then the GFR increases, and water loss increases. These changes in GFR work hand-in-hand with other mechanisms to limit water loss when blood pressure is low, and to increase water loss when blood pressure is high.

A peptide hormone; the posterior pituitary gland

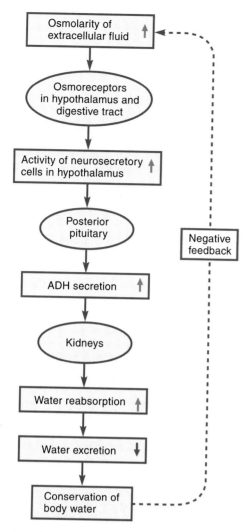

FIGURE 18.11 The pathway by which ADH secretion in response to changes in extracellular fluid osmolarity effects changes in water reabsorption and urine output. *This flowchart is available as an interactive exercise under Activities at www.physiologyplace.com.*

Quick Test 18.2

1. Name two stimuli for ADH release from the posterior pituitary. Where does ADH act in the kidneys, and what does it do?

2. Describe the effects of ADH on principal cells.

SODIUM BALANCE

Because sodium is the primary solute in extracellular fluid, it must be regulated if normal osmolarity is to be maintained. In addition, because the electrochemical gradient for sodium across plasma membranes is critical to the function of excitable cells, it is important that plasma sodium levels be regulated. An increase in plasma sodium levels above normal, called **hypernatremia,** is often accompanied by water retention and an increase in blood

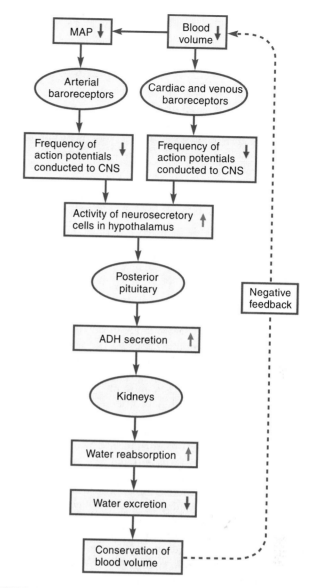

FIGURE 18.12 Effects of arterial and cardiac baroreceptors on ADH release. *A decrease in mean arterial pressure and/or blood volume stimulates ADH release.*

pressure. A decrease in plasma sodium levels below normal, called **hyponatremia,** is associated with low plasma volume and hypotension.

Sodium is freely filtered at the glomerulus and undergoes tubular reabsorption, but it is not secreted. Therefore, regulation can occur at the level of filtration or reabsorption. Regulation of GFR was described in the previous section and in Chapter 17; this section focuses on the regulation of sodium reabsorption. Two hormones function in the regulation of sodium reabsorption: aldosterone and atrial natriuretic peptide or ANP.

The Effects of Aldosterone

Aldosterone is a steroid hormone released from the adrenal cortex that regulates both the reabsorption of

You have likely encountered these famous lines from Samuel Taylor Coleridge's epic poem, *The Rime of the Ancient Mariner:*

> Water, water everywhere
> Nor any drop to drink

As you are probably well aware, the poet is referring to the fact that sailors cannot rely on the sea as a source of drinking water because if they do, they face certain death by dehydration. What you probably do not realize is that if our nephrons had longer loops of Henle, these lines never would have been written!

The danger in drinking seawater lies in the concentration of the solutes dissolved in it. Depending on the location, the solute concentration of seawater varies from 2000 mOsm to 2400 mOsm, most of which is due to dissolved sodium chloride. At 2400 mOsm, the solute concentration of seawater is eight times that of plasma. More importantly, seawater is much more concentrated than the most concentrated urine that our kidneys are capable of producing, which is about 1400 mOsm.

To see the problem with drinking seawater, suppose that you have lost the equivalent of one liter of pure water from your plasma as a result of sweating. (In reality, solutes are also lost in sweat.) Furthermore, suppose that you have tried to compensate for the loss by drinking one liter of seawater. After you have drunk the water, your plasma volume returns to normal because the volume of fluid drunk equals the volume of fluid lost from your body. But because the solute concentration of seawater is greater than that of plasma, ingesting seawater causes the osmolarity of your plasma to exceed normal. (If you had drunk one liter of pure water instead, your plasma osmolarity would have returned to normal.) The higher-than-normal plasma osmolarity stimulates the secretion of ADH, and as a result your kidneys excrete a small volume of highly concentrated urine, which helps your body conserve water while ridding itself of the excess solutes. However, there is a catch: Ideally the kidneys would excrete the total excess of ingested solute (all 2400 milliosmoles of it) while excreting only a negligible volume of water. This is not possible, however, because the highest possible urine concentration is 1400 mOsm. Consequently, to excrete 2400 milliosmoles of solute, the kidneys would have to excrete at least 1.7 liters of water (2400 milliosmoles/1.7 liter = 1400 mOsm), which is 0.7 liters more than the volume of fluid you drank! This volume of water—the minimum that must accompany a given quantity of excreted solutes—is called the *obligatory water loss.*

People who drink seawater soon find themselves in a predicament because the volume of water lost in the urine over a given time period invariably exceeds the volume of water drunk, resulting in a net loss of body water. Drinking more seawater to replace the lost volume only makes the problem worse because the quantity of excess solutes in the plasma becomes greater. The accumulation of excess solutes then leads to larger volumes of water lost in the urine. Eventually this water loss leads to circulatory collapse and death.

Certain animals could, unlike humans, survive on seawater indefinitely. One is the desert-dwelling kangaroo rat, which can produce urine that is much more concentrated than seawater (up to 9000 mOsm) because its kidneys possess nephrons with unusually long loops of Henle. Other things being equal, longer loops of Henle translate into higher solute concentrations in the renal medulla, which increase the kidneys' ability to reabsorb water from fluid in the collecting ducts before it emerges as urine. The ability to produce very concentrated urine conserves water, which is a precious commodity in the kangaroo rat's desert habitat.

---

sodium and secretion of potassium. Here we focus on the role of aldosterone in sodium reabsorption.

Aldosterone binds to cytosolic receptors in principal cells of the late distal tubules and collecting ducts, where it has several effects (Figure 18.13). Aldosterone increases the number of open sodium channels and potassium channels in the apical membrane, both by causing existing channels to open and by stimulating the synthesis of new channels. Aldosterone also stimulates the synthesis of Na^+/K^+ pumps, which increases the concentration of Na^+/K^+ pumps in the basolateral membrane. Through these actions, aldosterone increases sodium reabsorption and potassium secretion simultaneously; it cannot affect one without affecting the other.

Of the factors that control aldosterone release, the one that is most important in the control of sodium reabsorption is the renin-angiotensin-aldosterone system (RAAS).

The Renin-Angiotensin-Aldosterone System

Recall from Chapter 17 that where the distal tubule travels close to the afferent and efferent arterioles, these structures form the juxtaglomerular apparatus. Within the walls of the afferent arteriole are granular cells that secrete renin; within the walls of the distal tubule are the macula densa cells, which can detect changes in the flow and sodium and chloride concentrations of the tubular fluid. When sodium concentration in the tubular fluid decreases, renin secretion increases. Although often called a hormone, renin is actually a proteolytic enzyme.

Once renin is released from the granular cells into the bloodstream, it starts a series of reactions that lead to the release of aldosterone (Figure 18.14). Renin acts on another protein that is always present in the plasma, *angiotensinogen*, which like most plasma proteins is secreted by the liver. Renin cleaves off some amino acids from an-

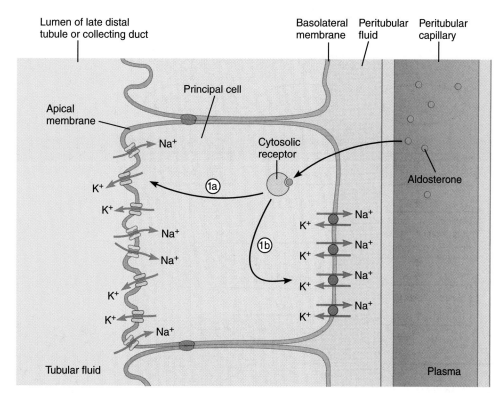

Lumen of late distal tubule or collecting duct

Basolateral membrane | Peritubular fluid | Peritubular capillary

Principal cell

Apical membrane

Na⁺

Cytosolic receptor

(1a)

K⁺

K⁺

Na⁺

Na⁺

K⁺

Na⁺

(1b)

K⁺ → Na⁺

K⁺ → Na⁺

K⁺ → Na⁺

Na⁺

Aldosterone

K⁺

K⁺

Na⁺

Tubular fluid

Plasma

FIGURE 18.13 **Effects of aldosterone on principal cells of the distal tubules and collecting ducts.** *After binding to its receptor, aldosterone* (1a) *stimulates both the opening of sodium channels and potassium channels and the synthesis of new channels on the apical membrane, and* (1b) *stimulates the synthesis and insertion of more Na⁺/K⁺ pumps on the basolateral membrane.*

What class of hormone is aldosterone, and in what location in principal cells are receptors for the hormone located?

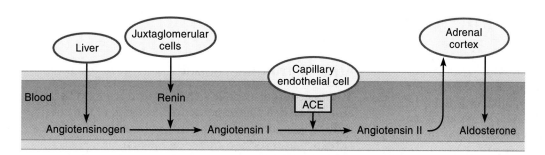

Liver | Juxtaglomerular cells | Capillary endothelial cell | Adrenal cortex

Blood

Renin

ACE

Angiotensinogen → Angiotensin I → Angiotensin II → Aldosterone

FIGURE 18.14 The renin-angiotensin-aldosterone system. *The liver and the juxtaglomerular cells secrete and release angiotensinogen and renin, respectively, into the blood, where renin cleaves amino acids from angiotensinogen to form angiotensin I. Angiotensin converting enzyme (ACE) located on certain endothelial cells in capillaries cleaves amino acids from angiotensin I to form angiotensin II, which travels in the bloodstream to the adrenal cortex, where it stimulates the release of aldosterone into the blood.*

giotensinogen, converting it to angiotensin I. As angiotensin I molecules circulate in the bloodstream, they encounter another proteolytic enzyme called *angiotensin converting enzyme,* or *ACE,* which is bound to the inner surfaces of capillaries throughout the body and is particularly abundant in the capillaries of the lungs. ACE cleaves off some amino acids from angiotensin I, converting it to angiotensin II. In addition to acting as a vasoconstrictor that is important in the regulation of mean arterial pressure (see Chapter 13), angiotensin II has another key role: the stimulation of aldosterone release from the adrenal cortex. (Angiotensin II also acts in the hypothalamus, where it stimulates ADH release and thirst.)

Figure 18.15 summarizes the four mechanisms whereby angiotensin II increases mean arterial pressure: (1) Angiotensin II stimulates vasoconstriction of systemic arterioles, which by increasing the total peripheral resistance increases mean arterial pressure. (2) Angiotensin II stimulates the adrenal cortex to secrete aldosterone, which by increasing sodium reabsorption causes water reabsorption to increase. (3) Angiotensin II stimulates the posterior pituitary to secrete ADH, which by increasing

A steroid hormone; in the cytosol

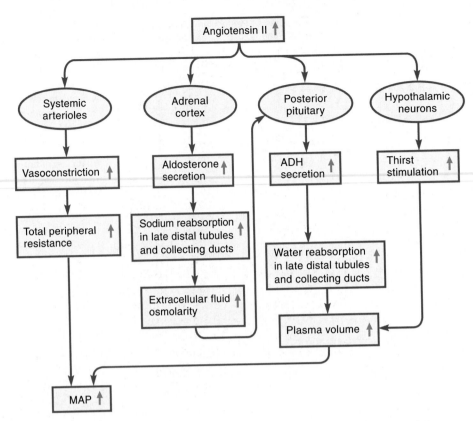

FIGURE 18.15 Mechanisms by which angiotensin II increases mean arterial pressure.

water reabsorption minimizes fluid loss and maintains plasma volume, thereby maintaining mean arterial pressure. (4) Angiotensin II activates hypothalamic neurons to stimulate thirst and fluid intake, which by increasing plasma volume increases mean arterial blood pressure.

Renin release is controlled by a variety of stimuli related to blood pressure (Figure 18.16). Because the RAAS tends to increase blood pressure, a decrease in blood pressure is a primary stimulus for renin release. Specifically, a decrease in afferent arteriolar pressure triggers renin release, because granular cells are directly sensitive to the degree of stretch of the afferent arteriole. An increase in renal sympathetic nerve activity also stimulates renin release by direct input to granular cells. The sympathetic nervous system is activated during the baroreceptor reflex response to a decrease in blood pressure (see Chapter 14). A large decrease in mean arterial pressure also decreases the glomerular filtration rate, which when coupled with the continual reabsorption of sodium and chloride in the proximal tubule and ascending loop of Henle, decreases sodium and chloride levels in the distal tubule. Recall from Chapter 17 that a decrease in sodium and chloride in the distal tubules is detected by macula densa cells of the tubule, which secrete a chemical signal that stimulates renin release from the juxtaglomerular cells.

Atrial Natriuretic Peptide

Atrial natriuretic peptide (ANP), is secreted by cells in the atria of the heart in response to distension of the atrial wall, which occurs when plasma volume has increased. ANP increases sodium excretion by increasing the glomerular filtration rate and by decreasing sodium reabsorption (Figure 18.17). ANP causes dilation of the afferent arteriole and constriction of the efferent arteriole, which by increasing glomerular capillary pressure increases the glomerular filtration rate and increases the filtered sodium load. ANP decreases sodium reabsorption directly by decreasing the number of open sodium channels in the apical membrane of the principal cells. In addition, ANP decreases secretion of both renin and aldosterone.

Quick Test 18.3

1. Briefly explain how an increase in the secretion of renin stimulates the secretion of aldosterone. How does an increase in aldosterone secretion affect sodium reabsorption?

2. Where are the cells that secrete renin located? Name three stimuli for renin secretion.

3. How does atrial natriuretic peptide affect sodium reabsorption?

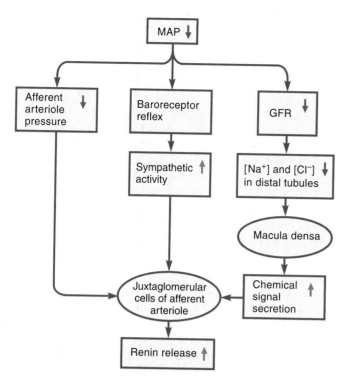

FIGURE 18.16 Mechanisms by which decreases in mean arterial pressure stimulate renin release.

What is renin's RAAS function in the plasma?

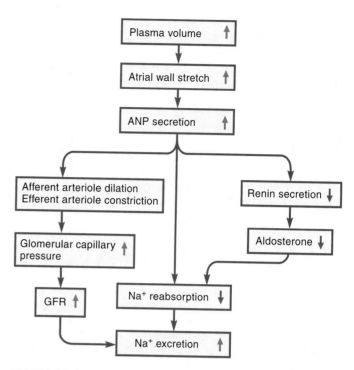

FIGURE 18.17 Mechanisms by which secretion of atrial natriuretic peptide increases sodium excretion in response to increased plasma volume.

4. Describe the effects of aldosterone on principal cells.

POTASSIUM BALANCE

The gradient that results from high potassium concentrations in the intracellular fluid and low potassium concentrations in the extracellular fluid is critical to the function of excitable cells. An increase in plasma potassium levels is called **hyperkalemia.** Some common symptoms of hyperkalemia include cardiac arrhythmias, muscle weakness and cramps, dizziness, nausea, and diarrhea. A decrease in plasma potassium levels is called **hypokalemia.** Common symptoms of hypokalemia include cardiac arrhythmias, muscle weakness and tenderness, hypotension, confusion, alkalosis, and shortness of breath.

Renal Handling of Potassium Ions

In the kidneys, potassium is freely filtered at the glomerulus and undergoes both reabsorption and secretion in the tubules. Normally, the amount of potassium reabsorbed is greater than the amount secreted; that is, the net effect is reabsorption. In fact, most of the potassium filtered is reabsorbed.

Unlike water and sodium, whose plasma levels are regulated by varying the amounts that are reabsorbed from the renal tubules, the plasma concentration of potassium is regulated by varying the amounts that are secreted into the renal tubules. Therefore, the net renal handling of potassium can be varied only by regulating the rate of secretion. Regulation of potassium secretion occurs in the late distal tubule and collecting duct.

Figure 18.18 compares the movement of potassium ions in the proximal tubule with that in the late distal tubule and collecting duct. In the proximal tubule (Figure 18.18a), potassium is reabsorbed by the following mechanisms: Potassium ions move from the peritubular fluid into the tubule epithelial cell via the Na^+/K^+ pump located on the basolateral membrane; potassium ions also move from the tubular fluid into the epithelial cell by some as yet unknown mechanism. Once inside the epithelial cell, potassium ions move through potassium channels in the basolateral membrane into the peritubular fluid. Therefore, most potassium entering the tubule epithelial cell (whether it originated in the peritubular fluid or in the tubular fluid) moves into the peritubular fluid and then into the plasma. In addition, potassium ions can move between cells from tubule lumen to the peritubular fluid and then into the plasma.

It cleaves amino acids off angiotensinogen to form angiotensin I.

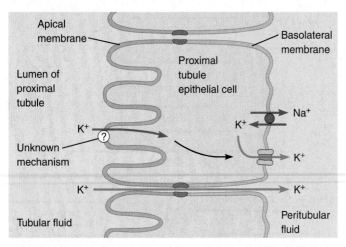

(a) Potassium reabsorption in the proximal tubule

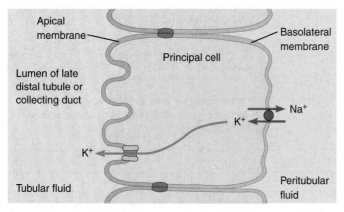

(b) Potassium secretion in the principal cells of the late distal tubule and collecting duct

FIGURE 18.18 Potassium transport in renal tubules. (a) *In the proximal tubule, potassium is reabsorbed because of the presence of potassium channels in the basolateral membrane.* **(b)** *In principal cells of the distal tubule and collecting duct, potassium is secreted because of the presence of potassium channels in the apical membrane.*

In principal cells of the late distal tubule and collecting duct (Figure 18.18b), potassium is secreted by the following mechanism: As in the proximal tubule, potassium ions move from the peritubular fluid into the epithelial cell via the Na$^+$/K$^+$ pump in the basolateral membrane. Unlike epithelial cells in the proximal tubule, however, in principal cells potassium channels are located in the apical membrane, allowing potassium ions to move out of the epithelial cell into the tubular fluid of the distal tubule and collecting duct.

Regulation of Potassium Secretion by Aldosterone

Potassium secretion is regulated by aldosterone. Recall that this hormone increases both the number of Na$^+$/K$^+$ pumps on the basolateral membrane in principal cells lin-

ing the late distal tubules and collecting ducts, and the number of potassium channels in the apical membrane. The increase in Na$^+$/K$^+$ pumps causes greater potassium movement into the epithelial cells, which is followed by greater movement of potassium ions through apical potassium channels and into the lumen of the tubules, resulting in greater excretion of potassium in the urine.

As discussed previously, aldosterone secretion is regulated by the renin-angiotensin-aldosterone system, whereby angiotensin II stimulates aldosterone release from the adrenal cortex. However, high plasma potassium levels also directly stimulate aldosterone secretion by acting on secretory cells in the adrenal cortex. The aldosterone released then increases potassium secretion, which brings plasma potassium levels toward normal.

CALCIUM BALANCE

Calcium is critical to the function of most cells: It triggers exocytosis of chemical messengers, stimulates secretion of various substances, stimulates muscle contraction, and increases the contractility of the heart and blood vessels. Calcium is also an important component of the bone and teeth. An increase in plasma calcium, called **hypercalcemia,** has widespread effects on the body, including muscle weakness and atrophy, lethargy, behavioral changes, hypertension, constipation, and nausea. A decrease in plasma calcium, called **hypocalcemia,** causes numbness and tingling sensations, muscle cramps and spasms, exaggerated reflexes, and hypotension.

Plasma calcium concentration is regulated through the interaction of a number of organs, including the kidneys, digestive tract, bone, and skin (Figure 18.19). Calcium can be added to the plasma from bone and absorbed via the digestive tract, and it can be removed from the plasma by bone and the kidneys. Even though most of the calcium in the body (99%) is located in the bones, this calcium is not permanently fixed in the bone. The bone actually provides a reservoir of calcium such that when plasma calcium levels are low, the plasma can obtain calcium via a process called *resorption.* Conversely, when plasma calcium levels are high, calcium can be deposited into bone. (When It Goes Wrong: Osteoporosis, www.physiologyplace.com, Challenge Yourself) The body also obtains calcium from ingested food. Although the absorption of most substances by the gastrointestinal tract is not regulated, calcium absorption is regulated according to the needs of the body.

Renal Handling of Calcium Ions

Calcium is transported in blood both bound to carrier proteins and free in the plasma. Calcium that is free in the plasma is freely filtered at the glomerulus. Normally, 99% of the filtered calcium is reabsorbed as the tubular fluid moves through the renal tubules. Approximately

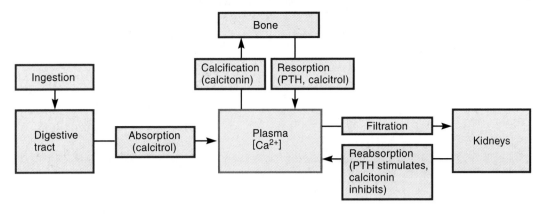

FIGURE 18.19 Routes of calcium exchange. *Calcium can enter the plasma by absorption from the digestive tract or resorption of bone. Calcium leaves the plasma by calcification of bone or excretion in the urine. The amount of calcium excreted in the urine is regulated by varying the rate of calcium reabsorption.*

70% of the filtered calcium is reabsorbed in the proximal tubules, 20% is reabsorbed in the thick ascending limbs of the loops of Henle, and the remaining 10% is reabsorbed in the distal tubules. Reabsorption in the loops of Henle and the distal tubules is regulated by hormones.

Hormonal Control of Plasma Calcium Concentrations

Several hormones regulate plasma calcium levels, including **parathyroid hormone (PTH), calcitrol,** and **calcitonin.** PTH is the primary regulator of plasma calcium levels.

Effects of Parathyroid Hormone

PTH is a peptide hormone produced in the parathyroid glands and secreted in response to a decrease in the plasma calcium concentration (Figure 18.20). Among its functions, PTH (1) stimulates calcium reabsorption in the ascending limb of the loop of Henle and the distal tubules, which decreases the excretion of calcium and sustains plasma calcium levels; (2) stimulates the activation in the kidneys of calcitrol, which stimulates calcium absorption in the digestive tract and calcium reabsorption in the kidneys; and (3) stimulates resorption of bone, which increases plasma calcium levels. PTH also directly causes a small increase in the absorption of calcium from the digestive tract.

Effects of Calcitrol

Calcitrol is a steroid hormone that acts to increase plasma calcium levels by stimulating calcium absorption from the digestive tract and calcium reabsorption in the distal tubules of the kidneys. Calcitrol is synthesized in several steps from vitamin D_3, which can either be synthesized from 7-dehydrocholesterol in the skin upon exposure to sunlight, or absorbed from the diet (Figure 18.21). Once in the plasma, vitamin D_3 travels to the liver, where it is converted to 25-hydroxyvitamin D_3, or 25-OH D_3. From the liver, 25-OH D_3 travels in the bloodstream to the kidneys, where, in response to low plasma calcium levels, PTH regulates the conversion of 25-OH D_3 to calcitrol.

Effects of Calcitonin

Unlike PTH and calcitrol, calcitonin decreases plasma calcium levels. Calcitonin is a peptide hormone secreted from C cells of the thyroid gland, which are distinct from the cells that secrete thyroid hormone. Calcitonin secretion is triggered by increases in plasma calcium levels. Although the primary action of calcitonin is to increase bone formation by depositing calcium, it also decreases the reabsorption of calcium by the kidneys, which leads to an increase in urinary excretion of calcium; both of these actions decrease plasma calcium levels.

Quick Test 18.4

1. What hormone regulates potassium secretion? What are some stimuli for the release of this hormone?
2. Name the three hormones that affect plasma calcium levels. Describe the effects of each.
3. Describe the function of bone as a reservoir of calcium ions.

INTERACTIONS BETWEEN FLUID AND ELECTROLYTE REGULATION

The previous sections described fluid and electrolyte balance as if water and ions were independently regulated. However, that is not the case, because reabsorption of any ion increases the osmotic gradient for water reabsorption. In addition, hormones often affect more than one system. Consider, for example, ADH, which in addition to increasing water reabsorption also stimulates production of sodium channels in the principal cells, thereby enhancing the movement of sodium ions from the tubule lumen into these cells. As more sodium enters the cell, the Na^+/K^+ pump works faster to rid the cell of sodium, increasing sodium reabsorption and potassium secretion. ADH also decreases water flow to the late distal tubules and collecting ducts because less water is left following its reabsorption. The decrease in water increases

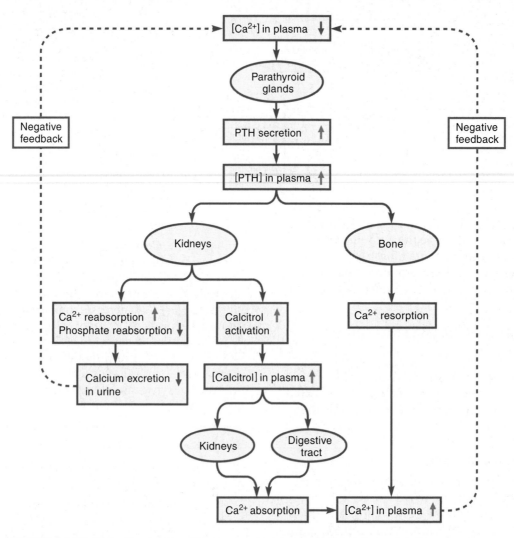

FIGURE 18.20 Role of parathyroid hormone in calcium balance. *PTH release is stimulated by a decrease in plasma calcium levels. PTH then acts on the kidneys to increase calcium reabsorption and activate calcitrol, and on the bone to increase calcium resorption, all of which lead to increases in plasma calcium levels.*

solute concentration in the tubules, which affects the reabsorption or secretion of any solutes present in these tubules. In other examples, angiotensin II and ANP affect water reabsorption in addition to their primary effects on sodium reabsorption. Whereas angiotensin II increases ADH secretion, which stimulates water reabsorption, ANP inhibits ADH secretion, which decreases water reabsorption.

We can illustrate the interactions between systems regulating fluid and electrolyte balance by considering the events following hemorrhage. When we first used this example to demonstrate control of mean arterial pressure (see Chapter 14, p. 453), neural control of mean arterial pressure was the focus, although hormonal control of blood volume also clearly had a role. Here we return to the restoration of blood volume following hemorrhage; Figure 18.22 is similar to Figure 14.9, except

that the cardiovascular effects of hemorrhage are abbreviated, and the renal effects are expanded.

Hemorrhage results in a decrease in blood volume, which decreases venous pressure and therefore decreases venous return. With less blood returning to the heart, stroke volume is decreased, causing a decrease in mean arterial pressure. This decrease in mean arterial pressure is detected by arterial baroreceptors, which (through the baroreceptor reflex) activate the sympathetic nervous system and inhibit the parasympathetic nervous system. As a result, heart rate and stroke volume increase, and total peripheral resistance increases, all of which produce an increase in mean arterial pressure. This neural control of blood pressure is very rapid, but it does not correct for the initial cause of the problem—the loss of blood volume.

Through its effects on blood volume, hormonal control of the kidneys is instrumental in long-term regulation of blood pressure. Several factors induced by hemorrhage contribute to renal control of water loss (see Figure 18.22):

1. The decrease in mean arterial pressure stimulates renin secretion both directly and indirectly as follows: Pressure in the afferent arteriole is decreased, which stimulates renin secretion. The baroreceptor reflex increases sympathetic input to juxtaglomerular cells, which also stimulates renin secretion. Increased renin secretion from the kidneys causes an increase in plasma angiotensin II levels, producing several effects that increase mean arterial pressure. Angiotensin II causes vasoconstriction, which increases the total peripheral resistance and thus mean arterial pressure. Angiotensin II also stimulates the adrenal cortex to release aldosterone, which increases sodium reabsorption in the kidneys, which in turn increases water reabsorption. Angiotensin II also acts in the hypothalamus to stimulate thirst, which increases fluid intake and thus plasma volume.

2. Decreased activity in venous and cardiac baroreceptors stimulates ADH release from the posterior pituitary gland. ADH increases water reabsorption in the kidneys, conserving plasma volume.

3. A decrease in mean arterial pressure decreases the glomerular filtration rate, which conserves water and sodium.

4. Although not shown in the figure, hemorrhage includes the loss of plasma and blood cells. The blood cells are replaced by the synthesis of new cells, which takes days. The decrease in mean arterial pressure results in a decrease in blood flow to the kidneys, so the oxygen supply to the kidneys is decreased, and erythropoietin secretion is stimulated. Recall from Chapter 13 that following its release by the kidneys, erythropoietin travels in the blood to the bone marrow, where it stimulates the production of new erythrocytes.

Clearly, organ systems must work together to maintain homeostasis. What also should be apparent is the critical role the kidneys have in maintaining homeostasis (see When It Goes Wrong: Kidney Disease, p.586). In the next section we see how acid-base balance of the plasma is maintained by interactions between two organ systems: the lungs and the kidneys.

Quick Test 18.5

1. Describe the relationship between blood volume and blood pressure.

2. Compare the short-term and long-term regulation of blood pressure.

3. Name the hormones affecting blood volume, and their major actions.

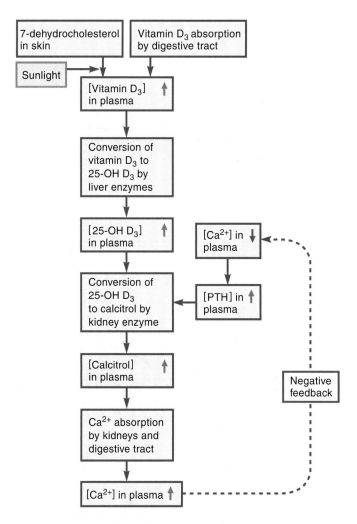

FIGURE 18.21 Activation of calcitrol. *Several steps are required to convert vitamin D_3 to 25-OH D_3; in response to low plasma calcium levels. PTH regulates the conversion of 25-OH D_3 to calcitrol.*

ACID-BASE BALANCE

The hydrogen ion concentration or pH of arterial blood is regulated by the combined actions of the lungs and the kidneys. It is essential that the arterial pH be tightly controlled to near or within its normal range of 7.35 to 7.45. A decrease in pH (increase in hydrogen ion concentration) to less than 7.35 is called acidosis; an increase in pH (decrease in hydrogen ion concentration) to greater than 7.45 is called alkalosis. Changes in pH of as little as a few tenths of a unit in either direction can have the following profound effects on the body:

1. Interactions between hydrogen ions and certain amino acids result in conformational changes in proteins, and thus in changes in the proteins' functions. For example, enzyme activity may increase or decrease with pH changes.

2. Activity of the nervous system changes, although the precise mechanisms are not fully understood. Acidosis

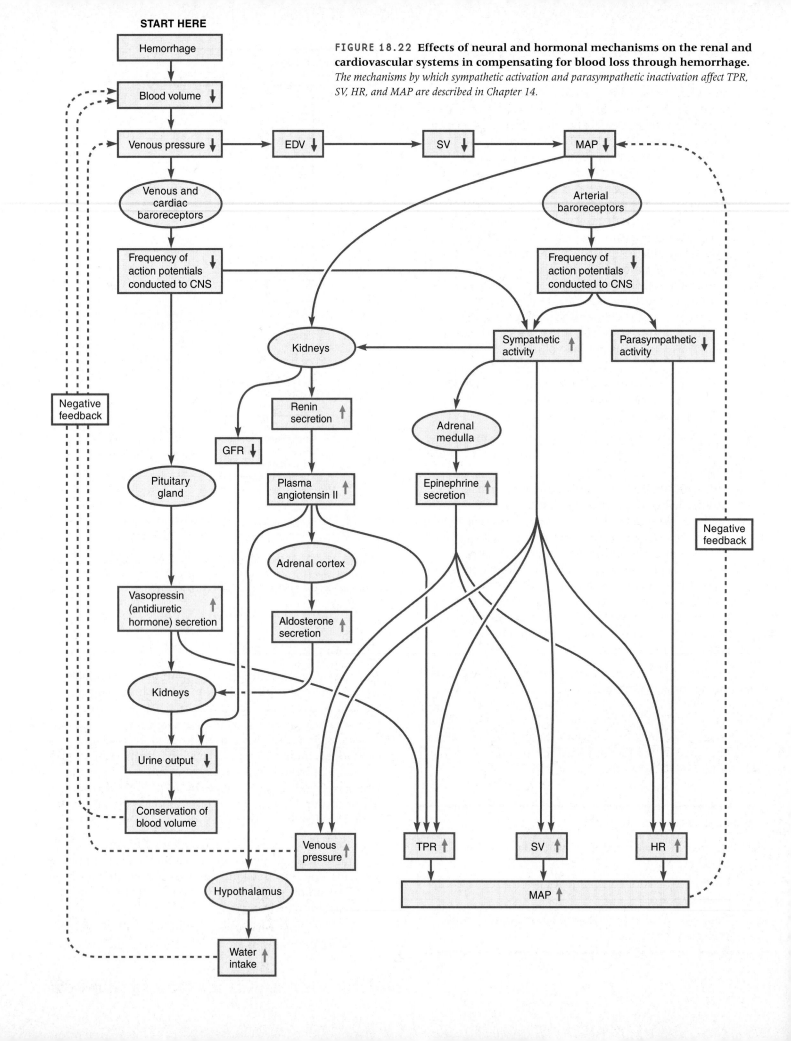

START HERE

FIGURE 18.22 Effects of neural and hormonal mechanisms on the renal and cardiovascular systems in compensating for blood loss through hemorrhage. *The mechanisms by which sympathetic activation and parasympathetic inactivation affect TPR, SV, HR, and MAP are described in Chapter 14.*

causes a decrease in the excitability of neurons, especially in the CNS; severe acidosis can lead to confusion, coma, and even death. Alkalosis, such as that produced by hyperventilation, causes an increase in the excitability of neurons. If you have ever hyperventilated, you may have noticed a tingling sensation particularly in the hands or feet that is caused by overactive afferent neurons that generate action potentials even in the absence of a stimulus. Motor neurons may also become hyperexcitable, resulting in muscle spasms and twitches.

3. Due to complex interactions in the movement of potassium and hydrogen ions across renal tubules (such as electrochemical interactions and competition for carrier proteins), acid-base disturbances are often coupled to potassium imbalances. Acidosis results in potassium retention (hyperkalemia), whereas alkalosis results in potassium depletion (hypokalemia).

4. Acidosis causes cardiac arrhythmias and vasodilation of blood vessels to the skin due to impaired activity of catecholamines.

Sources of Acid-Base Disturbances

Figure 18.23 shows the various inputs and outputs of acid (free hydrogen ions) to the blood. Inputs include dietary and metabolic sources. Protein and fats in the diet provide hydrogen ions in the forms of amino acids and fatty acids. Cellular metabolism produces several acids, including carbon dioxide, lactic acid, and ketoacids. Hydrogen ions can be removed from the blood by the kidneys or lungs. Whereas the lungs remove acid in the form of carbon dioxide during ventilation, the kidneys excrete hydrogen ions in the urine. As with many other systems in the body, these inputs and outputs must be balanced if normal blood pH is to be maintained. Because acid production occurs regularly in the body during metabolism, potential increases in blood acidity must be prevented.

Respiratory Disturbances

One source of acid is carbon dioxide (see Chapter 16). Recall that the enzyme carbonic anhydrase catalyzes the reaction converting carbon dioxide and water to carbonic acid, which dissociates to bicarbonate and hydrogen ions:

$$CO_2 + H_2O \underset{\text{Carbonic anhydrase}}{\rightleftharpoons} H_2CO_3 \rightleftharpoons HCO_3^- + H^+$$

Carbon dioxide is produced during cellular metabolism—in particular, in the Krebs cycle. As carbon dioxide increases in the blood, it is converted to bicarbonate and hydrogen ions. When blood rich in carbon dioxide reaches the lungs, the reaction reverses, and carbon dioxide is exhaled. The respiratory system contributes to acid-base balance by regulating carbon dioxide levels in the blood.

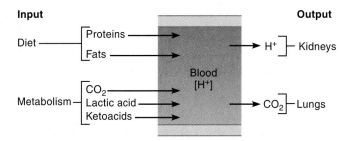

FIGURE 18.23 Inputs and outputs of acid to the blood. *Whereas diet is the primary contributor of acids to the blood, cellular metabolism also generates acids. To compensate for acid inputs, the kidneys excrete hydrogen ions and the respiratory system clears the blood of carbon dioxide.*

Normally, the P_{CO_2} of arterial blood is maintained at its normal value of 40 mm Hg by respiratory chemoreceptor reflexes, which adjust alveolar ventilation to keep pace with the rate of metabolic carbon dioxide production. Respiratory acidosis or respiratory alkalosis results from an excess or deficit, respectively, of carbon dioxide in the blood. Respiratory acidosis can result either from lung diseases that interfere with the exchange of carbon dioxide between the blood and alveolar air, or from hypoventilation, an alveolar ventilation rate that is inappropriately low. In either case, arterial P_{CO_2} rises, leading to an increase in carbonic acid levels and an increase in the blood's hydrogen ion concentration (a decrease in pH). Respiratory alkalosis, in contrast, can result from hyperventilation, an alveolar ventilation rate that is inappropriately high; this causes the P_{CO_2} to decrease, which leads to a decrease in the blood's carbonic acid concentration and the hydrogen ion concentration (an increase in pH).

Metabolic Disturbances

Metabolic acidosis and **metabolic alkalosis** are disturbances in blood pH caused by something other than an abnormal P_{CO_2}. Metabolic acidosis can be caused by excess elimination from the body of alkaline substances such as bicarbonate, excess production of acid in metabolism, or excess consumption of acids in the diet. Metabolic alkalosis can be caused by the excess elimination of acids from the body or the addition of alkaline substances to the blood. Among the factors that can produce metabolic disturbances in acid-base balance are the following:

1. *A high-protein diet.* Because protein catabolism produces phosphoric acid and sulfuric acid, diets high in protein can result in metabolic acidosis.

2. *A high-fat diet.* Because catabolism of fats or triglycerides produces fatty acids, high-fat diets can result in metabolic acidosis.

3. *Heavy exercise.* During heavy activity, the oxygen demands of tissues cannot be met, and the resulting anaerobic metabolism produces increased amounts of lactic acid, which can result in metabolic acidosis.

Diseases of the kidneys are responsible for approximately 35,000 deaths annually in the United States. Although this is a small number relative to the number of deaths caused by the leading killers—heart disease and cancer—the actual cost of kidney disease in monetary terms and in terms of lives disrupted is much higher because many of those affected require continual medical treatment for the rest of their lives.

The term *kidney disease* includes any condition that impairs the kidneys' ability to properly regulate the volume and composition of the plasma. As such, the term includes conditions that interfere with glomerular filtration, tubular function, or both. Although a bewildering number of kidney diseases have been identified and characterized, these fall into a relatively small number of general categories. Here, we look at some of the major ones, including *acute nephritic syndrome, nephrotic syndrome, renal failure,* and *specific tubular abnormalities* that interfere with the reabsorption of certain substances.

Acute nephritic syndrome is a condition that often occurs within one to three weeks following infection of the tonsils, skin, or other parts of the body with certain strains of streptococcal bacteria. As a result of such an infection, certain cells of the immune system secrete proteins called antibodies. The antibodies bind to foreign substances in the body, known as antigens, to form complexes.

Though the formation of antigen-antibody complexes is a normal part of the immune response to invading microorganisms, these complexes can trigger acute nephritic syndrome when they are present at high concentrations in the blood because they can become trapped in capillaries of the glomeruli. When this occurs, it can trigger the proliferation of epithelial cells and other cells in the glomeruli and the subsequent infiltration of the glomeruli with white blood cells, resulting in blockage of the slit pores in the glomerular membrane. As a consequence, the filtration of fluid from glomerular capillaries is reduced, or in some cases is stopped altogether. In other nephrons, antigen-antibody complexes trigger immune reactions that damage the glomeruli, making them permeable to proteins and possibly even red blood cells. This causes *proteinuria* and *hematuria,* which are the appearance of proteins and blood, respectively, in the urine. Within a few weeks or months after the condition develops, however, kidney function usually returns to normal.

In *nephrotic syndrome* there is a similar increase in glomerular permeability, resulting in proteinuria. However, in this condition the permeability increase is not caused by streptococcal infection, and the loss of proteins in the urine is usually more severe, exceeding 3.5 grams per day in some cases. Nephrotic syndrome is frequently caused by bacterial infection and subse-

quent *glomerulonephritis* (inflammation of the glomeruli), but there are other possible causes as well. One of the signs of nephrotic syndrome is edema of body tissues, which develops because the loss of protein from the blood causes a reduction in the plasma oncotic pressure, which leads to increased capillary filtration. This edema is exacerbated by the retention of salt and water in the body, which is due to a decrease in the excretion of these substances in the urine. Normally, you would expect the urinary excretion of salt and water to be *higher* than normal under these circumstances due to the increase in glomerular permeability, which should cause an increase in the GFR. However, excretion of salt and water in the urine is actually *lower* than normal for the following reason: The filtration of fluid from capillaries reduces the blood volume, leading to a fall in arterial and venous blood pressures. This reduction in pressure triggers a host of compensatory endocrine responses, including increased secretion of aldosterone and ADH and decreased secretion of atrial natriuretic peptide. As a result, reabsorption of salt and water is stimulated, which causes the inappropriate retention of these substances in the body.

Renal failure is characterized by an abnormally low rate of urine flow (*oliguria*) or the complete cessation of urine flow (*anuria*), depending on the severity of the condition. Renal failure can be acute, developing within a period of days, or chronic, developing over a period of many years. Furthermore, the condition can affect a relatively small percentage of nephrons, or it can affect entire kidneys. A person can function normally with few or no symptoms, with as little as one third of his or her nephrons working properly. In part, this is because the remaining undamaged nephrons can compensate for any loss by increasing the rate at which they filter the blood and the rate at which they reabsorb material from the tubular fluid. The precise mechanism of this compensatory response is unknown. If the number of functioning nephrons falls below 10–20% of normal, however, the condition is inevitably fatal in the absence of treatment. Death occurs within 8–14 days if renal shutdown is complete.

Renal failure can occur as a result of glomerulonephritis, if the condition impairs glomerular filtration, or it can result from conditions that interfere with the flow of blood through the kidneys. One such condition is *nephrosclerosis,* in which small renal blood vessels develop plaques that occlude blood flow. Renal failure can also be caused by *polycystic kidney disease,* a congenital condition in which the kidneys develop large cysts that can impair the function of neighboring nephrons by pressing against them. Another cause of renal failure is *tubular necrosis,* the death of renal tubule epithelial cells, which can cause cells to slough off into the tubular lumen, leading to the block-

age of fluid flow. Tubular necrosis can occur as a result of exposure to poisons such as certain heavy metal ions (for example, mercury or lead) or organic compounds (for example, carbon tetrachloride). Tubular necrosis can also be caused by severe ischemia of kidney tissue, which kills cells by depriving them of oxygen and nutrients. Such ischemia might occur as a result of blockage of blood vessels in the kidneys or as a result of a prolonged reduction in arterial pressure, such as in circulatory shock. When the arterial pressure drops, blood flow to the kidneys is further compromised because the fall in pressure triggers an increase in sympathetic nervous activity, which induces constriction of the afferent and efferent arterioles, thereby raising the kidneys' vascular resistance.

Because urine output is reduced in renal failure, the condition causes the body to retain materials that would ordinarily be excreted, such as excess sodium chloride and water, urea and other nitrogenous wastes, sulfates, phosphates, and potassium. The accumulation of these unwanted materials in the blood eventually triggers a constellation of symptoms that is sometimes referred to as *uremia.* The retention of salt and water raises the blood pressure, leading to increased capillary filtration and edema of body tissues. The rise in plasma potassium can lead to cardiac arrhythmias and possibly death by cardiac arrest. Renal failure also impairs the kidneys' ability to secrete hydrogen

ions and to transport bicarbonate ions into the bloodstream. Because of this, another sign of renal failure is acidosis and a reduction in plasma bicarbonate. It is possibly for this reason that renal failure invariably progresses to coma and death if left untreated. Fortunately, long-term survival of those with renal failure is possible through *dialysis* treatment. In this procedure, blood is withdrawn, circulated through a machine that removes unwanted materials, and then returned to the patient.

Specific tubular abnormalities are generally hereditary and are due to deficiencies in the number or type or certain transport proteins in tubule epithelial cells. Such a deficiency impairs the kidneys' ability to reabsorb a specific substance or group of related substances. For example, in *glycosuria,* the capacity of the kidneys for reabsorbing glucose is reduced, which causes glucose to appear in the urine even when plasma glucose levels are normal. In *aminoaciduria,* the kidneys' ability to reabsorb amino acids is reduced, leading to the appearance of these compounds in the urine. Another specific tubular abnormality is *nephrogenetic diabetes insipidus,* in which the renal tubules lack the normal responsiveness to ADH. This reduces the kidneys' capacity for reabsorbing water, thereby leading to inappropriately high rates of urine flow.

A hydrogen ion *buffer* is an acid-base pair that has the ability to minimize changes in the proton concentration or pH by taking up hydrogen ions (H^+) from a solution when acid is added and by releasing hydrogen ions into the solution when acid is removed. Recall from Chapter 3 that the pH of a solution declines as the proton concentration increases and is defined as the logarithm of the inverse of the proton concentration: $pH = \log(1/[H^+])$. Because a buffer must be able to release and bind hydrogen ions, it functions as both an acid and a base. Substances that can release and take up hydrogen ions are defined as *weak acids* because they have a tendency to give up some, but not all, of their bound hydrogen ions when dissolved in water. In other words, they only *partly dissociate* in solution. In contrast, a *strong acid,* such as hydrochloric acid (HCl), has little tendency to bind hydrogen ions and completely dissociates in solution.

To see how a buffer works, consider what happens when a hypothetical weak acid (HA) is dissolved in water. The acid releases some of its hydrogen ions into solution, such that some molecules go into the base form (A^-). Over a given time period, some of these bases take up other hydrogen ions from solution, and in so doing revert to the acid form. Over the same time period an equal number of acids release their hydrogen ions, thereby going to the base form. Therefore, the concentration of hydrogen ions in solution does not change, and the concentrations of the acid and base forms of the buffer remains constant. Under these conditions, the system is in equilibrium, which is represented as follows:

$$HA \rightleftharpoons H^+ + A^-$$

When acid (hydrogen ions) is added to water, the pH decreases because the concentration of hydrogen ions rises. If a buffer is present, this rise in concentration pushes the equilibrium to the left by mass action, as follows:

$$\text{add } H^+$$
$$\Downarrow$$
$$HA \underset{\longrightarrow}{\longleftarrow} H^+ + A^-$$

The combination of hydrogen ions with the base form of the buffer (A^-) removes some of the added hydrogen ions from solution. Consequently, the pH still falls, but not as much as it would if the buffer were not present. When acid is removed from a solution, the pH rises because the hydrogen ion concentration falls. If a buffer is present, the decrease in hydrogen ion concentration pulls the equilibrium to the right, as follows:

$$\text{remove } H^+$$
$$\Uparrow$$
$$HA \underset{\longleftarrow}{\longrightarrow} H^+ + A^-$$

4. *Excessive vomiting.* Because it results in the loss of hydrogen ions that are secreted in the stomach and are normally reabsorbed in the small intestine, excessive vomiting can produce metabolic alkalosis.

5. *Severe diarrhea.* Because it results in the loss of bicarbonate, which is produced in the upper small intestine and is normally reabsorbed in the lower small intestine, severe diarrhea can produce metabolic acidosis.

6. *Alterations in renal function.* Because the kidneys secrete hydrogen ions and reabsorb bicarbonate (as described shortly), changes in renal function can produce metabolic acidosis or metabolic alkalosis.

Defense Mechanisms Against Acid-Base Disturbances

Any alteration in the rate of acid or base production by the body changes the pH of arterial blood by upsetting the balance between the rate at which hydrogen ions are added to and removed from the blood. Fortunately, three "lines of defense" protect against such changes in pH: (1) *buffering of hydrogen ions,* which occurs through the binding or release of hydrogen ions by substances that are always present in the blood and other body fluid compartments; (2) *respiratory compensation,* which adjusts the rate at which carbon dioxide is cleared from the blood via elimination through the lungs; and (3) *renal compensation,* which adjusts the rate at which hydrogen ions are secreted and bicarbonate ions are reabsorbed by renal tubules. The term *compensation* is used because the respiratory and renal systems do not correct an acid-base disturbance; they only compensate for it by activating mechanisms that work to restore normal arterial pH.

Buffering of Hydrogen Ions

Buffers, the first line of defense against changes in pH, act immediately to compensate for disturbances in pH. A chemical buffer is a compound that minimizes changes in pH when either an acid or base is added or removed from

The release of hydrogen ions from the acid form of the buffer molecule (HA) adds hydrogen ions to the solution, which replaces some of the hydrogen ions removed initially. As a consequence, the pH rises, but not as much as it would have in the absence of the buffer.

From this discussion it is apparent that buffers limit increases or decreases in the pH of a solution when acid is added or removed. When acid is added to or removed from a given volume of pure water, the size of the resulting change in pH depends on the quantity of acid involved. When the same quantity of acid is added or removed in the presence of a buffer, the resulting change in pH is smaller, as seen in the figure at right. Notice that when the buffer is present, the pH is fairly constant over a certain range, but it changes more quickly as the pH becomes either very high or very low. This means that a buffer's ability to protect against changes in pH is restricted to a certain range of pH values (the flat portion of the curve). Outside this range, a buffer is relatively ineffective at limiting changes in pH. Different buffers are most effective over different ranges of pH, depending on how strongly they tend to dissociate.

To be useful in protecting the blood against changes in pH, a buffer's effective range must be within the physiological pH range. The buffer should also be present at sufficient concentration to enable it to consume or release significant quantities of hydrogen ions. Important physiological buffers include bicarbonate ions

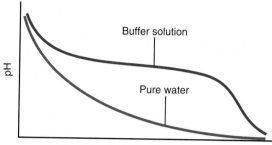

Quantity of acid (hydrogen ions) added

(HCO_3^-) and monoprotonated phosphates (HPO_4^{2-}). These buffers take up and release hydrogen ions as follows:

$$H^+ + HCO_3^- \rightleftharpoons H_2CO_3$$

$$H^+ + HPO_4^{2-} \rightleftharpoons H_2PO_4^-$$

Proteins are also important physiological buffers because they contain amino acids that possess carboxyl groups (—COOH) and amino groups (—NH_2), which can accept or give up hydrogen ions as follows:

$$H^+ + —COO^- \rightleftharpoons —COOH$$

$$H^+ + —NH_2 \rightleftharpoons —NH_3^+$$

---

a solution (Toolbox: Buffers). The most important buffer in the extracellular fluid (which includes plasma) is bicarbonate. The buffering ability of bicarbonate is shown in the following equation:

$$HCO_3^- + H^+ \rightleftharpoons H_2CO_3$$

Important buffers located primarily (but not exclusively) in intracellular fluid include proteins and phosphates, whose buffering abilities are shown in the following equations:

$$Protein^- + H^+ \rightleftharpoons H \cdot Protein$$

$$HPO_4^{2-} + H^+ \rightleftharpoons H_2PO_4^-$$

When acid is added to pure water, the pH decreases because the concentration of dissolved hydrogen ions increases. When the same quantity of acid is added to an equivalent volume of blood, the pH again decreases, but not to the same degree, because blood contains buffers that bind some of the added hydrogen ions.

When acid is removed from pure water, the pH increases because the concentration of dissolved hydrogen ions decreases. When the same quantity of acid is removed from blood, the pH again rises, but to a lesser degree, because buffers in the blood release hydrogen ions that replace some of those that were removed.

The law of mass action determines whether a buffer binds or releases hydrogen ions. Consider the following example, which shows hydrogen ions (H^+) binding reversibly to a buffer (A^-):

$$H^+ + A^- \rightleftharpoons HA$$

When acid is added to a solution, the resulting rise in the hydrogen ion concentration pushes the reaction to the right, so that the ionized form of the buffer (A^-) combines with hydrogen ions to produce the protonated (or acid) form of the buffer (HA):

$$\overset{\text{add}}{\Downarrow}$$
$$H^+ + A^- \underset{\longleftarrow}{\longrightarrow} HA$$

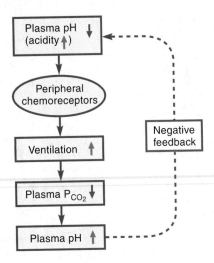

FIGURE 18.24 **The mechanism by which decreases in plasma pH increase ventilation.**

Where are the peripheral chemoreceptors located?

When acid is removed from a solution, the resulting fall in hydrogen ion concentration pulls the equilibrium to the left, so that the protonated form of the buffer releases hydrogen ions:

$$\overset{\text{remove}}{\underset{\displaystyle \Uparrow}{}} \quad H^+ + A^- \xleftarrow{\quad\longrightarrow\quad} HA$$

Of the three lines of defense against alterations in blood pH, buffering is the fastest; its response time is limited only by the time required for buffers to bind or release hydrogen ions. However, buffering can only *limit* changes in pH by binding or releasing hydrogen ions; it cannot *reverse* changes in pH unless more buffer molecules are added to or removed from the blood. Consider, for example, a situation in which a decrease in pH has occurred. Once a buffer molecule has bound a hydrogen ion to remove it from solution and thus minimize the reduction in pH, that molecule cannot bind another hydrogen ion. The excess hydrogen ions bound to buffers must eventually be eliminated from the body, or the buffering capacity of blood will be exceeded. Therefore, whereas buffering is the first mechanism of defense against pH changes, it cannot do the job alone. Once arterial pH has deviated from its normal value, it can be returned to normal only by respiratory or renal compensation.

Respiratory Compensation

The respiratory system, the second line of defense against changes in blood pH, usually acts within minutes. The respiratory system regulates pH by increasing or decreasing alveolar ventilation, which tends to raise or lower pH,

respectively. An increase in alveolar ventilation lowers arterial P_{CO_2}. Given that carbon dioxide is in equilibrium with hydrogen ions and bicarbonate, the law of mass action dictates that hydrogen ions and bicarbonate ions decrease, as follows:

Increased alveolar ventilation
removes CO_2
$$\Uparrow$$
$$CO_2 + H_2O \xleftarrow{\quad\rightarrow\quad} H_2CO_3 \xleftarrow{\quad\rightarrow\quad} H^+ + HCO_3^-$$

Conversely, a decrease in alveolar ventilation raises the arterial P_{CO_2}. By the law of mass action, hydrogen ions and bicarbonate ions increase, as follows:

Decreased alveolar ventilation
adds CO_2
$$\Downarrow$$
$$CO_2 + H_2O \xrightarrow{\quad\longrightarrow\quad} H_2CO_3 \xrightarrow{\quad\longrightarrow\quad} H^+ + HCO_3^-$$

Unlike simple hydrogen ion buffering, which minimizes but cannot reverse changes in pH, respiratory compensation is a true homeostatic regulatory mechanism that can reverse pH changes. Figure 18.24 shows what happens to ventilation when plasma pH decreases. The increase in hydrogen ions in the plasma activates the peripheral chemoreceptors, which reflexively increase ventilation. An increase in ventilation causes a decrease in arterial P_{CO_2}, which by the law of mass action causes the conversion of bicarbonate and hydrogen ions to carbon dioxide, thereby removing free hydrogen ions from solution and increasing the pH.

The respiratory system alone generally cannot completely restore the pH to normal. Usually, the last line of defense, renal compensation, must be called into action as well.

Renal Compensation

The third line of defense against changes in blood pH is the renal system, which takes hours or even days to compensate for changes in pH. The kidneys regulate the pH of arterial blood by regulating the renal excretion of hydrogen ions and bicarbonate, and by producing new bicarbonate, according to the following rules: *If the hydrogen ion concentration in the blood increases, the kidneys increase hydrogen ion secretion and bicarbonate reabsorption and synthesize new bicarbonate; if hydrogen ion concentration in the blood decreases, the kidneys decrease hydrogen ion secretion and bicarbonate reabsorption.* The secretion of hydrogen ions is coupled to the reabsorption or synthesis of bicarbonate ions, as described shortly.

Several substances critical to renal compensation for acid-base disturbances are filtered at the glomerulus, including CO_2, H^+, HCO_3^-, and $H_2PO_4^-$. The fates of these substances vary in the different segments of the renal tubules.

Carotid bodies

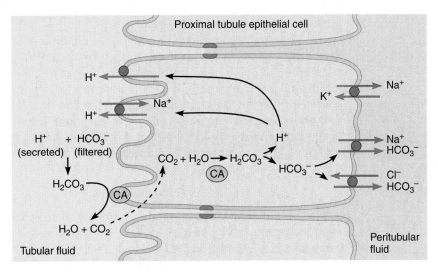

FIGURE 18.25 Bicarbonate reabsorption and hydrogen ion secretion in the proximal tubule. *Filtered bicarbonate ions combine with secreted hydrogen ions to form carbonic acid, which is converted to water and carbon dioxide by carbonic anhydrase on the apical membrane. The carbon dioxide diffuses into the epithelial cell, where intracellular carbonic anhydrase catalyzes the conversion of carbon dioxide and water to carbonic acid, which then dissociates into bicarbonate and hydrogen ions. The hydrogen ions are secreted by countertransport with sodium ions, whereas the bicarbonate ions are reabsorbed by cotransport with sodium ions and by countertransport with chloride ions.*

Renal Handling of Hydrogen and Bicarbonate Ions in the Proximal Tubule In the proximal tubule, bicarbonate reabsorption is coupled to hydrogen ion secretion (Figure 18.25). In epithelial cells of the proximal tubule, several carrier proteins required for movement of hydrogen or bicarbonate ions are located on either the basolateral or apical membrane. Which transporters are active depends on the pH of the extracellular fluid.

The basolateral membrane contains three transporters: (1) Na^+/K^+ pumps that transport sodium ions out of the cell and into the peritubular fluid while transporting potassium ions into the cell, (2) Na^+/HCO_3^- cotransporters that transport both sodium and bicarbonate ions out of the cell and into the peritubular fluid, and (3) HCO_3^-/Cl^- countertransporters that transport chloride ions into the cell and bicarbonate ions into the peritubular fluid.

The apical membrane contains two transporters: (1) Na^+/H^+ countertransporters that transport sodium ions into the cell and hydrogen ions out of the cell and into the tubular fluid, and (2) H^+ pumps that use ATP to transport hydrogen ions into the tubular fluid.

The enzyme carbonic anhydrase (CA), which is located in the cytosol and on the apical membrane (specifically, on the microvilli) of the epithelial cell, catalyzes the following reversible reaction:

$$CO_2 + H_2O \rightleftharpoons H_2CO_3$$

The membrane-bound carbonic anhydrase converts carbonic acid (which comes from filtered bicarbonate ions, as described shortly) to carbon dioxide in the lumen of the proximal tubule. The carbon dioxide then diffuses into the epithelial cell, where it is converted back to carbonic acid by carbonic anhydrase. The carbonic acid then dissociates by the following reversible reaction:

$$H_2CO_3 \rightleftharpoons H^+ + HCO_3^-$$

The hydrogen ion formed inside the epithelial cell by this reaction is secreted into the lumen of the tubules by either countertransport with sodium ions or active transport by the H^+ pumps. The intracellular concentration of sodium is kept low by the Na^+/K^+ pumps on the basolateral membrane. In the lumen of the tubule, hydrogen ions combine with filtered bicarbonate to form carbonic acid. The carbonic anhydrase located on microvilli catalyzes the conversion of carbonic acid to carbon dioxide and water. The carbon dioxide can then diffuse into the epithelial cell, as previously described.

The bicarbonate ion formed inside the epithelial cell by the carbonic anhydrase-catalyzed reaction moves from the epithelial cell into the peritubular fluid by either cotransport with sodium or countertransport with chloride. The Na^+/HCO_3^- cotransporter functions in reabsorption of both sodium and bicarbonate. The net effect for the bicarbonate ion is reabsorption, because bicarbonate is moved from the lumen of the tubules into the peritubular fluid, as follows: A bicarbonate ion in the lumen is converted to a carbon dioxide molecule that moves from the lumen into the epithelial cell, where it is converted back into a bicarbonate ion that moves into the peritubular fluid, where it can diffuse into the blood.

Overall, these actions in the proximal tubule produce three primary effects: (1) Under normal conditions, approximately 80–90% of the filtered bicarbonate is reabsorbed, (2) hydrogen ions are secreted, and (3) sodium is reabsorbed.

Renal Handling of Hydrogen and Bicarbonate Ions in the Late Distal Tubule and Collecting Duct In certain epithelial cells of the late distal tubule and collecting duct, called intercalated cells, secretion of hydrogen ions is coupled to the synthesis of new bicarbonate ions (Figure 18.26). The intercalated cells lining the distal tubules and collecting ducts have different membrane proteins than the epithelial cells lining the proximal tubules.

The basolateral membrane contains (1) HCO_3^-/Cl^- countertransporters that move bicarbonate out of the cell and into the peritubular fluid while moving chloride ions

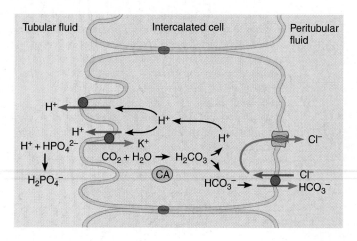

FIGURE 18.26 Bicarbonate synthesis and hydrogen ion secretion by intercalated cells of the distal tubule and collecting duct. *Carbonic anhydrase in the cytosol converts carbon dioxide (which is either metabolically produced or has diffused into the cell from the peritubular fluid) into carbonic acid, which dissociates into hydrogen ions and bicarbonate ions. The hydrogen ions are secreted by a H^+ pump or a K^+/H^+ countertransporter; the bicarbonate ions are transported into the peritubular fluid by a HCO_3^-/Cl^- countertransporter.*

into the cell, and (2) chloride channels that allow the chloride to diffuse back into the peritubular fluid. The apical membrane contains (1) H^+ pumps that utilize ATP to transport hydrogen ions out of the cell and into the tubular fluid, and (2) K^+/H^+ countertransporters that transport potassium ions into the cell and hydrogen ions into the tubular fluid.

The enzyme carbonic anhydrase is located in the cytosol of the intercalated cell. Carbon dioxide levels in the epithelial cell increase from either cellular metabolism or diffusion from the peritubular fluid into the cell. Inside the cell, carbonic anhydrase converts carbon dioxide and water to carbonic acid, which then dissociates into H^+ and HCO_3^-, as follows:

$$CO_2 + H_2O \rightleftharpoons H_2CO_3 \rightleftharpoons H^+ + HCO_3^-$$

The reaction is driven to the right by the removal of hydrogen ions and bicarbonate ions from the epithelial cell. Hydrogen ions are removed through transport by the H^+ pump or in exchange for K^+ by the countertransporter, resulting in secretion of hydrogen ions. Bicarbonate ions are removed by countertransport with chloride ions across the basolateral membrane into the peritubular fluid, and then into the plasma. However, this bicarbonate is not being reabsorbed; because it was never in the lumen of the renal tubules but instead was produced by the epithelial cell, this bicarbonate is considered *new* bicarbonate.

The secreted hydrogen ions decrease the pH of the tubular fluid. However, the pH of the tubular fluid, and therefore urine, is limited to a minimum of 4.5, at which

point hydrogen ion secretion stops. To minimize decreases in urine pH, secreted hydrogen ions are buffered. Recall that hydrogen ions secreted in the proximal tubule were buffered by filtered bicarbonate ions. In the distal tubule and collecting ducts, however, very little bicarbonate remains in the lumen of the tubules. Recall as well that phosphate ions are freely filtered by the glomerulus. In the lumen of the distal tubules and collecting ducts, hydrogen ions are buffered by phosphates according to the following equation:

$$HPO_4^{2-} + H^+ \rightleftharpoons H_2PO_4^-$$

Overall, these actions in the distal tubules and collecting ducts produce two primary effects: Newly formed bicarbonate ions are added to the plasma, and hydrogen ions are secreted into the tubular fluid.

Role of Glutamine in Renal Compensation During Severe Acidosis The mechanisms just described generally compensate for increases in hydrogen ion concentration produced by normal daily activities. However, these mechanisms are insufficient to compensate for large increases in the plasma hydrogen ion concentration. Under conditions of severe acidosis, a third renal mechanism contributes to compensation (Figure 18.27).

In the proximal convoluted tubule, glutamine is transported from both the tubular fluid and the peritubular fluid into the epithelial cells. Catabolism of glutamine in the epithelial cells generates bicarbonate ions and ammonia (NH_3), as follows:

$$Glutamine \rightarrow HCO_3^- + NH_3$$

The bicarbonate moves into the peritubular fluid by either cotransport with sodium or countertransport with a chloride ion. This bicarbonate is not being reabsorbed, however; because it was never in the tubular fluid, *new* bicarbonate is being added to the blood. The ammonia is converted to ammonium (NH_4^+) by the following reaction:

$$NH_3 + H^+ \rightarrow NH_4^+$$

This ammonium is transported into the tubular fluid by countertransport with sodium ions and is eventually excreted.

The overall effect of these actions is that a new bicarbonate ion is added to the blood, and a hydrogen ion is secreted in the form of ammonium.

Compensation for Acid-Base Disturbances

Recall from Chapter 16 that the Henderson-Hasselbalch equation describes the relationship of plasma pH to the ratio of bicarbonate and carbon dioxide levels in blood:

$$pH = 6.1 + \log [HCO_3^-]/[CO_2]$$

Given that blood pH must be maintained at 7.4, this equation can be solved for the ratio of bicarbonate to carbon dioxide, as follows:

$$7.4 = 6.1 + \log [HCO_3^-]/[CO_2]$$

$$1.3 = \log [HCO_3^-]/[CO_2]$$

$$[HCO_3^-]/[CO_2] = 20$$

Therefore, for plasma pH to be normal, the ratio of bicarbonate to carbon dioxide must be 20:1. The respiratory and renal systems work together to control this ratio; the respiratory system controls carbon dioxide levels, and the kidneys regulate bicarbonate levels. An increase in bicarbonate levels or a decrease in carbon dioxide levels increases the ratio; a decrease in bicarbonate levels or an increase in carbon dioxide levels decreases the ratio.

In acidosis, the ratio of bicarbonate to carbon dioxide decreases to less than 20:1, either because of a decrease in bicarbonate or an increase in carbon dioxide. In alkalosis, the ratio of bicarbonate to carbon dioxide is greater than 20:1, either because of an increase in bicarbonate or a decrease in carbon dioxide.

We turn now to a description of the four types of acid-base disturbances, and how the body compensates for them.

Respiratory Acidosis

Respiratory acidosis is caused by hypoventilation, in which ventilation is less than that needed by the body. Carbon dioxide increases in the plasma, decreasing the pH:

$$\overset{\overset{\text{add}}{\Downarrow}}{CO_2} + H_2O \underset{\leftarrow}{\longrightarrow} \uparrow H_2CO_3 \underset{\leftarrow}{\longrightarrow} \uparrow H^+ + \uparrow HCO_3^-$$

Hypoventilation can be caused by lung diseases, depression of the respiratory center in the brainstem, or diseases that affect respiratory muscles. As a result of hypoventilation, arterial P_{CO_2} increases, decreasing the ratio of bicarbonate to carbon dioxide. To bring the ratio (and therefore pH) back to normal, the kidneys compensate by increasing the secretion of hydrogen ions and the reabsorption of bicarbonate ions. The lungs cannot compensate because that is where the problem developed initially (unless the hypoventilation was voluntary and not pathological, in which case the lungs can compensate by returning ventilation to normal).

Respiratory Alkalosis

Respiratory alkalosis is caused by hyperventilation, in which ventilation is greater than that needed by the body. Causes of hyperventilaton include fever and anxiety. As a result of the hyperventilation, arterial P_{CO_2} de-

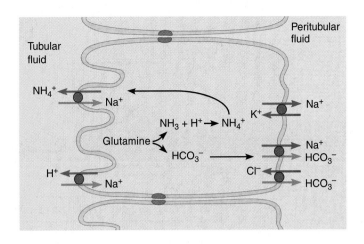

FIGURE 18.27 Bicarbonate production and hydrogen secretion by glutamine metabolism in the proximal tubule. *Glutamine is catabolized to bicarbonate ions and ammonia. The bicarbonate ions are transported into the peritubular fluid by a Na^+/HCO_3^- cotransporter. The ammonia binds a hydrogen ion to form ammonium, which is secreted by countertransport with sodium ions.*

creases, increasing the ratio of bicarbonate to carbon dioxide:

$$\overset{\overset{\text{remove}}{\Uparrow}}{CO_2} + H_2O \underset{\leftarrow}{\longrightarrow} \downarrow H_2CO_3 \underset{\leftarrow}{\longrightarrow} \downarrow H^+ + \downarrow HCO_3^-$$

To bring the ratio back to normal, the kidneys compensate by decreasing the reabsorption of bicarbonate ions and by secreting fewer hydrogen ions. The lungs cannot compensate because that is where the problem developed initially (unless the hyperventilation was voluntary).

Metabolic Acidosis

Metabolic acidosis is caused by an increase in acids in the plasma from sources other than carbon dioxide:

$$CO_2 + H_2O \underset{\longleftarrow}{\rightarrow} H_2CO_3 \underset{\longleftarrow}{\rightarrow} \overset{\overset{\text{add}}{\Downarrow}}{\uparrow H^+} + \downarrow HCO_3^-$$

Causes of metabolic acidosis include diarrhea, which results in loss of bicarbonate through elimination of the intestinal contents; diabetes mellitus, which by increasing fat metabolism causes a buildup of keto acids; strenuous exercise, which increases lactic acid production; and renal failure. Compensation for metabolic acidosis includes an increase in ventilation and, if the kidneys are not part of the initial problem, an increase in production of new bicarbonate by the kidneys and an increase in excretion of H^+.

In respiratory compensation, an increase in hydrogen ion levels in the plasma activates peripheral chemorecep-

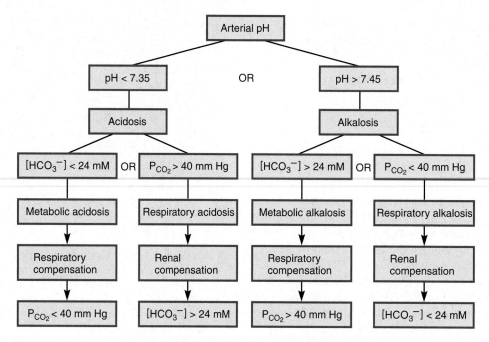

FIGURE 18.28 Summary of acid-base disturbances and compensation.

tors, which reflexively increase ventilation. The increase in ventilation decreases arterial P_{CO_2}, which drives the conversion of bicarbonate and hydrogen ions to more carbon dioxide, decreasing the free hydrogen ion levels in blood. However, the lungs cannot completely compensate because the increased ventilation decreases arterial P_{CO_2}, which decreases the stimulatory influence of carbon dioxide on central chemoreceptors.

In renal compensation, the production of new bicarbonate is essential to replace bicarbonate that was lost via two processes: (1) when bicarbonate was used to buffer excess acid, and (2) when plasma bicarbonate levels were decreased during respiratory compensation as described above.

Metabolic Alkalosis

Metabolic alkalosis is caused by a decrease in acids in the plasma from sources other than carbon dioxide:

$$CO_2 + H_2O \xrightleftharpoons{} H_2CO_3 \xrightleftharpoons[]{\text{remove} \Uparrow} \downarrow H^+ + \uparrow HCO_3^-$$

Causes of metabolic alkalosis include vomiting, which results in a loss of acidic gastric contents, and ingestion of alkaline drugs such as sodium bicarbonate (baking soda) or antacids. Compensation for metabolic alkalosis involves both the lungs and kidneys.

In respiratory compensation, decreases in hydrogen ion levels in the plasma remove a stimulatory effect on peripheral chemoreceptors, which reflexively decreases ventilation. The decrease in ventilation increases arterial P_{CO_2}, which combines with water to produce bicarbonate

and hydrogen ions, increasing free hydrogen ion levels in blood. As in metabolic acidosis, the lungs cannot completely compensate because the decreased ventilation increases arterial P_{CO_2}, which activates the central chemoreceptors and increases ventilation.

In renal compensation, the kidneys excrete more bicarbonate ions and fewer hydrogen ions. Ridding the blood of bicarbonate shifts the equilibrium to the right, causing more carbon dioxide to react with water to form carbonic acid, which then dissociates and increases the plasma concentration of hydrogen ions.

Evaluation of Acid-Base Disturbances

Diagnosis of acid-base disturbances involves measuring plasma levels of hydrogen ion concentration (pH), carbon dioxide (P_{CO_2}), and bicarbonate. The different acid-base disturbances can be diagnosed as follows (Figure 18.28):

- A decrease in pH coupled with a decrease in plasma bicarbonate levels indicates that a metabolic acidosis is occurring. Because the lungs compensate for a metabolic acidosis by increasing ventilation, P_{CO_2} levels decrease.

- A decrease in pH coupled with an increase in P_{CO_2} indicates that a respiratory acidosis is occurring. Because the kidneys compensate for a respiratory acidosis by increasing bicarbonate reabsorption, plasma bicarbonate levels increase.

- An increase in pH coupled with an increase in plasma bicarbonate levels indicates that a metabolic alkalosis is occurring. P_{CO_2} levels increase as part of respiratory compensation.

- An increase in pH coupled with a decrease in P_{CO_2} indicates that a respiratory alkalosis is occurring. Plasma bicarbonate levels decrease as part of renal compensation.

Quick Test 18.6

1. Describe the three lines of defense against changes in acid-base balance. Which most rapidly corrects changes in blood pH? Which takes the longest to compensate?

2. Explain how hydrogen ion secretion is coupled to bicarbonate reabsorption in the proximal tubule, and how it is coupled to new bicarbonate synthesis in the distal tubule and collecting duct.

3. Define the following terms: *metabolic acidosis, respiratory acidosis, metabolic alkalosis,* and *respiratory alkalosis.*

4. What is the normal plasma ratio of bicarbonate ions to carbon dioxide? Why?

SUMMARY OF FACTORS INFLUENCING PLASMA VOLUME AND SOLUTE CONCENTRATION

Figure 18.29 summarizes the interrelated actions of the urinary system and includes material covered in Chapters 17 and 18. Although various sources supply water and solutes to the plasma, the kidneys ultimately regulate plasma volume and plasma solute concentration. They regulate the plasma composition by filtering the blood, removing unneeded solutes and water and excreting them in the urine while retaining necessary water and solutes.

The water content of plasma is regulated by the endocrine system. The primary hormone controlling the amount of water excreted by the kidneys is antidiuretic hormone, which is released from the posterior pituitary in response to an increase in plasma osmolarity and stimulates water reabsorption. Water reabsorption is also affected by two other hormones, aldosterone and atrial natriuretic peptide. Aldosterone, which is released from the adrenal cortex in response to the renin-angiotensin-aldosterone system and high levels of potassium in the plasma, stimulates sodium reabsorption and potassium secretion. An increase in sodium reabsorption increases water reabsorption by osmosis. Atrial natriuretic peptide decreases water reabsorption indirectly by inhibiting sodium reabsorption.

The solute content of plasma is also regulated by the endocrine system. As previously mentioned, aldosterone stimulates sodium reabsorption, potassium secretion, and atrial natriuretic peptide decreases sodium reabsorption.

The rate of exchange between renal tubules and blood affects both the water and solute concentration of plasma. For a substance to be excreted in the urine, it must be filtered or secreted from blood into the renal tubules. Some of the filtered substance is reabsorbed into the blood to decrease the amount excreted.

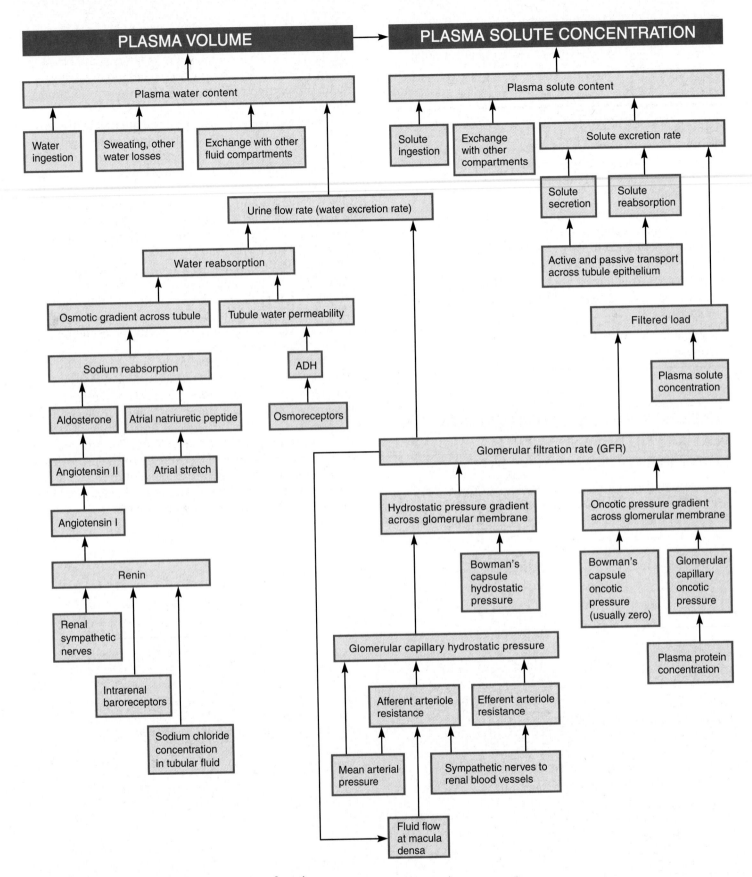

FIGURE 18.29 Summary of urinary system function.

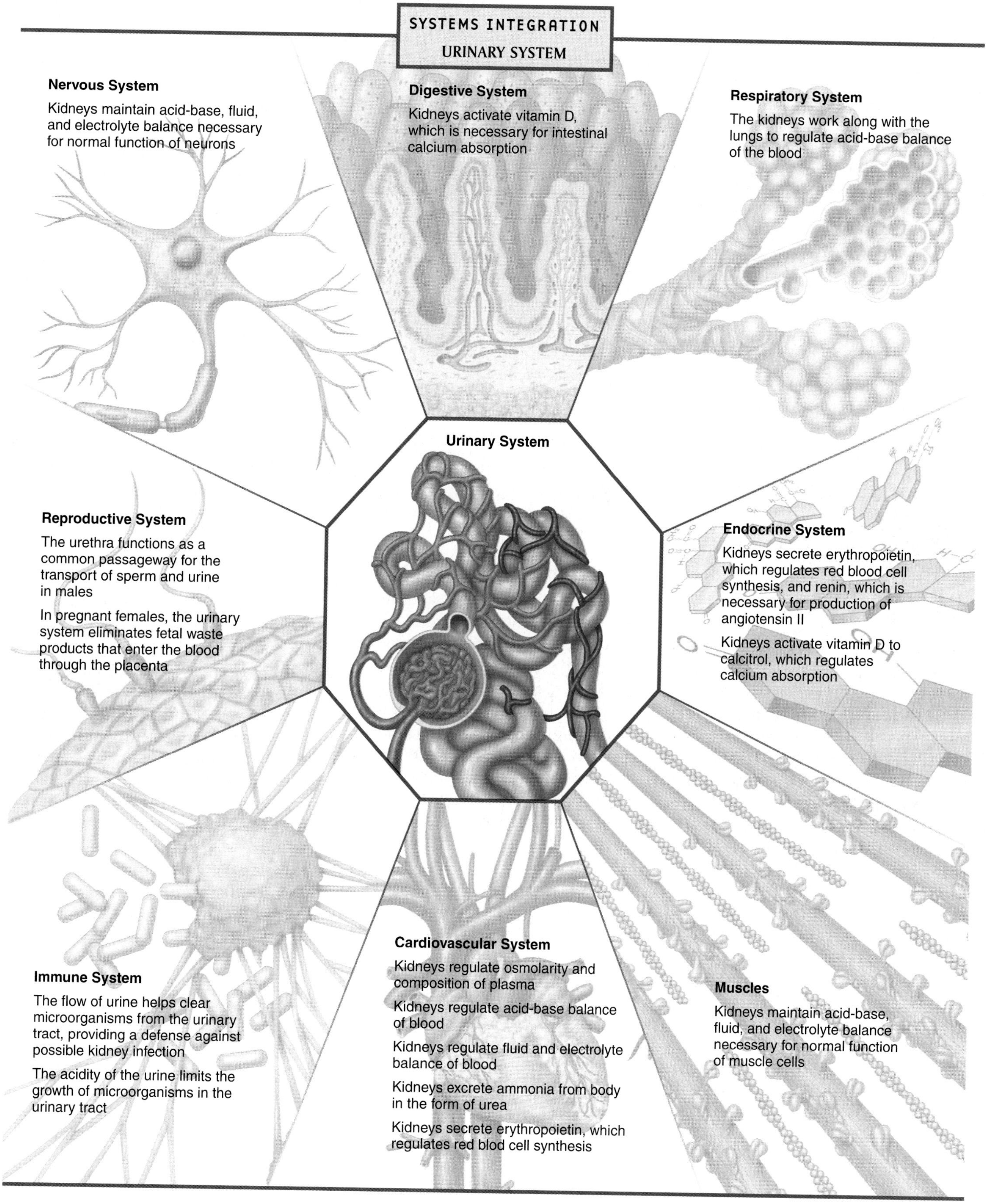
SYSTEMS INTEGRATION
URINARY SYSTEM
Nervous System
Kidneys maintain acid-base, fluid, and electrolyte balance necessary for normal function of neurons
Digestive System
Kidneys activate vitamin D, which is necessary for intestinal calcium absorption
Respiratory System
The kidneys work along with the lungs to regulate acid-base balance of the blood
Urinary System
Reproductive System
The urethra functions as a common passageway for the transport of sperm and urine in males
In pregnant females, the urinary system eliminates fetal waste products that enter the blood through the placenta
Endocrine System
Kidneys secrete erythropoietin, which regulates red blood cell synthesis, and renin, which is necessary for production of angiotensin II
Kidneys activate vitamin D to calcitrol, which regulates calcium absorption
Immune System
The flow of urine helps clear microorganisms from the urinary tract, providing a defense against possible kidney infection
The acidity of the urine limits the growth of microorganisms in the urinary tract
Cardiovascular System
Kidneys regulate osmolarity and composition of plasma
Kidneys regulate acid-base balance of blood
Kidneys regulate fluid and electrolyte balance of blood
Kidneys excrete ammonia from body in the form of urea
Kidneys secrete erythropoietin, which regulates red blod cell synthesis
Muscles
Kidneys maintain acid-base, fluid, and electrolyte balance necessary for normal function of muscle cells

The Concept of Balance, p. 565

To be in balance, the sum of the input and production of a substance must equal the sum of the output and utilization of that substance. The plasma can gain or lose materials by exchange with cells or extracellular connective tissue. The plasma can also gain or lose materials as a result of exchange between it and the external environment. When solutes and water enter and exit the plasma at the same rate, the plasma is in balance. When a substance enters the body faster than it exits, a state of positive balance is said to exist. When a substance leaves the body faster than it enters, a state of negative balance exists.

> **IP** Fluid, Electrolyte, and Acid/Base Balance, Introduction to Body Fluids, pages 1–3, 9–15

Water Balance, p. 567

For water to be in balance, the input from consuming food and fluids and through cellular metabolism must equal the output in urine, feces, and insensible losses. The control of water excretion by the kidneys regulates plasma volume and osmolarity. In the renal tubules, water reabsorption occurs via osmosis that is coupled to the active reabsorption of solutes. The medullary osmotic gradient creates a force for water reabsorption via osmosis as tubular fluid moves through the distal tubule and collecting duct. The osmolarity in this gradient varies from 300 mOsm at the edge of the medulla to approximately 1400 mOsm deep within the medulla; this gradient exists because of the countercurrent multiplier.

Most filtered water is reabsorbed in the proximal tubule. How much of the remaining 30% can be reabsorbed in the late distal tubule and collecting ducts depends on the plasma levels of ADH. The tubular fluid in the late distal tubule and collecting ducts is hypo-osmotic to the interstitial fluid, creating an osmotic force for water reabsorption. ADH increases the permeability of these tubules to water, thereby allowing water reabsorption to occur. ADH is released from the posterior pituitary in response to either increases in the osmolarity of the extracellular fluid or decreases in blood pressure and blood volume.

> **IP** Fluid, Electrolyte, and Acid/Base Balance, Introduction to Body Fluids, pages 4–9, 19–22

> **IP** Fluid, Electrolyte, and Acid/Base Balance, Water Homeostasis, pages 1–10, 12, 14–19*

> **IP** Fluid, Electrolyte, and Acid/Base Balance, Electrolyte Homeostasis, page 8

Sodium Balance, p. 575

Sodium regulation is critical to maintain normal extracellular fluid osmolarity and normal activity of excitable tissues. The hormones aldosterone and atrial natriuretic peptide regulate sodium reabsorption. Aldosterone release is controlled by plasma potassium levels and the renin-angiotensin-aldosterone system. Renin release is stimulated by an increase in sympathetic nerve activity, a decrease in afferent arteriole pressure, or a decrease in sodium and chloride levels in the distal tubule. Renin converts angiotensinogen to angiotensin I, which is converted by angiotensin converting enzyme to angiotensin II, which stimulates aldosterone release from the adrenal cortex. Aldosterone increases sodium reabsorption and potassium secretion. Atrial natriuretic peptide is secreted by cells in the atria in response to distension of the atrial wall caused by an increase in plasma volume. ANP decreases the glomerular filtration rate and sodium reabsorption, which increases sodium excretion.

> **IP** Fluid, Electrolyte, and Acid/Base Balance, Electrolyte Homeostasis, pages 1–6, pages 17–24

Potassium Balance, p. 579

Potassium balance is critical to the normal function of excitable cells. Potassium undergoes both reabsorption and secretion in the renal tubules. Although the net effect of potassium movement across renal tubules is reabsorption, it is the secretion of potassium that is regulated. Potassium secretion is increased by aldosterone. High plasma potassium levels stimulate aldosterone release.

> **IP** Fluid, Electrolyte, and Acid/Base Balance, Electrolyte Homeostasis, pages 25–32

Calcium Balance, p. 580

Calcium, which is critical to the function of most cells, can be added to the plasma from the bone and digestive tract, and removed from the plasma by the bone and kidneys. PTH stimulates resorption of bone, calcium absorption in the digestive tract, calcium reabsorption in the kidneys, and activation of calcitrol in the kidneys. Calcitrol stimulates calcium absorption in the digestive tract and reabsorption in the kidneys. Calcitonin decreases plasma calcium levels by increasing bone formation and decreasing the reabsorption of calcium by the kidneys.

> **IP** Fluid, Electrolyte, and Acid/Base Balance, Electrolyte Homeostasis, pages 33–37

Interactions Between Fluid and Electrolyte Regulation, p. 581

Considerable overlap exists in the regulation of fluid and electrolytes in that a single hormone often affects both water and electrolyte excretion by the kidneys. In addition, movement of solute affects the forces acting on water molecules, and movement of water affects the forces acting on solute molecules. Hemorrhage provides an example of how systems interact to maintain homeostasis. In this example, interactions between fluid and electrolyte regulatory systems restore blood pressure back to normal levels.

Acid-Base Balance, p. 583

Arterial pH is highly regulated to maintain the normal range of 7.35 to 7.45. A decrease in pH to below 7.35 is called acidosis, whereas an increase in pH to above 7.45 is called alkalosis. The respiratory system contributes to acid-base balance by regulating carbon dioxide levels in the blood. Carbon dioxide can be converted to carbonic acid by the enzyme carbonic anhydrase. Respiratory acidosis is caused by an increase in P_{CO_2}, whereas respiratory alkalosis is caused by a decrease in P_{CO_2}. Metabolic acidosis and metabolic alkalosis are disturbances in blood pH caused by something other than an abnormally high or low P_{CO_2}, respectively.

Three "lines of defense" protect against changes in blood pH: (1) buffering of hydrogen ions, (2) respiratory compensation, and (3) renal compensation. Buffering acts immediately because chemical buffers are always present in the blood. However, the blood has a limited buffering capacity, and when excess hydrogen ions are added to the plasma, buffered hydrogen ions must eventually be eliminated by the body. The respiratory system acts within minutes to eliminate hydrogen ions in the form of carbon dioxide. The renal system takes hours to days to synthesize new bicarbonate and to eliminate hydrogen ions.

IP Fluid, Electrolyte, and Acid/Base Balance, Acid/Base Homeostasis, pages 1–60

---

*This topic is available on the *InterActive Physiology*® *Sampler CD* that comes with the purchase of a new copy of this book.

EXERCISES

Multiple-Choice Questions

1. Which of the following would be expected to trigger a decrease in the secretion of renin?
   a) a fall in mean arterial pressure
   b) a fall in pressure inside the afferent arteriole
   c) a decrease in the activity of renal sympathetic nerves
   d) a decrease in the concentration of sodium chloride in tubular fluid

2. Which of the following would be expected to trigger a decrease in the secretion of ADH?
   a) ingestion of a large quantity of pure water
   b) sweating
   c) hemorrhage
   d) ingestion of a large quantity of salty food

3. Assume that fluid enters the distal tubule with an osmolarity of 100 mOsm, and that the maximum osmolarity of medullary interstitial fluid is 1100 mOsm. As plasma ADH levels rise, the osmolarity of the urine
   a) approaches 100 mOsm as a lower limit.
   b) approaches 1100 mOsm as an upper limit.
   c) eventually exceeds 1100 mOsm.
   d) approaches 300 mOsm, the normal osmolarity of plasma.

4. Which of the following tends to be accompanied by an increase in the rate at which bicarbonate is excreted in the urine? (Do not include compensatory changes that might be triggered as a result.)
   a) a decrease in hydrogen ion secretion
   b) a decrease in the plasma bicarbonate concentration
   c) an increase in the production of ammonia by renal tubule epithelial cells
   d) an increase in the urinary excretion of nonvolatile acids

5. Assuming that arterial P_{CO_2} is normal, metabolic acidosis promotes which of the following?
   a) decreased hydrogen ion secretion by the renal tubule
   b) decreased alveolar ventilation
   c) decreased transport of bicarbonate from renal tubule cells to the bloodstream
   d) increased bicarbonate reabsorption

6. Which of the following tends to promote an increase in sodium excretion?
   a) an increase in the glomerular filtration rate
   b) an increase in plasma renin concentration
   c) an increase in sodium reabsorption
   d) a decrease in the secretion of atrial natriuretic peptide

7. The osmolarity of tubular fluid increases as it flows through the descending limb of the loop of Henle because
   a) solutes are passively transported into the descending limb.
   b) solutes are actively transported into the descending limb.
   c) water moves passively into the descending limb.
   d) water moves passively out of the descending limb.

8. Because sweat is essentially a salt solution with an osmolarity lower than that of plasma, severe sweating leads to a reduction in plasma volume and an increase in plasma osmolarity.

How do these changes affect the secretion of ADH?

a) The decrease in plasma volume inhibits ADH secretion, but the increase in osmolarity stimulates it.

b) The decrease in plasma volume stimulates ADH secretion, but the increase in osmolarity inhibits it.

c) Both the decrease in plasma volume and the increase in osmolarity stimulate ADH secretion.

d) Both the decrease in plasma volume and the increase in osmolarity inhibit ADH secretion.

9. In the cytosol of intercalated cells, the carbon dioxide that is converted to carbonic acid can come from

a) the lumen of the distal tubule and the collecting duct.

b) metabolism inside the intercalated cell.

c) catabolism of glutamine.

d) all of the above

10. In the lumen of the proximal tubule, secreted hydrogen ions are primarily buffered by

a) bicarbonate.

b) phosphates.

c) proteins.

d) sulfates.

11. An increase in mean arterial pressure stimulates which of the following?

a) ADH release

b) angiotensin II production

c) aldosterone release

d) increased water excretion in urine

12. Which of the following does *not* stimulate aldosterone release?

a) atrial natriuretic peptide

b) an increase in plasma potassium

c) an increase in renin secretion

d) an increase in angiotensin II production

13. The Na^+/K^+ pump is located

a) on the basolateral membrane of proximal tubule cells, and on the apical membrane of principal cells.

b) on the apical membrane of proximal tubule cells, and on the basolateral membrane of principal cells.

c) on the basolateral membrane of both proximal tubule cells and principal cells.

d) on the apical membrane of both proximal tubule cells and principal cells.

14. In epithelial cells lining the proximal tubules, carbonic anhydrase is located

a) inside the cell and on the apical membrane.

b) inside the cell and on the basolateral membrane.

c) inside the cell only.

d) on the basolateral membrane only.

15. A person has the following symptoms: arterial pH = 7.48, P_{CO_2} = 44 mm Hg, plasma bicarbonate concentration = 27 mM. What is the diagnosis?

a) respiratory acidosis

b) respiratory alkalosis

c) metabolic acidosis

d) metabolic alkalosis

Objective Questions

1. An increase in the reabsorption of solutes (increases/decreases) water reabsorption.

2. Most solute and water are reabsorbed in the (proximal tubule/distal tubule and collecting duct).

3. Epithelial cells of the descending limb of the loop of Henle actively transport solutes from tubular fluid to the surrounding peritubular space. (true/false)

4. Urine flow rate increases as the plasma ADH level (increases/decreases).

5. Stretching of the atria of the heart promotes the secretion of _____ _____ _____ , a hormone that promotes sodium excretion.

6. If the plasma volume is below normal, an increase in water reabsorption will be sufficient to restore it to normal. (true/false)

7. _____ stimulates the insertion of Na^+/K^+ pumps into the plasma membrane of principal cells of the distal tubules and collecting ducts.

8. _____ stimulates the insertion of water pores into the plasma membrane of epithelial cells of the distal tubules and collecting ducts.

9. ADH increases water permeability of the loop of Henle. (true/false)

10. Potassium secretion is (stimulated/inhibited) by aldosterone.

11. Calcitonin (increases/decreases) plasma calcium levels.

12. Resorption of bone (increases/decreases) plasma calcium levels.

13. By adding new bicarbonate to the blood, the kidneys can bring about a compensatory (increase/decrease) in the plasma pH.

14. The kidneys can excrete urine that is pure water. (true/false)

15. There is no limit to the amount of hydrogen ions the kidneys can excrete. (true/false)

Essay Questions

1. Describe the cellular effects of ADH on principal cells in the distal tubule and collecting duct.

2. Describe how each of the following hormones affects water reabsorption, either directly or indirectly: ADH, angiotensin II, aldosterone, and ANP.

3. Describe the effects of aldosterone on renal handling of sodium and potassium.

4. What are the stimuli for renin release? Describe the pathway by which renin leads to aldosterone release.

5. Explain how the following hormones affect blood pressure: ADH, angiotensin II, aldosterone, and ANP.

6. Describe the changes in arterial P_{CO_2} and bicarbonate levels that occur during metabolic acidosis. Which change is compensatory?

Find the answers to these exercises, and additional study tools, at the Physiology Place (www.physiologyplace.com).

19

The Digestive System

OBJECTIVES

- Identify the major organs of the digestive system, and describe the functions of each.

- Identify the various tissue layers that make up the wall of the digestive tract.

- Describe the fundamental mechanisms involved in the absorption of carbohydrate, protein, and lipid digestion products, and explain how the mechanism of lipid absorption is related to the hydrophobic nature of fats.

- In general terms, describe the role of short reflex pathways, long reflex pathways, and gastrointestinal hormones in the control of digestive function.

- Describe the functions of saliva, stomach acid, pancreatic juice, and bile, and explain how the secretion of each of these substances is regulated.

- Define peristalsis, segmentation, migrating motility complex, haustration, mass movement, and basic electrical rhythm, and describe the role of each in digestion.

CHAPTER OUTLINE

Overview of Digestive System Function   602

Functional Anatomy of the Digestive System   602

Digestion and Absorption of Nutrients and Water   614

General Principles of Gastrointestinal Regulation   620

Gastrointestinal Secretion and its Regulation   624

Gastrointestinal Motility and its Regulation   628

Above: Electromicrograph of small intestine villus

The human body demands a continuous supply of organic nutrients to provide it with energy and the substrates of metabolism. Although the body can store some nutrients and synthesize others (glucose, for instance), all molecules used by the body ultimately come from the food we eat. In this chapter we turn our attention to the digestive system, which exists essentially for one purpose: to extract needed materials from the food we eat and the fluids we drink and to transport them into the bloodstream so that they can be distributed to cells throughout the body. These materials include not only organic nutrients, but also water and electrolytes, which are necessary for maintaining the normal volume and composition of extracellular fluid. We begin with an overview of the basic functions of the digestive system.

OVERVIEW OF DIGESTIVE SYSTEM FUNCTION

Most of the nutrient molecules in food are unusable in their original forms; most are too large to be transported into the bloodstream. For this reason, these molecules must be chemically broken down to smaller molecules by enzymes in the lumen of the digestive tract, a process called **digestion.** This process is aided by the mechanical breakdown of food as well—the physical reduction of food into smaller particles, which both enables it to move more easily through the tract and renders it more susceptible to the action of digestive enzymes. Once the larger nutrient molecules have been reduced to smaller digestive end-products, these molecules are transported into the bloodstream via a process called absorption. To aid in digestion and absorption, fluids containing enzymes and other substances are transported into the lumen of the tract via a process called secretion. As these processes are occurring, the contents of the lumen are mixed and slowly propelled from one end of the tract to the other by the contractile activity of muscle located in the wall of the tract itself, which gives digestive organs the ability to move, called **motility.** The four basic

processes of digestion, absorption, secretion, and motility are summarized in Figure 19.1

In the next section we explore the digestive system in more detail by seeing how its anatomy relates to its function. In subsequent sections we focus on the four basic digestive processes and how digestive function is regulated.

FUNCTIONAL ANATOMY OF THE DIGESTIVE SYSTEM

The **digestive system** (gastrointestinal system) comprises two major divisions: the **digestive tract** (also known as the *gastrointestinal tract* or *GI tract*), several organs joined in series to form a passageway through which food and digestion products are conducted, and **accessory glands,** a number of glands located outside the GI tract that secrete various fluids and enzymes into the lumen of the tract to aid the digestive process.

The Digestive Tract

The digestive tract is essentially a hollow tube approximately 4.5 meters (15 feet) long that runs through the body and opens to the outside at either end. The tract begins at the *mouth,* where food enters, and ends at the *anus,* where unabsorbed material exits. These and the other organs of the GI tract—the *pharynx, esophagus, stomach, small intestine, colon,* and *rectum*—are shown in Figure 19.2 and discussed later in the chapter. Although each organ has its own distinguishing structure and function, the wall of the GI tract has a relatively uniform structure throughout most of its length, which we examine next.

Generalized Structure of the Gastrointestinal Wall

Figure 19.3 shows structural features of the gastrointestinal wall, and its four major layers; (1) the **mucosa,** which lines the lumen of the digestive tract; (2) the **submucosa,** an underlying layer of connective tissue; (3) the **muscularis externa,** a layer made up primarily of smooth muscle fibers; and (4) the **serosa** (or *adventitia*), an outer layer composed mostly of connective tissue.

The Mucosa   The mucosa is composed of three layers: (1) an innermost layer of cells called the *mucous membrane,* (2) a middle layer called the *lamina propria,* and (3) an outer layer of smooth muscle called the *muscularis mucosae.*

The **mucous membrane** is a layer of epithelial cells of various types (collectively referred to as **enterocytes**) that lines the inside of the GI tract, forming a continuous barrier separating the lumen from the body's internal environment. Some enterocytes are classified as **absorptive cells** because they are specialized for *absorption,* the transport of nutrients and other materials from the lumen

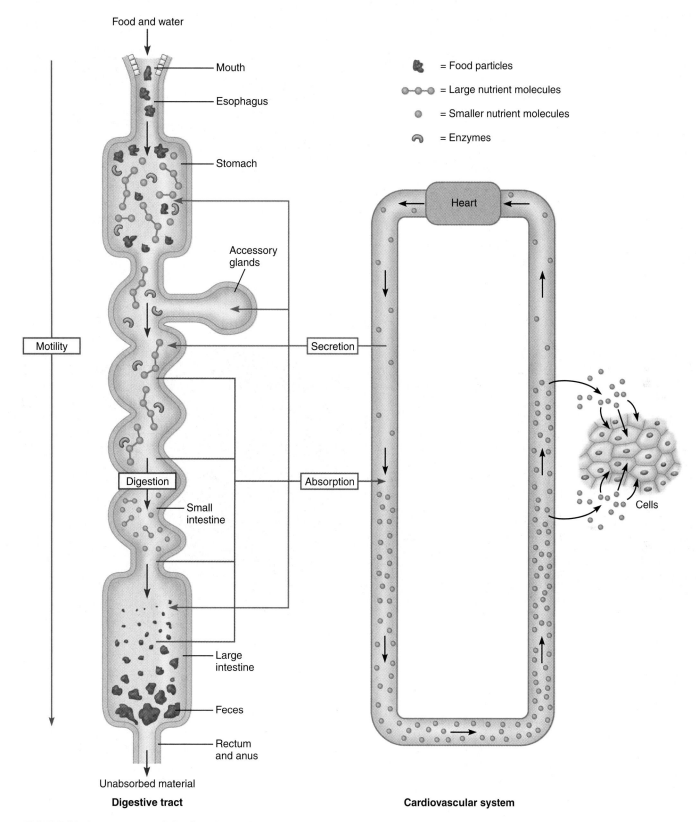

Food and water

Mouth

Esophagus

Stomach

Accessory glands

Motility

Secretion

= Food particles

= Large nutrient molecules

= Smaller nutrient molecules

= Enzymes

Heart

Absorption

Digestion

Small intestine

Cells

Large intestine

Feces

Rectum and anus

Unabsorbed material

Digestive tract

Cardiovascular system

FIGURE 19.1 **Overview of the four basic digestive processes: digestion, absorption, secretion, and motility.**

FIGURE 19.2 **Major structures of the digestive system.**

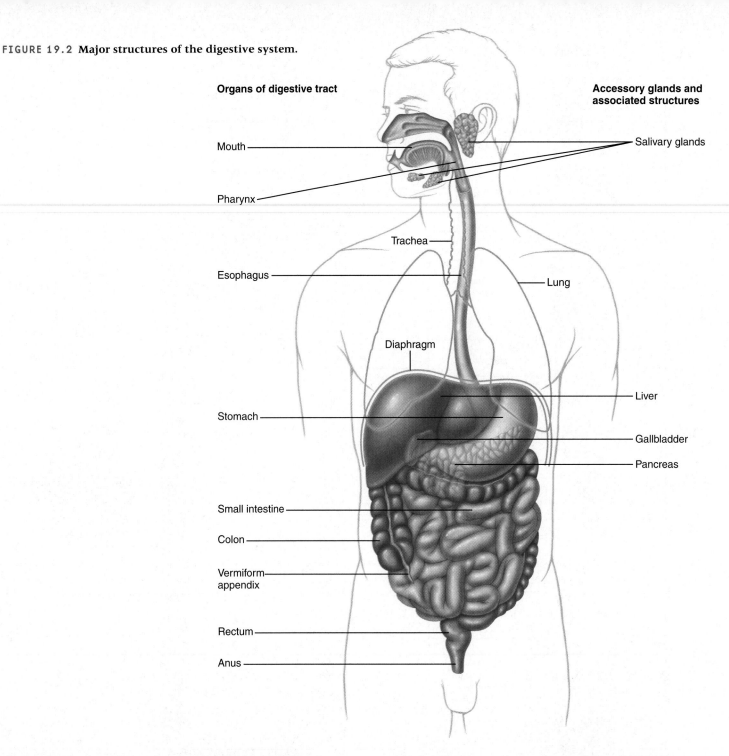

Organs of digestive tract

Accessory glands and associated structures

Mouth

Pharynx

Salivary glands

Trachea

Esophagus

Lung

Diaphragm

Liver

Stomach

Gallbladder

Pancreas

Small intestine

Colon

Vermiform appendix

Rectum

Anus

to the bloodstream. Other enterocytes are classified as **exocrine cells** because they secrete materials such as fluids and enzymes into the lumen (which is outside the body, as explained in Chapter 1). Among these cells are *goblet cells,* which secrete **mucus** (a sticky, viscous fluid containing glycoproteins called *mucins*) throughout the length of the GI tract; mucus forms a coating that protects the lining against abrasion and substances in the lumen that may attack tissue. Still other enterocytes are **endocrine cells,** which secrete hormones into the bloodstream. As we will see, these hormones play an important role in the regulation of digestive function.

The **lamina propria** is a layer of connective tissue underlying the mucous membrane. Contained within this layer are small blood vessels, nerves, and lymphatic vessels that communicate with larger nerves and vessels in still deeper tissue layers. The lamina propria also contains lymphoid tissue, including *lymph nodules* and *Peyer's patches,* that is important in defending the body against bacteria, which are very plentiful in the lumen of the intestines.

The **muscularis mucosae** is a thin layer of smooth muscle that serves primarily to contract the mucosa into folds, which stirs the lumenal contents and promotes

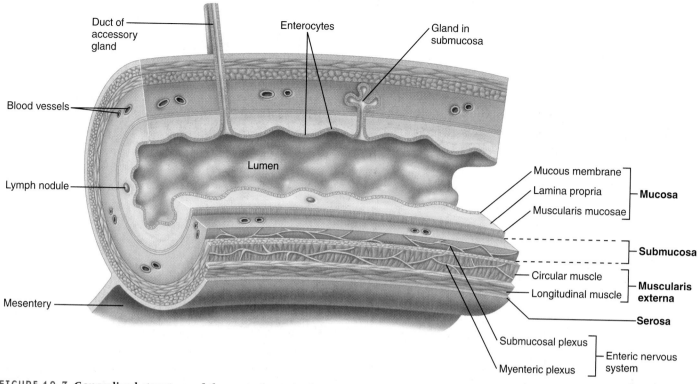

FIGURE 19.3 Generalized structure of the gastrointestinal wall, depicting the four major layers and the tissues within them.

contact with the mucosal surface. Within this layer are *longitudinal* muscle fibers, which run parallel to the tract's long axis, and *circular* muscle fibers, which run around the tract's circumference.

The Submucosa The submucosa is a thick layer of connective tissue that provides the GI tract with much of its distensibility and elasticity, enabling it to tolerate a large degree of stretch without sustaining damage. This layer also contains many of the tract's larger blood and lymphatic vessels. At its outer border is a network of nerve cells known as the **submucosal** (*Meissner's*) **plexus,** which communicates with another nerve cell network in the muscularis externa called the **myenteric** (*Auerbach's*) **plexus** (see Figure 19.3); together these nerve plexuses make up what is called the **enteric** (or *intrinsic*) **nervous system.** The enteric nervous system receives input from both autonomic nerves and sensory neurons located within the wall of the GI tract. Output from the enteric nervous system goes to effector cells located within the GI tract, including muscle cells, exocrine cells, and endocrine cells, and plays an important role in the control of digestive function. Although the enteric nervous system receives input from the central nervous system, it is capable of regulating many functions independently and thus serves as "the digestive system's brain."

The Muscularis Externa The muscularis externa is largely responsible for the motility of the GI tract and contains two separate layers of smooth muscle: an inner layer of circular muscle and an outer layer of longitudinal

muscle. Contraction of the circular muscle layer narrows the digestive tract's diameter, whereas contraction of the longitudinal muscle decreases its length.

The motility of the GI tract propels the lumenal contents from one organ to the next and mixes it, promoting contact between undigested food and digestive enzymes. Motility also promotes contact of digestive end-products with the mucosal epithelium, which is necessary for efficient absorption.

The Serosa The serosa, the outermost layer of the gastrointestinal wall, consists of an inner layer of fibrous connective tissue, which provides structural support, and an outer layer of epithelial tissue called the **mesothelium,** which secretes a watery lubricating fluid that makes it easier for organs to slide past one another. The mesothelium (along with a layer of underlying connective tissue) is continuous with the **mesenteries,** a system of clear, thin membranes that interconnects most of the abdominal organs and houses nerves and blood vessels running to them. The mesenteries help to anchor the organs in place and are continuous with the **peritoneum,** a membrane lining the inside of the abdominal cavity. A mesentery called the *greater omentum* connects the stomach to the spleen, whereas another, called the *lesser omentum,* connects the stomach to the liver and pancreas and contains ducts that carry secretions from these organs to the small intestine.

Although the generalized structure shown in Figure 19.3 pertains to *most* of the digestive tract, it does not

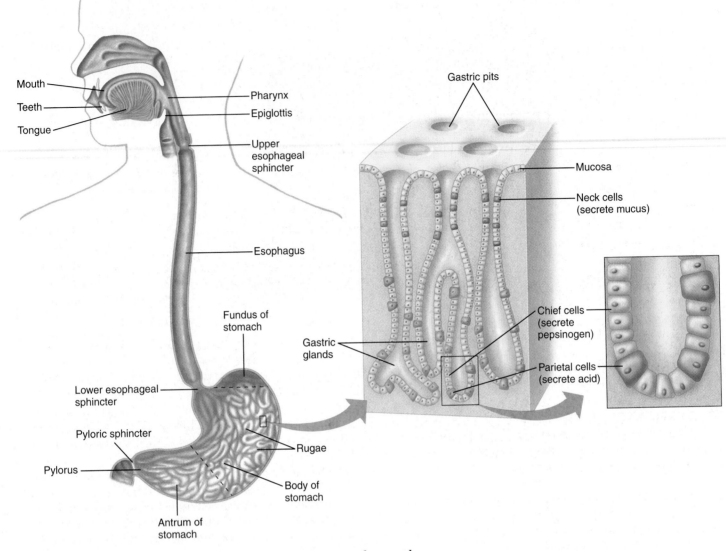

FIGURE 19.4 Anatomy of the mouth, pharynx, esophagus, and stomach.
Enlarged views show gastric pits and gastric glands.

pertain to all portions. The mouth, pharynx, upper third of the esophagus, and the externalmost portion of the anus contain *skeletal muscle* rather than smooth muscle, and the stomach contains a layer of obliquely-oriented smooth muscle in addition to the layers of circular and longitudinal muscle.

Quick Test 19.1

1. Name the four major layers that make up the wall of the digestive tract.

2. Name the three major cell types in the mucous membrane, and briefly describe the function of each.

3. Name the two major divisions of the enteric nervous system, and give their locations.

4. Define the following terms: *digestion, absorption, secretion, motility, digestive tract, accessory glands, longitudinal muscle, circular muscle, mesenteries, peritoneum,* and *enterocyte.*

Functional Anatomy of Digestive Tract Organs

Now we turn our attention to the individual organs of the GI tract, progressing in order from the upper end of the tract to the lower end. We begin with the mouth, pharynx, and esophagus.

The Mouth, Pharynx, and Esophagus The *mouth* or *oral cavity,* the beginning of the GI tract, is where food enters and where the processes of mechanical breakdown and digestion begin (Figure 19.4). In the mouth, food is chewed (a process called **mastication**) and mechanically broken down into smaller particles by the cutting and grinding actions of the teeth. The food is also mixed with a secretion called **saliva,** which lubricates it and contains an enzyme called **salivary amylase,** which begins the digestion of carbohydrates by breaking down starch and glycogen.

From the mouth, the food-saliva mixture is propelled by the tongue into the pharynx (commonly known as the *throat*), a common passageway for food and air. From the pharynx, the passageways for food and air diverge. Whereas air enters the larynx and trachea via the glottis and proceeds toward the lungs, food enters the esophagus, which runs parallel and dorsal to the trachea.

The **esophagus** is a muscular tube whose primary function is to conduct food from the pharynx to the stomach. Unlike the trachea, it is thin-walled and pliant, so that it can easily stretch to accommodate food as it is swallowed; when food is not present, however, it is normally collapsed. The esophagus is unusual among the organs of the GI tract in that its wall contains both skeletal muscle (in the upper third of its length) and smooth muscle. The movement of food from the pharynx to the esophagus is regulated by the **upper esophageal sphincter** (see Figure 19.4), a ring of skeletal muscle surrounding the esophagus at its upper end. (A **sphincter** is generally defined as a ring of muscle that surrounds an orifice and regulates the passage of material through it by altering its diameter.) At the esophagus' lower end is the **lower esophageal sphincter,** a ring of smooth muscle that regulates the flow of food from the esophagus to the stomach. Both of these sphincters are normally closed, and open only when food is being swallowed. The lower esophageal sphincter prevents the contents of the stomach, which are acidic, from entering the esophagus. However, backflow of stomach contents into the esophagus (*gastric reflux*) can on occasion occur and produce *heartburn,* a burning sensation in the chest caused by irritation of the esophageal lining.

The Stomach   An important function of the **stomach,** a J-shaped sac located beneath the diaphragm, is to store food after it is swallowed and to release it into the small intestine at a rate slow enough to permit its proper digestion and absorption. The lining of the stomach contains glands (called **gastric glands**) that secrete a watery fluid called **gastric juice** into the lumen. Contractile activity of smooth muscle in the stomach's wall pulverizes food into smaller particles and mixes it with gastric juice, forming a mixture called **chyme.**

The stomach has three major anatomical regions (see Figure 19.4): a domed upper portion called the **fundus,** which extends above the lower esophageal sphincter; a middle region called the **body,** which accounts for the bulk of the stomach's volume; and a lower region called the **antrum,** which is narrower and smaller in volume. Of these three regions, the antrum is the most muscular and generates the strongest contractions; for this reason,

mixing of the chyme is most vigorous in this region. Contractions of the antrum also propel the chyme from the stomach into the small intestine, a process called *gastric emptying.* As chyme exits the stomach, it passes through a narrow passage called the **pylorus** on its way to the small intestine. The flow of chyme through the pylorus is regulated by a surrounding ring of smooth muscle called the **pyloric sphincter,** which opens and closes with each cycle of stomach contraction, such that chyme exits the stomach in spurts.

The volume of the stomach is about 50 milliliters when empty but can increase to 1000 milliliters or greater (a factor of 20 or more) following the ingestion of an average-sized meal. To accommodate stretching, the gastric mucosa is thrown into folds called **rugae,** which flatten as the stomach expands. Over most of its surface the lining of the stomach is studded with openings called **gastric pits,** which lead to the gastric glands.

Gastric juice contains the following substances that play a direct role in the digestion and absorption of certain constituents of the food: (1) **pepsinogen,** an inactive precursor of the enzyme **pepsin,** which is secreted by specialized **chief cells** and acts in the lumen to begin the digestion of proteins; (2) *hydrogen ions,* which are secreted by **parietal cells** and act to acidify the stomach contents; and (3) **intrinsic factor,** a protein secreted by parietal cells that binds to vitamin B_{12} and is necessary for the absorption of this vitamin in the small intestine. The stomach lining also contains **G cells** (not shown), which secrete **gastrin,** a hormone that regulates many gastrointestinal functions, including acid secretion by the stomach.

The lumen of the stomach is the only locale in the GI tract where the contents are acidic. In fact, the pH of stomach contents can go as low as 2, which is equivalent to a 10 mM solution of hydrochloric acid! This acidity is necessary for converting pepsinogen into its active form, pepsin; it is also useful because it helps denature proteins in the food and kills many foodborne bacteria and thus protects against certain illnesses.

The stomach's lining is protected against the potentially harmful effects of acid and pepsin by a surface covering of mucus, which is secreted by *neck cells* in gastric glands and also by cells in the surface epithelium. (Because of its protective role, this mucus layer is often called the **gastric mucosal barrier.**) The epithelium also contains cells that secrete bicarbonate, which neutralizes acid near the surface of the stomach's lining.

The Small Intestine   From the stomach, chyme travels to the **small intestine,** a coiled tube about 2.5–3 meters (8–10 feet) long that is the primary site for the digestion of all nutrients in food. The small intestine is also where most of the ingested nutrients, water, vitamins, and minerals (inorganic ions such as sodium, potassium, and calcium) are absorbed.

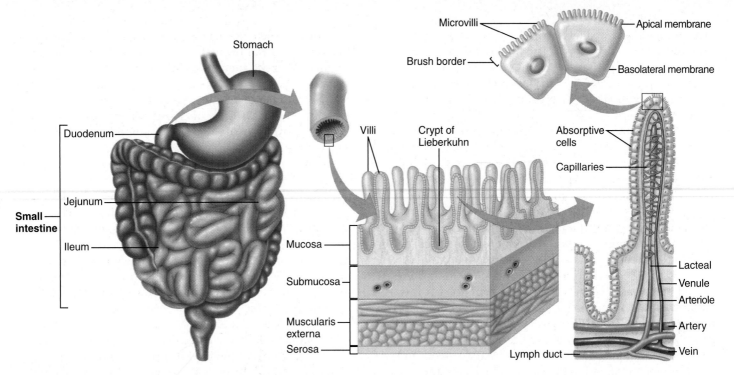

FIGURE 19.5 Anatomy of the small intestine. *The duodenum, jejunum, and ileum are shown in relation to other GI tract organs; the enlarged views show the structure of villi and microvilli in the duodenum.*

On the basis of subtle anatomical distinctions, the small intestine is divided into three major regions (Figure 19.5): an initial portion called the **duodenum,** which begins at the pylorus and extends for approximately 30 cm (12 in.); a middle portion called the **jejunum,** which extends for about another 1 meter (3–4 feet); and a terminal portion called the **ileum,** which extends approximately 1.5 meters (4–5 feet) and joins the colon.

In the duodenum, chyme is mixed with a watery secretion from the pancreas called **pancreatic juice,** which both contains a wide variety of digestive enzymes and is rich in bicarbonate, which neutralizes the acid in the chyme when it exits the stomach. This is important because the enzymes in pancreatic juice function at the normal pH of the small intestine (which is slightly basic) but not at acidic pHs. In addition to pancreatic juice, the duodenum also receives **bile,** a fluid secreted by the liver that contains bicarbonate and *bile salts,* which aid in the digestion of fats, as described later.

As nutrients in the chyme are broken down by enzymes, digestive end-products are released into solution and absorbed by cells in the mucosal epithelium. These simultaneous processes of digestion and absorption begin in the duodenum and continue to completion in the remainder of the small intestine. Unless an unusually large quantity of food has been ingested, absorption is typically completed within about the first 20% of the small intestine's length, or before the chyme has reached the ileum. Thus the small intestine has a large excess capacity for absorbing nutrients, indicating that its absorptive mechanisms are highly efficient.

The small intestine's absorptive efficiency is attributable in part to the folding of the mucosal surface into structures called **villi** (singular: *villus*), which facilitate the transport of materials by increasing the surface area of the epithelium (see Figure 19.5). Each villus houses structures crucial to the absorption of nutrients, including a network of capillaries and a blind-ended lymphatic vessel called a **lacteal.** After nutrients are absorbed from the lumen, they are transported across the mucosal epithelium into the interstitial fluid, where most absorbed nutrients diffuse into the capillaries and then are carried away from the intestine and eventually into the general circulation. (Absorbed *fats* are an exception to this rule, as we will see shortly.)

Another factor that increases the absorptive efficiency of the small intestine is the presence of a brush border on the mucosal surface, which increases surface area even further. This **brush border** is made up of *microvilli,* located on the apical surface of epithelial cells and numbering from 3000 to 6000 per cell. The brush border of microvilli (and indeed the villi themselves) are most prominent in the proximal small intestine (the duodenum, for example) and less abundant in the more distal portions of the small intestine.

Also located in the lining of the small intestine are pits known as *crypts of Lieberkühn* (see Figure 19.5), which contain cells that secrete copious amounts of bicarbonate-rich fluid into the lumen. This fluid is secreted mostly in the more proximal portions of the small intestine and is almost completely absorbed (along with fluid that is ingested) before the chyme reaches the colon.

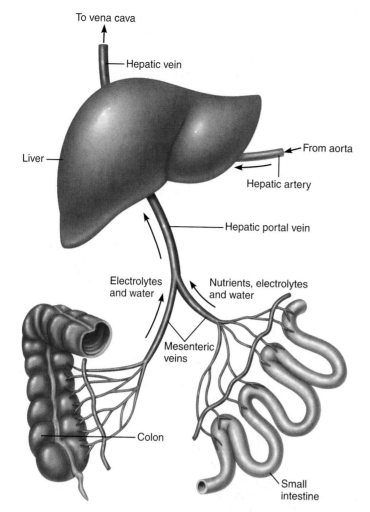

FIGURE 19.6 Circulatory route of materials absorbed in the small intestine and colon. *Materials absorbed in the intestines travel to the liver via the mesenteric veins and the hepatic portal vein, and then into the general circulation via the hepatic vein.*

Materials that are absorbed from the intestines are carried by the bloodstream to the liver, which extracts certain nutrients for further processing. (Nutrients that remain in the blood are carried into the general circulation.) Blood from intestinal capillaries drains into the *mesenteric veins* and is carried to the liver by the *hepatic portal vein* (Figure 19.6). Blood delivered by the hepatic portal vein is deoxygenated, but a supply of oxygenated blood (which is necessary for the liver's proper function) is delivered to the liver by the *hepatic artery*. Blood is carried from the liver to the general circulation by the *hepatic vein,* which drains into the inferior vena cava.

The Colon The **colon** is divided into four major regions on the basis of its anatomy (Figure 19.7): (1) the *ascending colon,* which runs upward on the right side of the body from the end of the small intestine toward the diaphragm; (2) the *transverse colon,* which runs across the abdominal cavity; (3) the *descending colon,* which runs downward on the left side, and (4) the *sigmoid colon,* an S-shaped segment leading to the rectum. The first three segments are specialized for absorbing water and ions

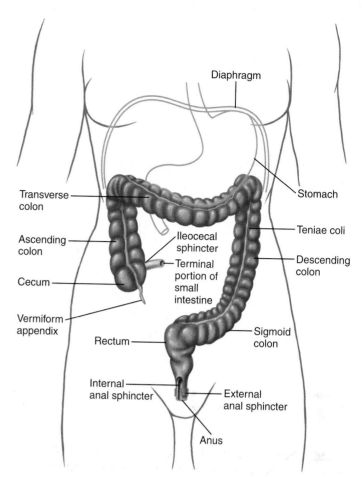

FIGURE 19.7 Major anatomical features of the colon.

from the chyme; the sigmoid colon serves primarily as a storage depot for whatever material remains in the lumen after absorption has occurred. Although the wall of the colon has the same fundamental structure as other parts of the GI tract, the longitudinal muscle layer of the muscularis externa is not continuous but is instead compressed into three relatively narrow bands called *teniae coli,* which run the colon's length.

At the junction between the ileum and colon, the flow of material is regulated by a ring of smooth muscle called the **ileocecal sphincter** (see Figure 19.7). Below this junction is a blind-ended bulb called the *cecum,* to which is attached the *vermiform appendix,* a wormlike appendage having no known function. On rare occasions the opening of the appendix can become blocked and then inflamed, a condition known as *appendicitis.* If the condition persists, the appendix may rupture, spilling the lumenal contents into the abdominal cavity. Rupture is a dangerous situation because it almost always leads to *peritonitis* (inflammation of the peritoneum), which is fatal in most cases if left untreated.

By the time chyme reaches the colon, it contains very few digestible nutrients because most of the materials present in the chyme have already been absorbed. What remains consists mainly of water, inorganic ions, indigestible material from food, and bacteria. The colon's primary function is to reduce the volume of the chyme by

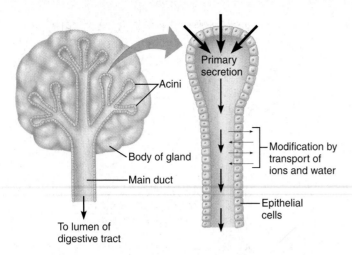

FIGURE 19.8 Generalized structure of an accessory gland.
The gland depicted here is typical of the salivary glands or the pancreas, in which the body of the gland contains acini and associated ducts. Heavy arrows at right indicate the formation of the primary secretion in an acinus; lighter arrows indicate modification of this secretion by the transport of ions and water within the ducts.

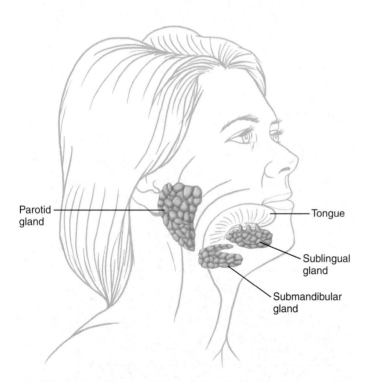

FIGURE 19.9 Location of the major salivary glands.

absorbing most of the water remaining in it, thereby transforming it into a semisolid material called **feces.** The colon then stores the feces until it is ready to be eliminated from the body. Together, the cecum, colon, and rectum constitute what is often referred to as the *large intestine.*

The Rectum and Anus   Intermittently, the colon contracts strongly, pushing fecal material into the rectum.

This material does not exit the body immediately because the movement of material through the anus is controlled by two sphincters: the **internal anal sphincter,** which is composed of smooth muscle, and the **external anal sphincter,** a ring of skeletal muscle that controls the opening to the outside. Relaxation of both sphincters, which are normally closed, allows fecal material to be eliminated from the body, a process called **defecation.**

Quick Test 19.2

1. Arrange the following terms in order such that they correctly describe the path of ingested material as it travels through the GI tract: *descending colon, esophagus, rectum, stomach, lower esophageal sphincter, pharynx, ascending colon, ileum, upper esophageal sphincter, transverse colon, duodenum, sigmoid colon, ileocecal sphincter, pyloric sphincter, jejunum.*

2. Define the following terms: *fundus, antrum, chyme, cecum, crypt of Lieberkuhn, villus.*

3. Describe the brush border. Where is it located and what is its function?

4. Where does most digestion and absorption occur in the GI tract?

The Accessory Glands

The accessory glands of the digestive system include the **salivary glands,** which secrete *saliva;* the pancreas, which secretes *pancreatic juice;* and the **liver,** which secretes *bile.* Despite the fact that the accessory glands look very different superficially and perform different functions, they share many structural and functional similarities. Secretions of these glands are carried to the GI tract via ducts lined by epithelial cells, and within the body of a gland the ducts branch extensively, terminating in an enclosed space that is completely surrounded by a layer of specialized secretory epithelial cells (Figure 19.8). In the salivary glands and pancreas, these epithelial cells are arranged in ball-like clusters called **acini** (singular: *acinus*). (In the liver, secretory cells are arranged differently, as we will see.) The acinar cells secrete a fluid (referred to as the *primary secretion*) containing water, inorganic ions, and other solutes whose nature depends on the gland in question. As this fluid flows through the ducts, the epithelial cells lining the ducts secrete or absorb ions and/or water, thereby modifying the fluid's composition.

The Salivary Glands

Saliva is produced by three pairs of major salivary glands (Figure 19.9): the **parotid glands,** located on both sides of the head at approximately ear level; the **sublingual**

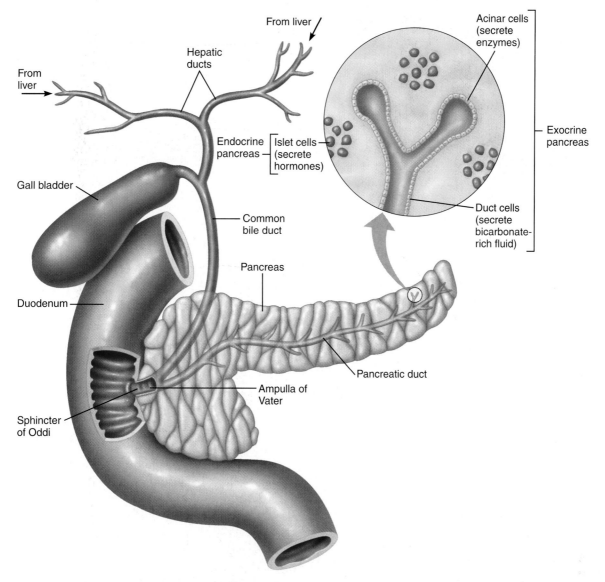

FIGURE 19.10 The pancreas and associated structures. *Enlarged view shows acini and ducts (exocrine pancreas) and islets (endocrine pancreas).*

glands, located beneath the tongue on either side; and the **submandibular glands,** located on each side beneath the lower jaw. Other smaller salivary glands are located in the wall of the mouth and pharynx.

Among the components of saliva are (1) *bicarbonate,* which makes the saliva alkaline and helps to neutralize acid; (2) *mucus,* which lubricates the food and protects the lining of the mouth from abrasion; (3) *salivary amylase,* a digestive enzyme that breaks down starch and glycogen; and (4) *lysozyme,* an enzyme that by destroying or *lysing* certain bacteria helps to prevent tooth decay.

The Pancreas

The pancreas, located behind and beneath the stomach (see Figure 19.2), is not only an exocrine organ of the digestive system but is also an endocrine organ that secretes

hormones that are important in the regulation of metabolism (see Chapter 20). The *exocrine pancreas* comprises the numerous acini and their associated ducts, whereas the *endocrine pancreas* consists of *pancreatic islets* scattered among the acini and ducts (Figure 19.10). Ducts from the acini converge to larger ducts, which eventually converge to the **pancreatic duct,** the main duct that carries pancreatic juice to the duodenum.

Pancreatic juice is rich in bicarbonate and also contains several digestive enzymes, including **pancreatic amylase,** which is similar to salivary amylase and breaks down starch and glycogen, and **pancreatic lipases,** which break down fats. Also present are a number of *proteases,* which break down proteins, and *nucleases,* which break down nucleic acids.

Relative to its weight, the pancreas secretes more protein than any other tissue in the body, most of it in the form of digestive enzymes. Because many of these

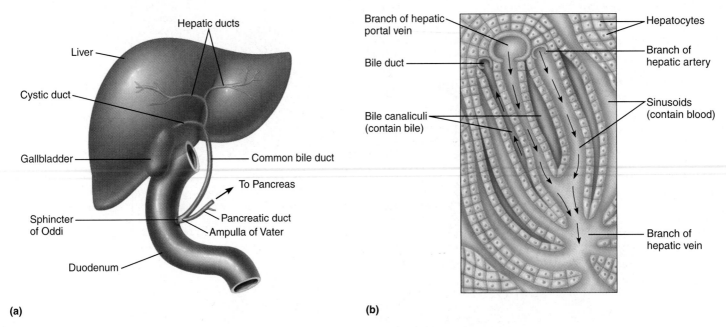

FIGURE 19.11 Structures of the biliary system. (a) *Gross anatomy of the liver, gallbladder, and their associated ducts.* **(b)** *Microscopic anatomy of the liver, showing hepatocytes, sinusoids, and bile canaliculi.*

Nutrients are carried from the small intestine to the sinusoids via branches of which blood vessel?

enzymes (the proteases in particular) are capable of breaking down molecules that make up the structure of the pancreas itself, they are stored within acinar cells in inactive forms. These molecules, collectively referred to as **zymogens,** are stored in vesicles called *zymogen granules* and are secreted by exocytosis. After secretion the zymogens are converted to their active forms in the lumen of the gastrointestinal tract, generally through the action of proteolytic enzymes located there.

The Liver

The liver, the largest organ in the abdominal cavity, is amazingly versatile. (Discovery: Liver Regeneration, www.physiologyplace.com, Challenge Yourself) Among its more important functions are the following:

1. *Secretion of bile.* As mentioned previously, the liver secretes bile, which contains bicarbonate, phospholipids, inorganic ions, and bile salts, which are derivatives of cholesterol.

2. *Metabolic processing of nutrients.* Following a meal, the liver converts some of the absorbed glucose to glycogen, and some absorbed amino acids to fatty acids; the liver also synthesizes triglycerides and cholesterol and uses them to synthesize *lipoprotein particles,* which it then secretes into the bloodstream. During periods in which nutrients are not being absorbed, the liver converts glycogen to glucose and fatty acids to ketones. It also produces glucose by gluconeogenesis and synthesizes urea from ammonia, which is generated as a by-product of amino acid catabolism (see Chapter 3).

3. *Removal of aged red blood cells from the blood.* The liver contains macrophages that remove old red blood cells and bacteria from the blood. Hemoglobin from old red blood cells is then broken down by the liver; some components (such as iron) are saved for reuse while others (such as *bilirubin*) are eliminated from the body.

4. *Elimination of wastes from the body.* Bilirubin and other breakdown products of hemoglobin are secreted in the bile and eliminated from the body in the feces. (Bilirubin and similar breakdown products are referred to as *bile pigments* because in combination they impart a greenish color to the bile.) Other substances eliminated in the bile include excess cholesterol, other organic compounds (including foreign compounds such as drugs or poisons), and trace metals. The liver also chemically transforms many hydrophobic compounds (including toxins) into more hydrophilic forms so that they can be more readily dissolved in the plasma and eliminated by the kidneys.

5. *Synthesis of plasma proteins.* The liver synthesizes most of the proteins that are present in the plasma, includ-

The hepatic portal vein

TABLE 19.1 GASTROINTESTINAL SYSTEM
ORGANS AND THEIR FUNCTIONS

ORGAN	FUNCTIONS
Mouth	Mechanical breakdown of food; mixing of food with saliva; initiation of chemical digestion of carbohydrates by salivary amylase
Pharynx	Conduction of food to esophagus
Esophagus	Conduction of food to stomach
Stomach	Mechanical breakdown of food; secretion of acid, pepsinogen, and intrinsic factor; initiation of chemical digestion of proteins by pepsin; secretion of gastrin into bloodstream; transformation of food into chyme
Small intestine	Chemical digestion of all nutrient classes by pancreatic enzymes and membrane-bound enzymes; absorption of digestive end-products, water, ions, and vitamins; secretion of enterogastrones into bloodstream; secretion of bicarbonate-rich fluid
Colon	Absorption of ions and water; transformation of chyme into feces; storage of feces
Rectum	Storage of feces prior to elimination
Anus	Control of defecation
Salivary glands	Secretion of saliva (contains amylase, mucus, bicarbonate, and lysozyme)
Pancreas	Secretion of pancreatic juice (contains digestive enzymes and bicarbonate)
Liver	Secretion of bile (contains bile salts and bicarbonate); processing of absorbed nutrients
Gallbladder	Storage and concentration of bile

ing albumin, steroid-binding and thyroid-hormone-binding proteins, clotting proteins, and angiotensinogen.

6. *Secretion and modification of hormones.* The liver participates with the kidney in the activation of vitamin D, and it secretes somatomedins, also known as *insulin-like growth factors* (discussed in Chapter 20). The liver also helps to clear many hormones from the body by metabolizing them.

Because it manufactures and secretes bile, the liver is part of the **biliary system,** which comprises all structures involved in synthesizing or storing bile and delivering it to the GI tract (Figure 19.11). Bile is manufactured and secreted by the liver continually, but it is released into the tract only when food is present. Between meals the bile secreted by the liver is stored in a small muscular sac called the **gallbladder,** which is located immediately adjacent to and beneath the liver (Figure 19.11a). During meals the gallbladder is stimulated to contract, which forces the stored bile into the **common bile duct,** which carries it to the duodenum.

At their juncture with the duodenum, the common bile duct and the pancreatic duct converge to form a common passageway (called the *ampulla of Vater*) for the flow of bile and pancreatic juice (see Figure 19.10). The flow of these two fluids is regulated by a ring of smooth muscle called the **sphincter of Oddi,** which is normally closed but opens when food is present. Between meals, when the sphincter is closed, the bile secreted by the liver backs up in the common bile duct and "spills over" into the gallbladder, where it is stored and concentrated.

The secretion of bile, like that of saliva and pancreatic juice, is a two-step process involving the formation of a primary secretion and its subsequent modification in ducts. However, the spongelike liver differs substantially in structure from the salivary glands and pancreas in that it contains numerous blood-filled cavities called **sinusoids,** which take the place of true capillaries (Figure 19.11b). The blood that fills these sinusoids is a mixture of oxygenated blood flowing from branches of the hepatic artery, and deoxygenated blood flowing from branches of the portal vein. (Blood leaves the sinusoids via branches of the hepatic vein.) The walls of the sinusoids consist of sheetlike layers of liver cells (**hepatocytes**), which are arranged in such a way that one side of each hepatocyte faces the blood while the other side faces a space called a **bile canaliculus.** The cells take from the blood the raw materials for making bile and then secrete the bile into the bile canaliculi. The canaliculi drain into **bile ducts** that eventually converge to form the **common hepatic duct,** which carries bile away from the liver.

A summary of the organs of the digestive system and their major functions is given in Table 19.1.

Quick Test 19.3

1. Name the three pairs of major salivary glands. What are the main secretory products found in saliva?

2. What is the distinction between the exocrine pancreas and the endocrine pancreas? Which secretes bicarbonate-rich fluid? digestive enzymes? hormones?

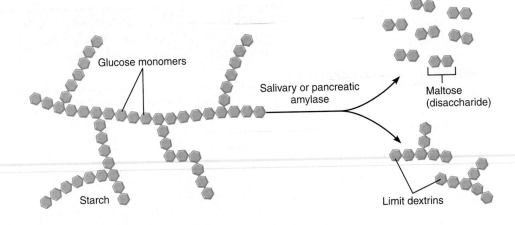

FIGURE 19.12 Digestion of starch by salivary or pancreatic amylase. *Shown are digestion products that would be obtained after digestion of starch by amylase enzymes is complete.*

Is starch digestion an *anabolic* reaction or a *catabolic* reaction?

3. Which organ synthesizes bile? Stores bile?

4. Define the following terms: *bile, common hepatic duct, common bile duct, gallbladder, sphincter of Oddi,* and *sinusoids.*

DIGESTION AND ABSORPTION OF NUTRIENTS AND WATER

Most of the nutrient molecules present in food are large polymeric molecules (primarily carbohydrates, proteins, and lipids) that must be enzymatically broken down into smaller molecules before they can be absorbed. In the next three sections we see how molecules in each of these major nutrient classes are digested and absorbed. Then we see how the digestive system handles vitamins, minerals, and water, components of food that are not digested but instead are simply absorbed in their original forms.

Carbohydrates

Between 250 to 800 grams of carbohydrates are ingested daily in a typical diet. Although glucose and other simple carbohydrates (monosaccharides) are present in the diet in small quantities, most dietary carbohydrates are in the form of disaccharides or larger molecules. These disaccharides include *sucrose* (table sugar), which is composed of glucose and fructose; *lactose* (milk sugar), which is composed of glucose and galactose; and *maltose*, which is composed of two glucose molecules. About two-thirds of dietary carbohydrates are in the form of *starch*, a mixture of straight and branched-chain glucose polymers found in plants such as corn and potatoes, and in plant-derived food products such as bread and pasta. A similar compound, *glycogen*, is found in certain animal products.

A catabolic reaction

Digestion

Because the small intestine cannot absorb carbohydrates larger than monosaccharides (glucose, galactose, and fructose), complex carbohydrates must first be broken down to monosaccharides. The chemical digestion of starch or glycogen begins in the mouth with the action of salivary amylase, which breaks these molecules down by breaking bonds between glucose monomers. However, this enzyme is not capable of reducing starch or glycogen completely to glucose monomers (more on this shortly). Salivary amylase is inactivated once it comes into contact with the acid environment of the stomach, but in the small intestine another enzyme, pancreatic amylase, resumes the task of carbohydrate digestion. Pancreatic amylase, which is present in pancreatic juice, is also incapable of reducing starch or glycogen to glucose monomers.

Because neither salivary nor pancreatic amylase can break bonds at the ends of chains or at branch points, they can only reduce starch or glycogen to either disaccharides composed of two glucose monomers (maltose) or short, branched polysaccharides called *limit dextrins* (Figure 19.12). Depending on the time available for the enzyme to act, longer fragments may be produced.

To be absorbed, the end-products of amylase digestion must be broken down further, which is accomplished by enzymes bound to the apical membranes of absorptive cells lining the small intestine. These enzymes (and others like them) are often referred to as *brush border enzymes* because they are located in the small intestine's brush border. The brush border enzymes include *dextrinase* and *glucoamylase*, which break limit dextrins and straight-chain glucose polymers down to glucose monomers; *sucrase*, which splits sucrose into fructose and glucose; *lactase*, which splits lactose into galactose and glucose; and *maltase*, which splits maltose into two glucose molecules. These enzymes are in close proximity to other membrane-bound proteins that transport the monosaccharides into the epithelial cells, the first step in absorption. To illustrate the action of brush border enzymes, the digestion and absorption of maltose is shown in Figure 19.13.

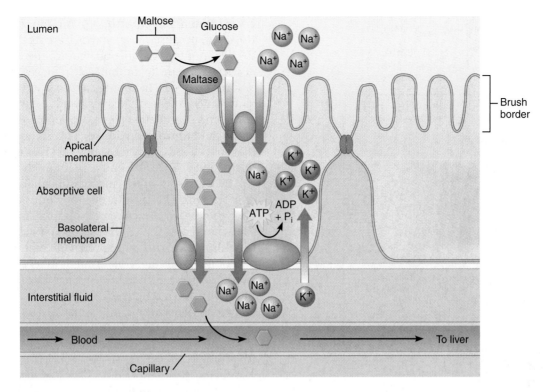

FIGURE 19.13 Digestion and absorption of maltose. *Maltose, a disaccharide consisting of two glucose monomers (hexagons), an end-product of starch digestion by amylase, is digested by maltase, a brush border enzyme. The end-products, glucose monomers, are transported across the apical membrane of mucosal epithelial cells by sodium-linked secondary active transport. Energy for this process is derived from the action of Na^+/K^+ pumps in the basolateral membrane, which create an inwardly directed electrochemical gradient for sodium ions. Glucose molecules exit cells by facilitated diffusion across the basolateral membrane and then diffuse into capillaries, which carry them into the general circulation.*

Humans cannot digest and absorb all carbohydrates. For example, *cellulose,* a glucose polymer found in plant products, cannot be digested because its glucose-glucose bonds are different from those in starch or glycogen and are not acted upon by the enzymes present in the GI tract. When cellulose is ingested, it remains in the intestine and is eventually eliminated. What nutritionists call *dietary fiber* is actually mostly cellulose.

Absorption

Once dietary carbohydrates have been reduced to monosaccharides by the action of digestive enzymes, they are transported across the epithelium that lines the small intestine. Whereas glucose and galactose enter the epithelial cells via sodium-linked secondary active transport across the apical membrane (see Figure 19.13), fructose enters by facilitated diffusion. Entry of these sugars is followed by their exit across the basolateral membrane via facilitated diffusion. Following transport across the epithelium, these molecules then diffuse into capillaries and are carried by the bloodstream into the general circulation.

Proteins

Although the average dietary intake of proteins is approximately 125 grams per day, normal adults of average weight require only about 40–50 grams of protein per day. For a person who is not growing or building muscle mass, this is the quantity required to replace proteins that are normally catabolized within the body. In addition, another 20–60 grams of protein enter the GI tract each day via secreted fluids (pancreatic juice and mucus, for example) and in cells shed from the mucosal lining.

Proteins are not absorbed in significant amounts unless they are first enzymatically broken down to smaller molecules. Collectively the digestive enzymes that act on proteins in the lumen of the GI tract are capable of breaking them down completely to individual amino acids, but this is not absolutely necessary because enterocytes in the small intestine can also absorb short peptides (dipeptides or tripeptides).

Digestion

The digestion of proteins begins in the stomach with the action of *pepsin* (Figure 19.14). Pepsin is secreted in the form of a zymogen called *pepsinogen,* which is inactive

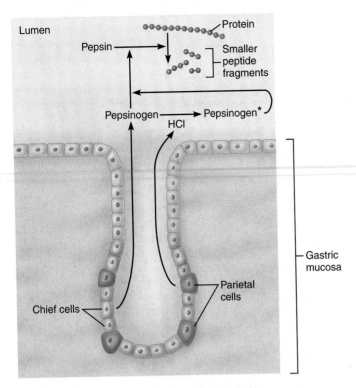

Lumen

FIGURE 19.14 Activation and activity of pepsin in the stomach. *Chief cells secrete pepsinogen, which is partially activated by hydrochloric acid (HCl) secreted by parietal cells. Partially activated pepsinogen molecules (pepsinogen*) then fully activate other pepsinogen molecules by cleaving off amino acids to form pepsin, which digests proteins to smaller fragments.*

Is pepsin capable of completely breaking down a protein to its constituent amino acids? Why or why not?

until it comes into contact with the acidic contents of the stomach. Contact of pepsinogen with this acidic environment induces in the protein a conformational change that partially activates the enzyme. The partially active pepsinogen acts on other pepsinogen molecules, cleaving off a small portion of the protein to yield the fully active pepsin.

Once activated, pepsin digests proteins to smaller peptide fragments, but it cannot break them down completely to amino acids because it can split bonds between *certain* amino acids only. Pepsin's actions are also limited by the length of time it stays in the stomach, because it requires an acidic environment and becomes inactive once it enters the more basic environment of the small intestine.

In the small intestine, the task of protein digestion is taken over by proteases secreted in pancreatic juice and by other proteases bound to the intestinal brush border. The pancreatic proteases include **trypsin and chymotrypsin,** which are secreted as the zymogens

trypsinogen and **chymotrypsinogen,** respectively. However, because these enzymes are like pepsin in that they cannot break all peptide bonds, the complete breakdown of proteins to amino acids requires the action of additional enzymes, including **carboxypeptidase,** a pancreatic protease that splits bonds between amino acids at the end of a peptide chain containing the carboxyl group, and **aminopeptidase,** a brush border enzyme that splits bonds between amino acids at the opposite end of a peptide chain (the end containing the amino group).

Because trypsin, chymotrypsin, and carboxypeptidase are secreted in inactive forms, they must be activated once they reach the lumen of the small intestine, as shown in Figure 19.15. Trypsinogen is acted upon by **enterokinase,** a brush border enzyme that clips off a portion of the molecule, converting it to its active form, trypsin. Once activated, trypsin acts on other proteins in the chyme, including chymotrypsinogen and procarboxypeptidase, causing these enzymes to be activated.

Absorption

As previously mentioned, protein digestion products are absorbed either as individual amino acids or as dipeptides or tripeptides. Many amino acids enter absorptive epithelial cells via sodium-linked secondary active transport across the apical membrane; others are transported into the cells by facilitated diffusion. Dipeptides and tripeptides are actively transported across the apical membrane and then broken down to individual amino acids by proteases within the cells. Amino acids inside the cells are transported by facilitated diffusion across the basolateral membrane and diffuse into capillaries and the general circulation.

Quick Test 19.4

1. The digestion of carbohydrates is complete when they have been reduced to what form? When are proteins digested completely?

2. Are salivary or pancreatic amylases capable of digesting starch completely? Why or why not?

3. Name the substrate for each of the following enzymes: sucrase, lactase, dextrinase, trypsin, chymotrypsin, pepsin, amylase, carboxypeptidase, and enterokinase.

Lipids

A typical daily diet contains between 25 and 160 grams of lipids, most of which are triglycerides. The digestion of lipids follows the same basic pattern as that for carbohydrates and proteins—large molecules are broken down by enzymes to smaller molecules before they are absorbed.

No: pepsin can break peptide bonds between certain amino acids only.

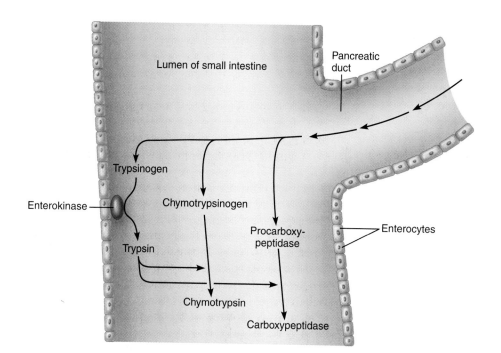

FIGURE 19.15 **Activation of proteases in the small intestine.** *Trypsinogen, a zymogen secreted by the pancreas, is converted to trypsin by the brush border enzyme enterokinase. Trypsin then converts other zymogens to their active forms in the lumen of the small intestine and also converts trypsinogen to its active form (not shown).*

However, the mechanism of lipid absorption is markedly different from what we have seen so far. In this section we place particular emphasis on the digestion and absorption of triglycerides, because they are the most abundant lipids in the diet.

Digestion

The digestion of most lipids does not begin until they reach the small intestine, where the chyme is mixed with **lipases** present in pancreatic juice. (The stomach secretes an enzyme that breaks down butterfat, however.) Although several different enzymes in pancreatic juice act on lipid substrates, we use the term *lipase* to refer to any of them.

Because lipids are hydrophobic, ingested fats do not mix readily with the rest of the stomach contents; instead, they coalesce to form large droplets that float on top of the chyme. As these fats leave the stomach, they do so as large globules that are practically indigestible because lipases are water soluble (like other digestive enzymes) and can act only on molecules near a globule's surface. Thus the vast majority of the lipid molecules in these globules cannot be reached by the enzymes.

The Action of Bile Salts Efficient digestion of lipids is made possible by the action of bile, which first comes into contact with fat globules in the duodenum. The bile does not actually digest lipids because it contains no enzymes; instead, it simply facilitates the action of lipases by breaking fat globules down into smaller droplets, a process called **emulsification.** By increasing the total surface area of the droplets, emulsification increases the amount of lipid that is exposed to water and therefore susceptible to enzymatic digestion.

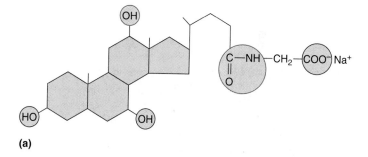

(a)

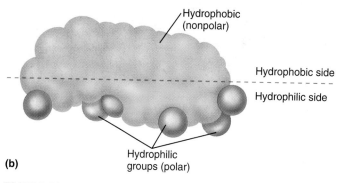

(b)

FIGURE 19.16 **A representative bile salt. (a)** *Structural formula, in which hydrophilic groups are highlighted by shading.* **(b)** *A space-filling model illustrating the amphipathic nature of the molecule.*

The emulsification of fats is due primarily to the action of **bile salts,** cholesterol derivatives synthesized by hepatocytes and secreted in the bile. Although cholesterol itself is very hydrophobic because it is composed almost entirely of nonpolar hydrocarbons, bile salts possess a number of oxygen-containing polar groups (such as hydroxyl groups or —OH, and carboxyl groups or —COOH) that are hydrophilic (Figure 19.16a). Because all these polar groups are located on one side of the molecule (the opposite side is nonpolar), bile salts are amphipathic.

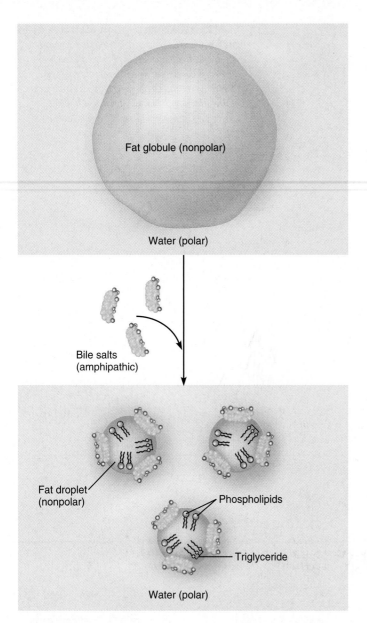

Fat globule (nonpolar)

Water (polar)

Bile salts
(amphipathic)

Fat droplet
(nonpolar)

Phospholipids

Triglyceride

Water (polar)

FIGURE 19.17 Emulsification of a fat globule by bile salts.

When bile salts come into contact with a fat globule, their hydrophobic sides face inward (toward the hydrophobic droplet), and their hydrophilic sides face outward (toward the water), as shown in Figure 19.17. In so doing, bile salts endow the droplets with a polar "coating" that allows them to mix more readily with water.

The Action of Pancreatic Lipase As fat is being emulsified in the duodenum, it is also mixed with pancreatic lipase, which begins to work on molecules located on the surfaces of lipid droplets (Figure 19.18). Lipases act on triglycerides to break the bonds linking fatty acids to the two carbons on either end of the glycerol backbone. As a result, the end-products of triglyceride digestion are two

free fatty acids and a monoglyceride (a glycerol molecule to which a single fatty acid is attached). Once fatty acids and monoglycerides are generated, they are released into solution. Some of these dissolved end-products are quickly absorbed into epithelial cells lining the small intestine; others remain in the chyme, aggregating into small particles called *micelles,* which readily exchange lipids with the surrounding solution. (Although not shown in Figure 19.18, these micelles also contain bile salts.)

Due to the continued action of pancreatic lipase and the release of fatty acids and monoglycerides into solution, fat droplets shrink during their transit through the small intestine, eventually disappearing by the time chyme reaches the colon. In the ileum, the bulk of the bile salts that were secreted into the duodenum (about 95%) are absorbed into the circulation; these bile salts are eventually recycled by the liver and secreted again in the bile via a pathway referred to as the *enterohepatic circulation* (Figure 19.19).

Absorption

The first step in the absorption of lipids is the entry of fatty acids and monoglycerides into absorptive intestinal epithelial cells (enterocytes), which occurs by simple diffusion. Inside the cells, these molecules enter the smooth endoplasmic reticulum, where they are acted upon by enzymes that *reassemble* them into triglycerides (Figure 19.20). (Other end-products of lipid digestion are also reassembled in a similar fashion.) These lipids are then packaged by the Golgi apparatus into large particles called **chylomicrons,** which belong to a general class of particles known as **lipoproteins** (Discovery: Lipoproteins and Plasma Cholesterol, p. 621).

Following their synthesis, chylomicrons are secreted by exocytosis across the basolateral membrane and into the interstitial fluid (see Figure 19.20). After reaching the interstitial fluid, they enter the lymphatic system via the lacteals, which like other lymphatic capillaries have openings in their walls that are large enough to allow such particles to pass through. (Chylomicrons cannot enter the bloodstream directly because they are too large to cross capillary walls.) The flow of lymphatic fluid eventually carries the chylomicrons to the bloodstream, where much of the lipid contained within them is released to cells for use, including cells in adipose tissue.

Vitamins

Fat-soluble vitamins (vitamins A, D, E, and K) are absorbed by epithelial cells in parallel with lipids because they dissolve readily in lipid droplets, micelles, and chylomicrons. Water-soluble vitamins are absorbed through

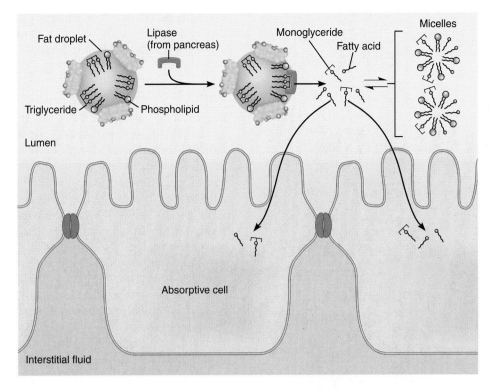

FIGURE 19.18 Liberation of fatty acids and monoglycerides from fat droplets by lipases. *Note that these end-products of fat digestion aggregate into micelles, which are in equilibrium with dissolved molecules. Some dissolved molecules cross the apical membranes of intestinal epithelial cells by simple diffusion.*

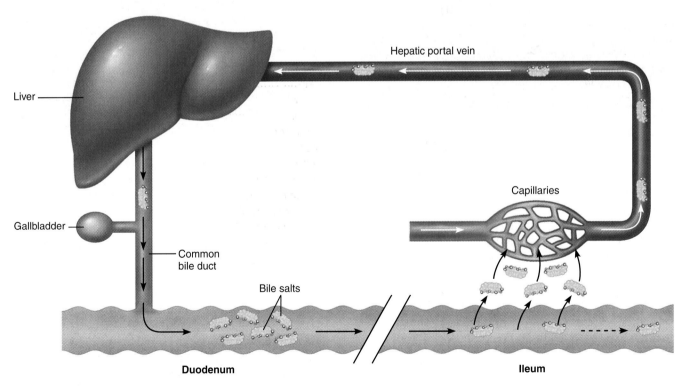

FIGURE 19.19 The enterohepatic circulation. *Bile salts are absorbed in the ileum, return to the liver via the circulation, and are eventually resecreted in the bile.*

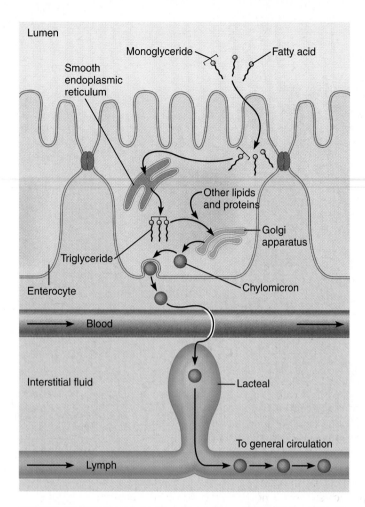

FIGURE 19.20 Events in the absorption of fats. *Fatty acids and monoglycerides absorbed into enterocytes are synthesized into triglycerides, which are then packaged into chylomicrons. After exocytosis into interstitial fluid, chylomicrons enter the lymphatic system via lacteals and eventually reach the general circulation.*

What prevents chylomicrons from entering the bloodstream directly via capillaries in intestinal villi?

the action of special transport proteins. One water-soluble vitamin, vitamin B_{12}, cannot be absorbed by itself; it can be absorbed only when bound to intrinsic factor, which is secreted into the lumen of the stomach. Intrinsic factor binds to vitamin B_{12} to form a complex that subsequently is absorbed in the ileum. Because vitamin B_{12} is necessary for the synthesis of hemoglobin, its deficiency leads to a particular form of anemia called *pernicious anemia,* which can result either from a lack of vitamin B_{12} in the diet or an inability to absorb it. Ulcers (When It Goes Wrong:

They are too large.

Ulcers, p. 622) and other damage to the gastric mucosa can lead to pernicious anemia by interfering with the production and secretion of intrinsic factor.

Minerals and Water

Events in the absorption of minerals and water by the digestive tract largely parallel the events that occur in the kidney tubule. Water absorption is passive and is driven by an osmotic gradient across the mucosal epithelium that is created by the transport of solutes (sodium and chloride in particular) from the lumen to the interstitial fluid. Usually, about 95% of the water that is initially present in the duodenum is absorbed by the time chyme reaches the colon.

Although the precise mechanism varies from region to region in the small intestine, the absorption of sodium across the mucosal epithelium is always active and is most often accompanied by the passive absorption of chloride. In the jejunum, bicarbonate is absorbed along with chloride, but in the ileum and colon, bicarbonate is usually secreted. In addition to sodium, minerals absorbed by the intestine include potassium, magnesium, calcium, iron, zinc, and various trace metals. Potassium is absorbed in the small intestine but may be absorbed or secreted in the colon, depending on its lumenal concentration. When its concentration in the lumen is low, potassium is secreted; when the concentration is higher, potassium is absorbed.

Quick Test 19.5

1. What are the end-products of triglyceride digestion?
2. Are absorbed fats transported directly into the bloodstream? Why or why not? What are chylomicrons, and what is their role in the absorption of fats?
3. What role do bile salts play in the digestion of fats?

GENERAL PRINCIPLES OF GASTROINTESTINAL REGULATION

The digestive system is unlike the other organ systems we have studied in that for the most part it does not act to maintain constancy of conditions in the body's internal environment—at least not directly—because generally the amount of material absorbed by the digestive tract is not determined by conditions inside the body. (This is not true for all substances, however.) Most of the nutrients ingested are absorbed completely, so virtually nothing with nutritive value is eliminated in the feces. Indeed, the

Chylomicrons belong to a general class of protein-and-lipid-containing particles known as *lipoproteins,* which range from 75–1500 μm in diameter and have an inner core of hydrophobic lipids surrounded by a hydrophilic coat. Lipoproteins are useful for transporting lipids in blood because their hydrophilic surfaces allow them to mix readily with water, which lipids cannot do on their own.

Chylomicrons are but one type of lipoprotein; the liver makes other types that are released into the bloodstream and distribute lipids to the body's cells. (These particles contain lipids that were either absorbed and stored in the liver or manufactured by the liver.) The inner core of a lipoprotein contains triglycerides, cholesterol, and other hydrophobic lipids and fat-soluble compounds. The outer coat contains phospholipids, other amphipathic lipids, and proteins called *apoproteins,* which both increase the particles' water solubility and serve as cofactors for various enzymes that act on the particles. Certain apoproteins bind to receptors on the surfaces of cells, triggering uptake by receptor-mediated endocytosis. Inside cells the lipoproteins are enzymatically degraded, enabling the cells to use their contents.

As lipoproteins circulate in the bloodstream, they are acted upon by an enzyme called *lipoprotein lipase,* which is found on the endothelium of capillaries throughout the body, particularly in adipose tissue and skeletal muscle. This enzyme hydrolyzes triglycerides within the lipoproteins, liberating glycerol and free fatty acids that can be used for energy by cells in the vicinity. As a result, the particles' size decreases and their density increases. (The density increases because the ratio of fat to protein decreases; fats are less dense than proteins.) Many of these particles are taken up by the liver and degraded, but a certain fraction remain in the bloodstream for long periods of time.

Plasma lipoproteins are classified into different groups according to their density, which differs depending on the ratio of protein to lipid. Particles with the lowest densities are relatively rich in lipid and are classified as *very-low-density lipoproteins* (VLDLs). Those with the highest densities are relatively rich in protein and are called *high-density lipoproteins* (HDLs). Other lipoproteins are classified as *low-density lipoproteins* (LDLs) or as *intermediate-density lipoproteins* (IDLs).

LDL particles have received much attention from the medical community in the past few decades because they have been implicated in the development of *atherosclerotic plaques,* cholesterol-rich deposits on the linings of blood vessels that can cause coronary artery disease and other cardiovascular problems. Because of the connection between cholesterol and atherosclerosis, physicians have long advised their patients (particularly those at risk for heart disease) to reduce their dietary intake of animal fats and egg yolk, which are rich in cholesterol. A plasma cholesterol level of 200 mg/dL (milligrams per deciliter) or lower is considered to be in the acceptable range; levels above 240 mg/dL are considered high and presumably put a person in the "high risk" category for heart disease. Recently, however, it has become apparent that the risk of heart disease depends not on *total* cholesterol level, but instead on the relative quantities of cholesterol carried by HDLs and LDLs. LDL cholesterol is often called "bad cholesterol" because higher concentrations of LDLs in the blood are associated with a higher risk for atherosclerosis (other things being equal). In contrast, HDL cholesterol is referred to as "good cholesterol" because higher HDL levels are associated with lower risks for the disease. As the ratio of HDL cholesterol to LDL cholesterol increases, the risk of atherosclerosis decreases.

Although the reason why HDL cholesterol is "good" is not yet known, it may be related to the ability of these particles to carry out *reverse cholesterol transport.* In this process, HDLs take up cholesterol from cells and from other lipoprotein particles through the action of enzymes that physically transfer molecules from one place to another. Subsequently the HDL particles are taken up by the liver and degraded, and much of the cholesterol contained within them is eliminated in the bile. In this manner, HDL particles help to clear cholesterol from the body by shuttling it to the liver for elimination.

To reduce their risk of atherosclerosis, people can take a number of measures to increase the ratio of HDL cholesterol to LDL cholesterol, including regular exercise, weight reduction (for those who are overweight), reduction of the dietary intake of saturated fats, and quitting smoking, all of which are known to raise plasma HDL levels. Total cholesterol levels can also be reduced by drugs that either suppress cholesterol synthesis or interfere with the absorption of cholesterol from the GI tract.

digestive system might be said to operate according to the principle of "waste not, want not"; it will absorb all the nutrients it can, regardless of the quantity ingested.

Although GI functions are not regulated according to the conditions in the internal environment, they are regulated by an impressive array of neural and hormonal regulatory mechanisms, many of which involve negative feedback control. But as we will see, many of these mechanisms work to control conditions *in the lumen of the GI tract,* not those in the internal environment. Now we take a brief look at general principles pertaining to the regulation of digestive function.

About 10% of the U.S. population is affected by *ulcers*, which are erosions of the lining of the GI tract that are deep enough to penetrate through the muscularis mucosae. Such lesions result from the combined action of acid and pepsin, which break down tissues. The most vulnerable areas are the lower regions of the stomach and the uppermost portion of the duodenum; duodenal ulcers occur approximately ten times more frequently than stomach ulcers. Affliction with either type of ulcer is termed *peptic ulcer disease*, the most common symptom of which is a chronic, rhythmic, and periodic gnawing or burning pain in the stomach area that is usually relieved by drinking milk (which contains proteins that buffer acid), eating, or taking antacids. For reasons that are not understood, ulcers appear more frequently in the spring and fall than at other times of the year!

If an ulcer is sufficiently deep, acid and pepsin can break down the walls of nearby blood vessels, causing bleeding. In extreme cases the lesion can penetrate through the wall of the stomach or small intestine, allowing the luminal contents to leak into the abdominal cavity. This condition, known as a *perforating ulcer*, is extremely serious because the leakage of bacteria-laden material almost invariably leads to inflammation of the peritoneum (*peritonitis*), which is often fatal.

Although ulcers are commonly associated with stress, they can be brought on by other risk factors. Chronic use of aspirin and other nonsteroidal anti-inflammatory drugs increases the risk of ulcer because these agents suppress the secretion of both mucus and bicarbonate, which normally protect the lining of the GI tract from the effects of acid and pepsin. The risk of ulcer is also increased by chronic alcohol use or the leakage of bile from the duodenum into the stomach, both of which can disrupt the mucus barrier. Surprisingly, ulcers are usually not associated with abnormally high rates of stomach-acid secretion; more often than not, acid secretion is normal or even below normal in most people with ulcers.

Researchers have recently discovered that many ulcers may result from infection with *Helicobacter pylori*, a bacterial species that is unusual in that it thrives in the acidic environment of the stomach. Although these bacteria are present in the stomachs of about 40% of the general population (most of whom experience no symptoms), they are present in over 90% of people with duodenal ulcers, and in 70% of those with gastric ulcers. These bacteria trigger ulcers not by destroying tissue directly, but instead by precipitating an immune response that destroys tissue, rendering it more vulnerable to attack by acid and pepsin.

Most ulcers can be successfully treated by drugs that suppress gastric acid secretion. These drugs include H_2 *receptor antagonists*, which block the binding of histamine to parietal cell H_2 receptors, and *pump blockers*, which directly suppress active transport of hydrogen ions in the stomach. Successful treatment of ulcers in those infected with *H. pylori* is usually accomplished by the administration of antibiotics in conjunction with acid-suppressing drugs.

Neural and Endocrine Pathways of Gastrointestinal Control

Most GI functions are controlled by reflexes coordinated by the enteric nervous system, but the central nervous system (CNS) also influences these functions. Many GI functions are also controlled by hormones secreted by endocrine cells in the stomach or small intestine.

The function of gastrointestinal organs is influenced by stimuli arising from within the GI tract, such as the presence or absence of food or changes in the acidity of the lumenal contents. Conditions in the lumen of the tract are monitored by three types of receptor neurons located within the gastrointestinal wall: mechanoreceptors, which detect the degree of distention of the wall; chemoreceptors, which monitor the concentrations in the lumen of specific substances such as hydrogen ions or fats; and osmoreceptors, which monitor the osmolarity of the lumenal contents. These receptors project to both the enteric nervous system and the CNS.

TABLE 19.2 GASTROINTESTINAL HORMONES
AND THEIR ACTIONS

HORMONE	SITE OF SECRETION	STIMULI FOR SECRETION	ACTIONS
Gastrin	Stomach	Proteins and protein digestion products in stomach; distention of stomach; parasympathetic input to stomach	Stimulates gastric secretion and motility; stimulates ileal motility and relaxes ileocecal sphincter; stimulates mass movement of colon
Cholecystokinin (CCK)	Duodenum and jejunum	Fat or protein digestion products in duodenum	Inhibits gastric secretion and motility; stimulates pancreatic bicarbonate secretion (potentiates actions of secretin); stimulates pancreatic enzyme secretion; stimulates bile secretion by liver; stimulates gallbladder contraction and relaxation of sphincter of Oddi
Secretin	Duodenum and jejunum	Acid in duodenum	Inhibits gastric secretion and motility; stimulates pancreatic bicarbonate secretion; stimulates pancreatic enzyme secretion (potentiates actions of CCK); stimulates bile secretion by liver
Glucose-dependent insulinotropic peptide (GIP)	Duodenum and jejunum	Glucose, fats, or acid in duodenum; distention of duodenum	Inhibits gastric secretion and motility; stimulates insulin secretion by pancreas

The enteric nervous system and CNS exert their control over GI function via neurons projecting to various types of effector cells in GI organs, usually smooth muscle cells or secretory cells (which may be exocrine or endocrine). Once endocrine cells are stimulated, the hormones they secrete circulate in the bloodstream and eventually return to the GI system to exert their effects. Of the four gastrointestinal hormones whose roles are well-established—gastrin, **cholecystokinin (CCK),** **secretin,** and **glucose-dependent insulinotropic peptide (GIP)**—the last three are secreted by cells in the duodenum and jejunum and are referred to collectively as *enterogastrones*. The sites of and the stimuli for the release of these hormones, and their actions, are tabulated in Table 19.2.

Short and Long Reflex Pathways

Neural and hormonal control of gastrointestinal function is summarized in Figure 19.21, which shows that a stimulus in the lumen of the digestive tract can trigger a response without any involvement of the CNS. In this case, signals may travel from receptors to the intrinsic nerve plexuses and then directly to the effectors, following what is called a **short reflex pathway;** alternatively, signals may follow a **long reflex pathway,** traveling from receptors to the CNS and then to the intrinsic nerve plexuses, which relay information to the effectors. (In some cases the CNS may relay signals directly to the effectors.)

As a general rule, long reflex pathways involve either the sympathetic nervous system, the parasympathetic nervous system, or both. Increased parasympathetic activity usually promotes an increase in gastrointestinal activity, which is manifested as an increase in muscle activity or fluid secretion; conversely, sympathetic activity generally has the opposite effect, promoting a reduction in gastrointestinal activity. (There are some exceptions to this rule, however.)

Phases of Gastrointestinal Control

A given region of the GI tract can respond to stimuli arising either within that region or in more remote regions. For example, the rate of stomach-acid secretion is influenced both by the degree of acidity of the stomach contents (secretion decreases with increasing acidity) and by the level of acidity in the duodenum. The function of GI organs can also be affected by stimuli arising from outside the GI tract; for instance, stomach acid secretion is influenced by the sight, taste, and smell of food.

Control of gastrointestinal function by stimuli arising in the head (such as the smell, taste, or thought of food) is referred to as **cephalic-phase control,** which is always effected by input from the CNS. Control of gastrointestinal function by stimuli arising in the stomach or the small intestine is referred to as **gastric-phase control** or **intestinal-phase control,** respectively. Gastric-phase and intestinal-phase stimuli exert their effects via long or short reflex pathways or by altering the secretion of gastrointestinal hormones.

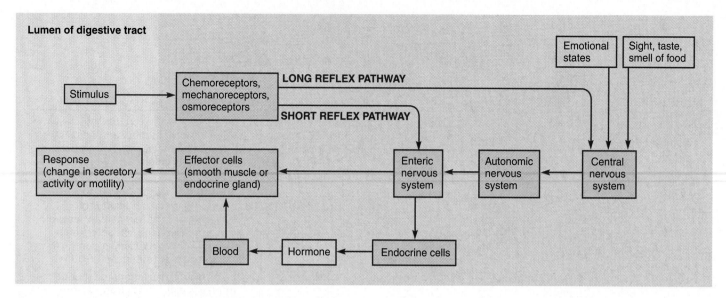

FIGURE 19.21 **Summary of neural and endocrine control of gastrointestinal function.**

Regulation of Food Intake

Even though food intake is largely independent of the body's need for new nutrients at a given time, food intake is strongly affected by other factors, including psychosocial influences. From your own experience you likely know that even when you are not particularly hungry, you can be persuaded to eat when you see an appealing dish, when you have a rare opportunity to eat a favorite food, or when you are with a group of friends who are eating.

Physiological regulation of food intake can occur over the short term or the long term. Eating a meal because you are hungry and then stopping because your hunger has disappeared is an example of short-term regulation; increases in your daily food consumption over the weeks since you began an exercise program is an example of long-term regulation. One factor that is apparently important in long-term regulation is **leptin,** a hormone secreted by adipose cells. When the dietary intake of calories in nutrients exceeds the body's demands, fat is deposited in adipose tissue, which secretes leptin in response. Leptin then acts on appetite-control centers in the hypothalamus to reduce the sensation of hunger. Leptin also promotes a general increase in the body's metabolic rate, which reduces fat storage.

Food intake is also influenced by a number of other physiological variables that are important in short-term regulation. The absorption of nutrients from a typical meal, for example, stimulates pancreatic islet cells to release the hormone insulin, which acts on the hypothalamus to reduce the sensation of hunger. Another hunger-suppressing hormone is CCK, which is released in

response to the presence of food in the duodenum. In response to the presence of food or digestion products in the lumen, signals from certain mechanoreceptors and chemoreceptors in the wall of the GI tract can also suppress hunger. Physiological signals having hunger-suppressing effects are referred to as *satiety signals.*

Quick Test 19.6

1. Name three major types of receptors that monitor conditions in the lumen of the GI tract.

2. What is the distinction between a short reflex pathway and a long reflex pathway?

3. To what do the terms *cephalic phase, gastric phase, and intestinal phase* refer?

4. Name the four confirmed gastrointestinal hormones, and identify their sites of release.

GASTROINTESTINAL SECRETION AND ITS REGULATION

In this section we examine the digestive system mechanisms of exocrine secretion in greater detail, with emphasis on how the mechanisms are regulated.

Saliva Secretion

The secretion of saliva is controlled by autonomic input to the salivary glands in response to stimuli arising in the mouth, such as the taste and texture of food. Information pertaining to the texture of food is relayed by mechano-receptors in the tongue and in other parts of the mouth to a *salivary center* in the medulla, which controls autonomic output to the salivary glands. Information pertaining to the taste of food is relayed to the salivary center by chemoreceptors in taste buds. The salivary center also receives input from the cerebral cortex pertaining to the sight and smell of food and other cephalic-phase stimuli.

Exercise Link

In the middle to late stages of the marathon, Bill and Jane noticed that their mouths felt dry and their saliva was thick, viscous, and harder to swallow. It might seem reasonable to assume this was related to evaporation as a result of their heavy breathing. However, it is most likely a result of the increased sympathetic activity during exercise, which reduces salivary flow and increases protein secretion.

Acid and Pepsinogen Secretion in the Stomach

The mechanism of acid secretion in the stomach is in some ways reminiscent of the mechanism of hydrogen ion secretion in the kidney tubule. The acid secreted by the stomach is generated inside parietal cells by the reaction of carbon dioxide with water, which is catalyzed by the enzyme *carbonic anhydrase* (CA) and generates hydrogen ions and bicarbonate (Figure 19.22). The hydrogen ions are actively transported into the lumen of the stomach by pumps in the apical membrane that use energy from ATP to transport hydrogen ions out of the cell in exchange for potassium ions. Bicarbonate ions exit the cell across the basolateral membrane in exchange for chloride ions, which then move into the lumen through channels in the apical membrane. The net result is that hydrogen and chloride ions are transported into the lumen while bicarbonate is transported into the interstitial fluid.

The secretion of acid by parietal cells is stimulated by parasympathetic nervous activity, gastrin, and histamine, which is secreted by cells in the stomach lining and acts as a paracrine agent. Acid secretion is controlled by cephalic-phase, gastric-phase, and intestinal-phase stimuli and is generally stimulated when food is present in the stomach and suppressed when the food leaves. Because the stimuli that affect acid secretion also tend to affect the secretion of pepsinogen in the same manner, the secretion of pepsinogen generally rises and falls in parallel with changes in acid secretion.

As shown in Figure 19.23a, cephalic-phase stimuli arising either from the sight, taste, and smell of food or

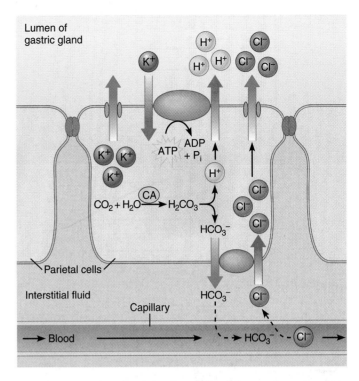

FIGURE 19.22 Mechanism of gastric acid secretion. *Note that because potassium ions are actively transported from the lumen into parietal cells and then passively leak back into the lumen, these ions are simply recycled across the apical membrane.*

Is the electrochemical driving force on hydrogen ions directed into or out of the cell?

from the acts of chewing and swallowing trigger increased activity in parasympathetic nerves to the stomach, which stimulates both parietal cells and chief cells to secrete acid and pepsinogen, respectively, and G cells to secrete gastrin. The gastrin enters the bloodstream and stimulates the parietal and chief cells, reinforcing the effects of parasympathetic input.

Once food reaches the stomach, gastric-phase stimuli come into play (Figure 19.23b). The presence of proteins and protein digestion products in the lumen stimulates chemoreceptors in the wall of the stomach, whereas the presence of food distends the stomach, which activates mechanoreceptors. As a result, signals are relayed to parietal cells and chief cells via short reflex and long reflex pathways, triggering the release of acid and pepsinogen, and also to G cells, triggering the release of gastrin. The proteins present in the lumen also exert a direct effect on G cells that stimulates them to secrete gastrin.

As food leaves the stomach, both gastric-phase and intestinal-phase stimuli work to *reduce* acid and pepsino-

Into the cell

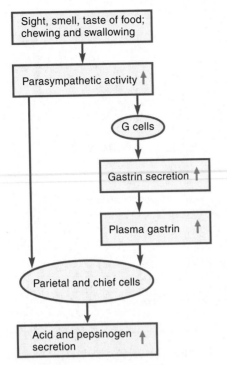

(a) Cephalic-phase control of gastric secretion

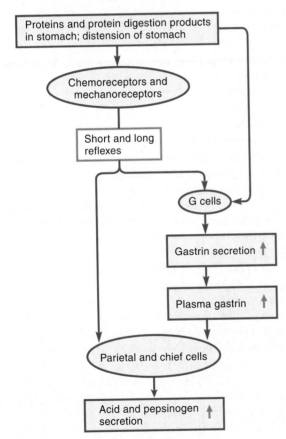

(b) Gastric-phase control of gastric secretion

FIGURE 19.23 Regulation of gastric secretion. *Shown are the pathways by which acid and pepsinogen are secreted in response to* **(a)** *cephalic-phase stimuli and* **(b)** *gastric-phase stimuli when food is being consumed and is present in the stomach. This flow chart is available as an interactive exercise under Activities at* www.physiologyplace.com.

gen secretion. The exit of food from the stomach reduces the degree of distention and lowers the concentration of proteins and protein digestion products in the lumen, causing a *withdrawal* of the stimuli that previously stimulated gastric secretion. The exit of proteins also tends to cause a rise in gastric acidity (because proteins normally buffer some of the secreted hydrogen ions), which acts directly on G cells to suppress gastrin secretion, thereby causing the withdrawal of another stimulus for acid secretion.

The reduction in gastric secretion that occurs during gastric emptying is also triggered by signals arising from the entry of food into the duodenum. As chyme leaves the stomach, the osmolarity of the duodenal contents rises, the concentrations of fats and acid in the contents rise, and the duodenum becomes more distended. These changes stimulate chemoreceptors, osmoreceptors, and mechanoreceptors, which relay signals via long reflex and short reflex pathways to parietal and chief cells in the stomach, inducing a decrease in acid and pepsinogen secretion. In addition, signals relayed to the endocrine cells in the small intestine that secrete CCK, secretin, and GIP suppress the secretory activity of parietal and chief cells.

Secretion of Pancreatic Juice and Bile

The secretion of pancreatic juice begins in the acini, where cells produce a relatively small volume of fluid containing water, electrolytes, and digestive enzymes. As this fluid flows through the ducts leading from the acini, a larger volume of bicarbonate-rich fluid is added to it. Even though both components of the pancreatic juice—enzyme-rich fluid and bicarbonate-rich fluid—are secreted together during a meal, the regulatory mechanisms that control their secretions are somewhat separate, and for this reason the composition of the pancreatic juice can vary.

The secretion of pancreatic juice is influenced by cephalic-phase, gastric-phase, and intestinal-phase stimuli, although the last predominate. The strongest influences on pancreatic secretion are the hormones CCK and secretin, which are released in response to the presence of food in the duodenum. CCK acts primarily on acinar cells to stimulate enzyme secretion, whereas secretin acts primarily on duct cells to stimulate the secretion of bicarbonate-rich fluid. Although CCK by itself is only a weak stimulus for bicarbonate secretion, its effect becomes stronger when secretin is present. Likewise, secretin by itself exerts only a weak stimulatory effect on enzyme secretion, but this effect becomes much stronger when CCK is present. Put another way, when CCK and secretin are both present, these hormones amplify each other's effects, a phenomenon known as **potentiation.**

The release of secretin and the subsequent secretion of bicarbonate-rich fluid is strongly affected by the acidity

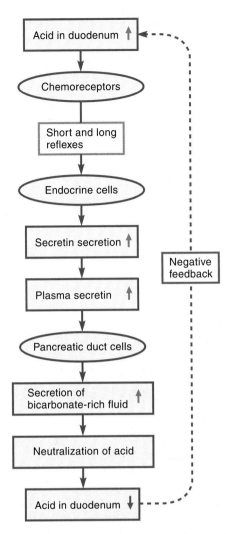

FIGURE 19.24 Stimulation of the secretion of bicarbonate-rich fluid by acidity in the duodenum. *This flow chart is available as an interactive exercise under Activities at* www.physiologyplace.com.

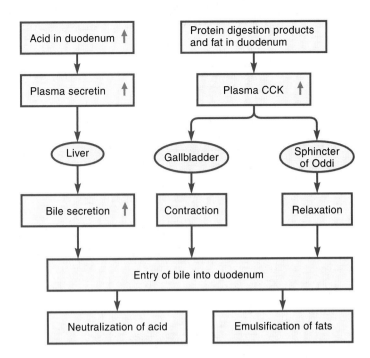

FIGURE 19.25 The mechanisms by which secretin and CCK regulate the entry of bile into the duodenum.

of the duodenal contents (Figure 19.24). The flow of chyme from the stomach into the duodenum raises the acidity of the duodenal contents, which stimulates chemoreceptors in the wall. As a result, signals are relayed via short reflex and long reflex pathways to endocrine cells in the duodenum and jejunum, which then release secretin in response. The resulting rise in plasma secretin levels acts on pancreatic duct cells to stimulate the secretion of bicarbonate-rich fluid, which combines with hydrogen ions in the duodenum to lower the acidity of the contents. (Note that these events constitute negative feedback control, because a rise in acidity triggers a series of events that culminate in a fall in acidity.) This reduction of acidity is necessary for the proper activity of pancreatic and other enzymes that work in the small intestine.

The secretion of CCK is regulated primarily by the concentrations of protein digestion products and fat in the duodenum, which rise as chyme leaves the stomach. This change stimulates chemoreceptors, which relay sig-

nals via short reflex and long reflex pathways to endocrine cells, triggering the release of CCK, which acts on acinar cells to stimulate enzyme secretion. Later, once digestion products have left the duodenum, these signals are turned off, so CCK secretion falls.

CCK and secretin are also responsible for regulating the entry of bile into the duodenum (Figure 19.25). In response to increased acidity in the duodenum, secretin acts on the liver to stimulate bile secretion; the presence of protein digestion products and fat in the duodenum stimulates the secretion of CCK, which by promoting gallbladder contraction and relaxation of the sphincter of Oddi, allows bile to flow into the duodenum.

Rates of Fluid Movement in the Digestive System

The volume of fluid that moves into and out of the digestive system each day is impressive, amounting to several times the normal plasma volume (Figure 19.26). The liver and pancreas combined secrete into the GI tract an average of 2 liters of fluid per day, and the salivary glands secrete another 1.5 liters. An additional 1.5 liters is secreted by glands in the wall of the small intestine, and about 2 liters enter the GI tract via ingested water. The great majority of the secreted and ingested fluid—some 8.5 liters each day—is absorbed by the small intestine; the colon

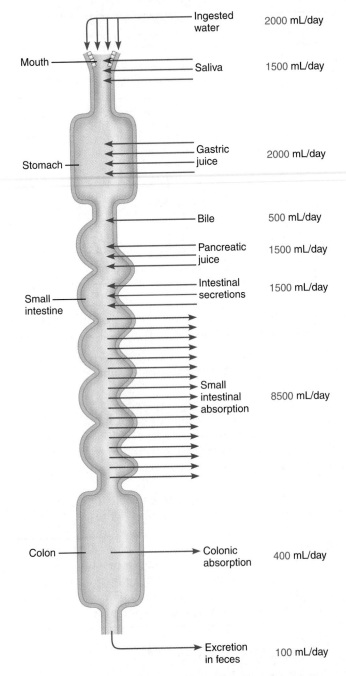

Ingested water		2000 mL/day
Mouth	Saliva	1500 mL/day
Stomach	Gastric juice	2000 mL/day
	Bile	500 mL/day
	Pancreatic juice	1500 mL/day
Small intestine	Intestinal secretions	1500 mL/day
	Small intestinal absorption	8500 mL/day
Colon	Colonic absorption	400 mL/day
	Excretion in feces	100 mL/day

FIGURE 19.26 Approximate daily fluid flows in the digestive system.

also absorbs some fluid. When the movements of fluid into and out of the GI tract have been accounted for, the difference between the amount that enters and the amount that is absorbed is quite small—only about 100 ml per day. This unabsorbed water is eliminated in the feces.

Quick Test 19.7

1. What is the role of carbonic anhydrase in stomach acid secretion? What other ions are secreted with hydrogen ions?
2. Of the following stimuli, choose those that would be expected to stimulate gastric acid secretion: distention of stomach, distention of duodenum, increased protein digestion products in stomach, increased acidity in duodenum, increased osmolarity in duodenum.
3. Of the following stimuli, choose those that would be expected to stimulate the secretion of enzymes or bicarbonate-rich fluid by the pancreas: increased acidity in the duodenum, a rise in plasma CCK levels, a rise in plasma secretin levels, decreased protein digestion products in the duodenum, increased fat in the duodenum.

GASTROINTESTINAL MOTILITY AND ITS REGULATION

The GI tract is capable of generating finely tuned patterns of motion that both propel forward and mix the lumenal contents. In fact, it is largely because of this motility that digestion and absorption are so efficient. In this section we examine the various patterns of motility in digestive organs, and how motility is regulated. We begin with the basis of gastrointestinal motility: electrical activity in smooth muscle.

Electrical Activity in Gastrointestinal Smooth Muscle

Contractions of gastrointestinal smooth muscle are triggered by electrical activity in pacemaker cells clustered in the muscle layers at various locations along the GI tract. These cells exhibit slow, spontaneous, graded depolarizations known as **slow waves,** which when large enough to bring the membrane potential to threshold are accompanied by action potentials (Figure 19.27). When action potentials do occur, they tend to occur in bursts coinciding with the peaks of the slow waves. Extensive electrical coupling between smooth muscle cells, due to the presence of gap junctions, allows propagation of this electrical activity (both slow waves and action potentials) to nearby nonpacemaker cells. The frequency of slow waves varies from region to region along the length of the tract because muscle cells in different regions are driven by different sets of pacemaker cells.

In any given region of the GI tract, slow waves occur at regular intervals at a fairly constant frequency, a pattern that is referred to as the **basic electrical rhythm**

(BER) and is affected by neural activity and hormones. In general, parasympathetic activity is excitatory and tends to promote increased contractile force; sympathetic activity has the opposite effect. These influences reflect changes in the pattern of slow wave activity.

Neural activity and hormonal signals generally affect the *height* of slow waves rather than their frequency. Excitatory stimuli shift the membrane potential upward, so that it still increases and decreases with each slow wave, but at a higher (more positive) average value; inhibitory stimuli shift the membrane potential downward. An upward shift in the membrane potential may trigger either the sudden appearance of action potentials (if none are occurring initially) or an increase in their frequency, both of which generally result in increases in contractile force (see Figure 19.27). Downward shifts in the membrane potential tend to reduce action potential frequency and contractile force.

Although contractile activity is related to the pattern of electrical activity in gastrointestinal smooth muscle, the nature of the relationship depends on the location of the muscle. In the stomach, smooth muscle generates force in a graded fashion that varies according to the degree of depolarization; larger depolarizations trigger stronger contractions. If the depolarizations are large enough to trigger action potentials, the force of contraction increases. In the intestines, no force is generated unless action potentials occur. When action potentials do occur, the strength of contraction varies in a graded fashion with changes in the action potential frequency, as occurs in gastric smooth muscle.

In the sections that follow we examine motility in greater detail, beginning with the mouth, pharynx, and esophagus, which are involved in chewing and swallowing. Because this part of the GI tract contains skeletal muscle rather than smooth muscle, the principles just outlined do not apply; those muscles, like skeletal muscles everywhere, contract only when commanded to do so by the central nervous system and do not exhibit slow-wave activity.

Chewing and Swallowing

Chewing is like breathing in that it is controlled both consciously and unconsciously; we can control the rate and force of chewing motions voluntarily, or we can chew without even thinking about it. Unconscious chewing is orchestrated by a **chewing reflex** that is activated by the presence of food in the mouth as follows: When food is not present in the mouth, the muscles that hold the lower jaw closed are tonically active and constantly exert force. However, the presence of food in the mouth stimulates pressure receptors and triggers inhibition of the jaw-closing muscles, allowing the jaw to drop in response to the pull of gravity. Opening of the jaw relieves the pressure of food against the receptors, which removes the in-

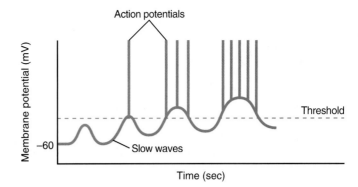

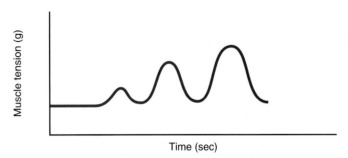

FIGURE 19.27 **Slow-wave electrical activity in gastrointestinal smooth muscle.** *Depolarizations that exceed threshold result in action potentials (top), inducing in gastrointestinal smooth muscle contractions whose strengths vary with the frequency of the action potentials (bottom).*

hibitory stimulus from the jaw-closing muscles, allowing them to close the jaw once again. When this happens, however, the pressure is restored, which triggers inhibition of the jaw-closing muscles and allows the jaw to open, and so on. Thus the operation of this reflex induces the alternating opening and closing motion of the jaw that constitutes chewing.

Chewing and the associated tongue movements both reduce the food mass to smaller particles and ensure that food is mixed thoroughly with saliva, transforming the food into a semisolid mass called a **bolus.** When the bolus becomes soft and moist enough to swallow, the tongue propels it to the back of the mouth and into the pharynx, where it stimulates mechanoreceptors that initiate the **swallowing reflex,** a series of muscle contractions coordinated by the *swallowing center* in the medulla. The swallowing reflex involves the following series of steps (Figure 19.28):

① As the bolus begins to descend from the pharynx, it presses downward on the epiglottis (a flap of tissue guarding the glottis), causing it to cover the glottis and preventing the bolus from entering the larynx and trachea. Closure of the glottis is aided by muscles of the neck, which raise the larynx. Reflex mechanisms also inhibit inspiratory muscles, which suppresses breathing motions.

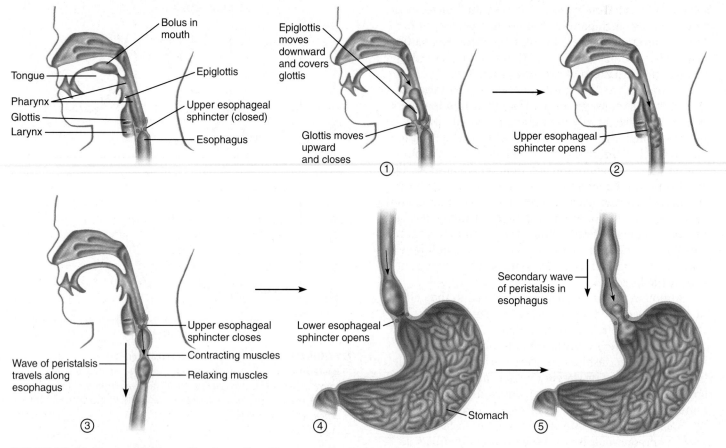

FIGURE 19.28 **Events of the swallowing reflex.** *The steps are described in greater detail in the text.*

② The upper esophageal sphincter relaxes, enabling the entry of the bolus into the esophagus. Once the bolus passes through, the sphincter closes again.

③ Entry of the bolus into the esophagus stimulates stretch receptors, triggering **peristalsis** (a wave of contraction that travels the length of a hollow vessel or organ). The peristaltic wave propels the bolus toward the stomach, which it reaches in about 9 seconds.

④ Upon arrival of the bolus at the lower end of the esophagus, the lower esophageal sphincter relaxes momentarily to allow the bolus to enter the stomach.

⑤ In the event that the initial or primary peristaltic wave is unsuccessful in delivering the bolus to the stomach, it is followed by additional secondary waves initiated by stimulation of mechanoreceptors in the esophagus.

Prior to the arrival of the bolus, the stomach makes preparations to accommodate it because the swallowing center in the medulla also triggers relaxation of smooth muscles in the upper portion of the stomach, which increases its volume (a phenomenon called *receptive relaxation*).

Gastric Motility

Following ingestion of a meal, the muscles of the stomach accomplish two tasks: mixing of the chyme to ensure that food and gastric juice are thoroughly combined, and regulation of gastric emptying such that chyme enters the small intestine at an appropriate rate. Both of these tasks are achieved by peristaltic waves in the stomach, which are coordinated by the enteric nervous system.

Gastric Motility Patterns

Peristaltic waves travel downward from the upper body of the stomach to the pylorus, normally at a rate of about three per minute. Each wave begins as a weak contraction but progressively increases in force as it advances toward the pylorus. Because the pyloric sphincter is normally closed (except for a brief period of opening that occurs upon arrival of the wave at the pylorus), the wave of contraction does not push the bulk of the chyme forward; instead, most of it is forced to flow backward, which mixes it.

Peristaltic waves in the stomach also are responsible for propelling the chyme forward into the duodenum, which occurs as the pyloric sphincter opens with each peristaltic wave. The rate of gastric emptying depends on several factors, including the composition of the chyme, the volume of chyme in the stomach, and the force of gastric contractions. In general, emptying is faster when the volume of chyme is larger and gastric contractions are stronger.

During periods of fasting, peristaltic contractions eventually cease, and the stomach becomes quiescent for an hour or two, after which time another pattern of activity appears. First, the antrum begins to undergo a series of intense contractions (called a **migrating motility complex**) that are accompanied by relaxation of the pyloric sphincter. After a while these contractions stop and the stomach goes into another period of quiescence, which is followed by another burst of contractions, and so on. This pattern of activity sweeps the stomach of its contents, including any particles that might have been too large to pass through the pyloric sphincter.

Regulation of Gastric Motility

Control of gastric motility is achieved primarily through changes in the force of smooth muscle contractions. (Changes in frequency do occur but are relatively minor.) The force of gastric contraction increases in response to gastrin and decreases in response to CCK, secretin, and GIP. (In fact, GIP was originally called *gastric inhibitory peptide* because of its inhibitory influence on gastric motility.)

The control of gastric motility is similar to the control of gastric secretion in that motility is influenced by cephalic-phase, gastric-phase, and intestinal-phase stimuli. Effective cephalic-phase stimuli include pain, fear, and depression, which normally inhibit gastric motility, and anger and aggression, which stimulate it. Gastric-phase stimuli (including distention of the stomach) and intestinal-phase stimuli (including distention of the duodenum and changes in lumenal acidity, osmolarity, and fat concentration) regulate gastric emptying such that chyme exits the stomach at a rate that is well matched to the small intestine's ability to process it.

Vomiting

On occasion, certain conditions can cause the contents of the stomach and sometimes of lower portions of the GI tract to be forcefully expelled through the mouth, a phenomenon known as **vomiting.** Vomiting can be triggered by a variety of stimuli, including illness (such as influenza), strong emotional states, severe pain, severe distention of the stomach or small intestine, rotational motion of the head (as in motion sickness), or the ingestion of certain substances (such as copper sulfate). Substances that stimulate vomiting are known as *emetics.* Given that many emetics are poisonous, vomiting acts as a protective mechanism that removes these substances from the GI tract before substantial quantities can be absorbed into the blood.

Vomiting involves a complex sequence of events (called the *vomiting reflex*) that is coordinated by a region in the medulla called the *vomiting center.* Prior to the actual act of vomiting, a person usually experiences a sensation of nausea, the skin becomes pallid, and heart rate and sweating increase. Eventually, a series of deep inspirations are followed by closure of the glottis. Abdominal muscles begin to contract strongly, causing the abdominal wall to move inward as the inspiratory movements are causing the diaphragm to move downward. The combination of motions raises the abdominal pressure substantially and effectively squeezes the stomach, raising the pressure inside. Finally, the lower esophageal sphincter relaxes, allowing stomach contents to enter the esophagus. If contractions are not strong enough to eject this material into and out of the mouth, it returns to the stomach; this can occur several times before abdominal contractions become strong enough to push the material through the upper esophageal sphincter and out through the mouth.

Motility of the Small Intestine

Like the stomach, the small intestine exhibits different patterns of motility that depend on whether food is present or absent. When food is present, motility in the small intestine takes two forms: peristalsis and segmentation, which is the more common of the two.

Motility Patterns in the Small Intestine

When chyme is present in the small intestine, brief periods during which peristaltic waves travel a short distance and propel the chyme forward (Figure 19.29a) are interspersed with relatively longer periods of **segmentation,** in which alternate contraction and relaxation of neighboring segments of intestine shuttles the chyme back and forth (Figure 19.29b). The result of segmentation is that the chyme is stirred, which thoroughly mixes it with bicarbonate, digestive enzymes, and bile. Stirring also brings products of digestion into contact with the mucosa, which promotes their absorption.

During fasting, the pattern of motility in the small intestine changes; peristalsis and segmentation cease and are replaced by migrating motility complexes that periodically move through the intestine, sweeping it of its contents.

Segmentation, peristalsis, and migrating motility complexes result from the contractile activity of the external muscle layers, but the innermost layer (the thin muscularis mucosae) also undergoes regular contractions when food is present. These contractions mix chyme near the surface of the mucosa and may also exert a "massaging" effect on the lacteals, that helps propel the lymphatic fluid forward.

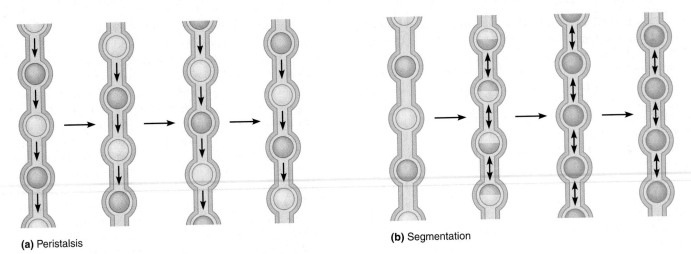

(a) Peristalsis

(b) Segmentation

FIGURE 19.29 Comparison of peristalsis and segmentation in the small intestine. (a) *In peristalsis, waves of contraction travel a short distance and then die out, propelling the chyme along the intestine's length.* **(b)** *In segmentation, waves of contraction cause neighboring segments of intestine to alternately contract and relax, mixing the chyme by shuttling it back and forth.*

Regulation of Motility of the Small Intestine

Contractions of the small intestine are influenced both by distention and by input from extrinsic nerves and hormones. Moderate distention of the intestine triggers an increase in contractile force, a response mediated by both short reflex and long reflex pathways that tends to relieve the distention by propelling the contents onward. Gastrin stimulates motility in the ileum and also promotes relaxation of the ileocecal sphincter.

The small intestine also exhibits a number of specialized reflexes that are orchestrated by the central nervous system and come into play only in certain situations. In the **intestino-intestinal reflex,** severe distention or injury to any portion of the small intestine inhibits contractile activity throughout the rest of the intestine, which helps protect the injured part from further stretching and additional injury. In the **ileogastric reflex,** distention of the ileum causes inhibition of gastric motility. In the **gastroileal reflex,** the presence of chyme in the stomach triggers increased motility in the ileum.

Motility of the Colon

In the colon, contractile activity serves to mix the chyme, to expose it to the mucosal surface, which facilitates the absorption of minerals and water, and to propel the lumenal contents toward the rectum for storage and eventual elimination.

Motility Patterns in the Colon

The more proximal portions of the colon exhibit a pattern of motility called **haustration,** which is similar to seg-

mentation in the small intestine, except that the segments *(haustra)* are delineated by permanent folds in the intestinal wall and are very regular in appearance (Figure 19.30). Haustration is also significantly slower than segmentation; contractions occur at a rate of about two per hour.

About three or four times a day, a different pattern of activity begins. This activity, called **mass movement,** is like a peristaltic wave, except that after a given portion of the intestine contracts, it remains contracted for a longer time before relaxing. These waves of contraction propel the luminal contents forward rapidly and sweep the colon clean.

Regulation of Motility of the Colon

Like the small intestine, the colon exhibits specialized reflexes that act only in certain circumstances. In the **colonocolonic reflex,** distention of one part of the colon induces relaxation of the remaining parts. In the **gastrocolic reflex,** the presence of a meal in the stomach triggers an increase in colonic motility and an increase in the frequency of mass movements.

Defecation

The elimination of feces from the body is controlled both unconsciously by a **defecation reflex** and voluntarily. The defecation reflex is triggered by distention of the rectum, which occurs as fecal material begins to enter the rectum from the colon, usually during mass movements of the colon. This distention stimulates stretch receptors and initiates several events. First, smooth muscle in the wall of the rectum is stimulated to contract, which raises the pressure inside. Peristaltic contractions of the sigmoid colon are also stimulated, which propels more fecal material into the rectum, further raising the pressure. At the same time, the internal anal sphincter relaxes while the external anal sphincter contracts, preventing the material from exiting the body. If the colonic contractions are strong enough to raise the pressure in the rectum to a certain critical level, the external anal sphincter relaxes, allowing defecation to proceed. In adults and children who are toilet trained, defecation may be postponed by voluntary contraction of the external anal sphincter.

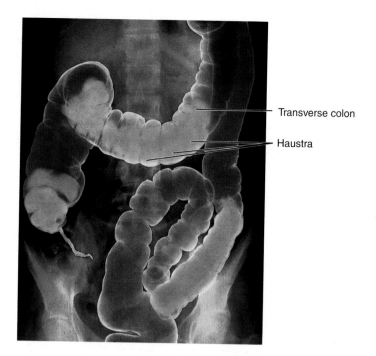

FIGURE 19.30 **An X-ray image of a human colon showing haustra.**

Quick Test 19.8

1. Define the following terms: *slow waves, basic electrical rhythm, peristalsis, segmentation, migrating motility complex, haustration, mass movement.*

2. Briefly describe what happens in the swallowing reflex.

3. Briefly describe what happens in the intestino-intestinal reflex, ileogastric reflex, gastroileal reflex, colonocolonic reflex, and gastrocolic reflex.

SUMMARY OF FACTORS INFLUENCING DIGESTION AND ABSORPTION OF NUTRIENTS

Figure 19.31 summarizes the various factors that influence the digestive system's ability to perform its primary function, which is to deliver nutrients to the bloodstream. Note that the regulation of secretion and motility ultimately influences digestion and absorption. Remember that this intricate, highly regulated system does not adjust the extent to which nutrients are digested and absorbed; instead, it works to digest and absorb whatever food is ingested! (Discovery: Gastrointestinal Facts, www.physiologyplace.com, Challenge Yourself)

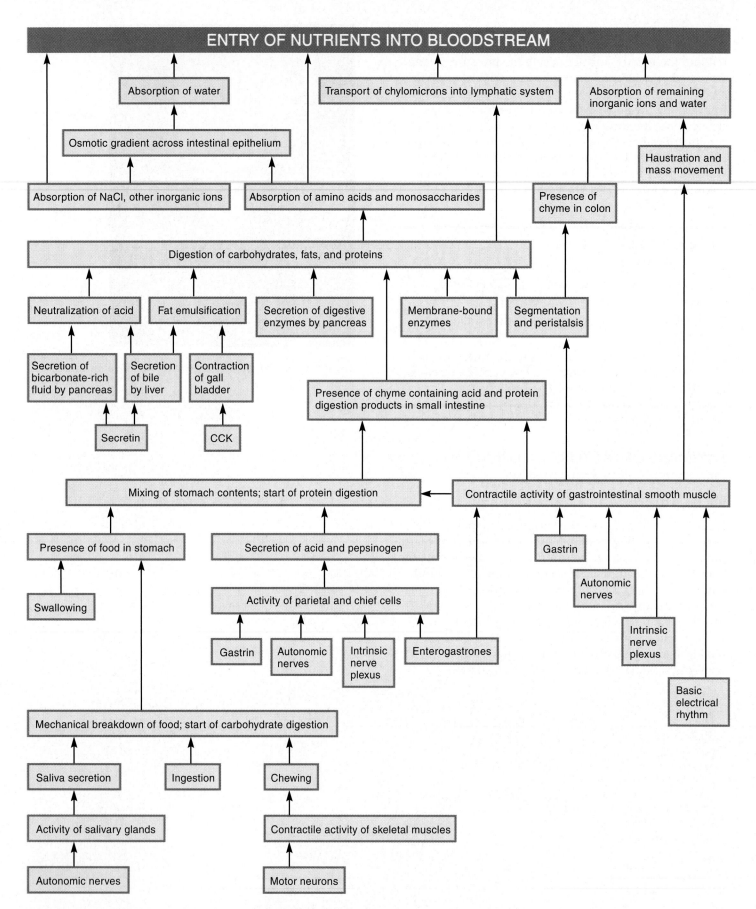

FIGURE 19.31 Summary of factors influencing digestive function.

SYSTEMS INTEGRATION

DIGESTIVE SYSTEM

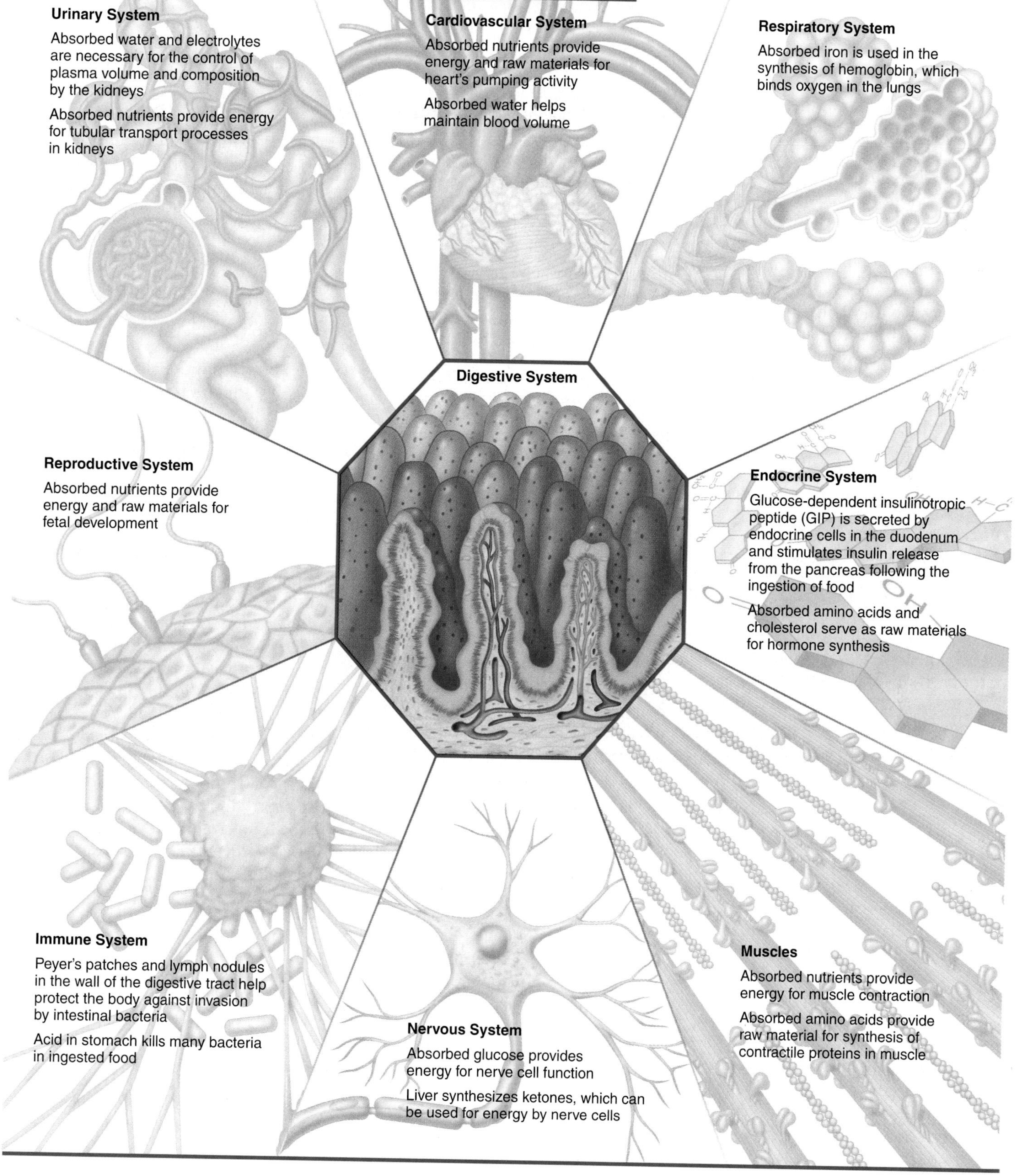

Urinary System
Absorbed water and electrolytes are necessary for the control of plasma volume and composition by the kidneys
Absorbed nutrients provide energy for tubular transport processes in kidneys
Cardiovascular System
Absorbed nutrients provide energy and raw materials for heart's pumping activity
Absorbed water helps maintain blood volume
Respiratory System
Absorbed iron is used in the synthesis of hemoglobin, which binds oxygen in the lungs
Digestive System
Reproductive System
Absorbed nutrients provide energy and raw materials for fetal development
Endocrine System
Glucose-dependent insulinotropic peptide (GIP) is secreted by endocrine cells in the duodenum and stimulates insulin release from the pancreas following the ingestion of food
Absorbed amino acids and cholesterol serve as raw materials for hormone synthesis
Immune System
Peyer's patches and lymph nodules in the wall of the digestive tract help protect the body against invasion by intestinal bacteria
Acid in stomach kills many bacteria in ingested food
Nervous System
Absorbed glucose provides energy for nerve cell function
Liver synthesizes ketones, which can be used for energy by nerve cells
Muscles
Absorbed nutrients provide energy for muscle contraction
Absorbed amino acids provide raw material for synthesis of contractile proteins in muscle

Overview of Digestive System Function, p. 602

The primary purpose of the digestive system is to extract needed materials from ingested food and deliver them to the bloodstream for distribution to cells throughout the body. Because most nutrient molecules are too large to be transported in the bloodstream, they must be broken down to smaller molecules by enzymes in the lumen of the digestive tract (digestion). The resultant digestion products are transported into the bloodstream (absorption). To aid these processes, fluids and enzymes are transported into the lumen of the tract (secretion), and muscular activity in the wall (motility) propels the lumenal contents from one digestive organ to the next.

Functional Anatomy of the Digestive System, p. 602

The digestive system comprises the digestive tract (GI tract) and accessory glands. Throughout most of its length, four layers make up the wall of the tract: (1) the mucosa, which lines the lumen; (2) the submucosa, an underlying layer of connective tissue containing numerous nerves and blood vessels; (3) the muscularis externa, which contains circular and longitudinal smooth muscle; and (4) the serosa, composed of connective tissue and an outermost mesothelium. The mucosa contains the mucous membrane, an epithelial layer containing secretory, absorptive, and endocrine cells (enterocytes). Within the wall is the enteric nervous system, consisting of two divisions: the submucosal plexus and the myenteric plexus. Organs of the GI tract include the mouth, esophagus (which conducts food to the stomach), stomach (which holds the food and mixes it with secretions to form chyme), small intestine (the primary site of digestion and absorption), colon (which ab-

sorbs water and electrolytes and stores feces), rectum, and anus (a passage leading to the outside). The flow of material between organs is regulated by sphincters. Accessory glands include the salivary glands (which secrete saliva), the pancreas (which secretes pancreatic juice containing enzymes and bicarbonate), and the liver (which secretes bile and processes absorbed nutrients). Pancreatic enzymes are generally secreted as inactive zymogens, which are activated upon entry into the GI tract.

Digestion and Absorption of Nutrients and Water, p. 614

Digestion of starch and glycogen begins in the mouth with the action of salivary amylase and is continued in the small intestine by pancreatic amylase. Brush border enzymes in the small intestine reduce carbohydrates to monosaccharides, which are transported across the mucosal epithelium and diffuse into the bloodstream. Protein digestion begins with the action of pepsin in the stomach and is continued in the small intestine by pancreatic enzymes (including trypsin, chymotrypsin, and carboxypeptidase) and membrane-bound enzymes (including aminopeptidase). Most proteins are reduced completely to amino acids, which are transported into the bloodstream. Dietary fats (mostly triglycerides) are reduced to fatty acids and monoglycerides by pancreatic lipases. This process is aided by bile salts, which emulsify fat droplets. Products of fat digestion enter epithelial cells by simple diffusion and then are reassembled into triglycerides, which are then transported (along with other lipids) into the lymphatic system in the form of chylomicrons, a type of lipoprotein. Water absorption is secondary to solute absorption and is driven by an osmotic gradient. Vitamins and minerals are absorbed chemically unaltered.

General Principles of Gastrointestinal Regulation, p. 620

Gastrointestinal regulatory mechanisms maximize the efficiency of digestion and absorption but generally do not act to maintain homeostasis. Digestive function is regulated by short reflex and long reflex pathways involving the enteric nervous system and hormones (including gastrin, secretin, cholecystokinin, and glucose-dependent insulinotropic peptide). The enteric nervous system receives input both from the autonomic nervous system and from mechano-receptors, chemoreceptors, and osmoreceptors that monitor conditions in the digestive tract.

Gastrointestinal Secretion and Its Regulation, p. 624

Saliva secretion is controlled by autonomic input to the salivary glands and is coordinated by the medullary salivary center. Gastric secretion of acid and pepsinogen is influenced by cephalic-phase, gastric-phase, and intestinal-phase stimuli and controlled by neural and hormonal reflexes. Pancreatic secretion is also controlled by neural and hormonal signals (primarily secretin and cholecystokinin). Bile secretion by the liver is stimulated by secretin and cholecystokinin, which also stimulates contraction of the gallbladder.

Gastrointestinal Motility and Its Regulation, p. 628

Gastrointestinal smooth muscle contractions are triggered by slow waves generated by pacemaker cells. Nerves and hormones generally influence the strength of contractions, but not the frequency. The stomach and intestines exhibit motility patterns that change depending on conditions in the lumen.

Multiple-Choice Questions

1. What do sodium, fatty acids, and vitamin A have in common?
   a) They are not enzymatically modified prior to absorption into the bloodstream.
   b) They cross the apical membranes of enterocytes by simple diffusion.
   c) They are transported into blood capillaries in the villi.
   d) They are all hydrophilic.

2. Blockage of the flow of bile into the duodenum interferes with the digestion of which of the following?
   a) starch
   b) proteins
   c) lipids
   d) disaccharides

3. Which of the following is an accurate statement regarding the various phases of gastrointestinal control?
   a) The autonomic nervous system is involved in cephalic-phase regulation only.
   b) The autonomic nervous system is involved in gastric-phase regulation only.
   c) The autonomic nervous system is involved in intestinal-phase regulation only.
   d) The autonomic nervous system is involved in all three phases of regulation.

4. Failure of the salivary glands to secrete amylase would make it impossible to digest which of the following?
   a) proteins
   b) fats
   c) starch
   d) none of the above

5. Which of the following tends to inhibit acid secretion by the stomach?
   a) neutralization of acid in the duodenum by bicarbonate in pancreatic juice
   b) entry of stomach acid into the duodenum
   c) the arrival of food in the stomach
   d) secretion of fluid into the duodenum by cells in the crypts of Lieberkuhn, which dilutes the concentration of nutrients in the duodenum.

6. Which of the following best illustrates the phenomenon of potentiation?
   a) Bile secretion is normally stimulated by both secretin and CCK.
   b) Secretin and CCK act primarily on different parts of the hepatic secretory apparatus.
   c) Secretin and CCK both stimulate bile duct cells to secrete fluid.
   d) The secretion of fluid by bile duct cells is greater when both CCK and secretin are present, compared to when either hormone is present alone (at the same concentration).

7. The enzyme enterokinase is directly or indirectly responsible for the proper functioning of
   a) bile salts.
   b) trypsin.
   c) chymotrypsin.
   d) both trypsin and chymotrypsin.

8. Which of the following is an example of a zymogen?
   a) enterokinase
   b) chymotrypsinogen
   c) salivary amylase
   d) cholecystokinin

9. Increases in gastric motility are generally accompanied by increases in ileal motility because of the
   a) gastroileal reflex.
   b) ileogastric reflex.
   c) gastrocolic reflex.
   d) colonocolonic reflex.

Objective Questions

1. Through the action of pancreatic amylase alone, starch could be broken down to a form that could be absorbed completely. (true/false)

2. The glands that secrete acid in the stomach are examples of accessory glands. (true/false)

3. The rate of gastric emptying tends to (increase/decrease) in response to the presence of acid in the duodenum.

4. Receptive relaxation is an example of (cephalic-phase/gastric-phase) control of stomach function.

5. The (submucosal/myenteric) nerve plexus is located within the muscularis externa.

6. The _____ is the outermost layer of the gastrointestinal wall, consisting of the mesothelium and an underlying layer of connective tissue.

7. In the stomach, food is mixed with gastrointestinal secretions to become a semisolid material called _____.

8. The phenomenon by which severe distention of one part of the small intestine induces inhibition of motility in the remainder of the intestine is known as the (intestino-intestinal/ileogastric) reflex.

9. When food is present in the small intestine, contraction of the gallbladder is stimulated by (secretin/cholecystokinin).

10. Bile salts are necessary for the proper functioning of pancreatic (proteases/lipases).

11. The lamina propria is located within the (mucosa/submucosa).

12. Gastrin (stimulates/inhibits) the secretion of acid by the stomach.

13. In the stomach, pepsinogen is secreted by (chief cells/parietal cells).

14. The small intestine is periodically swept of its contents by (segmentation/migrating motility complexes).

15. The term *enterohepatic circulation* refers to the conduction of blood from the intestine to the liver via the hepatic portal vein. (true/false)

16. Nerves and hormones exert their effects on gastrointestinal motility primarily by altering the (frequency/amplitude) of slow waves.

17. Chewing is under voluntary control but is also controlled by reflex neural pathways. (true/false)

18. Disaccharides are broken down to monosaccharides by (pancreatic enzymes/membrane-bound enzymes).

Essay Questions

1. Describe the process by which ingested triglycerides are digested, absorbed, and transported into the bloodstream. Indicate which of the steps are adaptations to triglycerides' hydrophobic nature, and explain why. Compare these steps to the corresponding processes pertaining to the digestion and absorption of hydrophilic substances such as proteins or carbohydrates.

2. When chyme moves from the stomach to the small intestine, a number of stomach functions are altered. Identify these functions, describe the changes that occur, and explain the regulatory mechanisms that bring about these changes. Explain why these changes make sense in light of the gastrointestinal system's ultimate function.

3. Describe the structural adaptations of the small intestine that increase its capacity for absorbing nutrients.

4. Describe the processes of segmentation, peristalsis, haustration, and mass movement, and explain their roles in digestive function. What is the apparent function of the muscularis mucosae?

5. Describe the various mechanisms that regulate the secretion of acid and pepsinogen by the stomach. Compare these mechanisms to those that regulate gastric motility.

Find the answers to these exercises, and additional study tools,
at the Physiology Place (www.physiologyplace.com).

20

The Endocrine System: Regulation of Energy Metabolism and Growth

OBJECTIVES

- Compare the metabolic pathways operating during energy mobilization to those operating during energy utilization.

- Explain the concepts of negative energy balance and positive energy balance.

- Describe the hormonal control of metabolism during absorptive and postabsorptive states.

- Explain how growth hormone regulates growth.

- Describe the synthesis and secretion of thyroid hormones. Distinguish between direct and permissive actions of thyroid hormones.

- Describe the effects of glucocorticoids on whole body metabolism. Compare the physiological effects of glucocorticoids to their pharmacological effects.

- Describe the stress response.

CHAPTER OUTLINE

An Overview of Whole Body Metabolism   640

Energy Intake, Utilization, and Storage   641

Energy Balance   643

Energy Metabolism During the Absorptive and Postabsorptive States   645

Regulation of Absorptive and Postabsorptive Metabolism   649

Hormonal Regulation of Growth   654

Thyroid Hormones   661

Glucocorticoids   663

Above: Photomicrograph of the thyroid gland

The next time you find yourself sitting down to eat a meal, ask yourself this question: Why am I eating? If you do this on several occasions, you will find that there are many possible reasons. Perhaps you smelled something cooking that stimulated your appetite, or you saw an advertisement for a food you like. Perhaps you decided to join some friends who were on their way to dinner, or maybe you just looked at your watch and decided it was mealtime. The number of possibilities is quite large. In fact, we have so many different motivations for eating that it is easy to forget that the ultimate reason—obtaining nutrition—is a biological necessity, because food is our sole source of energy and the raw materials from which our bodies are made.

Although our need for food is driven by biological necessity, our eating patterns are not constant and are influenced by other factors. (You have probably skipped meals to study for exams, or overeaten during the holidays.) Normally this is no cause for concern because the body has ways of maintaining the steady supply of energy cells need despite changes in the pattern of food intake. Between meals the body converts energy stores (including large carbohydrates, proteins, and lipids) into smaller molecules that cells can use for energy. When you eat, the body replenishes these stores by converting nutrients into energy storage molecules.

The way the body stores and utilizes energy—*energy metabolism*—is influenced not only by eating patterns, but also by such factors as growth, stress, and metabolic rate. In all cases, whether the body stores or utilizes energy is controlled primarily by endocrine signals. Two critical concepts drive the control of energy metabolism:

1. Because food intake is intermittent, the body must store nutrients during periods of intake and then break down these stores during periods between meals.

2. Because the brain depends on glucose as its primary energy source, blood glucose levels must be maintained at all times, even between meals.

In Chapter 5 we learned the basics about chemical messengers (including hormones) and some of the components of the endocrine system. In this chapter we continue our exploration of the endocrine system by focusing on a selected set of hormones that exert important influences on energy metabolism and growth.

As we study these hormones and their actions, we will find that their effects are varied and often overlap. To aid in the understanding of these hormones, the first section of this chapter presents general principles relating to energy metabolism and energy balance. A comparison of energy metabolism during and between meals follows. The chapter then describes the hormones regulating blood glucose level, especially insulin and glucagon. Then it discusses the hormones that regulate growth and the primary hormones that regulate whole body metabolism—thyroid hormones. Finally, the glucocorticoids and their role in adapting to stress are described.

AN OVERVIEW OF WHOLE BODY METABOLISM

In Chapter 3 we explored cellular metabolism, including some specific metabolic pathways and the role of enzymes in regulating metabolic pathways. In this chapter we discuss the coordinated regulation of metabolic pathways in different organs to maintain adequate energy supplies for all the body's cells. Fundamental in this whole body metabolism is the maintenance of plasma glucose levels. To fully appreciate the control of metabolism, we need to review some key concepts of cellular metabolism.

Adenosine Triphosphate (ATP)

ATP provides usable energy to cells. Energy from the hydrolysis of ATP is used for most cellular processes, including movement, active and vesicular transport, and anabolism. ATP is synthesized by two mechanisms, substrate-level phosphorylation and oxidative phosphorylation. Substrate-level phosphorylation involves the transfer of a phosphate group from a metabolic intermediate (X) to ADP to synthesize ATP, as in the following equation:

$$X - P + ADP \rightarrow X + ATP$$

Most ATP in body cells is produced by oxidative phosphorylation, in which energy released from electrons moving down the electron transport chain of the inner mitochondrial membrane is used to synthesize ATP. Important intermediates of oxidative phosphorylation include acetyl CoA and the reduced coenzymes NADH + H$^+$ and FADH$_2$. Acetyl CoA is produced during catabolism of carbohydrates, lipids, and proteins, and it is the initial substrate of the Krebs cycle, which reduces the coenzymes NAD$^+$ and FAD to NADH + H$^+$ and FADH$_2$, respectively. By supplying electrons to the electron transport chain, the reduced coenzymes provide the energy for ATP synthesis through oxidative phosphorylation.

Anabolism

An interesting aspect of metabolism is that the same small biomolecules that provide energy are also used to synthesize larger biomolecules. A good example is acetyl CoA, which can not only be catabolized for energy, as just described, but is also a substrate for triglyceride and cholesterol synthesis. Thus because carbohydrates, lipids, and proteins can all be catabolized to acetyl CoA, they can all eventually be converted to lipids. Many other metabolic intermediates of glycolysis and the Krebs cycle can be used to synthesize larger biomolecules. For example, some such metabolic intermediates can be converted to amino acids and used for protein synthesis, whereas other intermediates can be used in the synthesis of phospholipids.

Regulation of Metabolic Pathways

If anabolic and catabolic pathways have several of the same intermediates, what governs the direction of metabolism? Most important in determining which metabolic pathways are in operation is the number and activity of the enzymes involved in the pathways. The activity of enzymes can be regulated by changing their concentration through synthesis or degradation, or by changing the activity of individual enzyme molecules through allosteric or covalent regulation. Hormones that regulate metabolic pathways do so by regulating the activity of enzymes in one or more of these ways.

Metabolic pathways are also controlled by compartmentation. In Chapter 3 we examined an example of cellular compartmentation: whereas glycolysis occurs in the cytosol, the Krebs cycle occurs in the mitochondrial matrix. Compartmentation also occurs on the tissue level, because some enzymes are found in the cells of certain tissues only. In addition, hormones differentially affect tissues based on the types of receptors on the cells in that tissue. Tissues or organs that have special metabolic activities include the brain, skeletal muscle, the liver, and adipose tissue. We will explore how these tissues and organs affect whole body metabolism shortly; first we consider how the body handles different classes of biomolecules, from absorption, to cellular uptake, to utilization by the cell.

ENERGY INTAKE, UTILIZATION, AND STORAGE

When we eat, digestion breaks down the large molecules in food into smaller molecules, which are then absorbed into the bloodstream. Blood flow distributes them to tissues throughout the body, where they are eventually taken up by cells. Inside cells these molecules undergo three possible fates:

1. Absorbed biomolecules can be broken down to smaller molecules, in the process releasing energy that can be used for driving various cellular processes such as muscular contraction, transport, secretion, or anabolism.
2. Absorbed biomolecules can be used to synthesize other molecules needed by cells and tissues for normal function, growth, and repair.
3. Absorbed biomolecules in excess of those required for energy and synthesis of essential molecules are converted to energy storage molecules that provide energy during periods between meals. The two primary energy storage molecules are the polymer glycogen and triglyceride (fat).

The ultimate fate of consumed molecules depends on their chemical nature and the body's needs at the time of consumption. Next we examine the absorption, utilization, and storage of energy in carbohydrates, proteins, and fats.

Absorption, Utilization, and Storage of Energy in Carbohydrates

In the diet, carbohydrates come in a variety of forms, the most abundant of which are the polysaccharides starch and glycogen. Other major carbohydrates in the diet include the disaccharides (such as sucrose and lactose) and to a lesser degree the monosaccharides (such as glucose and fructose). The monosaccharides are the class of carbohydrates absorbed in the gastrointestinal tract.

Figure 20.1a illustrates how the most abundant absorbed carbohydrate, glucose, is absorbed, used, and stored. Polysaccharides are digested to monosaccharides, including glucose ①, which is then absorbed across the intestinal epithelium and into the bloodstream ②, as described in Chapter 19. Next ③ molecules of glucose are transported into cells throughout the body by *glucose transporters*. Inside cells, glucose can ④ be oxidized for energy, which generates carbon dioxide as a waste product; ⑤ provide substrates to other metabolic reactions; or

FIGURE 20.1 Digestion, absorption, and subsequent handling of biomolecules within cells. *Numbered steps are described in the text for the handling of* **(a)** *carbohydrates,* **(b)** *proteins, and* **(c)** *lipids. In (c), the interrupted arrow in the first panel indicates that the transport of triglycerides from the GI tract to the bloodstream is indirect.*

What three metabolic pathways are necessary for the handling of glucose shown in ④ in (a)?

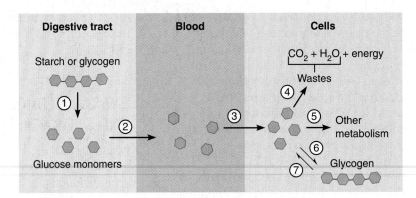

(a) Carbohydrates

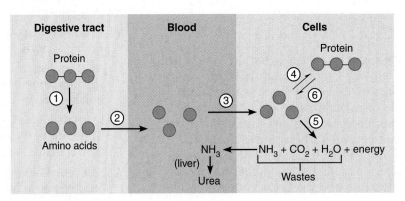

(b) Proteins

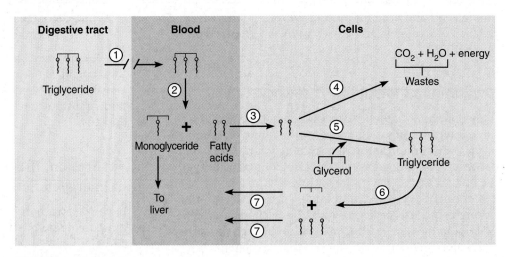

(c) Lipids

⑥ be converted to glycogen for storage. If glucose levels in the cell decrease, glycogen can be broken down to glucose by glycogenolysis ⑦.

Although this series of steps accurately describes what happens to glucose *in the body as a whole,* it may or may not describe what happens in individual cells. Most cells, for example, can oxidize glucose but have a limited ability to synthesize and store glycogen.

Absorption, Utilization, and Storage of Energy in Proteins

As depicted in Figure 20.1b, virtually all dietary protein is digested ① and absorbed into the bloodstream ② in the form of individual amino acids. Following uptake into cells ③, amino acids are used for the synthesis of proteins ④ or catabolized for energy by proteolysis ⑤. The proteins function as amino acid stores that subsequently can be broken down to amino acids ⑥, which can then be catabolized for energy or released into the bloodstream

Glycolysis, the Krebs cycle, and the electron transport chain

TABLE 20.1 SUMMARY OF CARBOHYDRATES, PROTEIN, AND LIPID PROCESSING

	FORM ABSORBED ACROSS GI TRACT	FORM CIRCULATING IN BLOOD	FORM STORED	STORAGE SITE	PERCENT OF TOTAL ENERGY STORED
Carbohydrates	Glucose	Glucose	Glycogen	Liver, skeletal muscle	1%
Proteins	Amino acids, some small peptides	Amino acids	Proteins	Skeletal muscle*	22%
Lipids	Monoglycerides and fatty acids (in chylomicrons)	Free fatty acids, lipoproteins	Triglycerides	Adipose tissue	77%

*Even though proteins are found in all cells of the body, most of the proteins mobilized for energy come from skeletal muscle cells.

for use by other cells. Catabolism of cellular proteins to generate energy occurs only during periods of starvation, and it generates ammonia (NH_3) and carbon dioxide as waste products. The ammonia is converted by the liver to *urea*, which is eventually eliminated in the urine.

Absorption, Utilization, and Storage of Energy in Fats

When the body uses dietary carbohydrates and proteins, they are taken into cells in the form of smaller components (glucose or amino acids), which can either be catabolized for energy or assembled into larger molecules. The same process occurs for fats, although the process is a little more complicated.

Figure 20.1c illustrates the body's handling of triglycerides, the predominant form in which fats are present in the diet. The absorption of triglycerides ① occurs via transport of chylomicrons into the lymphatic system (see Chapter 19), which eventually reach the blood when lymph drains into the bloodstream at the thoracic duct. After entering the bloodstream, the chylomicrons transport triglycerides to cells. The triglycerides at the outer surface of chylomicrons are broken down by an enzyme called **lipoprotein lipase** ②, which is located on the inside surface of capillaries throughout the body and is present in particularly high density in capillaries running through adipose tissue (body fat). This enzyme breaks down triglycerides into fatty acids and monoglycerides; the fatty acids are then taken up by nearby cells ③ while the monoglycerides remain in the bloodstream and are eventually metabolized by the liver.

Upon entry into cells, fatty acids may be oxidized for energy ④ or combined with glycerol to form new triglycerides ⑤, which are stored in fat droplets in the cytosol. This storage occurs mainly in **adipocytes,** adipose tissue cells that are specialized for fat storage. (The glycerol used

in triglyceride synthesis is not derived from absorbed triglycerides, but instead is synthesized within adipocytes.) Stored triglycerides can subsequently be broken down to glycerol and fatty acids ⑥, which can be catabolized for energy or released into the bloodstream for use by other cells ⑦. The catabolism of glycerol and fatty acids produces carbon dioxide as a waste product. The breakdown of triglycerides to fatty acids and glycerol, such as occurs in ② or ⑥, is called *lipolysis*.

The body's processing of carbohydrates, proteins, and lipids is summarized in Table 20.1. Fatty acids are included among the smaller nutrient molecules (even though they are not absorbed in this form) because they are the form in which fats are made available to most cells. Some small nutrients can be *interconverted;* for example, glucose can be synthesized from amino acids, and fatty acids can be synthesized from glucose or amino acids. These interconversions have a significant role in whole body metabolism, as we will see shortly.

ENERGY BALANCE

Chapter 18 describes endocrine control of kidney function in relation to fluid and electrolyte balance, but the endocrine system also functions in energy balance. As with fluids and electrolytes, for energy to be in balance, energy input must equal energy output.

The endocrine system regulates energy balance to ensure that a steady supply of small nutrients is always available to all body cells to meet their energy demands. As cells expend energy, they draw upon stores of nutrients within cells and in the bloodstream to provide more energy. This pool of nutrients must be continually replenished if uninterrupted energy expenditure is to occur. This replenishment can be accomplished in two ways: by absorption of more nutrients into the bloodstream, or by *mobilization* of energy stores—that is, the catabolism of

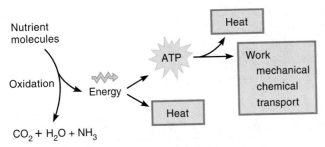

FIGURE 20.2 The forms of energy produced by the oxidation of nutrient molecules. *Some of the energy liberated during oxidation is used to generate ATP, which can perform various types of work within cells; the rest takes the form of heat.*

stored macromolecules into small nutrient molecules that are released into the bloodstream. The body mobilizes its energy stores when the rate of energy intake is insufficient to meet its energy needs.

Energy Input

Energy input into the body arrives in the form of absorbed nutrients. When a particular nutrient molecule (such as glucose) is oxidized in the body, a certain quantity of energy is liberated; this quantity represents the *energy content* of the molecule. A person's *energy intake* is the total energy content of all nutrients absorbed.

The energy content of a nutrient is generally determined by burning a known quantity of the substance in an instrument called a *calorimeter,* which measures the total amount of energy released in the form of heat as the substance burns. This amount is usually expressed in calories per gram of the substance burned.

Energy Output

In a calorimeter, all the energy contained in a molecule is converted to heat during oxidation, and thus the energy content of the molecule (input) is equal to the quantity of energy in the heat (output). Although the body also utilizes the oxidation of molecules to liberate energy, the energy released during oxidation or other catabolic reactions takes two forms: heat and work (Figure 20.2). Approximately 60% of the energy in consumed nutrients goes to heat production, which is necessary to maintain body temperature. Most of the remaining 40% of the energy is used to synthesize ATP, which is used to perform cellular work (a process that releases still more heat).

The energy-requiring processes of cells fall into three main categories: **Mechanical work** uses intracellular protein filaments to generate movement, such as occurs in muscle contraction or the beating of cilia lining the respiratory tract. **Chemical work** is used in the formation of bonds during chemical reactions, such as occurs when small molecules are used to synthesize large molecules.

Transport work utilizes energy to move a molecule from one side of a cell membrane to another, such as occurs in active transport (the Na^+/K^+ pump, for example) or in vesicular transport (exocytosis and endocytosis).

Metabolic Rate

When the body breaks down nutrients, it either releases energy as heat or uses it to perform work. The amount of energy so expended per unit time is the body's **metabolic rate.** A person's metabolic rate is influenced by a number of factors, including muscular activity, age, gender, body surface area, and environmental temperature. When sitting still, for example, the rate of energy expenditure is about 100 kilocalories per hour; when riding a bicycle at a moderate rate, it is over 300 kilocalories per hour.

The **basal metabolic rate** (BMR) is the rate of energy expenditure of a person who is awake, is lying down, is physically and mentally relaxed, and has fasted for at least 12 hours; under these conditions, both metabolic rate and work performed are minimal. The BMR is usually estimated by measuring a person's rate of oxygen consumption, which correlates with the rate at which nutrients are oxidized in the body.

The BMR represents the energy requirement of performing such necessary tasks as pumping blood and transporting ions. The BMR generally increases as body weight increases, because a larger mass of tissue requires greater energy expenditure for its upkeep; therefore, BMR is usually expressed as the rate of energy expenditure per unit body weight. For adults, the BMR averages 20–25 kilocalories per kilogram of body weight per day. Most of this expenditure is due to activity in the nervous system and skeletal muscles, which account for 40% and 20–30% of the BMR, respectively. The BMR varies from tissue to tissue; muscle tissue, for example, has a higher resting metabolic rate than adipose tissue. The BMR (per unit body weight) also varies with age: It is greater in growing children because of the energy expended in the synthesis of new tissue, and it is usually lower in the elderly than in young adults.

Negative and Positive Energy Balance

In today's society, virtually everyone is aware that connections exist among body weight, diet, and exercise. When people eat a lot of food but don't get much exercise, their body weight tends to increase; when they eat less and exercise more, their body weight decreases. These changes in body weight occur when energy input and output are not balanced.

First, we define energy output as follows:

Energy output = work performed + heat released

When the body is in energy balance, energy input equals energy output:

Energy input = energy output

= work performed + heat released

If the body is not in energy balance—that is, if energy input and energy output are not equal—then the difference between energy input and output determines whether the amount of stored energy increases or decreases (is reduced through the catabolism of energy stores):

Energy stored = energy input − (work performed + heat released)

When a person takes in energy at a rate greater than that at which it is expended as heat and work, the quantity of stored energy increases. This condition, *positive energy balance,* tends to be associated with increases in body weight; a net synthesis of macromolecules from absorbed nutrients occurs. (Later in this chapter we see that most excess nutrients are converted to lipids for storage.) When the rate of energy intake is less than the rate at which energy is expended as heat or work, the quantity of stored energy decreases. This condition, *negative energy balance,* tends to be associated with decreases in body weight. Under these conditions, a net breakdown of macromolecules (including lipid stores) provides energy for body functions.

When someone goes on a diet to lose weight, the idea is to decrease food intake and achieve negative energy balance. The same result can be attained through exercise, which increases work and heat production (energy output). When the increased energy output exceeds energy intake, the energy balance becomes negative.

Although the concept of energy balance is useful for explaining why diet and exercise affect body weight, note that a change in the body's energy content is not necessarily equivalent to a change in mass. If during a given time span 100 grams of glucose are absorbed and 100 grams of triglycerides are oxidized, then body weight does not change; however, the body's energy content decreases because 1 gram of glucose contains less energy than 1 gram of triglycerides.

Quick Test 20.1

1. Define the terms *metabolic rate* and *basal metabolic rate.*

2. What are the storage forms of carbohydrates and lipids?

3. Once energy is taken into the body, it is either stored or it appears in what two other forms?

ENERGY METABOLISM DURING THE ABSORPTIVE AND POSTABSORPTIVE STATES

We have seen that the maintenance of energy balance requires that energy input equals energy output. However, the body is generally not in energy balance at any given time, because the rate of energy input is determined by feeding, which is intermittent. For about 3–4 hours after a typical meal, nutrients are absorbed during the **absorptive state,** after which time absorption stops until the next meal. During this time the rate of energy input generally exceeds energy output, putting the body in positive energy balance. Nutrients in the blood are plentiful. Glucose serves as the primary energy source for cells while fats, amino acids, and excess glucose are taken up by liver, muscle, and fat cells and converted to energy storage molecules. The **postabsorptive state** corresponds to the time between meals when nutrients are not being absorbed; during this time the rate of energy expenditure is greater than the rate of energy intake. Energy stores are mobilized to provide the energy cells need. Whereas glucose serves as the energy source for cells in the central nervous system, other cells in the body utilize other energy sources (such as fatty acids), thereby *sparing* glucose for the central nervous system.

Energy metabolism during the absorptive and postabsorptive states can be summarized by the following rule: *During the absorptive state, energy is stored in macromolecules; during the postabsorptive state, these energy stores are mobilized.* This rule is important, because despite the fact that we eat different amounts at different times, the body provides a constant supply of nutrients to cells—a supply that cannot be interrupted for even a minute because the body must expend energy continuously just to stay alive. In the following sections we see how the body performs this necessary function.

Metabolism During the Absorptive State

The absorptive state is primarily an anabolic state; that is, the majority of reactions involve synthesis of macromolecules. However, cellular metabolism differs among cell types. Next we look at the typical absorptive state metabolic responses in body cells in general, in skeletal muscle cells, in liver cells, and in adipocytes (Figure 20.3).

Body Cells in General

The body's energy needs are supplied primarily by absorbed glucose, which is plentiful after a typical meal. Glucose is taken into the cells and catabolized as the primary fuel. Absorbed fatty acids and amino acids can also be catabolized for energy, particularly if the diet is rich in these nutrients but poor in carbohydrates. Fatty acids undergo oxidation to provide acetyl CoA subunits for the Krebs cycle, and amino acids are converted to keto acids

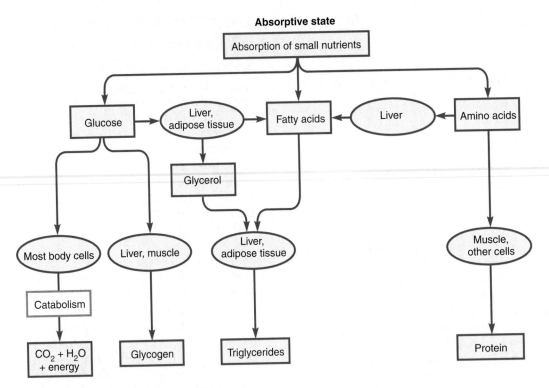

FIGURE 20.3 Major metabolic reactions of the absorptive state.

(organic acids with a carbonyl group, C=O), many of which are intermediates for the Krebs cycle. Amino acids can also be used to synthesize proteins. Note, however, that proteins are not synthesized as "storage molecules." Instead, most body proteins have important structural and functional roles in cells and are continuously turned over; that is, old proteins are degraded and replaced with new ones. For this reason, the body's protein mass is relatively stable and does not increase simply in response to the absorption of an excess of amino acids. Body cells catabolize proteins for energy only under extreme conditions, because they do so at the expense of losing functioning molecules.

Skeletal Muscle Cells

Like body cells in general, skeletal muscle cells take up glucose and amino acids from the blood for their own needs. However, unlike most body cells, skeletal muscle cells can convert glucose to glycogen for storage. Within individual muscle cells, these glycogen stores are limited, but taken together they constitute the majority (approximately 70%) of the body's total stored glycogen.

Liver Cells

The liver converts nutrient molecules to energy stores that subsequently can be mobilized to supply energy to most cells in the body. The liver converts glucose to glycogen or fatty acids, and fatty acids to triglycerides. The glycogen is stored in the liver (which contains ap-

proximately 24% of the body's glycogen stores), whereas the triglycerides are transported to adipose tissue for storage. Between liver and skeletal muscle, the body can only store approximately 500 grams of glycogen, which is only enough to meet the body's energy demands for a few hours. Any absorbed glucose that exceeds the quantity that is needed for energy or can be stored as glycogen is converted to fatty acids and then to triglycerides.

The liver also takes up amino acids. Although the liver uses some amino acids to synthesize proteins (including plasma proteins), most amino acids are converted to keto acids, many of which are intermediates of glycolysis or the Krebs cycle and can be used for energy. However, most of the keto acids are used to synthesize fatty acids, and therefore ultimately end up as triglycerides.

The triglycerides synthesized in the liver must be transported to adipose tissue, which is achieved by packaging the triglycerides into particles called **very-low-density lipoproteins** (VLDLs). (See Discovery: Lipoproteins and Plasma Cholesterol, p. 621.) Briefly, VLDLs transport triglycerides to the cells of the body (see Chapter 19). The plasma membranes of most cells contain the enzyme *lipoprotein lipase,* which catabolizes triglycerides at the outer surface of the VLDLs to fatty acids and monoglycerides. The fatty acids then diffuse into cells, where they can be used for energy (by most body cells) or converted back into triglycerides for storage (in adipocytes). Adipocytes have a high concentration of lipoprotein lipase on their plasma membranes and therefore take up most of the fatty acids transported in VLDLs.

	GLYCOGEN	TRIGLYCERIDES	PROTEINS (MOBILIZABLE)
		TABLE 20.2 ENERGY STORES (AS PERCENT OF TOTAL) IN A HEALTHY 70-KG MALE	
Skeletal muscle	71	<1	98
Liver	24	<1	2
Adipose tissue	5	99	<1
Brain	<1	0	0

Adipocytes

Adipocytes store energy in the form of triglycerides, or fat. Absorbed triglycerides are transported to adipocytes by chylomicrons. Lipoprotein lipase catabolizes the triglycerides in chylomicrons in the same manner as just described for VLDLs. Excess absorbed glucose enters the adipocytes and is converted to triglycerides. In addition, triglycerides synthesized in the liver are transported to adipocytes by VLDLs for storage.

Energy Reserves

Whereas the body is limited in its ability to store energy in the form of glycogen or protein, it is practically unlimited in its ability to store energy as fats. Triglyceride synthesis thus represents the "final common pathway" for all nutrients that are absorbed in excess of the body's needs. In most people, fat accounts for 20–30% of body weight, but in very overweight individuals it can account for as much as 80%. To those who are weight conscious, the body's propensity for storing nutrients as fats is at best an annoyance. However, given that triglycerides contain roughly 9 kilocalories per gram, as opposed to 4 kilocalories per gram for carbohydrates and 5–6 kilocalories per gram for proteins, triglyceride synthesis is clearly the best way to store the most energy in the least weight.

Under normal circumstances, glycogen stores account for 1% or less of the body's total energy reserves and can supply energy needs for only a few hours of quiet activity. Proteins account for about 20–25% of the total energy reserves. Although large amounts of protein can be mobilized for energy without serious consequences, particularly from skeletal muscle, continual use of protein for energy is harmful and potentially fatal because it eventually compromises cellular function. Thus a significant portion of the energy contained in protein stores is, in reality, unavailable for use. Fat stores represent about 75–80% of the total energy reserves and contain enough energy to last for about two months. For this reason, fats are absolutely essential to the body's ability to withstand prolonged periods of fasting. Table 20.2 tabulates the average energy stores of a 70-kg male.

Metabolism During the Postabsorptive State

Within a few hours after a typical meal, absorption of nutrients ceases. During this time, the body cannot rely on absorbed nutrients for its energy needs, and so it is forced to catabolize glycogen, proteins, and fats for energy (Figure 20.4). Thus the postabsorptive state is primarily a catabolic state. In addition, unlike most body cells, central nervous system cells rely on glucose as their sole energy source (cells of the central nervous system can also obtain energy from ketone bodies during extreme conditions such as starvation). Therefore, a primary function of the postabsorptive state is to maintain plasma glucose levels. Too large a reduction in plasma glucose can result in serious impairment of brain function, loss of consciousness, and even death.

Given the importance of maintaining a steady supply of glucose during the postabsorptive state, a question naturally arises: Because glucose is derived from the breakdown of glycogen, which is in relatively short supply (enough to last only a few hours), how is glucose made available for longer periods? The body synthesizes new glucose from amino acids, glycerol, and other breakdown products of catabolism, a process known as *gluconeogenesis*. Most tissues turn almost exclusively to other energy sources, primarily fatty acids, thereby conserving glucose for use by the central nervous system; this is called **glucose sparing.**

As in the absorptive state, cellular metabolism during the postabsorptive state differs among the types of cells.

Body Cells in General

Most cells utilize fatty acids for energy instead of glucose, sparing the glucose for the central nervous system.

Skeletal Muscle Cells

In a skeletal muscle cell, any glucose formed from glycogen during glycogenolysis can be used for energy only within that muscle cell. Glycogen is catabolized to glucose-6-P (a glucose molecule with a phosphate group attached to the sixth carbon, and an intermediate in glycolysis). The phosphate group cannot be removed

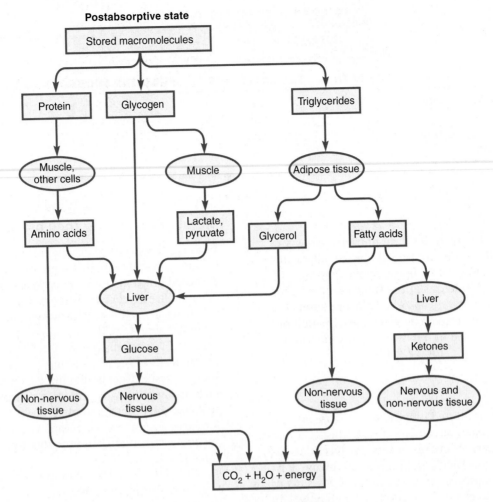

Postabsorptive state

FIGURE 20.4 **Major metabolic reactions of the postabsorptive state.**

When the liver converts amino acids, lactate, or pyruvate to glucose, what type of reaction is occurring?

from the glucose because skeletal muscle cells lack the enzyme (glucose-6 phosphatase) that catalyzes its removal. To transport glucose out of a cell, the glucose must be in its unphosphorylated form. Thus the glucose formed by glycogenolyis in skeletal muscle cells remains in the cell and is catabolized by glycolysis to form pyruvate or lactate. Any lactate produced then travels to the liver for further processing, as described shortly.

Skeletal muscle cells can also catabolize proteins to amino acids, which are then transported into the bloodstream to the liver for further processing.

Liver Cells

The liver is the primary source of plasma glucose during the postabsorptive state. The liver has glycogen stores that can be broken down by glycogenolysis to glucose-6-

P, and the liver also has the enzyme to convert glucose-6-P to glucose. Glucose can then be transported out of the liver cell and into the bloodstream. Therefore, liver glycogen stores, unlike skeletal muscle glycogen stores, can be mobilized to provide glucose to the blood.

The liver is also the primary site of gluconeogenesis. (Some gluconeogenesis also occurs in the kidneys.) Like the glucose produced by glycogenolysis, the newly synthesized glucose is transported from the liver into the bloodstream for use by other cells in the body.

During the postabsorptive state the liver converts some of the fatty acids entering it to ketone bodies, which are released into the bloodstream and eventually catabolized by most tissues. The production of ketones is important because during prolonged fasting, the central nervous system acquires the ability to use ketones for energy, thereby freeing it from some of its dependence on glucose.

Gluconeogenesis

TABLE 20.3 ENERGY METABOLISM DURING THE ABSORPTIVE AND POSTABSORPTIVE STATES

	ABSORPTIVE (ANABOLIC) STATE	POSTABSORPTIVE (CATABOLIC) STATE
Carbohydrates	In liver and skeletal muscle, glucose is converted to glycogen (glycogenesis) for storage. Liver also converts some glucose to triglycerides, which are transported to adipose tissue by lipoproteins. In adipose tissue, glucose is converted to triglycerides for storage. In body cells, glucose undergoes oxidation for energy.	In liver, glycogen is catabolized to glucose (glycogenolysis) and new glucose is formed from noncarbohydrate precursors (gluconeogenesis); glucose is then transported into the bloodstream. In muscle, glycogen is catabolized to glucose-6-P which can be used by the muscle cell for energy.
Lipids	Triglycerides are synthesized from ingested fats, proteins, and glucose by adipocytes; triglycerides synthesized in the liver are transported to adipose tissue for storage.	In adipose tissue, triglycerides are broken down to fatty acids and to glycerol, which enters the bloodstream and travels to the liver, where it is converted to glucose (gluconeogenesis). Fatty acids enter the bloodstream and provide the primary energy source for most body cells. In liver, fatty acids are converted to ketones.
Amino acids	In all cell types, amino acids are used for protein synthesis. In liver, amino acids can be converted to keto acids for energy production or triglyceride synthesis. Triglycerides are then transported to adipose tissue by lipoproteins.	In muscle, proteins are catabolized to amino acids, which are transported to the liver for gluconeogenesis.

Exercise Link

These mechanisms were important for ensuring that Jane and Bill's blood glucose concentrations remained relatively stable through most of their marathon run. A portion of the lactate produced by their exercising muscles was converted by their livers (and kidneys) back into glucose via gluconeogenesis. It has been estimated that perhaps 10% of the glucose released into the blood during moderate intensity exercise is derived from gluconeogenesis. Liver glycogen was also slowly broken down, providing about 15% of the total glucose consumed during the race. Ketone body production increased as well, but the contribution of these ketones to total energy consumption was small compared to glucose.

Adipocytes

In the postabsorptive state, adipose tissue supplies fatty acids to the bloodstream as energy sources for body cells, thereby sparing glucose for the central nervous system. Adipose tissue does this by catabolizing stored triglycerides into glycerol and free fatty acids. The glycerol is also released into the bloodstream, where it travels to the liver and is catabolized by glycolysis.

Table 20.3 summarizes the metabolism of energy molecules during the absorptive and postabsorptive states.

Quick Test 20.2

1. When glucose or amino acids are absorbed in excess of the quantities oxidized or stored as glycogen or proteins, what happens to them?

2. Where is most of the body's glycogen stored? What is the storage site of most of the glycogen that can supply glucose for cells throughout the body?

3. During the postabsorptive state, most tissues use fatty acids instead of glucose as their primary energy source. Why is this important in whole body metabolism?

4. Name two processes that help the body maintain blood glucose at normal levels during fasting.

REGULATION OF ABSORPTIVE AND POSTABSORPTIVE METABOLISM

The transitions between the absorptive and postabsorptive states are marked by profound alterations in the metabolic activity of tissues throughout the body. In this section we see how these metabolic changes are triggered primarily by endocrine signals involving the pancreatic

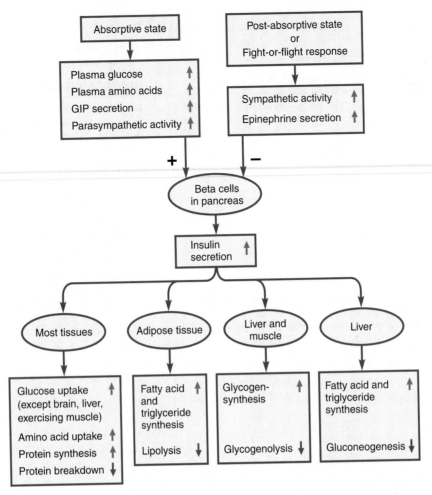

FIGURE 20.5 Factors affecting the secretion of insulin, and its actions on target tissues. *This flowchart is available as an interactive exercise under Activities at* www.physiologyplace.com.

hormones insulin and glucagon. In addition, epinephrine and sympathetic nervous activity play a role.

The Role of Insulin

The metabolic adjustments that occur as the body switches between the postabsorptive and absorptive states are largely triggered by changes in the plasma concentration of **insulin,** a peptide hormone secreted by beta cells located in the pancreatic islets of Langerhans (see Chapter 5). Even though the actions of insulin and the factors that affect its secretion are numerous, they all have a common thread: *Insulin promotes the synthesis of energy storage molecules and other processes characteristic of the absorptive state (Figure 20.5).* In other words, insulin is an anabolic hormone. Accordingly, its secretion is stimulated by signals associated with feeding and the absorption of nutrients into the bloodstream.

Factors Affecting Insulin Secretion

During the absorptive period, insulin secretion by beta cells increases, causing an increase in the plasma concen-

tration of insulin, which promotes many of the metabolic processes characteristic of the absorptive state. During the postabsorptive state, insulin secretion is decreased, causing a decrease in the plasma concentration of insulin, which helps turn off the absorptive processes. This raises a key question: How do beta cells know when to increase or decrease insulin secretion?

Figure 20.5 shows that insulin secretion is influenced by a variety of factors. Particularly important among these is plasma glucose concentration. During the absorptive period, plasma glucose levels increase as glucose is transported into the bloodstream from the gastrointestinal tract. This increase stimulates insulin secretion by a direct effect of glucose on beta cells, which are sensitive to the concentration of glucose in the fluid surrounding them. During the postabsorptive period, plasma glucose levels decrease, causing insulin secretion to fall. Insulin secretion is influenced in a similar fashion by the plasma amino acid concentration: Rising amino acid levels in plasma stimulate insulin secretion, whereas falling amino acid levels decrease insulin secretion.

Hormones and input from the autonomic nervous system also influence insulin secretion. Secretion is stim-

ulated by parasympathetic nervous activity and by glucose-dependent insulinotropic peptide (GIP), a hormone
secreted by cells in the wall of the gastrointestinal tract.
This is significant because parasympathetic activity and
secretion of GIP are both stimulated in response to the
presence of food in the gastrointestinal tract. Because
feeding occurs prior to the absorption of nutrients, these
signals prepare the body for transitions to the absorptive
state by triggering insulin secretion in advance. Insulin
secretion is inhibited by sympathetic nervous activity and
circulating epinephrine.

Actions of Insulin

Through its actions on a variety of target tissues, insulin
influences almost every major aspect of energy metabolism (see Figure 20.5). It promotes energy storage by
stimulating the synthesis of fatty acids and triglycerides in
the liver and adipose tissue, glycogen in liver and skeletal
muscle, and proteins in most tissues. At the same time, it
opposes the catabolism of energy stores by inhibiting the
breakdown of proteins, triglycerides, and glycogen, and
by suppressing gluconeogenesis by the liver. In short, insulin promotes reactions of the absorptive state and suppresses reactions of the postabsorptive state.

   Along with its metabolic actions, insulin affects the
transport of nutrients across the membranes of all body
cells except those in the liver and central nervous system.
In most tissues, insulin stimulates the uptake of amino
acids by cells, which facilitates the hormone's stimulatory
effect on protein synthesis. Insulin also stimulates the uptake of glucose in many tissues by increasing the number
of glucose transport proteins in cell plasma membranes.
The particular protein affected is a type of glucose transporter called GLUT 4. Many cells have pools of GLUT 4
transporters stored in vesicles in the cytoplasm. Insulin
can either trigger insertion of these *stored* transporters
into the plasma membrane by exocytosis, or stimulate the
synthesis of *new* transporters.

   The transport of glucose in the central nervous system and the liver is not affected by insulin. This is critical,
because when insulin levels are low during the postabsorptive state, glucose uptake by most cells is decreased,
sparing the glucose for use by cells of the central nervous
system, where glucose transport is not affected by insulin.

   Insulin also has little effect on the glucose permeability of cells in exercising muscle because exercise triggers
an increase in the number of glucose transporters by a
mechanism that operates independently of insulin. In
resting muscle, however, insulin stimulates increased glucose transport in the normal manner.

   In addition to the actions shown in Figure 20.5, insulin also has important growth-promoting effects. Although insulin by itself does not stimulate growth, it
must be present in the blood for growth hormone to exert its normal effects, a form of permissiveness discussed
in Chapter 5. This need results, at least in part, from

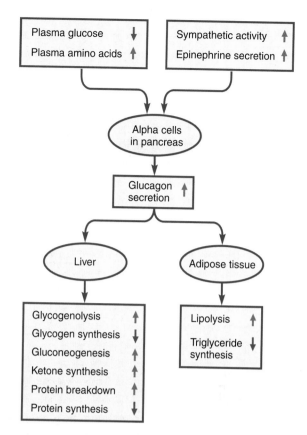

**FIGURE 20.6 Factors affecting the secretion of glucagon,
and its actions on target tissues.** *This flowchart is available as an
interactive exercise under Activities at* www.physiologyplace.com.

insulin's role in promoting protein synthesis, DNA synthesis, and cell division, all of which are essential to tissue
growth.

The Role of Glucagon

Insulin's actions to bring about the body's metabolic
adaptations to the absorptive and postabsorptive states
are reinforced by contrary changes in **glucagon,** a peptide hormone secreted by alpha cells of pancreatic islets of
Langerhans. Put another way, insulin and glucagon are
antagonists, hormones whose actions oppose each other:
Insulin promotes processes of the absorptive state;
glucagon promotes processes of the postabsorptive state.

   Glucagon secretion decreases during the absorptive
state and increases during the postabsorptive state. Because insulin levels are also changing with these states,
the metabolic adjustments from one state to the other are
orchestrated by contrary changes in plasma levels of insulin and glucagon.

Factors Affecting Glucagon Secretion

Most of the signals that stimulate the secretion of
glucagon (Figure 20.6) are the same signals that inhibit

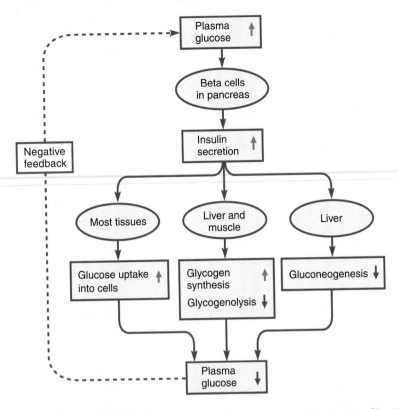

FIGURE 20.7 Regulation of plasma glucose concentration by insulin. *This flowchart is available as an interactive exercise under Activities at* www.physiologyplace.com.

the secretion of insulin. Decreases in blood glucose both stimulate glucagon secretion and inhibit insulin secretion. Glucagon secretion is also stimulated by sympathetic nervous activity and epinephrine, which have a suppressive effect on insulin secretion. Some studies suggest that glucagon and insulin function as paracrines in the islets of Langerhans, with insulin inhibiting the secretion of glucagon from alpha cells, and glucagon inhibiting the secretion of insulin from beta cells. Because of the opposite controls of these hormones, plasma glucagon levels tend to rise as insulin levels fall, and vice versa.

Actions of Glucagon

Figure 20.6 also shows that the actions of glucagon, though more limited than those of insulin, oppose insulin's actions. In the liver, glucagon promotes glycogenolysis and gluconeogenesis (which increase blood glucose levels), ketone synthesis, and breakdown of proteins, while at the same time inhibiting the opposing processes of glycogen and protein synthesis. In adipose tissue, glucagon stimulates lipolysis and suppresses triglyceride synthesis. These actions classify glucagon as a catabolic hormone. The overall effect of glucagon promotes mobilization of energy stores and synthesis of "new" energy sources (glucose and ketone bodies) that can be used by tissues; all of these actions are characteristic of the postabsorptive state.

Negative Feedback Control of Blood Glucose Levels by Insulin and Glucagon

Plasma glucose levels are normally tightly regulated by the antagonistic actions of insulin and glucagon to maintain stability. (Other hormones also regulate plasma glucose levels.) This stability is important because deviations too far from normal in either direction can have serious adverse effects on health. The normal fasting level of glucose in the blood is 70–110 mg/dL (blood glucose levels are usually measured clinically, not plasma glucose levels). Fasting blood glucose levels greater than 140 mg/dL constitute *hyperglycemia,* which is often indicative of **diabetes mellitus,** a serious and fairly common disease involving defects in insulin production or signaling (When It Goes Wrong: Diabetes Mellitus, p. 657). Fasting blood glucose levels below 60 mg/dL constitute *hypoglycemia,* which has widespread deleterious effects on nervous system function because the nervous system uses glucose almost exclusively as its source of energy.

Insulin and glucagon control plasma glucose concentration through negative feedback. Figure 20.7 diagrams insulin's negative feedback control of plasma glucose concentration. An increase in plasma glucose stimulates insulin secretion from beta cells of the pancreas, and the actions of insulin decrease plasma glucose. Insulin decreases plasma glucose concentration in three ways: (1) by promoting the uptake of glucose into cells by increasing the number of glucose transporters in the plasma membrane; (2) by reducing the concentration of free glucose within cells by converting it to glycogen, which promotes glucose uptake by increasing the size of the glucose concentration gradient; and (3) by suppressing gluconeogenesis, thereby reducing the rate at which new glucose is released into the bloodstream. If plasma glucose concentration decreases, insulin secretion decreases, causing an increase in plasma glucose.

Figure 20.8 depicts how together insulin and glucagon control plasma glucose concentration through negative feedback. An increase in plasma glucose concentration increases insulin secretion and decreases glucagon secretion from the pancreas, both of which cause a decrease in plasma glucose. Glucagon increases plasma glucose concentration by promoting gluconeogenesis and glycogenolysis in the liver, which directly increases plasma glucose concentration, and by stimulating lipolysis in adipose tissue, which provides fatty acids as an alternate energy source to glucose.

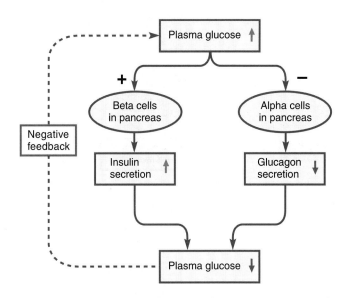

FIGURE 20.8 Regulation of plasma glucose concentration by insulin and glucagon.

Stimulation of Insulin and Glucagon Secretion by Amino Acids

Although insulin secretion and glucagon secretion are usually affected in opposite ways by a given stimulus, there is an exception to this pattern. Glucagon and insulin secretion are stimulated by an increase in certain plasma amino acids (see Figures 20.5 and 20.6). Although this might seem counterproductive, it is not. Consider what happens following a meal rich in proteins but low in carbohydrates: When nutrients are absorbed, plasma amino acid levels rise significantly, but glucose levels either do not change or rise only slightly. The rise in amino acids stimulates insulin secretion, which promotes increases in amino acid uptake by cells. At the same time, the rise in insulin tends to promote a *decrease* in plasma glucose. Because the plasma glucose level was already near normal, this change is inappropriate and potentially dangerous. However, the rise in plasma amino acids also stimulates the secretion of glucagon, which tends to promote an *increase* in plasma glucose. (Note that glucagon does not affect amino acid uptake.) When amino acids are absorbed with significant amounts of glucose, as in a typical diet, the effect of insulin prevails over that of glucagon because insulin secretion is stimulated by amino acids *and* glucose, whereas these two stimuli affect glucagon secretion in opposite ways.

Effects of Epinephrine and Sympathetic Nervous Activity on Metabolism

Figures 20.5 and 20.6 show that the sympathetic nervous system and epinephrine suppress insulin secretion and stimulate glucagon secretion, thereby indirectly promoting metabolic adjustments to the postabsorptive state.

These metabolic adjustments are also promoted by the direct actions of sympathetic neurons and epinephrine on certain target tissues.

The postabsorptive period is marked by decreased plasma glucose levels, which act directly on alpha and beta pancreatic cells to increase glucagon secretion and decrease insulin secretion. In similar fashion, a decrease in plasma glucose acts directly on *glucose receptors* in the central nervous system to raise the level of activity in sympathetic neurons, which trigger a rise in epinephrine secretion by the adrenal medulla (Figure 20.9); the resulting increase in plasma epinephrine acts on the liver to increase glycogenolysis and gluconeogenesis, on skeletal muscle to increase glycogenolysis, and on adipose tissue to increase lipolysis. Similar actions are promoted by sympathetic neural input to the liver and adipose tissue. (Skeletal muscle cells do not receive input from sympathetic neurons.)

Although the sympathetic control of metabolism plays a role in adapting the body to the postabsorptive state, under normal circumstances its influence is relatively minor compared to that of insulin and glucagon. The major importance of the sympathetic influence on metabolism is in the body's reaction to **stress,** a general term for any condition that actually or potentially poses a serious challenge to the body's ability to maintain homeostasis. (Thus stress includes physical conditions such as dehydration, hemorrhage, infection, exposure to temperature extremes, trauma, or severe exercise, as well as psychological states such as pain, fear, or anxiety.) In stress, the activation of the sympathetic nervous system that typically occurs triggers the familiar constellation of fight-or-flight responses (accelerated heart rate, generalized vasoconstriction, dilation of respiratory airways, and so on); it also elevates plasma glucose levels (due to

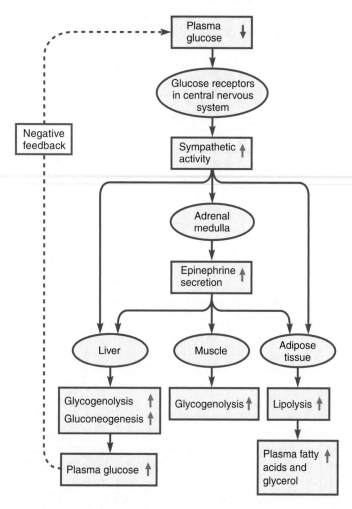

FIGURE 20.9 Regulation of energy metabolism by sympathetic nervous activity and circulating epinephrine in response to a fall in plasma glucose levels.

increased gluconeogenesis and glycogenolysis in the liver) and plasma levels of fatty acids and glycerol (due to increased lipolysis in adipocytes) (see Figure 20.9). These actions make fuel more readily available to cells, in the process helping prepare the body for the strenuous physical activity inherent in the fight-or-flight response. The increased availability of fuel also prepares the body for other activities requiring energy, such as tissue repair or fighting infections. Other components of the body's response to stress are described later in this chapter.

Quick Test 20.3

1. The concentration of which hormone, insulin or glucagon, is increased during the absorptive period?

2. Secretion of which hormone, insulin or glucagon, is stimulated in response to a decrease in plasma glucose concentration?

3. Sympathetic nervous activity and epinephrine promote metabolic reactions characteristic of which state, the absorptive state or the postabsorptive state?

4. For each of the following processes, indicate whether it is promoted by insulin or by glucagon: gluconeogenesis, glucose uptake by cells, glycogenolysis, glycogen synthesis, catabolism of energy stores, protein synthesis, a decrease in blood glucose levels, triglyceride synthesis, lipolysis.

HORMONAL REGULATION OF GROWTH

Although feeding (or the lack of it) is certainly a significant factor in the regulation of the body's overall metabolism, it is not the only one. In this section we explore the actions of hormones that play little (if any) role in everyday adjustments to feeding and fasting but nevertheless exert important influences on energy metabolism.

During their first two years of life humans experience a dramatic increase in height and body weight, a phenomenon called the *postnatal growth spurt* (Figure 20.10). After age 2, growth continues at a slower rate until the beginning of adolescence (about age 11 for girls and age 13 for boys), at which time another period of rapid growth, known as the *pubertal growth spurt,* begins. At the end of adolescence, which occurs in the late teens, growth stops, and individuals attain their full adult stature. Thereafter, no further increase in height is possible, although obviously it is possible for a person's weight to increase. Unless otherwise noted, in this text the term *growth* refers to the bodily changes that normally accompany increases in height only.

During periods of growth there is an increase in both the size and number of cells in the body's soft tissues (nonbony tissues such as skin and muscle) and in the length and thickness of bones. Observed increases in height are mostly attributable to increases in the length of bones in the legs and vertebral column. Lengthening of long bones in the limbs (the femur of the thigh or the humerus of the arm, for instance) are largely responsible for the changes in body proportion that accompany growth.

Body growth during childhood is regulated primarily by hormones, but it is also influenced by a person's genetic makeup, diet, and other factors such as disease or stress. Many of the bodily changes occurring during growth are attributable to the actions of **growth hormone** (GH), a peptide hormone secreted by the anterior pituitary (see Chapter 5). Other hormones that are essential for normal growth include insulin, the thyroid hormones, and the sex hormones (androgens and estrogens), which are especially important during the pubertal growth spurt. In addition, growth of various organs and tissues is influenced by numerous *growth factors* and *growth-inhibiting factors* that are usually specific to certain types of tissues and act locally as paracrine or autocrine agents. *Nerve growth factor,* for example, promotes elongation and proliferation of the axons and dendrites of neurons.

In the following section we concentrate on the actions of growth hormone and the factors influencing its secretion; the influence of other hormones on growth is discussed shortly.

Effects of Growth Hormone

In children, GH exerts several effects on bones and soft tissues that effect body growth. In adults it exerts many of the same effects, but instead of promoting growth it maintains bone mass and *lean body mass,* which is the proportion of body weight that is contributed by muscle (as opposed to fat).

Growth hormone directly promotes growth in two ways: It stimulates protein synthesis and increases cell size **(hypertrophy),** and it stimulates cell division, which results in increased cell number **(hyperplasia).** The results of these actions are *linear growth* (increased height) due to the elongation of bones, an increase in lean body mass due to the growth of muscle tissue, and an increase in the size of individual organs, including the heart, lungs, kidneys, and intestines.

Growth hormone also indirectly promotes a number of actions that affect growth. GH increases the plasma concentrations of glucose, fatty acids, and glycerol by inhibiting glucose uptake in adipose tissue and skeletal muscle, by stimulating lipolysis in adipose tissue, and by stimulating gluconeogenesis in the liver. Increasing plasma levels of these nutrients makes energy more readily available to tissues, which must expend energy to grow. GH also promotes increased uptake of amino acids by cells in muscle and other tissues, which facilitates protein synthesis.

Growth hormone alone cannot ensure normal growth; an adequate diet is clearly essential in providing the raw materials for growth. During growth, increases in the body's total protein mass resulting from protein synthesis requires an abundant supply of amino acids, which are most easily provided by eating food rich in proteins. Certain amino acids can be synthesized if they are lacking in the diet, but *essential amino acids* must be consumed in the diet. Many other raw materials for tissue growth, such as calcium for bones, must also be present in the diet in sufficient quantity. Finally, the *energy content* of the diet must provide enough calories to meet the heightened energy demands of growth. If the diet is inadequate in any of these respects, growth is inhibited. In younger children especially, the growth-stunting effects of poor diet are often irreversible.

Many of the growth-promoting effects of GH result from the actions of intermediary chemical messengers on

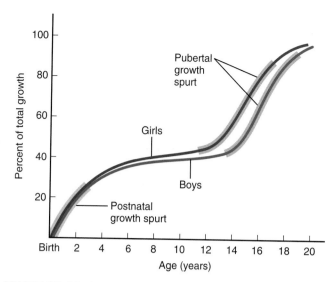

FIGURE 20.10 A representative human growth curve. *Note the postnatal and pubertal growth spurts.*

Based on the graph, who enters puberty at an earlier age, girls or boys?

target tissues, rather than a direct action of growth hormone itself. These messengers are peptides known as **somatomedins** or **insulinlike growth factors** (**IGFs**) because they bear some structural resemblance to insulin. To date, two IGFs (IGF-1 and IGF-2) have been positively identified. Growth hormone stimulates the production of somatomedins by the liver, which secretes them into the bloodstream for transport to target tissues throughout the body. In this respect, somatomedins function as hormones. Growth hormone also stimulates the production of somatomedins in other target tissues, where they act locally as paracrines.

Factors Affecting Growth Hormone Secretion

Secretion of growth hormone by the anterior pituitary is regulated by two hypothalamic hormones: **growth hormone releasing hormone** (GHRH), which stimulates growth hormone secretion, and **growth hormone inhibiting hormone** (GHIH) or **somatostatin,** which inhibits growth hormone secretion (Figure 20.11). Although the relative importance of these two hormones is unclear, variations in GH secretion are likely triggered primarily by GHRH, with GHIH playing a relatively minor role. Like other anterior pituitary hormones, growth hormone secretion is regulated through negative feedback loops (see Chapter 5). GH limits its own secretion via short loop negative feedback on the hypothalamus. Plasma somatomedins also exert long loop negative

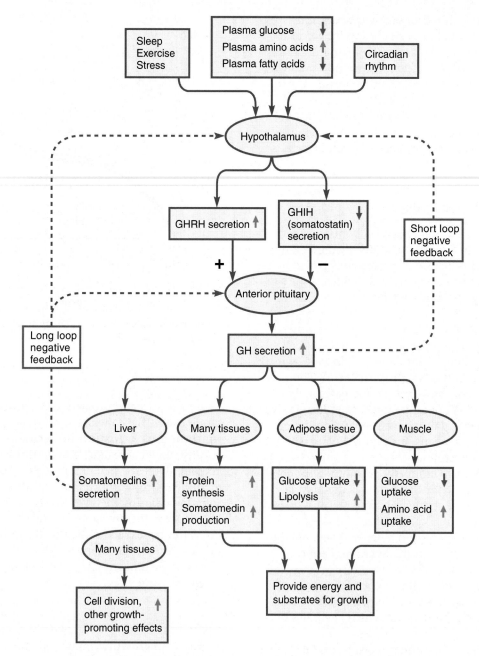

FIGURE 20.11 **Factors affecting the secretion of growth hormone, and its actions on target tissues.**

feedback controls on the hypothalamus and anterior pituitary to inhibit GHRH and GH secretion, respectively.

GHRH secretion is regulated by neural input of various types to the hypothalamus. Secretion is affected by plasma nutrient concentrations; decreases in plasma glucose or fatty acid levels, or increases in plasma amino acid levels, stimulate GHRH secretion (see Figure 20.11). Because it promotes changes in the opposite direction (by reducing glucose uptake and increasing lipolysis and amino acid uptake), growth hormone acts by negative feedback to limit variations in these nutrient concentra-

tions in plasma. Growth hormone secretion is also stimulated in response to sleep, exercise, or stress. Even though the significance of heightened GH secretion during sleep is not understood, a boost in GH during exercise or stress is useful because it tends to counteract reduced plasma levels of glucose and fatty acids, thereby helping to maintain a steady supply of these much-needed energy sources. Growth hormone secretion is also subject to a circadian rhythm that is mediated by neural input to the hypothalamus (see Figure 20.11). Experiments have shown that GH release exhibits a regular pattern, increas-

In *diabetes mellitus,* which affects an estimated 4–6% of the adult population and is the seventh leading cause of death in the United States, insulin regulation of energy metabolism and blood glucose levels is reduced or altogether absent. The disease has two basic forms: (1) *insulin-dependent diabetes mellitus* (IDDM), also known as *type 1* or *juvenile-onset diabetes,* which usually appears before age 20 and accounts for 10–15% of all cases, and (2) *non-insulin-dependent diabetes mellitus* (NIDDM), also known as *type 2* or *adult-onset diabetes,* which usually appears after age 40 and which accounts for the vast majority of cases. In IDDM, insulin secretion is reduced or absent, usually due to a reduction in the number of active pancreatic beta cells; in NIDDM the primary defect is a reduction in target cell responsiveness to insulin.

The primary sign of either form of diabetes is a persistent hyperglycemia, which is an expected consequence of reduced insulin activity. Hyperglycemia is due in part to reduced glucose uptake and utilization by many tissues, and also to increased glucose output by the liver, which results from increased gluconeogenesis and glycogenolysis and reduced glycogen synthesis. Frequently these effects are exacerbated by abnormally high plasma glucagon levels. Although hyperglycemia normally has a *suppressive* effect on glucagon secretion, glucagon secretion is often *elevated* in diabetes because the glucose permeability of alpha cells in the pancreas (which secrete glucagon) is insulin dependent. A lack of insulin hampers the ability of glucose to enter these cells, which "tricks" them into behaving as if the glucose level is lower than it actually is. (Recall that glucagon secretion is stimulated when plasma glucose levels fall.)

Often accompanying the changes in glucose metabolism in diabetes are other metabolic abnormalities that are usually more pronounced in IDDM than in NIDDM. Overstimulation of lipolysis and suppression of triglyceride synthesis (due to a lack of insulin or an excess of glucagon) can result in *hyperlipidemia,* an excess of fatty acids and other lipids in the blood. Excess utilization of fatty acids for energy can also lead to *ketosis,* elevated ketone levels in the blood. In-sulin lack also interferes with protein synthesis, resulting in excessive protein catabolism. This change hampers normal tissue repair and also causes muscle weakness and retardation of growth in children.

Adverse consequences of diabetes are many and varied and are secondary to the hyperglycemia and metabolic disturbances associated with the disease. Elevation of blood glucose results in *glucosuria* (the presence of glucose in urine) and excessive urine output caused by osmotic forces exerted by glucose in kidney tubules, as discussed in Chapter 17. In fact, the very name of the disease refers to these symptoms; the Greek roots of *diabetes mellitus* mean "sweet siphon." The high rate of water loss via the urine predisposes individuals to dehydration and loss of electrolytes (Na^+, K^+, and others) from the plasma. Unless compensated for by water intake (which also must be abnormally high) or other measures, dehydration can quickly lead to circulatory collapse and death. Electrolyte disturbances can result in neuromuscular problems. Excess ketone production, which is a greater

ing at night and falling during the day. (Secretion reaches its peak about 1–2 hours after the onset of sleep.)

Average daily plasma growth hormone levels also vary with a person's age. GH levels generally reach a maximum during puberty and then decline with age. Decreased GH levels are thought to be at least partially responsible for some signs of aging, such as decreased muscle mass and increased body fat.

Bone Growth

Because stimulation of bone growth is an important part of growth hormone's actions, it is appropriate to consider the nature of bone in some detail here. Moreover, as we saw in Chapter 18, bone is an important reserve for calcium, which can be liberated and moved into the bloodstream when plasma calcium levels decrease.

To support the weight of the body and to withstand forces placed on it by contracting muscles, bone must be strong but not brittle. Crystals of calcium phosphate in a form known as *hydroxyapatite* [$Ca_{10}(PO_4)_6(OH)_2$] give bone a mineral component that helps it withstand compressive forces (that is, "squeezing" or "crushing" forces). **Osteoid,** an organic component that consists of collagen fibers embedded in a gel-like substance, gives bone its

problem in IDDM than in NIDDM, is also potentially dangerous because some ketones are acids and can therefore cause a drop in blood pH—a particular form of metabolic acidosis called *ketoacidosis.* This condition depresses central nervous system function and can ultimately lead to coma and death.

The many complications of diabetes are capable of precipitating an acute crisis culminating in death within a few hours. Other complications are more chronic (often taking years to become apparent) but no less dangerous. Most often these problems are a consequence of vascular degeneration, which results in reduced blood flow to tissues. Diabetic patients are particularly prone to atherosclerosis, which narrows blood vessels, and to blood vessel weakening and breakage. For these and other reasons, diabetics are more likely to suffer from stroke, heart disease, and kidney failure. Reduced blood

flow to extremities also predisposes diabetics to infection and delayed wound repair, especially in the legs and feet. Gangrene is not uncommon and frequently necessitates amputation of limbs. Another vexing problem is *diabetic retinopathy*—degeneration of the retina caused by reduced blood flow—which is the fourth leading cause of blindness in the United States.

Precise causes of diabetes are not entirely known but are thought to involve both genetic and environmental factors. The fact that over 80% of NIDDM patients are obese is consistent with the current thinking that overeating can trigger this form of the disease. Overeating is likely to cause higher-than-normal rates of insulin secretion due to an excess of absorbed nutrients in the blood. One problem with this is that it can trigger downregulation of insulin receptors in target tissues, which reduces their responsiveness to

the hormone. Insulin excess may also precipitate changes in signal transduction mechanisms beyond the receptor, such as decreased numbers of glucose transporters. Overstimulation of beta cells to secrete insulin can also result in "burn-out" such that they can no longer secrete insulin at their normal rates. Typically, insulin secretion is adequate or increased in the early stages of NIDDM but becomes subnormal as the disease progresses.

Because beta cells are unable to secrete insulin in IDDM, the only available treatment is administration of insulin, which must be given by injection. In contrast, NIDDM can usually be controlled or even reversed through dietary weight reduction and exercise, which increases tissue insulin responsiveness. Use of drugs that stimulate insulin secretion may also be beneficial.

---

ability to withstand tensile or stretching forces, making it less prone to fracture.

Despite its nonliving organic and mineral components, bone is a dynamic living tissue that contains cells. The dynamic nature of bone is evident not only in its ability to grow during childhood, but also in the fact that bone can heal following a fracture and adapt its structure in response to forces placed on it. In a person who engages regularly in heavy lifting, for instance, the weight-bearing bones gradually increase in thickness and strength. In a person who is sedentary or bedridden, bone mass diminishes over time. Such restructuring of bone is called *remodeling,* and it is critical to the body's ability to regulate plasma calcium levels.

Central to the remodeling of bone are mobile cells known as osteoblasts and osteoclasts, both of which are found on the outer surfaces and inner cavities of bone tissue (Figure 20.12). **Osteoblasts** or "bone makers" are responsible for building up the mass of bone tissue, a

process called **deposition,** whereas **osteoclasts** or "bone breakers" are responsible for breaking down bone tissue, a process called **resorption.** When the activity of osteoblasts exceeds that of osteoclasts—when deposition exceeds resorption—bone growth occurs.

Osteoblasts initiate bone deposition by laying down the osteoid, which is followed by deposition of calcium phosphate, a process called **calcification.** The mechanism of calcification, which takes several days, is poorly understood, but interstitial calcium is attracted to the osteoid and calcifies it. As an osteoblast works to build bone, it eventually becomes immobilized within the surrounding tissue and is transformed into another type of cell called an **osteocyte,** which maintains the surrounding matrix but no longer actively lays down new bone tissue. An osteocyte is distinguishable from other bone cells in that it possesses long, filamentous processes that extend through channels in the bone tissue called *canaliculi.* These cell processes come into contact with processes be-

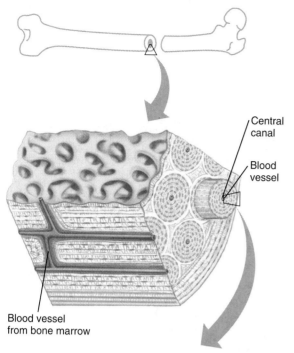

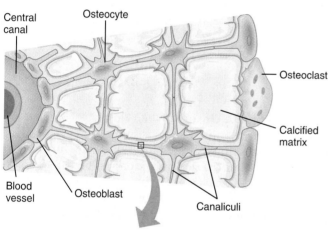

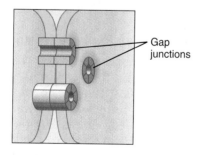

FIGURE 20.12 **The structure of bone.**

Which type of cell breaks down bone?

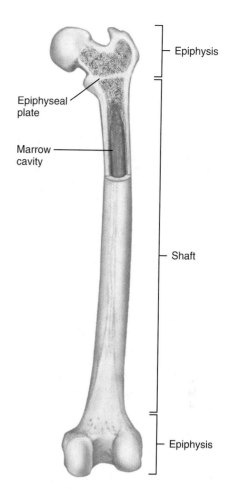

FIGURE 20.13 **Structure of a long bone (the femur).**

longing to other osteocytes or with nearby osteoblasts, such that the cells can communicate with each other through gap junctions. This communication is important because it allows cells in the interior of the bone tissue to exchange nutrients, wastes, and other materials with blood vessels, which run through bone cavities.

Osteoclasts effect resorption of bone tissue by secreting acid that dissolves calcium phosphate crystals, and enzymes that break down osteoid. Resorption releases calcium and phosphate into the bloodstream. Chapter 18 discusses the roles of bone deposition and resorption in calcium homeostasis.

Bone deposition is also necessary for increases in height, which occur through the growth of long bones. Figure 20.13 shows the structure of a typical long bone, which consists of a long, nearly cylindrical *shaft* capped at either end by a knob called an *epiphysis.* In the bones of growing children, the epiphyses are separated from the shaft by a thin layer of tissue called the **epiphyseal plate,** which is composed of **cartilage,** a soft material

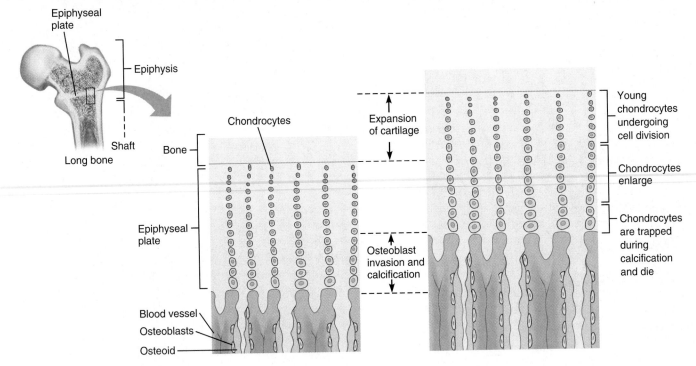

FIGURE 20.14 Long bone elongation. *Growth occurs at the epiphyseal plate. Chondrocytes lay down cartilage, which is invaded by osteoblasts. The osteoblasts cause calcification or bone formation.*

similar to uncalcified osteoid. The epiphyseal plate plays a key role in the elongation of bones during growth. Inside most bones is a central cavity containing red bone marrow and yellow bone marrow. *Red bone marrow* produces red and white blood cells; *yellow bone marrow* contains primarily adipocytes.

Under the influence of growth hormone, bones increase in circumference and length. The increase in circumference is brought about through the action of osteoblasts, which lay down new tissue on the outer bone surfaces. This increase in circumference is accompanied by resorption of bone by osteoclasts at the inner surface of the marrow cavity. As a consequence, the diameter of the marrow cavity increases as the outer diameter of the bone increases, which minimizes weight gain while increasing strength.

Increases in the length of a bone are brought about by the addition of new bone tissue to either end of the bone shaft (Figure 20.14). The process begins with the activity of cells in the epiphyseal plates called **chondrocytes,** which are similar to osteoblasts except that they produce cartilage rather than bone. Under the influence of growth hormone, chondrocytes increase in size and number, forming a layer of cartilage that causes the epiphyseal plate to become wider (elongating the bone). As new cartilage forms, chondrocytes located in the region adjacent to the shaft die, and osteoblasts from the nearby bone tissue replace them and begin converting the cartilage to bone. In this manner, the new bone is added to the end of the bone shaft.

In late adolescence the epiphyseal plates become completely filled in with bone tissue, a process called **epiphyseal plate closure.** When closure is complete, growth hormone can no longer stimulate bone elongation, and bones stop lengthening. This is why further increases in height are generally not possible after adolescence (even though changes in bone circumference and remodeling of bone continues). Plate closure is influenced by sex hormones, androgens and estrogens, during puberty.

Effects of Abnormal Growth Hormone Secretion

Deficient growth hormone secretion during childhood leads to a condition known as *dwarfism,* an irreversible stunting of growth, poor muscle development, and higher-than-normal amounts of body fat. Dwarfism is sometimes also caused by deficient tissue responsiveness to growth hormone, which can result from defective growth hormone receptors, insufficient production of somatomedins, or failure of tissues to respond to somatomedins. Abnormally low growth hormone secretion in adults produces few noticeable signs other than decreased muscle or bone mass.

When excessive secretion of growth hormone occurs before closure of the epiphyseal plates, the result can be *gigantism,* a condition in which stature is abnormally large but the body is normally proportioned. An excess of growth hormone that occurs following plate closure results in acromegaly. In this syndrome, no change in height occurs, but the overgrowth of soft tissues and an increase in bone circumference produces a characteristic pattern of disfiguration in certain body parts. Individuals with untreated acromegaly typically have an overly wide, protruding jaw (sometimes referred to as a "lantern jaw") and overly long limbs.

Other Hormones That Affect Growth

Normal body growth requires the actions of other hormones besides growth hormone, including thyroid hormones, sex hormones, and insulin. Thyroid hormones are required for the synthesis of growth hormone and are generally permissive for its actions; that is, thyroid hormones are required for growth hormone to exert its effects on target tissues. (Other actions of thyroid hormones are described later.) Insulin is also permissive for growth because it is required for secretion of IGF-1 and for normal protein synthesis in general.

Sex hormone levels are low until a few years before puberty and play little (if any) role in early childhood growth. In puberty, however, the dramatic rise in sex hormone secretion is essential for the growth spurt that normally occurs during this period. In contrast to insulin and the thyroid hormones, which are permissive for growth, the sex hormones actively promote growth by stimulating the secretion of growth hormone and IGF-1. In addition, the sex hormones stop bone elongation by virtue of their role in promoting epiphyseal plate closure. Androgens (testosterone in men and adrenal androgens in women) exert an additional growth-promoting effect by directly stimulating protein synthesis in many tissues, including skeletal muscle. The marked rise in muscle mass that occurs in boys during puberty is largely due to the rise in androgen levels during this period. Androgens also stimulate increased muscle mass in girls, but to a lesser extent because levels of these hormones are lower in girls than in boys.

In contrast to the hormones mentioned so far, which promote growth in one way or another, the glucocorticoids (such as cortisol) secreted from the adrenal cortex inhibit growth at high concentrations, in part because they promote bone resorption and protein catabolism. It is also worth noting that glucocorticoid secretion is stimulated by stress, which is one possible explanation for the observation that illness and other forms of stress can have a growth-retarding effect. Glucocorticoids and their role in stress are described later.

Quick Test 20.4

1. Define the following terms: *hypertrophy, hyperplasia, osteoid, osteoblast, osteoclast, osteocyte, chondrocyte, epiphysis, epiphyseal plate.*

2. Where are growth hormone releasing hormone and growth hormone inhibiting hormone (somatostatin) secreted? How do they influence growth hormone secretion?

3. What are somatomedins? What is their role with respect to body growth?

4. Why are further increases in body height not possible after adolescence?

THYROID HORMONES

Unlike the hormones we have studied so far, which show large changes in their rates of secretion throughout the course of a normal day, the thyroid hormones show little variation and their plasma levels are nearly steady. Consequently, the thyroid hormones do not normally "trigger" effects; instead they simply work to maintain the status quo.

Synthesis and Secretion of Thyroid Hormones

The thyroid gland contains numerous spherical follicles, each of which is composed of a single layer of secretory cells, called *follicular cells,* surrounding a glycoprotein the follicular cells secrete, called *colloid* (Figure 20.15a). Thyroid hormones are synthesized in the follicles. Located in the interstitial space between the follicles are C cells, which synthesize and secrete calcitonin (see Chapter 18). This section discusses the synthesis and secretion of thyroid hormones.

Thyroglobulin (TG), the primary substance found in the colloid, is a protein that functions as the precursor molecule for thyroid hormones. Also located in the colloid are the enzymes required for thyroid hormone synthesis and iodide (ionized form of iodine, I^-). The thyroglobulin and enzymes are synthesized in the follicular cells and secreted into the colloid by exocytosis; the iodide is actively transported by follicular cells from the blood into the colloid and is a necessary component of thyroid hormones. Thus all components of thyroid hormone synthesis are located in the colloid.

The steps of thyroid hormone synthesis and secretion are as follows (Figure 20.15b):

① Tyrosine residues of TG are iodinated. Addition of one iodide forms **monoiodotyrosine** (MIT), whereas addition of a second iodide to the same tyrosine residue forms **diiodotyrosine** (DIT).

② Two iodinated tyrosine residues (MIT or DIT) on the same TG molecule are coupled, at which time the two tyrosine residues are linked together by a covalent bond. If two DIT groups combine, the final product is 3,5,3',5'-tetraiodothyronine or T_4 (also called thyroxine); if a DIT and an MIT combine, the final product is 3,5,3'-triiodothyronine or T_3. T_3 and T_4 are

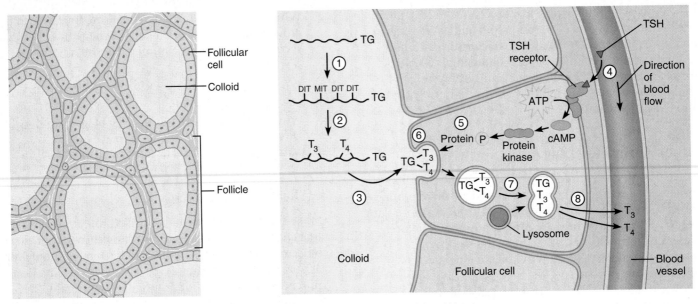

(a) Thyroid follicles

(b) Synthesis and secretion of thyroid hormones

FIGURE 20.15 Synthesis and secretion of thyroid hormones. (a) *Thyroid follicles, the sites of the synthesis of thyroid hormones.* **(b)** *Steps of thyroid hormone synthesis, which are described in detail in the text. Thyroid hormones can be stored in the colloid for months after their formation there, until binding of TSH stimulates the endocytosis into follicular cells of TG-thyroid hormone complexes, which are enzymatically degraded to release the thyroid hormones into the cells and ultimately into the bloodstream.*

the thyroid hormones, although at this step they are still attached to TG. Note that two MIT groups cannot combine.

③ Thyroid hormones are stored in the colloid bound to TG for up to three months until release.

④ Thyroid stimulating hormone (TSH) arriving via the bloodstream stimulates release of thyroid hormones. TSH first binds to receptors on the membrane of follicular cells, activating the cAMP second messenger system. This results in phosphorylation of a variety of follicular cell proteins necessary for the remaining steps.

⑤ The follicular cells take in iodinated TG molecules from the colloid by phagocytosis.

⑥ The phagosome containing the iodinated TG fuses with a lysosome.

⑦ Exposure of the TG molecule to lysosomal enzymes that break down the thyroglobulin causes the release of free T_3 and T_4 into the follicular cell.

⑧ Because T_3 and T_4 are lipophilic, they can diffuse across the plasma membrane and into the bloodstream, where they are selectively bound by protein carriers that include *thyroxine-binding globulin* and *transthyrethin,* and nonselectively bound by albumin.

T_4 is normally produced and secreted at a rate about ten times greater than T_3. However, T_3 is approximately

four times more potent at the target tissues. Most of the T_4 that is secreted into the plasma is eventually converted by the liver, kidneys, or target tissues to the more active T_3; in fact, the majority of T_3 in the plasma is synthesized from circulating T_4. Conversion of T_4 to T_3 is called *activation.*

Thyroid hormone levels are virtually constant under normal conditions because the primary control of its secretion occurs via negative feedback (Figure 20.16). As we have seen, thyroid hormone secretion is stimulated by TSH from the anterior pituitary. Secretion of TSH is, in turn, stimulated by **thyrotropin releasing hormone** (TRH) from the hypothalamus. Once thyroid hormones are released into the bloodstream, they feed back to the hypothalamus and the anterior pituitary to limit the secretion of TRH and TSH. Interestingly, T_4 provides stronger negative feedback than T_3.

The only known stimulus for TRH secretion, and thus for thyroid hormone secretion, is exposure to cold temperatures. This action is more pronounced in infants than in older children and is virtually absent in adults. In infants, the cold-stimulated TRH secretion is thought to promote heat production as the infant adapts to a colder environment outside the mother's body. TRH secretion, and thus thyroid hormone secretion, is inhibited by stress through neural input to the hypothalamus, although the significance of this inhibition is unknown.

Actions of Thyroid Hormones

Thyroid hormones are lipophilic and thus easily cross membranes, and the receptors for thyroid hormones are in the nuclei of target cells. Binding of thyroid hormone to receptors alters the rate of transcription of mRNA from DNA, thereby altering protein synthesis in the target cell. Such alterations take hours to days to exert an observable effect in the target cell; once induced, however, the effect generally lasts for days.

The primary action of the thyroid hormones is to raise the body's basal metabolic rate (see Figure 20.16)— that is, to increase the rate of oxygen consumption and energy expenditure at rest. As a result, heat generation also increases, a phenomenon termed a *calorigenic* effect. Thyroid hormone-stimulated increases in metabolism occur in most (but not all) tissues of the body; notable exceptions are the brain, spleen, and gonads. One way in which thyroid hormones increase metabolism is an increase in the rate of Na^+/K^+ pump activity in cells. As ATP is hydrolyzed during activity of the Na^+/K^+ pump, heat is liberated. Meanwhile, ATP is used up, necessitating higher rates of fuel oxidation and ATP production, which generates even more heat. Thyroid hormones also promote increased numbers of mitochondria in cells and increases in the concentrations of certain enzymes involved in oxidative phosphorylation.

In addition to stimulating energy *utilization,* the thyroid hormones also promote increased energy *mobilization* when present in *higher-than-normal* concentrations by promoting glycogenolysis, conversion of muscle proteins to amino acids, and lipolysis. They also promote gluconeogenesis and ketone synthesis. Conversely, at *lower-than-normal* concentrations, thyroid hormones have the opposite effect, promoting glycogenesis and protein synthesis.

Many effects of the thyroid hormones are permissive. Thyroid hormones promote the synthesis of beta adrenergic receptors, for example. Recall that adrenergic receptors bind epinephrine and norepinephrine, the chemical messengers of the sympathetic nervous system. Thus thyroid hormones *permit* many tissues to respond to sympathetic neural input and to circulating epinephrine.

Thyroid hormones are necessary for normal growth and development of many other tissues, and for maintaining normal function after growth has ceased. Many of these effects are mediated through stimulation of GH release (in synergism with glucocorticoids, discussed shortly) and permissiveness to growth hormone in target tissues. Developmental actions of thyroid hormones are especially important with respect to the nervous system. In infants, thyroid hormone deficiency can lead to a form of irreversible brain damage called *cretinism,* in which mental development is retarded and growth is stunted. Axons and dendrites of nerve cells are poorly developed, and myelination of axons is defective. Cretinism can be prevented by early diagnosis of hypothyroidism and initi-

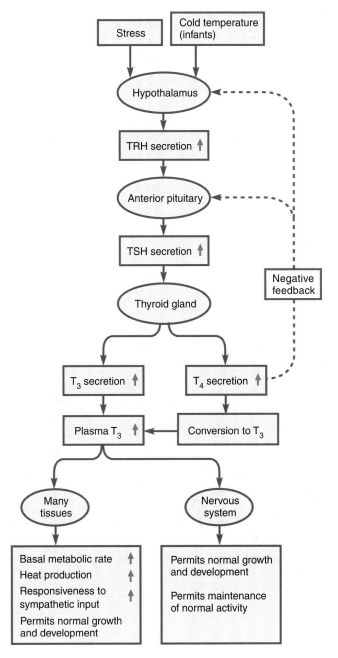

FIGURE 20.16 Factors affecting the secretion of thyroid hormones, and their actions on target tissues. *T_3 also provides negative feedback (not shown), but to a lesser extent than T_4.*

ation of T_3 replacement therapy. In the fully developed nervous system, thyroid hormones are essential for normal function. In adults, thyroid hormone deficiency can lead to impairment of mental function, but defects are fully reversible upon restoration of normal thyroid hormone levels.

GLUCOCORTICOIDS

At normal plasma concentrations, the glucocorticoids, which are steroid hormones secreted by the adrenal

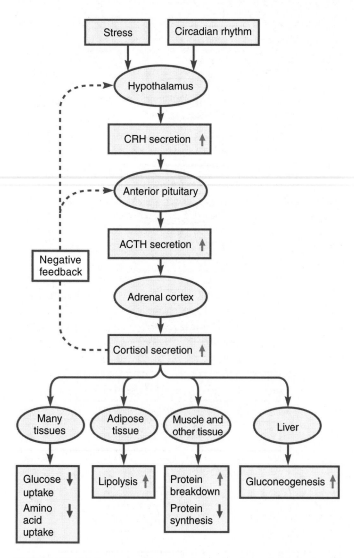

FIGURE 20.17 Factors affecting the secretion of cortisol, and its actions on target tissues.

cortex, maintain a wide variety of essential body functions. At higher concentrations they play a crucial role in the body's adaptation to stress.

Factors Affecting Secretion of Glucocorticoids

Secretion of glucocorticoids by the adrenal cortex is stimulated by **adrenocorticotropic hormone** (ACTH) from the anterior pituitary, which in turn is stimulated by **corticotropin releasing hormone** (CRH) from the hypothalamus (Figure 20.17). Because glucocorticoids are steroid hormones and thus are lipophilic, glucocorticoids diffuse out of the adrenal cortex and into the bloodstream immediately after synthesis. Plasma glucocorticoid levels are normally regulated by negative feedback on the hypothalamus and anterior pituitary, which limits the secretion of CRH and ACTH.

Cortisol is the primary glucocorticoid released from the adrenal cortex. Like growth hormone, cortisol is se-

creted in spurts and exhibits a circadian rhythm. Although the amount of hormone secreted per spurt is virtually constant, spurt *frequency* varies with the time of day (higher in the morning and lower at night). This pattern is tied to the sleep-wake cycle and reverses in people who are awake at night and sleep during the day. At lower spurt frequencies, plasma cortisol concentration rises and falls in a wavelike fashion because the added hormone is cleared from the plasma after each spurt. At higher spurt frequencies the concentration rises because insufficient time passes between spurts to permit complete clearance of the hormone from the plasma.

Stress, whether physical or emotional, is an important stimulus for cortisol secretion. Stressors that are most effective in stimulating cortisol secretion are usually noxious stimuli such as surgery, trauma, burns, infection, shock, and pain; other stressors include exposure to temperature extremes, strenuous exercise, and anxiety.

Actions of Glucocorticoids

Although the glucocorticoids do not *trigger* normal adjustments to the postabsorptive state, their presence is essential to the body's ability to mobilize fuels in response to signals from other hormones (for example, insulin and glucagon). The primary actions of glucocorticoids are to maintain normal concentrations of enzymes necessary for the breakdown of proteins, fats, and glycogen, and the conversion of amino acids to glucose in the liver. For this reason, the glucocorticoids are necessary for survival during prolonged fasting. In their absence, the resulting deficiencies in gluconeogenesis can lead to death by hypoglycemia once glycogen stores have been depleted.

Glucocorticoids are also required for growth hormone secretion (in synergism with thyroid hormones) and for maintaining the normal responsiveness of blood vessels to vasoconstrictive stimuli such as sympathetic nervous activity, epinephrine, and angiotensin II. In addition, glucocorticoids exert a variety of effects on the functions of the immune system, the nervous system, and the kidneys.

When plasma levels of glucocorticoids increase above resting levels, they exert a number of effects on metabolism that promote energy mobilization and glucose sparing. In many tissues they promote decreased uptake of glucose and amino acids. They stimulate lipolysis in adipose tissue, which raises plasma levels of fatty acids and glycerol. At the same time, glucocorticoids stimulate protein breakdown in muscle and other tissues, inhibit protein synthesis, and stimulate gluconeogenesis. As a consequence of all these actions, plasma concentrations of glucose, fatty acids, and amino acids rise. The metabolic actions of *cortisol*, the most abundant of the glucocorticoids, are illustrated in Figure 20.17.

Glucocorticoids are probably best known for their pharmacological effects when administered at doses that exceed normal physiological levels. At these dosages, glucocorticoids inhibit inflammation and allergic reactions. Glucocorticoids are given therapeutically to treat inflammation, such as occurs with arthritis, and to treat certain allergies. Glucocorticoids are also administered during tissue transplantation to decrease the likelihood of rejection, an immune response against foreign tissue. However, glucocorticoid administration must be done with caution because these hormones decrease the immune system's ability to defend the body against pathogens.

The Role of Cortisol in the Stress Response

For decades, cortisol has been considered the hormone of stress. Although it is generally acknowledged that enhanced cortisol secretion is important in helping the body adapt to stress, the reasons are poorly understood. Cortisol stimulates energy mobilization, which is useful in tissue repair. The ability to tolerate stress is poor in glucocorticoid-deficient individuals; mortality during recovery from surgery, for example, is significantly higher than in normal individuals.

However, cortisol is only one facet of the body's response to stress. In general, if a stimulus is effective in triggering increased cortisol secretion, it also triggers a consistent pattern of other neural and hormonal responses; for example, stress tends to promote increased activity of the sympathetic nervous system and secretion of epinephrine. These activities elicit the familiar fight-or-flight responses and stimulate gluconeogenesis, glycogenolysis, and lipolysis (see Figure 20.6), which augment cortisol's energy-mobilizing action. Other changes generally associated with stress include increased secretion of antidiuretic hormone by the posterior pituitary, increased renin release by the kidneys, and elevated plasma levels of angiotensin II. These responses help maintain blood pressure, and thus adequate blood flow to the heart and brain. This generalized, stereotypical pattern of stress responses is referred to as **general adaptation syndrome.**

Effects of Abnormal Glucocorticoid Secretion

An excess or deficiency of glucocorticoid secretion can result from either a defect originating in the adrenal cortex (a primary disorder) or a defect in the secretion of the tropic hormones CRH or ACTH (a secondary disorder).

Hypersecretion of cortisol is associated with a characteristic pattern of signs known as *Cushing's syndrome.* Signs include hyperglycemia (due to stimulation of gluconeogenesis and inhibition of glucose uptake) and protein depletion, which results in muscle wasting, weakness, and fragility in many tissues due to the breakdown of connective tissue. A frequent consequence of Cushing's syndrome is a tendency to bruise easily, which indicates weakened blood vessels. Although cortisol generally has a stimulatory effect on lipolysis, it also stimulates fat synthesis and proliferation of adipocytes in certain regions of the body, which promotes an unusual pattern of body fat distribution: Fat is deposited in the abdomen and above the shoulder blades, giving patients a protruding stomach and a hump-backed appearance; fat is also deposited in the face. Other regions, however, are not affected.

Hyposecretion of cortisol, known as *Addison's disease,* is characterized by hypoglycemia and poor tolerance of stress. In the primary form of the disease, which is usually a result of destruction of the adrenal cortex, there is often a defect in the secretion of aldosterone. Because aldosterone normally promotes sodium retention and potassium secretion by the kidneys, the disease is marked by excess sodium excretion and potassium retention, which by altering plasma sodium and potassium levels results in cardiac arrhythmias and other neuromuscular signs.

Quick Test 20.5

1. What does it mean to say that thyroid hormones have a calorigenic effect?

2. Describe the roles of thyrotropin releasing hormone and thyroid stimulating hormone in the regulation of thyroid hormone secretion.

3. Indicate which of the following actions are promoted by glucocorticoids: glycogen synthesis, glycogenolysis, an increase in plasma glucose levels, gluconeogenesis, protein synthesis.

4. What is the effect of stress on glucocorticoid secretion?

The metabolic effects of all the hormones discussed in this chapter are summarized in Table 20.4.

	SITE OF SECRETION	PRIMARY STIMULI FOR SECRETION (INDIRECT STIMULI IN PARENTHESES)	NET EFFECT ON CARBOHYDRATE METABOLISM	EFFECT ON PLASMA GLUCOSE	NET EFFECT ON LIPID METABOLISM	NET EFFECT ON PROTEIN METABOLISM
Insulin	Beta cells of islets of Langerhans in pancreas	↑Plasma glucose ↑Plasma amino acids	↑Glucose uptake into cells ↑Glycogen stores	↓Plasma glucose	↑Triglyceride stores	↑Amino acid uptake into cells ↑Protein synthesis
Glucagon	Alpha cells of islets of Langerhans in pancreas	↓Plasma glucose ↑Plasma amino acids	↑Glycogenolysis ↑Gluconeogenesis	↑Plasma glucose	↑Lipolysis	↑Proteolysis
Epinephrine	Adrenal medulla	Sympathetic nerve activity (stress, exercise)	↑Glycogenolysis	↑Plasma glucose	↑Lipolysis	None
Growth Hormone	Anterior pituitary	GHRH from hypothalamus (↓plasma glucose, ↑plasma amino acids, ↓fatty acids, sleep, stress, exercise)	↓Glucose uptake into cells	↑Plasma glucose	↑Lipolysis	↑Amino acid uptake into cells ↑Protein synthesis
Thyroid Hormones (T$_3$ and T$_4$)	Thyroid gland	TSH from anterior pituitary (TRH from hypothalamus, cold temperatures in infants)	↑Glycolysis	None	↑Lipolysis	↑Protein synthesis
Cortisol	Adrenal cortex	ACTH from anterior pituitary (CRH from hypothalamus, stress)	↓Glucose uptake into cells ↑Gluconeogenesis	↑Plasma glucose	↑Lipolysis	↓Amino acid uptake into cells ↑Proteolysis

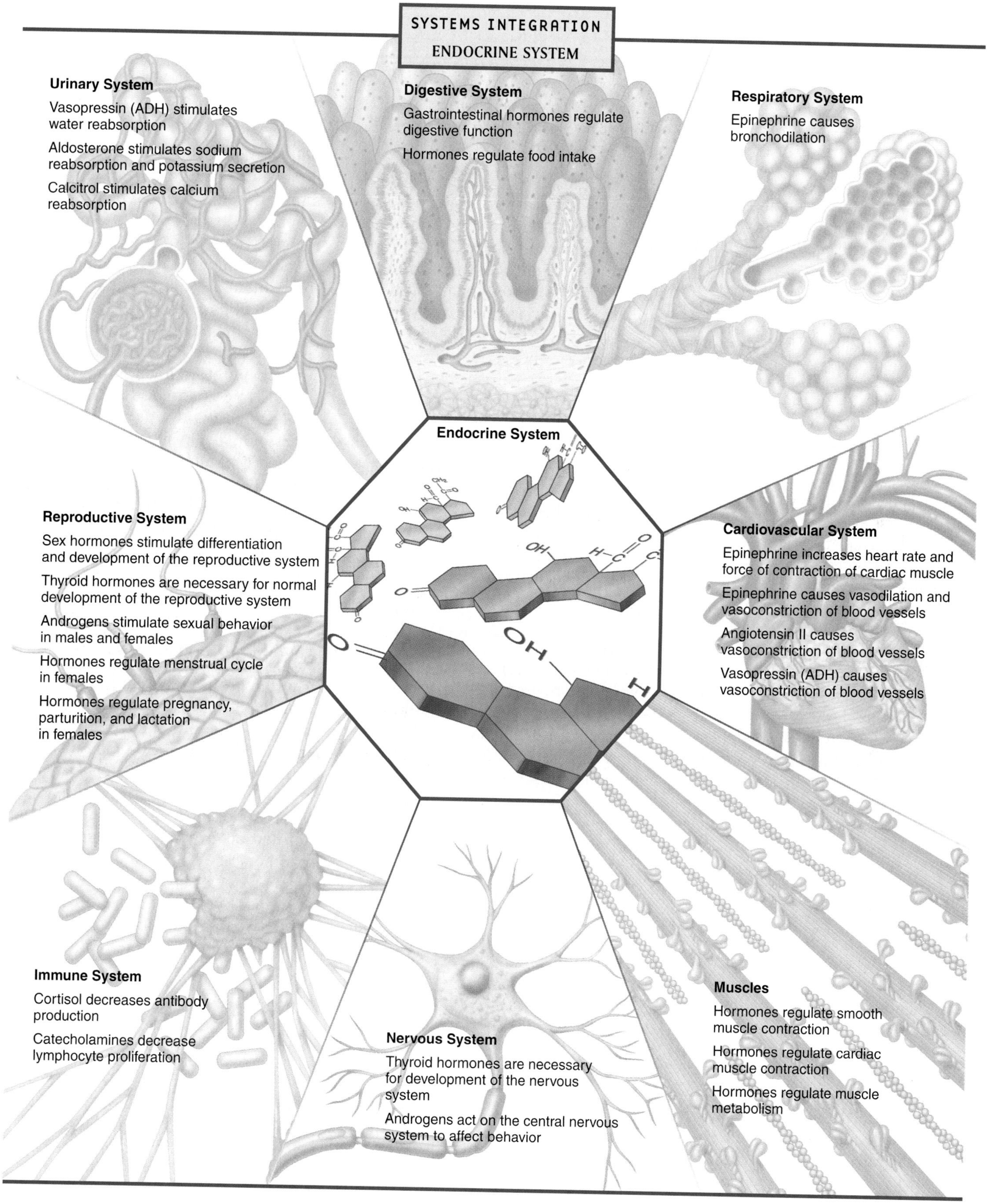

Endocrine System

Urinary System

Vasopressin (ADH) stimulates water reabsorption

Aldosterone stimulates sodium reabsorption and potassium secretion

Calcitrol stimulates calcium reabsorption

Digestive System

Gastrointestinal hormones regulate digestive function

Hormones regulate food intake

Respiratory System

Epinephrine causes bronchodilation

Reproductive System

Sex hormones stimulate differentiation and development of the reproductive system

Thyroid hormones are necessary for normal development of the reproductive system

Androgens stimulate sexual behavior in males and females

Hormones regulate menstrual cycle in females

Hormones regulate pregnancy, parturition, and lactation in females

Cardiovascular System

Epinephrine increases heart rate and force of contraction of cardiac muscle

Epinephrine causes vasodilation and vasoconstriction of blood vessels

Angiotensin II causes vasoconstriction of blood vessels

Vasopressin (ADH) causes vasoconstriction of blood vessels

Immune System

Cortisol decreases antibody production

Catecholamines decrease lymphocyte proliferation

Nervous System

Thyroid hormones are necessary for development of the nervous system

Androgens act on the central nervous system to affect behavior

Muscles

Hormones regulate smooth muscle contraction

Hormones regulate cardiac muscle contraction

Hormones regulate muscle metabolism

An Overview of Whole Body Metabolism, p. 640

Whole body metabolism requires the coordination of cellular metabolic activities. Cells use energy in the form of ATP, which they obtain from the oxidation of small nutrient molecules such as glucose, fatty acids, and amino acids. Cellular metabolism must be coordinated so that nutrients are provided to the appropriate cells when needed.

Energy Intake, Utilization, and Storage, p. 641

Energy is released in cells by the breakdown of nutrients into smaller molecules, as when glucose, amino acids, or fatty acids are oxidized to yield waste products. Energy mobilization is the breakdown of macromolecules into small nutrient molecules that are released into the bloodstream. Energy is stored by converting small nutrient molecules into macromolecules: Glucose is stored as glycogen in skeletal muscle and liver, fatty acids and glycerol are stored as triglycerides in adipose tissue, and amino acids are stored as proteins in all cells, but especially in skeletal muscle cells.

Energy Balance, p. 643

To maintain energy balance, energy input must equal energy output. Energy input comes from ingested nutrients, whereas energy output is the energy expended as heat or work. Positive energy balance occurs when energy input exceeds energy output; negative energy balance occurs when energy output exceeds energy input. The body's metabolic rate is the total amount of energy released per unit time as a result of nutrient oxidation. The metabolic rate at rest is the basal metabolic rate, or BMR.

Energy Metabolism During the Absorptive and Postabsorptive States, p. 645

In the absorptive state, glucose is used by most tissues as the primary fuel. Absorbed nutrients are also converted to glycogen, triglycerides, and proteins. Excess amino acids and glucose are mostly converted to fatty acids and stored as triglycerides. In the postabsorptive state, stored glycogen, triglycerides, and proteins are catabolized for energy. Fatty acids are used by most tissues as the primary fuel. An exception is the nervous system, which relies on a steady supply of glucose for its energy. Utilization of nonglucose fuels conserves glucose for use by the nervous system, a phenomenon called glucose sparing. The liver can also produce more glucose by gluconeogenesis.

Regulation of Absorptive and Postabsorptive Metabolism, p. 649

Metabolic adjustments to the absorptive state are promoted by insulin and include synthesis of energy stores (glycogen, proteins, fatty acids, and triglycerides) and uptake of glucose and amino acids by cells in many tissues. Insulin also suppresses gluconeogenesis and regulates plasma glucose levels via negative feedback control.

Metabolic adjustments to the postabsorptive state are promoted by glucagon and include glycogenolysis, protein breakdown by the liver, lipolysis, gluconeogenesis, and ketone synthesis. Glucagon also helps to regulate blood glucose levels. Postabsorptive metabolic adjustments are also promoted by increased epinephrine secretion and sympathetic nervous activity.

Hormonal Regulation of Growth, p. 654

Body growth during childhood is promoted by the actions of growth hormone, which is secreted by the anterior pituitary and acts to promote the growth of soft tissues and bones. In adulthood, growth hormone acts to maintain bone mass and lean body mass. Actions promoted by growth hormone include hypertrophy and hyperplasia in bones and soft tissues, protein synthesis, lipolysis, gluconeogenesis, and amino acid uptake by cells. Growth hormone also inhibits glucose uptake by adipose tissue and muscle. Combined metabolic actions work to raise plasma levels of glucose, fatty acids and glycerol, thereby making energy more readily available to growing tissues. Many of growth hormone's actions are mediated by somatomedins synthesized by the liver and other tissues.

Thyroid Hormones, p. 661

Thyroid hormones are normally secreted by the thyroid gland at near-constant rates and increase the metabolic rate in most tissues of the body. At high concentrations, thyroid hormones mobilize energy stores. Thyroid hormones are also necessary for normal growth, development, and maintenance of normal function in many tissues, particularly the nervous system. Thyroid hormones are secreted by thyroid follicles in two forms, T_3 and T_4. Whereas T_4 is the more abundant, T_3 is the more active.

Glucocorticoids, p. 663

Glucocorticoids are released by the adrenal cortex and are important in the body's response to stress. Glucocorticoids are also required for the body's ability to mobilize energy stores during postabsorptive periods.

Multiple-Choice Questions

1. Which of the following is an example of a permissive effect of a hormone?
   a) the effect of thyroid hormones on growth
   b) the effect of insulin on glucose uptake by cells
   c) the effect of sex hormones on the secretion of growth hormone
   d) all of the above

2. Which of the following is an example of a glucose-sparing effect of cortisol?
   a) inhibition of ACTH release
   b) stimulation of gluconeogenesis by the liver
   c) stimulation of lipolysis
   d) stimulation of glycogen breakdown

3. Which of the following cells of the pancreas secrete insulin?
   a) alpha cells
   b) beta cells
   c) delta cells
   d) exocrine cells

4. Stress stimulates secretion of which of the following hormones?
   a) growth hormone
   b) epinephrine
   c) thyroid hormones
   d) ACTH
   e) all of the above

5. Hypoglycemia inhibits secretion of which of the following?
   a) growth hormone
   b) insulin
   c) epinephrine
   d) glucagon

6. In the postabsorptive state, the central nervous system uses which of the following as its primary source of energy?
   a) fatty acids
   b) amino acids
   c) glucose
   d) glycerol

7. Which of the following cell types is directly responsible for building new bone material?
   a) osteoblasts
   b) osteoclasts
   c) osteocytes
   d) chondrocytes

8. Which of the following is true of adulthood?
   a) Growth hormone exerts no effects on body tissues.
   b) The secretion of growth hormone ceases altogether.
   c) Growth hormone cannot stimulate increases in the length of long bones.
   d) The structure of bone becomes permanently fixed.

9. Which form of thyroid hormone has greater activity at target cells?
   a) T_3
   b) T_4
   c) neither; T_3 and T_4 have equal activity

10. Which of the following hormones is a steroid?
    a) thyroid hormones
    b) insulin
    c) glucagon
    d) cortisol

Objective Questions

1. Energy mobilization is promoted by (insulin/glucagon).

2. Secretion of (insulin/glucagon) is increased during the absorptive period.

3. Insulin and glucagon both help regulate the plasma glucose concentration. (true/false)

4. Breakdown of triglycerides yields fatty acids and _____, which can be used by cells for energy.

5. Conversion of amino acids to fatty acids is more likely to occur in the (absorptive/postabsorptive) state.

6. Conversion of amino acids to glucose is more likely to occur in the (absorptive/postabsorptive) state.

7. An increase in plasma thyroid hormone levels tends to make the body's energy balance more (positive/negative).

8. Energy that is taken into the body is either stored or appears as work or _____.

9. Stress tends to (stimulate/inhibit) GHRH secretion.

10. Many of growth hormone's effects are due to the action on target tissues of other chemical messengers called _____.

11. Growth hormone secretion is not regulated by negative feedback. (true/false)

12. Closure of the epiphyseal plates is promoted by (growth hormone/sex hormones).

13. Thyroid hormones promote increased responsiveness of target tissues to (sympathetic/parasympathetic) nerve activity.

14. Glucocorticoids promote (increased/decreased) plasma glucose levels.

15. Stimulation of gluconeogenesis by glucagon is an example of a glucose-sparing effect. (true/false)

16. Plasma glucocorticoids have a(n) (stimulatory/inhibitory) effect on the secretion of ACTH.

Essay Questions

1. Describe the regulation of plasma glucose by insulin and glucagon. Include a description of the role of negative feedback.

2. Describe how insulin, glucagon, and the sympathetic nervous system work together to maintain adequate plasma glucose levels during fasting. Why is this important?

3. Describe the various factors that determine the body's energy balance. Be sure to describe what happens to energy that is liberated as a result of fuel oxidation.

4. Describe the similarities between the metabolic actions of thyroid hormones and glucocorticoids.

5. Describe the metabolic actions of growth hormone, and explain how these actions promote growth.

Find the answers to these exercises, and additional study tools, at the Physiology Place (www.physiologyplace.com).

The Reproductive System

OBJECTIVES

- Describe the events in meiosis, and explain its significance in gametogenesis and in the promotion of genetic diversity.

- Describe the role of sex chromosomes and sex hormones in the development of sexual characteristics.

- Describe the process of spermatogenesis and its hormonal regulation.

- Explain the cyclic variations in plasma hormone levels that occur during the menstrual cycle, and how these hormones regulate the ovarian and uterine changes that occur during this cycle.

- Describe the events that occur during fertilization, implantation, and early embryonic development.

- Describe the regulation of estrogen and progesterone secretion during pregnancy, and explain how these hormones help maintain pregnancy and prepare the body for parturition.

CHAPTER OUTLINE

An Overview of Reproductive Physiology 671

The Male Reproductive System 676

The Female Reproductive System 683

Fertilization, Implantation, and Pregnancy 695

Parturition and Lactation 701

Above: Scanning electron micrograph of a human sperm fertilizing an egg

STUDY HINTS

1. Hypothalamic tropic hormones, p. 157

2. Negative feedback, p. 9

3. Mitosis, p. 56

The reproductive system, the topic of this chapter, is unlike any other organ system in that its primary function is not to promote the survival of the immediate individual; instead, its function—to produce offspring—promotes the survival of the species as a whole. The reproductive system is also unique among organ systems because it differs between males and females, making it the only organ system that does not operate according to a single set of rules. We begin our study of the reproductive system with a general overview of reproductive physiology, with emphasis on similarities between the male and female systems, and how they work together to create offspring. In subsequent sections we concentrate on sex-specific aspects of reproductive function.

AN OVERVIEW OF REPRODUCTIVE PHYSIOLOGY

Reproduction in humans is *sexual,* meaning that offspring are produced as a result of the mating of parents of different sexes. Each offspring inherits a unique combination of genes from both parents and therefore develops its own unique set of characteristics. This ability to create new combinations of genes is a hallmark of sexual reproduction, and it is one reason for the great diversity of living things in nature.

The Role of Gametes in Sexual Reproduction

In sexual reproduction, each parent produces cells called **gametes,** which contain half of his or her genetic material. Male gametes are known as **spermatozoa** (or simply *sperm*); female gametes are **ova** (singular: **ovum**) or *eggs.* Such cells are described as **haploid** because they possess half the number of chromosomes found in most other cells of the body, which are described as **diploid.** (The chromosome number in haploid cells is designated by the symbol n, whereas the chromosome number in diploid cells is designated $2n$.) As a result of mating, gametes from each parent may fuse together to produce a new cell, a phenomenon known as **fertilization.** (Fertilization in humans is frequently referred to as *conception.*) This new cell, known as a **zygote,** has the potential to develop into a completely new individual and is diploid, having received half of its chromosomes from one parent and half from the other.

Gametes are generated from a pool of specialized, relatively undifferentiated precursor cells known as *germ cells* in a process called **gametogenesis.** In this process, the diploid germ cells undergo a series of cell divisions that ultimately reduces the number of chromosomes in each cell by half (that is, from $2n$ to n).

A single sperm or ovum contains 23 chromosomes—22 autosomes and one sex chromosome—each of which is distinguishable from the others on the basis of shape and possesses its own characteristic set of genes. **Sex chromosomes** are the chromosomes that determine an individual's sex and are of two types—the Y or male chromosome and the X or female chromosome. **Autosomes** are the chromosomes other than the sex chromosomes. During fertilization, genetic material from a sperm and an egg combines to yield a total of 46 chromosomes (44 autosomes plus two sex chromosomes), the normal complement of chromosomes found in most cells of the body. Unlike the sex chromosomes, autosomes always come in matching pairs, the members of which are similar in size and shape and possess genes governing the same characteristics. As a reflection of this, the members of each pair of autosomes are said to be *homologous.* In the zygote and all other diploid cells that arise from it, one member of each pair of homologous chromosomes is inherited from the egg and is therefore of *maternal* origin (that is, comes from the mother), whereas the other member of the pair is inherited from the sperm and is therefore of *paternal* origin (that is, comes from the father).

The fact that autosomes come in pairs means that for every autosomal gene, there is a corresponding gene on the homologous chromosome that governs the same trait or characteristic. Thus autosomal *genes* come in pairs, one each from the father and the mother. The genes in each pair, though similar in many important respects, are not necessarily identical because genes generally come in a variety of different versions or *alleles.* The allele inherited from the mother may be the same as or different from the allele inherited from the father. A familiar example involves a gene that determines eye color. One version of the gene (one allele) codes for blue eyes, whereas another allele codes for brown eyes. Because each person has two genes for eye color, it is possible for a person to have two "blue" alleles, two "brown" alleles, or one "blue" allele and one "brown" allele.

The rule that chromosomes come in matching pairs does not apply to the sex chromosomes. The Y chromosome is significantly smaller than the X chromosome and has a different shape. Furthermore, most of the genes

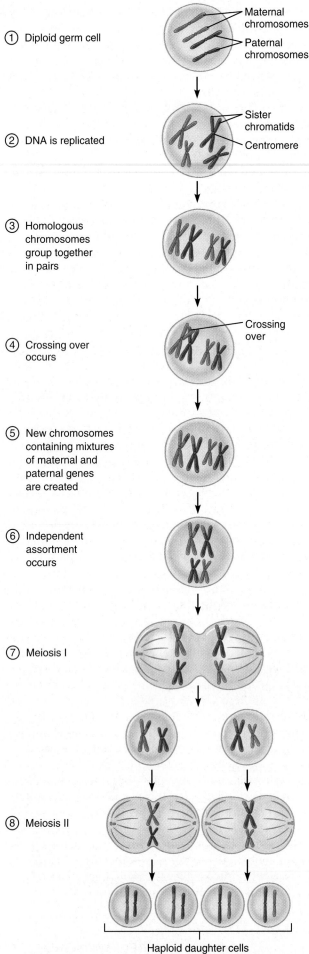

① Diploid germ cell

Maternal chromosomes
Paternal chromosomes

② DNA is replicated

Sister chromatids
Centromere

③ Homologous chromosomes group together in pairs

④ Crossing over occurs

Crossing over

⑤ New chromosomes containing mixtures of maternal and paternal genes are created

⑥ Independent assortment occurs

⑦ Meiosis I

⑧ Meiosis II

Haploid daughter cells

FIGURE 21.1 **Meiosis.** *Details of the steps of meiosis are described in text.*

present on the X chromosome are not matched by corresponding genes on the Y chromosome, and vice versa. Because many genes that are present on the X but absent from the Y chromosome are necessary for life, every individual must inherit at least one X chromosome. Females inherit two X chromosomes, whereas males inherit one Y and one X.

Gene Sorting and Packaging in Gametogenesis: Meiosis

In both males and females, all the gametes an individual will ever produce are ultimately derived from a relatively small set of diploid germ cells. At some point in life these cells undergo mitosis to produce a colony or *clone* of daughter cells, each possessing an exact copy of all 46 of the individual's chromosomes. Subsequently these cells undergo meiosis, a series of two cell divisions that generates daughter cells with half the normal chromosome number. Eventually these cells become mature sperm or ova.

In meiosis, an individual's chromosomes are sorted and packaged in such a way that each of the final daughter cells receives one sex chromosome plus one chromosome from each of the 22 pairs of autosomes. The sorting and packaging that occurs in meiosis proceeds in the following series of steps (Figure 21.1):

① The process begins with a diploid germ cell containing 46 chromosomes, which in the figure are represented by four bars of two different colors that indicate whether they are paternal or maternal. Homologous chromosomes are indicated by their similar lengths. (For simplicity, the germ cell's nuclear membrane is not shown.)

② In a process similar to what occurs prior to mitosis, the nuclear membrane breaks down, and the original DNA is replicated, yielding exact copies of all 46 chromosomes. The two copies of each chromosome (called **sister chromatids**) remain joined together at a structure called the centromere.

③ Homologous chromosomes begin to group together in pairs such that genes on paternal chromosomes line up opposite the corresponding genes on maternal chromosomes.

④ Portions of homologous chromosomes begin to overlap and exchange segments with one another, a phenomenon known as **crossing over.**

⑤ Crossing over results in new chromosomes that contain mixtures of maternal and paternal genes. Because crossing over occurs randomly, the number of possible new chromosomes is extremely large.

⑥ Pairs of homologous chromosomes line up along a plane bisecting the cell in such a way that a random mixture of maternal and paternal chromosomes is present on either side of the cell. This random grouping of maternal and paternal chromosomes is known as **independent assortment.**

⑦ The first meiotic division **(meiosis I)** occurs. The cell divides in two, and each daughter cell receives one chromosome from each homologous pair. Due to independent assortment, each daughter cell receives a different combination of maternal and paternal chromosomes. Because the chromosomes are duplicates consisting of two sister chromatids, each cell receives half the normal chromosome number but two copies of every chromosome. Thus each cell receives two copies of each autosome (either the maternal or paternal version) and two copies of either an X or a Y chromosome.

⑧ The second meiotic division **(meiosis II)** occurs. Each of the cells generated in meiosis I divides in two, yielding a total of four cells. In this process, sister chromatids separate, and each of the new haploid daughter cells receives one sister chromatid from each chromosome. Consequently, each of the four cells receives single copies of 23 chromosomes (22 autosomes plus either an X or a Y chromosome), which are eventually enclosed within a new nuclear membrane (not shown).

Because of crossing over and independent assortment, a given parent can produce a large number of genetically different gametes, even though they are all derived from the same gene pool. Accordingly, each child of a given set of parents normally inherits its own unique set of genes.

Components of the Reproductive System

The *reproductive system* of an individual encompasses all organs involved in mating, gametogenesis, or other functions directly involved in the production of offspring. Reproductive system organs include the gonads and the accessory reproductive organs.

The **gonads** (*testes* in the male and *ovaries* in the female) are considered the primary reproductive organs because they perform two functions that ultimately govern all reproductive activity: They produce gametes, and they secrete sex hormones, a variety of steroids that promote gametogenesis, growth and maintenance of reproductive organs, development of secondary sex characteristics, and various other effects throughout the body. The testes secrete a class of sex hormones known as **androgens** (notably *testosterone*); the ovaries secrete **estrogens** (such as *estradiol*) and **progesterone.**

Note that even though androgens and estrogens are commonly referred to as "male" and "female" sex hor-

mones, respectively, they are actually present in both sexes, but in different relative amounts. Androgens are more abundant in males, whereas estrogens are more abundant in females. As noted in Chapter 5, sex hormones are produced not only by the gonads, but also by the adrenal cortex. In women the adrenal cortex is the primary source of androgens, which are responsible for promoting the sex drive in either sex.

The **accessory reproductive organs** include structures specialized to perform additional functions required for proper reproductive activity, such as transporting gametes from one place to another and providing nourishment for gametes once they are produced. Accessory organs include (1) the organs of the **reproductive tract,** a system of interconnecting passageways through which gametes are transported after leaving the gonads, and (2) various glands that secrete fluids into the reproductive tract.

Events Following Fertilization

Sperm are transferred from the male to the female during the act of mating, known as **copulation.** Ova, in contrast, remain in the female's body, where they are fertilized by the arriving sperm. After fertilization, the ovum undergoes many cell divisions, eventually giving rise to billions of cells. As the cells increase in number, they also become specialized and begin to organize into distinct tissues according to instructions encoded in their genes. Over a period of approximately nine months, these cells and tissues gradually develop into a functional human organism. During this time the developing human is carried within its mother's body (a condition referred to as **pregnancy** or *gestation*), an arrangement that provides it with nourishment and protection from the outside environment. In the first two months after conception the developing human is called an **embryo;** from that point on it is referred to as a **fetus.** As the fetus develops it eventually acquires the capacity to live outside its mother's body and separates from it in a process known as **parturition** (birth). Next we examine the events in embryonic and fetal development that determine the offspring's gender.

Sex Determination

Because a female's germ cells possess two X chromosomes (and thus no Y chromosome), every egg she ever produces possesses an X chromosome. In contrast, a male produces some sperm with an X chromosome and others with a Y chromosome. For this reason, the sex of the offspring is determined entirely by the genetic makeup of the sperm, not the egg: When the sperm bears an X chromosome, the fertilized egg inherits two Xs (one from the mother and one from the father) and develops into a female; when the sperm bears a Y chromosome, the egg inherits one Y and one X and therefore develops into a male. That males and females are born in approximately equal numbers is a consequence of the fact that on average about 50% of all sperm cells possess a Y chromosome. Although a person's sex is normally determined by the genes he or she inherits, the genes themselves do not confer the full complement of male or female traits; instead they simply determine whether a fetus will develop ovaries or testes. The role of genes in deciding a person's sex is referred to as **sex determination.**

Whether a fetus develops ovaries or testes is determined by the presence or absence in the embryo of a gene located on the Y chromosome—the **srY gene**—which codes for a protein called *testis-determining factor.* (srY stands for *sex-determining region of the Y* chromosome.) In early life the embryo possesses primitive gonads that have the potential to become either ovaries or testes. When the embryo inherits a Y chromosome, the srY gene instructs the primitive gonads to become testes; when the embryo inherits two X chromosomes, the srY gene is absent, so the primitive gonads develop into ovaries. Thus the primitive gonads develop into ovaries "by default"—that is, unless they are instructed by the srY gene to do otherwise.

Sex Differentiation

The development of gonads in the embryo (sex determination) sets the stage for the development of other sexual characteristics. Whether these characteristics are male or female depends on the presence or absence of two hormones normally secreted by the testes—*testosterone* and another hormone called **Müllerian-inhibiting substance (MIS).** If these hormones are present (and target tissues are able to respond to them normally), the fetus develops as a male; if these hormones are absent, it develops as a female. The role of these hormones in the development of sexual characteristics is referred to as **sex differentiation.**

In the first few weeks of development, the embryo is said to be *sexually indifferent* and possesses rudimentary female and male reproductive systems, called **Wolffian ducts** *(mesonephric ducts)* and **Müllerian ducts** *(paramesonephric ducts)*, respectively (Figure 21.2). These structures have the potential to give rise to all reproductive organs except the gonads, the *external genitalia* (sex organs that are externally visible) and, in females, the vagina (the initial portion of the reproductive tract, which leads to the outside of the body).

When testes are present and functioning normally, embryonic tissues are exposed to both testosterone and MIS. Testosterone acts on the Wolffian ducts to promote the development of male reproductive organs, whereas MIS promotes the regression and eventual disappearance of the Müllerian ducts, thereby preventing the development of female organs. At the same time, testosterone acts on certain other embryonic tissues to promote the

development of male external genitalia (such as the penis and scrotum, which are discussed shortly). In many target tissues, however, it is not testosterone that triggers these changes; instead, an enzyme in these target tissues converts testosterone to dihydrotestosterone, which then binds to receptors and triggers the target tissue response.

When the embryo develops ovaries instead of testes, the absence of testosterone causes the Wolffian ducts to regress, thereby preventing the development of male organs. In addition, the absence of MIS allows the Müllerian ducts to develop into female organs. The lack of testosterone also eliminates the influences that would otherwise stimulate the development of male external genitalia and allows female external genitalia to develop instead. The roles of sex determination and sex differentiation in the development of males and females are depicted in Figure 21.3.

Patterns of Reproductive Activity Over the Human Life Span

Humans are not born with the ability to reproduce but instead acquire this ability during **puberty,** a period of sexual maturation that typically begins sometime between the ages of 10 and 14 and extends to the mid- to late teens. In the years before puberty the reproductive organs are immature and incapable of generating sperm or ova. In addition, a person's outward appearance is childlike, with little obvious distinction between the bodies of boys and girls. During puberty the reproductive organs mature, gametogenesis begins, and other physiological changes occur throughout the body. In the process, males and females develop various **secondary sex characteristics**—the external features that distinguish the sexes from each other, such as the growth of facial hair in men and the widening of the hips relative to the shoulders in women.

Exercise Link

Some other secondary sex characteristics may have affected Bill and Jane's marathon performance. Testosterone promoted increased muscle mass development in Bill, allowing him to run faster (at least for a short period of time). Estrogen influenced some of the enzymes that liberate and oxidize carbohydrates and fats for energy. As a result, Jane used her energy stores in a way that may have provided her with greater stamina.

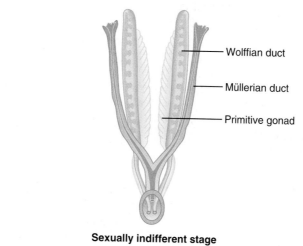

Wolffian duct

Müllerian duct

Primitive gonad

Sexually indifferent stage

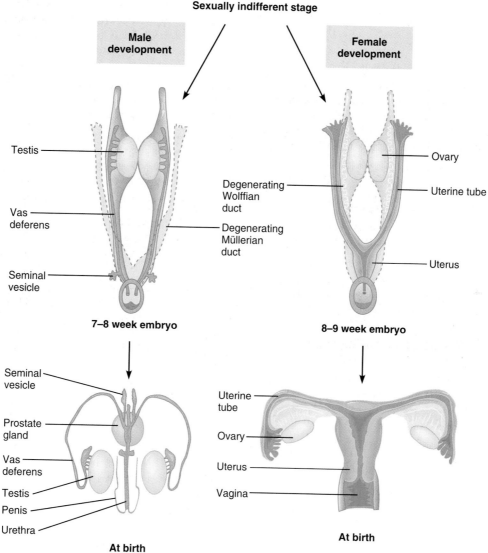

Male development

Female development

Testis

Vas deferens

Seminal vesicle

Degenerating Wolffian duct

Degenerating Müllerian duct

Ovary

Uterine tube

Uterus

7–8 week embryo

8–9 week embryo

Seminal vesicle

Prostate gland

Vas deferens

Testis

Penis

Urethra

Uterine tube

Ovary

Uterus

Vagina

At birth

At birth

FIGURE 21.2 **Development of male and female internal reproductive organs during embryonic life.** *Events shown include sex determination (development of primitive gonads into either testes or ovaries) and sex differentiation (development of either Wolffian ducts or Müllerian ducts into male or female reproductive structures, respectively). The penis and vagina do not develop from Wolffian or Müllerian ducts but are included for purposes of orientation.*

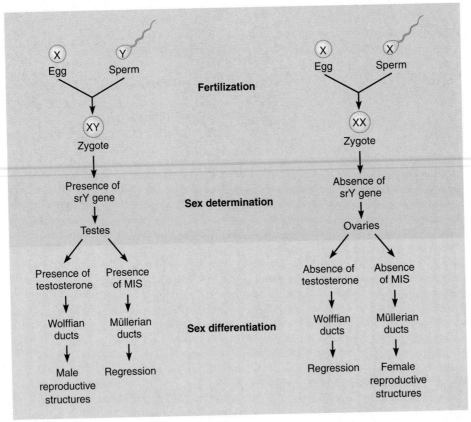

(a) Pivotal events in male development   **(b)** Pivotal events in female development

FIGURE 21.3 **The roles of sex determination and sex differentiation in the development of males and females.**

Although both men and women acquire full reproductive function during puberty, the subsequent patterns of reproductive activity are very different for men and women. Activity of the female reproductive system exhibits cyclic variations known as the *menstrual cycle,* which averages about one month (28 days) in length, during which usually only a single ovum is produced, whereas the male reproductive system is continuously active and normally produces millions of sperm every day. Furthermore, females produce only a limited number of ova in their lifetime and lose the ability to reproduce in middle age (around the age of 45 or 50). This loss of reproductive capacity marks the beginning of a period known as **menopause,** which is characterized by changes in hormonal secretory patterns and a host of other physiological changes. In contrast, men retain the ability to produce sperm—and thus the ability to reproduce—throughout their adult lives.

Quick Test 21.1

1. Define the following terms: *gamete, zygote, diploid, haploid, sperm, ovum, parturition, secondary sex characteristics, puberty,* and *reproductive tract.*

2. Define the following terms: *meiosis, sister chromatids, crossing over,* and *independent assortment.*

3. What are the male gonads called? the female gonads? Besides gametogenesis, what is the gonads' other primary function?

4. What is the distinction between sex determination and sex differentiation? What are the deciding factors in each?

THE MALE REPRODUCTIVE SYSTEM

The primary function of the male reproductive system is to produce sperm and deliver them into the female for fertilization. First we examine the anatomy of the male reproductive organs as it relates to their functions.

Functional Anatomy of the Male Reproductive Organs

Among the major structures of the male reproductive system (Figure 21.4) are the testes; the external genitalia, which include the penis and the scrotum; the reproduc-

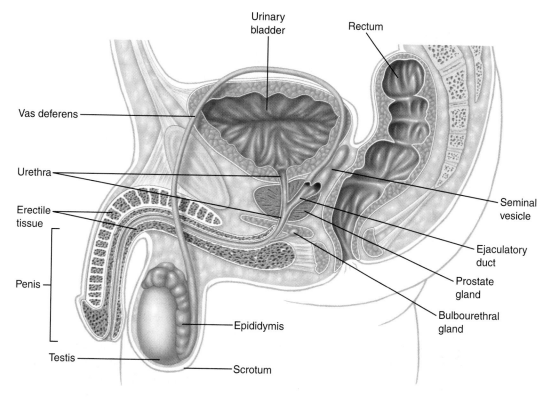

FIGURE 21.4 **The male reproductive system, as seen in a sagittal section.**

tive tract, a system of ducts that conveys sperm from the testes to the penis and outside the body; and several accessory glands (the paired seminal vesicles and bulbourethral glands and the single prostate gland) that secrete various fluids into the reproductive tract.

The Testes

The **testes,** the male gonads, are a pair of small egg-shaped organs that are encapsulated in a layer of fibrous connective tissue. Each testis (Figure 21.5a) is divided internally into 250–300 compartments, each containing a set of thin, highly coiled hollow tubes known as **seminiferous tubules,** where sperm are produced. In the spaces between the tubules (Figure 21.5b) are clusters of cells known as **Leydig cells** *(interstitial cells),* which are responsible for the synthesis and secretion of testosterone and other androgens.

The wall of a seminiferous tubule (Figure 21.5c) contains an outer layer of smooth muscle cells and an inner layer of epithelial cells called **Sertoli cells,** whose primary function is to nurture sperm and control their development. Sperm cells in various stages of development are located in the spaces between the Sertoli cells, and mature sperm are located in the lumen of the seminiferous tubules, which is filled with fluid. The smooth muscle layer, which is separated from the Sertoli cells by a basement membrane, exhibits peristaltic contractions that help propel the sperm and fluid through the tubule.

Each Sertoli cell is joined to its neighbors by tight junctions, which limit the diffusion of materials between cells. The cells and the tight junctions that connect them form a barrier (called the **blood-testis barrier**) that isolates the luminal fluid from the fluid bathing the cells on the other side (referred to as the *basal compartment*). The basal compartment communicates freely with the blood and is similar in composition to normal interstitial fluid, whereas the luminal fluid has a different composition. The significance of the blood-testis barrier is that by maintaining this difference in fluid composition, the barrier ensures that when sperm move from the basal compartment to the luminal compartment at a certain stage of their development, the new fluid environment they encounter in the lumen fulfills their developmental requirements. The blood-testis barrier is also important because it prevents possible attack of sperm cells by the male's immune system.

Sertoli cells perform a number of other functions crucial to sperm development and transport. They are responsible for secreting the luminal fluid, which serves as a medium for sperm development and transport, and for transporting nutrients to the developing sperm cells. The Sertoli cells also manufacture and secrete *androgen-binding protein,* which by binding androgens reversibly acts as an "androgen buffer" that helps maintain in the luminal fluid a steady concentration of androgens, which must be present for sperm development to take place. Additionally, Sertoli cells play a critical role in orchestrating germ

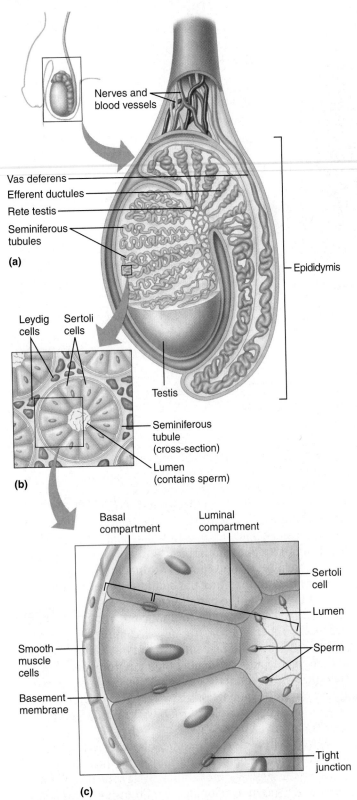

(a)

Nerves and blood vessels

Vas deferens

Efferent ductules

Rete testis

Seminiferous tubules

Epididymis

(b)

Leydig cells

Sertoli cells

Testis

Seminiferous tubule (cross-section)

Lumen (contains sperm)

(c)

Basal compartment

Luminal compartment

Sertoli cell

Lumen

Sperm

Smooth muscle cells

Basement membrane

Tight junction

FIGURE 21.5 The testis and associated structures. (a) *A sagittal section of the testis and epididymis showing the seminiferous tubules and other ducts that carry sperm away from the testis. **(b)** An enlargement showing seminiferous tubules in cross section and Leydig cells located in the spaces between the tubules. **(c)** A portion of a seminiferous tubule showing structure of the wall. This diagram is highly schematized to emphasize essential features.*

cells' responses to testosterone and *follicle-stimulating hormone* (FSH), the hormones that control their development. These hormones are both necessary for sperm production, but they do not target the developing sperm themselves; instead they target the Sertoli cells, which then secrete various chemical messengers that actually stimulate the germ cells to grow and develop. Sertoli cells are also responsible for the secretion of a hormone called *inhibin* (whose function is discussed later), and in embryonic life they secrete MIS.

The External Genitalia

The reproductive function of the **penis,** the male copulatory organ, is to penetrate into the vagina during copulation and deposit sperm inside the female reproductive tract, thereby initiating the steps that lead to fertilization. The penis is ordinarily flaccid and hangs downward (see Figure 21.4), but during sexual arousal it swells, elongates, and becomes much firmer and straighter, a process known as **erection.** These changes enable the penis to penetrate into the vagina so that sperm can be deposited as close to the site of fertilization as possible. At the height of sexual arousal, sperm are forcibly ejected through the urethra (an event called **ejaculation**), which runs the length of the penis and opens to the outside at its tip. As we saw in Chapter 17, the urethra also serves as the passageway that carries urine from the urinary bladder to the outside of the body.

Changes in the penis's shape during erection are possible because much of its volume is occupied by **erectile tissue** (see Figure 21.4), spongy masses of connective tissue and smooth muscle containing numerous interconnected vascular spaces. Ordinarily only a small volume of blood is present within the erectile tissue, but during erection the volume and pressure of this blood increases, producing the swelling and elongation characteristic of erection.

The **scrotum,** a sac of skin and underlying connective tissue that is suspended beneath the penis, houses each testis in separate right and left compartments divided by a septum. The scrotum provides an environment for the testes in which the temperature is slightly below core body temperature, which is important because even a small rise in testicular temperature can suppress sperm production. (Once produced, sperm can survive at normal body temperature, however.)

The Reproductive Tract

Within each testis the seminiferous tubules converge to a network of short interconnected tubules (called the *rete testis*) that lead to a set of small tubes *(efferent ductules)* that penetrate the fibrous outer covering of the testis (see Figure 21.5). Outside the testis these tubes lead to the **epididymis,** which is loosely attached to the testis's outer surface and contains a single highly coiled duct. This duct leads to a larger and more muscular thick-walled tube called the **vas deferens,** which enters the pelvic cavity.

In the pelvic cavity the vas deferens connects with a duct from one of the seminal vesicles, at which point it becomes known as an *ejaculatory duct* (see Figure 21.4). The two ejaculatory ducts (one from each testis) penetrate into the prostate gland, where they join the urethra. The urethra then emerges from the prostate, joins with ducts leading from the bulbourethral glands, and traverses the length of the penis.

Accessory Glands

The **seminal vesicles** (see Figure 21.4) are two elongated glands that secrete an alkaline fluid containing fructose and other nutrients (which the sperm use for energy) and other substances (including enzymes and prostaglandins). The alkalinity of the fluid helps to neutralize the acid environment the sperm encounter in the vagina, which otherwise would suppress the motility sperm require if they are to achieve fertilization of the egg.

The **prostate gland,** roughly the shape of an onion, surrounds the urethra, into which it empties its secretions via a number of small openings in the urethral wall. The prostate actually comprises a number of smaller glands that are surrounded by smooth muscle and connective tissue. During ejaculation the smooth muscle contracts and compresses the glands, forcing their secretions into the urethra. Prostate secretions contain several enzymes and citrate, which the sperm use for energy.

The **bulbourethral glands,** a pair of round, pea-sized glands located beneath the prostate, secrete a fluid containing a viscous, sticky mucus that is released prior to ejaculation and functions primarily as a lubricant.

In the following sections we take a closer look at how the male reproductive system performs its functions, and how these functions are regulated.

Hormonal Regulation of Reproductive Function in Males

Reproductive function in males is largely governed by the gonadotropins, by androgens (including testosterone), and by gonadotropin-releasing hormone via the processes diagrammed in Figure 21.6.

The **gonadotropins** are two protein hormones that are secreted by the anterior pituitary and act on the gonads: **follicle-stimulating hormone (FSH)** and **luteinizing hormone (LH).** (Later in the chapter we will see that these same two hormones are also involved in controlling reproductive function in females.) The primary function of FSH in males is to act on Sertoli cells to stimulate gametogenesis and other functions. LH, by con-

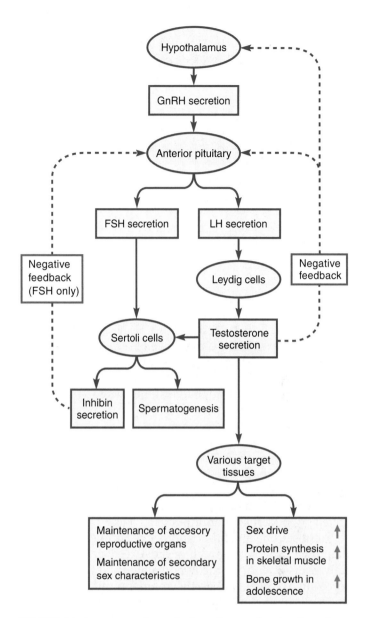

FIGURE 21.6 Hormonal regulation of reproductive function in males.

How does GnRH travel from the hypothalamus to the anterior pituitary?

trast, stimulates the secretion of androgens by Leydig cells. Because androgens are steroids, they are lipophilic and therefore readily diffuse from the Leydig cells to all testicular tissues, including the Sertoli cells. The androgens also enter the bloodstream, which delivers them to tissues all over the body.

In the testes, testosterone works with FSH to promote spermatogenesis. Testosterone also promotes the development and growth of accessory reproductive organs during puberty and is necessary for the maintenance and continued function of these structures in adult life. For this

Via hypothalamic-hypophyseal portal vessels

TABLE 21.1 ■ SELECTED ACTIONS OF ANDROGENS IN MALES
Stimulate spermatogenesis
Promote development of secondary sex characteristics during puberty and maintenance of these characteristics in adult life
Increase sex drive
Promote protein synthesis in skeletal muscle
Stimulate growth hormone secretion, which permits bone growth during adolescence
Promote development of male reproductive structures during embryonic life

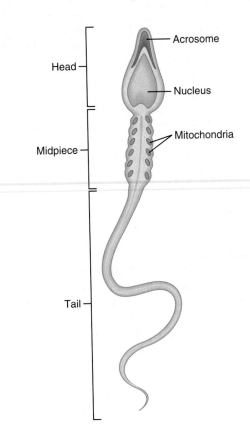

FIGURE 21.7 Anatomy of a sperm cell.

reason, a lack of testosterone leads to a decrease in the size of accessory organs and a decrease in smooth muscle and glandular activity. Testosterone also is responsible for the development and maintenance of secondary sex characteristics and acts on the brain to promote sex drive.

Gonadotropin-releasing hormone (GnRH) is a hypophysiotropic hormone that stimulates the secretion of gonadotropins and therefore tends to promote androgen secretion. (GnRH also stimulates gonadotropin secretion in females, as we will see.) As discussed in Chapter 5, most hypophysiotropic hormones stimulate the release of one particular hormone from the anterior pituitary, but GnRH is different because it stimulates the release of two hormones—FSH and LH—which are produced and secreted by the same anterior pituitary cells.

Testosterone and other androgens normally limit their own secretion through negative feedback. Specifically, they act both on the hypothalamus, to suppress the secretion of GnRH, and on the anterior pituitary, making it less responsive to GnRH. As a result, testosterone tends to suppress the release of FSH and LH, which in turn tends to suppress subsequent testosterone secretion. Sertoli cells also secrete a protein hormone called **inhibin,** which suppresses the release of FSH (but not the release of LH) by the anterior pituitary.

Rates of sex hormone secretion in males are fairly constant and are not subject to the cyclic variations that occur in females during the menstrual cycle. Throughout a male's reproductive life the hypothalamus releases GnRH in bursts occurring at approximately two-hour intervals, which causes rates of FSH and LH secretion to rise during bursts and fall between bursts. Despite these fluctuations, however, *average* plasma levels of FSH and LH (and hence levels of testosterone) are fairly constant from day to day. Significant variations over the long term do occur, however; testosterone levels rise dramatically during puberty, reach a peak sometime in the third decade of life, and then subsequently decline slowly.

In prepubescent males, testosterone levels are relatively low, reflecting low rates of GnRH and gonadotropin

secretion. However, GnRH secretion rises dramatically at the beginning of puberty in response to a change in brain activity that alters neural input to the hypothalamus. (The precise nature of this change in brain activity is presently unknown.) The increase in GnRH triggers a similar rise in gonadotropin secretion, which stimulates growth and maturation of the testes as well as testosterone secretion.

The rise in plasma testosterone levels that occurs during puberty stimulates further maturation of the reproductive system and other widespread changes throughout the body. For example, testosterone stimulates the secretion of growth hormone by the anterior pituitary, which in turn stimulates the bone growth responsible for the dramatic increase in stature that occurs during this period. Testosterone's actions also trigger the development of secondary sex characteristics during puberty, including growth of facial hair; growth of coarse body hair in the pelvic region and armpits; increased protein synthesis in muscle that leads to increased muscle size and strength, particularly in the upper body; growth of the larynx, which causes the voice to deepen; and secretion of a thick oil by sebaceous glands (oil glands) in the skin. Actions of androgens in males are summarized in Table 21.1.

Sperm and Their Development

Fully developed sperm cells possess three distinct segments (Figure 21.7): a bulblike *head,* a short cylindrical

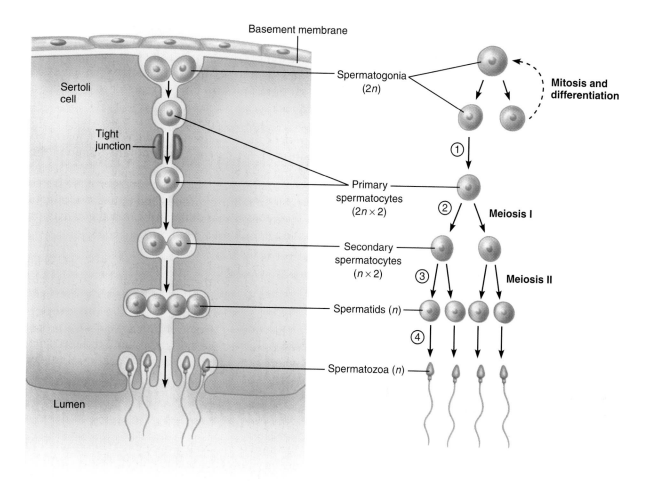

FIGURE 21.8 Spermatogenesis. *As germ cells migrate from the basal compartment to the lumen of a seminiferous tubule (depicted at left), they undergo various stages of development (depicted at right). The dashed arrow indicates that some spermatogonia do not differentiate following mitosis and remain as spermatogonia.*

Over a male's lifetime, does the number of spermatogonia normally increase, decrease, or stay the same?

midpiece, and a long, slender *tail* (or *flagellum*). Within the head are the chromosomes and a large vesicle called an **acrosome,** which contains enzymes and other proteins that enable the sperm to fuse with the egg during fertilization. The tail contains complex machinery made up of proteins that hydrolyze ATP and convert the released chemical energy into motion. The resulting whiplike movements of the tail propel the sperm forward in a swimming motion. The midpiece anchors the tail to the head and contains mitochondria, which produce the ATP required for motility.

Sperm production (**spermatogenesis**) starts near the basement membrane in the seminiferous tubules with undifferentiated germ cells called **spermatogonia** (Figure 21.8). As spermatogenesis proceeds, germ cells gradually migrate toward the lumen of the tubule in the spaces between Sertoli cells. At a certain point the developing germ cells move from the basal compartment to the luminal compartment by passing through the tight junctions, which open up temporarily to let the cells pass through.

As a result of this migration, cells at different stages of development are characteristically found at certain locations in the tubule; the older, more mature cells are nearer to the lumen, whereas the newer, less differentiated cells are nearer to the basement membrane.

Each male is born with a finite number of spermatogonia, but these cells undergo mitosis repeatedly, giving him the capacity to produce sperm for an indefinite period following puberty. One of the daughter cells produced by each mitotic division differentiates further and eventually becomes a mature sperm; the other daughter cell does not differentiate and remains a spermatogonium. Therefore, the total number of spermatogonia does not change, ensuring that this pool of potential sperm is never depleted.

To become a mature sperm, a spermatogonium goes through the following stages of development (see Figure

Stays the same

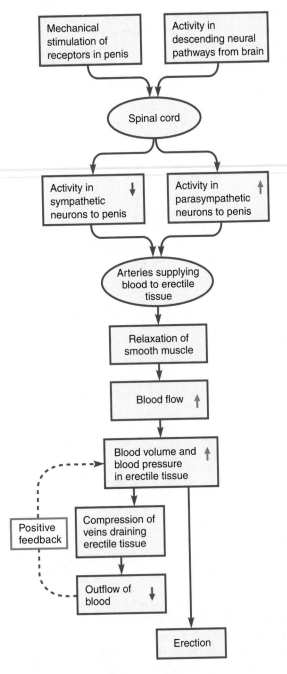

FIGURE 21.9 Events leading to erection.

When the activity of parasympathetic neurons increases, does the secretion of nitric oxide increase or decrease?

secondary *spermatocytes*, which possess 23 duplicated chromosomes ($n \times 2$). ③ Secondary spermatocytes undergo meiosis II to become *spermatids*, which have 23 single chromosomes (n) but have not developed the mature sperm's characteristic features. ④ Spermatids differentiate to become spermatozoa (n), which possess the characteristic head, midpiece, and tail. After spermatozoa are formed, they are released into the lumen of the seminiferous tubule.

At this stage, however, the sperm are immotile (incapable of self-propulsion) and remain so for about 20 days. Sperm acquire motility only after they have moved from the seminiferous tubules to the epididymis to undergo further maturation. Transport of sperm to the epididymis is driven by peristaltic contractions in the seminiferous tubules and by the flow of luminal fluid, a natural consequence of its continual secretion by Sertoli cells. Much of this fluid is absorbed in the epididymis, causing the sperm to become more concentrated. Peristalsis propels the sperm within the epididymis and to the vas deferens, where they are held until sexual arousal triggers ejaculation.

The Sexual Response in Males

Erection is controlled by spinal reflexes and influenced by activity in higher brain centers (Figure 21.9), as is evident in the fact that this event (and ejaculation as well) can be triggered by sights, smells, sexual thoughts, and even dreams. Mechanical stimulation of the penis or of certain other sensitive parts of the body (termed *erogenous zones*) causes impulses to travel via sensory neurons to the spinal cord, where systems of interneurons orchestrate changes in the activity of sympathetic and parasympathetic neurons that project to the penis. (The same responses can be triggered by activity in descending pathways from the brain.) In the sympathetic neurons that project to arteries supplying the erectile tissue with blood, activity decreases; in parasympathetic neurons that project to these same arteries, activity increases. These parasympathetic neurons are unusual not only in that they control arterial resistance, but also because they release nitric oxide, which has potent vasodilatory effects. (Recall from Chapter 14 that vascular resistance is controlled in most regions of the body by sympathetic, not parasympathetic, neurons.)

The combined result of the decrease in sympathetic activity and the increase in parasympathetic activity is arterial vasodilation, which increases blood flow to the erectile tissue. As a result, the vascular spaces inside the tissue fill with blood and expand, which brings about erection (When It Goes Wrong: Erectile Dysfunction, p. 685). The expansion of erectile tissue due to arterial dilation also compresses veins in the penis, which by reducing the outflow of blood from the vascular spaces promotes the accumulation of blood in the erectile tissue.

21.8, at right): ① The spermatogonium's chromosomes are replicated, and the cell differentiates to become a *primary spermatocyte*, which has 46 duplicated chromosomes possessing two sister chromatids apiece. (In the figure, the chromosome number is designated as $2n \times 2$ to indicate that the chromosomes are duplicated.) ② The primary spermatocyte goes through meiosis I to yield two

It increases.

As more blood accumulates, it further compresses the veins, leading to the accumulation of still more blood in a positive feedback loop that promotes expansion of the erectile tissue.

During the act of procreation, the male inserts the erect penis into the female's vagina and makes thrusting pelvic movements. The resulting mechanical stimulation of the penis stimulates sensory receptors, which in conjunction with psychological stimuli increases arousal and eventually triggers ejaculation. During ejaculation, the male experiences intense psychological stimulation accompanied by increases in heart rate and blood pressure and a host of other physiological changes, a phenomenon called *orgasm*.

Ejaculation, like erection, is triggered by mechanical stimulation of the penis and is controlled by a spinal reflex, but in this case the reflex causes an increase in the activity of sympathetic neurons projecting to the reproductive tract and accessory glands. This sympathetic activity triggers strong contractions in the epididymis, vas deferens, and ejaculatory ducts, and it also stimulates the secretion of fluids by the seminal vesicles and prostate gland. As a result, a mixture of sperm and fluid (called **semen**) moves into the urethra, an event called **emission.** Following emission, smooth muscle in the urethra and skeletal muscle at the base of the penis undergo a series of strong contractions that expel the semen through the urethral opening. The reflex also triggers closure of the urethral sphincter at the base of the bladder, which prevents urine from mixing with the semen during ejaculation.

Quick Test 21.2

1. Define the following terms: *penis, scrotum, seminiferous tubule, erection, ejaculation, semen, seminal vesicle, prostate gland, bulbourethral gland, blood-testis barrier, Sertoli cell, Leydig cell, acrosome, spermatogenesis, spermatogonium, primary spermatocyte, secondary spermatocyte, spermatid,* and *spermatozoa.*

2. For each of the following descriptions, indicate whether it pertains to Sertoli cells or to Leydig cells: are present in seminiferous tubules; are present outside seminiferous tubules; secrete testosterone; are target cells of FSH; are target cells of LH; secrete inhibin.

3. Arrange the following structures so that they correctly describe the path of sperm transport from the testes to the penis: ejaculatory duct, vas deferens, seminiferous tubule, urethra, epididymis, efferent ductule, rete testis.

THE FEMALE REPRODUCTIVE SYSTEM

Whereas the male reproductive system has the relatively uncomplicated job of producing sperm and delivering them to the female, the female reproductive system is responsible not only for the production and transport of ova, but also for everything else that must occur to enable a new human being to be brought into the world. The female reproductive system exhibits the following general characteristics:

1. *Cyclic changes in activity.* Each month the female reproductive system undergoes a series of dramatic structural and functional changes called the **menstrual cycle,** which is accompanied by regular, periodic changes in the secretion of hypothalamic, anterior pituitary, and ovarian hormones. The beginning of each cycle, which lasts for about 28 days, is marked by **menstruation,** the shedding of blood and tissue from the surface of the uterine lining.

2. *Restricted periods of fertility.* Ova mature at different rates and are typically released singly from the ovaries at intervals of approximately 28 days—once per menstrual cycle. The release of the egg, known as **ovulation,** is a prerequisite for fertilization and usually occurs in the middle of the cycle. A female is fertile (capable of having an egg fertilized) for only the few days per cycle that roughly coincide with ovulation.

3. *Limited gamete production.* Ova develop from a pool of germ cells whose number is fixed at birth at approximately 2–4 million. The number of potential ova declines steadily throughout life because most germ cells degenerate (a phenomenon called **atresia**) at some point in their development. Only about 400 eggs are released over a woman's lifetime.

Functional Anatomy of the Female Reproductive Organs

Our study of the functional anatomy of the female reproductive system (Figure 21.10) centers on the ovaries; the reproductive tract, a system of interconnected passages comprising the internal cavities of the uterus, uterine tubes, and vagina; and the external genitalia, a set of structures surrounding the entrance to the reproductive tract.

The Ovaries

The **ovaries,** the female gonads, are a pair of slightly flattened oval structures located in the pelvic region on either side of the uterus (see Figure 21.10a). Each ovary is composed mostly of dense connective tissue that is well supplied with blood vessels and enclosed within a layer of fibrous connective tissue. Embedded within this

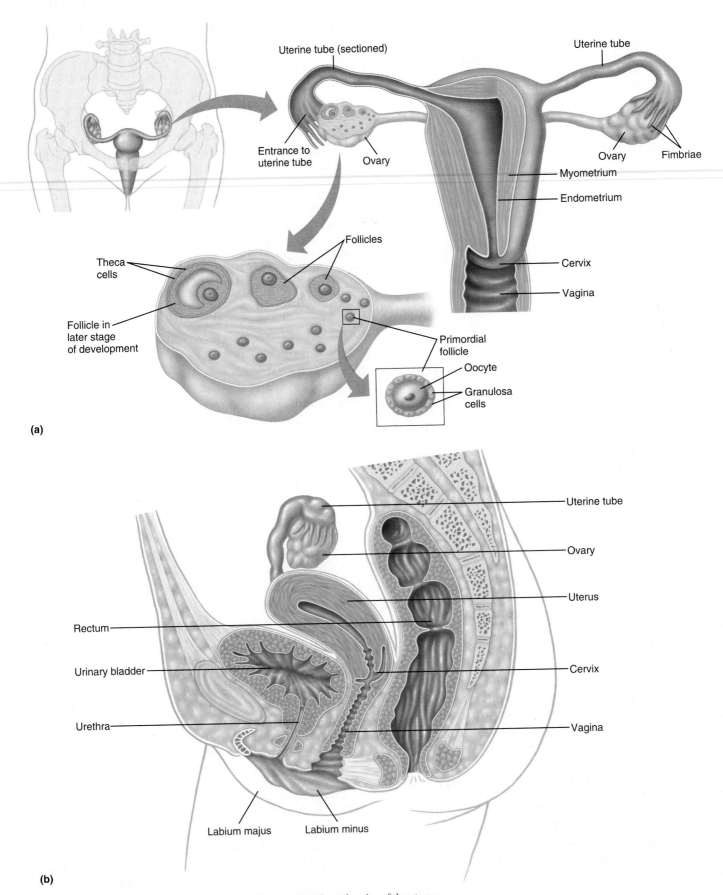

(a)

Uterine tube (sectioned)

Entrance to uterine tube

Ovary

Uterine tube

Ovary   Fimbriae

Myometrium

Endometrium

Cervix

Vagina

Follicles

Theca cells

Follicle in later stage of development

Primordial follicle

Oocyte

Granulosa cells

(b)

Uterine tube

Ovary

Uterus

Cervix

Vagina

Rectum

Urinary bladder

Urethra

Labium majus   Labium minus

FIGURE 21.10 The female reproductive system. (a) *A frontal section of the uterus and associated structures; the enlargement of an ovary shows ovarian follicles in different stages of development.* **(b)** *A sagittal section showing the major reproductive organs.*

Erectile dysfunction (ED), commonly known as *impotence*, is the inability to attain or maintain an erection sufficiently rigid to permit sexual intercourse. It affects about 10–15 million American men. Most often, ED results from damage to vascular or connective tissues in the penis, occurring as a normal consequence of aging, but it can also result from damage either to spinal tracts involved in the control of blood flow to erectile tissue or to afferent or efferent nerves to the penis. ED can be triggered by endocrine dysfunction (a lack of testosterone, for example) or other diseases, including kidney disorders, multiple sclerosis, atherosclerosis, or diabetes mellitus. The risk of ED increases with age and is also increased by

smoking, alcohol use, and various drugs such as antihypertensive agents, antidepressants, and tranquilizers. ED need not have a physical basis, however, and in 10–20% of cases the cause is thought to be purely psychological. Feelings of stress, anxiety, guilt, or depression are all known to increase the risk of ED.

Treatments for ED include oral testosterone replacement (if the primary problem is a lack of testosterone) and the injection into the penis of drugs that increase blood flow, which has the disadvantage that it induces erection almost immediately. Other treatments include surgery to rebuild damaged arteries or to insert into the penis prosthetic devices that allow simulation of

naturally occurring erections.

In recent years the treatment of ED has been revolutionized by a drug called Viagra (seldenifil) that enables erection to occur if taken an hour or so before intercourse. This drug works by mimicking the effects of nitric oxide, which induces arterial dilation in the penis. Nitric oxide normally works by stimulating guanylate cyclase in smooth muscle cells, which converts GTP to cyclic GMP (cGMP). cGMP promotes relaxation of arterial smooth muscle and an increase in blood flow to erectile tissue. Viagra mimics the effect of nitric oxide by inhibiting the phosphodiesterase enzyme that breaks down cGMP, thereby promoting a rise in intracellular cGMP levels.

---

connective tissue are numerous spherical structures called **follicles,** each of which contains a single developing ovum. As the ovum develops, the rest of the follicle develops along with it, increasing in size and structural complexity. Among the follicles present in an ovary at any given time, most are in the earliest stages of development; only a few are in later stages.

A follicle in the earliest stage of development (called a **primordial follicle**) is a simple structure consisting of a developing ovum or *oocyte* surrounded by a layer of specialized cells called **granulosa cells** (see Figure 21.10a). In later stages of development the granulosa cells proliferate, and the outermost layer is transformed into another cell type known as **theca cells.**

Granulosa cells perform a number of important functions that are analogous in many ways to those performed by Sertoli cells in males. Granulosa cells function as intermediaries between the oocyte and the hormones that control its development in that both estrogen and FSH stimulate granulosa cells to secrete chemical messengers that target the oocyte. In addition, the granulosa

cells secrete inhibin, which suppresses FSH secretion (as it does in males). Granulosa cells also transport nutrients to the oocyte's interior via cytoplasmic bridges that communicate with the oocyte's interior through gap junctions.

Granulosa cells are also responsible for secreting estrogens, but these hormones are actually derived from androgens that are first synthesized in theca cells. After synthesis, these androgens diffuse from the theca cells to the granulosa cells, where they are enzymatically modified and then secreted. The granulosa cells also manufacture and secrete progesterone.

The Reproductive Tract

The centerpiece of the female reproductive tract is the **uterus,** a hollow pear-shaped organ located in the center of the pelvic cavity that functions to house and nourish the developing human (see Figure 21.10a). During pregnancy the fetus is held in the upper portion of the uterus, called the *body.* In the lower, narrower portion of the uterus (called the **cervix**), a central canal leads directly to the

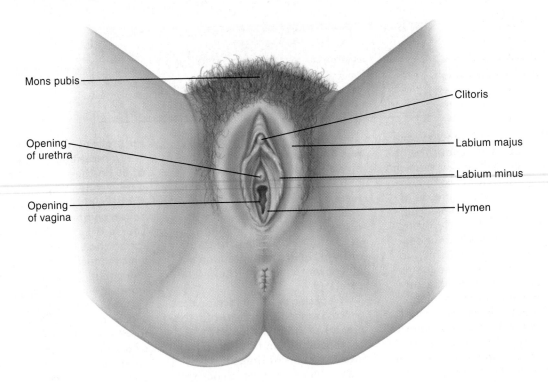

FIGURE 21.11 **Female external genitalia.**

Labels (clockwise from top left):
Mons pubis
Opening of urethra
Opening of vagina
Clitoris
Labium majus
Labium minus
Hymen

vagina. During parturition, the fetus passes through this canal and then through the vagina to exit the mother's body. For this reason, the cervical canal and vagina are often referred to as the *birth canal.*

To accommodate the growing fetus, the wall of the uterus must be strong and flexible. Beneath the outer layer of epithelial cells and connective tissue (called the *perimetrium*) is a layer of smooth muscle called the **myometrium,** which occupies most of the wall's thickness. Contractions of the myometrium help to expel the fetus from the uterus during birth. The innermost layer, called the **endometrium,** comprises a layer of epithelial cells and an underlying layer of thick connective tissue. The endometrium also contains numerous glands that secrete fluid that bathes the uterine lining.

The **vagina,** a canal about 8–10 cm long leading from the cervix to the outside of the body, receives the penis during mating and is thus the female organ of copulation (Figure 21.10). Its thin wall contains a middle layer of smooth muscle, and its inner surface is bathed by fluid that leaks from the uterus or is secreted by glands in the cervix. Due to the action of bacteria that normally live in the vagina, this fluid is acidic, which although potentially harmful to sperm also limits the growth of certain pathogenic microorganisms that could easily enter the vagina from the external environment.

Extending from the upper part of the uterus on either side are the **uterine tubes** (also *fallopian tubes* or *oviducts*), which function in egg transport and fertilization (see Figure 21.10a). Each tube, which is about 10 cm long, terminates in a funnellike structure (the *infundibu-*

lum) fringed by fingerlike projections *(fimbriae)* that partially envelop the ovary on that side. When an egg is released from the ovary, it is swept into the uterine tube by the motion of the surrounding fluid, which is drawn into the tube by the beating of cilia lining the tube's inner surface. For the first few minutes after ovulation, peristaltic contractions of the tube propel the egg toward the uterus; thereafter the egg is propelled solely by ciliary motion. The entire trip from the ovary to the uterus takes about four days.

Sperm deposited in the vagina migrate by self-propulsion through the uterus and into the uterine tubes, where they approach the advancing egg. (Sperm migration is also aided by uterine and vaginal contractions that occur during sexual arousal.) Fertilization, when it occurs, usually takes place in the uterine tubes. The fertilized egg undergoes several cell divisions and moves on to the upper portion of the uterus, where it adheres to and then embeds in the uterine wall (an event called **implantation**). If implantation is successful, further development of the fertilized egg continues. If fertilization does not occur, implantation does not occur, and the egg is eventually expelled from the uterus during menstruation.

The External Genitalia

The external genitalia of females are located in the single structure called the **vulva** (or *pudendum*), which encompasses the following structures (Figure 21.11): (1) the *mons pubis,* a mound of skin and underlying fatty tissue located centrally in the lower pelvic region; (2) the *labia* *majora* (singular: *labium majus*) and *labia minora* (singular: *labium minus*), a set of outer and inner skin folds that surround the entrance to the vagina; (3) the *vestibule*, an area enclosed by the labia minora that leads to the openings of the vagina and the urethra; (4) the *clitoris*, a small erectile organ located at the front of the vestibule; and (5) *vestibular glands*, a pair of small glands that secrete a lubricating fluid that facilitates entry of the penis into the vagina during copulation (not shown in figure). In young females the opening to the vagina is partially covered by a thin membrane called the *hymen*, which contains small blood vessels. This membrane is often ruptured by entry of the penis into the vagina during the first episode of copulation and thus is absent in most older females.

In the remainder of this section we examine in more detail the day-to-day functioning of the female reproductive system that occurs in the absence of fertilization. Thereafter we focus on processes that occur only when an egg becomes fertilized.

Ova and Their Development

In the process called **oogenesis,** ova develop from relatively undifferentiated germ cells called **oogonia.** The number of oogonia is fixed in number prior to birth, and they are not continuously regenerated, as are spermatogonia. Furthermore, the process of meiosis, which transforms oogonia into fully mature ova, begins in fetal life but is not completed until fertilization has occurred.

Oogenesis begins in the first three months of embryonic life, when oogonia (which are diploid) undergo mitosis to yield a clone of cells from which all ova are ultimately derived (Figure 21.12). These cells subsequently undergo further differentiation in the fetus to become *primary oocytes;* in this process the chromosomes are replicated, producing a total of 46, with two sister chromatids apiece ($2n \times 2$). The primary oocytes then go into a state of suspended development known as **meiotic arrest** and remain there until just before ovulation, when the primary oocyte undergoes meiosis I to yield two daughter cells possessing 23 replicated chromosomes each ($n \times 2$). One of the daughter cells, called a *secondary oocyte,* receives most of the cytoplasm during the cell division and continues to develop further; the other daughter cell, called the *first polar body,* degenerates and is lost. Only in the event that the secondary oocyte is fertilized does meiosis II occur. This second meiotic division yields an ovum, which receives most of the cytoplasm, and a *second polar body,* which degenerates. The fertilized ovum contains 23 single chromosomes that it inherits from the secondary oocyte plus an equal number that it inherits from the sperm, giving a total of 46 chromosomes ($2n$).

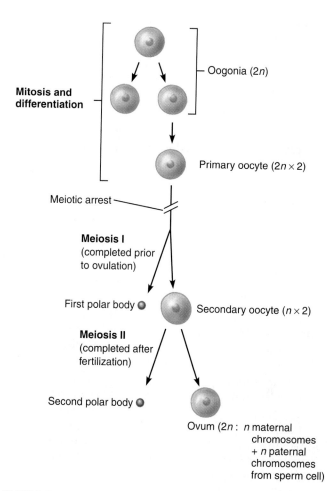

FIGURE 21.12 Oogenesis, the development of ova. *Note that the ovum, which develops after fertilization, contains 23 chromosomes of maternal origin plus an equal number from the sperm.*

How many sex chromosomes are present in an oogonium, and which are they?

The Sexual Response in Females

In females, sexual excitement produces a characteristic set of changes, including vascular congestion (increased blood volume) in certain tissues and muscular contraction in many areas of the body. For example, the clitoris begins to swell and elongate due to engorgement of its erectile tissue with blood. The breasts also become engorged, the nipples become erect, and the vagina begins to secrete a viscous lubricating fluid. The clitoris, which is equipped with numerous sensory endings, is stimulated by friction during sex acts, and this induces pleasurable sensations and raises the level of sexual arousal. When arousal reaches a certain level, it can trigger orgasm, which is accompanied by rhythmic contractions of the vagina and uterus and many other responses, including intense pleasure, a rise in blood pressure and heart rate, and widespread skeletal muscle contractions.

Two X chromosomes

1. Define the following terms: *uterus, uterine tubes, cervix, vulva, mons pubis, labia majora, labia minora, clitoris, vestibule, oogenesis, oogonium, primary oocyte, secondary oocyte, ovum, follicle, granulosa cells,* and *theca cells.*

2. How does oogenesis differ from spermatogenesis with regard to the number of gametes produced? the timing of gamete production?

3. When do meiosis I and meiosis II occur in oogenesis?

The Menstrual Cycle

Among the widespread changes in bodily function that occur during the menstrual cycle, which recurs throughout a female's reproductive life, are the following: cyclic changes in ovarian structure and function (referred to as the **ovarian cycle**), cyclic changes in uterine structure and function (referred to as the **uterine cycle**), and cyclic changes in the secretion of ovarian, hypothalamic, and pituitary hormones. In the next two sections we examine the ovarian and uterine cycles to see how they set the stage for possible fertilization and pregnancy, which are necessary for fulfillment of the reproductive system's ultimate function, the creation of new life. Then we link these events to the concomitant hormonal changes to understand the causes of the menstrual cycle.

The Ovarian Cycle

The ovarian cycle is divided into two phases: the **follicular phase,** which is variable in duration but 14 days long on average, and the **luteal phase,** which is always 14 days long. The follicular phase commences with the start of menstruation (day 1 of the menstrual cycle) and ends with ovulation; the luteal phase coincides with the remainder of the menstrual cycle.

The Follicular Phase   Through the course of an ovarian cycle (Figure 21.13), most of the follicles in the ovaries remain in the primordial stage ①, but a small fraction of follicles begin to develop, each proceeding independently of the others. As a follicle develops, the oocyte grows in size and the granulosa cells proliferate, forming multiple layers around the oocyte ②. A thick layer of noncellular material called the **zona pellucida** forms between the oocyte and the surrounding granulosa cells, and an outer layer of theca cells begins to form. In some follicles a fluid-filled cavity called an antrum forms within the granulosa layer and increases in size ③. Whereas some follicles progress only far enough to develop a small antrum and thus reach an *early antral* stage, others stop developing in a *preantral* stage (before the antrum has begun to form).

At the beginning of the follicular phase, about 10–25 follicles from the pool of preantral or early antral follicles are recruited to develop further. After about seven days, one of these follicles (referred to as the **dominant follicle**) is selected to develop to full maturity, thus setting the stage for ovulation. (On rare occasions, more than one follicle is selected.) The remaining "contenders" undergo atresia and are lost.

Follicular growth and development is stimulated by both FSH and the estrogens secreted by the follicles themselves. During the follicular phase, plasma FSH levels gradually fall (for reasons explained later), which tends to cause rates of estrogen secretion to decrease. Those follicles that cannot sustain adequate rates of estrogen secretion undergo atresia. Selection of the dominant follicle depends on its ability to secrete adequate levels of estrogen in the face of falling FSH levels. The dominant follicle is more sensitive to FSH than are the other follicles, and it also shows some responsiveness to LH, which does not appear until a follicle has reached a certain stage in development. Thus the dominant follicle ensures its dominance by developing more quickly than the other follicles.

As the dominant follicle continues to develop, the antrum grows and displaces some of the cellular tissue around the oocyte, causing the oocyte and a layer of surrounding granulosa cells (the *corona radiata*) to protrude from the wall ④. Eventually, meiosis I occurs, and the oocyte (now a *secondary* oocyte) detaches from the follicle wall and floats freely in the antral fluid, along with the surrounding corona radiata. The antrum continues to expand, and the follicle eventually grows to about 1.5 cm in diameter and reaches its fully mature form, referred to as a **Graafian follicle,** just before ovulation.

The Luteal Phase   In ovulation, which marks the beginning of the luteal phase, the wall of the Graafian follicle ruptures ⑤, causing a flow of antral fluid that carries the oocyte (with its surrounding cells) to the ovary's surface. The ruptured follicle is then transformed into a gland called the **corpus luteum** ⑥, which secretes estrogens and progesterone. Both ovulation and formation of the corpus luteum are triggered by the same event—an abrupt rise in plasma LH levels. In fact, its role in the formation of the corpus luteum is why this hormone is called luteinizing hormone.

After ovulation the released oocyte enters the uterine tube, and its fate ultimately determines that of the corpus luteum. If the oocyte is not fertilized, the corpus luteum reaches its maximum activity within 10 days of its formation ⑦ and then begins to degenerate ⑧. This degeneration causes a decline in plasma estrogen and progesterone levels that sets the stage for menstruation and the beginning of the next follicular phase. If the oocyte is fertilized, the corpus luteum does not degenerate; instead it persists well into the gestation period.

[handwritten margin note: best follicle develops, survival of the fittest.]

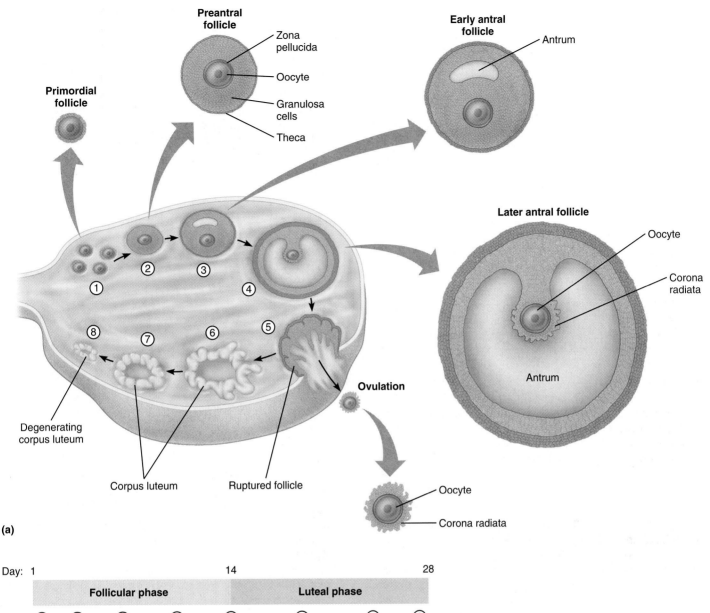

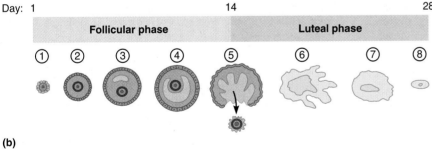

FIGURE 21.13 The ovarian cycle. (a) *Cross section of an ovary depicting the various stages of the ovarian cycle; events occurring during the stages are described in detail in the text.* **(b)** *A diagram relating the stages depicted in (a) to the follicular and luteal phases of the ovarian cycle. Note that ovulation marks both the end of the follicular phase and the beginning of the luteal phase.*

On rare occasions (about 1–2% of all ovarian cycles), two or more follicles are selected to become dominant. When this happens, two (or more) oocytes are released at ovulation. In the event that both are fertilized, the result will be two offspring referred to as *fraternal twins,* who do not look alike and inherit different sets of genes. These offspring are also called *dizygotic twins* because they originate from different zygotes.

The Uterine Cycle

The uterine cycle, which occurs coincidentally with the ovarian cycle, is divided into three phases: the **menstrual phase,** which begins on day 1 and lasts for three to five days; the **proliferative phase,** which lasts for the duration of the follicular phase (typically from the end of

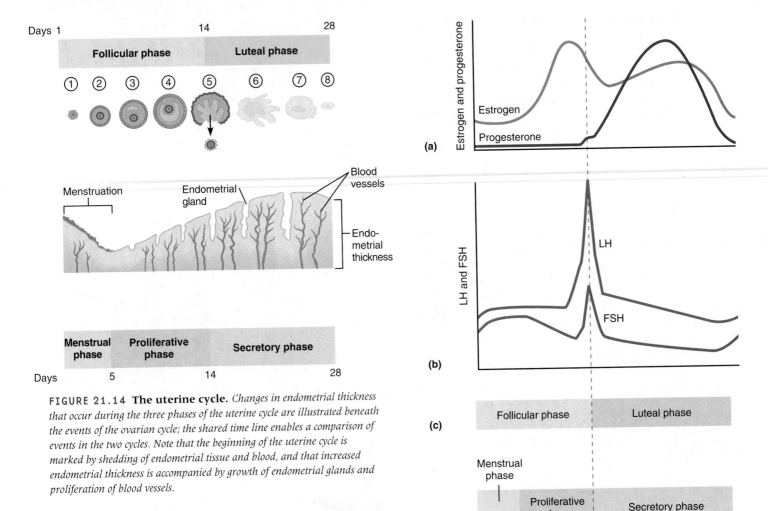

FIGURE 21.14 The uterine cycle. *Changes in endometrial thickness that occur during the three phases of the uterine cycle are illustrated beneath the events of the ovarian cycle; the shared time line enables a comparison of events in the two cycles. Note that the beginning of the uterine cycle is marked by shedding of endometrial tissue and blood, and that increased endometrial thickness is accompanied by growth of endometrial glands and proliferation of blood vessels.*

FIGURE 21.15 Summary of events in the menstrual cycle. *(a) Plasma estrogen and progesterone levels. (b) Plasma gonadotropin levels. (c) Phases of the ovarian cycle. (d) Phases of the uterine cycle. The vertical dashed line indicates the occurrence of ovulation.*

menstruation to day 14); and the **secretory phase,** which coincides with the luteal phase (Figure 21.14).

The Menstrual Phase The menstrual phase of the uterine cycle is so named because it corresponds to the period of menstruation, the shedding of the uterine lining. Menstruation is triggered in response to the fall in plasma estrogen and progesterone that occurs as the corpus luteum degenerates. At first, blood vessels in the outermost layer of the endometrium begin to constrict, which reduces blood flow to the tissue. As a result, these tissues die and start to separate from the underlying endometrial tissues, which remain intact. The dead tissue gradually sheds from the endometrial surface, which causes rupture of blood vessels and bleeding. Over the next few days a mixture of blood and sloughed tissue seeps into the vagina from the uterus and exits the body, a phenomenon called *menstrual flow.*

The Proliferative Phase During the proliferative phase of the uterine cycle, which begins with the end of menstruation, the uterus renews itself in preparation for possible pregnancy, which might occur following the next

ovulation. The endometrial tissues that were spared from destruction in the menstrual phase begin to grow, and the smooth muscle in the underlying myometrium thickens. The endometrial glands enlarge, and blood vessels increase in abundance. In the cervical canal, glands begin to secrete a thin mucus that bathes the inner surface; should sperm be deposited in the vagina, this mucus facilitates their migration through the uterus, a necessary prelude to fertilization. Uterine changes in the proliferative phase are promoted by estrogens, whose plasma levels rise due to the growth and increasing secretory activity of the dominant follicle.

The Secretory Phase During the secretory phase of the uterine cycle, the endometrium (which has been rebuilt during the proliferative phase) is transformed in such a way as to make it a favorable environment for implantation and subsequent housing and nourishment of the developing embryo. The blood supply of the endometrium becomes enriched as arteries branch. The endometrial glands enlarge further and begin to secrete fluids rich in glycogen, which the embryo uses as an energy source in its early stages of growth. Furthermore, the secretions of the cervical glands become stickier and more viscous, eventually blocking the cervical canal by forming a "plug" that effectively isolates the uterus from microorganisms in the outside environment that could possibly harm a developing embryo. These uterine changes are promoted by progesterone, whose plasma levels rise during the secretory phase (as do estrogen levels) due the action of the corpus luteum.

As the end of the secretory phase approaches, the corpus luteum degenerates, causing plasma estrogen and progesterone levels to fall. This decline causes a withdrawal of these hormones' growth-promoting influences on the endometrium, which triggers the previously described events of menstruation. With the onset of menstruation, the secretory phase ends and the next menstrual phase (and follicular phase of the ovarian cycle) begins. If fertilization and implantation have occurred, the corpus luteum does not degenerate, and estrogen and progesterone levels remain elevated. As a result, secretory-phase uterine conditions are maintained into pregnancy.

Hormonal Changes During the Menstrual Cycle

In this portion of our discussion we examine the hormonal changes that occur in the menstrual cycle relative to the phases of the ovarian cycle, because events in the follicular and luteal phases are directly or indirectly responsible for the hormonal changes. We begin by examining events occurring in the early and middle portions of the follicular phase.

Hormonal Changes in the Early to Mid-Follicular Phase

The hormonal changes occurring during the menstrual cycle are plotted in Figure 21.15a and b. As is apparent in Figure 21.15a, the early follicular phase is marked by short-lived declines in plasma estrogen and progesterone levels, which are actually the final stages of declines that began in the previous luteal phase as a result of degeneration of the corpus luteum. Because estrogens and progesterone tend to suppress the secretion of gonadotropins by the anterior pituitary (Figure 21.16), plasma levels of LH and FSH show a slight rise at this time, reflecting a withdrawal of this suppressive action (see Figure 21.15b). The rise in FSH stimulates a number of follicles to develop and grow.

FSH binds to receptors on granulosa cells and promotes their growth and proliferation, which causes follicles to increase in size (see Figure 21.16). Under the influence of this hormone, the outer layer of granulosa

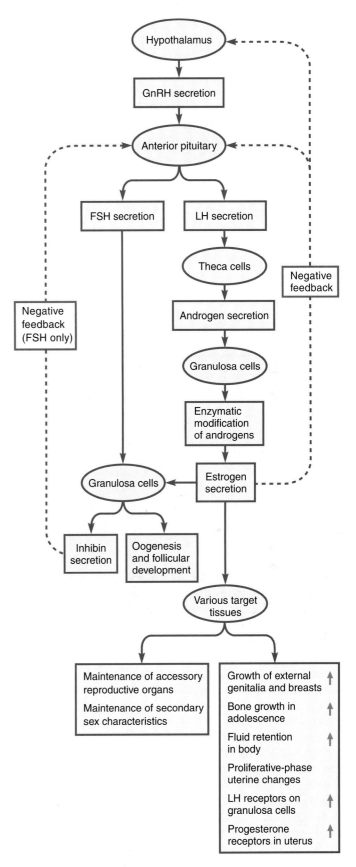

FIGURE 21.16 Regulation of hormone secretion during the early to mid-follicular phase.

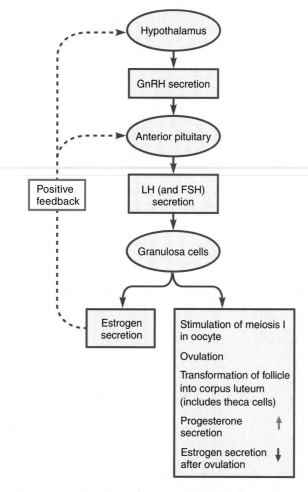

FIGURE 21.17 Regulation of hormone secretion during the late follicular phase.

cells differentiates into theca cells, which possess LH receptors. LH stimulates the theca cells to secrete androgens, which then travel to the granulosa cells and are converted to estrogens. Eventually, a dominant follicle emerges that secretes estrogens at a high rate, causing plasma levels of these hormones to rise rapidly. These estrogens feed back on the hypothalamus and anterior pituitary, suppressing LH and FSH secretion and halting the initial rise in plasma levels of these hormones. LH remains fairly steady, but FSH levels fall due to the influence of inhibin, which is secreted at ever-increasing rates by granulosa cells of the growing follicle. This drop in FSH is largely responsible for triggering atresia of nondominant follicles, as previously described.

During the early to mid-follicular phase, estrogens promote a variety of physiological changes throughout the body, including uterine changes characteristic of the proliferative phase, and estrogens also work in conjunction with FSH to promote oogenesis and follicular growth (see Figure 21.16). Estrogens also promote changes that prepare the body for subsequent events of the menstrual cycle. For instance, estrogens induce the expression of LH

receptors on granulosa cells, which renders these cells responsive to LH in preparation for the rise in LH that occurs just before the end of the follicular phase (see Figure 21.15b). Estrogens also induce the expression of progesterone receptors on endometrial cells, which primes the endometrium to respond to progesterone during the luteal phase.

Hormonal Changes in the Late Follicular Phase In the late follicular phase, rising levels of estrogens (see Figure 21.15a) trigger a fundamental change in the way these hormones affect secretory activity of the hypothalamus and anterior pituitary: Instead of suppressing the secretion of LH, estrogens *stimulate* the secretion of this hormone (Figure 21.17). As a consequence, LH levels rise (see Figure 21.15b, which stimulates more estrogen secretion, which stimulates more LH secretion, and so on. These events constitute a positive feedback loop that causes LH secretion to rise abruptly, bringing about a dramatic upswing in plasma LH levels called the **LH surge.** (FSH levels also rise, but not to the same degree.)

During the mid-follicular phase, estrogens stimulate granulosa cells to express LH receptors on their surfaces, making them responsive to LH. As a result, the rising tide of LH that occurs in the late follicular phase triggers the following changes in the dominant follicle: (1) Granulosa cells begin to secrete paracrines that stimulate the oocyte to complete meiosis I. (2) Estrogen secretion by granulosa cells falls, causing plasma estrogen levels to decrease. (3) Granulosa cells begin to secrete progesterone, which causes a small rise in plasma levels of this hormone. (4) Granulosa cells begin to secrete enzymes and paracrines that begin to break down the follicular wall. About 18 hours following the peak of the LH surge, the wall ruptures, triggering ovulation. (5) Granulosa cells and theca cells begin to differentiate, which eventually transforms the ruptured follicle into a corpus luteum. The appearance of the corpus luteum sets the stage for other events of the luteal phase, which are described next.

Hormonal Changes in the Luteal Phase Early in the luteal phase the corpus luteum is growing but not yet fully functional. Plasma estrogen levels continue to fall (see Figure 21.15a) because the rate of estrogen secretion by the corpus luteum is not yet high enough to compensate for the loss of the estrogen-secreting dominant follicle. This fall in estrogens removes the stimulus for LH secretion, thereby terminating the LH surge (see Figure 21.15b). Progesterone secretion by the corpus luteum causes plasma levels of this hormone to begin rising. As time passes, the corpus luteum grows and eventually attains full maturity. Estrogen and progesterone secretion rise, causing a corresponding dramatic increase in plasma levels that peak at about the middle of the luteal phase. Estrogens work to maintain many of the same effects ex-

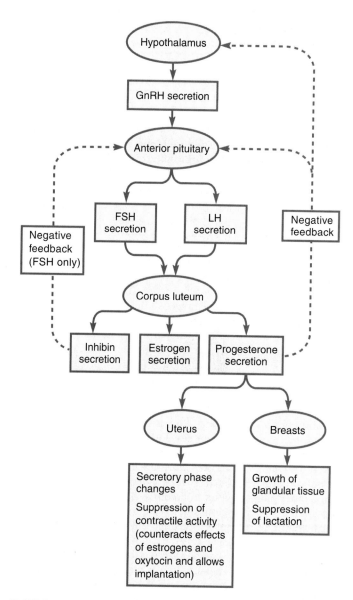

FIGURE 21.18 **Regulation of hormone secretion during the luteal phase.**

TABLE 21.2 SELECTED ACTIONS OF ESTROGENS AND PROGESTERONE IN FEMALES

Actions of estrogens

Stimulate oogenesis and follicular development

Promote development of secondary sex characteristics during puberty and maintenance of these characteristics in adult life

Promote growth of breasts during pregnancy

Stimulate growth hormone secretion, which promotes bone growth during adolescence

Promote proliferative-phase uterine conditions

Induce expression of uterine progesterone receptors and granulosa cell LH receptors

Stimulate decreased urine output by kidneys, which promotes fluid retention

Actions of progesterone

Promotes secretory-phase uterine conditions

Suppresses uterine contractile activity during pregnancy

Promotes growth of glandular tissue in breasts but suppresses milk production

---

erted during the follicular phase, whereas progesterone promotes secretory-phase uterine changes and other changes that generally adapt the body for possible pregnancy (described shortly). After about the tenth day of the luteal phase (in the absence of an implanted embryo), the corpus luteum begins to degenerate and its secretory activity falls, causing a drop in estrogen and progesterone levels that soon triggers menstruation. With the beginning of menstruation, the luteal phase ends and the next menstrual cycle begins with a new follicular phase.

As estrogen levels rise during the luteal phase, possible stimulatory effects of these hormones on LH secretion are blocked by progesterone, which strongly suppresses secretory activity of the hypothalamus and anterior pituitary (Figure 21.18). Thus further LH surges, which might otherwise be triggered by high estrogen levels, are pre-

vented. In fact, plasma LH (and FSH) levels decrease throughout the luteal phase due to this inhibitory influence of progesterone.

We have seen that the female sex hormones are responsible for regulating many of the events that occur during the menstrual cycle. In the next section we consider the long-term effects of these hormones, estrogens in particular.

Long-Term Hormonal Regulation of Female Reproductive Function

Estrogen and progesterone levels are low in early childhood and rise dramatically in puberty. The rising estrogen levels promote the development of female secondary sex characteristics, including widening of the hips relative to the shoulders; deposition of fat in the breasts, hips, and buttocks; secretion of skin oil (which is thinner than that of males); and growth of coarse body hair in the pattern characteristic of females. (Elongation of the hairs themselves is promoted by androgens, as it is in males.) Estrogens also promote the growth of bones during adolescence, which is similar to the action of testosterone in males. Actions of estrogens and progesterone in females are summarized in Table 21.2.

During a woman's reproductive life, average estrogen levels remain high and help maintain the secondary sex characteristics, but when she reaches menopause these levels decline as menstrual cycling comes to a halt. Over time this relative lack of estrogens causes a reversal of many of the changes that occurred during puberty and

Many sexually active couples use various methods of birth control to postpone or prevent pregnancy. Many birth control methods work by preventing fertilization (a process called *contraception*); others work by preventing implantation, and a few terminate pregnancy (called *abortion*).

One method of contraception, called *natural family planning* or the *rhythm method*, requires couples to refrain from sexual intercourse for a short time during each menstrual cycle—namely, for a few days around the time of ovulation, when the woman is fertile. Successful use of this method requires an accurate prediction of when the woman's next period of fertility is about to occur, which can be estimated from daily measurements of her body temperature or from other signs such as changes in vaginal secretions. Because these predictions are prone to error, however, a woman using the rhythm method faces a 20% chance of becoming pregnant over a year's time.

A more reliable method of contraception is *surgical sterilization*, which has a failure rate near zero. In men the most common method of sterilization is *vasectomy*, in which the vas deferens is cut and tied to prevent the passage of sperm from the testis to the urethra. In women the most common method is *tubal sterilization*, in which movement of eggs from the ovary to uterus is prevented by cauterizing the uterine tubes or by placing rings or clips around them (*tubal ligation*). Al-

though surgical sterilization is nearly 100% effective, it is very difficult to reverse and should be considered permanent.

Other contraceptive methods rely on the use of physical barriers to prevent sperm from reaching the egg. One barrier method involves placing a thin latex or rubber sheath called a *condom* over a man's penis; another involves insertion of a rubber *diaphragm* or *cap* into a woman's vagina prior to intercourse. (Caps or diaphragms are usually used in conjunction with spermicidal creams or jellies.) These barrier methods have a relatively high failure rate; about 15% of couples face an unexpected pregnancy over the course of a year.

Oral contraceptives, which are currently available for use by women only, are pills containing drugs that prevent ovulation. Some types contain a combination of synthetic estrogens and a progesteronelike substance; others contain only the latter. The combination pills are taken daily in a specific sequence, with each pill containing a different drug dosage; the other pills (often called *minipills*) all contain the same dosage and need not be taken in any sequence. The primary effect of these drugs is to suppress LH secretion—and hence ovulation—by mimicking the normal rise in plasma progesterone that occurs in the luteal phase. If ovulation should occur, these drugs still interfere

with implantation and sperm migration because they promote secretory-phase uterine conditions.

Contraceptive drugs can also be administered via intramuscular injection or through the use of *implants,* small capsules (about the size of a match stick) that are inserted under the skin, usually in the upper arm. Injections are effective for about three months, whereas implants retain their effectiveness for over five years. Another long-term treatment is the use of an *intrauterine device* (IUD), a small T- or Y-shaped piece of metal or plastic that is inserted into the uterine cavity. Some IUDs contain copper that is slowly released into the surrounding endometrial tissue, whereas others release a progesteronelike hormone. In either case, IUDs work by preventing implantation.

All the methods just described are intended for use before intercourse, but a recent development is a drug called RU 486 (mifepristone), sometimes referred to as the "morning-after pill" because it is effective when taken within a day or two after intercourse. This drug binds to progesterone receptors, thereby blocking progesterone's ability to promote secretory-phase uterine conditions and suppress uterine contractions. If fertilization and implantation have occurred, this drug can trigger detachment of the embryo from the endometrium and its subsequent expulsion from the uterus.

---

triggers many other changes throughout the body. For example, the breasts and reproductive organs, which grow during puberty, gradually shrink in size during menopause. Many women also experience so-called hot flashes—increases in body temperature and blood flow to the skin of the face and neck followed by profuse sweating—which appear suddenly for no apparent reason and last for a few seconds to minutes. Women in menopause also experience increased risks for coronary heart disease and bone fractures; the latter effect results from osteoporosis, a decrease in the mineral content of bones.

Quick Test 21.4

1. Define the following terms: *menstrual cycle, menstruation, ovulation, endometrium, atresia, dominant follicle, Graafian follicle, antrum, zona pellucida, follicular phase, luteal phase, corpus luteum, LH surge,* and *myometrium.*

2. Identify the two phases of the ovarian cycle, and briefly describe the events that occur in each.

3. Identify the three phases of the uterine cycle, and briefly describe the events that occur in each.

FERTILIZATION, IMPLANTATION, AND PREGNANCY

In order for fertilization to occur, sperm must be introduced into the female reproductive tract within five days before ovulation, because sperm can live in the tract for a maximum of five days and because the oocyte remains viable for only 12–24 hours after its release from the ovary.

When first deposited in the female reproductive tract, sperm are unable to fertilize an oocyte; they gain this capacity only after they have been in the tract for several hours, during which time they undergo a process called *capacitation*. One result of capacitation is a change in the pattern of tail movements, which permits sperm to move faster; another is that the plasma membrane is altered such that the sperm can fuse with the oocyte, which is necessary for fertilization to occur.

Sperm deposited in the vagina move through the cervical canal into the uterus, where they move along the uterine wall and eventually enter the opening to the uterine tube. (Of the millions of sperm that are deposited, only a few hundred typically reach the uterine tube; the others die along the way.) In the tube, sperm move toward the oocyte as it approaches them from the opposite direction. Fertilization occurs when a sperm contacts and is able to penetrate into the oocyte (Discovery: Birth Control Methods).

Events of Fertilization

To fertilize an oocyte, a sperm must penetrate through the surrounding corona radiata and reach the zona pellucida, where it binds to molecules of a specific sperm-binding protein. This binding then triggers an event called the **acrosome reaction,** which results in the contents of the acrosome being released to the outside. Among the acrosome's contents are enzymes capable of digesting the substance of the zona pellucida, thereby allowing the sperm access to the oocyte's surface. The first sperm to reach the oocyte binds to a receptor on the plasma membrane, which precipitates transport of the sperm into the egg's cytoplasm. This fusion also stimulates the oocyte to complete meiosis II and become an ovum. Inside the ovum the sperm's plasma membrane disintegrates and its chromosomes migrate toward the center of the cell, as do the chromosomes that were originally present in the oocyte (At this stage the chromosomes are contained in their nuclear membranes, which later disintegrate.) The chromosomes of the sperm and ovum combine, and the DNA is then replicated in preparation for the first mitotic division of the cell (now a *zygote*).

The fact that hundreds of sperm typically reach the uterine tube raises the distinct possibility that more than one sperm could fertilize an oocyte, a phenomenon called **polyspermy.** If this were to happen, the fertilized ovum would contain more than the normal number of chromosomes, and it would most likely fail to develop. Fortunately, polyspermy is averted by a number of mechanisms set in motion when a sperm first fuses with the oocyte's plasma membrane. Such fusion triggers the exocytosis of vesicles near the plasma membrane, which releases enzymes into the space between it and the surrounding zona pellucida. As a result, sperm-binding proteins in the zona pellucida become inactivated, and the zona pellucida hardens and pulls away from the plasma membrane. Because these changes create a barrier around the oocyte that prevents its fertilization by other sperm, they are referred to as the *block to polyspermy.*

Early Embryonic Development and Implantation

After fertilization has occurred, the ovum undergoes several mitotic divisions (Figure 21.19a), transforming within the next few days into a round ball of cells called a **morula.** Cell divisions at this stage are unlike subsequent cell divisions in that the total volume of cytoplasm does not increase. For this reason the number of cells increases with each division, but the size of each cell decreases. Cell division of this type is often referred to as *cell cleavage.* Three to four days after fertilization, repeated cell cleavages yield a total of 16–32 cells, each of which is **totipotent**—that is, has the potential to develop into a complete human being. By this time the morula has reached the uterus; for the next three to four days it floats in the intrauterine fluid and undergoes more cell divisions.

Eventually the morula develops into a more complex, hollow structure called a **blastocyst,** which has shed the zona pellucida and possesses the following components (see Figure 21.19a): a spherical outer cell layer called the *trophoblast;* a cluster of cells on the inside called the *inner cell mass,* which is attached to the trophoblast on one side and eventually gives rise to the embryo; and a fluid-filled cavity, called the *blastocoele,* adjacent to the inner cell mass. At this stage the cells have differentiated from one another; that is, they have lost their totipotency and are committed to eventual development into different tissues.

About six or seven days after fertilization the blastocyst attaches to the uterine wall (Figure 21.19b). For the next few weeks, cells of the trophoblast secrete enzymes that digest the adjacent endometrial cells, from which the embryo draws nourishment. The trophoblast also secretes paracrines that promote local changes in the endometrial tissue. Among these changes, collectively known as the *decidual response,* is an increase in the number of capillaries, which helps increase the delivery of oxygen and nutrients to the area. At the point of contact with the uterine wall, trophoblastic cells proliferate, differentiate,

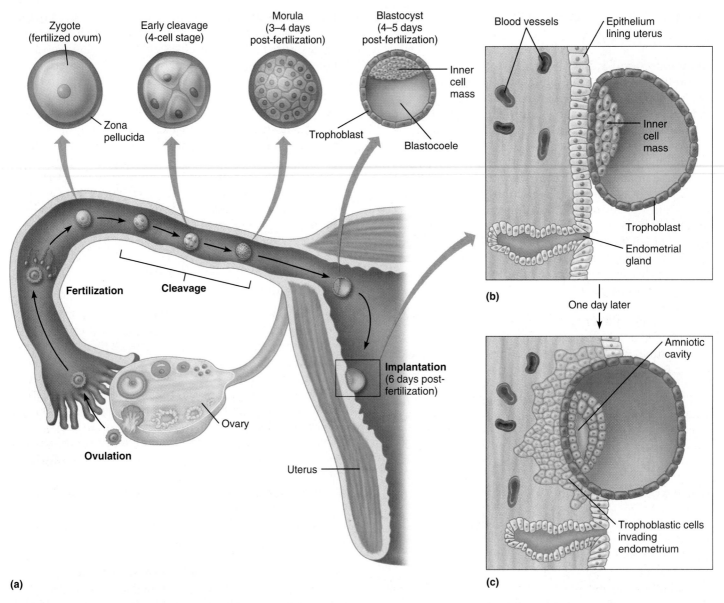

FIGURE 21.19 Early embryonic development and implantation. (a) *Stages of early embryonic development. Events shown begin with ovulation and proceed through implantation.* **(b)** *Early implantation, showing the blastocyst adhering to the surface of the endometrium.* **(c)** *A later stage of implantation, showing embryonic cells penetrating into the endometrial tissue. Note the appearance of the fluid-filled* amniotic cavity *within the inner cell mass.*

and begin to infiltrate into the endometrial tissue (Figure 21.19c). As described shortly, the trophoblast and underlying endometrium eventually develop into the *placenta,* a structure that allows efficient exchange of gases, nutrients, and wastes between the mother's circulatory system and the circulatory system of the developing embryo.

On rare occasions the embryonic cell mass splits into two (or occasionally more) parts in an early developmental stage, when the cells are still totipotent. When this occurs, each part can develop into a complete individual. The resulting offspring are *identical twins,* which look alike and have identical genes. The offspring are also called *monozygotic twins* because they originate from the same zygote.

Later Embryonic and Fetal Development

After implantation the embryo continues to grow in size and complexity. In the region of contact between the embryo and the uterine wall at about five weeks postfertilization, the trophoblast thickens and develops into the **chorion,** a tissue that eventually grows into a tough en-

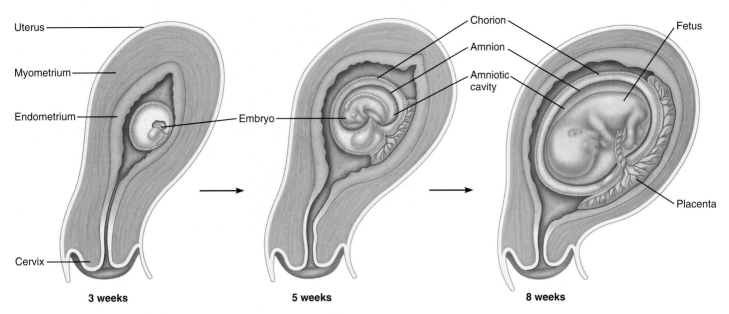

FIGURE 21.20 Growth and development of the embryo and the structures that support it. *Note that at about eight weeks the developing human is termed a fetus.*

velope that encapsulates the embryo, isolating it from its surroundings (Figure 21.20). Inside the embryo, other changes are occurring: A fluid-filled cavity called the **amniotic cavity** begins to form and grow in size within the inner cell mass (see Figure 21.19c). As the amniotic cavity expands, the surrounding cells develop into an epithelial tissue called the **amnion** (or *amniotic sac;* see Figure 21.20), which eventually fuses with the chorion to form a single membrane that surrounds the developing embryo. The fluid contained within the amnion (called **amniotic fluid**) is similar in composition to normal extracellular fluid; it bathes the embryo and forms a physical barrier between it and the uterine wall that cushions the developing embryo against possible physical trauma and protects it against sudden changes in temperature.

As development proceeds, the embryo undergoes a series of dramatic transformations in structure and function. Within the inner cell mass, cells begin to differentiate into distinct tissues that will ultimately give rise to all the organ systems found in adults. Indeed, the pace of change is so rapid that rudimentary organ systems are already discernible within the first few weeks after fertilization. Over time these primitive organ systems gradually acquire more and more of their adult characteristics; different systems develop at different rates. (When It Goes Wrong: Congenital Heart Disease, www.physiologyplace.com, Challenge Yourself) As this is occurring, the embryo's external appearance is also rapidly changing, so that by the end of the eighth week of life, it has a recognizably human form, complete with clearly discernible limbs as well as a head and face. From this point onward, the developing human is now a *fetus.*

In its earlier stages of development, the embryo is able to obtain nutrients through the breakdown of en-

dometrial tissue, and sufficient quantities of oxygen by diffusion from nearby blood vessels. However, as the embryo grows, its need for oxygen increases such that it can no longer rely on diffusion from the mother's tissues as a mechanism of delivery; likewise, its demand for nutrients increases such that it can no longer obtain sufficient amounts from the surrounding tissue.

To obtain essential materials (and to rid itself of carbon dioxide and other wastes), the growing embryo or fetus relies on the **placenta,** a structure that permits the ready exchange of gases, nutrients, and other materials between the fetus's bloodstream and the maternal circulatory system. The placenta develops in the first few weeks after implantation in the region of contact between the embryo and the endometrium, and it is made of specialized, closely interwoven fetal and maternal tissues (Figure 21.21). Placental development begins as the chorion sends out fingerlike projections called **chorionic villi.** The villi contain both capillaries of the fetal circulation and cells that secrete paracrines that alter the structure and function of the surrounding endometrial tissue, such that the villi become surrounded by sinuses containing maternal blood. As a result, maternal blood and fetal blood come into close proximity (which facilitates material exchange) without actually mixing. (Separation of maternal and fetal blood is important because it prevents attack of fetal tissues by the mother's immune system.) Maternal blood is delivered to the placental sinuses by the *uterine artery* and is returned to the mother's general circulation by the *uterine vein.* Fetal blood is carried to the placenta by the paired *umbilical arteries* and back to the fetus's general circulation by the *umbilical vein,* both of which are housed in a ropelike structure called the **umbilical cord,** which extends between the placenta and

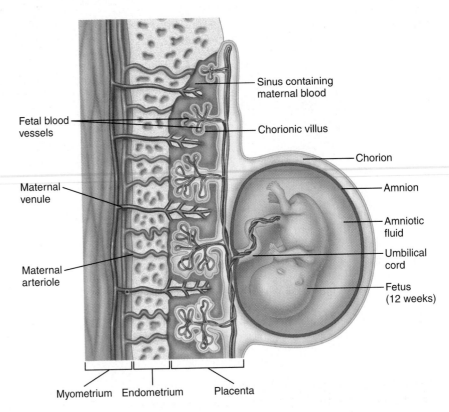

FIGURE 21.21 The placenta. *The interweaving of fetal and maternal tissues enables the exchange of materials between the circulatory systems of mother and fetus.*

Sinus containing maternal blood

Fetal blood vessels

Chorionic villus

Chorion

Amnion

Amniotic fluid

Maternal venule

Umbilical cord

Fetus (12 weeks)

Maternal arteriole

Myometrium   Endometrium   Placenta

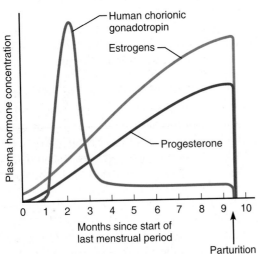

Human chorionic gonadotropin

Estrogens

Progesterone

Months since start of last menstrual period

Parturition

FIGURE 21.22 Plasma concentrations of estrogen, progesterone, and human chorionic gonadotropin during pregnancy.

creted by the chorionic portion of the placenta and exerts many of the same effects as LH. Under the influence of hCG, the corpus luteum continues to secrete large amounts of estrogens and progesterone, which exert several actions described later. Secretion of hCG begins at implantation, when trophoblast cells start to infiltrate the endometrial tissue, and then it rises for the next two months as the placenta grows (see Figure 21.22). At the end of this period the secretion of this hormone falls off dramatically, and as a result the corpus luteum begins to degenerate. Estrogen and progesterone levels remain high, however, and continue to rise until parturition because the placenta begins to secrete these hormones just as secretion by the corpus luteum is declining. Placental estrogens are actually manufactured from androgens (primarily dehydroepiandrosterone or DHEA) secreted by the adrenal cortex of the developing fetus. Placental progesterone is manufactured from cholesterol, which is delivered to the placenta by the maternal bloodstream.

During pregnancy, plasma levels of estrogen and progesterone, which rise after formation of the corpus luteum, continue to rise until parturition, at which time estrogen and progesterone levels fall precipitously (Figure 21.22). These elevated hormone levels are important for maintaining the pregnancy and for preparing the mother's body for the eventual delivery of the fetus. In particular, estrogen promotes the following effects:

1. *Growth of duct tissue in the breasts.* In the milk-producing glands of the breasts (called **mammary glands**), estrogens stimulate the development of tissue in ducts that carry milk to the nipples. This prepares the breasts for **lactation** (milk production), which begins

the fetus. As we will see shortly, the placenta is not just an organ of exchange; it also secretes several hormones that are important in maintaining pregnancy and preparing the mother's body for the eventual birth of the child.

Hormonal Changes During Pregnancy

During the first three months of pregnancy, the corpus luteum is maintained by the actions of a hormone called **human chorionic gonadotropin** (hCG), which is se-

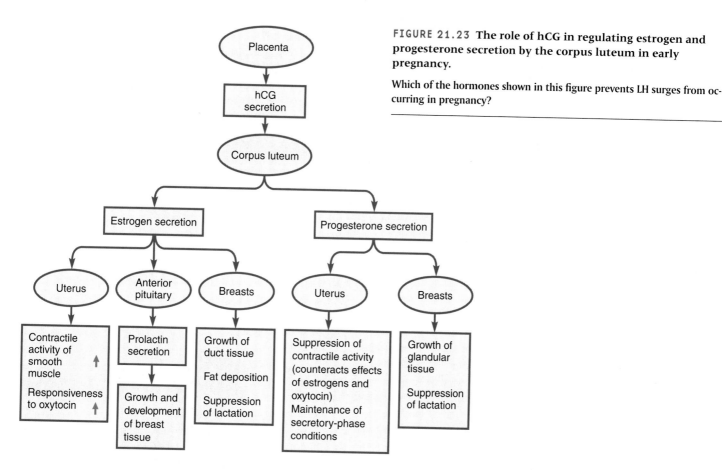

FIGURE 21.23 The role of hCG in regulating estrogen and progesterone secretion by the corpus luteum in early pregnancy.

Which of the hormones shown in this figure prevents LH surges from occurring in pregnancy?

after parturition. Estrogens also promote the deposition of fatty tissue in the breasts, which causes them to enlarge.

2. *Prolactin secretion.* Estrogen stimulates the secretion by the anterior pituitary of **prolactin,** which promotes breast growth and prepares the mammary glands for lactation. Although prolactin and estrogens both work to build up the milk-producing apparatus during pregnancy, actual milk production does not occur yet because the high plasma levels of estrogens and progesterone during pregnancy suppress lactation.

3. *Growth and enhanced contractile responsiveness of uterine smooth muscle.* Estrogens stimulate the growth of uterine smooth muscle, thereby enhancing the ability of the uterus to contract and expel the fetus during parturition. Estrogens also work to promote increased contractile activity in uterine smooth muscle and to increase its responsiveness to *oxytocin,* which stimulates uterine contractions during birth. Despite the fact that estrogens enhance contractile activity by themselves, the uterine contractions that occur during pregnancy are infrequent and weak, because the contraction-promoting actions of estrogens and oxytocin are blocked by the elevated levels of progesterone during this period.

While estrogens are working to promote these effects, progesterone is also working to maintain the pregnancy and prepare the mother's body for parturition. In particular, progesterone promotes the following effects:

1. *Growth of glandular tissue in the breasts.* In the breasts, progesterone stimulates the growth of glandular tissue (as opposed to ductal tissue, whose growth is stimulated by estrogens). This glandular tissue is responsible for secreting the milk that is produced after parturition.

2. *Suppression of contractile activity in uterine smooth muscle.* As previously mentioned, progesterone suppresses uterine contractions, thereby counteracting the effects of estrogens and oxytocin on uterine smooth muscle. Because contractions could prematurely expel an embryo or fetus from the uterus, this suppressive action is necessary for successful implantation and for the maintenance of pregnancy.

3. *Maintenance of secretory-phase uterine conditions.* Progesterone works during pregnancy to maintain the same secretory-phase uterine conditions that it promotes during the luteal phase of the ovarian cycle. This ensures hospitable conditions for embryonic and fetal growth.

Actions of estrogens and progesterone are summarized in Figure 21.23, which shows hormonal interactions occurring in early pregnancy, when the corpus luteum is still active.

Progesterone

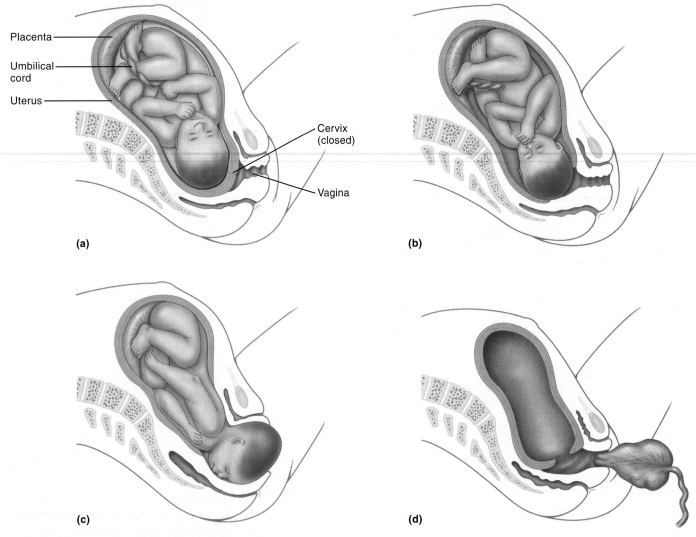

Placenta
Umbilical cord
Uterus

Cervix (closed)

Vagina

(a)

(b)

(c)

(d)

FIGURE 21.24 Events of parturition. (a) *Position of the fetus in the uterus before parturition.* (b) *Dilation of the cervix.* (c) *Movement of the fetus through the birth canal (delivery).* (d) *Expulsion of the placenta following delivery.*

In pregnancy, progesterone exerts negative feedback on the hypothalamus and anterior pituitary that keeps rates of LH and FSH secretion low. Consequently, no new dominant follicles appear and no LH surges occur, despite the fact that estrogen levels are elevated at this time. In addition, the placenta secretes a new hormone called **placental lactogen,** which has effects similar to some of the actions of growth hormone and prolactin. This hormone works hand in hand with estrogens and prolactin to promote not only breast growth, but also the mobilization within the mother of energy stores needed to meet the increased demands arising from the growth of breasts and fetal tissue.

Quick Test 21.5

1. Define the following terms: *capacitation, acrosome reaction, polyspermy, morula, blastocyst, inner cell mass, trophoblast, implantation, placenta, chorion, chorionic villi, amnion, amniotic fluid,* and *umbilical cord.*

2. Where does fertilization occur? implantation? What is the primary function of the placenta?

3. Estrogens and progesterone are secreted by what structure during the first three months of pregnancy? during the final six months?

4. For each of the following hormones, name the site from which it is secreted, and give a brief description of its function: human chorionic gonadotropin, prolactin, oxytocin, and placental lactogen.

PARTURITION AND LACTATION

Pregnancy normally lasts about nine months (40 weeks), during which time the fetus gains in size, weight, and maturity such that it acquires the ability to live outside its mother's body.

Events of Parturition

In the last few weeks of pregnancy, weak and infrequent uterine contractions begin to appear. Over time these contractions gradually increase in strength and frequency. Because estrogens are known to promote contractile activity in uterine smooth muscle, it is likely that these contractions are triggered, at least in part, by the increased levels of estrogens relative to progesterone that exist in late pregnancy. Moreover, uterine smooth muscle is capable of initiating its own contractions in response to stretch, and as the fetus grows and the uterus expands, the resulting stretch of the smooth muscle may initiate contractile activity.

Shortly before parturition the fetus, which is normally upside down, moves downward and comes into contact with the closed cervix (Figure 21.24a), so that the strong uterine contractions of parturition will push the fetus through the cervical canal headfirst. Moreover, the fetus's headfirst orientation enables the head, which is narrower than the rest of the body, to act as a wedge to force the cervical canal open. (Opening of the canal is called *cervical dilation*). In a small fraction of cases the legs or hips pass through the cervical canal first, a phenomenon called *breech birth*; because the hips are wider than the head, such a delivery is more difficult than normal and increases the time required. Breech birth also exposes the fetus to greater danger because it increases the risk that the umbilical cord may become compressed between the fetus and the wall of the birth canal, which chokes off the fetus's blood supply. For these reasons, *cesarean section* (surgical delivery through the abdominal wall) is often recommended when breech birth is imminent.

To facilitate delivery, the cervix undergoes a process called *ripening* in the weeks before parturition. During this process, the cervix becomes softer and more flexible as a result of the enzymatic breakdown of collagen fibers in its connective tissue. These changes make it easier for the cervical canal to dilate as the fetus begins to pass through it.

In the hours just before parturition, the amniotic sac typically ruptures, causing the leakage of amniotic fluid to the outside of the body. This event is often the first reliable signal to the mother that parturition is imminent. Soon afterward a series of strong uterine contractions (known as **labor**) begins. Each contraction starts at the top of the uterus and then travels downward through the smooth muscle as a wave. (Contractions are triggered by electrical signals that move from cell to cell through gap junctions.) The downward movement of these contractions pushes the contents of the uterus against the cervix, which eventually forces the cervical canal open. In labor, uterine contractions may be assisted by voluntary contractions of the mother's abdominal muscles, which increase pressure on the uterine contents.

Initially the contractions of labor are separated by intervals of 10–15 minutes, but contractions come closer together as labor progresses. The rising intrauterine pressure causes dilation of the cervical canal (Figure 21.24b), which eventually reaches a maximum diameter of about 10 centimeters. At that point the fetus moves into the canal and within a few minutes emerges from the mother's body (Figure 21.24c). Blood vessels in the umbilical cord begin to constrict, and the placenta then separates from the endometrium. (The umbilical cord is usually tied off by birth attendants and then cut on the maternal side of the knot.) Finally, a wave of strong uterine contractions expels the placenta (now referred to as the *afterbirth*) to outside the mother's body (Figure 21.24d).

Once the contractions of labor begin, the stretch-sensitivity of uterine smooth muscle tends to perpetuate them. As the contractions push the fetus downward, the lower portion of the uterus is stretched, which stimulates more contractile activity. Uterine contractions are also promoted by oxytocin, whose release is stimulated in labor. The pressure exerted by the fetus against the uterine wall excites stretch receptors located in the lower portion of the uterus, which triggers the release of oxytocin from the posterior pituitary. Oxytocin enters the bloodstream and then stimulates uterine contractions both by direct action on muscle cells and by stimulating other cells in the myometrium to secrete prostaglandins, which act locally to promote contractile activity. The contraction-promoting effects of oxytocin are normally blocked during pregnancy by progesterone, which helps to keep uterine contractile activity low.

Although the precise nature of the signals that trigger parturition are incompletely understood, evidence suggests that they may originate from the fetus itself. The fetal portion of the placenta is known to secrete corticoptropin-releasing hormone (CRH), the same hormone that is secreted by the hypothalamus in adults and stimulates the release of ACTH from the anterior pituitary. In the fetus, placental CRH stimulates the release of ACTH, which stimulates the adrenal gland to secrete DHEA. This hormone is then converted by the placenta to estrogens, which enter the maternal bloodstream. The rising rates of placental CRH secretion are believed to be largely responsible for the rapidly rising estrogen levels that occur toward the end of pregnancy and help to initiate the events of labor. Because some placental CRH leaks into the maternal blood, it is also possible that rising CRH levels act as a direct trigger for parturition.

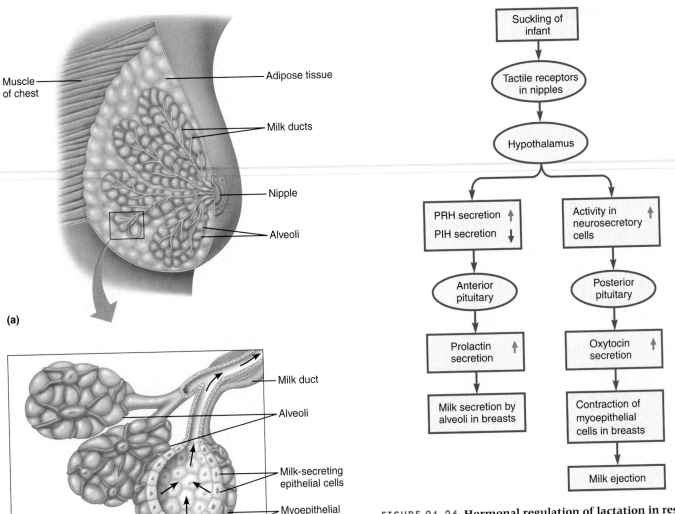

(a)

(b)

FIGURE 21.25 Mammary glands. (a) *A cutaway view of a breast, showing the alveoli and milk ducts that constitute the mammary glands.* **(b)** *An enlargement showing milk-secreting epithelial cells and myoepithelial cells in alveoli.*

FIGURE 21.26 Hormonal regulation of lactation in response to suckling.

Lactation

The decrease in estrogens and progesterone that occurs in parturition is important not only because it helps to promote uterine contractions in labor, but also because it allows lactation to begin. In the first months after birth the infant has no teeth and cannot eat solid foods. For this reason, the milk produced by the mother's breasts can serve as a valuable source of early nutrition. For the first one or two days after birth, this milk is little more than a watery fluid (called *colostrum*) containing a vast array of proteins but few other nutrients. Subsequently, however, the milk becomes enriched with a variety of additional constituents, including fat, lactose (the sugar present in cow's milk), growth factors and hormones (which promote tissue development in the infant), and antibodies, which confer on the infant some degree of immunity against bacteria and other pathogens. (In early life the infant is not able to make sufficient quantities of its own antibodies because its immune system is not yet developed.)

Milk is produced in the mammary glands by clusters of round, saclike glands called *alveoli* and is delivered by milk ducts to openings in the nipples (Figure 21.25a). To obtain this milk the infant simply sucks on the nipples (an act known as *suckling*), which triggers the flow of milk through the ducts (sometimes referred to as "let-down"). The alveoli are surrounded by *myoepithelial cells* (Figure 21.25b), which are flattened like ordinary epithelial cells but have the ability to contract like muscle cells. Suckling stimulates these cells to contract, which compresses the alveoli and forces the milk to flow through the ducts, a phenomenon called **milk ejection.**

The events by which suckling induces milk ejection are diagrammed in Figure 21.26. Suckling of the infant stimulates tactile receptors in the nipples, which project to the hypothalamus and excite neurosecretory cells that extend to the posterior pituitary and secrete oxytocin. Oxytocin travels in the bloodstream and stimulates my-oepithelial cells to contract. Suckling also stimulates the actual production of milk by stimulating the release of prolactin-releasing hormone (PRH) by the hypothalamus and inhibiting the release of prolactin-inhibiting hormone (PIH). Both of these changes stimulate the anterior pituitary to secrete prolactin, which induces milk production by cells in the alveoli.

During puberty, rising estrogen and progesterone levels stimulate growth of the breasts and development of the milk-producing apparatus. Estrogens promote not only the development of tissue in the ducts, but also the deposition of fatty tissue in the breasts, which results in visible breast enlargement; progesterone, by contrast, promotes development of the alveoli. In pregnancy, the breasts become enlarged and the milk-producing apparatus becomes fully developed as a result of the combined actions of increased plasma levels of estrogens, progesterone, prolactin, and placental lactogen. (Recall that prolactin release is stimulated by estrogen at this time.) No milk is actually produced, however, because the high estrogen and progesterone levels block the secretory activity of cells in the alveoli. After parturition the fall in estrogens and progesterone unblocks the stimulatory effect of prolactin on milk production and allows lactation to proceed. (Discovery: Reproductive Facts, www.physiologyplace.com, Challenge Yourself)

Quick Test 21.6

1. Define the following terms: *afterbirth, labor, colostrum, mammary glands,* and *alveoli.*

2. For each of the following descriptive phrases, indicate whether it pertains to oxytocin, prolactin, estrogens, or progesterone: (a) inhibits uterine contractions during pregnancy, (b) secretion is stimulated by estrogens, (c) promotes development of milk ducts and deposition of fat in breasts during puberty, (d) promotes development of alveoli during puberty, (e) is secreted in response to excitation of uterine stretch receptors, (f) is secreted in response to tactile stimulation of nipples, (g) stimulates milk production in response to suckling, (h) stimulates milk ejection in response to suckling.

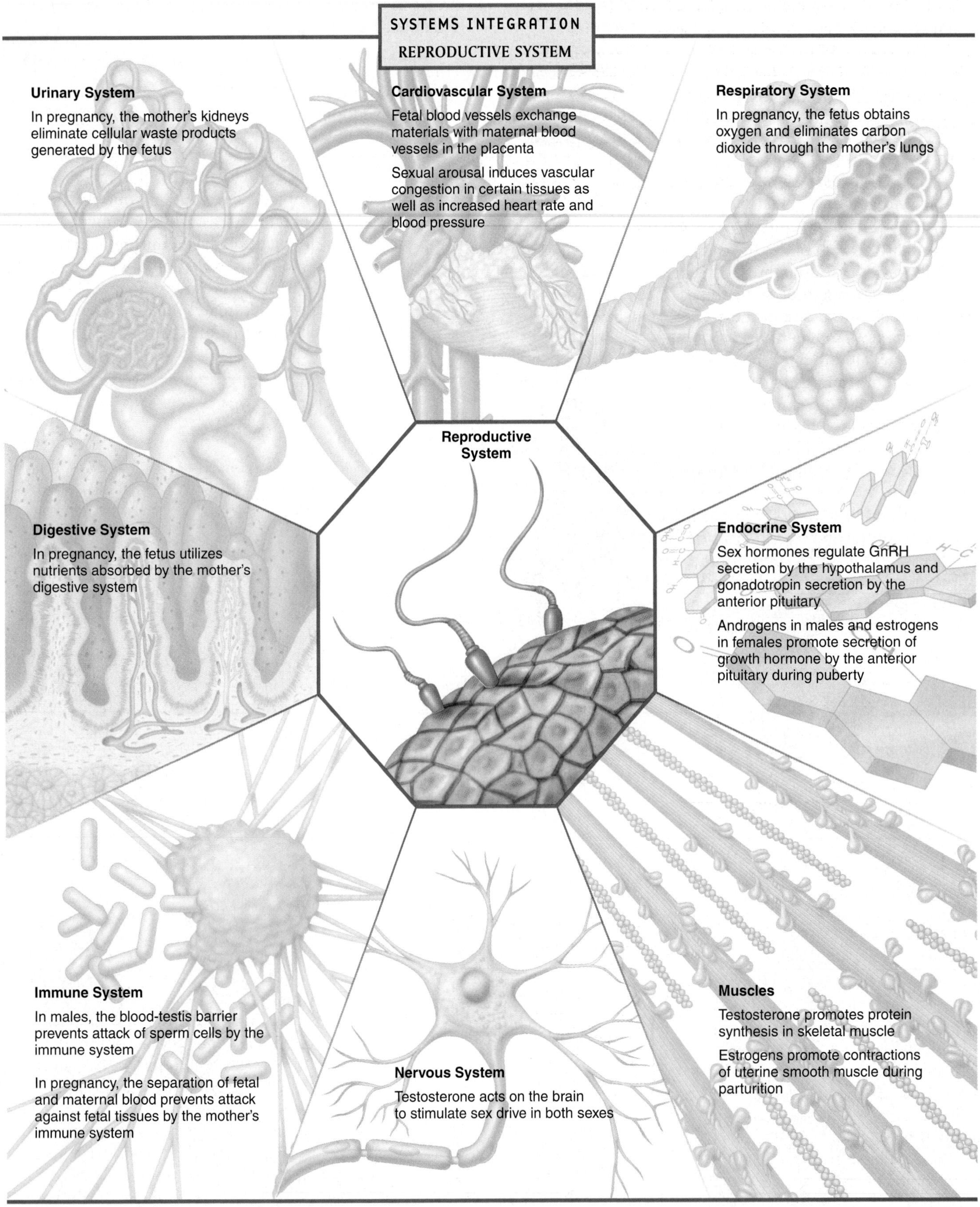
SYSTEMS INTEGRATION
REPRODUCTIVE SYSTEM
Urinary System
In pregnancy, the mother's kidneys eliminate cellular waste products generated by the fetus
Cardiovascular System
Fetal blood vessels exchange materials with maternal blood vessels in the placenta
Sexual arousal induces vascular congestion in certain tissues as well as increased heart rate and blood pressure
Respiratory System
In pregnancy, the fetus obtains oxygen and eliminates carbon dioxide through the mother's lungs
Reproductive System
Digestive System
In pregnancy, the fetus utilizes nutrients absorbed by the mother's digestive system
Endocrine System
Sex hormones regulate GnRH secretion by the hypothalamus and gonadotropin secretion by the anterior pituitary
Androgens in males and estrogens in females promote secretion of growth hormone by the anterior pituitary during puberty
Immune System
In males, the blood-testis barrier prevents attack of sperm cells by the immune system
In pregnancy, the separation of fetal and maternal blood prevents attack against fetal tissues by the mother's immune system
Nervous System
Testosterone acts on the brain to stimulate sex drive in both sexes
Muscles
Testosterone promotes protein synthesis in skeletal muscle
Estrogens promote contractions of uterine smooth muscle during parturition

An Overview of Reproductive Physiology, p. 671

Human reproduction involves the fundamental processes of gametogenesis, fertilization, pregnancy, and parturition. Reproductive ability is acquired during puberty, during which the reproductive organs mature, gametogenesis begins, and secondary sex characteristics develop. Males are able to reproduce continuously through adulthood, but in females reproductive capacity is cyclic and is lost at menopause.

The male and female reproductive systems include the gonads (testes in males and ovaries in females), which carry out gametogenesis and secrete sex hormones (androgens in males and estrogens and progesterone in females), and accessory reproductive organs, which include organs of the reproductive tract and various glands that secrete fluids into the tract.

The Male Reproductive System, p. 676

The male reproductive system includes the testes, external genitalia (the penis and scrotum), the reproductive tract (the epididymis, vas deferens, ejaculatory duct, and urethra), and accessory glands (seminal vesicles, bulbourethral glands, and the prostate gland). Sperm are formed in the testes in seminiferous tubules, which are lined by Sertoli cells. During sexual arousal a mixture of sperm and fluids (semen) is forcibly ejected from the penis through the urethra, an event called ejaculation.

Reproductive function in males is controlled by androgens and other hormones, including gonadotropins from the anterior pituitary (follicle-stimulating hormone, FSH, and luteinizing hormone, LH) and gonadotropin-releasing hormone (GnRH), a hypophysiotropic hormone. Spermatogenesis and other Sertoli cell functions are stimulated by FSH. LH stimulates androgen secretion by Leydig cells. During reproductive life, androgen levels are fairly steady because they limit their own secretion through negative feedback control of GnRH and gonadotropin secretion.

The Female Reproductive System, p. 683

The female reproductive system includes the ovaries, the reproductive tract (the uterus, uterine tubes, and vagina), and the external genitalia (the mons pubis, labia majora, labia minora, vestibule, clitoris, and vestibular glands). The menstrual cycle, which lasts about 28 days, is marked by cyclic changes in pituitary and ovarian hormone secretion and begins with menstruation, the shedding of tissue and blood from the endometrium. Ova develop from a set of germ cells whose number is fixed at birth and do not become fully mature until fertilization has occurred.

Each ovary contains numerous follicles, each of which contains one oocyte. Follicles also contain granulosa cells, which nourish the oocyte, regulate its development, and secrete estrogens, and (in later stages of development) theca cells. The ovarian cycle is divided into a follicular phase, during which a dominant follicle is selected and develops into a Graafian follicle, and a luteal phase, during which the follicle is transformed into a corpus luteum. The follicular phase ends with release of the oocyte (ovulation), which then enters the nearby uterine tube. Coinciding with the ovarian cycle is a uterine cycle consisting of a menstrual phase, a proliferative phase, and a secretory phase.

In the follicular phase, FSH targets granulosa cells to stimulate growth and estrogen secretion; LH stimulates theca cells to secrete androgens, which granulosa cells then convert to estrogens. The estrogens promote oogenesis (along with FSH) and proliferative-phase uterine changes. In the late follicular phase, rising estrogen levels trigger an LH surge, which is responsible for ovulation and development of the corpus luteum. In the luteal phase the corpus luteum secretes estrogen and progesterone, which suppresses LH and FSH secretion and promotes secretory-phase uterine changes. In the absence of fertilization the corpus luteum degenerates, causing a drop in estrogen and progesterone levels that triggers menstruation.

Fertilization, Implantation, and Pregnancy, p. 695

After fertilization, which normally occurs in the uterine tube, the zygote is eventually transformed into a blastocyst, which implants in the endometrium. At the point of contact, embryonic and endometrial tissues develop into a placenta, which permits exchange of materials between the mother and the developing embryo. During pregnancy, estrogens and progesterone (secreted by the corpus luteum at first and then by the placenta) promote many effects, including growth and development of the mammary glands, secretion of prolactin by the anterior pituitary (which promotes breast growth and eventually lactation), and maintenance of secretory-phase uterine conditions.

Parturition and Lactation, p. 701

Parturition typically occurs 40 weeks after fertilization and is accompanied by a wave of strong uterine contractions (labor), dilation of the cervix, expulsion of the fetus from the uterus, and separation of the placenta (afterbirth) from the uterine wall. After delivery, nourishment is provided to the infant by milk secreted by the mammary glands. Suckling of the infant triggers secretion of prolactin, which promotes milk production, and of oxytocin, which promotes milk ejection.

Multiple-Choice Questions

1. In both males and females, gonadotropin secretion by the anterior pituitary is stimulated by
   a) inhibin.
   b) androgens.
   c) GnRH.
   d) FSH.

2. In the embryo, testosterone promotes
   a) development of the primitive gonads into testes.
   b) regression of Müllerian ducts.
   c) development of Müllerian ducts into male reproductive organs.
   d) development of Wolffian ducts into male reproductive organs.

3. The testes are housed in a structure called the
   a) prostate gland.
   b) scrotum.
   c) penis.
   d) epididymis.

4. In the first step of spermatogenesis, spermatogonia differentiate into cells called
   a) spermatids.
   b) primary spermatocytes.
   c) spermatozoa.
   d) spermatophytes.

5. Cells in the ovaries secrete all of the following hormones *except*
   a) estrogens.
   b) progesterone.
   c) androgens.
   d) luteinizing hormone.

6. In oogenesis, meiosis I occurs
   a) in early fetal life.
   b) just before ovulation.
   c) after ovulation but before fertilization.
   d) after fertilization.

7. During the early to mid-follicular phase of the ovarian cycle, granulosa cell functions are stimulated by
   a) progesterone.
   b) FSH.
   c) LH.
   d) GnRH.

8. In the late luteal phase, estrogen and progesterone levels fall due to
   a) rupture of the dominant follicle.
   b) degeneration of the corpus luteum.
   c) an inhibitory effect of LH on secretory activity of the corpus luteum.
   d) the inhibitory effect of inhibin on the secretory activity of granulosa cells.

9. In the uterine cycle, the proliferative phase is immediately preceded by the
   a) menstrual phase.
   b) secretory phase.
   c) luteal phase.
   d) follicular phase.

10. The placenta serves not only as an organ of exchange; it also secretes all of the following hormones *except*
    a) prolactin.
    b) chorionic gonadotropin.
    c) placental lactogen.
    d) progesterone.

Objective Questions

1. In meiosis I, maternal and paternal chromosomes are segregated into separate daughter cells. (true/false)

2. The srY gene codes for (testosterone receptors/testis-determining factor), which determine(s) whether an embryo develops testes or ovaries.

3. In the absence of testosterone and MIS, the Müllerian ducts (persist/degenerate) in the embryo, and female structures eventually develop.

4. FSH and LH are classified as (sex hormones/gonadotropins).

5. GnRH, which is secreted by the hypothalamus, stimulates the secretion of both FSH and LH from the anterior pituitary. (true/false)

6. In the testes, androgens are secreted by (Sertoli cells/Leydig cells).

7. Spermatogenesis is stimulated by testosterone and (FSH/LH), which targets Sertoli cells.

8. The head of a sperm contains chromosomes and a(n) _____, a vesicle containing enzymes needed for fertilization.

9. Erection is accompanied by a(n) (increase/decrease) in the activity of sympathetic neurons projecting to blood vessels in the penis.

10. Once sperm are deposited in the female reproductive tract, they cannot fertilize the oocyte until they have undergone a process called _____.

11. Fertilization usually occurs in the (uterus/uterine tube).

12. The second half of the ovarian cycle is called the (luteal/follicular) phase.

13. In a follicle, the oocyte is surrounded by a layer of (granulosa/theca) cells that provide it nourishment and regulate its development.

14. (FSH/LH) stimulates theca cells to secrete androgens, which are converted to estrogens by granulosa cells.

15. The inner layer of the uterine wall is called the (endometrium/myometrium).

16. During the (proliferative/secretory) phase of the uterine cycle, the lining of the uterus thickens under the influence of rising estrogen levels.

17. Ovulation is triggered by (FSH/LH).

18. The corpus luteum secretes estrogens and (LH/progesterone).

19. In the late follicular phase, LH secretion is stimulated by (estrogens/progesterone).

20. Degeneration of the corpus luteum causes hormonal changes that trigger (ovulation/menstruation).

21. Before implantation, the morula develops into a _____, which consists of an inner cell mass contained within a hollow, fluid-filled outer cell layer.

22. Secretory-phase uterine conditions are promoted by (estrogen/progesterone), which inhibits gonadotropin secretion during the last half of the ovarian cycle.

23. During the last six months of pregnancy, estrogens and progesterone are secreted by the (ovaries/placenta).

24. During labor, strong uterine contractions are induced by (prolactin/oxytocin).

25. Suckling stimulates the release of _____, which promotes milk production by the breasts.

Essay Questions

1. Compare the hormonal regulation of reproductive function in males with that in females during the early to mid-follicular phase, drawing as many parallels as possible. Be sure to consider the actions of hormones as well as the mechanisms that regulate their secretion.

2. Compare the steps in spermatogenesis with those of oogenesis, drawing as many parallels as possible.

3. Describe all the events that must occur before a sperm deposited in the female reproductive tract can fertilize an oocyte released from an ovary.

4. Describe the events that occur during an LH surge, including the actions of LH on target cells. Be sure to include a description of the events that trigger the surge and prime the target tissue's responses to it.

5. Describe the processes that give rise to elevated rates of estrogen and progesterone secretion during pregnancy. Explain how these hormones promote the maintenance of pregnancy and prepare the body for the eventual birth and nourishment of the infant.

Find the answers to these exercises, and additional study tools,
at the Physiology Place (www.physiologyplace.com).

The Immune System

OBJECTIVES

- Name the five major leukocyte types and briefly describe their functions.

- Identify the lymphoid organs and briefly describe their functions.

- Explain events that occur during inflammation and how each contributes to nonspecific defense. Describe how the skin and mucous membranes, interferons, natural killer cells, and the complement system contribute to nonspecfic defense.

- Describe humoral immunity, or how B cells, through the production of antibodies, contribute to immune responses.

- Describe cell-mediated immunity, or how helper T cells and cytotoxic T cells contribute to immune responses.

- Explain how immunization can lead to protection from infectious disease.

- Discuss the major immunological issues regarding blood transfusion and organ transplantation.

- Explain how immune dysfunction can result in allergy, autoimmunity, or immunodeficiency.

CHAPTER OUTLINE

Anatomy of the Immune System   709

Organization of the Body's Defenses   713

Humoral Immunity   721

Cell-Mediated Immunity   724

Immune Responses in Health and Disease   728

Above: Scanning electron micrograph of an alveolar macrophage attacking E-coli bacteria

STUDY HINTS

1. *Phagocytosis, p. 47*

2. *Endocytosis and exocytosis, p. 46*

3. *Composition of blood, p. 404*

4. *Lymphatic system, p. 438*

Ming is a bright first-year college student who is a champion off-road bicyclist. In excellent health, Ming is surprised and dismayed when she develops a sore throat and a fever and feels so tired that she can't gather enough energy to so much as look at a textbook—or even at her bike! Her physician at the campus health center tells her she has a viral disease called *mononucleosis,* also known as "mono." She learns that there is no specific treatment for mono, but that in time she will feel better thanks to her **immune system**—the organs, tissues, circulating cells, and secreted molecules that resist and defeat infections. Like all of us, Ming understands the power of this system: When she had an infected cut, it healed; when she was ill with the flu, she got better; and when she was immunized against the bacterial infection diptheria, she became protected from that disease for life.

In the preceding chapters we have learned a great deal about how physiological systems function to maintain health. In this chapter we see how the body remains healthy even as disease-causing organisms and nonliving substances gain access to it via air, food, and water. **Immunity** refers to the immune system's capacity to protect individuals from disease by recognizing and eliminating potentially *pathogenic* (disease-generating) agents, including bacteria, bacterial toxins, viruses, parasites, and fungi. In addition to resisting foreign agents, the immune system disposes of unneeded components of the body, including aging cells and cellular debris present in diseased tissue; participates in healing wounds; and sometimes recognizes and eliminates mutant cells that may develop into cancer. As it seeks out and recognizes materials foreign to the body, the immune system also rejects tissues and cells that are not identical to "self" (that is, to the cells and tissues of the individual in question); such rejection poses the primary obstacle to organ transplantation. The presence in the body of such foreign and abnormal substances induces the immune system to develop an **immune response,** a complex series of physiological events that culminates in the destruction and elimination of these substances. Sometimes, however, immune responses are inappropriate and lead to disease. People with allergies, for example, experience exaggerated immune responses to otherwise benign foreign materials such as dust, pollen, or peanuts. In addition, some people suffer from *autoimmune diseases* in which a person's immune system attacks his or her own tissues and cells. Rheumatoid arthritis, diabetes mellitus, and multiple sclerosis are just a few examples of the many diseases that can result from autoimmune responses.

ANATOMY OF THE IMMUNE SYSTEM

The immune system consists of two components: leukocytes (white blood cells), which are responsible for producing a wide range of immune responses, and **lymphoid tissues,** such as the bone marrow, thymus, spleen, lymph nodes, and tonsils, in which leukocytes develop, reside, and come into contact with foreign materials. We discuss each of these components in detail in the following sections.

Leukocytes

There are five major types of leukocytes, each with a particular role in immunity (Table 22.1). Three of these cell types—neutrophils, eosinophils, and basophils—are called *granulocytes* because they have prominent protein-containing vesicles in their cytoplasm known as *cytoplasmic granules.* The others—monocytes and lymphocytes—do not have prominent granules and thus are sometimes called *agranulocytes.* Each leukocyte type is traditionally identified by its reaction with Wright's stain, which reveals its distinctive nuclear shape and cytoplasmic color.

Neutrophils

Neutrophils constitute about 60–80% of all leukocytes and are capable of one of the most important defense activities in the body: phagocytosis. As a phagocyte ("eating cell"), a neutrophil engulfs and digests microorganisms, abnormal cells, and foreign particles present in blood and tissues. Only a few other cell types (eosinophils, monocytes, and macrophages) are capable of phagocytosis. Newly produced neutrophils circulate in the blood for 7–10 hours and then migrate into the tissues, where they live for only a few days. During an infection the number of circulating neutrophils increases dramatically, providing not only assistance in defense but also a clinical indication of an ongoing infection.

TABLE 22.1	**LEUKOCYTES AND THEIR ROLES IN IMMUNITY★**

LEUKOCYTE MORPHOLOGY	FUNCTIONS IN IMMUNITY
Neutrophil	Engulf microorganisms, abnormal cells, and foreign particles by phagocytosis
Eosinophil	Secrete enzymes that kill parasites; contribute to tissue damage in allergic reactions
Basophil	Secrete chemical mediators of inflammation and allergic reactions
Monocyte ↓ Macrophage	Secrete cytokines; engulf microorganisms by phagocytosis
Lymphocyte	Plasma cells (mature form of B cells) secrete antibodies Helper T cells secrete cytokines that activate multiple cell types; cytotoxic T cells secrete factors that lead to the death of infected cells and tumor cells Null cells called natural killer cells secrete factors that lead to the death of infected cells and tumor cells

*Morphological features shown are characteristic of preparation with Wright's stain.

Other situations lead to increased numbers of circulating neutrophils as well. By the end of the marathon, Bill and Jane's circulating neutrophil counts increased over five-fold. This is because many of the neutrophils released from the bone marrow do not freely circulate, but rather reside in the vascular compartment in a stationary manner, loosely adhering to blood vessel walls via electrostatic forces. These cells congregate especially in large veins where blood flow is sluggish. Several elements of exercise liberate these cells so that they freely circulate. First, the increased blood flow (shear stress) sweeps the cells away from the vessel walls; second, epinephrine reduces the adherence forces; and third, cortisol promotes increased release of neutrophils from the bone marrow. These increases subside within 12 to 24 hours after the race and there is no evidence that such transient changes in circulating cell numbers have any important influence on overall immune function.

Eosinophils

About 1.5% of all leukocytes are **eosinophils.** Like neutrophils, eosinophils are phagocytic, but their main contribution in defense is in attacking parasitic invaders that are too large to be engulfed. Eosinophils mount an attack against these parasites by attaching to their bodies and discharging toxic molecules from their cytoplasmic granules. Unfortunately, eosinophil defense—and the overall immune response to such parasites—is weak. Eosinophil responses can sometimes be harmful because the toxic molecules they release can also damage normal tissues and may trigger allergic reactions.

Basophils

Basophils are nonphagocytic cells that are thought to defend against larger parasites; they may operate much like eosinophils by releasing toxic molecules that damage invaders. However, basophils also release histamine, heparin, and other chemicals that contribute significantly to allergic reactions such as hay fever. Although basophils constitute less than 1% of all leukocytes, they have a significant effect in people with allergies.

Monocytes

Monocytes, which make up about 5% of the leukocytes, are important in phagocytic defense. New monocytes circulate in the blood for only a few hours before they migrate into tissues, where they become 5–10 times larger and develop into very active phagocytic cells known as macrophages ("big eaters").

Some macrophages, called "wandering macrophages," migrate throughout body tissues, whereas others, known as "fixed macrophages," remain at particular sites. Macrophages are especially abundant in connective tissue, in the wall of the gastrointestinal tract, in the alveoli of the lung, and in the walls of certain blood vessels in the liver (where they are known as Kupffer cells) and in the spleen, two sites in which they phagocytose abnormal, dead, and dying erythrocytes. Macrophages are also prevalent in regions of the spleen and lymph nodes where they are most likely to contact infectious agents circulating in the blood and lymphatic fluid.

Lymphocytes

Lymphocytes constitute about 30% of all leukocytes in the blood, and about 99% of all cells found in interstitial fluid. Lymphocytes are of three major types: **B lymphocytes** (B cells), **T lymphocytes** (T cells), and **null cells,** so called because they lack cell membrane components that are characteristic of B cells and T cells. Most null cells are large, granular lymphocytes known as **natural killer (NK) cells** (described shortly).

When B cells contact foreign or abnormal molecules known as **antigens,** they develop into *plasma cells,* which secrete antibodies. **Antibodies,** also known as *immunoglobulins,* are proteins present in the plasma and interstitial fluid that target specific antigens for destruction. If, for example, a B cell contacts the bacterium *Staphylococcus aureus,* it develops into a plasma cell that secretes antibodies that bind only to *S. aureus.* These antibodies do not damage the bacteria by themselves but instead mark them for destruction by various mechanisms that are described later in the chapter.

T cells, in contrast, do direct damage to foreign cells. T cells contact infected cells, mutant cells, and transplanted cells and then take several days to develop into active *cytotoxic T cells* that destroy the infected or abnormal cells. T cells kill by secreting molecules that form pores in the target cell's membrane; once it has a perforated membrane, the target cell succumbs to lysis, a process in which it fills with fluid and bursts. The roles of B lymphocytes and T lymphocytes in the body's immune response are discussed in greater detail later in the chapter.

Even though natural killer cells constitute only a small proportion of circulating lymphocytes, they are very important in fighting viral infections. Viruses, unlike bacteria, must enter cells to reproduce, and by killing virus-infected cells NK cells limit the production of new viruses in the body. Whereas NK cells kill by a mechanism similar to that used by cytotoxic T cells, they differ from cytotoxic T cells in the manner in which they recognize their target cells. NK cells also differ from cytotoxic T cells in that they exhibit immediate readiness, enabling them to respond well before B cells and T cells and making them an essential part of early immune responses. NK cells also recognize and kill mutant cells that may develop into cancer.

Lymphoid Tissues

All leukocytes (and erythrocytes as well) develop from precursor cells called **hematopoietic** (blood-forming) **stem cells,** located in the bone marrow. Whereas B lymphocytes and most other leukocyte types come to full maturity in the bone marrow, T lymphocytes must migrate to the thymus gland (located in the thoracic cavity above the heart) before they develop to maturity. Because the bone marrow and the thymus (and the fetal liver as well) are the sites of lymphocyte maturation, they are called **central lymphoid tissues** (Figure 22.1). After B cells and T cells reach maturity, they migrate from central lymphoid tissues to the various sites in the body where they are most likely to contact foreign substances. These sites, known as **peripheral lymphoid tissues,** include the spleen, lymph nodes, tonsils, adenoids, appendix, and regions in the lining of the gastrointestinal tract called *Peyer's patches,* which are essentially collections of B cells, T cells, and macrophages.

Each of the peripheral lymphoid tissues contains a dense network of cells that trap microorganisms and foreign particles, and each is located where they can ensnare invaders soon after they enter the body. While the spleen is collecting worn out erythrocytes from the blood, it also is collecting blood-borne microorganisms and foreign particles. Once trapped in the spleen, these particles are eventually cleared by the actions of macrophages and lymphocytes. Similarly, microorganisms and particles carried in lymph are trapped by lymph nodes, which are found throughout the body where lymphatic vessels converge. Whereas the macrophage and lymphocyte networks of the spleen and lymph nodes filter blood and lymph, respectively, those of the tonsils and adenoids trap inhaled particles and microorganisms, and those of the appendix and Peyer's patches trap substances that enter the body in ingested food or water.

Quick Test 22.1

1. List the five major types of leukocytes, and describe one function of each.

2. Define phagocytosis. What types of cells are capable of phagocytosis?

3. Name the three major types of lymphocytes. Which type produces antibodies? Which types kill virus-infected cells, tumor cells, and transplanted cells?

4. Name the two central lymphoid tissues. Why are they known as *central* lymphoid tissues?

5. List six peripheral lymphoid tissues. What general role do these tissues play in immunity?

FIGURE 22.1 Lymphoid tissues.

Lymphocytes reach maturity in the central lymphoid tissues and interact with foreign antigen in the peripheral lymphoid tissues. These lymphoid tissues are interconnected by blood vessels (not shown) and lymphatic vessels through which lymphocytes circulate.

A few days after infection with chicken pox a person experiences viremia, a high level of viruses in the blood. In which lymphoid tissue are these viruses most likely to contact lymphocytes?

Central Lymphoid Tissues   **Peripheral Lymphoid Tissues**

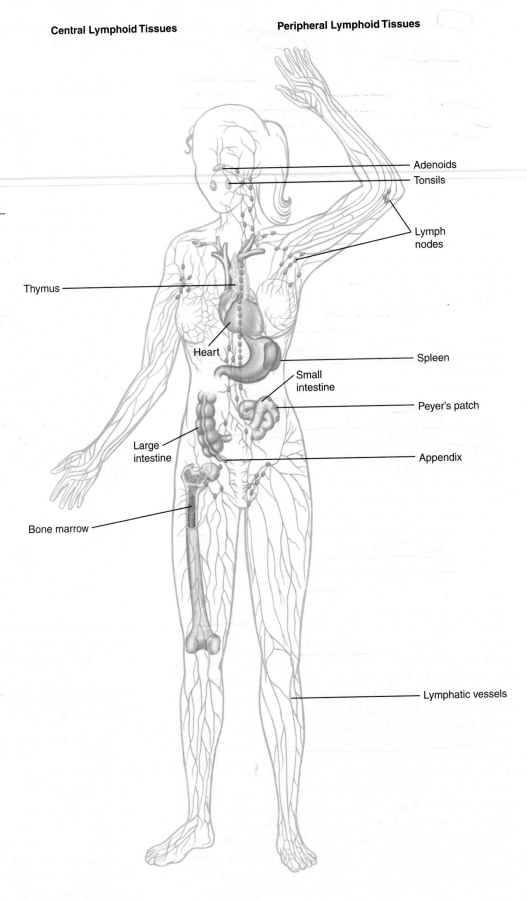

- Adenoids
- Tonsils
- Lymph nodes
- Thymus
- Heart
- Spleen
- Small intestine
- Peyer's patch
- Large intestine
- Appendix
- Bone marrow
- Lymphatic vessels

The spleen

ORGANIZATION OF THE BODY'S DEFENSES

The body's defenses are categorized as either nonspecific or specific, depending on their selectivity for the invader and when the response occurs. The mechanisms that constitute **nonspecific defenses** defend against potentially harmful substances without regard to their precise identity. These mechanisms operate even before foreign material enters the body, in the form of the skin and mucous membranes—the body's first line of defense against infection. If those barriers are broken, internal nonspecific defenses immediately begin to operate. Nonspecific mechanisms also clear wounds and damaged tissue of debris and contribute to healing. **Specific immune responses,** in contrast, are highly selective (meaning they target specific substances) and come into play after nonspecific responses have already begun. These responses are specific because they are mediated by lymphocytes, which are uniquely designed to recognize particular substances and aid in their destruction. Unlike nonspecific responses, which always operate with about the same speed and effectiveness, specific responses strengthen with each exposure to a particular offending agent.

Nonspecific Defenses

Intact skin and the mucous membranes that line the digestive, respiratory, urinary, and reproductive tracts provide excellent initial barriers to most bacteria and viruses. Mucous membranes also produce viscous mucus, which bathes the surfaces of exposed epithelia and can trap foreign matter and potential pathogens (disease-causing agents). Microorganisms that enter the upper respiratory tract, for example, are often caught in mucus and are then coughed into the mouth and swallowed, an action that is enhanced by ciliated epithelial cells that line the trachea. In addition to acting as physical barriers, the skin and mucous membranes provide chemical defenses: Secretions from sebaceous glands and sweat glands give the skin a pH ranging from 3 to 5, which is acidic enough to prevent colonization by many pathogens. (Bacteria that are normally found on the skin are adapted to its acidic, relatively dry environment.)

As soon as a potential pathogen or foreign matter enters the body, internal nonspecific defenses are quickly initiated, even if the body has not been previously exposed to that pathogen or material. The body's internal nonspecific defenses include (1) **inflammation,** a complex series of events that culminates in the accumulation of proteins, fluid, and phagocytic cells in an area of tissue that has been injured or invaded by microorganisms; (2) **interferons,** a family of related proteins that are secreted by leukocytes and virus-infected cells and can induce other cells to resist infection by the virus; (3) natural killer (NK) cells, which provide early defenses against virus-infected cells and cancer cells by recognizing and destroying them; and (4) the **complement system,** a set of plasma proteins that, once activated by a series of stepwise reactions, act to lyse foreign cells, especially bacteria. We consider each of these internal nonspecific defense mechanisms next.

Inflammation

Microbial invasion or damage to tissue triggers a complex series of events that rapidly lead to inflammation of the affected tissue. In fact, the tissue space, rather than the bloodstream, is the main site of defensive action against infection. Five major events occur in inflammation, typically in the following order: (1) nearby macrophages engulf debris and foreign matter, (2) nearby capillaries dilate and become more permeable to proteins and fluid, (3) foreign matter is contained, (4) additional leukocytes migrate into the region, and (5) recruited leukocytes continue to help clear the infection, mainly by phagocytosis. Note that these events are nonspecific; inflammation proceeds in much the same way, regardless of which bacteria, virus, or injury triggered it. In the following subsections we examine the major events in inflammation by considering the body's response to an injury in which the skin is cut and abraded.

Phagocytosis of Pathogens and Debris by Nearby Macrophages
Macrophages already present in affected tissues quickly detect bacteria introduced into the cut using receptor proteins on their surfaces that can bind to molecules on the surfaces of many different types of bacterial cells. The resulting attachment initiates phagocytosis (Figure 22.2a), and the macrophages begin to engulf the bacteria. Attaching to bacteria in this way also stimulates the macrophages to secrete cytokines, proteins that are secreted by cells in response to a stimulus (in this case, contact with bacteria) and affect the behavior of other nearby cells (see Chapter 5). As we will see, the cytokines secreted by such activated macrophages contribute to subsequent steps in inflammation (and to immune responses as well). The phagocytic activity of these macrophages is very important in limiting the spread of bacteria early in infection, but these cells are usually too few in number to eliminate all the foreign cells and debris present at an injury site.

Dilation and Increased Permeability of Capillaries
Within minutes of bacterial invasion, nearby blood vessels dilate (Figure 22.2b), which increases local blood flow. The capillary walls also become more permeable, allowing proteins and fluid normally contained in the plasma to move into tissue spaces. The increased blood flow brings additional leukocytes and defensive proteins into the local circulation, and the increased capillary permeability allows these proteins to move into the tissues where they are needed. In addition, the leukocytes that gather in these dilated vessels also migrate from the blood into the tissue spaces by a process we examine shortly.

Drugs known as antihistamines block the action of histamine. Explain how antihistamines diminish the symptoms of inflammation.

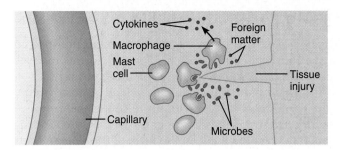

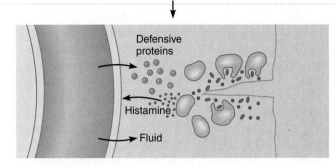

(a) Phagocytosis by nearby macrophages

Both the *vasodilation* and the increased capillary permeability are induced by *histamine* released from damaged **mast cells,** which are cells that are dispersed throughout the body's connective tissues and are similar to the basophils in the blood. Because mast cells are especially numerous in submucosal tissue and the dermis (a tissue layer in the skin), they are prone to damage when the skin is injured and are thus poised to stimulate the early events of inflammation by releasing histamine.

The vascular changes mediated by histamine are ultimately responsible for the four characteristic symptoms of inflammation (L. *inflammo,* "to set on fire"): redness, heat, edema (swelling), and pain. The increased blood flow both reddens the tissue (which is more or less apparent depending on skin tone) and makes it warmer. Both the histamine-induced increase in blood flow and capillary permeability contribute to the edema that follows. As the capillaries become engorged with blood, the resulting increase in hydrostatic pressure plus the increased interstitial osmotic pressure that accompanies the leakage of plasma proteins causes fluid to filter out of the capillaries and into the tissue spaces, resulting in edema. The edema exerts pressure against the surrounding tissue and skin, which contributes to pain, as does the production of pain-inducing chemicals such as *bradykinin* that stimulate nearby sensory neurons. Thus even though these vascular changes bring some discomfort, they also help to gather nonspecific defenses to the site of injury.

Containment of Foreign Matter Early in the process mast cells and basophils also release the anticoagulant heparin, which temporarily suspends blood clotting and allows leukocytes unimpeded access to the area of tissue injury. In time, however, clotting factors that have leaked from the plasma into the tissue become active and form clots around clusters of bacteria, thereby inhibiting their spread within the body (Figure 22.2c). With the aid of other plasma proteins and also proteins released from damaged tissues, the process of clot formation continues and effectively walls off the region of damage and infection. Eventually the portion of the clot that is at the skin's surface dries and hardens, forming a scab.

Leukocyte Migration and Proliferation As previously mentioned, macrophages already present in the damaged tissue are the first to phagocytose bacteria and debris, but

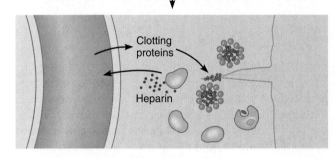

(b) Dilation and increased permeability of capillary

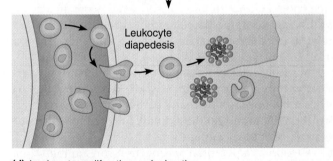

(c) Containment of bacteria and foreign matter

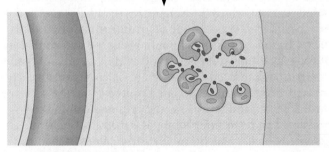

(d) Leukocyte proliferation and migration

Inhibition of capillary dilation reduces redness and heat; inhibition of increased capillary permeability reduces edema.

(e) Continued activity of recruited leukocytes

soon they are joined by other phagocytic cells. About an hour after the injury, neutrophils accumulate in great numbers within the affected tissue; about ten hours later, monocytes begin to move to the tissue, where they develop into large, active macrophages.

The signalling that tells leukocytes to move through blood vessel walls at the right location in the body is achieved by a process of regulated transit involving four events: margination, attachment, diapedesis, and chemotaxis (Figure 22.2d). **Margination,** the movement of leukocytes toward the blood vessel wall, begins when cytokines released from macrophages and mast cells at the site of injury induce nearby blood vessels to produce *adhesion molecules* called *selectins*. Selectins protrude from the vessel's interior wall and attach loosely to leukocytes as they pass by, stalling their forward progress. This gives leukocytes the chance to receive activating signals from other cytokines, which, in turn, signal them to form other adhesion molecules called *integrins* that bind the cells tightly to the blood vessel wall. **Attachment** is soon followed by the cell's transit across the wall, known as **diapedesis** ("jump across"). In this step the leukocyte essentially "crawls" between endothelial cells of the blood vessel wall and through the basement membrane beneath. Once in the tissues, leukocytes move steadily and directly toward the point of injury, attracted by chemicals released from bacteria and injured tissues themselves, in a process called **chemotaxis**. Phagocytic leukocytes are thus delivered to the place where they are most needed: the original site of injury and bacterial invasion. Once at the site, phagocytes contact and engulf additional bacteria and debris. The process whereby the regulated transit of leukocytes occurs is depicted in Figure 22.3.

The hereditary disease called *leukocyte adhesion deficiency* (LAD) clearly demonstrates that the migration of phagocytic leukocytes from blood to infected tissues is an essential component of defense. In people with LAD, leukocytes possess faulty integrins, so the cells are unable to bind tightly to the blood vessel wall and therefore cannot cross into the tissues. Without the aid of additional phagocytes, tissue-borne infections are liable to fester and spread, and for these reasons people with this disease suffer from frequent, severe bacterial infections.

Note that the mass movement of neutrophils and monocytes from blood to tissues during inflammation does not diminish the number of circulating leukocytes. In fact, a common sign of bacterial infection is *leukocytosis,* a four- to five-fold increase in the number of circulating neutrophils. The reason for this increase is that the cytokines secreted by macrophages eventually reach the bone marrow, where they stimulate the proliferation and release of neutrophils and, later, monocytes into the circulation.

Continued Activities of Recruited Leukocytes The inflammatory processes described so far have brought defensive proteins (such as clotting factors) and phagocytic

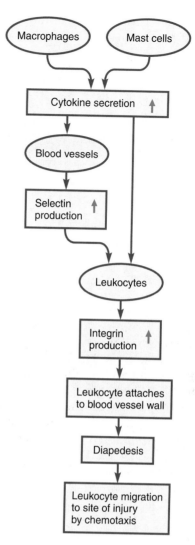

FIGURE 22.3 Events in the regulated transit of leukocytes from the blood stream to tissues.

leukocytes (neutrophils and macrophages) to the site of injury. We have seen that phagocytes perform two important functions: They engulf bacteria and debris, and they secrete cytokines that further regulate inflammation and other defensive mechanisms. Let's take a closer look at these two important tasks.

Phagocytosis is achieved in four steps: attachment, internalization, degradation, and exocytosis (Figure 22.4). The first step, *attachment,* enables the phagocyte to distinguish between substances that should be engulfed and those that should not. The selectivity of a phagocyte for its target material is mainly determined by how well the two adhere. Phagocytes tend to bind to damaged and dead cells with irregular, rough surfaces, but not to healthy cells. Phagocytes also attach to many types of bacteria. Sometimes, attachment is enhanced by **opsonins,** proteins (including antibodies) that bind tightly to the foreign material and make it easier for the phagocyte to engulf it. Phagocytosis is greatly enhanced by this

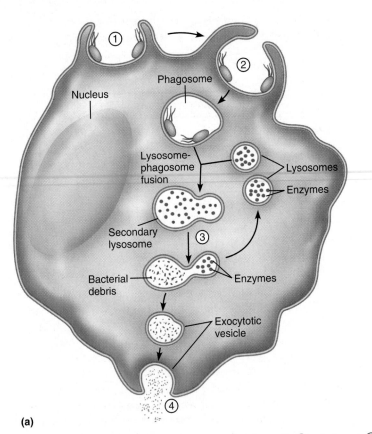

(a)

Nucleus

Phagosome

Lysosome-phagosome fusion

Lysosomes

Enzymes

Secondary lysosome

Bacterial debris

Enzymes

Exocytotic vesicle

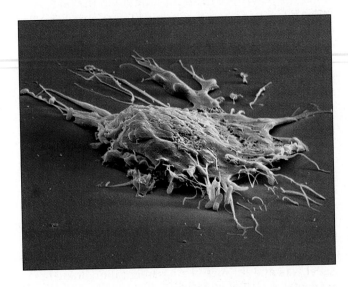

(b)

FIGURE 22.4 Phagocytosis. (a) *The steps of phagocytosis:* ① *attachment,* ② *internalization,* ③ *degradation, and* ④ *exocytosis.* **(b)** *Electron photomicrograph of a macrophage phagocytosing bacteria.*

binding of opsonins, a process known as *opsonization* ("to make tasty"); a macrophage can engulf and destroy a substance about 4000 times faster when it is coated with antibodies.

As soon as the phagocyte has attached to its prey, *internalization* occurs. In less than 0.01 second, the phagocyte's plasma membrane extends outward around the site of attachment and surrounds the material, enclosing it within a large intracellular vesicle called a *phagosome*. The phagosome moves toward the cell interior and fuses with a *lysosome,* which contains a variety of digestive enzymes. *Degradation* of the phagocytosed material then occurs within the enlarged, enzymatically active organelle, now called a *secondary lysosome*. Subsequently the phagocyte uses some of the harmless bacterial debris (amino acids, for example) and eliminates others by *exocytosis.*

Attachment of a phagocyte to foreign cells and debris not only initiates phagocytosis, but also induces the phagocyte to accomplish its other major task: the secretion of cytokines that help prolong and coordinate the body's responses to infection and injury. These cytokines are varied, and exert a wide range of effects on the body. Some of them are called *interleukins* because they act as

chemical messengers that send signals between (*inter-*) leukocytes (*-leukins*). We focus in particular on three cytokines secreted by activated macrophages: *interleukin-1 (IL-1), interleukin-6 (IL-6),* and *tumor necrosis factor-α (TNF-α)*. These cytokines act individually and collectively to induce a number of changes, including inducing blood vessel endothelial cells to synthesize more adhesion molecules and stimulating bone marrow to release neutrophils in greater numbers.

In addition to these actions, IL-1, IL-6, and TNF-α also act on the hypothalamus to raise body temperature; that is, they function as *endogenous pyrogens*. They achieve this function by stimulating the hypothalamus to release prostaglandins, which in turn adjust the body's "thermostat" setting, raising the temperature above normal. Although a high fever can be dangerous, a moderate one is generally thought to benefit host defense; elevated temperatures are thought to decrease the rates of bacterial and viral replication, and to accelerate phagocytosis and other defensive reactions.

The cytokines IL-1, IL-6, and TNF-α also stimulate liver cells to produce *acute phase proteins*, a group of proteins having a wide range of antibacterial and inflammatory effects and whose plasma concentrations increase soon after an infection begins. One of these is *C-reactive protein,* which binds to the surface of many types of bacteria. A bacterium bound by C-reactive protein molecules is opsonized, and therefore is more susceptible to phagocytosis.

Finally IL-1 helps induce the proliferation and differentiation of B lymphocytes and T lymphocytes. B cell differentiation leads to the production of antibodies that mark selected foreign substances for destruction, and T cell differentiation leads to *cell-mediated immunity,* in which certain types of T cells kill specific abnormal or infected body cells.

Interferons

A second nonspecific defense mechanism prevents the spread of viruses within the body by interfering with virus replication. This defense is provided by a group of related proteins appropriately named interferons. The secretion of two interferons, called *interferon-α* and *interferon-β,* from virus-infected cells signals to the surrounding cells the presence of the virus and induces the cells' resistance.

The viral nucleic acid that accumulates in a virus-infected cell during viral replication stimulates the cell to synthesize and secrete interferons. Even though the infected cell will likely die, the interferons it secretes bind to nearby healthy cells, initiating in them a series of intracellular changes that cause the cells to become more resistant to the virus. This virus-resistant state is conferred by the presence of RNA-degrading enzymes and protein synthesis inhibitors in the cytoplasm. Because these enzymes and inhibitors are potentially dangerous to the cell itself, they come into play only after viral infection and last only a short time. Interferon-induced cells are thus poised to block the production of new viruses, but they can do so for a limited time only. Both the production of interferon and the virus-resistant state it induces are nonspecific; virtually any viral nucleic acid can induce interferon production, and interferon-induced resistance can defeat virtually any viral infection.

A third type of interferon, *interferon-γ,* is secreted not by virus-infected cells, but by active T cells and NK cells, and it has a broader range of effects. In addition to inhibiting viral replication, interferon-γ enhances phagocytosis in macrophages, boosts antibody production in B cells, and helps to activate natural killer cells and cytotoxic T cells, both of which kill virus-infected cells and cancer cells. In addition, by inhibiting cell division, interferon-γ has a direct effect on cancer cells, suppressing the growth of tumors.

Natural Killer Cells

We have already encountered the third nonspecific defense mechanism: natural killer (NK) cells. NK cells function in early nonspecific defense by recognizing the general features of infected or abnormal cells and by causing their deaths via lysis. In the case of virus-infected cells, NK cells can act on them without detecting the virus itself, and NK cells are ready to act immediately. NK cells thus provide an essential, general defense in the early stages of viral infection, and of tumor growth as well. These cells are also brought into play during immune responses because their activities can be enhanced by the cytokine *interleukin-2 (IL-2)* and by antibodies, which are produced by T cells and plasma cells, respectively, during specific responses.

The Complement System

The fourth and final nonspecific defense mechanism, the complement system, is so-named because it completes or fulfills ("complements") the actions of specific antibodies. However, the system can also act in the absence of antibodies. The system consists of about 30 plasma proteins that act to destroy invading microorganisms, especially bacteria. The first in a series of complement protein reactions occurs in association with a bacterium and leads to a cascade of activation steps in which each component activates the next in the series (Figure 22.5). The cascade occurs at the bacterial surface and ends with the development of a *membrane attack complex* (MAC), a collection of pore-forming proteins that pierce the bacterial membrane. The insertion of these channels into the membrane causes the cell to lose its integrity, so it fills with fluid, swells, and then bursts (Figure 22.5). Such complement-mediated lysis is the primary way by which antibody-coated bacteria are killed.

The complement cascade can be activated in two ways: (1) by binding directly to carbohydrates present on the surface of a broad range of bacterial cells (known as the *alternative pathway*), or (2) by binding to antibodies that are already attached to bacterial cells (known as the *classical pathway*). The alternative pathway is important in nonspecific, early responses to infection because it occurs in the absence of antibodies (that is, before they are secreted in response to an infection). The classical pathway, by contrast, becomes important only in the later stages of an infection because it requires antibodies. Note that whereas the alternative pathway is nonspecific, the classical pathway is specific because it involves antibodies. This is another example in which nonspecific and specific mechanisms converge; here, complement is essential to effective antibody action because antibodies can target a microorganism for destruction but cannot directly kill it.

Although the complement cascade is often viewed in terms of its final outcome—lysis of bacteria—a number of activated complement proteins also contribute to the development of inflammation. Because complement proteins are activated in the vicinity of the bacteria and accumulate there, they are appropriately positioned to assist in antibacterial defense. Indeed, some complement proteins act in chemotaxis, guiding phagocytes into the area;

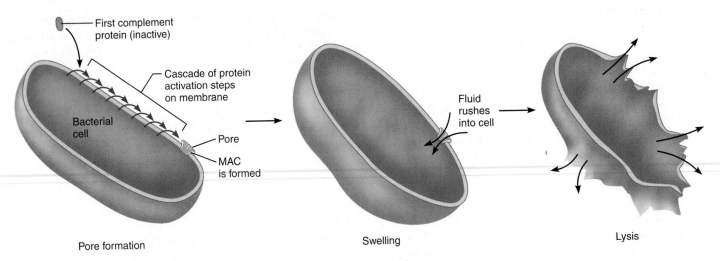

FIGURE 22.5 Actions of the complement system. *The complement system can be triggered when the first in a series of complement proteins binds to the surface of a bacterial cell. This mode of complement activation is nonspecific because it occurs on any of a broad range of bacterial types. After a series of activation steps, some of the complement proteins form a membrane attack complex (MAC), which inserts into the cell membrane as a protein-lined channel or pore. Fluid rushes into the cell, causing it to swell and burst (undergo lysis).*

others bind to nearby mast cells and induce them to release histamine, which causes vasodilation and increased capillary permeability. Finally, one specific protein of the complement system, known as C3b, is generated in great quantity in the cascade and coats bacterial surfaces; because it acts as an opsonin, it enhances phagocytosis of these bacteria.

Quick Test 22.2

1. Describe two ways in which nonspecific defenses differ from specific immune responses.

2. What are the characteristic symptoms of inflammation? What physiological events contribute to each?

3. List the four steps of phagocytosis. Which of these steps is improved by opsonins?

4. Name one nonspecific defense mechanism directed against viral infections, and one directed against bacterial infections.

Immune Responses

In the previous section we saw that the skin and mucous membranes constitute the body's first line of defense; that inflammation, interferons, NK cells, and the complement system form a second line of nonspecific defense; and that these mechanisms respond rapidly to injury or exposure to foreign material or infective agents, even if such exposure is the initial one. But nonspecific defenses are not always completely successful in eliminating foreign

materials; fortunately the body has an exquisite and powerful mechanism in reserve: the immune response.

Suppose that the body's nonspecific defenses were not completely effective in responding to the cut and abrasion we previously discussed. While these defenses continue to operate, bacteria, bacterial fragments, and other foreign molecules gain access to the bloodstream and become trapped in the netlike architecture of the spleen. Likewise, bacteria present in ISF are carried into lymphatic vessels and eventually into the netlike lymph nodes, which swell and become tender. In the spleen, lymph nodes, and other lymphoid tissues, the bacteria come into contact with B lymphocytes and T lymphocytes, thereby inducing these cells to generate efficient and selective immune responses that work throughout the body to eliminate the invaders.

B lymphocytes and T lymphocytes each generate a particular kind of immune response. B lymphocytes develop into plasma cells that secrete antibodies, the actions of which bring about **humoral immunity,** so called because antibodies circulate in the blood and lymph, body fluids long ago called "humors." In contrast, certain T lymphocytes develop into active cytotoxic T cells, which bind to and kill abnormal body cells. This type of immune response constitutes **cell-mediated immunity,** so called because cytotoxic T cells must come into direct contact with their targets in order to act on them.

The circulating antibodies of the humoral response defend mainly against bacteria, toxins, and viruses present in body fluids. In contrast, the T cells of the cell-mediated response are active against bacteria and viruses that are hidden within infected body cells. Moreover, the cell-mediated response operates in the body's reaction to transplanted tissue and cancer cells, both of which are perceived as foreign.

Features of Immune Responses

About 2400 years ago Thucydides of Athens described how those sick and dying of plague were attended by others who had recovered, "for no one was ever attacked a second time." We have all made similar observations: Those of us who experienced many childhood diseases do not worry about getting them again because exposure to each disease confers lifelong immunity to that disease. In this section we explore the nature of specific immune responses by addressing the following questions: Why does a measles infection generate a response to that disease but to no other disease (a phenomenon known as *specificity*)? How is it that the immune system is already poised to defend against a first exposure to chicken pox (which relates to a property known as *diversity*)? How do we acquire our lifelong immunity to chicken pox after the first exposure (which depends on the immune system's *memory*)? Finally, why does the immune system respond against foreign microbes but not against the body's own tissues (a phenomenon known as *self-tolerance*)? As we will see, these special features of the immune response—specificity, diversity, memory, and self-tolerance—derive from the nature of B lymphocytes and T lymphocytes.

Specificity B cells and T cells bind and respond to foreign or abnormal molecules known as *antigens*. Antigens (a term arising from *anti*body *gen*erators) are typically complex protein or polysaccharide components of viruses, bacteria, fungi, protozoa, parasitic worms, pollen, transplanted tissue, and tumor cells. Each antigen has a unique structure and contains different recognition sites called **epitopes** or *antigenic determinants,* each of which can be detected by specific lymphocytes, which then target that invader for destruction.

B cells and T cells are able to recognize specific antigens because they possess antigen-binding proteins called **antigen receptors** on their surfaces. All of the receptors on a particular B cell or T cell bind to certain antigens only and thus resemble other receptors that bind specific ligands, such as neurotransmitter receptors and hormone receptors.

The antigen receptors on B cells are similar to antibody molecules except that the receptors are bound to the plasma membrane, whereas antibodies are secreted into the extracellular fluid. For this reason, B cell antigen receptors are often called *membrane antibodies* or *membrane immunoglobulins*. A B cell's membrane antibodies have the same specificity as the antibodies it later secretes as a plasma cell.

A typical antibody is a Y-shaped molecule consisting of four protein chains: two identical **heavy chains** and two identical **light chains** that are joined by disulfide bridges (Figure 22.6a). Each antibody molecule possesses

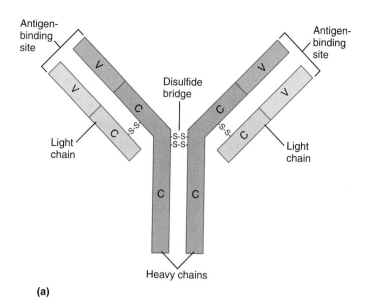

(a)

(b)

FIGURE 22.6 The basis of antigen-antibody specificity.
(a) *Each antibody molecule consists of two identical heavy chains and two identical light chains linked by disulfide bonds. The combined structures of the variable (V) regions form the antigen binding sites of the molecule, which vary from antibody to antibody.* (b) *A schematic representation of how the complementary physical structures of an antigen's epitope and the antibody's antigen-binding site constitute antigen-antibody specificity.*

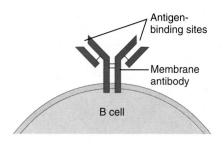

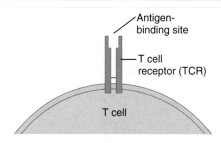

FIGURE 22.7 B cell and T cell antigen receptors. *Antigen receptors on B cells (left) are called membrane antibodies (or membrane immunoglobulins); antigen receptors on T cells (right) are called T cell receptors (TCRs). Each lymphocyte has about 100,000 identical receptors specific for a particular antigen.*

two identical antigen-binding sites and thus each antibody molecule is capable of binding two epitopes of the same kind. These antigen-binding sites are formed by the *variable regions* (V) of adjacent heavy and light chains, so called because the amino acid sequences in these regions vary extensively from antibody to antibody. The association of a heavy chain V region with a light chain V region forms the three-dimensional contours of the antigen-binding site. The interaction between the antigen-binding site and its antigen is similar to that between an enzyme and its substrate, which also shows specificity: The unique shape of the binding site allows for a close fit between the antibody and its antigen (Figure 22.6b), such that multiple noncovalent bonds can form between chemical groups on the respective molecules.

The antigen receptors on T cells, called *T cell receptors (TCR)* come from the same family of proteins that includes antibodies but are different in structure from antibodies (Figure 22.7). Moreover, unlike antibodies, T cell receptors act only as cell surface receptors for antigen. They are never secreted.

Diversity   A single T lymphocyte or B lymphocyte has about 100,000 antigen receptors, all with the same specificity. Therefore, each B cell and T cell can detect just a few of the millions of possible antigens that might gain entry to the body. The particular antigen receptor molecules a given lymphocyte produces are determined by random genetic events that occur early in the development of the lymphocyte. These early events, which are

unique to each lymphocyte and occur prior to any contact with foreign antigen, generate a phenomenal array of B lymphocytes and T lymphocytes in the body, each bearing antigen receptors of a particular specificity. With this diversity of lymphocytes, the immune system has the capacity to recognize and respond to millions of different antigens—even to antigens as yet unseen in the universe!

The specificity of B cell and T cell receptors explains the specificity of the B cell and T cell responses. When a particular microorganism invades the body, it interacts with and activates only those lymphocytes that have receptors specific for the antigens it possesses. When the virus that causes chicken pox invades the body, for example, only lymphocytes specific for chicken pox antigens are activated to proliferate (by successive cell divisions) and differentiate. In other words, the foreign antigen triggers an immune response against itself. This antigen-driven activation of lymphocytes is called **clonal selection,** and is absolutely necessary for immune responses (Figure 22.8). Lymphocyte differentiation gives rise to two populations (clones) of cells: **effector cells** (such as plasma cells), which are short-lived cells that combat the same antigen that stimulated their production, and **memory cells,** which are long-lived cells bearing membrane receptors specific for the same antigen. Each antigen, by binding to specific receptors, selectively activates a tiny fraction of cells from the body's diverse array of lymphocytes. This relatively small number of selected cells then gives rise to clones of thousands of cells, all specific for and dedicated to eliminating that particular microorganism.

Memory   The antigen-induced lymphocyte changes that occur when a person is first exposed to an antigen constitutes a **primary immune response.** In the primary response, antigen-selected B cells and T cells proliferate and differentiate into effector cells (antibody-producing plasma cells and cytotoxic T cells, respectively) about 10–17 days after exposure to the antigen. Often, a person becomes ill during this time, because it takes only a few days for most viruses or bacteria to cause symptoms. Eventually, however, the symptoms of illness diminish and disappear as antibodies and cytotoxic T cells help clear the offending agent from the body.

Once someone has suffered through an infection, he or she has likely become immune to further infection by that same microorganism. The basis of this so-called *acquired immunity* is that upon subsequent exposures to that same antigen, the response (called a **secondary immune response**) is much faster (only 2–7 days), greater in magnitude, and more prolonged than a primary response (Figure 22.9). The secondary response results from the existence of immunological memory, which is due to the fact that each exposure to an antigen gives rise not only to effector cells but also to clones of long-lived memory T cells and memory B cells. Upon subsequent exposure to the same antigen, memory cells are poised to

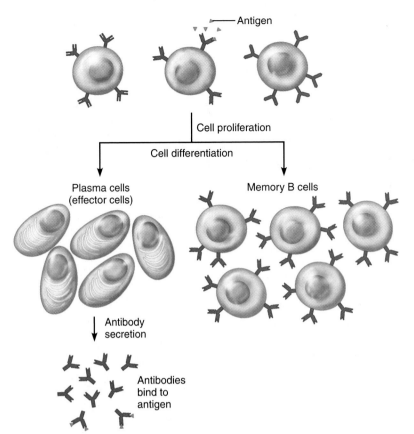

Antigen

Cell proliferation

Cell differentiation

Plasma cells
(effector cells)

Memory B cells

Antibody
secretion

Antibodies
bind to
antigen

FIGURE 22.8 Clonal selection in B lymphocytes. *Each lymphocyte produces a randomly generated set of identical antigen receptors. When a foreign antigen and a lymphocyte's receptors are sufficiently complementary in structure, binding induces the lymphocyte to proliferate and differentiate into a population (clone) of short-lived effector lymphocytes (in this case, plasma cells) and a clone of long-lived memory lymphocytes. Note that clonal selection also occurs in T lymphocytes.*

quickly proliferate and differentiate into still more memory and effector cells.

Self-Tolerance Given that the diverse repertoire of lymphocyte specificities is randomly generated, how is it that an individual's B cells and T cells do not react to the body's proteins? The answer is that as B cells and T cells mature in the bone marrow and thymus, their antigen receptors are in effect tested for their potential to recognize and react against self. In general, those lymphocytes whose receptors have the potential to react with self-molecules are either rendered nonfunctional or undergo **apoptosis,** the self-destructive events that accompany *programmed cell death.* As a result of these processes, only those lymphocytes that are reactive against foreign *(nonself)* molecules remain. This capacity to distinguish self from nonself, known as *self-tolerance,* continues to develop even as the lymphocytes migrate to lymphatic organs. Failure of self-tolerance can lead to various autoimmune diseases (discussed later).

Quick Test 22.3

1. What is humoral immunity? What lymphocyte type is responsible for this kind of immunity?

2. What is cell-mediated immunity? What lymphocyte type is responsible for this kind of immunity?

3. List the four key features of immune responses, and provide a brief explanation of each.

4. What is the basis for the stronger and more rapid immune response upon secondary exposure to an antigen?

HUMORAL IMMUNITY

Now that we've discussed the characteristics of immune responses, we examine in greater detail the first type of immunity: humoral immunity. We begin by considering the role of B lymphocytes.

The Role of B Lymphocytes in Antibody Production

Millions of B lymphocytes with different specificities normally circulate in the blood and lymph or reside in peripheral lymphoid tissues. When one of the B cells binds with an antigen, the cell responds in the two ways depicted in Figure 22.8. First, the B cell is stimulated to proliferate, which increases the number of B cells with the same specificity; second, the cells differentiate, such that some become long-lived memory B cells and others become short-lived antibody-synthesizing effector cells

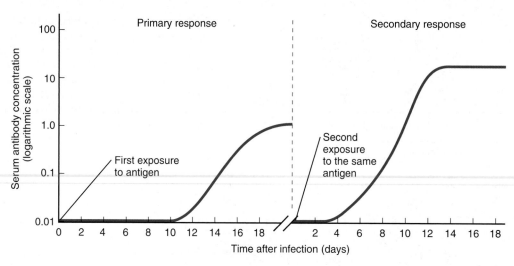

FIGURE 22.9 Primary and secondary responses to a given antigen. *In secondary responses, serum antibody concentrations rise sooner, reach a higher level, and remain elevated longer than those stimulated by the initial exposure to the antigen.*

If a person's initial exposure to antigen X coincided with a secondary exposure to antigen Y, what kind of response to antigen X would occur?

called *plasma cells*. A plasma cell secretes about 2000 antibody molecules per second over its life span of 4–7 days. These antibodies then circulate throughout the blood and lymph for several weeks, binding to the same antigens that stimulated their production, thereby marking them for destruction by phagocytosis or complement-mediated lysis.

The antigens that evoke the production of both plasma cells and memory B cells are known as *T-dependent antigens* because they do so only with help from a special kind of T cell called a **helper T cell.** We examine the precise functions of helper T cells shortly; we mention them here because they influence B cell activation. When helper T cells respond to specific antigens, they secrete many cytokines, including one called *interleukin-2 (IL-2)*. Together, IL-2 and T-dependent antigens (mainly proteins) induce the proliferation of B cells. By contrast, polysaccharide antigens, such as those found on many bacterial surfaces, can activate B cells without T cell help; these antigens are known as *T-independent antigens*. Because they contain long arrays of repeated subunits, these polysaccharides bind with many antigen receptors on the B cell surface, providing a strong enough stimulus to induce the B cell to proliferate even in the absence of IL-2. However, without IL-2, all of the proliferating cells differentiate into antibody-secreting plasma cells; none develop into memory B cells. Therefore, even repeated exposures to a particular T-independent antigen produce primary responses only, because the antigen alone never generates immunological memory. Thus, B cell responses to T-

independent antigens take 10–17 days to occur, and the quantity of antibodies produced is smaller than that produced in a secondary response. Specific events in B cell activation and the development of humoral immunity are discussed in the following sections.

Next we will see how antibodies bind to and mediate the disposal of microorganisms and foreign materials. Note that even though antibodies are generated in the immune response, they also enhance nonspecific defenses by binding to and thus marking specific microorganisms, thereby focusing these defenses on the invader at hand. A typical bacterium can be coated with as many as 4 million antibody molecules!

Antibody Function in Humoral Immunity

We have seen that an antibody has two functions: First it binds specifically to an antigen and then it aids in the inactivation or disposal of that antigen. Whereas the antigen-binding sites are responsible for the recognition of antigen, the tail of the Y-shaped antibody molecule is responsible for the mechanisms by which it mediates antigen disposal. The tail is made up of the constant regions (C) of the heavy chains (see Figure 22.6a). Because there are five kinds of heavy chains, there are five classes of antibodies: IgG, IgM, IgA, IgE, and IgD. (*Ig* stands for immunoglobulin.)

The way in which an antibody aids in antigen disposal depends on its class. The structures and functions of each of the five immunoglobulin classes are summarized in Table 22.2. All classes of immunoglobulins can mediate the simplest forms of antigen attack—neutralization and agglutination—whereas particular classes specialize in

A primary response

CLASS OF ANTIBODY	STRUCTURE	FEATURES	ROLES IN ANTIGEN DISPOSAL
IgM	J (joining) chain	The most common class of antibody produced in the primary response to antigen	Neutralizes antigen Agglutinates antigen; activates complement
IgD		Important as an antigen receptor on B cells	Neutralizes antigen Agglutinates antigen
IgG		The most common class of antibody in the blood, and the major class of antibody produced in secondary responses. Crosses the placenta, so is important in fetal and newborn immunity.	Neutralizes antigen Agglutinates antigen; activates complement; opsonizes antigen; enhances NK cell activity
IgE		Involved in allergies such as hay fever	Neutralizes antigen Agglutinates antigen Binds to mast cells and basophils, causing them to release histamine
IgA	J (joining) chain	Crosses epithelial cells, so is present on mucosal surfaces and in breast milk; is important in immunity in newborns	Neutralizes antigen Agglutinates antigen

opsonization of the antigen, activation of the complement system, and stimulation of natural killer cells. The functions of IgG are shown in Figure 22.10. In **neutralization** the antibody blocks an antigen's activity just by binding to it. For example, antibodies can neutralize a virus simply by attaching to the molecules that the virus must use to infect its host cell. Similarly, antibodies that coat bacterial toxins (such as the toxin produced by *Clostridium tetani,* which causes the disease *tetanus*) can render them inactive. Antigens are frequently neutralized and clumped together simultaneously by thousands of antibody molecules. This process, called **agglutination,** is possible because each antibody molecule has at least two antigen-binding sites. (Those belonging to the IgM and IgA classes have more than two). IgG, for example, can bind to equivalent epitopes on two separate pathogens, linking them together. With numerous IgG molecules likewise binding to two microbes apiece, a large complex forms.

Once bound to antibodies, an antigen is effectively opsonized and thus rendered more susceptible to phagocytosis. As previously mentioned, antibodies act as opsonins because they can bind to both antigens and phagocytic cells. IgG antibodies are specialized for opsonization because their tails bind to specific surface receptors on phagocytic cells. This binding triggers the phagocytes to engulf both the antibodies and their targeted prey. Likewise, both IgM and IgG antibodies can activate the complement system, which brings about the lysis of bacteria to which the antibodies are bound. Whereas the complement system can also be activated when certain complement proteins bind directly to some kinds of bacteria (the nonspecific, alternative pathway), the specific, classical pathway is effective against almost any bacterial cell that has been marked with antibodies. Regardless of the initiating event, activated complement proteins not only bring about the destruction of bacteria,

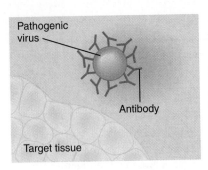

Neutralization

Antibodies block the activity of a pathogen.

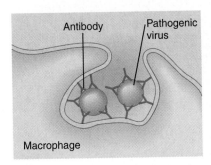

Agglutination

Multiple pathogens are aggregated by antibody molecules.

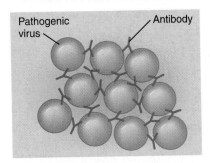

Opsonization

Pathogens bound by antibodies are more efficiently engulfed by phagocytes.

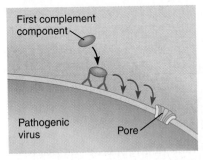

Complement activation

Antibodies bound to pathogens activate the complement cascade, resulting in lysis of the cell.

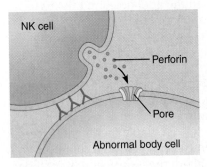

Enhanced NK cell activity

Abnormal body cells that are bound by antibodies are recognized by NK cells and are subsequently lysed.

FIGURE 22.10 Antibody-mediated mechanisms of antigen disposal.

but many of these proteins also help to advance and regulate inflammation, another essential nonspecific defense.

Finally, IgG antibodies can enhance the nonspecific killing action of natural killer cells. Recall that NK cells broadly recognize abnormal features of tumor cells and virus-infected cells and then produce membrane-perforating molecules that lead to the lysis of these cells. Often these abnormal body cells also possess abnormal surface molecules that can stimulate the production of antibodies. The antibodies mark the cells for death—in this case, death by NK cells. Like phagocytes, NK cells have surface receptors that bind with the constant regions of IgG. Antibodies thus provide a link between a specifically targeted, abnormal cell and the NK cell, serving once again to focus nonspecific responses upon a particular foreign substance.

Quick Test 22.4

1. Explain two differences between plasma cells and memory B cells.

2. What is the difference between T-dependent antigens and T-independent antigens? Why are no memory B cells generated in response to a T-independent antigen?

3. Draw an IgG molecule, and list the four other classes of antibody.

4. State five ways in which antibodies help eliminate antigen.

CELL-MEDIATED IMMUNITY

We have seen that whereas antibodies defend against invaders and other antigens floating free in the blood and lymph, T cells make contact with and respond to any body cells that are infected or otherwise abnormal. Because T cells require direct contact between them and their targets, their responses are described as being *cell-mediated*. We begin our discussion of cell-mediated responses by examining the roles of T cells in this type of immunity.

Roles of T Lymphocytes in Cell-Mediated Immunity

There are three major types of T lymphocytes: helper T cells, cytotoxic T cells, and suppressor T cells. Helper T cells, the primary regulators of immune responses, operate indirectly by secreting cytokines that enhance the activity of B cells, cytotoxic T cells, suppressor T cells, and helper T cells themselves. In addition, helper T cells produce cytokines that enhance the actions of macrophages and NK cells, which are essential to nonspecific defenses. Cytotoxic T cells, in contrast, are directly responsible for cell-mediated immunity in that they kill cells infected by viruses or intracellular bacteria, and cells that are otherwise abnormal (such as cancer cells and transplanted cells). Suppressor T cells are not well understood, but they are thought to produce cytokines that suppress the activity of B cells, helper T cells, and cytotoxic T cells.

All three types of T cells have antigen receptors (T cell receptors or TCRs) that detect foreign antigens on body cells, but they do so only when these antigens are associated with a special class of normal self-proteins known as **major histocompatibility (MHC) molecules.** This type of recognition is unlike the way in which B cells recognize antigens in that B cells and the antibodies they secrete are able to bind to epitopes in their native forms (for instance, as they exist on the surface of a bacterium). In order for T cells to be activated, their antigen receptors must contact an MHC molecule on the surface of some other body cell that is bound to a small fragment of antigen. Thus MHC molecules must first bind to a fragment of foreign antigen that is present within a body cell and then must transport it to the cell surface, where it can be detected by T cells. This process is called **antigen presentation**. Before we examine T cell responses in greater detail, let's take a closer look at MHC molecules and their roles in T cell maturation, antigen presentation, and T cell activation.

MHC Molecules: Markers of Self

Each body cell is identified as "self" by a set of MHC molecules (in humans, known as *human leukocyte antigens,* or HLA molecules), of which there are two classes. **Class I MHC** molecules are found on the surfaces of all nucleated cells—that is, on almost every cell of the body; **class II MHC** molecules are found on only a few specialized cell types, including macrophages, activated B cells and T cells, and the cells that make up the interior of the thymus.

Although the two classes of MHC molecules function the same way in all people, each person's MHC molecules are unique to himself or herself; it is virtually impossible for the tissues of any two people (except those of identical twins) to have the same set of MHC molecules, or *HLA tissue type.* That MHC molecules differ from person to person explains why skin grafts and organs are usually rejected when transplanted from one person to another; in fact, the existence of MHC molecules was discovered during investigations of graft rejection. These proteins play a major part in determining whether transplanted tissue is accepted as self (histocompatible) or rejected as foreign.

The discovery of MHC molecules and their role in graft rejection was somewhat puzzling; why would markers have evolved that prevent us from sharing tissues? The answer is that this diversity of MHC molecules is

adaptive to the survival of the human species as a whole. Each person's MHC molecules are capable of presenting some antigen fragments but not others. The existence of a range of different MHC molecules in the population ensures that if the population is infected with a given pathogenic microbe, at least some individuals will have MHC molecules capable of presenting its specific antigens, and those individuals will be able to mount an immune response to the invader and survive the infection. In reality, most of us are capable of mounting an attack against all the microbes we encounter in everyday life, because antigens are complex enough, and each person's set of MHC molecules are varied enough, such that each of us develops an effective immune response. Even if the cells of two different people present different portions of a given antigen, both people will be able to mount a strong immune response to that antigen.

The Role of MHC Molecules in T Cell Maturation

If individuals are to mount an effective immune response, their T cells must have antigen receptors capable of making specific contacts with their own MHC molecules. This poses a particular challenge to the development of T cells, for the following reason: Because T cell receptors (like B cell receptors or membrane antibodies) are randomly generated to provide a great array of different lymphocytes, each with a unique specificity, developing T cells differ in their capacity to bind to an individual's own MHC molecules.

In a healthy immune system, only "desirable" T cells reach maturity and leave the thymus. As young T cells are developing in the thymus, they come into contact with thymic cells bearing high levels of the body's own class I and class II MHC molecules. Only T cells bearing receptors with appropriate affinity for self-MHC molecules continue to develop. Furthermore, T cells develop into two different types, depending on the class of MHC molecules to which they bind: Those cells that bind to class I MHC molecules develop into cytotoxic T cells, whereas those that bind to class II MHC molecules develop into helper T cells.

The Role of MHC Molecules in Antigen Presentation and T Cell Activation

Each MHC molecule possesses a binding site that can bind to a variety of foreign antigen fragments. (In the absence of foreign antigen, MHC molecules bind to self-protein fragments, but the resulting combination does not normally induce an immune response.) As a newly synthesized MHC molecule makes its way to the surface of an infected or abnormal cell, it can capture an antigen fragment in its binding site and carry it out to the surface of the cell. Class I MHC molecules capture foreign or abnormal antigens synthesized within infected cells or tumor cells, whereas class II MHC molecules capture foreign

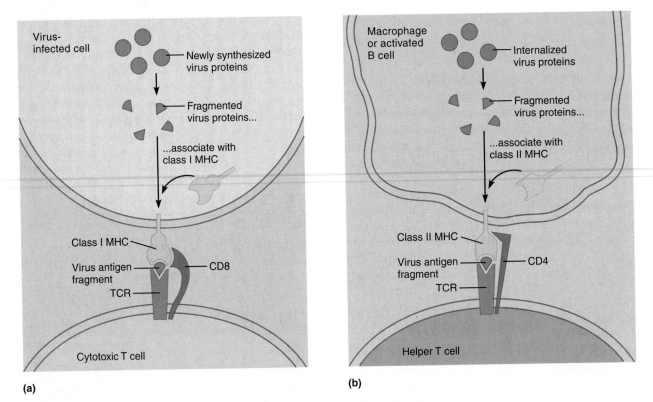

FIGURE 22.11 Presentation of antigens to T cells by major histocompatibility complex (MHC) molecules. (a) *Class I MHC molecules, made by all nucleated cells of the body, capture fragments of viral (or bacterial) antigens synthesized within an infected cell and transport them to the cell surface. A cytotoxic T cell then binds to the infected cell through its T cell receptors (TCR) and CD8 molecules.* (b) *Class II MHC molecules, made by macrophages and activated B cells, capture fragments of foreign antigens internalized by phagocytosis or receptor-mediated endocytosis, respectively, and transport them to the cell surface. A helper T cell then binds to the presenting cell through its TCR and CD4 molecules.*

antigens that have been taken into cells through phagocytosis or receptor-mediated endocytosis. Seen this way, the distribution of the two different classes of MHC molecules on body cells makes sense: Class I MHC molecules are found on all nucleated cells of the body (because *any* of these cells can be infected or become a tumor cell), whereas class II MHC molecules are found only on a few cell types, such as macrophages (which internalize foreign antigens by phagocytosis) and B cells (which internalize antigens by receptor-mediated endocytosis).

The two classes of MHC molecules differ in another way: Class I MHC molecules are recognized by cytotoxic T cells, whereas class II MHC molecules are recognized by helper T cells (Figure 22.11). Activated macrophages, after engulfing and degrading antigens early in infection, present fragments of these antigens to helper T cells, which in turn are stimulated to secrete cytokines that induce and regulate other immune responses. Fortunately, helper T cells do not kill the cells they contact, because they would then kill the macrophages. In contrast, infected cells or cancer cells present antigen fragments to cytotoxic T cells, which are then stimulated to kill the unhealthy cells. Helper T cells have a surface protein called

CD4 that binds to the class II MHC molecule and improves the association between the T cell and macrophage. Cytotoxic T cells have a similar protein called *CD8* that binds to a class I MHC molecule, improving its ability to associate with an abnormal target cell. For this reason, helper T cells and cytotoxic T cells are sometimes called CD4 cells and CD8 cells, respectively.

Quick Test 22.5

1. List the three major types of T cells, and explain their functions.

2. What are the two classes of MHC molecules? On which types of cells is each class found?

3. Why do MHC molecules pose limits to transplantation between people, except in the case of identical twins?

4. What type of T cell has CD4 on its surface, and with what class of MHC does this cell associate?

5. What type of T cell has CD8 on its surface, and with what class of MHC does this cell associate?

CYTOKINE	TARGET CELLS	EFFECTS UPON TARGET CELLS
Interleukin-2	Helper T cells and cytotoxic T cells	Stimulates proliferation
	B cells	Stimulates proliferation and plasma cell development
	NK cells	Enhances activity
Interleukin-4	B cells	Stimulates proliferation and plasma cell development; induces plasma cells to secrete IgE and IgG; increases numbers of surface class II MHC molecules
	Helper T cells	Stimulates proliferation
	Macrophages	Increases numbers of surface class II MHC molecules; enhances phagocytosis
	Mast cells	Stimulates proliferation
Interleukin-5	B cells	Stimulates proliferation; induces plasma cells to secrete IgA
	Hematopoietic stem cells	Induces proliferation and development of eosinophils
Interleukin-10	Macrophages	Inhibits cytokine production (helps downregulate the immune response)
Interferon-γ	Multiple cell types	Confers resistance to viruses
	Macrophages	Enhances phagocytosis
	B cells	Enhances antibody production
	Cancer cells	Inhibits proliferation
	Cytotoxic T cells and NK cells	Enhances killing capacity of cytotoxic T cells and NK cells

Helper T Cell Activation

Like B lymphocytes, millions of T lymphocytes with different specificities normally circulate in the blood and lymph or reside in peripheral lymphoid tissues. Activation of helper T cells involves two simultaneous events: Helper T cells first bind with class II MHC–foreign antigen complexes on the surfaces of macrophages and B cells, and then helper T cells receive from these cells an inducing signal in the form of IL-1. As a result, the helper T cells proliferate and differentiate. Some of the daughter cells begin to secrete cytokines, whereas a small proportion of them become long-lived memory T cells. As is the case with humoral responses, cell-mediated responses can take up to 17 days to develop, especially upon first exposure to an antigen. Once memory T cells are present in the body, however, subsequent responses to that same antigen are quicker and much more vigorous.

Activated helper T cells secrete several types of cytokines that help stimulate and regulate the activities of other helper T cells, B cells, cytotoxic T cells, macrophages, mast cells, NK cells, and hematopoietic stem cells

(Table 22.3). Helper T cells thus appear to be the central coordinators of immune responses—which is why the depletion of helper T cells, a primary characteristic of AIDS, is so devastating to the immune system (When It Goes Wrong, Acquired Immunodeficiency Syndrome, www.physiologyplace.com, Challenge Yourself).

Cytotoxic T Cell Activation: The Destruction of Virus-Infected Cells and Tumor Cells

A cytotoxic T cell becomes activated to kill when two events occur simultaneously: when it both binds with a class I MHC–foreign antigen complex on the surface of a virus-infected cell and receives an inducing signal (in the form of IL-2) from a helper T cell. Once activated, the cytotoxic T cell releases **perforins,** proteins that form pores in the infected cell's membrane, thereby increasing the membrane's permeability to ions and water. As ions and water flow into the cell, it swells and eventually lyses. The secretion of such lysis-inducing proteins is a common

FIGURE 22.12 Summary of the events in the immune response to a virus. *A helper T cell is activated both by specific contact with a macrophage presenting viral antigen via class II MHC and by IL-1, generating memory T cells and IL-2-secreting helper T cells. A cytotoxic T cell is activated by specific contact with an infected cell presenting viral antigen via class I MHC, and by IL-2 secreted by a helper T cell. The activated cytotoxic T cell produces perforins and fragmentins, each of which can induce target-cell death. A B cell is activated by specific contact with free-floating antigen and by IL-2 secreted by a helper T cell, generating memory B cells and plasma cells that secrete antibody specific for the antigen. The immune response to other types of pathogens is similar to that described here.*

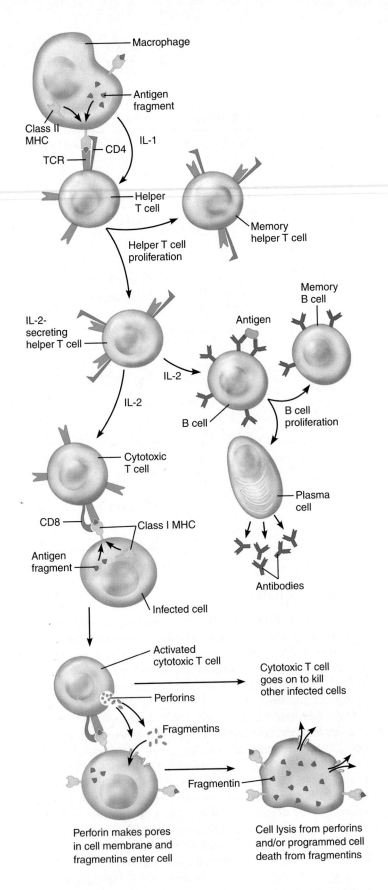

phenomenon in immune reactions; we have already encountered it in the complement cascade and in the way NK cells destroy their target cells. Cytotoxic T cells also release proteins called **fragmentins,** which first gain entry into target cells through the perforin-induced pores and then work inside them to bring about their death through apoptosis. Pathogens released from the dead cell are rapidly eliminated by nearby phagocytic cells or are targeted for destruction by antibodies.

In similar fashion, cytotoxic T cells also defend against tumor cells. These cells sometimes possess distinctive molecules, called *tumor antigens,* that are not present on normal cells. The class I MHC molecules that normally exist on tumor cells present fragments of these antigens to cytotoxic T cells, thereby initiating their killing response. Note that certain cancers and viruses (Epstein-Barr virus, for example) actively inhibit the production of class I MHC on affected cells, which enables such cancers and viruses to escape detection by cytotoxic T cells.

As we have seen, effective immune responses arise from multiple direct and indirect interactions among macrophages, helper T cells, B cells, and cytotoxic T cells, as depicted in Figure 22.12.

Quick Test 22.6

1. List four cytokines that are produced by helper T cells, and describe how each helps regulate immune responses.

2. Briefly describe how a cytotoxic T cell kills a virus-infected cell.

3. List the four cell types that are involved in immune responses, and briefly describe the functions of each.

IMMUNE RESPONSES IN HEALTH AND DISEASE

We have seen how effective immune responses depend on the interplay of many kinds of cells and molecules. In

728 CHAPTER 22 ■ *The Immune System*

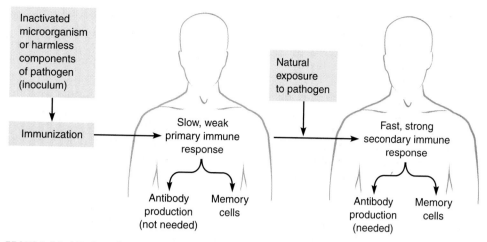

FIGURE 22.13 Acquisition of long-term immunity through vaccination (immunization). *Introduction into the body of the inoculum stimulates immune responses that both target the introduced materials for destruction and generate memory cells. Upon subsequent natural exposure to the pathogen, memory cells mount a rapid secondary response that prevents or diminishes the symptoms of the disease.*

Five-year old Ivan receives an immunization against chicken pox, and develops antibodies and memory cells specific for the virus. Which of these two components of his immune response is more important for Ivan, and why?

the sections that follow we explore how the immune system generates immunity, and responds to transplantation and transfusion. We'll also see how disease can result when the immune system malfunctions.

Generating Immunity: Immunization

In 1798, a physician named Edward Jenner was investigating the spread of smallpox when he noticed that milkmaids in Gloucestershire, England, who had previously been ill with cowpox (a mild disease contracted from cows) usually escaped smallpox infection, even when this disfiguring, often fatal disease was running rampant in their community. Suspecting that exposure to cowpox conferred some sort of protection against smallpox, Jenner began to deliberately inoculate people with cowpox, boldly predicting that this action would bring an end to this terrible scourge. This treatment came to be known as *vaccination* after the Latin word *vacca,* meaning "cow." Indeed, thanks to intensive worldwide vaccination programs in the 20th century, smallpox has been eradicated.

In vaccination, also known as *immunization,* a safe form of a microorganism or a collection of its components that are not expected to cause disease is introduced into the body, where the inoculum both stimulates an immune response to that pathogen and—even more significantly—induces immunological memory (Figure 22.13). A vaccinated person who then subsequently encounters the natural pathogen then mounts a strong immune response similar to a naturally occurring secondary immune response; a successfully vaccinated person is

immune to that pathogen. During the 20th century, routine immunization of infants and children has been extremely effective in preventing many infectious diseases, and such vaccine programs may result in the global eradication of polio by 2005. The term *immunization* is also used to refer to the conferral of immunity that results from a natural infection. Both artificial immunization and natural immunization confer a type of protection referred to as **active immunity** because it depends on the ability of the immunized person's immune system to mount a response. Unfortunately, not all infectious agents (for example, viruses that cause the common cold) are easily managed by vaccination.

Ready-made antibodies to a particular antigen can also be introduced into the body to provide another kind of protection called **passive immunity**. This type of immunization is passive because it does not require a response from the immunized person's immune system. Because no foreign antigens are introduced into the body in this procedure, the person's B cells do not make antibodies, and no memory B cells are made. Note, therefore, that although the introduced antibodies help fight an ongoing infection, passive immunization does not induce long-term immunity. The antibodies typically used for passive immunization are first isolated from people who are already immune to a particular disease; these

Memory cells. Antibodies function in the short term; they will help clear the harmless vaccine from Ivan's body. Memory cells, however, can survive a lifetime, and are poised to combat the chicken pox virus whenever he is exposed to it.

antibodies are then injected into another person's body, conferring in the process a short-lived but immediate protection from that disease. A person who had been bitten by a rabid animal, for example, may be injected with antibodies collected from other people who have been vaccinated against rabies. This measure is important because rabies may progress more rapidly than the time it takes for a person to mount an active immune response. Most individuals exposed to rabies are given both passive and active immunizations; the injected antibodies fight the virus for a few weeks, and then the person's own immune response, which is induced by both the immunization and the infection itself, takes over.

Passive immunity also occurs naturally when IgG antibodies in the blood of a pregnant woman cross the placenta and reach the fetus. IgA antibodies (see Table 22.2) are also passed from mother to nursing infant in breast milk, especially when the milk is in its early form, the colostrum. Passive immunity persists only as long as the transferred antibodies do—a few weeks to a few months—but it provides the infant protection from infections until the baby's own immune system has had a chance to mature.

Quick Test 22.7

1. How does an immunization such as the one against chicken pox confer lifelong immunity without causing the disease?

2. What is the difference between passive immunity and active immunity? Do the maternal antibodies a baby acquires through the placenta provide active immunity or passive immunity?

Roles of the Immune System in Transfusion and Transplantation

The immune system's ability to distinguish self from nonself, while essential to a healthy immune system, limits our ability to transfuse blood or transplant tissue from donor individuals. For this reason, material from the donor must be matched, as closely as possible, to the recipient in order to minimize immune reactions. In addition, transplant recipients are given medicines that suppress the ability of their immune systems to react against the foreign tissue. Note that the body's hostile response to an incompatible transfusion or transplant is not a disorder of the immune system, but instead a healthy response to foreign antigens.

Blood-Group Compatibility

Blood is classified into different types (designated by the letters A, B, AB, or O) according to the presence or absence of certain antigens on the surface of a person's red

blood cells. An individual who has *type A blood* possesses red blood cells with surface antigens known as A antigens. These antigens are perceived as self by the individual's immune system; they are not antigenic to their "owner." Similarly, B antigens are found on type B red blood cells, and both A antigens and B antigens are found on type AB red blood cells. Type O red blood cells possess neither antigen.

An individual with type B blood does not produce antibodies to B antigen because that antigen is recognized as "self." Surprisingly, however, this person will have antibodies to A antigen (called anti-A antibodies), *even if this individual has never been exposed to type A blood.* This totally unexpected presence of circulating antibodies specific for A antigen seems to suggest that the individual's B lymphocytes have detected and responded against type A blood. But what has in fact occurred is that antibodies to the A antigen have been produced in response to bacteria that are normally present in the body and possess epitopes that are very similar to the blood group antigens A and B. A person with type B blood does not make antibody to the B-like bacterial antigens—those are too much like self—but the individual does make antibodies to A-like bacterial antigens, which are perceived as foreign by his or her immune system. Thus in a type B individual, the anti-A antibodies that are constantly circulating in the blood induce an immediate and devastating **transfusion reaction** in the event that this individual receives a transfusion with type A or type AB blood, both of which contain the A antigen. For these reasons, a person with type O blood is considered to be a *universal donor* because any anti-A or anti-B antibodies that may be present in the recipient will find no target on the type O donor cells. In contrast, a person with type AB blood is considered to be a *universal recipient;* because such an individual lacks antibodies against A antigen or B antigen, transfusion of blood of any blood group into this individual will not induce a transfusion reaction.

One more issue must be considered with blood donations: When whole blood is transfused, it contains antibodies that could, in a mismatched transfusion, attack the *recipient's* red blood cells. This usually is not a problem because *packed* red blood cells, rather than whole blood, are typically used in transfusions.

Tissue Grafts and Organ Transplantation

As previously described, MHC molecules are present on almost every type of cell in the body, and more importantly they are present across the human population in many different forms. Therefore, it is unlikely that two randomly chosen people would have the same MHC (HLA) profile, and thus HLA molecules are responsible for stimulating the rejection that occurs in tissue grafts and organ transplants. Note that MHC molecules do *not* play a role in transfusion reactions because red blood cells do not have MHC molecules.

Autoimmune diseases such as multiple sclerosis (MS) occur when tolerance to self breaks down and the immune system wrongly identifies normal body components as foreign and mounts an attack. In multiple sclerosis, the tissue under attack is myelin, the material that wraps around and insulates the axons of neurons and is abundant in the white matter of the brain and spinal cord. The name of the disease refers to the many lesions or *scleroses* (from the Greek word for scarring or hardening) that result from the destruction of myelin (*demyelination*).

The symptoms of MS range from mild to severe and may appear in various combinations, depending on where the demyelination occurs. The symptoms may also come and go, or they may persist for the remainder of a person's life. The symptoms of MS typically include blurred vision (or blindness in one eye), muscle weakness, and trouble maintaining balance while walking (ataxia). These symptoms may worsen, leading to a complete inability to walk or stand. Other symptoms include muscle spasticity; tremors; impairment of pain, temperature, and touch sensa-

tions; speech disturbances; vertigo; and fatigue. MS is usually diagnosed between the ages of 20 and 40 and occurs more often in women than in men. Approximately 250,000–350,000 people currently live with this disease in the United States. In fact, MS is the most common cause of neurologic disability in developed countries.

Because the cerbrospinal fluid of MS patients has been shown to contain activated cytotoxic T cells (which are absent from the CSF of healthy people), it is likely that these cells are primarily responsible for the demyelination that occurs in MS. However, despite intensive research it is still not clear why the body's cytotoxic T cells turn against its own myelin. One possibility is that an infection of some sort either induces inappropriate immune responses to myelin or causes a breakdown in the blood-brain barrier that normally prevents lymphocytes from leaving capillaries within the CNS. Lymphocytes have no opportunity to develop tolerance to self-antigens (including myelin proteins) that are located in and around the brain and spinal cord; when lymphocytes are subsequently exposed to these anti-

gens, they perceive them as foreign and mount defensive responses, causing irreparable CNS damage. Scientists have studied a number of infectious agents (including measles, mumps, and rubella viruses) that may induce the autoimmune responses typical of MS but have been unable to implicate any particular causative agent.

Increasing evidence also suggests that heredity may play a role in determining a person's susceptibility to MS. Some ethnic groups, including gypsies and eskimos, never experience MS, and Native Americans, Japanese, and other Asian peoples have a very low incidence of the disease. Among the genes associated with susceptibility to the disease are those encoding HLA or MHC molecules. A higher incidence of MS is associated with people who have inherited a particular kind of class II MHC molecule.

Treatment and perhaps even prevention of this and other devastating autoimmune diseases requires better understanding of the precise mechanisms by which B cells and T cells respond against foreign antigens while remaining tolerant of the body's own tissues.

To minimize the chance of rejection, physicians attempt to match the HLA molecules of the donor and recipient as closely as possible, using a procedure called *tissue typing*. When a recipient has no identical twin (who would have exactly the same tissue type), siblings usually provide the closest HLA match. In addition to testing and matching HLA molecules, physicians also attempt to minimize the chance of rejection by prescribing drugs that

suppress the recipient's immune responses. The complication with this strategy is that this renders the recipient more susceptible to infections and cancer during the course of treatment. The main effect of drugs such as cyclosporin A and FK506, which have greatly improved the success of organ transplants, is to inhibit production of IL-2, which in turn inhibits B cell and T cell activation and the mounting of immune responses.

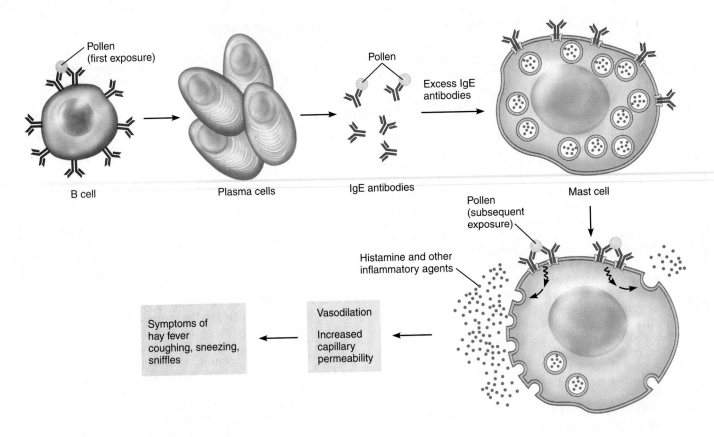

Pollen (first exposure)

B cell

Plasma cells

IgE antibodies

Excess IgE antibodies

Mast cell

Pollen (subsequent exposure)

Histamine and other inflammatory agents

Vasodilation

Increased capillary permeability

Symptoms of hay fever coughing, sneezing, sniffles

FIGURE 22.14 Events in hay fever, an allergic response. *In response to exposure to allergens such as pollen, an allergic person produces abundant IgE molecules; excess IgE binds to the surface of mast cells. Upon subsequent exposure to the same kind of pollen, the allergens and bound IgE bind, cross-linking the IgE and inducing rapid degranulation (histamine release). Histamine induces increased vascular permeability and the resulting symptoms of hay fever.*

In anaphylactic shock there is a drop in blood pressure due, in part, to widespread histamine release. What actions of histamine contribute to this loss of blood pressure?

Despite the risk of rejection, transplantation of bone marrow is often used successfully to treat leukemia and other cancers, as well as various blood-cell diseases. Before receiving the transplant, the recipient is typically treated with radiation to eliminate his or her own bone marrow cells. Such treatment both eliminates the abnormal cells and effectively inactivates that person's immune system, which minimizes the likelihood that the recipient will reject the graft. The great danger in this procedure is that the donated marrow, which contains lymphocytes, will mount immune responses against the recipient, an example of what is known as a **graft versus host reaction.** The intensity of such a reaction can be minimized if the HLA molecules of the donor and recipient are well matched, which is why bone marrow donor programs continually seek volunteer donors the world over. Because of the diverse array of HLA molecules within the human population, a diverse pool of potential donors is needed.

Quick Test 22.8

1. What cell surface molecules are tested in the process of tissue typing?

2. What is the name of the phenomenon in which a grafted tissue mounts an immune response against a recipient?

Immune Dysfunctions

The complex, highly regulated interplay of foreign substances with lymphocytes and other cells that make up

Histamine induces increased capillary permeability and vasodilation, resulting in the loss of fluid from blood vessels and decreased peripheral resistance, respectively. Both of these events contribute to a drop in mean arterial pressure.

the immune system provides us with extraordinary protection from infections. In addition, a growing body of evidence suggests that immune function is intimately associated with nervous system and endocrine system functions. When this delicately balanced network of interacting cells and molecules is disrupted, the effects upon the individual can range from the minor inconvenience of some allergies to the serious and sometimes devastating consequences of autoimmune and immunodeficiency diseases. In the following sections we examine what can happen when immune function goes awry.

Allergy

Allergies (also known as *hypersensitivity reactions*) are exaggerated responses to certain environmental antigens known as *allergens*. The most common allergies involve antibodies of the IgE class (see Table 22.2). Allergies occur in some people who are genetically predisposed to produce more than the usual amount of IgE when they are exposed to allergens. If an individual produces high levels of IgE in response to pollen, the result is an allergic reaction commonly known as *hay fever*. As shown in Figure 22.14, some of these IgE antibodies do not bind to the pollen, but instead attach by their tails to mast cells. When the person is subsequently exposed to pollen again, the pollen grains and the antigen-binding sites of these IgE antibodies bind, causing adjacent antibody molecules to become cross-linked and inducing the mast cell to degranulate—that is, to release histamine and other inflammatory agents into the surrounding fluid. Following its release, histamine causes dilation and increased permeability of small blood vessels in the immediate vicinity. These inflammatory events lead to the typical symptoms of hay fever: sneezing, runny nose, tearing eyes, and breathing difficulty, which can result from histamine-induced contraction of smooth muscle in respiratory airways. Drugs that act as antihistamines diminish allergy symptoms by blocking receptors located on blood vessel endothelial cells and smooth muscle cells that normally bind to histamine.

The most serious consequence of an acute allergic response is **anaphylactic shock,** a life-threatening reaction to injected or ingested allergens. Anaphylactic shock results from widespread degranulation of mast cells throughout the body, which triggers abrupt dilation of peripheral blood vessels. This dilation is not just confined to a restricted area, as in hay fever, but is widespread and therefore causes a precipitous drop in total peripheral resistance and mean arterial pressure. As a consequence of the drop in pressure, death may occur within a few minutes. Bee venom or penicillin are examples of allergens that can trigger anaphylactic shock in people who are extremely allergic to them. Some susceptible individuals carry syringes containing the hormone epinephrine as a prophylactic measure. When injected into the blood-

stream, this hormone counteracts the allergic response by stimulating increased cardiac output and constriction of blood vessels, both of which tend to raise blood pressure back toward normal levels.

Autoimmune Diseases

When the immune system loses tolerance to self and begins to react against normal molecules of the body, autoimmune disease can result. In a disease called *systemic lupus erythematosus* (often called *lupus*), for example, the immune system generates antibodies (known as *auto-antibodies*) against all sorts of self molecules, resulting in a widespread array of signs including skin rashes, fever, arthritis, and kidney dysfunction. *Rheumatoid arthritis,* another antibody-mediated autoimmune disease, causes painful inflammation of and eventual damage to the cartilage and bone of joints. In insulin-dependent diabetes mellitus, another autoimmune disease, the insulin-producing beta cells of the pancreas are targeted by cell-mediated immune responses. Another autoimmune disease is *multiple sclerosis* (MS), the most common chronic neurological disease in developed countries. In this disease, T cells that have a propensity to attack normal myelin are thought to infiltrate the central nervous system and cause demyelination of nerve fibers, thereby precipitating a number of serious neurological abnormalities (When It Goes Wrong: Multiple Sclerosis, p. 731).

The causes of autoimmunity are varied and complex. Although much remains to be learned about these diseases, we know that people who inherit particular MHC molecules also are more likely to develop certain autoimmune diseases. For example, individuals who inherit certain class II MHC molecules are at higher risk of developing insulin-dependent diabetes mellitus than are members of the general population.

Immunodeficiency Diseases

In allergies or autoimmune diseases, problems arise due to overactivity in the immune system; *immunodeficiency diseases* arise as a result of underactivity in the immune system. There are almost as many immunodeficiency diseases as there are functions within the immune system. Many congenital immune deficiencies affect the function of either humoral or cell-mediated immunity, but in *severe combined immunodeficiency disease* (SCID), both branches of the immune system fail to function. For people with this genetic disease, long-term survival may require a transplant of healthy bone marrow that will continue to supply functional B cells and T cells. One type of SCID is caused by deficiency of the enzyme adenosine deaminase (ADA), which plays a role in the breakdown of the DNA building blocks known as *purines*. ADA is particularly active in lymphocytes, and its deficiency results in the accumulation of lethal by-products. ADA has been treated with some success by gene therapy

As its name indicates, severe combined immunodeficiency disease (SCID) is a combination of deficiencies of both the humoral and cell-mediated branches of immunity. It is no surprise, therefore, that people with this disorder exhibit increased susceptibility to all types of infections. Infants with this congenital disease usually begin to develop recurrent infections at 3–6 months of age, and even normally harmless microbes can cause serious or even life-threatening infections. The most common diseases associated with SCID are pneumonia due to the bacterium *Pneumocystis carinii,* and (in infants) diarrhea caused by rotavirus.

About one in every million children are born with adenosine deaminase (ADA) deficiency, the first known cause of SCID. This enzyme plays a role in DNA metabolism, and although it is used by all cells of the body, the disruption of DNA synthesis its absence causes in B cells and T cells is particularly damaging. Because the antigen-driven proliferation of lymphocytes is the key to immunity, people with ADA deficiency have a markedly low lymphocyte count and therefore a virtually nonexistent immune response.

Although the frequent infections characteristic of SCID are treated by the administration of antiviral medicines,

antibiotics, and antibodies, ADA deficiency is fatal—usually by the time the child reaches 2 years of age—unless immune function is restored. Like other blood cell diseases, ADA deficiency can be cured by transplanting bone marrow from a donor with a similar HLA tissue type, but a well-matched donor is hard to find. Fortunately, ADA deficiency SCID has also been successfully treated by regular injections of the purified enzyme itself.

Now, with the advent of *gene replacement therapy,* or gene therapy, there is new hope for those born with genetic diseases such as ADA deficiency SCID. The concept is simple: If a gene is defective (that is, it produces an aberrant product or inadequate levels of it), then introducing a functional copy of the gene—one that directs normal synthesis of the product—should cure the problem. For gene therapy to be successful in treating ADA deficiency, a normal adenosine deaminase gene must be introduced into T cells and B cells, and the gene must also be replicated normally when the cells divide. Ideally, cells carrying the new gene should persist in the individual, so that repeated treatments are not needed.

Gene therapy is accomplished in four steps: (1) Lymphocytes or stem cells are collected from the blood, bone marrow, or

umbilical cord of an infant or child with the disease; (2) the harvested cells are treated with a chemical that causes them to proliferate in order to generate additional target cells for gene insertion; (3) a healthy ADA gene is introduced into the cells, often using a nonpathogenic virus as a carrier; and (4) the cells, now producing normal ADA, are infused back into the bloodstream of the affected child.

ADA deficiency was the first human disorder to be treated by gene therapy. Since 1990, a number of infants and children with the disease have been treated in this manner in the United States, Europe, and Japan. Reports indicate that recipients of this treatment are doing well. However, because most recipients are still receiving enzyme injections regularly, it is not yet possible to attribute their good health to gene therapy alone. Moreover, grave concerns about the viruses used to carry such genes, arose from the 1999 death of Jessie Gelsinger, a teenager who received gene therapy (for a different genetic disease) and had a severe inflammatory reaction to the virus used in the treatment. Much has been learned from this tragic event, and scientists continue to pursue this kind of treatment for debilitating and often deadly genetic diseases.

(Discovery: Gene Therapy for Severe Combined Immunodeficiency Disease); in this procedure an individual's own bone marrow cells are removed, provided with a functional ADA gene, and then returned to the body.

Immunodeficiency is not always a congenital condition; an individual may develop immune dysfunction later in life. For example, certain cancers can cause immunodeficiency by suppressing the immune system. A prime example is *Hodgkin's disease,* which damages the lymphatic system. Another well-known and devastating acquired immune deficiency is AIDS (see When It Goes Wrong: Acquired Immunodeficiency Syndrome, www.physiologyplace.com, Challenge Yourself).

The Role of Stress in the Immune Response

Clinical and personal observations have long suggested a correlation between psychosocial factors and immune function. Indeed, several studies have suggested that a positive outlook is associated with improved health in both cancer and AIDS survivors; conversely, people hospitalized for depression have been found to exhibit diminished immunity. The mind-immunity relationship arises from multiple complex interactions that exist among the immune system, the nervous system, and the endocrine system. The effects of the immune system on these and other systems is depicted in the Systems Integration Chart on page 737.

Many of these interactions are well established. Some steroid hormones are known to suppress immune responses; corticosteroids, for example, reduce the number and activity of immune cells and are potent anti-inflammatory agents. Likewise, research on wild animals has demonstrated that chemicals such as PCB and DDT, which mimic naturally occurring endocrine hormones, reduce the effectiveness of immune responses and increase the incidence of infection. By contrast, both growth hormone and thyroid hormone play important roles in T cell development and function.

The nervous, endocrine, and immune systems interact in a variety of ways. For example, autonomic nerves innervate lymphoid organs such as the bone marrow and the spleen, often terminating at B cell and T cell clusters, and lymphocytes bear receptors for epinephrine and acetylcholine. Additionally, lymphocytes and macrophages secrete cytokines that affect both the endocrine and central nervous systems. For example, IL-1 secreted by macrophages induces fever, suppresses appetite, inhibits thyroid function, and stimulates the release of pituitary hormones.

The correlation between stress and immune function continues to be demonstrated in various ways. One study found that when college students were tested for immune functions just after a vacation and again during final exams, their immune systems were impaired in various ways (for example, plasma interferon levels were lower) during exam week.

SUMMARY OF FACTORS INFLUENCING SPECIFIC AND NONSPECIFIC DEFENSES

Figure 22.15 graphically summarizes many of the concepts discussed in this chapter. It may help you to recall and understand the many components of immune function, and how they lead to the clearance of microorganisms from the body.

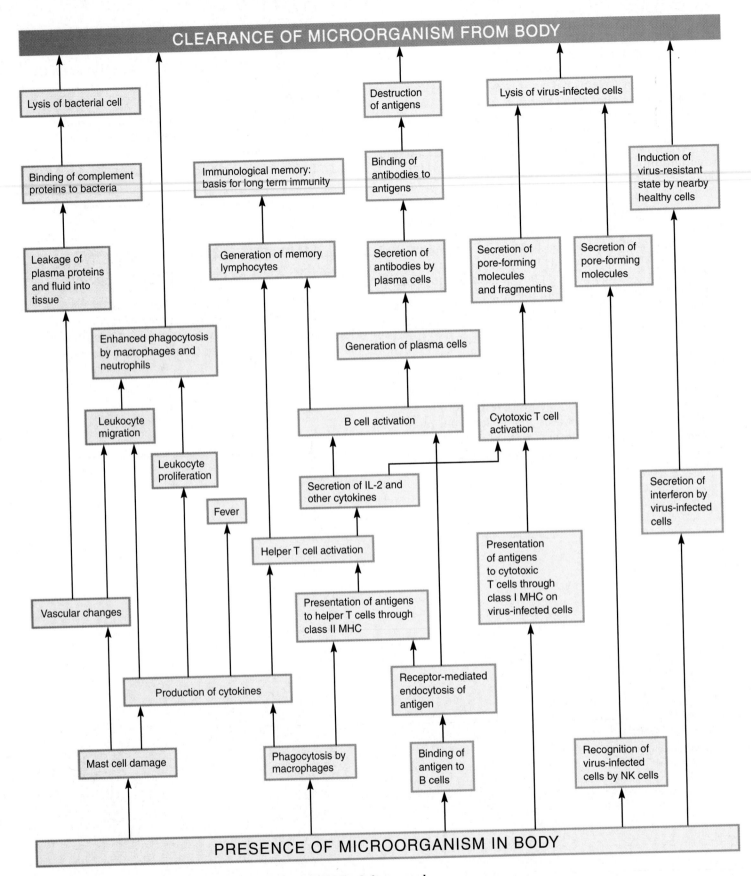

FIGURE 22.15 The events by which the body's nonspecific defenses and immune responses clear microorganisms from the body.

SYSTEMS INTEGRATION

IMMUNE SYSTEM

Urinary System
Lymphoid tissues in the wall of the urinary tract trap and help eliminate antigens
Cardiovascular System
The spleen traps antigens that are carried in the blood and destroys aging erythrocytes
Antibodies bind to and help eliminate foreign matter in the blood
Chemical mediators increase vascular permeability
Respiratory System
Lymphoid tissues in the walls of airways trap and help eliminate inhaled antigens
Chemical mediators induce tracheal and bronchiolar constriction
Macrophages in the alveoli engulf inhaled foreign matter
Immune System
Reproductive System
IgG antibodies cross the placenta and provide immunity to developing fetus
Lymphoid tissues in the wall of the reproductive tract trap and help eliminate antigens
Endocrine System
IL-1 induces the release of pituitary hormones
IL-2 helps regulate glucagon levels
Digestive System
Lymphoid cells in Peyer's patches trap and help eliminate ingested antigens
Chemical mediators induce intestinal smooth muscle contraction
Nervous System
Endogenous pyrogens act on the hypothalamus to raise the set point for body temperature
IL-2 and leukocyte-derived opioid peptides in inflamed tissue produce analgesia
Muscles
Chemical mediators induce pulmonary and intestinal smooth muscle contraction

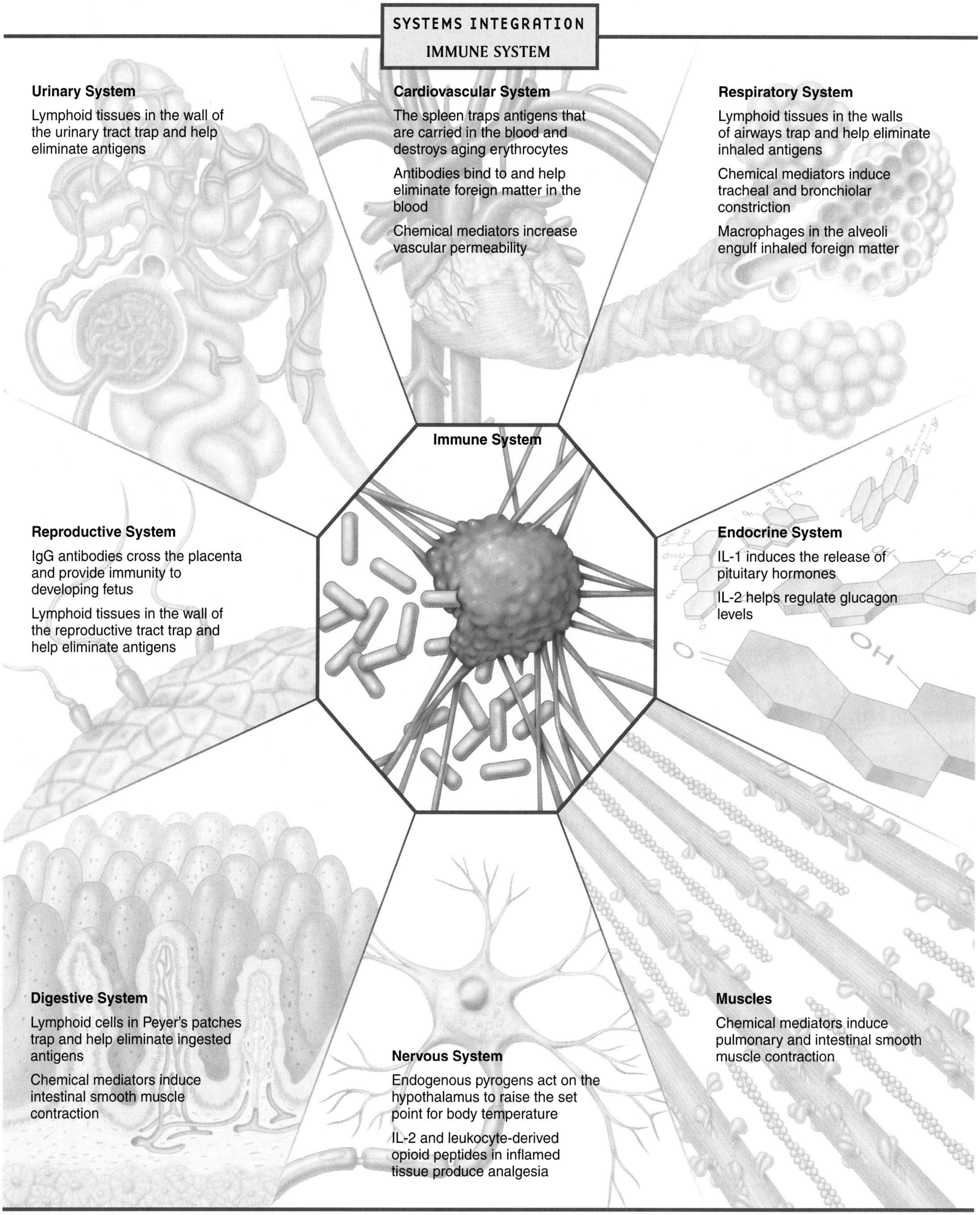

The components of the immune system—lymphoid tissues, leukocytes and the molecules they produce—work together to generate and regulate immune responses. The function of the immune response is two-fold: to clear foreign materials from the body, and to bring about long-term immunity to infectious diseases.

Anatomy of the Immune System, p. 709

There are five major types of leukocytes. Neutrophils, monocytes, and macrophages (which arise from monocytes) are phagocytic; they engulf and destroy foreign matter and debris. Eosinophils and basophils defend against large parasites, and are also involved in allergic reactions. The lymphocytes consist of B lymphocytes (B cells), T lymphocytes (T cells), and null cells. Whereas B cells and T cells exhibit specificity, null cells are nonspecific. Most null cells are large granular lymphocytes known as natural killer (NK) cells, which pose an important, early threat against viral infections. Leukocytes develop to maturity in the central lymphoid tissues (the bone marrow and, in the case of T cells, the thymus). The peripheral lymphoid tissues exhibit a netlike architecture that traps foreign matter present in the blood (spleen), lymphatic fluid (lymph nodes), air (tonsils, adenoids), and in food and water (appendix and Peyer's patches).

Organization of the Body's Defenses, p. 713

When an infectious agent breaks through the first line of defense (the skin and mucous membranes, for example), it faces both nonspecific defenses and immune responses. Nonspecific defenses provide the body's most rapid defenses against infection or injury. In inflammation, oxygen, nutrients, defensive molecules, and phagocytic cells are drawn to the affected region. Interferons, secreted from virus-infected cells, protect the surrounding healthy cells from infection, whereas NK cells can recognize and kill virus-infected cells. The complement system is activated when the first in a series of complement proteins binds to certain types of bacteria, leading to bacterial lysis. The complement cascade can also be activated by antibodies, so it contributes to both nonspecific defenses and immune responses.

B cells and T cells provide for the features of immune responses: specificity, diversity, memory, and self-tolerance. There are two types of immune responses: the humoral response and the cell-mediated response. The humoral response is the result of B cell activation. Upon contact with specific antigen, B cells proliferate and develop into long-lived memory B cells and short-lived plasma cells. Whereas memory cells provide for long-term immunity to the antigen, plasma cells secrete antibodies that bind to and target the antigen, and then recruit other defenses (such as phagocytic cells), to destroy it. The cell-mediated response occurs when cytotoxic T cells detect specific antigen presented by a class I MHC molecule (on a virus-infected cell or a tumor cell) and develop into active killers. Cytotoxic T cells destroy their targets in two ways: by releasing perforins that form pores in the stricken cell membrane, and fragmentins that enter the cell and induce apoptosis. Both humoral and cell-mediated responses are supported and regulated by cytokines secreted by activated helper T cells. Helper T cells are activated to proliferate and to secrete cytokines when they contact specific antigen presented by a class II MHC molecule (on a macrophage or a B cell). At the same time, long-lived memory T cells, are generated. Thus the responses of helper T cells, cytotoxic T cells, and B cells coordinate the specific disposal of, and long-term immunity to, offending antigens.

Immune Responses in Health and Disease, p. 728

The aim of vaccination, or immunization, is to provide protection from infection. Both immunization and natural infection induce what is known as active immunity because they depend on the response of a person's own immune system and generate memory against the agent. Passive immunity, in contrast, is generated when antibodies are transferred from one person to another. These ready-made antibodies immediately target and mediate the disposal of antigens for which they are specific. Passive immunization is used when a dangerous bacteria or virus has entered the body of a person who is not already immune to it.

The immune system's capacity to distinguish self from nonself limits our ability to share tissues through blood transfusion and transplantation. For example, a transfusion reaction results when antibodies induce lysis of mismatched red blood cells. Similarly, a graft rejection can result when HLA, or MHC molecules, which differ from person to person, are not well matched. The survival of a graft requires making the best possible HLA match and using immunosuppressive drugs to diminish the recipient's immune responses. Bone marrow transplantation poses a particular problem when tissues are not well matched: The donated marrow, which contains lymphocytes, can mount immune responses against the recipient, resulting in a type of rejection known as a graft versus host reaction.

Immune dysfunction can result in allergies, autoimmune diseases, or immunodeficiency diseases. An allergy results from an exaggerated response to environmental antigens (allergens). Autoimmune diseases occur when the immune system reacts against self, as in rheumatoid arthritis or multiple sclerosis. An immunodeficiency disease can result when a component of immunity is incapacitated by an inherited or acquired condition.

Immunodeficiency diseases may affect humoral or cell-mediated immune function, or both (as in severe combined immunodeficiency, or SCID).

Evidence suggests that the immune system, the nervous system, and the endocrine system are physiologically linked. Neuroendocrine mechanisms have been shown to regulate immune responses, and immune responses in turn can induce changes in both endocrine and neural function.

EXERCISES

Multiple-Choice Questions

1. Which of the following conditions would lead to the most serious immune deficiency disease?
   a) lack of IgG
   b) lack of neutrophils
   c) lack of B cells
   d) lack of cytotoxic T cells
   e) lack of helper T cells

2. Which of the following molecules can opsonize antigen?
   a) a T cell receptor
   b) interferon
   c) an antibody
   d) a perforin
   e) interleukin-2

3. Lymphocytes contact foreign antigen in all of the following tissues *except*
   a) bone marrow.
   b) spleen.
   c) lymph nodes.
   d) appendix.
   e) Peyer's patches.

4. Which of the following is *not* true about humoral immunity?
   a) It involves B cells.
   b) It involves antibody.
   c) It involves cytotoxic T cells.
   d) It can provide passive immunity when transferred from one person to another.

5. Macrophages
   a) have class I MHC molecules.
   b) have class II MHC molecules.
   c) are phagocytic.
   d) are indirectly inolved in specific immunity.
   e) all of the above

6. Activated cytotoxic T cells release pore-forming molecules called
   a) histamines.
   b) complement proteins.
   c) perforins.

d) immunoglobulins.
e) ready-porins.

7. An individual with type AB blood
   a) is considered a universal blood donor.
   b) is considered a universal blood recipient.
   c) produces antibodies to the B antigen.
   d) produces antibodies to the A antigen.
   e) is Rh-positive.

8. Which of the following events can result in lifelong immunity?
   a) passage of maternal antibodies to a developing fetus
   b) an inflammatory response to a splinter
   c) phagocytosis of bacteria by a neutrophil
   d) administration of the polio vaccine
   e) administration of antibodies against the rabies virus

9. Foreign antigens phagocytosed by macrophages are presented by
   a) class I MHC molecules to cytotoxic T cells.
   b) class II MHC molecules to helper T cells.
   c) class I MHC molecules to helper T cells.
   d) class II MHC molecules to CD8-bearing cells.
   e) class II MHC molecules to cytotoxic T cells.

10. Of the following events, which occurs earliest in the process of local inflammation?
    a) increased capillary permeability
    b) fever
    c) attack by cytotoxic T cells
    d) release of histamine
    e) lysis of microbes mediated by antibodies and complement

11. Which of the following is *not* true about helper T cells?
    a) They function in both cell-mediated and humoral immune responses.
    b) They secrete antibody.
    c) They bear surface CD4 molecules.
    d) They are subject to infection by HIV.
    e) When activated, they secrete IL-2 and other cytokines.

12. IL-2 is important for the activation of all of the following cell types *except*
    a) B cells.
    b) cytotoxic T cells.
    c) NK cells.
    d) T helper cells.
    e) macrophages.

13. Which of the following is an autoimmune disease in which myelinated neurons become the target of the immune response?
    a) myasthenia gravis
    b) multiple sclerosis
    c) diabetes mellitus
    d) rheumatoid arthritis

Objective Questions

1. Fill in the blank with the abbreviation for the cell type mediating the stated function: helper T cell (T_H), cytotoxic T cell (T_C), B cell (B), or macrophage (M)
   a. _____ Phagocytosis
   b. _____ Secretion of cytokines such as IL-2
   c. _____ Killing of virus-infected cells
   d. _____ Specific binding to free virus
   e. _____ Differentiation into antibody-secreting plasma cells

2. Fill in the blank with the letter that applies to the stated situation: humoral immune response (H), cell-mediated response (CM), both (B), or neither (N)

   a. _____ Occurs in a viral infection
   b. _____ Involves the production of antibodies
   c. _____ Involves the phagocytic activity of neutrophils
   d. _____ Involves killing of virus-infected cells
   e. _____ Involves T cells bearing CD8

3. A person who experiences life-threatening allergic reactions to bee sting venom might be given an experimental drug designed to (block/enhance) the binding of IgE to mast cells.

4. When a macrophage is infected by a virus, viral antigen will be presented by (class I/class II MHC molecules) to a (helper T cell/cytotoxic T cell).

5. Evidence exists of interactions among the immune system, the nervous system, and the endocrine system. (true/false)

6. A young girl who has never been immunized against tetanus cuts her foot on a rusty nail. In the emergency room her wound is cleaned, and she is given an injection of tetanus antitoxin (antibody to tetanus toxin). This is considered (active/passive) immunization.

7. Macrophages internalize foreign antigens by (endocytosis/phagocytosis), whereas B cells do so by (endocytosis/phagocytosis).

8. A person who has had a thymectomy as a treatment for a thymic tumor will likely experience a diminished (T cell/B cell) count.

Essay Questions

1. You notice that the area surrounding a paper cut on your hand has become red, warm, and swollen. Briefly describe the processes that lead to these symptoms of inflammation.

2. Through what lymphoid organs will a cell destined to be an active, mature T cell travel during its lifetime? Briefly state the function of each organ you mention.

3. Immune responses exhibit four features, all attributed to lymphocyte function. List these four features and discuss how they arise.

4. Draw a diagram showing the central role of helper T cells in both humoral and cell-mediated immune responses.

5. Consider a simple artificial antigen containing two distinct epitopes. Describe and diagram the process of clonal selection that results as B cells specific for these epitopes are selected by the antigen and are stimulated to proliferate and to differentiate into memory B cells and plasma cells. Also diagram how secreted antibodies interact with this antigen.

6. An infant who has experienced multiple bacterial infections since birth has been diagnosed with a macrophage deficiency. Describe the effect this deficiency has on the baby's nonspecific defenses and immune responses. Do you think that a neutrophil deficiency would present a more severe or a less severe state of immune deficiency?

7. A child needs a kidney transplant. Both of his parents and his two older siblings are willing to donate. Tissue typing reveals that his father has the HLA genotye "8,9" (one HLA gene copy is "8," the other is "9"), and his mother has an HLA genotype "2,6." Considering that a child inherits one copy of the HLA gene from each parent, which is the better potential donor—a parent or a sibling? Is it possible to have a 100% donor-recipient match of the HLA type?

Find the answers to these exercises, and additional study tools, at the Physiology Place (www.physiologyplace.com).

23

The Whole Body: Integrated Physiological Responses to Exercise

OBJECTIVES

- Describe the manner in which different physiological systems interact in the stressful, real-life situation of running a marathon.

- Provide an example of how some physiological responses are sensed by subjective experience.

- Describe how increased physiological function is accomplished during exercise.

- Describe the mechanisms that regulate physiological function during exercise, and explain how regulation of one system influences other systems, thus requiring coordination of multiple systems.

- Describe how resources are allocated to various tissues and organs under stressful conditions that cause conflicting physiological demands.

- Identify the physiological characteristics that limit exercise intensity and duration and thus influence performance.

CHAPTER OUTLINE

Principles of Physiological Integration 742

The Start: Transition from Rest to Exercise 742

The Long Haul: Almost Steady State 746

The Decline to the End: "The Wall" or the Finish Line? 753

The Aftermath 755

Now that we've seen how each physiological system operates on its own, let's revisit the marathon experiences of Bill and Jane and see how physiological systems interact with each other. Before reading the rest of this chapter, you may wish to return to the last part of Chapter 1 and read about Bill and Jane's marathons again (the key events are summarized in Figure 23.1).

PRINCIPLES OF PHYSIOLOGICAL INTEGRATION

Before we explore the physiological details at each stage of the race, we must cover a few general principles that determine how and why physiological systems interact.

The Metabolic Demands of Exercise

Changing posture, lifting weight, and moving from one place to another are accomplished through the voluntary actions of skeletal muscle. For very brief periods, the energy for muscle contraction can be provided by the energy stored within the muscle itself in molecules such as ATP, creatine phosphate, glucose, and glycogen. But more prolonged activity requires increased delivery of oxygen and metabolic fuels (such as glucose and fatty acids) to the muscle from other sources. In addition, contracting muscle cells produce heat and metabolic products (that is, metabolites, including lactic acid and carbon dioxide) that must be carried away in order for the muscle to continue functioning. If these metabolites are not removed, muscle pain and fatigue can result. Moreover, the increased metabolic demand of skeletal muscle must be satisfied without compromising the energy supply to the brain and other vital organs such as the heart.

How Physiological Function Is Increased

Locomotion, heart function, and respiration are all examples of physiological processes that operate in repeating cycles. The speed (or capacity) of these processes can be augmented by increasing the frequency of the cycles and/or the magnitude (or strength) of each cycle. During exercise, an individual can also tap into reserve anatomical capacity to increase function. For example, gas exchange in the lungs can be enhanced by inflating more alveoli, because at rest not all alveoli are inflated with each breath. Also, the contracting skeletal muscles that cause locomotion can aid cardiac and respiratory function.

The ability to achieve and maintain an increased level of physiological function depends on the availability of energy as well as the competing needs of different physiological systems. These needs are profoundly influenced by environmental conditions and the individual's behavior.

The changes leading to increased function within specific physiological systems are regulated by (1) local metabolites (intrinsic control or autoregulation), (2) hormonal signals, and (3) neural signals (both conscious and unconscious). Increasing function requires increased blood flow. Under nonstressful conditions, autoregulation (intrinsic control) is the dominant mechanism determining blood flow to skeletal muscle and many other individual organs. But when several organs begin competing for increased delivery of blood, regulation requires greater involvement of hormonal and neuronal control mechanisms (extrinsic control).

Quick Test 23.1

1. Name three fuels that can be used for energy by skeletal muscles and that are normally present within the muscle cells themselves.

2. Blood flow to muscles and other organs is regulated by intrinsic and extrinsic control. As the general demand for blood flow increases, which type of control assumes greater importance?

THE START: TRANSITION FROM REST TO EXERCISE

We have seen that the body's response to exercise is orchestrated by both intrinsic control mechanisms and extrinsic control mechanisms. Now let's look at how these mechanisms interacted while Bill and Jane ran their marathon.

Autonomic Nervous System Input: Fight or Flight

Marathon running and many other types of exercise are cultural substitutes for primitive activities such as hunting, fighting, and fleeing; activities that require rapid release of stored fuels and the distribution of these fuels (via the blood) to the muscles that need them. The autonomic nervous system activates and regulates this release and distribution of metabolic fuels. The subjective "heart-pounding" excitement Bill and Jane felt just before the start of the race was triggered by reduced parasympathetic nerve activity, increased sympathetic nerve activity and by epinephrine, the circulating hormone of the sympathetic autonomic nervous system. In modern society, many stresses still activate this sympathetic response (which can also engender fear and rage), but people tend to participate less and less in the physical activities for which the response evolved. The chronically elevated catecholamine concentrations that can result from such stress have been associated with anxiety, irritability, and depression. Exercise can provide an outlet for pent-up sympathetically mediated excitement, and increase clearance of catecholamines from the circulation. Perhaps this explains why Bill and Jane feel better after their daily exercise routine (although maybe not after a marathon).

The increased sympathetic activity (and decreased parasympathetic activity) that Bill and Jane experienced just before the marathon helped prepare them for the event. Sympathetic input to the SA node and myocardium causes increased cardiac output by increasing both heart rate and stroke volume, and as a result, mean arterial pressure rises in anticipation of the increased demand for blood flow to skeletal muscle. Sympathetically mediated increases in venomotor tone promote venous return, which also enhances cardiac output. Sympathetic input to the pancreas stimulates glucagon release and inhibits insulin release, causing a rise in blood glucose in anticipation of the increased demand for energy in skeletal muscle.

Energy Sources and Mobilization: Different Fuels for Different Needs

The proportional use of fuels (carbohydrate, fat, and even protein) during exercise depends upon the intensity and duration of the activity. In a marathon, the mix of fuels used at the beginning is different than the mix used later on. As we will see, the rate of carbohydrate expenditure at the beginning of the race has a profound effect on performance at the end. Although Bill and Jane could not directly adjust their fuel usage by force of will (the proportion of each fuel used is dictated by enzyme-mediated biochemical pathways), they could indirectly influence the relative depletion of carbohydrate by adjusting their behavior (exercise intensity).

We have seen that ATP is broken down to ADP during the crossbridge cycling that pulls thin filaments along thick filaments and causes muscle contraction. The instantaneous recharging of ADP back to ATP is accomplished by donation of high-energy phosphate from creatine phosphate. However, this "energy reservoir" by

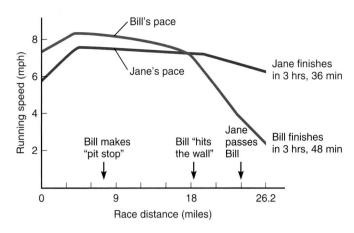

FIGURE 23.1 Bill and Jane's paces over the course of the marathon. *Both started slower than they wished because they were packed in tightly with other runners. Once the runners spread out a bit, Jane kept to a comfortable, even pace while Bill ran faster and darted through openings in the crowd.*

itself can only supply a few seconds of energy reserve. Glucose (whether free in the blood or stored as glycogen in muscle and liver) is another energy source that can be called on quickly because it can be broken down to produce ATP through glycolysis, which is anaerobic (that is, does not require oxygen). However, recall that whereas glycolysis liberates energy quickly, it is not a very efficient way of generating ATP. Exercising while this metabolic pathway is dominant (before oxygen delivery can be increased) means that a great deal of glucose will be broken down to produce rather limited amounts of usable energy. Therefore, this metabolic pathway is appropriate for sprinting, but it cannot support long-duration exercise such as a marathon.

The metabolic pathways involved in fat-burning, aerobic metabolism (that is, metabolic pathways that use oxygen, such as fatty acid oxidation and the Krebs cycle) require more time to respond to the changes in metabolic rate during exercise. Stated in more precise physiological terms, the activities of rate-limiting enzymes in these pathways change with the buildup or depletion of metabolic reaction products. Certain reaction products (including heat, H^+) cause vasodilation, which increases blood flow to exercising muscles, and these products also decrease the affinity of hemoglobin for oxygen, resulting in greater oxygen unloading in the muscle tissue. The time course for the transition from an anaerobic/glycolysis-dominant metabolism to an aerobic/lipolysis-dominant metabolism at the onset of moderate exercise is shown in Figure 23.2.

If a person begins exercising relatively gradually (at a low intensity), a greater proportion of ATP is generated through aerobic, lipolytic metabolism. This conserves blood glucose, which is normally the only fuel used by the brain, as well as muscle glycogen, which seems to be an important factor in limiting muscle function in the late

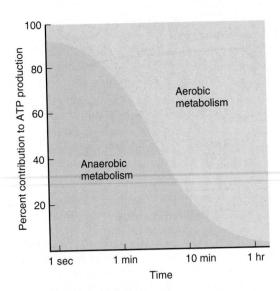

FIGURE 23.2 The relative contributions of anaerobic and aerobic metabolism to ATP production over the first hour of the marathon. *This time-compressed figure indicates the runners' metabolic status at the start, after 1 minute, after a little over 1 mile (10 minutes), and after about 7.5 miles (1 hour), running at approximately 70% of maximal aerobic capacity.*

Why is glycolysis said to be *anaerobic*?

stages of a marathon. Thus Jane's slower start conserved both liver and muscle glycogen and contributed to her ability to maintain her pace in the later stages of the marathon. Because Bill started at a higher intensity, he used relatively more glycogen at earlier stages of the race and thus depleted almost all of his muscle glycogen by about 20 miles into the race (when he "hit the wall").

Cardiovascular Adjustments: Anticipating the Needs

Support of aerobic metabolism in exercising muscle requires increased oxygen delivery, which in turn requires increased blood flow to the muscle. Both respiratory and cardiac function increase *immediately* at the onset of exercise. In fact, these functions, especially heart rate, may already be somewhat elevated in the anticipation of exercise as we just discussed. In addition to the "fight-or-flight" response, a mechanism that is unique to exercise—a feedforward mechanism called **central command**—is also involved. This feedforward signal originates in the motor cortex or other higher brain centers and is associated with Bill and Jane's conscious decisions to contract their muscles. An experiment that illustrates the effect of central command is shown in Figure 23.3.

When muscles are contracted *voluntarily*, heart rate increases with virtually the very next beat, far faster than possible if baroreceptor- or chemoreceptor-mediated autonomic negative feedback controls were involved. When muscle contraction is induced by *external electrical stimulation*, increases in heart rate are delayed and take longer to attain a steady state because negative feedback mechanisms activated in response to the involuntary muscle contractions are the only control signals involved. Central command is a mechanism that assures that blood pressure will not decrease at the onset of exercise. This theoretically would occur if a large proportion of blood volume suddenly rushed into exercising skeletal muscle.

Respiratory Responses: Recruiting Reserve Capacities

As exercise begins, active skeletal muscles consume greater amounts of oxygen and produce greater amounts of carbon dioxide, but arterial concentrations of these gases change very little (at least during mild-to-moderate exercise). This is because feedforward control circuits increase ventilation before the gas concentrations in arterial blood can change. One of these feedforward circuits is the central command associated with the intention to perform exercise (just described); another feedforward signal originates in mechanoreceptors and chemoreceptors in exercising skeletal muscles. Both of these feedforward signals stimulate ventilation.

In Chapter 16 we saw that changes in arterial oxygen and carbon dioxide concentrations regulate ventilation through negative feedback control. During exercise, these feedback mechanisms seem to *inhibit* ventilation in order to prevent hyperventilation. In contrast, the feedforward signals provide the stimulus for *increased* ventilation during exercise.

Ventilation increases during exercise as a consequence of both increased breathing frequency and increased tidal volume. At low-to-moderate exercise intensities, the dominant adjustment is increased tidal volume. As exercise intensity increases further, tidal volume stabilizes at about 65% of vital capacity, and additional increases in ventilation are due to increased frequency.

Recall that *all* of the cardiac output flows through the lungs. Resting cardiac output (5 L/min.) is easily accommodated by about one-third of pulmonary capillary capacity. Thus only one of every three capillaries is conducting blood at any given time, and more capillaries in the base of the lungs are carrying blood than in the apex. As cardiac output increases during exercise, more of the capillaries begin carrying blood continuously, and perfusion becomes more uniform throughout the lungs. (This constitutes one type of pulmonary reserve capacity.) Once cardiac output exceeds three times the resting value, all the reserve capillaries have been recruited, pul-

It does not require oxygen.

monary pressure rises modestly, and blood flows faster within individual capillaries. As a consequence, the amount of time each red blood cell spends exchanging gas with an alveolus is reduced. Transit time along the length of an alveolar capillary is normally about 1 second, but a red cell can fully exchange oxygen and carbon dioxide in one-third that time. (This constitutes a second type of pulmonary reserve capacity.) In combination, the two reserve mechanisms ensure that the pulmonary circuit can accommodate up to a six-fold increase in cardiac output during exercise, with no reduction in red cell oxygenation and carbon dioxide release. Bill and Jane's cardiac outputs never exceeded this limit while they ran.

World-class endurance athletes, by contrast, may attain rates of oxygen consumption and cardiac output that can exceed the reserves of the pulmonary circuit. That both skeletal muscle (the major consumer of oxygen) and the heart (the generator of cardiac output) can hypertrophy with training, but the lungs cannot, suggests to some researchers that the fixed anatomical capacity of the lungs may be the ultimate limitation in aerobic exercise performance. Other researchers, however, have argued that elite athletes may achieve high exercise intensities (in which *arterial* blood pH falls and body temperature rises extremely) such that the hemoglobin saturation curve shifts drastically to the right (see Figure 16.10), which means that less oxygen is loaded per hemoglobin molecule as it transits a lung capillary.

Temperature Regulation

Before they started running, Bill and Jane's core temperatures were approximately 37°C, and their leg muscle temperatures were about 2°C lower (at this time, body heat was generated primarily by their hearts and livers). When they began exercising, their muscles generated large amounts of heat (the mechanical efficiency of exercise is less than 25%). After 10 minutes of running, Jane and Bill's muscle temperatures reached a steady state (39°C), which were about one-half degree above core temperature. Under the mild environmental conditions at the start of the race (68°F/20°C), these increases in core temperature were proportional to the runners' exercise intensities and independent of ambient temperature. (But in the late stages of the race, when ambient temperature reached [86°F/30°C] and the hot sunshine was beating down on them, this would no longer be true.) The increase over resting temperature represents the error signal (actual temperature versus the regulated set-point temperature) in the negative feedback control system for thermoregulation.

The first physiological response implemented to maintain thermal homeostasis is an increase in blood flow to the skin. In this way, heat is carried by the blood from the muscles and the core to the body surface, where it is dissipated to the environment. Skin blood flow can increase from about 0.15 L/min. at rest to more than 7

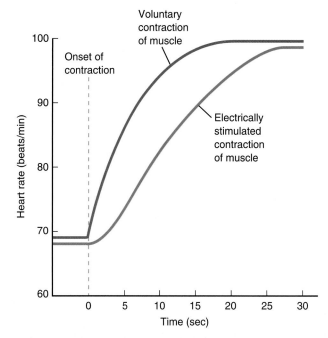

FIGURE 23.3 Central command. *Heart rate increases more rapidly in response to voluntary exercise than to muscle contraction in response to external electrical stimuli.*

Source: Redrawn from A. Krogh and J. Lindhard, "A comparison between voluntary and electrically induced muscular work in man." *Journal of Physiology* (London), 51(1917): 182–201.

Is central command an example of intrinsic control or extrinsic control of the heart?

L/min. during exercise. The heat is transferred to the external environment primarily by convection and evaporation (sweating); conduction to the air is a minor component because air has a low thermal conductance, and conduction to the ground is low because Bill and Jane's feet are insulated by shoes and socks.

At the onset of exercise, sympathetically mediated cutaneous vasoconstriction increases in proportion to exercise intensity (as part of the "fight-or-flight" response). As body temperature increases, vasoconstriction diminishes (because increased tissue temperatures reduce the sensitivity of venous adrenergic receptors), resulting in a two-fold increase in blood flow to the skin. But this increase is miniscule compared to the 30-fold increase caused by a special sympathetically mediated active vaso*dilation* mechanism in the skin. This active vasodilation is the efferent arm of the negative feedback response to an increase in core temperature. Core temperature has a far greater influence on blood flow to the skin than does either skin temperature or even the local metabolic needs of the skin. (There seem to be no local chemoreflex mechanisms in the skin.) The neurotransmitter and receptors involved in sympathetic cutaneous vasodilation have not been identified, but this mechanism seems somehow tied to the sweating response.

Extrinsic control

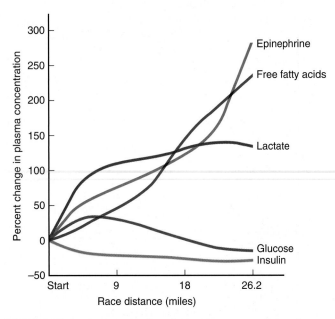

FIGURE 23.4 Changes in the plasma concentrations of some metabolic fuels and some hormones that regulate them over the course of a marathon.

Compiled from data reported by R. Cade et al., "Marathon running: Physiological and chemical changes accompanying late-race functional deterioration." *European Journal of Applied Physiology*, 65 (1992): 485–491, © Springer-Verlag; M.J. O'Brien et al., "Carbohydrate dependence during marathon running." *Medicine and Science in Sports and Exercise*, 25 (1993): 1009–1017; E.F. Coyle et al., "Muscle glycogen utilization during prolonged strenuous exercise when fed carbohydrate." *Journal of Applied Physiology*, 61 (1986): 165–172.

Quick Test 23.2

1. Prior to the start of exercise, there is often an anticipatory increase in sympathetic nervous activity and decrease in parasympathetic activity. What is the effect of these changes on cardiac output? blood glucose concentration? Explain.

2. A gradual increase in exercise intensity at the onset of exercise helps conserve a particular fuel that is present in the liver and in muscle. What is this fuel?

3. Define the term *central command,* as it relates to the cardiovascular and respiratory adjustments that occur at the start of exercise.

4. What are the two reserve mechanisms that enable the lungs to accommodate up to a six-fold increase in blood flow with little or no change in their ability to deliver oxygen to the blood or remove carbon dioxide?

THE LONG HAUL: ALMOST STEADY STATE

After the initial transition from primarily anaerobic to primarily aerobic metabolism in the first mile or so, the flow of energy (from glucose and fatty acids to ATP and then to heat and external work by exercising muscles)

has settled into a relatively "steady-state" condition. It is not a perfect steady state, however, because muscle glycogen is slowly but continuously being depleted, and blood glucose is taking up the slack. For the time being, Bill and Jane feel good; let us examine events occurring in their bodies at this time.

Hormonal Control of Energy Metabolism

Mobilization of glucose and fatty acids into the bloodstream from storage sites (liver and adipose tissue, respectively) is regulated by glucagon and epinephrine. A decline in insulin concentration also contributes to this mobilization. The changes in the circulating concentrations of some of these substrates and regulatory hormones over the course of a 26.2-mile marathon are illustrated in Figure 23.4. (Lactate is included in the graph because it is involved in gluconeogenesis.) Plasma glucagon concentration increases about 67% by the end of a marathon and cortisol, another important regulator of energy metabolism, increases about 95%.

The slight increase in glucose concentration during the first two-thirds of the race is due to the stimulatory effect of epinephrine on glycogenolysis and gluconeogenesis. Even though an increase in plasma glucose after a meal causes the pancreas to secrete insulin, plasma insulin concentrations decrease during exercise due to inhibition of insulin secretion by sympathetic nervous input to the pancreatic beta cells. Glucagon secretion is stimulated by sympathetic nerve impulses to the pancreatic alpha cells. Lactate (a product of anaerobic metabolism) rises quickly at the beginning of the race but levels off as aerobic metabolism and gluconeogenesis increase. Free fatty acid concentrations rise steadily as a result of epinephrine and a high glucagon-to-insulin ratio. Note that blood glucose levels fall below baseline, and that epinephrine concentrations rise precipitously in the final third of the race; these are signs of physiological distress (as we will see later in the chapter). For the time being (miles 4–18), blood glucose concentrations are slightly elevated, so Bill and Jane have a conscious, subjective feeling of well-being.

Some evidence suggests that the sex of a person influences how various energy substrates are used, but the relative importance of this influence during exercise and the mechanisms involved remain topics of debate. In some experiments, testosterone (higher in Bill) and estrogen and progesterone (both higher in Jane) directly alter key enzymes that control glucose and free-fatty-acid metabolism. These steroid hormones may also indirectly alter glucose and fat metabolism by influencing the secretion and action of other hormones such as epinephrine and insulin. Men tend to have higher circulating epinephrine concentrations during exercise, which would enhance the rate of glycogenolysis in muscle and liver. Whereas estrogens increase lipolysis in muscle and adi-

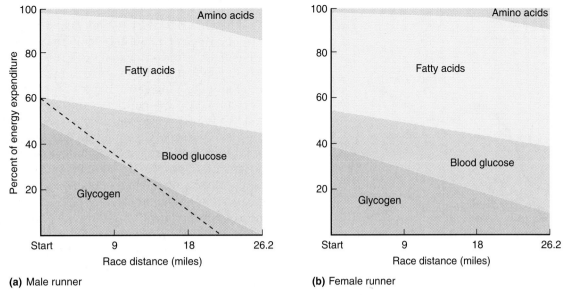

(a) Male runner

(b) Female runner

FIGURE 23.5 Use of nutritional fuels for energy by men and women during a marathon. (a) *The relative dependence of a typical male runner on each fuel source. The contribution from muscle glycogen declines from 50% at the start of the race to less than 10% at the end; this decline is offset by increased use of blood glucose (derived from gluconeogenesis) and fatty acid oxidation. In the late stages of the race, deaminated amino acids contribute significantly to gluconeogenesis. The dashed line, which represents Bill's rate of muscle glycogen consumption, indicates that by starting out too quickly, he consumed glycogen at an undesirably high rate that ultimately resulted in glycogen depletion before the race ended.* **(b)** *The relative dependence of a typical female runner on each fuel source. Compared to men, women rely more heavily on fatty acid oxidation, which results in less depletion of muscle glycogen and less oxidation of amino acids.*

Source: Adapted from Figure 5 in E.F. Coyle et al., "Muscle glycogen utilization during prolonged strenuous exercise when fed carbohydrate." *Journal of Applied Physiology* (1986): 165–173.

pose tissue, they inhibit gluconeogenesis and glycogenolysis; as a result, men burn a relatively greater proportion of carbohydrate for fuel, and women burn a relatively greater proportion of fat (Figure 23.5). Some researchers have estimated that over the course of a marathon, women may oxidize as much as 75% more fat and 40% less carbohydrate than men. However, other researchers have observed that the differences in substrate use between men and women decrease as the level of training and aerobic fitness increases.

Cardiovascular Control: Factors Influencing Blood Pressure

During the early and middle stages of the race, Bill and Jane's mean arterial pressure is approximately 20% higher than at rest. Why should this be? Why isn't peripheral resistance adjusted by baroreceptor-mediated negative feedback mechanisms to bring blood pressure back to normal? The answer is probably related, in part, to blood flow in muscles during exercise. Blood vessels are compressed each time a muscle contracts; in many muscles, blood flow stops altogether when muscle tension exceeds 70% of maximum. Various "metaboreceptors" that are sensitive to the mechanical stresses of compression, the transient buildup of metabolites such as potassium, or increases in temperature send afferent signals to integrative centers in the central nervous system; this information initiates a regulated rise in mean arterial pressure.

The results of a classic experiment that demonstrated this metabolic sensor system are shown in Figure 23.6. Blood pressure rises gradually during rhythmic contraction of skeletal muscle and then rapidly returns to resting value when the contractions cease. Part of this response is due to increases in metabolite concentrations in the muscle as blood flow is interrupted during each muscle contraction. The "metaboreceptors" send intermittent (but repetitive) signals to increase blood pressure. If the buildup of metabolites is exaggerated by completely blocking blood flow to the exercising muscle with a tourniquet, blood pressure rises faster and higher. Note that blood pressure remains elevated after exercise ceases and does not return to normal until after the tourniquet is removed and the metabolites are "washed out" by the restored flow of blood.

What is the benefit of this metabolite-regulated rise in blood pressure during exercise? After all, it means the heart must work harder. Because blood pressure is the driving force for perfusing the muscle with blood, a higher driving force ensures that flow is maintained through a greater portion of the skeletal muscle contraction cycle, and that flow will be enhanced between contractions to help wash out the metabolites that have built up with each momentary stoppage of flow.

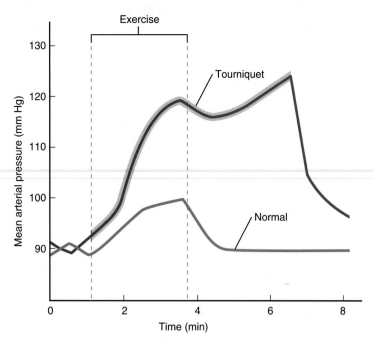

FIGURE 23.6 Regulation of blood pressure by metabolite buildup in exercising muscle. *When greater quantities of metabolites are trapped in an exercising arm by a tourniquet, blood pressure rises more quickly and remains elevated as long as the tourniquet is in place, even if exercise has ceased.*

Source: Redrawn from M. Alam and F.H. Smirk, "Observations in man upon a blood pressure raising reflex arising from the voluntary muscles." *Journal of Physiology* (London), 89 (1937): 372–383.

The muscle "metaboreceptors" are thought to contribute to cardiovascular control in humans in a meaningful way only at high exercise intensities; in contrast, central command signals arising from the motor cortex seem to be active during all exercise intensities (Figure 23.7). These signals keep Bill and Jane's blood pressure slightly elevated through at least the first two-thirds of the marathon; in the final stages of the race, blood volume diminishes (via sweating) to the point that adequate perfusion pressure to vital organs (especially the brain and heart) becomes threatened. At this point, the classical negative feedback system—baroreceptor-mediated autonomic control—assumes dominance in maintaining mean arterial pressure.

Respiration: The Effort of Breathing

During the race, Jane and Bill feel the effort involved in moving their bodies horizontally along the road and vertically up hills, but not the effort of breathing. However, moving air into and out of their lungs requires a significant amount of effort, and about 15% of cardiac output during exercise flows to respiratory muscles in order to accomplish this work. In this section we see how Bill and Jane have unconsciously learned to breathe with less effort, and how the various movements associated with exercise can provide a boost to respiratory movements.

First, it is necessary to understand how we can measure the work of breathing. Recall that inflation of the lungs starts with contraction of the diaphragm (with increasing help from the external intercostal muscles as ventilation rate increases), which causes a decrease in intrapleural pressure, which expands the lung tissue. At rest, expiration is due primarily to the passive recoil of the elastic lung tissue, but during exercise the contraction of internal intercostal and abdominal muscles increases intrapleural pressure, which helps "push" air out of the lungs. Measuring the magnitude of the decreases and increases in intrapleural pressure gives an indication of the effort (work) of breathing. Placing a pressure gauge directly in the pleural space would be difficult and downright dangerous; fortunately, the esophagus runs through the pleural cavity, and (so long as active swallowing is not occurring) the pressures measured there provide a reasonable approximation of intrapleural pressure.

If we were to ask Bill or Jane to hyperventilate under resting conditions in order to mimic their ventilation rates during the marathon, they would have to produce over four times as much pleural pressure during expiration to attain the same volume and flow rate. The explanation for this relates to airway resistance. During exercise, epinephrine relaxes airway smooth muscle (the opposite of what it does to vascular smooth muscle). But unconscious breathing during exercise is also influenced by receptors that monitor either the tension of respiratory muscles, the stretch of the lung tissue, or the pressure in the airways. These receptors send afferent signals to the respiratory control center that coordinate the overall breathing pattern by "fine tuning" both the recruitment order of respiratory muscles and the force output of each muscle. This results in an orderly compression of alveoli with minimal compression of the conducting airways, which would increase resistance. Such efficient ventilation mechanics during exhalation can be likened to what occurs when one squeezes a tube of toothpaste from the bottom (Figure 23.8a). Conscious hyperventilation at rest, however, overrides this control mechanism and results in haphazard compression of the lung structure accompanied by increased turbulence, airway resistance, and pockets of trapped air (Figure 23.8b). Exercise training seems to help refine the unconscious neural control mechanism, which may explain why untrained individuals "feel their lungs burning" (irritation due to air turbulence) when they exert themselves.

The muscular actions needed for running are not limited to the legs. The arms swing and the torso moves to counterbalance the leg movements and maintain balance. (This explains Bill's sore biceps muscles after the marathon.) Furthermore, abdominal organs bounce around, pushing on the diaphragm. These muscular contractions and intestinal movements contribute to the expansion and compression of the thoracic cavity and thus have some influence on ventilation. A runner can take advantage of these movements, because when breathing

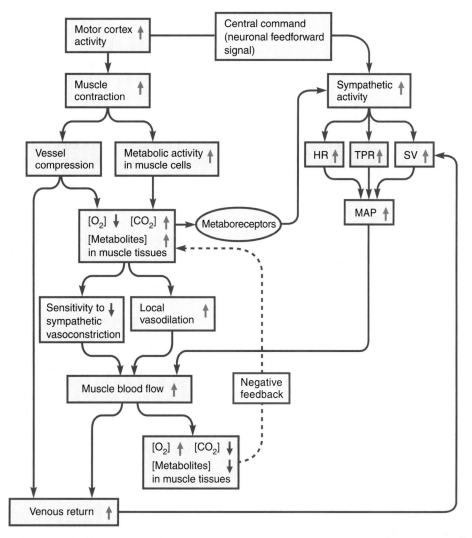

FIGURE 23.7 The roles of "metaboreceptors" and central command in control of mean arterial pressure during exercise. *The main benefit of increased mean arterial pressure is enhanced blood flow to muscles, which promotes delivery of oxygen and removal of metabolites from the active muscle tissue. Note that muscle contraction has two distinct influences on blood flow: By compressing veins running between muscles, venous return is enhanced (the skeletal muscle pump), but by compressing arterioles and capillaries within the muscle, blood flow and oxygen delivery are briefly interrupted, causing carbon dioxide and metabolite build up. These metabolites stimulate several mechanisms that promote skeletal muscle blood flow. They cause vascular smooth muscle relaxation, resulting in local vasodilation. They stimulate "metaboreceptors" that trigger sympathetically mediated increases in mean arterial pressure. Finally, they reduce the sensitivity of the vasculature in the exercising muscle to the sympathetic efferent signals that are causing vasoconstriction elsewhere in the body. This flowchart is available as an interactive exercise under Activities at* www.physiologyplace.com.

If an increase in muscle metabolism triggers local vasodilation and an increase in muscle blood flow, is it an example of active hyperemia or reactive hyperemia?

is synchronized with leg movements, 8% less oxygen is consumed compared to desynchronized running. An 8% energy savings compounded over several hours of marathon running is quite significant. People tend to synchronize these movements unconsciously, and trained runners get "in sync" faster than untrained runners.

The Gastrointestinal Tract: The Forgotten System

Because most runners have more than enough energy stores to support their metabolic needs through the completion of the marathon, they do not need to eat during

Active hyperemia

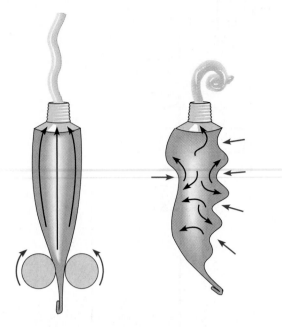

FIGURE 23.8 A model of ventilation mechanics during exhalation. (a) *Squeezing from the end of a toothpaste tube and moving the applied force toward the opening results in smooth, unidirectional, unimpeded flow.* (b) *Squeezing the tube in the middle with an open-fingered grasp results in turbulent, chaotic flow and pockets of trapped material.*

the event. Therefore, the energy needs of gastrointestinal tissues are relatively low, and blood flow through this region is "expendable," at least temporarily. Thus gastrointestinal function during exercise has received relatively little attention. However, as Bill discovered, sometimes the gastrointestinal tract cannot be ignored during exercise.

At rest, ingested food is propelled through the gastrointestinal tract by the slow, steady rippling action of the smooth muscle in the walls of the stomach and intestines (gastrointestinal motility). During and after a meal, motility and other digestive functions are stimulated by parasympathetic efferent signals and inhibited by sympathetic efferent signals. The regulated transit of digested food through the small intestine and colon allows adequate time for absorption of almost all of the water from the lumen.

Bill's gastrointestinal difficulties during the race are a common experience for runners. Diarrhea results from inadequate absorption of water due to excessively rapid transit of digested food through the gastrointestinal tract. Although intrinsic gastrointestinal motility is thought to be reduced during exercise (because parasympathetic efferent signals are reduced and sympathetic efferent signals are increased), the entire body is bouncing up and down with each stride. Therefore, the contents of the gastrointestinal tract are literally shaken, propelling them by forces that are outside the normal control of gastric and intestinal motility; this might lead to movement through

the small intestine and into the colon at a rate faster than appropriate for adequate absorption of water. The movement of radioactive-tracer-labeled food has been studied under controlled exercise conditions. Various studies have reported results ranging from no change to 20% increases in gastric emptying rate during exercise, and 10% increases in transit through the small intestine during exercise. Alternatively, psychological stress (at rest) increases transit through the small intestine by 25%. Runner's diarrhea, therefore, may be caused by a combination of stress and jarring movement. Furthermore, because caffeine stimulates secretion of fluids into the small intestine, Bill's coffee intake may have contributed to his distress at mile 8 in the race.

Some controversy exists regarding the influence of glucose contained in sports drinks on water absorption. As we will discuss shortly, fluid replacement during marathon running is a critical issue. *Hypo*osmotic carbohydrate-containing beverages are beneficial if runners find them more palatable than water and therefore drink more. However, *hyper*osmotic beverages may inhibit diffusion of water out of the intestinal lumen, leading to discomfort or diarrhea.

Gastrointestinal blood flow may possibly be constricted so drastically during heavy exercise in the heat that the epithelial cells are rendered somewhat ischemic, which can cause increased permeability of the epithelium to relatively large molecules. The consequences of this phenomenon are described in the section on postexercise fever.

Thermoregulation: Intensifying the Competition for Blood

Bill and Jane have been running for 2 hours, and the air temperature has increased to 86°F (30°C). The thermal gradient between their skin and the air is now smaller than it was at the beginning of the race, and thus heat flow due to convection is less effective, and heat loss via evaporation is increased. More blood is diverted to the skin, and increasing amounts of fluid are lost as sweat. The competition among organs for blood is becoming intense, and compromises must be made if both exercise performance and homeostasis are to be maintained.

The changes in the distribution of blood flow that result during heat exposure and exercise are depicted in Figure 23.9, in which a triangle represents the body. The vital organs (brain, heart, and lungs) are situated at the center of the triangle; the skin, skeletal muscle, and viscera (gastrointestinal tract) are situated at the corners. The relative amounts of blood flowing through the organs are represented by color: Deeper shades of red represent greater blood flow. The liver and kidneys are not easily categorized as belonging to either the "vital organs" or the viscera in this figure because even though they are vital organs, their blood flow can be reduced somewhat

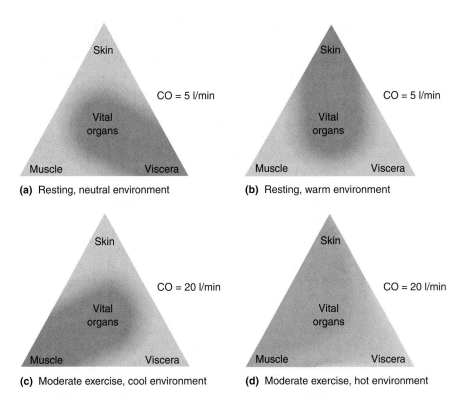

(a) Resting, neutral environment

(b) Resting, warm environment

(c) Moderate exercise, cool environment

(d) Moderate exercise, hot environment

FIGURE 23.9 Thermoregulatory-induced changes in blood flow distribution. (a) *During resting (thermoneutral) conditions, most of the cardiac output is distributed to the vital and visceral organs.* (b) *When environmental temperatures increase, more blood is directed to the skin at the expense of the viscera and muscles.* (c) *During exercise in a cool environment, cardiac output can increase by a factor of 4. Although the percentage of blood directed to the vital organs has decreased, the absolute flow rate to the heart actually increases and flow to the brain remains constant.* (d) *As environmental temperature increases, more blood is directed away from exercising muscle, which degrades performance, and from visceral organs, which can cause organ damage if flow reduction is of sufficient severity or duration.*

during exercise for the sake of working muscle and temperature regulation.

At rest (Figure 23.9a), about 60–75% of cardiac output flows through the vital organs and viscera, with the balance distributed through the skin and skeletal muscle. Over a modest range of environmental temperatures (called the *thermoneutral zone*), body core temperature can be regulated solely by regulating the amount of warm blood sent to the cooler skin surface. In warm conditions at rest more blood must be redirected to the skin in order to dissipate heat and maintain constant body temperature (Figure 23.9b). This adjustment can be made simply by changing blood flow distribution (no change in cardiac output is required), and it in no way interferes with the function of the other tissues. Muscle energy needs are low when a person is resting.

During exercise under cool conditions (Figure 23.9c), blood flow to working muscle increases dramatically. This is accomplished not only by redistribution of blood flow away from skin and viscera (including mild reductions in renal and hepatic blood flow), but also by an increase in cardiac output. Blood flow to all organs and tissues is still sufficient to meet their metabolic needs.

As core temperature rises, the cardiovascular system is faced with conflicting demands: providing increased blood flow to the skin for heat dissipation *and* providing increased blood flow to exercising muscle without ever sacrificing flow to the brain and heart. This was Bill and Jane's cardiovascular condition through the first two-thirds of the marathon. In spite of these conflicting needs, they were coping quite well because their premarathon

training had expanded their blood volumes. But as was the case with substrate metabolism, their cardiovascular systems were not quite in steady state. Blood volume slowly diminished as the rate of sweating increased. Eventually, blood volume was no longer sufficient to provide all tissues with adequate flow, so vasoconstriction of renal, hepatic, and splanchnic blood vessels (vessels supplying digestive organs) increased. Eventually, blood flow to vital visceral organs (kidney and liver) began to be compromised (Figure 23.9d); this is not a sustainable condition, and if it occurs, exercise performance cannot be maintained in the last stages of the race.

Renal and splanchnic vascular resistance varies in direct proportion to core temperature. A 1.5–2.0°C increase in core temperature causes decreases in blood flow of 25% to the liver, 30% to the kidneys, and over 50% to the stomach and intestines. Moderate, transient reductions in blood flow to the kidney and liver can be tolerated, but the function of these organs must be maintained above certain critical thresholds if they are to perform the necessary filtration and detoxification of the blood.

"Safety valves" exist to prevent blood flow reductions below these critical thresholds. Recall that at rest, renal blood flow is about 1 L/min. and represents 20% of cardiac output. The primary function of the kidneys—to filter waste products from the blood—is clearly a function of importance that cannot be discontinued during exercise. Furthermore, if fluid flow is not maintained through the renal tubules, they can become clogged with protein precipitates (casts). Thus renal blood flow cannot

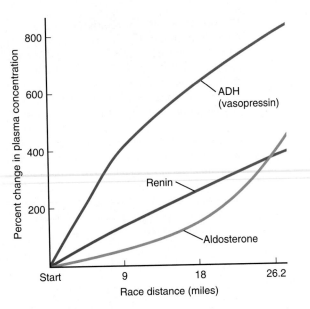

FIGURE 23.10 Changes in the plasma concentrations of some hormones that regulate fluid balance over the course of the marathon.

Compiled from data reported by J. Pastene et al., "Water balance during and after marathon running." *European Journal of Applied Physiology*, 73 (1996): 49–55; B.J. Freund et al., "Exaggerated ANF response to exercise in middle-aged vs. young runners." *Journal of Applied Physiology*, 69 (1990): 1607–1614; R.A. Irving et al., "The immediate and delayed effects of marathon running on renal function." *Journal of Urology*, 136 (1986): 1176–1180.

be interrupted and redirected to exercising muscle to the same extent as the flow to the stomach and intestines can be redirected. Although sympathetic activity reduces renal blood flow during exercise, it is counterbalanced by the strong influence of autoregulatory mechanisms within the kidney. In addition, prostaglandins (PGE_2 and PGI_2) produced locally in the kidney stimulate vasodilation that counteracts any excessive vasoconstriction due to sympathetic activity or increased plasma renin/angiotensin concentrations. Because aspirin and several other over-the-counter pain relievers work by blocking an enzyme involved in the synthesis of prostaglandins, taking such medication while exercising in the heat can inhibit an important safety mechanism.

Fluid Balance: The Weak Link in Steady State

Unlike energy stores, which, if managed properly, are more than adequate to complete a marathon, water stores must be replenished. Under warm conditions, rates of sweating can reach 30 mL/min. (1.8 L/hr). It is not unusual for runners to experience a net loss of 2 liters of body water during a marathon, even if conditions are cool and the runners are drinking during the race. Although sweat is a filtrate of plasma, only about 10% of the total fluid loss is evident as a change in plasma vol-

ume. In fact, running a marathon under mild conditions may result in very little change in plasma volume, because water (and solutes) redistribute from interstitial and intracellular locations to the plasma compartment.

One way the body counteracts fluid losses during exercise is to reduce urine production. At rest, urine production averages about 60 mL/hr, but this can vary enormously, depending upon fluid intake and ingestion of alcohol or caffeine. During exercise under cool conditions, urine flow rate may not change significantly, but sodium excretion decreases. Under warmer conditions such as those during the later stages of the marathon, both glomerular filtration rate and urine flow rate decrease.

Increased sympathetic nerve activity and reduced glomerular filtration rate cause the juxtaglomerular cells to release more renin. This enzyme is involved in the proteolytic cleavage process that ultimately produces angiotensin II, which in turn stimulates aldosterone secretion by the adrenal cortex. Aldosterone promotes sodium reabsorption in the renal tubules (and therefore helps conserve water). The time course for changes in the circulating concentrations of hormones that influence renal function are shown in Figure 23.10.

In the later stages of the race Bill and Jane's plasma became hyperosmotic due to their heavy sweating. Sweat glands are like miniature kidneys in the sense that iso-osmotic fluid from blood plasma is pumped into ducts, and then the sodium and other electrolytes are reabsorbed as the fluid moves through the ducts toward the skin surface. The loss of hypo-osmotic sweat results in hyperosmotic plasma, which is sensed by osmoreceptors in the hypothalamus. This triggers the release from the posterior pituitary gland of antidiuretic hormone (ADH, also known as vasopressin), which promotes water reabsorption by the renal tubules.

Although water conservation occurs via the physiological reflex mechanisms just described, replacing lost water requires a behavioral response: drinking. But the thirst mechanism in humans does not provide a tightly coupled negative feedback loop that keeps pace with fluid loss. After dehydrating exercise, people usually drink only about two-thirds of the water they lost as sweat, a situation called "voluntary dehydration." As a result, people must consciously try to drink more fluid than their thirst indicates. Because Jane drank whether she was thirsty or not, she kept her fluid volume more stable than Bill, who only drank to satisfy his thirst through most of the race. Voluntary dehydration is the major reason for flavoring and sweetening "sports drinks"—it makes them more palatable than plain water and thus encourages greater fluid intake. Although sports drink manufacturers and some researchers contend that adding sugar to the drink helps maintain blood glucose and therefore improves athletic performance, controlled experiments are equivocal on this matter.

1. Soon after the start of exercise, insulin secretion decreases despite an increase in plasma glucose levels. What is responsible for the decrease in insulin secretion?

2. What are muscle *metaboreceptors*? How are these involved in the control of cardiovascular function during exercise?

3. As core body temperature rises during exercise, the cardiovascular system is faced with the conflicting demands of providing increased blood flow to muscles and providing increased blood flow to what other organ or structure?

4. Define the term *voluntary dehydration* as it applies to exercise.

THE DECLINE TO THE END: "THE WALL" OR THE FINISH LINE?

Exercise diminishes energy stores, decreases blood volume, and places many other physiological stresses on the body that will ultimately cause performance to decline. The degree and speed of this decline are determined by how well the body is able to adapt to these stresses and by behavior, as we will see in this section.

The Prime Directive: Protect the Brain

The organ that directs Bill and Jane's motor control, guides their behavior, processes their objective and subjective sensations—and imposes the strictest, incontrovertible demands for blood flow—is the brain. The energy needs of the brain change very little under varying physiological conditions—"thinking as hard as you can" will not change the rates of total blood flow (about 750 mL/min.) or oxygen consumption (about 50 mL/min.). But depending on what type of information is being processed, the distribution of blood throughout the various structures of the brain can vary considerably. The energy is used to maintain ion gradients across neuronal membranes, which in turn allow neuronal depolarizations—the electrical activity of information processing and thought. Interrupting blood flow to the brain can cause unconsciousness within just a few seconds. Severe interruptions, such as those resulting from a blood clot lodging in a cerebral blood vessel (stroke) or loss of blood pressure (hemorrhage or cardiac arrest), can permanently affect memory and intellectual processes. By analogy, interrupting the power to a computer results in the loss of all information in memory that was not saved to disk. We do not seem to have any back-up system analogous to a disk drive for our memories, thus we cannot ever afford to have our brains power down. As such, all other organs (except the heart which delivers the "power" to the brain) can and will be sacrificed for the sake of the brain.

Fatigue: The Early Warning System

The fatigue that sets in during the later stages of a marathon is caused by a combination of three factors: (1) intrinsic factors within the exercising muscles, (2) systemic factors influenced by metabolic and environmental stresses, and (3) psychological factors that are still poorly understood.

Within the muscle, ATP, creatine phosphate, and glycogen are energy-related factors that could theoretically become depleted and cause muscle fatigue. In reality, ATP concentrations are reduced somewhat during exercise but attain a steady state. Creatine phosphate concentrations may fall progressively during high-intensity exercise, but they do not fall to zero during the moderate exercise intensity of a marathon. Most evidence indicates that the terminal fatigue experienced in an endurance event such as the marathon correlates with the depletion of glycogen. Although more and more of the energy used by muscle originates from beta oxidation of fatty acids as the race progresses, apparently some small energy contribution continues to be supplied by the carbohydrate stored as glycogen. Primarily small-diameter, slow-twitch, fatigue-resistant (oxidative) fibers are recruited in marathon running, but the glycogen stores in these smaller fibers can be depleted, resulting in recruitment of more large-diameter, fast-twitch fibers (which until then were relatively inactive). This recruitment allows the muscle as a whole to continue contracting.

Although lactic acid accumulates during fatigue, the lactate anion is not the cause of fatigue. In fact, lactate released by contracting muscle is recycled into glucose in the liver by gluconeogenesis. As shown previously (see Figure 23.4), plasma lactate concentrations rise moderately and remain relatively stable during a marathon. (In contrast, plasma lactate can increase over 1000% during exercise requiring maximal oxygen consumption.) The hydrogen ion, by contrast, contributes to the fatigue of maximal intensity exercise in several ways: (1) by inhibiting phosphofructokinase, an enzyme of glycolysis; (2) by displacing calcium ions from troponin, shutting down crossbridge cycling; (3) by stimulating local pain receptors; (4) by acting on the brain to cause pain and nausea; (5) and by interfering with *hormone-sensitive lipase,* an intracellular enzyme that mobilizes free fatty acids.

Are these H^+-mediated mechanisms of fatigue important during marathon running? As shown in Figure 23.11, the pH of arterial blood falls only modestly (and linearly) as exercise intensity increases up to the level attained during a marathon (about 70% of maximal aerobic capacity); at higher exercise intensities, plasma pH decreases more exponentially. In contrast, noninvasive NMR techniques have shown that muscle pH decreases linearly over the entire range of exercise intensities.

Few studies have examined the extent to which plasma pH might change over a long period of time at a contant exercise intensity. As shown in Figure 23.12, arterial pH falls quickly to a level determined by exercise

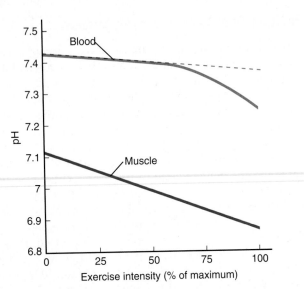

FIGURE 23.11 Influence of exercise intensity on arterial blood pH and intracellular pH of exercising leg muscle. *At moderate exercise intensities (up to about 65% of maximum), the rate of H^+ clearance from the circulation rises in proportion to its rate of release into the circulation, resulting in a linear relationship between arterial blood pH and exercise intensity (dashed line). At higher exercise intensities, H^+ is produced and released into the blood at much greater rates than it is cleared, resulting in the deviation from linearity. For practical reasons, experiments designed to investigate this relationship have maintained each exercise intensity for only a few minutes.*

Compiled from data reported by P.A. Mole et al., "Myoglobin desaturation with exercise intensity in human gastrocnemius muscle." *American Journal of Physiology*, 277 (1999): R173–R180; and J.A. Zoladz et al., "Changes in acid-base status of marathon runners during an incremental field test." *European Journal of Applied Physiology*, 67 (1993): 71–76.

As the kidneys compensate for acidosis, does bicarbonate excretion increase or decrease?

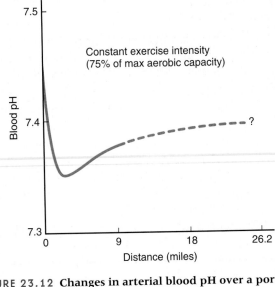

FIGURE 23.12 Changes in arterial blood pH over a portion of a marathon. *The dashed line is an extrapolation; measurements have not been made during later stages of the race, when blood volume and glucose stores are diminished.*

Source: Adapted from a study by A. Usaj and V. Starc, "Blood pH and lactate kinetics in the assessment of running endurance," *International Journal of Sports Medicine* 17 (1996): 34–40.

nase and displaces calcium ions from troponin. Calcium may accumulate in mitochondria, which interferes with oxidative phosphorylation. The significant increases in plasma potassium concentrations that occur during a marathon have been implicated as contributing factors in fatigue.

Even though blood glucose concentrations are usually fairly stable during exercise, even over the course of a marathon (see Figure 23.4), severe hyperthermia and dehydration can reduce renal and hepatic blood flow to the point that gluconeogenesis can no longer sustain the glucose needs of the brain. The resulting hypoglycemia is associated with profound psychological fatigue, disorientation, and possible loss of consciousness.

Tactical and Biological Influences on Fatigue

Based on Bill and Jane's physical characteristics as described in Chapter 1, Bill should have finished first. In this section we look at the reasons why he did not. His 2-minute "pit stop" was not a factor, because it was offset by Jane's eight 15-second stops to fully drink at the aid stations (but these two minutes of behavior on her part *were* a factor, as we will see).

When at the start Bill darted through the crowd while Jane remained trapped in the pack, he used up proportionally more of his muscle glycogen reserves than she did. Because he did not spend time at the aid stations, he covered the first 18 miles faster than Jane. Beyond this point, however, Bill paid for his early gains. He was in a much greater state of dehydration, which resulted in reduced venous return, higher heart rate, and reduced

intensity in subjects running at a typical marathon pace, then slowly rises. Why should it rise? Recall that the plasma concentration of any substance reflects the balance between its release into, and its removal from, the bloodstream. Many of the hydrogen ions released into the blood by exercising muscle are removed via respiration as they are combined with bicarbonate to yield carbon dioxide, which is subsequently exhaled. In addition, an increase in urinary hydrogen ion excretion during exercise is promoted by the rise in plasma aldosterone (see Figure 23.10), which also increases sodium reabsorption.

The evidence just presented suggests that hydrogen ion accumulation may play a contributing role in *local* muscle fatigue during a marathon but is probably not a major *systemic* factor (although we do not know much about how blood pH changes under the combined stresses of prolonged exercise, hyperthermia, and fluid depletion). Phosphate accumulation in muscle tissue, like accumulation of hydrogen ions, inhibits phosphofructoki-

It decreases.

ability to dissipate heat. His glycogen stores were depleted from smaller-diameter, slow-twitch fibers (which provide more controlled movement), which necessitated recruitment of large-diameter, fast-twitch fibers (which provide more explosive movement). The resulting loss of fine motor control led to stiff-legged, inefficient strides that resulted in greater energy expenditure per unit distance traveled—precisely the wrong development when energy stores were dwindling. Thus, in this dehydrated, glycogen-depleted, thermally stressed condition, Bill was unable to maintain his pace, and his final eight miles were run at a much slower pace. He lost whatever time he had gained by speeding past the aid stations. Thus, tactical errors impeded Bill's performance.

But Jane also had some physiological advantages that offset Bill's greater muscle mass and aerobic capacity. First, recall from Figure 23.5 that energy expenditure in a woman seems to differ from that in a man. A greater proportion of the energy Jane expended in the marathon was derived from fat, and thus muscle glycogen was used proportionally less. At the end of the marathon, Jane still had some glycogen remaining in her skeletal muscles. In addition, one research study found that women reported lower perception of leg muscle pain than men exercising at the same relative intensity. If confirmed, this type of pain tolerance might represent another reason Jane was better able to maintain her pace near the end of the marathon.

Conscious Decisions Versus Autonomic Control

Despite the existence of the "safety valve" mechanisms that protect the vital organs under extreme conditions, often the most effective homeostatic response is behavioral: A person ceases the activity or leaves the situation that is causing so much physiological stress. Afferent signals to the brain, including pain, acidosis, hypoglycemia, hypoxia, and hypotension, engender "psychological" fatigue. Therefore, fatigue can be considered as the efferent (and conscious) signal that prompts us to change our behavior (that is, stop running). A major element of marathon running requires that a person ignore this signal. However, the autonomic nervous system remains the ultimate arbiter: It can shut down excessively stressful activity no matter how hard runners will themselves to continue.

Quick Test 23.4

1. In an endurance event such as a marathon, terminal fatigue most likely results from depletion of what metabolic fuel?

2. In maximal intensity exercise, fatigue is associated with the accumulation of what chemical substance?

3. In marathon running, what type of skeletal muscle fiber is recruited initially?

THE AFTERMATH

In view of what we have learned about the physiological stresses of exercise, it should come as no surprise that a bout of severe and/or prolonged exercise, such as a marathon race, will have an impact on body function that can in some cases last for many hours after the exercise has stopped. In this section, we explore some of these post-exercise consequences.

Syncope: The Autonomic System's Final Veto Power

Even though Bill was hypovolemic in the last few miles of the race, he did not collapse until after he crossed the finish line. Why? Once Bill had come to a stop, the skeletal muscle pump activities of his legs ceased, and blood began to pool in his leg veins. Venous return to the heart decreased, and therefore ventricular filling pressure, stroke volume, cardiac output, and systemic blood pressure all fell precipitously. At this point, the autonomic nervous system used its final veto power: **syncope** (more commonly known as fainting).

In Bill's condition, syncope provided several homeostatic benefits: First, skeletal muscle activity ceased (except his diaphragm), which ended the competition for blood flow between the brain and muscle. Second, heat production was reduced, which also ended the competition for blood flow between the brain and the skin. Third, because 70% of Bill's blood volume was below the level of his heart when he was standing, venous return was impeded by gravity. Syncope brought Bill's heart, legs, and brain to the same level, allowing improved venous return unimpeded by gravity, better filling pressure, improved stroke volume, and improved cardiac output. However, syncope itself can pose hazards. Fortunately Bill was not hurt by the fall nor was he resting in a dangerous location.

Water Intoxication

Bill recovered from his fainting spell only to succumb later to a debilitating headache and vomiting. The cause was hypo-osmolarity and the resultant shifts of water between the intracellular fluid and the extracellular fluid (Figure 23.13).

Over the course of the marathon, Bill and Jane were gradually losing sodium, chloride, potassium, and other electrolytes ("salts") as well as water in their sweat. Their extracellular fluid volumes decreased steadily throughout the race. They both lost relatively more water than salts because some of the salts were reabsorbed along the sweat gland ducts. As a result, their plasma and interstitial fluid became hyperosmotic, which drew water out of cells by osmosis (Figure 23.13b). Their cell volumes

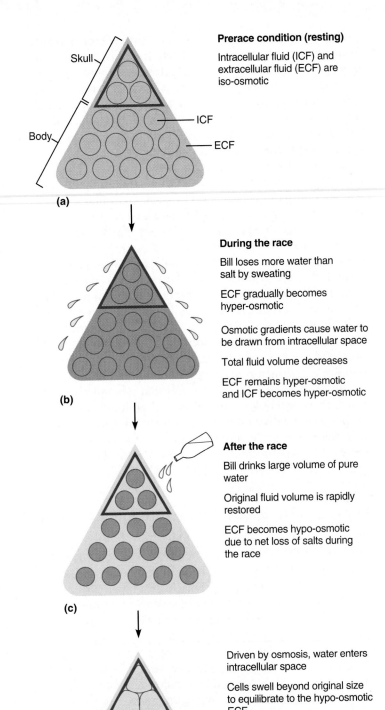

FIGURE 23.13 Fluid shifts during and after exercise. *In each diagram, the entire body (the large triangle) contains the extracellular fluid (ECF) and cells containing the intracellular fluid (ICF). Of the cells, three are encased in the skull (the small triangle). The darker the shade of blue, the greater the concentration of salts.*

Would you expect the decrease in ECF osmolarity that occurs after the race to stimulate ADH secretion by the posterior pituitary or inhibit it?

Prerace condition (resting)

Intracellular fluid (ICF) and extracellular fluid (ECF) are iso-osmotic

(a)

During the race

Bill loses more water than salt by sweating

ECF gradually becomes hyper-osmotic

Osmotic gradients cause water to be drawn from intracellular space

Total fluid volume decreases

ECF remains hyper-osmotic and ICF becomes hyper-osmotic

(b)

After the race

Bill drinks large volume of pure water

Original fluid volume is rapidly restored

ECF becomes hypo-osmotic due to net loss of salts during the race

(c)

Driven by osmosis, water enters intracellular space

Cells swell beyond original size to equilibrate to the hypo-osmotic ECF

Most tissues are compliant to cell expansion

Rigid skull restricts cell expansion, causing pressure buildup, crowding and neuronal dysfunction

(d)

shrank somewhat due to the fluid shifts; this shrinkage may have influenced certain cellular functions, which may be another factor that contributed to fatigue.

After the race, Jane drank a modest amount of water and ate some food, which replenished electrolytes as well as water. In contrast, after the race Bill drank a volume of pure water that was almost equal to the volume he lost through sweating during the race, but because he did not also take in any electrolytes, his plasma and interstitial fluid became hypo-osmotic (Figure 23.13c). Water moved into his cells by osmosis to equilibrate intracellular and extracellular osmolality. But due to the net loss of electrolytes during the race, more water entered Bill's cells than originally left them, causing the cells to swell (Figure 23.13d). Although this was not a problem for most of his cells, the critical exceptions were the central nervous system cells encased in his skull. The resultant intracranial pressure increase caused Bill's headache, vomiting, and convulsions—a condition termed **water intoxication.** We can now see that voluntary dehydration may be a protective mechanism that limits rehydration and guards against serious osmotic imbalances that can lead to disturbances in the central nervous system. (Voluntary dehydration is diminished if food is ingested with water.) Bill's water intoxication was actually a very serious condition and could have been life threatening. His friends probably should have taken Bill to the hospital for intravenous infusion of electrolytes, instead of letting him sleep.

Postexercise Fever

The elevated body temperature that occurs *during* exercise is characterized as hyperthermia—that is, a body temperature that is above the set-point temperature (an error signal). However, hours *after* long-duration exercise such as a marathon, some people feel chilled and wrap up in a blanket (as Bill did), even though their core temperatures are at or above "normal." The initiation of heat conservation mechanisms (vasoconstriction also occurs) at a time of elevated core temperature indicates that the thermoregulatory set point has been shifted upward (the definition of a fever).

The change in set point is due to the action of leukocyte-derived proteins (cytokines) on the thermoregulatory center in the hypothalamus. Production of specific cytokines, such as interleukin-1 and interleukin-6 (IL-1 and IL-6), is increased in the period 4–24 hours following

A decrease in ECF osmolarity will tend to inhibit ADH secretion.

exercise such as a marathon. This increased production occurs because fragments of extracellular matrix and damaged cells bind to "scavenger receptors" on macrophages, which initiates production of cytokines. These cytokines, in turn, stimulate the liver to synthesize plasma proteins that serve antioxidative functions and scavenge iron and heme. (After a marathon, the "blood" some runners see in their urine is actually heme from red cells damaged in the foot vasculature by the 42,000 impacts that occurred during the race.)

Another possible trigger for cytokine production after exercise is intestinal-derived endotoxin. We all have bacteria colonized in our lower gastrointestinal tracts. Fragments of the cell walls of some types of bacteria are a complex of carbohydrate and lipid known as endotoxin. If the epithelial barrier of the gastrointestinal tract becomes permeable to endotoxin (not necessarily to whole, viable bacteria), this endotoxin is taken up by the portal circulation and transported to the liver, which contains a high density of specialized macrophages called Kupffer cells. The endotoxin binds to specific receptors on the Kupffer cells, triggering cytokine production.

Exercising under conditions of severe heat stress can result in excessive vasoconstriction in the gastrointestinal tract, leading to ischemia. Ischemic damage to the epithelium allows large amounts of endotoxin to cross it, fully saturate the clearance mechanisms of the liver, and gain entry into the general circulation, activating monocytes and macrophages throughout the body. Overproduction of another cytokine, tumor necrosis factor (TNF), can lead to vascular permeability, peripheral vasodilation, and decreased cardiac contractility. At a time when maintenance of mean arterial pressure is already compromised by the fluid adjustments required to dissipate heat, TNF-induced decrements in cardiovascular function can lead to life-threatening situations. Severe cases of exertional heat stroke usually involve these mechanisms.

Delayed-Onset Muscle Soreness

Bill and Jane had somewhat sore and tired muscles after the race. In Bill's case, these subjective feelings were mild compared to the deep, diffuse aching, heaviness, and inflexibility he experienced upon waking the next morning. Microscopic examination of muscle samples taken after marathons reveals sarcomere damage, especially at the Z disks. The plasma membranes of muscle fibers become more permeable to large molecules, which is reflected by large increases in plasma concentrations of muscle cell enzymes such as creatine kinase. The changes in membrane permeability are thought to permit intracellular calcium concentration to increase, which disrupts muscle cell function.

Macrophages invade the affected muscle and clear out damaged fibers by phagocytosis. In addition, macrophages appear to deliver various cytokines that promote repair. In fact, evidence suggests that even though blocking the mild inflammatory response that normally follows a marathon by taking aspirin or other drugs alleviates acute symptoms of pain and swelling, taking these drugs can actually delay the long-term healing of contraction-induced skeletal muscle damage.

Jane's symptoms were less severe. Women exhibit much lower plasma creatine kinase concentrations following a marathon than do men. Various experimental approaches have shown that these reduced plasma creatine kinase concentrations are specifically due to circulating estrogens. The exact mechanism involved, and the functional significance in terms of susceptibility to muscle damage or rate of healing, are not presently known.

Effect of Marathon Running on Susceptibility to Sickness

Almost any exercise (even just running up a flight of stairs) leads to a transient increase in circulating white cells, primarily because of the mobilization of neutrophils. About half of the neutrophils in the vascular compartment do not circulate freely, but are instead loosely attached (electrostatically) to the inner walls of large-diameter veins, especially veins in the pulmonary circuit. The fluid shear stresses that accompany the increased cardiac output of exercise strip these cells free; epinephrine also triggers this "release." As a result, neutrophil counts can more than double following moderate exercise. The other populations of leukocytes can become activated via stress hormones or fragments of damaged extracellular matrix and cells, as previously mentioned. Once activated, these cells express adhesion molecules that bind to the vascular endothelium and help the leukocytes move out of the bloodstream into the interstitial space. Some subpopulations of lymphocytes may exhibit transient 50% (or greater) decreases in circulating numbers via this mechanism; other leukocyte subpopulations may increase as a result of increased lymphatic flow to the thoracic duct due to skeletal muscle pump activity.

These transient changes in circulating leukocytes have been misinterpreted as a change in overall readiness to resist infection, and anecdotal reports of chronic illness in elite athletes are offered as proof. However, little solid evidence exists to support this view. The changes in circulating leukocyte counts are small in magnitude and transient in nature. As we saw in Chapter 22, the main battlefield during infection is usually in the interstitial space, especially near the barriers to the external environment (skin, respiratory tract), not in the bloodstream. Only a tiny fraction of all leukocytes are actually in the bloodstream at any moment; a two-fold change in circulating cell numbers represents a shift in location of only 1% of the cells. Most controlled studies that have actually measured infection rates find no difference in incidence of infection between exercisers and sedentary people.

The resting (chronic) numbers and functioning of leukocytes are generally no different in highly trained versus untrained individuals. However, the spectrum of cytokines produced by the leukocytes of active people may be somewhat different than those of inactive people. Given that some of these cytokines are involved in the pathological mechanisms leading to atherosclerosis and other "inflammatory" diseases, the small, training-induced changes in the concentrations of these cytokines over long periods of time may contribute mechanistically to the well-established influence of exercise on cardiovascular health.

Quick Test 23.5

1. Explain how syncope can be benificial in restoring cardiac output following a drop in systemic blood pressure.

2. Explain why it is dangerous for a person to drink a large volume of pure water to replace fluid lost via sweating.

3. The elevation of thermoregulatory set point that occurs in post-exercise fever is triggered by what chemical substances acting on the brain?

CHAPTER SUMMARY

Principles of Physiological Integration, p. 742

Exercise places many demands on the body, which can be met only through the activities of different organ systems working together. One such demand is an increased need for energy to power skeletal muscle contractions, which for very brief periods can be supplied by molecules stored within the muscle cells themselves (ATP, creatine phosphate, glucose, and glycogen). More prolonged activity requires increased delivery of oxygen and fuels (glucose and fatty acids, for example) to the muscle cells from other sources. The body's ability to achieve and maintain increased levels of physiological function depends on the availability of energy and the often competing needs of different physiological systems. Within specific systems, adjustments are regulated by local metabolites (intrinsic control or autoregulation) and neural and hormonal signals (extrinsic control). Under most conditions, blood flow to muscles and other organs is controlled by autoregulatory mechanisms. When conditions become more stressful, organs begin to compete with each other for blood delivery, and neural and hormonal signals assume a greater role.

The Start: Transition from Rest to Exercise, p. 742

Often, exercise is preceded by an anticipatory increase in sympathetic nervous activity (and epinephrine secretion) and a decrease in parasympathetic activity. Cardiac ouput increases (due to increased stroke volume, heart rate, and venomotor tone), as does plasma glucose concentration (due to increased glucagon secretion and decreased insulin secretion by the pancreas). At the start of exercise, central command (a feedforward mechanism arising in the cerebral cortex) helps orchestrate the cardiovascular and respiratory adjustments that occur. The proportion of fuels utilized depends on exercise intensity and duration. Early on (before oxygen delivery can be increased), anaerobic glycolysis is the predominant ATP generating mechanism in exercising muscle cells; afterward aerobic metabolism (fatty acid oxidation and the Krebs cycle) becomes predominant.

Pulmonary ventilation increases due to central command and also in response to signals arising in mechanoreceptors and chemoreceptors in exercising muscles. In mild-to-moderate exercise arterial oxygen and carbon dioxide concentrations change very little despite a substantial increase in pulmonary blood flow. A sympathetically-mediated increase in blood flow to the skin helps dissipate the excess heat generated by exercising muscles.

The Long Haul: Almost Steady State, p. 746

After the initial shift from anaerobic to aerobic metabolism that occurs in prolonged exercise, fuel utilization settles into a relative steady state. However, glycogen (and body fluids) are continually being depleted. Mobilization of energy stores is regulated by changes in plasma levels of glucagon, epinephrine, and cortisol (which rise), and insulin (which falls). Plasma glucose rises initially due to epinephrine-stimulated glycogenolysis and gluconeogenesis but eventually begins to decline. Lactate rises due to the intial burst of anaerobic metabolism but levels off due to the rise in aerobic metabolism coupled with increased gluconeogenesis. Fatty acid levels rise steadily due to an increased glucagon to insulin ratio.

Central command signals appear to be important in maintaining an elevated mean arterial pressure (MAP) during exercise, which helps maintain increased blood flow to muscles. At higher exercise intensities, another regulatory mechanism comes into play that is triggered by signals arising in metaboreceptors that detect changing conditions in exercising muscles. Baroreceptor-mediated negative feedback becomes important in maintaining MAP only after blood volume has diminished (due to sweating) to the point where it threatens the cardiovascular system's ability to sustain adequate perfusion of vital organs. Increased pulmonary ventilation is aided by forced expiration and a decrease in airway resistance (due to epinephrine).

As core body temperature rises, the cardiovascular system becomes faced with the conflicting demands of providing increased blood flow to exercising muscles and to the skin (for heat dissipation). Vasoconstriction in renal, hepatic, and splanchnic vessels helps maintain MAP, especially as blood volume diminishes. Fluid losses may be counteracted through a decrease in urine production, a consequence of increased sympathetic activity, reduced GFR, and activation of the renin-angiotensin-aldosterone system. Water conservation is also promoted by ADH secretion, which is stimulated as sweating causes the plasma to become hyperosmotic.

The Decline to the End: "The Wall" or the Finish Line?, p. 753

The stresses of exercise ultimately cause a decline in performance; the degree and speed of this decline depend on how well the body is able to adapt to these stresses. Prolonged exercise ultimately leads to fatigue, which arises due to intrinsic factors within the exercising muscles, systemic factors influenced by metabolic and environmental stresses, and psychological factors. Glycogen depletion is most likely the pivotal event that triggers terminal fatigue in endurance exercise; in high-intensity exercise accumulation of hydrogen ions appears to be important.

The Aftermath, p. 755

At the end of exercise, cessation of skeletal muscle pump activity may result in a fall in blood pressure. In extreme situations, syncope may occur, which is actually beneficial in that it ends competition for blood flow between the muscles and the brain, reduces heat production by muscles, and increases venous return and cardiac output. Replacement of lost body fluids with large volumes of pure water after exercise can lead to cell swelling and water intoxication. Following prolonged exercise, a person may experience fever due to increased cytokine levels, which act on hypothalamic thermoregulatory centers to elevate the set point. A person may also experience delayed-onset muscle soreness, a symptom of inflammation triggered by contraction-induced muscle damage. This inflammatory response normally sets the stage for eventual muscle repair.

EXERCISES

Multiple-Choice Questions

1. Although water in sweat is derived from plasma, it is possible that a person could lose a significant volume of water via sweating while experiencing only a small decrease in blood volume. This is because:
   a) much of the blood volume is occupied by cells, not water.
   b) water moves into the plasma from other compartments, such as interstitial fluid.
   c) the kidneys divert water from the renal tubules to the peritubular capillaries.
   d) water moves from red blood cells to the plasma.

2. The skeletal muscle fibers that are first recruited in a marathon are primarily:
   a) slow glycolytic fibers.
   b) fast glycolytic fibers.
   c) slow oxidative fibers.
   d) fast oxidative fibers.

3. In the initial stages of a marathon plasma glucose levels rise and insulin secretion falls. This is because:
   a) pancreatic beta cells are sensitive to glucose levels in the fluid surrounding them and normally reduce their secretion of insulin in response to a rise in glucose concentration.
   b) the decrease in insulin secretion triggers the rise in glucose levels.
   c) increased sympathetic input to pancreatic beta cells suppresses insulin secretion despite the rise in plasma glucose.
   d) exercise reduces the responsiveness of many tissues to insulin.

4. Which of the following is evidence that negative feedback chemoreceptor control of ventilation is *not* responsible for the increase in pulmonary ventilation that occurs at the onset of moderate exercise?
   a) If negative feedback control were responsible, arterial P_{O_2} would fall and P_{CO_2} would rise at the beginning of exercise, triggering the increase in ventilation; these changes in partial pressure are not observed.
   b) If negative feedback control were responsible, arterial P_{O_2} would rise above normal during exercise and arterial P_{CO_2} would fall below normal; these changes in partial pressure are not observed.
   c) If negative feedback control were responsible, ventilation would not increase at all during exercise; negative feedback normally works to keep alveolar ventilation constant.
   d) If negative feedback control were responsible, ventilation would decrease at the start of exercise. Negative feedback almost always triggers responses that are inappropriate, which is why it is called *negative* feedback.

5. When core temperature rises, the resulting increase in blood flow to the skin conflicts with the cardiovascular system's efforts to maintain increased blood flow to exercising muscles. Why are these two processes said to be in conflict?
   a) Blood cools at it flows through the skin but warms as it flows through the muscles.
   b) Blood gains oxygen as it flows through the skin but loses oxygen as it flows through the muscles.
   c) The increase in blood flow to the skin is due to sympathetically mediated vasodilation, whereas the increase in muscle blood flow is due to parasympathetically mediated vasodilation.
   d) The vasodilation that allows skin blood flow to increase also lowers total peripheral resistance and thus tends to lower mean arterial pressure. Any reduction in mean arterial pressure would tend to decrease muscle blood flow.

6. Assume that the decrease in arterial pH that occurs during exercise is due to an increase in lactic acid production. If so, it is an example of:
   a) respiratory acidosis.
   b) metabolic acidosis.
   c) respiratory alkalosis.
   d) metabolic alkalosis.

7. Which of the following is an example of negative feedback control?
   a) stimulation of increased heart rate by central command signals at the onset of exercise
   b) stimulation of increased heart rate by signals arising in muscle metaboreceptors at the onset of exercise
   c) stimulation of pulmonary ventilation by signals arising in muscle mechanoreceptors and chemoreceptors at the onset of exercise
   d) vasodilation of skin blood vessels in response to a rise in core body temperature

8. If it were possible for a person to lose 1 L of *pure water* through sweating, how would it affect extracellular fluid (ECF) osmolarity *prior to any possible movement of water between body fluid compartments?* Assume that all intra- and extra-cellular solutes are impermeant and that ECF and intracellular fluid (ICF) osmolarities are initially at the normal value of 300 mOsm.
   a) ECF osmolarity will not change.
   b) ECF osmolarity will rise above 300 mOsm.
   c) ECF osmolarity will fall below 300 mOsm.
   d) More information is needed in order to determine how ECF osmolarity will be affected.

9. In the situation described in question 8, assume that any possible water movement between body fluid compartments has occurred and ECF and ICF have reached osmotic equilibrium. At this point, ECF osmolarity will be:
   a) normal at 300 mOsm.
   b) greater than 300 mOsm.
   c) less than 300 mOsm.
   d) More information is needed in order to determine how ECF osmolarity will be affected.

10. Starting with the situation described in question 9, suppose the person drinks 1 L of pure water. Assuming that this water has been completely absorbed into the ECF *but no movement of water between body fluid compartments has yet occurred,* how will ECF osmolarity be affected?
   a) It will decrease to normal (300 mOsm).
   b) It will decrease but remain above normal.
   c) It will decrease to below normal.
   d) More information is needed in order to determine how ECF osmolarity will be affected.

11. For the situation described in question 10, how will ECF volume compare to normal?
   a) ECF volume will be normal.
   b) ECF volume will be below normal.
   c) ECF volume will be above normal.
   d) More information is needed in order to determine how ECF volume will be affected.

12. For the situation described in question 10, how will ECF osmolarity compare to normal *after* any possible water movement between body fluid compartments has occurred and osmotic equilibrium has been attained?
   a) ECF osmolarity will be normal at 300 mOsm.
   b) ECF osmolarity will be below normal.
   c) ECF osmolarity will be above normal.
   d) More information is needed in order to determine how ECF osmolarity will be affected.

13. If the person described in question 9 lost 1 L of sweat containing solutes at a concentration of 50 mOsm and drank 1 L of pure water, how would ECF osmolarity compare to normal after the water has been absorbed and osmotic equilibrium has been attained?
   a) ECF osmolarity will be normal at 300 mOsm.
   b) ECF osmolarity will be below normal.
   c) ECF osmolarity will be above normal.
   d) More information is needed in order to determine how ECF osmolarity will be affected.

Objective Questions

1. Muscle mechanoreceptors send (feedforward/feedback) signals for respiratory control.

2. At the onset of exercise, pulmonary ventilation increases mostly as a result of an increase in (breathing frequency/tidal volume).

3. The velocity of blood flow through a typical alveolar capillary can increase to three times the normal resting value without affecting oxygen or carbon dioxide concentrations in systemic arterial blood. (true/false)

4. A gradual initial increase in exercise intensity helps conserve the body's stores of (glycogen/fatty acids).

5. During exercise core temperature normally rises due to elevation of the thermoregulatory set point. (true/false)

6. As the body's glycogen stores become depleted in prolonged exercise, blood glucose is maintained at or near its normal resting level through the conversion of (fatty acids/lactate) to glucose.

7. At the onset of low-intensity exercise, ATP production in exercising muscle cells occurs primarily through (aerobic/anaerobic) metabolism.

8. After exercising, people usually drink less fluid than they lost through sweating. This phenomenon is referred to as _____. (two words)

9. Sweat is (hypo-osmotic/hyperosmotic) relative to plasma.

10. Exercise has been shown to induce a significant increase in the number of leukocytes within the body. (true/false)

11. Under normal resting conditions, the distribution of blood flow to organs is controlled primarily by (intrinsic/extrinsic) regulatory mechanisms.

12. During prolonged exercise, plasma concentrations of renin would be expected to (increase/decrease).

13. As lactate begins to accumulate in the blood during prolonged exercise, some of it is converted to glucose. (true/false)

14. The central command signals that stimulate increased cardiac output at the onset of exercise are triggered in response to afferent signals arising in muscle metaboreceptors. (true/false)

15. As body temperature increases during exercise, skin blood flow increases due to (sympathetically/parasympathetically) mediated vasodilation.

Essay Questions

1. Many people who do not have use of their legs participate in marathons in wheelchairs. What physiological responses might be different in these individuals? How would someone who lacks motor neuron input to their legs differ from someone with absolutely no neuronal input to their legs?

2. Can heart transplant patients raise their cardiac output during exercise? If so, how?

3. Recall the discussion of the synchronized movement of our runners' leg muscles with their ventilatory muscles. What animals might have tightly coupled ventilatory-motor movements? Can you think of any disadvantages of having these processes too tightly coupled?

4. List the different physiological responses that are influenced by body movements or limb movements associated with locomotion.

5. Why is it more efficient to increase tidal volume instead of ventilatory frequency during moderate exercise?

6. As exercise intensity increases further, tidal volume stabilizes at about 65% of vital capacity, and additional increases in ventilation are due to increased frequency. Why?

7. How is an increase in body temperature during exercise different from a fever?

8. Aside from blocking pain, how does aspirin influence physiological function?

9. What might happen to free fatty acid mobilization during a marathon if a person ingested sports drinks or with too much sugar or ate orange slices along the way?

10. If sympathetic input generally promotes vasoconstriction, how is it that blood flow to skeletal muscles increases during exercise?

11. Make a list of all the physiological processes that are enhanced by epinephrine during exercise.

12. If you gave a rodent a chemical that inactivated all sympathetic efferent signaling (both neural and hormonal), could it still run? What mechanisms still exist during exercise under these conditions that promote physiological function?

13. If you gave the same drug mentioned in question 12 to human beings, they probably could not run but might be able to swim. Why?

Find the answers to these exercises, and additional study tools, at the Physiology Place (www.physiologyplace.com).

Answers to Odd-Numbered Multiple-Choice and Objective Questions

CHAPTER 1

Multiple-Choice Questions

1. b
3. c
5. d

Objective Questions

1. extracellular fluid
3. true
5. false
7. false
9. connective

CHAPTER 2

Multiple-Choice Questions

1. b
3. a
5. d
7. c
9. c
11. d
13. c

Objective Questions

1. polar
3. cholesterol
5. integral
7. false
9. connexons
11. UAG
13. leader sequence
15. cytokinesis

CHAPTER 3

Multiple-Choice Questions

1. b
3. d
5. c
7. b
9. a
11. d
13. a
15. b

Objective Questions

1. oxidation
3. negative
5. false
7. gluconeogenesis
9. chemiosmotic coupling
11. lowering
13. concentration
15. affinity

CHAPTER 4

Multiple-Choice Questions

1. b
3. c
5. b
7. a
9. c
11. a
13. c
15. c

Objective Questions

1. hydrophobic
3. true
5. false
7. true
9. true
11. true
13. true
15. false

CHAPTER 5
Multiple-Choice Questions

1. c
3. c
5. d
7. c
9. c
11. d
13. d
15. c

Objective Questions

1. secretory cell
3. hormone
5. adenylate cyclase
7. lipophilic
9. calmodulin
11. medulla
13. exocytosis
15. neurotransmitters

CHAPTER 6
Multiple-Choice Questions

1. d
3. d
5. d
7. d
9. a

Objective Questions

1. afferent and efferent divisions
3. glial cells
5. Schwann cells, oligodendrocytes
7. negative
9. dorsal root ganglia
11. false
13. saltatory
15. out of

CHAPTER 7
Multiple-Choice Questions

1. a
3. a
5. c
7. b
9. b

Objective Questions

1. gap junction
3. false
5. false
7. true
9. false
11. presynaptic inhibition
13. monoamine oxidase (MAO), catechol-o-methyltransferase (COMT)
15. aspartate, glutamate

Chapter 8
Multiple-Choice Questions

1. b
3. c
5. c
7. d
9. b
11. c
13. a

Objective Questions

1. choroid plexus
3. white
5. cerebellum
7. Broca's
9. thalamus, hypothalamus
11. dopamine
13. limbic
15. false

CHAPTER 9
Multiple-Choice Questions

1. c
3. e
5. b
7. a
9. b
11. b
13. d
15. b

Objective Questions

1. warm receptors; cold receptors
3. quickly
5. left
7. ganglion cells
9. false
11. opsin
13. false
15. false

CHAPTER 10
Multiple-Choice Questions

1. c
3. c
5. a
7. a
9. d
11. c

Objective Questions

1. parasympathetic
3. false
5. CN III, VII, IX, and X
7. all preganglionic neurons, postganglionic parasympathetic neurons, and motor neurons
9. phospholipase C
11. excitation
13. nicotinic
15. acetylcholinesterase

CHAPTER 11
Multiple-Choice Questions

1. d
3. a
5. c
7. b
9. b

Objective Questions

1. troponin
3. true
5. true
7. myosin head groups
9. myosin ATPase activity
11. oxidative

CHAPTER 12
Multiple-Choice Questions

1. b
3. c

5. b
7. a
9. a
11. a
13. d

Objective Questions

1. SA node
3. false
5. semilunar
7. systolic pressure
9. heart rate
11. increases
13. diastole
15. pulse

CHAPTER 13
Multiple-Choice Questions

1. c
3. b
5. c
7. d
9. c
11. a
13. a
15. a

Objective Questions

1. false
3. decreases
5. increases
7. larger
9. pressure
11. false
13. true
15. erythrocytes
17. thrombin

CHAPTER 14
Multiple-Choice Questions

1. d
3. b
5. c
7. a
9. b

Objective Questions

1. an increase
3. false
5. myogenic

7. decreases

9. heart rate

11. true

CHAPTER 15

Multiple-Choice Questions

1. d

3. a

5. c

7. b

9. d

Objective Questions

1. false

3. more

5. decreases

7. conducting zone

CHAPTER 16

Multiple-Choice Questions

1. c

3. d

5. a

7. b

9. d

11. c

13. d

Objective Questions

1. the same as

3. false

5. carbonic anhydrase

7. increases

9. carbaminohemoglobin ($HbCO_2$)

11. peripheral

13. irritant

15. higher

17. vasodilation

CHAPTER 17

Multiple Choice Questions

1. b

3. b

5. a

7. c

9. b

11. c

Objective Questions

1. urethra

3. afferent

5. higher

7. internal

9. basolateral

11. secretion

13. peritubular capillaries

15. proximal tubule

CHAPTER 18

Multiple Choice Questions

1. c

3. b

5. d

7. d

9. b

11. d

13. c

15. d

Objective Questions

1. increases

3. false

5. atrial natriuretic peptide

7. aldosterone

9. false

11. decreases

13. increase

15. false

CHAPTER 19

Multiple Choice Questions

1. b

3. d

5. b

7. d

9. a

Objective Questions

1. false

3. decrease

5. myenteric

7. chyme

9. cholecystokinin

11. mucosa

13. chief cells

15. false

17. true

CHAPTER 20

Multiple Choice Questions

1. a
3. b
5. d
7. a
9. a

Objective Questions

1. glucagon
3. true
5. absorptive
7. negative
9. stimulate
11. false
13. sympathetic
15. false

CHAPTER 21

Multiple Choice Questions

1. c
3. b
5. d
7. b
9. a

Objective Questions

1. false
3. persist
5. true
7. FSH
9. decrease
11. uterine tube
13. granulosa
15. endometrium
17. LH
19. estrogen
21. blastocyst
23. placenta
25. prolactin

CHAPTER 22

Multiple Choice Questions

1. e
3. a
5. e
7. b
9. b
11. b
13. b

Objective Questions

1. a) M   b) T_H c) T_C d) B   e) B
3. block
5. true
7. phagocytosis, endocytosis

CHAPTER 23

Multiple Choice Questions

1. b
3. c
5. d
7. d
9. b
11. c
13. b

Objective Questions

1. feedforward
3. true
5. false
7. anaerobic
9. hypo-osmotic
11. intrinsic
13. true
15. sympathetically

Credits

All illustrations by Precision Graphics.

Photos:

Chapter 1 Opener: Dr. Dennis Kunkel/Phototake

Chapter 2 Opener: Dr. Dennis Kunkel/Photo Researchers, Inc.
Exercise Link Photo: PhotoDisc, Inc.
2.16: Don W. Fawcett/Visuals Unlimited
2.17b: Visuals Unlimited
2.20b: Omnikron/Photo Researchers, Inc.
2.21: Biophoto Associates/Photo Researchers, Inc.
2.24: Don W. Fawcett/Visuals Unlimited
2.25: David M. Phillips/Visuals Unlimited

Chapter 3 Opener: Dr. Dennis Kunkel/Phototake

Chapter 4 Opener: David E. Scott/Phototake

Chapter 5 Opener: Dr. Dennis Kunkel/Phototake
5.27: John D. Cunningham/Visuals Unlimited
5.28: Don W. Fawcett/Visuals Unlimited

Chapter 6 Opener: David E. Scott/Phototake

Chapter 7 Opener: CNRI/Phototake

Chapter 8 Opener: Albert Tousson/Phototake

Chapter 8 Discovery: from: Damasio, H., T. Grabowski, R. Frank, A.M. Galaburba, A.R. Damasio: "The Return of the Phineas Gage: Clues about the Brain from a Famous Patient." Science, 264:11:02-1105, 1994. Department of Neurology and Image Analysis Facility, University of Iowa

Chapter 9 Opener: Insitit Pasteur/Phototake
9.21: Richard Megna/Fundamental Photographs
9.21: Richard Megna/Fundamental Photographs
9.39: Prof. P. Motta/ Dept. of Anatomy/University "La Sapienza". Rome/Science Photo Library/Photo Researchers, Inc.

Chapter 10 Opener: Carolina Biological Supply Co./Phototake

Chapter 11 Opener: John T. Hansen/Phototake
11.3: James E. Dennis/Phototake
11.3: Courtesy of H. Ris
11.5: Courtesy of John Heuser
11.6: James E. Dennis/Phototake
11.6: James E. Dennis/Phototake
11.24: Biophoto Associates
11.29: © Eric Graves/Photo Researchers
11.29: © Marion Rice

Chapter 12 Opener: CNRI/Phototake

Chapter 13 Opener: Dr. Dennis Kunkel/Phototake
13.1: Oliver Meckes/Eye of Science/Photo Researchers, Inc.
13.5: Meckes/Ottawa/Photo Researchers, Inc.

Chapter 14 Opener: Walter Rienhart/Phototake

Chapter 15 Opener: Institut Pasteur/CNRI/Phototake
15.4a: Alfred Pasieka/Peter Arnold, Inc.
15.4b: Biophoto Associates/Science Source/Photo Researchers, Inc.

Chapter 16 Opener: Dr. Dennis Kunkel/Phototake

Chapter 16 When It Goes Wrong: Stan Flegler/Visuals Unlimited

Chapter 17 Opener: Gopal Murti/Phototake

Chapter 18 Opener: Sovereign/IFM/Phototake

Chapter 19 Opener: Dr. Dennis Kunkel/Phototake
19.30: Science Photo Library/Photo Researchers, Inc.

Chapter 20 Opener: Carolina Biological Supply Co./Phototake

Chapter 21 Opener: Dr. Dennis Kunkel/Phototake

Chapter 22 Opener: Dr. Dennis Kunkel/Phototake
22.4: Meckers/Ottawa/Eye of Science/Photo Researchers, Inc.

Chapter 23 Opener: PhotoDisc, Inc.

Glossary

absolute refractory period period during and immediately following an action potential during which a second action potential cannot be generated in response to a second stimulus, regardless of the strength of that stimulus

absorption movement of substance from the external environment to the internal environment by transport across an epithelium

acetylcholine (ACh) (ass-ih-teel-koh'-leen) a neurotransmitter widely employed in both the central and peripheral nervous systems; the most abundant neurotransmitter in the peripheral nervous system, found in efferent neurons of both the somatic and autonomic nervous systems

acetylcholinesterase (assih-teel-koh-lin-es'-ter-ase) enzyme that degrades acetylcholine

acidosis condition in which arterial blood pH is 7.35 or lower

acrosome large vesicle within the head of the sperm; contains enzymes and other proteins that allow the sperm to fuse with the egg during fertilization

actin the contractile protein found in thin filaments in striated muscle

action potentials large changes in the membrane potential of excitable cells in which the inside of the cell becomes positive relative to the outside; function in transmitting information over long distances in axons

activation energy additional energy that molecules must acquire in order to react; the difference between the energy of the transition state and the energy of the reactants or products

active hyperemia (hy-per-ee'-me-ah) the increase in blood flow occurring in response to an increase in metabolic activity

active immunity a type of immune protection that depends on the ability of the immune system to mount a response

active transport any method of transport of molecules across a membrane that requires the use of energy

adenosine triphosphate (ATP) (ah-den'-oh-seen) compound that serves as primary direct energy source for cell activities; synthesized from adenosine diphosphate (or ADP) and inorganic phosphate (P_i)

adenylate cyclase (ad-den'-ah-late sy'-klase) an intracellular enzyme that catalyzes the conversion of ATP to cAMP

adequate stimulus the energy form or stimulus type to which the receptor responds best

adipose tissue (ad'-uh-poze) connective tissue that contains numerous fat cells (adipocytes) and stores energy in the form of triglycerides

adrenal glands primary endocrine gland located above the kidneys; divided into an outer cortex and an inner medulla

adrenergic pertaining to epinephrine (adrenaline) or norepinephrine (noradrenaline)

adrenocorticotropic hormone (ACTH) (ad-ren-oh-kor-tih-koh-troh'-pik) tropic hormone secreted from the anterior pituitary that stimulates secretion of glucocorticoids from the adrenal cortex

aerobic metabolism metabolism occurring under conditions of ready oxygen availability; normally includes the Krebs cycle, oxidative phosphorylation, and other pathways depending on them

afferent neurons neurons that transmit either sensory information or visceral information to the central nervous system for further processing

affinity a measure of how tightly substrate molecules bind to the active site of an enzyme, or ligand molecules bind to receptors

after-hyperpolarization third phase of an action potential during which the membrane potential is more negative than at rest due to high permeability to potassium

afterload the pressure that the ventricles have to work against as they pump blood

agglutination the linking together of many antigens through the specific binding of many antibodies

aldosterone (al-dos'-stir-own) a steroid hormone released from the adrenal cortex that regulates the reabsorption of sodium and secretion of potassium

alkalosis condition in which arterial blood pH is 7.45 or greater

allosteric regulation regulatory mechanism in which a modulator binds reversibly to the regulatory site on an enzyme, inducing a change in its conformation and activity

alveolar ventilation a measure of the volume of fresh air reaching the alveoli each minute, which is minute ventilation corrected for dead space volume

alveoli (al-vee-oh'-lie) terminal structures of the respiratory tract, where most gas exchange occurs; usually grouped in clusters

amines chemical messengers derived from amino acids

amino acid (ah-meen'-oh) biomolecule containing amine group, carboxyl group, hydrogen, and an R or residual group attached to a central carbon; found in proteins

amnion (am'-nee-on) a membrane that forms a fluid-filled sac around the embryo; also amniotic sac

amphipathic (am-fuh-path'-ick) having both polar and nonpolar regions, as in phospholipid molecules

anabolism (an-nab'-oh-lizm) synthesis of large molecules from smaller molecules, generally requiring an input of energy

anaerobic metabolism metabolism occurring under conditions of oxygen lack; normally includes glycolysis and lactic acid production

androgens a class of sex hormones secreted by the testes

angiotensin II (an'-gee-oh-ten-sin) a protein derived from a precursor called angiotensin I that stimulates aldosterone secretion by the adrenal cortex and other physiological responses that serve to maintain or increase blood volume and blood pressure

anterior pituitary anterior lobe of pituitary gland; primarily secretes tropic hormones

antibodies proteins present in the blood and interstitial fluids that target particular antigens for destruction; also known as immunoglobulins

antigens protein or polysaccharide components of viruses, bacteria, fungi, protozoa, parasitic worms, pollen, transplanted tissue, and tumor cells

aorta a major artery whose branches carry blood to all organs and tissues in the systemic circuit

aortic semilunar valve a valve located between the left ventricle and the aorta

apical membrane membrane on the side of an epithelial cell that faces the lumen of a body cavity; the lumen-facing membrane

arachidonic acid phospholipid found in the plasma membranes of cells and platelets and is the precursor molecule for synthesis of eicosanoids

arachnoid mater (ah-rak'-noyd) one of the three meninges, located between the dura mater and pia mater

arteries large vessels that carry blood away from the heart

arterioles small blood vessels that carry blood to the capillaries; walls contain smooth muscle that contracts and relaxes to regulate blood flow

aspartate (ah-spar'-tate) excitatory amino acid neurotransmitter

association areas areas of the cerebral cortex involved in complex processing that requires integrating different types of information

astrocytes (ass'-trow-sites) glial cells in the CNS that provide support to neurons and are critical to the formation of the blood-brain barrier by stimulating endothelial cells to develop tight junctions

ATP synthase the enzyme that synthesizes ATP during oxidative phosphorylation; located in the inner mitochondrial membrane

atria (ay'-tree-ah) the heart's two upper chambers, which receive blood carried to the heart in veins; singular, atrium

atrioventricular bundle a compact bundle of conducting fibers located in the interventricular septum; also known as the bundle of His

atrioventricular (AV) node (ay-tree-oh-ven-trik'-you-lar) a region located near the tricuspid valve

atrioventricular valve (AV valve) (ay-tree-oh-ven-trik'-you-lar) valves that separate the atrium and ventricle on either side of the heart and that permit blood to flow from the atrium to the ventricle, but not in the opposite direction

auditory cortex portion of the temporal lobe of the brain that processes auditory information

autocrine (au'-toh-krin) type of chemical messenger for which the secretory cell and target cell are the same

autonomic nervous system the division of the nervous system that encompasses efferent neurons that synapse with and regulate the function of internal organs and other structures not under voluntary control

autoreceptors receptors on an axon terminal that bind the same neurotransmitter that is released from the axon terminal, often enabling a neurotransmitter to modulate its own release

autoregulation when the function of an organ or tissue is regulated by factors originating from within the organ or tissue itself; also known as intrinsic control or local regulation

autorhythmicity the ability of the heart to generate signals that trigger its contractions on a periodic basis; the heart's ability to generate its own rhythm

axon branch that extends from the cell body of a neuron and sends information to other neurons or effector cells

axon hillock the site where the axon originates from the cell body of a neuron and the point of initiation of action potentials; trigger zone

axon terminal the end of the axon that forms a synapse with another neuron or an effector cell

B lymphocytes cells that produce antibodies that mark selected foreign substances for destruction; also called B cells

baroreceptor a type of sensory receptor neuron found in blood vessels and the heart; responds to changes in pressure within the cardiovascular system

basal metabolic rate the slowest metabolic rate of the body, indicative of the energy necessary to sustain vital functions; measured under conditions in which a person has fasted for at least 12 hours, is lying down, and is both physically and mentally relaxed

basal nuclei a particular group of nuclei located deep in the cerebrum that are important in controlling voluntary movements

basement membrane a layer of noncellular material that is relatively permeable to most substances; anchors the basolateral membrane and provides physical support for the epithelial layer

basic electrical rhythm (BER) a pattern of electrical activity in gastrointestinal smooth muscle, in which waves of depolarization occur at regular intervals at a fairly constant frequency

basilar membrane membrane in the cochlea of the inner ear that separates the scala tympani from the scala media

basolateral membrane membrane on the side of an epithelial cell that faces the internal environment and is in contact with interstitial fluid; blood-facing membrane

basophils (bay'-so-fils) leukocytes that defend against larger parasites; release histamine, heparin, and other chemicals

bicuspid valve the AV valve on the left side of the heart, which possesses two cusps; also called the mitral valve

bile fluid secreted by the liver that contains bicarbonate and bile salts, which emulsify fats

bile salts cholesterol derivatives manufactured by the liver and found in bile; function is to emulsify fats in the small intestine

biogenic amines a class of neurotransmitters derived from amino acids

blood-brain barrier physical barrier that exists between the blood and the interstitial fluid in the CNS, formed by tight junctions between endothelial cells of the cerebral capillaries

blood-testis barrier a physical barrier that isolates the luminal compartment from the basal compartment in a seminiferous tubule; consists of the Sertoli cells plus the tight junctions that link them together

bone marrow soft tissue located inside bones that contains cells dispersed within a porous matrix

Bowman's capsule in each nephron, a cup-shaped structure that surrounds the glomerulus and conducts filtrate into the renal tubule at the inflow end of the renal tubules; site where filtrate enters the renal tubules

brainstem bottom-most part of the brain that connects the forebrain and cerebellum to the spinal cord; consists of three main regions: midbrain, pons, and medulla oblongata

Broca's area area of association cortex devoted to language expression; located in the frontal lobe

bronchi branched tubes of the respiratory tract, located between the trachea and respiratory bronchioles of the lungs; includes smaller branches called secondary bronchi and bronchioles

bronchioles small tubules leading from the bronchi to the alveoli; less than 1 mm in diameter

brush border the collection of microvilli that are located on the apical membranes of epithelial cells lining the small intestine or renal tubule

bulbourethral glands (bul-bo-you-wreeth'-ral) accessory glands of the male reproductive system; the last to secrete fluid into the reproductive tract during sperm transport

calcitonin peptide hormone released from C cells of the thyroid gland that regulates plasma calcium levels

calcitrol steroid hormone derived from vitamin D that regulates plasma calcium levels

calmodulin (kal-mod'-you-lin) cytosolic calcium-binding protein; modulates the activity of intracellular proteins

calorie amount of energy that must be put into one gram (or one ml) of water to raise its temperature by one degree centigrade (°C) under a standard set of conditions

capillaries the smallest blood vessels in the body; possess thin walls that permit material exchange between blood and respiring tissues

carbaminohemoglobin (kar-bah-meen'-oh) hemoglobin with carbon dioxide bound to it

carbohydrates biomolecules composed of carbon, hydrogen, and oxygen in a ratio of 1:2:1

carbonic anhydrase enzyme that catalyzes the reversible reaction converting carbon dioxide and water to carbonic acid

cardiac cycle a series of mechanical and electrical events occurring within the heart during a single beat

cardiac muscle the type of muscle that is found in the heart; possesses striations and a sarcomere structure similar to that of skeletal muscle

cardiac output (CO) the volume of blood ejected from each ventricle per minute

carrier a transmembrane protein that binds molecules on one side of a membrane and transports them to the other side by means of a conformational change, or a change in shape

cartilage a connective tissue secreted by chondroblasts that is similar to uncalcified bone

catabolism breakdown of large molecules into smaller molecules; generally releases energy

catechol-O-methyltransferase (COMT) enzyme that degrades catecholamines

catecholamines amine compounds that contain a catechol group and are derived from the amino acid tyrosine

cell-mediated immunity reaction of certain types of T cells to kill abnormal or infected body cells

cellulose polysaccharide found in plants that humans are unable to digest or absorb

central chemoreceptors chemoreceptors located in the medulla oblongata that respond directly to changes in hydrogen ion concentration in the cerebrospinal fluid and indirectly to arterial P_{CO_2}; function in regulating ventilation

central lymphoid tissues tissues where leukocytes develop to maturity, including the bone marrow and thymus

central nervous system (CNS) the division of the nervous system that includes the brain and spinal cord; consolidates information received from the organs and develops commands to be sent to the organs; and is also the site of learning, memory, emotions, and cognition

central venous pressure (CVP) the pressure in the large veins in the thoracic cavity that lead to the heart

cerebellum (ser-ah-bel'-um) a bilaterally symmetrical brain structure, with an outer cortex and inner nuclei; located below the forebrain and posterior to the brainstem

cerebral cortex thin layer of gray matter that covers the cerebrum

cerebral hemispheres the two halves of the cerebrum

cerebrospinal fluid (CSF) clear, watery fluid that surrounds and protects the CNS and is similar in composition to plasma

cerebrum (seh-ree'-brum) largest structure of the brain, which contains both gray and white matter; gray matter areas include the cerebral cortex at the surface and deep subcortical nuclei

cervix the lower, narrower portion of the uterus, containing a central canal that leads directly to the vagina

channel a transmembrane protein that transports molecules by way of a passageway or pore that extends from one side of the membrane to the other

chemical synapses synapses where a neuron secretes a neurotransmitter into the extracellular fluid in response to an action potential arriving at the axon terminal

chemiosmotic coupling entire process that couples electron transport to ATP synthesis; the utilization of energy released during electron transport to transport hydrogen ions across the inner mitochondrial membrane up their concentration gradient

chemoreceptors receptors that monitor the concentrations of certain chemicals in various locations in the body

chemotaxis (kee-moh-tax'-iss) process by which leukocytes move to an injury, attracted by chemicals released from bacteria and wounded tissues

chief cells specialized cells in the gastric glands that secrete pepsinogen

chloride shift the movement of chloride ions into erythrocytes in exchange for the movement of bicarbonate into plasma

cholecystokinin (CCK) a hormone secreted by the duodenum and jejunum that inhibits gastric secretion and motility, and stimulates pancreatic enzyme secretion, gallbladder contraction, and relaxation of the sphincter of Oddi

choline acetyl transferase (CAT) enzyme that catalyzes the synthesis of acetylcholine

cholinergic (koh-lin-er'-jik) pertaining to acetylcholine

chondrocytes (kon'-droh-sites) cells that produce cartilage

chorion (kor'-ee-on) the outermost membrane that encapsulates the embryo, forming a tough envelope that isolates it from its surroundings

choroid plexus (kor'-oid) vascularized tissue lining the cerebral ventricles, synthesizes cerebrospinal fluid

chromatin the form in which DNA, along with its associated proteins, exists throughout most of the cell cycle; loosely-coiled DNA and proteins scattered throughout the nucleus

chromosome one complete molecule of DNA, along with its associated proteins; carries a specific set of genes

chylomicrons lipoprotein particles formed during the process of lipid absorption in the small intestine

chyme a mixture of food particles and gastrointestinal secretions; found in the stomach and intestines

cilia hairlike processes found on certain epithelial cells in the respiratory tract and oviduct

ciliary muscles muscles in the eye attached to the lens by zonular fibers; regulate the curvature of the lens for focusing light

circadian rhythm (sir-kay'-dee-an) endogenous fluctuations in body functions that occur on a 24-hour cycle

Class I MHC molecules found on the surfaces of all nucleated cells

Class II MHC molecules found on specialized cell types, including macrophages, activated B and T cells, and the cells of the interior of the thymus

clearance virtual measure of the volume of plasma from which a substance is completely removed or "cleared" by the kidneys per unit time

clonal selection antigen-driven activation of lymphocytes necessary for specific immune responses

coated pit an indentation of the plasma membrane that eventually forms an endocytotic vesicle; coated on its inner surface by specific proteins

cochlea a spiral-shaped structure in the inner ear that contains the receptor cells for hearing

codon complementary three-base sequence in mRNA that codes for a specific amino acid

coenzymes (koh-en'-zimes) molecules that do not themselves have catalytic activity but that are necessary for proper enzyme function and participate directly in reactions catalyzed by enzymes; often serve to transfer certain chemical groups from one reactant to another

cofactors nonprotein components of some enzymes necessary for them to hold normal conformation during metabolic reactions

collecting ducts ducts that collect fluid from several different renal tubules and carry it to the renal pelvis for eventual elimination

colon an organ of the digestive tract that absorbs water and ions from the chyme, and stores feces; comprises the ascending colon, descending colon, transverse colon, and sigmoid colon

complement system system that completes the actions of antibodies; 30 proteins that act to destroy invading microorganisms, especially bacteria

compliance a measure of the ability of blood vessels or other hollow structures to stretch as the pressure inside them rises

concentration gradient difference in the concentration of a substance between regions

conducting zone the upper part of the respiratory tract; conducts air from the larynx to the lungs

conduction system a set of specialized heart muscle cells that initiate and conduct action potentials

conduction the transfer of thermal energy from one object to another that occurs when the objects are in direct contact with each other

cones photoreceptors that enable visibility during relatively bright light and are responsible for color vision

connective tissue cells cells whose primary function is to provide physical support for other structures, to anchor them in place, or link them together

contractile proteins the proteins actin and myosin that make up the thin and thick filaments of the sarcomere; generate contractile force

contralateral referring to ascending and descending pathways that are on the side opposite their origin

convergence the transmission of information from two or more neurons to one cell

convex lens type of lens that causes parallel light waves to converge at a single point

core temperature temperature deep within the body

cornea transparent structure at the front of the eye that allows light waves to enter

corpus callosum (kor'-pus kal-loh'-sum) the primary band of nervous tissue that connects the two cerebral hemispheres

corpus luteum (kor'-pus loo'-tee-um) a gland that develops from the ruptured follicle following ovulation and secretes progesterone and estrogen

corticotropin releasing hormone (kor-tih-koh-troh'-pin) tropic hormone released by the hypothalamus that stimulates secretion of ACTH from the anterior pituitary

cotransport carrier-mediated transport by which two transported substances move in the same direction

countertransport carrier-mediated transport by which two transported substances move in opposite directions

covalent regulation regulatory mechanism in which changes in an enzyme's activity are brought about by the covalent bonding of a specific chemical group to a site on the enzyme molecule

cranial nerves 12 pairs of peripheral nerves that emanate directly from the brain

creatine phosphate a compound in muscle cells that can donate a high-energy phosphate to ADP to form ATP

crossbridge cycle the mechanism that drives muscle contraction

crossbridges protrusions on both ends of the thick filament that are responsible for generating the motion that causes muscle contraction

crypts of Lieberkühn pits located between villi in the small intestine containing cells that secrete copious volumes of bicarbonate-rich fluid into the lumen

cupula gelatinous area within the ampulla of the inner ear; contains hair cells and supporting cells

current the movement of electrical charges; in biological systems, currents are caused by movement of ions and expressed in units of microamps (10_{-6} amperes)

cytokines (sy'-toh-kines) proteins secreted by cells in response to a stimulus that affect the behavior of other nearby cells

cytoplasm everything inside the cell except the nucleus

cytoskeleton flexible lattice of fibrous proteins that gives structure and support to the cell

cytosol cytoplasm minus membrane-bounded organelles; synonymous with intracellular fluid

cytotoxic T cells (sy''-toh-tox'-ik) T cells that contact virus-infected cells, mutant cells, and transplanted cells, and gain the ability to destroy the infected or abnormal cells

dendrites branches that extend from the cell body of a neuron and receive information from other neurons

deoxyribonucleic acid (DNA) (dee-ox-see-ry-boh-noo-klay'-ik) a biomolecule consisting of two strands of nucleotides coiled together as a double helix; found in the nucleus and stores genetic information

dephosphorylation the removal of a phosphate group from a molecule during a metabolic reaction

depolarization any change in membrane potential in which the inside of the cell becomes more positive (less negative) than it is at rest

desmosome (dez'-moh-some) filamentous junction between two adjacent cells that provides a strong physical linkage between them; enables tissues to withstand stretching without cells being torn apart from one another

detrusor muscle smooth muscle fibers of the bladder

diabetes mellitus (die-ah-bee'-teez) a disease involving defects in insulin production or signaling

diacylglycerol (DAG) a second messenger released by the phosphatidyl inositol system

diaphragm primary inspiratory muscle for respiration; the muscular partition that separates the abdominal and thoracic cavities

diastole (dy-ass'-toh-lee) the period of ventricular relaxation during a cardiac cycle

diastolic pressure (DP) the minimum aortic pressure attained during the cardiac cycle; occurs during diastole

diencephalon (dy-en-sef'-ah-lon) the lowest portion of the forebrain; consists of the thalamus and hypothalamus

diffusion the passive movement of molecules from one location to another as a result of their own thermal motion

digestion the breakdown of nutrient molecules that are present in food to smaller molecules by enzymes in the lumen of the digestive tract

digestive tract a number of organs joined in series to form a passageway through which food and digestion products are conducted

diploid in reference to a cell's chromosome number, possessing the full complement of chromosomes

disaccharide (dy-sak'-er-ide) carbohydrate consisting of two monosaccharides covalently bonded together

distending pressure a difference in pressure across the wall of a hollow organ or vessel that tends to cause the organ or vessel to stretch

divergence the transmission of information from one neuron to two or more cells

dominant follicle a follicle that has been selected to go to full maturation during the course of a menstrual cycle

dorsal horn posterior half of the gray matter on either side of the spinal cord

dorsal respiratory group (DRG) respiratory control center located on the dorsal side of the medulla; contains primarily inspiratory neurons

dorsal root where a spinal nerve bifurcates before joining the spinal cord, the dorsalmost of the two resulting branches; contains afferent nerve fibers

dorsal root ganglia clusters of cell bodies of afferent neurons located outside the spinal cord

duodenum initial portion of the small intestine

dura mater outermost of the three meninges, closest to the bone

effector cells in the immune system, short-lived cells that work to combat the antigen that triggered the immune response

effectors cells, tissues, or organs that respond to neural or chemical signals; in homeostatic regulatory systems, cells, tissues, or organs that respond to output signals of the integrating center and bring about the final response

efferent neurons neurons that transmit information from the central nervous system to effector organs

eicosanoids modified fatty acids (all derived from arachidonic acid) that function in intercellular communication

ejection fraction (EF) the fraction of end-diastolic volume ejected during a heartbeat

electrocardiogram (ECG) a recording of the heart's electrical activity obtained through electrodes placed on the body's surface

electroencephalogram (EEG) noninvasive means of recording of electrical activity in the brain using electrodes placed on the scalp

electron transport chain a series of electron acceptors and other proteins in the inner mitochondrial membrane; involved in the synthesis of ATP by oxidative phosphorylation

electrotonic conduction the spread of a graded potential from the site of origin within a cell

embryo the developing human within the first two months after conception

emmetropia normal visual acuity

emulsification the process by which the action of bile salts breaks fat globules down into smaller droplets

end-plate potential (EPP) a depolarization of the motor end plate of a skeletal muscle fiber caused by acetylcholine binding to nicotinic cholinergic receptors

end-diastolic volume (EDV)   the volume of blood contained within each ventricle at the end of diastole

end-product inhibition   special case of feedback inhibition in which an enzyme is inhibited by the end-product of a metabolic pathway

end-systolic volume (ESV)   the volume of blood in each ventricle at the end of systole

endocrine glands   glands that secrete hormones into interstitial fluid

endocytosis (en-doh-sy-toh'-sis)   uptake of material into a cell via vesicles that pinch off from the plasma membrane; enables macromolecules and larger particles to enter cells

endolymph   fluid found in the scala media of the cochlea in the inner ear; similar in composition to intracellular fluid

endometrium (en-doh-mee'-tree-um)   the innermost layer of tissue in the uterus, made up of epithelial cells and an underlying layer of thick connective tissue

endoplasmic reticulum (en-doh-plas'-mik reh-tik'-you-lum) elaborate network of membranes inside cells that encloses a single interior compartment; includes rough endoplasmic reticulum and smooth endoplasmic reticulum

endorphins (en-dor'-fins)   morphinelike neuropeptides produced in the central nervous system

endothelium (en-doh-thee'-lee-um)   a layer of epithelial cells that line the interior surface of the walls of all blood vessels and the heart

enkephalins (en-kef-a'-lins)   morphinelike neuropeptides produced in the central nervous system

enteric nervous system   a system of neural networks in the wall of the GI tract that regulates many digestive functions; also called the intrinsic nervous system

enterocytes   epithelial cells of various types found in the mucous membrane of the stomach or intestine

enterogastrones   three gastrointestinal hormones that are secreted by the duodenum and jejunum: cholecystokinin (CCK), secretin, and glucose-dependent insulinotropic peptide (GIP)

enterokinase   brush border enzyme that converts trypsinogen to trypsin

enzymes (en'-zimes)   biomolecules, usually proteins, specialized to act as catalysts in metabolic reactions

eosinophils (ee-oh-sin'-oh-fils)   leukocytes that attack parasitic invaders too large to be engulfed

ependymal cells (ep-en'-dee-mal)   glial cells that line the cerebral ventricles of the brain and central canal of the spinal cord

epididymis (ep-ih-did'-eh-mus)   duct loosely attached to the testes' outer surface; serves as a site for sperm maturation

epiglottis   a flap of tissue over the glottis that prevents food or water from entering the larynx when swallowing

epinephrine   a hormone secreted by the adrenal medulla that exerts effects similar to those exerted by the sympathetic nervous system

epiphyseal plate (ep-ih-fiz'-ee-al)   a thin layer of tissue that separates the epiphyses of a long bone from the shaft; plays a key role in the elongation of bones during growth

epithelia (ep-ih-thee'-lee-ah)   continuous, sheetlike layers of cells found in the skin and linings of hollow organs; singular: epithelium

epitopes (ep'-ih-topes)   sites on an antigen that are recognized by antibodies

equilibrium potential   the membrane potential at which the electrical driving force is equal and opposite to the chemical driving force, giving an electrochemical driving force of zero

erectile tissue   spongy masses of connective tissue and smooth muscle containing numerous interconnected vascular spaces; found in the penis, clitoris, and certain other structures that expand when filled with blood

erection   process occurring in male sexual arousal, during which the penis swells, elongates, and becomes firmer and straighter to facilitate its entry into the vagina during copulation; also occurs in the clitoris during female sexual arousal

erythrocyte (eh-rith'-roh-site)   red blood cell

esophagus   a muscular tube whose primary function is to conduct food from the pharynx to the stomach; contains the upper esophageal sphincter and the lower esophageal sphincter

essential nutrient   any biomolecule necessary for proper body function that cannot be synthesized in cells and therefore must be obtained from dietary sources

estrogens   a class of sex hormones secreted by the ovaries

eustachian tube (you-stay'-shun)   a canal that connects the middle ear with the pharynx and allows equilibration of air pressure across the eardrum

excitable cells   cells capable of producing action potentials

excitation-contraction coupling   in a muscle cell, the sequence of events that links the action potential to the contraction

excitatory postsynaptic potential (EPSP)   a graded depolarization caused by neurotransmitter binding to receptors on the postsynaptic neuron

exocrine glands   glands that are specialized for the transport of materials from the body's internal environment to the external environment, including the lumen of the digestive tract

exocytosis (ex-oh-sy-toh'-sis)   transport of material out of a cell via vesicles that fuse with the plasma membrane; involved in the cellular secretion of hydrophilic molecules

expiration   the movement of air out of the lungs

expiratory reserve volume (ERV)   the maximum volume of air that can be expired from the end of a normal expiration

external auditory meatus   ear canal

external intercostals   inspiratory muscles of the chest wall

external respiration   the exchange of oxygen and carbon dioxide between the atmosphere and the tissues of the body; involves both the respiratory and cardiovascular systems

extracellular fluid (ECF)   fluid located outside cells, accounting for one-third of total body water

extrapyramidal tract   all motor control pathways outside the pyramidal system; indirect connections between the brain and motor neurons in the spinal cord

extrinsic control   regulation of the heart (or any other organ or tissue) by neural input, circulating hormones, or any other factor originating from outside the organ

facilitated diffusion   the passive movement of molecules across a membrane by way of a transport protein

fast pain   a sharp, pricking sensation that can be easily localized and is produced by activation of nociceptors; is transmitted by Aδ fibers

fast-twitch fibers skeletal muscle fibers that contract quickly and reach peak tension relatively quickly during a twitch

fatigue a decline in a muscle's ability to maintain a constant force of contraction in the face of long-term, repetitive stimulation

fatty acids long hydrocarbon chains with a carboxyl group (-COOH) at one end

feedback inhibition regulatory mechanism in which an enzyme in a metabolic pathway is inhibited by an intermediate appearing downstream

feedforward activation regulatory mechanism in which an enzyme in a metabolic pathway is stimulated by an intermediate appearing upstream

fertilization the process by which two gametes (one from each parent) fuse together to produce a new cell; known as conception in humans

fetus the developing human after two months following conception

fibrin the last active clotting factor in the coagulation cascade, which forms polymers that make up the actual clot

fight-or-flight response the group of physiological changes coordinated by the sympathetic nervous system that prepares the body to cope with threatening situations

filling time the time available for the ventricles to fill with blood during the cardiac cycle

filtered load the quantity of a certain solute that is filtered at the glomerulus per unit time; equals the product of the GFR and the plasma concentration of the solute

filtration the movement of fluid across capillary walls from plasma to the interstitium

flow autoregulation local regulation that tends to keep blood flow constant

flux the number of molecules crossing a membrane in a given length of time; a measure of the rate at which a substance is being transported

follicle spherical structure in the ovary containing a single developing ovum

follicle-stimulating hormone (FSH) a gonadotropic hormone that stimulates gametogenesis and regulates other gonadal functions in either sex

follicular phase (fuh-lik'-you-lar) the initial phase of the ovarian cycle, during which follicular maturation occurs; terminates with ovulation

forced expiratory volume a measure of the percentage of the forced vital capacity that can be exhaled within a certain time frame

forced vital capacity the maximum amount of air a person can forcefully expire following a maximum inspiration

forebrain largest and uppermost part of the brain, divided into left and right halves, or hemispheres; consists of the cerebrum and diencephalon

fovea (foh'-vee-ah) central point on the retina of the eye, in which visual acuity is greatest

fragmentin protein released from cytotoxic T cells that enters target cells and induces apoptosis

free nerve endings somatosensory receptors in the skin that lack identifiable specialized sensory structures; includes some mechanoreceptors, thermoreceptors, and nociceptors

frequency coding the coding of stimulus intensity by the frequency of action potentials in a neuron, in which a stronger depolarizing stimulus causes the action potential frequency to increase

frontal lobe one of four lobes of the cerebrum, located in the anterior portion of the cerebrum; important in voluntary motor control, behavior, and personality traits

functional residual capacity (FRC) the volume of air in the lungs at the end of a resting expiration

G proteins membrane proteins with the ability to bind guanosine nucleotides; function in coupling an extracellular messenger to a response inside the target cell

gallbladder small muscular sac located immediately adjacent to the liver; stores bile in between meals

gametes (gam-eets') cells of reproduction produced in meiosis that contain half of a person's genetic material; examples are sperm cells in males and ova in females

gametogenesis (gah-mee-toh-jen'-ih-sis) the production of gametes from a pool of germ cells

gamma-aminobutyric acid (GABA) inhibitory amino acid neurotransmitter

ganglion (gang'-glee-on) cluster of neuron cell bodies in the peripheral nervous system; plural ganglia

gap junctions areas where two adjacent cells are connected together by membrane proteins called connexons that form small channels between the cells, enabling ions and small molecules to move freely between them

gastric mucosal barrier mucus layer that protects the lining of the stomach from the effects of acid and pepsin

gastrin a hormone secreted by the stomach that regulates many gastrointestinal functions, including gastric acid secretion

gene section of DNA that codes for a particular protein or proteins

generator potential a change in membrane potential, also known as receptor potential

genetic code the correspondence between DNA triplets and specific amino acids that governs the expression of all genetic information

glial cells (glee'-al) cells in the nervous system that provide various types of support to the neurons, including structural and metabolic support

glomerular filtration the bulk flow of protein-free plasma from the glomerular capillaries into Bowman's capsule

glomerular filtration rate (GFR) the volume of the plasma filtered per unit time from all renal glomeruli combined

glomerulus (gloh-mer'-you-lus) in each nephron, a ball-like cluster of capillaries in the renal corpuscle; site of filtration

glottis opening to the larynx

glucagon a peptide hormone secreted by alpha cells of the pancreas; promotes metabolic processes of the postabsorptive state

glucocorticoids steroid hormones secreted from the adrenal cortex that regulate the body's response to stress, regulate protein, carbohydrate, and lipid metabolism in a variety of tissues, and regulate blood glucose levels; primary glucocorticoid is cortisol

gluconeogenesis (gloo-koh-nee-oh-jen'-ih-sis) process during which new glucose molecules can be synthesized from non-carbohydrate precursors by the liver

glucose most common monosaccharide; provides important source of cellular energy

glucose sparing process by which non-nervous tissues convert to using fatty acids for energy rather than glucose, which is spared for use by the central nervous system

glucose-dependent insulinotropic peptide (GIP) a hormone secreted by the duodenum and jejunum that stimulates insulin secretion by pancreas

glutamate excitatory amino acid neurotransmitter

glycerol (gliss'-er-ol) 3-carbon carbohydrate that functions as the "backbone" of a triglyceride or phospholipid

glycine (gly'-seen) inhibitory amino acid neurotransmitter

glycogen (gly'-coh-jen) a glucose polymer found in animal cells; functions as an energy store

glycogenesis (gly-koh-jen'-eh-sis) synthesis of glycogen from glucose monomers

glycogenolysis (gly-koh-jen-nol'-ih-sis) breakdown of glycogen to glucose monomers

glycolysis the first stage of glucose oxidation, occurring in the cytosol, in which each glucose molecule is broken down to two molecules of pyruvate

goblet cells epithelial cells in the respiratory tract and digestive tract that secrete mucus

Golgi apparatus (goal'-jee) an organelle consisting of membrane-bound flattened sacs called cisternae that process molecules synthesized in the endoplasmic reticulum and prepares them for transport

gonadotropin releasing hormone (GnRH) a hypophysiotropic hormone that stimulates the secretion of gonadotropins by the anterior pituitary

gonadotropins two hormones, follicle-stimulating hormone (FSH) and luteinizing hormone (LH), that are secreted by the anterior pituitary and regulate gonadal function in either sex

gonads the primary reproductive organs; serve as sites of gametogenesis and sex hormone secretion

Graafian follicle fully matured follicle, just prior to ovulation

graded potentials small changes in membrane potential that occur when ion channels open or close in response to a stimulus, such as the binding of neurotransmitters to receptors; magnitude of the potential change varies with the strength of the stimulus

granular cells specialized cells in the wall of the afferent and efferent arterioles that secrete renin; also called juxtaglomerular cells

granulosa cells specialized cells that surround a developing ovum and control its development

gray matter areas of the CNS consisting primarily of cell bodies, dendrites, and axon terminals of neurons; where synaptic transmission and neural integration occur

growth hormone inhibiting hormone (GHIH) tropic hormone released by the hypothalamus that inhibits growth hormone secretion from the anterior pituitary; also known as somatostatin

growth hormone peptide hormone secreted by the anterior pituitary, essential for normal growth

growth hormone releasing hormone (GHRH) tropic hormone released by the hypothalamus that stimulates growth hormone secretion from the anterior pituitary

gyri (jy'-ri) ridges in the highly convoluted gray matter of the cerebral cortex; singular: gyrus

haploid in reference to a cell's chromosome number, possessing half the number of chromosomes found in most other cells of the body

helper T cells T cells that secrete many different cytokines and thereby influence the action of other lymphocytes

hematocrit (heh-mat'-oh-krit) the fraction of the blood volume that is occupied by red blood cells

hemoglobin (hee'-moh-gloh-bin) a protein in red blood cells that carries oxygen and carbon dioxide

hemostasis process of stopping bleeding; includes vascular spasms, platelet plug formation, and blood clot formation

hepatocytes liver cells

histamine substance released by damaged mast cells; causes both vasodilation and increased capillary permeability

homeostasis (home-ee-oh-stay'-sis) maintenance of relatively constant conditions within the body's internal environment

hormones chemical messengers released from endocrine cells or glands into the interstitial fluid, where they then diffuse into the blood and travel to target cells

human chorionic gonadotropin (hCG) a hormone secreted by the chorion that maintains the corpus luteum during the first three months of pregnancy

humoral immunity a specific immune response generated by B lymphocytes and conferred by antibodies that circulate in the blood and lymph

hypercalcemia condition characterized by an increase in plasma calcium

hyperkalemia condition characterized by an increase in plasma potassium levels

hypernatremia (hy-per-na-tree'-mee-ah) condition characterized by an increase in plasma sodium levels above normal

hyperopia (hi-per-oh'-pee-ah) common visual defect of the eye causing far-sightedness

hyperosmotic a solution having an osmolarity that is higher than another

hyperplasia increase in cell number

hyperpnea (hy-perp-nee'-ah) an increase in alveolar ventilation in response to increased oxygen consumption and carbon dioxide production

hyperpolarization any change in membrane potential in which the inside of the cell becomes more negative than it is at rest

hypertension higher than normal arterial blood pressure

hyperthermia a condition of higher than normal body temperature

hypertonic a solution that draws water out of a cell causing the cell to shrink

hypertrophy increase in cell size

hyperventilation an increase in alveolar ventilation such that metabolic demands of the tissue are exceeded

hypo-osmotic a solution with a lower osmolarity than another

hypocalcemia   condition characterized by a decrease in plasma calcium

hypokalemia   condition characterized by a decrease in plasma potassium levels

hyponatremia   condition characterized by a decrease in plasma sodium levels

hypotension   lower than normal arterial pressure

hypothalamus   a region at the base of the brain that regulates autonomic functions and secretes several hormones, most of which regulate secretory activity of the pituitary gland

hypothermia   a condition of lower than normal body temperature

hypotonic   a solution that causes water to enter a cell causing the cell to swell

hypoventilation   a decrease in alveolar ventilation such that metabolic demands of the tissue are not met

ileum   terminal portion of the small intestine

immunoglobulins (Ig) (im-mun-oh-glob'-you-lins)   proteins present in the blood and interstitial fluids that target particular antigens for destruction; also known as antibodies.

implantation   process by which an embryo adheres to and embeds in the wall of the uterus

inclusions   cytosolic particles composed of triglycerides or glycogen; serve as energy stores for cellular metabolism

inflammation   a complex series of events that culminate in the accumulation of proteins, fluid, and phagocytic cells in an area where tissue has been injured or invaded by microorganisms

inhibin   a protein hormone secreted by the gonads that suppresses the release of FSH from the anterior pituitary

inhibitory postsynaptic potential (IPSP)   a graded hyperpolarization caused by neurotransmitter binding to receptors on the postsynaptic neuron

innervate   to make a synaptic connection with an effector organ

inositol triphosphate (IP3)   a second messenger released by the phosphatidyl inositol system that stimulates the release of calcium

inspiration   the movement of air into the lungs

inspiratory capacity (IC)   the maximum volume of air that can be inspired at the end of a resting expiration

inspiratory reserve volume (IRV)   the maximum volume of air that can be inspired from the end of a normal inspiration

insulin   a peptide hormone secreted by beta cells of the pancreas; promotes metabolic processes of the absorptive state

insulin-like growth factor   peptide hormone secreted by the liver in response to growth hormone; promotes protein synthesis and growth; also known as somatomedin

integral membrane proteins   proteins that span the lipid bilayer or are otherwise tightly embedded within it

interferons   a group of related proteins that interfere with virus replication

intermediate filaments   fibrous proteins with a diameter between that of microfilaments and microtubules; stronger and more stable than microfilaments

internal environment   fluid that surrounds the cells inside the body, including fluid in the bloodstream that surrounds blood cells; synonymous with extracellular fluid

internal respiration   cellular respiration that occurs in the mitochondria

interneurons   neurons located entirely in the central nervous system; account for 99% of all neurons in the body

internodal pathways   systems of conducting fibers that run from the SA node to the AV node through the walls of the atria

interphase period   period in the life cycle of the cell during which it is carrying on its normal physiological functions

interstitial fluid (ISF)   extracellular fluid that is present outside the blood, and that bathes most of the cells of the body

intra-alveolar pressure (P_{alv})   the pressure exerted by the air within the alveoli

intracellular fluid (ICF)   fluid located inside cells, accounting for two-thirds of total body water

intrapleural pressure (P_{ip})   the pressure of the fluid inside the pleural space

intrapleural space   a fluid-filled compartment located between the lungs and chest wall; is bounded by the visceral and parietal pleura

intrinsic control   regulation of an organ or tissue by factors originating from within the organ or tissue itself; also known as autoregulation or local regulation

intrinsic factor   a protein secreted by parietal cells of the stomach that binds to vitamin B_{12} and that is necessary for the absorption of this vitamin in the small intestine

ionotropic receptor (eye-oh-no-troh'-pik)   a receptor protein that also functions as an ion channel that opens or closes in response to the binding of a chemical messenger; also called a channel-linked receptor

ipsilateral (ip-sih-lat'-er-al)   referring to ascending and descending pathways that are on the same side as their origin

iris   pigmented smooth muscle in the eye that sits in front of the lens and regulates the diameter of the pupil to control the amount of light entering the eye

ischemia (iss-key'-me-ah)   a condition in which blood flow in a tissue is insufficient to keep up with metabolic demand

iso-osmotic   two solutions having the same osmolarity

isometric twitch (eye-soh-met'-rick)   a twitch during which a muscle generates force but is not allowed to shorten

isotonic   a solution that will not alter cell volume

isotonic twitch (eye-soh-tah'-nik)   a twitch during which a muscle is allowed to shorten and lift a constant load

isovolumetric contraction   contraction of the ventricles with all heart valves closed, such that the volume of blood contained within the ventricles is constant; occurs early in systole

isovolumetric relaxation   relaxation of the ventricles with all heart valves closed, such that the volume of blood contained within the ventricles is constant; occurs early in diastole

jejunum   middle portion of the small intestine

juxtaglomerular apparatus   a collection of specialized cells in the distal tubules and the afferent and efferent arterioles near where the three structures come together in the kidney; regulates glomerular filtration and renin secretion

juxtaglomerular cells (jux-tah-gloh-mer'-you-lar)   specialized cells in the wall of the afferent and efferent arterioles that secrete renin; also called granular cells

kinocilium large stereocilia projecting from the receptor cells for equilibrium; direction of bending in response to acceleration of the body determines direction of receptor potentials in receptor cells

Krebs cycle a cyclical metabolic pathway occurring in the mitochondrial matrix, in which acetyl coenzyme A is a primary reactant and carbon dioxide and reduced coenzymes are produced; also called the citric acid cycle, tricarboxylic acid cycle, or TCA cycle

lactation milk production that takes place in the mammary glands

lacteal blind-ended lymphatic vessel found within each villus in the lining of the small intestine

lactic acid compound produced from pyruvate in the event that pyruvate cannot proceed to the Krebs cycle; often produced under conditions of reduced oxygen availability; anionic form is called lactate

laminar flow a type of blood flow in which the blood in all parts of a vessel moves in the same direction

large intestine an organ of the digestive tract, made up of the cecum, colon, and rectum; absorbs water and electrolytes, and stores feces

larynx (lar'-inks) the initial passageway of the respiratory tract, which contains the vocal cords

latent period the lag of a few milliseconds that occurs between the action potential in a muscle and the start of contraction, when the muscle first begins to generate force

lateral horn a region of the gray matter of the spinal cord where certain autonomic preganglionic neurons originate; synonymous with intermediolateral cell column

lateral inhibition process during which a stimulus that strongly excites receptors in a certain location inhibits activity in the afferent pathways of other receptors located nearby

law of complementary base pairing states that whenever two strands of nucleic acids are held together by hydrogen bonds, G in one strand is always paired with C in the opposite strand, and A is always paired with T in DNA (or U in RNA)

law of mass action states that a reaction can be made to go either forward or reverse by raising or lowering the concentrations of reactants relative to products

law of specific nerve energies states that a given sensory receptor is specific for a particular energy form or stimulus type

leader sequence initial sequence of amino acids that is present in a newly synthesized polypeptide chain; synthesized during translation by ribosomes in the cytosol

leak channels ion channels that are always open; responsible for the resting membrane potential

leptin hormone secreted by adipose cells that is important in physiological regulation of food intake

leukocytes (loo'-koh-sites) white blood cells

Leydig cells the cells in the testes that secrete androgens

ligand-gated channels ion channels that open or close in response to a chemical messenger binding to a specific receptor

limbic system diverse collection of closely associated cortical regions, sub-cortical nuclei, and tracts in the forebrain; function in learning and emotions

lipases enzymes present in pancreatic juice that act on lipid substrates

lipids biomolecules that contain primarily carbon and hydrogen atoms linked together by nonpolar covalent bonds

lipogenesis (ly-poh-jen'-eh-sis) process by which fat is synthesized from non-lipid nutrients, such as proteins and carbohydrates

lipolysis (ly-pol'-ih-sis) the first stage of lipid breakdown; in regard to triglycerides, separation of the fatty acids from the glycerol backbone

lipophilic pertaining to a molecule that is lipid soluble and can readily cross the lipid bilayer of a cell membrane that is lipid soluble, or water insoluble; hydrophobic

lipophobic pertaining to a molecule that is water soluble (not lipid soluble) and therefore does not easily cross the lipid bilayer of a cell membrane

lipoprotein lipase enzyme found on the inside surface of capillaries throughout the body that breaks down triglycerides

lipoproteins protein- and lipid-containing particles that possess a hydrophobic interior surrounded by a hydrophilic coat; the primary vehicle of lipid transport in the blood or lymphatic system

liver accessory gland in the digestive system that secretes bile; also processes certain absorbed nutrients and performs many other functions

loop of Henle (hen'-lee) the portion of the renal tubule that forms a hairpin loop that dips into the medulla, including the descending limb, the thin ascending limb, and the thick ascending limb

lower esophageal sphincter a ring of smooth muscle at the lower end of the esophagus that regulates the flow of food from the esophagus to the stomach

lumen interior compartment of a hollow organ or vessel

luteal phase the final phase of the ovarian cycle during which the corpus luteum is active

luteinizing hormone (LH) a gonadotropic hormone that stimulates sex hormone secretion and regulates other gonadal functions in either sex

lymphatic fluid fluid that flows through lymphatic system

lymphatic system a network of vessels, or ducts, that courses throughout the body and contains lymphatic fluid

lysosomes small spherical membrane-bounded organelles containing numerous degradative enzymes; involved in the breakdown of unneeded intracellular material or foreign matter that has been taken into the cell

macrophages phagocytic cells that filter and remove particles in the lymph and body tissues, including bacteria or other foreign matter

macula densa specialized cluster of epithelial cells found in the distal convoluted tubule in the region adjacent to the afferent and efferent arterioles

macula lutea a depression in the center of the retina that contains the fovea; contains a high density of cones

major histocompatibility complex (MHC) a special class of normal self proteins on body cells that present foreign antigens so that T cell receptors can detect them

mammary glands   milk-producing glands in the female breasts

mast cells   cells dispersed throughout the body's connective tissues that help signal the early events of inflammation by releasing histamine

mean arterial pressure (MAP)   the average aortic pressure occurring during the cardiac cycle

mechanoreceptors   sensory receptors that detect physical forces such as pressure or vibration

medial geniculate body (je-nik'-you-late)   a nucleus of the thalamus that transmits information to the auditory cortex in the temporal lobe

mediated transport   transport of molecules across a membrane utilizing transmembrane proteins

medulla oblongata (meh-duhl'-ah ob-long-got'-ah)   lowest portion of the brainstem; connects to the spinal cord

meiosis (my-oh'-sis)   a series of two cell divisions that generate daughter cells (gametes) with half the normal chromosome number; involved in gametogenesis

membrane potential   a difference in electrical potential or voltage that appears across the membranes of most cells

memory cells   in the immune system, long-lived cells that bear membrane receptors specific for the antigen that triggered the immune response

meninges (men-in'-jees)   three membranes that separate the soft tissue of the CNS from the surrounding bone; dura mater, arachnoid mater, and pia mater

menopause   cessation of reproductive capacity that occurs in females in middle age

menstrual cycle   a sequence of events involving changes in reproductive function that occurs in females on a regular basis with a period of about one month

menstrual phase   phase of the uterine cycle during which the uterine lining is shed

mesenteries   a system of clear thin membranes that interconnects most of the abdominal organs and houses nerves and blood vessels running to them; helps anchor organs in place

mesothelium   a layer of epithelial tissue found on the outer surfaces of the organs of the GI tract

messenger RNA (mRNA)   molecule that carries genetic information from the nucleus to cytoplasm

metaboreceptors   various receptors in skeletal muscles that are sensitive to mechanical stresses, local metabolite concentrations, and temperature

metabotropic receptors   any receptor that works by triggering biochemical changes rather than triggering a direct change in membrane permeability

microfilaments   the smallest-diameter protein filaments of the cytoskeleton

microtubules   the largest-diameter protein filaments of the cytoskeleton, composed of long hollow tubes made of tubulin

microvilli   projections of the plasma membrane that increase its surface area; often found in epithelial cells specialized for transport

micturition (mik-chur-rish'-un)   the elimination of urine from the body, or urination

midbrain   uppermost portion of the brainstem; connects to the forebrain

mineralocorticoids   steroid hormones secreted from the adrenal cortex that regulate sodium reabsorption and potassium secretion by the kidneys; primary mineralocorticoid is aldosterone

minute ventilation   the total amount of air that flows into or out of the respiratory system in a minute

mitochondria   oval-shaped organelles bounded by two membranes; primary site of ATP synthesis in the cells

mitochondrial matrix   the innermost compartment in mitochondria, bounded by the inner mitochondrial membrane

mitosis (my-toh'-sis)   the type of cell division that yields two daughter cells containing the normal number of chromosomes; necessary for the growth and replacement of cells

modality   in the sensory system, the energy form of a stimulus

modulator   specific molecule that binds to regulatory site on an enzyme molecule and regulates its activity

monoamine oxidase (MAO) (moan-oh-am'-ine)   enzyme that degrades biogenic amines

monocytes   leukocytes capable of phagocytosis; circulate in the blood for only a few hours, and then migrate into tissues where they enlarge and develop into phagocytic cells called macrophages

monosaccharide (mah-no-sak'-er-ide)   simple sugar composed of a single unit

motor end plate   the specialized region of a skeletal muscle fiber's plasma membrane located at the neuromuscular junction

motor homunculus   map that indicates which areas of the primary motor cortex in the frontal lobe are devoted to particular regions of the body

motor neurons   efferent neurons of the somatic nervous system that synapse on skeletal muscle cells; originate in the spinal cord ventral horn or in analogous structures in the brainstem

motor unit   a motor neuron and all the muscle fibers it innervates

mucosa   the innermost of the four major layers that make up the wall of the digestive tract; composed of three layers including the mucous membrane, the lamina propria, and the muscularis mucosae

mucous membrane   a layer of epithelial cells that lines the inside of the GI tract and other hollow cavities in the body

Müllerian-inhibiting substance (MIS)   a hormone secreted by the testes that inhibits development of the Müllerian ducts

muscle fibers   muscle cells

muscularis externa   one of the four major layers of the wall of the digestive tract, located between the submucosa and serosa; composed primarily of smooth muscle fibers

myelin (my'-uh-lin)   layers of plasma membrane from oligodendrocytes and Schwann cells that form insulation around the axons of neurons

myenteric plexus   one of the two neural networks that make up the enteric nervous system; located in the muscularis externa

myocardium   the entire cardiac muscle mass

myofibrils   banded, rodlike elements that contain a muscle fiber's contractile machinery

myoglobin   an oxygen-binding protein found in certain muscle cells

myometrium a layer of smooth muscle in the uterine wall

myopia (my-oh'-pee-ah) common visual defect of the eye causing near-sightedness

myosin (my'-oh-sin) the contractile protein found in thick filaments in striated muscle

myosin kinase an enzyme involved in excitation-contraction coupling in smooth muscle; phosphorylates myosin crossbridges

natural killer cells (NK) large, granular lymphocytes, also known as null cells

negative feedback a type of feedback commonly employed in homeostatic regulatory systems in which the response of a system goes in a direction opposite to the change that set it in motion

nephrons functional units of the kidneys that filter the blood and form urine; each consists of a renal corpuscle (a glomerulus and Bowman's capsule) and renal tubule

nerve a bundle of axons in the peripheral nervous system; connects the central nervous system with organs in the periphery

net filtration pressure the force that determines the direction of fluid movement across capillary walls

neuroactive peptides short chains of amino acids that function as neurotransmitters in neurons; also neuropeptides

neurohormones special class of hormones released by neurosecretory cells

neuromodulators neurotransmitters that act through G proteins and generally produce long-lasting responses in postsynaptic neurons that affect other synaptic inputs

neuromuscular junction the synapse between a motor neuron and a skeletal muscle cell

neurons specialized cells in the nervous system that communicate via electrical and chemical signals; also known as nerve cells

neurotransmitter chemical messenger released from the axon terminal of a neuron

neutralization in the immune system, when an antibody blocks an antigen's activity by binding to it

neutrophils (noo'-troh-fil) leukocytes capable of phagocytosis

nitric oxide gas that functions as a chemical messenger

nociceptors (noh'-sih-sep-tors) sensory receptors on the surface of the body that detect tissue-damaging stimuli; include mechanical nociceptors that respond to intense mechanical stimuli, thermal nociceptors that respond to intense heat, and polymodal nociceptors

node of Ranvier (rahn'-vee-ay) gap in the myelin along axons; membrane in this region contains voltage-gated sodium and potassium channels that support the production of action potentials

nonspecific immunity mechanism that defends against potentially harmful substances without regard to their precise identity

nuclear envelope barrier that separates the nucleus from the cytoplasm; consists of two membranes

nuclear pores pores in the nuclear envelope that allow selective movement of molecules between the nucleus and cytoplasm

nucleic acids polymers of nucleotides that function in the storage and expression of genetic information; see deoxyribonucleic acid (DNA) and ribonucleic acid (RNA)

nucleotide a biomolecule containing one or more phosphate groups, a 5-carbon carbohydrate, and a nitrogenous base; involved in energy exchange and in the storage and transmission of genetic information in cells

nucleus in cells, membrane-bound structure that contains a cell's DNA; in the central nervous system, a collection of neuron cell bodies other than in the cerebral cortex; plural nuclei

occipital lobe (ok-sip'-ih-tul) one of four lobes of the cerebrum, located in the posterior and inferior portion of the cerebrum; important in processing visual information

odorants chemical substances that must bind to specific chemoreceptors in order to be smelled

olfactory epithelium the organ for smell within the nasal cavity

olfactory nerve the nerve that contains axons of olfactory receptor cells; cranial nerve I

oligodendrocytes (oh-lih-goh-den'-droh-sites) glial cells that form myelin around axons in the central nervous system; one oligodendrocyte provides myelin segments for many axons

oncotic pressure (ong-kot'-ik) osmotic pressure exerted by proteins in a fluid

oogenesis (oh-uh-jen'-ih-sis) the production of ova

opsonins (op-son'-ins) proteins generated in immune responses that bind tightly to foreign materials and make it easier for the phagocyte to engulf them later

optic chiasm (ki'-azm) portion of the visual neural pathway where some axons from both the right and left optic nerves cross to the opposite side of the brain

optic disk point on the retina of the eye that lacks photoreceptors; where the optic nerve and blood vessels that supply the eye join the retina

optic nerve the nerve that transmits visual information from the eye to the optic chiasm; contains axons of retinal ganglion cells

optic radiations pathways from the lateral geniculate body to the visual cortex on either side

optic tract tract containing axons of retinal ganglion cells that transmit visual information from the optic chiasm to the lateral geniculate nucleus of the thalamus

organ of Corti the sensory organ for sound; located on top of the basilar membrane in the cochlea of the inner ear

osmolarity total concentration of solute particles in a solution

osmoreceptors (oz'-moh-ree-sep-tors) receptors that detect the osmolarity of various body fluids

osmosis the passive movement of water across a membrane down its concentration gradient

osmotic pressure an indirect measure of a solute's concentration, expressed in ordinary units of pressure

ossicles three bones in the middle ear that transmit sound vibrations from the eardrum to the cochlea; includes the malleus, incus, and stapes

osteoblasts mobile cells within bone that secrete the extracellular matrix during bone formation

osteoclasts mobile cells that are responsible for breaking down bone tissue

osteoid (os'-tee-oyd) organic gel-like substance secreted by osteoblasts that forms the matrix of bone

otoliths   small calcium carbonate crystals within the gelatinous material found in the utricles and saccules

ovarian cycle   cyclic changes in ovarian structure and function; consists of two phases, the follicular phase and the luteal phase

ovaries   female gonads

ovulation   release of an oocyte from an ovarian follicle, which occurs at the conclusion of the follicular phase of the menstrual cycle

ovum   egg cell; plural ova

oxidative phosphorylation   the process of synthesizing ATP by harnessing the energy released when hydrogen atoms or electrons are transported through the electron transport chain in the inner mitochondrial membrane

oxytocin (ox-see-toh'-sin)   hormone secreted by the posterior pituitary that regulates contractions of the uterus and the flow of milk from the breasts; functions as a neurotransmitter in some neurons

pacemaker cells   cells that are capable of generating pacemaker potentials; in cardiac or smooth muscle, are responsible for triggering contractions

pancreas   a gland located in the abdominal cavity that performs endocrine and exocrine functions; secretes pancreatic juice and hormones, including insulin and glucagon

pancreatic juice   watery secretion from the pancreas that contains digestive enzymes and bicarbonate

paracrine (par'-ah-krin)   type of chemical messenger that communicates with neighboring cells by simple diffusion

parathyroid glands   primary endocrine glands located on the posterior surface of the thyroid gland; secretes parathyroid hormone (PTH)

parathyroid hormone (PTH)   peptide hormone released from the parathyroid glands that regulates plasma calcium levels

parietal cells   cells in the gastric glands that secrete hydrogen ions

parietal lobe   one of four lobes of the cerebrum, located immediately behind the frontal lobe of the cerebrum; important in processing somatic sensations and sensory integration

parturition   the process of birth

passive immunity   a type of immune protection that does not require a response from the immune system

passive transport   any method of transport of molecules across a membrane that does not require the use of energy

penis   the male copulatory organ

pepsin   the fully active form of pepsinogen; digests proteins to smaller peptide fragments

pepsinogen   an inactive precursor of the enzyme pepsin, secreted by specialized chief cells; acts in the lumen of the stomach to begin the digestion of proteins

peptides   short chains of amino acids, usually less than 50

perforin   protein that forms pores in a stricken cell's membrane and increases the membrane's permeability to ions and water, eventually causing lysis

perilymph   fluid found in the scala vestibuli and scala tympani of the cochlea in the inner ear; similar in composition to cerebrospinal fluid

peripheral chemoreceptors   chemoreceptors located in the carotid arteries that respond to changes in arterial P_{O_2}, P_{CO_2}, and pH and are involved in regulating ventilation

peripheral lymphoid tissues   tissues that trap foreign matter present in the blood, including the spleen, lymph nodes, tonsils, adenoids, appendix, and Peyer's patches

peripheral membrane proteins   proteins that are loosely bound to the lipid bilayer by associations with integral membrane proteins or phospholipids

peripheral nervous system   the division of the nervous system that contains nerve cells that provide communication between the central nervous system and organs of the body; includes afferent and efferent branches

peristalsis   a wave of momentary contraction that travels along the length of a hollow vessel or organ, such as the GI tract

peritoneum   a membrane lining the inside of the abdominal cavity

peritubular capillaries   capillary bed that branches off the efferent arterioles of cortical nephrons and is located close to the renal tubules; functions in exchange with renal tubules during reabsorption and secretion

permeability   measure of the ease with which molecules are able to move through the cell membrane

permissiveness   phenomenon in which one hormone is needed for another hormone to exert its actions

peroxisomes   spherical membrane-bounded organelles that function in the degradation of molecules such as amino acids, fatty acids, and toxic foreign matter

phagocytosis (fag-uh-sy-toh'-sis)   process by which a cell engulfs and digests microorganisms, abnormal cells, and foreign particles present in blood and tissues

pharynx (fair'-inks)   a passageway leading from the mouth to the esophagus or larynx that serves as a common passageway for food and air

phosphatase   an enzyme that catalyzes dephosphorylation

phosphatidyl inositol system (fos-fah-tide'-il in-os'-ih-tol)   a signal-transduction system that produces two second messengers, diacylglycerol and inositol triphosphate

phospholipid   amphipathic lipid molecule consisting of a glycerol backbone to which two fatty acids and a phosphorus-containing chemical group are attached

photopigment   the molecule in the photoreceptors that absorbs light, the first step of phototransduction

photoreceptors   cells located in the retina of the eye that detect light waves; includes rods, which detect dim light and are responsible for black-and-white vision, and cones, which detect bright light and are responsible for color vision

phrenic nerve (fren'-ik)   nerve that innervates the diaphragm

pia mater   innermost of the three meninges, adjacent to the nervous tissue

pineal gland   primary endocrine gland located in the brain; secretes the hormone melatonin

pinocytosis   a form of endocytosis in which a cell takes up fluid and dissolved molecules via endocytotic vesicles that pinch off the plasma membrane

pituitary gland   primary endocrine gland connected to the hypothalamus at the base of the brain; divided into the anterior pituitary and posterior pituitary

placenta a structure consisting of maternal and fetal tissues that allows exchange of gases, nutrients, and wastes between the mother's circulatory system and the circulatory system of the fetus

placental lactogen a hormone secreted by the placenta that promotes breast growth and the mobilization of energy stores in the mother

plasma liquid in the blood made up of water and dissolved solutes, including proteins; represents about 20% of the total volume of extracellular fluid

plasma cells mature form of B cells that secrete antibodies

plasma membrane consists of phospholipids, proteins, and cholesterol; separates the cell from its immediate environment

plasmin plasma protein that dissolves a blood clot

platelets cell fragments that play an important role in blood clotting

pleura (plur'-ah) the membrane that lines the chest wall and lung forming a pleural sac around each lung

pneumothorax condition in which air enters the pleural space causing the lungs to collapse and the chest wall to expand

polypeptide polymer containing amino acids joined together by peptide bonds

polysaccharide polymer composed of many monosaccharides joined together by covalent bonds

pons middle portion of the brainstem; connects to the cerebellum

positive feedback a type of feedback in which the response of a system goes in the same direction as the change that set it in motion

posterior pituitary posterior lobe of pituitary gland; secretes antidiuretic hormone (ADH) and oxytocin

postganglionic neuron the neurons of the autonomic nervous system that travel from autonomic ganglia to the effector organs

postsynaptic neuron at a synapse, the neuron that receives signals from another neuron

postsynaptic potential (PSP) a change in the membrane potential in a postsynaptic neuron that occurs in response to the binding of neurotransmitters to receptors

potentiation in endocrinology, the amplification of one hormone's effects by another hormone

preganglionic neuron the neurons of the autonomic nervous system that travel from the central nervous system to autonomic ganglia, where they communicate with postganglionic neurons

preload ventricular end-diastolic pressure

presynaptic facilitation a phenomenon occurring at an axo-axonic synapse such that activity in the presynaptic neuron enhances the release of neurotransmitter from the postsynaptic neuron

presynaptic inhibition a phenomenon occurring at an axo-axonic synapse such that activity in the presynaptic neuron decreases the release of neurotransmitter from the postsynaptic neuron

presynaptic neuron at a synapse, a neuron that transmits signals to a second neuron

primary active transport active transport of molecules utilizing a protein (pump) that uses ATP as the energy source

primary endocrine organs organs whose primary function is the secretion of hormones

primary immune response antigen-induced lymphocyte response that occurs when a person is exposed to an antigen for the first time

primary motor cortex region of the frontal lobe where voluntary movement is initiated

primary somatosensory cortex region of the parietal lobe of the cerebral cortex specialized for the processing of somatic sensory information

progesterone a sex hormone secreted by the ovaries, primarily during the luteal phase of the menstrual cycle

projection fibers axons in the CNS that connect the cerebral cortex with lower levels of the brain or the spinal cord

prolactin hormone that promotes breast growth and prepares the mammary glands for lactation

proliferative phase phase of the uterine cycle during which the uterus renews itself after the menstrual phase

promoter sequence a specific base sequence in DNA to which the enzyme RNA polymerase can bind, thereby initiating transcription

proprioception (pro-pree-oh-cep'-shun) the perception of the position of limbs and the body

prostacyclin eicosanoid released from healthy endothelial cells that inhibits platelet aggregation

prostaglandins substances with a wide range of effects, including the development of fever

prostate gland an accessory gland of the male reproductive system that secretes fluid into the urethra

protein a polymer containing amino acids joined together by peptide bonds; usually refers to chains containing more than 50 amino acids

protein C plasma protein that inhibits blood coagulation

protein kinase a type of enzyme that catalyzes phosphorylation of a target protein

proteolysis (proh-tee-oh-ly'-sis) process by which proteins are broken down to amino acids

proximal tubule portion of the renal tubule that includes the proximal convoluted tubule and the proximal straight tubule

pulmonary arteries arteries that carry blood to the lungs from the heart

pulmonary circuit the portion of the vasculature that encompasses all the blood vessels within the lungs and those connecting the lungs with the heart

pulmonary semilunar valve valve located between the right ventricle and pulmonary trunk

pulmonary surfactant detergent-like substance secreted by Type II alveolar cells; decreases the surface tension in the lungs

pulmonary veins veins that carry blood from the lungs to the heart

pulmonary ventilation the movement of air into and out of the lungs by bulk flow

pulse pressure (PP) the difference between systolic and diastolic pressure; represents the magnitude (height) of the arterial pressure wave

pumps proteins that actively transport molecules across a membrane utilizing ATP as the direct energy source

Purkinje fibers (purr-kin'-gee) an extensive network of conducting fibers that spread through the ventricular myocardium

pylorus a narrow passage between the stomach and the duodenum

pyramidal tract a direct pathway from the primary motor cortex to the spinal cord

pyrogens (py'-roh-jen) substances that act on the hypothalamus to raise body temperature

reabsorption in the kidneys, the transport of a substance from the lumen of a renal tubule to the surrounding interstitial fluid

reactive hyperemia a local increase in blood flow that occurs following termination of an occlusion of blood flow to the same area

receptive field the area over which an adequate stimulus can produce a response, either excitatory or inhibitory, in an afferent neuron or higher order neurons

receptor potential graded potential caused by the opening or closing of ion channels on sensory receptors, and triggered by sensory stimuli

receptor-mediated endocytosis a form of endocytosis in which endocytotic vesicles contain receptors that recognize and bind specific molecules in the extracellular fluid; enables cells to selectively take up certain molecules

receptors proteins on a target cell that recognize and bind a chemical messenger

recruitment an increase in the number of active motor units in a skeletal muscle

rectum the most distal portion of the large intestine, which functions in the storage of feces

referred pain the perception of a painful stimulus as originating at a site on the body distinct from the location of the stimulus

reflex arc simple reflex pathway made up of five components, including a sensory receptor, an afferent neuron, an integration center, an efferent neuron, and an effector organ

refractory period period of reduced membrane excitability; during and immediately after an action potential when the membrane is less excitable than it is at rest

regulatory site binding site on an enzyme molecule that is specific for molecules known as modulators

relative refractory period period immediately following the absolute refractory period during which it is possible to generate a second action potential, but only with a stimulus stronger than that needed to reach threshold under resting conditions

REM sleep sleep characterized by high frequency waves in the EEG and periodic rapid movements of the eyes; REM = rapid eye movement

renal arteries arteries that branch off the aorta and provide the kidneys with their blood supply

renal pelvis funnel-shaped passage forming the initial portion of the ureter

renal threshold the plasma concentration of solute at which the transport maximum is exceeded and excess solute appears in the urine

renal tubule a portion of a nephron, consisting of a long, coiled tube

renal veins transport blood from the kidneys back into general circulation

repolarization return of the membrane potential of a cell to the resting potential following a depolarization

residual volume (RV) the volume of air remaining in the lungs after a maximum expiration

resorption the breakdown of bone tissue

respiration the process of gas exchange within the body; includes internal respiration and external respiration

respiratory bronchioles small tubules of the respiratory tract located between terminal bronchioles and alveolar ducts

respiratory membrane the structure across which gas exchange occurs in the lungs; a barrier between blood and air consisting of capillary endothelial cells and their basement membranes and alveolar epithelial cells and their basement membranes

respiratory pump the blood-pumping action exerted by respiratory movements; increases the flow of blood to the heart

respiratory quotient the ratio of carbon dioxide produced by the body to the amount of oxygen consumed

respiratory zone the lower part of the respiratory tract; the site of gas exchange within the lungs

resting membrane potential the voltage that exists across a cell membrane when the cell is not transmitting electrical signals; polarity is such that the inside of the cell is negative with respect to the outside

reticular formation a diffuse network of nuclei located in the brainstem that is important in sleep-wake cycles, arousal of the cerebral cortex, and consciousness

retina innermost layer of the eye; consists of neural tissue and contains photoreceptors

reuptake active transport of neurotransmitter back into the presynaptic neuron that released it

ribonucleic acid (RNA) polynucleotide molecules found in a cell's nucleus and cytoplasm that are necessary in the expression of genetic information

ribosomes complexes of rRNA and proteins that function in protein synthesis; found in the cytosol and rough endoplasmic reticulum

rods photoreceptors that enable visibility during relatively low light and are responsible for black-and-white vision

saccule (sak'-yool) structure of the inner ear; detects up and down linear acceleration

salivary glands glands that secrete saliva, located in the mouth and pharynx

saltatory conduction a type of electrotonic conduction in axons that are sheathed in myelin

sarcolemma (sar-co-lem'-ah) a muscle fiber's plasma membrane

sarcomeres (sar'-kuh-meers) the fundamental repeating units that make up myofibrils

sarcoplasmic reticulum (SR) a saclike membranous network that surrounds the myofibrils and stores calcium

scala media fluid-filled duct in the cochlea; also called cochlear duct

scala tympani fluid-filled duct in the cochlea; also called tympanic duct

scala vestibuli fluid-filled duct in the cochlea; also called vestibular duct

Schwann cells glial cells that form myelin around axons in the peripheral nervous system; one Schwann cell provides myelin for one axon

sclera (sklee'-rah) tough connective tissue that makes up the white of the eye

scrotum a sac that houses the testes

second messenger an intracellular messenger molecule that is produced in response to the binding of an extracellular messenger (the first messenger) to a receptor

secondary active transport active transport of molecules utilizing a carrier that uses a concentration gradient or electrochemical gradient as a source of energy

secondary bronchi branches off the bronchi leading to the lungs; includes three secondary bronchi to the right lung and two secondary bronchi to the left lung

secondary endocrine organs organs whose secretion of hormones is secondary to another function

secondary immune response response caused by subsequent exposure to an antigen

secretin a hormone secreted by the duodenum and jejunum that inhibits gastric secretion and motility and stimulates pancreatic bicarbonate secretion

secretion movement of substance from the internal environment to the external environment by transport across an epithelium; movement of substance from inside a cell to outside the cell by movement across the plasma membrane

secretory phase phase of the uterine cycle during which the uterine endometrium prepares for the possible arrival and subsequent implantation of the fertilized egg

secretory vesicles intracellular vesicles containing molecules destined for secretion from the cell

self-tolerance the capacity of a lymphocyte to distinguish self from nonself so that the lymphocyte is reactive only against foreign cells

semicircular canals structures in the inner ear that contain the receptor cells for rotational acceleration

semilunar valves valves located between the ventricles and arteries on either side of the heart and that prevent blood from flowing back into the ventricles when the ventricles are relaxed

seminal vesicles accessory glands of the male reproductive system; the first to secrete fluid into the reproductive tract during sperm transport

seminiferous tubules (sem-ih-nif'-er-ous) thin, highly-coiled hollow tubes in the testes where sperm are produced

sensory homunculus map that indicates which areas of the primary somatosensory cortex are devoted to particular regions of the body

sensory receptors specialized neuronal structures that detect a specific form of energy in either the internal or external environment

sensory unit a single afferent neuron and all sensory receptors associated with it

serosa the outermost of the four major layers of the wall of the digestive tract; composed mostly of connective tissue

Sertoli cells epithelial cells lining the wall of a seminiferous tubule whose primary function is to nurture sperm and control their development

serum plasma with coagulation factors removed

set point the normal or desired value of the regulated variable in a homeostatic regulatory system

sex hormones steroid hormones including estrogens, progesterone, and androgens, secreted from the adrenal cortex and gonads that regulate reproductive function, promote gametogenesis, growth and maintenance of reproductive organs, and the development of secondary sex characteristics

signal transduction the process by which the binding of a chemical messenger to receptors brings about a response in a target cell

simple diffusion passive movement of molecules in response to chemical or electrical forces

sinoatrial (SA) node (sy-noh-ay'-tree-al) a region in the wall of the upper right atrium where pacemaker cells are concentrated; normally determines the heart rate

size principle the correspondence between the size of motor units and the order of recruitment

skeletal muscle the type of muscle that is generally connected to bone

skeletal muscle pump the blood-pumping action exerted by skeletal muscles as they contract and press against veins traveling between them, which forces blood to move toward the heart

slow ligand-gated ion channels channels regulated by G proteins that cause ion channels to open or close in response to the messenger binding to its receptor

slow pain a poorly localized, dull, aching sensation produced by activation of nociceptors; is transmitted by C fibers

slow-twitch fibers skeletal muscle fibers that contract slowly and take a longer time to reach peak tension during a twitch

slow-wave sleep (SWS) sleep characterized by multiple stages of low frequency waves in the EEG

small intestine an organ of the digestive tract, consisting of a coiled tube 8–10 feet long; the primary site for the digestion and absorption of all the nutrients in food

smooth muscle the type of muscle found in internal organs, blood vessels, and other structures that are not under voluntary control; lacks striations

sodium-potassium pump a protein that utilizes ATP to actively transport sodium ions out of the cell and potassium ions into the cell against their electrochemical gradients

somatic nervous system the division of the nervous system that encompasses nerve cells that regulate skeletal muscle contractions

somatomedin (soh-mat-oh-me'-din) peptide hormone secreted by the liver in response to growth hormone; promotes protein synthesis and growth; also known as insulinlike growth factor

somatosensory system (suh-mat-uh-sen'-suh-ree) branch of the nervous system associated with perception of sensations; associated with receptors in the skin and proprioception

somatostatin tropic hormone released by the hypothalamus that inhibits growth hormone secretion from the anterior pituitary; also known as growth hormone inhibiting hormone (GHIH)

somesthetic sensations sensations that arise from receptors in the skin

spatial summation in a postsynaptic cell, the addition of postsynaptic potentials generated at different synapses that occurs when the synapses are stimulated more or less simultaneously

specific immunity mechanism that targets and eliminates specific substances

spermatogenesis (sper-mah-toh-jen'-ih-sis) sperm production

spermatozoa (spur-ma-toh-zoh'-ah) newly formed sperm cells with characteristic head, midpiece, and tail

sphincter a ring of muscle that surrounds an orifice and regulates the passage of material through it by altering its diameter

sphincter of Oddi a ring of smooth muscle that regulates the flow of bile and pancreatic juice into the duodenum

spinal nerves a total of 31 pairs of nerves originating in the spinal cord and traveling to the periphery

spinothalamic tract somatosensory pathway that transmits information from thermoreceptors and nociceptors to the thalamus

starch polysaccharide found in plants

Starling's Law of the Heart law stating that when there is a change in the rate at which blood flows into the heart from the veins, the heart automatically adjusts its output to match the inflow

stereocilia (ster-ee-oh-sil'-ee-ah) hairlike projections on the upper surfaces of hair cells in the inner ear that move in response to sound vibrations or acceleration of the head

steroids lipids derived from cholesterol consisting of three 6-carbon rings and one 5-carbon ring, many of which function as hormones

stomach an organ of the digestive tract that stores food and releases it into the small intestine; major anatomical regions are the fundus, body, and antrum

striated muscle (stry'-ay-ted) muscle in which the cells have a striped appearance due to the presence of sarcomeres; includes skeletal and cardiac muscle

stroke volume (SV) the volume of blood ejected from each ventricle during a single heartbeat

subarachnoid space the space between the pia mater and arachnoid mater, filled with cerebrospinal fluid

submucosa one of the four major layers of the wall of the digestive tract, located between the mucosa and muscularis externa; composed mostly of connective tissue

submucosal plexus one of the two neural networks that make up the enteric nervous system; located in the submucosa

substance P hormone that decreases gastrointestinal motility; functions as a neurotransmitter in some neurons

substrate a reactant in an enzyme-catalyzed reaction; binds to the active site of an enzyme molecule

substrate-level phosphorylation a mechanism of ATP synthesis, in which an enzyme transfers a phosphate group from a substrate to ADP; occurs in steps 7 and 10 of glycolysis and step 5 of the Krebs cycle

sulci (sul'-sigh) grooves in the highly convoluted gray matter of the cerebral cortex; singular sulcus

summation in neurophysiology, the adding together of postsynaptic potentials that occurs within a neuron; in muscle physiology, the adding together of twitches that occurs when a muscle is stimulated at high frequency

sympathetic chains structures parallel to the spinal column on either side in which the sympathetic ganglia are linked together in rows; also called sympathetic trunks

synaptic cleft the extracellular space between the axon terminal of the presynaptic cell and the postsynaptic cell at a synapse

synaptic vesicles vesicles in the axon terminal that contain neurotransmitter molecules

syncope fainting

synergistic pertaining to a process in which the net effect is greater than the sum of the individual effects

systemic circuit the portion of the vasculature that encompasses all of the body's blood vessels, except those belonging to the pulmonary circuit

systole (sis'-toh-lee) the period of ventricular contraction during a cardiac cycle

systolic pressure (SP) the maximum aortic pressure attained during the cardiac cycle; occurs during systole

T cell receptor (TCR) receptors that detect foreign antigens on body cells

T lymphocytes cells that kill particular abnormal or infected body cells; also called T cells

T-dependent antigens antigens that evoke the production of plasma cells and memory B cells with help from a special kind of T cell

T-independent antigens antigens that can activate B cells without T cell help; plasma cells are produced, but no memory B cells are produced

tastants chemical substances that give foods their flavors

tectorial membrane (tek-tor'-ee-al) membrane in the organ of Corti in which the tips of stereocilia are embedded

temporal lobe one of four lobes of the cerebrum located at the side and separated from the frontal lobe by a very deep groove; important in processing auditory information and language

temporal summation in a postsynaptic cell, the addition of postsynaptic potentials generated at a particular synapse that occurs when the synapse is stimulated at a high frequency

tendons cords of elastic connective tissue that transmit force from skeletal muscles to bones

terminal bouton the axon terminal of a motor neuron, which stores and releases acetylcholine

terminal bronchioles bronchioles that lead directly to the airways of the respiratory zone of the respiratory tract; the last component of the conducting zone

testes male gonads

tetanus (tet'-ah-nus) in a muscle being stimulated at high frequency, the plateau phase of the contraction, during which the tension is relatively constant

thalamus cluster of nuclei located in the diencephalon; functions as a relay station for sensory information en route to the cerebral cortex

theca cells (thee'-ca) cells that surround the granulosa cells in a follicle

thermoneutral zone the range of environmental temperatures over which the body is able to regulate its temperature solely through changes in skin blood flow

thermoreceptors sensory receptors that detect temperature; include warm receptors that respond to temperatures between 30° and 43°C, and cold receptors that respond to temperatures between 35° and 20°C

thick filaments filaments composed of myosin that form part of the contractile machinery of a muscle cell

thin filaments filaments composed of actin that form part of the contractile machinery of a muscle cell; also contain troponin and tropomyosin in striated muscle cells

threshold in an excitable cell, the critical value of the membrane potential to which the cell must be depolarized in order to trigger an action potential

thrombin one of the active clotting factors in the coagulation cascade; reacts with fibrinogen to convert it to fibrin and activates factor XIII

thromboxane A_2 eicosanoid released from aggregated platelets that promotes hemostasis

thrombus blood clot

thymus primary endocrine gland located near the heart; secretes the hormone thymosin; also is site of T lymphocyte maturation

thyroid gland butterfly-shaped primary endocrine gland located on the ventral surface of the trachea; secretes tetraiodothyronine (T_4), triiodothyronine (T_3), and calcitonin

thyrotropin releasing hormone (TRH) tropic hormone secreted from the hypothalamus that stimulates secretion of TSH from the anterior pituitary

tidal volume (V_T) the volume of air that moves into and out of the lungs during a normal, unforced breath

tight junction a junction that connects two adjacent cells together, forming a nearly impenetrable barrier that limits the passage of molecules between them

tissue factor a protein found in subendothelial tissues that initiates the extrinsic clotting pathway when exposed to blood

total body water (TBW) the volume of water that is contained in all the body's compartments

total lung capacity (TLC) the volume of air in the lungs at the end of a maximum inspiration

total peripheral resistance (TPR) in the systemic circuit, the combined resistance of all the organs, including the blood vessels leading toward and away from them

trachea (tray'-key-ah) cartilaginous tube of the respiratory tract, located between the larynx and the bronchi

transcription the synthesis of messenger RNA using DNA as a template

transcytosis the transport of macromolecules across epithelial cells; involves endocytosis at one membrane, followed by exocytosis at the opposite membrane

translation the synthesis of polypeptides using messenger RNA as a template; the final step in the expression of genetic information

transmembrane proteins integral membrane proteins that span the lipid bilayer, with surfaces exposed to both the cytosol and interstitial fluid

transpulmonary pressure the difference between the intrapleural pressure and the intra-alveolar pressure, which represents the distending pressure acting on the lungs; $P_{alv}-P_{ip}$

transverse tubules structures that transmit action potentials from the sarcolemma into the cell's interior, triggering the release of calcium from the sarcoplasmic reticulum ; also called T tubules

TRH hormone secreted by the hypothalamus that regulates the release of TSH by the anterior pituitary; functions as a neurotransmitter in some neurons

tricuspid valve the AV valve on the right side of the heart, which has three cusps

trigger zone the initial portion of an axon, where action potentials are initiated

triglyceride a lipid consisting of three fatty acids linked to a glycerol backbone

tropic hormones hormones that regulate the secretion of another hormone; also called trophic hormones

tropomyosin (troh-poh-my'-oh-sin) one of the two regulatory proteins in striated muscle; a long fibrous molecule that acts to block myosin-binding sites on thin filaments when a muscle is not contracting

troponin (troh-poh'-nin) one of the two regulatory proteins in striated muscle; binds calcium reversibly and is responsible for starting the crossbridge cycle by moving tropomyosin out of its blocking position

trypsin a protease enzyme secreted by the pancreas

tubuloglomerular feedback autoregulatory mechanism in which a change in glomerular filtration rate is regulated by paracrines secreted from the macula densa, located downstream from the glomerulus

turbulent flow a type of blood flow in which the blood moves in different directions in different places within a vessel; increases the resistance to blood flow

twitch the mechanical response of an individual motor unit to a single action potential

tympanic membrane eardrum

umbilical cord a ropelike structure that extends from the placenta to the fetus; contains vessels that provide blood to the fetus

upper airways air passages in the head and neck; include the nasal cavity, oral cavity, and pharynx

urinary system organ system that consists of two kidneys, two ureters, the urinary bladder, and the urethra

urine a fluid produced by the kidneys and eliminated from the body

uterine cycle cyclic changes in uterine structure and function that occur during the menstrual cycle

uterine tubes tubes protruding from the upper part of the uterus on either side that function in egg transport and fertilization; also called fallopian tubes or oviducts

uterus   a hollow pear-shaped organ located in the center of the pelvic cavity that functions to house and nourish the developing human

utricle   structure of the inner ear; detects forward and backward linear acceleration

vagina   a canal about 8–10 cm long leading from the cervix to the outside of the body, receives the penis during intercourse; the female organ of copulation

vagus nerve (vay'-gus)   major parasympathetic nerve that originates in the medulla oblongata and innervates much of the viscera; cranial nerve X (CN X)

vas deferens (vas def'-er-enz)   a duct leading from each testis to the abdominal cavity that carries sperm and fluid

vasa recta   capillary bed that branches off the efferent arteriole of a juxtamedullary nephron and surrounds the loop of Henle; functions in maintaining medullary osmotic gradient

vasopressin   hormone secreted by the posterior pituitary that regulates urine output by the kidney; stimulates water reabsorption by the kidneys; functions as a neurotransmitter in some neurons; also known as antidiuretic hormone (ADH)

vaults   barrel-shaped cell organelles that may be involved in the transport of molecules across the nuclear envelope and may have other functions

veins   large blood vessels that carry blood toward the heart

vena cava (vee'-nah kay'-vah)   one of the two large veins that carry blood into the right atrium

venomotor tone   the degree of tension exerted by smooth muscle in the walls of veins; venous compliance decreases as venomotor tone increases

venous pooling   the accumulation of blood in the veins

venous return   blood flow into the heart

ventilation-perfusion ratio   relationship of ventilation to perfusion in alveoli, or $\dot{V}_A/\dot{Q}$

ventral horn   anterior half of the gray matter on either side of the spinal cord

ventral root   point at which a spinal nerve bifurcates before joining the spinal cord, the ventralmost of the two resulting branches; contains efferent nerve fibers

ventricles   the heart's two lower chambers, which pump blood into the arteries

ventricular ejection   the exit of blood from the ventricles

venules   blood vessels that carry blood from capillaries to veins

vertebral column   bony structure that surrounds and protects the spinal cord

very-low-density lipoproteins (VLDLs)   one of a group of lipoprotein particles consisting of lipids and proteins in various ratios; VLDLs contain a high ratio of lipids to proteins, thus its low density

vesicle   small, spherical membrane-bound sac

vestibular apparatus   structures of the inner ear that contain the receptor cells for equilibrium, including the semicircular canals, utricle, and saccule

vestibular membrane   membrane in the cochlea of the inner ear that separates the scala vestibuli from the scala media

vestibulocochlear nerve   nerve that contains the afferents for hearing and equilibrium; cranial nerve VIII

villi   folds in the mucosal surface of the small intestine that facilitate the transport of materials by increasing the surface area of the epithelium; singular, villus

visceral reflexes   automatic changes in the functions of organs that occur in response to changing conditions inside the body

visual cortex   portion of the cerebral cortex that processes visual information; located in the occipital lobe

vital capacity (VC)   the maximum volume of air that can be expired following a maximum inspiration

vitamins   organic molecules that are required in trace amounts in the body; often act as coenzymes

vitreous humor   jellylike material found in the vitreous chamber of the eye; maintains the spherical structure of the eye

voltage-gated channels   channels that open or close in response to a change in membrane potential

von Willebrand factor   protein that anchors platelets to collagen fibers during the platelet adhesion phase of hemostasis

vulva   external genitalia of the female; includes the mons pubis, labia majora, labia minora, the vestibule, and vestibular glands

Wernicke's area   area of association cortex devoted to language comprehension; located in the posterior and superior portion of the temporal lobe and the inferior parietal lobe

white matter   areas of the CNS consisting primarily of myelinated axons; specialized for the rapid transmission of information over relatively long distances in the form of action potentials

zona pellucida   a thick layer of noncellular material that forms between an oocyte and surrounding granulosa cells

zonular fibers   strands of connective tissue in the eye that connect the ciliary muscles to the lens and are involved in adjusting the shape of the lens to focus light

zygote (zy'-goat)   a diploid cell formed by the fusion of two gametes

zymogens   inactive precursor forms of pancreatic digestive enzymes that are stored in secretory cells and released by exocytosis

Index

Note: An *f* following a page number indicates a figure, a *t* indicates tabular material, and a *b* indicates boxed material. Boldface page numbers indicate definitions.

A band, muscle, **336**
Abdominal cavity, 371, 372*f*
Abortion, 694*b*
Absolute refractory period, **192**–93
Absorption, 6, **124**
    across capillaries, 437–38
    of carbohydrates, 614–15
    of lipids, 618, 620*f*
    of proteins, 615–16
    of vitamins, minerals, and water, 620
Absorption, digestive, 602, 603*f*
Absorptive cells, **602**
Absorptive state, **645**
    energy metabolism during, 645–47, 648*t*
    major metabolic reactions of, 646*f*
    regulation of metabolism during, 649–54
Accessory glands, digestive system, 602, 610–14
    generalized structure of, 610*f*
    liver, 612–13
    pancreas, 611–12
    salivary glands, 610–11
Accessory reproductive organs, **673**
    male, 677*f*, 679
Accommodation (eye), **281**, 282*f*, 323
A cells (alpha cells), 159, 161*f*
Acetylcholine (ACh), 134, **214**–16
    as autonomic neurotransmitter, 317
    excitation-contraction coupling and role of, 340, 341*f*
Acetylcholinesterase (AChE), **215**, 320, 321, 325
Acetylcoenzyme A (acetyl CoA), 73, 94, 95, 96, 215
    role of, in Krebs cycle, 82*f*, 83
Acetyl groups, 73
Achilles tendon, 357, 358*f*
Acid(s), 75*b*
    inputs and outputs of, into blood, 585*f*
    secretion of, in stomach, 607, 625–26
Acid-base homeostasis, 522–26, 583–95
    blood pH and, 522–23, 524*b*
    compensation for disturbances in, 592–95
    defense mechanisms against disturbances in, 588–92

role of respiratory system in, 524–26
    sources of disturbances in, 585–88
Acidosis, **523**, 583–85
    metabolic, 585–88, 593–94
    respiratory, **526**, 593
Acini, **610**
Acquired immunity, 720
Acromegaly, 164
Acrosome, **681**
Acrosome reaction, **695**
Actin, 334, **336**
    unbinding and binding of, to myosin, 339–40
Actin-binding site, 337
Action potentials, 173, **173**–74, 177*t*, 183, 187–96
    in cardiac muscle, 377–79, 380*f*, 381*f*, 382*f*, 383
    characteristics of neurons at rest and in phases of, 192*t*
    comparison of graded potentials and, 188*t*
    frequency coding, 194*f*
    ionic basis of, 187–88, 189*f*
    in muscle contraction, 340–43
    propagation of, 194–96, 197*t*
    refractory periods in, 192–94
    role of graded potentials in creation of, 184
    role of voltage-gated ion channels in, 188–92
Activation energy, **68**–69
Activation energy barrier, 68, 69*f*, 70*f*
    chemical reaction rates and height of, 71
Activation gates, **189**
Active hyperemia, **456**, 457*f*, 458*f*
Active immunity, **729**
Active site (enzyme), **72**
Active transport, **102**, 115–19
    characteristics of, 118*t*
    factors affecting rates of, 110
    passive transport versus, 101–2
    primary, 115, 116–17
    pumps and leaks and, 118–19
    pumps as mechanisms of (*see* Pump(s))
    renal reabsorption and, 546, 547*f*
    secondary, 115, 117–18
Acuity of location-stimulus perception, 268, 269, 270
Acute nephritic syndrome, 586*b*
Acute phase proteins, 716–17

Adaptation
    to light and dark, 286, 287–88
    of sensory receptors, **264**
Addison's disease, 665
Additive effects of hormones, **166**
Adenine, 32, 33*f*
Adenosine, sleep-wake cycles and, 252
Adenosine deaminase (ADA), 733, 734
Adenosine diphosphate (ADP), **77**
Adenosine monophosphate (AMP), 32
Adenosine triphosphate (ATP), 32, **77**–78, 640–41
    actions of, in glucose oxidation, 78–81
    differences in mode of production of, in muscle fibers, 354–55
    muscle contraction and production of, 343
    synthesis and hydrolysis of, 77, 78*f*, 79*f*
Adenylate cyclase, **146**, 152*b*, 153*f*, 306
Adequate stimulus, **262**
ADH. *See* Antidiuretic hormone (ADH)
Adhesion, cell-to-cell, 43–46
Adhesion molecules, inflammation and, 715
Adipocytes, **643**, 647, 649
Adipose tissue, 94
Adrenal cortex, **159**, 160*f*
Adrenal glands, 154*f*, **159**, 160*f*
    sympathetic innervation of, 314, 316*f*
Adrenaline, 141. *See also* Epinephrine (adrenaline)
Adrenal medulla, **159**, 160*f*, 314
Adrenergic neurons, **317**
Adrenergic receptors, 141, **217**
    effects of epinephrine and, 451–52
    types of, 317–18, 319*f*, 320*t*
Adrenocorticoids, **159**
Adrenocorticotropic hormone (ACTH), 157, 162, **664**
Aerobic exercise, 357. *See also* Exercise
Afferent arteriole, 536
    stretching of, 543
Afferent division of peripheral nervous system, **171**
Afferent neurons, **175**, 261
    receptive fields of, **264**, 265*f*, 268*f*–70
    sensory transduction and, 262–64
Affinity of enzymes, **73**
After-hyperpolarization, action potentials and phase of, 188, 189*f*
Afterload, influence of, on stroke volume, **398**
Agglutination as antibody function, **723**, 724*f*

Agonists, **142**
Agranulocytes, 709
AIDS, 727, 734
Air, gases in, 498. *See also* Gases; Gas exchange
Air embolisms, 499
Air hunger, 518
Airway resistance, **484**–87
Albumin, 140, 404, 405t
Aldosterone, 452, **575**, 752
  effects of, on principal cells of renal distal tubules and collecting tubules, 577f
  potassium levels and secretions of, 163–64, 165f, 580
  sodium balance and effects of, 575–78, 579f
Alkalosis, **523**, 583–85
  metabolic, **585**–88, 594
  respiratory, 525b, **526**, 593
Alleles, 671
Allergens, 487
Allergies, 136b, **733**
  asthma and, 486–87
  as immune system dysfunction, 732f, 733
  reduction in blood flow to brain and responses to, 443
All-or-none principle, action potentials and, **190**–92
Allosteric regulation of enzyme activity, **75**–76
Alpha [α] adrenergic receptors, 217, **317**–18, 319f, 320t
  epinephrine binding to, 451
Alpha cells, 159, 161f
Alpha waves (sleep), 252, 253f
Altitude, effects of, on oxygen blood levels, 525b
Alveolar ducts, **475**, 476f
Alveolar edema, 504b
Alveolar macrophages, **475**, 476f
Alveolar pores, **475**
Alveolar sacs, **475**, 476f
Alveolar ventilation, 490, **491**
Alveoli, **475**, 476f
  determinants of partial pressures in, 504–5
  fluid layer in, 484, 486–87b
  partial pressure in, 500, 501f, 502, 503f, 504–5
Amacrine cells, 285
Amines
  as chemical messengers, **135**, 137–38, 216–17
  synthesis and release of, 137–38
Amino acid(s), 22, **28**, 29f. *See also* Protein(s)
  as chemical messengers, 135, 137, 216, 217f
  deamination of, 95
  energy metabolism of, during absorptive and postabsorptive states, 649t
  essential, 96, 137, 655
  stimulation of insulin and glucagon secretions by, 653
  synthesis and release of, 137
Amino acid neurotransmitters, **217**
Aminoaciduria, 587b
Amino groups, 75b
Aminopeptidase, **616**
Amnion, **697**
Amniotic cavity, **697**
Amniotic fluid, **697**
Amphipathic molecules, **25**

Ampulla, **300**, 301f
Ampulla of Vater, 613
Amygdala, 241, 242f
Anabolic reactions, **63**
  as energy-requiring reactions, 65
  rates of, 46
Anabolism, **46**, 63, 641
Analgesia, **276**
Anaphase (mitosis), 58f
Anaphylactic shock, **733**
Anatomical dead space, **491**
Androgen-binding protein, 677
Androgens, 157, 160, **673**
  effects of, on growth, 661
Anemia, 406, 507, 510–11b, 620
Aneurysm, 230b
Angiotensin converting enzyme (ACE), 452, 577
Angiotensinogen, 452
  RAAS system and, 576–78
Angiotensin II, 533
  effects of, on mean arterial pressure, 452, 577, 578f
Anions, 31b
  negative electrical charge in, 103
Antagonism, hormonal, **165**
Antagonistic muscles, **358**
Antagonists, 142
Anterior cavity (eye), **278**, 279f
Anterior lobe, pituitary gland (adenohypophysis), **155**
  connection of, to hypothalamus, 156f
  hormone secretions from, 156f, 157, 158f
Anterograde transport, 206
Antibodies, **711**
  antigen-disposing mechanisms of, 724f
  function of, in humoral immunity, 722–24
  heavy and light chains of, 719, 720f, 722
  role of B lymphocytes in production of, 721–22
  structure and properties of major classes of, 723t
Anticodon, 51
Antidiuretic hormone (ADH), 48, 165, 452, 552, **572**. *See also* Vasopressin
  effects of, on water permeability in late distal tubule and collecting ducts, 572–73, 574f
  regulation of secretion of, 574, 575f
Antidiuretics, 573
Antigen(s), **711**
  antibody-mediated mechanisms of disposing of, 724f
  diversity of lymphocyte response to, 720, 721f
  memory and primary or secondary immune response to, 720–21, 722f
  role of MHC molecules in presentation of, 725–26
  specificity of B cell and T cell binding to, 719–20
  T-dependent, and T-independent, 722
  tumor, 728
Antigenic determinants, 719
Antigen presentation, **725**
Antigen receptors, **719**, 720f
  on T cells, 725

Antihistamines, 136b
Antrum of stomach, 606f, **607**
Anus, 610
Aorta, 372, 373f
  blood flow through, **375**, 376f
  pressure in, during cardiac cycle, 388, 389b, 390
Aortic arch, arterial baroreceptors on, 444
Aortic semilunar valve (aortic valve), **373**, 375f
Aphasia, 251
Apical membrane, 45, **124**
Aplastic anemia, 510b
Apoproteins, 621b
Apoptosis, **721**
Appendicitis, 609
Aquaporins, 571
  aquaporin-2, 573, 574f
Aqueous humor, **278**, 279f
Arachidonic acid, **408**
Arachnoid mater, **224**, 225f
Arachnoid villi, 225f, 226
Arcuate arteries, renal, 537, 539f
Arcuate fasciculus, 231
Arcuate veins, 538, 539f
Arrhythmias, cardiac, 385f, 386
Arterial baroreceptors
  cardiovascular response to change in mean arterial pressure and role of, 447f
  cardiovascular response to hemorrhage and role of, 453–55
  effects of, on release of antidiuretic hormone, 574, 575f
  response of, to changes in arterial pressure, 444f, 445
Arterial pressure, short-term and long-term regulation of, 443. *See also* Mean arterial pressure (MAP)
Arteries, **371**, 411–12
  blood pressure in (*see* Mean arterial pressure (MAP))
  brachial, 425b
  carotid, 444
  compliance of, 412, 415b
  coronary, 419
  hardening of, 389b
  hepatic, 609
  as pressure reservoirs, 390, 411–12
  pulmonary, 376
  renal, 534, 537, 539f
  structural characteristics of, 413f
  uterine, 697
Arteriolar smooth muscle, vascular smooth muscle response to stretching of, 458–60
Arterioles, **371**, 412–13
  afferent, 536, 543
  efferent, 536
  local controls on radius of pulmonary, 523t
  regulation of blood flow to organs, and role of, 419
  structural characteristics of, 413f
Arteriosclerosis, 230b
Ascending colon, 609
Ascending reticular activating system (ARAS), 252, 253f
Ascending tracts, spinal cord, **232**–33, 234f, 235f, 236–37

A site, ribosome, 51, 52*f*
Aspartate, 135, 137, **217**
Aspirin as anticoagulant, 411
Association areas, cerebral cortex, **239**
Association fibers, 229*f*, **231**
Associative learning, **255**
Asthma, 486–87
   treatment of, 318–20, 487
Astigmatism, 282
Astrocytes, 176, **227**
Atherosclerotic plaques, 621*b*
Atmospheric pressure (P$_{atm}$), **477**, 478*f*
Atomic number, 23
Atoms, 23*b*
ATP. *See* Adenosine triphosphate (ATP)
ATPases, 116
ATPase site, 337
ATP hydrolysis, 77, 78*f*
ATP synthase, **84**
Atresia, **683**
Atria (atrium), **371**, 373*f*
   pressure in, during cardiac cycle, 388*f*, 389*b*
Atrial fibrillation, 386
Atrial myocardium, 372
Atrial natriuretic peptide (ANP), 160
   sodium balance and role of, 578, 579*f*
Atrioventricular node (AV), 363, **377**
   action potentials in, 377
Atrioventricular valves (AV valves), **372**, 373*f*
   action of, 374*f*
   bicuspid (mitral), 372
   tricuspid, 372
Attachment
   inflammation and cell, **715**
   phagocytosis and, 715, 716*f*
Auditory cortex, 129, 240*f*, **298**
Auditory system, 291–99
Autocrines, 133*f*, **134**
Autoimmune disease, 709, 733
   multiple sclerosis as, 731*b*
Autonomic ganglia, **313**
Autonomic nervous system, 171*f*, **172**,
   312–23
   anatomy of, 313–17
   control of cardiovascular function by,
      444–46
   dual innervation in, 312, 313*f*
   effects of innervation by, 324–25*t*
   initiation of exercise and input from,
      742–46
   neuroeffector junctions, 320, 321*f*, 322*f*
   neurotransmitters and receptors of,
      317–20
   properties of, 329*t*
   regulation of, 321–23
   regulation of cardiac output by, 393, 394,
      395*t*, 399*f*, 429
   regulation of smooth muscle contractions
      by, 362–63
   veto power of, during exercise, 755
Autonomic reflexes, 244, 247
Autoreceptors, 136, **217**
Autorhythmicity, **377**
Autosomes, **671**
Axoaxonic synapse, **202**
   presynaptic modulation at, 212–13, 214*f*
Axodendritic synapses, **202**

Axon (nerve fiber), **173**
   bundles of, 175
   central, 174
   C fibers, 275
   conduction velocities in various types of,
      197*t*
   neurotransmitter release on terminal of,
      206, 207*f*
   peripheral, 174
   propagation of action potentials in
      myelinated, 196
   propagation of action potentials in
      unmyelinated, 195–96
Axonal transport, 206
Axon hillock, **174**
Axon terminal, 134, **174**
Axosomatic synapses, **202**

Bacteria
   attacks of, on proteins, 152–53*b*
   lysis of, 717, 718*f*
Balance, concept of, 565, 566*f*. *See also*
   Homeostasis
Bare zone of thick filament, 337
Baroreceptor(s), 261, **444**
   ADH release and effects, 574, 575*f*
   arterial (*see* Arterial baroreceptors)
   blood volume receptors, 452–53
   cardiac (*see* Cardiac baroreceptors)
   cardiovascular function and input from,
      445–46
Baroreceptor reflex, **446**–51
   drop in mean arterial pressure and response
      of, 448*f*, 453, 454*f*, 455
   hemorrhage and response of, 447, 449*f*
   hypertension, hypotension, and, 447–48,
      450*b*
Basal cells, **306**
Basal compartment, 677
Basal metabolic rate (BMR), **644**, 663
Basal nuclei (basal ganglia), **240**, 242*f*
   motor control and, 250–51
Basement membrane, 4, **124**
Bases, 75*b*
Bases in nucleic acids, 32, *33*
   complementary pairing of, 32, 34*f*
   in DNA, protein synthesis and, 49
Basic electrical rhythm (BER), **628**–29
Basilar membrane, 295*f*, **296**, 297
Basket cell, 3*f*
Basolateral membrane, 45, **124**
Basophils, **710**
B cells (beta cells), 159, 161*f*
B cells (B lymphocytes). *See* B lymphocytes
   (B cells)
Beta [β] adrenergic receptors, 217, **317**–18,
   319*f*, 320*t*
   epinephrine binding to, 451–52
Beta blockers, treatment of hypertension with,
   450
Beta cells, 159, 161*f*
Bicarbonate
   in saliva, 611
   secretion of fluid rich in, into stomach, 626,
      627*f*
Bicarbonate ions as buffer, 525–26, 589, 590–92
   renal handling of, 591–92

Biceps muscle, 358*f*
   force generated by, 359*f*, 360
Bicuspid valve (mitral), **372**
Bile, **608**
   liver secretion of, 612
   secretion of, 626–27
Bile canaliculus, 612*f*, **613**
Bile ducts, 612*f*, **613**
Bile pigments, 612
Bile salts, 608, **617**
   emulsification of fat by, 618*f*
   representative, 617*f*
Biliary system, **613**
   structures of, 612*f*
Bilirubin, 406
Binocular visual field, 290*f*, 291
Biogenic amines, **216**–17
   depression and deficient, 218
Biomolecules, **22**–33
   atoms and, 23*b*
   carbohydrates, 22–25
   common functional groups in, 23*t*
   energy intake, utilization, and storage of,
      641–43
   ions, ionic bonds, and, 31*b*
   lipids, 25–28
   nucleotides and nucleic acids, 32–33
   polar molecules and hydrogen bonds, 24*b*
   proteins, 28–32
Bipolar neuron, 174*f*
Birth canal, 686
Birth control methods, 694*b*
Blastocoele, 695, 696*f*
Blastocyst, **695**, 696*f*
Bleeding, mechanisms for stopping. *See*
   Hemostasis
Blind spot, **278**
Block to polyspermy, 695
Blood, **370**, 371. *See also* Plasma
   carbon dioxide transport in, 510–14
   cells of, 371
   central nervous system and supply of, 226,
      230*b*
   composition of, 376*t*, 404–7
   flow of (*see* Blood flow)
   hemostasis and platelets of, 407–11
   inputs and outputs of acids to, 585*f*
   oxygenated and deoxygenated, 375, 376*f*
   oxygen transport in, 505–9
   pH of, and acid-base homeostasis, 522–23
   pH of, and exercise, 754*f*
   resistance and viscosity of, 423–24
   transfusions and blood-group compatibility,
      730
   transport of oxygen in, 505–9
   vessels transporting, 411–17 (*see also* Blood
      vessels)
   volume (*see* Blood volume)
Blood-brain barrier, **226**–27, 228*f*
Blood clot
   aspirin and, 411
   factors limiting formation of, 410–11
   formation of, 408–9
   platelet plug leading to, 407, 408*f*
   role of coagulation factors in formation
      disorders, 411
   vascular spasm leading to, 407

Blood flow, 374–76. *See also* Blood supply
  baroreceptor reflex and shifts in, 449–51
  through cardiovascular system, 376f,
    417–19
  changes in central venous pressure
    affecting, 430–32
  changes in central venous pressure and, to
    organs, 430–32
  control of cardiovascular function and
    regulation of, 455–61
  during exercise, 750, 751f, 753
  extrinsic control of glomerular filtration and
    renal, 543–44
  factors affecting distribution of blood and, to
    organs, 427–29, 464f, 465
  to kidneys, 537–38, 539f, 556
  physical laws governing blood pressure and,
    420–26
  prime directive: brain protection, 753
  relationship of blood vessels to size and
    direction of, 412f
  resistance and manner of (laminar versus
    turbulent), 424
  response of vascular smooth muscle to
    changes in, 458, 459f
  role of arterioles in regulating, 413, 414f
  role of pressure gradients in driving, 421–22
Blood glucose
  balance of, 566
  clearance of, 556, 557f
  insulin secretions and levels of, 163, 165f,
    566, 652f, 653
  renal function and levels of, 534, 547, 548f
Blood pressure, 389–90
  in arteries (*see* Mean arterial pressure
    (MAP))
  autonomic control of, when body changes
    position, 322, 323f
  factors influencing, during exercise, 747,
    748f
  high (*see* Hypertension)
  physical laws governing blood flows and,
    420–26
  role of turbulence in measurement of, 425b
  treatment of, 115
  in veins (*see* Central venous pressure)
Blood supply. *See also* Blood flow
  to kidneys, 537–38, 539f, 556
Blood-testis barrier, **677**
Blood type, 730
Blood vessels, **370**, 371. *See also* Vasculature
  arteries, 411–12 (*see also* Arteries)
  arterioles, 412–13 (*see also* Arterioles)
  capillaries, 414, 415f, 416 (*see also*
    Capillaries)
  components of, 376t
  pressures in, 420f
  relationship of, to size and direction of
    blood flow, 412f
  resistance of individual, 422–24
  resistance of networks of, 424–26
  in skin, 13
  structure and function of, 411–17
  vasodilation and vasoconstriction of, **423**
    (*see also* Vasoconstriction; Vasodilation)
  veins, 416–17 (*see also* Veins)
  venules, 371, 416

Blood volume
  baroreceptors and detection of changes in,
    452–53
  distribution of, in cardiovascular system,
    417f
  influence of, on central venous pressure
    and blood flow to organs, 431–32
  water balance related to, 567
Blowfish, neurotoxins in, 185b
Blue (deoxygenated) blood, 375, 376f
B lymphocytes (B cells), **711**, 718
  clonal selection in, 720, 721f
  diversity of antigen receptors on, 720
  role of, in antibody production, 721–22
  specificity of antigen binding by, 718–20
Body, 2–14
  cells, tissues, organs, and organ systems,
    3–5
  growth of, 654–61
  homeostasis and, 8–14 (*see also*
    Homeostasis)
  overall body plan, 5–8
  perception of sensations in (*see*
    Somatosensory system)
  temperature (*see* Thermoregulation)
  two-point discrimination thresholds for
    select areas of, 270t
Body, muscle, **334**, 335f
Body fluid compartments, 7–8
Body of stomach, 606f, **607**
Bohr effect, **509**, 524
Bolus, **629**, 630f
Bonds, chemical, 23
Bone(s)
  attachment of muscles to, 334, 357–58
  elongation of long, 660f
  growth of, 657–60
  as levers and implications for muscle
    contraction, 358–60
  structure of, 659f
Bone marrow, 2, 660
  transplantation of, 732
Bony labyrinth, 299
Bowman's capsule, **536**, 540
Bowman's capsule hydrostatic pressure, 540–41
Bowman's capsule oncotic pressure, 540
Bowman's space, 540
Boyle's Law, **479**, 482b
Brachial artery, 390
  measurement of blood pressure and, 425b
Bradykinin, 272
  inflammation, pain, and, 714
  as vasoactive substance, 460, 461t
Brain, **224**, 225f, 237–43
  autonomic function and related regions of,
    322, 323f
  blood brain barrier, 226–27, 228f
  blood flow to, during exercise, 753
  breathing and, 515–17
  cerebrum and cerebral cortex, 238–39, 240f,
    241f
  forebrain, cerebellum, and brainstem of,
    236f, 237–38
  gray and white matter of, 228–31
  neurogenesis in, 173b
  reduction in blood flow to, 443
  subcortical nuclei, 230, 237, 239–43

thermoregulatory center, 11
  types of cells found in, 2, 3f
  ventricles of, 226, 227f
  water balance and cellular swelling in, 568
Brainstem, 217, 236f, **237**
  control of posture by, 249–50
  regulation of breathing rhythm by, 515–17
Brain waves, 251
Breathing
  motor neurons and neural control of, 514,
    515f
  peripheral inputs into respiratory centers,
    517–18
  quiet, model of respiratory control during,
    516, 517f
  quiet, versus active ventilation, 515f
  rhythm of, generated in brainstem, 515–17
Breathing, mechanics of, 479–84
  Boyle's Law and, 479, 482b
  expiration, 482–83
  inspiration, 480–82, 483f
  intra-alveolar pressure and, 479–80
  muscles of, 481f
  volume and pressure changes during, 483f
Broca's area, 240f, **251**
Bronchi, **471**, 472f
Bronchioles, **472–73**, 475, 476f
  local controls on radius of, 523t
  terminal, **473**, 476f
Bronchoconstriction, 485
Bronchodilation, 486
Bronchodilators, 487, 490b
Brush border, kidney, 551, 552f
Brush border, small intestine, **608**
Brush border enzymes, 614
Buffer(s), 588–89b
  bicarbonate ions as, 525–26, 589, 590–92
  buffering of hydrogen ions, 588–90
  hemoglobin as, 524–25
Bulbourethral glands, 677f, **679**
Bulk flow
  blood flow and, 421
  movement of molecules in blood by, 370
  permeability, resistance, and, 423b
Bundle of His, 377–78
Bypass reaction, 92

Cadherins, 45
Caffeine, 147
Calcification, bone, **658**
Calcitonin, 158
  plasma calcium concentrations and, **581**
Calcitrol, 28
  plasma calcium concentrations and, **581**, 583f
Calcium
  changing levels of, through ligand-gated
    channels, 144f
  electrical activity in cardiac muscle and role
    of, 380, 381f, 382, 383
  excitation-contraction coupling (muscles)
    and role of, 342–43, 361
  homeostasis and balance of, 580–81
  hormonal regulation of blood levels of, 158,
    581
  intracellular signaling and role of, 144–45
  neurotransmitter release and role of, 203,
    206

renal function and balance of 533, 580–81
   as second messenger, 146, 147t
   strokes and, 230b
Calcium channel blocker, 115
   for treatment of hypertension, 450
Calcium-induced calcium release, 383
Calmodulin, **145**, 361
Calorie, **66**
cAMP (cyclic adenosine monophosphate), 146, 306, 573
   as second messenger system, 147t, 148f
   signal amplification by, 149, 150f
cAMP phosphodiesterase, 147
Canaliculi, 658
Cancer
   as uncontrollable cell division, 59b
   vaults and chemotherapy for, 42b
Cap (mRNA), 50
Capillaries, 8, **371**, 411, 414
   brain, 226, 228f
   carbon dioxide exchange and transport in, 511–13
   gas exchange as function of pulmonary, 502, 503f
   inflammation and dilation and increased permeability of, 713, 714f
   movement of fluids across walls of, 432–38
   peritubular, 537, 539f, 546
   role of arterioles in regulating blood flow through, 413, 414f
   structural characteristics of, 413f
   transport across, 414, 416f
   two types of, 414, 415f
Capillary beds, 374–75
Carbamino effect, **509**
Carbaminohemoglobin, **509**
Carbohydrates, **22**–25
   absorption, utilization, and storage of energy in, 641–42, 643t
   digestion and absorption of, 614–15
   energy metabolism of, during absorptive and postabsorptive states, 649t
   metabolism of, 91–94 (see also Glucose oxidation)
   in plasma membrane, 36–37
Carbo-loading, 18, 93
Carbon, 22, 23
Carbon dioxide, 498
   acid-base disturbances caused by, 585
   affinity of hemoglobin for oxygen and partial pressure of, 509
   effect of, on bronchioles and airway resistance, 486
   gas exchange of oxygen and, 500–505, 511–13
   movements of oxygen and, in pulmonary and systemic tissues during rest, 496f
   relationship between acidity and, 524b
   solubility of, in water, 499, 500f
Carbon dioxide transport in blood, 510–14
   effect of oxygen on, 513, 514f
   in pulmonary capillaries and veins, 512f, 513
   role of carbonic anhydrase in, 510–11
   in systemic capillaries and veins, 511–14
Carbonic anhydrase
   bicarbonate reabsorption and hydrogen ion secretion catalyzed by, 591f, 592

carbon dioxide transport in blood and role of, **510**–11
   stomach acid and, 625
Carbon monoxide, affinity of hemoglobin for, 509
Carboxyl groups, 75b
Carboxypeptidase, **616**, 617f
Cardiac arrhythmias, 385f–86
Cardiac baroreceptors
   ADH release and effects of, 574, 575f
   control of blood volume, pressure, and, 452–53
   restoration of mean arterial pressure following hemorrhage and role of, 455
Cardiac cycle, **386**–92
   aortic pressure and, 388–90
   atrial and ventricular pressure and, 387f, 388
   diagram of, 387f
   heart sounds and, 391–92
   pump cycle and, 386–88
   ventricular volume and, 390–91
Cardiac function
   cardiac cycle, 386–92
   cardiac output, 392–99
   heart, 370, 371–73 (see also Heart)
   heart, electrical activity of, 377–86
   path of blood flow through heart and vasculature, 374–76
Cardiac function curve (Starling curve), 396–97
Cardiac muscle, 363–64, 371. See also Myocardium
   action potentials in, 377–79, 380f, 381f, 382f, 383
   desmosomes in, 45f, 377, 378f
   electrical connections between cells of, 377, 378f
   gap junctions in, 45f, 361f, 377, 378f
   ionic basis of electrical activity in, 379–83
   myogenic nature of, 377
   relative duration of contractions and action potentials in, 364f
   spread of electrical excitation through, 379, 380f
Cardiac output (CO), **392**–99
   autonomic input to heart, 393
   changes in heart rate and, 393–95
   changes in stroke volume and, 395–99
   during exercise, 394b, 744–45
   heart failure as reduced, 434–35b
   integration of factors affecting, 399
   mean arterial pressure and increases in, 428, 429f
Cardiovascular system, 5, **370**
   blood (see Blood; Blood pressure)
   blood flow through, 374, 375, 376f
   blood volume distribution in, 417f
   cardiac function in (see Cardiac function)
   components of, 370–73, 376t
   exercise and adjustments in, 744, 745f, 747–48
   microcirculation in, 412, 414f
   pressure gradients in, 421–22
   resistance in, 422–26
   systems integration chart for, 466f
Cardiovascular system regulation, 443–68
   additional regulatory processes, 461–65
   regulation of blood flow distribution to organs, 455–61

regulation of mean arterial pressure, 443–55
   summary of factors influencing organ blood flow, 464f, 465
Carotid arteries, arterial baroreceptors on carotid sinuses of, 444, 445f
Carotid bodies, peripheral chemoreceptors in, 518f, 519
Carrier proteins, 46, **113**–14
   globular proteins as, 31
   transport of chemical messengers by, 140
   transport of hormones bound to, 164
Cartilage, **659**–60
   in larynx, 471, 472f
Cascade of reactions, **149**
Catabolic reactions, **63**
   as energy-releasing reactions, 65
   rate of, 46
Catabolism, **46**, 63
Catalase, 40
Catalysts, 70
   enzymes as, 71–77
Catalytic rate, **73**
Cataracts, 282
Catecholamines, **135**, 216, 217
   synthesis of, 137f
Catechol-O-methyltransferase (COMT), **217**
Cations, 31b
   positive electrical charge in, 103
Cauda equina, 231
Caudate nucleus, 240, 242f
Cell(s), **2**, 21–61
   biomolecules in, 22–33 (see also Biomolecules)
   cell-to-cell adhesions, 43–46
   division of (see Cell division)
   effect of growth hormone on size and number of, 655
   energy metabolism in, during absorptive state, 645–46
   energy metabolism in, during postabsorptive state, 647
   four major types of, 3, 4f, 5
   general functions of, 44t, 46–48
   intercellular communication between, 48, 132–33 (see also Intercellular communication)
   long-distance communication between, 151 (see also Endocrine system; Nervous system)
   nervous system, 172–77 (see also Neuron(s))
   programmed death of, 721
   protein synthesis in, 48–56
   structure of, 34–43, 44t
   types of, found in brain, 2, 3f
   volume changes of, in hypotonic and hypertonic solutions, 123f
Cell body (soma) of neuron, **173**
Cell cleavage, 695, 696f
Cell cycle, 57–58
Cell division, 56–59
   cancer and faulty, 59b
   cell cycle and, 57–58
   DNA replication in, 56–57
   mitosis and meiosis in, 56, 57f, 58f
Cell-mediated immunity, **718**, 724–28
   cytotoxic T cell activation, 727–28
   helper T cell activation, 727
   role of T lymphocytes in, 724–26

Cell membrane(s), selective permeability of, 7.
  See also Plasma membrane
Cell membrane transport, 46–48, 100–130
  active transport, 115–19
  characteristics of, 118*t*
  epithelial transport, 124–27
  facilitated diffusion: passive transport,
    113–15
  factors affecting direction of, 101–7
  factors affecting rate of, 108–10
  osmosis, 119–23
  simple diffusion and, 111–12
Cell metabolism, 46, 62–99. *See also* Metabolism
  carbohydrate, fat, and protein metabolism,
    91–96
  energy and metabolic reactions, 65–70
  glucose oxidation in, 77–91
  rates of metabolic reactions, 70–77
  types of metabolic reactions, 63–65
Cell-to-cell adhesions, 43–46
  desmosomes, 45
  gap junctions, 45–46
  tight junctions, 43–45
Cellulose, **22**, 615
Central axon, 174
Central canal, brain, **226**, 227*f*
Central chemoreceptors, **519**
Central command mechanism, **744**, 745*f*
Central (psychological) fatigue, 261
Central lymphoid tissues, **711**, 712*f*
Central nervous system (CNS), **171**, 223–59
  brain, 224, 237–43
  diseases affecting, 230*b*
  functional classes of neurons in, 175*f*
  gastrointestinal control by, 622–23, 624*f*
  general anatomy of, 224–31
  integrated function: emotions and
    motivation, 254
  integrated function: language, 251
  integrated function: learning and memory,
    255–57
  integrated function: reflexes, 244–47
  integrated function: sleep, 251–54
  integrated function: voluntary motor
    control, 247–51
  plasticity in, 256–57
  spinal cord, 224, 231–37
Central pattern generator, **516**
Central sulcus, 238
Central thermoreceptors, 13
Central veins, 416
Central venous pressure, 398, **422**
  effects of changes in, on blood flow to
    organs, 430–32
  process chart of factors affecting, 433*f*
Centrioles, 35*f*, 40, 44*t*
Centromere, 58
Cephalic-phase control, **623**, 625, 626*f*
Cerebellum, 236*f*, **237**
  major pathways for information flow to and
    from, 250*f*
  motor coordination and role of, 250
Cerebral cortex, **228**, 230, 236*f*, 237, 238–39
  functional areas of, 235*f*, 239, 240*f*, 241*f*
  functional specializations of left and right,
    239
  organization of, 238*f*

sensory areas of, 266
somatosensory and motor areas of, 235*f*
  (*see also* Somatosensory cortex)
  voluntary movement and, 248, 249*f*
Cerebral hemispheres, **231**, 236*f*, 237, 238,
  239*f*
Cerebrospinal fluid (CSF), **224**, 226
  composition of, 226*t*
Cerebrovascular accident (CVR) (stroke), 230*b*
Cerebrum, 228, 236*f*, **237**. *See also* Cerebral
  cortex
  effects of injury in, 243*b*
  lobes of, 238, 239*f*
Cervical nerves, **231**, 232*f*
Cervix, 684*f*, **685**
C fibers, 275
cGMP (cyclic guanosine monophosphate), 145,
  146, 147*t*
Channel(s), ion, 15, 46
  funny, 381
  gated, 183, 184*f*
  G proteins affecting slow ligand-gated, 146
  leak, 174
  ligand-gated, 143–45, 174
  T-type, and L-type, 381
  voltage-gated, 174 (*see also* Voltage-gated
    channels)
  water (*see* Aquaporins)
Channel-linked receptors, **204**
  signal transduction mechanisms for, 143–45
Channel proteins, **114**
Charged tRNA, 51
Chemical driving forces, **102**–3
  actions of, on potassium ions crossing cell
    membrane, 106*f*
  production of resting membrane potential
    by actions of, 178–80
Chemical messengers, 133–40
  bacterial infections affecting, 152–53*b*
  chemical classification of, 135–37
  endocrine system and hormones as, 151,
    152–67
  functional classification of, 133–35
  globular proteins as, 31
  indirect intercellular communication via,
    132–33
  nervous system and, 151
  properties of, 150*t*
  responses of smooth vascular muscle to
    locally secreted, 460, 461*t*
  signal amplification by, 149, 150*f*
  signal transduction mechanisms of, 140–49
  synthesis and release of, 137–40
Chemical reactions
  energy and, 65–70
  rates of, 70–77
  types of metabolic, 63–65 (*see also* Metabolic
    reactions)
Chemical synapses, **202**–10. *See also*
  Neurotransmitters
  excitatory synapses, 205–9
  functional anatomy of, 202–3, 204*f*
  inhibitory synapses, 209–10
  signal transduction mechanisms at, 204,
    205*f*
Chemical work, **644**
Chemiosmotic coupling, 84, 86*f*, 87

Chemoreceptor(s), 11, 261, **517**
  central, 519
  control of pulmonary ventilation by, 517,
    518–20
  in digestive tract, 622
  neural control of cardiovascular function
    through input from, 445*f*, 446, 453–55
  peripheral, 518–19
Chemoreceptor reflexes, 462, 520
  as cardiovascular regulatory process, 462
Chemotaxis, **715**
Chemotherapy, 42*b*
Chest wall, **475**, 477*f*
Chewing reflex, **629**
Chief cells, **607**, 616*f*
Chloride (Cl-)
  cystic fibrosis and faulty membrane
    transport of, 127*b*
  inhibitory synapses and channels for, 209,
    210*f*
Chloride shift, **513**
Cholecalciferol (Vitamin D3), 160
Cholecystokinin (CCK), **623**, 624, 626, 627*f*
Cholera, 126, 152*b*
Cholesterol, 28, 29*f*, 35, 617
Choline, 215
Choline acetyl transferase (CAT), **215**
Cholinergic neurons, **317**
Cholinergic receptors, **215**
  muscarinic, 216, 317, 318*f*, 319*t*
  nicotinic, 216, 317, 318*f*, 319*t*
  signal transduction mechanisms of, 216*f*,
    317, 319*t*
Cholinergic synapses, neurotransmitter
  synthesis, action, and degradation at, 215*f*
Chondrocytes, **660**
Chordae tendineae, 372
Chorion, **696**–97
Chorionic villia, **697**
Choroid, **278**, 279*f*
Choroid plexus, **226**, 227*f*
Chromaffin cells, 159, 314, 316*f*
Chromatid, **58**
Chromatin, **37**, 56*f*
Chromosome, 56*f*
  sex, 671, 672–73
Chronic bronchitis, 490*b*
Chronic mountain sickness, 525*b*
Chronic obstructive pulmonary diseases
  (COPD), 487, 489, 490*b*
Chylomicrons, **618**, 620*f*, 621*b*
Chyme, **607**
Chymotrypsin, **616**
  activation of, 616, 617*f*
Chymotrypsinogen, **616**, 617*f*
Cilia
  microtubules of, 42*f*, 43
Ciliary body, **278**, 279*f*
Ciliary muscles, **278**
Ciliated cells, **473**, 474*f*
Cingulate gyrus, 241, 242*f*
Circadian rhythm, **158**, 164, 241
Circular muscle, (eye iris), **284**
Circulatory shock, 451
Circulatory system, 374. *See also* Cardiovascular
  system
Cis face, Golgi apparatus, 38

Cisternae, Golgi apparatus, 38, 39f
Citric acid cycle. *See* Krebs cycle
Class I MHC molecules, **725**, 726
Class II MHC molecules, **725**, 726
Clearance, **554**–56
    of glucose and PAH, 556, 557f
    of inulin, 555f, 556
    rates of, for common substances processed
      by kidneys, 558t
Clitoris, 686f, 687
Clonal selection in B lymphocytes, **720**, 721f
Clostridium tetani, 349b
Coagulation cascade, 408–9
Coagulation factors, 408–9
    extrinsic and intrinsic pathways to, 409, 410f
Coated pit, **47**
Coccygeal nerve, **231**, 232f
Cochlea, **292**
    functional anatomy of, 293–96
Cochlear duct (scala media), 295f, **296**
Cochlear nerve (cranial nerve VIII), 297
Codon, **49**
Coenzyme(s), **72**–73
Coenzyme 2-H, 91, 92f
Coenzyme A (CoA), 32, 73
Coenzyme Q, 85
Cofactors, **72**
Cold thermoreceptors, **272**
Collagen, 5, 31, 32f, 411
Collateral ganglia, **315**
Collaterals of neuron, **173**
Collecting ducts, renal, **535**
    effects of ADH on water permeability in late
      distal tubules and, 572–73, 574f
    effects of aldosterone on, 577f
    handling of hydrogen and bicarbonate ions
      in, 591–92
    reabsorption and secretion in, 551–52
Colloid, 661
Colon, 602, 604f, **609**–10
    motility of, 632–33
Colonocolonic reflex, **632**
Color vision and color blindness, 289b
Commissural fibers, 229f, **231**
Commissures, **175**
Common bile duct, 612f, **613**
Common hepatic duct, **613**
Complement system, **713**, 717–18, 724f
Compliance, 415b
    of arteries **412**, 415
    of lungs, 484
Concave surfaces, light and, 280
Concentration
    of enzymes, 73
    of solution, 69b
Concentration gradient ([delta]C), **102**–3
    movement of oxygen and carbon dioxide in
      respiratory system along, 497–505
    net flux changes, and changes in, 109f
    relationship of net flux and, for carrier-
      transported molecules, 114, 115f
Condensation reaction, 28, 29f, 64
Conditioned reflexes, 244
Condom, 694b
Conductance (g), **177**
Conducting zone of respiratory tract, **471**–74
    anatomical features, 473f

Conduction, **12**
Conduction fibers, 377
Conduction system, heart, **377**–78, 379f
    control of heartbeat by pacemakers,
      378
    initiation and conduction of impulse
      during heartbeats, 377–78
Cones (eye), 278, 284f, **285**
    characteristics of, 288t
Conformation, protein, 28, 30f
Congestive heart failure, 432, 435b
Connecting tubule, renal, 537
Connective tissue, 4–5
Connective tissue cells, **4**
Connexins, 45f, 46, 132
Connexons, 45f, 46, 132
Constrictor muscle, 284
Continuous capillaries, 414, 415f
Contraception, 694b
Contractile cells, cardiac, 377
    electrical activity in, 382–83
Contractile component, **346**
Contractile proteins, 334, **336**. *See also* Actin;
    Myosin
Contraction phase, muscle twitch, **344**
Contralateral pathway, **237**
Convergence, neural, **186**, 212, 213f, 288
Converting enzyme inhibitors, treatment of
    hypertension with, 450b
Convex surfaces, light and, 280, 281f
Copulation, **673**
Core temperature, **13**
Cornea, **278**, 279f
Coronary arteries, 419
Corpus callosum, 229f, **231**
Corpus luteum, **688**, 689f
Cortex, kidney, **535**
Cortical nephron, 537f
Corticospinal tracts, 230–31
Corticosteroid-binding globulin, 140
Corticosteroids, as treatment for pulmonary
    disease, 487, 490b
Corticotropin releasing hormone (CRH), 157,
    664, 701
Cortisol, 28
    stress and secretions of, 664, 665, 666t
Cotransport (symport), **117**
Countercurrent multiplier, 570f, 571
Countertransport (antiport), **117**
Covalent bonds, 23, 24
    in proteins, 28
Covalent regulation of enzyme activity, **76**
Cranial nerves, 231, **237**
    number, name, class, and function of,
      237t
    I (olfactory nerve), 307
    III (oculomotor nerve), 315
    VII (facial nerve), 315
    VIII (cochlear nerve), 297
    IX (glossopharyngeal nerve), 315
    X (vagus nerve), 315
Cranial reflexes, 244, 247
Craniosacral division, autonomic nervous
    system, 315
Cranium, **224**, 225f
C-reactive protein, 717
Creatine, **343**

Creatine/creatine phosphate system,
    production of ATP for muscle contraction
    and role of, 343
Creatine kinase, 343
Creatine phosphate, **343**
Creatinine, 556
Cretinism, 663
Cristae, mitochondria, **38**, 39f
Crossbridge(s), **336**, 337f
Crossbridge cycle, **339**–40
    production of ATP for operation of, 343–44
    in smooth muscle versus in skeletal muscle,
      361, 362f
Cross-extensor reflex, **246**
Crossing over, **672**
Crotoxin, 327
Crypts of Lieberkuhn, 608
Cupula, **300**, 301f
Curare, 329b
Current (l), **177**
Cushing's syndrome, 665
Cyanosis, heart failure and, 434b
Cyclic AMP (cAMP), 32, 33f
Cyclic GMP (cGMP), 32
Cyclic nucleotides, 32
Cyclooxygenase, 139
Cyclooxygenase pathway, 139
Cystic fibrosis, 126, 127b
Cytochromes, 84, 86
Cytokines, 133f, **134**, 713
    helper T cell secretion of, 727t
    interferon-[gamma], 727
    interleukin-1 (IL-1), 716
    interleukin-2 (IL-2), 722, 727
    interleukin-4 (IL-4), 727
    interleukin-5 (IL-5), 727
    interleukin-6 (IL-6), 716
    interleukin-10 (IL-10), 727
    tumor necrosis fact [alpha] (TNF-[alpha]),
      716
Cytokinesis, **58**
Cytoplasm, **34**
    protein synthesis in, 48–56
Cytoplasmic granules, 709
Cytosine, 32, 33f
Cytoskeleton, 36, 44t
    structure of, **40**, 41f
Cytosol, **34**, 44t
    contents of, 37
Cytotoxic T lymphocyte cells, 711
    activation of, 727–28
    CD8 surface protein on, 726
    cell-mediated immunity and, 718
    recognition of Class I MHC molecules by, 726

Dalton's Law, partial pressure of gases and,
    498b
Dark, adaptation to light and, 287–88
D cells (delta cells), 159, 161f
DDT, as mutagen, 59
Decibels, 292, 294b
Decidual response, 695
Declarative memory, **256**
Defecation, **610**
Defecation reflex, **632**–33
Degradation phase in phagocytosis, 716
Delayed rectifier channels, 383

[δ]C. See Concentration gradient ([δ] C)
Delta cells, 159, 161f
[δ]E (energy changes in reactions), 65–66
Dementia, 250
Dendrites of neuron, **173**
Dense bodies, 360
Deoxygenated blood, 375, 376f
Deoxyhemoglobin, 505
Deoxyribonucleic acid (DNA), **32**
  in cell nucleus, 37
  complementary base pairing in, 34f
  genes as sections of (see Gene(s))
  replication of, and cell division, 56, 57f
  transcription of, in protein synthesis, 48, 49–50
  triplet base pair sequences in, 49
Deoxyribose, 22
Dephosphorylation, **64**
Depolarization of membrane potentials, **183**
  action potentials, 188, 189f
  phototransduction and, 286, 287f
  spontaneous, in smooth muscle cells, 363
Deposition, bone, **658**
Depression, treatment for clinical, 218b
Depth perception, 290f, 291
Dermatomes, 231, 233f
Dermis, 272
Descending colon, 609
Descending limb, loop of Henle, **536**
Descending tracts, spinal cord, **232**–33, 234f, 235f, 236–37
Desmosomes, **45**, 377, 378f
Detrusor muscle, **556**, 558f
Dextrinase, 614
Diabetes insipidus, 574
Diabetes mellitus, 164, **652**, 657–58b
Diabetes nephrogenetic, 587
Diabetic retinopathy, 657b
Diacylglycerol (DAG), 146, **147**
Dialysis, 550–51b, 587b
Diapedesis, **715**
Diaphragm, 371, 372f, **475**, 477f
  phrenic nerve innervation of, 514
Diaphragm, contraception, 694b
Diastole, **386**, 387f
Diastolic pressure, **389**
Dicrotic notch, 389
Diencephalon, 236f, **237**
Diet
  acid-base disturbances and, 585
  growth and, 655
Dietary fiber, 615
Diffusion, **111**. See also Facilitated diffusion; Simple diffusion
  of gases in pulmonary circulation, 497–505
  renal reabsorption by, 546, 547f
Diffusional equilibrium, 108
Digestion, 5, **602**, 603f, 614–20
  of carbohydrates, 614–15
  of lipids, 616–18
  of proteins, 615–16
  vitamins, minerals, water, and, 620
Digestive system, **602**–38
  digestion and absorption of nutrients and water, 614–20
  digestion and absorption of nutrients and water, summary of factors influencing, 633, 634f

functional anatomy of, 602–14
gastrointestinal motility, regulation of, 628–33
gastrointestinal regulation, general principles of, 620–24
gastrointestinal secretion, regulation of, 624–28
overview of, 602, 603f
systems integration chart for, 635f
Digestive tract (gastrointestinal tract), 5, **602**–10
  digestion and absorption in, 614–20
  during exercise, 749–50
  gastrointestinal wall, generalized structure of, 602–6
  homeostasis and transport across walls of, 565
  motility in, 628–33
  organs of, 606–10, 613t
  regulation of, 620–24
  secretion in, 624–28
Dihydropyridine receptors, 342
Diiodotyrosine (DIT), **661**
Dilator muscle, 284
Dinoflagellates, 185b
Diploid cells, **671**
Direct intercellular communication, gap junctions and, 132
Direction of cell membrane transport, 101–6
Disaccharides, **22**, 25f
Discrimination threshold, 270
Disease. See Illnesses, diseases, and conditions
Distal convoluted tubule, **536**
  effects of ADH on water permeability in collecting ducts and late, 572–73, 574f
  effects of aldosterone on principal cells of, 577f
  epithelial cells of, 566
  regulation of reabsorption and secretion in, 551–52
  renal handling of hydrogen and bicarbonate ions in, 591–92
  sodium reabsorption in, 549f
  solute reabsorption in, 545–47
Distending pressure, 415b
Disulfide bridge, 28
Diuresis, 452, 573
Diuretics, treatment of hypertension with, 450
Divergence, neural, **186**
Diversity of immune response, 719, 720, 721f
Dizygotic twins, 689
DNA. See Deoxyribonucleic acid (DNA))
DNA polymerase, 56
Dominant follicle, **688**
Dopa decarboxylase, 137
Dopamine, 135, 217
Dorsal column-medial lemniscal pathway, **273**, 274f
Dorsal columns, spinal cord, 232
Dorsal horn, spinal cord gray matter, **231**, 234f
Dorsal respiratory group, **515**, 516f
Dorsal root ganglia, **231**
Dorsal roots, spinal cord, **231**, 234f
Double bonds, 23
Down regulation, 142
2,3-DPG (2,3-diphosphoglycerate), affinity of oxygen for hemoglobin and effects of, 509

Driving forces on molecules, 102–6
  directions of, acting on uncharged solutes and ions, 107t
  magnitude of, 109–10
Dual innervation of autonomic nervous system, 312, 313f
Duodenal ulcers, 622b
Duodenum, **608**
  secretions into, 626–27
Dura mater, **224**, 225f
Dwarfism, 660
Dynein arms, 42f, 43
Dyspnea (labored breathing), 486–87

Ear, equilibrium and, 299–303
  anatomy of vestibular apparatus, 299–300
  neural pathways for, 303
  semicircular canals and rotation transduction, 300, 301f
  utricle, saccule, and linear acceleration transduction, 300, 302f, 303
Ear, hearing and, 291–98
  ear anatomy, 291, 292f
  neural pathways for sound, 297–98
  signal transduction for sound, 293–97, 299f
  sound amplification in middle ear, 292–93
  sound waves and, 292, 293f, 294b
Edema, heart failure and, 435b
Effector(s), homeostasis and, **11**
Effector cells, immune response and, **720**
Effector organs, **171**, 202
Efferent arteriole, 536
Efferent division of peripheral nervous system, **171**–72
Efferent neurons, **174**, 175f
  motor, and somatic nervous system, 324–27
  neuroeffector junctions as synapse for, 320–21
Eggs (ova), 671
Eicosanoids, **28**
  as chemical messengers, 136–37, 139–40, 460
  synthesis and release, 139–40
Einthoven, Willem, 383–84
Einthoven's triangle, 384f
Ejaculation, **678**
Ejaculatory duct, 677f, 679
Ejection fraction (EF), **391**b
Elastin, 411
Electrical activity
  in cardiac muscle, 377–86
  in gastrointestinal smooth muscle, 628, 629f
Electrical driving forces, 103, **104**
  actions of, on potassium ions crossing cell membrane, 106f
  creation of, by membrane potential, 104
  factors affecting direction and magnitude of, 104, 105f
  production of resting membrane potential by actions of, 178–80
Electrical forces in polar molecules, 24b
Electrical potential (voltage), 103
  types of, in biological systems, 177t
Electrical potentials in neurons, 177–86
  action potentials and, 183, 187–96
  changes in membrane potential and, 183
  graded potentials and, 183, 184, 186

neural integration and, 186–87, 210–12
neurotoxins and attack on, 185*b*
resting membrane potential and, 178–82
Electrical signals through changes in membrane potential, 183
Electrical synapses, **202**
Electrocardiogram (ECG, EKG), recording electrical activity of heart with, 383–86
Electrochemical driving force, **104–6**
actions of, on potassium ions crossing cell membrane, 106*f*
determining direction of, 105–6
resting membrane potential and, 178–80
significance of, 106, 107*t*
Electrochemical gradient, **106**
Electroencephalograms (EEGs), **251**
study of sleep using, 251–52, 253*f*
Electrolytes, 31*b*
balancing body, 565–66, 581–83, 584*f* (*see also* Homeostasis, urinary system and)
Electrolytic solutions, 31*b*
Electromagnetic spectrum, 279*f*
Electromagnetic waves, light as, 278–81
Electron(s), 23
oxidation-reduction reactions and, 64–65
Electron transport chain, 38, **84–86**
Electrotonic conduction, **184**
Elements, 23
Embryo, **673**
early development and implantation of, 695, 696*f*
later development of, 696–98
Emetics, 631
Emission, male, **683**
Emmetropia, **281**, 283*f*
Emotions
autonomic nervous system and, 323
central nervous system function and, 254, 255*f*
Emphysema, 490*b*
Emulsification of fats, **617**, 618*f*
End-diastolic pressure (preload), 397–98
End-diastolic volume (EDV), **391***b*
effects of, on stroke volume, 396–98
Endocrine cells, **604**
Endocrine glands, **152–60**
hormones released by, 134
Endocrine hypertension, 450*b*
Endocrine pancreas, 159, 611
Endocrine system, 5, 151, 152–60, 162–63*t*
characteristics of, 151*t*
energy balance and, 643–45
glucocorticoids and, 663–65
hormones and, 134, 161–67 (*see also* Hormone(s))
hypothalamus as link between nervous system and, 240–41
primary endocrine organs, 153–60
regulation of absorptive and postabsorptive metabolism by, 649–54
regulation of growth by, 654–61
secondary endocrine organs, 160
systems integration chart, 667*f*
thyroid hormones and, 661–63 (*see also* Thyroid hormones)
Endocytosis, **46**, 47*f*
Endocytotic vesicle, 47

Endogenous analgesia systems, **276**, 278*f*
Endogenous opioids, **219**
Endogenous pyrogens, 716
Endolymph, **296**, 300, 301*f*
Endolysosome, 47
Endometrium, **686**
Endoplasmic reticulum, **38**, 44*t*
protein synthesis on, 52, 53*f*
Endorphins, **219**
Endothelin-1, as vasoactive substance, 460, 461*t*
Endothelium, 4
End-plate potential (EEP), **326**, 328*b*, 342
End-product inhibition of enzyme activity, 76
End-products of metabolic pathways, **64**
End-stage renal disease (ESRD), 550*b*
End-systolic volume (ESV), **391***b*
Energy, **65**
intake, utilization, and storage of, in food, 641–43
metabolic reactions and changes in, 65–68
metabolic reactions that release or require, 65
metabolism (*see* Energy metabolism)
Energy balance, 643–45
energy input and output, 644
metabolic rate, 644
negative and positive, 644–45
Energy exchange in glucose hydrolysis, ATP and, 78–81
Energy metabolism, **63**, 639–69
in absorptive and postabsorptive states, 645–49
energy balance and, 643–45
energy intake, utilization, and storage and, 641–43
exercise and demands on, 743–44, 746–47
glucocorticoids, 663–65
hormonal regulation of growth and, 654–61
overview of whole body metabolism and, 640–41
regulation of absorptive and postabsortive, 649–54
thyroid hormones and, 661–63
Energy of motion (kinetic), 66, 67*f*
Energy reserves, 647
Enkephalins, **219**, 276–77
Enteric (intrinsic) nervous system, 172, **605**
gastrointestinal control by, 622–23, 624*f*
Enterocytes, **602**, 604
Enterogastrones, 623
Enterohepatic circulation, 618, 619*f*
Enterokinase, **616**
Enzyme(s), **71**
as catalysts in chemical reactions, 70, 71–77
cofactors and coenzymes, 72–73
factors affecting rates of enzyme-catalyzed reactions, 73–74, 75*b*
globular proteins as, 31
mechanisms of actions by, 71–72
receptors linked to, 145
regulation of activity in, 75–77
substrates of, 71
substrate specificity of, 72
transport proteins as, 116
Enzyme-linked receptors, **145**
Eosinophils, **710**

Ependymal cells, 176, **226**
Epidermis, 272
Epididymis, **678**
Epidural, 231
Epiglottis, **471**, 472*f*
Epinephrine (adrenaline), 135, 141, 217, 666*t*, 733
binding to adrenergic receptors, 317–18, 319*f*
controls on mean arterial pressure by, 451–52
effects of, on metabolism, 653–54
heart rate and effects of, 394, 395*t*
stress and, 665
vasoconstriction and, 407–8
ventricular contractility and effects of, 396
Epiphyseal plate, **659**, 660
Epiphyseal plate closure, **660**
Epithelia (epithelium), **4**
general structure of, 124*f*
olfactory, 306*f*
of proximal tubule, 551, 552*f*, 566
of respiratory tract, 474*f*
selective permeability of, 7
transport across (*see* Epithelial transport)
Epithelial cells, 3, **4**. *See also* Epithelia (epithelium)
Epithelial transport, 45, 124–27
cystic fibrosis and faulty, 127*b*
epithelial solute transport, 124–26
epithelial structure and, 124
epithelial water transport, 126
Epitopes, **719**
Equilibrium (balance), 299–303
anatomy of vestibular apparatus, 299, 300*f*
neural pathways for, 303
semicircular canals and transduction of rotation, 300, 301*f*
utricle, saccule, and transduction of linear acceleration, 300–303
Equilibrium (chemical reactions), **67**
Equilibrium potential, **105**, 177*t*
Nernst equation and, 107*b*
Erectile dysfunction (ED), 685*b*
Erectile tissue, 677*f*, **678**
Erection, penile, **678**, 682–83
dysfunctional, 685*b*
events leading to, 682*f*
Erogenous zone, 682
Error signal, **11**
Erythrocytes, 2, **371**, 405–6
anemia and, 510*b*
liver and removal of aged, 612
Erythropoietin, 2, 160, 406, 525
kidney secretion of, 533
renal anemia and, 510*b*
Esophagus, 602, 604*f*, 606*f*, **607**
Essential nutrient, **96**, 655
Estradiol, 28
Estrogens, 11, 157, 160, **673**
actions of, in female reproduction, 693*t*
during pregnancy, 698–700
Eustachian tube, **291**, 292*f*
Evaporation, **12**
Exchangeable proteins, transport of, across capillary walls, 414, 416*f*
Excitable cells, **173**

Excitation-contraction coupling, **340**, 341*f*, 342
   calcium, troponin, and tropomyosin in skeletal muscle, 342–43
   in cardiac contractile cells, 383
   in smooth muscle, 360–62
Excitatory graded potentials, **184**
Excitatory postsynaptic potential (EPSP), **205**, 208–9, 326
Excitatory synapses, **205**–9
Excretion, 7, **538**, 553–59
   clearance and, 554–56
   micturition and, 556–59
   plasma glucose levels and, 548*f*
   rate of, 553–54
   regulation of glomerular filtration rate and water, 574
Exercise
   cardiac output during, 394*b*
   cardiovascular responses to, 463–65
   hierarchy of resource allocation during, 15–16
   long-term response of muscles to, 356–57
   marathon (*see* Exercise links; Marathon case study)
   muscle cell metabolism linked to intensity of, 343–44
   rate of anabolic and catabolic reactions during, 46
   response of muscle fiber types to, 355
   response to, and cardiovascular system regulation, 463–65
Exercise, integrated physiological response to, 741–61
   aftermath and post-exercise consequences, 755–58
   decline to end of exercise, 753–55
   principles of physiological integration and, 743
   steady-state condition and, 746–50
   transition from rest to exercise, 742–46
Exercise links
   acetylcholinesterase and muscle function, 325
   beta-endorphins and pain, 219
   blood flow shifts, 449
   blood glucose levels, 165–66
   brain swelling due to low plasma osmolarity, 568
   cardiac output, 392–93
   central venous pressure, 430
   central venous pressure and orthostatic hypotension, 432
   deaminated proteins, 95
   fatigue, 261
   gluconeogenesis, 649
   increased number of neutrophils, 710
   increased ventilation, 520
   muscles and potassium levels, 189
   physical characteristics of two runners, 16*t*
   pulmonary ventilation, 484
   salivary flow, 625
   secondary sex characteristics and, 674
   size principle and recruitment of motor units in late stages of, 352
   sweating, role of, 122, 240
   sweating, water loss from, 545

Exercise physiology, 15–18. *See also* Exercise, integrated physiological response to
Exocrine cells, **604**
Exocrine glands, 152
Exocrine pancreas, 159, 161*f*, 611
Exocytosis, **46**, 47, 48*f*
   of neurotransmitters, 203, 206, 207*f*
Exons, 50
Expiration, 482
   changes in alveolar pressure and breath volume during, 480*f*
   muscles of, 481*f*
Expiratory muscles, **480**, 481*f*
Expiratory neurons, **515**
Expiratory reserve volume (ERV), **488**
Expressive aphasia, 251
External anal sphincter, **610**
External auditory meatus, **291**, 292*f*
External ear, 291, 292*f*
External environment, **5**, 6*f*
   exchanges between internal and, 6–7
   mechanisms of heat transfer between body and, 12–13
External intercostals, **475**, 477*f*
External respiration, **470**, 471*f*
External urethral sphincter, **556**
Extracellular fluid (ECF), **7**–8. *See also* Internal environment
   concentrations of select solutes in, 101*t*
   homeostasis in composition, temperature, and volume of, 8
Extracellular matrix, 5
Extrapyramidal tracts, voluntary motor control and, **248**, 249*f*
Extrasystole, 385
Extrinsic control, **392**
   of cardiovascular system, 443–55
   of glomerular filtration and renal blood flow, 543–44
Eye
   anatomy of, 278, 279*f*
   diabetic retinopathy, 657*b*
   pupillary light reflex in, 247, 284
   vision in (*see* Vision)

Facial nerve (cranial nerve VII), 315
Facilitated diffusion, **113**–15
   in central nervous system, 227–28
   characteristics of, 118*t*
   factors affecting membrane permeability in, 114–15
   transport proteins in, 113–14
Factor XIII (Hageman factor), 409
FAD (flavin adenine dinucleotide), 32, 72
Fallopian tubes, 686
Fast axonal transport, 206
Fast glycolytic fibers, **355**, 356*t*
Fast ligand-gated channels, 143, 144*f*
Fast oxidative fibers, **355**, 356*t*
Fast pain, **274**–75
Fast response in postsynaptic neurons, **204**, 205*f*
Fast-twitch muscle fibers, 345, **354**
Fatigue, **356**
   muscle, 356, 357*f*
   physical and psychological experience of, 261

   tactical and biological influences on, 754–55
   as warning signal during exercise, 753–54
Fats. *See* Lipids
Fatty acids, **25**, 26*f*
   liberation of, from fat by lipases, 619*f*
F cells, 160
Feces, **610**
Feedback inhibition of enzyme activity, **76**
Feedback loop, 11*f*
   regulation of hypothalmic and anterior pituitary hormones by, 157, 158*f*
Feedforward activation of enzyme activity, **77**
Female(s)
   birth control methods used by, 694*b*
   sex determination and differentiation of, 673–74, 675*f*, 676*f*
Female reproductive system, 683–94
   fertilization, implantation of egg, and pregnancy, 695–700
   functional anatomy of organs of, 683–87
   long-term hormonal regulation of, 693–94
   menopause, 676
   menstrual cycle, 676, 683, 688–93
   parturition and lactation, 701–3
   sexual response, 687
Fenestrated capillaries, 414, 415*f*
Fertilization, **671**, 686, 695–700
   events of, 695
   sex determination and differentiation following, 673–74, 675*f*, 676*f*
Fetus, **673**
   development of, 696–98
Fever, 13
   post-exercise, 756–57
Fibrillations, 385–86
Fibrin, **408**
Fibrin clots, 408, 409*f*, 410*f*
Fibrinogen, 404, 405*t*, 409
Fibrous proteins, **31**, 32*f*
Fibrous skeleton, 372
Fick's Law, permeability and, 110, 112*b*
"Fight-or-flight" response, **312**, 742–43
Filaments, muscle, 336, 337*f*, 338*f*
Filtered load, **542**
Filtrate, renal, 536
Filtration, 6. *See also* Glomerular filtration
   across capillaries, **433**, 437–38
Filtration fraction, 541*f*, 542
Fimbriae, 686
First law of thermodynamics, 65
First-order neuron, **265**
First polar body, 687
Flagella, microtubules of, 42*f*, 43
Flavin adenine dinucleotide (FAD), 32, 72
Flavin mononucleotide (FMN), 85
Flow autoregulation, 460
Fluid balance, 565–66, 567–75, 581–83, 584*f*
   during exercise, 752, 756*f*
Fluid mosaic, 35, 36*f*
Fluid movement in digestive system, rates of, 627, 628*f*
Flux, **108**
Focal point, 280
Folic acid, 406, 510*b*
Follicles, ovary, 684*f*, **685**
Follicle stimulating hormone (FSH), 157, 678, **679**, 691

Follicular cells, thyroid, 661, 662f
Follicular phase, ovarian cycle, **688**, 689f
    hormonal changes in early to mid-, 690f, 691–92
    hormonal changes in later, 692
Food
    as energy source (*see* Energy metabolism)
    intake regulation of, 624
    swallowing bolus of, 629, 630f
Foot processes of glomerular capillaries, 540
Foot structures, protein, 342
Force (voltage), 177
Forced expiratory volume (FEV), **489**
Forced vital capacity (FVC), **489**
Force generation in muscles, 338–44. *See also* Tension, muscle
    crossbridge cycle, 339–40
    excitation-contraction coupling, 340–43
    frequency of muscle fiber stimulation and, 347–48
    muscle-cell metabolism and, 343–44
    muscle fiber diameter and, 348
    muscle fiber length and, 348–50
    sliding-filament model, 338
Forebrain, 236f, **237**
Foreign matter, inflammation and containment of, 714
Formed elements in blood, 404–7
Fornix, 241, 242f
Fovea, **278**
Fragmentins, **728**
Fraternal twins, 689
Free energy, 67
Free nerve endings, **272**
Frequency coding
    action potentials and, 194f, **212**
    for sensory stimulus intensity, 266–68
    for sound intensity and pitch, 297, 299f
Frequency of muscle stimulation and forces generated, 347–48
Frontal lobe, cerebrum, **238**, 239f
    effects of injury in, 243b
Fructose, 22, 25f
Functional groups in biomolecules, 23t
Functional residual capacity (FRC), **478**
Functional unit in nervous system, neuron as, 173
Fundus of stomach, 606f, **607**
Funny channels, 381

GABA (gamma-aminobutyric acid), **217**
G actin, 336
Gage, Phineas, 243b
Galactose, 22, 25f
Gallbladder, 612f, **613**
Gametes, 160, **671**–72
Gametogenesis, **671**, 672–73
Gamma-amino butyric acid (GABA), 135, 137
Ganglion (ganglia), **175**
    autonomic, 313
    collateral, 314f, 315
    dorsal root, 231
    receptive fields of, 288f
Gap junctions, **45**–46
    in cardiac muscle, 45f, 361f, 377, 378f
    direct intercellular communication through, 132

electrical synapses and, 202
    in smooth muscle, 361f, 363
Gases, 497–99
    Henry's Law and solubility of, 502b
    partial pressure and diffusion of, 498–99
    in respiratory system, 500–505
    solubility in liquids, 499–500, 502b
    transport of, in blood, 505–14
Gas exchange, 470, 500–505, 526, 527f
    in conducting zone of respiratory tract, 473–74
    determinants of alveolar P_{O_2} and P_{CO_2} and, 504–5
    in lungs, 500–503
    in respiring tissue, 503
Gas transport in blood, 505–14
    carbon dioxide, 510–14
    oxygen, 505–9
Gastric emptying, 607
Gastric glands, 606f, **607**
Gastric inhibitory peptide, 631
Gastric juice, **607**
Gastric motility, 630–31
Gastric mucosal barrier, **607**
Gastric-phase control, **623**, 625, 626f
Gastric pits, 606f, **607**
Gastric reflex, 607
Gastrin, **607**, 623t
Gastrocnemius muscle, 357, 358f
Gastrocolic reflex, **632**
Gastroileal reflex, **632**
Gastrointestinal tract. *See* Digestive tract (gastrointestinal tract)
Gastrointestinal wall, 602–6
    mucosa of, 602–5
    muscularis externa of, 605
    serosa of, 605–6
    submucosa of, 605
Gate-control theory of pain modulation, **276**, 277f
Gated channels, 183
    ligand-, 143–45, **174**
    neurotoxins and faulty, 185b
    voltage-, 174, 188–92
G cells, **607**
Gene(s), **49**, 671–72
    sorting and packaging of, in gametogenesis, 672–73
General adaptation syndrome, **665**
Generator potentials, **262**
Gene therapy, 734b
Genetic code, **48**
    role of, in protein synthesis, 49
Genitalia
    differentiation of, 674, 675f
    female external, 686f–87
    male external, 677f, 678
Genome, human, **56**
Germ cells, 671
Gestation. *See* Pregnancy
GFR. *See* Glomerular filtration rate (GFR)
GHK (Goldman, Hodgkin, Katz) equation, 182b
Gigantism, 660
Glaborous skin, 272
Glaucoma, 282
Glial cells, 151, 173, **176**–77
Globin, 505

Globular proteins, **31**, 32f
Globulin, 404, 405t
Globus pallidus, 240, 242f
Glomerular capillary hydrostatic pressure, 540
Glomerular filtration, **536**, 538–45
    clearance measurements of rate of, 555–56
    filtered load, 542
    filtration fraction, 541f, 542
    plasma glucose concentration and, 548f
    rates of, 541, 545t
    regulation of, and water excretion regulation, 574
    regulation of filtration rate, 542–44
    Starling forces and, 538, 540–41
Glomerular filtration pressure, **540**–41
Glomerular filtration rate (GFR), 541–45. *See also* Glomerular filtration
    clearance and estimates of, 555–56
    reabsorption of water and solutes related to normal, 545t
    regulation of, 542–44
Glomerular membrane (filtration barrier), 540
Glomerular oncotic pressure, 541
Glomeruli (neurons), 307
Glomerulonephritis, 586–87b
Glomerulus (kidney), **536**
Glossopharyngeal nerve (cranial nerve IX), 315
Glottis, **471**, 472f
Glucagon, 159, 161f, **651**–53
    actions of, 651f, 652, 666t
    blood glucose levels and, 652, 653f
    cardiac function and effects of, 394
    factors affecting secretion of, 651f, 652, 653
Glucoamylase, 614
Glucocorticoids, **159**, 663–65
    actions of, 664–65
    effects of, on growth, 661
    effects of abnormal secretions of, 665
    factors affecting secretion of, 664
    stress response and, 665
Gluconeogenesis, **93**–94, 534, 647, 652
    during exercise, 649
    in liver, 648
Glucose, 22, 25f
    absorption, utilization, and storage of energy in, 641, 642f
    in blood (*see* Blood glucose)
    in brain, 226
    clearance rates of, 556, 557f
    formation of new, 93–94
    synthesis of new (*see* Gluconeogenesis)
    transport of, across cell membrane by carrier protein, 113f
Glucose-dependent insulinotropic peptide (GIP), **623**
Glucose oxidation, 64, 77–91
    energy exchange in, and role of ATP, 77–81
    stages of, 81–91
    summary of, 88f, 89
Glucose receptors, 653
Glucose sparing, **647**
Glucosuria, 657b
Glutamate, 135, 137, **217**
    excitotoxicity of, 230b
    long-term potentiation mechanism involving, 256, 257f
    taste and, 305

Glutamic acid decarboxylase, 137
Glutamine, role of, in renal compensation
   during severe acidosis, 592, 593*f*
GLUT 4 glucose transporter, 651
Glycerol, **25**, 26*f*
Glycine, 135, **217**
Glycocalyx, 36, 37*f*
Glycogen, **22**, 25*f*, **92**, 614
   exercise and depletion of, 753
   metabolism of, 91, 92–93
   storage of, 646, 647
Glycogenesis, **92**, 93*f*
Glycogenolysis, **93**, 652
Glycolipids, 36
Glycolysis, **81**–83
   reactions of, 80*f*
   summary of, 81*f*
Glycolytic muscle fibers, **354**, 355*f*
   fast, 355–57
Glycoproteins, 31–32, 36
Glycosuria, 587*b*
Glycosylation, 53
Goblet cells, **473**, 474*f*, 604
Goldman, Hodgkin, Katz (GHK) equation, 182*b*
Golgi apparatus, **38**, 39*f*, 44*t*, 53
Gonad(s), 160, **673**. *See also* Ovaries; Testes
Gonadotropin-releasing hormone (GnRH), 157,
   680
Gonadotropins, **679**–80
G protein(s), 36, 143, **145**
   actions of, 146*f*
   actions of, on slow ligand-gated ion
      channel, 146*f*
   adrenergic receptors coupled to, 317, 319*f*
   aquaporins, antidiuretic hormone and
      actions of, 573, 574*f*
   attack of bacterial infections on, 152–53*b*
   enzymes regulated by, 147*f*
   slow responses in postsynaptic neurons and
      role of, 204, 205*f*
G-protein-linked receptors, **145**–49
G-protein-regulated enzymes, **146**, 147*f*
Graafian follicle, **688**
Graded potentials, 177*t*, 183, **184**, 186
   comparison of action potentials and, 188*t*
   excitatory and inhibitory postsynaptic
      potentials as, 205–9
   integration (summation) of, 186, 211–12
   properties of, 184*f*
Graft versus host reaction, **732**
Granular cells, **537**, 538*f*
Granulocytes, 709
Granulosa cells, 684*f*, **685**
Gray matter
   brain, **228**, 229*f*
   spinal cord, 231–32, 234*f*
Gray ramus, 314, 315*f*
Greater omentum, 605
Growth, hormonal regulation of, 654–61
   effects of insulin and, 651, 661
   glucocorticoids and, 661
   growth hormone and, 654–61
   sex hormones and, 661
   thyroid hormones and, 661
Growth hormone (GH), 31, 157, **654**–61, 666*t*
   bone growth and, 657–60
   effects of abnormal secretions of, 660–61
   factors affecting secretion of, 655–57

Growth hormone inhibiting hormone (GHIH),
   157, **655**
Growth hormone releasing hormone (GHRH),
   157, **655**
GTP (guanosine triphosphate), 145
Guanine, 32, 33*f*
Guanylate cyclases, 145
Gustatory cortex, 305
Gustatory nucleus, 305
Gustducin, 305
Gyri (gyrus), **238**

H₂ receptor antagonists, 622*b*
Habituation, **255**
Hair cells (organ of Corti), 295*f*, **296**
   sound transduction by, 296, 297*f*, 298*f*
Hair follicle receptors, 272
Haldane effect, **513**
Half-life of blood-born chemical messengers,
   140
Haploid cells, **671**
Haustration, **632**, 633*f*
Hay fever, immune response and, 732*f*, 733
Hearing, 291–98
   ear anatomy and, 291–92
   nature of sound waves and, 292, 293*f*
   neural pathway for sound and, 297–98
   signal transduction for sound and, 293–97,
      299*f*
   sound amplification in middle ear and,
      292–93
Heart, **370**, 371–73, 376*t*
   atria, ventricles, and atrioventricular valves
      of, 373*f*, 374*f*
   blood flow through, 374–76, 417, 418*f*,
      419
   cardiac cycle of, 386–92
   electrical activity (*see* Heart, electrical
      activity of)
   enlargement of, 397
   failure of, 432, 434–35*b*
   location of, in thoracic cavity, 372*f*
   muscle of (*see* Cardiac muscle)
   output of, 392–99
   path of blood flow through, 374–76
   rate (*see* Heartbeat; Heart rate (HR))
   sounds of, 391–92
   stroke volume (*see* Stroke volume (SV),
      cardiac)
Heart, electrical activity of, 377–86
   conduction system and, 377–78
   ionic basis of, 379–83
   recording, with electrocardiogram, 383–86
   spread of excitation through muscles of,
      379, 380*f*
Heartbeat, 372
   autorhythmicity of, 377
   control of, by pacemakers, 378
   initiation and conduction of impulse during,
      377–78, 379*f*
   rate of, 392, 393–95
   rate of, mean arterial pressure and,
      427–29
Heartblock, 434*b*
Heartburn, 607
   histamines and, 136
Heart failure, 432, 434–35*b*
Heart muscle. *See* Cardiac muscle

Heart rate (HR), **392**
   cardiac output and changes in, 393–95
   integration of controls on, 394–95
Heart sounds, 391, 392*f*
Heat (thermal energy), 66
   release of, in ATP synthesis, 79*f*
Heat exhaustion, 9*b*
Heat load, 463
Heat stroke, 9*b*
Heavy chains, antibody, **719**, 720*f*, 722
Helicobacter pylori bacteria, ulcers and 622*b*
Helicotrema, 295*f*, **296**
Helium dilution method, calculation of residual
   lung volume by, 488
Helper T cell, **722**
   activation of, 727
   CD4 surface protein on, 726
   recognition of Class II MHC molecules by,
      726
Hematocrit, **406**, 407*f*
Hematopoietic (blood-forming) stem cells, **711**
Hematuria, 586*b*
Heme groups, 84, 405, 406*f*, 505
Hemodialysis, 550*b*
Hemoglobin, 28, 371, 405, 406*f*, **505**–9
   additional factors affecting affinity of
      oxygen and, 508–9
   affinity of carbon dioxide for, 509
   as buffer, 524–25
   as carrier of oxygen in blood, 505–7
   decreased oxygen-carrying capacity of blood
      and, 507, 510–11*b*
   dissociation curve with oxygen and, 507,
      508*f*
   saturation of, 505, 507*f*
Hemoglobin-oxygen dissociation curve, 507,
   508*f*
Hemolytic anemia, 511*b*
Hemophilia, 411
Hemorrhage
   baroreceptor response to, 447, 449*f*
   control of glomerular filtration rate and
      renal vascular resistance in response to, *f*
   fluid and electrolyte balance following,
      582–83, 584*f*
   integrated neural and hormonal controls on
      cardiovascular function in response to,
      453–55
Hemorrhagic anemia, 510–11*b*
Hemorrhagic stroke, 230*b*
Hemostasis, **407**–11
   blood clot formation, 408–11
   platelet plug, 407, 408*f*
   vascular spasm, 407
Henderson-Hasselbalch equation, 524*b*, 525
Henry's Law, 499, 502*b*
Hepatic artery, 609
Hepatic portal vein, 419, 609
Hepatic vein, 419
Hepatocytes, 612*f*, **613**
Hertz, 292
High-density lipoproteins (HDLs), 621*b*
High-energy phosphate bond, 78
Hippocampus, 241, 242*f*
Histamine, 134, 135, 272, 460
   antihistamines to combat effects of, 136*b*
   bronchoconstriction caused by, 486
   as neurotransmitter, 217

release of, as allergic response, 443, 732f
release of, by injured cells, 437–38
release of, by mast cells during inflammation, 714
Histones, 56
HLA tissue type, 725
Hodgkin's disease, 734
Homeostasis, **8**–14
  acid-base, 522–26, 583–95
  heat exhaustion and stroke as malfunction of, 9b
  kidneys and (see Homeostasis, urinary system and)
  negative feedback control in, 9–12
  respiratory system and, 522–26, 585
  role of autonomic nervous system in, 312
  thermoregulation and, 12–13
Homeostasis, urinary system and, 564–600
  acid-base balance, 583–95
  calcium balance, 580–81
  fluid and electrolyte interactions, regulation of, 581–83
  plasma composition and, 565, 566f
  potassium balance, 579–80
  sodium balance, 575–78
  solute and water balance, overview of, 566
  water balance, 567–75
Homeothermic animals, 12
Homologous chromosomes, 671, 672
Horizontal cells, 285
Hormone(s), **134**, 161–67
  abnormal secretion of, 164–65, 166f
  as chemical messengers, 133f, 134, 142, 143f, 151
  control of blood levels of, 161–65
  control of energy metabolism during exercise by, 746–47
  control of gastrointestinal tract by, 622, 623t
  control of heart rate by, 394, 395t
  control of mean arterial pressure by, 451–55
  interactions of, 165–67
  lactation and, 702f
  liver and secretion of, 613
  in menstrual cycle, 691–93
  metabolism rate of, 164
  pregnancy and, 698–700
  regulation of growth by, 654–61
  regulation of male reproductive function by, 679–80
  secretion of specific, by endocrine organs, 162–63t
  secretion rate of, 161–64
  transport of carrier protein-bound, 164
  tropic, 155–57, 158f
Hormone response element (HRE), 143
Hormone sensitive lipase, 753
Human body. See Body
Human chorionic gonadotropin (hCG), pregnancy and, **698**, 699f
Human genome, **56**
Human life span, patterns of reproductive activity over, 674–76
Humoral immunity, **718**, 721–24
  antibody structure and function in, 723t
  B lymphocytes and production of antibodies in, 721–22
  function of antibodies in, 722–24

Humoral signals affecting hormone secretions, 162–64
Huntington's chorea, 250
Hydrated carbohydrates, 22
Hydrochloric acid (HCl), 75b
  stomach secretions of, 616f
Hydrogen, 23
  buffering ions of, 588–90
  oxidation-reduction reactions and, 64–65
  renal handling of, 591
  secretion of, and stomach acidification, 607
  sour taste and presence of, 304
Hydrogen bonds, 24b
  in proteins, 28, 30f
Hydrolysis, **64**
Hydrophilic (polar) bonds, 31b
Hydrophobic (nonpolar) bonds, 24, 31b
Hydrostatic column, 421
Hydrostatic pressure gradient, movement of fluids in and out of capillaries and, 434–35, 436f
Hydroxyapatite, 657
Hymen, 686f, 687
Hypercalcemia, **580**
Hyperemia
  active, 456, 457f, 458f
  reactive, 458, 459f
Hyperglycemia, 652
Hyperkalemia, **579**
Hypernatremia, **575**
Hyperopia, **281**–2, 283f
Hyperosmotic solutions, **121**
Hyperplasia, **655**
Hyperpnea, **505**
Hypersecretion of hormones, 164–65, 166f
Hypertension, 450b
  baroreceptor reflex and, 447–48
  blood volume and regulation of, 431
  heart failure and, 434b
  strokes and, 230b
  treatment of, 320, 450b
Hyperthermia, **12**
Hypertonic solutions, **122**, 123f
Hypertrophy, **655**
Hyperventilation, **505**
  effects of, on minute ventilation, 521f
Hypervolemia, 567
Hypocalcemia, **580**
Hypoglycemia, 652
Hypokalemia, **579**
Hyponatremia, **575**
Hypo-osmotic solutions, **121**
Hyposecretion of hormones, 164–65
Hypotension, 449
  orthostatic, 432
Hypothalamus, 13, **153**–57, 236f, 237, 240–41
  connection between pituitary gland and, 155f, 156f
  hormone secretions from, 155, 157, 158f
  neural control of cardiovascular function and, 445f, 446
  role of, in regulating autonomic function, 322, 323f
Hypothalmic-pituitary portal system, 156f, **157**, 158f
Hypothermia, **12**, 70
Hypotonic solutions, 122, 123f

Hypoventilation, **505**
  effects of, on minute ventilation, 521f
Hypovolemia, 567
Hypoxemia, 490, 525b
Hypoxia, 456, 504, 525b
H zone, muscle, **336**

I band, muscle, **336**
Ideal gas law, 482b, 498b
Identical twins, 696
Idioventricular pacemakers, 378
Ileocecal sphincter, **609**
Ileogastric reflex, **632**
Ileum, **608**
Illnesses, diseases, and conditions
  abnormal secretion of hormones, 164–65, 166f
  affecting motor control, 250–51
  anemia, 406, 507, 510–11b, 620
  asthma, 318–20, 486–87
  autoimmune disease, 709, 731b, 733
  bacterial infections, 152–53b
  blood clotting disorders, 411
  cancer (see Cancer)
  cholera, 126, 152b
  cystic fibrosis, 126, 127b
  depression, 218b
  diabetes, 164, 574, 587, 652, 657–58b
  effect of marathon running on susceptibility to, 757–58
  effects of neurotoxins, 185b
  heart failure, 432, 434–35b
  heat stroke and heat exhaustion, 9b
  hypertension (see Hypertension)
  immune response in health and, 728–35
  immunodeficiency disease, 733–34
  kidney disease, 586–87b
  lactose intolerance, 90b
  leukocyte adhesion deficiency (LAD), 715
  myasthenia gravis, 327, 328b
  obstructive pulmonary disease, 487, 489, 490b
  Parkinson's, 250–51
  pertussis (whooping cough), 152–53b
  pulmonary edema, 504b
  renal failure, 550–51b, 586–87b
  stroke, 230b
  tetanus, 349b
  ulcers, 622b
  vision defects, 281–82, 283f
Immune response, **709**
  diversity of, 720, 721f
  dysfunctional, 732–35
  memory of, 720–21
  primary, and secondary, 720–21, 722f
  specificity of, 719–20
  stress and, 734–35
  in transfusion and transplantation, 730–32
  types of (see Cell-mediated immunity; Humoral immunity
  to virus, 728f
Immune system, 5, **709**–40
  anatomy of, 709–12
  cancer and, 59b
  cell-mediated immunity, 718, 724–28
  health, disease, and responses of, 728–35
  humoral immunity, 718, 721–24

Immune system, *continued*
  nonspecific defenses, 713–18, 736f
  specific immune responses, 718–21, 736f
  systems integration chart, 737f
Immunity, **709**
  acquired, 720
  active, and passive, 729
  immunization for generating, 729–30
Immunization, 729–30
Immunodeficiency disease, 733–34
  gene therapy for, 734b
Immunoglobulins, 711, 722–24. *See also*
  Antibodies
Implantation of fertilized egg, **686**, 695, 696f
Implants, contraceptive, 694b
Inactivation gates, **189**
Inclusions, cytosol, **37**
Incus, 291, 292f
Independent assortment, **673**
Indirect intercellular communication, chemical
  messengers and, 132–33. *See also* Chemical
  messengers
Induced fit model of enzyme and substrate, 72
Inflammation, **713**–17
  capillary dilation and increased permeability
    and, 713–14
  containment of foreign matter and, 714
  continued activities of recruited leukocytes
    and, 715–17
  leukocyte migration and proliferation and,
    714–15
  major events in local, 714f
  phagocytosis and, 713, 714f, 715, 716f
  release of histamine and, 134, 714
Infundibulum, 153, 686
Inhibin, 678, **680**
Inhibiting hormones, 155
Inhibitory graded potentials, **184**
Inhibitory post-synaptic potential (IPSP), **209**
Inhibitory synapse, **209**–10
Initiation factors, 51
Initiation of translation, 51f
Initiator codon, 49
Innate reflexes, 244
Inner cell mass, 695
Inner ear, 291, 292f
Innervation of effector organs, **171**
Inorganic ions, 31b
Inositol triphosphate (IP3), 146, **147**
Input, homeostasis and, **11**
Insensible loss, 567
Insensible water loss, 12
Insertion, skeletal muscle, **357**, 358f
Inspiration, 480–82
  changes in alveolar pressure and breath
    volume during, 480f
  events in process of, 483f
Inspiratory capacity (IC), **488**
Inspiratory muscles, **480**, 481f
Inspiratory neurons, **515**, 517f
Inspiratory reserve volume (IRV), **488**
Insulin, 115, 134, **650**–51
  actions of, 650f, 651, 666t
  beta cell secretion of, 159, 161f
  blood glucose levels and secretions of, 163,
    165f, 566, 652f, 653f
  cardiac function and effects of, 394

controls on secretions of, 163, 165f
diabetes and, 657–58b
factors affecting secretion of, 650–51, 653
Insulin-dependent diabetes mellitus, 164, 657b
Insulin-like growth factors (IGF), 160, **655**
Integral membrane proteins, **36**
Integrating center, **11**
Integrator neurons, 186
Integrins, 715
Intention tremor, 250
Interatrial septum, 371
Intercalated cells, renal tubule 566
  bicarbonate synthesis and hydrogen ion
    secretion by, 591, 592f
Intercalated disks, 377, 378f
Intercellular communication, 48, 132–33
  direct, through gap junctions, 132
  hormone actions at target cell and, 161–67
  indirect, through chemical messengers,
    132–50 (*see also* Chemical messengers)
  signal transduction and, 140–49
Intercostal nerves, **514**
Interferons, **713**, 717
  α, and β, 717, 727t
  γ, 717, 727t
Interleukin, 716
Interleukin-1 (IL-1), 716
Interleukin-2 (IL-2), 722, 727
Interleukin-4 (IL4), 727
Interleukin-5 (IL5), 727
Interleukin-6 (IL-6), 716
Interleukin-10 (IL-10), 727
Interlobar arteries, renal, 537, 539f
Interlobar veins, 538, 539f
Interlobular arteries, 537, 539f
Interlobular veins, 538, 539f
Intermediate-density lipoproteins (IDLs), 621b
Intermediate filaments, **42**
Intermediates in metabolic pathways, **64**
Intermediolateral cell column, 231, 314
Intermembrane space, mitochondria, **38**, 39f
Internal anal sphincter, **610**
Internal environment, 5, **6**. *See also*
  Extracellular fluid (ECF)
  exchanges between external and, 6–7
  homeostasis and, 8–14
Internal intercostals, **475**, 477f
Internalization process in phagocytosis, 716
Internal respiration, **470**, 471f
Internal urethral sphincter, **556**, 558f
Interneurons, **175**
Interphase, **57**, 58f
Interstitial edema, 504b
Interstitial fluid (ISF), 7f, **8**
  regulation of, by renal function, 533
Interstitium, 432
Interventricular septum, 371
Intestinal-phase control, **623**
Intestino-intestinal reflex, **632**
Intra-alveolar pressure (P_{alv}), **477**, 478f
  determinants of, 479–80
Intracellular fluid (ICF), **7**
  concentrations of selected solutes in, 101t
Intracellular receptor-mediated responses,
  signal transduction mechanisms for, 142–43
Intrapleural fluid, **475**, 477f
Intrapleural pressure (P_{ip}), **477**, 478f

Intrapleural space, **475**, 477f
  air trapped in (pneumothorax), 479f
Intrauterine device (IUD), 694b
Intrinsic control, **392**
  of cardiovascular function, 455–61
  of glomerular filtration, 542–43
Intrinsic factors (protein), **607**
Intrinsic neurons, 314
Introns, 50
Inulin, clearance of, 555f, 556
Involuntary nervous system. *See* Autonomic
  nervous system
Ion(s), 31b
  directions of driving forces acting on
    uncharged, 107t
  electrochemical driving force and, 104–6
  membrane potential and, 103, 104
  transport of, across cell membrane by
    channel protein, 114f
Ion channels. *See* Channel(s), ion
Ionic bonds, 31b
  in proteins, 28
Ionotropic receptor (channel-linked), **204**
Ipsilateral pathway, **237**
Iris, **278**, 279f
  muscles of, 284f
Iron-deficiency anemia, 406
Iron-sulfur proteins, 84
Irritant receptors, 518
Ischemia, **456**
Islets of Langerhans, 159, 161f
Isoelectric line, 384
Isometric twitch, **345**
  duration of, 347f
Iso-osmotic solutions, **121**
Isotonic solution, **122**
Isotonic twitch, **345**–46
  effect of load on peak tension in, 346f
Isovolumetric contraction, **386**
Isovolumetric relaxation, **388**

Jejunum, **608**
Juvenile-onset diabetes, 657b
Juxtaglomerular apparatus, **537**, 538f
Juxtaglomerular cells, **537**, 538f
Juxtamedullary nephrons, 537f

Keratin, 42
Ketoacidosis, 658b
Ketones, 95, 658b
Kidney(s), 534. *See also* Urinary system
  basic renal exchange processes and, 538–48,
    553f
  blood supply to, 537–38, 539f
  diseases of, 586–87b
  excretion by, 553–59
  function of, 533–34
  macroscopic anatomy of, 535
  microscopic anatomy of, 535–37
  regional specialization of renal tubules of,
    549–53
  renal hypertension and malfunctioning,
    450b
  water balance and, 567–74
Kinesins, 206
Kinetic energy, **66**, 67f

Kinocilium, **300**, 301*f*, 303
Korotkoff sounds, 425*b*
Krebs cycle, **83**–84
   reactions of, 82*f*
   summary of, 83*f*

Labeled lines, **265**
Labia majora, 686*f*, 687
Labia minora, 686*f*, 687
Labor, parturition and, **701**
Lactase, 90*b*, 614
Lactate, exercise and levels of, 746, 753
Lactate dehydrogenase, 89
Lactation, **698**–99
Lacteal, **608**
Lactic acid, 83
   conversion of pyruvate to, 89–91
   exercise and build up of, 343, 356
Lactose, 22, 25*f*, 614
Lactose intolerance, 90*b*
Lamina properia, **604**
Laminar (streamlined) blood flow, 424*f*
Language, central nervous system function
   and, 251
Laplace's Law, 398*b*, 486*b*
Large intestine, 610
Larynx, **471**, 472*f*
Latent phase, muscle twitch, **344**
Lateral geniculate body, **291**
Lateral horn, **314**
Lateral inhibition, 268, **269**
Lateral reticular formation, 276
Lateral sulcus (Sylvian fissure), 239
Latroxin, 327
Law of complementary base pairing, 32, 34*f*
Law of LaPlace, 398*b*, 486–87*b*
Law of mass action, **67**, 68*f*
Law of specific nerve energies, **262**
Law of thermodynamics, first, 65
Leader sequence, translation of, **52**, 53*f*
Leak channels, **174**
Leaks
   pumps and membrane, 118–19
   sodium and potassium, and resting
     membrane potential, 180
Lean body mass, 655
Learning, **255**
Left heart, **371**, 376
Length-tension curve, **348**, 350*f*
Lens (eye), **278**, 279*f*
   accommodation of, 281, 282*f*
Leptin, **624**
Lesser omentum, 605
Leukocyte(s), 13, **371**, 407, 709–11
   cytokine release by, 134
   inflammation and activities of recruited,
     715–17
   inflammation and migration and
     proliferation of, 714–15
   morphology and functions of, 710*t*
   phagocytosis by, 47, 713, 714*f*, 715, 716*f*
Leukocyte adhesion deficiency (LAD), 715
Leukocytosis, 715
Leukotrienes, 28, 139
Leydig cells (interstitial cells), **677**, 678*f*
LH surge, 11, **692**
Ligaments, 4

Ligand(s), 132
   agonistic and antagonistic, 142
Ligand-gated channels, **143**–45, **174**
Light adaptation, 286, 287–88
Light chains, antibody, **719**, 720*f*
Light waves
   nature and behavior of, 278–81
   regulating amount of, entering eye, 282–84
Limbic system, **241**, 242*f*, 243
Limit dextrins, 614
Linear acceleration, 299
   utricle, saccule, and transduction of, 300–303
Lipases, **617**, 753
Lipids, 22, **25**–28
   absorption, utilization, and storage of
     energy in, 642*f*, 643
   digestion and absorption of, 616–19
   eicosanoids, 28
   energy metabolism of, during absorptive
     and postabsorptive states, 649*t*
   metabolism of, 94–95
   phospholipids, 27
   steroids, 28, 29*f*
   triglycerides, 25–26
Lipogenesis, **95**
Lipolysis, **94**–95, 643
Lipophilic molecules, **135**
   as chemical messengers, 142–43
Lipophobic molecules, **135**
Lipoprotein lipase, 621*b*, **643**, 646
Lipoproteins, 32, 612, **618**, 621*b*
Lipoxygenase, 139
Lipoxygenase pathway, 139
Liquids, solubility of gases in, 499–500, 502*b*
Liver, **610**, 621
   energy metabolism in, during absorptive
     state, 646
   energy metabolism in, during
     postabsorptive state, 648
   functions of, 612–13
Loads, muscle, 345
   effect of, on muscle shortening, 353*f*
   effect of, on peak tension in isotonic twitch,
     346*f*
Load-velocity curve, 353*f*
Lock-and-key model of enzyme and substrate,
   72
Long reflex pathway, **623**, 624*f*
Long-term memory, **256**
Long-term potentiation (LTP), 256, 257*f*
Loop of Henle, **536**
   countercurrent multiplier and, 570–71
   water conservation in, 552, 572
Low-density lipoproteins (LDLs), 621*b*
Lower esophageal sphincter, 606*f*, **607**
Lower motor neurons, **247**–48
L-type channels, 381
Lumbar nerves, **231**, 232*f*
Lumen, **4**, 45, 411
   regulation of, 621–23
Lumen, stomach, 607
Lungs, 470, 477*f*
   anatomical dead space, effects on alveolar
     ventilation of, 491*f*
   compliance of, 484
   gas exchange in, 500–503
   pressures affecting, 477, 478*f*, 479*f*

Lung volumes, **488**
   calculating residual, 488
   factors affecting, 485*f*
   total lung capacity and, 488, 489*f*
Luteal phase, ovarian cycle, **688**–89
   hormonal changes in, 692–93
Luteinizing hormone (LH), 11, 157, **679**
   surge, 11, **692**
Lymphatic capillaries, 438
Lymphatic fluid (lymph), 433, **438**
Lymphatic system, **438**
   capillary filtration and absorption and, 437
   diagram of, 438*f*
Lymph nodes, 438
Lymph nodules, 604
Lymphocytes, 710*t*, **711**
   B, 711
   null and natural killer (NK), 711, 724*f*
   T, 158, 711
Lymphoid tissues, **709**, 711, 712*f*
Lysosomes, **39**, 40*f*, 44*t*, 716
   secondary, 716
Lysozyme, 611

Macromolecules, transport of, 46–48
Macrophages, **438**, 474, 710–11
   alveolar, 475, 476*f*
   cytokines secreted by, 716–17
   phagocytosis by, 713, 714*f*, 715, 716*f*
   wandering, 711
Macula densa, **537**, 538*f*
Macula lutea, 284*f*, **285**
Major histocompatibility (MHC) molecules, **725**
   antigen presentation, T cell activation, and
     role of, 725, 726*f*
   as markers of self, 725
   T cell maturation, and role of, 725
Major vault protein (MVP), 42
Male(s), sex determination and differentiation
   of, 673–74, 675*f*, 676*f*
Male reproductive system, 676–83
   accessory glands, 679
   hormonal regulation of, 679–80
   organs of, 676–79
   sexual response, 682–83, 685*b*
   sperm, 680–82
Malleus, 291, 292*f*
Maltase, 614
Maltose, digestion and absorption of, 614,
   615*f*
Mammary glands, **698**, 702*f*
   lactation and, 702–3
Marathon case study, 16–18. *See also* Exercise
   links
   carbo-loading prior to, 18, 93
   physiological integration during (*see*
     Physiological integration during
     exercise)
   training and, 16–17
Margination of leukocytes, **715**
Mass absorber, proximal tubule as, 551
Mass action, law of, 67, 68*f*
Mass motion, colon, **632**
Mass of solute, 69*b*
Mast cells, 134, **714**
Mastication, **606**–7
Maximum tetanic tension, **348**

Mean arterial pressure (MAP), **389**b
  blood flow, blood distribution to organs, and, 427, 428f
  determinants of, 427–29
  drop in, upon standing, 432
  integrated response to restore, following hemorrhage, 453–55, 582–83, 584f
  metaboreceptors and central command controlling, during exercise, 748, 749f
  pressure gradients and, 422f
  renin release stimulated by decrease in, 579f
Mean arterial pressure (MAP), regulation of, 443–55
  effects of angiotensin II on, 577, 578f
  hormonal control of, 451–55
  integrated control of, 453–55
  neural control of, 444–51, 453–55
  plasma volume and, 565
Mechanical nociceptors, **272**
Mechanical work, **644**
Mechanoreceptors, 261, **271**
  in digestive tract, 622
  in mouth and tongue, 625
  in skin, 272
Medial geniculate body, **298**
Mediated transport, **101**. See also Active transport; Facilitated diffusion
Medulla (kidney), **535**
  medullary osmotic gradient in, 569–71
  salivary center in, 625
Medulla oblongata (brain), 236f, **237**
  central chemoreceptors in, 519f
  mean arterial pressure controlled by, 444–46
  respiratory control centers of, 515–16
Medullary osmotic gradient, **569**–71
  countercurrent multiplier and, 570–71
  urea and, 571
  vasa recta and, 571, 572f
Medullary pyramids, 248
Megakaryocytes, 407
Meiosis, **56**, 672f
  gene sorting and packing during, 672–73
Meiosis I, **673**
Meiosis II, **673**
Meiotic arrest, **687**
Meissner's corpuscles, 272
Melatonin, 158, 164, 241
Membrane antibodies, 719
Membrane attack complex (MAC), 717, 718f
Membrane-bound receptor-mediated responses, signal transduction mechanisms for, 143–49
Membrane immunoglobulins, 719
Membrane permeability, 7, 109
  factors affecting, in facilitated diffusion, 114–15
  factors affecting, in simple diffusion, 112
  Fick's law and, 110, 112b
  magnitude of driving force and, 109–10
Membrane potential (Vm), **103**–4, 177t
  action, 187–96 (see also Action potentials)
  effects of ion valence and, on electrical driving force, 105f
  electrical forces created by, 104, 105f
  electrical signaling through changes in, 183
  graded, 183, 184, 186

resting, 178–82
summary of factors influencing, 198f
threshold, 184
Membrane proteins. See also Transport proteins
  cell adhesion, 43
  classes of, 36
Membrane surface area, molecule transport and, 110
Membrane transport. See Cell membrane transport
Membranous labyrinth, 299
Membranous organelles, structure of, 37–40, 44t
Memory, **255**–56
Memory cells, immune response, **720**
Memory of immune response, 719, 720–21
Meninges, **224**, 225f
Menopause, **676**
Menstrual cycle, 676, **683**, 688–93
  hormonal cycles during, 691–93
  ovarian cycle, 688–89
  uterine cycle, 689–91
Menstrual phase, uterine cycle, **689\90**
Menstruation, **683**, 690–91
Merkel's disks, 272
Mesangial cell contraction, 542, 543
Mesenteries, **605**
Mesonephric ducts (Wolffian ducts), 674
Mesothelium, **605**
Messenger RNA (mRNA), **48**
  codon base-pair sequence in, 49
  DNA transcription and formation of, 48, 49
  translation of, and protein synthesis, 51–52
Mesenteric veins, 609
Metabolic acidosis, **585**–88, 593–94
Metabolic alkalosis, **585**–88, 594
Metabolic pathway(s), **63**–64
  of glucose oxidation, 80f, 82f
  of protein, glycogen, and fat metabolism, 92f, 93f, 94f, 95f
  regulation of, 641
Metabolic rate, **644**
Metabolic reactions
  of absorptive state, 646f
  of carbohydrate, fats, and protein metabolism, 91–96
  coupled, in ATP synthesis, 79f
  energy and, 65–70
  of glucose oxidation, 77–81
  of glucose oxidation, stages of, 81–91
  of postabsorptive state, 648f
  rates of, 70–77
  response of vascular smooth muscle to changed rate of, 456–58
  types of, 63–65
Metabolic wastes, removal of, 533
Metabolism, **46**, **63**–99
  acid-base disturbances caused by problems in, 585–88
  of carbohydrates, fats, and proteins, 91–96
  demands of exercise on, 742
  energy and reactions of, 65–70 (see also Energy metabolism)
  of glucose (glucose oxidation), 77–91
  of hormones, 164
  overview of whole body, 640–41
  rates of metabolic reactions, 70–77
  types of reactions in, 63–65

Metaboreceptors, 747–48, 749f
Metabotropic receptors, **204**–5
Metaphase (mitosis), 58f
Metarterioles, 413, 414f
MHC. See Major histocompatability (MHC) molecules
Micelle, 27f, 618
Microcirculation, 412
Microfilaments, **41**–42
Microglia, 176
Microtubules, **42**–43
Microvilli, 41f, **42**
  in small intestine, 608
Micturition, **556**–59
  anatomy of bladder and urethra, 556, 557f
Micturition reflex, 558, 559f
Midbrain, 236f, **237**
Middle ear, 291, 292f
  sound amplification in, 292–93
Migrating motility complex, **631**
Milk ejection, **702**
Mineralocorticoids, **159**
Minerals, absorption of, 620
Minor calyces, kidney, 535
Minute ventilation, **490**–91
  effects of hypoventilation and hyperventilation on, 521f
  effects of respiration rate and tidal volume on, 491t
  summary of factors affecting, 492f
Mitochondria, **38**, 39f, 44t
Mitochondrial matrix, **38**, 39f
Mitosis, **56**, 57f, 58f
Mitotic spindle, 40
Mitral cells, 307
Mixed nerves
  autonomic nerves as, 317
  spinal nerves as, **232**
Mixed neurons, respiratory, 516
Mixed venous blood, **503**
M lines, muscle, **336**
Modality of stimuli, **262**, 263t
Modulating neuron, 213
Modulators, **75**, 76
Molecule(s). See also Biomolecules
  amphipathic, 25
  atoms and, 22b
  diffusion of individual versus populations of, 111
  driving forces on, 102–6
  polar, 24b
  transport of, across membranes, 46 (see also Cell membrane transport)
Monoamine oxidase (MAO), **217**
Monoamine oxidase inhibitors, 218
Monocytes, **710**–11
Monoglycerides, liberation of, from fat by lipases, 619f
Monoiodotyrosine (MIT), **661**
Mononucleosis, 709
Monosaccharides, **22**, 25f
Monosynaptic reflexes, 244
Monounsaturated fatty acid, 26
Monozygotic twins, 696
Mons pubis, 686
Morula, **695**, 696f

Motility, digestive, **602**, 603f, 628–33
  chewing, swallowing, and, 629–30
  colon motility, 632–33
  electrical activity in smooth muscle, 628, 629f
  gastric motility, 630–31
  small intestine motility, 631–32
Motion sickness, 323
Motivation, central nervous system function and, 254
Motor control, voluntary, 247–51
Motor cortex, 239, 240f, 241f
  voluntary movement and, 248, 249f
Motor end plate, **325**, 327f, 341–42
Motor homunculi, **239**, 241f
Motor neurons, **172**. *See also* Motor unit
  control of breathing by, 514, 515f
  release of acetylcholine by, 340, 341f
  size principle, order of recruitment, and, 351, 352f
  voluntary motor control and lower, 247–48
  voluntary motor control and upper, 248
Motor unit, **324**, 326f, **344**
  recruitment as increase in active, 351–52
  twitch event in, 344–47
Mouth (oral cavity), 606f–7, 625
Mucins, 604
Mucosa, **602**–5, 606f
Mucous membrane, **602**, 605f
  as nonspecific chemical defense, 713
Mucus, 473, **604**
  in saliva, 611
Mucus escalator, 474
Müllerian (paramesonephric ducts), **674**
Müllerian-inhibiting substance (MIS), **674**
Multidrug resistance, 42
Multiple sclerosis, 731b, 733
Multipolar axon, 174f
Multi-unit smooth muscle, **363**
Muscarinic cholinergic receptors, **216**, 317, 318f, 319t
Muscle(s), 333–68
  cardiac, 363–64
  of chest wall, 475, 477f
  delayed on-set soreness in, post-exercise, 757
  inhibition of, 246
  inspiratory and expiratory, 480, 481f
  micturition and urinary bladder, 556, 558f
  myogenic versus neurogenic, 377
  skeletal (*see* Skeletal muscles)
  smooth, 360–63 (*see also* Smooth muscle)
Muscle cells, **3**, 4f
  metabolism in, and production of ATP, 343–44
Muscle contraction, 344–53
  bones as levers and implications for, 358–60
  in cardiac muscle, 363, 364f
  crossbridge cycle and, 339–40
  determining velocity of shortening in, 352–53
  factors affecting force generated by muscle fibers, 347–50
  neural regulation of smooth, 362–63
  regulation of force generate by whole muscles, 350–52, 353f
  role of ATP and muscle cell metabolism in, 343–44

sliding-filament model of, 338
  turning on and off (excitation-contraction coupling), 340–43
  twitch event in, 344
Muscle fibers, **324**, **334**, 353–57
  differences in ATP production mode in, 354–55
  factors affecting force generated by individual, 347–50
  fast-twitch, and slow-twitch, 345, 354
  long-term responses of, to exercise, 356–57
  properties of different, 356t
  slow oxidative, fast oxidative, and fast glycolytic, 355–57
  stretch sensitive, 458
  structure of, 335f
Muscle spindle, 245
Muscle spindle stretch reflex, **245**
Muscularis externa, **602**, 605
Muscularis mucosae, **604**–5
Mutagens, 57, 59b
Myasthenia gravis, 327, 328b
Myelin, **176**–77
  multiple sclerosis and, 731b, 733
  in white matter, 228
Myelin sheaths, formation and origins of, 176f
Myenteric (Auerbach's) plexus, **605**
Myocardial infarction, 434b, 450
Myocardium, **371**, 372. *See also* Cardiac muscle
Myoepithelial cells, 702f
Myofibrils, **334**, 335f
  thick and thin filaments of, 336–38
Myogenic muscle contractions, 364, 377
Myogenic regulation of glomerular filtration rate, **542**, 543f
Myogenic response, vascular resistance, changes in perfusion pressure, and, **458**, 460f
Myoglobin, 31, 32f, **354**
Myometrium, **686**
Myopia, **281**–2, 283f
Myosin, 31, 32f, 42, 334, **336**, 337
  fast, and slow, 354
  unbinding and binding of, to actin, 339–40
Myosin kinase, **361**

NAD (Nicotinamide adenine dinucleotide), 32, 73
NADH, 73
  glucose oxidation and role of, 81, 82f, 83, 85, 86
Natural family planning, 694b
Natural gas, 23
Natural killer (NK) cells, **711**, 717, 724f
Neck cells, 607
Negative balance, 566
Negative feedback control, **9**–12
  of blood glucose levels, 652–53
  in homeostasis, 11–12
  loop, 11f
  of mean arterial pressure, 444
  of regulated variable, 10f
  in secretion of hypothalmic and anterior pituitary hormones, 157, 158f
Nephrogenetic diabetes insipidus, 587b
Nephrons, **535**
  cortical, and juxtamedullary, 537f

Nephrosclerosis, 587b
Nephrotic syndrome, 586b
Nernst equation, equilibrium potentials and, 107b
Nerve(s), **175**
  cranial, 231, 237t
  spinal, 231, 232f
Nerve cells. *See* Neuron(s)
Nerve fibers, **230**
Nervous system, 151, 170–97
  action potentials in tissue membranes and, 187–96
  autonomic (*see* Autonomic nervous system)
  basis of neural stability in, 187–98
  central (*see* Central nervous system (CNS))
  cells of, 172–77
  characteristics of, 151t
  electrical signals in neurons of, 177–86
  enteric, 172, 605
  hypothalamus as link between endocrine system and, 240–41
  integration chart for, 365f
  organization of, 171f
  overview of, 171–72
  sensory systems of (*see* Sensory systems)
  somatic (*see* Somatic nervous system)
Net filtration pressure, **436**–37, 540
Net flux, **108**
  concentration gradient changes, and changes in, 109f
  relationship of concentration gradient and, for carrier-transported molecules, 114, 115f
Net rate (reaction rates), **71**
Neural integration, 186, 187f, **210**–12
  frequency coding and, 212
  summation as, 186, 211–12
Neural pathways
  ascending, and descending, in spinal cord, 232–33, 234f, 235f, 236
  autonomic nervous system, 312, 313f, 314f
  axon, **175**
  for equilibrium, 303
  for hearing, 297–98
  for information flow to and from cerebellum, 250f
  ipsilateral and contralateral, 237
  for olfaction, 307
  parasympathetic nervous system, 316f
  sensory, 264–65, 266f
  short and long reflex, and gastrointestinal function, 623, 624f
  somatosensory, 273, 274f
  for taste, 305
  for vision, 290–91
Neural processes, 173
Neural signals altering hormone secretions, 162, 164f
Neuroactive peptides (neuropeptides), **217**, 219
Neuroeffector junctions, 202, 320–21, 322f
Neurogenesis, 173b
Neurogenic muscle, 377
  contractions, 364
Neurohormones, 133f, **134**–35, 155
Neuromodulators, **205**
Neuromuscular fatigue, 356

Neuromuscular junction, **324**–29
    diagram of skeletal muscle and, 361*f*
    excitation-contraction coupling in muscles
        and role of, 340–42
    functional anatomy of, 327*f*
    malfunctions in, 327, 328*b*, 329*b*
Neuron(s), 2, **3**, 4*f*, 151, 173–75
    characteristics of, at rest and during action
        potential, 192*t*
    communication and integration of, 186
    functional classification of, 174–75
    inspiratory, and expiratory, 515
    mechanisms in, for moving products in,
        206
    motor (*see* Motor neurons)
    preganglionic and postganglionic, 313, 314*f*,
        315*f*, 393
    presynaptic and postsynaptic, 202
    production of new, 173*b*
    release of neurotransmitters by, 134
    at rest, 181–82
    signal transmission in, 151*f*
    structural classification of, 174*f*
    structural organization of, 175
    structure of, 172*f*
    temporal and spatial summation in, 186,
        187*f*, 211–12
Neuron-to-neuron chemical synapses, 202,
    203*f*
Neurosecretory cells, 134
Neurotoxins, 185*b*
Neurotransmitters, 133*f*, **134**, 214–19
    acetylcholine as, 214–16
    amino acids as, 135, 137, 217
    of autonomic nervous system, 317–20
    biogenic amines as, 216–17
    chemical synapses and actions of, 202–10
    neuroactive peptides, 217–19
    other, 219
    select members of classes of, 215*t*
    varicosities and release of, 320–21, 322*f*
Neutralization as antibody function, **723**, 724*f*
Neutrons, 23
Neutrophils, **709**, 710*t*
    phagocytosis by, 715, 716*f*
Nicotinamide adenine dinucleotide (NAD+),
    32, 73
Nicotinic cholinergic receptors, **216**, 317, 318*f*,
    319*t*
Nitric oxide, 214, **219**
    as vasoactive sustance, 460, 461*t*
Nitric oxide synthetase, **219**
Nitrogen, 23
Nitrogenous bases in nucleic acids, 32, 33*f*
    in DNA, protein synthesis and, 49
    law of complementary pairing of, 32, 34*f*
NK (natural killer) cells, **711**, 717, 724*f*
Nociceptors, 245, 246*f*, **271**
    pain response and activation of, 274
    in skin, 272
Nodes of Ranvier, **177**
Nonassociative learning, **255**
Non-insulin dependent diabetes mellitus
    (NIDDM), 657*b*
Nonmembranous organelles, structure of,
    40–43, 44*t*
Nonpolar (hydrophobic) bonds, 24*b*, 31*b*

Nonspecific defenses, **713**–18, 736*f*
    complement system, 717–18
    inflammation, 713–17
    interferons, 717
    natural killer cells, 717
Norepinephrine (noradrenaline), 135, 141,
    142, 217, 218
    as autonomic system neurotransmitter,
        317
    binding of, to adrenergic receptors, 317–18,
        319*f*
Normovolemia, 567
Nuclear envelope, 34, **37**
Nuclear pores, **37**
Nucleases, 611
Nucleic acid(s), **32**
    DNA (*see* Deoxyribonucleic acid (DNA))
    RNA (*see* Ribonucleic acid (RNA))
    structure of, 34*f*
Nucleolus, 37, 44*t*
Nucleotide(s), 22, **32**
    structure of, 33*f*
Nucleotide diphosphate, 32, 33*f*
Nucleotide monophosphate, 32, 33*f*
Nucleotide triphosphate, 32, 33*f*
Nucleus (atom), 23
Nucleus (cell), **34**, 44*t*
    structure of, 37
Nucleus (neuron cell body), **175**
Nucleus raphe magnus, 276
Null cells, **711**
Nutrients
    entry of, into blood stream, 634*f*
    essential, 96, 655
    liver and metabolic processing of, 612
Nutrition, 640. *See also* Energy metabolism
Nutritional anemias, 510*b*

Obesity, diabetes and, 658*b*
Obligatory water loss, **572**, 576*b*
Obstructive pulmonary diseases, 487, 489,
    490*b*
Occipital lobe, cerebrum, **238**, 239*f*
Occludins, 43
Occlusive stroke, 230*b*
Oculomotor nerve (cranial nerve III), 315
Odorants, **306**
Ohm's law (I = E/R), **177**
Olfaction, 306–7
    anatomy of olfactory system, 306
    neural pathway for, 307
    signal transduction in, 306–7
Olfactory binding proteins, **306**
Olfactory epithelium, **306**
Olfactory nerve (cranial nerve I), **307**
Olfactory receptors, 264, **306**
Olfactory tract, 307
Olfactory tubercle, 307
Oligodendrocytes, **176**–77
Oncotic pressure, **436**
Oogenesis, **687**
Oogonia, **687**
Opponent process theory, 289*b*
Opsin, 286
Opsonins, **715**
Opsonization, 724*f*
Optic chiasm, **290**–91

Optic disk, **278**, 279*f*
Optic nerve, **288**
Optic radiations, **291**
Optic tract, **291**
Organ(s), **5**
    blood flow to, 418*f*, 419 (*see also* Blood flow;
        Blood supply)
    cardiovascular function and regulation of
        blood flow distribution to, 455–61
    effects of changes in central venous
        pressure on blood flow to, 430–32
    mean arterial pressure, blood flow, and
        resistance of, 427, 428*f*
    of respiratory system, 470–75
    summary of factors influencing blood flow
        to, 464*f*, 465
Organelles, **34**
    structure of membranous, 37–40
    structure of nonmembranous, 40–43
Organ of corti, anatomy of, 295*f*, **296**
Organ systems, **5**
Organ transplantation and immune response,
    730–32
Orgasm, male, 683
Origin, skeletal muscle, **357**, 358*f*
Orthostatic hypotension, 432
Osmolarity, **121**, 122*f*
    tonicity versus, 122*f*, 123*t*
    water balance and, 567–69
Osmoreceptors, **574**, 575*f*
    in digestive tract, 622
Osmosis, **119**–23
    osmolarity and, 121, 122*f*, 123*t*
    tonicity and, 121–23
Osmotic equilibrium, 567
Osmotic pressure, 120*b*, **121**
Osmotic pressure gradient, movement of fluids
    in and out of capillaries and, 434, 436
Ossicles, **291**, 292*f*, 293
Osteoblasts, **658**, 659*f*
Osteoclasts, **658**, 659*f*
Osteocyte, **658**
Osteoid, **657**–58
Otoliths, **300**
Output, homeostasis and, **11**
Ova (ovum), **671**
    development of, 687
    fertilization of (*see* Fertilization)
    limited number of, 683
Oval window (ear), **291**, 292*f*
Ovarian cycle, **688**–89
Ovaries, 673, **683**–85
Oviducts, 686
Ovulation, 11, **683**
Oxidation, **64**
Oxidation-reduction reactions, 64–65
Oxidative muscle fibers, **354**, 355*f*
    myoglobin in, 354
    slow, and fast, 355–57
Oxidative phosphorylation, 65, **84**–89
    chemiosmotic coupling, 86*f*, 87
    electron transport chain and, 84–86
    summary of, 87–89
Oxygen, 23
    in air, 498
    altitude and blood levels of, 525*b*
    in brain, 226

carbon dioxide transport and effect of, 513, 514f
conversion of pyruvate to lactic acid and, 89–91
gas exchange of carbon dioxide and, 500–505
movements of carbon dioxide and, in pulmonary and systemic tissues during rest, 496f, 497
role of, in oxidative phosphorylation, 84–89
solubility of, in water, 499, 500f
Oxygenated blood, 375, 376f
Oxygen transport in blood, 505–9
addition factors affecting affinity of hemoglobin for oxygen, 508–9
anemia as decreased capacity for, 507, 510–11b
hemoglobin as carrier of, 505, 506f, 507
hemoglobin-oxygen dissociation curve, 507, 508f
Oxyhemoglobin, 505
formation of, 506f, 507
Oxytocin, 155, **219**, 701, 703

Pacemaker cells, heart, **363**, **377**
control of heartbeat by, 378
electrical activity in, 379–82
Pacemaker potentials, **363**, **380**
Pacinian corpuscles, 264, 272
Pain perception, 274–77
brain and sensations of, 239
inflammation, bradykinin, and, 714
insensitivity to, 275b
modulation of pain signals and, 219, 236, 275–77, 278f
pain response and, 274–75
visceral, and referred pain, 275, 276f
Pancreas, **159**–60, 161f, 610, 611–12
Pancreatic amylase, **611**
digestion of starch by, 614f
Pancreatic duct, **611**
Pancreatic islets, 611
Pancreatic juice, **608**, 610
secretion of, 626–27
Pancreatic lipase, **611**
action of, 618, 619f
Papillae of renal pyramids, 535
Papillary muscles, 372
Para-aminohippuric acid (PAH), clearance of, 556, 557f
Paracellular movement, 45
Paracellular spaces, 124
Paracellular transport, 551
Paracrines, 133f, **134**
Paradoxical sleep, 253
Parallel flow, blood, 417, 418f, 419
Parallel processing, **248**
in visual system, **291**
Parasympathetic nervous system, 171f, 172, 312
anatomy of, 315–17
control of cardiovascular function and input from, 446
control of heart rate by, 394, 395t
neural pathway of, 313f
Parathyroid glands, 154f, **158**, 159f

Parathyroid hormone (PTH), 158, **581**
plasma calcium concentrations and effects of, 581, 582f
Paraventricular nucleus, 155
Parietal cells, 606f, **607**, 616f, 625
Parietal lobe, cerebrum, **238**, 239f
Parkinson's disease, 250–51
Parotid glands, **610**
Partial pressure of gases, **498**–99
alveolar oxygen and carbon dioxide, 500, 501f, 502, 503f, 504–5, 520f
binding of oxygen to hemoglobin and, 506
Dalton's Law and, 498b
gas exchange in respiratory system along gradient of, 500, 501f
high altitude and, 525b
of oxygen and carbon dioxide in air, alveolar air, and various body sites, 501f, 503b
Parturition, **673**
events of, 700f, 701
Passive immunity, **729**–30
Passive transport, **102**
active transport versus, 101–2
factors affecting membrane permeability in facilitated diffusion and, 114–15
factors affecting rates of, 108–10
osmosis of water as, 119–23
role of transport proteins in facilitated diffusion and, 113–14
Pathogenic agents, 709
Pathways. See Metabolic pathway(s); Neural pathways
Pavlov, Ivan, 245
Penis, **678**
Pepsin, 72, **607**
activation and activity of, in stomach, 615, 616f
Pepsinogen, **607**, 615–16
secretion of, in stomach, 625–26
Peptic ulcer disease, 622b
Peptide(s), **28**, 29f
as chemical messengers, 135–36, 138–39
synthesis and release of, 138–39
Peptide bond, 28, 29f
Percent saturation of enzymes, **74**
Perception, **261**. See also Sensory systems
fallibility of, 262f
of pain (see Pain perception)
Perforating ulcer, 622b
Perforins, **727**
Perfusion, local regulation of ventilation and, 521–22, 523f
Perfusion pressure, changes in myogenic response to changes in, 458, 460f
Periaqueductal gray matter, 276
Pericardium, 371
Perilymph, **296**
Perimetrium, 686
Peripheral axon, 174
Peripheral chemoreceptors, **518**–19
Peripheral lymphoid tissues, **711**
Peripheral membrane proteins, **36**
Peripheral nervous system, **171**
functional classes of neurons in, 175f
input of, into respiratory centers, 517–18
Peripheral thermoreceptors, 13

Peripheral veins, 416
Peristalsis, **630**, 631
segmentation compared to, 632f
Peritoneal dialysis, 550b
Peritoneum, 534, **605**
Peritonitis, 609, 622
Peritubular capillaries, **537**, 539f, 546
Peritubular fluid, 546
Peritubular space, 546
Permeability of membranes, 7, **109**, 110
factors affecting, in facilitated diffusion, 114–15
factors affecting, in simple diffusion, 112
Fick's law and, 110, 112b, 423
resistance and, 423b
Permissiveness, hormones and process of, **167**
Pernicious anemia, 406, 510b, 620
Peroxisomes, **39**–40, 44t
Pertussis (whooping cough), 152–53b
Peyer's patches, 604, 711
pH, 75b
affinity of hemoglobin for oxygen and effects of, 509
effects of changes in, on ventilation, 520f
exercise and, of blood and muscle, 754f
of plasma, renal regulation of, 533, 592–93
Phagocytic vesicle, 47
Phagocytosis, 46, **47**, 713, 714f
steps of, 715, 716f
Phagolysosome, 47
Phagosome, 47, 716
Pharynx (throat), 291, **470**, 602, 604f, 606f, **607**
Phasic (rapidly adapting) receptors, **264**
Phenoxybenzamine, 142
Phenylephrine, 142
Phenylethanolamine N-methyl transferase (PNMT), 138
Phosphatase, **76**
Phosphate, renal regulation of levels of, 533
Phosphate bond, 64
Phosphatidylinositol 4, 5-biphosphate (PIP2), 149
Phosphatidyl inositol second messenger system, **147**, 148f
Phosphodiesterase, 286
Phospholipase A_2, 139
Phospholipase C, 149
Phospholipid(s), **27**
Phospholipid bilayer, 27f, 35, 36f
Phosphoprotein phosphatases, 147
Phosphorylation, **64**
oxidative, 84–89
substrate-level, 84
Photopigment, **286**
Photoreceptors, 262, **278**, 284f, **285**
absorbance spectra for different, 287f
morphology of, 285f
Phototransduction, 278, **285**–87
Phrenic nerve, **514**
Physiological integration during exercise, 742–61
decline to end of exercise and, 753–55
increased physiological function and, 742
metabolic demands of exercise and, 742
post-exercise consequences and, 755–58
steady state conditions and, 746–53
transition from rest to exercise and, 742–46

Physiology, 1–20
    defined, **2**
    exercise, 15–18 (*see also* Exercise;
        Physiological integration during exercise)
    homeostasis as central principle of, 8–14
    organization of human body and, 2–8
Pia mater, **224**, 225*f*
Pineal gland, 154*f*, **158**, 241
Pinna, 291, 292*f*
Pinocytosis, 46, **47**
Pituitary gland, **153**–57
    connection between hypothalamus and,
        155*f*, 156*f*
    positive feedback systems and, 11
Placenta, 696, **697**, 698*f*
    as endocrine gland, 160
Placental lactogen, **700**
Plaque
    atherosclerotic, 621*b*
    desmosome, 45
Plasma, 7*f*, **8**, 404
    cholesterol in, 621*b*
    composition of, 226*t*, 404, 565, 566*f*
    glucose levels and renal function, 534,
        547–48, 556, 557*f* (*see also* Blood
        glucose)
    renal regulation of pH of, 533, 592–93
    renal regulation of volume and composition
        of, 533–34, 560*f*
Plasma cells (B lymphocytes), 711, 722
Plasma membrane, 28, **34**, 44*t*
    carbohydrates in, 36–37
    permeability of (*see* Membrane
        permeability)
    proteins in, 36
    structure of, 34–37
    transport across (*see* Cell membrane
        transport)
Plasma osmolarity, effects of low, 567, 568
Plasmapheresis, 328
Plasma proteins, liver and synthesis of, 612–13
Plasma volume, mean arterial pressure and,
    565
Plasmin, **410**
Plasminogen, 410
Plasminogen activators, 410
Plasticity, **256**
Platelet adhesion, 407
Platelet aggregation, 408
Platelet factor 3, **409**
Platelet plug, 407, 408*f*
Platelets, **371**, 407–11
    blood clot formation and, 408–11
    plug created by, 407, 408*f*
    vascular spasm and response of, 407
Pleura, **475**, 477*f*
Pleural sac, **475**, 477*f*
Pneumothorax, **479**
Podocytes, 540
Poikilothermic animals, 12
Poiseuille's Law, 424*b*
Poisons, 185*b*
Polar (hydrophilic) bonds, 31*b*
Polar molecules and hydrogen bonds, 24*b*
Polycystic kidney disease, 587*b*
Polycythemia, 424, 525*b*
Polymers, 22

Polymodal nociceptors, **272**
Polypeptides, **28**, 29*f*
    synthesis of, 51–52 (*see also* Protein synthesis)
Polyribosome, 52
Polysaccharides, **22**, 25*f*
Polyspermy, **695**
Polysynaptic reflexes, 244
Polyunsaturated fatty acid, 26
Pons, 236*f*, **237**
    respiratory control centers of, 516
    role of, in regulating autonomic function,
        322, 323*f*
Pontine respiratory group, **516**
Population coding, sensory stimuli, 266, 268*f*
Portal circulation, 419
Positive balance, 566
Positive cooperativity, 507
Positive feedback, **11**
Postabsorptive state, **645**
    energy metabolism in, 647–49
    metabolic reactions of, 648*f*
    regulation of metabolism during, 649–54
Posterior lobe, pituitary gland
    (neurohypophysis), **155**
Post-exercise consequences, 755–58
    delayed-onset muscle soreness, 757
    fever, 756–57
    susceptibility to sickness, 757–58
    syncope and autonomic system veto, 755
    water intoxication, 755–56
Postganglionic neurons, **313**, 314*f*, 315*f*, 393,
    446
Postmitotic muscle fibers, 357
Postnatal growth spurt, 654, 655*f*
Postsynaptic cell, **134**
Postsynaptic neuron, **202**
    fast and slow responses of, 204, 205*f*
Postsynaptic potential (PSP), **204**
Post-transcriptional processing, 50*f*
Post-translational processing, **53**
Posture, control of, by brainstem, 249–50
Potassium (K+)
    aldosterone secretion linked to levels of,
        163–64, 165*f*, 580
    electrical activity in cardiac muscles and role
        of, 380, 381*f*
    homeostasis and balance of, 579–80
    resting membrane potential and, 178–80
Potassium ion channels
    action potentials and, 187–88, 189*f*, 190,
        191*f*
    inhibitory post-synaptic potentials and, 209*f*
Potential difference (E), 177*t*
Potential energy, **66**, 67*f*
Potentiation, **626**
Preganglionic neurons, **313**, 314*f*, 315*f*, 393, 446
Pregnancy, **673**, 695–700
    hormonal changes during, 698–700
Preload (end-diastolic pressure), **397**–98
Presbyopia, 282
Pressure drop, 421
Pressure gradients, **420**
    in cardiovascular system, 421–22, 426*t*
    movement of fluids into and out of
        capillaries and, 434–36
    relating resistance to, in systemic
        circulation, 426

Pressure reservoirs, arteries as, 390, 411–12
Presynaptic cell, **134**
Presynaptic facilitation, **213**, 214*f*
Presynaptic inhibition, **213**, 214*f*
Presynaptic modulation, **212**–14
Presynaptic neuron, **202**
Primary active transport, **115**
Primary endocrine organs, **152**, 153–60,
    162–63*t*
Primary (essential) hypertension, 450*b*
Primary immune response, **720**, 722*f*
Primary motor cortex, **239**, 240*f*
    voluntary movement and, 248, 249*f*
Primary oocytes, 687
Primary protein structure, 28, 30*f*
Primary secretion, acinar cells, 610
Primary somatosensory cortex, **239**, 240*f*, 241*f*,
    273
Primary spermatocyte, 682
Primordial follicle, 684*f*, **685**
Principal cells, renal tubules, 566
    effects of aldosterone on, 577*f*
Procedural memory, **256**
Products, chemical reaction, **63**
    reactant and concentrations of, 71
Progesterone, 157, 160, **673**, 691
    actions of, in female reproduction, 693*t*
    during pregnancy, 698–700
Programmed cell death, 721
Projection fibers, 229*f*, **230**
Prolactin, **699**
Prolactin inhibiting factor (PIF), 703
Prolactin inhibiting hormone (PIH), 157
Prolactin releasing factor (PRF), 703
Prolactin releasing hormone (PRH), 157
Prolapse of AV valve cusps, 372
Proliferative phase, uterine cycle, **690**
Prometaphase (mitosis), 58*f*
Promoter sequence, **50**
Prophase (mitosis), 58*f*
Proprioception, 239, **262**, 271
Prostacyclins, 139, **408**
    as vasoactive sustance, 460, 461*t*
Prostaglandins, 28, 139, 272
Prostate gland, 677*f*, **679**
Proteases, 56
    activation of, in small intestine, 617*f*
    in saliva, 611
Protein(s), 22, **28**–32
    absorption, utilization, and storage of
        energy in, 642*f*, 643*t*
    degradation of, 56
    destination of, following synthesis, 52–53
    digestion and absorption of, 615–16
    energy metabolism of, during absorptive
        and postabsorptive states, 649*t*
    fibrous versus globular, 31, 32*f*
    G (*see* G protein(s))
    metabolism of, 95–96
    osmotic pressure exerted by, 436
    in plasma membrane (*see* Membrane
        proteins)
    post-translational processing and packaging
        of, 53
    structure of, 28, 30*f*
    transport of exchangeable, across capillary
        walls, 414, 416*f*

Protein C, **410**
Protein kinase, **76**
Protein kinase A, 146, 573
Protein kinase C, 149
Protein kinase G, 147
Protein synthesis, 48–56, 138–39
  destination of proteins following, 52–53
  post-translational protein processing and packaging, 53
  regulation of, 53–56
  ribosomes and, 40, 51–52
  role of genetic code in, 49
  summary of, 55f
  transcription in, 49–50
  translation in, 51–52
Proteinuria, 586b
Proteolysis, **95**–96
Proteosome, 56
Protons, 23, 75b
Proto-oncogenes, 57, 59b
Proximal convoluted tubule, **536**
  reabsorption of solutes in, 545–47
Proximal straight tubule, **536**
Proximal tubule, **536**. See also Proximal convoluted tubule
  nonregulated reabsorption in, 550–51, 552f
  renal handling of hydrogen and bicarbonate ions in, 591
  sodium reabsorption in, 549f, 568f
  solutes reabsorption in, 545–47
  water reabsorption in, 568f, 569
Pseudo-unipolar neurons, 174f
P site, ribosome, 51, 52f
Pubertal growth spurt, 654
Puberty, **674**
Pudendum, 686
Pulmonary arteries, **376**
  mixed venous blood in, 503
Pulmonary capillaries
  carbon dioxide exchange and transport in, 512f, 513
  gas exchange as function of length of, 502, 503f
Pulmonary circuit, **374**, 376f
  blood flow through, 417, 418f
  pressure gradients across, 422
Pulmonary circulation, 495–531
  acid-base homeostasis and, 522–26
  central regulation of ventilation and, 514–18
  chemoreceptor control of ventilation and, 518–20
  exchange of oxygen and carbon dioxide and, 500–505, 526, 527f
  gas diffusion and, 497–99
  local regulation of ventilation and perfusion and, 521–22, 523f
  overview of, 496–97
  transport of gases in blood and, 505–14
Pulmonary edema, 108, 502–3, 504b
  heart failure and, 434b
Pulmonary epithelium, 108
Pulmonary function tests, 489
Pulmonary semilunar valve (pulmonary valve), **373**
Pulmonary stretch receptors, 517–18

Pulmonary surfactant, **484**, 486–87b
  lung compliance and, 485f
Pulmonary trunk, 372, 373f, 376
Pulmonary veins, carbon dioxide exchange and transport in, 512f, 513
Pulmonary ventilation, **470**
  acid-base homeostasis and, 522–26
  central regulation of, 514–18
  control of, by chemoreceptors, 518–20
  exercise and response of, 744–45
  factors affecting, 484–87
  forces for, 475–84
  local regulation of perfusion and, 521–22, 523f
  mechanics of breathing and, 479–84
  pulmonary pressures affecting, 477–79
  respiratory system anatomy and, 470–75
  respiratory volumes, air flows, and, 488–91
  summary of factors affecting minute alveolar ventilation and, 492f
Pulse pressure, 389b
Pump(s), **102**, 115
  leaks and, 118–19
  primary active transport by sodium potassium, 116f
  transport rate of, 110
Pump blockers, 622b
Pump cycle, cardiac, **386**–88
Pupil (eye), **278**, 279f
  regulating light exposure by varying size of, 282–84
Pupillary light reflex, 247, 322
Purines, 32, 33f, 733
Purkinje cell, 3f, 378
Putamen, 240, 242f
P wave, **384**, 385
Pyloric sphincter, 606f, **607**
Pylorus of stomach, 606f, **607**
Pyramidal cell, 3f
Pyramidal tracts, spinal cord, 232, 235f
  voluntary movement and role of, **248**, 249f
Pyrimidines, 32, 33f
Pyrogens, 13
Pyruvate
  conversion of, to lactic acid, 89–91
  glycolysis and formation of, 81
Pyruvic acid, 81

QRS complex, **384**
Quantum, **206**
Quaternary protein structure, 28, 30f

Radial muscle (eye iris), **284**
Radiation, **12**
Rapidly adapting (phasic) receptors, **264**
Rate of cell membrane transport, 108–10
Rate of change, muscle shortening, 352–53
Rate of chemical reactions, **70**–71
Reabsorption, renal, 7, **538**, 545–48
  barriers to, 546f
  mechanism of, 546, 547f
  of sodium, in renal tubule, 549–50
  of solutes, 545–47
  transport maximum, 547–48
  of water, 545, 547, 569–75
Reactants, **63**
  product concentrations and, 71

Reactions. See Chemical reactions; Metabolic reactions
Reactive hyperemia, **458**, 459f
Receptive aphasia, 251
Receptive fields of afferent neurons, **264**, 265f, 268f, 288f
Receptive relaxation, 630
Receptor(s), 48, **133**
  agonists and antagonists, 142
  of autonomic nervous system, 317–20
  channel-linked, 143–45
  in digestive tract, 622–23, 624f
  enzyme-linked, 145
  globular proteins as, 31, 47
  G-protein linked, 145–49
  properties of, 141–42
  sensory, 175 (see also Sensory receptors)
  specificity of, 141
  visual, 175
Receptor binding, target cell response related to, 141–42
Receptor-mediated endocytosis, 46, **47**
Receptor-mediated responses
  signal transduction mechanisms for intracellular, 142–43
  signal transduction mechanisms for membrane-bound, 143–49
Receptor potentials, **262**
Recruitment, **351**
  size principle and, 351, 352f
Rectum, 602, 604f, 610
Red (oxygenated) blood, 375, 376f
Red blood cells. See Erythrocytes
Red muscle, 354
Red nuclei, 249
Reduction, **64**
Redundancy versus parallel processing, 248
Referred pain, **275**, 276f
Reflex, 243, **244**–47
  chewing, 629
  classes of, 244t
  colonocolonic, 632
  defecation, 632
  digestive tract, 623, 624f
  gastrocolic, 632
  gastroileal, 632
  ileogastric, 632
  intestino-intestinal, 632
  micturition, 558, 559f
  pupillary light, 247, 284, 322
  stretch, 245
  swallowing, 322–23, 629, 630f
  visceral, 244, **321**–23
  withdrawal and crossed-extensor, 245–47
Reflex arc, 244, **245**
Reflexion of light, **279**
Refraction of light, **279**, 280f, 281f
Refractory periods
  action potentials and, **192**, 193f, 194
  in atrioventricular nodes, 378
Regenerative mechanism, **190**
Regulated variable, **9**
  negative feedback control of, 10f
Regulation
  of autonomic nervous system function, 321–23
  of digestive tract, 620–24

Regulation, *continued*
    of energy metabolism during absorptive and
        postabsorptive states, 649–54
    of enzyme activity, 75–77
    of female reproductive function, 693–94
    of glomerular filtration rate, 542–44
    of male reproductive function, 679–80
    of protein synthesis, 53–56
    of renal reabsorption and secretion, 551–52
    of ventilation, 514–22
Regulatory proteins in muscle filaments,
    336–37
Regulatory site, enzyme, **75**
Relative refractory period, **193**
Relaxation phase, muscle twitch, **344**
REM sleep, **252**, 253
Renal anemia, 510*b*
Renal arteries, **534**, 537, 539*f*
Renal compensation as defense mechanism
    against acid-base disturbances, 588, 590–92,
    593*f*, 594*f*
Renal corpuscle, 535
    anatomy of, 536, 540*f*
    Starling forces in, 540–41
Renal exchange processes, 538–48
    dialysis as treatment for failed, 550–51*b*, 587*b*
    excretion, 553–59
    glomerular filtration, 536, 538–45
    reabsorption, 545–48
    renal tubule specialization and, 549–53
    schematic representation of four basic, 553*f*
    secretion, 548
Renal failure, 533, 586–87*b*
    dialysis for, 550–51*b*, 587*b*
Renal function
    regulation of plasma volume and
        composition by, 533–34, 560*f*
    removal of metabolic wastes by, 533
Renal hilus, 534
Renal hypertension, 450*b*
Renal pelvis, **535**
Renal pyramids, 535
Renal threshold, **547**
Renal tubules, **535**, 536–37, 549–53
    abnormalities of, 587*b*
    anatomy of, 536–37
    clearance measurements and determination
        of fate of solutes in, 556, 557*f*
    distal (*see* Distal convuluted tubule)
    epithelia of, 551, 552*f*
    homeostasis and transport across walls of,
        565
    mechanisms of sodium reabsorption in,
        549–50
    nonregulated reabsorption in proximal,
        550–51, 552*f*
    potassium transport in, 579, 580*f*
    proximal (*see* Proximal tubule)
    regulated reabsorption and secretion in
        distal tubule and collecting duct, 551–52
    three exchange processes in, 538, 539*f*
    water conservation in the loop of Henle,
        552, 572
Renal veins, **534**, 538, 539*f*
Renin, 533, 537
    release of, in response to drop in arterial
        pressure, 452

Renin-angiotensin-aldosterone (RAAS) system,
    576–78
Replication of DNA, **56**, 57*f*
Repolarization of membrane potentials, **183**
    action potentials and, 188, 189*f*
Reproductive organs
    development of, during embryonic life, 675*f*
    functional anatomy of female, 683–87
    functional anatomy of male, 676–79
Reproductive system, 670–707
    components of, 673
    female, 683–94
    fertilization, implantation, and pregnancy,
        695–70
    male, 676–83
    overview of physiology of, 671–76
    parturition and lactation, 701–3
    systems integration chart, 704*f*
Reproductive tract, **673**
    female, 684*f*, 685–86
    male, 678–79
Resistance (R), **177**
    permeability and, 423*b*
    Poiseuille's Law on, 424*b*
Resistance in cardiovascular system, 422–26
    in individual blood vessels, 422–24
    local factors controlling vascular, 456–60
    in networks of blood vessels (total
        peripheral resistance), 424–26
    physical laws applied to, 426*t*
    relating pressure gradients and, to systemic
        circulation, 426
Resistance in renal vascular system, 544*f*
Resistance in respiratory tract, 484–87
    factors affecting air flow due to, 521, 523*f*
Resistance of organs, blood flow, blood
    distribution, and, 427, 428*f*
Resistance vessels, **426**
Resorption of bone tissue, **658**
Resource allocation, activity and hierarchy of,
    15–16
Respiration, **470**
    chemoreceptors and control of, 462
    internal, and external, 470, 471*f*
Respiration rate, **490**–91
    effect of tidal volume and, on minute
        alveolar ventilation, 491*t*
Respiratory acidosis, **526**, 593
Respiratory alkalosis, 525*b*, **526**, 593
Respiratory bronchioles, **475**, 476*f*
Respiratory compensation as defense
    mechanism against acid-base disturbances,
    588, 590, 594*f*
Respiratory membrane, **475**, 476*f*, 497*f*
Respiratory pump, 416
    central venous pressure, blood flow to
        organs, and, 430
Respiratory quotient, **496**–97
Respiratory sinus arrhythmia, as cardiovascular
    regulatory process, 462
Respiratory system
    acid-base homeostasis and, 522–26, 585
    anatomy of, 470–75
    circulation and gas exchange in (*see*
        Pulmonary circulation)
    exercise and response of, 744–45
    function of, overview of, 470, 471*f*

systems integration chart for, 528*f*
terminology used in physiology of, 505*t*
ventilation in (*see* Pulmonary ventilation)
Respiratory tract, **470**–75
    anatomy of, 471, 472*f*
    conducting zone of, 471–74
    cystic fibrosis and, 127*b*
    respiratory zone of, 471, 473*f*, 474–75, 476*f*
    thoracic cavity structures and, 475
Respiratory zone of respiratory tract, **471**
    anatomical features of, 473*f*
    anatomy of, 476*f*
Resting membrane potential (Vm), 177*t*,
    178–80
    in cardiac muscle, 383
    determinants of, 180, 181*f*
    GHK equation and, 182*b*
    neurons at rest and, 181–82
    production of, 178–80
Reticular formation, **237**, 249
Retina, **278**, 279*f*, 285
    anatomy of, 284*f*
    diabetes and degeneration of, 658*b*
    distribution of rods and cones in, 285*f*
    neural processing in, 288–90
Retinal, 286
Retrograde transport, 206
Reuptake, neurotransmitter, **203**
Reverse cholesterol transport, 621*b*
Rheumatoid arthritis, 733
Rhodopsin, 286
Rhythm method of contraception, 694*b*
Ribonucleic acid (RNA), **32**, 34*f*
    DNA transcription and formation of, 49–50
    messenger (mRNA), 48
    ribosomal (rRNA), 37
    transfer (tRNA), 49, 51*f*
Ribose, 22
Ribosomal RNA (rRNA), 37
    DNA transcription and formation of, 49–50
Ribosomes, **38**, 44*t*
    protein synthesis and, 40, 51–52
    structure of, 40
Right heart, **371**, 374
RNA. *See* Ribonucleic acid (RNA)
RNA polymerase, 50
Rods (eye), 278, 284*f*, **285**
    characteristics of, 288*t*
    components of, 286*f*
    distribution of, in retina, 285*f*
Rotational acceleration, 299
Rough endoplasmic reticulum, 38, 44*t*
Round window, **291**, 292*f*
Ruffini's endings, 272
RU 486, 694*b*
Rugae, 606*f*, **607**
Ryanodine receptors, 342

Saccule, 300–303
    functional anatomy of, 302*f*
Sacral nerves, **231**, 232*f*
Saliva, **607**, 610
    components of, 611
    regulation of secretion of, 624–25
Salivary amylase, **607**, 611
    digestion of starch by, 614*f*
Salivary center in medulla, 625

Salivary glands, **610**–11
    location of major, 610*f*
Salivation reflex, 245
Saltatory conduction, **196**
Sarcolemma, **334**, 335*f*
    motor end plate of, 341–42
Sarcomere, **336**
    contractile component of, 346
    contraction and relaxation in, 338*f*
    structure of, 337*f*
Sarcoplasmic reticulum, **334**, 335*f*
Satellite cells, 357
Satiety signals, 624
Saturated fatty acids, **26**
Saturation of hemoglobin, 505, 507*f*
Saxitoxin (STX), 185*b*
Scala media (cochlear duct), 295*f*, **296**
Scala tympani (tympanic duct), 295*f*, **296**
Scala vestibuli (vestibular duct), 295*f*, **296**
Schwann cells, **176**–77
Sclera, **278**, 279*f*
Scrotum, **678**
Seawater, effects of drinking, 576*b*
Secondary active transport, **115**, 117–18
Secondary bronchi, **471**, 472*f*
Secondary endocrine organs, **152**, 154*f*, 160, 163*t*
Secondary hypertension, 450*b*
Secondary immune response, **720**, 722*f*
Secondary lysosome, 716
Secondary oocyte, 687
Secondary polar body, 687
Secondary protein structure, 28, 30*f*
Secondary sex characteristics, **674**
Secondary spermatocytes, 682
Second messenger, **145**
    G-protein-regulated enzymes and production of, 146, 147*f*
    systems of, 147*t*, 148*f*, 205*f*
Second order neurons, **265**
Secretin, **623**, 626, 627*f*
Secretion, 6, 7, **124**, 548
    from cytosol, **37**
Secretion, digestive tract, **602**, 603*f*
    regulation of, 624–28
Secretion, renal, **538**
Secretory phase, uterine cycle, **690**–91
Secretory vesicles, **37**
Segmental arteries, renal, 537, 539*f*
Segmentation, **631**
    peristalsis compared to, 632*f*
Selectins, 715
Selective permeability of cell membranes and epithelial tissues, 7
Self markers, MHC molecules as, 725
Self-tolerance, immune response and, 719, 721
Semen, **683**
Semicircular canals, 292
    functional anatomy of, 301*f*
    transduction of rotation and, 300
Semiconservative DNA replication, 56, 57*f*
Semilunar valves, **372**–73
    operation of, 375*f*
Seminal vesicles, 677*f*, **679**
Seminiferous tubules, **677**, 678*f*
Sensitization, **255**
Sensors, **11**

Sensory coding, 265–70
    for stimulus intensity and duration, 266–68
    for stimulus location, 268–70, 271*f*
    for stimulus type, 266
Sensory division of peripheral nervous system, **171**
Sensory homunculi, **239**, 241*f*
Sensory pathways, 264–65, 266*f*
Sensory receptors, 175, 261, **262**–64
    adaptation of, 264
    cardiovascular function and input to medulla from, 445–46
    characteristics of, 263*t*
    in skin, 271*t*, 272*f*
    structure and function of, 263*f*
    transduction in, 262, 263*f*
Sensory systems, 260–310
    ear and equilibrium, 299–303
    ear and hearing, 291–98
    general principles of, 261–70
    olfaction, 306–7
    somatosensory system (body sensation), 270–77
    taste, 303–5
    vision, 277–91
Sensory transduction, **262**
Sensory unit, **264**, 265*f*
Separation of charge, 103*f*
Series elastic component, **346**
Serosa, **602**, 605–6
Serotonin, 135, 217, 218
    vasoconstriction and, 407–8
Serotonin reuptake inhibitors (SSRIs), 218*b*
Sertoli cells, **677**
Serum, **411**
Set point, **11**
Severe combined immunodeficiency disease (SCID), 733
    gene therapy for, 734*b*
Sex chromosomes, **671**
    meiosis, gametogenesis, and, 672–73
Sex determination, 673, **674**, 676*f*
Sex differences in energy metabolism during exercise, 746, 747*f*
Sex differentiation, **674**, 675*f*, 676*f*
Sex hormones, **159**, 160, 673. *See also* Androgens; Estrogens; Progesterone; Testosterone
    effects of, on growth, 661
    female, 693, 698–700
    male, 679–80
Sexual reproduction, 671. *See also* Reproductive system
    gene sorting and packaging in gametogenesis of, 672–73
    role of gametes in, 671–72
Sexual response
    female, 687
    male, 682–83, 685*b*
Shivering, 13
Short reflex pathway, **623**, 624*f*
Short-term memory, 256
Shunts, 413
Sickle cell anemia, 511*b*
Sigmoid colon, 609
Signal amplification by chemical messenger systems, 149, 150*f*
Signal recognition proteins, 52–53

Signal transduction mechanisms, **133**, 140–49, 202
    at chemical synapses, 202, 204–5
    at cholinergic receptors, 216*f*
    for intracellular receptor-mediated responses, 142–43
    for membrane-bound receptor-mediated responses, 143–49
    at neuromuscular junction, 326, 327*f*
    for olfaction, 306–7
    phototransduction, 285–87
    properties of chemical messengers and, 150*t*
    receptor properties and, 141–42
    of rotation and acceleration, 300–303
    for select chemical messengers, 149*t*
    sensory transduction, 262
    for sound, 293–97
    sympathetic nervous system, 318, 319*f*
    for taste, 304*f*, 305
Simple diffusion, **101**
    basis for, 111
    characteristics of, 118*t*
    factors affecting membrane permeability in, 112
    permeability, resistance, and, 423*b*
Single-unit smooth muscle, **363**
Sinoatrial node (SA node), 363, **377**
    initiation of action potentials in, 377
Sinus arrhythmia, respiratory, as cardiovascular regulatory process, 462
Sinusoids, 612*f*, **613**
Sister chromatids, **672**
Size principle, **351**, 352*f*
Skeletal muscle pump, central venous pressure and blood flow to, 430, 431*f*
Skeletal muscles, **334**–60
    attachment of, to bones, 357–58
    bones as levers and contraction of, 358–60
    of digestive tract, 606
    energy metabolism in cells of, during absorptive state, 646
    energy metabolism in cells of, during postabsorptive state, 647–48
    inhibition of, 246
    long-term responses of, to exercise, 356–57
    mechanics of contraction in, 344–53
    mechanism of force generation in, 338–44
    micturition and, 558, 559*f*
    neurogenic nature of, 377
    neuromuscular junctions in, 361*f*
    regulation of force generation in whole, 350–52, 353*f*
    shivering and thermoregulation, 13
    somatic nervous system effects on, 323–27
    structure of, 334–38
    types of fibers in, 353–57
Skin
    as nonspecific defense, 713
    sensory receptors in, 271*t*, 272*f*
Skull, 224, 225*f*
Sleep, 251–54
    characteristics of slow wave and REM, 252*t*
    electrical activity during wakefulness and, 252–54
    functions of, 251–52
    sleep-wake cycles, 252, 253*f*
    stages of, 253, 254*f*
Sleep-wake cycles, 252, 253*f*

Sliding-filament model of muscle contraction, **338**
Slit diaphragms, 540
Slit pores, 540
Slow axonal transport, 206
Slow ligand-gated ion channels, **146**
Slowly adapting (tonic) receptors, **264**, 266
Slow oxidative fibers, **355**, 356t
Slow pain, **275**
Slow response in postsynaptic neurons, **204**, 205f
Slow-twitch muscle fibers, 345, **354**
Slow waves, gastrointestinal motility and, **628**, 629f
Slow-wave sleep (SWS), **252**
Small intestine, 602, 604f, **607**–9
    anatomy of, 608f
    circulatory route of materials absorbed in, 609f
    motility of, 631–32
Smell. *See* Olfaction
Smooth endoplasmic reticulum, 38, 44t
Smooth muscle, **360**–63
    arteriolar, 458–60
    in blood vessels, 411, 413f
    factors affecting vascular, related to cardiovascular regulation, 456–60
    gap junctions in, 361f
    in gastrointestinal tract, 628, 629f
    mechanism of excitation-contraction coupling in, 360–62
    neural regulation of contraction in, 362–63
    single-unit, and multi-unit, 363
Sodium (Na)
    action potentials and, 187–88, 189f
    electrical activity in cardiac muscles and role of, 380, 381f, 382
    homeostasis and balance of, 575–78
    reabsorption of, in renal tubule, 549–50, 568f
    resting membrane potential and, 178–80
    salty taste and presence of, 305
Sodium chloride (NaCl), 31b
Sodium ion channels
    action potentials and, 187–88, 189f
    action potentials and three conformations of, 189, 190f
    effect of neurotoxins on voltage-gated, 185b
Sodium-linked glucose transport, 117
Sodium potassium (Na+/K+) pump, 31, **116**–17, 124
    basis of neural stability and, 197, 198f
    renal handling of potassium by way of, 579, 580f
    renal reabsorption of sodium by way of, 549–50
    resting membrane potential and, 178, 180, 181f
Sodium-proton exchange, 117
Solute(s), 31b
    in blood plasma, 404, 405t
    cell membrane transport of, 111–19
    clearance measurements and determination of rate of solutes in renal tubules, 556, 557f
    clearance rate for common, 558t
    concentrations of select, in intracellular and extracellular fluids, 101t

directions of driving forces acting on, 107t
    epithelium transport of, 124–26
    homeostasis and balance in, 566, 575–81
    normal rates of glomerular filtration and reabsorption of water and, 545t
    reabsorption of, 545–47
Solution(s), 69b
    acidic, basic, neutral, and pH of, 75b
    concentration and mass of, 69b
    electrolytic, 31b
    iso-osmotic, hyperosmotic, and hypo-osmotic, 121
    isotonic, hypertonic, and hypotonic, 122
    osmotic pressure of, 120b
Solvent, 31b
Somatic nervous system, 171f, **172**, 312, 323–29
    anatomy of, 324, 326f
    faulty function of, 328b
    neuromuscular junctions and, 324–29
    properties of, 329t
Somatic reflexes, 244
Somatomedins, 157, 160, **655**
Somatosensory cortex, 239, 240f, 241t, 273
Somatosensory pathways, 273, 274f
Somatosensory receptors, 271–72
Somatosensory system, **262**, 270–77
    pathways, 273, 274f
    receptors, 271–72
    somatosensory cortex and, 273
Somatostatin, 159, 161f, **655**
Somatotropic organization, 239
Somesthetic sensations, **262**
Sound waves
    amplification of, in middle ear, 292–93
    coding for intensity and pitch of, 297, 299f
    conduction of, in ear, 297f
    nature of, 292, 293f, 294b
    neural pathways for, 297–98
    transmission of, in middle ear, 294f
Spatial summation, **186**, 187f, 211f, 212, 213f
Special sensory systems, **262**
    balance and equilibrium, 299–303
    hearing, 291–98
    olfaction, 306–7
    taste, 303–5
    vision, 277–91
Specific immune responses, **713**, 718–21, 736f. *See also* Cell-mediated immunity; Humoral immunity
Specificity
    of immune responses, 719–20
    of receptors for chemical messengers, **141**
Sperm, 671, 680–82
    anatomy of, 680f
    migration of, in vagina, 686
Spermatids, 682
Spermatogenesis, **681**f
Spermatogonia, **681**–82
Spermatozoa, **671**
Sphincter, **607**
Sphincter of Oddi, 612f, **613**
Sphygmomanometer, 425b
Spinal cord, **224**, 225f, 231–37
Spinal nerves, **231**, 232f, 233f
Spinal reflexes, 244, 245–46, 323
Spinal tap, 228, 231
Spindle fibers, 43

Spinothalamic tract, 273, 274f
Spirometer, **488**
Spirometry, 488f
Spontaneous pneumothorax, 479
srY gene, **674**
Stapes, 291, 292f
Starch, **22**
    digestion of, 614f
Starling curve, 396f–97
Starling effect, 396, 434b
Starling forces, 433, 434
    glomerular filtration and, 538, 540–41
Starling's Law of the Heart, **396**
Steady state conditions, 76, 746–50
    cardiovascular control and blood pressure in, 747–48
    fluid balance in, 752
    gastrointestinal tract in, 749–50
    hormonal control of energy metabolism in, 746–47
    respiration and effort of breathing in, 748–49
    thermoregulation in, 750–52
Stellate cell, 3f
Stereocilia, 295f, **296**
    linear acceleration and, 300
    role of, in sound transduction by hair cells, 298f
Steroids, **28**, 29f
    as chemical messengers, 136, 139, 140
    synthesis and release of, 139
Stimulating hormones, 155
Stomach, 602, 604f
    acid and pepsin secretion in, 625–26
    anatomy of, 606f, **607**
    pancreatic juice and bile in, 626–27
    ulcers, 622b
Stress
    exercise as (*see* Exercise; Exercise links)
    hormone secretions in response to, 162, 164f, 664, 665
    immune response and, 734–35
    sympathetic nervous system influence on metabolism and reaction to, 653–54
Stretch reflex, 245
Striated muscle, **334**–36
Stroke, 226, 230b, 450
Stroke volume (SV), cardiac, **391**b, 392
    influence of afterload on, 398
    influence of end-diastolic volume on, 396–98
    influence of ventricular contractility on, 395–96
    mean arterial pressure and, 427–29
    summary of factors affecting, 398f
Strong acid, 75b
Subarachnoid space, **224**, 225f
Subcortical nuclei, 230, 237, 239–43
    hypothalamus, 240–41
    limbic system, 241–43
    thalamus, 240
Subendothelial tissue, 407
Sublingual glands, **610**–11
Submandibular glands, **611**
Submucosa, **602**, 605
Submucosal (Meissner's) plexus, **605**
Substance P, **219**, 275, 276
Substantia nigra, 251

Substrate, **71**, 72*f*
    concentration of, 73, 74*f*
Substrate-level phosphorylation, **84**
Substrate specificity, **72**
Subthreshold stimulus, **190**
Sucrase, 614
Sucrose, 22, 25*f*, 614
    sweet taste and presence of, 305
Sulci (sulcus), **238**
Summation
    integration of graded potentials, 186,
        211–12
    muscle twitch, **347**, 348*f*
    spatial, **186**, 187*f*, 211*f*, 212, 213*f*
    temporal, **186**, 187*f*, 211–12
Supporting cells, **306**
Suprachiasmatic nucleus, 164
Supraoptic nucleus, 155
Suprathreshold stimulus, **190**
Surgical sterilization, 694*b*
Swallowing reflex, 322–23, **629**, 630*f*
Sweat, 12, 13, 122
Sweat glands, 12, 13
Sweating
    control of glomerular filtration rate and
        renal vascular resistance in response to
        excessive, 544*f*
    thermoregulation and, 240
Sympathetic chains, **314**, 315*f*
Sympathetic nervous system, 171*f*, 172, 312
    anatomy of, 314–15, 316*f*
    control of heart rate by, 394, 395*b*
    control of ventricular contractility by,
        395–96
    effects of, on metabolism, 653–54
    neural pathway of, 313*f*
Sympathetic trunks, **314**, 315*f*
Sympathetic vasoconstrictor nerves, 429
Synapse(s), 134, 173, 202–19
    chemical, 202-10 (*see also*
        Neurotransmitters)
    electrical, 202
    neuroeffector junctions as, 320–21
    presynaptic modulation and, 212–14
Synaptic cleft, **202**, 204*f*
Synaptic delay, **204**
Synaptic potential, 177*t*, 186
Synaptic signaling, **134**
Synaptic vesicles, **203**, 206*b*, 207*f*
Syncope, **755**
Synergistic effects of hormones, **166**
Systemic circuit, **374**, 376*f*
    blood flow through, 417, 418*f*, 419
    pressure gradients, resistance, and flow
        applied to, 426*t*
    pressure gradients across, 422
Systemic lupus erythematosus, 733
Systole, **386**, 387*f*
Systolic pressure (SP), **389**

Target cells, 48, **132**–33
    action of lipophilic hormones on, 143*f*
    hormone actions at, 161–67
    relationship between receptor binding and
        response of, 141–42
Targeting of cytosol-translated proteins, 53*f*,
    55*f*
Tastants, **304**

Taste, 303–5
    anatomy of taste buds, 303–4
    neural pathway for, 305
    signal transduction in, 304–5
Taste buds, anatomy of, **303**–4
T cell receptors (TCR), 720
T cells. *See* T lymphocytes (T cells)
T-dependent antigens, 722
Tectorial membrane, 295*f*, **296**
Teleophase (mitosis), 58*f*
Temperature
    affinity of hemoglobin for oxygen and
        effects of, 508, 509
    chemical reaction rates and, 71
    inflammation and higher body, 716
    regulation of body (*see* Thermoregulation)
Temporal lobe, cerebrum, **238**, 239*f*
Temporal summation, **186**, 187*f*, 211–12
Tendons, 4, **334**, 357
Teniae coli, 609
Tension, muscle, 344
    effect of load on peak, in isotonic twitch, 346*f*
    increase in peak (treppe), 347–48
    maximum tetanic, 348
Terminal boutons, **324**, 327*f*
Terminal bronchioles, **473**, 476*f*
Termination codon, 49
Tertiary bronchi, 472
Tertiary protein structure, 28, 30*f*
Testes, 673, **677**, 678*f*
Testis-determining factor, 674
Testosterone, 28, 29*f*, 160
    male reproductive function and, 679–80
    sex differentiation and, 674
    as treatment for erectile dysfunction, 685*b*
Tetanus (infectious disease), 349*b*
Tetanus, muscle, **347**, 348*f*
Tetraiodothyronine (T_4), 158
Tetrodotoxin (TTX), 185*b*
Thalamus, 236*f*, 237, 240
Theca cells, 684*f*, **685**
Thermal energy (heat), 66
Thermal motion, 66
Thermal nociceptors, **272**
Thermodynamics, first law of, 65
Thermoneutral zone, 13, 751
Thermoreceptors, 11, **271**, 462
    in skin, 272
Thermoregulation, 11, **12**–13
    components of body system for, 13
    exercise and, 745–46, 750–52
    heat exhaustion and heat stroke as failure
        of, 9*b*
    mechanisms of heat transfer between body
        and external environment, 12–13
    post-exercise fever, 756–57
    response of, to drop in temperature, 13, 14*f*
    responses of, as cardiovascular regulatory
        process, 462–63
    sweating and, 240
Thermoregulatory center, brain, 11, 462–63
Thermoregulatory responses and regulation of
    cardiovascular system, 462–63
Thick ascending limb, loop of Henle, **536**
Thick filaments, muscle, **336**–38
    oblique arrangement of, in smooth muscle,
        362*f*
    structure of skeletal, 337*f*

Thin ascending limb, loop of Henle, **536**
Thin filaments, muscle, **336**–38
    oblique arrangement of, in smooth muscle,
        362*f*
Third-order neurons, **265**
Thoracic cavity, 371, 372*f*
    structures of, 475, 477*f*
Thoracic nerves, **231**, 233*f*
Thoracolumbar division, autonomic nervous
    system, 314
Threshold, membrane potential, **184**
Thrombin, **409**
Thrombocytes, 407. *See also* Platelets
Thrombomodulin, 410
Thromboxanes, 28, 139
    thromboxane A_2 (TXA_2), **408**
Thrombus, 407. *See also* Blood clot
Throphoblast, 695, 696*f*
Thymine, 32, 33*f*
Thymosin, 158
Thymus, 154*f*, **158**, 328
Thyroglobulin (TG), 661
Thyroid gland, 154*f*, **158**, 159*f*
Thyroid hormones, 135, 140, 157, 158*f*,
    661–63
    abnormal secretion of, 165, 166*f*
    actions of, 663, 666*t*
    affects of, on growth, 661
    cardiac function and effects of, 394
    synthesis and secretion of, 661–63, 663*f*
Thyroid stimulating hormone (TSH), 157, 662
Thyrotropin releasing hormone (TRH), 157,
    158*f*, **219**, 662
Thyroxine-binding globulin, 662
Tidal volume (V_T), **488**
    effects of respiration rate and, on minute
        alveolar ventilation, 491*t*
Tight junctions, **43**–45, 124
    blood-brain barrier due to, 227
    in renal tubules, 546
T-independent antigens, 722
Tissue, **5**
    gas exchange in respiring, 501*f*, 503
    typing of, 731
Tissue factor (factor III), 409, 410*f*
Tissue factor pathway inhibitor, **410**–11
Tissue grafts, immune response and, 730–32
Tissue plasminogen activator (TPA), 410
T lymphocytes (T cells), 158, **711**, 718
    cytotoxic, 711, 718, 724–25, 726
    diversity of antigen receptors on, 720
    helper, 722, 724–25, 726, 727
    role of, in cell-mediated immunity, 724–26
    role of MHC molecules in antigen
        presentation and activation of, 725,
        726*f*
    role of MHC molecules in maturation of,
        725
    specificity of antigen binding by, 718–20
    suppressor, 724–25
Tone, resting muscle tension, 363
Tonicity, 121–23
    hypertonicity and hypotonicity, 122, 123*f*
    osmolarity versus, 122*f*, 123*t*
Tonic (slowly adapting) receptors, **264**, 266
Tonotopic organization, **298**
Total body water (TBW), **7**
Total lung capacity, 488, **489**

Total peripheral resistance, 424, **426**, 484
    mean arterial pressure and, 427–29
Totipotent, **695**
Toxins
    attacks on neuromuscular junctions by, 327, 329b
    neurotoxins, 185b
Trachea, **471**, 472f
Tractive forces, 485
Tracts, axon, **175**
Training, exercise, 16–17
    anatomical and functional changes resulting from, 17t
    upward spiral of, 16f
Transcription, 49, 50f
    regulation of, 53–56
Transcutaneous electrical nerve stimulation, 276
Transcytosis, **48**, 49f, 414, 416f
Transducin, 286, 305
Transduction, **262**. See also Signal transduction mechanisms
Trans face, Golgi apparatus, 38
Transfer RNA (tRNA), 49, 51f
    charged, 51
    free, 52
    translation and, 52f
Transfusion reaction, **730**
Transfusions, immune system and blood, 730
Translation, 51, 52f
    regulation of, 56
Transmembrane proteins, **36**
Transmural pressure, 415b
Transplantation, immune response and organ, 730–32
Transport
    across capillary walls, 414, 416f
    of carbon dioxide in blood, 510–14
    across cell membrane (see Cell membrane transport)
    of oxygen in blood, 505–9
    across wall of gastrointestinal tract, 565
    across walls of renal tubules, 565
Transport maximum, **547**–48
Transport proteins
    as enzymes, 116
    in facilitated diffusion, 113–14
    as pumps, 102, 116–17
Transport vesicles, 53, 54f
Transport work, **644**
Transpulmonary pressure, **482**, 483f
Transthyrethin, 662
Transverse colon, 609
Transverse tubules (T tubules), **334**, 335f, 342
Treppe, muscle, 347–48
TRH (thyrotropin releasing hormone), 157, 158f, **219**, 662
Triceps muscle, 358f
Tricuspid valve, **372**
Tricyclics, 218
Triglycerides, **25**, 26f
    metabolism of, 94–95
    synthesis of, 618, 620f, 646, 647
Triiodothyronine (T_3), 158
Triplet (DNA base pair sequence), **49**
Tropic hormones, **155**–57, 158f

Tropomyosin, 31, **336**, 337f
    excitation-contraction coupling (muscles) and role of, 342–43
Troponin, **336**, 337f
    excitation-contraction coupling (muscles) and role of, 342–43
Trypsin, 74, **616**
    activation of, 616, 617f
Trypsinogen, **616**, 617f
T-type channels, 381
Tubal sterilization, 694b
Tubular necrosis, 587b
Tubulin, 42f, 43
Tubuloglomerular feedback, 542, **543**
Tumor antigens, 728
Tumor necrosis fact [alpha] (TNF-[alpha]), 716
Turbulent blood flow, 424f
    blood pressure measurement and role of, 425b
T wave, **384**
Twitch, **344**–47
    isometric and isotonic, 345–47
    phases of, 344–45
Two-point discrimination, **270**
    thresholds of, in select body areas, 270t
Tympanic duct (scala tympani), 295f, **296**
Tympanic membrane (eardrum), 291, 292f
Type I alveolar cells, **475**, 476f
Type II alveolar cells, 475, **484**
Tyrosine [beta]-hydroxylase, 137
Tyrosine kinases, 145

Ubiquitin, 56
Umami, 305
Umbilical artery and vein, 697
Umbilical cord, **697**, 698f
Universal blood donors and recipients, 730
Unsaturated fatty acids, **26**
Upper airways, **470**, 472f
Upper esophageal sphincter, 606f, **607**
Upper motor neurons, voluntary motor control and, **248**
Up-regulation, 142
Uracil, 32, 33f
Urea, 95
    role of, in medullary osmotic gradient, 571
Uremia, 587b
Urethra, 534
Urethral sphincter, 556, 558f
Urinary bladder, 534
    anatomy of, 556, 558f
    micturition and, 556–59
Urinary system, 532–600
    acid-base balance and, 583–95
    anatomy of, **534**–38
    basic renal exchange processes of, 538–48
    calcium balance and, 580–81
    concept of balance and homeostasis related to, 565–66
    excretion function in, 553–59
    functions of, 533–34
    interactions between fluid and electrolyte regulation and, 581–83
    potassium balance and, 579–80
    regional specialization of renal tubules in, 549–53
    sodium balance and, 575–78

summary of factors in determining plasma volume and composition, 560f, 595, 596f
    systems integration chart for, 597f
    water balance and, 567–75
Urine, **533**, 535
    excretion of, 553–59
    output rate of, 444, 452, 553–54
Uterine artery and vein, 697
Uterine cycle, **688**, 689–91
Uterine tubes, 684f, **686**
Uterus, 684f, **685**
Utricle, 300–303
    functional anatomy of, 302f

Vaccination, 729–30
Vagina, 684f, **686**
Vagus nerve (cranial nerve X), **315**
Van der Waals forces, 28
Varicosities of neuroeffector junctions, **320**, 321f
Vasa recta, **537**
    role of, in prevention of medullary osmotic gradient dissipation, 571, 572f
Vascular smooth muscle, 456
    changes in blood flow and response of, 458, 459f
    changes in metabolic activity and response of, 456–58
    locally secreted chemical messengers and response of, 460, 461t
    stretch of arteriolar smooth muscle and response of, 458–60
Vascular spasm, 407
Vasculature, **371**. See also Blood vessels
    pressures in, 420f
Vas deferens, **679**
Vasoconstriction, **423**, 426
    angiotensin II and, 452, 577–78
    blood clot formation and, 407–8
    changes in metabolic activity and, 456–60
Vasodilation, **423**, 426
    changes in metabolic activity and, 456–60
    inflammation and increased, 714
    in response to chemical messengers, 460, 461t
    as response to epinephrine, 451–52
Vasopressin, 135, 155, **219**. See also Antidiuretic hormone (ADH)
    effects of, on mean arterial pressure, 452
Vaults, **40**
    chemotherapy and, 42b
Veins, **371**, 416–17
    carbon dioxide exchange and transport in, 511–13
    hepatic portal, and hepatic, 419, 609
    pulmonary, 376
    renal, 538, 539f
    structural characteristics of, 413f
    uterine, 697
    venomotor tone and constriction of, 432
    as volume reservoirs, 416, 417f
Velocity of shortening, muscle, 352, 353f
Venae cavae, blood flow through, **375**, 376f
Venomotor tone, **432**
Venous baroreceptors
    control of blood volume, pressure, and, 452–53

restoration of mean arterial pressure following hemorrhage and role of, 455

Venous pooling, **432**

Venous return, **386**, 404
central venous pressure and, 398

Ventilation. *See* Pulmonary ventilation

Ventilation-perfusion ratio, **521**, 522*f*

Ventral horn, spinal cord gray matter, **231**, 234*f*

Ventral respiratory group (VRG), **515**, 516*f*

Ventral roots, spinal cord, **232**, 234*f*

Ventricles, brain, **226**, 227*f*

Ventricles, heart, 371
differences in muscle thickness of, 373*f*
pressure of, during cardiac cycle, 388*f*
volume changes in, during cardiac cycle, 390*f*

Ventricular contractility
heart failure and reduced, 434*b*
stroke volume and influence of, **395–96**

Ventricular ejection, **386**

Ventricular fibrillation, 386

Ventricular filling, **388**

Ventricular myocardium, 372
hypertrophy of, 434*b*

Ventricular volume, 390–91

Venules, **371**, 416
structural characteristics of, 413*f*

Veriform appendix, 609

Vertebral column, **224**, 225*f*

Very-low-density lipoproteins (VLDLs), 621*b*, **646**

Vestibular apparatus, 292, **299**
anatomy of, 299, 300*f*

Vestibular duct (scala vestibuli), 295*f*, **296**

Vestibular glands, 687

Vestibular membrane, 295*f*, **296**

Vestibular nuclei, 249, 303

Vestibular system, 291

Vestibule, 687

Vestibulocochlear nerve, **292**

Villi, small intestine, **608**

Viruses, 711
events in immune response to, 728*f*

Visceral afferents, 261

Visceral information, **171**

Visceral pain, 275, 276*f*

Visceral receptors, 175, 261

Visceral reflexes, 244, **321–23**

Vision, 277–91
accommodation, 281, 282*f*
adaptation to light and dark, 287–88
clinical defects in, 281–82, 283*f*
color, 289*b*
depth perception, 290*f*, 291

eye anatomy, 278, 279*f*
light waves and, 278–81
neural pathways for, 290–91
neural processing in retina, 288–90
parallel processing in visual system, 291
phototransduction, 285–87
regulating amount of light entering eye, 282–84
retina and, 284*f*, 285

Visual cortex, 239, 240*f*,

Visual field, 280, 281*f*

Vital capacity (VC), **488**

Vitamins, **72**
absorption of, 620
B12, 406, 510, 607, 620
D3, 160, 533
K, 411

Vitreous chamber, **278**, 279*f*

Vitreous humor, **278**, 279*f*

Voltage-gated channels, **174**
action potential and, 188–92
cardiac muscle and, 380–83
effect of neurotoxins on sodium, 185*b*
model for operation of, 190*f*
muscle contraction and calcium, 342
neurotransmitter release and calcium, 203, 206, 326, 327*f*

Volume reservoirs, veins as, 416–17

Voluntary motor control, 247–51
basal nuclei and, 250–51
brainstem control of posture, 249–50
cerebellum and, 250
cortical control of voluntary movement, 248, 249*f*
four processes in, 247, 248*f*
neural components for, 247–48

Voluntary nervous system, 324

Vomiting, **631**
reflex, 322, 631

Vomiting center, 631

Von Willebrand factor (vWf), **407**

Von Willebrand's disease, 411

Vulva, **686**

Wakefulness, electrical activity during, 252, 253*f*, 254

Warm thermoreceptors, **272**

Waste products, liver and removal of, 612

Water
absorption of, 620
balance (*see* Water balance and homeostasis)
in blood plasma, 405*t*
body fluid compartments and, 7–8
effects of drinking salty seawater, 576*b*

epithelial transport of, 126, 127*b*
excretion of, 574 (*see also* Excretion
fluid movement in digestive system and, 627–28
movement of, and osmolarity, 567–69
normal rates of glomerular filtration and reabsorption of solutes and, 545*t*
osmosis of, across membranes, 119–23
as polar molecule, 24
reabsorption of, 545, 547, 569–75
solubility of carbon dioxide and oxygen in, 499, 500*f*
thermoregulation and loss of, 12

Water balance and homeostasis, 566, 567–75
during exercise, 752
osmolarity and water movement, 567–69
reabsorption of water, and role of medullary osmotic gradient, 569–71
reabsorption of water in distal tubule and collecting duct, 571–74
reabsorption of water in proximal tubule, 569

Water conservation
during exercise, 752
by kidneys and loop of Henle, 552, 572

Water intoxication, 119, 755, **756**

Weak acid, 75*b*

Wernicke's area, 231, 240*f*, **251**

White blood cells. *See* Leukocyte(s)

White matter
brain, **228**, 229*f*, 230–31
white matter, 232–33, 234*f*

White muscle, 354

White ramus, 314, 315*f*

Withdrawal reflex, **245**, 246*f*

Wolffian ducts, **674**

X chromosome, 671–72
sex determination and, 673–74

Y chromosome, 671–72
sex determination and, 673–74

Z lines, muscle, **336**

Zona fasciculata, adrenal cortex, 159, 160*f*

Zona glomerulosa, adrenal cortex, 159, 160*f*

Zona pellucida, **688**, 689*f*, 695

Zona reticularis, adrenal cortex, 159, 160*f*

Zonular fibers, **278**

Zygote, **671**, 695
early development and implantation of, 695, 696*f*

Zymogen granules, 612

Zymogens, **612**

Dear Student:

As the authors of *Principles of Human Physiology,* we wanted to make the complex topics in physiology easy to understand and integrate. We would appreciate hearing about your experiences with this textbook and its multimedia support, and we invite your suggestions for improvements!

Many thanks,

William J. Germann

Cindy Lou Stanfield

1. What did you like most about *Principles of Human Physiology*? Please provide examples in order of priority. _____

_____

_____

_____

_____

2. Which study tool(s)—e.g. Study Hints, Quick Tests, Chapter Summaries, End-of-Chapter Exercises — did you find most useful? _____

_____

_____

_____

_____

3. Do you have any ideas about how this book could be improved? Please write your specific suggestions, citing page numbers if appropriate. _____

_____

_____

_____

_____

4. Did you use the **Interactive Physiology® Sampler CD-ROM** that was included with the book?
☐ Yes ☐ No
If so, which topic(s) did you find most useful? _____

_____

_____

_____

5. Did you subsequently use the full **Interactive Physiology® 7-System Suite CD-ROM** or an online subscription to **IPweb™**? ☐ Yes ☐ No

6. Are there topics that should be added to **Interactive Physiology**®? How can we improve **Interactive Physiology**®? _____

_____

_____

_____

_____

7. Did you use **The Physiology Place** web site? ☐ Yes ☐ No

If so, what did you like or use most? Was it easy to use, or did you experience problems? How can we improve **The Physiology Place**? _____

_____

_____

_____

BUSINESS REPLY MAIL

FIRST-CLASS MAIL   PERMIT NO. 275   SAN FRANCISCO CA

POSTAGE WILL BE PAID BY ADDRESSEE

BENJAMIN CUMMINGS
PEARSON EDUCATION
1301 SANSOME STREET
SAN FRANCISCO CA 94111-9328

NO POSTAGE
NECESSARY
IF MAILED
IN THE
UNITED STATES

School: _____

Instructor's Name: _____

Optional

Your name: _____

Email: _____

Date: _____

May Benjamin Cummings have permission to quote your comments in promotions for ***Principles of Human Physiology?*** ☐ Yes ☐ No

FREQUENTLY USED ABBREVIATIONS AND SYMBOLS

ACh acetylcholine

ACTH adrenocorticotropic hormone

ADH antidiuretic hormone

ADP adenosine diphosphate

ATP adenosine triphosphate

AV atrioventricular

CA carbonic anhydrase

cAMP cyclic AMP

CCK cholecystokinin

cGMP cyclic GMP

CNS central nervous system

CO cardiac output

CRH corticotropin-releasing hormone

CVP central venous pressure

C_X clearance of X

DAG diacylglycerol

DNA deoxyribonucleic acid

DP diastolic pressure

ECF extracellular fluid

EDV end-diastolic volume

EPSP excitatory postsynaptic potential

ESV end-systolic volume

E_X equilibrium potential of X

FAD flavin adenine dinucleotide

FRC functional residual capacity

FSH follicle-stimulating hormone

GDP guanosine diphosphate

GFR glomerular filtration rate

GH growth hormone

GnRH gonadotropin-releasing hormone

GTP guanosine triphosphate

HR heart rate

ICF intracellular fluid

IP_3 inositol triphosphate

IPSP inhibitory postsynaptic potential

ISF interstitial fluid

LH luteinizing hormone

MAP mean arterial pressure

mm Hg millimeters of mercury

mV millivolts

NAD^+ nicotinamide adenine dinucleotide

NFP net filtration pressure

π osmotic pressure

P_{alv} intra-alveolar pressure

P_{atm} atmospheric pressure

P_{ip} intrapleural pressure

PIP_2 phophatidylinositol 4,5-biphosphate

PSP postsynaptic potential

P_X permeability of X; partial pressure of X; plasma concentration of X

RNA ribonucleic acid

SA sinoatrial

SP systolic pressure

SR sarcoplasmic reticulum

SV stroke volume

TRH thyrotropin-releasing hormone

TSH thyroid-stimulating hormone

TPR total peripheral resistance

V_m membrane potential

[X] concentration of X

THE METRIC SYSTEM

PREFIXES

PREFIX (ABBREVIATION)	STANDARD NOTATION	SCIENTIFIC NOTATION
Mega- (M)	multiply by 1,000,000	multiply by 1×10^6
kilo- (k)	multiply by 1,000	multiply by 1×10^3
deci- (d)	multiply by 0.1	multiply by 1×10^{-1}
centi- (c)	multiply by 0.01	multiply by 1×10^{-2}
milli- (m)	multiply by 0.001	multiply by 1×10^{-3}
micro- (μ)	multiply by 0.000001	multiply by 1×10^{-6}
nano- (n)	multiply by 0.000000001	multiply by 1×10^{-9}
pico- (p)	multiply by 0.000000000001	multiply by 1×10^{-12}

CONVERSION FACTORS

METRIC UNIT (ABBREVIATION)	ENGLISH UNIT	CONVERSION: METRIC TO ENGLISH	CONVERSION: ENGLISH TO METRIC
Lengths			
millimeters (mm)	inches	mm $\times$ 0.03937 = inches	inches $\times$ 25.4 = mm
centimeters (cm)	inches	cm $\times$ 0.3937 = inches	inches $\times$ 2.54 = cm
meters (m)	feet	m $\times$ 3.2808 = feet	feet $\times$ 0.3048 = m
meters (m)	yards	m $\times$ 1.0936 = yards	yards $\times$ 0.9144 = m
Weights			
grams (gm)	ounces	gm $\times$ 0.0353 = ounces	ounces $\times$ 28.3495 = gm
grams (gm)	pounds	gm $\times$ 0.0022 = pounds	pounds $\times$ 453.6 = gm
kilograms (kg)	pounds	kg $\times$ 2.2046 = pounds	pounds $\times$ 0.4536 = kg
Volumes			
milliliters (ml)	ounces	ml $\times$ 0.0338 = ounces	ounces $\times$ 29.574 = ml
milliliters (ml)	quarts	ml $\times$ 0.00106 = quarts	quarts $\times$ 946.3 = ml
liters (L)	gallons	L $\times$ 0.2643 = gallons	gallons $\times$ 3.784 = L
Temperature			
Celsius (°C)	Fahrenheit (°F)	(°C $\times$ 9/5) + 32 = °F	(°F − 32) $\times$ 5/9 = °C